List of the Elements with Their Symbols and Atomic Masses[a]

Element	Symbol	Atomic number	Atomic mass	Element	Symbol	Atomic number	Atomic mass
Actinium	Ac	89	(227)[b]	Neon	Ne	10	20.18
Aluminum	Al	13	26.98	Neptunium	Np	93	(237)
Americium	Am	95	(243)	Nickel	Ni	28	58.69
Antimony	Sb	51	121.8	Niobium	Nb	41	92.91
Argon	Ar	18	39.95	Nitrogen	N	7	14.01
Arsenic	As	33	74.92	Nobelium	No	102	(253)
Astatine	At	85	(210)	Osmium	Os	76	190.2
Barium	Ba	56	137.3	Oxygen	O	8	16.00
Berkelium	Bk	97	(247)	Palladium	Pd	46	106.4
Beryllium	Be	4	9.012	Phosphorus	P	15	30.97
Bismuth	Bi	83	209.0	Platinum	Pt	78	195.1
Boron	B	5	10.81	Plutonium	Pu	94	(242)
Bromine	Br	35	79.90	Polonium	Po	84	(210)
Cadmium	Cd	48	112.4	Potassium	K	19	39.10
Calcium	Ca	20	40.08	Praseodymium	Pr	59	140.9
Californium	Cf	98	(249)	Promethium	Pm	61	(147)
Carbon	C	6	12.01	Protactinium	Pa	91	(231)
Cerium	Ce	58	140.1	Radium	Ra	88	(226)
Cesium	Cs	55	132.9	Radon	Rn	86	(222)
Chlorine	Cl	17	35.45	Rhenium	Re	75	186.2
Chromium	Cr	24	52.00	Rhodium	Rh	45	102.9
Cobalt	Co	27	58.93	Rubidium	Rb	37	85.47
Copper	Cu	29	63.55	Ruthenium	Ru	44	101.1
Curium	Cm	96	(247)	Samarium	Sm	62	150.4
Dysprosium	Dy	66	162.5	Scandium	Sc	21	44.96
Einsteinium	Es	99	(254)	Selenium	Se	34	78.96
Erbium	Er	68	167.3	Silicon	Si	14	28.09
Europium	Eu	63	152.0	Silver	Ag	47	107.9
Fermium	Fm	100	(253)	Sodium	Na	11	22.99
Fluorine	F	9	19.00	Strontium	Sr	38	87.62
Francium	Fr	87	(223)	Sulfur	S	16	32.07
Gadolinium	Gd	64	157.3	Tantalum	Ta	73	180.9
Gallium	Ga	31	69.72	Technetium	Tc	43	(99)
Germanium	Ge	32	72.59	Tellurium	Te	52	127.6
Gold	Au	79	197.0	Terbium	Tb	65	158.9
Hafnium	Hf	72	178.5	Thallium	Tl	81	204.4
Helium	He	2	4.003	Thorium	Th	90	232.0
Holmium	Ho	67	164.9	Thulium	Tm	69	168.9
Hydrogen	H	1	1.008	Tin	Sn	50	118.7
Indium	In	49	114.8	Titanium	Ti	22	47.88
Iodine	I	53	126.9	Tungsten	W	74	183.9
Iridium	Ir	77	192.2	Unnilennium	Une	109	(266)
Iron	Fe	26	55.85	Unnilhexium	Unh	106	(263)
Krypton	Kr	36	83.80	Unniloctium	Uno	108	(265)
Lanthanum	La	57	138.9	Unnilpentium	Unp	105	(260)
Lawrencium	Lr	103	(257)	Unnilquadium	Unq	104	(257)
Lead	Pb	82	207.2	Unnilseptium	Uns	107	(262)
Lithium	Li	3	6.941	Uranium	U	92	238.0
Lutetium	Lu	71	175.0	Vanadium	V	23	50.94
Magnesium	Mg	12	24.31	Xenon	Xe	54	131.3
Manganese	Mn	25	54.94	Ytterbium	Yb	70	173.0
Mendelevium	Md	101	(256)	Yttrium	Y	39	88.91
Mercury	Hg	80	200.6	Zinc	Zn	30	65.39
Molybdenum	Mo	42	95.94	Zirconium	Zr	40	91.22
Neodymium	Nd	60	144.2				

[a] All atomic masses have four significant figures. These values are recommended by the Committee on Teaching of Chemistry, International Union of Pure and Applied Chemistry.

[b] Approximate values of atomic masses for radioactive elements are given in parentheses.

GENERAL CHEMISTRY

GENERAL CHEMISTRY

RAYMOND CHANG

WILLIAMS COLLEGE

RANDOM HOUSE

NEW YORK

First Edition

9 8 7 6 5 4 3 2 1

Copyright © 1986 by Random House, Inc.

Library of Congress Cataloging-in-Publication Data

Chang, Raymond.
  General chemistry.

  Includes index.
  1. Chemistry. I. Title.
QD31.2.C3735 1986   540   85-24467
ISBN 0-394-34122-8

This book contains revised material from *Chemistry*, second edition, by Raymond Chang © 1981, 1984 by Random House, Inc.

Cover photograph: Jerome Kresch
Book and Cover Design: Lorraine Hohman

Manufactured in the United States of America

Photograph and other acknowledgments appear on pages 991–92, which constitute a continuation of this copyright page.

PREFACE

General Chemistry is written for students taking a full-year introductory chemistry course. In recent years, a growing number of instructors have felt the need to introduce more descriptive chemistry in such a beginning course. However, teaching the chemistry of the elements early in the course before the fundamental principles have been presented or arbitrarily inserting a chapter on the chemistry of hydrogen and oxygen between the properties of solutions and chemical equilibrium, say, seems inappropriate. In writing this book, I have tried to strike a balance between theory and application, and I have attempted to solve the problem of "what to do with descriptive chemistry in a general chemistry course" by emphasizing the use of the periodic table. In many ways the periodic table is the single most important and useful source of information regarding the elements. It correlates the chemical behavior of the elements in a systematic manner and helps us to remember and understand many facts. Most of the topics presented in an introductory course can be expanded upon by reference to the periodic table, which allows descriptive chemistry to be taught and learned in a more natural way.

No ideal single organizational scheme exists for a general chemistry course. The chapters in this text have been arranged to follow the mainstream sequence, but the organization also allows for flexibility. Chapter 1 presents the basic vocabulary and tools needed for chemistry and Chapter 2 introduces the basic concepts of atoms and molecules and the nomenclature of inorganic compounds. As a group, Chapters 3 through 5 are devoted to the nature and types of chemical reactions and the mass and energy relationships accompanying chemical changes. That a fair amount of descriptive chemistry is presented in Chapter 3 through the discussion of chemical reactions enables students to gain an early appreciation of the reactivity of some common substances. After a chapter on the physical properties of gases (Chapter 6), the structure and quantum mechanical treatment of atoms are presented in Chapters 7 and 8. The general properties of elements and periodic trends are the topic of Chapter 9.

A very important topic, chemical bonding, is the subject of four chapters (Chapters 10–13). Chapters 14 and 15 deal with the physical properties of liquids and solids and solutions, and Chapter 16 discusses one important type of chemical reaction, oxidation–reduction reactions. As a group, Chapters 17 through 24 deal with the more quantitative aspects of chemistry. Acid-base reactions, which are among the most common and important of chemical processes, are discussed in three chapters (Chapters 18–20). After Chapter 25, which studies nuclear reactions, are four chapters that deal with descriptive chemistry in a systematic manner. The last chapter is on the chemistry of organic compounds.

Several features of the text deserve special mention.

- At the end of a number of chapters are box features called **An Aside on the Periodic Table.** These are short, elementary discussions related to the chapter materials and they serve to illustrate the periodic relationships among the elements. These box features enable students to appreciate descriptive chemistry in a more meaningful way.

- The text has two **Color Plate** sections. The first group of Color Plates (1–10) shows the appearance of many of the elements, their minerals, and the mining processes and methods used for their extraction. The second group of Color Plates (11–24) illustrates various physical and chemical processes and the properties of a number of substances. These Color Plates serve to bring the living color of chemistry to beginning students and enhance their appreciation for descriptive chemistry.

- Every **important term** appears in boldface when it is introduced and defined. For quick reference, these key words are also listed alphabetically at the end of each chapter and defined in a glossary at the end of the book.

- There is a **summary at the end of each chapter** to serve as a review of the important concepts introduced in the chapter.

- Numerous **marginal notes** that serve as reminder of facts already presented and additional comments are provided throughout the text.

The best way to test one's understanding of chemical concepts is by solving problems. The many **worked examples** within each chapter demonstrate problem-solving techniques. In addition, there are over 1,500 **end-of-chapter problems.** These problems are grouped according to specific topics in the chapters. **Answers to selected problems** are given at the end of the book.

Most of the units used in this book are SI units. For practical laboratory reasons, I have retained the use of atmosphere and mmHg for pressure and liter and milliliter for volume.

Supplements available for use with this text are:
    Philip C. Keller and Jill L. Keller, *Study Guide*
    Raymond Chang, *Solutions Manual*

These supplements contain many ideas and insights that are helpful to understanding chemical concepts, as well as problem-solving techniques. An *Instructor's Manual*, written by the author, and a *Test Bank*, written by Kenneth W. Watkins, are available to instructors upon request to the publisher.

Your comments and suggestions on the text and its ancillaries will be greatly appreciated.

Raymond Chang
Department of Chemistry
Williams College
Williamstown, Mass. 01267

ACKNOWLEDGMENTS

It is a pleasure for me to thank the following individuals whose comments and suggestions have helped to improve the text significantly:

David L. Adams, *North Shore Community College*
Ismail Y. Ahmed, *The University of Mississippi*
John L. Bonte, *Clinton Community College*
LeRoy P. Breimeier, *Vincennes University*
Arthur C. Breyer, *Beaver College*
David W. Brooks, *University of Nebraska, Lincoln*
Diane Bunce, *Catholic University*
Gordon J. Ewing, *New Mexico State University*
Dorothy Gabel, *Indiana University*
Julanna V. Gilbert, *Metropolitan State College*
L. Peter Gold, *Pennsylvania State University*
Edwin S. Gould, *Kent State University*
Philip C. Keller, *University of Arizona*
Russell Larsen, *Texas Tech University*
Tamar Y. Susskind, *Oakland Community College*
Kenneth W. Watkins, *Colorado State University*
David Weill, *Shady Side Academy*
Milton J. Wieder, *Metropolitan State College*

In particular, I would like to thank Henry W. Heikkinen (University of Maryland) whose detailed critique helped to make the text more student oriented.

I have benefited from helpful discussions with my colleagues at Williams, and I would like to thank James F. Skinner for letting me use the apparatus that appear in Figure 6.19 and Color Plate 23, and Bud Wobus for his advice on minerals.

I greatly appreciate the enthusiastic support and assistance given to me by the following members of Random House's College Department: Seibert Adams, Jane Bess, Edith Beard Brady, Patricia Chu, Sam Fussell, Lorraine Hohman, Thomas Holton, Anita Kann, Della Mancuso, Nancy Messing, Laurel Miller, Dorothy Sparacino, Doug Thompson, and Suzanne Thibodeau. In particular, I would like to mention Barry Fetterolf who encouraged me to write this book, Mary Shuford, the project editor who expertly assisted me through the critical stages of production, and Kathy Bendo, the competent and resourceful photo editor who is patient beyond the call of duty. Finally my thanks are due Harry Spector who did his usually careful job of copyediting and Ken Karp for his marvelous black and white and color photographs.

CONTENTS

GENERAL CHEMISTRY

1 INTRODUCTION

Every field of human activity has its specialized vocabulary and particular "working style." This observation holds true for such diverse enterprises as automotive engineering, musical composition, and athletics. In this opening chapter you will be introduced to the distinctive language and procedures associated with chemistry—the vigorous and productive branch of science that is the focus of both this course and this textbook.

Such an orientation to fundamental terminology and techniques is essential to your detailed studies later. And since chemistry is a quantitative science, some of the numerical aspects of "doing chemistry" are also covered and explained in this chapter.

1.1   THE NATURE OF CHEMISTRY

Chemistry is *the science that studies the properties of substances and how substances react with one another.* It plays a central role in almost all aspects of our lives. In fact, without the contributions of chemistry, the great advances in medicine, food production, and the manufacture of numerous clothing items and household objects that provide comfort and simplify our daily chores would have been impossible. But chemistry is not without its problems. For instance, sometimes the mishandling of chemicals has led to the contamination of our environment. Even here, though, our knowledge of chemistry can help us clean up the pollution.

The word "chemistry" is derived from the Greek word "chemia," meaning "the art of metalworking." This definition obviously needs to be broadened to describe chemistry today. In the truest sense of the word, chemistry has become an *interdisciplinary* science. Chemistry has been fruitfully applied to the study of biology, geology, physics, and many other areas, and the development of other fields has also made great advances in chemistry possible. An introductory chemistry course is essential for anyone intending to major in the subject. However, many of you take this course not because you intend to be professional chemists, but because your proposed fields of study require you to have one or more years of college chemistry. Whether you plan to become a biologist, a geologist, a physicist, an engineer, or a health-related professional, or to enter one of many other related careers, at least an elementary knowledge of chemistry is necessary. Thus the study of chemistry can be extremely important to you. It is an intellectual enterprise with worlds of practical applications. With reasonable effort on your part, you will soon begin to understand the basic principles of chemistry and find the subject interesting and enjoyable.

1.2   SCIENCE AND ITS METHODS

All sciences, including the social sciences, employ variations on what is called the *scientific method*—a systematic approach to the generation of new knowledge. After scientists have defined a goal, their next step is to make careful observations and collect bits of information about the system. The bits of information are called *data. System* here denotes that part of the universe that is under investigation. The information obtained may be both **qualitative,** *consisting of general observations about the system,* and **quantitative,** *comprising numbers obtained by various measurements of the system.*

When enough information has been gathered, a **hypothesis**—*a tentative explanation for a set of observations*—can be formulated. Further experiments then are devised to test the validity of the hypothesis in as many ways as possible.

After a large amount of data has been collected, it is often desirable to summarize the information in a concise way. A **law** is *a concise verbal or mathematical statement of a relationship between phenomena that is always the same under the same conditions.* As mentioned earlier, hypotheses provide only tentative explanations that must be tested by many experiments. If they survive such tests, hypotheses may develop into theories. A ***theory*** is *a unifying principle that explains a body of facts and those laws that are based on them.* Theories too are constantly being tested. If a theory is proven incorrect by experiment, then it must be discarded or modified so that it becomes consistent with experimental observations.

An Example of the Scientific Method

Perhaps the best way to appreciate the scientific method is to follow the steps of scientists in solving a particular problem. One example is the problem of the extinction of dinosaurs—a fascinating study that began in 1977 and is still being investigated at the present time.

The dinosaurs, which dominated Earth for millions of years, disappeared very suddenly (Figure 1.1). By studying the fossils and skeletons in rocks in various layers of Earth's crust, paleontologists are able to map out which species existed on Earth during specific geologic periods. These studies show that in the rocks that were formed immediately after the Cretaceous period, which dates back some 65 million years, not one dinosaur skeleton has been found. It is therefore assumed that the dinosaurs became extinct 65 million years ago. What happened?

Among the many hypotheses put forward to account for their disappearance are disruptions of the food chain and a dramatic change in climate caused by violent volcanic eruptions. However, none of these phenomena would seem by itself to be a convincing cause of the dinosaur extinction. Then, in 1977, a group of paleontologists working in Italy obtained some very puzzling data at a site near Gubbio. From the chemical analysis of a layer of clay that is *above* the sediments formed during the Cretaceous period (and therefore this layer recorded events that occurred *after* the Cretaceous period) they found a surprisingly high content of a substance called iridium. Iridium is rarely present in Earth's crust but is comparatively abundant in asteroids.

This investigation led scientists to hypothesize the extinction of dinosaurs as follows. To account for the quantity of iridium they found, they assume that a large asteroid several miles in diameter must have hit Earth at that time. Presumably the impact of the asteroid on Earth's surface was so tremendous that it literally vaporized a large quantity of surrounding rocks, soils, and other objects. The resulting dust and debris floated around in the air and blacked out the sun for months or perhaps even years. Without ample sunlight most plants could not grow, and eventually animals, whether they were herbivorous or carnivorous, gradually perished. This catastrophe obviously affected larger animals more than smaller ones.

The hypothesis about dinosaur extinction is both interesting and provocative. It can of course be further tested. If the hypothesis is correct then we should find similar high iridium content in corresponding layers at different locations on Earth. Moreover, we would expect the simultaneous extinction of other large species in addition to dinosaurs. Both of these predictions are

FIGURE 1.1 *Where have the dinosaurs gone, long time passing! The study of dinosaur extinction illustrates what we mean by the scientific method.*

strongly supported by recently collected data. In fact, the evidence has become so convincing that scientists now refer to the explanation as the theory of dinosaur extinction.

The Study of Chemistry

The dinosaur story illustrates the scientific approach in general. We now focus on how chemists study their subject.

Chemistry is largely an experimental science. Most chemists work in a laboratory of one kind or another. In a broad sense we can view chemistry on three levels. The first level is the *observational*. This is the level at which the chemist makes observations of what actually takes place in an experiment; for example, temperature rise, color change, gas evolution, and so on. The second level is the *representational*. A chemist records and describes an experiment in scientific language using special symbols and equations. Such a language helps to simplify the description and establishes a common base for chemists to communicate with one another. The third level is the *interpretational*, which deals with the explanation of observed phenomena (Figure 1.2).

All of us have witnessed the rusting of iron at one time or another. This is a process that takes place in the *macroscopic world*, in which we deal with things that can be seen, touched, weighed, and so on. If we were studying the rusting of iron as a chemistry project, our next step would be to describe this process with a "chemical equation" that tells us how rust is formed from iron, oxygen gas, and water, under a given set of conditions. Lastly, we would need to address questions like "What actually happens when iron rusts?" and "Why does iron rust but gold does not, under similar conditions?" To answer these and other related questions, we have to know the behavior of the fundamental units of substances, which are atoms and molecules. Because atoms and molecules are extremely small compared to macroscopic objects, the interpretation of an observed phenomenon takes us into the *microscopic world*.

The study of chemistry requires that we pay attention to both the macroscopic and the microscopic worlds. The data for chemical investigations

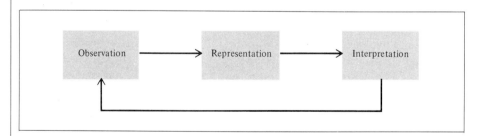

FIGURE 1.2 *The three levels of studying chemistry and their relationships. Observation deals with events in the macroscopic world. Representation is a scientific shorthand for describing an experiment in special symbols and chemical equations. Chemists use their knowledge of atoms and molecules to explain an observed phenomenon. The extremely small atoms and molecules constitute the microscopic world.*

most often come from large-scale phenomena and observations. But the testable hypotheses, theories, and explanations, which make chemistry an experimental science, are frequently expressed in terms of the notations of the unseen and partially imagined microscopic world of atoms and molecules.

Part of the difficulty some of you may feel in the beginning stage of learning chemistry is due to the frequent shifts that occur as the instructor or textbook moves back and forth between these two distinctly different ways of considering our physical universe. It has been said that often a chemist *sees* one thing (in the macroscopic world) and *thinks* another (in the microscopic world). Welcome to this fascinating "dual world" of chemistry!

1.3 SOME BASIC DEFINITIONS

Words that we use in everyday life often have altered meanings in scientific language. In this course you will soon learn how a number of terms, both familiar and unfamiliar, are used in chemistry. These terms are usually best introduced and explained at the time they are needed, in their proper context. However, a few of them are so important and fundamental that they have direct application to nearly all the subjects discussed in this book. Such terms are briefly introduced and defined here.

Matter

Anything that occupies space and possesses mass is called **matter.** Matter is all around us. Things we can see and touch (like water, earth, and trees) as well as things we cannot see and touch (like air) are matter.

Mass and Weight

Mass is *a measure of the quantity of matter contained in an object.* The terms "mass" and "weight" are often used interchangeably, although, strictly speaking, they refer to different quantities. In scientific language, **weight** refers to *the force that gravity exerts on an object.* An apple that falls from a tree is pulled downward by Earth's gravity. The mass of the apple is constant and does not depend on its location, but its weight does. For example, the apple's weight on the surface of the moon would be only one-sixth that on Earth, since the moon's gravity is only one-sixth that of Earth. This is the reason that astronauts are able to jump about rather freely on the moon's surface despite their massive suits and equipment. The mass of an object can be determined readily with a balance, and this process, oddly, is called *weighing.*

Substances and Mixtures

A **substance** is *a form of matter that has a definite or constant composition* (the number and type of basic units present) *and distinct properties.* Examples include water, ammonia, table sugar (sucrose), gold, oxygen, and so on. Substances differ from one another in composition and can be identified by their appearance, smell, taste, and other properties. At present, the number of known substances exceeds 5 million and the list of new substances is growing rapidly.

A **mixture** is *a combination of two or more substances in which the substances retain their identity.* Some familiar examples are air, soft drinks, milk, and cement. Mixtures do not have constant composition; samples of air collected over two different cities will most probably have different compositions, as a result of their differences in altitude, pollution, and so on.

Mixtures are either homogeneous or heterogeneous. When a spoonful of sugar dissolves in water, *the composition of the mixture, after sufficient stirring, is the same throughout the solution.* This solution is a **homogeneous mixture.** Now, if the sugar is mixed with sand instead, we obtain a heterogeneous mixture. In a **heterogeneous mixture** *the individual components remain physically separated and can be seen as separate components.* Any mixture, be it homogeneous or heterogeneous, can be put together and then separated into pure components without any change in the identity of the components; this is done by physical means. Thus, sugar can be removed from the homogeneous mixture we have described by evaporating the solution to dryness. Condensing the water vapor that comes off will give us back the water component. The sand and sugar can be separated by dissolving the sugar in water, reclaiming the sugar as just described, and then drying the sand. Thus, in this physical separation process, there has been *no* change in the composition of the substances making up the mixture.

Physical and Chemical Properties

Substances are characterized by their individual and sometimes unique properties. The color, melting point, boiling point, and density of a substance are examples of its physical properties. A **physical property** *can be measured and observed without changing the composition or identity of a substance.* For example, we can measure the melting point of ice by heating a block of ice in a special apparatus and recording the temperature at which the ice is converted to water. But since water differs from ice only in appearance and not in composition, this is a physical change; we can freeze the water to recover the original ice. Therefore, the melting point of a substance is a physical property. Similarly, when we say that hydrogen gas is lighter than air, we are referring to a physical property. On the other hand, the statement "Hydrogen gas burns in oxygen gas to form water" describes a **chemical property** of hydrogen because *in order to observe this property we must carry out a chemical reaction,* in this case burning. After the reaction, the original hydrogen and oxygen gases will have vanished, and all that will be left is water. We *cannot* recover the hydrogen from the water by a physical change such as boiling or freezing of the water.

Every time we hard-boil an egg for breakfast, we are causing chemical changes. When subjected to a temperature of about 100°C, the yolk and the egg white undergo reactions that alter not only their physical appearance but their chemical makeup as well. When eaten, the egg is altered again by substances in the stomach called *enzymes*. This digestive action provides additional examples of chemical reactions. The specific way such a process is carried out depends on the chemical properties of the enzymes and of the food involved.

All measurable properties of matter fall into two categories: extensive properties and intensive properties. *The measured value of an* **extensive property** *depends on how much matter is being considered.* Mass, length, and volume are extensive properties. More matter means more mass. Values of an extensive property can be added together. For example, two identical blocks each with the same mass will have a combined mass that is the sum of the two separate masses, and the volume occupied by any two beakers of water is the sum of the volumes of the water in the two individual beakers. The value of an extensive quantity depends on the size of matter.

The measured value of an **intensive property** *does not depend on how much matter is being considered.* Temperature is an intensive property. Suppose we have two beakers of water at the same temperature. If we combine them to make a single quantity of water in a larger beaker, the temperature of the larger quantity of water remains the same. Unlike mass and volume, temperature is not additive.

Atoms and Molecules

Formal definitions of atoms and molecules will be given in Chapter 2. Here you simply need to know that *all* matter is composed of atoms of different kinds, combined in many different ways. *Atoms* are the smallest possible units of a substance. *Molecules* are made up of atoms held together by special forces.

Elements and Compounds

Substances can be either elements or compounds. An **element** is *a substance that cannot be separated into simpler substances by chemical means.* At present, 109 elements have been positively identified. Eighty-three of them occur naturally on Earth. The others have been created by scientists in nuclear reactions (about which you will learn more later).

All matter on Earth is composed of various combinations of these known elements. Thus elements play as central a role in chemistry as, say, letters of the alphabet play in the study of language. Because of this importance, chemists have devised several related ways of tabulating information regarding the known elements.

The names of the elements are listed alphabetically inside the front cover of this book. This listing is convenient if you want to look up the symbol or some property of an element. However, if you visit a chemical laboratory or if you simply look at the front wall of your lecture hall, chances are that you will see an "element symbol list" in the form of a periodic table,

TABLE 1.1 Some Common Elements and Their Symbols

Name	Symbol	Name	Symbol	Name	Symbol
Aluminum	Al	Fluorine	F	Oxygen	O
Arsenic	As	Gold	Au	Phosphorus	P
Barium	Ba	Hydrogen	H	Platinum	Pt
Bismuth	Bi	Iodine	I	Potassium	K
Bromine	Br	Iron	Fe	Silicon	Si
Calcium	Ca	Lead	Pb	Silver	Ag
Carbon	C	Magnesium	Mg	Sodium	Na
Chlorine	Cl	Manganese	Mn	Sulfur	S
Chromium	Cr	Mercury	Hg	Tin	Sn
Cobalt	Co	Nickel	Ni	Tungsten	W
Copper	Cu	Nitrogen	N	Zinc	Zn

similar to the one inside the front cover of this book. More will be said of the periodic table in Section 1.4.

The first letter of the symbol for an element is *always* capitalized, but the second and third letters are *never* capitalized. For example, Co is the symbol for the element cobalt, whereas CO is the formula for a molecule of carbon monoxide. Table 1.1 shows some common elements and their symbols. The symbols of some elements are derived from Latin names—for example, Au from "aurum" (gold), Fe from "ferrum" (iron), and Na from "natrium" (sodium). Special three-letter element symbols have been proposed for the most recently synthesized elements, as inspection of the periodic table reveals.

As mentioned earlier, water can be formed by burning hydrogen in oxygen. The atoms of hydrogen and oxygen combine to form water, which has properties that are distinctly different from those of the starting materials. Thus, water is an example of a **compound,** *a substance composed of atoms of two or more elements chemically united in fixed proportions.* In every water unit, there are two H atoms and one O atom. This composition does not change whether we are dealing with water found in the United States, in Outer Mongolia, or on Mars. Unlike mixtures, compounds cannot be separated into their pure components, which are the atoms of the elements present, except by chemical methods.

The forms of matter defined here and earlier in this chapter are summarized in Figure 1.3.

1.4 CHEMICAL ELEMENTS AND THE PERIODIC TABLE

As their knowledge of chemistry grew, chemists in the nineteenth century began to realize that many elements show very strong similarities to one another and that there are regularities in their physical and chemical behavior. These facts led to the development of the **periodic table**—*a tabular arrangement of the elements* (Figure 1.4). *The elements in a column of the*

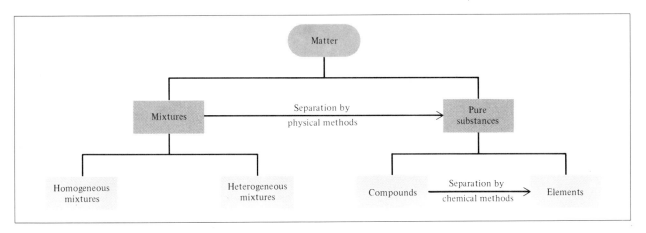

FIGURE 1.3 *Classification of matter and relationships between mixtures and pure substances and between compounds and elements.*

	1A																	8A
1	1 H	2A											3A	4A	5A	6A	7A	2 He
2	3 Li	4 Be											5 B	6 C	7 N	8 O	9 F	10 Ne
3	11 Na	12 Mg	3B	4B	5B	6B	7B	8B		1B	2B		13 Al	14 Si	15 P	16 S	17 Cl	18 Ar
4	19 K	20 Ca	21 Sc	22 Ti	23 V	24 Cr	25 Mn	26 Fe	27 Co	28 Ni	29 Cu	30 Zn	31 Ga	32 Ge	33 As	34 Se	35 Br	36 Kr
5	37 Rb	38 Sr	39 Y	40 Zr	41 Nb	42 Mo	43 Tc	44 Ru	45 Rh	46 Pd	47 Ag	48 Cd	49 In	50 Sn	51 Sb	52 Te	53 I	54 Xe
6	55 Cs	56 Ba	57 La	72 Hf	73 Ta	74 W	75 Re	76 Os	77 Ir	78 Pt	79 Au	80 Hg	81 Tl	82 Pb	83 Bi	84 Po	85 At	86 Rn
7	87 Fr	88 Ra	89 Ac	104 Unq	105 Unp	106 Unh	107 Uns	108 Uno	109 Une									

Period

58 Ce	59 Pr	60 Nd	61 Pm	62 Sm	63 Eu	64 Gd	65 Tb	66 Dy	67 Ho	68 Er	69 Tm	70 Yb	71 Lu
90 Th	91 Pa	92 U	93 Np	94 Pu	95 Am	96 Cm	97 Bk	98 Cf	99 Es	100 Fm	101 Md	102 No	103 Lr

Metals

Metalloids

Nonmetals

FIGURE 1.4 *The positions of metals, nonmetals, and metalloids in a modern version of the periodic table. With the exception of hydrogen (H), nonmetals appear at the far right of the table. The elements are arranged according to the numbers above their symbols. These numbers are based on a fundamental property of the elements which will be discussed in Chapter 7.*

periodic table are known as a **group** or **family.** *Each horizontal row of the periodic table* is called a **period.**

The elements can be divided into three categories—metals, nonmetals, and metalloids. A **metal** is *a good conductor of heat and electricity.* With the exception of mercury (which is a liquid), all metals are solids at room temperature (usually designated as 25°C). A **nonmetal** is *usually a poor conductor of heat and electricity, and it has more varied physical properties than a metal.* A **metalloid** *has properties that fall between those of metals and nonmetals.*

For convenience, some element groups have special names. *The Group 1A elements (Li, Na, K, Rb, Cs, and Fr)* are called **alkali metals,** and *the Group 2A elements (Be, Mg, Ca, Sr, Ba, and Ra)* are called **alkaline earth metals.** *Elements in Group 7A (F, Cl, Br, I, and At)* are known as **halogens,** and *those in Group 8A (He, Ne, Ar, Kr, Xe, and Rn)* are called **noble gases** (or **rare gases**). The names of other groups or families will be introduced later.

Looking at Figure 1.4 we see that two horizontal rows of elements are set aside at the bottom of the chart. Actually, cerium (Ce) should come right after lanthanum (La), and thorium (Th) should come right after actinium (Ac). These two rows of metals are set aside from the main body of the periodic table simply to avoid making the table too wide.

Figure 1.4 shows that the majority of known elements are metals, only seventeen elements are nonmetals, and eight elements are metalloids. The characteristic shape of the periodic table was originally devised to place elements with related physical and chemical properties near each other. Elements belonging to the same group resemble one another in chemical properties. In moving across any period from left to right, the physical and chemical properties of the elements change gradually from metallic to nonmetallic.

The periodic table is the single most important and useful source of information regarding the elements in the study of chemistry. It correlates the chemical behavior of the elements in a systematic way and helps us remember and understand many facts. Thus a major mission of this book is to demonstrate the utility and power of the periodic table in regularizing and simplifying your chemistry learning. Interestingly, the first periodic table was devised by the Russian chemist Dimitri Mendeleev (1836–1907) to aid students' learning in the general chemistry textbook he wrote over a century ago.

1.5   THE MEANING OF MEASUREMENT

As mentioned earlier, chemistry is largely an experimental science and deals with things that can be measured. The measurements we make are often used in calculations to obtain other related quantities. Our ability to measure properties depends to a large degree on the current state of technology. For instance, radiant energy from the sun could not be measured before the invention of devices that could detect it. The scope of chemistry continually expands as new instruments increase the range and precision of possible measurements.

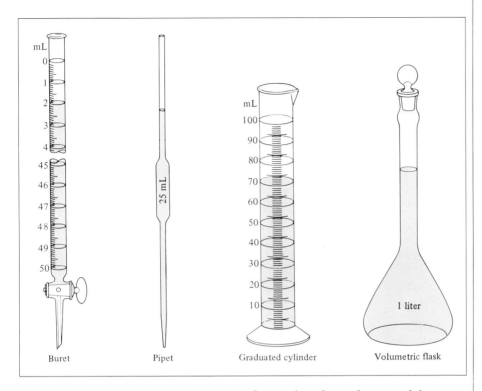

FIGURE 1.5   *Some common measuring devices found in a chemistry laboratory. These devices are not drawn to scale relative to one another. We will discuss the uses of these measuring devices in Chapter 4.*

A number of common devices available in the laboratory enable us to make simple measurements. For measuring length, there is the meter stick or scale; for volume, there are the buret, the pipet, the graduated cylinder, and the volumetric flask (Figure 1.5); for mass, there is the balance; for temperature, there is the thermometer.

Now suppose we want to know the mass of a single iron atom. Since atoms are extremely small, we cannot handle individual atoms nor do we have balances sensitive enough to weigh lone atoms. We must therefore take an indirect approach. We can, for example, measure the mass of a piece of iron metal. If we know the total number of iron atoms present in the metal, we can calculate the mass of one iron atom as follows:

$$\text{mass of one iron atom} = \frac{\text{mass of iron metal sample}}{\text{total number of iron atoms present}}$$

This approach is indeed plausible, as we shall see in the next chapter.

When we speak of measurements, we may be referring either to a **macroscopic property,** which is *determined directly,* or to a **microscopic property** on the atomic or molecular scale, which must be *determined by less direct methods.* In most cases a measured quantity is written as a number with an appropriate unit. To say that the distance between New York and San Francisco by car along a certain route is 5166 is meaningless. We must say that the distance is 5166 kilometers. The same is true in chemistry; units are important in stating a measurement correctly.

TABLE 1.2 SI Base Units

Base Quantity	Name of Unit	Symbol
Length	Meter	m
Mass	Kilogram	kg
Time	Second	s
Electrical current	Ampere	A
Temperature	Kelvin	K
Amount of substance	Mole	mol
Luminous intensity	Candela	cd

1.6 UNITS OF MEASUREMENT

For many years, the units used in science (including chemistry) were, in general, *metric units,* which were developed in France in the eighteenth century. Metric units are related decimally, that is, by powers of 10. This relationship is usually indicated by a prefix attached to the unit.

SI Units

In 1960 the General Conference of Weights and Measures, the international authority on units, proposed a revised and modernized metric system called the *International System of Units* (abbreviated *SI* from the French Le Système International d'Unités). Table 1.2 shows the seven SI base units. As we shall see shortly, all other units needed for measurement can be derived from these base units. Like metric units, SI units are modified in decimal fashion by a series of prefixes like those shown in Table 1.3. Most of these prefixes will be used often in this book, and you should become familiar with their meanings.

Length. The SI base unit of length is the *meter (m).* Other common units used are centimeter (cm) and decimeter (dm):

$$1 \text{ cm} = 0.01 \text{ m} = 1 \times 10^{-2} \text{ m}$$
$$1 \text{ dm} = 0.1 \text{ m} = 1 \times 10^{-1} \text{ m}$$

TABLE 1.3 Some Prefixes Used with SI Units

Prefix	Symbol	Meaning	Example
Tera-	T	1,000,000,000,000, or 10^{12}	1 terameter (Tm) $= 1 \times 10^{12}$ m
Giga-	G	1,000,000,000, or 10^9	1 gigameter (Gm) $= 1 \times 10^9$ m
Mega-	M	1,000,000, or 10^6	1 megameter (Mm) $= 1 \times 10^6$ m
Kilo-	k	1,000, or 10^3	1 kilometer (km) $= 1 \times 10^3$ m
Deci-	d	1/10, or 10^{-1}	1 decimeter (dm) $= 0.1$ m
Centi-	c	1/100, or 10^{-2}	1 centimeter (cm) $= 0.01$ m
Milli-	m	1/1,000, or 10^{-3}	1 millimeter (mm) $= 0.001$ m
Micro-	μ	1/1,000,000, or 10^{-6}	1 micrometer (μm) $= 1 \times 10^{-6}$ m
Nano-	n	1/1,000,000,000, or 10^{-9}	1 nanometer (nm) $= 1 \times 10^{-9}$ m
Pico-	p	1/1,000,000,000,000, or 10^{-12}	1 picometer (pm) $= 1 \times 10^{-12}$ m

Mass. The SI base unit of mass is the *kilogram* (*kg*). In chemistry, the most frequently used unit for mass is the gram (g):

$$1 \text{ kg} = 1000 \text{ g} = 1 \times 10^3 \text{ g}$$

Temperature. The SI base unit of temperature is the *kelvin* (*K*). (Note the absence of "degree" in the name and in the symbol.) The degree Celsius (°C), formerly termed the degree centigrade, is also allowed with SI. We will consider the relationship of the Celsius temperature scale to the nonmetric Fahrenheit scale later in this chapter. The Kelvin scale will play a major role when we study gas behavior and thermodynamics.

Time. The SI base unit of time is the *second* (*s*). Greater time intervals can be expressed either with suitable prefixes, such as in kiloseconds (ks), or with the familiar units called minutes (min) and hours (h).

Of the remaining base units, the mole (mol) and the ampere (A) play important measurement roles in chemistry. They will be defined and explained in context later in this book. The seventh unit, the candela (cd), is not involved in the development of general chemistry.

Derived SI Units

As mentioned, there are a number of quantities whose units can be derived from the SI base units. For example, from the base unit of length, we can define volume, and from the base units of length, mass, and time, we can define energy. Following are some of the quantities frequently encountered in the study of chemistry.

Volume. Since *volume* is length cubed, its SI derived unit is m^3. Related units are the cubic centimeter (cm^3) and the cubic decimeter (dm^3):

$$1 \text{ cm}^3 = (1 \times 10^{-2} \text{ m})^3 = 1 \times 10^{-6} \text{ m}^3$$
$$1 \text{ dm}^3 = (1 \times 10^{-1} \text{ m})^3 = 1 \times 10^{-3} \text{ m}^3$$

Another common (but non-SI) unit of volume is the liter (L). A **liter** is *the volume occupied by 1 cubic decimeter.* One liter of volume is equal to 1000 milliliters (mL), and one milliliter of volume is equal to one cubic centimeter:

$$1 \text{ L} = 1 \text{ dm}^3$$
$$1 \text{ mL} = 1 \times 10^{-3} \text{ L}$$
$$1 \text{ L} = 1000 \text{ mL}$$
$$1 \text{ mL} = 1 \text{ cm}^3$$

Figure 1.6 compares the relative sizes of two volumes. This book mainly uses the units L and mL for volume.

Velocity and Acceleration. By definition, *velocity* is the change in distance with time, that is

$$\text{velocity} = \frac{\text{change in distance}}{\text{time}}$$

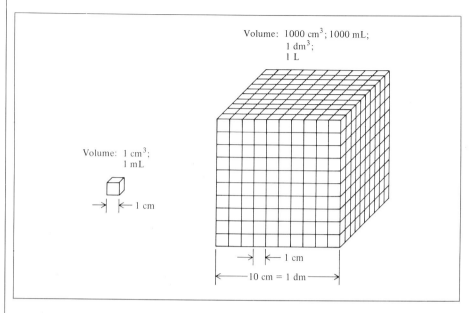

Volume: 1000 cm^3; 1000 mL;
1 dm^3;
1 L

Volume: 1 cm^3;
1 mL

1 cm

1 cm

10 cm = 1 dm

Acceleration is the change in velocity with time, that is

$$\text{acceleration} = \frac{\text{change in velocity}}{\text{time}}$$

Therefore, velocity has the units of m/s (or cm/s), and acceleration has the units of m/s^2 (or cm/s^2). Although velocity itself does not have much relevance in chemistry, it is needed to define acceleration, which in turn is required to define force and hence energy. Both force and energy play important roles in many areas in chemistry.

Force.   According to Newton's second law of motion (Sir Isaac Newton, 1642–1726)

$$\text{force} = \text{mass} \times \text{acceleration}$$

In common language, a force is often regarded as a push or a pull. Chemistry is concerned mainly with electrical forces that exist among atoms and molecules. The nature of these forces will be explored in later chapters. *The derived SI unit for force is the* **newton (N),** where

$$1 \text{ N} = 1 \text{ kg m/s}^2$$

Pressure.   *Pressure* is defined as *force applied per unit area,* that is

$$\text{pressure} = \frac{\text{force}}{\text{area}}$$

The force experienced by any area exposed to Earth's atmosphere is equal to *the weight of the column of air above it. The pressure exerted by this column of air is called* **atmospheric pressure.** The actual value of atmospheric pressure depends on location, temperature, and weather conditions. A common reference pressure of *one atmosphere* (1 atm) represents the atmospheric pressure exerted by a column of dry air at sea level and 0°C. The derived SI unit of pressure is obtained by applying the derived force unit of one newton over one square meter, which is the derived unit of area. *A pressure of one newton per square meter (1 N/m²)* is called a **pascal** (**Pa**). One atmosphere is now defined by the exact relation

$$1 \text{ atm} = 101,325 \text{ Pa}$$
$$= 101.325 \text{ kPa}$$

Energy.   ***Energy*** is *the capacity to do work or to produce change.* In chemistry we are mainly interested in energy effects that result from chemical or physical changes. In mechanics, work is defined as "force × distance." Since energy can be measured as work, we can write

$$\text{energy} = \text{work done}$$
$$= \text{force} \times \text{distance}$$

Thus *the SI derived unit of energy has the units of newtons × meters* (*N m*) or kg m²/s². This SI derived energy unit is more commonly called a **joule** (***J***):

$$1 \text{ J} = 1 \text{ kg m}^2/\text{s}^2$$
$$= 1 \text{ N m}$$

Sometimes energy is expressed in kilojoules (kJ):

$$1 \text{ kJ} = 1000 \text{ J}$$

Until recently, chemists expressed energy in *calories* (*cal*). The calorie is defined by the relationship

$$1 \text{ cal} = 4.184 \text{ J}$$

In this book we will mainly use joules and kilojoules as the units for energy.

Density.   The ***density*** of an object is *the mass of the object divided by its volume:*

$$\text{density} = \frac{\text{mass}}{\text{volume}}$$

or

$$d = \frac{m}{V}$$

where d and V denote density and volume, respectively. Note that the density of a given material does not depend on the quantity of mass present. This is because V increases as m does, so that the ratio of the two quantities always remains the same for the given material. Thus density is an intensive quantity.

The SI derived unit for density is the kilogram per cubic meter (kg/m³).

This unit is awkwardly large for most chemical applications. The g/cm^3 unit and its equivalent, the g/mL, are more commonly used for solid and liquid densities. Because gas densities are often very low, we use the units of g/L:

$$1 \text{ g/cm}^3 = 1 \text{ g/mL} = 1000 \text{ kg/m}^3$$
$$1 \text{ g/L} = 0.001 \text{ g/mL}$$

EXAMPLE 1.1

An iron bar has a mass of 64.2 g and its volume is 8.16 cm³. What is the density of the element iron (Fe) based on these data?

Answer

The density of the iron is given by

$$d = \frac{m}{V}$$
$$= \frac{64.2 \text{ g}}{8.16 \text{ cm}^3}$$
$$= 7.87 \text{ g/cm}^3$$

Similar example: Problem 1.23.

EXAMPLE 1.2

The density of a certain soft drink is 1.16 g/mL. Calculate the mass of 28.8 mL of the soft drink.

Answer

The mass of the soft drink is given by

$$m = d \times V$$
$$= (1.16 \text{ g/mL})(28.8 \text{ mL})$$
$$= 33.4 \text{ g}$$

Similar example: Problem 1.22.

Temperature Scales

Three temperature scales are currently in use. Their units are K (kelvin), °C (degree Celsius), and °F (degree Fahrenheit). The Fahrenheit scale defines the normal freezing and boiling points of water to be exactly 32°F and 212°F, respectively. The Celsius scale divides the range between the freezing point (0°C) and boiling point (100°C) of water into 100 degrees (Figure 1.7). The size of a degree on the Fahrenheit scale is only 100/180, or 5/9 times that on the Celsius scale. To convert degrees Fahrenheit to degrees Celsius, we write

$$?°C = (°F - 32°F) \times \frac{5°C}{9°F}$$

To convert degrees Celsius to degrees Fahrenheit, we write

$$?°F = \frac{9°F}{5°C} \times (°C) + 32°F$$

The kelvin temperature scale is discussed in Chapter 6.

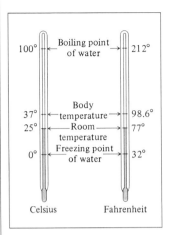

FIGURE 1.7 *Comparison of the Celsius and Fahrenheit temperature scales. Note that there are 100 divisions, or 100 degrees, between the freezing point and boiling point of water on the Celsius scale, and there are 180 divisions, or 180 degrees, between the same two temperature limits on the Fahrenheit scale. The Celsius scale was formerly called the centigrade scale.*

EXAMPLE 1.3

Convert (a) 37.0°C into °F and (b) 65°F into °C.

Answer

(a) The normal body temperature is 37.0°C or

$$\frac{9°F}{5°C} \times 37.0°C + 32°F = 98.6°F$$

(b) Here we have

$$(65°F - 32°F) \times \frac{5°C}{9°F} = 18°C$$

Similar examples: Problems 1.25, 1.26.

1.7 HANDLING NUMBERS

Having surveyed some of the units used in chemistry, we are now ready to take a closer look at the techniques of handling numbers. We are concerned with two particular techniques here: scientific notation and significant figures.

Scientific Notation

In chemistry, we often deal with numbers that are either extremely large or extremely small. For example, in one gram of the element hydrogen there are roughly

602,200,000,000,000,000,000,000

hydrogen atoms. Each hydrogen atom has a mass of only

0.00000000000000000000000166 g

These numbers are cumbersome to handle, and it is easy to make mistakes when using them in arithmetic computations. Consider the following multiplication:

0.0000000056 × 0.00000000048 = 0.000000000000000002688

It would be easy for us to miss one zero or add one more zero after the decimal point. To handle these large and small numbers, we use a system called *scientific notation*. Regardless of their magnitude, all numbers can be expressed in the form

$$N \times 10^n$$

where N is a number between 1 and 10 and n is an exponent that can be a positive or negative *integer* (whole number). Any number expressed in this way is said to be written in scientific notation.

Suppose we are given a certain number and asked to express it in scientific notation. Basically, this amounts to finding n. We count the number of places that the decimal point must be moved to give the number N (which is between 1 and 10). If the decimal point has to be moved to the left, then n is a positive integer; if it has to be moved to the right, then n is a negative integer. The following examples illustrate the use of scientific notation:

(a) Express 568.762 in scientific notation:

$$568.762 = 5.68762 \times 10^2$$

Note that the decimal point is moved to the left by two places and $n = 2$.

(b) Express 0.00000772 in scientific notation:

$$0.00000772 = 7.72 \times 10^{-6}$$

Note that the decimal point is moved to the right by six places and $n = -6$.

Next, we consider how scientific notation is handled in arithmetic operations.

Addition and Subtraction. In addition and subtraction, each quantity must first be written with the same exponent n. Then the addition or subtraction is performed on the N parts of the numbers; the exponent parts remain the same. Consider the following examples:

$$(7.4 \times 10^3) + (2.1 \times 10^3) = 9.5 \times 10^3$$

$$(4.31 \times 10^4) + (3.9 \times 10^3) = (4.31 \times 10^4 + 0.39 \times 10^4)$$
$$= 4.70 \times 10^4$$

$$(2.22 \times 10^{-2}) - (4.10 \times 10^{-3}) = (2.22 \times 10^{-2}) - (0.41 \times 10^{-2})$$
$$= 1.81 \times 10^{-2}$$

Multiplication and Division. In multiplication, the integers expressed as exponents are added; in division, the integers expressed as exponents are subtracted. Consider the following examples:

$$(8.0 \times 10^4) \times (5.0 \times 10^2) = (8.0 \times 5.0)(10^{4+2})$$
$$= 40 \times 10^6$$
$$= 4.0 \times 10^7$$

$$(4.0 \times 10^{-5}) \times (7.0 \times 10^3) = (4.0 \times 7.0)(10^{-5+3})$$
$$= 28 \times 10^{-2}$$
$$= 2.8 \times 10^{-1}$$

$$\frac{4.5 \times 10^7}{3.0 \times 10^5} = \left(\frac{4.5}{3.0}\right) \times 10^{7-5}$$
$$= 1.5 \times 10^2$$

$$\frac{2.0 \times 10^{-8}}{5.0 \times 10^9} = \left(\frac{2.0}{5.0}\right) \times 10^{-8-9}$$
$$= 0.40 \times 10^{-17}$$
$$= 4.0 \times 10^{-18}$$

Significant Figures

Except where integers are involved (for example, in counting the number of students in a class), it is often impossible to obtain the exact value of the quantity under investigation. For this reason, it is important to indicate the margin of error in a measurement by clearly indicating the **significant figures**, which are *the meaningful digits in a measured or calculated quantity*. When significant figures are counted, the last digit is understood to be uncertain. For example, we might measure the volume of a given amount of liquid using a graduated cylinder (see Figure 1.5) that is known to be in error by 1 mL. If the volume is found to be 6 mL, then the actual volume is in the range of 5 mL to 7 mL. We represent the volume of the liquid as (6 ± 1) mL. In this case, there is only one significant figure (the digit 6) that is uncertain by either plus or minus 1 mL. As an improvement, we might use a graduated cylinder that has finer divisions, so that the volume we measure is now uncertain by only 0.1 mL. If the volume of the liquid is now found to be 6.0 mL, we may express the quantity as (6.0 ± 0.1) mL, and the actual value is somewhere between 5.9 mL and 6.1 mL. We can further improve the measuring device and obtain more significant figures. In each case, the last digit is always uncertain; the amount of this uncertainty depends on the particular measuring device we use.

Guidelines for Using Significant Figures.   The previous discussion shows that we must always be careful in scientific work to write the proper number of significant figures. In general, it is fairly easy to determine how many significant figures are present in a number by following these rules:

- Any digit that is not zero is significant. Thus 845 cm has three significant figures, 1.234 kg has four significant figures, and so on.
- Zeros between nonzero digits are significant. Thus 606 m contains three significant figures, 40,501 J contains five significant figures, and so on.
- Zeros to the left of the first nonzero digit are not significant. These zeros are used to indicate the placement of the decimal point. Thus 0.08 L contains one significant figure, 0.0000349 g contains three significant figures, and so on.
- If a number is greater than 1, then all the zeros written to the right of the decimal point count as significant figures. Thus 2.0 mg has two significant figures, 40.062 mL has five significant figures, 3.040 dm has four significant figures. If a number is less than 1, then only the zeros that are at the end of the number and the zeros that are between

nonzero digits are significant. Thus 0.090 kg has two significant figures, 0.3005 J has four significant figures, 0.00420 min has three significant figures, and so on.

- For numbers that do not contain decimal points, the trailing zeros may or may not be significant. Thus 400 cm may have one significant figure (the digit 4), two significant figures (40), or three significant figures (400). We cannot know which is correct without more information. By using scientific notation, however, we avoid this ambiguity. In this particular case, we can express the number 400 as 4×10^2 for one significant figure, 4.0×10^2 for two significant figures, or 4.00×10^2 for three significant figures.

EXAMPLE 1.4

Determine the number of significant figures in the following measured quantities: (a) 478 cm, (b) 6.01 g, (c) 0.825 m, (d) 0.043 kg, (e) 1.310×10^{22} atoms, and (f) 7000 mL.

Answer

(a) Three. (b) Three. (c) Three. (d) Two. (e) Four. (f) This is an ambiguous case. The number of significant figures may be four (7.000×10^3), three (7.00×10^3), two (7.0×10^3), or one (7×10^3).

Similar examples: Problems 1.31, 1.32.

Our next step is to see how significant figures are handled in calculations. We can adhere to the following rules:

- In addition and subtraction, the number of significant figures to the right of the decimal point in the final sum or difference is determined by the lowest number of significant figures to the right of the decimal point in any of the original numbers. Consider the examples:

$$\begin{array}{r} 89.332 \\ + 1.1 \\ \hline 90.432 \end{array}$$ ⟵ one significant figure after the decimal point

round off to 90.4

$$\begin{array}{r} 2.097 \\ -0.12 \\ \hline 1.977 \end{array}$$ ⟵ two significant figures after the decimal point

round off to 1.98

The rounding off procedure is as follows. If we wish to round off a number at a certain point, we simply drop the digits that follow if the first of them is less than 5. Thus 8.724 rounds off to 8.72 if we want only two figures after the decimal point. If the first digit following the point of round off is equal to or greater than 5, we add 1 to the preceding digit. Thus 8.727 rounds off to 8.73 and 0.425 rounds off to 0.43.

- In multiplication and division, the number of significant figures in the final product or quotient is determined by the original number that has the smallest number of significant figures. The following examples illustrate this rule:

$$2.8 \times 4.5039 = 12.61092 \longleftarrow \text{round off to 13}$$

$$\frac{6.85}{112.04} = 0.0611388789 \longleftarrow \text{round off to 0.0611}$$

- Keep in mind that *exact numbers* obtained from definitions or by counting numbers of objects can be considered to have an infinite number of significant figures. If an object has the mass 0.2786 g, then the mass of eight such objects is

$$0.2786 \text{ g} \times 8 = 2.229 \text{ g}$$

We do *not* round off this product to one significant figure, because the number 8 is actually 8.00000 . . . , by definition. Similarly, to take the average of the two measured lengths 6.64 cm and 6.68 cm, we write

$$\frac{6.64 \text{ cm} + 6.68 \text{ cm}}{2} = 6.66 \text{ cm}$$

because the number 2 is actually 2.00000 . . . , by definition.

EXAMPLE 1.5

Carry out the following arithmetic operations: (a) 11,254.1 + 0.1983, (b) 66.59 − 3.113, (c) 8.16 × 5.1355, and (d) 0.0154/883.

Answer

(a)
$$\begin{array}{r} 11,254.1 \\ + \quad 0.1983 \\ \hline 11,254.2983 \end{array} \longleftarrow \text{round off to 11,254.3}$$

(b)
$$\begin{array}{r} 66.59 \\ - \quad 3.113 \\ \hline 63.477 \end{array} \longleftarrow \text{round off to 63.48}$$

(c) $8.16 \times 5.1355 = 41.90568 \longleftarrow$ round off to 41.9

(d) $\dfrac{0.0154}{883} = 0.0000174405436 \longleftarrow$ round off to 0.0000174, or 1.74×10^{-5}

Similar example: Problem 1.33.

Accuracy and Precision. In discussing measurements and significant figures it is useful to distinguish two terms: *accuracy* and *precision*. **Ac-curacy** *tells us how close a measurement is to the true value of the quantity that was measured.* **Precision** *refers to how closely two or more measurements of the same quantity agree with one another.* Suppose three students

accuracy = closeness of measurement

Precision — how closely two or measurements of the same quantity agree.

are asked to determine the mass of a piece of copper wire of mass 2.000 g. The results of two successive weighings by each student are

	Student A	Student B	Student C
	1.964 g	1.972 g	2.000 g
	1.978 g	1.968 g	2.002 g
Average value	1.971 g	1.970 g	2.001 g

Student B's results are more *precise* than those of Student A (1.972 g and 1.968 g deviate less from 1.970 g than 1.964 g and 1.978 g from 1.971 g). Neither set of results is very *accurate*, however. Student C's results are not only *precise* but also the most *accurate*, since the average value is closest to the true value. Highly accurate measurements are usually precise too. On the other hand, highly precise measurements do not necessarily guarantee accurate results. For example, an improperly calibrated meter stick or a faulty balance may give precise readings that are in error.

1.8   THE FACTOR-LABEL METHOD OF SOLVING PROBLEMS

The procedure we will use in solving problems is called the *factor-label method* (also called the *dimensional analysis method*). A simple and powerful technique that requires little memorization, the factor-label method is based on the relationship between different units that express the same physical quantity. The following example illustrates this method. Suppose we want to convert a certain measured length, say 57.8 meters, into centimeters. This problem may be expressed as

$$? \text{ cm} = 57.8 \text{ m}$$

By definition

$$1 \text{ cm} = 1 \times 10^{-2} \text{ m}$$

Dividing both sides by 1 cm, we obtain

$$\frac{1 \text{ cm}}{1 \text{ cm}} = \frac{1 \times 10^{-2} \text{ m}}{1 \text{ cm}} = 1$$

This ratio can be read as 1×10^{-2} m per centimeter. The factor (fraction) 1×10^{-2} m/1 cm is a *unit factor* (equal to 1) because the numerator and denominator describe the same length. On the other hand, if we divide the equation 1 cm = 1×10^{-2} m by 1×10^{-2} m we obtain

$$\frac{1 \text{ cm}}{1 \times 10^{-2} \text{ m}} = \frac{1 \times 10^{-2} \text{ m}}{1 \times 10^{-2} \text{ m}} = 1$$

This ratio reads as 1 cm per 1×10^{-2} m, and the factor 1 cm/1×10^{-2} m is also a unit factor. We see that the reciprocal of any unit factor is also a unit factor. Stated another way, dividing one by one always yields one. Multiplication of a quantity by a unit factor will therefore change only its

unit, not its value. In our example we proceed using the unit factor that
will give us the answer in centimeters, as was desired:

$$? \text{ cm} = 57.8 \text{ m} \times \frac{1 \text{ cm}}{1 \times 10^{-2} \text{ m}}$$
$$= 5780 \text{ cm}$$
$$= 5.78 \times 10^3 \text{ cm}$$

In the calculation, the unit m, present in both the numerator and denominator, cancels out, leaving the desired unit cm. Scientific notation is used
to indicate that the number has three significant figures. Note that the unit
factor 1 cm/1 $\times$ 10^{-2} m contains exact numbers; therefore, it does not
affect the number of significant figures.

The advantage of the factor-label method is that if the equation is set up
correctly, then all the units will cancel except for the desired one in the
final answer. If this is not the case, then an error must have been made
somewhere, and it can usually be spotted by inspection.

EXAMPLE 1.6

Convert 5.6 dm to meters.

Answer

The problem is

$$? \text{ m} = 5.6 \text{ dm}$$

By definition

$$1 \text{ dm} = 1 \times 10^{-1} \text{ m}$$

The unit factor is

$$\frac{1 \times 10^{-1} \text{ m}}{1 \text{ dm}} = 1$$

Therefore we write

$$? \text{ m} = 5.6 \text{ dm} \times \frac{1 \times 10^{-1} \text{ m}}{1 \text{ dm}} = 0.56 \text{ m}$$

EXAMPLE 1.7

A person weighs 162 pounds (lb). What is his mass in milligrams (mg)?

Answer

The problem can be expressed as

$$? \text{ mg} = 162 \text{ lb}$$

(Continued)

The conversion factors are

$$1 \text{ lb} = 453.6 \text{ g}$$

or

$$\frac{453.6 \text{ g}}{1 \text{ lb}} = 1$$

and

$$1 \text{ mg} = 1 \times 10^{-3} \text{ g}$$

so that

$$\frac{1 \text{ mg}}{1 \times 10^{-3} \text{ g}} = 1$$

Thus

$$? \text{ mg} = 162 \text{ lb} \times \frac{453.6 \text{ g}}{1 \text{ lb}} \times \frac{1 \text{ mg}}{1 \times 10^{-3} \text{ g}} = 7.35 \times 10^7 \text{ mg}$$

Similar examples: Problems 1.40(c), 1.40(d).

Note that unit factors may be squared or cubed because $1^2 = 1^3 = 1$. The use of such factors is illustrated with the following examples.

EXAMPLE 1.8

Calculate the number of cubic centimeters in 6.2 m³.

Answer

The problem can be stated as

$$? \text{ cm}^3 = 6.2 \text{ m}^3$$

By definition

$$1 \text{ cm} = 1 \times 10^{-2} \text{ m}$$

The unit factor is

$$\frac{1 \text{ cm}}{1 \times 10^{-2} \text{ m}} = 1$$

It follows that

$$\left(\frac{1 \text{ cm}}{1 \times 10^{-2} \text{ m}}\right)^3 = 1^3 = 1$$

Therefore we write

$$? \text{ cm}^3 = 6.2 \text{ m}^3 \times \left(\frac{1 \text{ cm}}{1 \times 10^{-2} \text{ m}}\right)^3 = 6{,}200{,}000 \text{ cm}^3$$
$$= 6.2 \times 10^6 \text{ cm}^3$$

Similar example: Problem 1.40(i).

EXAMPLE 1.9

The density of gold is 19.3 g/cm^3. Convert the density to units of kg/m^3.

Answer

The problem can be stated as

$$? \text{ kg/m}^3 = 19.3 \text{ g/cm}^3$$

We need two conversion factors. First

$$1 \text{ kg} = 1000 \text{ g}$$
$$\frac{1 \text{ kg}}{1000 \text{ g}} = 1$$

Second, from Example 1.8

$$\left(\frac{1 \text{ cm}}{1 \times 10^{-2} \text{ m}}\right)^3 = 1$$

Thus we write

$$? \text{ kg/m}^3 = \frac{19.3 \text{ g}}{1 \text{ cm}^3} \times \frac{1 \text{ kg}}{1000 \text{ g}} \times \left(\frac{1 \text{ cm}}{1 \times 10^{-2} \text{ m}}\right)^3 = 19,300 \text{ kg/m}^3$$
$$= 1.93 \times 10^4 \text{ kg/m}^3$$

Similar example: Problem 1.35.

AN ASIDE ON THE PERIODIC TABLE
Distribution of Elements in Earth's Crust and in Living Systems

Before we embark on the study of the elements, let's first look at how these elements are distributed in Earth's crust and which of them are essential to living systems.

By Earth's crust, we mean the layer measured from the surface of Earth to a depth of about 40 km (about 25 miles). Because of technical difficulties, scientists have not been able to study the inner portions of Earth as easily as the crust. It is believed that there is a solid core made up mostly of iron and nickel at the center of Earth. Surrounding the core is a very hot fluid called the *mantle,* which is made up of iron, carbon, silicon, and sulfur (Figure 1.8).

Of the eighty-three elements that occur naturally, twelve of them make up 99.7 percent of Earth's crust by mass. They are, in decreasing order of natural abundance, oxygen (O), silicon (Si), aluminum (Al), iron (Fe), calcium (Ca), magnesium (Mg), sodium (Na), potassium (K), titanium (Ti), hydrogen (H), phosphorus (P), and manganese (Mn) (Figure 1.9). In discussing the natural abundance of the elements, we should keep in mind that (1) the elements are not evenly distributed throughout Earth's crust and (2) most elements occur in combined forms. These features are the basis of methods used to obtain pure elements from their compounds, as we shall see in later chapters.

Only twenty-six elements are believed to be essential to living systems, shown in Figure

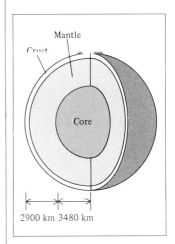

FIGURE 1.8 Structure of Earth's interior.

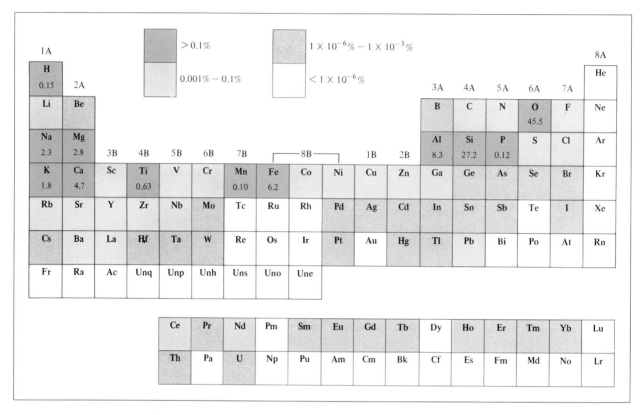

FIGURE 1.9 Natural abundance of the elements according to their positions in the periodic table in percent by mass. For example, oxygen's abundance is 45.5 percent. This means that in a 100 g of Earth's crust there are, on the average, 45.5 g of element oxygen. Percentages are shown only for the twelve most abundant elements.

1.10. By comparing Figure 1.10 with Figure 1.9, you can see that many of the abundant elements in living systems are also among the most abundant elements in Earth's crust. Table 1.4 lists the essential elements in the human body.

TABLE 1.4   Essential Elements in the Human Body

Element	Percent by Mass*	Element	Percent by Mass*
Oxygen	65	Sodium	0.1
Carbon	18	Magnesium	0.05
Hydrogen	10	Iron	<0.05
Nitrogen	3	Cobalt	<0.05
Calcium	1.5	Copper	<0.05
Phosphorus	1.2	Zinc	<0.05
Potassium	0.2	Iodine	<0.05
Sulfur	0.2	Selenium	<0.01
Chlorine	0.2	Fluorine	<0.01

* *Percent by mass* gives the mass of the element present in grams in a 100 g sample.

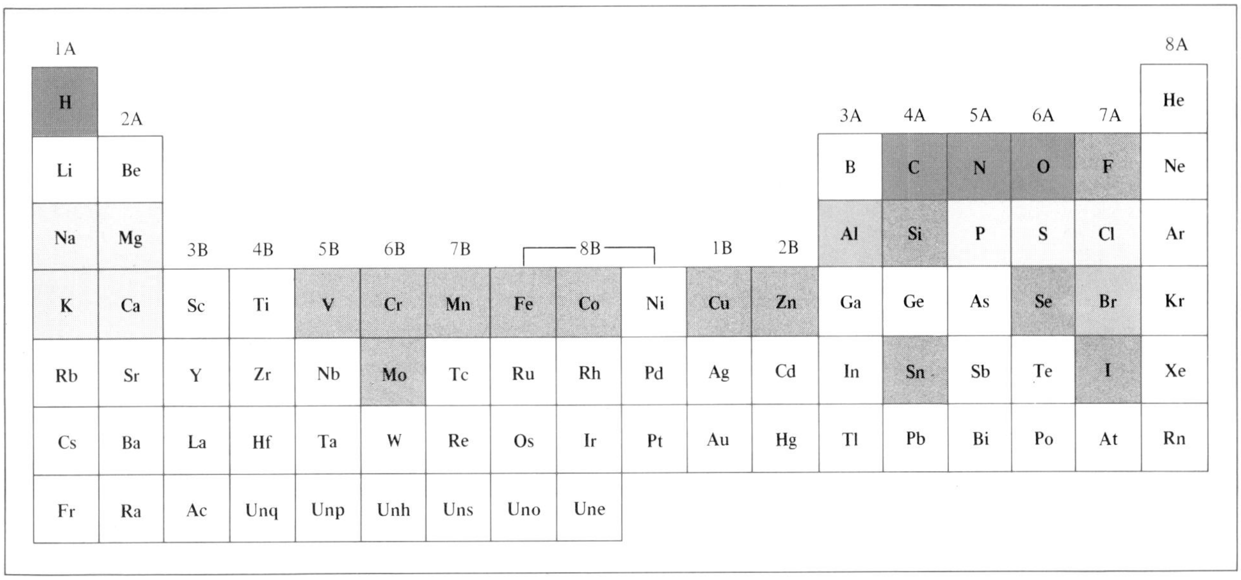

FIGURE 1.10 The twenty-six elements essential to living systems. The four most abundant elements (in dark color) are H (hydrogen), C (carbon), N (nitrogen), and oxygen (O). The seven next most abundant elements (light color) are Na (sodium), Mg (magnesium), P (phosphorus), S (sulfur), Cl (chlorine), K (potassium), and Ca (calcium). The elements needed only in trace amounts (gray color) are F (fluorine), Si (silicon), V (vanadium), Cr (chromium), Mn (manganese), Fe (iron), Co (cobalt), Cu (copper), Zn (zinc), Se (selenium), Mo (molybdenum), Sn (tin), and I (iodine). The specific roles of Al (aluminum) and Br (bromine) in living systems have not been established (medium color).

SUMMARY

1. The scientific method involves gathering information by making observations and measurements. In the process, hypotheses, laws, and theories are devised and tested.
2. The study of chemistry involves three basic steps: observation, representation, and interpretation. Observations refer to measurements in the macroscopic world; representations involve the use of special symbols and equations for communication; interpretations are based on atoms and molecules, which belong to the microscopic world.
3. Substances have unique physical properties that can be observed without changing the identity of the substances, and unique chemical properties that, when they are demonstrated, change the identity of the substances.
4. The simplest substances in chemistry are elements. Compounds are formed by the combination of atoms of different elements.
5. Elements can be grouped together according to their related properties in a pattern called a periodic table. The periodic table is a central source of chemical information.
6. SI units are used to express physical quantities in all sciences, including chemistry.

7. Numbers expressed in scientific notation in the form $N \times 10^n$, where N is between 1 and 10 and n is a positive or negative integer, help us handle very large and very small quantities.
8. The degree of certainty in a measured number is indicated by expressing only the significant figures and by rounding off answers to calculations involving that value to correct numbers of significant figures.

KEY WORDS

Accuracy, p. 21
Alkali metals, p. 10
Alkaline earth metals, p. 10
Atmospheric pressure, p. 14
Chemical property, p. 6
Chemistry, p. 2
Compound, p. 8
Density, p. 15
Element, p. 7
Energy, p. 15
Extensive property, p. 7
Family (in periodic table), p. 10
Group (in periodic table), p. 10
Halogens, p. 10
Heterogeneous mixture, p. 6
Homogeneous mixture, p. 6
Hypothesis, p. 2
Intensive property, p. 7
Joule, p. 15
Law, p. 2
Liter, p. 13
Macroscopic property, p. 11

Mass, p. 5
Matter, p. 5
Metal, p. 10
Metalloid, p. 10
Microscopic property, p. 11
Mixture, p. 6
Newton, p. 14
Noble gases, p. 10
Nonmetal, p. 10
Pascal, p. 15
Period (in periodic table), p. 10
Periodic table, p. 8
Physical property, p. 6
Precision, p. 21
Pressure, p. 14
Qualitative, p. 2
Quantitative, p. 2
Rare gases, p. 10
Significant figures, p. 19
Substance, p. 5
Theory, p. 3
Weight, p. 5

PROBLEMS

More challenging problems are marked with an asterisk.

The Scientific Method

1.1 Define the following terms: (a) hypothesis, (b) law, (c) theory.
1.2 Referring to the dinosaur extinction story, answer the following questions. (a) What is the hypothesis? (b) What kind of data are collected to test the hypothesis? Are they quantitative or qualitative? (c) Why is the hypothesis now called the theory?
1.3 Classify the following as quantitative or qualitative statements, giving your reasons. (a) The sun is approximately 93 million miles from Earth.

(b) Leonardo da Vinci was a better painter than Michelangelo. (c) Lead is denser than copper. (d) Butter tastes better than margarine. (e) A stitch in time saves nine.
1.4 Classify each of the following statements as a hypothesis, a law, or a theory. (a) Beethoven's contribution to music would have been much greater if he had married. (b) An autumn leaf gravitates toward the ground because there is an attractive force between the leaf and Earth. (c) All matter is composed of very small particles called atoms.
1.5 Discuss whether the methods of investigation used by your favorite fictional detective are scientific or not.

Basic Definitions

1.6 Does each of the following describe a physical change or a chemical change? Explain your answers. (a) Black coffee is sweetened by adding sugar to it. (b) In a blast furnace iron ore is smelted with coke (a form of carbon) to produce cast iron. (c) Frozen orange juice is reconstituted by adding water to it. (d) A lake's surface freezes over in winter. (e) A flashlight beam slowly gets dimmer and finally goes out.

1.7 Do the following statements describe chemical or physical properties? (a) Mothballs repel not only moths, but sometimes people as well. (b) Water boils at 78°C on top of a 18,500 ft mountain. (c) The helium gas inside a balloon tends to leak out after a few hours. (d) Water is denser than ice.

*1.8 Give one quantitative and one qualitative statement about each of the following: (a) water, (b) carbon, (c) iron, (d) hydrogen gas, (e) sucrose (cane sugar), (f) table salt (sodium chloride), (g) mercury, (h) gold, (i) air, (j) sulfuric acid.

1.9 Give the names of the elements represented by the chemical symbols Li, F, Sc, Cu, As, Cd, Os, Pt, Bi, Xe, U.

1.10 Give the chemical symbols for the following elements: (a) chromium, (b) selenium, (c) krypton, (d) zirconium, (e) silver, (f) tungsten, (g) potassium, (h) sulfur.

1.11 Classify each of the following substances as an element, a compound, or a mixture: (a) seawater, (b) helium inside a balloon, (c) mud, (d) distilled water, (e) rainwater, (f) cement, (g) blue ink.

1.12 Classify each of the following, with a brief explanation, as an element, a compound, a homogeneous mixture, or a heterogeneous mixture: (a) lake water, (b) chlorine, (c) cast iron, (d) milk, (e) mercury, (f) baking soda, (g) wine, (h) brass.

1.13 Which of the following properties are intensive, which are extensive? (a) length, (b) area, (c) color, (d) temperature, (e) mass.

*1.14 How is it that an astronaut of mass 75 kg achieves a weightless condition when aboard a Skylab orbiting space laboratory?

The Periodic Table

1.15 Briefly explain the significance of the periodic table.

1.16 Give two differences between a metal and a nonmetal.

1.17 Write names and symbols of four elements in each of the following categories: nonmetal, metal, metalloid.

1.18 Define, with examples, the following terms: (a) alkali metals, (b) alkaline earth metals, (c) halogens, (d) noble gases.

1.19 The elements in Group 8A of the periodic table are called noble gases. Can you guess the meaning of "noble" in this context?

Units

1.20 Give the SI units for expressing the following: (a) length, (b) area, (c) volume, (d) mass, (e) time, (f) energy, (g) temperature, (h) force.

1.21 Write numbers for the following prefixes: (a) mega-, (b) kilo-, (c) deci-, (d) centi-, (e) milli-, (f) micro-, (g) nano-, (h) pico-.

1.22 The density of mercury is 13.6 g/cm³. How many grams of mercury will occupy a volume of 95.8 mL?

1.23 The density of lead is 11.4 g/cm³. What is the volume occupied by 1.20×10^3 g of lead?

*1.24 Consult a handbook of chemical and physical data (ask your instructor where you can locate a copy of the handbook) to find (a) two metals less dense than water, (b) two metals more dense than mercury, (c) the densest known solid element, (d) the least dense solid element, (e) the element with the highest melting point.

1.25 Calculate the temperature in °C of the following: (a) a hot summer day of 95°F, (b) a cold winter day of 12°F, (c) the healthy human body, 98.6°F, (d) a furnace at 1852°F.

1.26 (a) Normally our bodies can endure a temperature of 105°F for only short periods of time without permanent damage to the brain and other vital organs. What is that temperature in degrees Celsius? (b) Ethylene glycol is a liquid organic compound that is used as an antifreeze in car radiators. It freezes at -11.5°C. Calculate its freezing point in degrees Fahrenheit. (c) The temperature on the surface of our sun is about 6.3×10^3 °C. What is this temperature in degrees Fahrenheit?

*1.27 At what temperature does the numerical reading on a Celsius thermometer equal that on a Fahrenheit thermometer?

Scientific Notation

1.28 Express the following numbers in scientific notation: (a) 0.000000027, (b) 356, (c) 47,764, (d) 0.096.

1.29 Express the following numbers as decimals: (a) 1.52×10^{-2} and (b) 7.78×10^{-8}.

1.30 Express the answers to the following in scientific notation:

(a) $145.75 + (2.3 \times 10^{-1})$
(b) $79,500 \div (2.5 \times 10^2)$
(c) $(7.0 \times 10^{-3}) - (8.0 \times 10^{-4})$
(d) $(1.0 \times 10^4) \times (9.9 \times 10^6)$
(e) $0.0095 + (8.5 \times 10^{-3})$
(f) $653 \div (5.75 \times 10^{-8})$
(g) $850,000 - (9.0 \times 10^5)$
(h) $(3.6 \times 10^{-4}) \times (3.6 \times 10^6)$

Significant Figures

1.31 What is the number of significant figures in each of the following numbers? (a) 4867, (b) 25, (c) 50,002, (d) 5200, (e) 80.1, (f) 0.0000001, (g) 0.9, (h) 1.2×10^{17}.

1.32 How many significant figures are there in each of the following? (a) 0.009, (b) 0.0903, (c) 90.3, (d) 903.3, (e) 9.0×10^{-3}, (f) 9, (g) 90?

1.33 Carry out the following operations as if they were calculations of experimental results, and express each answer in the correct units and with the correct number of significant figures:

(a) 5.6792 m + 0.6 m + 4.33 m
(b) 3.70 g − 2.9133 g
(c) 4.51 cm × 3.6666 cm
(d) 7.310 km ÷ 5.70 km

Factor-Label Method

1.34 The explosion of a small atomic bomb releases about 5.0×10^{13} J of energy. Convert this quantity into calories.

1.35 The density of aluminum is 2.70 g/cm³. What is its density in kg/m³?

1.36 The density of ammonia gas under certain conditions is 0.625 g/L. Calculate its density in g/cm³.

1.37 The average price of gold in September of 1984 was $342 per ounce. How much did 1.00 g of gold cost at that time? (1 ounce = 28.4 g)

1.38 Carry out the following conversions. (a) A 6.0 ft person weighs 162 lb. Express this person's height in meters and the mass in kilograms. (b) The current speed limit on U.S. highways is 55 mph, that is, 55 miles per hour. What is the speed limit in kilometers per hour? (c) The speed of light is 3.0 × 10¹⁰ cm/s. How many miles does light travel in one hour? (d) The "normal" lead content in human blood is about 0.40 ppm (that is, 0.40 g of lead per million grams of blood). A value of 0.80 ppm is considered to be dangerous. How many grams of lead are contained in 24.5 lb of blood if the lead content is 0.62 ppm?

1.39 In a city with heavy automobile traffic such as Los Angeles or New York, it is estimated that about 9.0 tons of lead from exhausts are deposited on or near the highways every day. What is this amount in kilograms per month? (1 ton = exactly 2000 lb; 1 lb = 453.6 g; 1 month = 31 days)

1.40 Carry out the following conversions: (a) 1.42 light-years to miles (a light-year is an astronomical measure of distance—the distance traveled by light in a year, or 365 days), (b) 32.4 yd to cm, (c) 155 lb to kg, (d) 25.4 mg to kg, (e) 3.0 × 10¹⁰ cm/s to ft/s, (f) 47.4°F to °C, (g) 27.3°C to °F, (h) −273.15°C to °F, (i) 68.3 cm³ to m³, (j) 7.2 m³ to liters.

1.41 How many seconds are in a solar year (365.24 days)?

1.42 A jogger does a mile in 13 minutes. Calculate the average speed in (a) in/s, (b) m/min, (c) km/h.

1.43 How many minutes does it take light from the sun to reach Earth? (Distance from the sun to Earth = 93 million miles; the speed of light = 3.00×10^8 m/s)

Miscellaneous Problems

1.44 Define the following terms: (a) matter, (b) mass, (c) force, (d) pressure, (e) energy, (f) density, (g) substance, (h) homogeneous mixture, (i) heterogeneous mixture, (j) element, (k) compound, (l) physical properties, (m) chemical properties.

*1.45 In a determination of the density of the material of a metal bar of rectangular form, a student made the following measurements: length, 8.53 cm; width, 2.4 cm; height, 1.0 cm; mass, 52.7064 g. Calculate the density of the material, giving your answer with the correct number of significant figures.

*1.46 Calculate the volume of each of the following: (a) a rod of magnesium of mass 210.6 g (density of magnesium = 1.74 g/cm³), (b) a crystal of silicon weighing 15.1 mg (density of silicon = 2.33 g/cm³), (c) 102 g of sulfuric acid (density of sulfuric acid = 1.832 g/cm³).

*1.47 Calculate the mass of each of the following: (a) a sphere of gold of radius 10.0 cm (volume of a sphere of radius r, $V = \frac{4}{3}\pi r^3$; density of gold = 19.3 g/cm³), (b) a cube of platinum of edge length 0.040 mm (density of platinum = 21.4 g/cm³), (c) 50.0 mL of ethanol (density of ethanol = 0.798 g/mL).

1.48 Calculate the density in g/cm³ for each of the following: (a) a brick occupying a volume of 12.0 mL and having a mass of 151 g, (b) a piece of metal occupying a volume of 4.6 mL and having a mass of 89 g, (c) a liquid with a volume of 0.324 L and a mass of 415 g. (Give your answers with the correct number of significant figures.)

*1.49 A cylindrical glass tube 12.7 cm in length is filled with mercury. The mass of mercury needed to fill the tube is found to be 105.5 g. Calculate the inner diameter of the tube. (Density of mercury = 13.6 g/cm^3)

1.50 The density of tungsten is 19.3 g/cm^3 and that of osmium is 22.6 g/cm^3. Which of the following weighs more, a cube of tungsten measuring 1.0 cm on a side or a cube of osmium measuring 0.030 ft on a side?

1.51 The following procedure was carried out to determine the volume of a flask. The flask was weighed dry and then filled with water. If the masses of the empty and filled flasks were 56.12 g and 87.39 g and the density of water is 0.9976 g/cm^3, calculate the volume of the flask in cm^3.

1.52 The speed of sound in air at room temperature is about 343 m/s. Calculate this speed in miles per hour.

1.53 Which of the following describe physical and which describe chemical properties? (a) Iron has a tendency to rust. (b) Rainwater in industrialized regions tends to be acidic. (c) Hemoglobin molecules have a red color. (d) When a glass of water is left out in the sun, the water gradually disappears. (e) Carbon dioxide in the air is converted to more complex molecules by plants during photosynthesis.

*1.54 Describe a chemical property exhibited by each of the following: (a) air, (b) water, (c) alcohol, (d) wax, (e) bread, (f) sodium.

*1.55 Suppose a new temperature scale has been established on which the melting point of ethanol (−117.3°C) and boiling point of ethanol (78.3°C) are taken as 0° and 100°, respectively. Derive an equation relating a reading on this scale to a reading on the Celsius scale. What would be the reading of this thermometer at room temperature (25°C)?

1.56 Group the following elements in pairs that you would expect to show similar physical and chemical properties: F, K, P, Na, Cl, and As. (Hint: See Figure 1.4.)

2

CHEMICAL FORMULAS AND THE NOMENCLATURE OF INORGANIC COMPOUNDS

The knowledge concerning the structure and composition of matter is based on careful experiments and observations of the behavior of matter, that is, on macroscopic data. However, the detailed descriptions that physicists and chemists have given us about such entities as elements, compounds, atoms, and molecules all involve microscopic concerns.

This chapter mainly focuses on microscopic knowledge about the composition of matter: The symbolic representations (chemical formulas) and names (chemical nomenclature) that chemists conventionally associate with various substances are featured. Such formulas and names provide us with convenient ways of communicating detailed knowledge of the makeup of chemical substances.

2.1 THE ATOMIC THEORY—FROM EARLY IDEAS TO JOHN DALTON

In the fifth century B.C. the Greek philosopher Democritus expressed the belief that all matter is composed of very small indivisible particles, which he named "atomos" (Greek word, meaning uncuttable or indivisible). Today we call these particles *atoms*. Although Democritus' idea of "atomism" was not accepted by many philosophers of his day (notably Plato and Aristotle), his suggestion persisted through the centuries. Gradually, his idea gained support from scientists and gave rise to the modern definitions of elements and compounds. However, it took almost two thousand years before an English scientist and school teacher, John Dalton (1766–1844), formulated a more precise definition of atoms, which could be supported by experimental evidence.

In 1808 Dalton presented the atomic theory that started the modern era of chemistry. The hypotheses about the nature of matter on which Dalton based his theory can be summarized as follows:

- Elements are composed of extremely small particles, called atoms. All atoms of a given element are identical, having the same size, shape, mass, and chemical properties. The atoms of one element differ from the atoms of other elements.
- Compounds are composed of atoms of more than one element. In any compound, the ratio of the numbers of atoms of any two of the elements present is either an integer or a simple fraction.
- A chemical reaction involves only the separation, combination, or rearrangement of atoms; it does not result in their creation or destruction.

Figure 2.1 is a schematic representation of the first two hypotheses.

As you can see, Dalton's concept of an atom was far more detailed and specific than the earlier view held by Democritus. The first hypothesis

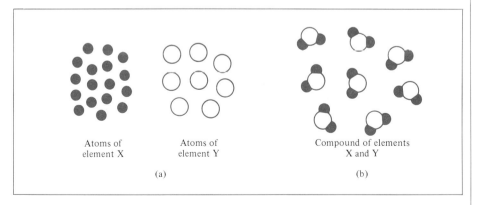

Atoms of
element X

Atoms of
element Y

Compound of elements
X and Y

(a)

(b)

FIGURE 2.1 *(a) According to Dalton's atomic theory, atoms of the same element are identical, but atoms of one element differ from atoms of other elements. (b) Compound formed from atoms of elements X and Y. In this case, the ratio of the atoms of element X to the atoms of element Y is two to one.*

states that atoms, whatever their structure or composition, are different for different elements. You should realize that at the time Dalton presented his theory he had no idea what an atom is really like. However, he recognized that the different properties shown by elements such as hydrogen and oxygen, for example, can be explained by assuming that hydrogen atoms are not the same as oxygen atoms.

The second hypothesis suggests that to form a certain compound we need not only atoms of the right kinds of elements but the correct numbers of these atoms as well. The last hypothesis is another way of stating the *law of conservation of mass,* which says that *matter can neither be created nor destroyed.* Since matter is made of atoms that are unchanged in a chemical reaction, it follows that mass must be conserved as well. Dalton's brilliant insight into the nature of matter was the main cause of the rapid progress of chemistry in the nineteenth century.

2.2 CHEMICAL FORMULAS

On the basis of Dalton's atomic theory we can now define atoms and molecules as follows. An ***atom*** is *the basic unit of an element that can enter into chemical combination,* and a ***molecule*** is *an aggregate of at least two atoms in a definite arrangement held together by special forces.* The structure of atoms and molecules will be discussed in later chapters. For the present we will deal with the problem of how to represent atoms and molecules.

Chemists represent compounds by chemical formulas. A ***chemical formula*** *expresses the composition of a compound in terms of the symbols for the atoms of the elements involved.* By *composition* we mean not only the elements present in the compound but also the ratios in which atoms occur in the compound.

The chemical formula for a single atom is the same as the symbol for the element. Thus C, the symbol for the element carbon, also represents a single carbon atom; and Fe, the symbol for the element iron, also represents a single iron atom. Chemical formulas used to represent molecules are called molecular formulas and empirical formulas.

The names and symbols of the elements are given in the inside front cover of the book.

Molecular Formula

A ***molecular formula*** *shows the exact number of atoms of each element in a molecule. The simplest type of molecule contains only two atoms and is called a **diatomic molecule.*** Elements that exist as diatomic molecules under atmospheric conditions are hydrogen (H_2), nitrogen (N_2), oxygen (O_2), as well as the Group 7A halogens—fluorine (F_2), chlorine (Cl_2), bromine (Br_2), and iodine (I_2). Figure 2.2 shows models for some of these diatomic molecules. In each case the subscript (2) in the formula indicates the number of atoms in the molecule. Of course, a diatomic molecule can contain atoms of two different elements, as in hydrogen chloride (HCl) and carbon monoxide (CO). These formulas have no subscripts because when the num-

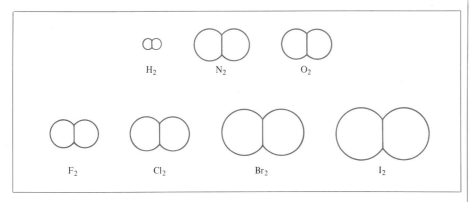

FIGURE 2.2 *Relative sizes of the diatomic molecules H_2, N_2, O_2, and the halogens (F_2, Cl_2, Br_2, and I_2).*

ber of a particular type of atom present is one, the number is not shown as a subscript.

Sometimes chemists are rather sloppy in terminology. For example, when a chemist says "hydrogen" it is not always clear whether he or she means atomic hydrogen (H) or a hydrogen molecule (H_2). To avoid such confusion, we will adhere to the practice, whenever appropriate, of using the term "atomic hydrogen" for hydrogen atoms, "molecular hydrogen" for hydrogen molecules, and "the hydrogen element" when we are discussing the properties of the element hydrogen. The same practice also applies to other substances.

A molecule may contain more than two atoms either of the same type, as in ozone (O_3), or of different types, as in water (H_2O) and ammonia (NH_3). Figure 2.3 shows models for these three molecules. *Molecules containing more than two atoms* are called ***polyatomic molecules.*** Note that both oxygen (O_2) and ozone (O_3) are elemental forms of the same element oxygen. *Different forms of the same element* are called ***allotropes.*** Two allotropic forms of the element carbon—diamond and graphite—present dramatic differences not only in properties but also in their relative cost.

You should be aware of the important differences between "molecule" and "compound," two terms that are often used interchangeably but do not

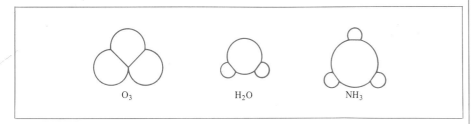

FIGURE 2.3 *Relative sizes and approximate shapes of the polyatomic molecules O_3, H_2O, and NH_3.*

necessarily have the same meaning. A compound is a *substance* composed of the atoms of two or more *elements,* whereas a molecule is a *unit* of a substance composed of two or more *atoms* of the same or different elements. Thus the symbol F_2 represents a molecule but not a compound because there is only one type of element present. On the other hand, the symbol NH_3 represents both a molecule (because there are four atoms present) and a compound (because there are two different elements present). (See Figures 2.2 and 2.3.)

Empirical Formula

The molecular formula of hydrogen peroxide, a substance used both as an antiseptic and as a bleaching agent for textiles and hair, is H_2O_2. This formula indicates that each hydrogen peroxide molecule consists of two hydrogen atoms and two oxygen atoms. The ratio of hydrogen to oxygen atoms in this molecule is 2:2 or 1:1. The empirical formula of hydrogen peroxide is written as HO. Thus the **empirical formula** *tells us which elements are present and the ratio of their atoms,* but not necessarily the actual number of atoms present in the molecule. As another example, consider the compound hydrazine (N_2H_4), which is used as a rocket fuel. The empirical formula of hydrazine is NH_2. Although the ratio of nitrogen to hydrogen is 1:2 in both the molecular formula (N_2H_4) and the empirical formula (NH_2), only the molecular formula tells us the actual number of N atoms (two) and H atoms (four) present in a hydrazine molecule. Empirical formulas are therefore the simplest chemical formulas; they are always written so that the subscripts in the molecular formulas are converted to the *lowest* integral numbers. Molecular formulas are the *true* formulas of molecules.

For many molecules, the molecular formulas and empirical formulas are one and the same. Some examples are water (H_2O), ammonia (NH_3), carbon dioxide (CO_2), and methane (CH_4).

The word *empirical* means "derived from experiment." As we will see shortly, empirical formulas are formulas determined experimentally.

EXAMPLE 2.1

Write the empirical formulas for the following molecules: (a) phosphorus (P_4), (b) dinitrogen tetroxide (N_2O_4), (c) glucose $(C_6H_{12}O_6)$, (d) diiodine pentoxide (I_2O_5).

Answer

(a) Elemental phosphorus exists as a molecule containing four phosphorus atoms. The empirical formula of P_4 is P.
(b) In dinitrogen tetroxide there are two nitrogen atoms and four oxygen atoms. Dividing the subscripts by 2, we obtain the empirical formula NO_2.
(c) In glucose there are six carbon atoms, twelve hydrogen atoms, and six oxygen atoms. Dividing the subscripts by 6, we obtain the empirical formula CH_2O. Note that if we had divided the subscripts by 3, we would have obtained the formula $C_2H_4O_2$. Although the ratio of carbon to hydrogen to oxygen atoms in $C_2H_4O_2$ is the same as that in $C_6H_{12}O_6$ (1:2:1), $C_2H_4O_2$ is not the simplest

formula because its subscripts have not been converted to their lowest integral numbers.

(d) Since the subscripts in I_2O_5 are already the lowest integral numbers (there is no number by which both 2 and 5 are divisible), the empirical formula for iodine pentoxide is the same as its molecular formula.

Similar example: Problem 2.15.

2.3   ATOMIC MASSES

As we have seen, the molecular formula of a compound tells us about the number and types of atoms present. However, to have a complete picture of the composition of the compound, we also need to know the masses of the atoms present. Atoms are extremely small particles—even the smallest speck of dust that our unaided eyes can see contains as many as 1×10^{16} atoms. Since atoms are so tiny, we cannot directly determine their individual masses. Fortunately there is a way of dealing with this seemingly very difficult situation. However, in order to discuss the masses of atoms, we need to have some idea about their composition. The structure of the atom will be discussed in detail in Chapter 7. Here we will present only the bare facts that will help us grasp the concept of atomic mass.

All atoms contain one or more positively charged particles called *protons* and one or more negatively charged particles called *electrons*. The protons are in a very small volume at the center of the atom called the *nucleus*, and the electrons are outside the nucleus. With the exception of hydrogen, the nucleus of an atom of any element also contains electrically neutral particles called *neutrons*. The identity of an atom of any element is determined solely by the number of protons present, but its mass is determined by the total number of electrons, protons, and neutrons in the atom. Atoms of the same element that contain different numbers of neutrons are called *isotopes*. Isotopes of the same element have identical chemical properties but their atoms have different masses.

By international agreement, an atom of the isotope of carbon that contains six protons and six neutrons (called carbon-12) was assigned a mass of exactly 12 **atomic mass units (amu).** Therefore, 1 amu is defined as *a mass exactly equal to 1/12th the mass of one carbon-12 atom*:

$$\text{mass of one C-12 atom} = 12 \text{ amu}$$

$$1 \text{ amu} = \frac{\text{mass of one carbon-12 atom}}{12}$$

One atomic mass unit is also called one dalton.

Experiments have shown that, on average, a hydrogen atom is only 8.400 percent as massive as the standard carbon-12 atom. Thus if we accept the mass of one carbon-12 atom to be exactly 12 amu, then the **atomic mass** (that is, *the mass of the atom in atomic mass unit*) of hydrogen must be $0.08400 \times 12 = 1.008$ amu. Similar calculations show that the atomic mass of oxygen is 16.00 amu and that of iron is 55.85 amu. Note that

The term *atomic weight* has also been used to mean atomic mass.

although we do not know just how much an average iron atom's mass is, we know that it is approximately fifty-six times as massive as a hydrogen atom. Note also that the atomic mass of the element carbon is given in the tables as 12.01 amu instead of 12.00 amu, as you might expect. The reason is that, like most other elements (for example, hydrogen, oxygen, and iron), carbon is composed of several isotopes. Its atomic mass is therefore an average value that includes the atomic masses of each of its isotopes. A table of atomic masses of all the elements is given inside the front cover of this book.

2.4 MOLAR MASS OF AN ELEMENT AND AVOGADRO'S NUMBER

We have seen that atomic mass units provide a relative scale for the elements. In any real situation (for example, in the laboratory) we deal with samples of substances containing enormous numbers of atoms. Therefore it would be convenient to have a special unit to describe a very large number of atoms. The idea of a unit to describe a particular number of objects is not new. For example, the pair (2 items), the dozen (12 items) and the gross (144 items) are all familiar units.

The adjective formed from the noun "mole" is "molar."

The unit defined by the SI system is the **mole (mol),** which is *the amount of substance that contains as many elementary entities (atoms, molecules, or other particles) as there are atoms in exactly 12 grams (or 0.012 kilograms) of the carbon-12 isotope.* Notice that this definition specifies only the method by which the number of elementary entities in a mole may be found. The actual number itself is an experimentally determined value. Over the years chemists have determined and refined this number, and the currently accepted value is

$$1 \text{ mole} = 6.022045 \times 10^{23} \text{ particles}$$

This number is called **Avogadro's number,** in honor of the Italian scientist Amedeo Avogadro (1776–1856). In most of our calculations, we will round this number to 6.022×10^{23}. The term "mole" and its symbol "mol" are used to represent Avogadro's number. Just as one dozen oranges contains twelve oranges, one mole of hydrogen atoms contains 6.022×10^{23} H atoms. On the other hand, one mole of molecular hydrogen (H_2) contains two moles of H atoms, that is, $2(6.022 \times 10^{23})$ or 1.204×10^{24} H atoms. Figure 2.4 shows one mole each of several common substances.

The term *gram molecular weight* has also been used to mean molar mass.

We have seen that one mole of carbon-12 atoms has a mass of exactly 12 grams and contains 6.022×10^{23} atoms. This quantity is called the **molar mass** of carbon-12, *the mass (in grams or kilograms) of one mole of units* (such as atoms or molecules) of the substance. Since each carbon-12 atom has an atomic mass of exactly 12 amu, it is useful to observe that the molar mass of an element (in grams) is numerically equal to its atomic mass in amu. Thus the atomic mass of sodium (Na) is 22.99 amu and its molar mass is 22.99 grams; the atomic mass of copper (Cu) is 63.55 amu and its molar mass is 63.55 grams; and so on. If we know the atomic mass of an element, we also know its molar mass.

The units of molar mass are g/mol (or kg/mol). However, it is also acceptable to say that the molar mass of Na is 22.99 g rather than 22.99 g/mol.

FIGURE 2.4 *One mole each of several common substances.*

We can now calculate the mass (in grams) of a single atom of the carbon-12 isotope:

$$\text{mass in grams of one carbon-12 atom} = \frac{\text{molar mass of carbon-12}}{\text{Avogadro's number}}$$

$$= \frac{12.00 \text{ g/mol}}{6.022 \times 10^{23} \text{ atoms/mol}}$$

$$= 1.993 \times 10^{-23} \text{ g/atom}$$

Furthermore, we can find the relationship between amu and gram by noting that since the mass of every carbon-12 atom is exactly 12 amu, the number of grams equivalent to 1 amu is

$$\frac{\text{gram}}{\text{amu}} = \frac{1.993 \times 10^{-23} \text{ g}}{\text{one carbon-12 atom}} \times \frac{\text{one carbon-12 atom}}{12 \text{ amu}}$$

$$= 1.661 \times 10^{-24} \text{ g/amu}$$

Thus

$$1 \text{ amu} = 1.661 \times 10^{-24} \text{ g}$$

and

$$1 \text{ g} = 6.022 \times 10^{23} \text{ amu}$$

This example shows that Avogadro's number can be used to convert from the atomic mass unit to the mass in grams and vice versa.

Look up atomic masses in the inside front cover. Atomic masses accurate to four significant figures are used in most calculations in this textbook.

EXAMPLE 2.2

How many moles of helium (He) atoms are in 6.46 g of He?

Answer

First we find that the molar mass of He is 4.003 g. This can be expressed by the equality 1 mol of He = 4.003 g of He. To convert the amount of He in grams to He in moles we write

$$6.46 \text{ g He} \times \frac{1 \text{ mol He}}{4.003 \text{ g He}} = 1.61 \text{ mol He}$$

Thus there are 1.61 moles of He atoms in 6.46 g of He.

EXAMPLE 2.3

How many grams of zinc (Zn) are there in 0.356 mole of Zn?

Answer

Since the molar mass of Zn is 65.39 g, the mass of Zn in grams is given by

$$0.356 \text{ mol Zn} \times \frac{65.39 \text{ g Zn}}{1 \text{ mol Zn}} = 23.3 \text{ g Zn}$$

Thus there are 23.3 g of Zn in 0.356 mole of Zn.

EXAMPLE 2.4

The atomic mass of silver (Ag) is 107.9 amu. What is the mass (in grams) of one silver atom?

Answer

The molar mass of silver (Ag) is 107.9 g. Since there are 6.022×10^{23} Ag atoms in one mole of Ag, the mass of one Ag atom is

$$\frac{107.9 \text{ g Ag}}{1 \text{ mol Ag}} \times \frac{1 \text{ mol Ag}}{6.022 \times 10^{23} \text{ Ag atoms}} = 1.792 \times 10^{-22} \text{ g Ag/Ag atom}$$

Similar examples: Problems 2.5, 2.6.

EXAMPLE 2.5

How many atoms are in 0.500 g of boron (B)?

Answer

Solving this problem requires two steps. First, we need to find the number of moles of B in 0.500 g of B (as in Example 2.2). Next, we need to calculate the number of B atoms from the known number of moles of B. We can combine the two steps as follows:

$$0.500 \text{ g B} \times \frac{1 \text{ mol B}}{10.81 \text{ g B}} \times \frac{6.022 \times 10^{23} \text{ B atoms}}{1 \text{ mol B atoms}} = 2.79 \times 10^{22} \text{ B atoms}$$

Similar example: Problem 2.8.

EXAMPLE 2.6

What is the mass in grams of exactly 1 trillion (1×10^{12}) aluminum (Al) atoms?

Answer

Since the molar mass of Al is 26.98 g, the mass of 1 trillion Al atoms is

$$1 \times 10^{12} \text{ Al atoms} \times \frac{1 \text{ mol Al}}{6.022 \times 10^{23} \text{ Al atoms}} \times \frac{26.98 \text{ g Al}}{1 \text{ mol Al}}$$
$$= 4.480 \times 10^{-11} \text{ g Al}$$

Similar example: Problem 2.7.

2.5 MOLECULAR MASS

Once we know the atomic masses we can proceed to calculate the masses of molecules. The **molecular mass** is *the sum of the atomic masses (in amu) present in the molecule.* For example, the molecular mass of H_2O is

$$2(\text{atomic mass of H}) + \text{atomic mass of O}$$

or

$$2(1.008 \text{ amu}) + 16.00 \text{ amu} = 18.02 \text{ amu}$$

The term *molecular weight* has also been used to mean molecular mass.

EXAMPLE 2.7

Calculate the molecular masses of the following compounds: (a) sulfur dioxide (SO_2); (b) ascorbic acid, or vitamin C $(C_6H_8O_6)$.

Answer

(a) From the atomic masses of S and O we get

$$\text{molecular mass of } SO_2 = 32.07 \text{ amu} + 2(16.00 \text{ amu})$$
$$= 64.07 \text{ amu}$$

(b) From the atomic masses of C, H, and O we get

$$\text{molecular mass of } C_6H_8O_6 = 6(12.01 \text{ amu}) + 8(1.008 \text{ amu}) + 6(16.00 \text{ amu})$$
$$= 176.12 \text{ amu}$$

Similar example: Problem 2.19.

The **molar mass of a chemical compound** is *the mass (in grams or kilograms) of one mole of the compound*. It is useful to remember that the molar mass of a compound (in grams) is numerically equal to its molecular mass (in amu). For example, the molecular mass of water is 18.02 amu, so its molar mass is 18.02 g. Both the molecular mass and the molar mass of a compound have the same number, but they differ from each other in their units (amu versus g or kg). The principle here is similar to that regarding atomic mass and molar mass of an atom discussed in Section 2.4. One mole of water weighs 18.02 g and contains 6.022×10^{23} H_2O molecules.

EXAMPLE 2.8

Methane (CH_4) is the principal component of natural gas. How many moles of CH_4 are present in 6.07 g of CH_4?

Answer

First we calculate the molar mass of CH_4:

$$\text{molar mass of } CH_4 = 12.01 \text{ g} + 4(1.008 \text{ g})$$
$$= 16.04 \text{ g}$$

We then follow the procedure in Example 2.2:

$$6.07 \text{ g } CH_4 \times \frac{1 \text{ mol } CH_4}{16.04 \text{ g } CH_4} = 0.378 \text{ mol } CH_4$$

Similar examples: Problems 2.20, 2.22(a).

EXAMPLE 2.9

How many hydrogen atoms are present in 25.6 g of sucrose, or table sugar ($C_{12}H_{22}O_{11}$)? The molar mass of sucrose is 342.3 g.

Answer

There are twenty-two hydrogen atoms in every sucrose molecule; therefore, the total number of hydrogen atoms is

$$25.6 \text{ g } C_{12}H_{22}O_{11} \times \frac{1 \text{ mol } C_{12}H_{22}O_{11}}{342.3 \text{ g } C_{12}H_{22}O_{11}} \times \frac{6.022 \times 10^{23} \text{ molecules } C_{12}H_{22}O_{11}}{1 \text{ mol } C_{12}H_{22}O_{11}}$$

$$\times \frac{22 \text{ H atoms}}{1 \text{ molecule } C_{12}H_{22}O_{11}} = 9.91 \times 10^{23} \text{ H atoms}$$

We could calculate the number of carbon and oxygen atoms by the same procedure. However, there is a shortcut. Note that the ratio of carbon to hydrogen atoms in sucrose is 12/22, or 6/11 and that of oxygen to hydrogen atoms is 11/22 or 1/2. Therefore the number of carbon atoms in 25.6 g of sucrose is $(6/11)(9.91 \times 10^{23})$, or 5.41×10^{23} atoms. The number of oxygen atoms is $(1/2)(9.91 \times 10^{23})$, or 4.96×10^{23} atoms.

2.6 IONS AND IONIC COMPOUNDS

The structure of atoms was briefly discussed in Section 2.3; we will now look more closely inside an atom. The positively charged protons in the nucleus of an atom remain there during ordinary chemical changes (also called chemical reactions), but the negatively charged electrons in atoms are readily gained or lost. *When electrons are removed from or added to a neutral atom, a charged particle called an* **ion** *is formed. An ion that bears a net positive charge* is called a **cation;** *an ion whose net charge is negative* is called an **anion.** For example, a sodium atom (Na) can readily lose an electron to become the cation represented by Na^+ (called the sodium cation):

Na ATOM	*Na$^+$ ION*
11 protons	11 protons
11 electrons	10 electrons

A chlorine atom (Cl) can gain an electron to become the anion represented by Cl^- (called the chloride ion):

Cl ATOM	*Cl$^-$ ION*
17 protons	17 protons
17 electrons	18 electrons

Sodium chloride (NaCl), ordinary table salt, is a compound formed from Na^+ cations and Cl^- anions.

Of course, an atom can lose or gain more than one electron, as in Mg^{2+}, Fe^{3+}, S^{2-}, and N^{3-}. Furthermore, a group of atoms may join together as in a molecule, but may also form an ion that has a net positive or a net negative charge, as in OH^- (hydroxide ion), CN^- (cyanide ion), NH_4^+ (ammonium ion), NO_3^- (nitrate ion), SO_4^{2-} (sulfate ion), and PO_4^{3-} (phosphate ion).

A solid sample of sodium chloride (NaCl) consists of equal numbers of Na^+ and Cl^- ions arranged in a three-dimensional network (Figure 2.5). In such a compound there is a one-to-one ratio of cations to anions, so that

In a neutral atom, the number of electrons is equal to the number of protons.

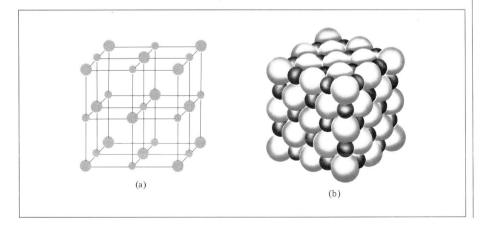

FIGURE 2.5 *(a) Structure of solid NaCl. (b) In reality, the cations are in contact with the anions. In both (a) and (b), the smaller spheres represent Na$^+$ ions and the larger spheres Cl$^-$ ions.*

(a)

(b)

the compound is electrically neutral. *Any neutral compounds containing cations and anions* are called **ionic compounds.**

In most cases, ionic compounds contain a metallic element as the cation and a nonmetallic element as the anion. As mentioned in Section 1.4, metals are good conductors of heat and electricity; they also tend to form cations in compounds. On the other hand, nonmetallic elements are generally poor conductors of heat and electricity, and they form anions when combined with metals. There is one important exception to this description of ionic compounds, however. The ammonium ion (NH_4^+), which is made up of two nonmetallic elements (N and H), is the cation in a number of ionic compounds.

As you can see from Figure 2.5, a particular Na^+ ion in NaCl is not associated with just one particular Cl^- ion. In fact, each Na^+ ion is equally held by six Cl^- ions surrounding it, and vice versa. The NaCl ionic network does *not* contain discrete molecular NaCl units. For other ionic compounds, the actual structure may be different but the arrangement between cations and anions is similar to that shown for NaCl. For this reason, we normally do not call ionic compounds molecules. Thus the formula NaCl represents the empirical formula for sodium chloride. Nevertheless, it is appropriate to use the term "mole" for such compounds. One mole of NaCl, then, refers to 6.022×10^{23} NaCl *formula units*. The molar mass of an ionic compound is the sum of the molar masses of the cations and anions present. Thus the molar mass of NaCl is given by

$$\begin{aligned} \text{molar mass of NaCl} &= \text{molar mass of } Na^+ \text{ ion} + \text{molar mass of } Cl^- \text{ ion} \\ &= 22.99 \text{ g} + 35.45 \text{ g} \\ &= 58.44 \text{ g} \end{aligned}$$

The atomic mass of Na is 22.99 amu; the mass of an electron is only about 0.0006 amu.

Because the mass of an electron is very small compared to that of the entire atom, the molar mass of an ion is virtually the same as that of the uncharged atom from which it is derived. Thus Na and the Na^+ ion are regarded as having the same mass; and Cl and the Cl^- ion also have essentially the same mass.

EXAMPLE 2.10

How many formula units of calcium fluoride (CaF_2) are present in 146.4 g of CaF_2? (The molar mass of CaF_2 is 78.08 g/mol.)

Answer

The number of CaF_2 formula units in 146.4 g of CaF_2 is

$$146.4 \text{ g } CaF_2 \times \frac{1 \text{ mol } CaF_2}{78.08 \text{ g } CaF_2} \times \frac{6.022 \times 10^{23} \text{ formula units } CaF_2}{1 \text{ mol } CaF_2}$$

$$= 1.129 \times 10^{24} \ CaF_2 \text{ formula units}$$

Similar example: Problem 2.29.

2.7 PERCENT COMPOSITION BY MASS OF COMPOUNDS

As we have seen, the formula of a compound tells us the composition of the compound. The composition of a compound is conveniently expressed as the **percent composition by mass** (also known as the **percentage composition by weight**), which is *the percent by mass of each element in a compound.* It is obtained by dividing the mass of each element in one mole of the compound by the molar mass of the compound and multiplying by 100 percent. In one mole of hydrogen peroxide (H_2O_2), for example, there are two moles of H atoms and two moles of O atoms. The molar masses of H_2O_2, H, and O are 34.02 g, 1.008 g, and 16.00 g, respectively. Therefore, the percent composition of H_2O_2 is calculated as follows:

$$\%H = \frac{2.016 \text{ g}}{34.02 \text{ g}} \times 100\% = 5.926\%$$

$$\%O = \frac{32.00 \text{ g}}{34.02 \text{ g}} \times 100\% = 94.06\%$$

The sum of the percentages is 5.926 percent + 94.06 percent = 99.99 percent. The small discrepancy from 100 percent is due to the way we rounded off the molar masses of the elements. If we had used the empirical formula HO for the calculation, we would have written

$$\%H = \frac{1.008 \text{ g}}{17.01 \text{ g}} \times 100\% = 5.926\%$$

$$\%O = \frac{16.00 \text{ g}}{17.01 \text{ g}} \times 100\% = 94.06\%$$

Since both the molecular formula and the empirical formula tell us the composition of the compound, it is not surprising that they give us the same percent composition by mass.

One reason for wanting to know the percent composition by mass of a compound is that by comparing the calculated value with that found experimentally we can verify the purity of the compound.

EXAMPLE 2.11

Sodium nitrite ($NaNO_2$) is a food preservative that is added to ham, hot dogs, and bologna. In recent years its use has come under attack because it has been shown to cause cancer in certain animals. Calculate the percent composition by mass of Na, N, and O in this compound.

Answer

The molar mass of $NaNO_2$ is given by

$$22.99 \text{ g} + 14.01 \text{ g} + 2(16.00 \text{ g}) = 69.00 \text{ g}$$

Therefore, the percent by mass of the elements in $NaNO_2$ is

$$\%Na = \frac{22.99 \text{ g}}{69.00 \text{ g}} \times 100\% = 33.32\%$$

(Continued)

$$\%N = \frac{14.01\ g}{69.00\ g} \times 100\% = 20.30\%$$

$$\%O = \frac{32.00\ g}{69.00\ g} \times 100\% = 46.38\%$$

Similar examples: Problems 2.30, 2.32.

The procedure in the example can be reversed if necessary. Suppose we are given the percent composition by mass of a compound. We can determine the empirical formula of the compound, as the following examples show.

EXAMPLE 2.12

Ascorbic acid (vitamin C) cures scurvy and may help prevent the common cold. It is composed of 40.92 percent carbon (C), 4.58 percent hydrogen (H), and 54.50 percent oxygen (O) by mass. Determine its empirical formula.

Answer

Because we are given percent values here, it is convenient to assume we are dealing with exactly 100 g of the substance. Therefore, in 100 g of ascorbic acid there are 40.92 g of C, 4.58 g of H, and 54.50 g of O. Next, we need to calculate the number of moles of each element in the compound. Let n_C, n_H, and n_O be the number of moles of elements present. Using the molar masses of these elements, we write

$$n_C = 40.92\ g\ C \times \frac{1\ mol\ C}{12.01\ g\ C} = 3.407\ mol\ C$$

$$n_H = 4.58\ g\ H \times \frac{1\ mol\ H}{1.008\ g\ H} = 4.54\ mol\ H$$

$$n_O = 54.50\ g\ O \times \frac{1\ mol\ O}{16.00\ g\ O} = 3.406\ mol\ O$$

Thus, we arrive at the formula $C_{3.407}H_{4.54}O_{3.406}$, which gives the identity and the ratios of atoms present. However, since chemical formulas are written with whole numbers we cannot have 3.407 C atoms, 4.54 H atoms, and 3.406 O atoms. Therefore we divide each number by the smallest subscript (3.406), and obtain $CH_{1.33}O$. The smallest integer that will convert 1.33 into a whole number is 3. Multiplying the subscripts by 3, we get $C_3H_4O_3$, which is the empirical formula of ascorbic acid.

Similar example: Problem 2.35.

EXAMPLE 2.13

The major air pollutant in coal-burning countries is a colorless, pungent gaseous compound containing only sulfur and oxygen. Chemical analysis of a 1.078 g sample of this gas showed that it contained 0.540 g of S and 0.538 g of O. What is the empirical formula of this compound?

Answer

Our first step is to calculate the number of moles of each element present in the sample of the compound. Let these numbers be n_S and n_O. The molar masses of S and O are 32.07 g and 16.00 g, respectively, so we proceed as follows:

$$n_S = 0.540 \text{ g S} \times \frac{1 \text{ mol S}}{32.07 \text{ g S}} = 0.0168 \text{ mol S}$$

$$n_O = 0.538 \text{ g O} \times \frac{1 \text{ mol O}}{16.00 \text{ g O}} = 0.0336 \text{ mol O}$$

From these results we could say that the formula is $S_{0.0168}O_{0.0336}$, but we know that we must convert the subscripts to whole numbers. Therefore we divide each number by the smaller of the two subscripts, that is, 0.0168, so that the subscript of S becomes 1. This procedure gives us $SO_{2.0}$. We know immediately that the empirical formula is SO_2, since the subscript of O must be an integer.

Chemists often want to know the actual mass of an element in a certain mass of a compound. Since the percent composition by mass of the element in the substance can be readily calculated, such a problem can be solved in a rather direct way.

EXAMPLE 2.14

Calculate the mass of Al (aluminum) in a 25.0 g sample of Al_2O_3 (aluminum oxide).

Answer

The molar masses of Al_2O_3 and Al are 102.0 g and 26.98 g, respectively, so that the percent composition by mass of Al is

$$\%\text{Al} = \frac{2 \times 26.98 \text{ g}}{102.0 \text{ g}} \times 100\% = 52.9\%$$

To calculate the mass of Al in the 25.0 g sample of Al_2O_3, we need to convert the percentage to a fraction (that is, convert 52.9 percent to 0.529) and write

$$\text{mass of Al in } Al_2O_3 = 0.529 \times 25.0 \text{ g} = 13.2 \text{ g}$$

This calculation can be simplified by combining the above two steps as follows:

$$\text{mass of Al in } Al_2O_3 = 25.0 \text{ g } Al_2O_3 \times \frac{2 \times 26.98 \text{ g Al}}{102.0 \text{ g } Al_2O_3}$$

$$= 13.2 \text{ g Al}$$

Similar examples: Problems 2.37, 2.38.

Determination of Empirical Formulas

The empirical formula of a compound can be obtained by an experimental procedure. For example, consider the compound ethanol. When ethanol is burned in an apparatus such as that shown in Figure 2.6, carbon dioxide (CO_2) and water (H_2O) are given off. Since neither carbon nor hydrogen was in the inlet gas, we can conclude that both carbon (C) and hydrogen (H) were present in ethanol and that oxygen (O) may also be present. (Molecular oxygen was used in the combustion process, but some of the oxygen may also have been in the original ethanol sample.)

The mass of CO_2 and of H_2O produced can be determined by measuring the increase in mass of the CO_2 and H_2O absorbers, respectively. Suppose that in one experiment the combustion of 11.5 g of ethanol produced 22.0 g of CO_2 and 13.5 g of H_2O. We can calculate the mass of C and H in the original 11.5 g sample of ethanol as follows:

$$\text{mass of C} = 22.0 \text{ g } CO_2 \times \frac{1 \text{ mol } CO_2}{44.01 \text{ g } CO_2} \times \frac{1 \text{ mol C}}{1 \text{ mol } CO_2} \times \frac{12.01 \text{ g C}}{1 \text{ mol C}}$$
$$= 6.00 \text{ g C}$$

$$\text{mass of H} = 13.5 \text{ g } H_2O \times \frac{1 \text{ mol } H_2O}{18.02 \text{ g } H_2O} \times \frac{2 \text{ mol H}}{1 \text{ mol } H_2O} \times \frac{1.008 \text{ g H}}{1 \text{ mol H}}$$
$$= 1.51 \text{ g H}$$

Thus, in 11.5 g of ethanol, 6.00 g are carbon and 1.51 g are hydrogen. The remainder must be oxygen, whose mass is

$$\text{mass of O} = \text{mass of sample} - (\text{mass of C} + \text{mass of H})$$
$$= 11.5 \text{ g} - (6.00 \text{ g} + 1.51 \text{ g})$$
$$= 4.0 \text{ g}$$

The number of moles of each element present in 11.5 g of ethanol is

$$\text{moles of C} = 6.00 \text{ g C} \times \frac{1 \text{ mol C}}{12.01 \text{ g C}} = 0.500 \text{ mol C}$$

$$\text{moles of H} = 1.51 \text{ g H} \times \frac{1 \text{ mol H}}{1.008 \text{ g H}} = 1.50 \text{ mol H}$$

$$\text{moles of O} = 4.0 \text{ g O} \times \frac{1 \text{ mol O}}{16.00 \text{ g O}} = 0.25 \text{ mol O}$$

FIGURE 2.6 *Apparatus for determining the empirical formula of ethanol.*

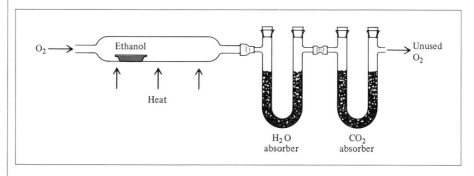

The formula of ethanol is therefore $C_{0.50}H_{1.5}O_{0.25}$ (we round off the number of moles to two significant figures). Since the number of atoms must be an integer, we divide the subscripts by 0.25 and obtain for the empirical formula C_2H_6O.

Now we can better understand the word "empirical," which literally means "based only on observation and measurement." The empirical formula of ethanol is determined from analysis of the compound in terms of its component elements. No knowledge of the actual structure of the molecule is required.

Determination of Molecular Formulas

The formula calculated from data about the percent composition by mass is always the empirical formula. To calculate the actual or molecular formula we must know the approximate molar mass of the compound in addition to its empirical formula. The following example illustrates the point.

EXAMPLE 2.15

The empirical formula of acetic acid (the important ingredient of vinegar) is CH_2O. What is the molecular formula of the compound, given that its molar mass is approximately 60 g?

Answer

First we calculate the molar mass that corresponds to the empirical formula CH_2O:

$$12.01 \text{ g} + 2(1.008 \text{ g}) + 16.00 \text{ g} = 30.03 \text{ g}$$

Then we divide the estimated molecular molar mass by the empirical molar mass:

$$\frac{60 \text{ g}}{30.03 \text{ g}} = 2$$

The answer is rounded off to 2 because molecular molar mass is an integral multiple of empirical molar mass.

Thus there is twice the mass and therefore twice as many atoms in the molecular formula as in the empirical formula. The molecular formula must be $(CH_2O)_2$, or $C_2H_4O_2$, a more conventional way to write it.

Similar example: Problem 2.34.

EXAMPLE 2.16

A compound of oxygen (O) and nitrogen (N) has the following composition: 1.52 g of N and 3.47 g of O. The molar mass of this compound is known to be between 90 g and 95 g. Determine its molecular formula and the molecular molar mass of the compound to three significant figures.

(Continued)

Answer

We first determine the empirical formula as outlined in Example 2.13. Let n_N and n_O be the number of moles of nitrogen and oxygen. Then

$$n_N = 1.52 \text{ g N} \times \frac{1 \text{ mol N}}{14.01 \text{ g N}} = 0.108 \text{ mol N}$$

$$n_O = 3.47 \text{ g O} \times \frac{1 \text{ mol O}}{16.00 \text{ g O}} = 0.217 \text{ mol O}$$

Thus, the formula of the compound is $N_{0.108}O_{0.217}$. Dividing the subscripts by 0.108 and rounding off, we get NO_2 as the empirical formula. The molecular formula will be equal to the empirical formula or to some integral multiple of it (for example, two, three, four, or more times the empirical formula). The molar mass of the empirical formula NO_2 is calculated as 46.0 g. Since the molar mass is between 90 g and 95 g, roughly twice that of the empirical molar mass, the molecular formula must be $(NO_2)_2$, or N_2O_4. And the compound's molar mass must be 2(46.0 g) or 92.0 g.

2.8 LAWS OF CHEMICAL COMBINATION

Having discussed chemical formulas of molecules and compounds, we will now consider two important laws that played a major role in the early steps toward understanding chemical compounds.

The **law of definite proportions** states that *different samples of the same compound always contain its constituent elements in the same proportions by mass.* This law is generally attributed to Joseph Proust (1754–1826), a French chemist who published it in 1799, eight years before Dalton's atomic theory was advanced. The law says that if, for example, we analyze samples of carbon dioxide (CO_2) gas obtained from different sources, we would find in each sample the same ratio by mass of carbon to oxygen. This statement seems obvious today, for we normally expect all molecules of a given compound to have the same composition, that is, to contain the same numbers of atoms of its constituent elements. If the ratios of different types of atoms are fixed, then the ratios of the masses of these atoms must also be fixed.

The other fundamental law is the **law of multiple proportions,** which states that *if two elements can combine to form more than one compound, the masses of one element that combine with a fixed mass of the other element are in ratios of small whole numbers.* For example, carbon forms two stable compounds with oxygen, namely, CO (carbon monoxide) and CO_2 (carbon dioxide). Experiments show that the masses of oxygen that combine with 12 g of carbon to form those two compounds are 16 g and 32 g, respectively. The masses 16 g and 32 g are in the ratio 1:2, a ratio of small whole numbers. As another example, consider the three compounds formed by oxygen and nitrogen: NO (nitric oxide), NO_2 (nitrogen dioxide), and N_2O (nitrous oxide). Here the masses of nitrogen that combine with

16 g of oxygen are found to be 14 g, 7 g, and 28 g. These masses are in the ratios 14:7:28 or 2:1:4.

Dalton's atomic theory explains the law of multiple proportions quite simply. Compounds differ in the number of atoms of each kind that combine. In the case of the two compounds formed between carbon and oxygen, measurements suggest that one atom of carbon combines with one atom of oxygen in one compound (that is, in CO) and that one carbon atom combines with two oxygen atoms in the other compound (that is, CO_2).

EXAMPLE 2.17

For many years the Group 8A noble gas elements (He, Ne, Ar, Kr, and Xe) were called "inert gases" because no one had succeeded in synthesizing compounds containing them. Since 1962, however, a number of stable xenon (Xe) compounds have been prepared. Fluorine (F) forms two compounds with xenon. In one of them, 0.312 g of F is combined with 1.08 g of Xe; in the other compound, 0.426 g of F is combined with 0.736 g of Xe. Prove that these data are consistent with the law of multiple proportions.

Answer

We need to calculate the mass of F (called x) that combines with 1.00 g of Xe in each of these two compounds.

First compound

First we set up the ratios

$$\frac{x \text{ g F}}{1.00 \text{ g Xe}} = \frac{0.312 \text{ g F}}{1.08 \text{ g Xe}}$$

Thus

$$x \text{ g F} = \frac{(0.312 \text{ g F})(1.00 \text{ g Xe})}{1.08 \text{ g Xe}} = 0.289 \text{ g F}$$

Second compound

Similarly

$$\frac{x \text{ g F}}{1.00 \text{ g Xe}} = \frac{0.426 \text{ g F}}{0.736 \text{ g Xe}}$$

Thus

$$x \text{ g F} = \frac{(0.426 \text{ g F})(1.00 \text{ g Xe})}{0.736 \text{ g Xe}} = 0.579 \text{ g F}$$

The ratios of F to Xe in these two compounds are compared as follows:

$$\frac{0.579 \text{ g}}{0.289 \text{ g}} = \frac{2}{1}$$

The 2:1 ratio is consistent with the law of multiple proportions.

Similar examples: Problems 2.45, 2.46, 2.47.

You can arrive at the same conclusion by calculating the number of grams of Xe that combines with 1.00 g of F in each of these two compounds.

2.9 NAMING INORGANIC COMPOUNDS

One challenge faced by chemists early in the development of this science was that of devising clear, systematic ways of naming chemical substances. Fortunately for the advancement of chemistry, such naming schemes were developed and accepted worldwide, facilitating communication among chemists and providing useful ways of dealing with the overwhelming variety of substances currently identified. The number of known compounds is well above 5 million, and there is no need to memorize their names, even if it were possible to do so. However, to study the nomenclature, or the naming of compounds, we first need to distinguish inorganic compounds from organic compounds.

Organic compounds *contain carbon, usually in combination with elements such as hydrogen, oxygen, nitrogen, and sulfur. All other compounds are classified as* **inorganic compounds.** Some carbon-containing compounds such as carbon monoxide (CO), carbon dioxide (CO_2), carbon disulfide (CS_2), compounds containing the cyanide group (CN^-), and carbonate (CO_3^{2-}) and bicarbonate (HCO_3^-) groups are considered for convenience to be inorganic compounds. Although the nomenclature of organic compounds will not be discussed until Chapter 30, we will use some organic compounds to illustrate chemical principles throughout this textbook.

To organize and simplify our venture into naming compounds, we can divide inorganic compounds into four categories: ionic compounds, molecular compounds, acids and bases, and hydrates.

Ionic Compounds

Section 2.6 explained that ionic compounds are made up of cations (positive ions) and anions (negative ions). The names of many cations are simply the names of the single elements of which the ions consist. For example:

	Element		*Name of Cation*
Na	sodium	Na^+	sodium ion (or sodium cation)
K	potassium	K^+	potassium ion (or potassium cation)
Mg	magnesium	Mg^{2+}	magnesium ion (or magnesium cation)
Al	aluminum	Al^{3+}	aluminum ion (or aluminum cation)

Figure 2.7 shows the ionic charges of some metals according to their positions in the periodic table.

Many ionic compounds are **binary compounds,** or *compounds formed from just two elements.* For binary ionic compounds such as $NaCl$ or CaF_2, the first element you write is customarily the metallic cation, followed by the nonmetallic anion. The rules for naming such substances follow similar guidelines. It is conventional to name the metallic element first and to end the name of the second (nonmetallic) element with the suffix "-ide." Table 2.1 shows the "-ide" nomenclature of some common nonmetals according to their positions in the periodic table.

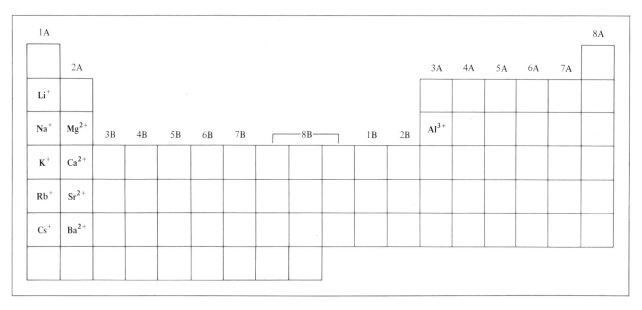

FIGURE 2.7 *Ionic charges of some metals according to their positions in the periodic table. Ions from Group 1A metals all bear +1 charges, those from Group 2A metals bear +2 charges, and those from Group 3A metals bear +3 charges. Note that these are the only stable charges that these metals can assume. Most other metals can form more than one cation, each with a different charge.*

TABLE 2.1 The "-ide" Nomenclature of Some Common Non-metals According to Their Positions in the Periodic Table

Group 4A	Group 5A	Group 6A	Group 7A
C Carbide $(C^{4-})^*$	N Nitride (N^{3-})	O Oxide (O^{2-})	F Fluoride (F^-)
Si Silicide (Si^{4-})	P Phosphide (P^{3-})	S Sulfide (S^{2-})	Cl Chloride (Cl^-)
		Se Selenide (Se^{2-})	Br Bromide (Br^-)
		Te Telluride (Te^{2-})	I Iodide (I^-)

*The word "carbide" is also used for the anion C_2^{2-}.

Some examples of binary ionic compounds are

KBr	potassium bromide
ZnI_2	zinc iodide
MgO	magnesium oxide
Ca_3N_2	calcium nitride

If an element can form two different cations, we must have a way to distinguish them. An older method that still finds limited use assigns the ending "-ous" to the cation with fewer positive charges and the ending "-ic" to the cation with more positive charges:

Fe^{2+}	ferrous ion
Fe^{3+}	ferric ion

The names of the compounds that these iron ions form with chlorine would thus be

$FeCl_2$ ferrous chloride
$FeCl_3$ ferric chloride

This method of naming has some distinct limitations. First, the "-ous" and "-ic" suffixes do not provide information regarding the actual charges of the two cations involved. Thus ferric ion is Fe^{3+}, but the cation of copper named cupric has the formula Cu^{2+}. In addition, the "-ous" and "-ic" designations provide element cation names for only two different positive charges. Some metallic elements can assume three or more different positive charges. Therefore, it has become increasingly common to designate different cations with Roman numerals (I for one positive charge, II for two positive charges, and so on). For example, manganese (Mn) atoms can assume several different positive charges:

Mn^{2+}: MnO   manganese(II) oxide
Mn^{3+}: Mn_2O_3 manganese(III) oxide
Mn^{4+}: MnO_2 manganese(IV) oxide

These compound names are pronounced manganese-two oxide, manganese-three oxide, and manganese-four oxide. Using the Stock system, we can express the ferrous ion and the ferric ion as iron(II) and iron(III), respectively; ferrous chloride becomes iron(II) chloride; and ferric chloride is called iron(III) chloride. In keeping with modern practice, we will favor the use of the Stock system of naming compounds in this textbook.

The "-ide" ending is also used for certain anion groups (containing two different elements), such as hydroxide (OH^-) and cyanide (CN^-). Thus the compounds LiOH and KCN are named lithium hydroxide and potassium cyanide. These and a number of other such ionic substances are called **ternary compounds,** meaning *compounds consisting of three elements.* Another example of a ternary ionic compound is ammonium chloride (NH_4Cl). In this case the cation (NH_4^+) is made up of two different elements. Table 2.2 lists alphabetically the names of a number of common inorganic cations and anions.

An important guideline for writing correct formulas of ionic compounds is that each compound must be electrically neutral. This means that the sum of charges on the cation and anion must add up to zero. The following examples illustrate this rule:

- *Potassium bromide.* The cation is K^+ and the anion is Br^-. The formula is KBr. The sum of the charges is $+1 + (-1) = 0$.
- *Calcium carbonate.* The cation is Ca^{2+} and the anion is CO_3^{2-}. The formula is $CaCO_3$. The sum of the charges is $+2 + (-2) = 0$.
- *Iron(III) nitrate.* Iron(III) means Fe^{3+}. Since Fe^{3+} has three positive charges and nitrate (NO_3^-) has only one negative charge, the formula should be $Fe(NO_3)_3$. In this formula the sum of the charges is $+3 + 3(-1) = 0$. The expression $(NO_3)_3$ indicates there are three nitrate ions present in each unit of iron(III) nitrate.
- *Magnesium nitride.* The cation is Mg^{2+} and the anion is N^{3-}. To have

This method of naming compounds is called the *Stock system* (after the German chemist Alfred Stock, 1876–1946).

TABLE 2.2 Names and Formulas of Some Common Inorganic Cations and Anions

Cation	Anion
Aluminum (Al^{3+})	Bromide (Br^-)
Ammonium (NH_4^+)	Carbonate (CO_3^{2-})
Barium (Ba^{2+})	Chlorate (ClO_3^-)
Calcium (Ca^{2+})	Chloride (Cl^-)
Cesium (Cs^+)	Chromate (CrO_4^{2-})
Chromium(III) or chromic (Cr^{3+})	Cyanide (CN^-)
Cobalt(II) or cobaltous (Co^{2+})	Dichromate ($Cr_2O_7^{2-}$)
Copper(I) or cuprous (Cu^+)	Dihydrogen phosphate ($H_2PO_4^-$)
Copper(II) or cupric (Cu^{2+})	Fluoride (F^-)
Hydrogen (H^+)	Hydride (H^-)
Iron(II) or ferrous (Fe^{2+})	Hydrogen carbonate or bicarbonate
Iron(III) or ferric (Fe^{3+})	(HCO_3^-)
Lead(II) or plumbous (Pb^{2+})	Hydrogen phosphate (HPO_4^{2-})
Lithium (Li^+)	Hydrogen sulfate or bisulfate
Magnesium (Mg^{2+})	(HSO_4^-)
Manganese(II) or manganous (Mn^{2+})	Hydroxide (OH^-)
Mercury(I) or mercurous (Hg_2^{2+})*	Iodide (I^-)
Mercury(II) or mercuric (Hg^{2+})	Nitrate (NO_3^-)
Potassium (K^+)	Nitride (N^{3-})
Silver (Ag^+)	Nitrite (NO_2^-)
Sodium (Na^+)	Oxide (O^{2-})
Strontium (Sr^{2+})	Permanganate (MnO_4^-)
Tin(II) or stannous (Sn^{2+})	Peroxide (O_2^{2-})
Zinc (Zn^{2+})	Phosphate (PO_4^{3-})
	Sulfate (SO_4^{2-})
	Sulfide (S^{2-})
	Sulfite (SO_3^{2-})
	Thiocyanate (SCN^-)

*Mercury(I) exists as a pair as shown.

the sum of the charges equal to zero, we can triple the $+2$ charge on Mg and double the -3 charge on N:

$$3(+2) + 2(-3) = 0$$

Thus the formula for magnesium nitride should be Mg_3N_2; the factors 3 and 2 appear as the subscripts in this formula.

When writing the formula of ionic compounds, remember the following very useful rule: The subscript of the cation is *numerically* equal to the charge on the anion, and the subscript of the anion is *numerically* equal to the charge on the cation. For example, in aluminum oxide the cation is Al^{3+} and the anion is O^{2-}, so that we write

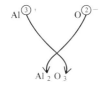

You may wish to refer to Table 2.2 in the following examples and end-of-chapter problems.

This rule applies only if the numerical charges on the cation and anion are different. If the charges are numerically equal, then the subscripts for both the cation and the anion will be 1, as in MgO and AlN.

EXAMPLE 2.18

Name the following ionic compounds: (a) CuI_2, (b) KH_2PO_4, (c) NH_4ClO_3, (d) Na_2O_2.

Answer

(a) Since the iodide ion (I^-) bears one negative charge, the copper ion must have two positive charges. Therefore, the compound is copper(II) iodide.
(b) The cation is K^+ and the anion is $H_2PO_4^-$ (dihydrogen phosphate). The compound is potassium dihydrogen phosphate.
(c) The cation is NH_4^+ (ammonium ion) and the anion is ClO_3^- (chlorate ion). The compound is ammonium chlorate.
(d) Since each sodium ion bears a positive charge and there are two Na^+ ions present, the total negative charge must be -2. Thus the anion is the peroxide ion O_2^{2-} and the compound is sodium peroxide.

Similar example: Problem 2.49.

EXAMPLE 2.19

Write chemical formulas for the following compounds: (a) mercury(I) nitrite, (b) cesium sulfide, (c) calcium phosphate, (d) potassium dichromate.

Answer

(a) The mercury(I) ion is diatomic, namely, Hg_2^{2+}, and the nitrite ion is NO_2^-. Therefore, the formula is $Hg_2(NO_2)_2$.
(b) Each sulfide ion bears two negative charges, and each cesium ion bears one positive charge (cesium is in Group 1A, as is sodium). Therefore, the formula is Cs_2S.
(c) Each calcium ion (Ca^{2+}) bears two positive charges, and each phosphate ion (PO_4^{3-}) bears three negative charges. To have the sum of charges equal to zero, the numbers of cations and anions must be adjusted:

$$3(+2) + 2(-3) = 0$$

Thus the formula is $Ca_3(PO_4)_2$.
(d) Each dichromate ion $(Cr_2O_7^{2-})$ has two negative charges and each potassium ion carries one positive charge, so the formula is $K_2Cr_2O_7$.

Similar example: Problem 2.50.

Refer to Figure 2.7 to determine the charges on some metal cations.

Molecular Compounds

Unlike ionic compounds, molecular compounds contain discrete molecular units. They are usually composed of nonmetallic elements. Naming binary molecular compounds is similar to naming binary ionic compounds, that is, we name the first element first and end the name of the second element with the suffix "-ide." Some examples are

HCl   hydrogen chloride
HBr   hydrogen bromide

H_2S	hydrogen sulfide
SiC	silicon carbide

Quite often we find that a pair of elements can form several different compounds. In these cases, confusion in naming the compounds is avoided by using Greek prefixes to denote the number of atoms of each element present (Table 2.3). Consider the following examples:

CO	carbon monoxide
CO_2	carbon dioxide
SO_2	sulfur dioxide
SO_3	sulfur trioxide
PCl_3	phosphorus trichloride
PCl_5	phosphorus pentachloride
NO_2	nitrogen dioxide
N_2O_4	dinitrogen tetroxide
Cl_2O_7	dichlorine heptoxide

This was the basis for the law of multiple proportions discussed in Section 2.8.

The following guidelines are helpful when you are naming compounds with prefixes:

- The prefix "mono-" may be omitted for the first element. For example, SO_2 is named sulfur dioxide, not monosulfur dioxide. Thus the absence of a prefix for the first element usually implies there is only one atom of that element present in the molecule.
- For oxides, the ending "a" in the prefix is sometimes omitted. For example, N_2O_4 may be called dinitrogen tetroxide rather than dinitrogen tetraoxide.

TABLE 2.3 Greek Prefixes Used in Naming Compounds

Prefix	Meaning
Mono-	1
Di-	2
Tri-	3
Tetra-	4
Penta-	5
Hexa-	6
Hepta-	7
Octa-	8
Nona-	9
Deca-	10

EXAMPLE 2.20

Name the following compounds: (a) BF_3, (b) $SiCl_4$, (c) IF_7, (d) P_4O_{10}.

Answer

(a) Since there are three fluorine atoms present, the compound is boron trifluoride.

(Continued)

(b) There are four chlorine atoms present, so the compound is silicon tetrachloride.

(c) Since there are seven fluorine atoms present, the compound is iodine heptafluoride.

(d) There are four phosphorus atoms and ten oxygen atoms present, so the compound is tetraphosphorus decoxide.

Similar example: Problem 2.49.

EXAMPLE 2.21

Write chemical formulas of the following compounds: (a) nitrogen trichloride, (b) carbon disulfide, (c) disilicon hexabromide, and (d) tetranitrogen tetrasulfide.

Answer

(a) Since there are one nitrogen atom and three chlorine atoms present, the formula is NCl_3.

(b) There are one carbon atom and two sulfur atoms present, so the formula is CS_2.

(c) Since there are two silicon atoms and six bromine atoms present, the formula is Si_2Br_6.

(d) There are four nitrogen atoms and four sulfur atoms present, so the formula is N_4S_4.

Similar example: Problem 2.50.

Note that the prefixes listed in Table 2.3 are normally not required for naming ionic compounds. The reason is that we can usually deduce the subscripts for the cation and anion from the known charges on these ions. For example, we know that tin(II) bromide must be $SnBr_2$ because Sn(II) means Sn^{2+} and the bromide ion *always* carries a -1 charge (that is, Br^-). Therefore, there is no need to call the compound tin(II) dibromide. Similarly, Na_3PO_4 should be called sodium phosphate and not trisodium phosphate because the sodium ion *always* carries a $+1$ charge (that is, Na^+) and the phosphate ion *always* carries a -3 charge (that is, PO_4^{3-}).

Acids and Bases

Naming Acids. For present purposes an ***acid*** is defined as *a substance that yields hydrogen ions* (H^+) *when dissolved in water*. Formulas for inorganic acids contain one or more hydrogen atoms as well as an anionic group. Anions whose names end in "-ide" have associated acids with a

The H^+ ion is also called the *proton.*

TABLE 2.4 Some Simple Acids

Anion	Corresponding Acid
F^- (fluoride)	HF (hydrofluoric acid)
Cl^- (chloride)	HCl (hydrochloric acid)
Br^- (bromide)	HBr (hydrobromic acid)
I^- (iodide)	HI (hydroiodic acid)
CN^- (cyanide)	HCN (hydrocyanic acid)
S^{2-} (sulfide)	H_2S (hydrosulfuric acid)

"hydro-" prefix and an "-ic" ending, as shown in Table 2.4. You may have noticed that in some instances two different names seem to be assigned to the same chemical formula. For example

HCl	hydrogen chloride
HCl	hydrochloric acid

The name assigned to the compound depends on its physical state. When HCl exists in the gaseous or pure liquid state, it is a molecular compound and we call it hydrogen chloride. When it is dissolved in water, the molecules break up into H^+ and Cl^- ions; in this state, the substance is called hydrochloric acid.

Acids containing hydrogen, oxygen, and another element (the central element) are called **oxyacids.** The formulas of oxyacids are usually written with the H first, followed by the central element, then O. When two oxyacids have the same central atom, we use the suffix "-ous" to indicate the oxyacid with fewer oxygen atoms and the suffix "-ic" to indicate the oxyacid with more oxygen atoms (Table 2.5). *The anions derived from oxyacids are called* **oxyanions.** Their names end with "-ite" and "-ate."

When three or four oxyacids have the same central atom, the prefixes "hypo-" and "per-" are used with the suffixes "-ous" and "-ic" to indicate different numbers of oxygen atoms. The prefix "hypo-" with the suffix "-ous" indicates the lowest number of oxygen atoms, while "per-" with "-ic" pertains to the acid having the highest number of oxygen atoms (Table 2.6). For intermediate cases we use the "-ous" and "-ic" suffixes as we have described.

TABLE 2.5 Oxyanions and the Oxacids

Oxyanion	Corresponding Oxyacid
NO_2^- (nitrite)	HNO_2 (nitrous acid)
NO_3^- (nitrate)	HNO_3 (nitric acid)
SO_3^{2-} (sulfite)	H_2SO_3 (sulfurous acid)
SO_4^{2-} (sulfate)	H_2SO_4 (sulfuric acid)

TABLE 2.6 The Names of a Series of Oxyacids

Oxyanion	Corresponding Oxyacid
ClO^- (hypochlorite)	$HOCl$ (hypochlorous acid)
ClO_2^- (chlorite)	$HClO_2$ (chlorous acid)
ClO_3^- (chlorate)	$HClO_3$ (chloric acid)
ClO_4^- (perchlorate)	$HClO_4$ (perchloric acid)

Naming Bases.   For present purposes, a ***base*** is defined as *a substance that yields hydroxide ions (OH⁻) when dissolved in water.* Some examples are

NaOH	sodium hydroxide
KOH	potassium hydroxide
$Ba(OH)_2$	barium hydroxide
$Al(OH)_3$	aluminum hydroxide
$Fe(OH)_2$	iron(II) hydroxide
$Fe(OH)_3$	iron(III) hydroxide

Ammonia (NH_3), a molecular compound in the gaseous or pure liquid state, is also classified as a common base. At first glance this may seem to be an exception to the definition for base that we have given. But note that all that is required is that a base *yields* hydroxide ions when dissolved in water. It is not necessary for the original base to contain hydroxide ions in its structure. In fact, when ammonia dissolves in water, NH_3 reacts at least partially with water to yield NH_4^+ and OH^- ions. Thus it is properly classified as a base.

Hydrates

Compounds that have a specific number of water molecules attached to them are called **hydrates.** For example, in its normal state (say, from a reagent bottle) each unit of copper(II) sulfate has five water molecules associated with it. The formula of copper(II) sulfate pentahydrate is $CuSO_4 \cdot 5H_2O$. These water molecules can be driven off by heating. When this occurs, the resulting compound is $CuSO_4$, which is sometimes called *anhydrous* copper(II) sulfate, where "anhydrous" means that the compound no longer has water molecules associated with it. Some examples of hydrates are

$BaCl_2 \cdot 2H_2O$	barium chloride dihydrate
$LiCl \cdot H_2O$	lithium chloride monohydrate
$MgSO_4 \cdot 7H_2O$	magnesium sulfate heptahydrate
$Sr(NO_3)_2 \cdot 4H_2O$	strontium nitrate tetrahydrate

Pentahydrate here means five water molecules associated with each formula unit of $CuSO_4$.

Familiar Inorganic Compounds

Some compounds are better known by their common names rather than by systematic chemical names. The names of some familiar inorganic compounds are given in Table 2.7.

TABLE 2.7 Common and Systematic Names of Some Compounds

Formula	Common Name	Systematic Name
H_2O	Water	Dihydrogen oxide
NH_3	Ammonia	Trihydrogen nitride
CO_2	Dry Ice	Solid carbon dioxide
NaCl	Table salt	Sodium chloride
N_2O	Laughing gas	Dinitrogen oxide (nitrous oxide)
$CaCO_3$	Marble, chalk, limestone	Calcium carbonate
CaO	Quicklime	Calcium oxide
$Ca(OH)_2$	Slaked lime	Calcium hydroxide
$NaHCO_3$	Baking soda	Sodium hydrogen carbonate
$MgSO_4 \cdot 7H_2O$	Epsom salt	Magnesium sulfate heptahydrate
$Mg(OH)_2$	Milk of magnesia	Magnesium hydroxide
$CaSO_4 \cdot 2H_2O$	Gypsum	Calcium sulfate dihydrate

AN ASIDE ON THE PERIODIC TABLE
Binary Ionic Compounds

The discussions of ionic compounds enable us to use the periodic table in a simple and informative way. In the vast majority of cases, binary ionic compounds are formed by the combination of the most reactive metals and the most reactive nonmetals, shown in Figure 2.8. (As we said earlier, the ammonium ion, NH_4^+, is an important exception.) The most reactive metals are the alkali metals (the Group 1A elements) and,

FIGURE 2.8 Positions of the most reactive metallic and non-metallic elements in the periodic table. By "reactive," we mean the strong tendency of the elements to react with other substances.

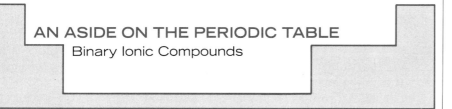

with the exception of beryllium, the alkaline earth metals (the Group 2A elements). Among the most reactive nonmetallic elements are the halogens (the Group 7A elements) and to a lesser extent oxygen (O) and sulfur (S). Molecular nitrogen (N_2) is normally a very stable molecule. However, under suitable conditions it does react with metals to form the nitride ion (N^{3-}).

Depending on their group positions, these metals have a strong tendency to lose one or more electrons, and the nonmetals have a strong tendency to accept one or more electrons in chemical combination. A number of other metals, especially those found in Group 1B through 8B [for example, chromium (Cr), manganese (Mn), iron (Fe), and zinc (Zn)], also form ionic compounds. As a useful exercise for the present, you should write the formulas and names of *all* possible binary ionic compounds that can form from the metals and nonmetals shown in Figure 2.8.

SUMMARY

1. Modern chemistry began with Dalton's atomic theory, which states that all matter is composed of tiny, indivisible particles called atoms; that all atoms of the same element are identical; that compounds contain atoms of different elements combined in whole number ratios; and that atoms are neither created nor destroyed in chemical reactions (the law of conservation of mass).

2. Chemical formulas combine the symbols for the constituent elements with whole number subscripts to show the type and number of atoms contained in the smallest unit of a compound.

3. The molecular formula shows the specific number and type of atoms combined in each molecule of a compound. The empirical formula shows the simplest ratios of the atoms combined in a molecule.

4. Atomic masses can be expressed in atomic mass units (amu), a relative unit based on assigning exactly 12 amu to one atom of the carbon-12 isotope.

5. A mole is an Avogadro's number (6.022×10^{23}) of atoms, molecules, or other particles. The molar mass (in grams) of an element or a compound is numerically equal to the mass of the atom, molecule, or formula unit (in amu), and contains an Avogadro's number of atoms (in the case of elements), molecules, or simplest units of cations and anions (in the case of ionic compounds).

6. The molecular mass of a molecule is the sum of the atomic masses of the atoms in the molecule.

7. Ionic compounds are made up of cations and anions, formed when atoms lose and gain electrons, respectively. Chemical compounds are either molecular compounds (in which the smallest units are discrete, individual molecules) or ionic compounds (in which positive and negative ions are held together by mutual attraction).

8. The percent composition by mass of a compound is the percent by mass of each element present. If we know the percent composition by mass of a compound we can deduce the empirical formula of the

compound and also the molecular formula of the compound if the approximate molar mass is known.

9. Atoms of constituent elements in a particular compound are always combined in the same proportions by mass (law of definite proportions). When two elements can combine to form more than one type of compound, the masses of one element that combine with a fixed mass of the other element are in the ratio of small whole numbers (law of multiple proportions).

10. The names of many inorganic compounds can be deduced from a set of simple rules.

KEY WORDS

Acid, p. 58
Allotrope, p. 35
Anion, p. 43
Atom, p. 34
Atomic mass, p. 37
Atomic mass unit, p. 37
Avogadro's number, p. 38
Base, p. 60
Binary compound, p. 52
Cation, p. 43
Chemical formula, p. 34
Diatomic molecule, p. 34
Empirical formula, p. 36
Hydrate, p. 60
Inorganic compound, p. 52
Ion, p. 43
Ionic compound, p. 44
Law of conservation of mass,
   p. 34

Law of definite proportions, p. 50
Law of multiple proportions, p. 50
Molar mass, p. 38
Molar mass of a chemical compound,
   p. 42
Mole, p. 38
Molecular formula, p. 34
Molecular mass, p. 41
Molecule, p. 34
Organic compound, p. 52
Oxyacid, p. 59
Oxyanion, p. 59
Percent composition by mass,
   p. 45
Percentage composition by weight,
   p. 45
Polyatomic molecule, p. 35
Ternary compound, p. 54

PROBLEMS

More challenging problems are marked with an asterisk.

Atomic Mass and Avogadro's Number

2.1 What is the difference between mass and weight?

2.2 Explain clearly what is meant by the statement "The atomic mass of gold is 197.0 amu."

2.3 Earth's population is about 4.0 billion. Suppose every person on Earth participates in a process of counting identical particles at the rate of two particles per second. How many years would it take to count 6.0×10^{23} particles? Assume that there are 365 days in a year.

2.4 Avogadro's number has often been described as a conversion factor between the units of amu and g. Use the fluorine atom (19.00 amu) as an example to show the relation between these two units.

2.5 What is the mass in grams of 2.6 amu?

2.6 How many amu are there in 3.4 g?

2.7 What is the mass in grams of a single atom of each of the following elements? (a) Hg, (b) Ne, (c) As, (d) Pb.

2.8 How many atoms are present in 3.14 g of copper?

2.9 Which of the following has more atoms: 1.10 g of hydrogen atoms or 14.7 g of chromium atoms?

Chemical Formulas

2.10 What does a chemical formula represent?

2.11 What are the similarities and differences between the empirical formula and molecular formula of a compound?

2.12 What does the expression S_8 signify? How does this differ from $8S$?

2.13 Give two examples each of a diatomic molecule and a polyatomic molecule.

2.14 What are allotropes?

2.15 What are the empirical formulas of the following compounds? (a) $C_6H_{12}O_6$, (b) C_6H_6, (c) C_8H_8, (d) $Na_2S_2O_4$, (e) Al_2Cl_6, (f) B_2H_6, (g) $K_2Cr_2O_7$.

Molecular Mass

2.16 What is the difference between atomic mass and molecular mass?

2.17 Why do chemists prefer not to express the masses of individual atoms and molecules in grams in practical work?

2.18 Calculate the molecular mass and molar mass of a substance if 0.89 mole of it has a mass of 79 g.

2.19 Calculate the molecular mass (in amu) of each of the following compounds: (a) CH_4, (b) H_2O, (c) H_2O_2, (d) $Co(NH_3)_5Cl$, (e) $C_{2952}H_{4664}N_{812}O_{832}S_8Fe_4$ (hemoglobin).

2.20 The molecular mass of hemoglobin is about 6.5×10^4 amu. How many molecules are present in 74.3 g of hemoglobin?

2.21 Calculate the molecular mass and molar mass of the following substances: (a) S_8, (b) CS_2, (c) $CHCl_3$ (chloroform), (d) $C_6H_8O_6$ (ascorbic acid or vitamin C), (e) $C_{55}H_{72}MgN_4O_5$ (chlorophyll).

2.22 How many (a) moles of $K_2Cr_2O_7$ and (b) atoms of K, Cr, and O are represented by 45.7 g of $K_2Cr_2O_7$?

2.23 The density of water is 1.00 g/mL at 4°C. How many water molecules are present in 2.56 mL of water at this temperature?

*2.24 What mole ratio of molecular chlorine (Cl_2) to molecular oxygen (O_2) would result from the decomposition of the compound Cl_2O_7 into its constituent elements?

2.25 Calculate the numbers of C, H, and O atoms in 1.50 g of sucrose ($C_{12}H_{22}O_{11}$).

Ionic Compounds

2.26 Why are ionic compounds normally not called molecules?

2.27 Give the numbers of protons and electrons in each of the following ions: K^+, Mg^{2+}, Al^{3+}, Fe^{2+}, I^-, F^-, S^{2-}, and O^{2-}.

2.28 Which of the following statements are true and which are false? (a) One mole of KCl is more massive than one mole of NaBr. (b) If F is lighter than Cl, then F^- must be heavier than Cl^-. (c) If we are given the molecular mass of NH_3, we can deduce the atomic masses of N and H without further information.

2.29 Most cyanides are deadly poisonous compounds. For example, ingestion of a quantity as small as 1.00×10^{-3} g of potassium cyanide (KCN) may prove fatal. How many KCN units does this quantity of KCN contain?

Percent Composition and Chemical Formulas

2.30 Calculate the percent by mass of all the elements in each of the following compounds: (a) NaF, (b) NaCl, (c) NaBr, (d) NaI.

2.31 What are the empirical formulas of the compounds with the following compositions? (a) 39.3 percent Na, 60.7 percent Cl; (b) 2.1 percent H, 65.3 percent O, 32.6 percent S; (c) 20.2 percent Al, 79.8 percent Cl; (d) 40.1 percent C, 6.6 percent H, 53.3 percent O; (e) 18.4 percent C, 21.5 percent N, 60.1 percent K.

2.32 For many years chloroform ($CHCl_3$) was used as an inhalation anesthetic in spite of the fact that it is also a toxic substance that may cause severe liver, kidney, and heart damage. Calculate the percent composition by mass of this compound.

2.33 Cinnamic alcohol is mainly used in perfumery, particularly soaps and cosmetics. Its molecular formula is $C_9H_{10}O$. (a) Calculate the percent compositions by mass of C, H, and O in cinnamic alcohol. (b) How many molecules of cinnamic alcohol are contained in a sample weighing 0.469 g?

2.34 The molar mass of caffeine is 194.19 g. Is the molecular formula of caffeine $C_4H_5N_2O$ or $C_8H_{10}N_4O_2$?

2.35 Peroxyacylnitrate (PAN) is one of the components of smog. It is a compound of C, H, N, and O. Determine the percentage of oxygen and the empirical formula from the following percent compositions by mass: C 19.8 percent; H 2.50 percent; N 11.6 percent.

*2.36 Platinum forms two different compounds with chlorine. One contains 26.7 percent Cl by mass and the other contains 42.1 percent Cl by mass. Determine the empirical formulas of the two compounds.

2.37 Rust is iron oxide (Fe_2O_3). How many moles of Fe are present in 24.6 g of the compound?

2.38 Tin(II) fluoride (SnF_2) is often added to toothpaste as an ingredient to prevent tooth decay. What is the mass of F in grams in 24.6 g of the compound?

2.39 Some scientists suspect monosodium glutamate (MSG) to be the cause of "Chinese restaurant syndrome" because they believe that this food-flavor enhancer can induce headaches, chest pains, and other symptoms. (Note that the Chinese people have used MSG for at least a thousand years with no ill effects.) MSG has the following percent composition by mass: 35.51 percent C, 4.77 percent H, 37.85 percent O, 8.29 percent N, and 13.60 percent Na. What is its molecular formula if its molar mass is 169.1 g?

*2.40 A compound containing only C and H is burned in air to form 4.40 g of CO_2 and 0.900 g of H_2O. What is the empirical formula of this compound?

*2.41 A substance is known to contain hydrogen. In a combustion experiment a sample of the substance weighing 0.03600 g is found to yield 0.01890 g of water. Calculate the percent composition by mass of hydrogen in the substance.

2.42 Pheromones are a special type of compound secreted by the females of many insect species to attract the males for mating. One pheromone has the molecular formula $C_{19}H_{38}O$. (a) Calculate the percent composition by mass of this compound. (b) Normally, the amount of this pheromone secreted by a female insect is on the order of 1.0×10^{-12} g. How many molecules of $C_{19}H_{38}O$ are contained in this quantity?

*2.43 When 0.880 g of an organic compound containing C, H, and O is burned completely in oxygen, 1.760 g of CO_2 and 0.720 g of H_2O are produced. (a) Calculate the empirical formula of the compound. (b) The molar mass of the compound is found to be 88.0 g. What is the molecular formula of the compound?

2.44 All of the materials listed below are fertilizers that contribute nitrogen to the soil. Which of these is the richest source of nitrogen on a mass percentage basis?

(a) urea, $(NH_2)_2CO$
(b) ammonium nitrate, NH_4NO_3
(c) guanidine, $HNC(NH_2)_2$
(d) ammonia, NH_3

Law of Multiple Proportions

2.45 Phosphorus forms two compounds with chlorine. In one compound, 1.94 g of P combines with 6.64 g of Cl; in another, 0.660 g of P combines with 3.78 g of Cl. Are these data consistent with the law of multiple proportions?

2.46 Nitrogen forms a number of compounds with oxygen. It is found that in one compound, 0.681 g of N is combined with 0.778 g of O; in another compound, 0.560 g of N is combined with 1.28 g of O. Prove that these data support the law of multiple proportions.

2.47 Elemental analysis of two iron chlorides shows their composition to be 44.06 g of Fe and 55.94 g of Cl; 34.43 g of Fe and 65.57 g of Cl. Are these data consistent with the law of multiple proportions?

*2.48 The following phosphorus sulfides are known: P_4S_3, P_4S_7, and P_4S_{10}. For each compound calculate (a) the molar mass, and (b) the mass of sulfur combined with 1.00 g of phosphorus. Are your results in (b) consistent with the law of multiple proportions?

Naming Compounds

2.49 Name the following compounds: (a) KH_2PO_4, (b) $Sn_3(PO_4)_2$, (c) HBr (gas), (d) HBr (in water), (e) Li_2CO_3, (f) $K_2Cr_2O_7$, (g) NH_4NO_2, (h) PF_3, (i) PF_5, (j) P_4O_6, (k) CdI_2, (l) $SrSO_4$, (m) $Al(OH)_3$, (n) $KClO$, (o) Ag_2CO_3, (p) $FeCl_2$, (q) $KMnO_4$, (r) $CsClO_3$, (s) KNH_4SO_4, (t) FeO, (u) Fe_2O_3, (v) $TiCl_4$, (w) NaH, (x) Li_3N, (y) Na_2O, (z) Na_2O_2.

2.50 Write the formulas for the following compounds: (a) rubidium nitrite, (b) potassium sulfide, (c) sodium hydrogen sulfide, (d) magnesium phosphate, (e) calcium hydrogen phosphate, (f) potassium dihydrogen phosphate, (g) iodine heptafluoride, (h) ammonium sulfate, (i) silver perchlorate, (j) iron(III) chromate, (k) cuprous cyanide, (l) strontium chlorite, (m) perbromic acid, (n) hydroiodic acid, (o) disodium ammonium phosphate, (p) lead(II) carbonate, (q) tin(II) fluoride, (r) tetraphosphorus decasulfide, (s) mercuric oxide, (t) mercury(I) iodide, (u) epsom salt, (v) milk of magnesia, (w) bismuth hydroxide, (x) baking soda, (y) laughing gas, (z) limestone.

2.51 Name the following anions: (a) BrO^-, (b) BrO_2^-, (c) BrO_3^-.

Miscellaneous Problems

2.52 What is wrong or ambiguous with each of the following statements? (a) One mole of hydrogen. (b) The molecular mass of NaCl is 58.5 amu.

2.53 What is the difference between a compound and a molecule?

2.54 Which of the following are elements and which are compounds? (a) SO_2, (b) P_4, (c) Cs, (d) N_2O_5, (e) O, (f) O_2, (g) O_3, (h) CH_4.

2.55 Which of the following has the greater mass,

0.72 g of O_2 or 0.0011 mole of chlorophyll (molecular formula $C_{55}H_{72}MgN_4O_5$)?

2.56 Arrange the following in order of increasing mass: (a) 16 water molecules, (b) 2 atoms of lead, (c) 5.1×10^{-23} mole of helium.

2.57 The atomic mass of an element X is 33.42 amu. A 27.22 g sample of X combines with 84.10 g of another element Y to form a compound XY. Calculate the atomic mass of Y in amu.

2.58 In the formation of carbon monoxide, CO, it is found that 2.445 g of carbon combines with 3.257 g of oxygen. What is the atomic mass of oxygen if the atomic mass of carbon is 12.01 amu?

2.59 Give two examples each of (a) an acid that contains one or more oxygen atoms and (b) an acid that does not.

2.60 Give (a) two examples of a base that contains a hydroxide (OH^-) group and (b) one example of a base that does not.

2.61 Label each of the following compounds as inorganic or organic: (a) NH_4Br, (b) CCl_4, (c) $KHCO_3$, (d) $NaCN$, (e) CH_4, (f) $CaCO_3$, (g) CO, (h) CH_3COOH.

3
CHEMICAL REACTIONS I: CHEMICAL EQUATIONS AND TYPES OF CHEMICAL REACTIONS

A considerable amount of practical information regarding chemical change was known even by ancient civilizations. Evidence of fermentation processes (producing, for example, alcoholic beverages), soap making, metalworking, and other activities involving chemical change has been found among the artifacts of early civilizations in all parts of the world. Such macroscopic knowledge has been of continuing importance to humankind.

As we begin our study of Chapter 3, we will explore how modern chemists represent chemical changes by means of equations and major classifications of types of reactions. Fundamentally, their efforts simply involve faithfully describing the known information regarding specific chemical changes. In this sense the chemist serves as a "reporter" of the observed details of particular chemical systems—clearly a macroscopic concern. In later chapters, however, we will see that the explanation of such chemical changes involves careful attention to a variety of microscopic details at the atomic and subatomic levels.

3.1 CHEMICAL EQUATIONS

A chemical change is called a *chemical reaction.* In order to communicate clearly with one another about chemical reactions, chemists need a standard way to represent them. This section deals with two related aspects of describing chemical reactions by use of chemical symbols—writing chemical equations and balancing chemical equations.

Writing Chemical Equations

Consider what happens when hydrogen gas (H_2) burns in air (which contains molecular oxygen, O_2) to form water. This reaction can be represented as

$$H_2 + O_2 \longrightarrow H_2O \tag{3.1}$$

where the + sign means "reacts with" and the $\longrightarrow$ sign means "to yield." Thus, this symbolic expression can be read: "Hydrogen gas reacts with oxygen gas to yield water." The reaction is assumed to proceed from left to right as written.

The expression (3.1) is not complete, however, because the total number of oxygen atoms on the left side of the arrow (two) is twice that on the right side (one). To conform with the law of conservation of mass, the number of each type of atom must be the same at the start and at the end of the reaction, that is, on both sides of the arrow. Therefore, we must *balance* this expression by placing an appropriate coefficient (2 in this case) in front of H_2 and H_2O:

$$2H_2 + O_2 \longrightarrow 2H_2O$$

This *balanced chemical equation* shows that "two hydrogen molecules can combine (or react) with one oxygen molecule to form two water molecules" (Figure 3.1). Alternatively, the equation can be read as "two moles of hydrogen molecules react with one mole of oxygen molecules to produce two

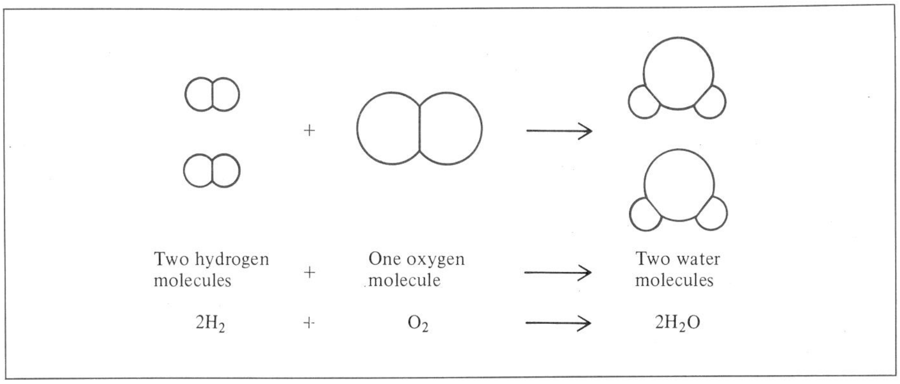

A detailed discussion of balancing chemical equations begins on p. 70.

Two hydrogen molecules   +   One oxygen molecule $\longrightarrow$ Two water molecules

$$2H_2 \quad + \quad O_2 \quad \longrightarrow \quad 2H_2O$$

FIGURE 3.1 *The balanced chemical equation for the combustion of H_2. In accordance with the law of conservation of mass, the number of each type of atom must be the same on both sides of the equation.*

moles of water molecules." Because we know the mass of a mole of each of these substances, we can also interpret the equation to mean "4.04 g of H_2 react with 32.00 g of O_2 to give 36.04 g of H_2O." These three ways of reading the equation are summarized in Table 3.1.

TABLE 3.1 Interpretation of a Chemical Equation

$$2H_2 \qquad + O_2 \qquad \longrightarrow 2H_2O$$

Two molecules + one molecule $\longrightarrow$ two molecules

Two moles + one mole $\longrightarrow$ two moles

$2(2.02 \text{ g}) = 4.04 \text{ g} + 32.00 \text{ g} \qquad \longrightarrow 2(18.02 \text{ g}) = 36.04 \text{ g}$

$\underbrace{\qquad\qquad\qquad\qquad}_{\text{36.04 g Reactants}} \qquad \underbrace{\qquad\qquad\qquad}_{\text{36.04 g Products}}$

In the above equation, we refer to H_2 and O_2 as **reactants,** which are *the starting substances in a chemical reaction.* H_2O is the **product,** which is *the substance formed as a result of a chemical reaction.* A chemical equation, then, can be thought of as the chemist's shorthand description of a reaction. In chemical equations the reactants are conventionally written on the left and the products on the right of the arrow:

$$\text{reactants} \longrightarrow \text{products}$$

It is important to keep in mind that to balance an equation we can change only the coefficients in front of the formulas. If we changed subscripts in the formulas we would change the identity of the compound. For example, $2NO_2$ means "two molecules of nitrogen dioxide," but if we doubled the subscripts, we would have N_2O_4, the formula of dinitrogen tetroxide, which is a completely different molecule.

In writing chemical equations, chemists commonly indicate the physical states of the reactants and products by using the abbreviations *g, l, s,* and *aq* in parentheses to denote gas, liquid, solid, and the aqueous (that is, water) environment, respectively. For example

$$2H_2(g) + O_2(g) \longrightarrow 2H_2O(l)$$

$$CH_4(g) + 2O_2(g) \longrightarrow CO_2(g) + 2H_2O(l)$$

$$2HgO(s) \longrightarrow 2Hg(l) + O_2(g)$$

$$Br_2(l) \longrightarrow Br_2(aq)$$

The last equation describes what happens when liquid bromine dissolves in water. These abbreviations help remind us of the state of the reactants and products, but they are not essential in writing and balancing equations.

Sometimes the condition under which the reaction proceeds is indicated above the arrow. For example, if a reaction is initiated by heating the reactants, the symbol Δ to denote the addition of heat is sometimes used:

$$2KClO_3(s) \xrightarrow{\Delta} 2KCl(s) + 3O_2(g)$$

EXAMPLE 3.1

Describe in words what can be deduced from the following chemical equation:

$$2ZnS(s) + 3O_2(g) \xrightarrow{\Delta} 2ZnO(s) + 2SO_2(g)$$

Answer

1. Solid ZnS (zinc sulfide) reacts with O_2 (oxygen gas) to form solid ZnO (zinc oxide) and SO_2 (sulfur dioxide) gas.
2. Two zinc sulfide units (not molecules, because ZnS is an ionic compound) react with three oxygen molecules to produce two zinc oxide units (ZnO is an ionic compound) and two sulfur dioxide molecules.
3. Two moles of zinc sulfide react with three moles of oxygen molecules to produce two moles of zinc oxide and two moles of sulfur dioxide molecules.
4. 194.9 g of zinc sulfide react with 96.00 g of oxygen to produce 162.8 g of zinc oxide and 128.1 g of sulfur dioxide.
5. Heat is necessary to sustain the reaction.

Similar examples: Problems 3.2, 3.3.

A chemical equation is far from a complete description of what actually occurs in a chemical reaction. It generally describes only the *overall* change, that is, the number and types of atoms, molecules, or ions before and after a reaction, and nothing more. In the earlier example involving water formation, the equation showed that the overall change in the réaction is the disappearance of two H_2 molecules and one O_2 molecule and the appearance of two H_2O molecules. The equation gives no details about how the reaction occurs. For example, it is not at all clear whether the reaction proceeds by the simultaneous attack of two H_2 molecules on an O_2 molecule or by the reaction of a H_2 molecule with an O_2 molecule first, followed by the attack of another H_2 molecule, or by some other route.

Furthermore, a chemical equation does not tell us how long it will take for the change to occur. Do we have to wait for days or months for the reaction to be completed, or will the reaction be finished in a few seconds? In fact, a single equation may represent a very fast reaction as well as a very slow reaction, depending on conditions. The oxidation of iron (that is, the combination of iron with oxygen) can take years if it involves the gradual rusting of an iron nail in an old house, or it can be very rapid if red-hot iron filings react with pure oxygen gas. Yet both of these reactions can be represented by the same chemical equation:

Oxygen gas is gaseous molecular oxygen (O_2).

$$4Fe(s) + 3O_2(g) \longrightarrow 2Fe_2O_3(s)$$

In general, therefore, we must avoid reading into an equation something that is not there.

Balancing Chemical Equations

We will see later that one of the most useful advantages of a chemical equation is that it allows us to predict the quantities of products formed if we know the quantities of reactants used in the reaction. But this prediction is valid only if we are dealing with a *balanced* equation. We will now examine the basic steps involved in balancing simple chemical equations:

- From the identities of the reactants and the products, write a correct formula for each substance on the appropriate side of the equation (that is, on the left or the right side of the arrow). At this stage, you are not expected to predict the actual products if only the reactants are given. Later in your study of chemistry you will be able to make intelligent guesses.
- Once you have written correct formulas for the reactants and the products, balance the equation by inserting suitable coefficients to precede the formulas. The goal is to make the number of atoms of each element the same on both sides of the equation. (Under no circumstances do you change any of the subscripts in the formulas because that would change the identity of the substances.)
- Look for elements that appear only once on each side of the equation and with equal numbers (of atoms) on each side. The compounds containing these elements must have the same coefficients.
- If the preceding step does not apply, look for elements that appear only once on each side of the equation but in unequal numbers (of atoms). Balance these elements first.
- The remaining elements can usually be balanced by simple inspection.

Consider the reaction of iron with hydrochloric acid to produce iron(II) chloride and hydrogen gas. From this information we write

$$Fe + HCl \longrightarrow FeCl_2 + H_2$$

We see that all three elements (Fe, H, and Cl) appear only once on each side of the equation, but only Fe appears in equal numbers of atoms on both sides. Thus Fe and $FeCl_2$ must have the same coefficient. The next step is to equalize either the number of Cl atoms or the number of H atoms on both sides of the equation. To balance Cl, we place the coefficient 2 in front of HCl:

$$Fe + 2HCl \longrightarrow FeCl_2 + H_2 \tag{3.2}$$

We see now that the equation is balanced for all the elements. As a final check, an atom inventory for each element may be tallied:

Reactants	Products
Fe (1)	Fe (1)
H (2)	H (2)
Cl (2)	Cl (2)

where the numbers in parentheses indicate the numbers of atoms. Of course, this equation could be balanced in many other ways, for example

$$2Fe + 4HCl \longrightarrow 2FeCl_2 + 2H_2$$
$$100Fe + 200HCl \longrightarrow 100FeCl_2 + 100H_2$$

For simplicity, the physical states of reactants and products will be omitted.

**Ethane is a component of
natural gas.**

In practice, however, we normally use the *smallest* possible set of whole number coefficients to balance an equation. Equation (3.2) conforms with this conventional practice.

Now let us consider the combustion (that is, burning) of ethane (C_2H_6) in oxygen or air, which yields carbon dioxide and water. We write

$$C_2H_6 + O_2 \longrightarrow CO_2 + H_2O$$

We find that the number of atoms is not the same on both sides of the equation for any of the elements (C, H, and O). We see that C and H appear only once on each side of the equation; O appears in two compounds on the right side (CO_2 and H_2O). To balance the C atoms, we place a 2 in front of CO_2:

$$C_2H_6 + O_2 \longrightarrow 2CO_2 + H_2O$$

To balance the H atoms, we place a 3 in front of H_2O:

$$C_2H_6 + O_2 \longrightarrow 2CO_2 + 3H_2O$$

At this stage, C and H atoms are balanced, but O atoms are not balanced, because there are seven O atoms on the right-hand side and only two O atoms on the left-hand side of the equation. This inequality of O atoms can be eliminated by writing $\frac{7}{2}$ in front of the O_2 on the left-hand side.

$$C_2H_6 + \tfrac{7}{2}O_2 \longrightarrow 2CO_2 + 3H_2O$$

The "logic" of the $\frac{7}{2}$ factor can be clarified as follows. There were seven oxygen atoms on the right-hand side of the equation, but only a pair of oxygen atoms (O_2) on the left. To balance them we ask how many *pairs* of oxygen atoms are needed to equal seven oxygen atoms. Just as 3.5 pairs of shoes equal seven shoes, $\frac{7}{2}O_2$ molecules equal seven O atoms. As the following tally shows, the equation is now completely balanced:

Reactants	Products
C (2)	C (2)
H (6)	H (6)
O (7)	O (7)

However, we normally prefer to express the coefficients as whole numbers rather than as fractions. Therefore, we multiply the entire equation by 2 to convert $\frac{7}{2}$ to 7:

$$2C_2H_6 + 7O_2 \longrightarrow 4CO_2 + 6H_2O$$

The final tally is

Reactants	Products
C (4)	C (4)
H (12)	H (12)
O (14)	O (14)

Note that the coefficients used in balancing the equation are the smallest possible set of whole numbers.

EXAMPLE 3.2

Write a balanced equation for each of the following reactions: (a) copper(II) oxide (CuO) reacts with ammonia (NH_3) to yield copper (Cu), water (H_2O), and nitrogen gas (N_2), (b) ammonia reacts with oxygen gas (O_2) to produce nitric oxide (NO) and water.

Answer

(a)
$$CuO + NH_3 \longrightarrow Cu + H_2O + N_2$$

We see that both Cu and O appear only once on each side of the equation and in equal numbers. Therefore, the coefficient for CuO must be the same as that for Cu and H_2O. We note that both H and N appear only once on each side of the equation but with unequal numbers of atoms. To balance N, we must use the coefficient 2 for NH_3; to balance H, we must use the coefficient 3 for H_2O:

$$CuO + 2NH_3 \longrightarrow Cu + 3H_2O + N_2$$

Since the coefficients for CuO, Cu, and H_2O must all be the same, we write

$$3CuO + 2NH_3 \longrightarrow 3Cu + 3H_2O + N_2$$

The final tally shows

Reactants	*Products*
Cu (3)	Cu (3)
N (2)	N (2)
H (6)	H (6)

Therefore the equation is completely balanced.

(b)
$$NH_3 + O_2 \longrightarrow NO + H_2O$$

Of these three elements (N, H, and O), only N appears once on each side of the equation and in equal number. Therefore, the coefficient for NH_3 must be the same as that for NO. We note that H appears once on each side of the equation but in unequal numbers of atoms (three on the left and two on the right). To balance H, we need to place a coefficient in front of NH_3 such that, when it is multiplied by 3 (the subscript in NH_3), the product will be equal to the product of the coefficient we place in front of H_2O and 2 (the subscript in H_2O); that is, we must have

$$3x = 2y$$

where x and y are the coefficients for NH_3 and H_2O, respectively. By simple inspection, we find that $x = 2$ and $y = 3$, so we write

$$2NH_3 + O_2 \longrightarrow NO + 3H_2O$$

We next balance the N atoms by placing a 2 in front of NO:

$$2NH_3 + O_2 \longrightarrow 2NO + 3H_2O$$

There are two O atoms on the left and five O atoms on the right, so we must place the coefficient $\frac{5}{2}$ in front of O_2:

$$2NH_3 + \tfrac{5}{2}O_2 \longrightarrow 2NO + 3H_2O$$

(Continued)

The equation is now balanced, but to convert all the coefficients into whole numbers, we multiply the equation by 2:

$$4NH_3 + 5O_2 \longrightarrow 4NO + 6H_2O$$

The final tally shows

Reactants	Products
N (4)	N (4)
H (12)	H (12)
O (10)	O (10)

Similar examples: Problems 3.4, 3.5.

3.2 PROPERTIES OF AQUEOUS SOLUTIONS

Now that we have learned the fundamentals for writing and balancing chemical equations, we are ready to study chemical reactions in a systematic way. However, even before we begin to classify reactions into categories, it will be useful to have some knowledge about the medium in which reactions take place. Because many important chemical reactions and virtually all biological processes take place in aqueous media, we will digress for a moment to survey the properties of water solutions. We begin by defining some terms that will be used frequently.

A **solution** is *a homogeneous mixture of two or more substances. The substance present in smaller proportion is called the* **solute,** and *the substance that is present in a larger amount is called the* **solvent.** Solutions may be gaseous (such as air), solid (such as a metal alloy), or liquid (a soft drink, for example). In this section we will discuss only cases in which the solute is a liquid or a solid and the solvent is water—that is, aqueous solutions. An aqueous solution can of course contain more than one kind of solute. Seawater is a solution that contains more than sixty different substances.

Electrolytes versus Nonelectrolytes

All solutes in aqueous solutions can be divided into two categories: electrolytes and nonelectrolytes. An **electrolyte** is *a substance that, when dissolved in water, results in a solution that can conduct electricity.* A **nonelectrolyte** *does not conduct electricity when dissolved in water.* The arrangement shown in Figure 3.2 provides an easy and straightforward method of distinguishing between electrolytes and nonelectrolytes. A pair of platinum electrodes is immersed in a beaker containing water. To light the bulb, electric current must flow from one electrode to the other, thus completing the circuit. Pure water is a very poor conductor of electricity. However, if we add a small amount of an ionic compound, such as sodium chloride (NaCl), the bulb will glow as soon as the salt dissolves in the water. When solid NaCl dissolves in water, it breaks up into Na^+ and Cl^-

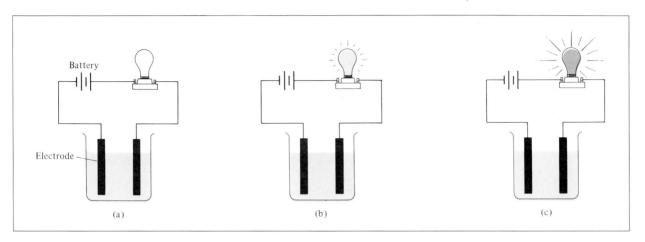

FIGURE 3.2 *An arrangement for distinguishing between electrolytes and non-electrolytes. A solution's ability to conduct electricity depends on the number of ions it contains. (a) A nonelectrolyte solution does not contain ions and the light bulb is not lit. (b) A weak electrolyte solution contains a small number of ions, and the light bulb is dimly lit. (c) A strong electrolyte solution contains a large number of ions, and the light bulb is brightly lit. The molar amounts of the dissolved solutes are equal in all three cases.*

ions. The movement of Na^+ ions toward the negative electrode and of Cl ions toward the positive electrode is equivalent to the flow of electrons along a metal wire. In this way a solution containing an electrolyte becomes able to conduct electricity.

Comparison of the light bulb's brightness for the *same molar amounts* of dissolved substances helps us distinguish between strong and weak electrolytes. Ionic compounds such as sodium fluoride (NaF), lithium chloride (LiCl), and potassium iodide (KI) and certain acids and bases such as hydrochloric acid (HCl), nitric acid (HNO_3), and sodium hydroxide (NaOH) are all strong electrolytes. They share the common characteristic of ionizing completely when dissolved in water:

$$HCl\ (aq) \longrightarrow H^+(aq)\ +\ Cl^-(aq)$$

In other words, *all* the dissolved HCl molecules give H^+ and Cl^- ions in solution. On the other hand, weak electrolytes such as acetic acid (CH_3COOH), which is found in vinegar, ionize much less. We represent the ionization of acetic acid as

$$CH_3COOH(aq) \rightleftharpoons CH_3COO^-(aq)\ +\ H^+(aq)$$

where CH_3COO^- is called the acetate ion.

The double arrow $\rightleftharpoons$ signifies that the reaction is ***reversible,*** that is, *the reaction can occur in both directions.* Initially, a number of CH_3COOH molecules break up to yield CH_3COO^- and H^+ ions. As time goes on, some of the CH_3COO^- and H^+ ions recombine to give CH_3COOH molecules. Eventually a state is reached in which the acid molecules break up as fast as its ions recombine. Such *a chemical state, in which no net change can be observed,* is called ***chemical equilibrium.*** Acetic acid, then, is a

The term *ionization* refers to the breakup of acids and bases into ions in solution; the term *dissociation* applies to the breakup of ionic compounds (other than acids and bases) into ions in solution.

Chemical equilibrium is an extremely important concept, to which we will return in later chapters.

weak electrolyte because its ionization in water is incomplete. By contrast, in a hydrochloric acid solution the H^+ and Cl^- ions have no tendency to recombine to form molecular HCl. Therefore, we use the single arrow to represent the ionization.

Substances that dissolve in water as neutral molecules rather than ions are nonelectrolytes because their solutions do not conduct electricity. Various sugars and alcohols are examples of nonelectrolytes.

Solubility Rules of Electrolytes in Aqueous Solutions

The term *solubility* indicates *the maximum amount of solute that can be dissolved in a given quantity of solvent at a specific temperature.* For example, at 20°C, we can dissolve up to 34.7 g of potassium chloride (KCl) in 100 mL of water. Thus we can represent the solubility of potassium chloride (KCl) as 34.7 g/100 mL H_2O. With a few exceptions, the solubility of electrolytes increases with increasing temperature.

Knowledge about the solubility of substances in water is useful in the study of the physical properties of solutions and the reactions the solutes undergo. Therefore, we will examine briefly some rules that apply to solubility of common electrolytes in water. We divide compounds into three categories called "soluble," "slightly soluble," and "insoluble." For convenience, we define soluble compounds as those having solubilities of 1 gram or more per 100 mL of water; slightly soluble compounds as having solubilities less than 1 gram but more than 0.1 gram per 100 mL of water; and insoluble compounds as having solubilities of less than 0.1 gram per 100 mL of water. In the following rules we assume the temperature to be 25°C.

These rules will be frequently referred to in later chapters. Gradually you will develop a feeling for which compounds are soluble and which are insoluble.

1. All common inorganic acids (see p. 58) are soluble in water.
2. All Group 1A (alkali metal) hydroxides (LiOH, NaOH, KOH, RbOH, and CsOH) are soluble. Of the Group 2A (alkaline earth metal) hydroxides, only barium hydroxide [Ba(OH)$_2$] is soluble. Calcium hydroxide [Ca(OH)$_2$] is slightly soluble. All other hydroxides are insoluble.
3. Most compounds containing chlorides (Cl^-), bromides (Br^-), or iodides (I^-) are soluble. The exceptions are those containing Ag^+, Hg_2^{2+}, and Pb^{2+} ions.
4. Most sulfates (SO_4^{2-}) are soluble. Calcium sulfate ($CaSO_4$) and silver sulfate (Ag_2SO_4) are slightly soluble. Barium sulfate ($BaSO_4$), mercury(II) sulfate ($HgSO_4$), and lead(II) sulfate ($PbSO_4$) are insoluble.
5. All compounds containing nitrate (NO_3^-), chlorate (ClO_3^-), perchlorate (ClO_4^-), and acetate (CH_3COO^-) are soluble.
6. All carbonates (CO_3^{2-}), phosphates (PO_4^{3-}), sulfides (S^{2-}), and sulfites (SO_3^{2-}) are insoluble, except those of ammonium (NH_4^+) and alkali metals.
7. All ammonium (NH_4^+) compounds are soluble.
8. All alkali metal compounds are soluble.

Note that a substance can be both a strong electrolyte *and* insoluble in water. We can take silver chloride (AgCl) as an example. At 10°C, the solubility of AgCl is only 0.000089 g/100 mL H_2O. Thus for all practical purposes, AgCl is a virtually insoluble substance. Yet because the very small amount of solid AgCl that does enter into solution is completely dissociated into Ag^+ and Cl^- ions, AgCl is a strong electrolyte:

$$AgCl(s) \xrightarrow{H_2O} Ag^+(aq) + Cl^-(aq)$$

Writing H_2O above the arrow indicates the physical process of dissolving a substance in water.

EXAMPLE 3.3

Classify the following ionic compounds as soluble, slightly soluble, and insoluble: (a) silver sulfate (Ag_2SO_4), (b) calcium carbonate ($CaCO_3$), (c) sodium phosphate (Na_3PO_4).

Answer

(a) According to solubility rule 4 we see that Ag_2SO_4 is slightly soluble.
(b) Calcium is an alkaline earth metal (a member of the Group 2A elements). According to solubility rule 6, $CaCO_3$ is insoluble.
(c) Sodium is an alkali metal (a member of the Group 1A elements). According to solubility rule 8, Na_3PO_4 is soluble.

Similar example: Problem 3.11.

At this stage we are ready to study various types of chemical reactions. There is no way to be sure, but the number of chemical reactions known to occur is probably in the tens of millions. Many of these, although certainly not all, can be classified into five types of reactions: combination reactions, decomposition reactions, displacement reactions, metathesis (or double displacement) reactions, and neutralization reactions. Classification in this manner gives us an overall perspective of chemical reactions and provides us with a better understanding of the nature of reactions in general.

3.3 COMBINATION REACTIONS

In a ***combination reaction,*** *two or more substances react to produce one product.* This type of reaction can be represented as

$$A + B \longrightarrow AB$$

Figure 3.3 shows three kinds of combination reactions: between the atoms of two elements to form a compound, between the atom of an element and a molecule to form a new compound, and between two compounds to form another compound.

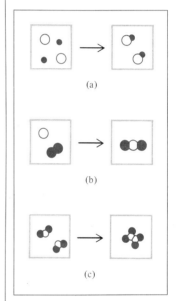

FIGURE 3.3 *Three types of combination reaction. (a) Two atoms of different elements combine to form a compound. (b) An atom and a molecule combine to form a compound. (c) Two compounds combine to form a different compound.*

Combination of Two Elements to Form a Compound

Because oxygen is involved in many chemical processes, the reactions between molecular oxygen and metallic and nonmetallic elements will be discussed first, followed by the reactions between metallic elements and nonmetallic elements other than oxygen.

Combination with oxygen is sometimes called *oxidation*.

Metals with Oxygen. Many metallic elements react directly with molecular oxygen to form metallic oxides. In the alkali metal family (the Group 1A metals; see Figure 1.4), we find that lithium reacts with molecular oxygen to form lithium oxide:

$$4Li(s) + O_2(g) \longrightarrow 2Li_2O(s)$$

whereas sodium reacts with molecular oxygen to form sodium peroxide:

$$2Na(s) + O_2(g) \longrightarrow Na_2O_2(s)$$

and potassium reacts with molecular oxygen to form potassium superoxide:

$$K(s) + O_2(g) \longrightarrow KO_2(s)$$

The alkaline earth metals (Group 2A) all react with molecular oxygen to form oxides (BeO, MgO, CaO, SrO, and BaO). The rapid reaction between magnesium and molecular oxygen

$$2Mg(s) + O_2(g) \longrightarrow 2MgO(s)$$

gives off much light (see Color Plate 11) and is widely used in photographic flash bulbs.

Less reactive metals, such as iron, react more slowly with molecular oxygen:

$$4Fe(s) + 3O_2(g) \longrightarrow 2Fe_2O_3(s)$$

This is the process that leads to the corrosion of the metal (that is, the formation of rust). Metals such as gold and platinum are *inert* toward oxygen, meaning that no combination reaction occurs between these elements and molecular oxygen.

Nonmetals with Oxygen. A number of nonmetallic elements combine directly with molecular oxygen to form oxides. Sometimes the product formed depends on the amount of oxygen available. For example, if the oxygen gas supply is limited, coal or coke burns to form carbon monoxide, a poisonous gas:

$$2C(s) + O_2(g) \longrightarrow 2CO(g)$$

However, if the oxygen gas supply is plentiful, carbon dioxide is formed:

$$C(s) + O_2(g) \longrightarrow CO_2(g)$$

Elemental sulfur burns in air to form sulfur dioxide (see Color Plate 12):

$$S(l) + O_2(g) \longrightarrow SO_2(g)$$

Under ordinary conditions (say, at 25°C) molecular nitrogen does not react with molecular oxygen. The following reaction occurs in a car's engine operating at very high temperatures (above 1000°C):

$$N_2(g) + O_2(g) \xrightarrow{\Delta} 2NO(g)$$

The Group 7A halogens [fluorine (F_2), chlorine (Cl_2), bromine (Br_2), and iodine (I_2)] do not react directly with molecular oxygen.

Compounds containing the O^{2-} ions are called *oxide*, the O_2^{2-} ions are called *peroxide*, and the O_2^- ions are called *superoxide*.

Coal is a black solid formed from fossilized plants. Coke is the product that results from heating coal in an oven or closed chamber.

Metals with Nonmetals. Metals combine with a number of nonmetals other than oxygen. These reactions are usually carried out at elevated temperatures. Some examples are

$$2Na(s) + Cl_2(g) \xrightarrow{\Delta} 2NaCl(s)$$

$$2Cs(l) + I_2(g) \xrightarrow{\Delta} 2CsI(s)$$

$$3Mg(s) + N_2(g) \xrightarrow{\Delta} Mg_3N_2(s)$$

$$Zn(s) + S(l) \xrightarrow{\Delta} ZnS(s)$$

Nonmetals with Nonmetals. A number of nonmetals react with each other to form compounds:

$$P_4(s) + 6Br_2(l) \longrightarrow 4PBr_3(l)$$

$$N_2(g) + 3H_2(g) \longrightarrow 2NH_3(g)$$

$$S(s) + 3F_2(g) \longrightarrow SF_6(g)$$

Combination of an Element and a Compound to Form a New Compound

These reactions are usually observed between an element and a *molecular* compound. Some examples are

$$Cl_2(g) + PCl_3(l) \longrightarrow PCl_5(s)$$

$$F_2(g) + SF_4(g) \longrightarrow SF_6(g)$$

$$Ni(s) + 4CO(g) \longrightarrow Ni(CO)_4(g)$$

Combination of Two Compounds to Form a New Compound

An example of reactions in this category is the combination of gaseous ammonia and hydrogen chloride to form ammonium chloride (Figure 3.4):

$$NH_3(g) + HCl(g) \longrightarrow NH_4Cl(s)$$

Many metallic oxides, especially those of the alkali and alkaline earth metals, react with water to form the corresponding hydroxides:

$$Li_2O(s) + H_2O(l) \longrightarrow 2LiOH(aq)$$

$$MgO(s) + H_2O(l) \longrightarrow Mg(OH)_2(s)$$

Nonmetallic oxides, on the other hand, usually react with water to form the corresponding acids:

$$CO_2(g) + H_2O(l) \longrightarrow H_2CO_3(aq)$$

$$SO_2(g) + H_2O(l) \longrightarrow H_2SO_3(aq)$$

$$P_4O_{10}(s) + 6H_2O(l) \longrightarrow 4H_3PO_4(l)$$

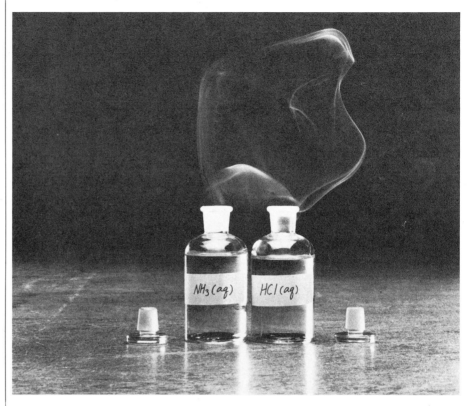

FIGURE 3.4 *HCl gas (from a bottle containing hydrochloric acid) combines rapidly with NH_3 gas (from a bottle containing aqueous ammonia) to form solid NH_4Cl.*

At high temperatures, some metal oxides combine with nonmetal oxides:

$$CaO(s) + CO_2(g) \xrightarrow{\Delta} CaCO_3(s)$$

$$CaO(s) + SiO_2(s) \xrightarrow{\Delta} CaSiO_3(l)$$

$$6Li_2O(s) + P_4O_{10}(l) \xrightarrow{\Delta} 4Li_3PO_4(s)$$

3.4 DECOMPOSITION REACTIONS

Most decomposition reactions are promoted by heat.

In a ***decomposition reaction,*** *one substance undergoes a reaction to produce two or more substances.* Therefore, this type of reaction is the opposite of a combination reaction. It can be represented by

$$AB \longrightarrow A + B$$

Figure 3.5 shows diagrams of two kinds of decomposition reactions. Decomposition reactions are usually carried out by heating the reactant. It is convenient to classify decomposition reactions according to the type of product formed, as follows.

Formation of Molecular Oxygen

When heated, most oxides decompose to form molecular oxygen:

$$2HgO(s) \xrightarrow{\Delta} 2Hg(l) + O_2(g)$$

$$2MgO(s) \xrightarrow{\Delta} 2Mg(s) + O_2(g)$$

$$2N_2O(g) \xrightarrow{\Delta} 2N_2(g) + O_2(g)$$

In the presence of heat or light, hydrogen peroxide also decomposes:

$$2H_2O_2(aq) \longrightarrow 2H_2O(l) + O_2(g)$$

In fact, water itself can be decomposed to form hydrogen and oxygen gas. This reaction is best carried out by *electrolysis*, a process in which decomposition is achieved by passing an electric current through the reactant:

$$2H_2O(l) \longrightarrow 2H_2(g) + O_2(g)$$

On heating, both potassium chlorate ($KClO_3$) and potassium nitrate (KNO_3) give off oxygen gas (Figure 3.6):

$$2KClO_3(s) \xrightarrow{\Delta} 2KCl(s) + 3O_2(g)$$

$$2KNO_3(s) \xrightarrow{\Delta} 2KNO_2(s) + O_2(g)$$

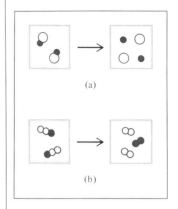

FIGURE 3.5 *Two types of decomposition reaction. (a) A compound decomposes to form atoms of two elements. (b) Two molecular compounds decompose to form three diatomic molecules. Decomposition reactions are the reverse of combination reactions.*

a

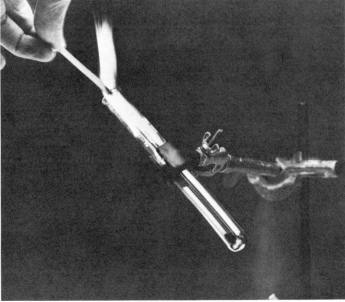

b

FIGURE 3.6 *(a) A piece of burning wood splint over unheated $KClO_3$. (b) Heating $KClO_3$ produces oxygen gas, which further facilitates the combustion process.*

Formation of Carbon Dioxide

All hydrogen carbonates (also called bicarbonates) are decomposed by heat to give carbon dioxide. For example, sodium hydrogen carbonate or sodium bicarbonate ($NaHCO_3$), which is commonly called baking soda, decomposes on heating:

$$2NaHCO_3(s) \xrightarrow{\Delta} Na_2CO_3(s) + H_2O(g) + CO_2(g)$$

The carbon dioxide gas that is released is responsible for the rising of bread, muffins, pancakes, and so on. Sodium bicarbonate is also used as a fire extinguisher because the carbon dioxide formed in the decomposition smothers flames.

Many carbonates also produce carbon dioxide when heated:

$$CaCO_3(s) \xrightarrow{\Delta} CaO(s) + CO_2(g)$$

$$NiCO_3(s) \xrightarrow{\Delta} NiO(s) + CO_2(g)$$

The exceptions are sodium carbonate (Na_2CO_3), potassium carbonate (K_2CO_3), and barium carbonate ($BaCO_3$), which are unaffected even by the highest temperatures attainable in the laboratory.

Carbonate compounds usually decompose at much higher temperatures (600°C–800°C) than bicarbonate compounds (about 200°C).

Formation of Ammonia

All ammonium compounds decompose when heated. Only ammonium chloride (NH_4Cl) and ammonium sulfate [$(NH_4)_2SO_4$] give off ammonia (NH_3) on heating:

$$NH_4Cl(s) \xrightarrow{\Delta} NH_3(g) + HCl(g)$$

$$(NH_4)_2SO_4(s) \xrightarrow{\Delta} 2NH_3(g) + H_2O(g) + SO_3(g)$$

Formation of Water from a Hydrate

When a hydrate (see p. 60) is heated, we can reasonably predict that water is given off first, since it is loosely bound to the parent compound. This is indeed the case for many hydrates:

$$BaCl_2 \cdot 2H_2O(s) \xrightarrow{\Delta} BaCl_2(s) + 2H_2O(g)$$

$$CuSO_4 \cdot 5H_2O(s) \xrightarrow{\Delta} CuSO_4(s) + 5H_2O(g)$$

$$CaCl_2 \cdot 6H_2O(s) \xrightarrow{\Delta} CaCl_2(s) + 6H_2O(g)$$

$$CoCl_2 \cdot 6H_2O(s) \xrightarrow{\Delta} CoCl_2(s) + 6H_2O(g)$$

Color Plate 13 shows the colors of copper(II) sulfate pentahydrate and cobalt(II) chloride hexahydrate and their anhydrous forms.

Note that anhydrous cobalt(II) chloride is frequently used as an indicator for detecting water in desiccators. A *desiccator* is an apparatus for keeping chemicals and small flasks, such as weighing bottles, dry. The drying agent (called the *desiccant*) in the desiccator is anhydrous calcium sulfate, which is mixed with a few small particles of the more expensive anhydrous $CoCl_2$. Because anhydrous $CaSO_4$ has a greater tendency to bind water than anhydrous $CoCl_2$, the former will be used up first. Then, in the presence of moist air, the blue $CoCl_2$ will be converted to the red $CoCl_2 \cdot 6H_2O$. When this happens, it is time to replace the desiccant with a fresh sample. The used desiccant can be regenerated by heating it in an oven to drive off the water, as shown in the above equations.

Both $CaCl_2$ and $CaCl_2 \cdot 6H_2O$ are colorless.

Miscellaneous Decomposition Reactions

Many other decomposition reactions do not fit the categories we have discussed. Some of them are

$$N_2O_4(g) \xrightarrow{\Delta} 2NO_2(g)$$

$$2HCl(g) \xrightarrow{\Delta} H_2(g) + Cl_2(g)$$

$$2NaH(s) \xrightarrow{\Delta} 2Na(l) + H_2(g)$$

These reactions are initiated by heating the reactants.

3.5 DISPLACEMENT REACTIONS

In a **displacement reaction,** *an atom or an ion in a compound is replaced by an atom of another element.* This type of reaction can be represented as

$$A + BC \longrightarrow AC + B$$

where A is an atom and B is either an atom or an ion. Figure 3.7 is a diagram of a displacement reaction. Displacement reactions can be conveniently divided into three major categories.

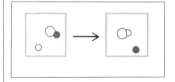

FIGURE 3.7 *A displacement reaction. An atom of one element displaces an atom or an ion of a different element from a compound.*

Hydrogen Displacement

Alkali metals and some alkaline earth metals (Ca, Sr, Ba), which are the most reactive of the metallic elements, will displace hydrogen from cold water (Figure 3.8):

$$2Na(s) + 2H_2O(l) \longrightarrow 2NaOH(aq) + H_2(g)$$

$$Ca(s) + 2H_2O(l) \longrightarrow Ca(OH)_2(s) + H_2(g)$$

Part of the H in the original H_2O appears in the metal hydroxide.

Less reactive metals, such as aluminum (Al, Group 3A) and other metals such as iron (Fe) and chromium (Cr) react with steam to give hydrogen gas:

FIGURE 3.8 *Reactions of Na (left beaker) and Ca (right beaker) with cold water. Note that the reaction is more violent with Na than with Ca.*

$$2Al(s) + 3H_2O(g) \longrightarrow Al_2O_3(s) + 3H_2(g)$$

$$3Fe(s) + 4H_2O(g) \longrightarrow Fe_3O_4(s) + 4H_2(g)$$

$$2Cr(s) + 3H_2O(g) \longrightarrow Cr_2O_3(s) + 3H_2(g)$$

Note that hydrogen gas is formed in each case, but the more active metals form hydroxides and the less active ones form oxides.

Many metals, including those that do not react with water, are capable of displacing hydrogen from acids. For example, cadmium (Cd) and tin (Sn) do not react with water but do react with hydrochloric acid, as follows:

$$Cd(s) + 2HCl(aq) \longrightarrow CdCl_2(aq) + H_2(g)$$

$$Sn(s) + 2HCl(aq) \longrightarrow SnCl_2(aq) + H_2(g)$$

On the other hand, copper and silver do not react even with hydrochloric acid. An easy way to predict whether a displacement reaction will actually occur is to refer to an ***activity series*** (sometimes called the ***electromotive series***), shown in Figure 3.9. Fundamentally, an activity series such as this can be considered simply *a convenient summary of the results of many possible displacement reactions* similar to those already described. According to this series, any metal above hydrogen will displace it from water or from an acid, but metals below hydrogen will not react with either water or an acid. Figure 3.10 shows the reactions between HCl and the metals Fe, Zn, and Mg.

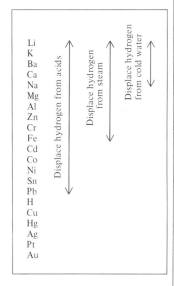

FIGURE 3.9 *The activity series for metals. The metals are arranged according to their ability to displace hydrogen from an acid or water. Li (lithium) is the most reactive metal listed and Au (gold) the least reactive.*

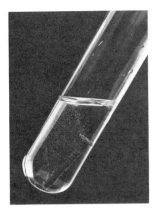 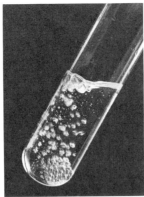 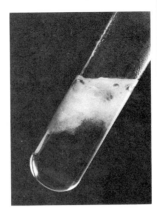

FIGURE 3.10 *From left to right: the reactions of iron (Fe), zinc (Zn), and magnesium (Mg) with hydrochloric acid, to form hydrogen gas and the corresponding metal chlorides ($FeCl_2$, $ZnCl_2$, $MgCl_2$) in solution. The reactivity of these metals is reflected in the rate of hydrogen gas evolution, which is slowest for the most unreactive metal Fe and fastest for the most reactive metal Mg.*

Metal Displacement

The activity series can also be used to predict the likelihood of a displacement reaction of one metal in a compound by another. A metal can displace from a compound any metal listed below it, but it cannot displace any metal above it. For example, when metallic zinc is added to a solution containing copper(II) sulfate, the zinc metal displaces copper(II) ions from the solution (see Color Plate 14):

$$Zn(s) + CuSO_4(aq) \longrightarrow ZnSO_4(aq) + Cu(s)$$

Similarly, metallic copper displaces silver ions from a solution containing silver nitrate:

$$Cu(s) + 2AgNO_3(aq) \longrightarrow Cu(NO_3)_2(aq) + 2Ag(s)$$

Reversing the roles of these metals would result in no reactions. Thus copper metal will not displace zinc from zinc sulfate, and silver metal will not displace copper from copper nitrate:

$$Cu(s) + ZnSO_4(aq) \longrightarrow \text{no reaction}$$

$$Ag(s) + Cu(NO_3)_2(aq) \longrightarrow \text{no reaction}$$

The equations representing displacement reactions are called ***molecular equations*** because *the formulas of the compounds are written as though all species existed as molecules or whole units.* This is not true in reality. As pointed out earlier, when ionic compounds dissolve in water, they break apart into their component cations and anions. Therefore these equations, to be more faithful to reality, should be written to indicate the dissociation of dissolved ionic compounds into ions. Returning to the reaction between zinc and copper(II) sulfate we would thus write

$$Zn(s) + Cu^{2+}(aq) + SO_4^{2-}(aq) \longrightarrow Cu(s) + Zn^{2+}(aq) + SO_4^{2-}(aq)$$

In chemistry, the word "species" is used to mean atoms, molecules, ions, or atomic particles such as electrons, protons, and neutrons.

Such an equation, which *shows dissolved ionic compounds in terms of their free ions*, is called an **ionic equation**. *Ions that are not involved in the overall reaction*, in this case the sulfate ions (SO_4^{2-}), are called **spectator ions**. The **net ionic equation**—that is, *the equation that indicates only the species that actually take part in the reaction*—is given by

$$Zn(s) + Cu^{2+}(aq) \longrightarrow Zn^{2+}(aq) + Cu(s)$$

For each hydrogen displacement reaction discussed earlier, we can write a similar net ionic equation. For example, in considering the reaction between magnesium and hydrochloric acid, we note that HCl is completely ionized into H^+ and Cl^- ions. However, the actual reaction occurs only between Mg and H^+ ions. We can thus write the net ionic reaction as

$$Mg(s) + 2H^+(aq) \longrightarrow Mg^{2+}(aq) + H_2(g)$$

Halogen Replacement

An analogous activity series also applies to the halogens. The reactivity of these nonmetallic elements decreases as we move down the group from fluorine to iodine in Group 7A, so molecular fluorine can replace chloride, bromide, and iodide in solution. In fact, molecular fluorine is so reactive that it also attacks water; thus these reactions cannot be carried out in aqueous solutions. On the other hand, molecular chlorine can displace bromides and iodides in solution (see Color Plate 15):

$$Cl_2(g) + 2KBr(aq) \longrightarrow 2KCl(aq) + Br_2(l)$$
$$Cl_2(g) + 2NaI(aq) \longrightarrow 2NaCl(aq) + I_2(s)$$

and molecular bromine can displace iodide in solution:

$$Br_2(l) + 2KI(aq) \longrightarrow 2KBr(aq) + I_2(s)$$

Reversing the roles of the halogens leads to no reaction:

$$Br_2(l) + 2LiCl(aq) \longrightarrow \text{no reaction}$$
$$I_2(s) + 2KBr(aq) \longrightarrow \text{no reaction}$$

F_2 attacks water, Cl_2 reacts with water to a very small extent, and Br_2 and I_2 do not react with water.

EXAMPLE 3.4

Write balanced molecular and net ionic equations for the following displacement reactions: (a) cadmium + tin(II) sulfate, (b) lithium + water.

Answer

(a) In the activity series we see that cadmium (Cd) appears above tin (Sn). Therefore, Cd can displace tin(II) ions (Sn^{2+}) from $SnSO_4(aq)$. The balanced molecular equation is

$$Cd(s) + SnSO_4(aq) \longrightarrow CdSO_4(aq) + Sn(s)$$

and the net ionic equation is

$$Cd(s) + Sn^{2+}(aq) \longrightarrow Cd^{2+}(aq) + Sn(s)$$

The sulfate ions serve as spectator ions in this reaction.

(b) Since lithium appears above hydrogen in the activity series, it can displace hydrogen from water to form lithium hydroxide (LiOH) and hydrogen (H_2) gas. The molecular equation is

$$2Li(s) + 2H_2O(l) \longrightarrow 2LiOH(aq) + H_2(g)$$

Because water is not ionized to any appreciable extent (it is a very weak electrolyte), the reactants do not appear in ionic form. However, the product LiOH can be expressed as ionized species, so that the net ionic equation is

$$2Li(s) + 2H_2O(l) \longrightarrow 2Li^+(aq) + 2OH^-(aq) + H_2(g)$$

Similar example: Problem 3.18.

3.6 METATHESIS REACTIONS

A **metathesis reaction** is *a double displacement reaction*. It can be represented by

$$AD + BC \longrightarrow AC + BD$$

Figure 3.11 is a diagram of a metathesis reaction. In many metathesis reactions, one product formed is a **precipitate,** *an insoluble solid that separates from the solution.* For example, consider what happens when an aqueous solution of barium chloride is mixed with an aqueous solution of sodium sulfate (Figure 3.12):

$$BaCl_2(aq) + Na_2SO_4(aq) \longrightarrow 2NaCl(aq) + BaSO_4(s)$$

In Section 3.2 we noted that barium sulfate ($BaSO_4$) is insoluble (solubility rule 4, p. 76). Therefore, it forms a precipitate in solution. To focus on the change that actually occurs, we write the net ionic equation

$$Ba^{2+}(aq) + SO_4^{2-}(aq) \longrightarrow BaSO_4(s)$$

The Na^+ and Cl^- ions remain in solution as spectator ions.

Another metathesis reaction involves potassium chloride and silver nitrate:

$$KCl(aq) + AgNO_3(aq) \longrightarrow KNO_3(aq) + AgCl(s)$$

From solubility rule 3 we note that silver chloride (AgCl) is an insoluble compound and should therefore precipitate from the solution. The net ionic equation is

$$Ag^+(aq) + Cl^-(aq) \longrightarrow AgCl(s)$$

In general, then, we can use solubility rules to predict the outcome of ionic metathesis reactions.

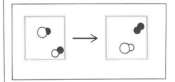

FIGURE 3.11 *A metathesis reaction. We can think of a metathesis reaction as a double displacement reaction that usually takes place between two ionic compounds. The cations and anions "change partners" in such a reaction.*

FIGURE 3.12 *When an aqueous solution of barium chloride ($BaCl_2$) is mixed with an aqueous solution of sodium sulfate (Na_2SO_4), a finely divided white precipitate, barium sulfate ($BaSO_4$), is formed.*

EXAMPLE 3.5

Predict the products of the following metathesis reaction, and write a net ionic equation for the reaction.

$$K_3PO_4(aq) + Ca(NO_3)_2(aq) \longrightarrow ?$$

Answer

Both reactants are soluble salts, but according to solubility rule 5, calcium ions $[Ca^{2+}(aq)]$ and phosphate ions $[PO_4^{3-}(aq)]$ can form an insoluble compound, calcium phosphate $[Ca_3(PO_4)_2]$. Therefore this is a precipitation reaction. The other product, potassium nitrate (KNO_3), is a soluble compound and therefore remains in solution. The molecular equation is

$$2K_3PO_4(aq) + 3Ca(NO_3)_2(aq) \longrightarrow 6KNO_3(aq) + Ca_3(PO_4)_2(s)$$

and the net ionic equation is

$$3Ca^{2+}(aq) + 2PO_4^{3-}(aq) \longrightarrow Ca_3(PO_4)_2(s)$$

3.7 NEUTRALIZATION REACTIONS

A **neutralization reaction** is *a reaction between an acid and a base.* Among the acids commonly used in the laboratory are hydrochloric acid (HCl), nitric acid (HNO_3), acetic acid (CH_3COOH), and sulfuric acid (H_2SO_4). The first three acids are **monoprotic acids,** that is, *each unit of the acid yields one hydrogen ion:*

$$HCl(aq) \longrightarrow H^+(aq) + Cl^-(aq)$$

$$HNO_3(aq) \longrightarrow H^+(aq) + NO_3^-(aq)$$

$$CH_3COOH(aq) \rightleftharpoons CH_3COO^-(aq) + H^+(aq)$$

As we mentioned earlier, because the ionization of CH_3COOH is incomplete (note the use of double arrows), it is a weak electrolyte. For this reason it is called a weak acid. On the other hand, both HCl and HNO_3 are strong acids because they are strong electrolytes (note the use of single arrows).

Sulfuric acid (H_2SO_4) is an example of a **diprotic acid** because *each unit of the acid yields two H^+ ions,* in two separate steps:

$$H_2SO_4(aq) \longrightarrow H^+(aq) + HSO_4^-(aq)$$

$$HSO_4^-(aq) \rightleftharpoons H^+(aq) + SO_4^{2-}(aq)$$

H_2SO_4 **is a strong acid, and HSO_4^- is a weak acid.**

H_2SO_4 is a strong electrolyte (note the first step of ionization is complete), but HSO_4^- is a weak electrolyte and we need to use the double arrow sign to represent its incomplete ionization.

Bases such as sodium hydroxide (NaOH) and barium hydroxide $[Ba(OH)_2]$ are strong electrolytes and hence strong bases:

$$NaOH(s) \xrightarrow{H_2O} Na^+(aq) + OH^-(aq)$$

$$Ba(OH)_2(s) \xrightarrow{H_2O} Ba^{2+}(aq) + 2OH^-(aq)$$

We noted in Section 2.9 that ammonia (NH_3), which does not contain a hydroxide group, is classified as a base because it ionizes partially to form OH^- ions when it dissolves in water:

$$NH_3(aq) + H_2O(l) \rightleftharpoons NH_4^+(aq) + OH^-(aq)$$

Ammonia is a weak electrolyte and is therefore termed a weak base because not all dissolved NH_3 molecules are converted to NH_4^+ and OH^- ions.

The most common bases used in the laboratory are sodium hydroxide and aqueous ammonia solution. The Group 2A metal hydroxides, with the exception of barium hydroxide [$Ba(OH)_2$], are all insoluble.

An acid-base neutralization reaction can be represented as

$$acid + base \longrightarrow salt + water$$

Neutralization reactions are identified by the following characteristics: (1) The reactants always include an acid and a base, and (2) the products are a salt and usually water. A **salt** is *an ionic compound made up of a cation other than H^+ and an anion other than OH^- or O^{2-}*. Some acid-base neutralization reactions are

$$HCl(aq) + NaOH(aq) \longrightarrow NaCl(aq) + H_2O(l)$$

$$2HCl(aq) + Ba(OH)_2(aq) \longrightarrow BaCl_2(aq) + 2H_2O(l)$$

$$H_2SO_4(aq) + 2NaOH(aq) \longrightarrow Na_2SO_4(aq) + 2H_2O(l)$$

Note that these neutralization reactions can be considered a special case of metathesis reactions. In each case the net ionic equation is simply the combining of hydrogen ions and hydroxide ions to form water:

$$H^+(aq) + OH^-(aq) \longrightarrow H_2O(l)$$

Since water is a very weak electrolyte and does not ionize to any appreciable extent, this reaction is assumed to go essentially to completion from left to right, that is, all of the reactants are converted to products.

Aqueous ammonia is sometimes erroneously called ammonium hydroxide. There is no evidence to support the existence of the NH_4OH species.

3.8 HOW CAN WE TELL IF A REACTION WILL OCCUR?

You have now been introduced to five major types of chemical reactions. As we observed earlier, these reaction types include most, but certainly not all, the reactions that we will study. When we consider any chemical system, a fundamental question is whether a reaction will occur when the reactants are brought together at specified conditions (that is, at a certain temperature, pressure, and with specific reactant amounts).

In theory it should be possible to predict the products of any reaction using the concepts (which we will study later) of chemical bonding, chemical kinetics, and laws of thermodynamics. In practice, most reactions are too complicated for us to treat precisely by applying those principles. Although such principles provide valuable insights into the behavior of chemical systems, chemists conduct many reactions with no clear ideas of what specific products may form. This is not to say that intelligent guesses cannot be made. As your knowledge of chemistry grows, you will gradually develop an informed sense for judging which reactions are more likely to occur than others.

In predicting the outcome of a chemical reaction, it is necessary to consider the "driving force" of the reaction. A **driving force** is *needed for reactants to be largely or totally converted to products.* We will describe three types of driving forces, using the reaction types introduced in this chapter.

Formation of a Gaseous Product

In a decomposition reaction one of the products may be a gas. If the gas formed is allowed to escape from the reaction flask, then a reaction like the one shown here can be carried to completion:

$$2HgO(s) \longrightarrow 2Hg(l) + O_2(g)$$

The driving force here is the escape of the oxygen gas from the reaction flask. Of course, in order for the reaction to begin, the reactant must first be heated to the appropriate temperature.

Formation of a Precipitate

The driving force in a metathesis reaction is the formation of a precipitate that separates from the solution. For example

$$Pb(NO_3)_2(aq) + 2KI(aq) \longrightarrow 2KNO_3(aq) + PbI_2(s)$$

or, written as a net ionic equation

$$Pb^{2+}(aq) + 2I^-(aq) \longrightarrow PbI_2(s)$$

Thus whenever a precipitate, such as lead(II) iodide [$PbI_2(s)$], is formed, we can usually conclude that the reaction will proceed from left to right to a large extent.

Formation of Water in a Neutralization Reaction

In acid-base neutralization reactions, H^+ and OH^- ions combine to form water:

$$H^+(aq) + OH^-(aq) \longrightarrow H_2O(l)$$

Because H_2O is largely nonionized, the driving force is the formation of water molecules. Thus acid-base neutralizations usually proceed essentially to completion.

AN ASIDE ON THE PERIODIC TABLE
Reactivity of Metallic Elements

The activity series introduced in Section 3.5 is a useful reference for us to compare the reactivity of metals. The discussion there is focused on the relative ease with which a metal can displace hydrogen from water or an acid. Although the activity series is useful in predicting the outcome of displacement reactions, it is difficult to remember the series in detail. As a handy guide we can use the periodic table to help us study the trends and variations in the reactivity of metals. In many instances it is sufficient just to see in the table approximately where a metallic element is, in relation to others, in order to have some idea of how this element will react with other substances. In addition, since a large number of metals do not react with water or acid, but do combine with molecular oxygen, a good general comparison of the reactivity of metallic elements is to examine how easily they form oxides.

Figure 3.13 shows how metals vary in the ease of their combination with molecular oxygen, according to their positions in the periodic table. The metals can be divided into three categories, as shown. The Group 1A alkali metals, the Group 2A alkaline earth metals (with the exception of beryllium), and aluminum (in Group 3A) are the most reactive metals. A number of metals such as chromium (Cr), manganese (Mn), iron (Fe), and zinc (Zn) are intermediate in reactivity. Metals like gold and platinum show little tendency to combine with molecular oxygen. These two metals are often described as being chemically inert.

Because the alkali and alkaline earth metals are so reactive, they are never found in the uncombined state in nature. Similarly, aluminum is highly reactive and is always

FIGURE 3.13 Variation in the ease of combination of a metal with molecular oxygen, according to its position in the periodic table.

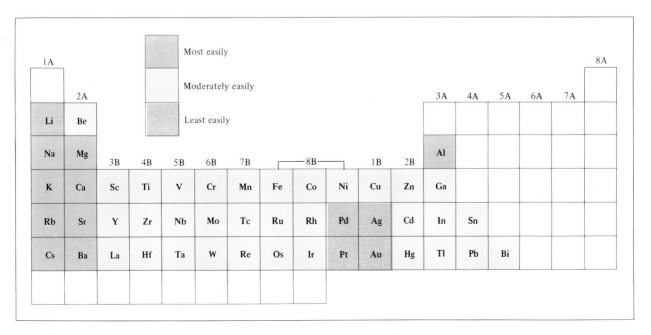

found in nature combined with other elements. Yet we often see aluminum metal that appears quite stable although it is exposed to air and water. (For example, a discarded soft drink can may withstand months of harsh weather without visible change.) This is because aluminum readily forms the oxide Al_2O_3 when exposed to air:

$$4Al(s) + 3O_2(g) \longrightarrow 2Al_2O_3(s)$$

It is a tenacious film of this oxide that protects the aluminum metal underneath from further reaction. On the other hand, the rust (Fe_2O_3) formed when iron is exposed to air and water is quite porous in texture and lacks the ability to protect the metal underneath, as in the case of aluminum. Finally, we note that the inertness of metals like gold and platinum is one of the reasons they find widespread use in jewelry.

SUMMARY

1. Chemical changes, called chemical reactions, are represented by chemical equations. Substances that undergo change—the reactants—are written on the left and substances formed—the products—appear on the right of the arrow.
2. Chemical equations must be balanced, in accordance with the law of conservation of mass. The number of atoms of each type of element in the reactants and products must be equal.
3. Aqueous solutions are electrically conducting if the solutes are electrolytes. If the solutes are nonelectrolytes, the solutions do not conduct electricity.
4. Five major types of chemical reactions are combination, decomposition, displacement, metathesis, and neutralization reactions.
5. An acid yields H^+ ions in solution and a base yields OH^- ions in solution. The reaction of an acid with a base is called neutralization. Acid-base neutralization is a special type of metathesis reaction.
6. In some cases, we can predict the outcome and extent of a reaction by recognizing a driving force in the reaction. The three types of driving forces we have identified so far are the formation of a gas, the formation of a precipitate, and the formation of water (in an acid-base neutralization reaction).

KEY WORDS

Activity series, p. 84
Chemical equilibrium, p. 75
Combination reaction, p. 77
Decomposition reaction, p. 80
Diprotic acid, p. 88
Displacement reaction, p. 83

Driving force, p. 90
Electrolyte, p. 74
Electromotive series, p. 84
Ionic equation, p. 86
Metathesis reaction, p. 87
Molecular equation, p. 85

PROBLEMS

More challenging problems are marked with an asterisk.

Chemical Equations

3.1 What is the difference between a chemical reaction and a chemical equation?

3.2 Describe in words the meaning of the following equation:

$$CaC_2(s) + 2H_2O(l) \longrightarrow Ca(OH)_2(s) + C_2H_2(g)$$

  calcium   water   calcium   acetylene
  carbide   hydroxide

3.3 From the chemical equation

$$N_2(g) + 3H_2(g) \longrightarrow 2NH_3(g)$$

which of the following characteristics or quantities can be deduced? (a) The reaction is started by heating. (b) Three moles of molecular hydrogen react with one mole of molecular nitrogen to form two moles of ammonia. (c) The reaction is essentially completed in a few minutes after mixing the reactants. (d) All substances are gases in this reaction. (e) This reaction takes place by one N_2 molecule and three H_2 molecules colliding with one another. (f) Six hundred molecules of H_2 would react with two hundred N_2 molecules.

3.4 Balance the following equations:

(a) $H_2 + I_2 \longrightarrow HI$
(b) $Na + H_2O \longrightarrow NaOH + H_2$
(c) $CO + O_2 \longrightarrow CO_2$
(d) $KClO_3 \longrightarrow KCl + O_2$
(e) $KNO_3 \longrightarrow KNO_2 + O_2$
(f) $NaOH + H_2SO_4 \longrightarrow Na_2SO_4 + H_2O$
(g) $HCl + CaCO_3 \longrightarrow CaCl_2 + H_2O + CO_2$
(h) $C_2H_6 + O_2 \longrightarrow CO_2 + H_2O$
(i) $Mg + O_2 \longrightarrow MgO$
(j) $P_4O_{10} + H_2O \longrightarrow H_3PO_4$
(k) $Al + H_2SO_4 \longrightarrow Al_2(SO_4)_3 + H_2$
(l) $CO_2 + KOH \longrightarrow K_2CO_3 + H_2O$
(m) $Zn + AgCl \longrightarrow ZnCl_2 + Ag$
(n) $Cl_2 + NaBr \longrightarrow NaCl + Br_2$
(o) $Ni(NO_3)_2 + NaOH \longrightarrow Ni(OH)_2 + NaNO_3$
(p) $H_2O_2 \longrightarrow H_2O + O_2$
(q) $Cu + HNO_3 \longrightarrow Cu(NO_3)_2 + NO_2 + H_2O$
(r) $O_3 \longrightarrow O_2$
(s) $N_2 + H_2 \longrightarrow NH_3$
(t) $KOH + H_3PO_4 \longrightarrow K_3PO_4 + H_2O$

3.5 Balance the following equations:

(a) $CH_4 + O_2 \longrightarrow CO_2 + H_2O$
(b) $CH_4 + Br_2 \longrightarrow CBr_4 + HBr$
(c) $Fe_2O_3 + CO \longrightarrow Fe + CO_2$
(d) $S_8 + O_2 \longrightarrow SO_2$
(e) $HNO_3 + S \longrightarrow H_2SO_4 + H_2O + NO_2$
(f) $Be_2C + H_2O \longrightarrow Be(OH)_2 + CH_4$

Aqueous Solutions

3.6 What is the difference between an electrolyte and a nonelectrolyte?

3.7 Water, as we know, is a nonelectrolyte and therefore cannot conduct electricity. Yet we are often cautioned not to operate electrical appliances when our hands are wet. Why?

3.8 Identify each of the following substances as a strong electrolyte, a weak electrolyte, or a nonelectrolyte: (a) H_2O, (b) $RbBr$, (c) H_2SO_4, (d) CH_3COOH, (e) $C_6H_{12}O_6$, (f) $MgCl_2$, (g) N_2, (h) NH_3, (i) $NaOH$.

3.9 The passage of electricity through an electrolyte solution is caused by the movement of (a) electrons only, (b) cations only, (c) anions only, (d) both cations and anions.

3.10 Predict and explain which of the following systems are electrically conducting: (a) solid NaCl, (b) molten NaCl, (c) an aqueous solution of NaCl.

Solubility of Ionic Compounds

3.11 Classify the following compounds as soluble or insoluble in water: (a) $Ca_3(PO_4)_2$, (b) $Mn(OH)_2$, (c)

$AgClO_3$, (d) K_2S, (e) $CaCO_3$, (f) $Mg(CH_3COO)_2$, (g) $Hg(NO_3)_2$, (h) $HgSO_4$, (i) NH_4ClO_4, (j) $BaSO_3$.

3.12 Name two soluble metal hydroxides and two insoluble metal hydroxides.

*3.13 Fluorides, that is, salts containing F^- ions, are not mentioned in the solubility rules in Section 3.2. Look up the solubilities of some metal fluorides in the *Handbook of Chemistry and Physics*, including LiF, CaF_2, and AgF. Does F^- fit with the other halides (that is, salts of Cl^-, Br^-, and I^-) in the solubility rules? How would you place fluorides in the solubility rules?

*3.14 On the basis of the solubility rules given in this chapter, suggest one method by which you might separate (a) K^+ from Ag^+, (b) Ag^+ from Pb^{2+}, (c) NH_4^+ from Ca^{2+}, (d) Ba^{2+} from Cu^{2+}. All cations are assumed to be in aqueous solution and the common anion is the nitrate ion.

3.15 Which of the following processes will result in a precipitation reaction? (a) mixing a $NaNO_3$ solution with a $CuSO_4$ solution, (b) mixing a $BaCl_2$ solution with a K_2SO_4 solution.

3.16 Give an example of a strong electrolyte that is insoluble and a weak electrolyte that is soluble in water.

Ionic Equations

3.17 What is the difference between an ionic equation and a molecular equation? What is the advantage of writing a net ionic equation?

*3.18 Write ionic and net ionic equations for the following reactions:

(a) $2AgNO_3(aq) + Na_2SO_4(aq) \longrightarrow$
(b) $BaCl_2(aq) + ZnSO_4(aq) \longrightarrow$
(c) $(NH_4)_2CO_3(aq) + CaCl_2(aq) \longrightarrow$
(d) $Na_2S(aq) + ZnCl_2(aq) \longrightarrow$
(e) $2K_3PO_4(aq) + 3Sr(NO_3)_2(aq) \longrightarrow$

(*Hint:* Use the solubility rules in Section 3.2 to help you predict the products.)

Types of Reactions

3.19 Balance the following equations and classify each reaction as one of the following types: combination, decomposition, displacement, metathesis, neutralization.

(a) $B_2H_6(g) \longrightarrow B(s) + H_2(g)$
(b) $CaCl_2(aq) + Na_2CO_3(aq) \longrightarrow$
$$CaCO_3(s) + NaCl(aq)$$
(c) $Al(s) + O_2(g) \longrightarrow Al_2O_3(s)$
(d) $Cd(s) + AgNO_3(aq) \longrightarrow Cd(NO_3)_2(aq) + Ag(s)$
(e) $CH_3SnH_3(g) \longrightarrow CH_4(g) + Sn(s) + H_2(g)$
(f) $Ba(OH)_2(aq) + HCl(aq) \longrightarrow$
$$BaCl_2(aq) + H_2O(l)$$

3.20 Which of the following metals can react with water? (a) Au, (b) Li, (c) Hg, (d) Ca, (e) Pt.

3.21 Predict the outcome of the reactions represented by the following equations by using the activity series, and balance the equations:

(a) $Cu(s) + HCl(aq) \longrightarrow$
(b) $I_2(s) + NaBr(aq) \longrightarrow$
(c) $Mg(s) + CuSO_4(aq) \longrightarrow$
(d) $Cl_2(g) + KBr(aq) \longrightarrow$

*3.22 Write a balanced equation for the preparation of (a) molecular oxygen, (b) ammonia, (c) carbon dioxide, (d) molecular hydrogen. Indicate the physical state of the reactants and products in each equation.

4

CHEMICAL REACTIONS II: MASS RELATION- SHIPS

A properly written and balanced chemical equation clearly specifies the identities of all reactants and products, as Chapter 3 illustrated. However, even when all the substances in a chemical system have been identified, a major macroscopic concern still remains to be addressed. The concern can be summed up in a simple two-word question: How much?

How much reactant is used up? How much product is formed? How much (if any) of a reactant remains when the reaction is finished? These are important questions that apply to theoretical considerations (What do these quantities tell us about microscopic behavior of atoms and molecules?) as well as to practical problems (Can a chemical industry produce enough product to make the chemical reaction profitable? Do I have sufficient fuel to heat the water to boiling?). Thus, this chapter is a "how much?" look at chemical reactions—a macroscopic view of great interest and importance not only to chemists but to anyone who encounters chemical changes in daily life. And that includes all of us!

95

4.1 AMOUNTS OF REACTANTS AND PRODUCTS

The mass relationships among reactants and products in a chemical re-action represent the **stoichiometry** of the reaction. To interpret such a reaction quantitatively, we need to apply our knowledge of molar masses and the mole concept. The basic question posed in many stoichiometric calculations is, "If we know the quantities of the starting materials (that is, the reactants) in a reaction, how much product will be formed?" There are, of course, various ways to do such calculations, since the quantities of interest may be expressed in moles, grams, liters (for gases), or other units. One useful *approach to determine the amount of product formed in a reaction* is called the **mole method.** It is based on the fact that *the stoichiometric coefficients in a chemical equation can be interpreted as the number of moles of each substance.* Referring to the formation of water from hydrogen and oxygen gases discussed in Chapter 3

$$2H_2 + O_2 \longrightarrow 2H_2O$$

The equation and the stoichiometric coefficients can be "read" as two moles of hydrogen gas combining with one mole of oxygen gas to form two moles of water. The mole method as commonly practiced consists of the following steps:

1. Obtain correct formulas for all reactants and products, and balance the equation under study.
2. Convert all the given or known quantities of substances (usually reactants) into moles.
3. Use the coefficients of substances in the balanced equation to calculate the moles of the sought or unknown quantities (usually products) in the problem.
4. From the calculated moles and the molar masses, convert the unknown quantities to whatever units are required (typically grams).

Step 1 is obviously a prerequisite to any stoichiometric calculation. We must know the identities of the reactants and the products, and the mass relationships among them must not violate the law of conservation of mass. Step 2 is the critical step of converting grams (or other units) of substances to number of moles. This conversion then allows us to analyze the actual reaction in terms of moles only.

To complete step 3 we need the balanced equation, already furnished by step 1. The key point here is that the coefficients in a balanced equation provide us with the ratio in which moles of one substance react with or form moles of another substance. Step 4 is similar to step 2, except that now we are dealing with the quantities sought in the problem. Figure 4.1 shows three types of stoichiometric calculations commonly encountered.

In discussing quantitative relationships it is useful to introduce the symbol ⇌, which means "stoichiometrically equivalent to" or simply "equivalent to." In the balanced equation for the formation of water, we see that

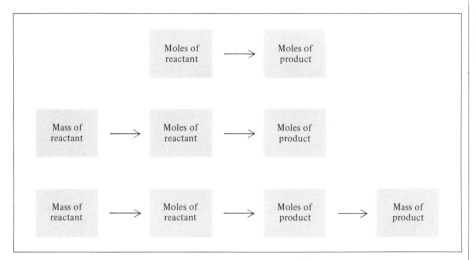

FIGURE 4.1   *Three types of stoichiometric calculations based on the mole method.*

two moles of H_2 react with one mole of O_2, so two moles of H_2 are equivalent to one mole of O_2:

$$2 \text{ mol } H_2 \backsimeq 1 \text{ mol } O_2$$

Similarly, since two moles of H_2 (or one mole of O_2) produce two moles of H_2O, we can say that two moles of H_2 (or one mole of O_2) are equivalent to two moles of H_2O:

$$2 \text{ mol } H_2 \backsimeq 2 \text{ mol } H_2O$$

$$1 \text{ mol } O_2 \backsimeq 2 \text{ mol } H_2O$$

Note that we generally use molecular equations in stoichiometric calculations because we are usually interested in the masses of whole units and not just the masses of separate cations or anions.

We will now demonstrate the utility of the four-step mole method in solving some typical stoichiometry problems.

This kind of equivalence forms the basis for using the factor-label method (see Section 1.8) in the following examples.

EXAMPLE 4.1

Magnesium (Mg) reacts with hydrochloric acid (HCl) to produce magnesium chloride ($MgCl_2$) and hydrogen gas (H_2). (a) How many moles of H_2 can be formed by the reaction of 2.46 moles of HCl with sufficient magnesium? (b) How many grams of H_2 can be formed by the reaction of 0.568 mole of Mg with sufficient HCl?

Answer

(a)

Step 1

$$Mg(s) + 2HCl(aq) \longrightarrow MgCl_2(aq) + H_2(g)$$

(Continued)

Refer to Figure 4.1 for the conversions.

Step 2

No conversion is needed here because the amount of the starting material, HCl, is given in moles.

Step 3

Since two moles of HCl produce one mole of H_2, or 2 mol HCl $\approx$ 1 mol H_2, we calculate moles of H_2 as follows:

$$\text{moles of } H_2 \text{ produced} = 2.46 \text{ mol HCl} \times \frac{1 \text{ mol } H_2}{2 \text{ mol HCl}}$$

$$= 1.23 \text{ mol } H_2$$

Step 4

This step is not required.

(b)

Step 1

The reaction is the same as in (a).

Step 2

No conversion is needed here because the amount of the starting material, Mg, is given in moles.

Step 3

Since one mole of Mg produces one mole of H_2, or 1 mol Mg $\approx$ 1 mol H_2, we calculate moles of H_2 as follows:

$$\text{moles of } H_2 \text{ produced} = 0.568 \text{ mol Mg} \times \frac{1 \text{ mol } H_2}{1 \text{ mol Mg}}$$

$$= 0.568 \text{ mol } H_2$$

Step 4

From the molar mass of H_2 (2.016 g), we calculate the mass of H_2 produced:

$$\text{mass of } H_2 \text{ produced} = 0.568 \text{ mol } H_2 \times \frac{2.016 \text{ g } H_2}{1 \text{ mol } H_2}$$

$$= 1.15 \text{ g } H_2$$

Similar example: Problem 4.8.

EXAMPLE 4.2

The food we eat is degraded, or broken down, in our bodies to provide energy for growth and function. A general overall equation for this very complex process represents the degradation of glucose ($C_6H_{12}O_6$) to carbon dioxide (CO_2) and water (H_2O):

$$C_6H_{12}O_6 + 6O_2 \longrightarrow 6CO_2 + 6H_2O$$

If 856 g of $C_6H_{12}O_6$ is consumed by the body over a certain period, what is the mass of CO_2 produced?

Answer

Step 1

The balanced equation is given.

Step 2

The molar mass of $C_6H_{12}O_6$ is 180.2 g, so the number of moles of $C_6H_{12}O_6$ in 856 g of $C_6H_{12}O_6$ is

$$\text{moles of } C_6H_{12}O_6 = 856 \text{ g } C_6H_{12}O_6 \times \frac{1 \text{ mol } C_6H_{12}O_6}{180.2 \text{ g } C_6H_{12}O_6}$$
$$= 4.75 \text{ mol } C_6H_{12}O_6$$

Step 3

From the balanced equation we see that 1 mol $C_6H_{12}O_6 \approx 6$ mol CO_2; thus the number of moles of CO_2 produced is given by

$$\text{moles of } CO_2 \text{ produced} = 4.75 \text{ mol } C_6H_{12}O_6 \times \frac{6 \text{ mol } CO_2}{1 \text{ mol } C_6H_{12}O_6}$$
$$= 28.5 \text{ mol } CO_2$$

Step 4

The molar mass of CO_2 is 44.01 g. To convert moles of CO_2 to grams of CO_2 we write

$$\text{mass of } CO_2 \text{ produced} = 28.5 \text{ mol } CO_2 \times \frac{44.01 \text{ g } CO_2}{1 \text{ mol } CO_2}$$
$$= 1.25 \times 10^3 \text{ g } CO_2$$

Similar example: Problem 4.4.

After some practice, you may find it convenient to combine steps 2, 3, and 4 in a single factor-label equation, as the following example shows.

EXAMPLE 4.3

From the following balanced equation calculate the number of grams of sodium phosphate needed to prepare 45.9 g of sodium chloride:

$$3CaCl_2 + 2Na_3PO_4 \longrightarrow Ca_3(PO_4)_2 + 6NaCl$$

| calcium chloride | sodium phosphate | calcium phosphate | sodium chloride |

Answer

Step 1

The balanced equation is given.

(Continued)

Steps 2, 3, and 4

From the balanced equation we see that 2 mol $Na_3PO_4 \leftrightharpoons$ 6 mol NaCl. The molar masses of Na_3PO_4 and NaCl are 163.9 g and 58.44 g, respectively. We combine all of these data into one factor-label equation as follows:

$$\text{mass of sodium phosphate required} = 45.9 \text{ g NaCl} \times \frac{1 \text{ mol NaCl}}{58.44 \text{ g NaCl}} \times \frac{2 \text{ mol } Na_3PO_4}{6 \text{ mol NaCl}}$$

$$\times \frac{163.9 \text{ g } Na_3PO_4}{1 \text{ mol } Na_3PO_4}$$

$$= 42.9 \text{ g } Na_3PO_4$$

Similar examples: Problems 4.1, 4.2, 4.3.

4.2 LIMITING REAGENTS

When a chemist carries out a reaction, the reactants are usually not present in exact **stoichiometric amounts,** that is, *in the proportions indicated by the balanced equation. The reactant used up first in a reaction* is called the **limiting reagent**, since the maximum amount of product formed depends on how much of this reactant was originally present (Figure 4.2). When this reactant is used up, no more product can be formed. Thus *one or more of the other reactants will often be present in quantities greater than those needed to react with the quantity of the limiting reagent present.* These reactants are called **excess reagents.**

The concept of the limiting reagent is analogous to the relationship between the number of stamps available and the number of letters to be mailed. If there are nine letters and only six stamps, then the maximum number of letters that can be sent is six. The number of stamps thus *lfmits*

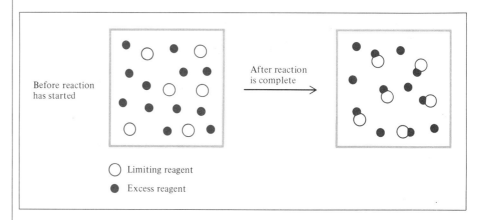

FIGURE 4.2 *Diagram showing the complete use of a limiting reagent (open circles) in a reaction.*

the number of letters that can be mailed, and there is an *excess* of envelopes.

Consider the reaction between lithium and molecular oxygen to produce lithium oxide:

$$4Li(s) + O_2(g) \longrightarrow 2Li_2O(s)$$

This equation tells us that four moles of Li react with one mole of O_2 to produce two moles of Li_2O. Suppose that in a certain reaction six moles of Li are exposed to one mole of O_2. Since 4 mol Li $\backsimeq$ 1 mol O_2, at the end of the reaction there will be two moles of Li_2O as product and two moles of Li left over. All the O_2 will be used up. Thus the reactant O_2 is the limiting reagent and Li is the excess reagent. The amount of Li_2O produced depends only on how much O_2 was originally present.

In stoichiometric calculations involving limiting reagents, the first step is to decide which reactant is the limiting reagent. After the limiting reagent has been identified, the rest of the problem can be solved as outlined in Section 4.1. The following examples show the approach used in this type of calculation.

EXAMPLE 4.4

At high temperatures, sulfur combines with iron to form iron(II) sulfide:

$$Fe(s) + S(l) \xrightarrow{\Delta} FeS(s)$$

In one experiment 7.62 g of Fe are allowed to react with 8.67 g of S. (a) Which of the two reactants is the limiting reagent? (b) Calculate the mass of FeS formed. (c) How much of the excess reagent (in grams) is left at the end of the reaction?

Answer

(a) Since we cannot tell by inspection which of the two reactants is the limiting reagent, we have to proceed by first converting their masses into number of moles. The molar masses of Fe and S are 55.85 and 32.07 g, respectively. Thus

$$\text{moles of Fe} = 7.62 \text{ g Fe} \times \frac{1 \text{ mol Fe}}{55.85 \text{ g Fe}}$$
$$= 0.136 \text{ mol Fe}$$

$$\text{moles of S} = 8.67 \text{ g S} \times \frac{1 \text{ mol S}}{32.07 \text{ g S}}$$
$$= 0.270 \text{ mol S}$$

From the balanced equation we see that 1 mol Fe $\backsimeq$ 1 mol S. We now divide the actual number of moles of Fe and of S by the coefficients shown in the equation:

$$\frac{0.136 \text{ mol Fe}}{1 \text{ mol Fe}} = 0.136 \qquad \frac{0.270 \text{ mol S}}{1 \text{ mol S}} = 0.270$$

(Continued)

Thus Fe must be the limiting reagent because a smaller proportionate amount of that substance was supplied in the reaction (0.136 is smaller than 0.270). (b) The number of moles of FeS produced is

$$\text{moles of FeS} = 0.136 \text{ mol Fe} \times \frac{1 \text{ mol FeS}}{1 \text{ mol Fe}}$$
$$= 0.136 \text{ mol FeS}$$

The mass of FeS produced is given by

$$\text{mass of FeS} = 0.136 \text{ mol FeS} \times \frac{87.92 \text{ g FeS}}{1 \text{ mol FeS}}$$
$$= 12.0 \text{ g FeS}$$

(c) The number of moles of the excess reagent (S) is $0.270 - 0.136$, or 0.134 mole, and therefore

$$\text{mass of S left over} = 0.134 \text{ mol S} \times \frac{32.07 \text{ g S}}{1 \text{ mol S}}$$
$$= 4.30 \text{ g S}$$

Similar examples: Problems 4.11, 4.12.

EXAMPLE 4.5

At high temperatures ammonia reacts with copper(II) oxide according to the equation:

$$3CuO(s) + 2NH_3(g) \xrightarrow{\Delta} 3Cu(s) + 3H_2O(g) + N_2(g)$$

In a certain experiment 236.1 g of CuO is treated with 64.38 g of NH_3. (a) Which compound is the limiting reagent? (b) How many grams of Cu would be produced? (c) Calculate the mass (in grams) of the excess reagent remaining at the end of the reaction.

Answer

(a) As in Example 4.3, our first step is to convert the masses of the reactants into molar amounts. The molar masses of CuO and NH_3 are 79.55 g and 17.03 g, respectively. Thus

$$\text{moles of CuO} = 236.1 \text{ g CuO} \times \frac{1 \text{ mol CuO}}{79.55 \text{ g CuO}}$$
$$= 2.968 \text{ mol CuO}$$

$$\text{moles of NH}_3 = 64.38 \text{ g NH}_3 \times \frac{1 \text{ mol NH}_3}{17.03 \text{ g NH}_3}$$
$$= 3.780 \text{ mol NH}_3$$

From the balanced equation we see that 3 mol CuO $\backsimeq$ 2 mol NH_3. We now divide the actual number of moles of CuO and of NH_3 by the coefficients shown in the equation:

$$\frac{2.968 \text{ mol CuO}}{3 \text{ mol CuO}} = 0.989 \qquad \frac{3.780 \text{ mol NH}_3}{2 \text{ mol NH}_3} = 1.890$$

Thus CuO must be the limiting reagent because a smaller proportionate amount of this substance was supplied in the reaction (0.989 is smaller than 1.890). (b) The mass of Cu produced is calculated as follows:

$$\text{mass of Cu produced} = 2.968 \text{ mol CuO} \times \frac{1 \text{ mol Cu}}{1 \text{ mol CuO}} \times \frac{63.55 \text{ g Cu}}{1 \text{ mol Cu}}$$

$$= 188.6 \text{ g Cu}$$

(c) The number of moles of NH_3 that reacted is then calculated:

$$\text{moles of } NH_3 \text{ reacted} = 2.968 \text{ mol CuO} \times \frac{2 \text{ mol } NH_3}{3 \text{ mol CuO}}$$

$$= 1.979 \text{ mol } NH_3$$

The number of moles of NH_3 left over is

$$\text{moles of } NH_3 \text{ left over} = 3.780 \text{ mol} - 1.979 \text{ mol}$$

$$= 1.801 \text{ mol}$$

Therefore, the mass of the excess reagent NH_3 is

$$\text{mass of } NH_3 \text{ left over} = 1.801 \text{ mol } NH_3 \times \frac{17.03 \text{ g } NH_3}{1 \text{ mol } NH_3}$$

$$= 30.67 \text{ g } NH_3$$

Similar example: Problem 4.14.

4.3   YIELDS: THEORETICAL, ACTUAL, AND PERCENT

The amount of limiting reagent present at the start of a reaction is related to *the quantity of product we can obtain from the reaction.* This quantity is called the **yield of the reaction.** There are three types of yields that concern us in the quantitative study of chemical reactions.

The **theoretical yield** of a reaction is *the amount of product predicted by the balanced equation when all of the limiting reagent has reacted.* The theoretical yield, then, is the *maximum* obtainable yield. In practice, *the amount of product obtained,* called **actual yield,** is almost always less than the theoretical yield. There are many reasons for this. For instance, many reactions are reversible, and so they do not proceed 100 percent from left to right. Even when a reaction is 100 percent complete, it may be difficult to recover all the product from the reaction medium (say, from an aqueous solution). Therefore chemists often use the term **percent yield (% yield),** which describes *the proportion of the actual yield to the theoretical yield.* It is defined as follows:

$$\% \text{ yield} = \frac{\text{actual yield}}{\text{theoretical yield}} \times 100\%$$

Many reactions have percent yields that are considerably below 90 percent. In addition to the complications listed above, some reactions are complex in the sense that the product(s) formed may further react among themselves or with the reactants to produce other undesired products. These reactions will further reduce the yield of the reaction. For these reasons the percent yields of many reactions may fall in the range between 5 percent and 50 percent.

EXAMPLE 4.6

Iron (Fe) can be obtained from its ore, iron(III) oxide (Fe_2O_3), by reaction with coke (C) at high temperature according to the equation

$$Fe_2O_3(s) + 3C(s) \longrightarrow 2Fe(l) + 3CO(g)$$

In a certain operation 2.86×10^4 kg of Fe_2O_3 is reacted with 9.82×10^3 kg of C. (a) Calculate the theoretical yield of Fe in kilograms. (b) Calculate the percent yield if 1.52×10^4 kg of Fe is actually obtained.

Answer

(a) The molar masses of Fe_2O_3 and C are 159.7 g and 12.01 g, respectively. Using the conversion factor 1 kg = 1000 g we write

$$\text{moles of } Fe_2O_3 = 2.86 \times 10^7 \text{ g } Fe_2O_3 \times \frac{1 \text{ mol } Fe_2O_3}{159.7 \text{ g } Fe_2O_3}$$
$$= 1.79 \times 10^5 \text{ mol}$$

$$\text{moles of C} = 9.82 \times 10^6 \text{ g C} \times \frac{1 \text{ mol C}}{12.01 \text{ g C}}$$
$$= 8.18 \times 10^5 \text{ mol C}$$

Next, we must determine which of the two substances is the limiting reagent. From the balanced equation we see that 1 mol Fe_2O_3 ⇆ 3 mol C. We now divide the actual number of moles of Fe_2O_3 and C by the stoichiometric coefficients shown in the equation

$$\frac{1.79 \times 10^5 \text{ mol } Fe_2O_3}{1 \text{ mol } Fe_2O_3} = 1.79 \times 10^5$$

$$\frac{8.18 \times 10^5 \text{ mol C}}{3 \text{ mol C}} = 2.73 \times 10^5$$

Thus Fe_2O_3 must be the limiting reagent because a smaller proportionate amount of this compound was supplied in the reaction. Since 1 mol Fe_2O_3 ⇆ 2 mol Fe, the theoretical amount of Fe formed is

$$\text{mass of Fe produced} = 2.86 \times 10^4 \text{ kg } Fe_2O_3 \times \frac{1 \text{ mol } Fe_2O_3}{159.7 \text{ g } Fe_2O_3}$$
$$\times \frac{2 \text{ mol Fe}}{1 \text{ mol } Fe_2O_3} \times \frac{55.85 \text{ g Fe}}{1 \text{ mol Fe}}$$
$$= 2.00 \times 10^4 \text{ kg Fe}$$

(b) To find the percent yield, we write

$$\% \text{ yield} = \frac{\text{actual yield}}{\text{theoretical yield}} \times 100\%$$

$$= \frac{1.52 \times 10^4 \text{ kg}}{2.00 \times 10^4 \text{ kg}} \times 100\%$$

$$= 76.0\%$$

Similar examples: Problems 4.16, 4.17.

Knowing how to calculate the percent yield of a reaction allows us also to calculate the purity of a substance. Most substances we work with are not 100 percent pure. When an impure substance is used in a chemical reaction, the percent yield is always less than 100 percent even if the reaction goes to completion, because there is less of the reactant present than we think. The following example shows how we can experimentally determine the percent purity of a sample.

EXAMPLE 4.7

An impure sample of zinc (Zn) is treated with an excess of hydrochloric acid (HCl) to form zinc chloride ($ZnCl_2$) and hydrogen gas (H_2) according to the equation

$$Zn(s) + 2HCl(aq) \longrightarrow ZnCl_2(aq) + H_2(g)$$

If 0.146 g of H_2 is obtained from a 5.10 g of the impure Zn sample, calculate the percent purity of the sample.

Answer

In this case we are given the mass of a product. Our first step is to calculate the mass of Zn contained in the impure sample that would give 0.146 g of H_2. Thus we write

$$\text{mass of Zn reacted} = 0.146 \text{ g } H_2 \times \frac{1 \text{ mol } H_2}{2.016 \text{ g } H_2} \times \frac{1 \text{ mol Zn}}{1 \text{ mol } H_2} \times \frac{65.39 \text{ g Zn}}{1 \text{ mol Zn}}$$

$$= 4.74 \text{ g Zn}$$

The percent purity of the Zn sample is given by

$$\text{percent purity} = \frac{\text{mass of pure component}}{\text{total mass of impure sample}} \times 100\%$$

$$= \frac{4.74 \text{ g}}{5.10 \text{ g}} \times 100\%$$

$$= 92.9\%$$

Similar example: Problem 4.46.

A cautionary note: In calculating percent purity of a substance according to the procedure shown in the above example, we must make the following assumptions. First, the reaction is assumed to go to completion. Second, we assume that the impurities in the sample do not react with hydrochloric acid to give hydrogen gas.

Many chemical reactions occur in solutions, particularly in water solutions or, more properly, aqueous solutions. Certainly most, or perhaps all, of the experiments you will be carrying out in the laboratory in this introductory chemistry course will involve the use of aqueous solutions. Thus in the next section concentration and dilution of solutions will be discussed.

4.4 CONCENTRATION AND DILUTION OF SOLUTIONS

Concentration of Solutions

See Section 3.2 for definitions of solute, solvent, and solution.

The **concentration of a solution** is *the amount of solute dissolved in a given quantity of solvent.* (In our present discussion we will assume the solute to be a liquid or a solid and the solvent to be a liquid.) One of the most common units of concentration in chemistry is **molarity,** symbolized by M and also called **molar solution.** Molarity is *the number of moles of solute in one liter of solution (soln).* Molarity is defined by the equation

$$M = \text{molarity} = \frac{\text{moles of solute}}{\text{liters of soln}}$$

Thus, a 3.40 molar potassium nitrate (KNO_3) solution, expressed as 3.40 M KNO_3, contains 3.40 moles of the solute (KNO_3) in one liter of the solution; a 1.66 molar glucose ($C_6H_{12}O_6$) solution, expressed as 1.66 M $C_6H_{12}O_6$, contains 1.66 moles of $C_6H_{12}O_6$ (the solute) in one liter of solution; and so on. Of course, we do not always need (or desire) to work with solutions of exactly one liter. Thus, a 500 mL solution containing 1.70 moles of KNO_3 still has the same concentration of 3.40 M:

$$M = \text{molarity} = \frac{\text{moles of solute}}{\text{liters of soln}}$$
$$= \frac{1.70 \text{ mol}}{0.500 \text{ L}}$$
$$= 3.40 \text{ mol/L} = 3.40 \ M$$

As you can see, molarity actually has units of mol/L, so that 3.40 M is equivalent to 3.40 mol/L.

It is important to keep in mind that molarity refers only to the *amount of solute originally dissolved* in water and does not attempt to reflect any subsequent processes (such as the ionization of an acid or the dissociation of a salt). Let us consider the following two solutions: 1 M NaBr (sodium bromide) and 1 M MgI$_2$ (magnesium iodide). Since both compounds are strong electrolytes, they are completely dissociated into ions:

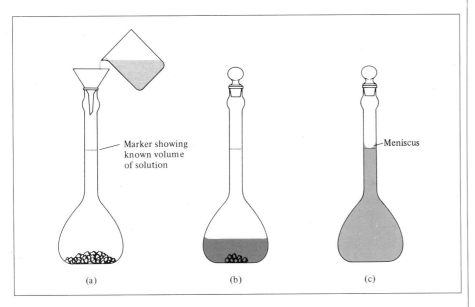

FIGURE 4.3 *Preparation of a solution of known molarity. (a) A known amount of substance (the solute) is put into the volumetric flask; then water is added through a funnel. (b) The solid is slowly dissolved by gently shaking the flask. (c) After the solid has completely dissolved, more water is added to bring the level of the solution to the mark. Knowing the volume of the solution and the amount of solute dissolved in it, we can calculate the molarity of the prepared solution.*

$$NaBr(s) \xrightarrow{H_2O} Na^+(aq) + Br^-(aq)$$

$$MgI_2(s) \xrightarrow{H_2O} Mg^{2+}(aq) + 2I^-(aq)$$

The concentrations of the ions in the 1 M NaBr solution are $[Na^+] = 1\ M$ and $[Br^-] = 1\ M$, and in the 1 M MgI_2 solution they are $[Mg^{2+}] = 1\ M$ and $[I^-] = 2\ M$. The situation is more complicated if the solute is a weak electrolyte. We will consider that case in a later chapter.

One advantage of expressing concentration in units of molarity is that solutions of known concentration can be conveniently prepared in a *volumetric flask*. A volumetric flask is designed to contain an exact volume of liquid when the bottom of the *meniscus*—the curved surface of the liquid— just reaches the etched line on the neck of the flask. The steps for preparing a solution of a specified molarity are illustrated in Figure 4.3.

The square brackets [ ] indicate that the concentration is expressed in molarity.

In general it is easier to measure the volume of a liquid than to determine its mass.

EXAMPLE 4.8

A sample of 6.98 g of sucrose ($C_{12}H_{22}O_{11}$) is dissolved in enough water to form 67.8 mL of solution. What is the molarity of this solution? (The molar mass of sucrose is 342.3 g.)

(Continued)

Answer

To change 6.98 g of sucrose to moles of sucrose we write

$$\text{moles of } C_{12}H_{22}O_{11} = 6.98 \text{ g } C_{12}H_{22}O_{11} \times \frac{1 \text{ mol } C_{12}H_{22}O_{11}}{342.3 \text{ g } C_{12}H_{22}O_{11}}$$

$$= 0.0204 \text{ mol } C_{12}H_{22}O_{11}$$

The molarity of the sucrose solution is given by

$$\text{molarity} = \frac{\text{moles of } C_{12}H_{22}O_{11}}{\text{liters of soln}}$$

$$= \frac{0.0204 \text{ mol } C_{12}H_{22}O_{11}}{67.8 \text{ mL } C_{12}H_{22}O_{11} \text{ soln}} \times \frac{1000 \text{ mL soln}}{1 \text{ L soln}}$$

$$= \frac{0.301 \text{ mol } C_{12}H_{22}O_{11}}{1 \text{ L soln}}$$

$$= 0.301 \ M$$

Similar examples: Problems 4.19, 4.23, 4.25.

EXAMPLE 4.9

How many grams of sodium chloride (NaCl) are present in 50.0 mL of a 2.45 M NaCl solution?

Answer

The first step is to find out the number of moles of NaCl present in 50.0 mL of the solution:

$$\text{moles of NaCl} = 50.0 \text{ mL NaCl soln} \times \frac{2.45 \text{ mol NaCl}}{1000 \text{ mL NaCl soln}}$$

$$= 0.123 \text{ mol NaCl}$$

The molar mass of NaCl is 58.44 g, so we write

$$\text{mass of NaCl} = 0.123 \text{ mol NaCl} \times \frac{58.44 \text{ g NaCl}}{1 \text{ mol NaCl}}$$

$$= 7.19 \text{ g NaCl}$$

Similar example: Problem 4.22.

EXAMPLE 4.10

How many grams of potassium nitrate (KNO_3) are required to prepare exactly 250 mL of a solution whose concentration is 0.700 M?

Answer

The first step is to determine the number of moles of KNO_3 in 250 mL of solution:

$$\text{moles of KNO}_3 = 250 \text{ mL KNO}_3 \text{ soln} \times \frac{0.700 \text{ mol KNO}_3}{1000 \text{ mL KNO}_3 \text{ soln}}$$

$$= 0.175 \text{ mol KNO}_3$$

The molar mass of KNO_3 is 101.1 g, so we write

$$\text{mass of KNO}_3 = 0.175 \text{ mol KNO}_3 \times \frac{101.1 \text{ g KNO}_3}{1 \text{ mol KNO}_3}$$

$$= 17.7 \text{ g KNO}_3$$

Similar example: Problem 4.27.

Dilution of Solutions

We frequently find it convenient *to prepare a less concentrated solution from a more concentrated solution.* The procedure for this preparation is called ***dilution.*** Suppose, for example, that we want to prepare one liter of 0.20 *M* $CuSO_4$ solution from a solution of 1.0 *M* $CuSO_4$. This requires the use of 0.20 mole of $CuSO_4$ from the 1.0 *M* $CuSO_4$ solution. Since there is 1.0 mole of $CuSO_4$ in 1 L or 1000 mL of a 1.0 *M* $CuSO_4$ solution, there is 0.20 mole of $CuSO_4$ in 200 mL of the same solution. Therefore we must withdraw 200 mL from the 1.0 *M* $CuSO_4$ solution (using a pipet) and dilute it to 1000 mL by adding water (in a one-liter volumetric flask). This method gives us one liter of the desired solution of 0.20 *M* $CuSO_4$. In carrying out a dilution process, it is useful to remember that adding more solvent to a given amount of solution changes (decreases) the concentration of the solution without changing the number of moles of solute present in the solution (Figure 4.4), that is

moles of solute before dilution = moles of solute after dilution

Since molarity is moles of solute/liters of soln, we see that the number of moles of solute before or after dilution is given by

$$\text{moles of solute} = \frac{\text{moles of solute}}{\text{liters of soln}} \times \text{volume of solution (in liters)}$$

$$= MV$$

The concentrated solution is often called the *stock solution.*

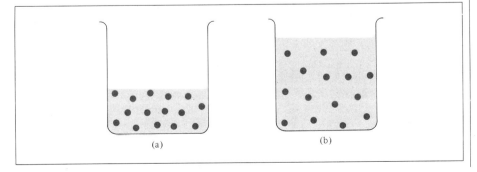

(a)

(b)

FIGURE 4.4 *The dilution of (a) a more concentrated solution to (b) a less concentrated one. Note that the total number of solute particles in (b) is the same as that in (a).*

we can conclude that

$$M_{initial}V_{initial} = M_{final}V_{final} \qquad (4.1)$$

moles of solute   moles of solute

before dilution   after dilution

where $M_{initial}$ and M_{final} are the initial and final concentrations of the solution in molarity, and $V_{initial}$ and V_{final} are the initial and final volumes of the solution, respectively. Of course, the *units* of $V_{initial}$ and V_{final} must be the same in the calculation.

EXAMPLE 4.11

Describe how you would prepare 5.00×10^2 mL of 2.20 M NH$_3$ solution, starting with a 5.70 M NH$_3$ solution.

Answer

Since the concentration of the final solution is less than that of the original one, this is a dilution process. We prepare for the calculation by tabulating our data:

$M_{initial} = 5.70\ M$ $M_{final} = 2.20\ M$

$V_{initial} = ?$ $V_{final} = 5.00 \times 10^2$ mL

Substituting in Equation (4.1)

$$(5.70\ M)(V_{initial}) = (2.20\ M)(5.00 \times 10^2\ \text{mL})$$

$$V_{initial} = \frac{(2.20\ M)(5.00 \times 10^2\ \text{mL})}{5.70\ M}$$

$$= 193\ \text{mL}$$

Thus, we must dilute 193 mL of the 5.70 M solution to a final volume of 5.00×10^2 mL in a 500 mL volumetric flask to obtain the desired concentration.

Similar examples: Problems 4.29, 4.31.

Having discussed the concentration and dilution of solutions, we are now ready to look at some reactions that take place in aqueous solutions. The next two sections will focus on two types of reactions that you are likely to encounter soon in the laboratory, namely, gravimetric analysis and acid-base titrations.

4.5 GRAVIMETRIC ANALYSIS

Gravimetric analysis is *an analytical procedure that involves the measurement of mass.* A sample substance of unknown composition is dissolved in water and then converted into an insoluble compound (that is, a precipitate) by allowing it to react with another substance. The precipitate formed is filtered off, dried, and weighed. Knowing the mass of the starting sample and the mass and chemical formula of the precipitated compound, we can calculate the percent by mass of a particular chemical component of the original sample. This information can help us identify the original substance or (if it was a mixture) determine the sample's composition.

A reaction that is studied often in gravimetric analysis is

$$AgNO_3(aq) + NaCl(aq) \longrightarrow NaNO_3(aq) + AgCl(s)$$

This is an example of a metathesis reaction, discussed in Section 3.6. The precipitate formed in this reaction is silver chloride (AgCl). If we want to determine the quantity of Cl in NaCl, we must add enough $AgNO_3$ to a solution of NaCl to cause as much Cl^- as possible to precipitate out of the solution as AgCl. Hence, NaCl must be the limiting reagent and $AgNO_3$ the excess reagent in this reaction. Figure 4.5 shows the basic steps in a gravimetric analysis experiment involving the precipitation of AgCl.

EXAMPLE 4.12

A sample of 0.5662 g of an unknown ionic compound containing chloride ions is dissolved in water and treated with an excess of $AgNO_3$. If the mass of the AgCl precipitate that forms is 1.0882 g, what is the percent of Cl (by mass) in the original compound?

Answer

We need to find the mass of Cl present in the unknown sample. To do this, let us first calculate the percent by mass of Cl in AgCl:

$$\% Cl = \frac{\text{molar mass of Cl}}{\text{molar mass of AgCl}} \times 100\%$$
$$= \frac{35.45 \text{ g}}{143.4 \text{ g}} \times 100\%$$
$$= 24.72\%$$

The mass of Cl in AgCl is then given by

$$\text{mass of Cl} = \% Cl \text{ in AgCl} \times \text{mass of AgCl}$$
$$= 0.2472 \times 1.0882 \text{ g}$$
$$= 0.2690 \text{ g}$$

The percent by mass of Cl in the unknown sample is given by

$$\% Cl \text{ by mass} = \frac{\text{mass of Cl}}{\text{mass of sample}} \times 100\%$$
$$= \frac{0.2690 \text{ g}}{0.5662 \text{ g}} \times 100\%$$
$$= 47.51\%$$

As an exercise, you should compare this value with that of Cl in KCl.

Similar example: Problem 4.37.

Remember to convert percentages to decimals and vice versa as needed in problems like Example 4.12.

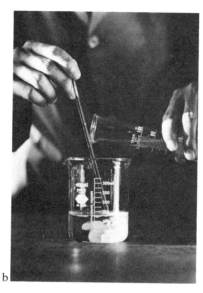

a   b   c

FIGURE 4.5 *(a) A solution containing a known amount of NaCl (or some other compound containing chloride ions) in a beaker. (b) The precipitation of AgCl upon the addition of $AgNO_3$ solution from a graduated cylinder. In this reaction, $AgNO_3$ is the excess reagent and NaCl is the limiting reagent. (c) The solution containing the AgCl precipitate is filtered through a preweighed sintered-disc crucible, which allows the liquid (but not the solid precipitate) to pass through. The crucible is then removed from the apparatus, dried in an oven, and weighed again. The difference between this mass and the mass of the empty crucible gives the mass of the AgCl precipitate.*

In cases where the identity of the original compound is known, the gravimetric analysis technique allows us to determine the concentration of the solution containing the compound.

EXAMPLE 4.13

The concentration of lead ions (Pb^{2+}) in a sample of polluted water that also contains nitrate ions (NO_3^-) is determined by adding solid sodium sulfate (Na_2SO_4) to exactly 500 mL of the water. (a) Write the molecular and net ionic equations for the reaction. (b) Calculate the molar concentration of Pb^{2+} if 0.450 g of Na_2SO_4 was needed for the complete precipitation of Pb^{2+} ions.

Answer

(a) The molecular equation for the reaction is

$$Pb(NO_3)_2(aq) + Na_2SO_4(aq) \longrightarrow 2NaNO_3(aq) + PbSO_4(s)$$

From the solubility rules listed in Section 3.2, we see that lead sulfate ($PbSO_4$) is insoluble. The net ionic equation is

$$Pb^{2+}(aq) + SO_4^{2-}(aq) \longrightarrow PbSO_4(s)$$

(b) From the molecular equation we see that 1 mol $Na_2SO_4 \cong$ 1 mol $Pb(NO_3)_2$ or 1 mol $Na_2SO_4 \cong$ 1 mol Pb^{2+}. The number of moles of Na_2SO_4 (molar mass = 142.1 g) in 0.450 g of the compound is

$$0.450 \text{ g } Na_2SO_4 \times \frac{1 \text{ mol } Na_2SO_4}{142.1 \text{ g } Na_2SO_4} = 3.17 \times 10^{-3} \text{ mol } Na_2SO_4$$

This quantity of Na_2SO_4 is needed to precipitate Pb^{2+} ions in a 500 mL solution. Thus, the concentration of Pb^{2+} ions in one liter of the solution is given by

$$[Pb^{2+}] = \frac{3.17 \times 10^{-3} \text{ mol}}{500 \text{ mL}} \times \frac{1000 \text{ mL}}{1 \text{ L}}$$
$$= 6.34 \times 10^{-3} \text{ mol/L}$$
$$= 6.34 \times 10^{-3} \ M$$

Similar example: Problem 4.39.

4.6   ACID-BASE TITRATIONS

Quantitative studies of acid-base neutralization reactions are most conveniently carried out by a procedure known as **titration.** In a titration experiment, *a solution of accurately known concentration,* called a **standard solution,** *is added gradually to another solution of unknown concentration, until the chemical reaction between the two solutions is complete.* If we know the volumes of the standard and unknown solutions used in the titration, and the concentration of the standard solution, we can calculate the concentration of the unknown solution.

Suppose we have at our disposal a certain base, say, sodium hydroxide (NaOH), of accurately known concentration, and we are asked to determine the concentration of a hydrochloric acid (HCl) solution by titration. First, a known volume of the acid solution (measured with a pipet) is transferred to an Erlenmeyer flask (Figure 4.6). Next, a buret is filled with the NaOH solution. The NaOH solution is carefully added to the acid solution until we reach the **equivalence point,** that is, *the point at which the acid is completely reacted with or neutralized by the base.* This point is usually signaled by a color change of an indicator that has been added to the acid solution. In acid-base titrations, **indicators** are usually *substances that have distinctly different colors in acidic and basic media.* One commonly used indicator is phenolphthalein, which is colorless in acidic and neutral solutions but reddish pink in basic solutions.

At the equivalence point, all of the HCl present has been neutralized by the added NaOH:

$$HCl(aq) + NaOH(aq) \longrightarrow NaCl(aq) + H_2O(l)$$

and the solution is still colorless. However, upon the addition of another drop of NaOH solution from the buret, the solution will turn reddish pink because the solution is now basic. The actual arrangement of this titration is shown in Figure 4.7.

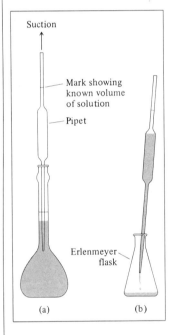

FIGURE 4.6 *Preparation for titration. (a) A pipet is used to draw a known volume of acid from the volumetric flask. This is done by drawing the solution up into the pipet until it reaches the mark on the pipet. (Never put a pipet in your mouth.) (b) This volume of solution is then transferred to an Erlenmeyer flask, ready for titration.*

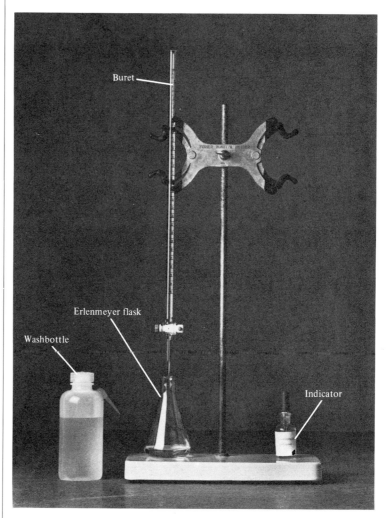

FIGURE 4.7 *A practical arrangement for acid-base titration. The base in the buret is carefully added to the acid in the Erlenmeyer flask until the equivalence point is reached.*

The acid-base neutralization reaction between NaOH and HCl is one of the simplest types of neutralization known. Suppose, though, that instead of HCl a diprotic acid such as sulfuric acid (H_2SO_4) is used. The reaction is represented by

$$H_2SO_4(aq) + 2NaOH(aq) \longrightarrow Na_2SO_4(aq) + 2H_2O(l)$$

Since 2 mol NaOH $\backsimeq$ 1 mol H_2SO_4, we would need twice the amount of NaOH to react completely with an H_2SO_4 solution of the same concentration as an HCl solution. On the other hand, we would need twice the amount of HCl to completely neutralize with a $Ba(OH)_2$ solution having the same concentration as a solution of NaOH:

One mole of Ba(OH)$_2$ yields two moles of OH$^-$ ions.

$$2HCl(aq) + Ba(OH)_2(aq) \longrightarrow BaCl_2(aq) + 2H_2O(l)$$

In calculations involving acid-base titrations, regardless of the acid or base involved, keep in mind that the total moles of H^+ ions that have reacted when the equivalence point has been reached must be exactly equal to the total moles of OH^- ions that have reacted.

EXAMPLE 4.14

In a titration experiment, a student finds that 37.42 mL of 0.1078 M NaOH solution is needed to completely neutralize 25.00 mL of an HCl solution. What is the concentration (in molarity) of the acid solution?

Answer

The number of moles of NaOH in 37.42 mL solution is

$$\text{moles of NaOH} = 37.42 \text{ mL NaOH soln} \times \frac{1 \text{ L soln}}{1000 \text{ mL soln}} \times \frac{0.1078 \text{ mol NaOH}}{1 \text{ L soln}}$$
$$= 4.034 \times 10^{-3} \text{ mol NaOH}$$

From the neutralization reaction shown previously we know that 1 mol NaOH $\rightleftharpoons$ 1 mol HCl. Therefore, the number of moles of HCl reacted is also 4.034 × 10⁻³ mole. (Keep in mind that 4.034 × 10⁻³ mole of NaOH supplies 4.034 × 10⁻³ mole of OH^- ions, and 4.034 × 10⁻³ mole of HCl supplies 4.034 × 10⁻³ mole of H^+ ions.) Now we can calculate the concentration of the HCl solution, as follows:

$$\text{molarity of HCl} = \frac{\text{mol HCl}}{1 \text{ L soln}} = \frac{4.034 \times 10^{-3} \text{ mol HCl}}{25.00 \text{ mL soln}} \times \frac{1000 \text{ mL soln}}{1 \text{ L soln}}$$
$$= 0.1614 \ M \text{ HCl}$$

Similar examples: Problems 4.41, 4.43.

Actually, the reaction occurs as soon as the solutions are mixed. For the purpose of calculation, we pretend that they can be mixed first and then allowed to react.

EXAMPLE 4.15

A volume of 17.8 mL of a 0.344 M H_2SO_4 solution is required to completely neutralize 20.0 mL of a KOH solution. Calculate the concentration (in molarity) of the KOH solution.

Answer

The equation for the neutralization reaction is

$$2KOH(aq) + H_2SO_4(aq) \longrightarrow K_2SO_4(aq) + 2H_2O(l)$$

The number of moles of H_2SO_4 in 17.8 mL solution is given by

$$\text{moles of } H_2SO_4 = 17.8 \text{ mL } H_2SO_4 \text{ soln} \times \frac{1 \text{ L soln}}{1000 \text{ mL soln}} \times \frac{0.344 \text{ mol } H_2SO_4}{1 \text{ L soln}}$$
$$= 6.12 \times 10^{-3} \text{ mol } H_2SO_4$$

(Continued)

6.12×10^{-3} mole of H_2SO_4 supplies $2(6.12 \times 10^{-3})$ mole or 1.22×10^{-2} mole of H^+ ions, and 1.22×10^{-2} mole of KOH supplies 1.22×10^{-2} mole of OH^- ions.

Since 1 mol $H_2SO_4 \approx 2$ mol KOH, the number of moles of KOH reacted must be

$$\text{moles of KOH} = 6.12 \times 10^{-3} \text{ mol } H_2SO_4 \times \frac{2 \text{ mol KOH}}{1 \text{ mol } H_2SO_4}$$
$$= 1.22 \times 10^{-2} \text{ mol KOH}$$

Now we can calculate the concentration of the KOH solution:

$$\text{molarity of KOH} = \frac{\text{mol KOH}}{1 \text{ L soln}} = \frac{1.22 \times 10^{-2} \text{ mol}}{20.0 \text{ mL soln}} \times \frac{1000 \text{ mL soln}}{1 \text{ L soln}}$$
$$= 0.610 \; M \text{ KOH}$$

Similar examples: Problems 4.41, 4.43.

EXAMPLE 4.16

What volume of a $0.5622 \; M$ HNO_3 solution is needed to react completely with 50.00 mL of $0.02844 \; M$ $Ba(OH)_2$?

Answer

The equation for the neutralization reaction is

$$2HNO_3(aq) + Ba(OH)_2(aq) \longrightarrow Ba(NO_3)_2(aq) + 2H_2O(l)$$

The number of moles of $Ba(OH)_2$ in 50.00 mL solution is

$$\text{moles of } Ba(OH)_2 = 50.00 \text{ mL } Ba(OH)_2 \text{ soln} \times \frac{1 \text{ L soln}}{1000 \text{ mL soln}}$$
$$\times \frac{0.02844 \text{ mol } Ba(OH)_2}{1 \text{ L soln}}$$
$$= 1.422 \times 10^{-3} \text{ mol } Ba(OH)_2$$

1.422×10^{-3} mole of $Ba(OH)_2$ supplies $2(1.422 \times 10^{-3})$ mole or 2.844×10^{-3} mole of OH^- ions, and 2.844×10^{-3} mole of HNO_3 supplies 2.844×10^{-3} mole of H^+ ions.

Since 1 mol $Ba(OH)_2 \approx 2$ mol HNO_3, the number of moles of HNO_3 reacted must be

$$\text{moles of } HNO_3 = 1.422 \times 10^{-3} \text{ mol } Ba(OH)_2 \times \frac{2 \text{ mol } HNO_3}{1 \text{ mol } Ba(OH)_2}$$
$$= 2.844 \times 10^{-3} \text{ mol } HNO_3$$

Now we can calculate the volume of the HNO_3 solution as follows. Since

$$\text{molarity of } HNO_3 \text{ soln} = \frac{\text{mol } HNO_3}{\text{liters of } HNO_3 \text{ soln}}$$

we have

$$\text{liters of } HNO_3 \text{ soln} = \frac{\text{mol } HNO_3}{M \; HNO_3 \text{ soln}}$$
$$= 2.844 \times 10^{-3} \text{ mol } HNO_3 \times \frac{1 \text{ L soln}}{0.5622 \text{ mol } HNO_3}$$
$$= 5.059 \times 10^{-3} \text{ L } HNO_3 \text{ soln}$$

To convert liters to mL, we write

$$5.059 \times 10^{-3} \text{ L} \times \frac{1000 \text{ mL}}{1 \text{ L}} = 5.059 \text{ mL HNO}_3 \text{ soln}$$
$$= 5.06 \text{ mL HNO}_3 \text{ soln}$$

We have rounded off the volume as shown because buret readings usually give only two decimal places.

SUMMARY

1. Stoichiometric calculations involve the amounts of products and reactants in chemical reactions. The calculations are best done by expressing both the known and unknown quantities in terms of moles and then converting to other units if necessary.
2. A limiting reagent is the reactant that is present in the smallest stoichiometric amount. It limits the amount of product that can be formed.
3. The amount of product obtained in a reaction (the actual yield) may be less than the maximum possible amount (the theoretical yield). The ratio of the two is expressed as the percent yield.
4. The concentration of a solution is the amount of solute dissolved in a given amount of solution. Molarity expresses concentration as the number of moles of solute in one liter of solution.
5. Adding solvent to a solution, a process known as dilution, decreases the concentration (molarity) of the solution without changing the total moles of solute present in the solution.
6. Gravimetric analysis is a technique for determining the identity of a compound and/or the concentration of a solution by measurement of mass. Precipitation reactions are involved in gravimetric experiments.
7. In acid-base titration, a solution of known concentration (say, a base) is added gradually to a solution of unknown concentration (say, an acid) with the goal of determining the unknown concentration. The point at which the reaction in a titration is exactly complete is called the equivalence point.

KEY WORDS

Actual yield, p. 103
Concentration of a solution, p. 106
Dilution, p. 109
Equivalence point, p. 113
Excess reagent, p. 100
Gravimetric analysis, p. 110
Indicator, p. 113
Limiting reagent, p. 100
Molar solution, p. 106

Molarity, p. 106
Mole method, p. 96
Percent yield, p. 103
Standard solution, p. 113
Stoichiometric amount, p. 100
Stoichiometry, p. 96
Theoretical yield, p. 103
Titration, p. 113
Yield of the reaction, p. 103

PROBLEMS

More challenging problems are marked with an asterisk.

Amounts of Reactants and Products

4.1 The annual production of sulfur dioxide from burning coal and fossil fuels, auto exhaust, and other sources is about 26 million tons. The equation for the reaction is

$$S(s) + O_2(g) \longrightarrow SO_2(g)$$

How much sulfur, present in the original materials, would result in that quantity of SO_2?

4.2 When baking soda (sodium bicarbonate or sodium hydrogen carbonate, $NaHCO_3$) is heated, it releases carbon dioxide gas, which is responsible for the rising of cookies, donuts, and bread. (a) Write a balanced equation for the decomposition of the compound (the products are Na_2CO_3, H_2O, and CO_2). (b) Calculate the mass of $NaHCO_3$ required to produce 20.5 g of CO_2.

4.3 When potassium cyanide (KCN) reacts with acids, a deadly poisonous gas, hydrogen cyanide (HCN), is given off. Here is the equation:

$$KCN(aq) + HCl(aq) \longrightarrow KCl(aq) + HCN(g)$$

If a sample of 0.140 g of KCN is treated with an excess of HCl, calculate the amount of HCN formed, in grams.

4.4 Fermentation is a complex chemical process in which glucose is converted into ethanol and carbon dioxide:

$$\underset{\text{glucose}}{C_6H_{12}O_6} \longrightarrow \underset{\text{ethanol}}{2C_2H_5OH} + 2CO_2$$

Starting with 500.4 g of glucose, what is the maximum amount of ethanol in grams and in liters that can be obtained by this process? (Density of ethanol = 0.789 g/mL)

4.5 Each copper(II) sulfate unit is associated with five water molecules in crystalline copper(II) sulfate pentahydrate ($CuSO_4 \cdot 5H_2O$). When this compound is heated in air above 100°C, it loses the water molecules and also its blue color:

$$CuSO_4 \cdot 5H_2O \longrightarrow CuSO_4 + 5H_2O$$

If 9.60 g of $CuSO_4$ is left after heating 15.01 g of the blue compound, calculate the number of moles of H_2O originally present in the compound.

4.6 For many years the recovery of gold (that is, the separation of gold from other materials) involved the treatment of gold by isolation from other substances, using potassium cyanide:

$$4Au + 8KCN + O_2 + 2H_2O \longrightarrow$$
$$4KAu(CN)_2 + 4KOH$$

What is the minimum amount of KCN in moles needed in order to extract 29.0 g (about an ounce) of gold?

4.7 How many grams of potassium are needed to react completely with 19.2 g of molecular bromine (Br_2) to produce KBr?

4.8 How many moles of sulfuric acid would be needed to produce 4.80 moles of molecular iodine (I_2), according to the following balanced equation:

$$10HI + 2KMnO_4 + 3H_2SO_4 \longrightarrow$$
$$5I_2 + 2MnSO_4 + K_2SO_4 + 8H_2O$$

4.9 Limestone ($CaCO_3$) is decomposed by heating to quicklime (CaO) and carbon dioxide. Write a balanced equation for the reaction and calculate how many grams of quicklime can be produced from 1.0 kg of limestone.

4.10 Nitrous oxide (N_2O) is also called "laughing gas." It can be prepared by the thermal decomposition of ammonium nitrate (NH_4NO_3). The other product is H_2O. (a) Write a balanced equation for this reaction. (b) How many grams of N_2O are formed if 0.46 mole of NH_4NO_3 is used in the reaction?

Limiting Reagents

4.11 Balance the equation

$$SF_4 + I_2O_5 \longrightarrow IF_5 + SO_2$$

and calculate the maximum number of grams of IF_5 that can be obtained from 10.0 g of SF_4 and 10.0 g of I_2O_5.

4.12 The depletion of ozone (O_3) in the stratosphere has been a matter of great concern among scientists in recent years. It is believed that ozone can react with nitric oxide (NO) that is discharged from the high-altitude jet plane, the SST. The reaction is

$$O_3(g) + NO(g) \longrightarrow O_2(g) + NO_2(g)$$

If 0.740 g of O_3 reacts with 0.670 g of NO, how many grams of NO_2 would be produced? Which compound is the limiting reagent? Calculate the number of moles of the excess reagent remaining at the end of the reaction.

4.13 Propane (C_3H_8) is a component of natural gas and is used in domestic cooking and heating. (a) Balance the following equation representing the combustion of propane in air:

$$C_3H_8(g) + O_2(g) \longrightarrow CO_2(g) + H_2O(l)$$

(b) How many grams of carbon dioxide can be produced by burning 3.65 moles of propane? Assume that oxygen is the excess reagent in this reaction.

4.14 Consider the reaction

$$MnO_2 + 4HCl \longrightarrow MnCl_2 + Cl_2 + 2H_2O$$

If 0.86 mole of MnO_2 and 48.2 g of HCl react, which reagent will be used up first? How many grams of Cl_2 will be produced?

Yield of the Reaction

4.15 Suggest reasons why the actual yield of a reaction is almost always less than the theoretical yield.

4.16 The reaction

$$Cr_2O_3 + 3CCl_4 \longrightarrow 2CrCl_3 + 3CCl_2O$$

is used to make $CrCl_3$. In one experiment 6.37 g of Cr_2O_3 was treated with excess CCl_4 and yielded 8.75 g of $CrCl_3$. Calculate the percent yield of $CrCl_3$.

4.17 Nitroglycerin $(C_3H_5N_3O_9)$ is a powerful explosive. Its decomposition may be represented by

$$4C_3H_5N_3O_9 \longrightarrow 6N_2 + 12CO_2 + 10H_2O + O_2$$

This reaction generates a large amount of heat and many gaseous products. It is the sudden formation of these gases, together with their rapid expansion, that produces the explosion. (a) What is the maximum amount of O_2 in grams that can be obtained from 2.00×10^2 g of nitroglycerin? (b) Calculate the percent yield in this reaction if the amount of O_2 generated is found to be 6.55 g.

Concentration of Solutions

4.18 Calculate the mass of NaOH in grams required to prepare a 5.00×10^2 mL solution of concentration 2.80 M.

4.19 A quantity of 5.25 g of NaOH is dissolved in a sufficient amount of water to make up exactly one liter of solution. What is the molarity of the solution?

4.20 How many moles of $MgCl_2$ are present in 60.0 mL of 0.100 M $MgCl_2$ solution?

4.21 Calculate the molarity of a phosphoric acid (H_3PO_4) solution containing 1.50×10^2 g of the acid in 7.50×10^2 mL of solution.

4.22 How many grams of KOH are present in 35.0 mL of a 5.50 M solution?

4.23 Calculate the molarity of each of the following solutions: (a) 29.0 g of ethanol (C_2H_5OH) in 545 mL of solution, (b) 15.4 g of sucrose $(C_{12}H_{22}O_{11})$ in 74.0 mL of solution, (c) 9.00 g of sodium chloride (NaCl) in 86.4 mL of solution.

4.24 Calculate the number of moles of solute present in (a) 75.0 mL of 1.25 M HCl, (b) 100.0 mL of 0.35 M H_2SO_4.

4.25 Calculate the molarity of each of the following solutions: (a) 6.57 g of methanol (CH_3OH) in 1.50 $\times 10^2$ mL of solution, (b) 10.4 g of calcium chloride $(CaCl_2)$ in 2.20×10^2 mL of solution, (c) 7.82 g of naphthalene $(C_{10}H_8)$ in 85.2 mL of benzene solution.

4.26 Calculate the volume in mL of a solution required to provide the following: (a) 2.14 g of sodium chloride from a 0.270 M solution, (b) 4.30 g of ethanol from a 1.50 M solution, (c) 0.85 g of acetic acid (CH_3COOH) from a 0.30 M solution.

4.27 Determine how many grams of each of the following solutes would be needed to make 2.50×10^2 mL of a 0.100 M solution: (a) cesium iodide (CsI); (b) sulfuric acid (H_2SO_4); (c) sodium carbonate (Na_2CO_3); (d) potassium dichromate $(K_2Cr_2O_7)$; (e) potassium permanganate $(KMnO_4)$.

4.28 A 325 mL sample of solution contains 25.3 g of $CaCl_2$. (a) Calculate the molar concentration of Cl^- in this solution. (b) How many grams of Cl^- are there in 0.100 L of this solution?

Dilution of Solutions

4.29 Describe how to prepare 1.00 liter of 0.646 M HCl solution, starting with a 2.00 M HCl solution.

4.30 A quantity of 25.0 mL of a 0.866 M KNO_3 solution is poured into a 500 mL volumetric flask and water is added until the volume of the solution is exactly 500 mL. What is the concentration of the final solution?

4.31 How would you prepare 60.0 mL of 0.200 M HNO_3 from a stock solution of 4.00 M HNO_3?

4.32 You have 505 mL of a 0.125 M HCl solution and you want to dilute it to exactly 0.100 M. How much water should you add?

4.33 Describe how you would prepare 2.50×10^2 mL of a 0.450 M KOH solution, starting with an 8.40 M KOH solution.

4.34 A 46.2 mL 0.568 M calcium nitrate $[Ca(NO_3)_2]$ solution is mixed with a 80.5 mL 1.396 M calcium nitrate solution. Calculate the concentration of the final solution.

Gravimetric Analysis

4.35 The volume 30.0 mL of 0.150 M $CaCl_2$ is added to 15.0 mL of 0.100 M $AgNO_3$. What is the mass in grams of AgCl precipitate formed?

4.36 Distilled water must be used in the gravimetric analysis of chlorides. Why?

4.37 A sample of 0.6760 g of an unknown compound containing barium ions (Ba^{2+}) is dissolved in water and treated with an excess of Na_2SO_4. If the mass of the $BaSO_4$ precipitate formed is 0.4105 g, what is the percent by mass of Ba in the original unknown compound?

4.38 How many grams of NaCl are required to precipitate practically all the Ag^+ ions from 2.50×10^2 mL of 0.0113 M $AgNO_3$ solution? Write the net ionic equation for the reaction.

4.39 The concentration of Cu^{2+} ions in the water (which also contains sulfate ions) discharged from a certain industrial plant is determined by adding excess sodium sulfide (Na_2S) solution to 0.800 L of the water. The molecular equation is

$$Na_2S(aq) + CuSO_4(aq) \longrightarrow Na_2SO_4(aq) + CuS(s)$$

Write the net ionic equation and calculate the molar concentration of Cu^{2+} in the water sample if 0.0177 g of solid CuS is formed.

Acid-Base Titrations

4.40 Explain clearly what is meant by titration. What is the function of an indicator in an acid-base titration?

4.41 Calculate the volume in milliliters (mL) of a 1.420 M NaOH solution required to titrate (a) 25.00 mL of a 2.430 M HCl solution, (b) 25.00 mL of a 4.500 M H_2SO_4 solution, (c) 25.00 mL of a 1.500 M H_3PO_4 solution.

4.42 Acetic acid (CH_3COOH) is an important ingredient of vinegar. A sample of 50.0 mL of a commercial vinegar is titrated against a 1.00 M NaOH solution. What is the concentration (in M) of acetic acid present in the vinegar if 5.75 mL of the base was required for the titration? (Acetic acid is monoprotic.)

4.43 What volume of a 0.50 M KOH solution is needed to neutralize completely each of the following? (a) 10.0 mL of a 0.30 M HCl solution, (b) 10.0 mL of a 0.20 M H_2SO_4 solution, (c) 15.0 mL of a 0.25 M H_3PO_4 solution.

Miscellaneous Problems

4.44 Define the following terms: (a) stoichiometry, (b) percent composition by mass, (c) limiting reagent, (d) theoretical yield, (e) actual yield, (f) percent yield.

*4.45 A quantity of 2.40 g of the oxide of the metal X (molar mass of X = 55.9 g/mol) was heated in carbon monoxide (CO). The products are the pure metal and carbon dioxide. The mass of the metal formed is 1.68 g. From the data given show that the simplest formula of the oxide is X_2O_3 and write a balanced equation for the reaction.

4.46 An impure sample of zinc (Zn) is treated with an excess of sulfuric acid (H_2SO_4) to form zinc sulfate ($ZnSO_4$) and molecular hydrogen (H_2). (a) Write a balanced equation for the reaction. (b) If 0.0764 g of H_2 is obtained from 3.86 g of the sample, calculate the percent purity of the sample. (c) What assumptions must you make in (b)?

4.47 One of the reactions that occurs in a blast furnace, where iron ore is converted to cast iron, is

$$Fe_2O_3 + 3CO \longrightarrow 2Fe + 3CO_2$$

Suppose that 1.64×10^3 kg of Fe is obtained from a 2.62×10^3 kg sample of Fe_2O_3. Assuming the reaction goes to completion, what is the percent purity of Fe_2O_3 in the original sample?

*4.48 A student took an unknown sample mass of a compound of Ti and Cl and put it into some water. The Ti formed TiO_2, which was removed, dried, and found to weigh 0.777 g. $AgNO_3$ was then added to the solution until all the Cl^- ions were converted to 5.575 g of AgCl. Determine the empirical formula of the unknown compound of Ti and Cl.

4.49 When 0.273 g of Cr is heated in an atmosphere of Cl_2 gas, a chemical reaction occurs and a solid compound containing only Cr and Cl forms that weighs 0.832 g. Calculate the empirical formula of the compound.

4.50 When 0.273 g of Mg is heated strongly in a nitrogen (N_2) atmosphere, a chemical reaction occurs. The product of the reaction weighs 0.378 g. Calculate the empirical formula of the compound containing Mg and N.

4.51 A sample of a compound of Cl and O reacts with excess H_2 to give 0.233 g of HCl and 0.403 g of H_2O. Determine the empirical formula of the compound.

*4.52 The formula of a hydrate of barium chloride is $BaCl_2 \cdot xH_2O$. If 1.936 g of the compound gives 1.864 g of anhydrous $BaSO_4$ on treatment with sulfuric acid, calculate the value of x.

*4.53 A 0.8870 g sample of a mixture of NaCl and KCl yielded 1.913 g of AgCl. Calculate the percent by mass of each compound in the mixture.

*4.54 A 1.00 g sample of a metal X (that is known to form X^{2+} ions) was added to 0.100 liter of 0.500 M H_2SO_4. After all of the metal had reacted, the remaining acid required 0.0334 liter of 0.500 M NaOH solution for neutralization. Calculate the molar mass of the metal.

4.55 Write balanced equations for the following: (a) the reaction on heating calcium carbonate, (b) the reaction of calcium carbonate with hydrochloric acid. Calculate the masses of carbon dioxide obtained from 15.3 g of calcium carbonate in (a) and in (b).

4.56 Sulfuric acid, which is used in metallurgical processes, is often discharged into lakes and rivers. To minimize water pollution, the acid is treated with slaked lime, $Ca(OH)_2$, to form calcium sulfate and water. (a) Write a balanced equation for this reaction. (b) If slaked lime costs 1.5¢ per pound, calculate the cost of treating 6.8×10^3 liters of 2.5 M sulfuric acid. (1 lb = 453.6 g)

*4.57 A quantity of 2.000 g of a mixture of CuO and Cu was heated in an atmosphere of molecular hydrogen. (H_2 reacts with CuO to form Cu and H_2O but does not react with Cu.) The residue containing only Cu weighed 1.750 g. Write an equation for the reaction and calculate the percent by mass of Cu and CuO in the original mixture.

4.58 What is the volume (in mL) of 0.50 M HCl that is required to react completely with 10.4 g of iron according to the following equation?

$$Fe(s) + 2HCl(aq) \longrightarrow FeCl_2(aq) + H_2(g)$$

*4.59 A student carried out a titration by adding NaOH solution from a buret into an Erlenmeyer flask containing a HCl solution. The equivalence point was carefully determined by the change of the phenolphthalein indicator from colorless to reddish pink. However, after the solution was left standing on the bench for a few minutes, she noticed that the faint reddish pink color had disappeared. What do you suppose happened to the solution? (*Hint:* Air contains a small amount of carbon dioxide gas.)

5
CHEMICAL REACTIONS III: ENERGY RELATION-SHIPS

Every chemical reaction obeys two fundamental laws: the law of conservation of mass and the law of conservation of energy. The mass relationships between reactants and products were discussed in Chapter 4; here we will take a close look at the energy changes that accompany chemical reactions.

Most of this chapter focuses on a macroscopic view of the thermal (or heat) effects associated with chemical reactions. Even though such temperature and energy concerns are governed by atomic and molecular behavior, we do not need to invoke any microscopic models to deal with these concepts—heat effects can be easily measured in the laboratory. You may be interested to know that much of the important work in this branch of chemistry took place in the nineteenth century, long before any detailed model for atomic structure or chemical bonding was available.

5.1 SOME DEFINITIONS

In Chapter 4 we focused on the mass relationships associated with chemical changes. However, the energy changes involved with chemical reactions are often of equal or greater practical interest. For example, combustion reactions involving fuels such as natural gas and coal are carried out in everyday life more for the thermal energy they release than for the particular quantities of combustion products (whether expressed in grams or in moles!) they produce. But before we examine energy changes in chemical reactions more closely, several fundamental terms associated with this topic need to be introduced and clarified.

A **system** is *any specific part of the universe that is of interest to us.* For chemists, systems usually include substances involved in chemical and physical changes. For example, in an acid-base neutralization experiment, the system may be a beaker containing 50 mL HCl and 50 mL NaOH. *The rest of the universe outside the system* is called the **surroundings.**

There are three types of systems. An **open system** *can exchange mass and energy (usually in the form of heat) with its surroundings.* For example, an open system may be a quantity of water in an open container, as shown in Figure 5.1(a). If we close the flask, as in Figure 5.1(b), so that no water vapor can escape from or condense into the container, we create a **closed system,** which *allows the transfer of energy (heat) but not mass.* By placing the water in a totally insulated container, we construct an **isolated system,** which *does not allow the transfer of either mass or energy,* as shown in Figure 5.1(c).

"Energy" is a much-used term, although it represents a rather abstract concept. Unlike matter, energy cannot be seen, touched, smelled, or weighed. Energy is known and recognized by its effects. It is usually defined as *the capacity to do work.* "Work" has a special meaning to scientists. In

The unit of energy is the joule (J). We will frequently use kilojoule (kJ) in our calculations. 1 kJ = 1000 J.

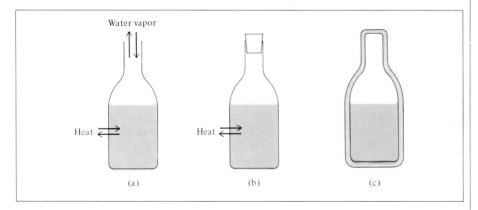

FIGURE 5.1 *Three types of systems represented by water in a flask: (a) an open system, which allows both energy and mass transfer; (b) a closed system, which allows energy transfer but not mass transfer; and (c) an isolated system, which allows neither energy nor mass transfer (here the flask is enclosed by a vacuum jacket).*

the study of mechanics, work is "force $\times$ distance," but we will see later that there are other kinds of work. All forms of energy are capable of doing work (that is, of exerting a force over a distance), but not all of them are equally relevant to chemistry. The energy contained in tidal waves, for example, can be harnessed to perform useful work, but the importance of tidal waves to chemistry is minimal. We will discuss some forms of energy that are of particular interest to chemists.

Radiant energy from the sun (solar energy) is Earth's primary energy source. Solar energy is responsible for heating the atmosphere and Earth's surface, for the growth of vegetation through the process known as photosynthesis, and for global climate patterns.

Heat (or *thermal energy*) is the energy associated with the random motion of atoms and molecules. In general, thermal energy can be calculated from temperature measurements—the more vigorous the motion of the atoms and molecules in a sample of matter, the hotter the sample is and the greater its thermal energy. However, we need to distinguish carefully between thermal energy and temperature. A cup of coffee at 70°C has a higher temperature than a bathtub filled with warm water at 40°C, but much more thermal energy is stored in the bathtub water because it has a much larger volume and greater mass than the coffee and therefore more water molecules and more molecular motion.

Chemical energy is a form of energy stored within the structural units of chemical substances; its quantity is determined by the type and arrangement of atoms in the substance being considered. When substances participate in chemical reactions, chemical energy is released, stored, or converted to other forms of energy.

Energy is also available by virtue of an object's position. This form of energy is called **potential energy.** For instance, because of its altitude, a rock at the top of a cliff has more potential energy and will make a bigger splash in the water below than a similar rock located part way down. Chemical energy can be considered a form of potential energy; it is associated with the relative positions and arrangements of atoms within the substances of interest.

Energy available because of the motion of an object is called **kinetic energy.** The kinetic energy of a moving object depends on both the mass and the velocity of the object.

All forms of energy can be changed (at least in principle) from one form to another. We feel warm when we stand in sunlight because radiant energy from the sun is converted to thermal energy on our skin. When we exercise, stored chemical energy in our bodies is used to produce kinetic energy of motion. When a ball starts to roll downhill, its potential energy is converted to kinetic energy. You can undoubtedly think of many other examples. Scientists have reached the conclusion that although energy can assume many different forms that are interconvertible, energy can neither be destroyed nor created. When one form of energy disappears, some other form of energy (of equal magnitude) must appear, and vice versa. *The total quantity of energy in the universe is thus assumed to remain constant.* This statement is generally known as the **law of conservation of energy.**

5.2 ENERGY CHANGES IN CHEMICAL REACTIONS

Almost all chemical reactions absorb or produce (release) energy. Heat (thermal energy) is the form of energy most commonly absorbed or released in chemical reactions. *The study of heat changes in chemical reactions* is called **thermochemistry.**

The combustion of hydrogen gas in oxygen is one of many familiar chemical reactions that release considerable quantities of energy (Figure 5.2):

$$2H_2(g) + O_2(g) \longrightarrow 2H_2O(l) + \text{energy}$$

In discussing energy changes of this type, we can label the reacting mixture (hydrogen, oxygen, and water molecules) the *system* and the rest of the universe the *surroundings.* Since energy cannot be created or destroyed, any energy lost by the system must be gained by the surroundings. Thus the heat generated by the combustion process is transferred from the system to its surroundings. *Any process that gives off heat* (that is, *transfers thermal energy to the surroundings*) is called an **exothermic process.** Figure 5.3(a) shows the energy change for the combustion of hydrogen gas.

Color Plate 16 shows the combustion of hydrogen gas in air.

FIGURE 5.2 *The Hindenburg disaster. The Hindenburg, a German airship filled with hydrogen gas, was destroyed by fire in Lakehurst, New Jersey, in 1937.*

FIGURE 5.3 *(a) An exothermic process. (b) An endothermic process. The scales in (a) and (b) are not the same. Therefore, the heat released in the formation of H_2O from H_2 and O_2 is not equal to the heat absorbed in the decomposition of HgO.*

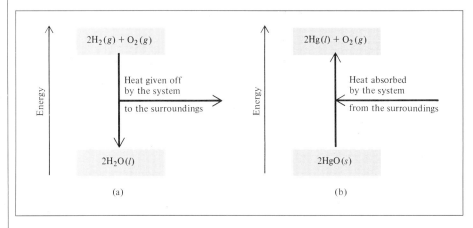

Now consider another reaction, the decomposition of mercury(II) oxide (HgO) at high temperatures:

$$\text{energy} + 2HgO(s) \longrightarrow 2Hg(l) + O_2(g)$$

This is an example of an **endothermic process,** *in which heat has to be supplied to the system* (that is, to HgO) *by the surroundings* [Figure 5.3(b)]

From Figure 5.3 you can see that in exothermic reactions, the total energy of the products is less than the total energy of the reactants. The difference in the energies is the thermal energy supplied by the system to the surroundings. Just the opposite happens in endothermic reactions. Here, the difference in the energies of the products and reactants is equal to the thermal energy supplied to the system by the surroundings.

5.3 ENTHALPY

Most physical and chemical changes, including those in living systems, occur in the constant-pressure conditions of our atmosphere. In the laboratory, for example, reactions are generally carried out in beakers, flasks, or test tubes that remain open to their surroundings and hence to a pressure of approximately 1 atm. When expressing the thermal energy released or absorbed in a constant-pressure process, chemists use a quantity called the **heat content,** or **enthalpy,** represented by the symbol H. The change in enthalpy of a system during a process at·constant pressure, represented by ΔH ("delta H," where the symbol Δ denotes change), is equal to the heat given off or absorbed by the system during the process. The **enthalpy of reaction** is *the difference between the enthalpies of the products and the enthalpies of the reactants:*

$$\Delta H = H(\text{products}) - H(\text{reactants}) \tag{5.1}$$

Enthalpy of reaction can be positive or negative, depending on the process. For an endothermic process (heat absorbed by the system from the surroundings), ΔH is positive (that is, $\Delta H > 0$). For an exothermic process (heat released by the system to the surroundings), ΔH is negative (that is, $\Delta H < 0$). Now let us apply the idea of enthalpy changes to two common processes—the first involving a physical change, the second a chemical change.

At 0°C and a constant pressure of 1 atm, ice can melt to form liquid water. Measurements show that for every mole of ice converted to liquid water under these conditions, 6.01 kilojoules (kJ) of energy are absorbed by the system (ice). Thus we can write the equation for this physical change as

$$H_2O(s) \longrightarrow H_2O(l) \qquad \Delta H = 6.01 \text{ kJ}$$

where

$$\begin{aligned}\Delta H &= H(\text{products}) - H(\text{reactants}) \\ &= H(\text{ice}) - H(\text{liquid water}) \\ &= 6.01 \text{ kJ}\end{aligned}$$

Since ΔH is a positive value, this is an endothermic process, which is what we would expect for the energy-absorbing change of melting ice (Figure 5.4).

As another example, consider the combustion of methane (CH_4):

$$CH_4(g) + 2O_2(g) \longrightarrow CO_2(g) + 2H_2O(l) \qquad \Delta H = -890.4 \text{ kJ}$$

where

$$\begin{aligned}\Delta H &= H(\text{products}) - H(\text{reactants}) \\ &= [H(CO_2, g) + 2H(H_2O, l)] - [H(CH_4, g) + 2H(O_2, g)] \\ &= -890.4 \text{ kJ}\end{aligned}$$

From experience we know that burning natural gas releases heat to its surroundings, so it is an exothermic process and ΔH must have a negative value (Figure 5.5).

The equations representing the melting of ice and the combustion of methane not only represent the mass relationships involved but also show the enthalpy changes. Equations showing both the mass and enthalpy re-

Methane is the principal component of natural gas.

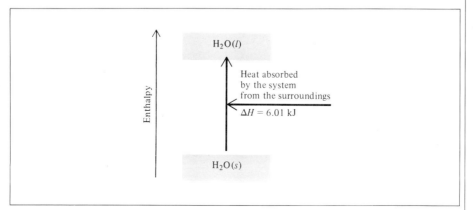

FIGURE 5.4 *Melting one mole of ice at 0°C results in an enthalpy increase in the system of 6.01 kJ.*

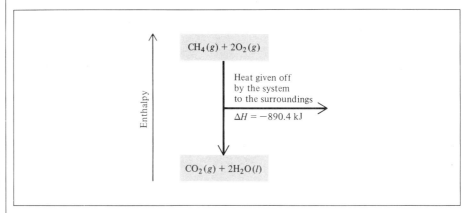

FIGURE 5.5 *Burning one mole of methane in oxygen gas results in an enthalpy decrease in the system of 890.4 kJ.*

lations are called *thermochemical equations.* The following guidelines are helpful in writing and interpreting thermochemical equations:

- The stoichiometric coefficients always refer to the number of moles of each substance. Thus, the equation representing the melting of ice may be "read" as follows: When one mole of liquid water is formed from one mole of ice at 0°C, the enthalpy change is 6.01 kJ. For the combustion of methane, we interpret the equation this way: When one mole of gaseous methane reacts with two moles of gaseous oxygen to form one mole of gaseous carbon dioxide and two moles of liquid water, the enthalpy change is −890.4 kJ.

- If we reverse a reaction, the magnitude of ΔH for the equation remains the same but its sign changes. This will seem reasonable if you consider the processes involved. For example, if a reaction consumes thermal energy from its surroundings (that is, if it is endothermic), then the reverse of that reaction must release thermal energy back to its surroundings (that is, it must be exothermic). This means that the enthalpy change expression must also change its sign. Thus, reversing the melting of ice and the combustion of methane, the thermochemical equations are

$$H_2O(l) \longrightarrow H_2O(s) \qquad \Delta H = -6.01 \text{ kJ}$$

$$CO_2(g) + 2H_2O(l) \longrightarrow CH_4(g) + 2O_2(g) \qquad \Delta H = 890.4 \text{ kJ}$$

and what was an endothermic process now becomes exothermic, and vice versa.

Enthalpy is an extensive property (see p. 7).

- If we multiply both sides of a thermochemical equation by a factor n, then ΔH must also change by the same factor. Thus, for the melting of ice, if $n = 2$

$$2H_2O(s) \longrightarrow 2H_2O(l) \qquad \Delta H = 2(6.01 \text{ kJ}) = 12.0 \text{ kJ}$$

and if $n = \frac{1}{2}$, then

$$\tfrac{1}{2}H_2O(s) \longrightarrow \tfrac{1}{2}H_2O(l) \qquad \Delta H = \tfrac{1}{2}(6.01 \text{ kJ}) = 3.01 \text{ kJ}$$

- When writing thermochemical equations, we must always specify the physical states of all reactants and products, because they help determine the actual enthalpy changes. For example, in the equation for the combustion of methane, if we show water vapor rather than liquid water as a product

$$CH_4(g) + 2O_2(g) \longrightarrow CO_2(g) + 2H_2O(g)$$

the enthalpy change would be -802.4 kJ rather than -890.4 kJ because 88.0 kJ of energy are needed to convert two moles of liquid water to water vapor; that is

$$2H_2O(l) \longrightarrow 2H_2O(g) \qquad \Delta H = 88.0 \text{ kJ}$$

- The enthalpy of a substance increases with temperature, and the enthalpy change for a reaction also depends on the specified temperature. We will see shortly that enthalpy changes are usually expressed for a reference temperature of 25°C. This convention may seem to present a problem for reactions such as combustion. During the burning of methane, for example, the temperature of the reacting system is considerably higher than 25°C. However, temperature changes during such a reaction should present no problem. We are concerned only with the enthalpy change when we convert one mole of methane and two moles of oxygen gas at 25°C to one mole of carbon dioxide and two moles of water at the same temperature. If the products form at a temperature higher than 25°C, they eventually cool down to 25°C, and the heat evolved on cooling simply becomes part of the overall enthalpy change.

5.4 STANDARD ENTHALPIES OF FORMATION AND REACTION

So far we have seen that the enthalpy change of a reaction can be obtained from the heat absorbed or released in the reaction (at constant pressure). From Equation (5.1) we see that ΔH can also be calculated if we know the actual enthalpies of all reactants and products. Unfortunately there is no way to determine the absolute value of the enthalpy of any substance. Only values relative to an arbitrary reference can be given. This is similar to the problem geographers faced in expressing the elevations of specific mountains or valleys. Rather than trying to devise some type of "absolute" elevation scale (perhaps based on distance from the center of Earth?), by common agreement all geographic heights and depths are expressed relative to sea level, an arbitrary reference with a defined elevation of "zero" meters or feet. A procedure for determining such relative values will now be introduced, which will permit the calculation of enthalpy changes for any chemical reactions.

The "sea level" reference point for all enthalpy expressions is based on implications of the term *enthalpy of formation* (also called the *heat of formation*). The enthalpy of formation of a compound is the heat change (in kJ) when one mole of the compound is synthesized from its elements under constant pressure conditions. This quantity can vary with experimental conditions (for example, the temperature and pressure at which the process is carried out). Therefore, we define the **standard enthalpy of formation** of a compound (ΔH_f°) to be *the heat change that results when one mole of the compound is formed from its elements in their standard states. Standard state refers to the condition of 1 atm.* The superscript $^\circ$ denotes that the measurement was carried out under standard state conditions (1 atm), and the subscript $_f$ denotes formation. Although the standard state does not specify what the temperature should be, we will always use ΔH_f° values measured at 25°C.

By convention, *the standard enthalpy of formation of any element in its most stable form is zero.* Take oxygen as an example. Of the three elemental forms (atomic oxygen, O; molecular oxygen, O_2; and ozone, O_3), O_2 is the most stable form at 1 atm and 25°C. Thus we can write $\Delta H_f^\circ(O_2) = 0$, but $\Delta H_f^\circ(O) \neq 0$ and $\Delta H_f^\circ(O_3) \neq 0$. Similarly, graphite is a more stable allotropic form of carbon than diamond at 1 atm and 25°C, so $\Delta H_f^\circ(C, graphite) = 0$ and $\Delta H_f^\circ(C, diamond) \neq 0$.

Consider the following exothermic reaction:

$$C(graphite) + O_2(g) \longrightarrow CO_2(g) \qquad \Delta H^\circ = -393.5 \text{ kJ}$$

This equation represents the synthesis of carbon dioxide from its elements. The **standard enthalpy of reaction, ΔH°,** is *the enthalpy change when the reaction is carried out under standard state conditions (1 atm), that is, reactants in their standard states are converted to products in their standard states.* The quantity ΔH° can be expressed as the standard enthalpy of formation of the product minus the sum of the standard enthalpies of formation of the reactants, that is

$$\Delta H^\circ = \Delta H_f^\circ(CO_2, g) - [\Delta H_f^\circ(C, graphite) + \Delta H_f^\circ(O_2, g)] \qquad (5.2)$$

Since both graphite and O_2 are stable forms, it follows that $\Delta H_f^\circ(C, graphite)$ and $\Delta H_f^\circ(O_2)$ are zero. Therefore

$$\Delta H^\circ = \Delta H_f^\circ(CO_2)$$

that is, the standard enthalpy of reaction is also the standard enthalpy of formation for CO_2. Note that ΔH° is given in kilojoules (kJ) and ΔH_f° is expressed in kilojoules per mole (kJ/mol). Thus, if $\Delta H^\circ = -393.5$ kJ for the combustion of graphite to form CO_2, then 393.5 kJ of thermal energy is given off for every mole of CO_2 formed from its elements. Therefore, $\Delta H_f^\circ(CO_2) = -393.5$ kJ/mol. In Equation (5.2) all ΔH_f° terms are multiplied by the stoichiometric coefficient of 1 mol so that the units (kJ) are the same on both sides of the equation.

This example shows how we can determine the standard enthalpy of formation of compounds in general. Note that our arbitrarily assigning zero ΔH_f° for each element in its most stable form at the standard state does not affect our calculations in any way. In thermochemistry we are interested only in the *changes* of enthalpies because they can be determined experimentally but their absolute values cannot. The choice of a zero "reference level" for enthalpy makes calculations easier to handle.

Table 5.1 lists standard enthalpies of formation of a number of elements and compounds. (A more complete list of ΔH_f° values is given in Appendix 1.) Note that in Table 5.1, as you may have expected, the ΔH_f° value for one form of each element is zero. As a good rule of thumb, compounds with positive standard enthalpies of formation are usually less stable (have a greater tendency to react) than those with negative standard enthalpies of formation.

With a set of the ΔH_f° values in hand, we can calculate the standard enthalpy changes of a large number of reactions. For a hypothetical reaction of the type

$$a\text{A} + b\text{B} \longrightarrow c\text{C} + d\text{D}$$

we can express the standard enthalpy of reaction as

$$\Delta H^\circ = [c\Delta H_f^\circ(\text{C}) + d\Delta H_f^\circ(\text{D})] - [a\Delta H_f^\circ(\text{A}) + b\Delta H_f^\circ(\text{B})] \qquad (5.3)$$

where a, b, c, and d, the stoichiometric coefficients, all have the unit mol. We can generalize Equation (5.3) as

$$\Delta H^\circ = \Sigma n\Delta H_f^\circ(\text{products}) - \Sigma m\Delta H_f^\circ(\text{reactants}) \qquad (5.4)$$

where n and m are the stoichiometric coefficients and Σ (sigma) means "the sum of."

In all calculations based on ΔH°, we assume that the temperature is 25°C.

Equation (5.1) is a simplified form of Equation (5.4).

TABLE 5.1   Standard Enthalpies of Formation of Some Inorganic Substances at 25°C

Substance	State	ΔH_f° (kJ/mol)	Substance	State	ΔH_f° (kJ/mol)
Ag	s	0	H_2O_2	l	−187.6
AgCl	s	−127.04	Hg	l	0
Al	s	0	I_2	s	0
Al_2O_3	s	−1669.8	HI	g	25.94
Br_2	l	0	Mg	s	0
HBr	g	−36.2	MgO	s	−601.8
C (graphite)	s	0	$MgCO_3$	s	−1112.9
C (diamond)	s	1.90	N_2	g	0
CO	g	−110.5	NH_3	g	−46.3
CO_2	g	−393.5	NO	g	90.4
Ca	s	0	NO_2	g	33.85
CaO	s	−635.6	N_2O_4	g	9.66
$CaCO_3$ (calcite)	s	−1206.9	N_2O	g	81.56
Cl_2	g	0	O	g	249.4
HCl	g	−92.3	O_2	g	0
Cu	s	0	O_3	g	142.2
CuO	s	−155.2	S (rhombic)	s	0
F_2	g	0	S (monoclinic)	s	0.30
HF	g	−268.61	SO_2	g	−296.1
H	g	218.2	SO_3	g	−395.2
H_2	g	0	H_2S	g	−20.15
H_2O	g	−241.8	ZnO	s	−347.98
H_2O	l	−285.8			

EXAMPLE 5.1

Calculate the standard enthalpy change for the decomposition of magnesium oxide (MgO):

$$2MgO(s) \longrightarrow 2Mg(s) + O_2(g)$$

Answer

Substituting in Equation (5.4) we write

$$\Delta H^\circ = [2\Delta H_f^\circ(Mg) + \Delta H_f^\circ(O_2)] - [2\Delta H_f^\circ(MgO)]$$

To avoid cumbersome notation, the physical states of the reactants and products in the equation have been omitted. In Table 5.1 we see that $\Delta H_f^\circ(Mg) = 0$, $\Delta H_f^\circ(O_2) = 0$, and $\Delta H_f^\circ(MgO) = -601.8$ kJ/mol. Therefore

$$\Delta H^\circ = [(2 \text{ mol})(0) + (1 \text{ mol})(0)] - [(2 \text{ mol})(-601.8 \text{ kJ/mol})]$$
$$= 1204 \text{ kJ}$$

The decomposition is an endothermic process. This is a reasonable conclusion, since the reverse of this process is the reaction of Mg with oxygen gas, an exothermic reaction that releases much heat and light. (Recall that reversing a reaction also reverses the sign of the enthalpy change expression.) This ΔH° value corresponds to the decomposition of two moles of MgO. Thus for one mole of MgO, we get $\Delta H^\circ = 602$ kJ.

Similar example: Problem 5.8.

EXAMPLE 5.2

The overall metabolism of glucose ($C_6H_{12}O_6$) in our bodies can be represented by the equation

$$C_6H_{12}O_6(s) + 6O_2(g) \longrightarrow 6CO_2(g) + 6H_2O(l)$$

Calculate the standard enthalpy of this reaction. The standard enthalpy of formation of $C_6H_{12}O_6$ is -1274.5 kJ/mol.

Answer

Substituting in Equation (5.4) we write

$$\Delta H^\circ = [6\Delta H_f^\circ(CO_2) + 6\Delta H_f^\circ(H_2O)] - [\Delta H_f^\circ(C_6H_{12}O_6) + 6\Delta H_f^\circ(O_2)]$$

The ΔH_f° values for O_2, CO_2, and H_2O are listed in Table 5.1. The value for glucose is known:

$$\Delta H^\circ = [(6 \text{ mol})(-393.5 \text{ kJ/mol}) + (6 \text{ mol})(-285.8 \text{ kJ/mol})]$$
$$- [(1 \text{ mol})(-1274.5 \text{ kJ/mol}) - (6 \text{ mol})(0)]$$
$$= -2801 \text{ kJ}$$

We see that a large quantity of energy is given off by this overall chemical change. Part of this energy is used by the body to synthesize complex biological molecules and part of it to sustain various activities such as muscle movements.

Similar example: Problem 5.12.

What if we are interested in the $\Delta H°$ value for the metabolism of glucose at 1 atm and 37°C, the body temperature? Since standard enthalpy of formation values increase with temperature, the $\Delta H°$ value at 37°C will very likely be different from that at 25°C. Equations that allow us to calculate $\Delta H°$ at 37°C (or any other temperature) are available. However, as temperature increases the enthalpies of formation of *both* products and reactants also increase. These effects tend to cancel to a certain extent [see Equation (5.4)], so $\Delta H°$ itself may remain fairly constant over a moderate temperature range (about 50°C). Thus, except for very accurate work, we can often use the $\Delta H_f°$ values at 25°C for reactions carried out at other temperatures.

5.5 HESS'S LAW

Chemists sometimes encounter experimental difficulties when they try to determine the enthalpies of formation of compounds in the laboratory. Undesired side reactions often produce substances other than the one of interest. Application of Hess's law (Germain Hess, 1802–1850) permits indirect determination of the desired enthalpy change in such cases. **Hess's law** can be stated as follows: *The enthalpy change for a reaction is equal to the sum of enthalpy changes for the individual reactions that can be added together to give the desired reaction.*

Suppose we want to determine the $\Delta H_f°$ value for carbon monoxide (CO). Theoretically this quantity could be obtained by measuring the heat given off by the following reaction at constant pressure:

(a) $C(graphite) + \frac{1}{2}O_2(g) \longrightarrow CO(g)$ $\Delta H_f° = ?$

However, it is extremely difficult (if not impossible) to burn carbon in oxygen gas without forming some CO_2 in addition to the desired product, CO. If the reaction is not "clean," that is, if side reactions yield one or more other products, then we cannot use the experimentally determined enthalpy change to calculate $\Delta H_f°$ for any particular compound.

Hess's law provides a way to avoid this difficulty. We know that the following two reactions can be cleanly and easily carried out to completion in the laboratory:

(b) $CO(g) + \frac{1}{2}O_2(g) \longrightarrow CO_2(g)$ $\Delta H° = -283.0$ kJ

(c) $C(graphite) + O_2(g) \longrightarrow CO_2(g)$ $\Delta H° = -393.5$ kJ

Reversing Equation (b) we write

(d) $CO_2(g) \longrightarrow CO(g) + \frac{1}{2}O_2(g)$ $\Delta H° = +283.0$ kJ

We can now add (c) and (d) to obtain the desired equation for the formation of CO.

$C(graphite) + O_2(g) \longrightarrow CO_2(g)$ $\Delta H° = -393.5$ kJ
+ $\quad\quad CO_2(g) \longrightarrow CO(g) + \frac{1}{2}O_2(g)$ $\Delta H° = +283.0$ kJ
(e) $C(graphite) + \frac{1}{2}O_2(g) \longrightarrow CO(g)$ $\Delta H° = -110.5$ kJ

Equation (e) is the same as Equation (a).

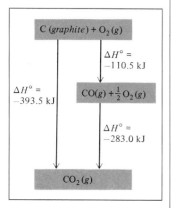

FIGURE 5.6 *Enthalpy changes in the formation of carbon dioxide from graphite and molecular oxygen. The overall enthalpy change [Equation (c), see text] is equal to the sum of the individual steps [Equations (a) + (b)], in accordance with Hess's law.*

Notice that in adding the two thermochemical equations, we treated the formulas like algebraic expressions. Since one mole of CO_2 would appear on both the left and right sides of the final equation, it can be removed (canceled out) from this expression. Likewise, the $\frac{1}{2}$ mol of O_2 on the right side of the second equation cancels out $\frac{1}{2}$ mol of O_2 on the left side of the first equation, leading to the final thermochemical expression, that is, Equation (e). Since the two equations add up to the desired equation, it follows that the two enthalpy changes can be added to give the desired enthalpy change.

Thus, the standard enthalpy of formation of carbon monoxide is

$$\Delta H_f^\circ(CO) = -110.5 \text{ kJ/mol}$$

Looking back, we see that Equation (c) represents the complete combustion of graphite to CO_2. Equations (a) and (b) can be regarded as two separate steps leading to the overall combustion process. Figure 5.6 shows the relationship between the two individual steps and the overall formation of CO_2.

EXAMPLE 5.3

Graphite is the more stable form of carbon at 1 atm and 25°C. Calculate the standard enthalpy change for the process

$$C(graphite) \longrightarrow C(diamond)$$

given that

(a) $C(graphite) + O_2(g) \longrightarrow CO_2(g)$ $\Delta H^\circ = -393.5 \text{ kJ}$

(b) $C(diamond) + O_2(g) \longrightarrow CO_2(g)$ $\Delta H^\circ = -395.4 \text{ kJ}$

Answer

First we reverse (b).

(c) $CO_2(g) \longrightarrow C(diamond) + O_2(g)$ $\Delta H^\circ = +395.4 \text{ kJ}$

Next, we add (a) and (c) to get the desired equation.

$$
\begin{aligned}
C(graphite) + O_2(g) &\longrightarrow CO_2(g) & \Delta H^\circ &= -393.5 \text{ kJ} \\
+ \quad\quad CO_2(g) &\longrightarrow C(diamond) + O_2(g) & \Delta H^\circ &= +395.4 \text{ kJ} \\
\hline
C(graphite) &\longrightarrow C(diamond) & \Delta H^\circ &= +1.9 \text{ kJ}
\end{aligned}
$$

The result shows that the conversion of graphite to diamond is endothermic, that is, energy has to be supplied to graphite. Since the standard enthalpy of formation of graphite is zero, the enthalpy of reaction here is also the enthalpy of formation for diamond:

$$\Delta H_f^\circ(diamond) = 1.9 \text{ kJ/mol}$$

Similar example: Problem 5.24.

The ΔH_f° values of graphite and diamond show that diamond is the less stable allotropic form of carbon.

EXAMPLE 5.4

From the following equations and the enthalpy changes

(a) $C(graphite) + O_2(g) \longrightarrow CO_2(g)$ $\Delta H° = -393.5$ kJ

(b) $H_2(g) + \frac{1}{2}O_2(g) \longrightarrow H_2O(l)$ $\Delta H° = -285.8$ kJ

(c) $2C_2H_2(g) + 5O_2(g) \longrightarrow 4CO_2(g) + 2H_2O(l)$ $\Delta H° = -2598.8$ kJ

calculate the enthalpy change for the reaction

$$2C(graphite) + H_2(g) \longrightarrow C_2H_2(g)$$

Answer

Since we want to obtain one equation containing only C, H_2, and C_2H_2, we need to eliminate O_2, CO_2, and H_2O from the first three equations. We note that (c) contains 5 moles of O_2, 4 moles of CO_2, and 2 moles of H_2O. First we reverse (c) to get C_2H_2 on the product side:

(d) $4CO_2(g) + 2H_2O(l) \longrightarrow 2C_2H_2(g) + 5O_2(g)$ $\Delta H° = +2598.8$ kJ

Next, we multiply (a) by 4 and (b) by 2 and carry out the addition of 4(a) + 2(b) + (d):

$$
\begin{array}{ll}
4C(graphite) + 4O_2(g) \longrightarrow 4CO_2(g) & \Delta H° = -1574.0 \text{ kJ} \\
+ \quad 2H_2(g) + O_2(g) \longrightarrow 2H_2O(l) & \Delta H° = -571.6 \text{ kJ} \\
+ \quad 4CO_2(g) + 2H_2O(l) \longrightarrow 2C_2H_2(g) + 5O_2(g) & \Delta H° = +2598.8 \text{ kJ} \\
\hline
4C(graphite) + 2H_2(g) \longrightarrow 2C_2H_2(g) & \Delta H° = +453.2 \text{ kJ}
\end{array}
$$

or

$$2C(graphite) + H_2(g) \longrightarrow C_2H_2(g) \qquad \Delta H° = +226.6 \text{ kJ}$$

This equation represents the synthesis of C_2H_2 (acetylene) from its component elements; therefore, the standard enthalpy of formation of acetylene is given by $\Delta H_f°(C_2H_2) = 226.6$ kJ/mol. This example should convince you of the usefulness of Hess's law. It is normally not possible to form acetylene by heating graphite in hydrogen gas, so there is no way of *measuring* the enthalpy change. However, by taking an indirect route, that is, using data on reactions that can be readily studied in the laboratory, we are able to calculate the enthalpy of formation of acetylene. The same procedure can be applied to a large number of other compounds.

Similar examples: Problems 5.25, 5.26.

5.6   HEAT OF SOLUTION AND DILUTION

So far we have focused on thermal energy effects resulting from chemical reactions. Many physical processes (for example, phase changes such as the melting of ice or the vaporization of a liquid) also involve the absorption or release of heat. Enthalpy changes are also involved when a solute dissolves in a solvent or when a solution is diluted. We will examine these

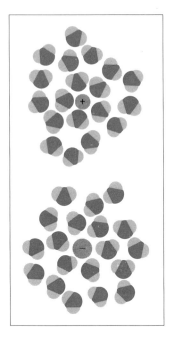

FIGURE 5.7 *Ion–solvent interactions of Na^+ and Cl^- ions in water. Each ion is surrounded by a number of water molecules in a specific way and is said to be hydrated.*

two closely related physical processes, involving heat of solution and heat of dilution, in greater detail here.

Heat of Solution

In the vast majority of cases, dissolving a solute in a solvent produces measurable heat effects. At constant pressure, the quantity of heat change is equal to the enthalpy change. The **heat of solution** (also called **enthalpy of solution**), **ΔH_{soln}**, is *the heat generated or absorbed when a certain amount of solute dissolves in a certain amount of solvent.* The quantity ΔH_{soln} represents the difference between the enthalpy of the final solution and the enthalpies of its original components (that is, solute and solvent) before they are mixed. Thus

$$\Delta H_{soln} = H_{soln} - H_{components}$$

Neither H_{soln} nor $H_{components}$ can be measured, but their difference, ΔH_{soln}, can be readily determined. Like other enthalpy changes, ΔH_{soln} is positive for endothermic (heat-absorbing) processes and negative for exothermic (heat-generating) processes.

Consider the heat of solution involving ionic compounds as solute and water as solvent. For example, what happens when solid NaCl dissolves in water? While NaCl is a solid, the Na^+ and Cl^- ions are held together by strong positive–negative (electrostatic) forces, but when a small crystal of NaCl dissolves in water, the three-dimensional network of ions breaks into its individual units. (The structure of solid NaCl is shown in Figure 2.5.) The separated Na^+ and Cl^- ions are stabilized in solution by their interaction with water molecules (Figure 5.7). These ions are said to be *hydrated*. **Hydration** describes the *process in which an ion or molecule is surrounded by water molecules arranged in a specific manner.* Here water plays a role similar to that of a good electrical insulator. Water molecules shield the ions (Na^+ and Cl^-) from each other and effectively reduce the electrostatic attraction that held them together in the solid state. The heat of solution is defined by the process

$$NaCl(s) \xrightarrow{H_2O} Na^+(aq) + Cl^-(aq) \qquad \Delta H_{soln} = ?$$

Dissolving an ionic compound such as NaCl in water involves complex interactions among the solute and solvent species. At first glance it might seem such an impossibly involved process that no clear account could be given in terms of the energy involved. However, for the sake of energy analysis we can imagine the solution process takes place in only two separate steps. First, the Na^+ and Cl^- ions in the solid crystal are separated from each other in a gas phase (Figure 5.8):

$$energy + NaCl(s) \longrightarrow Na^+(g) + Cl^-(g)$$

Lattice energy is a positive quantity.

The energy required for this process is called **lattice energy (U)**, *the energy required to completely separate one mole of a solid ionic compound into gaseous ions.* The lattice energy of NaCl is 774 kJ/mol. In other words, we would need to supply 774 kJ of energy to break one mole of solid NaCl into one mole of Na^+ ions and one mole of Cl^- ions.

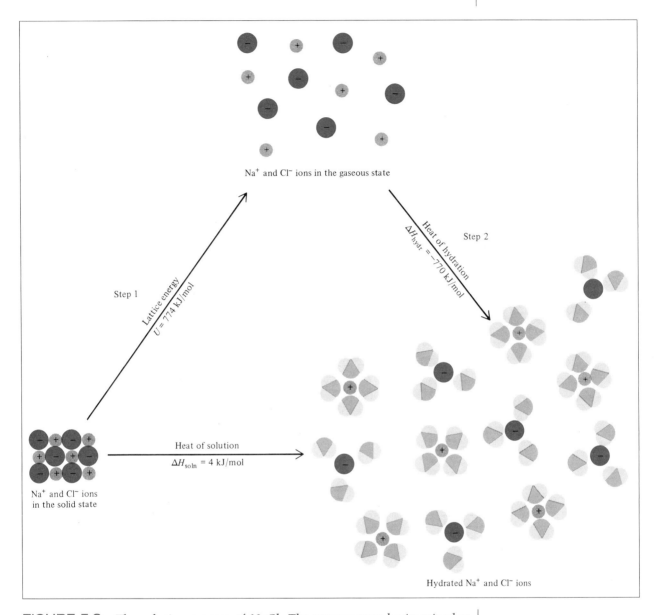

Na$^+$ and Cl$^-$ ions in the gaseous state

Step 1

Lattice energy
$U = 774$ kJ/mol

Heat of hydration
$\Delta H_{hydr} = -770$ kJ/mol

Step 2

Heat of solution
$\Delta H_{soln} = 4$ kJ/mol

Na$^+$ and Cl$^-$ ions
in the solid state

Hydrated Na$^+$ and Cl$^-$ ions

FIGURE 5.8   *The solution process of NaCl. The process may be imagined to occur in two separate steps. Step 1: separation of ions from the crystal state to the gaseous state. Step 2: hydration of the gaseous ions. The heat of solution is equal to the energy changes for these two steps,* $\Delta H_{soln} = U + \Delta H_{hydr}.$

Next, the "gaseous" Na$^+$ and Cl$^-$ ions enter the water and become hydrated:

$$Na^+(g) \xrightarrow{H_2O} Na^+(aq) + \text{energy}$$

$$Cl^-(g) \xrightarrow{H_2O} Cl^-(aq) + \text{energy}$$

The sum of these two equations is

$$Na^+(g) + Cl^-(g) \xrightarrow{H_2O} Na^+(aq) + Cl^-(aq) + energy$$

The enthalpy change associated with the hydration process is called the **heat of hydration** (ΔH_{hydr}). Applying Hess's law, it is possible to consider ΔH_{soln} as the sum of two related quantities: lattice energy (U) and heat of hydration (ΔH_{hydr}):

$$
\begin{array}{ll}
NaCl(s) \longrightarrow Na^+(g) + Cl^-(g) & U = 774 \text{ kJ} \\
+ \; Na^+(g) + Cl^-(g) \xrightarrow{H_2O} Na^+(aq) + Cl^-(aq) & \Delta H_{hydr} = -770 \text{ kJ} \\
\hline
NaCl(s) \xrightarrow{H_2O} Na^+(aq) + Cl^-(aq) & \Delta H_{soln} = 4 \text{ kJ}
\end{array}
$$

or summing up the enthalpy changes

$$
\begin{aligned}
\Delta H_{soln} &= U + \Delta H_{hydr} \\
&= 774 \text{ kJ/mol} - 770 \text{ kJ/mol} \\
&= 4 \text{ kJ/mol}
\end{aligned}
$$

We see that the solution process for NaCl is slightly endothermic. When one mole of NaCl dissolves in water, 4 kJ of heat will be absorbed from the immediate surroundings. We would observe this effect by noting that the beaker containing the solution becomes slightly colder. Table 5.2 lists the ΔH_{soln} of several ionic compounds. Depending on the nature of the cation and anion involved, ΔH_{soln} for an ionic compound may be either negative (exothermic) or positive (endothermic).

An interesting application of the heat of solution is the use of instant cold and hot packs as first-aid devices to treat injuries. A typical pack consists of a water-tight plastic bag containing a pouch of water and a water-soluble solid substance (usually an ionic compound). Striking the pack causes the inner pouch to break and the substance to dissolve. The temperature of the pack will either rise or fall, depending on whether the heat of solution of the solid substance is exothermic or endothermic (Figure 5.9).

TABLE 5.2 Heat of Solution of Some Ionic Compounds

Compound	ΔH_{soln} (kJ/mol)
LiCl	−37.1
NaCl	4.0
KCl	17.2
NH₄Cl	15.2
NH₄NO₃	26.2
CaCl₂	−82.8

Heat of Dilution

When a previously prepared solution is *diluted,* that is, when more solvent is added to lower the overall concentration of the solute, additional heat is usually given off or absorbed. *The heat change associated with the dilution process is called* **heat of dilution.** If a certain solution process had been endothermic, and the solution is diluted again, *more* heat would be absorbed by the same solution from the surroundings. The converse holds true for an exothermic solution process—more heat would be liberated if additional solvent were added to dilute the solution. Therefore, always be cautious when working on a dilution procedure in the laboratory. Due to its highly exothermic heat of dilution, concentrated sulfuric acid (H_2SO_4) poses a particularly hazardous problem if its concentration must be reduced by mixing it with additional water. Concentrated H_2SO_4 is composed of 98 percent acid and 2 percent water by volume. When it is diluted with water, considerable heat is released to the surroundings. This process is so exo-

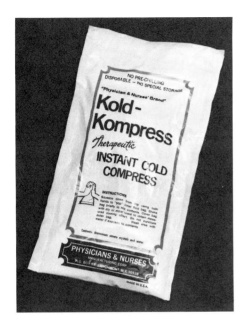 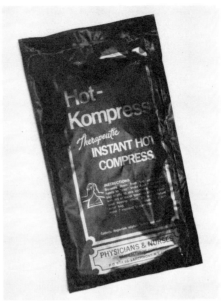

FIGURE 5.9 *Commercially available cold pack and hot pack. The cold pack contains ammonium nitrate (NH_4NO_3) and water, and the hot pack contains magnesium sulfate ($MgSO_4$) and water.*

thermic that you must *never* attempt to dilute the concentrated acid by adding water to it. The heat generated could easily cause the acid solution to boil and splatter. The recommended procedure is to add the concentrated acid slowly into the water (while constantly stirring).

Generations of chemistry students have been urged to remember this procedure by the venerable saying, "Do as you oughter, add acid to water."

5.7 CALORIMETRY

In this section we will consider experimental methods for measuring the heat of reaction. Two concepts that we need to understand in order to talk meaningfully about **calorimetry** (that is, *the measurement of heat changes*) are specific heat and heat capacity. Let us consider them first.

Specific Heat and Heat Capacity

The **specific heat (ρ)** of a substance is *the amount of heat required to raise the temperature of one gram of the substance by one degree Celsius*. The **heat capacity (C)** of a substance is *the amount of heat required to raise the temperature of a given quantity of the substance by one degree Celsius*. The relationship between the heat capacity and specific heat of a substance is

$$C = m\rho \tag{5.5}$$

where m is the mass of the substance in grams. For example, the specific heat of water is 4.184 J/g·°C, and the heat capacity of 60.0 g of water is

$$(60.0 \text{ g})(4.184 \text{ J/g·°C}) = 251 \text{ J/°C}$$

Note that specific heat has the units J/g·°C and heat capacity has the units J/°C. Table 5.3 shows the specific heats of some common substances.

The dot between g and °C reminds us that both g and °C are in the denominator.

TABLE 5.3 Specific Heats of Some Common Substances

Substance	Specific Heat $(J/g \cdot °C)$
Al	0.900
Au	0.129
C(graphite)	0.720
C(diamond)	0.502
Cu	0.385
Fe	0.444
Hg	0.139
H_2O	4.184
C_2H_5OH (ethanol)	2.46

From a knowledge of the specific heat or the heat capacity, the heat absorbed or released in a given process, q, can be calculated from the following equations:

$$q = mp\Delta t \qquad (5.6)$$

$$q = C\Delta t \qquad (5.7)$$

where m is the mass of the system, and Δt is the temperature change:

$$\Delta t = t_{final} - t_{initial}$$

The sign convention for q is the same as that for enthalpy change: q is positive for endothermic processes and negative for exothermic processes. Combining Equations (5.6) and (5.7) gives

$$C = mp$$

which is Equation (5.5).

EXAMPLE 5.5

A 466 g sample of water is heated from 8.50°C to 74.60°C. Calculate the amount of heat absorbed by the water.

Answer

Using Equation (5.5) we write

$$q = mp\Delta t$$
$$= (466 \text{ g})(4.184 \text{ J/g} \cdot °C)(74.60°C - 8.50°C)$$
$$= 1.29 \times 10^5 \text{ J}$$
$$= 129 \text{ kJ}$$

Similar examples: Problems 5.33, 5.34.

Constant-Volume Calorimetry

Constant volume refers to the volume of the container, which does not change during the reaction. Note that the container remains intact after the measurement. The term *bomb calorimeter* refers to the explosive nature of the reaction (on a small scale) in the presence of excessive oxygen gas.

Heats of combustion are usually measured by placing a known mass of the compound under study in a steel container, called a *constant-volume bomb calorimeter*, which is filled with oxygen at about 30 atm of pressure. The closed bomb is immersed in a known amount of water, as shown in Figure 5.10. The sample is ignited electrically, and the heat produced by the combustion can be calculated accurately after noting the rise in temperature of the water. The heat given off by the sample is absorbed by the water and the calorimeter. The special design of the bomb calorimeter allows us to assume that no heat (or mass) is lost to the surroundings during the time it takes to make measurements. Therefore we can call the bomb calorimeter and the water in which it is submerged an isolated system. Because there

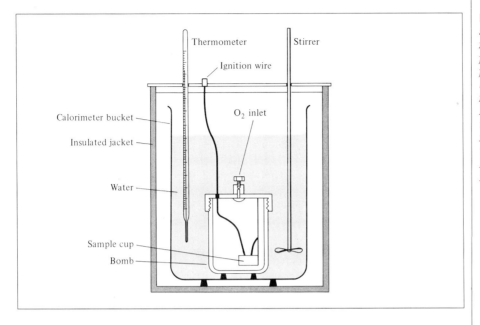

FIGURE 5.10 *A constant-volume bomb calorimeter. The calorimeter is filled with oxygen gas before it is placed in the bucket. The sample is ignited electrically and the heat produced by the reaction can be accurately determined by measuring the temperature increase in the known amount of surrounding water.*

is no heat entering or leaving the system throughout the process, we can write

$$q_{\text{system}} = q_{\text{water}} + q_{\text{bomb}} + q_{\text{reaction}}$$
$$= 0 \tag{5.8}$$

where q_{water}, q_{bomb}, and q_{reaction} are the heat changes for the water, the bomb, and the reaction, respectively. Thus

$$q_{\text{reaction}} = -(q_{\text{water}} + q_{\text{bomb}}) \tag{5.9}$$

The quantity q_{water} is obtained by

$$q = m\rho\Delta t$$

$$q_{\text{water}} = (m_{\text{water}})(4.184 \text{ J/g·°C})\Delta t$$

The product of the mass of the bomb and its specific heat is the heat capacity of the bomb, which remains constant for all experiments carried out in the bomb calorimeter.

$$C_{\text{bomb}} = m_{\text{bomb}} \times \rho_{\text{bomb}}$$

Hence

$$q_{\text{bomb}} = C_{\text{bomb}}\Delta t$$

We must know the heat capacity of the bomb before we can use it to measure heat of combustion.

Note that because reactions in a bomb calorimeter occur under constant volume rather than constant pressure conditions, the heat changes do not correspond to the enthalpy change ΔH (see Section 5.3). It is possible to correct the measured heat changes so that they correspond to ΔH values, but the corrections usually are quite small, so we will not concern ourselves with the details of the correction procedure.

EXAMPLE 5.6

A quantity of 1.435 g of naphthalene ($C_{10}H_8$) was burned in a constant-volume bomb calorimeter. Consequently, the temperature of the water rose from 20.17°C to 25.84°C. If the quantity of water surrounding the calorimeter was exactly 2000 g and the heat capacity of the bomb calorimeter was 1.80 kJ/°C, calculate the heat of combustion of naphthalene on a molar basis, that is, find the molar heat of combustion.

Answer

First we calculate the heat changes for the water and the bomb calorimeter.

$$q = mp\Delta t$$

$$q_{water} = (2000 \text{ g})(4.184 \text{ J/g·°C})(25.84°C - 20.17°C)$$
$$= 4.74 \times 10^4 \text{ J}$$

$$q_{bomb} = (1.80 \times 1000 \text{ J/°C})(25.84°C - 20.17°C)$$
$$= 1.02 \times 10^4 \text{ J}$$

(Note that we changed 1.80 kJ/°C to 1.80 × 1000 J/°C.) Next, from Equation (5.9) we write

$$q_{reaction} = -(4.74 \times 10^4 \text{ J} + 1.02 \times 10^4 \text{ J})$$
$$= -5.76 \times 10^4 \text{ J}$$

The molar mass of naphthalene is 128.2 g, and so the heat of combustion of one mole of naphthalene is

$$\text{molar heat of combustion} = \frac{-5.76 \times 10^4 \text{ J}}{1.435 \text{ g C}_{10}H_8} \times \frac{128.2 \text{ g C}_{10}H_8}{1 \text{ mol C}_{10}H_8}$$
$$= -5.15 \times 10^6 \text{ J/mol}$$
$$= -5.15 \times 10^3 \text{ kJ/mol}$$

Similar examples: Problems 5.36, 5.37.

Interestingly, the fuel values of food are determined in a bomb calorimeter, as are the heats of combustion of naphthalene and other substances.

Constant-Pressure Calorimetry

A simpler device for determining the heats of reactions other than combustion reactions is a *constant-pressure calorimeter*. In its most basic form, it can be constructed from two Styrofoam coffee cups, as shown in Figure 5.11. Such a calorimeter can be used to measure the heat effect for a variety of reactions, such as acid-base neutralization reactions as well as heats of solution and heats of dilution. Because the measurements are carried out under constant atmospheric pressure conditions, the heat change for the process ($q_{reaction}$) is equal to the enthalpy change (ΔH). The measurements

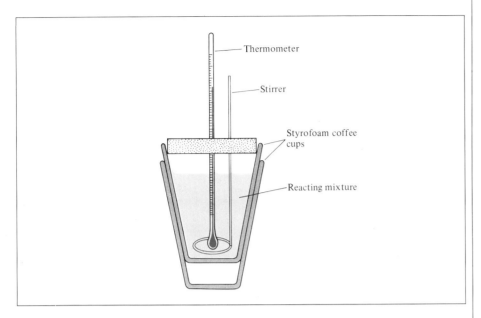

FIGURE 5.11 *A constant-pressure calorimeter made up of two Styrofoam coffee cups. The outer cup helps to insulate the reacting mixture from the surroundings. Two solutions of known volume containing the reactants initially at the same temperature are carefully mixed in the calorimeter. The heat produced or absorbed by the reaction can be determined by measuring the change in temperature. The cover on top is for insulation. Since it is not air-tight, the reaction occurs under atmospheric conditions.*

are similar to those of a constant-volume calorimeter—we need to know the heat capacity of the calorimeter, as well as the temperature change of the solution.

EXAMPLE 5.7

A quantity of 1.00×10^2 mL of 0.500 M HCl is mixed with 1.00×10^2 mL of 0.500 M NaOH in a constant-pressure calorimeter having a heat capacity of 335 J/°C. The initial temperature of the HCl and NaOH solutions is the same, 22.50°C, and the final temperature of the mixed solution is 24.90°C. Calculate the heat change for the neutralization reaction

$$NaOH(aq) + HCl(aq) \longrightarrow NaCl(aq) + H_2O(l)$$

Assume that the densities and specific heats of the solutions are the same as for water (1.00 g/mL and 4.184 J/g·°C, respectively).

Answer

From Equation (5.9) we write

$$q_{reaction} = -(q_{solution} + q_{calorimeter})$$

(Continued)

Because the density of the solution is 1.00 g/mL, the mass of a 100 mL solution is 100 g.

where

$$q_{solution} = (1.00 \times 10^2 \text{ g} + 1.00 \times 10^2 \text{ g})(4.184 \text{ J/g·°C})(24.90°C - 22.50°C)$$
$$= 2.01 \times 10^3 \text{ J}$$
$$q_{calorimeter} = (335 \text{ J/°C})(24.90°C - 22.50°C)$$
$$= 804 \text{ J}$$

Hence

$$q_{reaction} = -(2.01 \times 10^3 \text{ J} + 804 \text{ J})$$
$$= -2.81 \times 10^3 \text{ J}$$
$$= -2.81 \text{ kJ}$$

From the molarities given, we know there are 0.0500 mole of HCl in 1.00×10^2 g of the HCl solution and 0.0500 mole of NaOH in 1.00×10^2 g of the NaOH solution. Therefore the heat of neutralization when 1.00 mole of HCl reacts with 1.00 mole of NaOH is

$$\text{heat of neutralization} = \frac{-2.81 \text{ kJ}}{0.0500 \text{ mol}} = -56.2 \text{ kJ/mol}$$

Note that because the reaction takes place at constant pressure, the heat given off is equal to the enthalpy change.

AN ASIDE ON THE PERIODIC TABLE
Standard Enthalpies of Atomization of Elements and Chemical Reactivity

In Section 5.4 we introduced the quantity called standard enthalpy of formation of the elements. By arbitrarily assigning a zero value for ΔH_f° for an element in its most stable form, it is possible to obtain the ΔH_f° values of many compounds. Just as atomic masses are used to calculate molecular and molar masses of compounds, the tabulated ΔH_f° values can be used to calculate the enthalpies of reactions. As a rule, substances with positive ΔH_f° values are usually less stable (and therefore chemically more reactive) than those with negative ΔH_f° values.

We can study the reactivity of many elements by comparing the energy required to convert each element to a monatomic gas (a process called *atomization*). The standard enthalpy of atomization of an element is the energy required to convert one mole of the element in its most stable form at 25°C to one mole of monatomic gas, also at 25°C. Take the element hydrogen as an example. We know that the stable form of hydrogen at 25°C is gaseous molecular hydrogen. Experimentally we find that

$$H_2(g) \longrightarrow H(g) + H(g) \qquad \Delta H^\circ = 436.4 \text{ kJ}$$

Therefore, to form one mole of gaseous atomic hydrogen at 25°C we need to supply $\frac{1}{2}(436.4 \text{ kJ})$ or 218.2 kJ of energy. On the other hand, atomization of metallic sodium involves the conversion of one mole of Na atoms in the solid state at 25°C to one mole of Na atoms in the gas phase at 25°C:

$$Na(s) \longrightarrow Na(g) \qquad \Delta H^\circ = 108.4 \text{ kJ}$$

1A	2A	3B	4B	5B	6B	7B	8B	8B	8B	1B	2B	3A	4A	5A	6A	7A	8A
H 218.2																	**He** 0
Li 155.2	**Be** 320.5											**B** 407	**C** 718	**N** 470.7	**O** 249.4	**F** 75.3	**Ne** 0
Na 108.4	**Mg** 150.2											**Al** 314	**Si** 368	**P** 315	**S** 223	**Cl** 121.4	**Ar** 0
K 89.6	**Ca** 176.6	**Sc** 343	**Ti** 471	**V** 502	**Cr** 337	**Mn** 286	**Fe** 405	**Co** 439	**Ni** 424	**Cu** 341	**Zn** 131	**Ga** 276	**Ge** 328	**As** 254	**Se** 203	**Br** 96.3	**Kr** 0
Rb 82.0	**Sr** 163.6	**Y** 427	**Zr** 527	**Nb** 743	**Mo** 651	**Tc** ?	**Ru** 577	**Rh** 577	**Pd** 393	**Ag** 289	**Cd** 113	**In** 244	**Sn** 301	**Sb** 254	**Te** 199	**I** 75.5	**Xe** 0
Cs 78.2	**Ba** 174.5	**La** 368	**Hf** 703	**Ta** 782	**W** 843	**Re** 777	**Os** 728	**Ir** 690	**Pt** 564	**Au** 344	**Hg** 60.7	**Tl** 181	**Pb** 194	**Bi** 208			**Rn** 0

Figure 5.12 shows the standard enthalpy of atomization of many elements according to their positions in the periodic table. Since energy is needed for atomization, all the standard enthalpy values are positive—that is, all atomizations are endothermic processes. Reactions of many elements require atomization; therefore, those elements having lower standard enthalpies of atomization (easier to break up into gaseous atoms) are expected to be more reactive. This is particularly evident for the Group 1A elements (the alkali metals), the Group 2A elements (the alkaline earth metals with the exception of Be), and the Group 7A elements (the halogens).

In studying Figure 5.12 you should keep in mind that a small standard enthalpy of atomization is only a prerequisite for an element to exhibit chemical reactivity and does not guarantee it. For example, because mercury is a liquid at room temperature, it is relatively easy to atomize (60.7 kJ/mol). However, mercury is *not* a very reactive element; it reacts neither with water nor with hydrochloric acid. Finally, note that because the Group 8A elements (the noble gases) are already monatomic gases at 25°C, they all have zero standard enthalpy of atomization. The fact that these elements exist as monatomic gases in their normal state indicates their great stability. In fact, as a group they are among the least reactive of the elements—no known compounds containing He, Ne, or Ar have ever been prepared.

FIGURE 5.12 Standard enthalpies of atomization (in kJ/mol) of a number of elements according to their positions in the periodic table.

SUMMARY

1. Energy is the capacity to do work. There are many forms of energy that are interconvertible. The law of conservation of energy states that although energy can be converted from one form to another, it cannot be created or destroyed.

2. Any process that gives off heat to the surroundings is called an exothermic process; any process that absorbs heat from the surroundings is called an endothermic process.

3. The change in enthalpy (ΔH, usually given in kilojoules) is a measure of the heat of reaction (or any other process) at constant pressure.
4. Hess's law states that the overall enthalpy change in a reaction is equal to the sum of enthalpy changes for the individual steps that make up the overall reaction.
5. The heat of solution of an ionic compound in water is the sum of the lattice energy and the heat of hydration. The relative magnitudes of these two quantities determine whether the solution process is endothermic or exothermic. The heat of dilution is the heat absorbed or evolved when a solution is diluted.
6. Constant-volume and constant-pressure calorimeters are used to measure heat changes of physical and chemical processes.

KEY WORDS

Calorimetry, p. 139
Closed system, p. 123
Endothermic process, p. 126
Enthalpy, p. 126
Enthalpy of reaction, p. 126
Enthalpy of solution, p. 136
Exothermic process, p. 125
Heat capacity, p. 139
Heat content, p. 126
Heat of dilution, p. 138
Heat of hydration, p. 138
Heat of solution, p. 136
Hess's law, p. 133
Hydration, p. 136

Isolated system, p. 123
Kinetic energy, p. 124
Lattice energy, p. 136
Law of conservation of energy, p. 124
Open system, p. 123
Potential energy, p. 124
Specific heat, p. 139
Standard enthalpy of formation, p. 130
Standard enthalpy of reaction, p. 130
Standard state, p. 130
Surroundings, p. 123
System, p. 123
Thermochemistry, p. 125

PROBLEMS

More challenging problems are marked with an asterisk.

Law of Conservation of Energy

5.1 A truck initially traveling at 50 kilometers per hour is brought to a complete stop at a traffic light. Does this change violate the law of conservation of energy? Explain.
5.2 The following are various forms of energy: chemical, heat, light, mechanical, and electrical. Suggest ways of interconverting these forms of energy.
5.3 Describe the interconversions of forms of energy occurring in the following processes: (a) You throw a softball up in the air and catch it. (b) You switch on a flashlight. (c) You take the ski lift to the top of the hill and then ski down. (d) You strike a match and let it burn down.

Enthalpy Changes

5.4 Explain the meaning of the following thermochemical equation:

$$4NH_3(g) + 5O_2(g) \longrightarrow 4NO(g) + 6H_2O(g)$$
$$\Delta H° = -905 \text{ kJ}$$

5.5 Under what condition is the heat of a reaction equal to the enthalpy change of the same reaction?
5.6 In writing thermochemical equations, why is it important to indicate the physical state (that is,

liquid, solid, gaseous, or aqueous) of each substance?

5.7 What is meant by standard-state condition?

5.8 Calculate the heat of decomposition for this process at constant pressure and 25°C:

$$CaCO_3(s) \longrightarrow CaO(s) + CO_2(g)$$

(Look up the standard enthalpy of formation of the reactant and products in Table 5.1.)

5.9 The standard enthalpies of formation of H^+ and OH^- ions in water are zero and -229.6 kJ/mol, respectively, at 25°C. Calculate the enthalpy of neutralization when one mole of a strong monoprotic acid (such as HCl) is titrated by one mole of a strong base (such as KOH) at 25°C.

5.10 Calculate the heats of combustion of the following from the standard enthalpies of formation listed in Appendix 1:

(a) $2H_2(g) + O_2(g) \longrightarrow 2H_2O(l)$
(b) $2C_2H_2(g) + 5O_2(g) \longrightarrow 4CO_2(g) + 2H_2O(l)$
(c) $C_2H_4(g) + 3O_2(g) \longrightarrow 2CO_2(g) + 2H_2O(l)$
(d) $2H_2S(g) + 3O_2(g) \longrightarrow 2H_2O(l) + 2SO_2(g)$

5.11 The standard enthalpy change for the reaction

$$H_2(g) \longrightarrow H(g) + H(g)$$

is 436.4 kJ. Calculate the standard enthalpy of formation of atomic hydrogen (H).

5.12 From the standard enthalpies of formation, calculate $\Delta H°$ for the reaction

$$C_6H_{12}(l) + 9O_2(g) \longrightarrow 6CO_2(g) + 6H_2O(l)$$
$$\Delta H_f° \text{ for } C_6H_{12}(l) = -151.9 \text{ kJ/mol}$$

5.13 Methanol, ethanol, and n-propanol are three common alcohols. When 1.00 g of each of three alcohols is burned in air, heat is liberated as follows:

(a) methanol (CH_3OH): 22.6 kJ
(b) ethanol (C_2H_5OH): 29.7 kJ
(c) n-propanol (C_3H_7OH): 33.4 kJ

Calculate the heat of combustion of these alcohols in kJ/mol.

5.14 How much heat (in kJ) is given off when 1.26×10^4 g of ammonia is produced according to the equation

$$N_2(g) + 3H_2(g) \longrightarrow 2NH_3(g) \qquad \Delta H° = -92.6 \text{ kJ}$$

Assume the reaction to take place under standard-state conditions and 25°C.

*5.15 Dissolving 1.00 g of anhydrous copper(II) sulfate in a large amount of water liberates 418 kJ of heat. Dissolving 5.00 g of $CuSO_4 \cdot 5H_2O$ in a large amount of water requires the input of 231 kJ of heat. From these data, calculate the standard enthalpy change for the process

$$CuSO_4(s) + 5H_2O(l) \longrightarrow CuSO_4 \cdot 5H_2O(s)$$

5.16 Industrially, the first step in the recovery of zinc from the zinc sulfide ore is roasting, that is, the conversion of ZnS to ZnO by heating:

$$2ZnS(s) + 3O_2(g) \xrightarrow{\Delta} 2ZnO(s) + 2SO_2(g)$$
$$\Delta H° = -879 \text{ kJ}$$

Calculate the heat evolved (in kJ) per gram of ZnS roasted.

5.17 At 850°C, $CaCO_3$ undergoes substantial decomposition to yield CaO and CO_2. Assuming that the $\Delta H_f°$ values of the reactant and products at 850°C to be the same as those at 25°C, calculate the enthalpy change (in kJ) if 66.8 g of CO_2 is produced in one reaction.

5.18 Decomposition reactions are usually endothermic, whereas combination reactions are usually exothermic. Give a qualitative explanation for these trends.

Standard Enthalpy of Formation

5.19 Which is a more negative quantity: $\Delta H_f°$ for $H_2O(l)$ or $\Delta H_f°$ for $H_2O(g)$? Explain.

5.20 The standard enthalpy of formation of $Fe_2O_3(s)$ is -823 kJ/mol. How much heat (in kJ) is evolved when 762 g of Fe_2O_3 is formed in its standard state from its elements?

5.21 Suggest ways (with appropriate equations) that would allow you to measure the $\Delta H_f°$ values of Ag_2O and PCl_5. No calculation is necessary.

*5.22 In general, compounds with negative $\Delta H_f°$ values are more stable than those with positive $\Delta H_f°$ values. From Table 5.1 we see that H_2O_2 has a negative $\Delta H_f°$. Why, then, does H_2O_2 have a tendency to decompose to H_2O and O_2?

*5.23 The convention of arbitrarily assigning a zero enthalpy value for the most stable form of each element in the standard state is a convenient way of dealing with enthalpies of reactions. Explain why this convention cannot be applied to nuclear reactions. (In a nuclear reaction, the identity of elements may be altered.)

Hess's Law

5.24 From these data

$$S(rhombic) + O_2(g) \longrightarrow SO_2(g)$$
$$\Delta H° = -296.06 \text{ kJ}$$
$$S(monoclinic) + O_2(g) \longrightarrow SO_2(g)$$
$$\Delta H° = -296.36 \text{ kJ}$$

calculate the enthalpy change for the transformation

$$S(rhombic) \longrightarrow S(monoclinic)$$

(Monoclinic and rhombic are different forms of elemental sulfur.)

5.25 From the following data

$$C(graphite) + O_2(g) \longrightarrow CO_2(g)$$
$$\Delta H° = -393.5 \text{ kJ}$$
$$H_2(g) + \tfrac{1}{2}O_2(g) \longrightarrow H_2O(l)$$
$$\Delta H° = -285.8 \text{ kJ}$$
$$2C_2H_6(g) + 7O_2(g) \longrightarrow 4CO_2(g) + 6H_2O(l)$$
$$\Delta H° = -3119.6 \text{ kJ}$$

calculate the enthalpy change for the reaction

$$2C \ (graphite) + 3H_2(g) \longrightarrow C_2H_6(g)$$

5.26 From the following heats of combustion

$$CH_3OH(l) + \tfrac{3}{2}O_2(g) \longrightarrow CO_2(g) + 2H_2O(l)$$
$$\Delta H° = -726.4 \text{ kJ}$$
$$C(graphite) + O_2(g) \longrightarrow CO_2(g)$$
$$\Delta H° = -393.5 \text{ kJ}$$
$$H_2(g) + \tfrac{1}{2}O_2(g) \longrightarrow H_2O(l)$$
$$\Delta H° = -285.8 \text{ kJ}$$

calculate the enthalpy of formation of methanol (CH_3OH) from its elements

$$C(graphite) + 2H_2(g) + \tfrac{1}{2}O_2(g) \longrightarrow CH_3OH(l)$$

5.27 Calculate the standard enthalpy change for the reaction

$$2Al(s) + Fe_2O_3(s) \longrightarrow 2Fe(s) + Al_2O_3(s)$$

given that

$$2Al(s) + \tfrac{3}{2}O_2(g) \longrightarrow Al_2O_3(s) \qquad \Delta H° = -1601 \text{ kJ}$$
$$2Fe(s) + \tfrac{3}{2}O_2(g) \longrightarrow Fe_2O_3(s) \qquad \Delta H° = -821 \text{ kJ}$$

5.28 Calculate the standard enthalpy of formation of carbon disulfide (CS_2) from its elements given

$$C(s) + O_2(g) \longrightarrow CO_2(g) \qquad \Delta H° = -393.5 \text{ kJ}$$
$$S(s) + O_2(g) \longrightarrow SO_2(g) \qquad \Delta H° = -296.1 \text{ kJ}$$
$$CS_2(l) + 3O_2(g) \longrightarrow CO_2(g) + 2SO_2(g)$$
$$\Delta H° = -1072 \text{ kJ}$$

*5.29 The standard enthalpy of formation at 25°C of HF(aq) is -320.1 kJ/mol; of OH$^-$(aq), it is -229.6 kJ/mol; of F$^-$(aq), it is -329.1 kJ/mol; and of H$_2$O(l), it is -285.8 kJ/mol.

(a) Calculate the standard enthalpy of neutralization of HF(aq):

$$HF(aq) + OH^-(aq) \longrightarrow F^-(aq) + H_2O(l)$$

(b) Using the value of -56.2 kJ as the standard enthalpy change for the reaction

$$H^+(aq) + OH^-(aq) \longrightarrow H_2O(l)$$

calculate the standard enthalpy change for the reaction

$$HF(aq) \longrightarrow H^+(aq) + F^-(aq)$$

Specific Heat and Heat Capacity

5.30 What is the difference between specific heat and heat capacity?

5.31 A piece of silver of mass 362 g has a heat capacity of 85.7 J/°C. What is the specific heat of silver?

5.32 Explain the cooling effect experienced when ethanol is rubbed on your skin. Given that

$$C_2H_5OH(l) \longrightarrow C_2H_5OH(g) \qquad \Delta H° = 42.2 \text{ kJ}$$

5.33 A piece of copper metal of mass 6.22 kg is heated from 20.5°C to 324.3°C. Calculate the heat absorbed (in kJ) by the metal.

5.34 Calculate the amount of heat liberated (in kJ) from 366 g of mercury when it cools from 77.0°C to 12.0°C.

*5.35 A sheet of gold weighing 10.0 g and at a temperature of 18.0°C is placed flat on a sheet of iron weighing 20.0 g and at a temperature of 55.6°C. What is the final temperature of the combined metals? Assume that no heat is lost to the surroundings. (*Hint:* The heat gained by gold must be equal to the heat lost by iron.)

Calorimetry

5.36 A 0.1375 g sample of solid magnesium is burned in a constant-volume bomb calorimeter that has a heat capacity of 1769 J/°C. The calorimeter contains exactly 300 g of water and the temperature increases by 1.126°C. Calculate the heat given off by the burning Mg, in kJ/g and in kJ/mol.

*5.37 The enthalpy of combustion of benzoic acid (C_6H_5COOH) is commonly used as the standard for calibrating constant-volume bomb calorimeters; its value has been accurately determined to be -3226.7 kJ/mol. When 1.9862 g of benzoic acid is burned, the temperature rises from 21.84°C to 25.67°C. What is the heat capacity of the calorimeter? (Assume that the quantity of water surrounding the calorimeter is exactly 2000 g.)

6
PHYSICAL PROPERTIES OF GASES

We recognize and deal with the three states of matter (solids, liquids, and gases) in everyday life without great confusion or difficulty. However, there are important differences among them. *Solids* do not conform to the shape of the container in which they are placed— that is, they have definite shapes. On the other hand, liquids and gases (both termed *fluids*) do take a container's shape. A gas completely fills any closed container, while a liquid, possessing a well-defined volume, may not necessarily fill a given container.

Of the three states of matter, gases involve the greatest degree of disorder among the component units (molecules). You might thus conclude that gases would be the most difficult state of matter to study. However, the macroscopic properties of gases (volume, temperature, pressure, and amount) are linked by relatively simple relationships. These relationships represent a major focus of this chapter. In addition, since a discussion of gases leads our attention to the microscopic behavior of gaseous molecules themselves, we will also consider the gaseous state at that level.

149

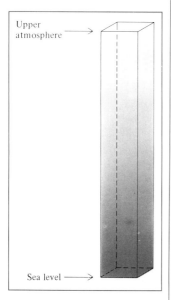

FIGURE 6.1   *A column of air extending from sea level to the upper atmosphere.*

6.1   PRESSURE OF A GAS

We live at the bottom of an ocean of air whose composition by volume is roughly 78 percent N_2, 21 percent O_2, and 1 percent other gases, such as Ar and CO_2. The molecules (and atoms) of these gases, like those of all other matter, are subject to Earth's gravitational pull. If we could see air, we would note that it is much denser near the surface of the Earth. Moving upward, we would find the air density rapidly decreasing (Figure 6.1). Measurements show that about 50 percent of the atmosphere is within 4 miles (6.4 km) of Earth's surface, 90 percent within 10 miles (16 km), and 99 percent within 20 miles (32 km).

One of the most readily measurable properties of a gas is its pressure. Gases exert pressure on any surface with which they come into contact because gas molecules are constantly in motion and they collide with the surface. We humans have adapted so well physiologically to the pressure of the air around us that we are usually unaware of its existence, perhaps as fish are unconscious of the water's pressure on them. There are many ways to demonstrate the existence of atmospheric pressure. One example is our ability to drink a liquid through a straw. Sucking air out of the straw creates a vacuum, which is filled quickly as the fluid in the container is pushed up into the straw by atmospheric pressure.

An instrument that measures atmospheric pressure is the simple **barometer.** This instrument can be constructed by filling a long glass tube, closed at one end, with mercury, and then carefully inverting the tube in a dish of mercury, making sure that no air enters the tube. Some mercury in the tube will flow down into the dish, creating a vacuum at the top (Figure 6.2). The weight of the mercury column remaining in the tube is supported by atmospheric pressure acting on the surface of the mercury in the dish. The *standard* atmospheric pressure (1 atm) is equal to the pressure that supports a column of mercury exactly 760 mm (or 76 cm) high at 0°C at sea level. The standard atmosphere thus equals a pressure of 760 mmHg, where mmHg represents the pressure exerted by a column of mercury 1 mm high. The mmHg unit is also called the *torr*, after the Italian scientist Evangelista Torricelli (1608–1647), who invented the barometer. Thus

$$1 \text{ torr} = 1 \text{ mmHg}$$

and

$$1 \text{ atm} = 760 \text{ mmHg}$$
$$= 760 \text{ torr}$$

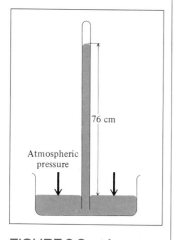

FIGURE 6.2   *A barometer for measuring atmospheric pressure. Above the mercury in the tube is a vacuum. The column of mercury is supported by the atmospheric pressure.*

EXAMPLE   6.1

The pressure outside a jet plane flying at high altitude falls considerably below atmospheric pressure at sea level. The air inside the cabin must therefore be pressurized to protect the passengers. What is the pressure in atmospheres in the cabin if the barometer reading is 688 mmHg?

Answer

The pressure in atmospheres is calculated as follows:

$$\text{pressure} = 688 \text{ mmHg} \times \frac{1 \text{ atm}}{760 \text{ mmHg}}$$

$$= 0.905 \text{ atm}$$

In SI units, pressure is measured in *pascals (Pa)*, defined as one newton per square meter:

See Section 1.6 (p. 14) to review the definition of a newton.

$$\text{pressure} = \frac{\text{force}}{\text{area}}$$

$$1 \text{ Pa} = 1 \frac{\text{N}}{\text{m}^2}$$

The relation between atmospheres and pascals (see Appendix 2) is

$$1 \text{ atm} = 101{,}325 \text{ Pa}$$
$$= 1.01325 \times 10^5 \text{ Pa}$$

This equation gives the SI definition of 1 atm.

and since 1000 Pa = 1 kPa (kilopascal)

$$1 \text{ atm} = 1.01325 \times 10^2 \text{ kPa}$$

EXAMPLE 6.2

The atmospheric pressure in San Francisco on a certain day was 732 mmHg. What was the pressure in kPa?

Answer

The conversion equations show that

$$1 \text{ atm} = 1.01325 \times 10^5 \text{ Pa} = 760 \text{ mmHg}$$

which allows us to calculate the atmospheric pressure in kPa:

$$\text{pressure} = 732 \text{ mmHg} \times \frac{1.01325 \times 10^5 \text{ Pa}}{760 \text{ mmHg}}$$

$$= 9.76 \times 10^4 \text{ Pa}$$
$$= 97.6 \text{ kPa}$$

EXAMPLE 6.3

Calculate the height in meters of the column in a barometer for 1 atm pressure if water were used instead of mercury. The density of water is 1.00 g/cm³ and that of mercury is 13.6 g/cm³.

(Continued)

Answer

The difference in heights here is the result of the different densities. Therefore

$$\text{height of } H_2O = \text{height of Hg} \times \frac{\text{density of Hg}}{\text{density of } H_2O}$$

$$= 760 \text{ mm} \times \frac{13.6 \text{ g/cm}^3}{1.00 \text{ g/cm}^3}$$
$$= 1.03 \times 10^4 \text{ mm}$$
$$= 10.3 \text{ m}$$

The calculation in Example 6.3 shows that there is a limit to how deep an old-fashioned farmyard well can be. Drawing water from such a well by using a suction pump is analogous to drinking soda with a straw. Strictly speaking, only the air in the pipe is pumped up. Water is then pushed up by atmospheric pressure. However, since this operation depends entirely on atmospheric pressure, we cannot get water from more than 10.3 m (or 34 ft) below the surface. Motorized pumps must be used to get water from greater depths.

Measurements of changing atmospheric pressure are of great value to meteorologists. Forecasts of rain, for example, are associated with "low-pressure fronts." Since water vapor has a lower density than the air it dilutes, wet air exerts a lower pressure than dry air. This occurs because the molar mass of water is 18 g, whereas the *average* molar mass of air is about 29 g. Therefore, an approaching front of low pressure means the approach of moisture-laden air.

As we most often use the terms, the difference between a "gas" and a "vapor" is: A *gas* is a substance that is *normally* in the gaseous state at ordinary temperatures and pressures; a *vapor* is the gaseous form of any substance that is a liquid or a solid at normal temperatures and pressures. Thus, at 25°C and 1 atm pressure, we speak of water *vapor* and oxygen *gas*.

6.2 SUBSTANCES THAT EXIST AS GASES UNDER ATMOSPHERIC CONDITIONS

Before we explore the physical behavior of gases, we will briefly survey the gaseous "cast of characters"—representative substances that might be expected to occur as gases (rather than liquids or solids) under normal conditions of pressure and temperature (that is, 1 atm and 25°C). Once we are able to recognize substances that act as gases we can infer their characteristic physical behavior from our understanding of this distinct state of matter.

In Chapter 2 we learned that compounds can be classified into two categories: ionic and molecular. Ionic compounds do not exist as gases at 25°C, because cations and anions in an ionic solid are held together by very strong electrostatic forces. To overcome these attractions we must apply a large amount of energy, which in practice means strongly heating the solid. Under normal conditions, all we can do is melt the solid; for example, NaCl melts at the rather high temperature of 801°C. In order to boil NaCl to form "gaseous NaCl," the temperature would have to be raised well above 1000°C.

The behavior of molecular compounds is more varied. Some of them, for example, CO, CO_2, HCl, NH_3, and CH_4 (methane), are gases, but the majority of molecular compounds are liquids or solids at room temperature. However, by heating they can be converted to gases much more easily than ionic compounds can. In other words, molecular compounds usually boil at much lower temperatures than ionic compounds. There is no simple rule to help us determine if a certain molecular compound is a gas under atmospheric conditions. To make such a determination we would need to understand the nature and magnitude of the attractive forces among the molecules. In general, the stronger these attractions, the less likely that the compound can exist as a gas at ordinary temperatures.

Figure 6.3 shows the elements that are gases under normal atmospheric conditions. Note that the elements hydrogen, nitrogen, oxygen, fluorine, and chlorine exist as gaseous diatomic molecules: H_2, N_2, O_2, F_2, and Cl_2. An allotrope of oxygen, ozone (O_3), is also a gas at room temperature. All the elements in Group 8A, the noble gases, are monatomic gases: He, Ne, Ar, Kr, Xe, and Rn.

Table 6.1 lists the common gases we will use in our study of the physical behavior of gases. Of these, only O_2 is essential for our survival. Hydrogen

When sodium chloride boils, the particles that enter the vapor phase are made of individual NaCl units.

FIGURE 6.3 *Elements that exist as gases at 25°C and 1 atm. The noble gases (the Group 8A elements) are monatomic species; the other elements exist as diatomic molecules. Ozone (O_3) is also a gas at room temperature.*

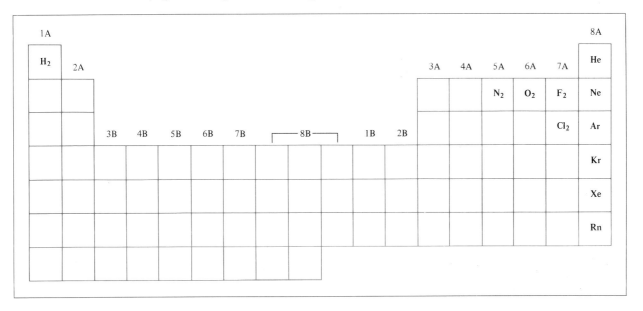

TABLE 6.1 Some Substances Found as Gases at 1 atm and 25°C

Elements	Compounds
H_2 (molecular hydrogen)	HF (hydrogen fluoride)
N_2 (molecular nitrogen)	HCl (hydrogen chloride)
O_2 (molecular oxygen)	HBr (hydrogen bromide)
O_3 (ozone)	HI (hydrogen iodide)
F_2 (molecular fluorine)	CO (carbon monoxide)
Cl_2 (molecular chlorine)	CO_2 (carbon dioxide)
He (helium)	NH_3 (ammonia)
Ne (neon)	NO (nitric oxide)
Ar (argon)	NO_2 (nitrogen dioxide)
Kr (krypton)	N_2O (nitrous oxide)
Xe (xenon)	SO_2 (sulfur dioxide)
Rn (radon)	H_2S (hydrogen sulfide)
	HCN (hydrogen cyanide)*
	CH_4 (methane)

* The boiling point of HCN is 26°C, but it is close enough to qualify as a gas at ordinary atmospheric conditions.

sulfide (H_2S) and hydrogen cyanide (HCN) are deadly poisons. Several others, such as CO, NO_2, O_3, and SO_2, are somewhat less toxic. The gases He, Ne, and Ar are chemically inert, that is, they do not react with any other substance. Most gases are colorless. Of the gases listed in Table 6.1, only NO_2 is dark brown; its color is sometimes visible in polluted air.

6.3 THE GAS LAWS

The study of gas behavior has occupied scientists for many centuries. Their observations and analyses led to the formulation of several useful and well-known laws dealing with the relationships among pressure, volume, and temperature of a sample of gas.

The Pressure-Volume Relationship: Boyle's Law

In the seventeenth century, Robert Boyle (1627–1691) studied gas behavior systematically and quantitatively. In one series of studies, Boyle investigated the pressure-volume relationship of a gas sample using an apparatus such as that shown in Figure 6.4. In Figure 6.4(a) the pressure exerted on the gas is equal to atmospheric pressure. In Figure 6.4(b) there is an increase in pressure due to the added mercury and the resulting unequal levels in the two columns; consequently, the volume of the gas decreases. Boyle noticed that when temperature is held constant, the volume (V) of a given amount of a gas decreases as the total applied pressure (P)—atmospheric pressure plus the pressure due to the added mercury—is increased. This pressure-volume relationship is shown pictorially in Figure 6.4(b), (c), and (d). Conversely, if the applied pressure is decreased, the gas volume becomes larger. The results of several measurements are given in Table 6.2.

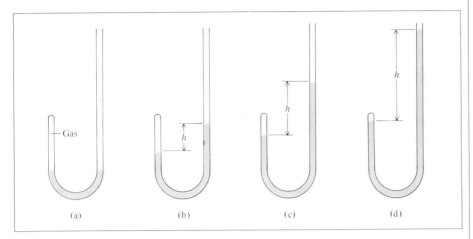
(a)   (b)   (c)   (d)

The *P-V* data actually recorded for such an experiment are consistent with these mathematical expressions showing an inverse relationship:

$$V \propto \frac{1}{P}$$

or

$$V = C \times \frac{1}{P}$$

where the symbol $\propto$ means *proportional to* and C is the proportionality constant. The above equation is called **Boyle's law,** which states that *the volume of a fixed amount of gas maintained at constant temperature is inversely proportional to the gas pressure.* Rearranging the last equation, we obtain

$$PV = C \qquad\qquad (6.1)$$

Equation (6.1) is another form of Boyle's law.

The conditions are constant temperature and constant amount of gas.

The concept of one quantity being proportional to another, and the use of a proportionality constant, can be clarified through the following analogy. The daily income of a movie theater depends on the number of people buying tickets. Thus we say that the income is proportional to the number of tickets sold:

$$\text{income} \propto \text{number of tickets sold}$$

However, since income is measured in dollars, it cannot be directly equated to the number of tickets sold. Instead, we must write

$$\text{income} = C \times \text{number of tickets sold}$$

TABLE 6.2 Typical Pressure-Volume Relationships Obtained by Boyle

P (mmHg)	724	869	951	998	1230	1893	2250
V (arbitrary units)	1.50	1.33	1.22	1.16	0.94	0.61	0.51
PV	1.09×10^3	1.16×10^3	1.16×10^3	1.16×10^3	1.2×10^3	1.2×10^3	1.1×10^3

FIGURE 6.5 *Graphs showing variation of the volume of a gas sample with the pressure exerted on the gas, at constant temperature. (a) P versus V; (b) P versus 1/V. Since P is proportional to 1/V, Equation (6.1) can be written as P = C(1/V). Therefore, a plot of P versus 1/V yields a straight line.*

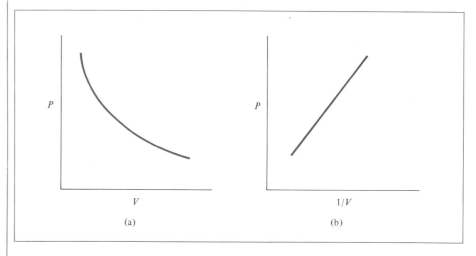

where C is the proportionality constant. It should be clear that C is the price (in dollars) per ticket, that is, it has the units of dollar/ticket. (We assume here that the theater has only one price for all tickets.)

Figure 6.5 shows Boyle's findings graphically. Although the individual values of pressure and volume can vary greatly for a given sample of gas, according to Equation (6.1) we can write

$$P_1 V_1 = P_2 V_2 \qquad (6.2)$$

where V_1 and V_2 are the volumes at pressures P_1 and P_2, respectively. As long as the temperature is held constant and the amount of the gas does not change, P times V is always equal to the same constant.

EXAMPLE 6.4

An inflated balloon has a volume of 0.55 L at sea level (1.0 atm) and is allowed to rise to a height of 6.5 km, where the pressure is about 0.40 atm. Assuming the temperature remains constant, what is the final volume of the balloon?

Answer

We use Equation (6.2)

$$P_1 V_1 = P_2 V_2$$

where

$$P_1 = 1.0 \text{ atm} \qquad P_2 = 0.40 \text{ atm}$$
$$V_1 = 0.50 \text{ L} \qquad V_2 = ?$$

Therefore

$$V_2 = V_1 \times \frac{P_1}{P_2}$$
$$= 0.55 \text{ L} \times \frac{1.0 \text{ atm}}{0.40 \text{ atm}}$$
$$= 1.4 \text{ L}$$

As we expected, the balloon expanded as it rose because the atmospheric pressure decreased.

Similar example: Problem 6.7.

EXAMPLE 6.5

A sample of chlorine gas occupies a volume of 946 mL at a pressure of 726 mmHg. Calculate the pressure of the gas if the volume is reduced to 154 mL. Assume the temperature remains constant.

Answer

Since

$$P_1 V_1 = P_2 V_2$$

where

$$P_1 = 726 \text{ mmHg} \qquad P_2 = ?$$
$$V_1 = 946 \text{ mL} \qquad V_2 = 154 \text{ mL}$$

we continue with

$$P_2 = P_1 \times \frac{V_1}{V_2}$$
$$= 726 \text{ mmHg} \times \frac{946 \text{ mL}}{154 \text{ mL}}$$
$$= 4.46 \times 10^3 \text{ mmHg}$$

Similar examples: Problems 6.10, 6.11.

The Temperature-Volume Relationship: Charles and Gay-Lussac's Law

We turn our attention now to the effect of temperature on gas volume. The earliest investigators of this relationship were French scientists, Jacques Charles (1746–1823) and Joseph Gay-Lussac (1778–1850). Their studies showed that, at constant pressure, the volume of a gas sample expands when heated and contracts when cooled (Figure 6.6). This conclusion may seem almost self-evident, and the quantitative relations involved in these changes in temperature and volume turn out to be remarkably consistent. For example, we observe an interesting phenomenon when we study the temperature-volume relationship at various pressures. At any given pressure, the plot of volume versus temperature yields a straight line. By extending the line to zero volume, we find the intercept on the temperature axis to be $-273.15°C$. At any other pressure, we obtain a different straight line for the volume-temperature plot, but we get the *same* zero-volume temperature intercept (Figure 6.7). (In practice, we can measure the volume

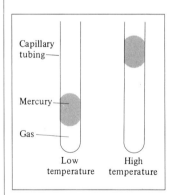

Capillary tubing—

Mercury—

Gas—

Low temperature   High temperature

FIGURE 6.6 *Variation of the volume of a gas sample with temperature, at constant pressure. The pressure exerted on the gas is the sum of the atmospheric pressure and the pressure due to the weight of mercury.*

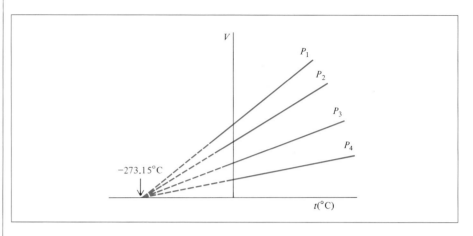

FIGURE 6.7   *Variation of the volume of a gas sample with temperature, at constant pressure. Each line represents the variation at a certain pressure. These pressures increase from P_1 to P_4. All gases ultimately condense (become liquids) if they are cooled to sufficiently low temperatures; the solid portions of the lines represent the temperature region above the condensation point. When these lines are extrapolated, or extended (the dashed portions), they all intersect at the point representing zero volume and temperature of $-273.15°C$.*

of a gas over only a limited temperature range because all gases condense at low temperatures to form liquids.)

In 1848 Lord Kelvin (1824–1907) realized the significance of this behavior. He identified the temperature $-273.15°C$ as *theoretically the lowest attainable temperature*, called **absolute zero.** With *absolute zero as a starting point*, he set up an **absolute temperature scale,** now called the **Kelvin temperature scale.** One degree Celsius is equal *in magnitude* to one kelvin (K). (Note that the absolute temperature scale has no degree sign, so that 25 K is called twenty-five kelvins.) The only difference between the absolute temperature scale and the Celsius scale is that the zero position is shifted. Important points on the scales match up as follows:

$$\text{absolute zero:} \quad 0\text{ K} = -273.15°C$$
$$\text{freezing point of water: } 273.15\text{ K} = 0°C$$
$$\text{boiling point of water: } 373.15\text{ K} = 100°C$$

The relationship between °C and K is

$$T(\text{K}) = t(°C) + 273.15°C \qquad (6.3)$$

Equation (6.3) shows that to convert a temperature reading in degrees Celsius to kelvins we have to add 273.15 to the reading. This equation can be rewritten as

$$\text{K} = (°C + 273.15°C)\left(\frac{1\text{ K}}{1°C}\right)$$

The unit factor (1 K/1°C) is there to make the units consistent on both sides of the equation. In most calculations we will use 273 instead of 273.15 as the term relating K and °C. By convention we use T to denote absolute (kelvin) temperature and t to indicate temperature on the Celsius scale.

Scientists have succeeded approaching absolute zero under experimental conditions within a small fraction of a kelvin.

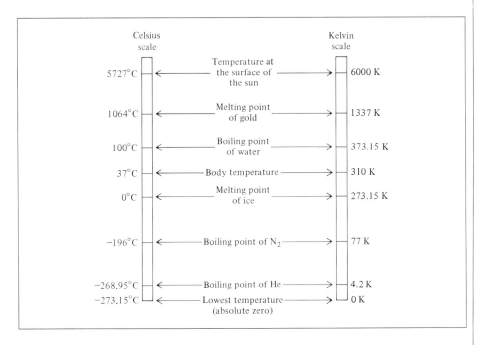

FIGURE 6.8 *Relationship between the Celsius temperature scale and the Kelvin temperature scale. The distances between temperature marks are not drawn accurately.*

Figure 6.8 shows corresponding temperatures on the Celsius and Kelvin scales for a number of systems.

With the development of an absolute temperature scale, it becomes possible to express the temperature-volume relationship in simple mathematical terms. For a given amount of gas at constant pressure

$$V \propto T$$
$$V = CT$$

or

$$\frac{V}{T} = C \qquad (6.4)$$

The conditions are constant pressure and constant amount of gas.

where C is the proportionality constant. Equation (6.4) is known as **Charles and Gay-Lussac's law**, or simply **Charles' law**. Charles' law states that *the volume of a fixed amount of gas maintained at constant pressure is directly proportional to the absolute temperature of the gas.* Note that although the same symbol C is used for the proportionality constant, it represents a different value than it did in Equation (6.1). The same is true for other equations in which it appears later.

From Equation (6.4) we can write

$$\frac{V_1}{T_1} = \frac{V_2}{T_2} \qquad (6.5)$$

where V_1 and V_2 are the volumes of the gas at temperatures T_1 and T_2 (both in kelvins), respectively. Figure 6.9 shows a straight line plot of V versus T at constant pressure.

In all subsequent calculations we assume that the temperatures given in °C are exact so that they do not affect the significant figures.

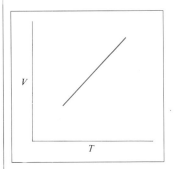

FIGURE 6.9 *Volume of a gas sample plotted versus temperature, at constant pressure.*

EXAMPLE 6.6

A gas whose volume is 452 mL is heated from 22°C to 187°C at constant pressure. What is its final volume?

Answer

We use Equation (6.5)

$$\frac{V_1}{T_1} = \frac{V_2}{T_2}$$

where

Remember to convert °C to K when solving gas problems.

$V_1 = 452$ mL $\qquad V_2 = ?$
$T_1 = (22 + 273)$ K $= 295$ K $\qquad T_2 = (187 + 273)$ K $= 460$ K

Hence

$$V_2 = V_1 \times \frac{T_2}{T_1}$$

$$= 452 \text{ mL} \times \frac{460 \text{ K}}{295 \text{ K}}$$

$$= 705 \text{ mL}$$

Similar example: Problem 6.13.

EXAMPLE 6.7

A sample of carbon monoxide gas occupies 3.20 L at 125°C. Calculate the temperature at which the gas will occupy 1.54 L if the pressure remains constant.

Answer

Since

$$\frac{V_1}{T_1} = \frac{V_2}{T_2}$$

where

$V_1 = 3.20$ L $\qquad V_2 = 1.54$ L
$T_1 = (125 + 273)$ K $= 398$ K $\qquad T_2 = ?$

Hence

$$T_2 = T_1 \times \frac{V_2}{V_1}$$

$$= 398 \text{ K} \times \frac{1.54 \text{ L}}{3.20 \text{ L}}$$

$$= 192 \text{ K}$$

or, from Equation (6.3)

$$°C = T_2 - 273 = 192 - 273$$
$$= -81°C$$

The Volume-Amount Relationship: Avogadro's Law

The work of the Italian scientist Amedeo Avogadro complemented the studies of Boyle, Charles, and Gay-Lussac. In 1811 he published a hypothesis that stated that at the same temperature and pressure, equal volumes of different gases contain the same number of molecules (or atoms if the gas is monatomic). It follows that the volume of any given gas must be proportional to the number of molecules present, that is

$$V \propto n$$
$$V = Cn \tag{6.6}$$

where n represents the number of moles and C is the proportionality constant.

Equation (6.6) is the mathematical expression of **Avogadro's law,** which states that *at constant pressure and temperature, the volume of a gas is directly proportional to the number of moles of the gas present.* From Avogadro's law we learn that when two gases react with each other their reacting volumes have a simple ratio to each other. If the product is a gas, its volume is related to the volume of the reactants by a simple ratio (a fact demonstrated earlier by Gay-Lussac). For example, consider the synthesis of ammonia from molecular hydrogen and molecular nitrogen:

$$3H_2(g) \;+\; N_2(g) \;\longrightarrow\; 2NH_3(g)$$
$$\text{3 mol} \qquad \text{1 mol} \qquad \text{2 mol}$$

Since at the same temperature and pressure, the volumes of gas are directly proportional to the number of moles of the gases present, we can now write

$$3H_2(g) \qquad + \; N_2(g) \qquad \longrightarrow \; 2NH_3(g)$$
$$\text{3 volumes} \qquad \text{1 volume} \qquad \text{2 volumes}$$

The ratio of volumes for molecular hydrogen and molecular nitrogen is 3:1, and that for ammonia (the product) and molecular hydrogen and molecular nitrogen (the reactants) is 2:4, or 1:2 (Figure 6.10).

Amedeo Avogadro also gave us Avogadro's number (see Section 2.4).

Because the hypothesis has been verified experimentally, it is now called the law.

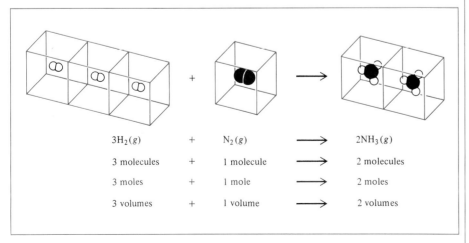

$$3H_2(g) \qquad + \qquad N_2(g) \qquad \longrightarrow \qquad 2NH_3(g)$$

3 molecules	+	1 molecule	⟶	2 molecules
3 moles	+	1 mole	⟶	2 moles
3 volumes	+	1 volume	⟶	2 volumes

FIGURE 6.10 *Volume relationship of gases in a chemical reaction. The ratio of volumes for molecular hydrogen and molecular nitrogen is 3:1 and that for ammonia (the product) and hydrogen and nitrogen combined (the reactants) is 2:4 or 1:2.*

6.4   THE IDEAL GAS EQUATION

Let us now summarize the various gas laws we have discussed so far:

$$\text{Boyle's law: } V \propto \frac{1}{P} \text{ (at constant } n \text{ and } T)$$
$$\text{Charles' law: } V \propto T \text{ (at constant } n \text{ and } P)$$
$$\text{Avogadro's law: } V \propto n \text{ (at constant } P \text{ and } T)$$

Combining all three equations, we obtain

$$V \propto \frac{nT}{P}$$
$$= R\,\frac{nT}{P}$$

or

$$PV = nRT \tag{6.7}$$

where R, *the proportionality constant*, is called the **gas constant** and n is the number of moles of the gas present. Equation (6.7) is called the **ideal gas equation.** It *describes the relationship in an ideal gas among four experimental variables—P, V, T, and n.* An **ideal gas** is *a hypothetical gas whose pressure-volume-temperature behavior can be completely accounted for by the ideal gas equation.* Although there is no such thing in nature as an ideal gas, discrepancies in the behavior of real gases over reasonable temperature and pressure ranges do not significantly affect calculations. Thus we can safely use the ideal gas equation to solve many gas problems.

Before we can apply Equation (6.7) to a real system, we must evaluate the gas constant R. At 0°C (273.15 K) and 1 atm pressure most real gases

The molecules of an ideal gas do not attract or repel one another, and their volume is negligible compared to the volume of the container.

FIGURE 6.11 *The volume of one mole of an ideal gas at 0°C and 1 atm (STP). A basketball is shown for comparison.*

behave like an ideal gas. Experiments show that under these conditions, one mole of gas occupies 22.414 L (Figure 6.11). *The conditions 0°C and 1 atm are called **standard temperature and pressure,** often abbreviated **STP.*** From Equation (6.7) we can write

$$R = \frac{PV}{nT}$$

$$= \frac{(1 \text{ atm})(22.414 \text{ L})}{(1 \text{ mol})(273.15 \text{ K})} = 0.082057 \frac{\text{L} \cdot \text{atm}}{\text{K} \cdot \text{mol}}$$

$$= 0.082057 \text{ L} \cdot \text{atm/K} \cdot \text{mol}$$

The dots (between L and atm and between K and mol) remind us that both L and atm are in the numerator and both K and mol are in the denominator. For most calculations, we will round off the value of R to three significant figures (0.0821 L·atm/K·mol).

The gas constant can also be expressed in other units (see Appendix 2).

EXAMPLE 6.8

Calculate the volume (in liters) occupied by 7.40 g of CO_2 at STP.

Answer

This problem can be solved in two different ways. We will show both approaches here.

Approach 1

The molar mass of CO_2 is 44.01 g; thus the number of moles of CO_2 in 7.40 g is 7.40 g/(44.01 g/mol) = 0.168 mol. From Equation (6.7) we write

$$V = \frac{nRT}{P}$$

$$= \frac{(0.168 \text{ mol})(0.0821 \text{ L} \cdot \text{atm/K} \cdot \text{mol})(273 \text{ K})}{1.00 \text{ atm}}$$

$$= 3.77 \text{ L}$$

Approach 2

Recognizing that one mole of an ideal gas occupies 22.414 L at STP, we write

$$V = 7.40 \text{ g } CO_2 \times \frac{1 \text{ mol } CO_2}{44.01 \text{ g}} \times \frac{22.414 \text{ L}}{1 \text{ mol } CO_2}$$

$$= 3.77 \text{ L}$$

Equation (6.7) is useful for problems that do not involve changes among the variables P, V, T, and n for a gas sample. At times, however, we need to deal with changes in pressure, volume, and temperature, or even the amount of a gas. When pressure, volume, temperature, and amount change, we must employ a modified form of Equation (6.7) that involves initial and final conditions. We derive it as follows. From Equation (6.7)

$$R = \frac{P_1 V_1}{n_1 T_1} \quad \text{(before change)}$$

and

$$R = \frac{P_2 V_2}{n_2 T_2} \quad \text{(after change)}$$

so that

$$\frac{P_1 V_1}{n_1 T_1} = \frac{P_2 V_2}{n_2 T_2}$$

If $n_1 = n_2$, as is usually the case because the amount of gas normally does not change, the equation then becomes

$$\frac{P_1 V_1}{T_1} = \frac{P_2 V_2}{T_2} \tag{6.8}$$

The subscripts 1 and 2 denote the initial and final states of the gas.

EXAMPLE 6.9

A small bubble rises from the bottom of a lake, where the temperature and pressure are 8°C and 6.4 atm, to the water's surface, where the temperature is 25°C and pressure is 1.0 atm. Calculate the final volume (in mL) of the bubble if its initial volume was 2.1 mL.

Answer

The initial and final conditions are

$$P_1 = 6.4 \text{ atm} \qquad P_2 = 1.0 \text{ atm}$$
$$V_1 = 2.1 \text{ mL} \qquad V_2 = ?$$
$$T_1 = (8 + 273)\text{K} = 281 \text{ K} \qquad T_2 = (25 + 273)\text{K} = 298 \text{ K}$$

The amount of the gas in the bubble remains constant, so that $n_1 = n_2$. By rearranging Equation (6.8) we write

$$V_2 = V_1 \times \frac{P_1}{P_2} \times \frac{T_2}{T_1}$$

$$= 2.1 \text{ mL} \times \frac{6.4 \text{ atm}}{1.0 \text{ atm}} \times \frac{298 \text{ K}}{281 \text{ K}}$$

$$= 14 \text{ mL}$$

Thus, the bubble's volume increases from 2.1 mL to 14 mL because of the decrease in water pressure and the increase in temperature.

Similar examples: Problems 6.16, 6.18.

We can use any appropriate units for volume (or pressure) as long as we use the same units on both sides of the equation.

EXAMPLE 6.10

A steel bulb contains xenon (Xe) gas at 2.64 atm and 25°C. Calculate the pressure of the gas when the temperature is raised to 436°C. Assume that the volume of the bulb remains constant.

Answer

The volume and amount of the gas are unchanged, so that $V_1 = V_2$ and $n_1 = n_2$. Equation (6.8) can therefore be written as

$$\frac{P_1}{T_1} = \frac{P_2}{T_2}$$

The initial and final conditions are

$P_1 = 2.64 \text{ atm}$ $P_2 = ?$
$T_1 = (25 + 273)\text{K} = 298 \text{ K}$ $T_2 = (436 + 273)\text{K} = 709 \text{ K}$

The final pressure is given by

$$P_2 = P_1 \times \frac{T_2}{T_1}$$

$$= 2.64 \text{ atm} \times \frac{709 \text{ K}}{298 \text{ K}}$$

$$= 6.28 \text{ atm}$$

We see that at constant volume, the pressure of a given amount of gas is directly proportional to its absolute temperature.

Similar example: Problem 6.17.

One practical consequence of this relationship is that automobile tire pressures should be checked only when the tires are at normal temperatures. After a long drive (particularly in the summer) tires become quite hot, leading to a higher measured air pressure.

Density Calculations

If we rearrange the ideal gas equation, we can use it to calculate the density of a gas:

$$\frac{n}{V} = \frac{P}{RT}$$

The number of moles of the gas, n, is given by

$$n = \frac{m}{\mathcal{M}}$$

where m is the mass of the gas in grams and $\mathcal{M}$ is its molar mass. Therefore

$$\frac{m}{\mathcal{M}V} = \frac{P}{RT}$$

Since density, d, is mass per unit volume, we can write

$$d = \frac{m}{V} = \frac{P\mathcal{M}}{RT} \tag{6.9}$$

Unlike molecules in condensed matter (that is, in liquids and solids), gaseous molecules are separated by distances that are large compared to their size. Consequently, the density of gases is very low under atmospheric conditions. For this reason, the units for gas density are usually expressed as g/L rather than g/mL.

EXAMPLE 6.11

Calculate the density of ammonia (NH_3) in grams per liter at 752 mmHg and 55°C.

Answer

To convert the pressure to atmospheres, we write

$$P = 752 \text{ mmHg} \times \frac{1 \text{ atm}}{760 \text{ mmHg}}$$

$$= \left(\frac{752}{760}\right) \text{ atm}$$

Using Equation (6.9) and $T = 273 + 55 = 328$ K, we have

$$d = \frac{P\mathcal{M}}{RT}$$

$$= \frac{(752/760) \text{ atm } (17.0 \text{ g/mol})}{(0.0821 \text{ L·atm/K·mol})(328 \text{ K})}$$

$$= 0.625 \text{ g/L}$$

Similar example: Problem 6.29.

In units of g/mL, the calculated density would be 6.25 $\times$ 10^{-4} g/mL.

The Molar Mass of a Gaseous Substance

Up to this point in your study of chemistry you may have the impression that the molar mass of a substance is found by examining its formula and summing the molar masses of its component atoms. However, this procedure works only if the actual formula of the substance is known. In practice, chemists often deal with substances of unknown or only partially defined composition. If the unknown substance is gaseous, its molar mass can nevertheless be found, thanks to the ideal gas equation. All that is needed is an experimentally determined density value (or mass and volume data) for the gas at a known temperature and pressure. By rearranging Equation (6.9) we get

$$\mathcal{M} = \frac{dRT}{P} \tag{6.10}$$

In a typical experiment, a bulb of known volume is filled with the gaseous substance under study. The temperature and pressure of the contained gas sample are recorded, and the total mass of the bulb plus gas sample is determined (Figure 6.12). The bulb is then evacuated (emptied) and weighed again. The difference in mass is the mass of the gas. The density of the gas is equal to its mass divided by the volume of the bulb. Then we can calculate the molar mass of the substance using Equation (6.10).

Under atmospheric conditions, 200 mL of air weighs about 0.24 g, an easily measured quantity.

EXAMPLE 6.12

The gas sulfur hexafluoride has a density of 20.8 g/L measured at 3.57 atm and 32°C. What is the molar mass of this substance?

Answer

We substitute in Equation (6.10):

$$\mathcal{M} = \frac{dRT}{P}$$

$$= \frac{(20.8 \text{ g/L})(0.0821 \text{ L}\cdot\text{atm/K}\cdot\text{mol})(32 + 273)\text{ K}}{3.57 \text{ atm}}$$

$$= 146 \text{ g/mol}$$

The molecular formula for sulfur hexafluoride (as its name implies) is SF_6. The actual molar mass of this compound, which you can calculate from the known molar masses of S and F, is 146.1 g/mol. Thus the experimental and actual values are in good agreement.

Similar example: Problem 6.26.

Since Equation (6.10) is derived from Equation (6.7), we can also calculate the molar mass of a gaseous substance using the ideal gas equation. The following two examples illustrate this approach.

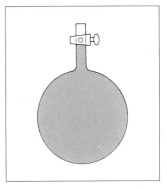

FIGURE 6.12 *An apparatus for measuring the density of a gas. The bulb of known volume is filled with the gas under study at a certain temperature and pressure. First the bulb is weighed, and then it is emptied (evacuated) and weighed again. The difference in masses gives the mass of the gas. With the volume of the bulb known, we can calculate the density of the gas as described in the text.*

EXAMPLE 6.13

Calculate the molar mass of methane (CH_4) if 279 mL of the gas measured at 31°C and 492 mmHg has a mass of 0.116 g.

Answer

From Equation (6.7) we write

$$n = \frac{PV}{RT}$$

where

$$P = 492 \text{ mmHg} \times \frac{1 \text{ atm}}{760 \text{ mmHg}} = 0.647 \text{ atm}$$

$$V = 279 \text{ mL} = 0.279 \text{ L}$$

$$T = (31 + 273)\text{ K} = 304 \text{ K}$$

Therefore

$$n = \frac{(0.647 \text{ atm})(0.279 \text{ L})}{(0.0821 \text{ L}\cdot\text{atm/K}\cdot\text{mol})(304 \text{ K})} = 7.23 \times 10^{-3} \text{ mol}$$

(Continued)

Since 7.23×10^{-3} mol of CH_4 weighs 0.116 g, the molar mass of CH_4 must be

$$\mathcal{M} = \frac{0.116 \text{ g } CH_4}{7.23 \times 10^{-3} \text{ mol } CH_4}$$
$$= 16.0 \text{ g/mol}$$

Similar examples: Problems 6.24, 6.31.

EXAMPLE 6.14

Chemical analysis of a gaseous compound showed that it contained 33.0 percent silicon and 67.0 percent fluorine by mass. At 35°C, 0.210 L of the compound exerted a pressure of 1.70 atm. If the mass of 0.210 L of the compound was 2.38 g, calculate the molecular formula of the compound.

Answer

First we determine the empirical formula of the compound.

$$n_{Si} = 33.0 \text{ g Si} \times \frac{1 \text{ mol Si}}{28.09 \text{ g Si}} = 1.17 \text{ mol Si}$$

$$n_F = 67.0 \text{ g F} \times \frac{1 \text{ mol F}}{19.00 \text{ g F}} = 3.53 \text{ mol F}$$

Therefore, the formula is $Si_{1.17}F_{3.53}$, or SiF_3. Next, we calculate the number of moles contained in 2.38 g of the compound. From Equation (6.7)

$$n = \frac{PV}{RT}$$

$$= \frac{(1.70 \text{ atm})(0.210 \text{ L})}{(0.0821 \text{ L·atm/K·mol})(308 \text{ K})} = 0.0141 \text{ mol}$$

Therefore, the molar mass of the compound is

$$\mathcal{M} = \frac{2.38 \text{ g}}{0.0141 \text{ mol}}$$
$$= 169 \text{ g/mol}$$

The actual molar mass of SiF_3 (empirical formula) is 85.09 g. Therefore the molecular formula of the compound must be Si_2F_6 because 2×85.09, or 170.2 g is very close to 169 g.

Similar example: Problem 6.30.

6.5 STOICHIOMETRY INVOLVING GASES

In Chapter 4 we used relationships between amounts (in moles) and masses (in grams) of reactants and products to solve stoichiometry problems. When the reactants and products are gases, we can also use the relationships between amounts (moles, n) and volume (V) to solve such problems. The following examples show how the gas laws are used in these calculations.

EXAMPLE 6.15

Calculate the volume of O_2 (in liters) at STP required for the complete combustion of 2.64 L of acetylene (C_2H_2) at STP:

$$2C_2H_2(g) + 5O_2(g) \longrightarrow 4CO_2(g) + 2H_2O(l)$$

Answer

Since both C_2H_2 and O_2 are gases measured at the same temperature and pressure, their reacting volumes are related to their coefficients in the balanced equation, that is, 2 L of C_2H_2 react with 5 L of O_2. Knowing this ratio we can calculate the volume of O_2 that will react with 2.64 L of C_2H_2.

$$\begin{aligned} \text{volume of } O_2 &= 2.64 \text{ L } C_2H_2 \times \frac{5 \text{ L } O_2}{2 \text{ L } C_2H_2} \\ &= 6.60 \text{ L } O_2 \end{aligned}$$

Similar examples: Problems 6.14, 6.41.

EXAMPLE 6.16

Calculate the volume of CO_2 (in liters) measured at 0°C and 1 atm that could be obtained by allowing 45.0 g of $CaCO_3$ to react with an excess of hydrochloric acid:

$$CaCO_3(s) + 2HCl(aq) \longrightarrow CaCl_2(aq) + H_2O(l) + CO_2(g)$$

The molar mass of $CaCO_3$ is 100.1 g.

Answer

From the balanced equation we see that 1 mol $CaCO_3$ ≎ 1 mol CO_2. The number of moles of $CaCO_3$ present in 45.0 g is

$$\begin{aligned} \text{moles of } CaCO_3 &= 45.0 \text{ g } CaCO_3 \times \frac{1 \text{ mol } CaCO_3}{100.1 \text{ g } CaCO_3} \\ &= 0.450 \text{ mol } CaCO_3 \end{aligned}$$

Therefore 0.450 mole of CO_2 is produced. Since one mole of CO_2 occupies 22.4 L at STP, the volume occupied by 0.450 mole is

$$\begin{aligned} \text{volume of } CO_2 &= 0.450 \text{ mol } CO_2 \times \frac{22.4 \text{ L } CO_2}{1 \text{ mol } CO_2} \\ &= 10.1 \text{ L} \end{aligned}$$

Similar examples: Problems 6.34, 6.40.

EXAMPLE 6.17

The first step in the industrial preparation of nitric acid (HNO_3) is the production of nitric oxide (NO) from ammonia (NH_3) and oxygen (O_2):

$$4NH_3(g) + 5O_2(g) \longrightarrow 4NO(g) + 6H_2O(g)$$

If 3.00 L of NH_3 at 802°C and 1.30 atm completely reacts with oxygen, how many liters of steam measured at 125°C and 1.00 atm are formed?

Answer

Our first step is to find the volume that would be occupied by the NH_3 at 125°C and 1 atm. Since $n_1 = n_2$, we can use Equation (6.8):

$$\frac{P_1 V_1}{T_1} = \frac{P_2 V_2}{T_2}$$

where

$$P_1 = 1.30 \text{ atm} \qquad\qquad P_2 = 1.00 \text{ atm}$$
$$V_1 = 3.00 \text{ L} \qquad\qquad V_2 = ?$$
$$T_1 = (802 + 273) \text{ K} = 1075 \text{ K} \qquad T_2 = (125 + 273) \text{ K} = 398 \text{ K}$$

Rearrangement of Equation (6.8) gives

$$V_2 = V_1 \times \frac{P_1}{P_2} \times \frac{T_2}{T_1}$$

$$= 3.00 \text{ L} \times \frac{1.30 \text{ atm}}{1.00 \text{ atm}} \times \frac{398 \text{ K}}{1075 \text{ K}} = 1.44 \text{ L}$$

Following the procedure in Example 6.15, the volume of gaseous H_2O (steam) is given by

$$\text{volume of } H_2O = 1.44 \text{ L } NH_3 \times \frac{6 \text{ L } H_2O}{4 \text{ L } NH_3}$$

$$= 2.16 \text{ L } H_2O$$

EXAMPLE 6.18

A flask of volume 0.85 L is filled with carbon dioxide (CO_2) gas at a pressure of 1.44 atm and a temperature of 312 K. A solution of lithium hydroxide (LiOH) of negligible volume is introduced into the flask. Eventually the pressure of CO_2 is reduced to 0.56 atm because some CO_2 is consumed in the reaction

$$CO_2(g) + 2LiOH(aq) \longrightarrow Li_2CO_3(aq) + H_2O(l)$$

How many grams of lithium carbonate are formed by this process? Assume that the temperature remains constant.

Answer

First we calculate the number of moles of CO_2 consumed in the reaction. The drop in pressure, which is 1.44 atm − 0.56 atm, or 0.88 atm, corresponds to the consumption of CO_2. From Equation (6.7) we write

The ability of LiOH to react with CO_2 makes it useful as an air purifier in space vehicles and submarines.

$$n = \frac{PV}{RT}$$

$$= \frac{(0.88 \text{ atm})(0.85 \text{ L})}{(0.0821 \text{ L·atm/K·mol})(312 \text{ K})} = 0.029 \text{ mol}$$

From the equation we see that 1 mol $CO_2 \rightleftharpoons$ 1 mol Li_2CO_3, so the amount of Li_2CO_3 formed is also 0.029 mole. Then, with the molar mass of Li_2CO_3 (73.89 g) we calculate its mass:

$$\text{mass of } Li_2CO_3 \text{ formed} = 0.029 \text{ mol } Li_2CO_3 \times \frac{73.89 \text{ g } Li_2CO_3}{1 \text{ mol } Li_2CO_3}$$

$$= 2.1 \text{ g } Li_2CO_3$$

6.6 DALTON'S LAW OF PARTIAL PRESSURES

Thus far we have concentrated on the behavior of pure gaseous substances. However, experimental studies are very often based on mixtures of gases. For example, we may be interested in the pressure-volume-temperature relationship of a sample of air, which contains several gases. In considering a gaseous mixture, we need to understand how the total gas pressure is related to the *pressures of individual gas components in the mixture*, called **partial pressures.** In 1801 Dalton formulated a law, now known as **Dalton's law of partial pressures,** which states that *the total pressure of a mixture of gases is just the sum of the pressures that each gas would exert if it were present alone.*

Consider a particularly simple case in which two gaseous substances, A and B, are in a container of volume V. The pressure exerted by gas A, according to Equation (6.7), is

> As we mentioned earlier, gas pressure is a result of the impact of gas molecules against the walls of the container.

$$P_A = \frac{n_A RT}{V}$$

where n_A is the number of moles of A present. Similarly, for gas B its pressure is

$$P_B = \frac{n_B RT}{V}$$

Now, in a mixture of gases A and B, the total pressure P_T is the result of the collisions of both types of molecules, A and B, with the walls. Thus, according to Dalton's law

$$P_T = P_A + P_B \tag{6.11}$$

$$= \frac{n_A RT}{V} + \frac{n_B RT}{V}$$

$$= \frac{RT}{V}(n_A + n_B)$$

$$= \frac{nRT}{V}$$

> This result shows that P_T depends only on the total number of moles of gases present, but not on the types of molecules.

where n, the total number of moles of gases present, is given by $n = n_A + n_B$, and P_A and P_B are the partial pressures of gases A and B, respectively. In general, the total pressure of a mixture of gases is given by

$$P_T = P_1 + P_2 + P_3 + \ldots$$

where $P_1, P_2, P_3, \ldots$ are the partial pressures of components 1, 2, 3, ...

There is a simple relation between total pressure and individual partial pressures. Consider again the case of a mixture of gases A and B. Dividing P_A by P_T, we obtain

$$\frac{P_A}{P_T} = \frac{n_A RT/V}{(n_A + n_B)RT/V}$$

$$= \frac{n_A}{n_A + n_B}$$

$$= X_A$$

where X_A is called the mole fraction of gas A. The **mole fraction** is *a dimensionless quantity that expresses a ratio of the number of moles of one component to the number of moles of all components present*. It is always smaller than 1, except when A is the only component present. In that case, $n_B = 0$ and $X_A = n_A/n_A = 1$. We can now express the partial pressure of A as

$$P_A = X_A P_T$$

Similarly

$$P_B = X_B P_T$$

Note that the sum of mole fractions must of course be unity:

$$X_A + X_B = \frac{n_A}{n_A + n_B} + \frac{n_B}{n_A + n_B} = 1$$

If a system contains more than two gases, then the partial pressure of the ith component is related to the total pressure by

$$P_i = X_i P_T \tag{6.12}$$

where X_i is the mole fraction of substance i.

EXAMPLE 6.19

Assume that 1 mole of air molecules contains 0.78 mole of molecular nitrogen, 0.21 mole of molecular oxygen, and 0.010 mole of argon. Calculate the partial pressures of these gases when the total air pressure is 1.0 atm.

Answer

The total pressure of the air is given by

$$P_{air} = \frac{nRT}{V} = (n_{N_2} + n_{O_2} + n_{Ar})\frac{RT}{V}$$

where n_{N_2}, n_{O_2}, and n_{Ar} are the numbers of moles of the gases present. The partial pressure of nitrogen gas is

$$P_{N_2} = n_{N_2}\frac{RT}{V}$$

Taking the ratio of P_{N_2}/P_{air}, we get

$$\frac{P_{N_2}}{P_{air}} = \frac{n_{N_2}}{n_{N_2} + n_{O_2} + n_{Ar}} = X_{N_2} = \frac{0.78 \text{ mol}}{1 \text{ mol}} = 0.78$$

where X_{N_2} is the mole fraction of nitrogen gas in the air. Thus

$$P_{N_2} = 0.78 \times 1.0 \text{ atm}$$
$$= 0.78 \text{ atm}$$

Similarly

$$P_{O_2} = X_{O_2} \times 1.0 \text{ atm}$$
$$= 0.21 \text{ atm}$$

and

$$P_{Ar} = X_{Ar} \times 1.0 \text{ atm}$$
$$= 0.010 \text{ atm}$$

Similar examples: Problems 6.44, 6.46.

Dalton's law of partial pressures has a practical application in calculating volumes of gases collected over water, for example, oxygen. When potassium chlorate ($KClO_3$) is heated, the products are KCl and O_2:

$$2KClO_3(s) \xrightarrow{\Delta} 2KCl(s) + 3O_2(g)$$

The evolved oxygen gas can be collected over water, as shown in Figure 6.13. Initially, the inverted bottle is completely filled with water. As oxygen gas is generated, the gas bubbles rise to the top and water is pushed out of the bottle. Note that this water displacement method for collecting a gas is based on the assumptions that the gas does not react with water and that it is not appreciably soluble in it. These assumptions are valid for oxygen gas, but the procedure will not work for gases such as NH_3, which dissolves readily in water. The oxygen gas collected in this way is not pure, however, because water vapor is also present in the same space. The total gas pressure is equal to the sum of the pressures exerted by the oxygen gas and the water vapor:

$$P_T = P_{O_2} + P_{H_2O}$$

Consequently, we must allow for the pressure caused by the presence of water vapor when we calculate the amount of O_2 generated. Table 6.3 shows the pressure of water vapor at various temperatures. These data are plotted in Figure 6.14.

TABLE 6.3
Pressure of Water Vapor at Various Temperatures

Temperature (°C)	Pressure (mmHg) of Water Vapor
0	4.58
5	6.54
10	9.21
15	12.79
20	17.54
25	23.76
30	31.82
35	42.18
40	55.32
45	71.88
50	92.51
55	118.04
60	149.38
65	187.54
70	233.7
75	289.1
80	355.1
85	433.6
90	525.76
95	633.90
100	760.00

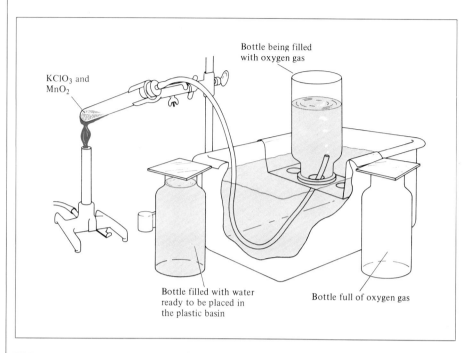

FIGURE 6.13 *An apparatus for collecting gas over water. The oxygen generated by heating potassium chlorate in the presence of a small amount of manganese dioxide (to speed up the reaction) is bubbled through water and collected in a bottle as shown. Water originally present in the bottle is pushed into the trough by the oxygen gas.*

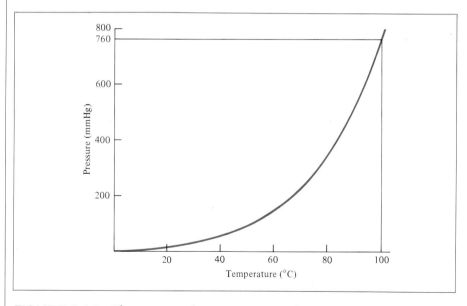

FIGURE 6.14 *The pressure of water vapor as a function of temperature. Note that at the boiling point of water (100°C), the pressure is 760 mmHg, which is exactly equal to one atmosphere of pressure.*

EXAMPLE 6.20

Oxygen gas generated in the decomposition of potassium chlorate is collected as shown in Figure 6.13. The volume of the gas collected at 24°C and atmospheric pressure of 762 mmHg is 128 mL. Calculate the mass (in grams) of oxygen gas obtained. The pressure of the water vapor at 24°C is 22.4 mmHg.

Answer

Our first step is to calculate the partial pressure of O_2. Since

$$P_T = P_{O_2} + P_{H_2O}$$

therefore

$$
\begin{aligned}
P_{O_2} &= P_T - P_{H_2O} \\
&= 762 \text{ mmHg} - 22.4 \text{ mmHg} \\
&= 740 \text{ mmHg} \\
&= 740 \text{ mmHg} \times \frac{1 \text{ atm}}{760 \text{ mmHg}} \\
&= 0.974 \text{ atm}
\end{aligned}
$$

From the ideal gas equation we write

$$PV = nRT = \frac{m}{\mathcal{M}} RT$$

where m and $\mathcal{M}$ are the mass of O_2 collected and the molar mass of O_2, respectively. Rearranging the equation we get

$$
\begin{aligned}
m &= \frac{PV\mathcal{M}}{RT} = \frac{(0.974 \text{ atm})(0.128 \text{ L})(32.00 \text{ g/mol})}{(0.0821 \text{ L·atm/K·mol})(273 + 24) \text{ K}} \\
&= 0.164 \text{ g}
\end{aligned}
$$

Similar example: Problem 6.48.

6.7 THE KINETIC MOLECULAR THEORY OF GASES

The gas laws we have surveyed in this chapter can be regarded as useful shorthand summaries of the results of countless experiments on the physical properties of gases over several centuries. The gas laws are important generalizations regarding the *macroscopic* behavior of gaseous substances. They played a major role in the development of many ideas about chemistry, but they did not explain what happens at the *molecular* level to account for changes in the volume of a gas caused by changes in temperature and pressure. For example, why *does* the volume of a gas expand on heating? Starting in the 1850s, a number of physicists, notably Ludwig Boltzmann (1844–1906) in Germany and James Clerk Maxwell (1831–1879) in England, found that the physical properties of gases could be satisfactorily explained in terms of the motions of individual molecules. Their work subsequently laid the foundation for the kinetic molecular theory of gases.

The *kinetic molecular theory of gases* (sometimes called simply the *kinetic theory of gases*) is based on the following assumptions:

1. A gas is composed of molecules that are separated from each other by distances far greater than their own dimensions. The molecules can be considered to be "points," that is, they possess mass but have negligible volume.
2. Gas molecules are in constant motion in random directions and they frequently collide with one another. Although energy may be transferred from one molecule to another as a result of a collision, the total energy of all the molecules remains unchanged.
3. The *average* kinetic energy of a collection of gas molecules is equal to the total kinetic energy of the molecules divided by the number of molecules. The average kinetic energy of a molecule is written

$$\overline{KE} = \tfrac{1}{2}m\overline{c^2}$$

where m is the mass of the molecule (we assume that all the gas molecules have the same mass), and c is its speed. The horizontal bar denotes an average value. According to the kinetic molecular theory, the average kinetic energy is proportional to the absolute temperature of the gas, that is

$$\overline{KE} \propto T$$
$$\tfrac{1}{2}m\overline{c^2} \propto T$$
$$\tfrac{1}{2}m\overline{c^2} = CT \tag{6.13}$$

where C is the proportionality constant. The quantity $\overline{c^2}$ is called the *mean square speed*. Suppose we have N molecules and that the speeds of molecules $1, 2, \ldots$ are $c_1, c_2, \ldots$, then $\overline{c^2}$ is given by

$$\overline{c^2} = \frac{c_1^2 + c_2^2 + \cdots + c_N^2}{N}$$

4. Gas molecules exert neither attractive nor repulsive forces on one another.

Application to the Gas Laws

Although the kinetic theory of gases is based on a rather simple model, the mathematical details involved are too complex to be dealt with here. However, on a qualitative basis, it is possible to apply the theory to account for the general properties of substances in the gaseous state. The following examples illustrate the range of its usefulness and applicability:

- *Compressibility of gases.* Since molecules in the gas phase are separated by large distances (Assumption 1), gases can be compressed easily to occupy smaller volumes.
- *Boyle's law.* The pressure exerted by a gas results from the impact of its molecules on the walls of the container. The rate, or the number of molecular collisions with the walls per second, is proportional to the density of the gas. Decreasing the volume of a given amount of gas increases its density and hence its collision rate. For this reason, the pressure of a gas is inversely proportional to the volume it occupies.

- *Charles' law.* Since the average kinetic energy of gas molecules is proportional to the sample's absolute temperature (Assumption 3), raising the temperature increases the average kinetic energy. Consequently, molecules will collide with the walls of the container more frequently and with greater impact if the gas is heated, and thus the pressure increases. The volume of gas will expand until the gas pressure is balanced by the constant external pressure (see Figure 6.6).

- *Avogadro's law.* We have shown that the pressure of a gas is directly proportional to both the density of the gas (m/V) and the temperature of the gas. Since the mass of the gas is directly proportional to the number of moles (n) of the gas, we can represent density by n/V. Therefore

$$P \propto \frac{n}{V} T$$

For two gases, 1 and 2, we write

$$P_1 \propto \frac{n_1 T_1}{V_1} = C \frac{n_1 T_1}{V_1}$$

$$P_2 \propto \frac{n_2 T_2}{V_2} = C \frac{n_2 T_2}{V_2}$$

Thus, for two gases under the same conditions of pressure, volume, and temperature (that is, when $P_1 = P_2$, $T_1 = T_2$, and $V_1 = V_2$), it follows that $n_1 = n_2$, which is a mathematical expression of Avogadro's law.

- *Dalton's law of partial pressures.* If molecules do not attract or repel one another (Assumption 4), then the pressure exerted by one type of molecule is unaffected by the presence of another gas. Consequently, the total pressure is given by the sum of individual gas pressures.

> Another way of stating Avogadro's law is that at the same pressure and temperature, equal volumes of different gases contain the same number of molecules.

Distribution of Molecular Speeds

The kinetic theory of gases helps us investigate molecular motion in detail. Suppose we have a large number of molecules of a gas, say, one mole, in a container. As you might expect, the motion of the molecules is totally random and unpredictable. As long as we hold temperature constant, however, the average kinetic energy and the average molecular speed (c) remain unchanged as time passes. The characteristic that is of interest to us here is the spread, or distribution, of molecular speeds. At a given instant, how many molecules are moving at a particular speed? An equation to solve this problem was formulated by Maxwell in 1860. The equation, which incorporates the assumptions we have mentioned (p. 176), is based on a statistical analysis of the behavior of the molecules.

Figure 6.15 shows typical *Maxwell speed distribution curves* for nitrogen gas at two different temperatures. At a given temperature, the distribution curve tells us the number of molecules moving at a certain speed. The peak of each curve gives the *most probable speed*, that is, the speed of the largest number of molecules. Note that at the higher temperature the most probable speed is greater than at the lower temperature. Comparing parts (a)

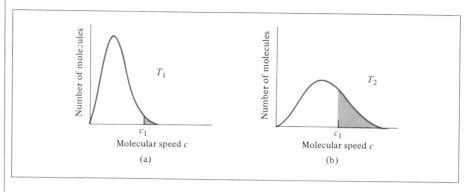

FIGURE 6.15 *Maxwell's speed distribution for N_2 gas at (a) a temperature T_1 and (b) a higher temperature T_2. Note that the curve flattens out at the higher temperature. The shaded areas represent the number of molecules traveling at a speed equal to or greater than a certain speed c_1. The higher the temperature, the greater the number of molecules moving at high speed.*

and (b) of Figure 6.15, we see that as temperature increases, not only does the peak shift toward the right, but the curve also flattens out, indicating that larger numbers of molecules are moving at greater speeds.

The *average speed, $\bar{c}$,* is defined as

$$\bar{c} = \frac{\text{sum of all molecular speeds}}{\text{total number of molecules present}}$$

$$= \frac{c_1 + c_2 + \cdots + c_N}{N}$$

Maxwell showed that

$$\bar{c} = \sqrt{\frac{2.55\ RT}{\mathcal{M}}}$$

(6.14)

where c is in meters per second, R is the gas constant, T the temperature in kelvins, and $\mathcal{M}$ the molar mass of the gas. From Equation (6.14) you can see that the average speed of a gas increases with the square root of its absolute temperature. Because $\mathcal{M}$ appears in the denominator, it follows that the heavier the gas, the more slowly its molecules move. If we use 8.314 J/K·mol for R (see Appendix 2) and convert the molar mass to kg/mol, then c will be calculated in m/s. This procedure is used in Example 6.21.

EXAMPLE 6.21

Calculate the average speeds of helium atoms and nitrogen molecules (in meters per second) at 25°C.

Answer

We need Equation (6.14) for this calculation. For He the molar mass is 4.00 g/mol, or 4.00×10^{-3} kg/mol:

$$\bar{c} = \sqrt{\frac{2.55\ RT}{\mathcal{M}}}$$

$$= \sqrt{\frac{(2.55)(8.314\ \text{J/K}\cdot\text{mol})(298\ \text{K})}{4.00 \times 10^{-3}\ \text{kg/mol}}}$$

$$= \sqrt{1.58 \times 10^6\ \text{J/kg}}$$

Using the conversion factor (see Section 1.6)

$$1\ \text{J} = 1\ \text{kg m}^2/\text{s}^2$$

we get

$$\bar{c} = \sqrt{1.58 \times 10^6\ \text{kg m}^2/\text{kg}\cdot\text{s}^2}$$

$$= \sqrt{1.58 \times 10^6\ \text{m}^2/\text{s}^2}$$

$$= 1.26 \times 10^3\ \text{m/s}$$

Similarly for N_2

$$\bar{c} = \sqrt{\frac{(2.55)(8.314\ \text{J/K}\cdot\text{mol})(298\ \text{K})}{2.80 \times 10^{-2}\ \text{kg/mol}}}$$

$$= \sqrt{2.25 \times 10^5\ \text{m}^2/\text{s}^2}$$

$$= 475\ \text{m/s}$$

Because of its smaller mass, a helium atom, on the average, moves about $2\frac{1}{2}$ times as fast as a nitrogen molecule at the same temperature $(1260 \div 475 = 2.65)$.

Similar examples: Problems 6.51, 6.53.

The calculation in Example 6.21 has an interesting relationship to the composition of Earth's atmosphere. Earth, unlike Jupiter, does not have appreciable amounts of gases such as hydrogen or helium in its atmosphere. A smaller planet, Earth has a weaker attraction for these lighter molecules. A fairly straightforward calculation shows that to escape Earth's gravitational field, a molecule must possess a speed equal to or greater than 1.1×10^3 m/s. This is usually called the *escape velocity*. Because the average speed of helium is considerably greater than that of molecular nitrogen or molecular oxygen, more helium atoms escape from Earth's atmosphere into outer space. Consequently, only a trace amount of helium is present in our atmosphere. Jupiter, on the other hand, with a mass about 320 times greater than that of Earth, is able to retain both heavy and light gases in its atmosphere.

6.8   GRAHAM'S LAW OF DIFFUSION AND EFFUSION

Diffusion

Gas **diffusion,** *the gradual mixing of molecules of one gas with the molecules of another by virtue of their kinetic properties,* provides a direct demonstration of random molecular motion. Without gas diffusion, the

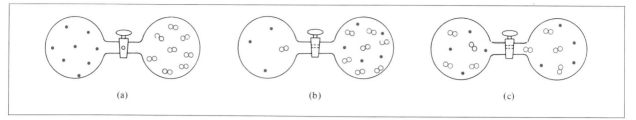

FIGURE 6.16   *Gaseous diffusion. (a) He and N_2 gases in two separate flasks. (b) Shortly after the stopcock is opened there are more He atoms in the right flask than there are N_2 molecules in the left flask, because He atoms move faster than N_2 molecules. (c) Given enough time, the two gases distribute themselves evenly between the two flasks. Situation (c) holds only for a large number of molecules, for example, one mole of He and one mole of N_2.*

perfume industry could not exist, and skunks would be much less feared. Consider the arrangement shown in Figure 6.16. Initially, the helium and nitrogen gases are separated from each other by the stopcock [Figure 6.16(a)]. After the stopcock is opened, the number of helium atoms that move into the bulb containing nitrogen molecules is greater than the number of nitrogen molecules moving in the opposite direction [Figure 6.16(b)]. We saw in Example 6.21 that the average speed of He is greater than that of N_2. Since the motion of any molecule is completely random, a lighter gas will diffuse a greater distance in a given time than will a heavier one. Consequently, more He will move, or diffuse, from left to right, resulting in the unequal distribution shown in Figure 6.16(b). This process does not continue in one direction for very long because the helium also diffuses from right to left. Nitrogen gas follows the same pattern of movement in the opposite direction, but more slowly. Eventually the gases are equally distributed between the two bulbs [Figure 6.16(c)], but only if the numbers of He atoms and N_2 molecules are very large, say, on the order of one mole of each gas. If we have only two He atoms and two N_2 molecules to start with, then all four might be in the same bulb or distributed in any combination in the two bulbs at any time.

In 1832 the Scottish chemist Thomas Graham (1805–1869) found that *under the same conditions of temperature and pressure, rates of diffusion for gaseous substances are inversely proportional to the square roots of their molar masses.* This statement, now known as **Graham's law of diffusion,** is expressed mathematically as

$$\frac{r_1}{r_2} = \sqrt{\frac{\mathcal{M}_2}{\mathcal{M}_1}} \qquad (6.15)$$

where r_1 and r_2 are the diffusion rates of gases 1 and 2, and $\mathcal{M}_1$ and $\mathcal{M}_2$ are the molar masses, respectively. Equation (6.15) is based on experimental observation, but it can also be deduced from the kinetic theory of gases as follows. Consider two gases, 1 and 2. We apply Equation (6.14) as follows:

$$\bar{c}_1 = \sqrt{\frac{2.55\ RT}{\mathcal{M}_1}} \qquad \bar{c}_2 = \sqrt{\frac{2.55\ RT}{\mathcal{M}_2}}$$

The ratio of the two average speeds is

$$\frac{\bar{c}_1}{\bar{c}_2} = \frac{\sqrt{\dfrac{2.55\ RT}{\mathcal{M}_1}}}{\sqrt{\dfrac{2.55\ RT}{\mathcal{M}_2}}}$$

$$= \sqrt{\frac{2.55\ RT}{\mathcal{M}_1}} \times \sqrt{\frac{\mathcal{M}_2}{2.55\ RT}}$$

$$= \sqrt{\frac{\mathcal{M}_2}{\mathcal{M}_1}}$$

Since the rate of diffusion is proportional to the average speed, this equation is equivalent to Equation (6.15).

Effusion

Whereas diffusion is a process by which one gas gradually mixes with another, **effusion** is *the process by which a gas under pressure escapes from one compartment of a container to another by passing through a small opening.* Figure 6.17 shows the effusion of a gas into a vacuum. Although effusion differs from diffusion in nature, the rate of effusion of a gas is also given by Graham's law of diffusion [(Equation (6.15)]. As in the case of diffusion, we see that at a given temperature lighter gases effuse faster than heavier gases (Figure 6.18). A practical demonstration of gaseous effusion is shown in Figure 6.19.

In practice, the rate of diffusion is *inversely* proportional to the time it takes for a gas to diffuse—a lighter gas takes less time to diffuse through a certain space than does a heavier gas through the same space.

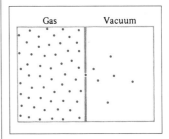

FIGURE 6.17 *Gaseous effusion. Gas molecules move from a high-pressure region (left) to a low-pressure region through a pinhole.*

EXAMPLE 6.22

Compare the effusion rates of helium and air at the same temperature and pressure. The molar mass of air may be taken as 29.0 g/mol.

Answer

From Equation (6.15) we write

$$\frac{r_{He}}{r_{air}} = \sqrt{\frac{29.0\ \text{g/mol}}{4.00\ \text{g/mol}}} = 2.69$$

Thus helium effuses 2.69 times faster than air.

Similar example: Problem 6.58.

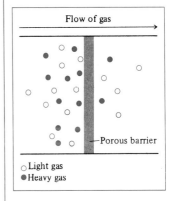

FIGURE 6.18 *Rates of effusion of a heavy gas (solid circle) and a light gas (open circles) through a porous barrier. A porous barrier contains many tiny holes suitable for effusion. Light gases effuse more rapidly than heavy gases.*

a 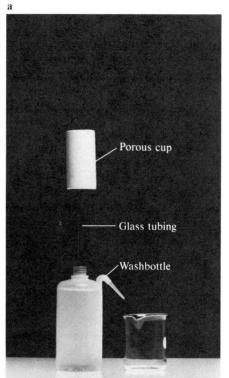   b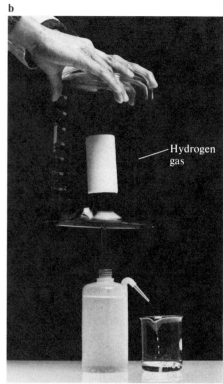

FIGURE 6.19   *(a) A porous cup fitted with glass tubing and stopper is attached to a washbottle filled with water. (b) An inverted beaker filled with hydrogen gas is placed over the porous cup. Because hydrogen is lighter than air, it effuses through the cup into the bottle faster than the air in the bottle can effuse out. The difference in the effusion rates results in a greater gas pressure inside the bottle. Consequently, water is squirted out of the bottle.*

EXAMPLE 6.23

A flammable gas made up only of carbon and hydrogen is generated by certain anaerobic bacterium cultures in marshlands and areas where sewage drains. A pure sample of this gas was found to effuse through a certain porous barrier in 1.50 min. Under identical conditions of temperature and pressure, it takes an equal volume of bromine gas (the molar mass of Br_2 is 159.8 g) 4.73 min to effuse through the same barrier. Calculate the molar mass of the unknown gas, and suggest what this gas might be.

Answer

The rate of effusion of a gas is inversely proportional to the time the gas takes to effuse through a barrier. Therefore, we can write

$$\frac{t_1}{t_2} = \frac{r_2}{r_1} = \sqrt{\frac{\mathcal{M}_1}{\mathcal{M}_2}}$$

or

$$\frac{1.50 \text{ min}}{4.73 \text{ min}} = \sqrt{\frac{\mathcal{M}}{159.8 \text{ g/mol}}}$$

where $\mathcal{M}$ is the molar mass of the unknown gas. Solving for $\mathcal{M}$, we obtain

$$\mathcal{M} = \left(\frac{1.50 \text{ min}}{4.73 \text{ min}}\right)^2 \times 159.8 \text{ g/mol}$$

$$= 16.1 \text{ g/mol}$$

Since the molar mass of carbon is 12.01 g and that of hydrogen is 1.008 g, the gas is probably methane (CH_4).

Similar examples: Problems 6.56, 6.57.

6.9 DEVIATION FROM IDEAL BEHAVIOR

So far our discussion has assumed that molecules in the gaseous state do not exert any force, either attractive or repulsive, on one another. Further, we have assumed that the volume of the molecules is negligibly small compared to that of the container. A gas that satisfies these two conditions is said to exhibit *ideal behavior*.

These would seem to be fair assumptions, although we cannot expect them to hold under all conditions. For example, without intermolecular forces, gases could not condense to form liquids. The important question is: Under what conditions will gases most likely exhibit nonideal behavior?

Figure 6.20 shows the PV/RT versus P plots for three real gases at a given temperature. These plots offer a test of ideal gas behavior. According to the ideal gas equation (for one mole of the gas), PV/RT equals one, regardless of the actual gas pressure. For real gases, this is true only at moderately low pressures ($\lesssim$ 10 atm); significant deviations are observed as pressure increases. The attractive forces among molecules operate at relatively short distances. For a gas at atmospheric pressure, the molecules are relatively far apart and these attractive forces are negligible. At high pressures, the density of the gas increases; the molecules are much closer to one another.

When $n = 1$, $PV = nRT$ becomes $PV = RT$ or $PV/RT = 1$.

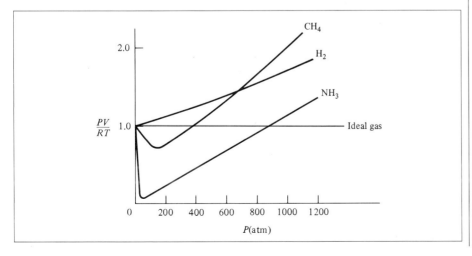

FIGURE 6.20 *Plot of PV/RT versus P of one mole of a gas at 0°C. For one mole of an ideal gas, PV/RT is equal to 1, no matter what the pressure of the gas is. For real gases, we observe various deviations from ideality at high pressures. Note that at very low pressures, all gases exhibit ideal behavior, that is, their PV/RT values all converge to 1 as P approaches zero.*

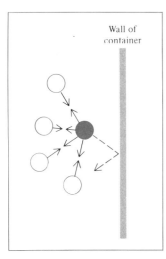

Wall of
container

FIGURE 6.21 *Effect of intermolecular forces (that is, forces that act between molecules or atoms) on the pressure exerted by a gas sample. The speed of a molecule that is moving toward the container wall (solid circle) is reduced by the attractive forces exerted by its neighbors (open circles). Consequently, the impact this molecule makes with the wall is less than it would be if no intermolecular forces were present. In general, the measured gas pressure is always lower than the pressure the gas would exert if it behaved ideally.*

The term *nb* represents the volume occupied by *n* moles of the gas.

Then intermolecular forces can become significant enough to affect the motion of the molecules, and the gas will no longer behave ideally.

Another way to observe the nonideality of gases is to lower the temperature. Cooling a gas decreases the molecules' average kinetic energy, which in a sense deprives molecules of the drive they need to break away from their mutual attractive influences.

The nonideality of gases can be dealt with mathematically by modifying Equation (6.7), taking into account intermolecular forces and finite molecular volumes. Such an analysis was first made by the Dutch physicist J. D. van der Waals (1837–1923) in 1873. Besides being mathematically simple, van der Waals' treatment provides us with an interpretation of real gas behavior at the molecular level.

Consider the approach of a particular molecule toward the wall of a container (Figure 6.21). The intermolecular attractions exerted by its neighbors tend to soften the impact made by this molecule against the wall. The overall effect is a lowered gas pressure, as compared to a gas with no such attractive forces present. Van der Waals suggested that the pressure exerted by an ideal gas, P_{ideal}, is related to the experimentally measured pressure, P_{real}, by

$$P_{ideal} = P_{real} + \frac{an^2}{V^2}$$

where a is a constant and n and V are the number of moles and volume of the gas, respectively. The correction term for pressure (an^2/V^2) can be understood as follows. The interaction between molecules that gives rise to nonideal behavior depends on how frequently any two molecules approach each other closely. The number of such "encounters" increases as the square of the number of molecules per unit volume, $(n/V)^2$, because the presence of each of the two molecules in a particular region is proportional to n/V. The quantity P_{ideal} is the pressure we would measure if there were no intermolecular attractions. The quantity a, then, is just a proportionality constant in the correction term for pressure.

Another correction concerns the volume occupied by the gas molecules. The quantity V in Equation (6.7) represents the entire volume occupied by the gas. However, each molecule does occupy a finite, although small, intrinsic volume, so the effective volume of the gas becomes $(V - nb)$ where n is the number of moles of the gas and b is a constant.

Having taken into account the corrections for pressure and volume, we can rewrite Equation (6.7) as

$$\left(P + \frac{an^2}{V^2}\right)(V - nb) = nRT \qquad (6.16)$$

Equation (6.16) is known as the *van der Waals equation*. The van der Waals constants a and b are selected for each gas to give the best possible agreement between the equation and actually observed behavior.

Table 6.4 lists the values of a and b for a number of gases. The value of a is an expression of how strongly the gas molecules attract one another. We see that the helium atoms have the weakest attraction for one another, because helium has the smallest a value. There is also a rough correlation between molecular size and b. Generally, the larger the molecule (or atom), the greater b is, but the relationship between b and molecular (or atomic) size is not a simple one.

TABLE 6.4 Van der Waals Constants of Some Common Gases

Gas	a (atm·L²/mol²)	b (L/mol)
He	0.034	0.0237
Ne	0.211	0.0171
Ar	1.34	0.0322
Kr	2.32	0.0398
Xe	4.19	0.0266
H_2	0.244	0.0266
N_2	1.39	0.0391
O_2	1.36	0.0318
Cl_2	6.49	0.0562
CO_2	3.59	0.0427
CH_4	2.25	0.0428
CCl_4	20.4	0.138
NH_3	4.17	0.0371
H_2O	5.46	0.0305

EXAMPLE 6.24

A quantity of 3.50 moles of NH_3 occupies 5.20 L at 47°C. Calculate the pressure of the gas (in atm) using (a) the ideal gas equation and (b) the van der Waals equation.

Answer

(a) We have the following data:

$$V = 5.20 \text{ L}$$

$$T = (47 + 273) \text{ K} = 320 \text{ K}$$

$$n = 3.50 \text{ mol}$$

$$R = 0.0821 \text{ L·atm/K·mol}$$

which we substitute in the ideal gas equation:

$$P = \frac{nRT}{V}$$

$$= \frac{(3.50 \text{ mol})(0.0821 \text{ L·atm/K·mol})(320 \text{ K})}{5.20 \text{ L}}$$

$$= 17.7 \text{ atm}$$

(b) from Table 6.4, we have

$$a = 4.17 \text{ atm·L}^2/\text{mol}^2$$

$$b = 0.0371 \text{ L/mol}$$

(Continued)

It is convenient to calculate the correction terms first, for Equation (6.16). They are

$$\frac{an^2}{V^2} = \frac{(4.17 \text{ atm} \cdot \text{L}^2/\text{mol}^2)(3.50 \text{ mol})^2}{(5.20 \text{ L})^2} = 1.89 \text{ atm}$$

$$nb = (3.50 \text{ mol})(0.0371 \text{ L/mol}) = 0.130 \text{ L}$$

Finally, substituting in the van der Waals equation, we write

$$(P + 1.89 \text{ atm})(5.20 \text{ L} - 0.130 \text{ L}) = (3.50 \text{ mol})(0.0821 \text{ L} \cdot \text{atm/K} \cdot \text{mol})(320 \text{ K})$$

$$P = 16.2 \text{ atm}$$

Note that the actual pressure measured under these conditions is 16.0 atm. Thus, the pressure calculated by the van der Waals equation (16.2 atm) is closer to the actual value than that calculated by the ideal gas equation (17.7 atm).

Similar example: Problem 6.62.

AN ASIDE ON THE PERIODIC TABLE
Physical States of the Elements Under Atmospheric Conditions

FIGURE 6.22 The physical states of the elements in the stable forms at 1 atm and 25°C. The lanthanides and actinides are all solids.

In Section 6.2 we considered those substances that are more likely to exist as gases. But what determines the physical state—solid, liquid, or gas—of the elements?

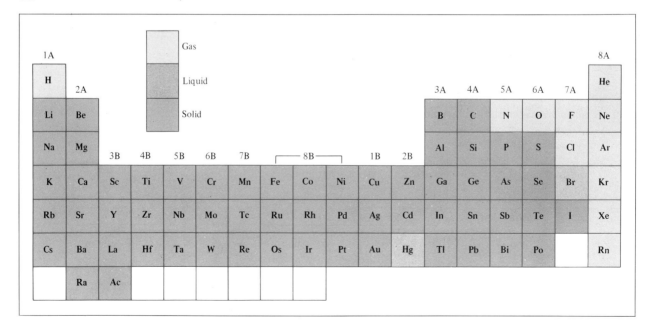

Whether a substance exists as a solid, a liquid, or a gas under atmospheric conditions is determined primarily by the forces that hold the basic units (atoms, molecules, or ions) of the substance together. If the forces are strong compared to the energy associated with molecular motion at 25°C, then the substance is most likely to exist as a condensed state of matter (that is, liquid or solid). If the attractive forces among the basic units are weak, then these units will break up and enter the gas phase, that is, the substance becomes a gas.

Figure 6.22 shows the elements in the periodic table according to the three physical states. As you can see, most elements are solids, eleven are gases (also see Figure 6.3), and interestingly, only two are liquids. The two liquids are the metallic element mercury (melting point: $-38.9°C$) and the nonmetallic element bromine (melting point: $-7.2°C$). Two other elements come close to qualifying as liquids at 25°C—gallium (melting point: 29.8°C) and cesium (melting point: 28.4°C).

SUMMARY

1. Gases exert pressure because their molecules collide with any surface with which they make contact. Gas pressure units include millimeters of mercury (mmHg), torr, pascals, and atmospheres. One atmosphere is equal to 760 mmHg, or 760 torr.

2. Under atmospheric conditions, ionic compounds exist as solids rather than as gases. The behavior of molecular compounds is more varied. A number of elemental substances occur as gases in elemental form: H_2, N_2, O_2, O_3, F_2, Cl_2, and the Group 8A elements (the noble gases).

3. The pressure-volume relationships of ideal gases are governed by Boyle's law: volume is inversely proportional to pressure (at constant T and n).

4. The temperature-volume relationships of ideal gases are described by Charles and Gay-Lussac's law: volume is directly proportional to temperature (at constant P and n).

5. Absolute zero ($-273.15°C$) is the lowest theoretically attainable temperature. The Kelvin temperature scale takes 0 K as absolute zero. In all gas law calculations, temperature must be expressed in kelvins.

6. The amount-volume relationships of ideal gases are described by Avogadro's law: equal volumes of gases contain equal numbers of molecules (at the same T and P).

7. The ideal gas equation, $PV = nRT$, is the combined expression of the laws of Boyle, Charles, and Avogadro. This equation describes the behavior of an ideal gas.

8. Dalton's law of partial pressures states that in a mixture of gases each component exerts the same pressure as it would if it were alone and occupied the same volume.

9. The kinetic molecular theory, a mathematical way of describing the behavior of gas molecules, is based on the following assumptions: Gas molecules are separated by distances far greater than their own dimensions, they possess mass but have negligible volume, they are in constant motion, and they frequently collide with one another. The molecules neither attract nor repel one another.

10. A Maxwell speed distribution curve shows how many gas molecules are moving at various speeds at a given temperature. As temperature increases, more molecules move at greater speeds.
11. In diffusion, two gases gradually mix with each other. In effusion, gas molecules move through a small opening under pressure. Both processes demonstrate random molecular motion and are governed by the same mathematical laws (Graham's laws of diffusion and effusion).
12. The van der Waals equation is a modification of the ideal gas equation that takes into account the nonideal behavior of real gases. It corrects for the fact that real gas molecules do exert forces on each other and that they do have volume. The van der Waals constants are determined experimentally for each gas.

KEY WORDS

Absolute temperature scale, p. 158
Absolute zero, p. 158
Avogadro's law, p. 161
Barometer, p. 150
Boyle's law, p. 155
Charles' law, p. 159
Charles and Gay-Lussac's law, p. 159
Dalton's law of partial pressures, p. 171
Diffusion, p. 179

Effusion, p. 181
Gas constant, p. 162
Graham's law of diffusion, p. 180
Ideal gas, p. 162
Ideal gas equation, p. 162
Kelvin temperature scale, p. 158
Mole fraction, p. 172
Partial pressure, p. 171
Standard temperature and pressure (STP), p. 163

PROBLEMS

More challenging problems are marked with an asterisk.

Atmospheric Pressure

6.1 Would it be easier to drink water with a straw on the top or at the foot of Mt. Everest? Explain.
6.2 Is the atmospheric pressure in a mine that is 500 m below sea level greater or less than 1 atm?
6.3 If the maximum distance that water may be brought up a well by using a suction pump is 34 ft (10.3 m), explain how it is possible to obtain water and oil from hundreds of feet below the surface of the earth.
6.4 Calculate the pressure of the gas in the arrangements shown in the illustration to the right (the liquid is mercury).
6.5 How does a vacuum cleaner work?
6.6 Why do astronauts have to wear protective suits when they are on the surface of the moon?

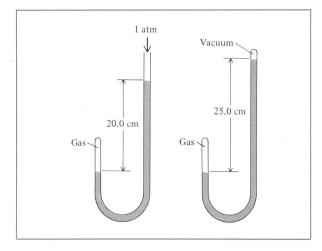

Gas Laws

6.7 A gas occupying a volume of 725 mL at a pressure of 0.970 atm is allowed to expand at constant temperature until its pressure becomes 0.541 atm. What is its final volume?

6.8 From the following data collected at 0°C, comment on whether carbon dioxide behaves as an ideal gas. (*Hint:* Determine *PV*. Is it constant?)

P (atm)	0.0500	0.100	0.151	0.202	0.252
V (L)	448.2	223.8	148.8	110.8	89.0

*6.9 A diver ascends quickly to the surface of the water from a depth of 4.08 m without exhaling gas from his lungs. By what factor would the volume of his lungs increase by the time he reaches the surface? Assume the density of seawater to be 1.03 g/cm^3 and the temperature to remain constant. (1 atm = 1.01325×10^5 N/m^2 and acceleration due to gravity is 9.8067 m/s^2.) (*Hint:* See Appendix 2.)

6.10 The volume of a gas is 5.80 L, measured at standard pressure. What is the pressure of the gas in mmHg if the volume is changed to 9.65 L? (The temperature remains constant.)

6.11 A sample of air occupies 3.8 L when the pressure is 1.2 atm. (a) What volume does it occupy at 6.6 atm? (b) What pressure is required in order to compress it to 0.075 L? (The temperature is kept constant.)

6.12 A dented (but not punctured) Ping-Pong ball can often be restored to its original shape by immersing it in very hot water. Why?

6.13 A quantity of 36.4 L of methane gas is heated from 25°C to 88°C at constant pressure. What is its final volume?

*6.14 Ammonia burns in oxygen gas to form nitric oxide (NO) and water vapor. How many volumes of nitric oxide are obtained from one volume of ammonia at the same temperature and pressure?

The Ideal Gas Equation

6.15 A certain amount of gas at 25°C and at a pressure of 0.800 atm is contained in a glass vessel. Suppose the vessel can withstand a pressure of 2.0 atm. How high can you increase the temperature of the gas without bursting the vessel?

6.16 A gas-filled balloon having a volume of 2.50 L at 1.2 atm and 25°C is allowed to rise to the stratosphere, where the temperature and pressure are −23°C and 3.00×10^{-3} atm, respectively. Calculate the final volume of the balloon.

6.17 The temperature of 2.5 L of a gas initially at STP is increased to 250°C at constant volume. Calculate the final pressure of the gas in atm.

6.18 A gas evolved during the fermentation of glucose (wine making) has a volume of 0.78 L when measured at 20.1°C and 1.00 atm. What was the volume of this gas at the fermentation temperature of 36.5°C and 1.00 atm pressure?

6.19 An ideal gas originally at 0.85 atm and 66°C was allowed to expand until its final volume, pressure, and temperature were 94 mL, 0.60 atm, and 45°C. What was its initial volume?

6.20 The volume of a gas at STP is 488 mL. Calculate its volume at 22.5 atm and 150°C.

6.21 A gas at 772 mmHg and 35.0°C occupies a volume of 6.85 L. Calculate its volume at STP.

6.22 Dry Ice is solid carbon dioxide. A 0.050 g sample of Dry Ice is placed in an evacuated vessel of volume 4.6 L at 30°C. Calculate the pressure inside the vessel after all the Dry Ice has been converted to CO_2 gas.

6.23 A sample of neon gas occupies 566 cm^3 at 10°C and 772 mmHg pressure. Calculate its volume at 15°C and 754 mmHg.

6.24 A quantity of gas weighing 7.10 g at 741 torr and 44°C occupies a volume of 5.40 L. What is its molar mass?

*6.25 The ozone molecules present in the stratosphere absorb much of the harmful radiation from the sun. Typical temperature and pressure of ozone in the stratosphere are 250 K and 1.0×10^{-3} atm, respectively. How many ozone molecules are present in 1.0 L of air under these conditions?

6.26 A 2.10 L vessel contains 4.65 g of gas at 1.00 atm and 27.0°C. (a) Calculate the density of the gas in g/L. (b) What is the molar mass of the gas?

6.27 It is quite easy to achieve a pressure as low as 1.0 $\times 10^{-6}$ mmHg using a diffusion pump. How many molecules of an ideal gas are present in 1.0 L at 25°C under this condition?

6.28 Calculate the volume of the following at STP: (a) 0.681 g of N_2, (b) 7.4 mole of SO_2, (c) 0.58 kg of CH_4.

6.29 Calculate the density of hydrogen bromide (HBr) gas in g/L at 733 mmHg and 46°C.

6.30 A certain anesthetic contains 64.9 percent C, 13.5 percent H, and 21.6 percent O by mass. 1.00 L of the gaseous compound measured at 120°C and 750 mmHg weighs 2.30 g. What is the molecular formula of the compound?

6.31 A volume of 0.280 L of a gas at STP weighs 0.400 g. Calculate the molar mass of the gas.

6.32 Assuming air contains 78 percent N_2, 21 percent O_2, and 1 percent Ar, all by volume, how many molecules of each type of gas are present in 1.0 L of air at STP?

Gases In Chemical Reactions

6.33 Propane (C_3H_8) burns in oxygen to produce carbon dioxide gas and water vapor. (a) Write a balanced equation for this reaction. (b) Calculate the number of liters of carbon dioxide measured at STP that could be produced from 7.45 g of propane.

6.34 Some commercial drain cleaners contain two components: sodium hydroxide and aluminum powder. When the mixture is poured down a clogged drain, the following reaction occurs:

$$2NaOH(aq) + 2Al(s) + 6H_2O(l) \longrightarrow$$
$$2NaAl(OH)_4(aq) + 3H_2(g)$$

The heat generated in this reaction helps melt away obstructions such as grease, and the hydrogen gas released stirs up the solids clogging the drain. Calculate the volume of H_2 formed at STP if 3.12 g of Al is treated with excess NaOH.

6.35 The equation for the metabolic breakdown of glucose $(C_6H_{12}O_6)$ is the same as that for the combustion of glucose in air:

$$C_6H_{12}O_6(s) + 6O_2(g) \longrightarrow 6CO_2(g) + 6H_2O(l)$$

Calculate the volume of CO_2 produced at 37.0°C and 1.00 atm when 5.60 g of glucose is used up in the reaction.

*6.36 A compound of P and F was analyzed as follows: Heating 0.2324 g of the compound in a 378 cm^3 container turned all of it to gas, which had a pressure of 97.3 mmHg at 77°C. Then the gas was mixed with calcium chloride solution, which turned all of the F to 0.2631 g of CaF_2. Determine the molecular formula of the compound.

*6.37 The volume of a sample of pure HCl gas was 189 mL at 25°C and 108 mmHg. It was completely dissolved in about 60 mL of water and titrated with a NaOH solution; 15.7 mL of the NaOH solution was required to neutralize the HCl. Calculate the molarity of the NaOH solution.

*6.38 A quantity of 0.225 g of a metal M (molar mass = 27.0 g/mol) liberated 0.303 L of molecular hydrogen (measured at 17°C and 741 mmHg) from an excess of hydrochloric acid. Deduce from these data the corresponding equation and write formulas for the oxide and sulfate of M.

6.39 A quantity of 73.0 g of NH_3 is mixed with an equal mass of HCl. What is the mass of the solid NH_4Cl formed? What is the volume of the gas remaining, measured at 14.0°C and 752 mmHg? What gas is it?

6.40 Dissolving 3.00 g of an impure sample of calcium carbonate in hydrochloric acid produced 0.656 L of carbon dioxide (measured at 20.0°C and 792 mmHg). Calculate the percent by mass of calcium carbonate in the sample.

6.41 A volume of 5.6 L of molecular hydrogen measured at STP is reacted with an excess of molecular chlorine gas. Calculate the mass in grams of hydrogen chloride produced.

6.42 A piece of sodium metal undergoes complete reaction with water as follows:

$$2Na(s) + 2H_2O(l) \longrightarrow 2NaOH(aq) + H_2(g)$$

The hydrogen gas generated is collected over water at 25.0°C. The volume of the gas is 246 mL measured at 1.00 atm. Calculate the number of grams of sodium used in the reaction. (Vapor pressure of water at 25°C = 0.0313 atm)

*6.43 Ethanol (C_2H_5OH) burns in air:

$$C_2H_5OH(l) + O_2(g) \longrightarrow CO_2(g) + H_2O(l)$$

Balance the equation and determine the volume of air in liters at 35.0°C and 790 mmHg required to burn 227 g of ethanol. Assume air to be 21.0 percent O_2 by volume.

Dalton's Law of Partial Pressures

6.44 A mixture of gases contains CH_4, C_2H_6, and C_3H_8. If the total pressure is 1.50 atm and the number of moles of the gases present are 0.31 mole for CH_4, 0.25 mole for C_2H_6, and 0.29 mole for C_3H_8, calculate the partial pressures of the gases.

6.45 A 2.5 L flask at 15°C contains a mixture of three gases, N_2, He, and Ne, at partial pressures of 0.32 atm for N_2, 0.15 atm for He, and 0.42 atm for Ne. (a) Calculate the total pressure of the mixture. (b) Calculate the volume in liters at STP occupied by He and Ne if the N_2 is removed selectively.

6.46 Dry air near sea level has the following composition by volume: N_2, 78.08 percent; O_2, 20.94 percent; Ar, 0.93 percent; CO_2, 0.05 percent. The atmospheric pressure is 1.00 atm. Calculate (a) the partial pressure of each gas in atm and (b) the concentration of each gas in mol/L at 0°C. (*Hint:* Since volume is proportional to the number of moles present, mole fractions of gases can be expressed as ratios of volumes at the same temperature and pressure.)

6.47 A mixture of helium and neon gases is collected over water at 28.0°C and 745 mmHg. If the partial pressure of helium is 368 mmHg, what is the partial pressure of neon? (Vapor pressure of water at 28°C = 28.3 mmHg)

6.48 A sample of zinc metal is allowed to react completely with an excess of hydrochloric acid:

$$Zn(s) + 2HCl(aq) \longrightarrow ZnCl_2(aq) + H_2(g)$$

The hydrogen gas produced is collected over water at 25.0°C using an arrangement similar to that shown in Figure 6.13. The volume of the gas is 7.80 L and the pressure is 0.980 atm. Calculate the amount of zinc metal in grams consumed in the reaction. (Vapor pressure of water at 25°C = 23.8 mmHg)

Kinetic Molecular Theory of Gases

6.49 What prevents the molecules in the atmosphere from escaping into outer space?

6.50 Which of the following two statements is correct? (a) Heat is produced by the collision of gas molecules against one another. (b) When a gas is heated, the molecules collide with one another more often.

6.51 Compare the average molecular speeds of O_2 and UF_6 at 25°C.

6.52 What is the Maxwell distribution of speeds? Does Maxwell's theory work for a sample of 150 molecules? Explain.

6.53 The temperature and pressure in the stratosphere are −23°C and 3.00×10^{-3} atm, respectively. Calculate the average speeds of N_2, O_2, and O_3 molecules in this region.

6.54 As we know, UF_6 is a much heavier gas than helium. Yet at a given temperature, the average kinetic energies of the samples of the two gases are the same. Explain.

*6.55 Two vessels are labeled A and B. Vessel A contains NH_3 gas at 70°C and vessel B contains Ne gas at the same temperature. If the average kinetic energy of NH_3 is 7.1×10^{-21} J/molecule, calculate the mean square speed of Ne atoms in m^2/s^2.

Diffusion and Effusion

6.56 It requires 57 seconds for 1.4 liters of an unknown gas to effuse through a porous wall, and it takes 84 seconds for the same volume of N_2 gas to effuse at the same temperature and pressure. What is the molar mass of the unknown gas?

6.57 A sample of the gas discussed in Problem 6.18 is found to effuse through a porous barrier in 15.0 min. Under the same conditions of temperature and pressure, it takes N_2 12.0 min to effuse through the same barrier. Calculate the molar mass of the gas and suggest what gas it might be.

6.58 List the following gases in order of increasing diffusion rates: PH_3, ClO_2, Kr, NH_3, and HI. Calculate the ratio of diffusion rates of fastest to slowest.

6.59 Under certain conditions of temperature and pressure, 60.0 cm^3 of a gas X effused through a pinhole in 10.0 seconds. Under the same conditions, 480 cm^3 of H_2 gas effused in 20.0 seconds. Calculate the molar mass of X.

*6.60 Consider the arrangement shown in the illustration.

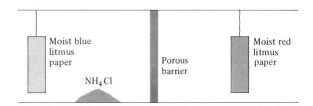

The ammonium chloride sample is being heated. (a) Write a balanced equation for the thermal decomposition of ammonium chloride. (b) Predict the subsequent changes in the color of the blue and the red litmus papers.

Nonideal Gas Behavior

*6.61 The temperature of a real gas that is allowed to expand into a vacuum usually drops. Explain.

6.62 Using the data shown in Table 6.4, calculate the pressure exerted by 2.50 moles of CO_2 confined in a volume of 5.00 L at 450 K. Compare the pressure with that calculated using the ideal gas equation.

6.63 Under the same conditions of temperature and pressure, which of the following gases would behave most ideally: Ne, N_2, or CH_4? Explain.

Miscellaneous Problems

6.64 Explain the following terms: atmospheric pressure, barometer, ideal gas equation, average speed, gas diffusion, gas effusion.

6.65 (a) Convert the following temperatures to kelvins: 0°C, 37°C, 100°C, −225°C. (b) Convert the following temperatures to degrees Celsius: 77 K, 4.2 K, 6.0×10^3 K.

6.66 Discuss the following phenomena in terms of the gas laws: (a) the pressure in an automobile tire increasing on a hot day, (b) the "popping" of a paper bag, (c) the expansion of a weather balloon as it rises in the air, (d) the loud noise heard when a light bulb shatters.

6.67 Nitroglycerin, an explosive, decomposes according to the equation

$$4C_3H_5(NO_3)_3(s) \longrightarrow$$
$$12CO_2(g) + 10H_2O(g) + 6N_2(g) + O_2(g)$$

Calculate the total volume of gases produced when

collected at 1.2 atm and 25°C from 2.6×10^2 g of nitroglycerin. What are the partial pressures of the gases under these conditions?

6.68 Explain why a balloon filled with helium gas generally deflates faster than a balloon (of similar size) filled with air. (*Hint:* A balloon has many invisible pinholes in the taut rubber, through which a gas can escape by effusion.)

*6.69 Consider the apparatus shown to the right. When a small amount of water is introduced into the flask by squeezing the bulb of the medicinal dropper, water is squirted upward out of the long glass tubing. Explain this observation. (*Hint:* Hydrogen chloride gas is soluble in water.)

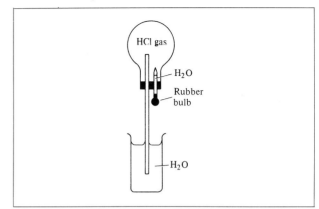

7

ATOMS I: SUBATOMIC PARTICLES AND NUCLEAR STABILITY

D alton's atomic theory, discussed in Chapter 2, accounted for a wide range of early macroscopic observations regarding the behavior of matter. These led to the theory's general acceptance by chemists. However, Dalton's detailed view of the nature of atoms began to be challenged by the middle of the nineteenth century. The challenge, which extended into the current century, came in the form of new observations that were not explained by assuming that atoms were indivisible, indestructible entities.

In this chapter we will describe the work that led to the unmistakable idea that an atom is composed of even smaller structural units, now termed *subatomic particles.* We will consider in detail three of these subatomic particles—electrons, protons, and neutrons.

7.1   THE ELECTRON

Electrons are normally associated with atoms. However, they can also be studied individually.

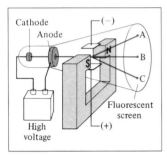

FIGURE 7.1 *A cathode ray tube with perpendicular (to the direction of cathode rays) electric field and external magnetic field. The symbols N and S denote the north and south poles of the magnet. The cathode rays will strike the end of the tube at A in the presence of a magnetic field; at C in the presence of an electric field; at B when there are no external fields present or when the effects of the electric field and magnetic field cancel each other.*

The discovery of electrons and the first detailed study of their behavior came about with the invention of the cathode ray tube, which was the forerunner of today's television tube. Figure 7.1 shows a schematic diagram of a cathode ray tube. *Negatively charged particles,* or ***electrons,*** emitted from the cathode, are accelerated by a positively charged plate, the *anode,* toward itself. The anode has a hole in it that allows electrons to pass through. The stream of electrons forms what early investigators named a *cathode ray.* The cathode ray goes on to strike the inside surface of the end of the tube. The surface is coated with a fluorescent material, such as ZnS, so that a strong fluorescence, or emission of light, is observed when the surface is bombarded by the electrons.

In some experiments two electrically charged plates and a magnet were added to the cathode ray tube, as shown in Figure 7.1. When the magnetic field is on and the electric field is off, the cathode ray strikes point A. When only the electric field is on, the ray strikes point C. When both the magnetic and the electric fields are off or when they are both on but balanced so as to cancel each other's influence, the ray strikes point B. Such behavior is consistent with the fact that electrons possess a negative charge. Electromagnetic theory tells us that a moving charged body behaves like a magnet and can interact with electric and magnetic fields through which it passes. Since the cathode ray is attracted by the plate bearing positive charges and repelled by the plate bearing negative charges, it is clear that it must consist of negatively charged particles.

From a knowledge of the effects of electrical and magnetic forces on a negatively charged particle, J. J. Thomson (1856–1940) was able to obtain the ratio of electric charge to mass for an electron. He found the ratio to be -1.76×10^8 C/g, where C stands for coulomb, which is the unit of electric charge. It was left to R. A. Millikan (1868–1953) to further study the electron. In a series of experiments carried out between 1908 and 1917, Millikan deduced the charge of an electron to be -1.60×10^{-19} C. From these data we can calculate the mass of an electron:

$$\begin{aligned} \text{mass of an electron} &= \frac{\text{charge}}{\text{charge/mass}} \\ &= \frac{-1.60 \times 10^{-19} \text{ C}}{-1.76 \times 10^8 \text{ C/g}} \\ &= 9.09 \times 10^{-28} \text{ g} \end{aligned}$$

The mass of an electron is so small that even a mole of these particles would have a mass of less than 0.0006 g. You can verify this by a simple calculation.

which is an exceedingly small mass.

7.2   X RAYS AND RADIOACTIVITY

In the 1890s many scientists became caught up in the study of cathode rays and other kinds of rays. Some of these rays were associated with the newly discovered phenomenon of ***radioactivity,*** which is *the spontaneous break-*

down of an atom by emission of particles and/or radiation. **Radiation** is the term used to describe *the emission and transmission of energy through space in the form of waves.* A radioactive substance *decays,* or breaks down, spontaneously. Information gained by studying various rays and their effects on other materials contributed greatly to the growing understanding of the structure of the atom.

In 1895 Wilhelm Röntgen (1845–1923) noticed that when cathode rays struck glass and metals, new and very unusual rays were emitted. These rays were highly energetic and could penetrate matter. They also darkened covered photographic plates and could produce fluorescence in various substances. Since these rays could not be deflected by a magnet, they did not consist of charged particles as did cathode rays. Röntgen called them *X rays.* They were later identified as a type of high-energy radiation.

Not long after Röntgen's discovery, Antoine Becquerel (1852–1908), a professor of physics in Paris, began to study fluorescent properties of substances. Purely by accident, he noticed that a certain compound containing uranium, placed nearby, was able to darken photographic plates that were wrapped in thick papers or even in thin metal sheets, without the stimulation of cathode rays. The nature of the radiation that was doing this was not known, although it seemed to resemble X rays in that both types of radiation were highly energetic and did not consist of charged particles. One of Becquerel's students, Marie Curie (1867–1934), suggested the name "radioactivity" for this phenomenon. Any element, such as uranium, that exhibits radioactivity is said to be radioactive. Marie Curie and her husband, Pierre, later studied and identified many radioactive elements.

Further investigation showed that three types of rays can be emitted by radioactive elements. These rays were studied by using an arrangement similar to that shown in Figure 7.2. It was found that two of the three types of rays could be deflected when they passed between two oppositively charged metal plates.

Depending on the deflection, these two rays are called alpha (α) rays and beta (β) rays. The third type of ray, which is unaffected by charged plates, is called a gamma (γ) ray. **α rays** or **α particles** were found to be *helium ions with a positive charge of +2.* They are therefore attracted by the negatively charged plate. The opposite holds true for **β rays** or **β particles**—they consist of negatively charged electrons, which are drawn to the positively charged plate. Because **γ rays** *do not consist of charged particles,* their movement is unaffected by an external electric field. *They are high-energy radiation.*

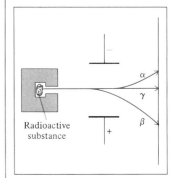

FIGURE 7.2 *Three types of rays emitted by radioactive elements. β rays consist of negatively charged particles (electrons), and are therefore attracted by the positively charged plate. The opposite holds true for α rays—they are positively charged (He²⁺) and are drawn to the negatively charged plate. Because γ rays do not consist of charged particles, their movement is unaffected by an external electric field.*

Some indication of the initial uncertainty regarding the identity of these three types of rays is seen in the fact that their original names are based on the first three letters of the Greek alphabet—alpha, beta, and gamma.

7.3 THE PROTON AND THE NUCLEUS

By the early 1900s, two features of atoms had become clear: They contain electrons, and they are electrically neutral. Since it is neutral, every atom must contain an equal number of positive and negative charges, to maintain the electrical neutrality. Around the turn of the century, the accepted model for atoms was the one that was proposed by J. J. Thomson. According to

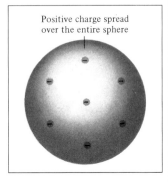

Positive charge spread
over the entire sphere

FIGURE 7.3 *Thomson's model of atoms, sometimes described as the "raisin muffin" model. The electrons are embedded in a uniform, positively charged sphere.*

his description, an atom could be thought of as a uniform, positive sphere of matter in which electrons are embedded (Figure 7.3).

In 1910 Ernest Rutherford (1871–1937), who had earlier studied under Thomson at Cambridge University, decided to use particles to probe the structure of atoms. Together with his associate Hans Geiger (1882–1945) and an undergraduate named Ernest Marsden (1889–1970), Rutherford carried out a series of experiments in which very thin foils of gold and other metals were used as targets for α particles emitted from a radioactive substance (Figure 7.4). They observed that the majority of the particles penetrated the foil either undeflected or with only a slight deflection. They also noticed that every now and then an α particle would be scattered (or deflected) at a large angle. In some instances, an α particle would even be turned back in the direction from which it had come! This was a most surprising finding, for in Thomson's model the positive charge of the atom was so diffuse that the positive α particles were expected to pass through with very little deflection. To quote Rutherford's initial reaction when told of this discovery: "It was as incredible as if you had fired a 15-inch shell at a piece of tissue paper and it came back and hit you."

Rutherford was later able to explain the results of the α-scattering experiment, but he had to abandon Thomson's model and propose a new model for the atom. According to Rutherford, most of the atom must be empty space. This explains why the majority of α particles passed through the gold foil with little or no deflection. The atom's positive charges, Rutherford proposed, are all concentrated in *a central core within the atom*, which he called the **nucleus.** Whenever an α particle came close to a nucleus in the scattering experiment, it experienced a large repulsive force and therefore a large deflection. If an α particle traveled directly toward a nucleus, it would experience an enormous repulsion that could completely reverse the direction of the moving particle.

The positively charged particles in the nucleus are called **protons,** and

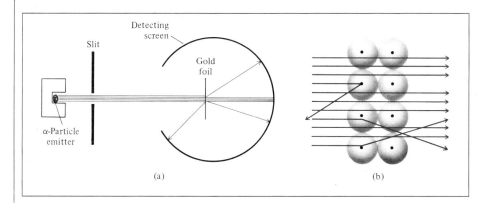

FIGURE 7.4 *(a) The experimental design for Rutherford's measurement of the scattering of α particles by a piece of gold foil. Most of the α particles passed through the gold foil with little or no deflection. A few were deflected at wide angles. Occasionally an α particle was turned back. (b) Diagram showing magnified view of α particles passing through and being deflected by atoms.*

each has a mass of 1.6725×10^{-24} g. In separate experiments, it was found that each proton carries the same *quantity* of charge as an electron and is about 1840 times heavier than the oppositely charged electron.

At this stage of investigation, scientists perceived the atom as follows. The mass of a nucleus comprises most of the mass of the entire atom, but the nucleus occupies only about $1/10^{13}$ of the volume of the atom. For atomic (and molecular) dimensions, we will express lengths in terms of the SI unit called the *picometer* (*pm*) where

$$1 \text{ pm} = 1 \times 10^{-12} \text{ m}$$

A typical atomic radius is about 100 pm, whereas the radius of an atomic nucleus is only about 5×10^{-3} pm. You can appreciate the relative sizes of an atom and its nucleus by imagining that if an atom were the size of the Houston Astrodome, the volume of its nucleus would be comparable to that of a small marble. While the protons are confined to the nucleus of the atom, the electrons are conceived of as spreading out about the nucleus at some distance from it.

7.4 THE NEUTRON

Our description of the atom is not yet complete. For example, it is not clear how a number of positively charged particles can be confined in a tiny volume—the nucleus. The laws of electrostatics tell us that like charges repel one another and unlike charges attract one another. Therefore, we would expect the protons to repel one another and the nucleus to be unstable. Although nuclear disintegration does occur in radioactive elements, many of the atomic nuclei are actually quite stable. Thus attractive forces of some kind must be present in the nucleus to hold the protons together.

In the early 1930s, Rutherford and others postulated the presence in the atomic nucleus of another type of subatomic particles called ***neutrons.*** Convincing experimental evidence was provided in 1932 by one of Rutherford's students, James Chadwick (1891–1972). When beryllium was bombarded with α particles, a very penetrating radiation was emitted that seemed to have properties similar to those of γ rays but had much more energy. Chadwick was able to show that these properties could be explained by assuming that the radiation was a beam of *neutral particles having a mass slightly greater than that of protons.* The neutron had been discovered.

7.5 ATOMIC NUMBER, MASS NUMBER, AND ISOTOPES

The subatomic particles we have introduced can help us better understand the properties of individual atoms. All atoms can be identified by the number of protons and neutrons they contain.

A common non-SI unit for atomic lengths is the *angstrom* (Å): $1 \text{ Å} = 1 \times 10^{-10}$ m. The conversion factor is $1 \text{ Å} = 100$ pm.

Reference to the radius of an atom may incorrectly imply that atoms have well-defined boundaries or surfaces similar to the wooden or plastic models used in chemistry classrooms. Even though atomic radius values are quite useful, we will learn later that the outer regions of atoms are relatively "fuzzy" and that an atom has no specific boundary.

In recent years physicists have discovered a large number of different kinds of subatomic particles. Atoms release these particles when bombarded by extremely energetic particles under specially created conditions in "atom smashers." Chemists deal only with electrons, protons, and neutrons because most chemical reactions are carried out under normal conditions.

The **atomic number** (**Z**) is *the number of protons in the nucleus of each atom of an element.* In a neutral atom the number of protons is equal to the number of electrons, so that the atomic number also indicates the number of electrons present in the atom. The chemical identity of an atom can be determined solely by its atomic number. For example, the atomic number of nitrogen is 7; this means that each neutral nitrogen atom has 7 protons and 7 electrons. Or, viewing it another way, every atom in the universe that contains 7 protons is correctly named "nitrogen."

The **mass number** (**A**) is *the total number of neutrons and protons present in the nucleus of an atom of an element.* Except for the atom of the most common form of hydrogen, which has one proton and no neutrons, all atomic nuclei contain both protons and neutrons. In general the mass number is given by

$$\text{mass number} = \text{number of protons} + \text{number of neutrons}$$
$$= \text{atomic number} + \text{number of neutrons}$$

The number of neutrons in an atom is equal to the difference between the mass number and the atomic number or $(A - Z)$. For example, the mass number of fluorine is 19 and the atomic number is 9 (indicating 9 protons in the nucleus). Thus the number of neutrons present is $19 - 9 = 10$. Note that all three quantities (atomic number, number of neutrons, and mass number) must be positive integers, or whole numbers.

In most cases atoms of a given element do not all have the same mass. For example, there are three types of hydrogen atoms, which differ only in their number of neutrons. They are hydrogen, with one proton and no neutrons; deuterium, with one proton and one neutron; and tritium, with one proton and two neutrons. As we noted in Section 2.3, *atoms having the same atomic number but different mass numbers* are called **isotopes.**

The accepted way to denote the atomic number of an atom of element X is as follows:

$$\text{mass number} \longrightarrow {}^{A}_{Z}X$$
$$\text{atomic number} \longrightarrow$$

Thus, for the isotopes of hydrogen, we write

$$\underset{\text{hydrogen}}{{}^{1}_{1}H} \qquad \underset{\text{deuterium}}{{}^{2}_{1}H} \qquad \underset{\text{tritium}}{{}^{3}_{1}H}$$

As another example, consider two common isotopes of uranium with mass numbers of 235 and 238, respectively:

$$^{235}_{92}U \qquad ^{238}_{92}U$$

The first isotope is used in nuclear reactors and atomic bombs, whereas the second isotope lacks the properties to be utilized in these respects. It is customary to identify isotopes by mass numbers only. In this case we can speak of uranium-235 and uranium-238.

The chemical properties of an element are determined primarily by the number of protons and electrons in its atoms, not by the mass number. For this reason, isotopes of the same element are chemically alike, forming the same types of compounds and displaying similar reactivities.

Atomic number is actually proton number.

To distinguish the isotope hydrogen from the element name hydrogen (which includes all three isotopes), the name *protium* has been assigned to the isotope of hydrogen containing one proton and no neutrons.

Knowing the element always enables us to know its atomic number but not the mass number.

EXAMPLE 7.1

Give the number of protons, neutrons, and electrons for each of the following species: (a) $^{25}_{12}Mg$, (b) $^{195}_{78}Pt$, (c) $^{198}_{78}Pt$.

Answer

(a) The atomic number is 12, so there are 12 protons. The mass number is 25, so the number of neutrons is $25 - 12 = 13$. The number of electrons is the same as the number of protons, that is, 12.

(b) The atomic number is 78, so there are 78 protons. The mass number is 195, so the number of neutrons is $195 - 78 = 117$. The number of electrons is 78.

(c) Here the number of protons is the same as that in (b), or 78. The number of neutrons is $198 - 78 = 120$. The number of electrons is also the same as that in (b), 78. The species in (b) and (c) are two isotopes of platinum.

Similar examples: Problems 7.14, 7.15.

7.6 STABILITY OF THE ATOMIC NUCLEUS

Density of the Nucleus

Table 7.1 summarizes the masses and electrical charges of the three subatomic particles of importance to chemistry—the electron, the proton, and the neutron. As we saw earlier, the nucleus occupies a very small portion of the total volume of an atom. However, the nucleus contains most of the atom's mass because both the protons and the neutrons reside there. We can easily estimate the density of the nucleus itself. For example, assume that a nucleus has a radius of 5×10^{-3} pm and a mass of 1×10^{-22} g. These figures correspond roughly to a nucleus containing 50 protons and 50 neutrons. Density is mass/volume, and we can calculate the volume from the known radius (the volume of a sphere is $\frac{4}{3}\pi r^3$, where r is the radius of the sphere). First we convert the pm units to cm. Then we calculate the density in g/cm^3:

$$r = 5 \times 10^{-3} \text{ pm} \times \frac{1 \times 10^{-12} \text{ m}}{1 \text{ pm}} \times \frac{100 \text{ cm}}{1 \text{ m}} = 5 \times 10^{-13} \text{ cm}$$

TABLE 7.1 Masses and Charges of Subatomic Particles

Particle	Mass		Charge	
	(g)	(amu)	coulomb (C)	Charge Unit
Electron*	9.1095×10^{-28}	0.000548	-1.6022×10^{-19}	-1
Proton	1.67265×10^{-24}	1.00729	1.6022×10^{-19}	$+1$
Neutron	1.67495×10^{-24}	1.008665	0	0

* More refined experiments have given us a more accurate value of an electron's mass than the one originally determined by Millikan.

$$\text{density} = \frac{\text{mass}}{\text{volume}} = \frac{1 \times 10^{-22} \text{ g}}{\frac{4}{3}\pi r^3} = \frac{1 \times 10^{-22} \text{ g}}{\frac{4}{3}\pi(5 \times 10^{-13} \text{ cm})^3}$$
$$= 2 \times 10^{14} \text{ g/cm}^3$$

This is an exceedingly high density. The highest density known for an element is only 22.6 g/cm³, for osmium (Os). Thus the average atomic nucleus is roughly 9×10^{12} (or 9 trillion) times more dense than the densest element known!

To dramatize the almost incomprehensibly high density of the nucleus, it has been suggested that it is equivalent to packing the mass of all the world's automobiles into one thimble.

Nuclear Binding Energy

The enormously high density of the nucleus prompts us to wonder what holds the particles together so tightly. From *Coulomb's law* we learn that like charges repel and unlike charges attract one another. We would thus expect the protons to repel one another strongly. This indeed is so. However, in addition to the repulsion, there are also short-range attractions between proton and proton, proton and neutron, and neutron and neutron. The study of these interactions is beyond the scope of this book. What you should understand is that the stability of any nucleus is determined by the difference between coulombic repulsion and this attraction. If the repulsion outweighs attraction, the nucleus disintegrates, emitting particles and/or radiation. This is the phenomenon of radioactivity that we discussed earlier. If the attraction prevails, the nucleus will be stable. Our concern here is to estimate the overall nuclear stability by considering the mass relationships in the formation of given atoms.

A quantitative measure of nuclear stability is the ***nuclear binding energy,*** which is *the energy required to break up a nucleus into its component protons and neutrons.* It was noticed long ago in the study of nuclear properties that the masses of nuclei are always less than the sums of the masses of the *nucleons.* (The term "nucleon" refers to both protons and neutrons in a nucleus.) For example, the $^{19}_{9}\text{F}$ isotope has an atomic mass of 18.9984 amu. The nucleus has 9 protons and 10 neutrons and therefore a total of 19 nucleons. The mass of 9 $^{1}_{1}\text{H}$ atoms (that is, the mass of 9 protons and 9 electrons) is

$$9 \times 1.007838 \text{ amu} = 9.070542 \text{ amu}$$

and the mass of 10 neutrons is (see Table 7.1)

$$10 \times 1.008665 \text{ amu} = 10.08665 \text{ amu}$$

Therefore, the atomic mass of a $^{19}_{9}\text{F}$ atom calculated from the known numbers of electrons, protons, and neutrons is

$$9.070542 \text{ amu} + 10.08665 \text{ amu} = 19.15719 \text{ amu}$$

which is larger than 18.9984 amu (the measured mass of $^{19}_{9}\text{F}$) by 0.1588 amu. *This difference between the mass of the atom and the sum of the masses of its protons, neutrons, and electrons* is called the ***mass defect.*** Where did the mass go?

In the formation of the fluorine nucleus, a large quantity of energy is released. Relativity theory tells us that a loss of energy is always accompanied by an equivalent loss of mass. According to *Einstein's mass-energy equivalence relationship* ($E = mc^2$, where E is energy, m is mass, and c is the velocity of light), we can calculate the energy released as follows. We start by writing

$$\Delta E = (\Delta m)c^2$$

where ΔE and Δm are defined as follows:

$$\Delta E = \text{energy of product} - \text{energy of reactants}$$

$$\Delta m = \text{mass of product} - \text{mass of reactants}$$

Here we have

$$\Delta m = 18.9984 \text{ amu} - 19.15717 \text{ amu}$$
$$= -0.1588 \text{ amu}$$

Because $^{19}_{9}\text{F}$ has a mass that is less than the mass calculated from the number of electrons and nucleons present, Δm is a negative quantity. Consequently, ΔE is also a negative quantity; that is, energy is released to the surroundings as a result of the formation of the fluorine-19 nucleus. So we calculate ΔE as follows:

$$\Delta E = (-0.1588 \text{ amu})(3.00 \times 10^8 \text{ m/s})^2$$
$$= -1.43 \times 10^{16} \text{ amu m}^2/\text{s}^2$$

With the conversion factors

$$1 \text{ kg} = 6.022 \times 10^{26} \text{ amu}$$

$$1 \text{ J} = 1 \text{ kg m}^2/\text{s}^2$$

we obtain

$$\Delta E = \left(-1.43 \times 10^{16} \frac{\text{amu m}^2}{\text{s}^2}\right) \times \left(\frac{1.00 \text{ kg}}{6.022 \times 10^{26} \text{ amu}}\right) \times \left(\frac{1 \text{ J}}{1 \text{ kg m}^2/\text{s}^2}\right)$$

$$= -2.37 \times 10^{-11} \text{ J}$$

This is the energy released (note the negative sign, which indicates that this is an exothermic process) when one fluorine-19 nucleus is formed from 9 protons and 10 neutrons. The nuclear binding energy of the nucleus has the value of 2.37×10^{-11} J, which is the amount of energy needed to decompose the nucleus into separate protons and neutrons. In general, the larger the mass defect, the greater is the nuclear binding energy and the more stable is the nucleus. In the formation of one mole of fluorine nuclei, for instance, the energy released is

$$\Delta E = (-2.37 \times 10^{-11} \text{ J})(6.022 \times 10^{23}/\text{mol})$$
$$= -1.43 \times 10^{13} \text{ J/mol}$$
$$= -1.43 \times 10^{10} \text{ kJ/mol}$$

The nuclear binding energy therefore is 1.43×10^{10} kJ for one mole of fluorine-19 nuclei, which is a tremendously large quantity when we consider that the enthalpies of ordinary chemical reactions are of the order of

Mass defect applies only to the loss of mass of the nucleus. The mass of electrons remains unchanged.

Note that on the basis of one mole of fluorine nuclei, a mass defect of only 0.1588 g is associated with a nuclear binding energy of over 10^{10} kJ. The fact that an incredibly large energy effect is related to an extremely small nuclear mass change is the basis for both nuclear power and nuclear weapons technology, as we will learn later.

FIGURE 7.5 *Plot of nuclear binding energy per nucleon versus mass number.*

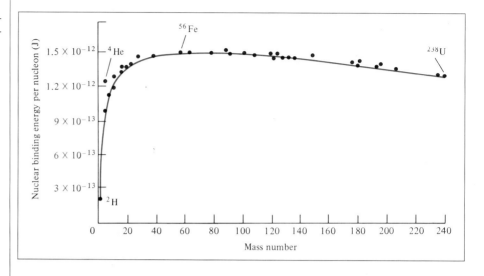

only 200 kJ. The procedure we have followed can be used to calculate the nuclear binding energy of any nucleus.

As we have noted, nuclear binding energy is an indication of the stability of a nucleus. However, in comparing the stability of any two nuclei we must account for the fact that they have different numbers of nucleons. For this reason, it is more meaningful to use the quantity *nuclear binding energy per nucleon*, defined as

$$\text{nuclear binding energy per nucleon} = \frac{\text{nuclear binding energy}}{\text{number of nucleons}}$$

For the fluorine-19 nucleus

$$\text{nuclear binding energy per nucleon} = \frac{2.37 \times 10^{-11} \text{ J}}{19 \text{ nucleons}}$$
$$= 1.25 \times 10^{-12} \text{ J/nucleon}$$

In this manner, we can compare the stability of all nuclei on a common basis. Figure 7.5 shows the variation of nuclear binding energy per nucleon plotted against mass number. As you can see, the curve rises rather steeply; the highest binding energies per nucleon belong to elements with intermediate mass numbers—between 40 and 100—and are greatest for elements in the iron, cobalt, and nickel region (the Group 8B elements) of the periodic table. This result shows that the *net* attractive forces among the particles (protons and neutrons) are greatest for the nuclei of these elements.

EXAMPLE 7.2

The atomic mass of $^{127}_{53}\text{I}$ is 126.9004 amu. Calculate the nuclear binding energy of this nucleus and the corresponding nuclear binding energy per nucleon.

Answer

There are 53 protons and 74 neutrons in the nucleus. The mass of 53 ^1_1H atoms is

$$53 \times 1.007838 \text{ amu} = 53.41541 \text{ amu}$$

and the mass of 74 neutrons is

$$74 \times 1.008665 \text{ amu} = 74.64121 \text{ amu}$$

Therefore, the predicted mass for $^{127}_{53}I$ is $(53.41541 + 74.64121) = 128.05662$ amu, and the mass defect is

$$\Delta m = 126.9004 \text{ amu} - 128.05662 \text{ amu}$$
$$= -1.1562 \text{ amu}$$

The energy released is

$$\begin{aligned}
\Delta E &= \Delta mc^2 \\
&= (-1.1562 \text{ amu})(3.00 \times 10^8 \text{ m/s})^2 \\
&= -1.04 \times 10^{17} \text{ amu m}^2/\text{s}^2 \\
&= \left(-1.04 \times 10^{17} \frac{\text{amu m}^2}{\text{s}^2}\right) \times \left(\frac{1.00 \text{ kg}}{6.022 \times 10^{26} \text{ amu}}\right) \times \left(\frac{1 \text{ J}}{1 \text{ kg m}^2/\text{s}^2}\right) \\
&= -1.73 \times 10^{-10} \text{ J}
\end{aligned}$$

Thus the nuclear binding energy is 1.73×10^{-10} J. The nuclear binding energy per nucleon is obtained as follows:

$$\frac{1.73 \times 10^{-10} \text{ J}}{127 \text{ nucleons}} = 1.36 \times 10^{-12} \text{ J/nucleon}$$

Similar examples: Problems 7.24, 7.25.

Note that in mass defect calculations we always use as many significant figures for the masses as are available because even a small mass defect is equivalent to an appreciably large change in energy.

AN ASIDE ON THE PERIODIC TABLE
Atomic Number and the Periodic Table

Nineteenth century chemists constructed the periodic table using their knowledge of atomic masses. At that time, they only had a vague idea about atoms and molecules and did not know of the existence of electrons and protons. On the other hand, accurate measurements of atomic masses of many elements had already been made. Arranging elements according to their atomic masses in the periodic table seemed logical to those chemists, who felt that chemical behavior should somehow be related to atomic mass. However, this kind of correlation led to serious discrepancies between predictions and observed facts in a few cases. For example, the atomic mass of argon (39.95 amu) is greater than that of potassium (39.10 amu). If the elements were arranged by increasing atomic mass, this would put argon in the position occupied now by potassium (see Figure 1.4). But of course no chemist would place argon, an inert gas, in the same group as lithium and sodium, two extremely reactive metals. Such discrepancies suggested that some fundamental property other than atomic mass is the cause of the observed periodicity. This property turned out to be associated with atomic number.

On the basis of α-scattering experiments, Rutherford was able to estimate the number of positive charges in the nucleus of a few elements, but there was no general procedure

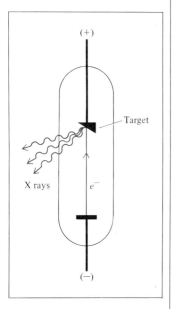

FIGURE 7.6 An X-ray tube. The target is made of the element under study. It emits X rays of characteristic frequencies when bombarded by high energy electrons.

for determining the atomic numbers of the elements. In 1913 a young English physicist, Henry Moseley (1887–1915), was able to provide the necessary measurements. When high-energy cathode rays (that is, high-energy electrons) are focused on a target made up of the element under study, X rays were generated (Figure 7.6). Moseley noticed that the frequencies of X rays emitted from elements could be correlated by the equation

$$\sqrt{\nu} = a(Z - b)$$

where ν is the frequency of the emitted X ray and a and b are constants that are the same for all the elements. (X rays are a form of radiation possessing wave properties. We will discuss the properties of waves in Chapter 8. For the present, just keep in mind that the energy of radiation is directly proportional to the frequency of the waves.) A plot of $\sqrt{\nu}$ versus Z yields a straight line (Figure 7.7). The atomic number of an element can therefore be determined. (From the measured frequency and hence $\sqrt{\nu}$ of emitted X rays, we can determine the atomic number of the element from the plot.)

With a few exceptions, Moseley found that the order of increasing atomic number is the order of increasing atomic mass. For example, calcium is the twentieth element in the increasing order of atomic mass and it has an atomic number of 20. The discrepancies mentioned earlier are now explained: The atomic number of argon is 18 and that of potassium is 19, so that potassium should follow argon.

In any modern periodic table the atomic number is usually listed with the element. As you already know, the atomic number also indicates the number of electrons in the atoms of an element. In the next few chapters we will see how this information is used to study the physical and chemical properties of the elements.

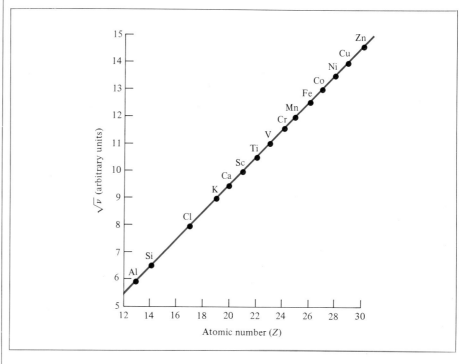

FIGURE 7.7 A plot of the square root of the frequency of X rays versus the atomic number of the target element.

SUMMARY

1. An atom consists of a very dense central nucleus containing protons and neutrons, with electrons moving about the nucleus at relatively large distances from it.
2. Protons are positively charged, neutrons have no charge, and electrons are negatively charged. Protons and neutrons have roughly the same mass, which is about 1840 times greater than the mass of an electron.
3. Electrons were discovered in experiments with cathode rays. Protons and the nucleus were discovered in scattering experiments using α particles from radioactive elements to bombard gold foil. Neutrons were discovered in the rays produced by α-particle bombardment of beryllium.
4. The atomic number of an element is the number of protons in the nucleus of an atom of the element; it is the identifying characteristic of an element. The mass number is the sum of the number of protons and the number of neutrons in the nucleus.
5. Isotopes are atoms of the same element, with the same number of protons, but different numbers of neutrons.
6. Nuclear binding energy is equivalent (in accordance with $E = mc^2$) to the mass lost when protons and neutrons combine to form nuclei.
7. In a modern periodic table, elements are arranged according to their increasing atomic number, not according to their increasing atomic mass.

KEY WORDS

α particles, p. 195
α rays, p. 195
Atomic number, p. 198
β particles, p. 195
β rays, p. 195
Electron, p. 194
γ rays, p. 195
Isotope, p. 198

Mass defect, p. 200
Mass number, p. 198
Neutron, p. 197
Nuclear binding energy, p. 200
Nucleus, p. 196
Proton, p. 196
Radiation, p. 195
Radioactivity, p. 194

PROBLEMS

More challenging problems are marked with an asterisk.

X Rays and Radioactivity

7.1 Define the following terms: (a) alpha particle, (b) beta particle, (c) gamma rays, (d) X rays.
7.2 List the types of radiation that are known to be emitted by radioactive elements.
7.3 A sample of a radioactive element is found to be losing mass gradually. Explain what is happening to the sample.
7.4 Compare the properties of the following: alpha particles, cathode rays, protons, neutrons, electrons.

The Atomic Nucleus

7.5 What was Thomson's contribution to our knowledge about the structure of the atom?

7.6   What was Rutherford's contribution to our knowledge about the structure of the atom?

7.7   Which of the following are subatomic particles? alpha particle, electron, hydrogen atom, proton, gamma rays, neutron.

7.8   The diameter of a neutral hydrogen atom is about 1×10^2 pm. Suppose we could line up hydrogen atoms side by side in contact with one another. Approximately how many atoms would it take to make the distance between the first and last atoms one inch?

7.9   Roughly speaking, the radius of an atom is about 10,000 times greater than that of its nucleus. If an atom were magnified so that the radius of its nucleus became 10 cm, what would be the radius of the atom in miles? (1 mi = 1609 m)

7.10 The radius of a uranium-235 nucleus is about 7.0×10^{-3} pm. Calculate the density of the nucleus in g/cm^3. (Assume the atomic mass to be 235 amu.)

7.11 Describe the experimental basis for believing that the nucleus occupies a very small fraction of the volume of the atom.

7.12 Describe the observations that would be made in an α-particle scattering experiment (Rutherford, Geiger, Marsden) if (a) the nucleus of an atom were negatively charged and the protons occupied the empty space outside of the nucleus; (b) the electrons were embedded in a positively charged sphere (Thomson's model); (c) a beam of electrons instead of α particles was shot at the target of gold foil.

Atomic Number, Mass Number, and Isotopes

7.13 Define the following terms: (a) atomic number, (b) mass number. Why does a knowledge of atomic number enable you to deduce the number of electrons present in an atom?

7.14 Explain the meaning of each term in the symbol $^A_Z X$.

7.15 Why do all atoms of a chemical element have the same atomic number, although they may have different mass numbers? What do we call atoms of the same elements with different mass numbers?

7.16 For each of the following isotopes determine the number of protons and the number of neutrons in the nucleus:

$$^{14}_{7}N \qquad ^{15}_{7}N \qquad ^{35}_{17}Cl \qquad ^{37}_{17}Cl \qquad ^{51}_{23}V \qquad ^{72}_{32}Ge \qquad ^{231}_{91}Pa$$

7.17 Indicate the number of protons, neutrons, and electrons in each of the following isotopes:

$$^{6}_{3}Li \quad ^{27}_{13}Al \quad ^{31}_{15}P \quad ^{64}_{30}Zn \quad ^{79}_{35}Br \quad ^{127}_{53}I \quad ^{196}_{80}Hg \quad ^{235}_{92}U$$

7.18 For the noble gas atoms

$$^{4}_{2}He \qquad ^{20}_{10}Ne \qquad ^{40}_{18}Ar \qquad ^{84}_{36}Kr \qquad ^{132}_{54}Xe$$

(a) determine the number of protons and neutrons in the nucleus of each atom, and (b) determine the ratio of neutrons to protons in the nucleus of each atom. Describe any general trend you discover in the way this ratio changes with increasing atomic number.

7.19 Write the appropriate symbol for each of the following isotopes: (a) $Z = 11$, $A = 23$, (b) $Z = 28$, $A = 64$, (c) $Z = 74$, $A = 186$, (d) $Z = 80$, $A = 201$.

7.20

Atom or Ion of Element	A	B	C	D	E	F	G
Number of electrons	5	7	9	12	7	6	9
Number of protons	5	7	10	10	7	5	9
Number of neutrons	5	7	10	10	8	6	10

The table gives numbers of electrons, protons, and neutrons in atoms or ions of a number of elements. Answer the following: (a) Which of the species are neutral? (b) Which are negatively charged? (c) Give the conventional symbols for B, D, and F.

7.21 One isotope of a metallic element has mass number 65 and 35 neutrons in the nucleus. The cation derived from the isotope has 28 electrons. Write the symbol for this cation.

7.22 In which one of the following pairs do the two species resemble each other most closely in chemical properties? (a) $^{129}_{53}I$ and $^{131}_{53}I^-$, (b) $^{35}_{17}Cl$ and $^{37}_{17}Cl$, (c) $^{7}_{3}Li$ and $^{8}_{3}Li^+$.

Nuclear Stability and Nuclear Binding Energy

*7.23 Given that

$$H(g) + H(g) \longrightarrow H_2(g) \qquad \Delta H° = -436.4 \text{ kJ}$$

Calculate the change in mass (in kg) per mole of H_2 formed.

7.24 Why is it impossible for the isotope ^2_2He to exist?

7.25 Estimates show that the total energy output of the sun is 5×10^{26} J/s. What is the corresponding mass loss in kg/s of the sun?

*7.26 Calculate the nuclear binding energy (in J) and the binding energy per nucleon of the following isotopes: (a) $^{7}_{3}Li$ (7.01600 amu), (b) $^{35}_{17}Cl$ (34.95952 amu), (c) $^{4}_{2}He$ (4.0026 amu), (d) $^{209}_{83}Bi$ (208.9804 amu).

*7.27 Calculate the nuclear binding energies, in J/nucleon, for the following species: (a) ^{10}B (10.0129 amu), (b) ^{11}B (11.00931 amu), (c) ^{14}N (14.00307 amu), (d) ^{56}Fe (55.9349 amu), (e) ^{184}W (183.9510 amu).

7.28 The atomic masses of $^{35}_{17}Cl$ (75.53 percent) and $^{37}_{17}Cl$ (24.47 percent) are 34.968 amu and 36.956 amu, respectively. Calculate the atomic mass of natural chlorine (that is, the average atomic mass of chlorine). The percentages in parentheses denote the relative abundances.

8

ATOMS II: THE QUANTUM MECHANICAL DESCRIPTION

F rom the perspective of modern chemistry, the chemical properties of elements (and, by extension, the chemistry of all substances) are determined largely by the numbers and arrangements of electrons in the atoms involved. Although the submicroscopic electrons cannot be observed directly in the same way that we observe macroscopic entities such as marbles or sand grains, the properties and characteristics of electrons in atoms can be inferred from numerous experiments, some of which will be summarized in this chapter.

The central notion we want to emphasize is that "thinking chemistry" in these late years of the twentieth century often reduces down to consideration of such fundamental questions as "Where are the electrons?" and "What are they doing?" The refinements that have been made in our understanding of matter since the days of John Dalton have all been based on the interplay between what we observe (the macroscopic, large-scale phenomena involving matter) and what we choose to infer from these observations—namely, the theories, models, and hypotheses that energize advancements in chemistry. In this chapter you will find part of the record of a major revolution in science, a revolution in ideas called the *quantum theory*. Its impact on our modern views of atoms and the structure of matter is a major theme of the story that follows.

8.1 TRANSITION FROM CLASSICAL PHYSICS TO QUANTUM THEORY

Attempts by nineteenth-century physicists to understand atoms and molecules by using a "tennis ball" model met with only limited success. By assuming that molecules behave like little rebounding balls, those early physicists were able to predict and explain some macroscopic phenomena, such as the pressure of a gas. However, when the same model was applied to explain the stability of molecules, for instance, the results were unsatisfactory. It took a long time to realize—and an even longer time to accept—that the properties of atoms and molecules are *not* governed by the same laws that work so well for larger objects.

It all started with a young German physicist in 1900. Max Planck (1858–1947) took the first bold departure from the established ideas in physics by introducing the *quantum theory*. Physics was never the same again. This theory accounted for the energy emitted by solids heated to various temperatures. According to Planck, when a solid emits energy it does so in a very specific manner—energy is emitted by atoms and molecules only in certain discrete amounts. It does not vary continuously, as had always been assumed in physics. This idea was completely alien to the thinking of Planck's contemporaries. The scientific community greeted it with skepticism, and in fact the theory was so revolutionary that Planck himself was not totally convinced of its validity. He tried for years to find alternative ways to explain the nature of radiation by solids.

In the development of science, a single major experimental discovery or the formulation of one important theory often sets off an avalanche of activity. Thus, in the thirty years that followed Planck's introduction of the quantum theory, a flurry of investigations transformed physics and altered our concept of nature.

To understand Planck's quantum theory, we must first know something about the nature of *radiation*, which is the emission and transmission through space of energy in the form of waves. We will thus start with a discussion of the properties of waves.

> Physics before the advent of the quantum theory is generally referred to as classical physics.

Properties of Waves

A ***wave*** can be thought of as *a vibrating disturbance by which energy is transmitted.* The speed of a wave depends on the type of wave and the nature of the medium through which the wave is traveling. The fundamental properties of a wave can be illustrated by considering a familiar example—water waves. Figure 8.1 shows a seagull floating on the ocean. Water waves are generated by pressure differences in various regions in the surface of water. If we carefully observe the motion of a water wave as it affects the motion of the seagull, we find it is periodic in character, that is, the wave form repeats itself at regular intervals.

The distance between identical points on successive waves is called the ***wavelength*** (λ). *The number of times per second the seagull moves through a complete cycle of upward and downward motion* indicates the ***frequency***

FIGURE 8.1 *Properties of water waves. The distance between corresponding points on successive waves is called the wavelength, and the number of times the seagull rises up and down per unit of time is called the frequency. It is assumed that the seagull does not move horizontally.*

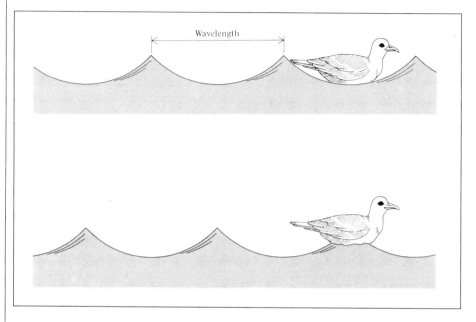

(ν) of the wave. The **amplitude** is *the vertical distance from the midline of a wave to the peak or trough* [Figure 8.2(a)]. Figure 8.2(b) shows two waves that have the same amplitude but different wavelengths and frequencies.

An important property of a wave traveling through space is its speed (c). The speed of a wave depends on the medium through which it is traveling.

FIGURE 8.2 *(a) The wavelength and amplitude of an ordinary wave. (b) Two waves having different wavelengths and frequencies. The wavelength of the top wave is three times that of the lower wave, but its frequency is only one-third that of the lower wave. Both waves have the same amplitude.*

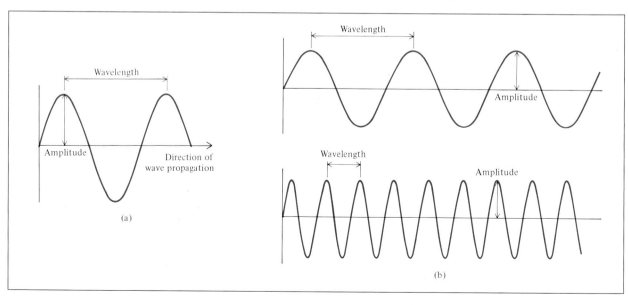

It also depends on the number of cycles of the wave passing through a given point per second (that is, on the frequency) and on the wavelength. In fact, the speed of a wave is given by the product of wavelength and frequency:

$$c = \lambda \nu \qquad (8.1)$$

The inherent "sensibility" of Equation (8.1) can be seen by analyzing the physical dimensions involved in the three terms. The wavelength (λ) expresses the length of a wave, or distance/wave. The frequency (ν) indicates the number of these waves that pass any reference point per unit of time, or waves/time. Thus the product of these terms results in dimensions of distance/time, which is speed:

$$\frac{\text{distance}}{\text{wave}} \times \frac{\text{waves}}{\text{time}} = \frac{\text{distance}}{\text{time}}$$
$$\lambda \qquad \times \quad \nu \quad = \quad c$$

Wavelength is usually expressed in units of centimeters or nanometers, and frequency is measured in units of hertz (Hz), where

$$1 \text{ Hz} = 1 \text{ cycle/s}$$

The word "cycle" may be left out and the frequency expressed as, for example, 25/s.

EXAMPLE 8.1

Calculate the speed of a wave whose wavelength and frequency are 17.4 cm and 87.4 Hz, respectively.

Answer

From Equation (8.1)

$$c = 17.4 \text{ cm} \times 87.4 \text{ Hz}$$
$$= 17.4 \text{ cm} \times 87.4/\text{s}$$
$$= 1.52 \times 10^3 \text{ cm/s}$$

Electromagnetic Radiation

Radiation, as we said earlier, is the emission and transmission of energy through space in the form of waves. There are many kinds of waves, such as water waves, sound waves, and light waves—an example of radiant energy. In 1873 James Maxwell showed theoretically that visible light consists of **electromagnetic waves.** According to Maxwell's theory, an electromagnetic wave *has an electric field component and a magnetic field component.* These two components have the same wavelength and frequency, and hence the same speed, but they travel in mutually perpendicular planes (Figure 8.3). The significance of Maxwell's theory is that it provides a mathematical description of the general behavior of light. In particular, his model accurately describes how energy in the form of radiation can be

Sound waves and water waves are not electromagnetic waves.

FIGURE 8.3 *The electric field and magnetic field components of an electromagnetic wave. These two components have the same wavelength, frequency, and amplitude, but they vibrate in two mutually perpendicular planes.*

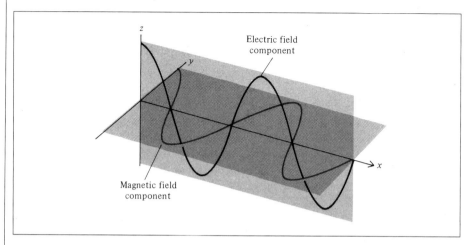

A more accurate value of the speed of light is given in the back inside cover of the book.

The wavelength of electromagnetic waves is given in nanometers (nm).

propagated through space in the form of vibrating electric and magnetic fields. We now know that light is but one form of *electromagnetic radiation,* which is the emission of energy in the form of electromagnetic waves.

A feature common to all electromagnetic waves is the speed with which they travel: 3.00×10^8 meters per second, or 186,000 miles per second, which is the speed of light in a vacuum. Although the speed differs from one medium to another, the variations are small enough to allow us to use 3.00×10^8 m/s as the speed of light in our calculations.

EXAMPLE 8.2

The wavelength of the green light from a traffic signal is centered at 522 nm. What is the frequency of this radiation?

Answer

Rearranging Equation (8.1) we get

$$\nu = \frac{c}{\lambda}$$

Because we are dealing with electromagnetic waves, c is 3.00×10^8 m/s. Recalling that 1 nm = 1×10^{-9} m (see Table 1.3), we write

$$\nu = \frac{3.00 \times 10^8 \text{ m/s}}{522 \text{ nm} \times \dfrac{1 \times 10^{-9} \text{ m}}{1 \text{ nm}}}$$

$$= 5.75 \times 10^{14}/\text{s}$$

Similar example: Problem 8.1.

Figure 8.4 shows various types of electromagnetic radiation, which differ from one another in wavelength and frequency. The long radio waves are emitted by large antennas, such as those used by broadcasting stations. The

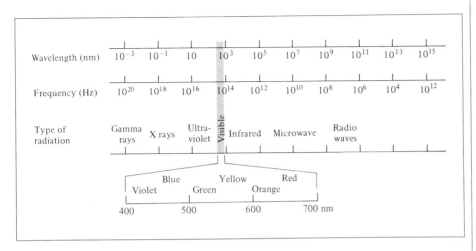

FIGURE 8.4   *Types of electromagnetic radiation. Gamma rays have the shortest wavelength and highest frequency; radio waves have the longest wavelength and lowest frequency. Each type of radiation is spread over a specific range of wavelengths (and frequencies). The visible region ranges from 400 nm (violet) to 700 nm (red).*

shorter, visible light waves are produced by the motions of electrons within atoms and molecules. The shortest waves, which also have the highest frequency, are those associated with γ (gamma) rays (see Chapter 7), which result from changes within the nucleus of the atom. As we will see shortly, the higher the frequency, the more energetic the radiation. Thus, ultraviolet radiation, X rays, and γ rays are high-energy radiation.

Planck's Quantum Theory

When solids are heated, they emit radiation over a wide range of wavelengths. The dull red glow of an electric heater and the bright white light of a tungsten light bulb are examples of radiation from solids heated to different temperatures.

Measurements taken in the latter part of the nineteenth century showed that the amount of radiation energy emitted depends on its wavelength. Attempts to account for this dependence in terms of established wave theory and thermodynamic laws were only partially successful. One theory explained short-wavelength dependence but failed to account for the longer wavelengths. Another theory accounted for the longer wavelengths but failed for short wavelengths. It seemed that something fundamental was missing from the laws of classical physics.

In 1900, Planck solved the problem with an assumption that departed drastically from accepted concepts. Classical physics had assumed that atoms and molecules could emit (or absorb) any arbitrary amount of radiant energy. Planck said that atoms and molecules could emit (or absorb) energy only in discrete quantities, like small packages or bundles. Planck gave the name **quantum** to *the smallest quantity of energy that can be emitted (or absorbed) in the form of electromagnetic radiation.* The energy E of an emitted single quantum of energy is proportional to the frequency of the radiation:

$$E \propto \nu$$

The proportionality constant for this relationship, symbolized as h, is now called *Planck's constant*:

$$E = h\nu \tag{8.2}$$

where h has the value of 6.63×10^{-34} J s.

According to Planck's quantum theory, energy is always emitted in multiples of $h\nu$; for example, $h\nu$, $2h\nu$, $3h\nu$, . . . but never, for example, $1.67\ h\nu$ or $4.98\ h\nu$. At the time Planck presented his theory, he could not explain why energies should be fixed or quantized in this manner. Starting with this hypothesis, however, he had no trouble correlating the experimental data for emission by solids over the *entire* range of wavelengths; they all supported the quantum theory.

The idea that energy should be quantized or "bundled" in this manner may seem strange at first, but the concept of quantization has many analogies. For example, an electric charge is also quantized; there can be only whole number multiples of e, the charge of one electron. Matter itself is quantized, for the numbers of electrons, protons, and neutrons and the numbers of atoms in a sample of matter must also be integers. Our money system is based on a "quantum" of value called a penny. Even processes in living systems involve quantized phenomena. The eggs laid by hens are quantized, and pregnant cats give birth to an integral number of kittens, not to one-half or three-quarters of a kitten.

8.2 THE PHOTOELECTRIC EFFECT

In 1905, only five years after Planck presented the quantum theory, Albert Einstein (1879–1955) used the theory to solve another mystery in physics, the *photoelectric effect*. Experiments had already demonstrated that electrons are ejected from the surface of certain metals exposed to light (Figure 8.5). According to the wave theory of light, both the number of electrons ejected and their energies should depend on the intensity, or brightness, of the light shining on the metal. The more intense the light, the theory suggests, the larger the number of ejected electrons and the higher their velocities. However, in practice it was found that while the *number* of electrons ejected does depend on the intensity of the incident light, the *energies* of the electrons do not. The energy depends only on the frequency of the light.

Einstein was able to explain the photoelectric effect by making an extraordinary assumption. He suggested that we should not think of a beam of light as wavelike but rather as consisting of *a stream of particles* which he called **photons.** Using Planck's quantum theory of radiation as a starting point, Einstein deduced that each photon must possess energy E, given by the equation

$$E = h\nu$$

where ν is the frequency of light. Electrons are held in a metal by attraction, and so to remove them from the metal we must employ light of a sufficiently high frequency (which corresponds to high energy). The process of

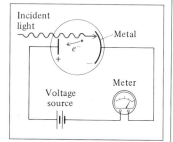

FIGURE 8.5 *An apparatus for studying the photoelectric effect. Light of a certain frequency falls on a clean metal surface. Ejected electrons are attracted toward the positive electrode. The flow of electrons is indicated by a detecting meter.*

This equation has the same form as Equation (8.2) because, as we will see shortly, radiation is emitted as well as absorbed in the form of photons.

shining a beam of light onto a metal surface can be viewed as shooting a beam of particles—the photons—at the metal atoms. If a photon strikes an electron with sufficient energy, the electron may be knocked out of the atom. The more energetic the photon (that is, the higher its frequency), the more energy it transfers to the ejected electron.

Now consider two beams of light having the same frequency but different intensities. The more intense beam of light consists of a larger number of photons; consequently, the number of electrons ejected from the metal's surface is greater than the number of electrons produced by the weaker beam of light. Thus the more intense the light, the greater the number of electrons emitted by the target metal; the higher the frequency of the light, the greater the energy of the emitted electrons.

EXAMPLE 8.3

The maximum wavelength of light from a certain light source is 588 nm. Calculate the energy (in joules) of a photon with this wavelength.

Answer

We use Equation (8.2):

$$E = h\nu$$

From Equation (8.1) $\nu = c/\lambda$, therefore

$$E = \frac{hc}{\lambda}$$

$$= \frac{(6.63 \times 10^{-34}\ \text{J s})(3.00 \times 10^{8}\ \text{m/s})}{(588\ \text{nm})\left(\dfrac{1 \times 10^{-9}\ \text{m}}{1\ \text{nm}}\right)}$$

$$= 3.38 \times 10^{-19}\ \text{J}$$

This is the energy possessed by a single photon of wavelength 588 nm.

Similar example: Problem 8.6.

Einstein's theory of light has posed a dilemma for scientists. On the one hand, it explains the photoelectric effect satisfactorily. On the other hand, the particle theory of light is not consistent with the known wave behavior of light. The only way to resolve the dilemma is to accept the idea that light possesses both particle- and wavelike properties. Depending on the experiment, we find that light behaves either as a wave or as a stream of particles. This concept was totally alien to the way physicists thought about matter and radiation, and it took a long time for them to become convinced of its validity. It turns out that the property of dual nature (particles and waves) is not unique to light but is also characteristic of submicroscopic particles like electrons, as we will see in Section 8.4.

8.3 BOHR'S THEORY OF THE HYDROGEN ATOM

Emission Spectra

Einstein's work paved the way for the solution of yet another nineteenth-century "mystery" in physics: the emission spectra of atoms. Ever since the seventeenth century, when Newton first passed a beam of sunlight through a glass prism to show that sunlight is composed of various color components, chemists and physicists have studied the characteristics of **emission spectra** of various substances, that is, *continuous or line spectra of radiation emitted by the substances*. The emission spectrum of a substance is obtained by energizing a sample of the material either with thermal energy or with some other energy form (such as a high-voltage electrical discharge if the substance is gaseous). A "red hot" or "white hot" iron bar freshly removed from a high-temperature source glows in a characteristic way. This visible glow is the portion of its emission spectrum that is sensed by eye. The warmth felt at a distance from the same iron bar is another portion of its emission spectrum—this portion in the infrared region. A feature common to the emission spectra of the sun and of a heated solid is that both are continuous, that is, all wavelengths of light are represented in the spectra (see Color Plate 17).

The emission spectra of atoms in the gas phase, on the other hand, do not show a continuous spread of wavelengths from red to violet; rather, the atoms *emit light only at specific wavelengths*. Such spectra are called **line spectra** because the radiation is identified by the appearance of bright lines in the spectra. Figure 8.6 is a schematic diagram of a discharge tube that is used to study emission spectra. Color Plate 18 shows the color emitted by

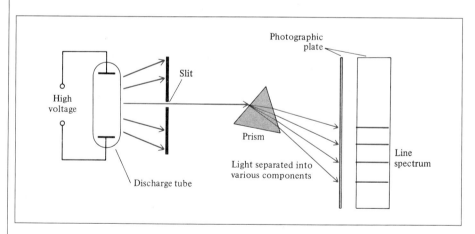

FIGURE 8.6 *An experimental arrangement for studying the emission spectra of atoms and molecules. The gas under study is in a discharge tube containing two electrodes. As electrons flow from the negative electrode to the positive electrode, they collide with the gas. This collision process eventually leads to the emission of light by the atoms (and molecules). The emitted light is separated into its components by a prism. Each component color is focused at a definite position, according to its wavelength, and forms a colored image of the slit on the photographic plate. The colored images are called spectral lines.*

hydrogen atoms in a discharge tube, and the portion of its line spectrum in the visible region is shown in Color Plate 17.

Every element has a unique emission spectrum. The characteristic lines in atomic spectra can be used in chemical analysis to identify unknown atoms, much as fingerprints are used to identify people. When the lines of the emission spectrum of a known element exactly match the lines of the emission spectrum of an unknown sample, the identity of the latter is quickly established. Although it was immediately recognized that this procedure is useful, the origin of these lines was unknown until early in this century.

Emission Spectrum of the Hydrogen Atom

In 1913, not too long after Planck's and Einstein's discoveries, Niels Bohr (1885–1962) offered a theoretical explanation of the emission spectrum of the hydrogen atom. Bohr's treatment is very complex, and it is not strictly correct. Thus, we will concentrate only on his important assumptions and final results, which do account for the positions of the spectral lines.

When Bohr first tackled this problem, physicists already knew that the atom contains electrons and protons (see Chapter 7). They thought of an atom as an entity with its electrons whirling around the nucleus in circular orbits at high velocities. This was an appealing model because it resembled the well-understood motions of the planets around the sun. In the hydrogen atom, it was believed that the electrostatic attraction between the "solar" proton and the "planetary" electron, which tends to pull the electron inward, is balanced exactly by the acceleration due to the circular motion of the electron.

Bohr's model for the atom included the idea of electrons moving in circular orbits, but he imposed a rather severe restriction: The single electron in the hydrogen atom could be located only in certain orbits. Since each orbit has a particular energy associated with it, Bohr's restriction meant that energies associated with electron motion in the permitted orbits are fixed in value, that is, they are quantized.

The emission of radiation by an energized hydrogen atom could then be explained in terms of the electron dropping from a higher-energy orbit to a lower one and giving up a quantum of energy (a photon) in the form of light (Figure 8.7). Using arguments based on electrostatic interaction and Newton's laws of motion, Bohr showed that the energies that the electron in the hydrogen atom can possess are given by

$$E_n = -R_H\left(\frac{1}{n^2}\right) \qquad (8.3)$$

where R_H, the *Rydberg constant*, has the value 2.18×10^{-18} J. The number n is an integer called the *principal quantum number*; it has the values $n = 1, 2, 3, \ldots$.

The negative sign in Equation (8.3) may seem strange, for it implies that all the allowable energies of the electron are negative. Actually, this sign is nothing more than an arbitrary convention; it says that the energy of the electron in the atom is *lower* than the energy of a *free electron*, or an

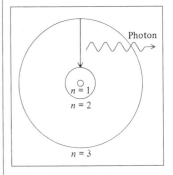

FIGURE 8.7 *The emission process in an excited hydrogen atom, according to Bohr's theory. An electron originally in a higher-energy orbit (n = 3) falls back to a lower-energy orbit (n = 2). As a result, a photon with energy hν is given off. The quantity hν is equal to the difference in energies of the two orbits occupied by the electron in the emission process. For simplicity, only three orbits are shown.*

electron that is infinitely far from the nucleus. The energy of a free electron is arbitrarily assigned a value of zero. Mathematically, this corresponds to setting n equal to infinity in Equation (8.3) so that $E_\infty = 0$. As the electron gets closer to the nucleus (as n decreases), E_n becomes larger in absolute value, yet also more negative. The most negative value, then, is reached when $n = 1$, which corresponds to the most stable orbit. We call this the **ground state**, or the **ground level**, which refers to *the lowest energy state of a system* (which is an atom in our discussion). The stability of the electron diminishes for $n = 2, 3, \ldots$, and each of these is called an **excited state**, or **excited level**, which is *higher in energy than the ground state*. An electron in a hydrogen atom that occupies an orbit with n greater than 1 is said to be in an excited state. The radius of each circular orbit depends on n^2. Thus as n increases from 1 to 2 to 3, the orbit radius increases in size very rapidly. The higher the excited state, the farther away the electron is from the nucleus.

Bohr's theory of the hydrogen atom enables us to explain the line spectra of that atom. Radiant energy absorbed by the atom causes the electron to move from a lower-energy orbit (characterized by a smaller n value) to a higher-energy orbit (characterized by a larger n value). Conversely, radiant energy is emitted when the electron moves from a higher-energy orbit to a lower-energy orbit. The quantized movement of the electron from one orbit to another has its analogy in the movement of a tennis ball either up or down a set of stairs. The ball can reside at a variety of different stair levels but never between stairs. Its journey from a lower stair to a higher one is an energy-requiring process, whereas that from a higher stair to a lower stair is an energy-releasing process. And like the electron in a Bohr atom, the quantity of energy involved in either type of change is determined by the difference in the levels of the initial and final states (or stairs).

Let us now apply Equation (8.3) to the emission process in a hydrogen atom. Suppose that the electron is initially in an excited state characterized by the principal quantum number n_i. During emission, the electron drops to a lower energy state characterized by the principal quantum number n_f (the subscripts i and f denote the initial and final states, respectively). This lower energy state may be either another excited state or the ground state. The difference between the energies of the initial and final states is ΔE (delta E), where

$$\Delta E = E_f - E_i$$

From Equation (8.3)

$$E_f = -R_H\left(\frac{1}{n_f^2}\right)$$

and

$$E_i = -R_H\left(\frac{1}{n_i^2}\right)$$

Therefore

$$\Delta E = \left(\frac{-R_H}{n_f^2}\right) - \left(\frac{-R_H}{n_i^2}\right)$$

$$= R_H\left(\frac{1}{n_i^2} - \frac{1}{n_f^2}\right)$$

Because this transition results in the emission of a photon of frequency v and energy hv (Figure 8.7), we can write

$$\Delta E = hv = R_H\left(\frac{1}{n_i^2} - \frac{1}{n_f^2}\right) \qquad (8.4)$$

When a photon is emitted, $n_i > n_f$. Consequently the term in parentheses is negative and ΔE is negative (energy is lost to the surroundings). When energy is absorbed, $n_i < n_f$ and the term in parentheses is positive, so ΔE is positive. Each spectral line in the emission spectrum corresponds to a particular transition in a hydrogen atom. When we study a a large number of hydrogen atoms, we observe all possible transitions and hence the corresponding spectral lines. The brightness of a spectral line depends on how many photons of the same wavelength are emitted.

The emission spectrum of hydrogen covers a wide range of wavelengths from the infrared to the ultraviolet. Table 8.1 shows the series of transitions in the hydrogen spectrum; they are named after their discoverers. The Balmer series was particularly easy to study because a number of its lines fall in the visible range.

TABLE 8.1 The Various Series in Atomic Hydrogen Emission Spectrum

Series	n_f	n_i	Spectrum Region
Lyman	1	2, 3, 4, . . .	Ultraviolet
Balmer	2	3, 4, 5, . . .	Visible and ultraviolet
Paschen	3	4, 5, 6, . . .	Infrared
Brackett	4	5, 6, 7, . . .	Infrared

Figure 8.7 shows a single transition. However, it is more informative to express transitions as shown in Figure 8.8. Each horizontal line is called an *energy level*. The position of the energy level, as measured on the energy scale, shows the energy associated with the particular orbit. The orbits are labeled with their principal quantum numbers.

EXAMPLE 8.4

What are the frequency and wavelength of a photon emitted during a transition from the $n_i = 5$ state to the $n_f = 2$ state in the hydrogen atom?

Answer

Since $n_f = 2$, this transition gives rise to a spectral line in the Balmer series (see Table 8.1). From Equation (8.4) we write

$$\Delta E = R_H\left(\frac{1}{n_i^2} - \frac{1}{n_f^2}\right)$$

$$= 2.18 \times 10^{-18} \text{ J}\left(\frac{1}{5^2} - \frac{1}{2^2}\right)$$

$$= -4.58 \times 10^{-19} \text{ J}$$

(Continued)

The negative sign is in accord with our convention that energy is given off to the surroundings.

The negative sign indicates that this is energy associated with an emission process. Since

$$\Delta E = h\nu$$

the frequency of the photon is given by

$$\nu = \frac{\Delta E}{h}$$
$$= \frac{4.58 \times 10^{-19} \text{ J}}{6.63 \times 10^{-34} \text{ J s}}$$
$$= 6.91 \times 10^{14}/\text{s}$$

Note that in calculating the frequency we have omitted the minus sign for ΔE because the energy of the photon must be positive. The wavelength of the photon is

$$\lambda = \frac{c}{\nu}$$
$$= \frac{3.00 \times 10^8 \text{ m/s}}{6.91 \times 10^{14}/\text{s}} = 4.34 \times 10^{-7} \text{ m}$$
$$= 434 \text{ nm}$$

Similar examples: Problems 8.25, 8.27.

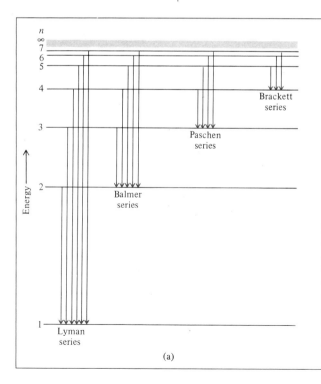

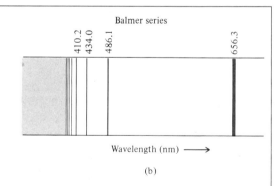

FIGURE 8.8 *(a) The energy levels in the hydrogen atom and the various emission series. Each energy level corresponds to the energy associated with the motion of an electron in an orbit, as postulated by Bohr and shown in Figure 8.7. The emission lines are labeled acccording to the scheme in Table 8.1. (b) Spectral lines in the Balmer series. Each line can be explained as a transition from a higher-energy level to the n = 2 energy level, as shown in (a). In theory, n increases from 1 to infinity (∞).*

8.4   THE DUAL NATURE OF THE ELECTRON

Physicists were mystified but intrigued by Bohr's theory. The question they asked about it was: Why are the energies of the hydrogen electron quantized? Or, phrasing the question in a more concrete way, Why is the electron in a Bohr atom restricted to orbiting the nucleus at certain fixed distances? For a decade no one, not even Bohr himself, had a logical explanation. In 1924 Louis de Broglie (1892–1977) provided a solution to this puzzle. De Broglie reasoned as follows: If light waves can behave like a stream of particles (photons), then perhaps particles such as electrons can possess wave properties. The exact nature of these "electron waves" was unclear, but de Broglie presented convincing mathematical support for his idea.

We said earlier that the radius of a Bohr model orbit depends on n^2. Thus the distances of successive electron orbits from the nucleus are in the ratios of the squares of whole numbers, that is, $n_1^2, n_2^2, n_3^2, \ldots$ or $1^2, 2^2, 3^2, \ldots$. These numbers indicate the relative sizes or circumferences of the circular orbits. De Broglie argued that if an electron does behave like a wave, the length of the wave must fit the circumference of the orbit exactly (Figure 8.9). Otherwise the wave would partially cancel itself on each successive orbit; eventually the amplitude of the wave would be reduced to zero, and the wave would not exist.

The relation between the circumference of an allowed orbit $(2\pi r)$ and the wavelength (λ) of the electron is given by

$$2\pi r = n\lambda \tag{8.5}$$

where r is the radius of the orbit, λ the wavelength of the electron wave, and $n = 1, 2, 3, \ldots$. Because n is an integer, it follows that r can have only certain values as n increases from 1 to 2 to 3 and so on. And since the energy of the electron depends on the size of the orbit (or the value of r), we see that its value must be quantized.

De Broglie's reasoning led to the conclusion that waves can behave like particles and particles can exhibit wave properties. De Broglie deduced that the particle and wave properties are related by the expression

$$\lambda = \frac{h}{mu} \tag{8.6}$$

Note that the left side of Equation (8.6) involves the wavelike property of wavelength, whereas the right side makes reference to mass, a distinctly particlelike property.

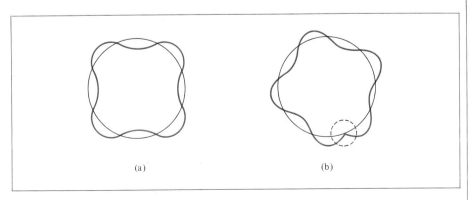

(a)   (b)

FIGURE 8.9   *(a) The circumference of the orbit is equal to an integral number of wavelengths. This is an allowed orbit. (b) The circumference of the orbit is not equal to an integral number of wavelengths. As a result, the electron wave does not close in on itself. This is a nonallowed orbit.*

where λ, m, and u are the wavelength associated with a moving particle, its mass, and its velocity, respectively. Equation (8.6) implies that a particle in motion can be treated as a wave, and a wave can exhibit the properties of a particle.

EXAMPLE 8.5

Calculate the wavelength of the "particle" in the following two cases: (a) The fastest serve in tennis is about 140 miles per hour, or 62 m/s. Calculate the wavelength associated with a 6.0×10^{-2} kg tennis ball traveling at this speed. (b) Calculate the wavelength associated with an electron moving at 62 m/s.

Answer

(a) Using Equation (8.6) we write

$$\lambda = \frac{h}{mu}$$

$$= \frac{6.63 \times 10^{-34} \text{ J s}}{6.0 \times 10^{-2} \text{ kg} \times 62 \text{ m/s}}$$

The conversion factor is 1 J = 1 kg m^2/s^2 (see Section 1.6). Therefore

$$\lambda = 1.8 \times 10^{-34} \text{ m}$$

This is an exceedingly small wavelength, since the size of an atom itself is on the order of 1×10^{-10} m. For this reason, the wave properties of such a tennis ball are completely undetectable by any existing measuring device.

(b) In this case

$$\lambda = \frac{h}{mu}$$

$$= \frac{6.63 \times 10^{-34} \text{ J s}}{9.1095 \times 10^{-31} \text{ kg} \times 62 \text{ m/s}}$$

where 9.1095×10^{-31} kg is the mass of an electron. Proceeding as in (a) we obtain

$$\lambda = 1.2 \times 10^{-5} \text{ m}$$
$$= 1.2 \times 10^{4} \text{ nm}$$

A wavelength of 1.2×10^{4} nm falls in the infrared region.

Similar examples: Problems 8.31, 8.32, 8.33.

Example 8.5 shows that although de Broglie's equation can be applied to diverse systems, the wave properties become observable only for submicroscopic objects. This distinction is brought about by the smallness of Planck's constant, h, which appears in the numerator in Equation (8.6).

Shortly after de Broglie advanced his equation, Clinton Davisson (1881–1958) and Lester Germer (1896–1972) in the United States and G. P. Thomson (1892–1975) in England demonstrated that electrons do indeed possess

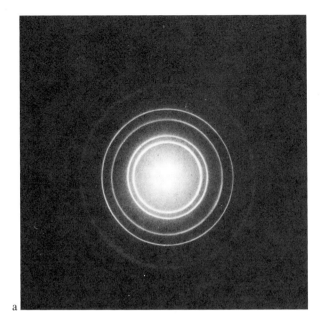
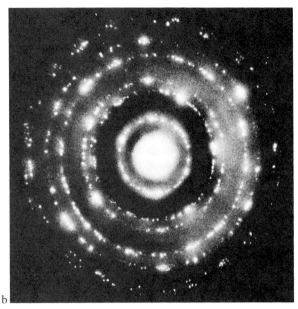

FIGURE 8.10   *Patterns made by a beam of electrons passing through (a) aluminum and (b) graphite. The patterns are similar to those obtained with X rays, which are known to be waves. The similarity provides strong support for the idea that electrons have wave properties.*

wavelike properties. By passing a beam of electrons through a thin piece of gold foil, Thomson obtained a set of concentric rings on a screen, similar to the pattern observed when X rays (which are waves) were used. Figure 8.10 shows such patterns for aluminum and graphite.

8.5   QUANTUM MECHANICS

The spectacular initial success of Bohr's theory was followed by a series of disappointments. For example, Bohr's approach could not account for the emission spectra of atoms containing more than one electron, such as atoms of helium and lithium. The theory also could not explain the appearance of the extra lines in the hydrogen emission spectrum that are observed when a magnetic field is applied. Another problem arose with the discovery that electrons are wavelike: How can the "position" of a wave be specified? We can speak of the amplitude at a certain point in the wave (see Figure 8.2), but we cannot define its precise location because a wave extends in space. Thus, although Bohr's suggestion that the energy of the electron in a hydrogen atom is quantized remained unchallenged, the assumption that the electron orbits the nucleus in well-defined paths was in error. If it did, we could predict its precise position relative to the nucleus at any moment, contrary to our knowledge of its wavelike behavior.

When scientists realized that the usefulness of Bohr's theory was limited to the hydrogen atom only, it became necessary to find a fundamental

In reality, Bohr's theory was able to account for the observed emission spectra of He^+ and Li^{2+} ions, as well as that of hydrogen. However, all three systems have one common feature—each contains a single electron. Thus the Bohr model worked successfully only for the hydrogen atom and for "hydrogenlike ions."

equation describing the behavior and energies of submicroscopic particles in general, an equation analogous to Newton's laws of motion for macroscopic objects. In 1926 Erwin Schrödinger (1887–1961) formulated the desired equation. The Schrödinger equation is too complex to discuss the details here. It is important for us simply to note that the equation incorporates both particle properties, in terms of mass m, and wave properties, in terms of a wave function ψ (psi) of the system (such as an electron in a confined region of space).

The wave function itself has no direct real physical meaning. However, the square of the wave function, ψ^2, is related to the probability of finding the electron in a certain region in space. We can think of ψ^2 as the probability per unit volume such that the product of ψ^2 and a small volume (called a *volume element*) yields the probability of finding the electron within that volume. (The reason for specifying a small volume is that ψ^2 varies from one region of space to another, but its value can be assumed constant within a small volume.) The total probability of locating the electron in a given volume (for example, around the nucleus of an atom) is then given by the sum of all the products of ψ^2 and the corresponding volume elements.

The idea of relating ψ^2 to the notion of probability stemmed from a wave theory analogy. According to wave theory, the intensity of light is proportional to the square of the amplitude of the wave, or ψ^2. The most likely place to find a photon is where the intensity is greatest, that is, where the value of ψ^2 is greatest. A similar argument thus followed for associating ψ^2 with the likelihood of finding an electron in regions surrounding the nucleus.

Schrödinger's equation began a new era in physics and chemistry, for it launched a new field, *quantum mechanics* (also called *wave mechanics*). We now refer to the developments in quantum theory from 1913—the time Bohr presented his analysis for the hydrogen atom—to 1926 as "old quantum theory."

8.6 APPLYING THE SCHRÖDINGER EQUATION TO THE HYDROGEN ATOM

We can now consider the quantum mechanical view of the hydrogen atom, which is the simplest atom, containing only one proton and one electron. Solving the Schrödinger equation for the hydrogen atom provides us with two types of information: It specifies the possible energy states the electron can occupy and identifies the corresponding wave functions (ψ) of the electron associated with each energy state. These energy states and wave functions are characterized by a set of quantum numbers (to be discussed shortly). Recall that the probability of finding an electron in a certain region is given by the square of the wave function, ψ^2. Thus once we know the values of ψ and the energies, we can calculate ψ^2 and construct a comprehensive view of the hydrogen atom.

This kind of information about the hydrogen atom is useful but not quite enough. The world of chemical substances and reactions involves systems considerably more complex than simple atomic species like hydrogen. How-

The wave function can be thought of as the variation of the amplitude of the wave with distance. The wave function tells us that in one cycle of the wave, the amplitude changes from its maximum value to zero and then reaches maximum value in the negative direction.

ever, it turns out that the Schrödinger equation cannot be solved exactly for any atom containing more than one electron! Thus even in the case of helium, which contains two electrons, the mathematics becomes too complex to handle. It would appear, therefore, that the Schrödinger equation suffers from the same limitation as Bohr's original theory—it can be applied in practice only to the hydrogen atom. However, the situation is not hopeless. Chemists and physicists have learned to get around this kind of difficulty by using methods of approximation. For example, although the behavior of electrons in helium and **many-electron atoms** (that is, *atoms containing three or more electrons*) is not the same as that in the hydrogen atom, we assume that the difference is probably not too great. On the basis of this assumption, we can use the energies and the wave functions obtained for the hydrogen atom as good approximations of the behavior of electrons in more complex atoms. In fact, we find that this approach provides fairly good descriptions of electron behavior in complex atoms.

Because the hydrogen atom serves as a starting point or model for all other atoms, we need a clear idea of the quantum mechanical description of this system. The solution of the Schrödinger equation shows that the energies an electron can possess in the hydrogen atom are given by the same expression as that obtained by Bohr [see Equation (8.3)]. Both Bohr's theory and quantum mechanics, therefore, show that the energy of an electron in the hydrogen atom is quantized. They differ, however, in their description of the electron's behavior with respect to the nucleus.

Since an electron has no well-defined position in the atom, we find it convenient to use terms like *electron density, electron charge cloud,* or simply *charge cloud* to represent the probability concept (these terms have essentially the same meaning). Basically, **electron density** *gives the probability that an electron will be found at a particular region in an atom.* Regions of high electron density represent a high probability of locating the electron while the opposite holds for regions of low electron density (Figure 8.11).

To distinguish the quantum mechanical description from Bohr's model, we replace "orbit" with the term **orbital** or **atomic orbital.** An orbital can be thought of as *the probability function (ψ^2) that defines the distribution of electron density in space around the atomic nucleus.* When we say that an electron (in a given allowed energy state) is in a certain orbital, we mean that the distribution of the electron density or the probability of locating the electron in space is described by the square of the wave function associated with that energy state. An atomic orbital therefore has a characteristic energy, as well as a characteristic distribution of electron density.

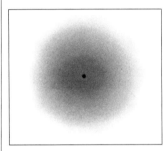

FIGURE 8.11 *A representation of the electron density distribution surrounding the nucleus in the hydrogen atom.*

The term *orbital* characterizes the quantum mechanical model of electron behavior, whereas the term *orbit* is used in connection with the Bohr model.

8.7 QUANTUM NUMBERS

Quantum mechanics tells us that three quantum numbers are required to describe the distribution of electrons in hydrogen and other atoms. These numbers are derived from the mathematical solution of the Schrödinger equation for the hydrogen atom. They are called the principal quantum number, the angular momentum quantum number, and the magnetic quantum number. These quantum numbers will be used to describe atomic orbitals and to label electrons that reside in them. A fourth quantum number that describes the behavior of a specific electron—the spin quantum number—completes the description of electrons in atoms.

The Principal Quantum Number (n)

Equation (8.3) holds only for the hydrogen atom.

The principal quantum number (n) can have integral values 1, 2, 3, and so forth; it corresponds to the quantum number in Equation (8.3). In a hydrogen atom, the value of n determines the energy of an orbital. As we will see shortly, this is not the case for a many-electron atom. The principal quantum number also relates to the average distance of the electron from the nucleus in a particular orbital. The larger the n, the greater the average distance of an electron in the orbital from the nucleus and therefore the larger the orbital.

The Angular Momentum Quantum Number (ℓ)

The shapes of orbitals are discussed in Section 8.8.

The angular momentum quantum number, ℓ, tells us the "shape" of the orbitals. The values of ℓ depend on the value of the principal quantum number, n. For a given value of n, ℓ has integral values from 0 to $(n - 1)$. If $n = 1$, there is only one value of ℓ, that is, $\ell = n - 1 = 1 - 1 = 0$. If $n = 2$, there are two values of ℓ, given by 0 and 1. If $n = 3$, there are three values of ℓ, given by 0, 1, and 2. The value of ℓ is generally designated by the letters s, p, d, . . . as follows:

ℓ	0	1	2	3	4	5
Name of orbital	s	p	d	f	g	h

Thus if $\ell = 0$ we have an s orbital, if $\ell = 1$ we have a p orbital, and so on.

The unusual sequence of letters (s, p, and d) has a historical origin. Physicists who studied atomic emission spectra tried to correlate the observed spectral lines with the particular energy states involved in the transitions. They noted that some of the lines were **s**harp; some were rather spread out, or **d**iffuse; and some were very strong and hence referred to as **p**rincipal lines. Subsequently, the initial letters of each adjective were assigned to those energy states. However, after the letter d and starting with the letter f, the orbital designations follow alphabetical order.

A collection of orbitals with the same value of n is frequently called a *shell*. One or more orbitals with the same n and ℓ values is referred to as a *subshell*. For example, the shell with $n = 2$ is composed of two subshells, $\ell = 0$ and 1 (the allowed values for $n = 2$). These subshells are called the 2s and 2p subshells.

The Magnetic Quantum Number (m_ℓ)

The magnetic quantum number, m_ℓ, describes the orientation of the orbital in space (to be discussed in Section 8.8). Within a subshell, the value of m_ℓ

depends on the value of the angular momentum quantum number, ℓ. For a certain value of ℓ, there are $(2\ell + 1)$ integral values of m_ℓ, as follows:

$$-\ell, (-\ell + 1), , , , 0, , , , (+\ell - 1), +\ell$$

If $\ell = 0$, then $m_\ell = 0$. If $\ell = 1$, then there are $[(2 \times 1) + 1]$ or three values of m_ℓ, namely, -1, 0, and 1. If $\ell = 2$, there are $[(2 \times 2) + 1]$, or five values of m_ℓ, namely, -2, -1, 0, 1, and 2. The number of m_ℓ values indicates the number of orbitals in a subshell with a particular ℓ value.

To summarize our discussion of these three quantum numbers, let us consider the situation in which we have $n = 2$ and $\ell = 1$. The values of n and ℓ indicate that we have a $2p$ subshell, and in this subshell we have *three* $2p$ orbitals (because there are three values of m_ℓ, given by -1, 0, and 1).

The Electron Spin Quantum Number (m_s)

Experiments on the emission spectra of hydrogen and sodium atoms indicated the need for a fourth quantum number to describe the electron in an atom. The experiments revealed that lines in the emission spectra could be split by the application of an external magnetic field. The only way physicists could explain these results was to assume that electrons act like tiny magnets. If electrons are thought to spin on their own axes, as Earth does, their magnetic properties can be accounted for. According to electromagnetic theory, a spinning charge generates a magnetic field, and it is this motion that causes the electron to behave like a magnet. Figure 8.12 shows the two possible spinning motions of an electron, one clockwise and the other counterclockwise. To take the electron spin into account, it is necessary to introduce a fourth quantum number, called the *electron spin quantum number* (m_s), which has the value of $+\frac{1}{2}$ or $-\frac{1}{2}$. These values correspond to the two possible spinning motions of the electron, shown in Figure 8.12.

Conclusive proof of electron spin was provided by Otto Stern (1888–1969) and Walther Gerlach (1889–   ) in 1924. Figure 8.13 shows the basic

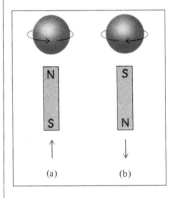

FIGURE 8.12 *The (a) counterclockwise and (b) clockwise spins of an electron. The magnetic fields generated by these two spinning motions are analogous to those from the two magnets. The upward and downward arrows are used to denote the two directions of spin.*

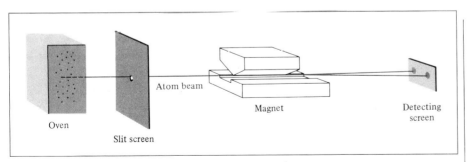

FIGURE 8.13 *Experimental arrangement for demonstrating the spinning motion of electrons. A beam of atoms is directed through a magnetic field. For example, when a hydrogen atom with a single electron passes through the field, it is deflected in one direction or the other, depending on the direction of the spin. In a stream consisting of many atoms, there will be equal distributions of the two kinds of spins, so that two spots of equal intensity are detected on the screen.*

experimental arrangement. A beam of atoms generated in a hot furnace is passed through a nonhomogeneous magnetic field. The interaction between an electron and the magnetic field causes the atom to be deflected from its straight-line path. Because the spinning motion is completely random, half of the atoms will be deflected one way and half the other, and two spots of equal intensity are observed on the detecting screen.

8.8 ATOMIC ORBITALS

Types of Atomic Orbitals

With the introduction of quantum numbers, we are now better able to discuss atomic orbitals in hydrogen, helium, and many-electron atoms more fully. Table 8.2 shows the relation between quantum numbers and atomic orbitals. We see that when $\ell = 0$, there is only one value of m_ℓ, and thus we have an s orbital. When $\ell = 1$, there are three values of m_ℓ and hence there are three p orbitals, labeled p_x, p_y, and p_z. When $\ell = 2$, there are five values of m_ℓ, and the corresponding five d orbitals are labeled with more elaborate subscripts. The following sections examine s, p, and d orbitals separately.

s Orbitals. In studying the properties of atomic orbitals, one of the important questions we ask is, "What are the shapes of the orbitals?" Strictly speaking, an orbital does not have a well-defined shape because the wave function characterizing the orbital extends from the nucleus to infinity. In that sense it is difficult to say what an orbital looks like. On the other hand, it is certainly convenient to think of orbitals in terms of specific shapes, particularly in discussing the formation of chemical bonds between atoms, as we will do in Chapters 10 and 11.

Although in principle an electron can be found anywhere, we know that it spends most of its time quite close to the nucleus. Figure 8.14(a) shows a plot of the electron density in a hydrogen atomic $1s$ orbital versus the distance from the nucleus. As you can see, the electron density falls off rapidly as the distance from the nucleus increases. Roughly speaking, there

An s subshell has one orbital, a p subshell three orbitals, and a d subshell five orbitals.

TABLE 8.2 Relation Between Quantum Numbers and Atomic Orbitals

n	ℓ	m_ℓ	Number of Orbitals	Designations of Atomic Orbitals
1	0	0	1	$1s$
2	0	0	1	$2s$
	1	$-1, 0, 1$	3	$2p_x$, $2p_y$, $2p_z$
3	0	0	1	$3s$
	1	$-1, 0, 1$	3	$3p_x$, $3p_y$, $3p_z$
	2	$-2, -1, 0, 1, 2$	5	$3d_{xy}$, $3d_{yz}$, $3d_{xz}$, $3d_{x^2-y^2}$, $3d_{z^2}$
$\vdots$	$\vdots$	$\vdots$	$\vdots$	$\vdots$

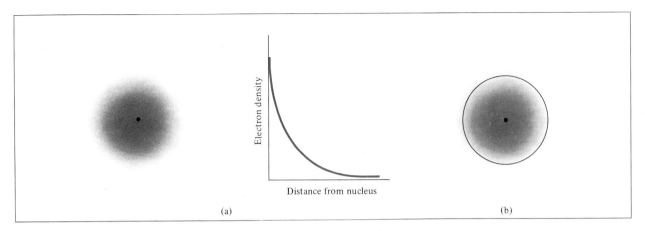

FIGURE 8.14 *(a) Plot of electron density in the hydrogen 1s orbital as a function of the distance from the nucleus. The electron density falls off rapidly as the distance from the nucleus increases. (b) Boundary surface diagram of the hydrogen 1s orbital.*

is about a 90 percent probability of finding the electron within a sphere of radius 100 pm surrounding the nucleus. Thus, we can represent the 1s orbital by drawing a ***boundary surface diagram*** which *encloses about 90 percent of the total electron density in an orbital,* as shown in Figure 8.14(b). A 1s orbital represented in this manner is merely a sphere.

Figure 8.15 shows boundary surface diagrams for the 1s, 2s, and 3s hydrogen atomic orbitals. All s orbitals are spherical in shape but differ in size, which increases as the principal quantum number increases. Although the details of electron density variation within each boundary surface are lost, there is no serious disadvantage. For us the most important features of atomic orbitals are their shapes and *relative* sizes, which are adequately represented by boundary surface diagrams.

***p* Orbitals.** It should be clear that the *p* orbitals start with the principal quantum number $n = 2$. If $n = 1$, then the angular momentum quantum number ℓ can assume only the value of zero; therefore, there is only a 1s orbital. As we saw earlier, when $\ell = 1$, the magnetic quantum number m_ℓ

Since the wave function for an orbital has no theoretical outer limit as one moves outward from the nucleus, this raises interesting philosophical questions regarding the sizes of atoms. Rather than pondering these issues, chemists have agreed to define an atom's size operationally, as we will see in later chapters.

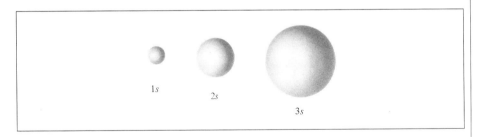

FIGURE 8.15 *Boundary surface diagrams of the hydrogen 1s, 2s, and 3s orbitals. Each sphere contains about 90 percent of the total electron density. All s orbitals are spherical. Roughly speaking, the size of an orbital is proportional to n^2, where n is the principal quantum number.*

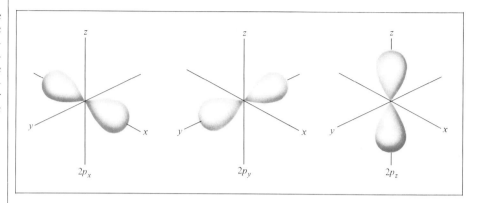

can have values of -1, 0, 1. Starting with $n = 2$ and $\ell = 1$, we therefore have three 2p orbitals: $2p_x$, $2p_y$, and $2p_z$ (Figure 8.16). The letter subscripts indicate the axes along which the orbitals are oriented. These three p orbitals are identical in size, shape, and energy; they differ from one another only in their orientations. Note, however, that there is no simple relation between the values of m_ℓ and the x, y, and z directions. For our purpose, you need only remember that because there are three possible values of m_ℓ, there are three p orbitals with different orientations.

The boundary surface diagrams of p orbitals in Figure 8.16 show that each p orbital can be thought of as two lobes; the nucleus is at the center of the p orbital. Like s orbitals, p orbitals increase in size from 2p to 3p to 4p orbital and so on.

d Orbitals and Other Higher-Energy Orbitals.   When $\ell = 2$, there are five values of m_ℓ, which correspond to five d orbitals. The lowest value of n for a d orbital is 3. Because ℓ can never be greater than $n - 1$, when $n = 3$ and $\ell = 2$, we have five 3d orbitals ($3d_{xy}$, $3d_{yz}$, $3d_{xz}$, $3d_{x^2-y^2}$, and $3d_{z^2}$), shown in Figure 8.17. As in the case of the p orbitals, the different orientations of the d orbitals correspond to the different values of m_ℓ, but again there is no direct correspondence between a given orientation and a particular m_ℓ value. All the 3d orbitals in an atom are identical in energy. The d orbitals for which n is greater than 3 (4d, 5d, . . .) have similar shapes.

As we saw in Section 8.7, beyond the d orbitals are f, g, . . . orbitals. The f orbitals are important in accounting for the behavior of elements with atomic numbers greater than 57, although their shapes are difficult to represent. We will have no need to concern ourselves with orbitals having ℓ values greater than 3 (the g orbitals and beyond).

EXAMPLE 8.6

List the values of n, ℓ, and m_ℓ for orbitals in the 4d subshell.

Answer

As we saw earlier, the number given in the designation of the subshell is the principal quantum number, so in this case $n = 4$. Since we are dealing with d

orbitals, $\ell = 2$. The values of m_ℓ can vary from $-\ell$ to ℓ. Therefore, m_ℓ can be $-2, -1, 0, 1, 2$ (which corresponds to the five d orbitals).

Similar example: Problem 8.37.

EXAMPLE 8.7

What is the total number of orbitals associated with the principal quantum number $n = 3$?

Answer

For $n = 3$, the possible values of ℓ are 0, 1, and 2. Thus there is one $3s$ orbital ($n = 3$, $\ell = 0$, and $m_\ell = 0$); there are three $3p$ orbitals ($n = 3$, $\ell = 1$, and $m_\ell = -1, 0, 1$); there are five $3d$ orbitals ($n = 3$, $\ell = 2$, and $m_\ell = -2, -1, 0, 1, 2$). Therefore the total number of orbitals is $1 + 3 + 5 = 9$.

Similar example: Problem 8.43.

The Energies of Orbitals

Despite the theoretical difficulties Bohr experienced in extending the usefulness of his model of the atom, his fundamental notion of electronic energy levels has been maintained and incorporated in the quantum mechanical concepts we are reviewing here. Thus along with the shapes and sizes of atomic orbitals, we are now ready to inquire into their relative energies and how these energy levels help determine the actual electronic arrangements found in atoms.

According to Equation (8.3), the energy of an electron in a hydrogen atom is determined solely by its principal quantum number. Thus the energies of hydrogen orbitals increase as follows (Figure 8.18):

$$1s < 2s = 2p < 3s = 3p = 3d < 4s = 4p = 4d = 4f < \cdots$$

Although the electron density distributions are different in the $2s$ and $2p$ orbitals, hydrogen's electron has the same energy whether it is in the $2s$ orbital or a $2p$ orbital. The $1s$ orbital in a hydrogen atom, as we have said, corresponds to the most stable condition and is called the ground state, and an electron residing in this orbital is most strongly held by the nucleus. An electron in the $2s$, $2p$, or higher orbitals in a hydrogen atom is in an excited state.

The energy picture is different for many-electron atoms. The energy of an electron in a many-electron atom, unlike that of the hydrogen atom, depends not only on its principal quantum number but also on its angular

FIGURE 8.17 *Boundary surface diagrams of the five 3d orbitals. Although the $3d_{z^2}$ orbital looks different, it is equivalent to the other four orbitals in all other respects. The d orbitals of higher principal quantum numbers have similar shapes.*

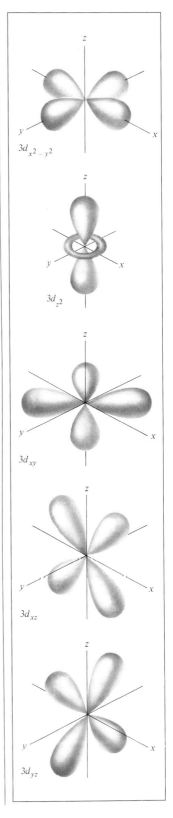

FIGURE 8.18 *Orbital energy levels in the hydrogen atom. Each short horizontal line here represents one orbital. Orbitals with the same principal quantum number (n) all have the same energy.*

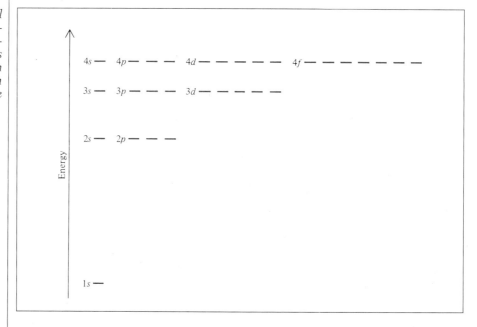

The arrangement of electrons in a many-electron atom can be understood by examining the atomic orbitals they reside in. As we will see shortly, each orbital can accommodate up to two electrons.

momentum quantum number. The stability of an electron in a many-electron atom reflects both the attraction between the electron and the nucleus and the repulsion between the electron and the rest of the electrons present. Attraction as well as repulsion depends on the shape of the orbital in which the electron resides. Consequently, the order of increasing orbital energy for a many-electron atom is more complicated and is usually determined by experiment. Figure 8.19 is a helpful diagram of the order in which atomic orbitals are filled by electrons in a many-electron atom. We will consider specific examples in the next section and in Chapter 9.

FIGURE 8.19 *The order in which atomic subshells are filled in a many-electron atom. Start with the 1s orbital and move downward, following the direction of the arrows. Thus the order goes as follows: 1s → 2s → 2p → 3s → 3p → 4s → 3d → . . . .*

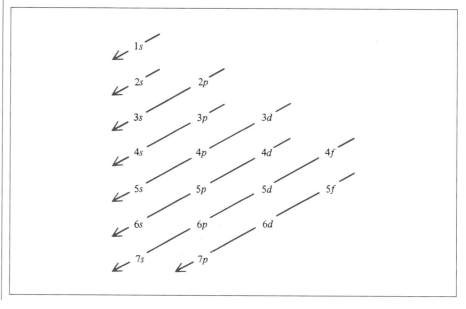

8.9 ELECTRON CONFIGURATION

The four quantum numbers n, ℓ, m_ℓ, and m_s enable us to label completely an electron in any orbital in any atom. For example, the four quantum numbers for a 2s orbital electron are either $n = 2$, $\ell = 0$, $m_\ell = 0$, and $m_s = +\frac{1}{2}$ or $n = 2$, $\ell = 0$, $m_\ell = 0$, and $m_s = -\frac{1}{2}$. In practice it is inconvenient to write out all the individual quantum numbers, and so we use the simplified notation (n, ℓ, m_ℓ, m_s). For the example above, the quantum numbers are either $(2, 0, 0, +\frac{1}{2})$ or $(2, 0, 0, -\frac{1}{2})$. The value of m_s has no effect on the energy or size and shape of an orbital, but it plays a profound role in determining how electrons are arranged in an orbital.

In a sense, we can regard the set of four quantum number values as the full "address" of an electron in an atom, somewhat in the same way that a postal Zip Code specifies the geographic region of an individual.

EXAMPLE 8.8

List the different ways to write the four quantum numbers that designate an electron in a 3p orbital.

Answer

To start with, we know that the principal quantum number n is 3 and the angular momentum quantum number ℓ must be 1 (because we are dealing with a p orbital). For $\ell = 1$, there are three values of m_ℓ given by -1, 0, 1. Since the electron-spin quantum number m_s can be either $+\frac{1}{2}$ or $-\frac{1}{2}$, we conclude that there are six possible ways to designate the electron:

$$(3, 1, -1, +\tfrac{1}{2}) \qquad (3, 1, -1, -\tfrac{1}{2})$$
$$(3, 1, 0, +\tfrac{1}{2}) \qquad (3, 1, 0, -\tfrac{1}{2})$$
$$(3, 1, 1, +\tfrac{1}{2}) \qquad (3, 1, 1, -\tfrac{1}{2})$$

The hydrogen atom is a particularly simple system because it contains only one electron. At any given instant, the electron may reside in the 1s orbital (the ground state) or it may be found in some higher orbital (an excited state). The situation is different for many-electron atoms. To understand electronic behavior in a many-electron atom, we must first know the **_electron configuration_** of the atom. The electron configuration of an atom tells us _how the electrons are distributed among the various atomic orbitals_. We will use the first ten elements (hydrogen to neon) to illustrate the basic rules for writing the electron configurations for the _ground states_ of the atoms. (Section 8.10 will describe how these rules can be applied to the remainder of the elements in the periodic table.) Recall that the number of electrons in a neutral atom is equal to its atomic number Z.

Figure 8.18 indicates that the electron in a ground state hydrogen atom must be in the 1s orbital, so its electron configuration is $1s^1$:

Hydrogen ($Z = 1$).

denotes the principal quantum number n ⟶ $1s^1$ ⟵ denotes the number of electrons in the orbital or subshell

denotes the angular momentum quantum number ℓ

The electron configuration can be represented also by an *orbital diagram* that shows the spin of the electron (see Figure 8.12):

$$\text{H} \quad \boxed{\uparrow}$$
$$1s^1$$

where the upward arrow denotes one of the two possible spinning motions of the electron. (We could have equally represented the electron with the downward arrow.) The box represents an atomic orbital.

The Pauli Exclusion Principle

We are now prepared to consider electron arrangements in atoms containing more than a single electron. Helium, with two electrons, is a reasonable choice to consider first. The $1s$ orbital in helium holds both electrons in its ground state. However, a problem arises when we try to indicate the spin of the two electrons in the $1s$ orbital because there are three possible choices, as shown by the following orbital diagrams:

$$\text{He} \quad \boxed{\uparrow\uparrow} \quad \boxed{\downarrow\downarrow} \quad \boxed{\uparrow\downarrow}$$
$$\quad\quad 1s^2 \quad 1s^2 \quad 1s^2$$
$$\quad\quad (a) \quad\; (b) \quad\; (c)$$

Diagrams (a) and (b) are ruled out by the **Pauli exclusion principle** (Wolfgang Pauli, 1900–1958), which states that *no two electrons in an atom can have the same four quantum numbers.* In diagram (a) both electrons would have the quantum numbers $(1, 0, 0, +\frac{1}{2})$; in diagram (b) both would have the same quantum numbers $(1, 0, 0, -\frac{1}{2})$. Only the configuration in (c) is physically acceptable, since one electron has the quantum numbers $(1, 0, 0, +\frac{1}{2})$ and the other has $(1, 0, 0, -\frac{1}{2})$.

We find that the two electrons of helium have the quantum numbers $(1, 0, 0, +\frac{1}{2})$ and $(1, 0, 0, -\frac{1}{2})$. Thus the atom has the electron configuration

$$\text{He} \quad \boxed{\uparrow\downarrow}$$
$$1s^2$$

Note that $1s^2$ is read "one s two," not "one s squared."

Diamagnetism and Paramagnetism

The Pauli exclusion principle is one of the fundamental principles of quantum mechanics. It can be tested by a simple observation. If the two electrons in the $1s$ orbital of a helium atom had the same, or parallel, spins ($\uparrow\uparrow$ or $\downarrow\downarrow$), their net magnetic fields would reinforce each other. Such an arrangement would make the helium atom **paramagnetic** [Figure 8.20(a)]. Paramagnetic substances are those that are *attracted by a magnet.* On the other hand, if the electron spins are paired, or antiparallel to each other ($\uparrow\downarrow$ or $\downarrow\uparrow$), the magnetic effects cancel out and the atom is **diamagnetic** [Figure 8.20(b)]. Diamagnetic substances are those that are *repelled by a magnet.*

Remember that the direction of electron spin has no effect on the energy of the electron.

In practice, the Pauli exclusion principle limits the capacity of any orbital to a maximum of two electrons.

Helium ($Z = 2$).

Electrons having opposite spins are said to be paired. In helium, $m_s = +\frac{1}{2}$ for one electron and, for the other, $m_s = -\frac{1}{2}$.

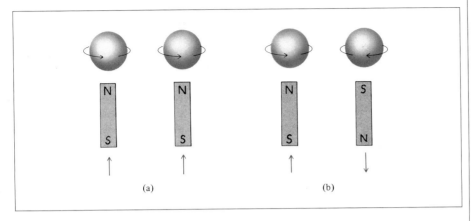

(a)   (b)

FIGURE 8.20 *The (a) parallel and (b) antiparallel spins of two electrons. In (a) the two magnetic fields reinforce each other and the atom is paramagnetic. In (b) the two magnetic fields cancel each other, and the atom is diamagnetic.*

By experiment we find that the helium atom is diamagnetic in the ground state, in accord with the Pauli exclusion principle. A useful general rule to keep in mind is that any atom with an odd number of electrons *must be* paramagnetic because we need an even number of electrons for complete pairing. On the other hand, atoms containing an even number of electrons may be either diamagnetic or paramagnetic. We will see the reason for this behavior shortly.

As another example, consider the lithium atom, which has three electrons. The third electron cannot go into the $1s$ orbital because it would inevitably have the same four quantum numbers as one of the first two electrons. Therefore, this electron "enters" the next (energetically) higher orbital, which is the $2s$ orbital (see Figure 8.19). The electron configuration of lithium is $1s^2 2s^1$, and its orbital diagram is

Li ↑↓ ↑
$1s^2$ $2s^2$

The lithium atom contains one unpaired electron and is therefore paramagnetic.

Lithium ($Z = 3$)

Experimentally we find that the $2s$ orbital lies lower in energy than the $2p$ orbital in a many-electron atom.

The Shielding Effect in Many-Electron Atoms

But why is the $2s$ orbital lower than the $2p$ orbital on the energy scale in a many-electron atom? In comparing the electron configurations of $1s^2 2s^1$ and $1s^2 2p^1$, we note that, in both cases, the $1s$ orbital is filled with two electrons. Because the $2s$ and $2p$ orbitals are larger than the $1s$ orbital, an electron in either of these orbitals will spend more time away from the nucleus (on the average) than an electron in the $1s$ orbital. Thus, we can speak of a $2s$ or $2p$ electron being partly "shielded" from the attractive force of the nucleus by the $1s$ electrons. The important consequence of the shielding effect is that it *reduces* the electrostatic attraction between protons in the nucleus and the electron in the $2s$ or $2p$ orbital.

The manner in which the electron density varies as we move outward from the nucleus differs for the $2s$ and $2p$ orbitals. The density for the $2s$

electron near the nucleus is greater than that for the $2p$ electron. In other words, a $2s$ electron spends more time near the nucleus than a $2p$ electron does (on the average). For this reason, the $2s$ orbital is said to be more "penetrating" than the $2p$ orbital, and it is less shielded by the $1s$ electrons. In fact, for the same principal quantum number n, the penetrating power decreases as the angular momentum quantum number ℓ increases, or

$$s > p > d > f > \cdots$$

Since the stability of an electron is determined by how strongly the electron is attracted by the nucleus, it follows that a $2s$ electron will be lower in energy than a $2p$ electron. To put it another way, less energy is required to remove a $2p$ electron than a $2s$ electron because a $2p$ electron is not quite as strongly held by the nucleus. The hydrogen atom has only one electron and, therefore, is without such a shielding effect.

Beryllium (Z = 4)

Continuing with our look into atoms of the first ten elements, the electron configuration of beryllium is $1s^2 2s^2$, or

$$\text{Be} \quad \boxed{\uparrow\downarrow} \quad \boxed{\uparrow\downarrow}$$
$$\qquad\quad 1s^2 \quad\; 2s^2$$

and beryllium atoms are diamagnetic as we would expect.

Boron (Z = 5)

The electron configuration of boron is $1s^2 2s^2 2p^1$, or

$$\text{B} \quad \boxed{\uparrow\downarrow} \quad \boxed{\uparrow\downarrow} \quad \boxed{\uparrow\;\;}$$
$$\qquad\quad 1s^2 \quad\; 2s^2 \qquad 2p^1$$

Note that the unpaired electron may be in the $2p_x$, $2p_y$, or $2p_z$ orbital. The choice is completely arbitrary. This follows from the fact that the three p orbitals are equivalent in energy. As the diagram shows, boron atoms are paramagnetic.

Hund's Rule

Carbon (Z = 6)

The electron configuration of carbon is $1s^2 2s^2 2p^2$. Here we have three choices for placing two electrons among three p orbitals, as follows:

$$\boxed{\uparrow\downarrow\;\;} \qquad \boxed{\uparrow\;\;\downarrow\;} \qquad \boxed{\uparrow\;\;\uparrow\;}$$
$$2p_x\, 2p_y\, 2p_z \qquad 2p_x\, 2p_y\, 2p_z \qquad 2p_x\, 2p_y\, 2p_z$$
$$\text{(a)} \qquad\qquad \text{(b)} \qquad\qquad \text{(c)}$$

None of the three arrangements violates the Pauli exclusion principle, so we must determine which one will give the greatest stability. The answer is provided by **_Hund's rule_** (Frederick Hund, 1896–　　), which states that _the most stable arrangement of electrons in subshells is the one with the greatest number of parallel spins._ The arrangement shown in (c) satisfies this condition. In both (a) and (b) the two spins cancel each other. Thus the electron configuration of carbon is $1s^2 2s^2 2p^2$, and its orbital diagram is

$$\text{C} \quad \boxed{\uparrow\downarrow} \quad \boxed{\uparrow\downarrow} \quad \boxed{\uparrow\ |\ \uparrow\ |\ \ }$$
$$\qquad\quad 1s^2 \qquad 2s^2 \qquad 2p^2$$

Qualitatively, we can understand why (c) is preferred to (a). In (a), the two electrons are in the same $2p_x$ orbital, and their proximity results in a greater mutual repulsion than when they occupy two separate orbitals, say $2p_x$ and $2p_y$. The choice of (c) over (b) is more subtle but can be justified on theoretical grounds.

Measurements of magnetic properties provide the most direct evidence supporting specific electron configurations of elements. Advances in instrument design during the last twenty years or so enable us to determine not only whether an atom is paramagnetic but also how many unpaired electrons are present. The fact that carbon atoms are paramagnetic, each containing two unpaired electrons, is in accord with Hund's rule.

Continuing, the electron configuration of nitrogen is $1s^2 2s^2 2p^3$, and the electrons are arranged as follows:

Nitrogen ($Z = 7$)

$$\text{N} \quad \boxed{\uparrow\downarrow} \quad \boxed{\uparrow\downarrow} \quad \boxed{\uparrow\ |\ \uparrow\ |\ \uparrow}$$
$$\qquad\quad 1s^2 \qquad 2s^2 \qquad 2p^3$$

Again, Hund's rule dictates that all three $2p$ electrons have spins parallel to one another; the nitrogen atom is therefore paramagnetic, containing three unpaired electrons.

The electron configuration of oxygen is $1s^2 2s^2 2p^4$. An oxygen atom is paramagnetic because it has two unpaired electrons:

Oxygen ($Z = 8$)

$$\text{O} \quad \boxed{\uparrow\downarrow} \quad \boxed{\uparrow\downarrow} \quad \boxed{\uparrow\downarrow\ |\ \uparrow\ |\ \uparrow}$$
$$\qquad\quad 1s^2 \qquad 2s^2 \qquad 2p^4$$

The electron configuration of fluorine is $1s^2 2s^2 2p^5$. The nine electrons are arranged as follows:

Fluorine ($Z = 9$)

$$\text{F} \quad \boxed{\uparrow\downarrow} \quad \boxed{\uparrow\downarrow} \quad \boxed{\uparrow\downarrow\ |\ \uparrow\downarrow\ |\ \uparrow}$$
$$\qquad\quad 1s^2 \qquad 2s^2 \qquad 2p^5$$

The fluorine atom is thus paramagnetic, having one unpaired electron.

In neon the $2p$ orbitals are completely filled. The electron configuration of neon is $1s^2 2s^2 2p^6$, and *all* the electrons are paired, as follows:

Neon ($Z = 10$)

$$\text{Ne} \quad \boxed{\uparrow\downarrow} \quad \boxed{\uparrow\downarrow} \quad \boxed{\uparrow\downarrow\ |\ \uparrow\downarrow\ |\ \uparrow\downarrow}$$
$$\qquad\quad 1s^2 \qquad 2s^2 \qquad 2p^6$$

The neon atom is thus predicted to be diamagnetic, which is in accord with experimental observation.

General Rules for Assigning Electrons to Atomic Orbitals

Based on the preceding examples we can formulate some general rules for determining the maximum number of electrons that can be assigned to the various subshells and orbitals for a given value of n:

- Each shell or principal level of quantum number n contains n subshells. For example, if $n = 2$, then there are two subshells (two values of ℓ) of angular momentum quantum numbers 0 and 1.
- Each subshell of quantum number ℓ contains $2\ell + 1$ orbitals. For example, if $\ell = 1$, then there are three p orbitals.
- No more than two electrons can be placed in each orbital. Therefore, the maximum number of electrons is simply twice the number of orbitals that are employed.

EXAMPLE 8.9

What is the maximum number of electrons that can be present in the principal level for which $n = 3$?

Answer

When $n = 3$, then $\ell = 0$, 1, and 2. The number of orbitals for each value of ℓ is given by

Value of ℓ	Number of Orbitals ($2\ell + 1$)
0	1
1	3
2	5

This result can be generalized by the formula $2n^2$. Here we have $n = 3$ so that $2(3^2) = 18$.

The total number of orbitals is nine. Since each orbital can accommodate two electrons, the maximum number of electrons that can reside in the orbitals is 2×9, or 18.

Similar examples: Problems 8.44, 8.45, 8.46.

EXAMPLE 8.10

An oxygen atom has a total of eight electrons. Write the four quantum numbers for each of the eight electrons in the ground state.

Answer

We start with $n = 1$, so that $\ell = 0$, a subshell corresponding to the 1s orbital. This orbital can accommodate a total of two electrons. Next, $n = 2$, and ℓ may be either 0 or 1. The $\ell = 0$ subshell contains one 2s orbital, which can accommodate a total of two electrons. The remaining four electrons are placed in the $\ell = 1$ subshell, which contains three $2p$ orbitals. The results are summarized in the following table:

O   ↑↓   ↑↓   ↑↓ ↑ ↑

     $1s^2$ $2s^2$ $2p^4$

Electron	n	ℓ	m_ℓ	m_s	Orbital
1	1	0	0	$+\frac{1}{2}$	1s
2	1	0	0	$-\frac{1}{2}$	
3	2	0	0	$+\frac{1}{2}$	2s
4	2	0	0	$-\frac{1}{2}$	
5	2	1	-1	$+\frac{1}{2}$	
6	2	1	0	$+\frac{1}{2}$	$2p_x, 2p_y, 2p_z$
7	2	1	1	$+\frac{1}{2}$	
8	2	1	1	$-\frac{1}{2}$	

Of course, the placement of the eighth electron in the orbital labeled $m_\ell = 1$ is completely arbitrary. Other suitable assignments would be $m_\ell = 0$ and $m_\ell = -1$.

Similar example: Problem 8.62.

At this point let's summarize what our examination has revealed about ground-state electron configurations and the properties of electrons in atoms:

- No two electrons in the same atom can have the same four quantum numbers. This is the Pauli exclusion principle.
- Each orbital can be occupied by a maximum of two electrons. They must have opposite spins, or different electron-spin quantum numbers.
- The most stable arrangement of electrons in a subshell is the one that has the greatest number of parallel spins. This is Hund's rule.
- Atoms in which one or more electrons are unpaired are paramagnetic. Atoms in which all of the electron spins are paired are diamagnetic.
- In a hydrogen atom, the energy of the electron depends only on its principal quantum number n. In a many-electron atom, the energy of an electron depends on both its principal quantum number n and its angular momentum quantum number ℓ.
- In a many-electron atom the subshells are filled in the order shown in Figure 8.19.
- For electrons of the same principal quantum number, their penetrating power, or proximity to the nucleus, decreases in the order $s > p > d > f$. This means that, for example, more energy is required to separate a $3s$ electron from an atom than is required for a $3p$ electron.

8.10 THE BUILDING-UP PRINCIPLE OF THE PERIODIC TABLE

In this final section of the chapter we will extend the rules used in writing the electron configurations for the first ten elements to apply to the rest of the elements in the periodic table. This process is based on the *building-up principle,* or *Aufbau principle*—the German word Aufbau means "building up." This principle enables us to build up the periodic table by increasing the number of electrons (and hence the atomic number) one at a time, starting with hydrogen and ending with the element having atomic number 109. Through this process we gain a detailed knowledge about the ground-state electron configurations of the elements. As we shall see later, a knowledge of the electron configurations helps us understand and predict the properties of the elements; it also explains why the periodic table works so well.

Table 8.3 gives the ground-state electron configurations of all the known elements from H ($Z = 1$) through Une ($Z = 109$). We have already discussed the electron configurations from hydrogen through neon in some detail. As you can see, the electron configurations of the elements from sodium ($Z = 11$) through argon ($Z = 18$) follow a pattern similar to those of lithium ($Z = 3$) through neon ($Z = 10$). (We compare only the highest filled subshells in the outermost shells.)

Starting with potassium ($Z = 19$), we begin to encounter some irregularities in filling the subshells. Potassium has nineteen electrons. There is no ambiguity about the configuration of the first eighteen electrons, which must be $1s^2 2s^2 2p^6 3s^2 3p^6$. Since this is also the electron configuration of argon, we can simplify this and similar notations by writing [Ar], which denotes the "argon core." But where does the nineteenth electron go? In the hydrogen atom, the $3d$ level has less energy than the $4s$ level (see Figure 8.18). But with potassium we are dealing with a many-electron atom whose energy depends on both n and ℓ. It turns out that the overall stability of potassium is greater if its last electron is in the $4s$ orbital rather than in the $3d$ orbital. Therefore the electron configuration of potassium is [Ar] $4s^1$ rather than [Ar] $3d^1$. The same argument holds for calcium ($Z = 20$), whose electron configuration is [Ar] $4s^2$. The placement of the outermost electrons in the $4s$ orbitals in potassium and calcium is strongly supported by experimental evidence.

The elements from scandium ($Z = 21$) to copper ($Z = 29$) belong to the transition metals. **Transition metals** either *have incompletely filled d subshells* or *readily give rise to cations that have incompletely filled d subshells.* Consider the first transition metal series, from scandium through copper. Along this series the added electrons are placed in the $3d$ orbitals according to Hund's rule. However, there are two irregularities. The electron configuration of chromium ($Z = 24$) is [Ar]$4s^1 3d^5$ and not [Ar]$4s^2 3d^4$, as we might expect. A similar break in the pattern is observed for copper, whose electron configuration is [Ar]$4s^1 3d^{10}$ rather than [Ar]$4s^2 3d^9$. The reason for these irregularities is that a slightly greater stability is associated with the half-filled ($3d^5$) and completely filled ($3d^{10}$) subshells. Electrons in the same subshells (in this case, the d orbitals) have equivalent but

Appendix 3 explains the derivations of the names and symbols of the elements.

The "argon core" notation introduced here is an accepted and useful simplification of the full electron configuration notation. Since the chemical properties of an element are largely determined by the behavior of the outer-shell electrons, these electrons remain fully specified. Any noble gas (Group 8A) element can serve as the "core." In practice, chemists ordinarily select the noble gas element that immediately precedes the element being considered.

TABLE 8.3   The Ground-State Electron Configurations of the Elements*

Atomic Number	Symbol	Electron Configuration	Atomic Number	Symbol	Electron Configuration	Atomic Number	Symbol	Electron Configuration
1	H	$1s^1$	37	Rb	[Kr] $5s^1$	73	Ta	[Xe] $6s^24f^{14}5d^3$
2	He	$1s^2$	38	Sr	[Kr] $5s^2$	74	W	[Xe] $6s^24f^{14}5d^4$
3	Li	[He] $2s^1$	39	Y	[Kr] $5s^24d^1$	75	Re	[Xe] $6s^24f^{14}5d^5$
4	Be	[He] $2s^2$	40	Zr	[Kr] $5s^24d^2$	76	Os	[Xe] $6s^24f^{14}5d^6$
5	B	[He] $2s^22p^1$	41	Nb	[Kr] $5s^1d^4$	77	Ir	[Xe] $6s^24f^{14}5d^7$
6	C	[He] $2s^22p^2$	42	Mo	[Kr] $5s^14d^5$	78	Pt	[Xe] $6s^14f^{14}5d^9$
7	N	[He] $2s^22p^3$	43	Tc	[Kr] $5s^24d^5$	79	Au	[Xe] $6s^14f^{14}5d^{10}$
8	O	[He] $2s^22p^4$	44	Ru	[Kr] $5s^14d^7$	80	Hg	[Xe] $6s^24f^{14}5d^{10}$
9	F	[He] $2s^22p^5$	45	Rh	[Kr] $5s^14d^8$	81	Tl	[Xe] $6s^24f^{14}5d^{10}6p^1$
10	Ne	[He] $2s^22p^6$	46	Pd	[Kr] $4d^{10}$	82	Pb	[Xe] $6s^24f^{14}5d^{10}6p^2$
11	Na	[Ne] $3s^1$	47	Ag	[Kr] $5s^14d^{10}$	83	Bi	[Xe] $6s^24f^{14}5d^{10}6p^3$
12	Mg	[Ne] $3s^2$	48	Cd	[Kr] $5s^24d^{10}$	84	Po	[Xe] $6s^24f^{14}5d^{10}6p^4$
13	Al	[Ne] $3s^23p^1$	49	In	[Kr] $5s^24d^{10}5p^1$	85	At	[Xe] $6s^24f^{14}5d^{10}6p^5$
14	Si	[Ne] $3s^23p^2$	50	Sn	[Kr] $5s^24d^{10}5p^2$	86	Rn	[Xe] $6s^24f^{14}5d^{10}6p^6$
15	P	[Ne] $3s^23p^3$	51	Sb	[Kr] $5s^24d^{10}5p^3$	87	Fr	[Rn] $7s^1$
16	S	[Ne] $3s^23p^4$	52	Te	[Kr] $5s^24d^{10}5p^4$	88	Ra	[Rn] $7s^2$
17	Cl	[Ne] $3s^23p^5$	53	I	[Kr] $5s^24d^{10}5p^5$	89	Ac	[Rn] $7s^26d^1$
18	Ar	[Ne] $3s^23p^6$	54	Xe	[Kr] $5s^24d^{10}5p^6$	90	Th	[Rn] $7s^2d^2$
19	K	[Ar] $4s^1$	55	Cs	[Xe] $6s^1$	91	Pa	[Rn] $7s^25f^26d^1$
20	Ca	[Ar] $4s^2$	56	Ba	[Xe] $6s^2$	92	U	[Rn] $7s^25f^36d^1$
21	Sc	[Ar] $4s^23d^1$	57	La	[Xe] $6s^25d^1$	93	Np	[Rn] $7s^25f^46d^1$
22	Ti	[Ar] $4s^23d^2$	58	Ce	[Xe] $6s^24f^15d^1$	94	Pu	[Rn] $7s^25f^6$
23	V	[Ar] $4s^23d^3$	59	Pr	[Xe] $6s^24f^3$	95	Am	[Rn] $7s^25f^7$
24	Cr	[Ar] $4s^13d^5$	60	Nd	[Xe] $6s^24f^4$	96	Cm	[Rn] $7s^25f^76d^1$
25	Mn	[Ar] $4s^23d^5$	61	Pm	[Xe] $6s^24f^5$	97	Bk	[Rn] $7s^25f^9$
26	Fe	[Ar] $4s^23d^6$	62	Sm	[Xe] $6s^24f^6$	98	Cf	[Rn] $7s^25f^{10}$
27	Co	[Ar] $4s^23d^7$	63	Eu	[Xe] $6s^24f^7$	99	Es	[Rn] $7s^25f^{11}$
28	Ni	[Ar] $4s^23d^8$	64	Gd	[Xe] $6s^24f^75d^1$	100	Fm	[Rn] $7s^25f^{12}$
29	Cu	[Ar] $4s^13d^{10}$	65	Tb	[Xe] $6s^24f^9$	101	Md	[Rn] $7s^25f^{13}$
30	Zn	[Ar] $4s^23d^{10}$	66	Dy	[Xe] $6s^24f^{10}$	102	No	[Rn] $7s^25f^{14}$
31	Ga	[Ar] $4s^23d^{10}4p^1$	67	Ho	[Xe] $6s^24f^{11}$	103	Lr	[Rn] $7s^25f^{14}6d^1$
32	Ge	[Ar] $4s^23d^{10}4p^2$	68	Er	[Xe] $6s^24f^{12}$	104	Unq	[Rn] $7s^25f^{14}6d^2$
33	As	[Ar] $4s^23d^{10}4p^3$	69	Tm	[Xe] $6s^24f^{13}$	105	Unp	[Rn] $7s^25f^{14}6d^3$
34	Se	[Ar] $4s^23d^{10}4p^4$	70	Yb	[Xe] $6s^24f^{14}$	106	Unh	[Rn] $7s^25f^{14}6d^4$
35	Br	[Ar] $4s^23d^{10}4p^5$	71	Lu	[Xe] $6s^24f^{14}5d^1$	107	Uns	[Rn] $7s^25f^{14}6d^5$
36	Kr	[Ar] $4s^23d^{10}4p^6$	72	Hf	[Xe] $6s^24f^{14}5d^2$	108	Uno	[Rn] $7s^25f^{14}6d^6$
						109	Une	[Rn] $7s^25f^{14}6d^7$

* The symbol [He] is called the *helium core* and represents $1s^2$.
   [Ne] is called the *neon core* and represents $1s^22s^22p^6$.
   [Ar] is called the *argon core* and represents [Ne]$3s^23p^6$.
   [Kr] is called the *krypton core* and represents [Ar]$4s^23d^{10}4p^6$.
   [Xe] is called the *xenon core* and represents [Kr]$5s^24d^{10}5p^6$.
   [Rn] is called the *radon core* and represents [Xe]$6s^24f^{14}5d^{10}6p^6$.

different spatial distributions. Consequently, their shielding of one another is relatively small, and the electrons are more strongly attracted by the nucleus when they have the $3d^5$ configuration. According to Hund's rule, the orbital diagram for Cr is

Cr [Ar] $4s^1$ $3d^5$

Thus, Cr has a total of six unpaired electrons. The orbital diagram for copper is

Cu   [Ar]   [↑]   [↑↓|↑↓|↑↓|↑↓|↑↓]

$4s^1$ $3d^{10}$

Note that extra stability is gained in this case by having the $3d$ orbitals completely filled.

For elements Zn $(Z = 30)$ through Kr $(Z = 36)$ the $4s$ and $4p$ subshells fill in a straightforward manner. With rubidium $(Z = 37)$, electrons begin to enter the $n = 5$ energy level.

The electron configurations in the second transition metal series [yttrium $(Z = 39)$ to silver $(Z = 47)$], on the whole, follow a pattern similar to those of the first transition metal series.

The sixth period of the periodic table begins with cesium $(Z = 55)$ and barium $(Z = 56)$, whose electron configurations are $[Xe]6s^1$ and $[Xe]6s^2$, respectively. Next we come to lanthanum $(Z = 57)$. From Figure 8.19 we would expect that after filling the $6s$ orbital we would place the additional electrons in $4f$ orbitals. In reality, the energies of the $5d$ and $4f$ orbitals are very close; in fact, for lanthanum $4f$ is slightly higher in energy than $5d$. Thus lanthanum's electron configuration is $[Xe]6s^25d^1$ and not $[Xe]6s^24f^1$. Following lanthanum are the fourteen elements [cerium $(Z = 58)$ to lutetium $(Z = 71)$] that form the **lanthanide,** or **rare earth, series.** The rare earth metals *have incompletely filled 4f subshells or readily give rise to cations that have incompletely filled 4f subshells.* In this series, the added electrons are placed in $4f$ orbitals. After the $4f$ subshells are completely filled, the next electron enters the $5d$ subshell which occurs with lutetium. Note that the electron configuration of gadolinium $(Z = 64)$ is $[Xe]6s^24f^75d^1$ rather than $[Xe]6s^24f^8$. Again, in this instance extra stability is gained by having half-filled subshells $(4f^7)$, as was the case for chromium.

Following completion of the lanthanide series, the third transition metal series, including lanthanum and hafnium $(Z = 72)$ through gold $(Z = 79)$, is characterized by the filling of the $5d$ orbitals. The $6s$ and $6p$ subshells are filled next, which takes us to radon $(Z = 86)$.

The last row of elements belongs to the **actinide series,** which starts at thorium $(Z = 90)$. *Most of these elements are not found in nature but have been synthesized.*

With few exceptions, you should now be able to write the electron configuration of any element, using Figure 8.19 as a guide. Elements that require particular care are those belonging to the transition metals, the lanthanides, and the actinides. As we noted earlier, at larger values of the principal quantum number n, the order of subshell filling may reverse from one element to the next.

EXAMPLE 8.11

Write the electron configurations of sulfur, mercury, and palladium.

Answer

Sulfur $(Z = 16)$

1. Sulfur has 16 electrons.
2. It takes 18 electrons to complete the third-period elements. Thus, the electron configuration of S is

$$1s^2 2s^2 2p^6 3s^2 3p^4$$

or $[Ne]3s^2 3p^4$.

Mercury $(Z = 80)$

1. Mercury has 80 electrons.
2. It takes 54 electrons to complete the fifth-period elements. Starting with the sixth period, we need 2 electrons for the $6s$ orbital, 14 electrons for the $4f$ orbitals, and 10 more for the $5d$ orbitals. The total number of electrons adds up to 80; thus the electron configuration of Hg is

$$1s^2 2s^2 2p^6 3s^2 3p^6 4s^2 3d^{10} 4p^6 5s^2 4d^{10} 5p^6 6s^2 4f^{14} 5d^{10}$$

or simply $[Xe]6s^2 4f^{14} 5d^{10}$.

Palladium $(Z = 46)$

1. Palladium has 46 electrons.
2. As in the previous cases, we need 36 electrons to complete the fourth-period elements, leaving us with 10 more electrons to distribute among the $5s$ and $4d$ orbitals. The three choices are (i) $4d^{10}$, (ii) $4d^9 5s^1$, and (iii) $4d^8 5s^2$. Unfortunately, there is no easy way to know whether the extra stability attained with the completely filled $4d$ subshell can outweigh the energy needed to promote the electrons from the $5s$ to the $4d$ orbital. However, it turns out that atomic palladium is diamagnetic, so its electron configuration must be

$$1s^2 2s^2 2p^6 3s^2 3p^6 4s^2 3d^{10} 4p^6 4d^{10}$$

or simply $[Kr]4d^{10}$.

SUMMARY

1. Classical physics, as expressed in Newton's laws of motion, failed to explain the behavior of atoms and molecules. A new theory was needed to account for the behavior of submicroscopic matter.
2. Since 1873, radiation had been known to possess the properties of waves—which we characterize by wavelength, frequency, and amplitude. One failure of classical physics, however, was in explaining the emission of radiation by heated solids. Max Planck solved this mystery in 1900. His theory, known as the quantum theory, revolutionized physics.
3. Planck's quantum theory states that radiant energy is emitted by atoms and molecules in small discrete amounts (quanta), rather than over a continuous range. This behavior is governed by the relationship $E = h\nu$, where E is the energy of the radiation, h is Planck's constant, and

ν is the frequency of the radiation. Energy is always emitted in whole number multiples of $h\nu$ ($1h\nu$, $2h\nu$, $3h\nu$, . . .).

4. By using quantum theory and, like Planck, making a radically new assumption, Albert Einstein in 1905 solved another mystery of physics—the photoelectric effect. Einstein proposed that light can behave like a stream of particles (photons).

5. The line spectrum of hydrogen, yet another mystery to nineteenth-century physics, was also explained by application of the quantum theory. Niels Bohr developed a model of the hydrogen atom in which the energy of the single electron moving in various circular orbits about the nucleus was quantized—limited to certain energy values determined by an integer, the principal quantum number.

6. An electron in its most stable energy state is said to be in the ground state, and an electron at a higher energy than its most stable state is said to be in an excited state. In the Bohr model, an electron emits a photon when it drops from a higher-energy orbit (an excited state) to a lower-energy orbit (the ground state or another, less excited state). The specific energies released as shown by the lines in the hydrogen emission spectrum can be accounted for in this way.

7. The dual behavior of light in exhibiting both particle and wave properties was extended by de Broglie to all matter in motion. The wavelength of a moving particle of mass m and velocity u is given by the de Broglie equation $\lambda = h/mu$.

8. The Schrödinger equation, formulated by Erwin Schrödinger in 1926, is a fundamental equation of quantum theory that describes the motions and energies of submicroscopic particles. This equation launched quantum mechanics and a new era in physics.

9. From the Schrödinger equation we find the possible energy states of the electron in a hydrogen atom and the probability of its location in a particular region surrounding the nucleus. These results can be applied with reasonable accuracy to many-electron atoms.

10. An atomic orbital is a probability function that defines the distribution of electron density in space. Orbitals are represented by electron density diagrams or boundary surface diagrams.

11. Four quantum numbers are needed to completely characterize each electron in an atom: the principal quantum number n, which determines the main energy level, or shell, of the orbital; the angular momentum quantum number ℓ, which determines the shape of the orbital; the magnetic quantum number m_ℓ, which determines the orientation of the orbital in space; and the electron spin quantum number m_s, which relates only to the electron's spin on its own axis.

12. The single s orbital in each energy level is spherical and centered on the nucleus. The three p orbitals each have two lobes, and the pairs of lobes are arranged at right angles to one another. There are five d orbitals, of more complex shapes and orientations.

13. The energy of the electron in a hydrogen atom is determined solely by its principal quantum number. In many-electron atoms, both the principal quantum number and the angular momentum quantum number determine the energy of an electron.

14. No two electrons in the same atom can have the same four quantum numbers (the Pauli exclusion principle).
15. The most stable arrangement of electrons in a subshell is the one that has the greatest number of parallel spins (Hund's rule). Atoms with one or more unpaired spins are paramagnetic. Atoms with all electron spins paired are diamagnetic.
16. The periodic table classifies the elements according to their atomic numbers and thus also by the electronic configurations of their atoms.

KEY WORDS

Actinide series, p. 242
Amplitude, p. 210
Atomic orbital, p. 225
Boundary surface diagram, p. 229
Diamagnetic, p. 234
Electromagnetic wave, p. 211
Electron configuration, p. 233
Electron density, p. 225
Emission spectra, p. 216
Excited level, p. 218
Excited state, p. 218
Frequency, p. 209
Ground state, p. 218
Ground level, p. 218

Hund's rule, p. 236
Lanthanide series, p. 242
Line spectra, p. 216
Many-electron atom, p. 225
Orbital, p. 225
Paramagnetic, p. 234
Pauli exclusion principle, p. 234
Photon, p. 214
Quantum, p. 213
Rare earth series, p. 242
Transition metals, p. 240
Wave, p. 209
Wavelength, p. 209

PROBLEMS

More challenging problems are marked with an asterisk.

Wave Properties

8.1 Convert 8.6×10^{13} Hz to nm, and 566 nm to Hz.

8.2 The average distance between Mars and Earth is about 1.3×10^8 miles. How long would it take TV pictures transmitted from the *Viking* space vehicle on Mars' surface to reach Earth? (1 mile = 1.61 km)

8.3 How long would it take a radio wave of frequency 5.5×10^5 Hz to travel from the planet Venus to Earth? (Average distance from Venus to Earth = 28 million miles)

8.4 (a) What is the frequency of light of wavelength 456 nm? (b) What is the wavelength (in nm) of radiation of frequency 2.20×10^9 Hz? (c) How much energy (in joules) is carried by a mole of photons of frequency 8.60×10^{10} Hz? (d) How much energy (in joules) is carried by a mole of photons of wavelength 752 nm?

8.5 Our eyes are sensitive to light in the approximate frequency range of 4.0×10^{14} to 7.5×10^{14} Hz. Calculte the wavelengths in nm that correspond to these two frequencies.

8.6 A photon has a wavelength of 624 nm. Calculate the energy of the photon in joules.

8.7 The blue color of the sky results from the scattering of sunlight by air molecules. The blue light has a frequency of about 7.5×10^{14} Hz. (a) Calculate the wavelength associated with this radiation, and (b) calculate the energy in joules of a single photon associated with this frequency.

8.8 A photon has a frequency of 6.0×10^4 Hz. (a) Convert this frequency into wavelength (nm). Does this frequency fall in the visible region? (b) Calculate the energy (in joules) of this photon, and (c) calculate the energy (in joules) of one mole of photons all with this frequency.

8.9 What is the wavelength in nanometers of radiation that has an energy content of 1.0×10^3 kJ/mol? In which region of the electromagnetic spectrum is this radiation found?

8.10 The basic SI unit of time is the second, which is defined as 9,192,631,770 cycles of radiation associated with a certain emission process in the cesium atom. Calculate the wavelength of this radiation (to three significant figures). In which region of the electromagnetic spectrum is this wavelength found?

8.11 The basic SI unit of length is the meter, which is defined as the length equal to 1,650,763.73 wavelengths of the light emitted by a particular transition in krypton atoms. Calculate the frequency of the light to three significant figures.

8.12 When copper is bombarded with high-energy electrons, X rays are emitted. Calculate the energy (in joules) associated with the photons if the wavelength of the X rays is 0.154 nm.

8.13 A particular electromagnetic radiation has a frequency of 8.11×10^{14} Hz. (a) What is its wavelength in nm? in m? (b) To what region of the electromagnetic spectrum would you assign it? (c) What is the energy (in joules) of one quantum of this radiation?

8.14 Explain what is meant by the phrase "quantization of energies."

The Photoelectric Effect

8.15 Explain the photoelectric effect. What important role did it play in the development of the particle–wave interpretation of the nature of electromagnetic radiation?

*8.16 In a photoelectric experiment a student uses a light source whose frequency is greater than that needed to eject electrons from a certain metal. However, after continuously shining the light on the same area of the metal for a long period of time the student notices that the maximum kinetic energy of ejected electrons begins to decrease, even though the frequency of the light is held constant. How would you account for this behavior?

Bohr's Theory of the Hydrogen Atom

8.17 Briefly describe Bohr's theory of the hydrogen atom. How does it differ from concepts of classical physics? What are its inadequacies?

8.18 Explain what is meant when we say that an atom is excited. Does this atom gain or lose energy?

8.19 Explain why elements produce their own characteristic colors when undergoing emission of photons.

8.20 When ionic compounds containing Cu^{2+} ions are heated in a flame, green light is emitted. How would you determine whether the light is of one wavelength or a mixture of two or more wavelengths?

8.21 Would it be possible for a fluorescent material to emit radiation in the ultraviolet region after absorbing visible light? Explain your answer.

8.22 Explain how astronomers are able to tell which elements are present in distant stars by analyzing the electromagnetic radiation emitted by the stars.

8.23 What is the physical significance of the minus sign in Equation (8.3)?

8.24 Consider the following energy levels of a hypothetical atom:

$$E_4 \underline{\hspace{2cm}} -1.0 \times 10^{-19} \text{ J}$$
$$E_3 \underline{\hspace{2cm}} -5.0 \times 10^{-19} \text{ J}$$
$$E_2 \underline{\hspace{2cm}} -10 \times 10^{-19} \text{ J}$$
$$E_1 \underline{\hspace{2cm}} -15 \times 10^{-19} \text{ J}$$

(a) What is the wavelength of the photon needed to excite an electron from E_1 to E_4? (b) What is the value (in joules) of the quantum of energy of a photon needed to excite an electron from E_2 to E_3? (c) When an electron drops from the E_3 level to the E_1 level, the atom is said to undergo emission. Calculate the wavelength of the photon emitted in this process.

8.25 Calculate the wavelength of a photon emitted by a hydrogen atom when its electron drops from the $n = 5$ state to the $n = 3$ state.

8.26 The first line of the Balmer series occurs at a wavelength of 656.3 nm. What is the energy difference between the two energy levels involved in the emission that results in this spectral line?

8.27 Calculate the frequency and wavelength of the emitted photon when an electron undergoes a transition from the $n = 4$ to the $n = 2$ level in a hydrogen atom.

8.28 Careful spectral analysis shows that the familiar yellow light of sodium lamps (such as street lamps) is made up of photons of two wavelengths, 589.0 nm and 589.6 nm. What is the difference in energy (in joules) between photons possessing these wavelengths?

8.29 An electron in an orbit of principal quantum number n_i in a hydrogen atom makes a transition to an orbit of principal quantum number 2. The photon emitted has a wavelength of 434 nm. Calculate n_i.

Particle–Wave Duality

8.30 Explain what is meant by the statement that matter and radiation have a "dual nature."

8.31 Calculate the wavelength (in nm) associated with a beam of neutrons moving at 4.00×10^3 cm/s. (Mass of a neutron $= 1.675 \times 10^{-27}$ kg)

8.32 What is the de Broglie wavelength in cm of a 12.4 g hummingbird flying at 1.20×10^2 mph? (1 mile = 1.61 km)

8.33 What is the de Broglie wavelength associated with a 2.5 g Ping-Pong ball traveling at 35 mph?

8.34 What properties of electrons are used in the operation of an electron microscope?

Quantum Numbers

8.35 An electron in a certain atom is in the $n = 2$ quantum level. List the possible values of ℓ and m_ℓ that it can have.

8.36 An electron in an atom is in the $n = 3$ quantum level. List the possible values of ℓ and m_ℓ that it can have.

8.37 Give the values of the quantum numbers associated with the following orbitals: (a) $2p_x$, (b) $3s$, (c) $4d_{z^2}$.

Atomic Orbitals

8.38 What is an atomic orbital?

8.39 Discuss the similarities and differences between a $1s$ and $2s$ orbital.

8.40 What is the difference between a $2p_x$ and a $2p_y$ orbital?

8.41 Draw the shapes (boundary surfaces) of the following orbitals: (a) p_y, (b) d_{z^2}, (c) $d_{x^2-y^2}$. (Show coordinate axes in your sketches.)

8.42 For the following subshells give the values of the quantum numbers (n, ℓ, and m_ℓ) and the number of orbitals in each subshell: (a) $4p$, (b) $3d$, (c) $3s$, (d) $5f$.

8.43 List all the possible subshells and orbitals associated with the principal quantum number n, if $n = 6$.

8.44 Calculate the total number of electrons that can occupy (a) one s orbital, (b) three p orbitals, (c) five d orbitals, (d) seven f orbitals.

*8.45 What is the total number of electrons that can be held in all orbitals having the same principal quantum number n?

8.46 What is the maximum number of electrons that can be found in each of the following subshells? $3s$, $3d$, $4p$, $4f$, $5f$.

8.47 Indicate the total number of (a) p electrons in N ($Z = 7$); (b) total s electrons in Si ($Z = 14$); and (c) $3d$ electrons in S ($Z = 16$).

8.48 Make a chart of all allowable orbitals in the first four principal energy levels of the hydrogen atom. Designate each by type (for example, s, p) and indicate how many orbitals of each type there are.

8.49 Which of the four quantum numbers (n, ℓ, m_ℓ, m_s) determine the energy of an electron in (a) a hydrogen atom and (b) a many-electron atom?

8.50 Why do the $3s$, $3p$, and $3d$ orbitals have the same energy in a hydrogen atom but different energies in a many-electron atom?

8.51 For each of the following pairs of hydrogen orbitals indicate which is higher in energy. (a) $1s$, $2s$; (b) $2p$, $3p$; (c) $3d_{xy}$, $3d_{yz}$; (d) $3s$, $3d$; (e) $5s$, $4f$.

8.52 Which orbital in each of the following pairs is lower in energy in a many-electron atom? (a) $2s$, $2p$; (b) $3p$, $3d$; (c) $3s$, $4s$; (d) $4d$, $5f$.

Pauli Exclusion Principle

8.53 Comment on the correctness of the following statement: The probability of finding two electrons with the same four quantum numbers in an atom is zero.

8.54 Which of the following sets of quantum numbers are unacceptable? Explain your answers. (a) $(1, 0, \frac{1}{2}, -\frac{1}{2})$, (b) $(3, 0, 0, +\frac{1}{2})$, (c) $(2, 2, 1, +\frac{1}{2})$, (d) $(4, 3, -2, +\frac{1}{2})$, (e) $(3, 2, 1, 1)$.

Diamagnetism and Paramagnetism

8.55 Predict the number of spots you would observe on the detecting screen in a Stern-Gerlach experiment for each of the following atoms: He, Li, N, Ne.

8.56 The atomic number of an element is 73. Are the atoms of this element diamagnetic or paramagnetic?

*8.57 Indicate the number of unpaired electrons present in each of the following atoms: B, Ne, P, Sc, Mn, Se, Zr, Ru, Cd, I, W, Pb, Ce, Ho.

8.58 Which of the following species has the most unpaired electrons: S^+, S, or S^-? Explain how you arrive at your answer.

Electron Configuration

8.59 Write the ground-state electron configurations for the following elements: B, V, Ni, As, I, and Au.

8.60 Write the ground-state electron configurations for the following elements: Ge, Fe, Zn, Ru, W, and Tl.

*8.61 The ground-state electron configurations listed here are *incorrect*. Explain specifically what mistakes have been made in each and write the correct electron configurations.

Al $1s^22s^22p^43s^23p^3$

B $1s^22s^22p^6$

F $1s^22s^22p^6$

8.62 The electron configuration of a neutral atom is $1s^22s^22p^63s^2$. Write a complete set of quantum numbers for each of the electrons. Name the element.

8.63 Draw orbital diagrams for atoms with the following electron configurations: (a) $1s^22s^22p^5$, (b) $1s^22s^22p^63s^23p^3$, (c) $1s^22s^22p^63s^23p^64s^23d^7$.

*8.64 The electron configurations described in this chapter all refer to atoms in their ground states. An atom may absorb a quantum of energy and promote one of its electrons to a higher-energy orbital. When this happens, we say that the atom is in an excited state. The electron configurations of some excited atoms are given. Identify these atoms and write their ground-state configurations. (a) $1s^22s^22p^23d^1$, (b) $1s^12s^1$, (c) $1s^22s^22p^64s^1$, (d) $[Ar]4s^13d^{10}4p^4$, (e) $[Ne]3s^23p^43d^1$.

Miscellaneous

8.65 Distinguish carefully between the following terms: (a) wavelength and frequency, (b) wave properties and particle properties, (c) quantization of energy and continuous variation in energy.

8.66 The term "quantum jump" (or "quantum leap") is sometimes used in ordinary conversation. Does this term have any resemblance to what physicists and chemists mean when they discuss the behavior of electrons in atoms?

8.67 Identify the following individuals and their contributions to the development of quantum theory: de Broglie, Einstein, Bohr, Planck, Schrödinger.

8.68 List the similarities and differences between the quantum mechanical and the Bohr treatments of the hydrogen atom.

8.69 Explain the meaning of each of the following terms: (a) wavelength, (b) emission process, (c) spectral lines, (d) photon, (e) energy level.

8.70 Explain the following terms or concepts: (a) electromagnetic spectrum, (b) ground state, (c) excited state.

8.71 Discuss the current view of the correctness of the following statements. (a) The electron in the hydrogen atom is in an orbit that never brings it closer than 100 pm to the nucleus. (b) Atomic absorption spectra result from transitions of electrons from lower to higher energy levels. (c) A many-electron atom behaves somewhat like a solar system that has a number of planets.

9
PERIODIC RELATION-SHIPS AMONG THE ELEMENTS

The periodic table stands as one of the most powerful conceptualizations of modern chemistry. It also provides a tangible and extremely useful link between the world of large-scale, macroscopic phenomena in chemistry and the related world of submicroscopic structure and behavior. In this chapter you will have the opportunity to see the periodic table "work" at both levels. Our tour of the periodic table will be organized horizontally across selected periods and vertically down selected element groups. As you will soon see, major physical and chemical properties of the elements are associated with the periodic table's special arrangement. And this special arrangement, in turn, is made understandable by patterns in the electron configurations in atoms of the elements being considered. Thus both macroscopic and molecular concerns are mirrored in the modern periodic table.

249

9.1 DEVELOPMENT OF THE PERIODIC TABLE

In the nineteenth century chemists began to realize that certain groups of elements display very strong similarities to one another, that is, there are regularities in their physical and chemical behavior. In 1864 the English chemist John Newland (1838–1898) arranged the known elements in order of increasing atomic mass and found that every eighth element had similar properties. Newland referred to this peculiar relationship as the *Law of Octaves*. However, this "law" turned out to be inadequate for elements beyond calcium ($Z = 20$), and Newland's work was not accepted by the scientific community.

In 1869 the Russian chemist Dimitri Mendeleev (1836–1907) and the German chemist Julius Meyer (1830–1895) independently proposed a much more extensive tabulation of the elements, based on regular, periodic recurrence of properties. Table 9.1 shows an early version of the periodic table drawn up by Mendeleev. Mendeleev's classification was a great improvement over Newland's in two important ways. First, it grouped the elements together more accurately, according to their properties. Equally important, it made possible the prediction of the properties of several ele-

TABLE 9.1   The Periodic Table as Drawn by Mendeleev*

THE ATOMIC MASSES OF THE ELEMENTS
Distribution of the Elements in Periods

Groups	Higher Salt-forming Oxides	Typical or 1st Small Period	Large Periods				
			1st	2nd	3rd	4th	5th
I.	R_2O	Li = 7	K 39	Rb 85	Cs 133	—	—
II.	RO	Be = 9	Ca 40	Sr 87	Ba 137	—	—
III.	R_2O_3	B = 11	Sc 44	Y 89	La 138	Yb 173	—
IV.	RO_2	C = 12	Ti 48	Zr 90	Ce 140	—	Th 232
V.	R_2O_5	N = 14	V 51	Nb 94	—	Ta 182	—
VI.	RO_3	O = 16	Cr 52	Mo 96	—	W 184	Ur 240
VII.	R_2O_7	F = 19	Mn 55	—	—	—	—
VIII.			Fe 56	Ru 103	—	Os 191	—
			Co 58.5	Rh 104	—	Ir 193	—
			Ni 59	Pd 106	—	Pt 196	—
I.	R_2O	H = 1. Na = 23	Cu 63	Ag 108	—	Au 198	—
II.	RO	Mg = 24	Zn 65	Cd 112	—	Hg 200	—
III.	R_2O_3	Al = 27	Ga 70	In 113	—	Tl 204	—
IV.	RO_2	Si = 28	Ge 72	Sn 118	—	Pb 206	—
V.	R_2O_5	P = 31	As 75	Sb 120	—	Bi 208	—
VI.	RO_3	S = 32	Se 79	Te 125	—	—	—
VII.	R_2O_7	Cl = 35.5	Br 80	I 127	—	—	—
		2nd Small Period	1st	2nd	3rd	4th	5th

Large Periods

* The groups are arranged horizontally, rather than vertically as in the modern periodic table. (From M. E. Weeks, *Discovery of the Elements*, 6th edition, Chemical Education Publishing Company, Easton, Pennsylvania, 1956.)

TABLE 9.2 Comparison of the Properties for Eka-Aluminum Predicted by Mendeleev and the Properties Found for Gallium*

Eka-Aluminum (Ea)	Gallium (Ga)
Atomic mass about 68.	Atomic mass 69.9.
Metal of specific gravity 5.9; melting point low; nonvolatile; unaffected by air; should decompose steam at red heat; should dissolve slowly in acids and alkalis.	*Metal* of specific gravity 5.94; melting point 30.15°C; nonvolatile at moderate temperatures; does not change in air; action of steam unknown; dissolves slowly in acids and alkalis.
Oxide: formula Ea_2O_3; specific gravity 5.5; should dissolve in acids to form salts of the type EaX_3. The hydroxide should dissolve in acids and alkalis.	*Oxide:* Ga_2O_3; specific gravity unknown; dissolves in acids, forming salts of the type GaX_3. The hydroxide dissolves in acids and alkalis.
Salts should have tendency to form basic salts; the sulfate should form alums; the sulfide should be precipitated by H_2S or $(NH_4)_2S$. The anhydrous chloride should be more volatile than zinc chloride.	*Salts* hydrolyze readily and form basic salts; alums are known; the sulfide is precipitated by H_2S and by $(NH_4)_2S$ under special conditions. The anhydrous chloride is more volatile than zinc chloride.
The element will probably be discovered by spectroscopic analysis.	Gallium was discovered with the aid of the spectroscope.

* From M. E. Weeks, *Discovery of the Elements*, 6th edition, Chemical Education Publishing Company, Easton, Pennsylvania, 1956.

ments that had not yet been discovered. For example, Mendeleev proposed the existence of an unknown element that he called eka-aluminum. "Eka" is the Sanskrit word meaning first; thus eka-aluminum is the first element under aluminum in the same group. When gallium was discovered four years later, predicted properties of eka-aluminum were found to match the observed properties of gallium very closely (Table 9.2).

The concept of the modern periodic table has been discussed in some detail in previous chapters. You will recall that Moseley showed that elements should be arranged according to their increasing atomic number, rather than atomic mass (see "An Aside on the Periodic Table" in Chapter 7). Electron configurations of elements, introduced in Chapter 8, help explain the periodic recurrence of physical and chemical properties (to be discussed shortly). The importance and usefulness of the periodic table can be seen in the fact that once we have learned the information contained in it, we do not need to study individual elements in isolation. We can use our understanding of the general properties and trends within a group or a period to predict with considerable accuracy the properties of any element, even though that element may be unfamiliar to us.

9.2 PERIODIC CLASSIFICATION OF THE ELEMENTS

Figure 9.1 shows the periodic table together with the ground-state electron configurations of the elements. (The electron configurations of all the elements are also given in Table 8.3.) Starting with hydrogen, we see that

	1A	2A	3B	4B	5B	6B	7B	8B	8B	8B	1B	2B	3A	4A	5A	6A	7A	8A
1	1 H $1s^1$																	2 He $1s^2$
2	3 Li $2s^1$	4 Be $2s^2$											5 B $2s^2 2p^1$	6 C $2s^2 2p^2$	7 N $2s^2 2p^3$	8 O $2s^2 2p^4$	9 F $2s^2 2p^5$	10 Ne $2s^2 2p^6$
3	11 Na $3s^1$	12 Mg $3s^2$											13 Al $3s^2 3p^1$	14 Si $3s^2 3p^2$	15 P $3s^2 3p^3$	16 S $3s^2 3p^4$	17 Cl $3s^2 3p^5$	18 Ar $3s^2 3p^6$
4	19 K $4s^1$	20 Ca $4s^2$	21 Sc $4s^2 3d^1$	22 Ti $4s^2 3d^2$	23 V $4s^2 3d^3$	24 Cr $4s^1 3d^5$	25 Mn $4s^2 3d^5$	26 Fe $4s^2 3d^6$	27 Co $4s^2 3d^7$	28 Ni $4s^2 3d^8$	29 Cu $4s^1 3d^{10}$	30 Zn $4s^2 3d^{10}$	31 Ga $4s^2 4p^1$	32 Ge $4s^2 4p^2$	33 As $4s^2 4p^3$	34 Se $4s^2 4p^4$	35 Br $4s^2 4p^5$	36 Kr $4s^2 4p^6$
5	37 Rb $5s^1$	38 Sr $5s^2$	39 Y $5s^2 4d^1$	40 Zr $5s^2 4d^2$	41 Nb $5s^1 4d^4$	42 Mo $5s^1 4d^5$	43 Tc $5s^2 4d^5$	44 Ru $5s^1 4d^7$	45 Rh $5s^1 4d^8$	46 Pd $4d^{10}$	47 Ag $5s^1 4d^{10}$	48 Cd $5s^2 4d^{10}$	49 In $5s^2 5p^1$	50 Sn $5s^2 5p^2$	51 Sb $5s^2 5p^3$	52 Te $5s^2 5p^4$	53 I $5s^2 5p^5$	54 Xe $5s^2 5p^6$
6	55 Cs $6s^1$	56 Ba $6s^2$	57 La $6s^2 5d^1$	72 Hf $6s^2 5d^2$	73 Ta $6s^2 5d^3$	74 W $6s^2 5d^4$	75 Re $6s^2 5d^5$	76 Os $6s^2 5d^6$	77 Ir $6s^2 5d^7$	78 Pt $6s^1 5d^9$	79 Au $6s^1 5d^{10}$	80 Hg $6s^2 5d^{10}$	81 Tl $6s^2 6p^1$	82 Pb $6s^2 6p^2$	83 Bi $6s^2 6p^3$	84 Po $6s^2 6p^4$	85 At $6s^2 6p^5$	86 Rn $6s^2 6p^6$
7	87 Fr $7s^1$	88 Ra $7s^2$	89 Ac $7s^2 6d^1$	104 Unq $7s^2 6d^2$	105 Unp $7s^2 6d^3$	106 Unh $7s^2 6d^4$	107 Uns $7s^2 6d^5$	108 Uno $7s^2 6d^6$	109 Une $7s^2 6d^7$									

58 Ce $6s^2 4f^1 5d^1$	59 Pr $6s^2 4f^3$	60 Nd $6s^2 4f^4$	61 Pm $6s^2 4f^5$	62 Sm $6s^2 4f^6$	63 Eu $6s^2 4f^7$	64 Gd $6s^2 4f^7 5d^1$	65 Tb $6s^2 4f^9$	66 Dy $6s^2 4f^{10}$	67 Ho $6s^2 4f^{11}$	68 Er $6s^2 4f^{12}$	69 Tm $6s^2 4f^{13}$	70 Yb $6s^2 4f^{14}$	71 Lu $6s^2 4f^{14} 5d^1$
90 Th $7s^2 6d^2$	91 Pa $7s^2 5f^2 6d^1$	92 U $7s^2 5f^3 6d^1$	93 Np $7s^2 5f^4 6d^1$	94 Pu $7s^2 5f^6$	95 Am $7s^2 5f^7$	96 Cm $7s^2 5f^7 6d^1$	97 Bk $7s^2 5f^9$	98 Cf $7s^2 5f^{10}$	99 Es $7s^2 5f^{11}$	100 Fm $7s^2 5f^{12}$	101 Md $7s^2 5f^{13}$	102 No $7s^2 5f^{14}$	103 Lr $7s^2 5f^{14} 6d^1$

FIGURE 9.1 *The ground-state electron configurations of the elements. For simplicity, only the configurations of the outer electrons are shown.*

subshells are filled in the order shown in Figure 8.19. According to the type of subshells being filled, the elements can be divided into categories—the representative elements, the noble gases, the transition elements (or transition metals), the lanthanides, and the actinides. Referring now to Figure 9.1, the **representative elements** are *the elements in Groups 1A through 7A, all of which have incompletely filled s or p subshell of highest principal quantum number.* With the exception of helium, the **noble gases** *(the Group 8A elements) all have completely filled p subshell. (The electron configurations are $1s^2$ for helium and ns^2np^6 for the other noble gases, where n is the principal quantum number for the outermost shell.)* The transition metals are the elements in Groups 1B and 3B through 8B, which have incompletely filled d subshell or readily produce cations with incompletely filled d subshell. Both the lanthanides and actinides have incompletely filled f subshell.

A clear pattern emerges when we examine the electron configurations of the elements in a particular group. First we will consider the representative elements. For Group 1A and 2A elements, the electron configurations are

Group 1A		Group 2A	
Li	$[He]2s^1$	Be	$[He]2s^2$
Na	$[Ne]3s^1$	Mg	$[Ne]3s^2$
K	$[Ar]4s^1$	Ca	$[Ar]4s^2$
Rb	$[Kr]5s^1$	Sr	$[Kr]5s^2$
Cs	$[Xe]6s^1$	Ba	$[Xe]6s^2$
Fr	$[Rn]7s^1$	Ra	$[Rn]7s^2$

We see that all members of the Group 1A alkali metals have similar outer electron configurations; each has a noble gas core and an ns^1 configuration of the outer electron. Similarly, the Group 2A alkaline earth metals have a noble gas core and an ns^2 configuration of the outer electrons. *The outer electrons of an atom, which are those involved in chemical bonding,* are often called the **valence electrons.** The similarity of the outer electron configurations (that is, they have the same number of valence electrons) is what makes the elements in the same group resemble one another in chemical behavior. This observation holds true for the other representative elements. Thus, for instance, the halogens (the Group 7A elements), all with outer electron configurations of ns^2np^5, have very similar properties as a group. We must be careful, however, in predicting properties when elements change from nonmetals through metalloids to metals. For example, the elements in Group 4A all have the same outer electron configuration, ns^2np^2, but there is much variation in chemical properties among these elements, since carbon is a nonmetal, silicon and germanium are metalloids, and tin and lead are metals.

As a group, the noble gases behave very much alike. With the exception of krypton and xenon, the rest of these elements are totally inert chemically. The reason is that these elements all have completely filled outer ns and np subshells, a condition that represents great stability. Although the outer electron configuration of the transition metals is not always the same within a group, and there is no regular pattern in the change of the electron configuration from one metal to the next in the same period, all transition metals share many characteristics that set them aside from other elements.

For the representative elements, the valence electrons are simply those electrons at the highest energy level n.

See Figure 1.4 for a classification of elements.

There have been some recent reports about reactions involving radon. However, all isotopes of radon are radioactive so it is both expensive and difficult to study radon in the laboratory.

This is because these metals all have an incompletely filled d subshell. Likewise, the lanthanide (and the actinide) elements resemble one another within the series because they have incompletely filled f subshells.

EXAMPLE 9.1

A neutral atom of a certain element has 15 electrons. Without consulting a periodic table, answer the following questions: (a) What is the electron configuration of the element? (b) Classify the element. (c) Are the atoms of this element diamagnetic or paramagnetic?

Answer

(a) Using the Aufbau principle and knowing the maximum capacity of s and p subshells, we can write the electron configuration of the element as $1s^2 2s^2 2p^6 3s^2 3p^3$.

(b) Since the $3p$ subshell is not completely filled, this is a representative element. Based on the information given, we cannot say whether it is a metal, a nonmetal, or a metalloid.

(c) According to Hund's rule, the three electrons in the $3p$ orbitals have parallel spins. Therefore, the atoms of this element are paramagnetic, with three unpaired spins. (Remember, we saw in Chapter 8 that any atom that contains an odd number of electrons *must be* paramagnetic.)

Similar example: Problem 9.13.

Representing Free Elements in Chemical Equations

Having classified the elements according to their electron configurations, we can now look at the way chemists represent metals, metalloids, and nonmetals that appear in chemical equations as free elements. Because metals do not exist in discrete molecular units, we always use their empirical formulas in chemical equations. The empirical formulas are, of course, the same as the symbols that represent the elements. For example, the empirical formula for iron is Fe, the same as the symbol for the element.

For nonmetals there is no single rule. Carbon, for example, exists as an extensive three-dimensional network of its atoms, and so we use its empirical formula (C) to represent it in chemical equations. Because hydrogen, nitrogen, oxygen, and the halogens exist as diatomic molecules, we use their molecular formulas (H_2, N_2, O_2, F_2, Cl_2, Br_2, I_2). The stable form of phosphorus exists as P_4 molecules, and so we use P_4. Although the stable form of sulfur is S_8, chemists often use its empirical formula in chemical equations. Thus, instead of writing the equation for the combustion of sulfur as

$$S_8(s) + 8O_2(g) \longrightarrow 8SO_2(g)$$

we often write simply

$$S(s) + O_2(g) \longrightarrow SO_2(g)$$

The structures of P_4 and S_8 will be discussed in Chapter 12.

Note that these two contrasting equations for the combustion of sulfur involve identical stoichiometry. This should not be surprising, since both equations describe the same chemical system. In both cases a number of sulfur atoms react with twice that number of oxygen atoms.

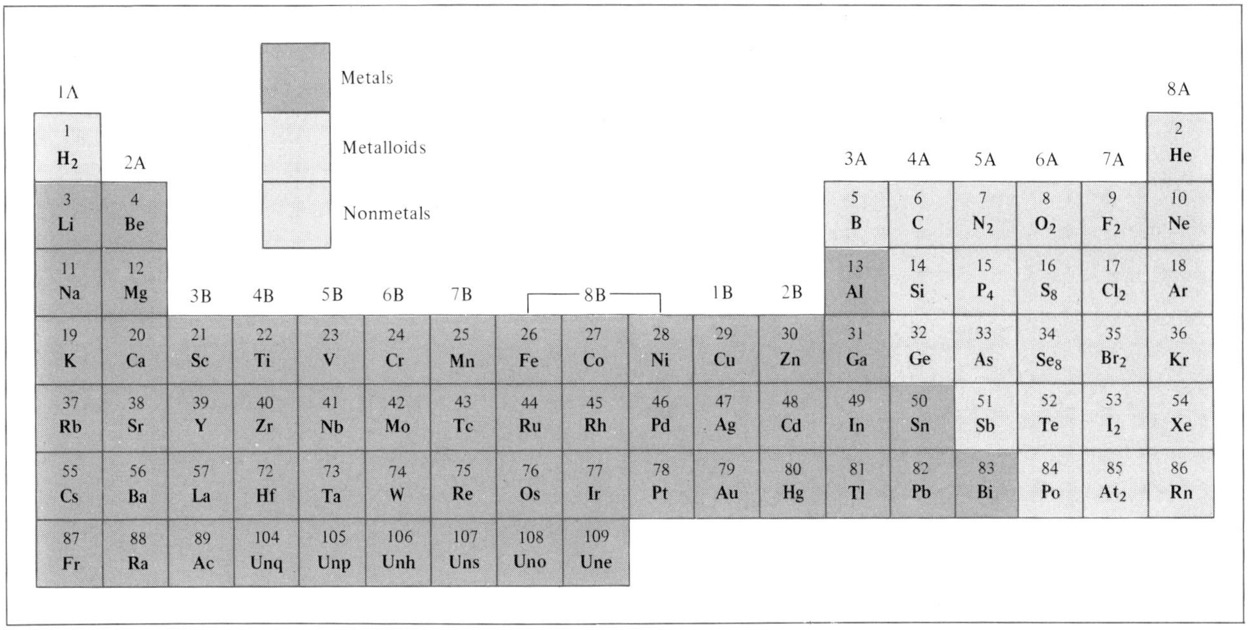

1A																	8A
1 H_2	2A											3A	4A	5A	6A	7A	2 He
3 Li	4 Be											5 B	6 C	7 N_2	8 O_2	9 F_2	10 Ne
11 Na	12 Mg	3B	4B	5B	6B	7B	8B		1B	2B		13 Al	14 Si	15 P_4	16 S_8	17 Cl_2	18 Ar
19 K	20 Ca	21 Sc	22 Ti	23 V	24 Cr	25 Mn	26 Fe	27 Co	28 Ni	29 Cu	30 Zn	31 Ga	32 Ge	33 As	34 Se_8	35 Br_2	36 Kr
37 Rb	38 Sr	39 Y	40 Zr	41 Nb	42 Mo	43 Tc	44 Ru	45 Rh	46 Pd	47 Ag	48 Cd	49 In	50 Sn	51 Sb	52 Te	53 I_2	54 Xe
55 Cs	56 Ba	57 La	72 Hf	73 Ta	74 W	75 Re	76 Os	77 Ir	78 Pt	79 Au	80 Hg	81 Tl	82 Pb	83 Bi	84 Po	85 At_2	86 Rn
87 Fr	88 Ra	89 Ac	104 Unq	105 Unp	106 Unh	107 Uns	108 Uno	109 Une									

Metals

Metalloids

Nonmetals

FIGURE 9.2 *Formulas of the stable forms of elements under atmospheric conditions. All metals and metalloids and the nonmetal carbon exist in giant three-dimensional structures; thus we use their empirical formulas (that is, their symbols) to represent them. The noble gases exist as monatomic gases, and so their formulas are the same as their symbols. The rest of the nonmetals exist as discrete molecular units. For simplicity, we omit the lanthanides and actinides.*

All the noble gases exist as monatomic species; thus we use their symbols: He, Ne, Ar, Kr, Xe, and Rn. The metalloids, like the metals, all exist in complex three-dimensional networks, and we represent them, too, with their empirical formulas, that is, their symbols: B, Si, Ge, and so on. Figure 9.2 shows the formulas of the stable forms of elements under atmospheric conditions.

9.3   PERIODIC VARIATION IN PHYSICAL PROPERTIES

As we have seen, the electron configurations of the elements show a periodic variation with increasing atomic number. Consequently, the elements also display periodic variations in their physical and chemical behavior. This section will examine some physical properties of elements in a group and across a period. The following three sections will discuss properties that influence the chemical behavior of elements.

Atomic Radius

A number of physical properties, such as density, melting point, and boiling point, are related to the sizes of atoms, but atomic size is difficult to define.

Atomic radius also influences chemical behavior. Here we will concentrate only on the periodic variation of atomic radii.

As we saw in Chapter 8, the electron density in an atom extends far beyond the nucleus. In practice, we normally think of atomic size as the volume containing about 90 percent of the total electron density around the nucleus.

Several techniques allow us to estimate the size of an atom. First consider the metallic elements. The structure of metals is quite varied, but they all share a common characteristic: Their atoms are linked to one another in an extensive three-dimensional network. Thus the **atomic radius** of a metal is *one-half the distance between the two nuclei in two adjacent atoms* [Figure 9.3(a)]. *For elements that exist as simple diatomic units*, for example, iodine (I_2), *the atomic radius is one-half the distance between the nuclei of the two atoms in a particular molecule* [Figure 9.3(b)].

Figure 9.4 shows the general trends in atomic radii of many elements according to their positions in the periodic table. Figure 9.5 plots the atomic radii of these elements against their atomic numbers. Periodic trends are clearly evident. Consider the second-period elements from Li to F. The Be atom is smaller than the Li atom because the two $2s$ electrons in Be do not shield each other effectively. Thus, the increase in nuclear charge (from Li to Be) results in a shrinkage in atomic radius. (That is, the electron density is pulled inward.) For this reason, the atomic radius of each alkaline earth metal (Group 2A) is smaller than that of the alkali metal (Group 1A) in the same period (see Figure 9.4).

Although the $2s$ electrons do shield the $2p$ electrons from the nucleus more effectively than they shield each other, the effect is not sufficient to offset increased nuclear charge. Consequently, the atomic radius decreases again from Be to B. Then the atomic radius decreases only slightly from C to F, indicating that the shielding effect and increased nuclear attraction must be nearly balanced in the carbon, nitrogen, oxygen, and fluorine atoms.

As we go down a group—say, Group 1A—we find that atomic radius increases with increasing atomic number. We can see this trend in Figure 9.5 by comparing the positions of Li, Na, K, Rb, and Cs. Each alkali metal atom has a single s electron outside a completely filled s (for lithium only) or p subshell (for the remaining metals). Since the orbital size increases with the increasing principal quantum number n, the size of the metal atoms increases from lithium to cesium, as Figure 9.5 illustrates.

The two opposing factors that affect the atomic radius are the shielding of the outer electron(s) by the inner electrons and the increase in nuclear charge that pulls the electron density inward. The former increases while the latter decreases the atomic radius.

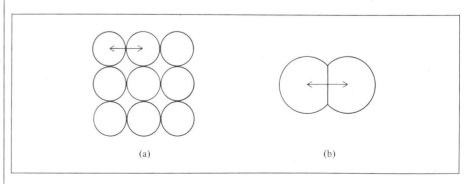

FIGURE 9.3 *(a) In a metal, the atomic radius is defined as one-half the distance between the centers of two adjacent atoms. (b) For elements such as iodine, which exist as discrete molecular units, the radius of the atom is defined as one-half the distance between the centers of the atoms in a molecule.*

(a)   (b)

Increasing atomic radius

1A												3A	4A	5A	6A	7A	8A
1 **H** 32	2A																2 **He** 93
3 **Li** 155	4 **Be** 112											5 **B** 98	6 **C** 91	7 **N** 92	8 **O** 73	9 **F** 72	10 **Ne** 71
11 **Na** 190	12 **Mg** 160	3B	4B	5B	6B	7B		8B		1B	2B	13 **Al** 143	14 **Si** 132	15 **P** 128	16 **S** 127	17 **Cl** 99	18 **Ar** 98
19 **K** 235	20 **Ca** 197	21 **Sc** 162	22 **Ti** 147	23 **V** 134	24 **Cr** 130	25 **Mn** 135	26 **Fe** 126	27 **Co** 125	28 **Ni** 124	29 **Cu** 128	30 **Zn** 138	31 **Ga** 141	32 **Ge** 137	33 **As** 139	34 **Se** 140	35 **Br** 114	36 **Kr** 112
37 **Rb** 248	38 **Sr** 215	39 **Y** 178	40 **Zr** 160	41 **Nb** 146	42 **Mo** 139	43 **Tc** 136	44 **Ru** 134	45 **Rh** 134	46 **Pd** 137	47 **Ag** 144	48 **Cd** 154	49 **In** 166	50 **Sn** 162	51 **Sb** 159	52 **Te** 160	53 **I** 133	54 **Xe** 131
55 **Cs** 267	56 **Ba** 222	57 **La** 187	72 **Hf** 167	73 **Ta** 149	74 **W** 141	75 **Re** 137	76 **Os** 135	77 **Ir** 136	78 **Pt** 139	79 **Au** 146	80 **Hg** 157	81 **Tl** 171	82 **Pb** 175	83 **Bi** 170	84 **Po** 176	85 **At** ?	86 **Rn** ?
87 **Fr** ?	88 **Ra** ?	89 **Ac** 188	104 **Unq** ?	105 **Unp** ?	106 **Unh** ?	107 **Uns** ?	108 **Uno** ?	109 **Une** ?									

Increasing atomic radius (left vertical axis)

FIGURE 9.4   *Trends in atomic radii (in picometers) of a number of elements according to their positions in the periodic table.*

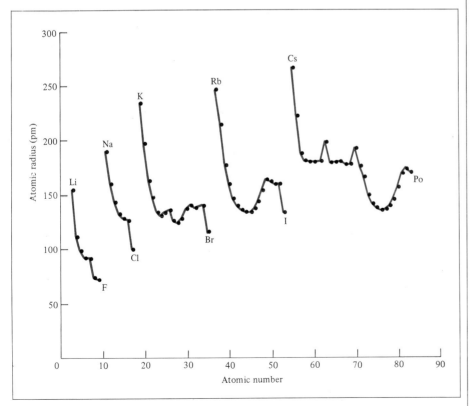

FIGURE 9.5   *Plot of atomic radii (in picometers) of elements against their atomic numbers.*

Note that the periodic trends in atomic radii (that is, the increase in atomic radii as we move down a group and the decrease in atomic radii as we move from left to right across a period) do not apply as well to the transition metals as they do to the representative elements (see Figure 9.4). Across a period of the transition metals there are only relatively small changes in the size of atomic radii. The reason is that the outer s electrons are shielded quite effectively from the increasing nuclear charge by the electrons being added to the d subshell. In the first-row transition metals, we see that the atomic radius first decreases from Sc to Cr, then rises slightly at Mn, then decreases slightly to Ni, and finally rises at Cu. Going down a column no clear trends are apparent in changes in atomic radii. From the first row to the second row the size of the atomic radius increases. However, from the second row to the third row there is only a slight increase in the size of radii.

Variation of Physical Properties Across a Period

As we move from left to right across a period, elements change from metals to metalloids to nonmetals, and therefore their physical properties change. Although these changes can be accounted for by the differing electron configurations of the elements, this section will look at the changes more broadly, that is, in terms of the characteristic properties of metallic and nonmetallic elements.

Table 9.3 shows a number of physical properties of the third-period elements. The molar heats of fusion and vaporization of a substance are the energies (in kJ) needed to melt and vaporize one mole of the substance at its melting point and boiling point, respectively. The terms electrical conductivity and thermal conductivity are used qualitatively here to indicate an element's ability to conduct electricity and heat. Sodium, the first element in the third period, is a very reactive metal, whereas chlorine, the second to last element of that period, is a very reactive nonmetal. In between, the elements show a gradual transition from metallic properties to

TABLE 9.3   Variations in Some Physical Properties for the Third-Period Elements

	Na	*Mg*	*Al*	*Si*	*P*	*S*	*Cl*	*Ar*
Type of element	←——— Metal ———→			Metalloid	←——— Nonmetal ———→			
Structure	← Extensive three-dimensional →				P_4	S_8	Cl_2	Atomic species
Density (g/cm³)	0.97	1.74	2.70	2.33	1.82	2.07	1.56*	1.40*
Melting point (°C)	98	649	660	1410	44	119	−101	−189
Molar heat of fusion (kJ/mol)	2.6	9.0	10.8	46.4	0.6	1.4	3.2	1.2
Boiling point (°C)	883	1090	2467	2355	280	445	−34.6	−186
Molar heat of vaporization (kJ/mol)	89.0	128.7	293.7	376.7	12.4	9.6	10.2	6.5
Electrical conductivity	Good	Good	Good	Fair	Poor	Poor	Poor	Poor
Thermal conductivity	Good	Good	Good	Fair	Poor	Poor	Poor	Poor

* The densities of Cl_2 and Ar are those of the liquids at their boiling points.

nonmetallic properties. Note that the melting point, molar heat of fusion, boiling point, and molar heat of vaporization rise from Na to Si and then fall to low values at Ar. Sodium, magnesium, and aluminum all have extensive three-dimensional atomic networks, and these atoms are held together by forces characteristic of the metallic state. Silicon is a metalloid; it has a giant three-dimensional structure in which the Si atoms are held together very strongly. That is why the heats of fusion and vaporization reach a maximum at silicon. (Recall that the stronger the force of attraction between the atoms, the greater the amount of energy needed to bring about the transition from solid to liquid or from liquid to vapor.) Starting with phosphorus, the elements exist in simple, discrete molecular units (P_4, S_8, Cl_2, and Ar). These molecules or atoms (for argon) are held together by relatively weak forces called *intermolecular forces* that operate among molecules. Therefore, they have lower heats of fusion and vaporization than do the metals and silicon.

The density of an element depends on three quantities: the atomic mass, the size of the atoms, and the way in which the atoms are packed together in the condensed state. (Recall that density is mass/volume.) We see that the density, like some of the other properties, rises as we move to the elements in the middle of the period, and then falls. The electrical conductivity and thermal conductivity are high for the metals and then fall rapidly as we move across the table. This is consistent with the fact that metals are good conductors of heat and electricity, whereas nonmetals are poor conductors in both respects. Note that silicon, a metalloid, has intermediate properties.

Similar trends are observed for representative elements in other periods. As we noted earlier, the transition metals, the lanthanides, and the actinides do not show this kind of periodic variation.

Predicting Physical Properties

One benefit of knowing the periodic trends in physical properties is that we can use this knowledge to predict properties of elements. Suppose we want to predict the boiling point of an element. One approach is to make use of the known boiling points of the element's immediate neighbors in the *same group.* In some cases we find that the boiling point of the element is fairly close to the average of the boiling points of the elements immediately above and below it in the group. The following example illustrates this approach.

EXAMPLE 9.2

The melting points of chlorine (Cl_2) and iodine (I_2) are $-101.0°C$ and $113.5°C$, and their boiling points are $-34.6°C$ and $184.4°C$, respectively. From these data estimate the melting point and boiling point of bromine (Br_2).

(Continued)

Answer

We note that chlorine, bromine, and iodine belong to Group 7A, the halogens. Since they are all nonmetals, we expect a gradual change in the melting point and boiling point of the elements as we move down the group. Bromine is located between chlorine and iodine; we therefore expect that many of its properties fall between those of chlorine and iodine. Thus we can estimate the melting point and boiling point of bromine by taking the average of those for chlorine and iodine. It is convenient mathematically to convert the temperatures to the Kelvin scale first by setting up the following table:

	Melting Point (K)	Boiling Point (K)
Cl_2	172.2	238.6
Br_2	?	?
I_2	386.7	457.6

$$\text{melting point of } Br_2 = \frac{172.2 \text{ K} + 386.7 \text{ K}}{2} = 279.5 \text{ K} = 6.4°C$$

and

$$\text{boiling point of } Br_2 = \frac{238.6 \text{ K} + 457.6 \text{ K}}{2} = 348.1 \text{ K} = 75°C$$

The actual melting point and boiling point of bromine are $-7.2°C$ and $58.8°C$, respectively, and so these estimates can be considered good "ballpark" values.

Note that predicting the properties of an element by examining the properties of its immediate neighbors in the same *period* is less reliable than predicting properties from *group* trends because the properties change more drastically across a period. This is particularly true when elements change from metal to metalloid and then to nonmetal.

9.4 IONIZATION ENERGY

Before we examine the periodic variations found in chemical properties, we need to discuss two quantities that play important roles in determining whether atoms of the elements will preferentially form ionic or molecular compounds: ionization energy and electron affinity (Section 9.5).

Ionization energy is *the minimum energy required to remove an electron from an isolated atom in its ground state.* The magnitude of ionization energy is a measure of the effort required to force an atom to give up an electron, or of how "tightly" the electron is held in the atom. The higher the ionization energy, the more difficult it is to remove the electron.

Ionization Energies of Many-Electron Atoms

For a many-electron atom, the amount of energy required to remove the first electron from the atom in its ground state

$$\text{energy} + X(g) \longrightarrow X^+(g) + e^- \tag{9.1}$$

is called the *first ionization energy* (I_1). In Equation (9.1) X represents an atom of any element, e is an electron, and g, as you know, denotes the gaseous state. Unlike an atom in the condensed phases (liquid and solid), an atom in the gaseous phase is virtually uninfluenced by its neighbors. The *second ionization energy* (I_2) and the *third ionization energy* (I_3) are shown in the following equations:

$$\text{energy} + X^+(g) \longrightarrow X^{2+}(g) + e^- \quad \text{second ionization}$$
$$\text{energy} + X^{2+}(g) \longrightarrow X^{3+}(g) + e^- \quad \text{third ionization}$$

The pattern continues for subsequent electrons.

After an electron is removed from a neutral atom, the repulsion among the remaining electrons decreases. Since the nuclear charge remains constant, more energy is needed to remove another electron from the positively charged ion. Thus, ionization energies always increase in the following order:

$$I_1 < I_2 < I_3 < \cdots$$

Figure 9.6 shows the first ionization energy of a number of elements, and the trends, according to their positions in the periodic table. Figure 9.7 shows the variation of the first ionization energy with atomic number. The plot clearly exhibits the periodicity in the stability of the most loosely held electron. Note that, apart from small irregularities, the ionization energies of elements in a period increase with increasing atomic number. Most notable are the peaks, which correspond to the noble gases. Since ionization energy is a measure of how strongly the electrons are held by the nucleus, we can see why most noble gases are chemically unreactive. In fact, helium has the highest first ionization energy of all the elements.

FIGURE 9.6 *Trends in the first ionization energies (in kJ/mol) of some elements according to their positions in the periodic table. Note that the ionization energy increases from left to right across a period and decreases from top to bottom down a group. These trends do not apply to the transition metals as well as they do to the representative elements.*

Ionization energies are measured in kJ/mol, that is, the amount of energy in kilojoules needed to remove one mole of electrons from one mole of atoms (or ions).

The resistance to chemical change is credited to the stability gained by the $1s^2$ configuration for helium and the completely filled outer s and p subshells of the other noble gases.

Increasing ionization energy

1A												3A	4A	5A	6A	7A	8A
H 1312	2A																**He** 2373
Li 520	**Be** 899											**B** 801	**C** 1086	**N** 1400	**O** 1314	**F** 1680	**Ne** 2080
Na 496	**Mg** 738	3B	4B	5B	6B	7B	⌐——8B——⌐			1B	2B	**Al** 578	**Si** 786	**P** 1012	**S** 1000	**Cl** 1251	**Ar** 1521
K 419	**Ca** 590	**Sc** 631	**Ti** 658	**V** 650	**Cr** 653	**Mn** 717	**Fe** 760	**Co** 759	**Ni** 737	**Cu** 745	**Zn** 906	**Ga** 579	**Ge** 760	**As** 947	**Se** 941	**Br** 1143	**Kr** 1350
Rb 403	**Sr** 549	**Y** 616	**Zr** 660	**Nb** 664	**Mo** 685	**Tc** 703	**Ru** 711	**Rh** 720	**Pd** 804	**Ag** 731	**Cd** 868	**In** 558	**Sn** 709	**Sb** 834	**Te** 870	**I** 1009	**Xe** 1170
Cs 376	**Ba** 503	**La** 541	**Hf** 676	**Ta** 760	**W** 770	**Re** 760	**Os** 820	**Ir** 870	**Pt** 869	**Au** 890	**Hg** 1007	**Tl** 589	**Pb** 716	**Bi** 703	**Po** 814	**At** 916	**Rn** 1037

Increasing ionization energy

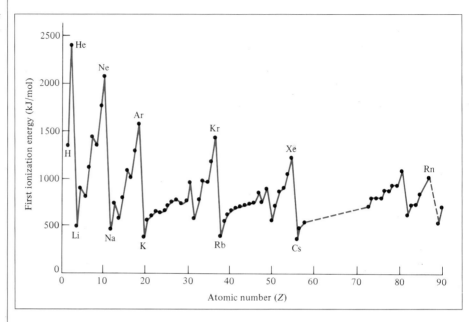

The Group 1A elements (the alkali metals) at the bottom of the graph have the lowest ionization energies. Each of these metals has one valence electron (the outermost electron configuration is ns^1) that is effectively shielded by the completely filled inner shells. Consequently, it is energetically easy to remove an electron from the atom of an alkali metal to form a *unipositive ion,* that is, an ion having a single positive charge (Li^+, Na^+, K^+, . . .). Significantly, the electron configurations of these ions are identical to those of the atoms of noble gases just preceding them in the periodic table (Table 9.4). *Ions, or atoms and ions, that possess the same number of electrons, and hence the same ground-state electron configuration,* are said to be **isoelectronic.** Thus the Li^+ ion is isoelectronic with He, the Na^+ ion is isoelectronic with Ne, and so on.

The Group 2A elements (the alkaline earth metals) have higher first ionization energies than the alkali metals do. The alkaline earth metals have two valence electrons (the outermost electron configuration is ns^2), which are more strongly held in the atom as a result of the increase in nuclear charge (for example, the increases from Li to Be and from Na to Mg). Most alkaline earth metal compounds contain *dipositive ions,* that is,

TABLE 9.4   Electron Configurations of the Alkali Metal Ions

Ion	Isoelectronic Noble Gas	Electron Configuration
Li^+	He	$1s^2$
Na^+	Ne	$[He]2p^6$
K^+	Ar	$[Ne]3s^23p^6$
Rb^+	Kr	$[Ar]4s^23d^{10}4p^6$
Cs^+	Xe	$[Kr]5s^24d^{10}5p^6$

ions containing two positive charges. The Be^{2+} ion is isoelectronic with Li^+ and with He, Mg^{2+} is isoelectronic with Na^+ and with Ne, and so on.

From the preceding discussion and Figure 9.6 we see that metals have relatively low ionization energies, whereas nonmetals possess much higher ionization energies. The ionization energies of the metalloids usually fall between those of metals and nonmetals. The difference in ionization energies suggests why metals always form cations and nonmetals form anions in ionic compounds. For a given group, the ionization energy decreases with increasing atomic number (that is, as we move down the group). Elements in the same group have similar outer electron configurations. However, as the principal quantum number n increases, so does the average distance of a valence electron from the nucleus. A greater separation between the electron and the nucleus means a weaker attraction, so that the electron becomes increasingly easier to remove as we go from element to element down a group. Thus the metallic character of the elements within a group increases from top to bottom. This trend is particularly noticeable for elements in Groups 3A to 7A. For example, in Group 4A we note that carbon is a nonmetal, silicon and germanium are metalloids, and tin and lead are metals.

Although the general trend in the periodic table is that of ionization energies increasing from left to right, some irregularities do exist. The first occurs in going from Group 2A to 3A (for example, from Be to B and from Mg to Al). The Group 3A elements all have a single electron in the outermost p subshell (ns^2np^1), which is well shielded by the inner electrons and the ns^2 electrons. Therefore, less energy is needed to remove a single p electron than to remove a paired s electron from the same principal energy level. This explains the *lower* ionization energies in Group 3A elements compared with those in Group 2A in the same period. The second irregularity occurs between Groups 5A and 6A (for example, going from N to O and from P to S). In the Group 5A elements (ns^2np^3) the p electrons are in three separate orbitals according to Hund's rule. In Group 6A (ns^2np^4) the additional electron must be paired with one of the three p electrons. The proximity of two electrons in the same orbital results in greater electrostatic repulsion, which makes it easier to ionize an atom of the Group 6A element, even though the nuclear charge has increased by one unit. Thus the ionization energies in Group 6A are *lower* than those in Group 5A in the same period.

The importance of ionization energy lies in the close correlation found between electron configuration (a microscopic property) and chemical behavior (a macroscopic property). As we shall see throughout the book, the chemical properties of any atom are related to the configuration of the atom's valence electrons, and the stability of these electrons is reflected directly in the atom's ionization energies.

Ionization Energies and Evidence for Shell Structure

In addition to revealing information about the relative stability of valence electrons, ionization energy measurements also provide strong evidence for the existence of electron shells in atoms. Consider the element sodium,

The only important nonmetallic cation is the ammonium ion NH_4^+.

FIGURE 9.8 *Plot of the log of ionization energy against the number (N) of ionization energy for sodium $(1s^2 2s^2 2p^6 3s^1)$. The first point (I_1) corresponds to the removal of the 3s electron; the next eight points $(I_2$ to $I_9)$, to the 2p and 2s electrons; and the last two points $(I_{10}$ and $I_{11})$, to the 1s electrons. The sharp breaks strongly suggest the existence of shell structure.*

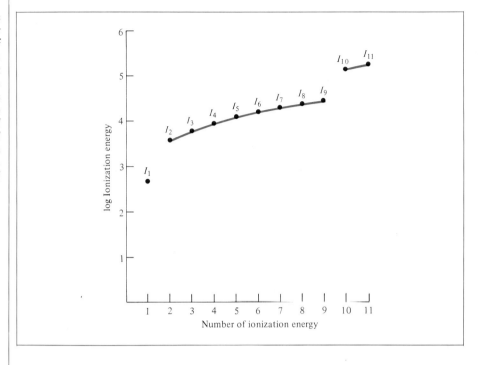

For example, for $N = 3$ we have I_3, which is the ionization energy for the third ionization process, that is

$$X^{2+}(g) \longrightarrow X^{3+}(g) + e^-$$

whose electron configuration is $1s^2 2s^2 2p^6 3s^1$. Figure 9.8 is a plot of the log of ionization energy against the number (N) of ionization energy (I_N). (We use a log scale here because ionization energies increase very rapidly as electrons are dislodged one after the other. A normal scale would make the y axis much too long.) It is clear from Figure 9.8 that sodium's eleven electrons are arranged in three distinct groups with different stabilities. After removing the first electron (I_1) there is a big jump between I_1 and I_2. Then the ionization energies increase more steadily from I_2 to I_9. Another jump occurs from I_9 to I_{10}. The three groups of electrons clearly correspond to the three shells characterized by the principal quantum numbers $n = 1$, 2, and 3. And I_1 corresponds to the $3s^1$ electron, I_2 through I_9 correspond to the $2s^2 2p^6$ electrons, and I_{10} and I_{11} correspond to the $1s^2$ electrons. Note that while $I_1 = 495.9$ kJ/mol, $I_{11} = 2.0 \times 10^5$ kJ/mol. This means that it takes about 400 times more energy to remove the last electron from the Na^{10+} ion

$$\text{energy} + Na^{10+}(g) \longrightarrow Na^{11+}(g) + e^-$$

than to remove the first electron from a neutral Na atom

$$\text{energy} + Na(g) \longrightarrow Na^+(g) + e^-$$

9.5   ELECTRON AFFINITY

Another property of atoms that greatly influences their chemical behavior is their ability to accept one or more electrons. This ability is measured by

1A	2A	3B	4B	5B	6B	7B	8B	1B	2B	3A	4A	5A	6A	7A	8A
H −77															He (21)
Li −58	Be (241)									B −23	C −123	N 0	O −142	F −333	Ne (29)
Na −53	Mg (230)									Al −44	Si −120	P −74	S −200	Cl −348	Ar (35)
K −48	Ca (154)									Ga (−35)	Ge −118	As −77	Se −195	Br −324	Kr (39)
Rb −47	Sr (120)									In −34	Sn −121	Sb −101	Te −190	I −295	Xe (40)
Cs −45	Ba (52)									Tl −48	Pb −101	Bi −100	Po ?	At ?	Rn ?

FIGURE 9.9 *Electron affinities (kJ/mol) of selected elements. The values in parentheses are estimates. By convention, energy released has a negative sign and energy absorbed has a positive sign. Note that the more negative the value, the greater the tendency for an atom of the element to accept an electron.*

electron affinity, which is *the energy change when an electron is accepted by an atom in the gaseous state.* The equation is

$$X(g) + e^- \longrightarrow X^-(g)$$

where X is an atom of an element. In accord with the convention used in thermochemistry (see Chapter 5), we assign a negative value to the electron affinity when energy is released. The more negative the electron affinity, the greater the tendency of the atom to accept an additional electron. Figure 9.9 shows the electron affinity values of some representative elements arranged according to their positions in the periodic table. No clear-cut periodic trends are apparent among electron affinities, in contrast to the trends found among ionization energies (Figure 9.10). In general, electron affinity values become more negative as we move from left to right across a period, but this trend is not observed among the Group 2A elements. The electron affinities of metals are generally more positive (or less negative) than those of nonmetals. The values differ little within each group. The halogens (Group 7A) have the most negative electron affinity values. This is not surprising when we realize that by accepting an electron each halogen atom assumes the electron configuration of a noble gas. For example, the configuration of F^- is $1s^2 2s^2 2p^6$ or [Ne]; for Cl^- it is [Ne]$3s^2 3p^6$ or [Ar]; and so on.

The electron affinity of oxygen has a negative value, which means that the process

$$O(g) + e^- \longrightarrow O^-(g)$$

is favorable. On the other hand, the electron affinity of the O^- ion

$$O^-(g) + e^- \longrightarrow O^{2-}(g)$$

is positive (708 kJ/mol) even though the O^{2-} ion is isoelectronic with the noble gas Ne. This process is unfavorable in the *gas phase* because the

FIGURE 9.10 *Plot of electron affinity against atomic number for the first twenty elements in the periodic table.*

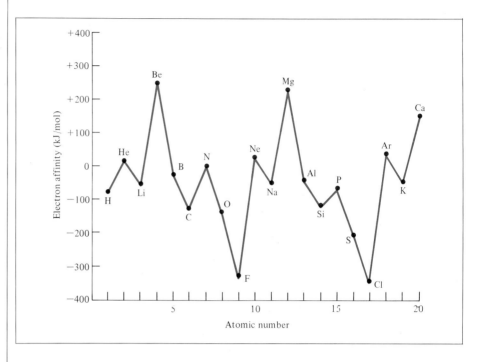

resulting increase in electron–electron repulsion outweighs the stability gained in achieving a noble gas configuration. However, note that ions such as O^{2-} are common in ionic compounds (for example, Li_2O and MgO); in solids, these ions are stabilized by the neighboring cations.

EXAMPLE 9.3

Explain why the electron affinities of the alkaline earth metals are all positive.

Answer

The outer-shell electron configuration of the alkaline earth metals is ns^2. For the process

$$M(g) + e^- \longrightarrow M^-(g)$$

where M denotes a member of the Group 2A family, the extra electron must enter the np subshell, which is effectively shielded by the two ns electrons (the np electrons are farther away from the nucleus than the ns electrons). Consequently, alkaline earth metals have no tendency to pick up an extra electron.

These two fundamental quantities are related in a simple way conceptually: Ionization energy indexes the attraction of an atom for its own outer electrons, whereas electron affinity expresses the attraction of an atom for an additional electron from some other source. Together, they give us insight into the general attraction of an atom for electrons or electron density.

9.6 VARIATIONS IN CHEMICAL PROPERTIES

Ionization energy and electron affinity are two characteristics that help chemists understand the types of reactions that elements undergo and the nature of the elements' compounds. Now that you are familiar with these characteristics, we can survey the chemical behavior of the elements systematically, paying particular attention to the relationship between chemical properties and electron configuration.

We have seen that as we move across a period from left to right in the periodic table, the metallic character of the elements decreases. And as we move down a group the metallic character of the elements increases. On the basis of these trends and the knowledge that metals usually have low ionization energies while nonmetals usually have high (more negative) electron affinities, we can frequently predict the chemical outcome when atoms of some of these elements react with one another. The following discussion focuses only on the representative elements, since variation in chemical behavior of transition metals across a period is not as pronounced as that of the representative elements. This is so because as we move across a period containing transition metals, electrons are being added to the $(n-1)d$ subshells, where n denotes the highest principal quantum number containing electrons. Since it is the outermost electrons that have the greatest influence on the properties of elements, adding an electron to an inner d orbital results in less striking changes in properties than does adding an electron to an outer s or p orbital. We will consider the chemistry of transition metals in a later chapter.

General Trends in Chemical Properties

Before we study the elements in individual groups, let us look at some overall trends. We have said that elements in the same group resemble one another in chemical behavior because they have similar outer electron configurations. This statement, although correct in the general sense, must be applied with caution. Chemists have long known that the first member of each group (that is, the element in the second period from lithium to fluorine) differs from the rest of the members of the same group. Thus lithium, while exhibiting many of the properties characteristic of the alkali metals, differs in some ways from the rest of the metals in Group 1A. Similarly, beryllium is somewhat atypical of the members of Group 2A, and so on.

Another trend in chemical behavior of the representative elements is the **diagonal relationship.** Diagonal relationship refers to *similarities that exist between pairs of elements in different groups and periods of the periodic table.* Although elements in the second period show differences from other elements in the same group, the first three members of the second period (Li, Be, and B) exhibit many similarities to those elements located diagonally below them in the periodic table (Figure 9.11). The chemistry of lithium resembles that of magnesium in some ways; the same holds for beryllium and aluminum and for boron and silicon. Thus each pair of these elements is said to exhibit a diagonal relationship. We will see a number of examples illustrating this relationship later.

In comparing the properties of elements in the same group, bear in mind that the comparison is most valid if we are dealing with elements of the same type. This guideline applies to the elements in Groups 1A and 2A, which are all metals, and to the elements in Group 7A, which are all

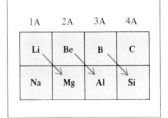

FIGURE 9.11 *Diagonal relationship in the periodic table. The relationship is most pronounced for lithium and magnesium, beryllium and aluminum, and boron and silicon.*

nonmetals. In Groups 3A through 6A, where the elements change either from nonmetals to metals or from nonmetals to metalloids, it is natural to expect greater variations in chemical properties even though the members of the same group have similar outer electron configurations.

We will now survey the chemical properties of hydrogen and the elements in Groups 1A through 8A. (Color Plates 1–8 show the physical appearance of and other facts about a number of elements discussed in this book.)

Chemical Properties in Individual Groups

Hydrogen ($1s^1$).   There is no totally suitable position for hydrogen in the periodic table. It resembles the alkali metals in having a single s valence electron and forming the H^+ ion, which is hydrated in solution [similar, for example, to the $Na^+(aq)$ ions]. On the other hand, hydrogen also forms the hydride ion (H^-), which is too unstable to exist in water but does exist in some ionic compounds. In this respect, hydrogen resembles the halogens, since they all form halide ions (F^-, Cl^-, Br^-, and I^-). Traditionally hydrogen is shown in Group 1A in the periodic table, but you should *not* think of it as a member of that group. The most important compound of hydrogen is, of course, water, which is formed when hydrogen burns in air:

$$2H_2(g) + O_2(g) \longrightarrow 2H_2O(l)$$

Group 1A Elements (ns^1, $n \geq 2$).   In Figure 9.6 we see that the alkali metals all have low ionization energies and therefore a great tendency to lose their single valence electron. In fact, in the vast majority of their compounds they appear as unipositive ions. These metals are so reactive that they are never found in the free, uncombined state in nature. They react with water to produce hydrogen gas and the corresponding metal hydroxide:

$$2M(s) + 2H_2O(l) \longrightarrow 2MOH(aq) + H_2(g)$$

where M denotes an alkali metal. When exposed to air, they gradually lose their shiny appearance as they combine with oxygen gas to form several different types of oxides or oxygen compounds. Lithium differs from the rest of the alkali metals in that it forms the normal oxide:

$$4Li(s) + O_2(g) \longrightarrow 2Li_2O(s)$$

The other alkali metals all form peroxides. For example

$$2Na(s) + O_2(g) \longrightarrow Na_2O_2(s)$$

Potassium, rubidium, and cesium also form superoxides:

$$K(s) + O_2(g) \longrightarrow KO_2(s)$$

Group 2A Elements (ns^2, $n \geq 2$).   As a group, the alkaline earth metals too are reactive metals, but less so than the alkali metals. As we saw earlier, the alkaline earth metals have a tendency to lose both valence electrons and to exist as dipositive ions in most of their compounds. In Figure 9.6

An oxide contains the O^{2-} ion, a peroxide the O_2^{2-} ion, and a superoxide the O_2^- ion.

we see that beryllium has the highest first ionization energy of the group. (The same relationship holds for their second ionization energies.) As a result, most beryllium compounds are molecular rather than ionic in nature. The reactivities of alkaline earth metals toward water differ quite markedly. Beryllium does not react with water; magnesium reacts slowly with steam; calcium, strontium, and barium are reactive enough to attack cold water:

$$M(s) + 2H_2O(l) \longrightarrow M(OH)_2(aq) + H_2(g)$$

where M denotes Ca, Sr, or Ba. The reactivities of the alkaline earth metals toward oxygen also increase from Be to Ba. Beryllium and magnesium form normal oxides (BeO and MgO) only at elevated temperatures, whereas CaO, SrO, and BaO are formed at room temperature.

Magnesium reacts with acids to liberate hydrogen gas:

$$Mg(s) + 2H^+(aq) \longrightarrow Mg^{2+}(aq) + H_2(g)$$

Calcium, strontium, and barium also react with acids to produce hydrogen gas. However, because these metals also attack water, two different reactions will occur simultaneously.

Group 3A Elements (ns^2np^1, $n \geq 2$). The first member of this group, boron, is a metalloid; the rest of the members are metals. Boron does not form binary ionic compounds and is unreactive toward oxygen gas and water. The next element, aluminum, also does not react with water, but it readily forms aluminum oxide when exposed to air:

$$4Al(s) + 3O_2(g) \longrightarrow 2Al_2O_3(s)$$

It is a protective layer of the oxide that renders aluminum unreactive. The metallic elements of Group 3A tend to lose all three valence electrons in reactions to form *tripositive ions*, that is, cations bearing three positive charges (Al^{3+}, Ga^{3+}, In^{3+}, Tl^{3+}). For example, aluminum reacts with hydrochloric acid to produce hydrogen gas:

$$2Al(s) + 6H^+(aq) \longrightarrow 2Al^{3+}(aq) + 3H_2(g)$$

However, these metallic elements also form many molecular compounds. Thus as we move from Group 1A to Group 3A, we begin to see a gradual shift from metallic character to nonmetallic character.

Group 4A Elements (ns^2np^2, $n \geq 2$). The first member of this group, carbon, is a nonmetal, and the next two members, silicon and germanium, are metalloids. These three elements do not form ionic compounds. They are stable toward water, although they do form oxides. For example, carbon forms carbon dioxide and carbon monoxide when it burns. The metallic elements of this group, tin and lead, do not react with water but form the oxides SnO, SnO_2, PbO, and PbO_2. SnO_2 and PbO_2 have little ionic character. Tin and lead react with acids (hydrochloric acid, for example) to liberate hydrogen gas:

$$Sn(s) + 2H^+(aq) \longrightarrow Sn^{2+}(aq) + H_2(g)$$

$$Pb(s) + 2H^+(aq) \longrightarrow Pb^{2+}(aq) + H_2(g)$$

Note that the reaction with water is similar to that found for the Group 1A elements. In both cases the metallic hydroxide is formed along with hydrogen gas.

Li resembles Mg in that both form the normal oxide. This is an example of the diagonal relationship.

Group 5A Elements (ns^2np^3, $n \geqslant 2$). In this group nitrogen and phosphorus are nonmetals, arsenic and antimony are metalloids, and bismuth is a metal. Thus we expect a larger variation in properties as we move down the group. Nitrogen exists as a diatomic gas (N_2). It forms a number of oxides (NO, N_2O, NO_2, N_2O_4, and N_2O_5), of which only N_2O_5 is a solid; the others are gases. Nitrogen has a tendency to accept three electrons to form the nitride ion, N^{3-} (thus achieving the electron configuration $2s^22p^6$, isoelectronic with neon). Most metallic nitrides (Li_3N and Mg_3N_2, for example) are ionic compounds. Phosphorus exists as P_4 molecules. It forms two solid oxides with the formulas P_4O_6 and P_4O_{10}. Arsenic, antimony, and bismuth have extensive three-dimensional structures. Bismuth is a far less reactive metal than those in the earlier groups.

Group 6A Elements (ns^2np^4, $n \geqslant 2$). The first three members of this group (oxygen, sulfur, and selenium) are nonmetals, and the last two members (tellurium and polonium) are metalloids. Oxygen is a diatomic gas; sulfur and selenium exist as S_8 and Se_8 units; tellurium and polonium have more extensive three-dimensional structures. Oxygen has a tendency to accept two electrons to form the oxide ion (O^{2-}) in many ionic compounds. (The electron configuration of O^{2-} is $2s^22p^6$, isoelectronic with neon.) Sulfur, selenium, and tellurium also form anions of the type S^{2-}, Se^{2-}, and Te^{2-}. The elements in this group (especially oxygen) form a large number of molecular compounds with nonmetals from the other groups.

Polonium is a radioactive element that is difficult to study in the laboratory.

Group 7A Elements (ns^2np^5, $n \geqslant 2$). All the halogens are nonmetals with the general formula X_2, where X denotes a halogen element. Because of their great reactivity, the halogens are never found in the elemental form in nature. Fluorine is so reactive that it attacks water to generate oxygen:

The last member of Group 7A is astatine, a synthetic element. Little is known about its properties.

$$2F_2(g) + 2H_2O(l) \longrightarrow 4HF(aq) + O_2(g)$$

The reaction between fluorine gas and water is quite complex; the products formed depend on reaction conditions. The reaction shown here is one of several possible changes.

The halogens have high ionization energies and large negative electron affinities. These facts suggest that they would preferentially form anions of the type X^-. Anions derived from the halogens (F^-, Cl^-, Br^-, and I^-) are called *halides.* They are isoelectronic with the noble gases. For example, F^- is isoelectronic with Ne, Cl^- with Ar, and so on. The vast majority of the alkali metal and alkaline earth metal halides are ionic compounds. The halogens also form many molecular compounds among themselves (such as ICl and BrF_3) and with nonmetallic elements in other groups (such as PCl_5 and NF_3). The halogens react with hydrogen to form hydrogen halides:

$$H_2(g) + X_2(g) \longrightarrow 2HX(g)$$

Here is an example of how the first member of a group differs from the other members.

When this reaction involves fluorine it is explosive, but it becomes less and less violent from chlorine to iodine. This is one indication that fluorine is considerably more reactive than the rest of the halogens. The hydrogen halides dissolve in water to form hydrohalic acids. Of this series, hydrofluoric acid (HF) is a weak acid (that is, it is a weak electrolyte), but the other acids (HCl, HBr, and HI) are strong acids (or strong electrolytes).

Group 8A Elements (ns^2np^6, $n \geqslant 2$). All noble gases exist as monatomic species. This fact suggests that they are probably very unreactive and have little or no tendency to combine among themselves or with other elements. That is indeed the case. The electron configurations of the noble gases show that their atoms have completely filled outer ns and np subshells, indicating great stability. The Group 8A ionization energies are among the highest of all elements (see Figure 9.6), and they have no tendency to accept extra electrons. For a number of years these elements were called inert gases due to their observed lack of chemical reactivity. Since 1962, however, a number of compounds involving krypton and xenon have been synthesized.

Comparison of Group 1A and Group 1B Elements

When we compare the Group 1A (alkali metals) and the Group 1B elements (copper, silver, and gold), we arrive at an interesting conclusion. Although the elements in these two groups have similar outer electron configurations, with one electron in the outermost s orbital, their chemical properties are quite different.

The first ionization energies of Cu, Ag, and Au are 745 kJ/mol, 731 kJ/mol, and 890 kJ/mol, respectively. Since these values are considerably larger than those of the alkali metals (see Figure 9.6), the Group 1B elements are much less reactive. The higher ionization energies of the Group 1B elements result from incomplete shielding of the nucleus by the inner d electrons (compared to the more effective shielding of the completely filled noble gas cores). Consequently the outer s electrons of these elements are more strongly attracted by the nucleus. In fact, copper, silver, and gold are so unreactive that they are usually found in the uncombined state in nature. The inertness and rarity of these metals have made them valuable in the manufacture of coins and in jewelry. For this reason, these metals are also called coinage metals. The difference in chemical properties between the Group 2A elements (the alkaline earth metals) and the Group 2B metals (zinc, cadmium, and mercury) can be similarly explained.

Properties of Oxides Across a Period

One way to compare the properties of the representative elements across a period is to examine the properties of a series of similar compounds. Since oxygen combines with almost all elements, we will briefly compare the properties of oxides of the third-period elements to see how metals differ from metalloids and nonmetals. Table 9.5 lists a few general characteristics of the oxides. We observed earlier that oxygen has a tendency to form the oxide ion. This tendency is greatly favored when oxygen combines with metals that have low ionization energies, namely, those in Groups 1A and 2A and aluminum. Thus Na_2O, MgO, and Al_2O_3 are ionic compounds as indicated by their high melting points and boiling points. They have extensive three-dimensional structures in which each cation is surrounded by a specific number of anions, and vice versa. As the ionization energies of the elements increase from left to right, so does the molecular nature of the oxides that are formed. Silicon is a metalloid; its oxide (SiO_2) also has a huge three-dimensional network, although no ions are present. The oxides of phosphorus, sulfur, and chlorine are molecular compounds composed of

The electron configuration of helium is $1s^2$.

Some elements in this period (P, S, and Cl) form several types of oxides. We consider only oxides in which the number of oxygen atoms in the compound is a maximum.

TABLE 9.5 Some Properties of Oxides of the Third-Period Elements

	Na_2O	MgO	Al_2O_3	SiO_2	P_4O_{10}	SO_3	Cl_2O_7
Type of compound	← Ionic →			← Molecular →			
Structure	← Extensive three-dimensional →			← Discrete molecular units →			
Melting point (°C)	1275	2800	2045	1610	580	16.8	−91.5
Boiling point (°C)	?	3600	2980	2230	?	44.8	82
ΔH_f° (kJ/mol)	−416	−602	−1670	−859	−2984	−395	+265
ΔH_f° per mole of O atoms*	−416	−602	−557	−430	−298	−132	+38
Acid–base nature	Basic	Basic	Amphoteric	← Acidic →			

* These values are obtained by dividing ΔH_f° by the number of O atoms in each unit of the oxide.

small discrete units. The weak attractions among these molecules result in relatively low melting points and boiling points.

The stability of the oxides is indicated by their standard enthalpies of formation. You will recall from Chapter 5 that substances with negative ΔH_f° values tend to be more stable than those with positive ΔH_f° values. The first row of ΔH_f° values in Table 9.5 does not show any distinct pattern. However, when we compare them with the (second row) standard enthalpies of formation per mole of oxygen atoms (a value that represents the stability of combination of one mole of the element with one mole of oxygen atoms), we see that the values become more positive as we move across the period from left to right. This tells us that the stability of the oxides decreases from metallic oxides to oxides formed between nonmetallic elements. (The only break in this trend is from Na_2O to MgO.)

In Section 3.7 we discussed some aspects of acid–base neutralization reactions. An important characteristic of oxides is their acid–base properties because it turns out that oxide substances can be classified as acidic or basic. (A third type of oxide, which has both acidic and basic properties, is called *amphoteric oxide*, and will be discussed shortly.) The first two oxides of the third period, Na_2O and MgO, are basic oxides. For example, Na_2O reacts with water to form metallic hydroxide:

$$Na_2O(s) + H_2O(l) \longrightarrow 2NaOH(aq)$$

Magnesium oxide is quite insoluble, and so it does not react with water to any appreciable extent. However, it does react with acids in a manner that resembles an acid–base reaction:

$$MgO(s) + 2HCl(aq) \longrightarrow MgCl_2(aq) + H_2O(l)$$

Note that the products of this reaction are a salt ($MgCl_2$) and water, the usual products from an acid–base neutralization. Aluminum oxide is even less soluble than magnesium oxide, so that it too does not react with water. However, it shows basic properties by reacting with acids:

$$Al_2O_3(s) + 6HCl(aq) \longrightarrow 2AlCl_3(aq) + 3H_2O(l)$$

and acidic properties by reacting with bases:

$$Al_2O_3(s) + 2NaOH(aq) + 3H_2O(l) \longrightarrow 2NaAl(OH)_4(aq)$$

Thus Al_2O_3 is classified as an ***amphoteric oxide,*** which is *an oxide that exhibits both acidic and basic properties.*

Silicon dioxide is insoluble and does not react with water. It has acidic properties, however, because it reacts with very concentrated bases:

$$SiO_2(s) + 2NaOH(aq) \longrightarrow Na_2SiO_3(aq) + H_2O(l)$$

The remaining third-period oxides are acidic as indicated by their reactions with water to form phosphoric acid (H_3PO_4), sulfuric acid (H_2SO_4), and perchloric acid ($HClO_4$):

$$P_4O_{10}(s) + 6H_2O(l) \longrightarrow 4H_3PO_4(aq)$$

$$SO_3(g) + H_2O(l) \longrightarrow H_2SO_4(aq)$$

$$Cl_2O_7(g) + H_2O(l) \longrightarrow 2HClO_4(aq)$$

This brief examination of oxides of the third-period elements tells us that as the metallic character of the elements decreases from left to right across the period, their oxides change from basic to amphoteric to acidic. Normal metallic oxides are usually basic, and oxides of nonmetals are always acidic. The intermediate properties of the oxides (as shown by the amphoteric oxides) are exhibited by elements whose positions are intermediate within the period. Note also that since the metallic character of the elements increases as we move down a particular group of the representative elements, we would expect that oxides of elements with larger atomic numbers would be more basic than the lighter elements. This is indeed the case.

Although the product is only a salt and no water is formed, the reaction qualifies as an acid–base reaction.

EXAMPLE 9.4

Classify the following oxides as acidic, basic, or amphoteric: (a) Rb_2O, (b) BeO, (c) As_2O_5.

Answer

(a) Since rubidium is an alkali metal, we would expect Rb_2O to be a basic oxide. This is indeed true, as shown by rubidium oxide's reaction with water to form rubidium hydroxide:

$$Rb_2O(s) + H_2O(l) \longrightarrow 2RbOH(aq)$$

(b) Beryllium is an alkaline earth metal. However, because it is the first member of Group 2A, we expect that it may differ somewhat from the other members in the group. Furthermore, beryllium and aluminum exhibit diagonal relationship, so that BeO may resemble Al_2O_3 in properties. It turns out that BeO, like Al_2O_3, is an amphoteric oxide, as shown by its reactions with acids and bases:

$$BeO(s) + 2H^+(aq) + 3H_2O(l) \longrightarrow Be(H_2O)_4^{2+}(aq)$$

$$BeO(s) + 2OH^-(aq) + H_2O(l) \longrightarrow Be(OH)_4^{2-}(aq)$$

(Continued)

(c) Since arsenic is a nonmetal, we expect As_2O_5 to be an acidic oxide. This prediction is correct, as shown by the formation of arsenic acid when As_2O_5 reacts with water:

$$As_2O_5(s) + 3H_2O(l) \longrightarrow 2H_3AsO_4(aq)$$

AN ASIDE ON THE PERIODIC TABLE
The Third Liquid Element

O f the 109 known elements, eleven are gases under atmospheric conditions. Six of these are the Group 8A elements (the noble gases He, Ne, Ar, Kr, Xe, and Rn), and the other five are hydrogen (H_2), nitrogen (N_2), oxygen (O_2), fluorine (F_2), and chlorine (Cl_2). Curiously, only two elements are liquids at 25°C: mercury (Hg) and bromine (Br_2).

We do not know the properties of all the known elements because some of them have never been prepared in quantities large enough for investigation. In these cases we must rely on periodic trends to predict their properties. What are the chances, then, of discovering a third liquid element?

Francium (Fr) is the last member of the Group 1A elements (the alkali metals). All of francium's isotopes are radioactive. The most stable isotope is francium-223, which has a half-life of 21 minutes. (*Half-life* is the time it takes for one-half of the atoms in any given amount of a radioactive substance to disintegrate.) This short half-life means that only very small traces of francium could possibly exist on Earth. And although it is feasible to prepare francium in the laboratory, we could never accumulate enough of it to study—the energy released during nuclear decay in all but the tiniest amount would vaporize the material instantly. Thus we know very little about francium's physical and chemical properties. Yet we can use the group periodic trends to predict some of those properties.

Take francium's melting point as an example. Figure 9.12 shows how the melting

FIGURE 9.12 Plot of the melting points of the alkali metals against their atomic numbers.

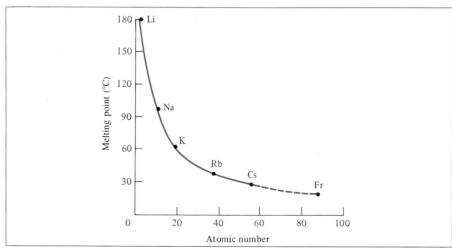

points of the alkali metals vary with atomic number. From lithium to sodium, the melting point drops 81.4 degrees; from sodium to potassium, 34.6 degrees; from potassium to rubidium, 24 degrees; from rubidium to cesium, 11 degrees. On the basis of this trend, we can predict that the drop from cesium to francium would be about 5 degrees. If so, the melting point of francium would be 23°C, which would qualify it as a liquid under atmospheric conditions.

SUMMARY

1. Electron configuration directly influences the properties of the elements. The periodic table classifies the elements according to their atomic numbers, and thus also by their electron configurations. The configuration of the outermost electrons (called valence electrons) directly affects the properties of the atoms of the representative elements.
2. As we move across a period, periodic variations are found in physical properties of the elements. These variations can be understood in terms of the changes in their structures brought about as the elements progress from metals to metalloids and then to nonmetals. The metallic character of elements increases as we move down a particular group containing representative elements.
3. The size of an atom, indicated by atomic radius, varies periodically with the arrangement of the elements in the periodic table.
4. Ionization energy is a measure of the tendency of an atom to lose an electron. The higher the ionization energy, the more strongly the nucleus holds the electron. Electron affinity is a measure of the tendency of an atom to gain an electron. The more negative the electron affinity, the greater the tendency for the atom to gain an electron. Metals usually have low ionization energies, and nonmetals usually have high (large negative) electron affinities.
5. Noble gases are very stable because their outer ns and np subshells are completely filled. The metals of the representative elements (in Groups 1A, 2A, and 3A) tend to lose electrons until the cations of these metals become isoelectronic with the noble gases preceding them in the periodic table. The nonmetals of the representative elements (in Groups 5A, 6A, and 7A) tend to accept electrons until the anions of these nonmetals become isoelectronic with the noble gases following them in the periodic table.

KEY WORDS

Amphoteric oxide, p. 273
Atomic radius, p. 256
Diagonal relationship, p. 267
Electron affinity, p. 265
Ionization energy, p. 260

Isoelectronic, p. 262
Noble gases, p. 253
Representative elements, p. 253
Valence electrons, p. 253

PROBLEMS

More challenging problems are marked with an asterisk.

Periodic Classification of the Elements

9.1 Give two examples each of a metal, a nonmetal, and a metalloid.

9.2 Compare the physical and chemical properties of metals and nonmetals.

9.3 Draw a rough sketch of a periodic table (no details are required). Indicate regions where metals, nonmetals, and metalloids are located.

9.4 What is a representative element? Give names and symbols of four representative elements.

9.5 What is a noble gas core?

9.6 For each of the following elements, describe its stable form of aggregation under atmospheric conditions (that is, whether the element exists as a monatomic gas, a diatomic gas, in molecular form in the liquid or solid state, or in extensive three-dimensional structure in the solid state): H, C, O, P, S, Ar, Br, Ag, I, Xe.

9.7 What would you consider to be the most important feature of the periodic table?

9.8 In the periodic table, the element hydrogen is sometimes grouped with alkali metals (as in this book) and sometimes with the halogens. Explain why hydrogen can resemble the Group 1A and the Group 7A elements.

9.9 Without referring to a periodic table, write the name and give the symbol for an element in each of the following: Group 1A, Group 2A, Group 3A, Group 4A, Group 5A, Group 6A, Group 7A, Group 8A, transition metals, lanthanides.

Electron Configuration and the Periodic Table

9.10 Write the outer electron configurations for (a) the alkali metals, (b) the alkaline earth metals, (c) the halogens, (d) the noble gases.

9.11 Use the first-row transition metals (Sc to Cu) as an example to illustrate the characteristics of the electron configurations of transition metals.

9.12 Group the following electron configurations in pairs that would represent similar chemical properties of their atoms:

(a) $1s^2 2s^2 2p^6 3s^2 3p^5$
(b) $1s^2 2s^2 2p^6 3s^2$
(c) $1s^2 2s^2 2p^3$
(d) $1s^2 2s^2 2p^6 3s^2 3p^6 4s^2 3d^{10} 4p^6$
(e) $1s^2 2s^2$
(f) $1s^2 2s^2 2p^6$

(g) $1s^2 2s^2 2p^6 3s^2 3p^3$
(h) $1s^2 2s^2 2p^5$

9.13 Without referring to a periodic table, write the electron configuration of elements with the following atomic numbers: (a) 9, (b) 20, (c) 26, (d) 33. Classify the elements.

9.14 Specify in what group of the periodic table each of the following elements is found: (a) $[Ne]3s^1$, (b) $[Ne]3s^2 3p^3$, (c) $[Ne]3s^2 3p^6$, (d) $[Ar]4s^2 3d^8$.

9.15 An ion M^{2+} derived from a metal in the first transition metal series has four and only four electrons in the $3d$ subshell. What element might M be?

Atomic Radius

9.16 Explain what the size of an atom means.

9.17 On the basis of their positions in the periodic table, select the atom with the larger atomic radius in each of the following pairs: (a) Na, Cs; (b) Be, Ba; (c) N, Sb; (d) F, Br; (e) Ne, Xe.

9.18 Why does the helium atom have a smaller atomic radius than the hydrogen atom?

9.19 Why is the radius of the lithium atom considerably larger than the radius of the hydrogen atom?

9.20 Arrange the following atoms in order of decreasing atomic radius: Na, Al, P, Cl, Mg.

9.21 Which is the smallest atom in Group 7A?

9.22 Use the second period of the periodic table as an example to show that the sizes of atoms decrease as we move from left to right. Explain the trend.

Ionization Energy

9.23 Ionization energy measurements are usually carried out with atoms in the gaseous state. Why?

9.24 Use the third period of the periodic table as an example to illustrate the change in first ionization energies of the elements as we move from left to right. Explain the trend.

*9.25 Ionization energy usually increases from left to right across a given period. Aluminum, however, has a lower ionization energy than magnesium. Explain.

9.26 The first and second ionization energies of K are 419 kJ/mol and 3052 kJ/mol, and those of Ca are 590 kJ/mol and 1145 kJ/mol, respectively. Compare their values and comment on the differences.

9.27 Why does potassium have a lower first ionization energy than lithium?

9.28 Two atoms have the electron configurations $1s^2 2s^2 2p^6$ and $1s^2 2s^2 2p^6 3s^1$. The first ionization energy of one is 2080 kJ/mol, and that of the other is

496 kJ/mol. Pair up each ionization energy with one of the given electron configurations. Justify your choice.

9.29 Arrange the following species in isoelectronic pairs: O^+, Ar, S^{2-}, Ne, Zn, Cs^+, N^{3-}, As^{3+}, N, Xe.

9.30 Which one of the following processes requires the greatest input of energy? Why?

$$P(g) \longrightarrow P^+(g) + e^-$$
$$P^+(g) \longrightarrow P^{2+}(g) + e^-$$
$$P^{2+}(g) \longrightarrow P^{3+}(g) + e^-$$

*9.31 A hydrogenlike ion is an ion containing only one electron. The energies of the electron in a hydrogenlike ion are given by

$$E_n = -(2.18 \times 10^{-18} \text{ J})Z^2(1/n^2),$$

where n is the principal quantum number and Z the atomic number of the element. Calculate the ionization energy (in kJ/mol) of the He^+ ion.

*9.32 Use the equation given in Problem 9.31 to calculate the following ionization energies (kJ/mol):

(a) $C^{5+}(g) \longrightarrow C^{6+}(g) + e^-$
(b) $Hg^{79+}(g) \longrightarrow Hg^{80+}(g) + e^-$

Electron Affinity

9.33 Electron affinity measurements are usually carried out with atoms in the gaseous state. Why?

9.34 Which of the following elements would you expect to have the greatest electron affinity? He, K, Co, S, Cl.

9.35 Which of the following species has a more negative electron affinity: S or S^-? Why?

Periodic Variation in Chemical and Physical Properties

9.36 Explain, with examples, the meaning of the diagonal relationship.

9.37 Use the alkali metals and alkaline earth metals as examples to show how we can predict the chemical properties of elements simply from their electron configurations.

9.38 Based on your knowledge of the chemistry of the alkali metals, predict some of the chemical properties of francium, the last member of the group.

9.39 As a group, the noble gases are very stable chemically (only Kr and Xe are known to form some compounds). Why?

9.40 Why are the Group 1B elements more stable than the Group 1A elements even though they seem to have the same outer electron configuration ns^1?

9.41 Which elements are more likely to form acidic oxides? Basic oxides? Amphoteric oxides?

9.42 How do the chemical properties of oxides change as we move across a period from left to right? As we move down a particular group?

9.43 Predict (and give balanced equations) the reactions between each of the following oxides with water: (a) Li_2O, (b) CaO, (c) SO_2.

9.44 Which oxide is more basic, MgO or BaO?

9.45 Write formulas and give names for the binary hydrogen compounds of the second-period elements (Li to F). Describe the changes in physical and chemical properties of these compounds as we move across the period from left to right.

Miscellaneous Problems

9.46 With reference to the periodic table, name (a) a halogen element in the fourth period, (b) an element similar to phosphorus in chemical properties, (c) the most reactive metal in the fifth period, (d) an element that has an atomic number smaller than 20 and is similar to strontium.

9.47 Define the following terms: (a) alkali metals, (b) alkaline earth metals, (c) coinage metals, (d) transition metals, (e) halogens.

9.48 How do ionization energy measurements provide evidence for electron shell structure?

9.49 Elements that have high ionization energies usually have more negative electron affinities. Why?

*9.50 The energies an electron can possess in a hydrogen atom or a hydrogenlike ion are given by $E_n = (-2.18 \times 10^{-18} \text{ J})Z^2(1/n^2)$. This equation cannot be applied to many-electron atoms. One way to modify it for the more complex atoms is to replace Z with $(Z - \sigma)$, where Z is the atomic number and σ is a positive dimensionless quantity called the shielding constant. Consider the helium atom as an example. The physical significance of σ is that it represents the extent of shielding that the two $1s$ electrons exert on each other. Thus the quantity $(Z - \sigma)$ is appropriately called the "effective nuclear charge." Calculate the value of σ if the first ionization energy of helium is 3.94×10^{-18} J per atom. (Ignore the minus sign in the given equation in your calculation.)

9.51 Match each of the elements on the right with a description on the left:

(a) A dark red liquid.
(b) A colorless gas that burns in oxygen gas.
(c) A reactive metal that attacks water.
(d) A shiny metal that is used in jewelry.
(e) A totally inert gas.

Calcium (Ca)
Gold (Au)
Hydrogen (H_2)
Bromine (Br_2)
Argon (Ar)

10
CHEMICAL BONDING I: IONIC COMPOUNDS

Much of our detailed discussion of atoms and atomic structure has been based on consideration of isolated, individual atoms somehow suspended in space. Thus properties such as ionization energy and electron affinity relate to such "detached" gaseous atoms of the elements. Although we can properly talk about the electron configuration or ionization energy of an isolated atom of, say, oxygen, fluorine, or sodium, we do not find these elements in the form of individual atoms in the natural world. Instead (with the exception of the noble gases) atoms are found linked or bonded to atoms of other elements (such as in H_2O, Na_2SO_4, or KF) or to atoms of the same element (such as in O_2).

In short, our macroscopic world is composed of matter involving various *aggregates* of atoms—of atoms linked or bonded together in particular ways. Curiously enough, ionization energy and electron affinity, properties that are based on the behavior of isolated atoms, play significant roles in determining the ways in which atoms of various elements interact to form entities such as molecules and ionic crystals. With this chapter we begin a detailed examination of the major ways in which these interactions take place.

10.1 LEWIS DOT SYMBOLS

When atoms interact to form a chemical bond, only their outer parts come into contact. For this reason, as we saw in Chapter 9, elements with similar outer electron configurations behave alike chemically. Therefore, when we study chemical bonding we are usually concerned only with valence electrons.

To show the valence electrons of atoms and to keep track of them in a chemical reaction, chemists use Lewis dot symbols (G. N. Lewis, 1875–1946). A **Lewis dot symbol** *consists of the symbol of an element and one or more dots that represent the number of valence electrons in an atom of the element.* Figure 10.1 shows the Lewis dot symbols of the representative elements. Note that with the exception of helium the number of valence electrons in each atom is the same as the group number of the element. For example, Li belongs to Group 1A and has one valence electron (one dot); Be belongs to Group 2A and has two valence electrons (two dots); and so on. Elements in the same group have similar outer electron configurations and hence similar Lewis dot symbols.

Note that the number of unpaired electrons in the Lewis dot symbol for an atom does not necessarily agree with the pairings you would make in writing the atom's electron configuration. For example, the electron configuration of B is $1s^2 2s^2 2p^1$, which indicates that the atom has one unpaired electron (see p. 236). However, the Lewis dot symbol for B in Figure 10.1 shows that it has three unpaired electrons. Lewis dot symbols are written in this manner because when atoms form bonds, they frequently behave as though they have the number of unpaired electrons shown by their Lewis dot symbols.

The transition metals, lanthanides, and actinides all have incompletely filled inner shells, and in general we cannot write simple Lewis symbols for them.

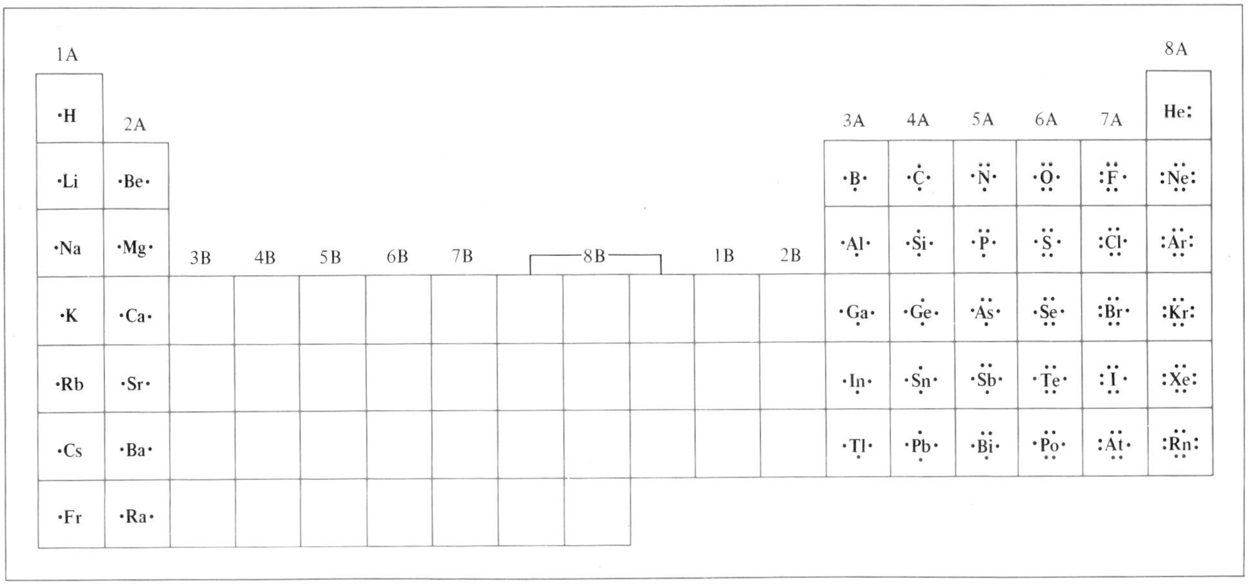

In a Lewis dot symbol the symbol for the element actually represents the nucleus of an atom of that element plus all the inner-shell electrons.

FIGURE 10.1 *Lewis dot symbols of the representative elements and the noble gases.*

10.2 TYPES OF ELEMENTS THAT FORM IONIC COMPOUNDS

Previous chapters have discussed some aspects of ionic compounds in detail (for example, their nomenclature and some of their general properties). Here we will examine factors that determine whether atoms of different elements will react to form an ionic compound.

In Chapter 9 we learned that atoms of elements with low ionization energies tend to form cations, while those with high negative electron affinities tend to form anions. This general rule tells us that elements most likely to form cations in ionic compounds are the alkali metals and alkaline earth metals, and that elements most likely to form anions are the halogens and oxygen. Consider the formation of the ionic compound lithium fluoride from lithium and fluorine. The electron configuration of lithium is $1s^2 2s^1$, and that of fluorine is $1s^2 2s^2 2p^5$. When lithium and fluorine atoms come into contact with each other, the outer $2s^1$ valence electron of lithium is transferred to the fluorine atom. The following are Lewis dot symbols and the corresponding electron configurations for the reaction:

$$\cdot\text{Li} \;+\; :\ddot{\underset{\cdot\cdot}{\text{F}}}\cdot \;\longrightarrow\; \text{Li}^+ :\ddot{\underset{\cdot\cdot}{\text{F}}}:^- \;(\text{or LiF}) \tag{10.1}$$

$$1s^2 2s^1 \qquad 1s^2 2s^2 2p^5 \qquad 1s^2\; 1s^2 2s^2 2p^6$$

For convenience, we can imagine this reaction as occurring in separate steps: the ionization of Li and the acceptance of an electron by F:

$$\cdot\text{Li} \longrightarrow \text{Li}^+ + e^-$$

and

$$:\ddot{\underset{\cdot\cdot}{\text{F}}}\cdot \;+\; e^- \longrightarrow :\ddot{\underset{\cdot\cdot}{\text{F}}}:^-$$

Then we can imagine the two separate ions joining to form an LiF unit:

$$\text{Li}^+ \;+\; :\ddot{\underset{\cdot\cdot}{\text{F}}}:^- \longrightarrow \text{Li}^+ :\ddot{\underset{\cdot\cdot}{\text{F}}}:^-$$

Note that the sum of these three reactions is

$$\cdot\text{Li} \;+\; :\ddot{\underset{\cdot\cdot}{\text{F}}}\cdot \longrightarrow \text{Li}^+ :\ddot{\underset{\cdot\cdot}{\text{F}}}:^-$$

which is the same as the direct transfer of an electron from lithium to fluorine to form the Li^+F^- unit as represented by Equation (10.1).

The Li^+ and F^- ions are held together by forces that result from electrostatic attraction between ions of opposite charges. These attractive forces give rise to the ionic bond. An ***ionic bond,*** then, is *the electrostatic force that holds ions together in an ionic compound.* According to Coulomb's law, the energy of interaction (E) between these two ions is given by

$$E \propto -\frac{Q_{\text{Li}^+}Q_{\text{F}^-}}{r}$$

$$= -k\frac{Q_{\text{Li}^+}Q_{\text{F}^-}}{r} \tag{10.2}$$

where Q_{Li^+} and Q_{F^-} are the charges on the Li^+ and F^- ions, r is the distance between the centers of the two ions, and k is the proportionality constant. The negative sign in Equation (10.2) means that the formation of an ionic

In the laboratory, LiF can be prepared by passing a stream of fluorine gas over heated lithium:

$$2\text{Li}(s) \;+\; \text{F}_2(g) \longrightarrow 2\text{LiF}(s)$$

Industrially, LiF (as well as many other ionic compounds) is obtained by recovering and purifying minerals containing the substance.

We normally write the empirical formulas of ionic compounds without showing the charges. The + and − signs are shown here to emphasize the transfer of electrons.

bond from a cation and an anion is an exothermic process. A pair of Li^+ and F^- ions together is therefore more stable than two separate Li^+ and F^- ions.

Following are additional examples of reactions that lead to the formation of ionic compounds. For instance, calcium burns in oxygen to form calcium oxide:

$$2Ca(s) + O_2(g) \longrightarrow 2CaO(s)$$

or in Lewis dot symbols

$$\cdot Ca \cdot \ + \ \cdot \ddot{O} \cdot \ \longrightarrow Ca^{2+} : \ddot{O} :^{2-}$$
$$[Ar]4s^2 \quad 1s^2 2s^2 2p^4 \quad [Ar] \ [Ne]$$

There is a transfer of two electrons from the calcium atom to the oxygen atom. Note that the resulting calcium ion (Ca^{2+}) has the argon core (or is isoelectronic with argon), and the oxide ion (O^{2-}) has the neon core (or is isoelectronic with neon).

In many cases the cations and anions do not carry the same charges. For instance, when lithium burns in air to form lithium oxide (Li_2O), the balanced equation is

$$4Li(s) + O_2(g) \longrightarrow 2Li_2O(s)$$

Using Lewis dot symbols we write

$$2 \cdot Li \ + \ \cdot \ddot{O} \cdot \ \longrightarrow 2Li^+ : \ddot{O} :^{2-} \text{ (or } Li_2O)$$
$$1s^2 2s^1 \quad 1s^2 2s^2 2p^4 \quad [He] \ [Ne]$$

In this process the oxygen atom receives two electrons (one from each of two lithium atoms) to form the oxide ion. The Li^+ ion is isoelectronic with helium.

When magnesium reacts with nitrogen at elevated temperatures, a white solid compound, magnesium nitride (Mg_3N_2), is formed:

$$3Mg(s) + N_2(g) \longrightarrow Mg_3N_2(s)$$

or

$$3 \cdot Mg \cdot \ + \ 2 \cdot \ddot{N} \cdot \ \longrightarrow 3Mg^{2+} 2 : \ddot{N} :^{3-} \text{ (or } Mg_3N_2)$$
$$[Ne]3s^2 \quad 1s^2 2s^2 2p^3 \quad [Ne] \quad [Ne]$$

There is a transfer of six electrons (two from each Mg atom) to two nitrogen atoms. The magnesium ion (Mg^{2+}) and the nitride ion (N^{3-}) are isoelectronic with neon.

This Lewis dot symbol equation is based on the assumption that the diatomic O_2 molecule has been divided into separate oxygen atoms to clarify this equation. Later we will focus in greater detail on the energetics of the required O_2 to O step.

EXAMPLE 10.1

Use Lewis dot symbols to describe the formation of aluminum oxide (Al_2O_3).

Answer

According to Figure 10.1, the Lewis dot symbols of Al and O are

$$\cdot \underset{\cdot}{Al} \cdot \qquad \cdot \ddot{O} \cdot$$

(Continued)

We use electroneutrality as our guide in writing formulas for ionic compounds.

Since aluminum tends to form the cation (Al^{3+}) and oxygen the anion (O^{2-}), the transfer of electrons is from Al to O. There are three valence electrons in each Al atom; each O atom needs two electrons to form the O^{2-} ion, which is isoelectronic with neon. Thus the simplest neutralizing ratio of Al^{3+} to O^{2-} is 2:3; two Al^{3+} ions have a total charge of $6+$, three O^{2-} ions have a total charge of $6-$. Thus the empirical formula of aluminum oxide is Al_2O_3 and the reaction is

$$2 \cdot \overset{\cdot}{Al} \cdot \quad + \quad 3 \cdot \overset{\cdot\cdot}{\underset{\cdot\cdot}{O}} \cdot \quad \longrightarrow 2Al^{3+} 3 : \overset{\cdot\cdot}{\underset{\cdot\cdot}{O}} :^{2-} \text{ (or } Al_2O_3)$$
$$[Ne]3s^2 3p^1 \quad 1s^2 2s^2 2p^4 \qquad [Ne] \quad [Ne]$$

Similar examples: Problems 10.12, 10.13.

Polyatomic Ions

So far we have concentrated on *monatomic ions,* that is, ions containing only one atom. Many ionic compounds contain *polyatomic cations* and/or *polyatomic anions,* which are ions containing more than one atom. At present, the only polyatomic cations you need to know are the mercury(I) ion (Hg_2^{2+}) and the ammonium ion (NH_4^+), mentioned in Chapter 2. On the other hand, there are many important polyatomic anions with which you do need to become familiar. These anions are listed in Table 2.2.

10.3 ELECTRON CONFIGURATIONS OF CATIONS AND ANIONS

The procedure for writing electron configurations of ions is only a slight extension of that for neutral atoms. We have already seen electron configurations of a number of cations and anions in the preceding section and in Chapter 9; in this section we will group the ions into a few categories for discussion.

Just as for neutral atoms, we use the Pauli exclusion principle and Hund's rule in writing the electron configurations of cations and anions.

Ions Derived from Representative Elements

In the formation of a cation from the neutral atom of a representative element, one or more electrons are removed from the highest filled or partially filled atomic orbitals. Following are the electron configurations of some neutral atoms and their corresponding cations:

Na	$[Ne]3s^1$	Na^+	$[Ne]$
Ca	$[Ar]4s^2$	Ca^{2+}	$[Ar]$
Al	$[Ne]3s^2 3p^1$	Al^{3+}	$[Ne]$

Note that each ion that is formed has a noble gas core; in other words, each ion is isoelectronic with a noble gas.

In the formation of an anion, electrons are added to the lowest vacant or highest partially filled atomic orbitals. Consider the following examples:

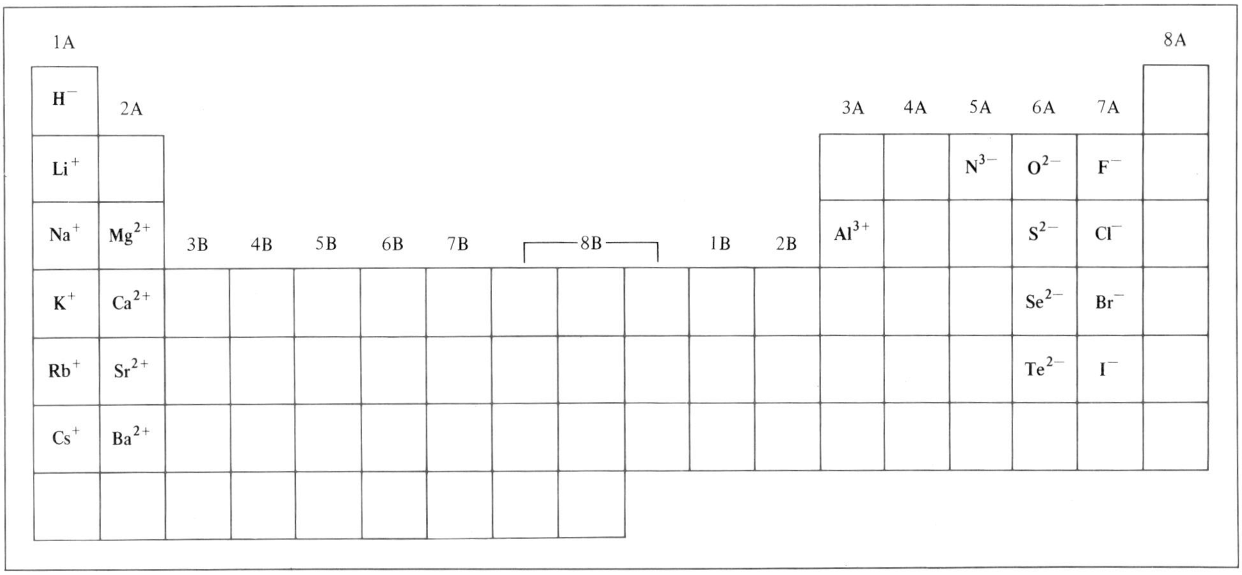

1A																	8A
H⁻	2A											3A	4A	5A	6A	7A	
Li⁺														N³⁻	O²⁻	F⁻	
Na⁺	Mg²⁺	3B	4B	5B	6B	7B		8B		1B	2B	Al³⁺			S²⁻	Cl⁻	
K⁺	Ca²⁺														Se²⁻	Br⁻	
Rb⁺	Sr²⁺														Te²⁻	I⁻	
Cs⁺	Ba²⁺																

FIGURE 10.2 *Cations and anions derived from representative elements. With the exception of H^- and Li^+, all the ions shown here have the noble gas electron configuration ns^2np^6; H^- and Li^+ have the helium electron configuration $1s^2$. Beryllium is not shown because it is an atypical element of Group 2A in that most of its compounds are covalent rather than ionic.*

H	$1s^1$	H⁻	$1s^2$ or [He]	
F	$1s^22s^22p^5$	F⁻	$1s^22s^22p^6$ or [Ne]	
O	$1s^22s^22p^4$	O²⁻	$1s^22s^22p^6$ or [Ne]	
N	$1s^22s^22p^3$	N³⁻	$1s^22s^22p^6$ or [Ne]	

As you can see, all the anions have noble gas cores. Thus a characteristic of most representative elements is that ions derived from their neutral atoms have the noble gas outer electron configuration ns^2np^6. Figure 10.2 shows the common ions of the representative elements that fit this description. (The exceptions, Ga, In, Tl, Sn, Pb, Sb, Bi, will be discussed later.)

Cations Derived from Transition Metals

Section 8.10 pointed out that in the first-row transition metals (Sc to Cu) the $4s$ orbital is always filled before the $3d$ orbitals. Consider manganese, whose electron configuration is $[Ar]4s^23d^5$. When the Mn^{2+} ion is formed, we might expect the two electrons to be removed from the $3d$ orbitals, to yield $[Ar]4s^23d^3$. In fact, the electron configuration of Mn^{2+} is $[Ar]3d^5$! We can understand this rather unexpected sequence if we realize that the electron–electron and electron–nucleus interactions in a neutral atom can be quite different in its ion. Thus, whereas the $4s$ orbital is always filled before the $3d$ orbitals in Mn, electrons are removed from the $4s$ orbital in forming Mn^{2+}. Therefore, in the formation of a cation from an atom of a

This configuration is confirmed by experimental measurements that show that the ground-state Mn^{2+} ion has five unpaired spins. The configuration $[Ar]4s^23d^3$ accounts for only three unpaired spins.

3B	4B	5B	6B	7B	8B			1B
Sc^{3+} [Ar]	Ti^{4+} [Ar]	V^{5+} [Ar]	Cr^{2+} [Ar]$3d^4$ Cr^{3+} [Ar]$3d^3$	Mn^{2+} [Ar]$3d^5$ Mn^{3+} [Ar]$3d^4$	Fe^{2+} [Ar]$3d^6$ Fe^{3+} [Ar]$3d^5$	Co^{2+} [Ar]$3d^7$ Co^{3+} [Ar]$3d^6$	Ni^{2+} [Ar]$3d^8$	Cu^{2+} [Ar]$3d^9$
					Ru^{2+} [Kr]$4d^6$ Ru^{3+} [Kr]$4d^5$		Pd^{2+} [Kr]$4d^8$	
							Pt^{2+} [Xe]$4f^{14}5d^8$ Pt^{4+} [Xe]$4f^{14}5d^6$	Au^{3+} [Xe]$4f^{14}5d^8$

FIGURE 10.3 *Cations derived from some of the more familiar transition metals. Note that Sc^{3+}, Ti^{4+}, and V^{5+} have the noble gas electron configuration, whereas the other ions have incompletely filled d subshells.*

This rule can be remembered easily if you recognize that electrons are removed from the highest filled subshells first. The major idea to grasp is that the order of electron filling does not determine or predict the order of electron removal for these transition metals.

transition metal, electrons are always removed first from the ns orbital and then from the $(n - 1)d$ orbitals.

Figure 10.3 shows the outer electron configurations of some common transition metal ions. Note that some of the metals can form more than one type of cation, and that frequently the cation is not isoelectronic with the noble gas.

Cations with Outer $ns^2np^6nd^{10}$ and $ns^2(n - 1)d^{10}$ Electron Configurations

Another outer electron configuration often found in cations can be illustrated by the element zinc ($Z = 30$). In Table 8.3 we see that the electron configuration of Zn is [Ar]$4s^2 3d^{10}$. Loss of the two $4s$ electrons results in the Zn^{2+} ion with the electron configuration [Ar]$3d^{10}$ or, written another way, [Ne]$3s^2 3p^6 3d^{10}$. This configuration, with eighteen electrons in the outer shell, is relatively stable; it is sometimes called a *pseudo-noble-gas electron configuration*. Figure 10.4 shows some stable cations that have the $ns^2 np^6 nd^{10}$ outer electron configuration.

Note that some of the commonly occurring cations have the [noble gas] $ns^2(n - 1)d^{10}$ configurations, which also represents a relatively stable configuration. Consider the element antimony ($Z = 51$). From Table 8.3 we write its electron configuration as [Kr]$5s^2 4d^{10} 5p^3$. Loss of three electrons (the $5p$ electrons) yields the Sb^{3+} ion (Figure 10.5) with the electron configuration [Kr]$5s^2 4d^{10}$. Both In^+ and Sn^{2+} have the same electron configuration as Sb^{3+}. On the other hand, the electron configurations of the Tl^+, Pb^{2+}, and Bi^{3+} ions are [Xe]$6s^2 4f^{14} 5d^{10}$. (The appearance of the $4f$ electrons here does not alter the general pattern for stability.)

Writing the electron configurations of ions helps us understand some of

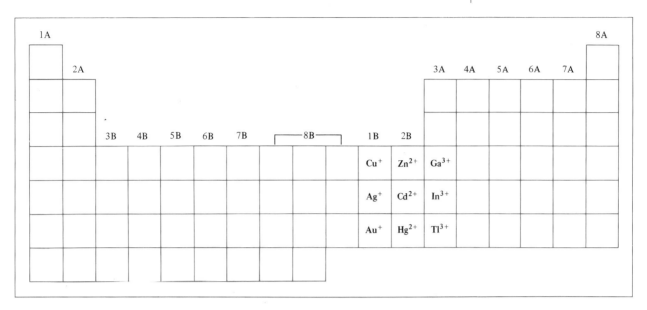

FIGURE 10.4 *Ions that have the pseudo-noble-gas electron configuration* $ns^2np^6nd^{10}$.

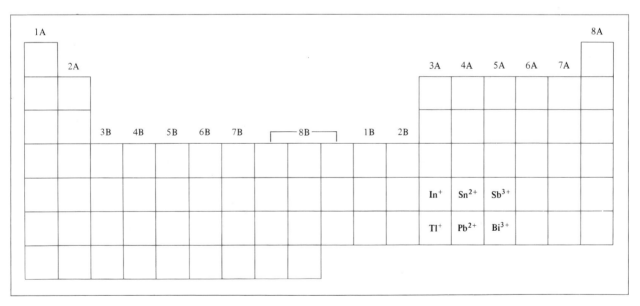

their macroscopic properties. For example, just as for neutral atoms, we can predict the magnetic properties of an ion (is it diamagnetic or paramagnetic?) from its electron configuration. This information is particularly important as we study the transition metals because their cations contain incompletely filled d subshells that give rise to the characteristic colors of the ions. And, as we have noted, sometimes we can use the electron configuration of a cation to predict its stability. For example, cations derived

FIGURE 10.5 *Ions that have the outer electron configuration* $ns^2(n-1)d^{10}$.

This generalization does not apply to ions derived from the transition metals, the lanthanides, and the actinides.

from the representative elements with the noble gas configuration, pseudo-noble-gas electron configuration, or electron configuration of [noble gas] $ns^2(n-1)d^{10}$ are usually more stable than cations without these configurations.

EXAMPLE 10.2

Compare the relative stabilities of the ions derived from thallium $(Z = 81)$: Tl^+, Tl^{2+}, and Tl^{3+}.

Answer

From Table 8.3 we see that the electron configuration of Tl is $[Xe]6s^24f^{14}5d^{10}6p^1$. From Figures 10.4 and 10.5 we find that the order of removing electrons is from the $6p$ subshell to the $6s$ subshell. Thus the predicted electron configurations of the three ions are

$$\begin{array}{ll} Tl^+ & [Xe]6s^24f^{14}5d^{10} \\ Tl^{2+} & [Xe]6s^14f^{14}5d^{10} \\ Tl^{3+} & [Xe]4f^{14}5d^{10} \end{array}$$

We see that the electron configuration of Tl^+ is similar to the [noble gas] $ns^2(n-1)d^{10}$ configuration. (As mentioned above, the presence of $4f$ electrons does not change the stability consideration.) Therefore we predict the Tl^+ ion to be stable.

The electron configuration of the Tl^{2+} ion does not fit any of the stable configuration patterns above. Thus we predict that Tl^{2+} would be relatively unstable (compared to Tl^+).

Since the electron configuration of Xe $(Z = 54)$ is $[Kr]5s^24d^{10}5p^6$, we can rewrite the electron configuration of Tl^{3+} as

$$[Kr]4d^{10}4f^{14}\ 5s^25p^65d^{10}$$
$$ns^2np^6nd^{10}$$

Thus the Tl^{3+} ion has a pseudo-noble-gas outer electron configuration and is expected to be stable.

In reality, Tl^+ and Tl^{3+} ions are known to exist, but compounds containing the Tl^{2+} ion have not been prepared.

10.4　IONIC RADII

Chapter 9 discussed atomic radii and their periodic variations. *Ionic radii, the radii of cations and anions,* contribute to the physical and chemical properties of an ionic compound. For example, the three-dimensional structure of an ionic compound depends on the relative sizes of its cation and anion.

When a neutral atom is converted to an ion, we naturally expect a change in size. If the atom forms an anion, its size (or radius) will increase, since the repulsion resulting from the additional electron(s) will enlarge the do-

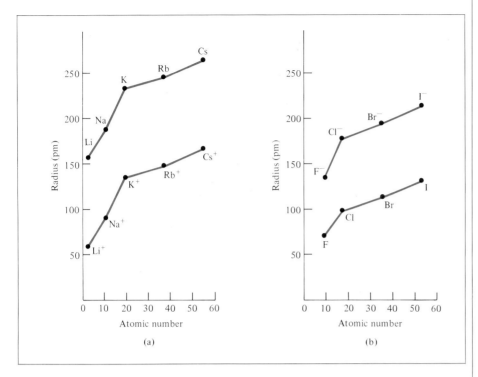

FIGURE 10.6 Comparison of atomic radii with ionic radii. (a) Alkali metals and alkali metal cations. (b) Halogens and halide ions.

main of the electron cloud. On the other hand, if a cation is formed, it will be smaller than the neutral atom, since removal of one or more electrons will shrink the electron cloud. Figure 10.6 shows the changes in size when alkali metals are converted to cations, and halogens to anions; Figure 10.7 shows the changes in size when a lithium atom reacts with a fluorine atom to form LiF.

Figure 10.8 shows the radii of ions derived from the more familiar elements, arranged according to their positions in the periodic table. We see that in some places there are parallel trends between atomic radii and ionic radii. For example, as we move down a group, the ionic radius increases. For ions derived from elements in different groups, the comparison in size is meaningful only if the ions are isoelectronic, that is, if they have the same number of electrons. If we examine isoelectronic ions, we find that anions are larger than cations. For example, look at the ions in the second period; note that O^{2-} and F^- are larger than Li^+ and Be^{2+}. Focusing on isoelectronic cations, we see that the radii of tripositive ions are smaller

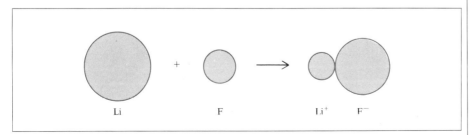

FIGURE 10.7 Changes in size when Li reacts with F to form LiF.

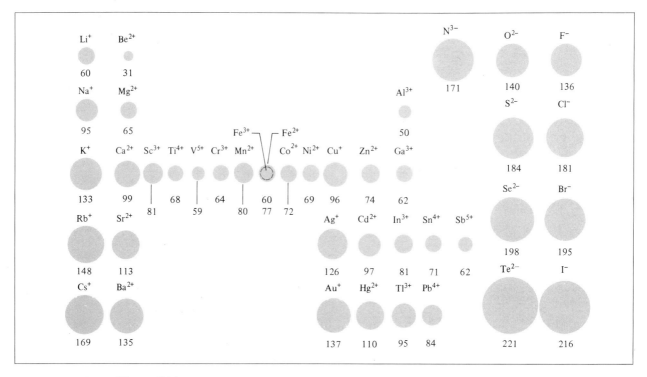

FIGURE 10.8 *The radii (pm) of some ions of more familiar elements shown in their approximate positions in the periodic table.*

than those of dipositive ions, which in turn are smaller than unipositive ions. This trend is nicely illustrated by comparing (in Figure 10.8) the sizes of three isoelectronic ions in the third period: Al^{3+}, Mg^{2+}, and Na^+. The Al^{3+} ion has the same number of electrons as Mg^{2+}, but it has one more proton. Thus the electron cloud in Al^{3+} is pulled inward more than that in Mg^{2+}. The smaller radius of Mg^{2+} compared to that of Na^+ can be similarly explained. Turning to isoelectronic anions, we find the radius increases as we go from ions with uninegative charge (that is, $-$) to dinegative charge (that is, $2-$) and so on. Thus the oxide ion is larger than the fluoride ion because oxygen has one fewer proton than fluorine; the electron cloud is spread out to a greater extent in O^{2-}.

Figure 10.9 compares the sizes of four isoelectronic species involving anions and cations.

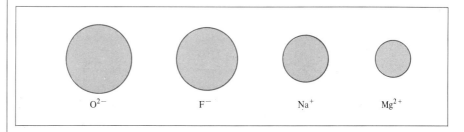

FIGURE 10.9 *Relative sizes of four isoelectronic species. Each species contains ten electrons. The size decreases from anions to cations.*

10.5 PROPERTIES OF IONIC COMPOUNDS

Physical Properties

Ionic compounds share a number of similar physical properties. Table 10.1 lists some of the common properties of several alkali metal and alkaline earth metal chlorides. In the solid state the ions are arranged in a highly ordered three-dimensional structure called the *crystal lattice* (see Figure 2.5). Because the ions are held together by strong electrostatic forces, any changes that disrupt the arrangement of the ions require a large energy input. Consequently ionic compounds all have comparatively high melting points and molar heats of fusion. When an ionic compound melts, it exists in the molten or fused state, which is sometimes called the *melt*. (For example, NaCl in the molten state is called the NaCl melt.) A great amount of energy is needed to vaporize ionic compounds; thus they generally have very high boiling points (see Table 10.1) and molar heats of vaporization. Ionic solids are usually quite hard and brittle. Unlike metals, ionic solids tend to shatter rather than deform.

The density of an ionic compound depends on the mass of the ions, the size of the ions, and the charge each ion carries. (Recall that density is mass/volume.) The charges on the cation and the anion determine the ratio of these ions in the crystal. The ratio of the ions and the ionic sizes together determine how the cations and the anions are arranged and packed together in a solid.

The formula of an ionic compound tells us the ratio of cations to anions in a crystal. In other words, it is an empirical formula.

In the solid state the cations and anions in an ionic compound are held rigidly in position, with little or no freedom of movement. Thus, solid ionic compounds are generally poor conductors of electricity. The situation

TABLE 10.1 Some Properties of Alkali Metal and Alkaline Earth Metal Chlorides

	$LiCl$	$NaCl$	KCl	$RbCl$	$MgCl_2$	$CaCl_2$	$SrCl_2$	$BaCl_2$
Melting point (°C)	605	801	770	718	714	782	875	960
Molar heat of fusion (kJ/mol)	13.4	30.2	26.8	18.4	33.9	25.5	17.2	22.5
Boiling point (°C)	1325	1413	>1500	1390	1412	>1600	1250	1560
Density (g/cm³)	2.07	2.17	1.98	2.80	2.32	2.15	3.05	3.86
Electrical conductivity								
Solid	Poor	Poor	Poor	Poor	Poor	Poor	Poor	Poor
Melt	Good	Good	Good	Good	Good	Good	Good	Good
Solubility*								
In water	Sol.	Sol.	Sol.	Sol.	Sol.	Sol.	Sol.	Sol.
In benzene	Insol.	Insol.	Insol.	Insol.	Insol.	Insol.	Insol.	Insol.

* "Sol." stands for soluble; "insol." for insoluble.

changes when an ionic compound melts (see Table 10.1). Then the cations and anions are able to move relatively freely, so that if an electric potential is applied across two electrodes in a NaCl melt, for example, the Na^+ ions migrate toward the negative electrode and the Cl^- ions toward the positive electrode. The movement of ions under applied voltage constitutes an electric current between the two electrodes, and the medium (that is, the melt) is said to be electrically conducting. Figure 10.10 shows the change in electrical conductivity of sodium chloride from the solid to the liquid state (that is, the melt).

For the same reason that ionic compounds in the solid state are poor conductors of electricity, they are also poor conductors of heat. When a solid ionic compound is heated, vibration of the ions increases. However, because the ions are confined to rather small domains, they cannot readily pass along their increased kinetic energy to other ions. Consequently solid ionic compounds have only limited ability to conduct heat.

Many ionic compounds are soluble in water but insoluble in solvents such as benzene and carbon tetrachloride, which lack the ability to interact with ions (that is, to *solvate* the ions). (The solubility rules of ionic compounds were discussed in Section 3.2.) When an ionic compound dissolves in water, the three-dimensional network is destroyed and the individual ions are stabilized by hydration (see Section 5.6). Because all ionic compounds are strong electrolytes, an aqueous solution containing a moderately soluble or soluble ionic compound is electrically conducting. On the other hand, an aqueous solution of an ionic compound with very low solubility, such as AgCl, is not able to conduct electricity because there is an insufficient number of ions in solution.

The vast majority of ionic compounds formed from representative elements are colorless. For example, all the alkali metal and alkaline earth metal halides are white solids. Ionic compounds containing either transition metal cations or anion groups containing a transition metal (for ex-

The relationship between color and electron configuration will be discussed in Chapter 29.

FIGURE 10.10. *The change in electrical conductivity of NaCl from the solid state to the liquid (melt) state. The melting point of NaCl is 801°C.*

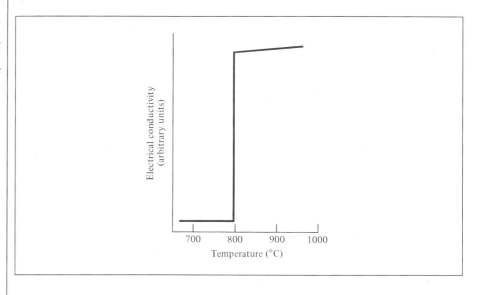

ample, CrO_4^{2-}, $Cr_2O_7^{2-}$, MnO_4^-) have distinctive colors as solids and in aqueous solution (see Color Plates 19 and 21). The color of transition metal ions results from the partially filled inner d subshells. Upon absorbing a photon in the visible region, a d electron may be promoted to a higher empty d orbital. This process gives the ion a characteristic color. Ions with completely filled d subshells do not absorb visible light, and therefore they usually appear colorless to us.

Chemical Properties

Since many different elements can form ionic compounds, the chemical properties of ionic compounds are rather varied and difficult to systematize. However, most of these compounds undergo displacement reactions and a number of them take part in precipitation reactions. You will recall that both of these kinds of reaction were discussed in Chapter 3.

In a displacement reaction, the cation of an ionic compound is replaced by an atom of another metallic element. For example

$$Mg(s) + NiSO_4(aq) \longrightarrow MgSO_4(aq) + Ni(s)$$

or, written as a net ionic equation

$$Mg(s) + Ni^{2+}(aq) \longrightarrow Mg^{2+}(aq) + Ni(s)$$

In general we can use the activity series shown in Figure 3.9 to help us predict the outcome of a displacement reaction. Apart from relatively inert metals such as silver and gold, most metals exist in nature in combination with other elements. To extract the pure metal from its ore, a displacement reaction involving the ionic compound that contains the metal ion is usually employed. Displacement reactions generally do not work if an alkali metal is to be extracted, because alkali metals have properties that put them at the top of the activity series. The ions of alkali metals in compounds *can* be converted to metals by a process called *electrolysis*.

Note that the displacement reaction is not restricted to the cations in an ionic compound. As we saw in Section 3.5, the halide ion in an ionic compound can also be displaced, by reaction with a more reactive halogen element:

$$Cl_2(g) + 2KBr(aq) \longrightarrow 2KCl(aq) + Br_2(l)$$

In fact, this process describes how the less reactive halogens such as bromine and iodine are prepared on an industrial scale.

Another common type of reaction that ionic compounds undergo is the precipitation reaction (which is an example of the metathesis reaction). The outcome of a precipitation reaction depends on the solubility of the compounds involved (see solubility rules in Section 3.2). For example, the reaction between two aqueous solutions containing the ionic compounds $Pb(NO_3)_2$ and KI results in the PbI_2 precipitate:

$$Pb(NO_3)_2(aq) + 2KI(aq) \longrightarrow PbI_2(s) + 2KNO_3(aq)$$

> The material obtained from a mineral deposit that is concentrated enough to allow economical recovery of a desired metal is known as an *ore*.

10.6   LATTICE ENERGY OF IONIC COMPOUNDS

In this final section of the chapter we will take a closer look at the stability of solid ionic compounds. Section 10.2 discussed the formation of ionic compounds in terms of the ionization energy and electron affinity of the elements. These important quantities do help us predict which elements are most likely to form ionic compounds, but they do not tell the whole story. They are limited because both the ionization process and the acceptance of an electron are defined as taking place in the gas phase. In the solid state each cation is surrounded by a specific number of anions, and vice versa. Thus the overall stability of an ionic compound must depend on the interactions of all these ions and not merely on the interaction of a single cation with a single anion. A quantitative measure of the stability of any ionic solid is its lattice energy, first introduced in Section 5.6.

Earlier we saw that the ionization energy values of an element increase rapidly as electrons are removed from its atom in succession. For example, the first ionization energy of magnesium is 738 kJ/mol, whereas the second ionization energy is 1450 kJ/mol, almost twice the first value. We might ask why, from the standpoint of energy, magnesium does not preferentially form unipositive ions in its compounds. For example, why doesn't magnesium chloride have the formula MgCl (containing the Mg^+ ion) rather than $MgCl_2$ (containing the Mg^{2+} ion)? Admittedly, only the Mg^{2+} ion has the noble gas configuration [Ne], which represents stability because of its completely filled shells. But the stability gained through the filled shells does not, in fact, outweigh the energy input needed to remove an electron from the Mg^+ ion. The solution to this puzzle lies in the extra stability gained when magnesium chloride is formed in the solid state. The attractive interactions between each Mg^{2+} ion and its neighboring Cl^- ions (and the interactions between each Cl^- ion and its neighboring Mg^{2+} ions) greatly stabilize the solid network. Thus when one mole of solid $MgCl_2$ is formed from its ions

$$Mg^{2+}(g) + 2Cl^-(g) \longrightarrow MgCl_2(s) + energy$$

2527 kJ of heat is given off to the surroundings. This highly exothermic process tells us that solid $MgCl_2$ is energetically much more stable than gaseous Mg^{2+} and Cl^- ions. Putting it another way, it takes 2527 kJ of energy to convert one mole of solid $MgCl_2$ into one mole of gaseous Mg^{2+} and two moles of gaseous Cl^- ions. As we noted in Section 5.6, the energy input needed to complete this separation process is called the *lattice energy*. Thus the larger the lattice energy, the more stable the solid. In fact, the lattice energy of $MgCl_2$ is more than enough to supply the energy needed to remove the first two electrons from a Mg atom (738 kJ/mol + 1450 kJ/mol = 2188 kJ/mol).

What about sodium chloride? Why is the formula for sodium chloride NaCl and not $NaCl_2$ (containing the Na^{2+} ion)? We might expect the latter, since Na^{2+} has a higher charge and therefore the hypothetical $NaCl_2$ should have a greater lattice energy. Again the answer is provided by the balance between energy input (that is, ionization energies) and stability gained. The sum of the first two ionization energies of sodium is

$$496 \text{ kJ/mol} + 4560 \text{ kJ/mol} = 5056 \text{ kJ/mol}$$

Since $NaCl_2$ does not exist, we do not know its lattice energy value. However, if we assume that the value is roughly the same as that for $MgCl_2$ (2527 kJ/mol), we see that it is far too small an energy yield to compensate for the energy required to produce the Na^{2+} ion.

What has been said about the cations applies also to the anions. In Section 9.5 we observed that the electron affinity of oxygen is a negative value, meaning the following process releases energy (and is therefore favorable):

$$O(g) + e^- \longrightarrow O^-(g)$$

As we would expect, adding another electron to the O^- ion

$$O^-(g) + e^- \longrightarrow O^{2-}(g)$$

would be unfavorable because of the increase in electrostatic repulsion. Indeed, the electron affinity of O^- is positive ($+708$ kJ/mol). Yet ionic compounds containing the oxide ion (O^{2-}) do exist and are very stable, whereas ionic compounds containing the O^- ion are not known. Again, the high lattice energy realized by the presence of O^{2-} ions in compounds such as Na_2O or MgO far outweighs the energy input needed to produce the O^{2-} ion.

The Born-Haber Cycle for Determining Lattice Energies

Lattice energy cannot be measured directly. However, if we know the structure and composition of an ionic compound, we can calculate the compound's lattice energy by using Coulomb's law. We can also determine the lattice energy indirectly, by assuming that the formation of an ionic compound takes place in a series of steps known as the ***Born-Haber cycle.*** The Born-Haber cycle *relates lattice energies of ionic compounds to ionization energies, electron affinities, heats of sublimation and formation, and bond dissociation energies.* This approach is based on Hess's law (see Section 5.5).

Developed by Max Born (1882–1970) and Fritz Haber (1868–1934), the Born-Haber cycle depicts the vaporization, ionization, bond-breaking, and electron-accepting steps, followed by the formation of an ionic solid. We will illustrate its use in finding the lattice energy of lithium fluoride.

Consider the reaction between lithium and fluorine gas:

$$Li(s) + \tfrac{1}{2}F_2(g) \longrightarrow LiF(s)$$

The standard enthalpy change for this reaction is found to be -594.1 kJ. (Note that -594.1 kJ is also the standard enthalpy of formation of one mole of LiF from its elements.) Applying the Born-Haber cycle, we can consider the reaction as comprising five separate steps, as follows:

1. Convert solid lithium to lithium vapor (the direct conversion of a solid to a gas is called *sublimation*):

$$\text{Li}(s) \longrightarrow \text{Li}(g) \qquad \Delta H_1^\circ = 155.2 \text{ kJ}$$

2. Dissociate one-half mole of F_2 gas into separate gaseous F atoms:

$$\tfrac{1}{2}F_2(g) \longrightarrow F(g) \qquad \Delta H_2^\circ = 75.3 \text{ kJ}$$

3. Ionize one mole of gaseous Li atoms (see Figure 9.6):

$$\text{Li}(g) \longrightarrow \text{Li}^+(g) + e^- \qquad \Delta H_3^\circ = 520 \text{ kJ}$$

4. Add one mole of electrons to one mole of gaseous F atoms (see Figure 9.9):

$$F(g) + e^- \longrightarrow F^-(g) \qquad \Delta H_4^\circ = -323 \text{ kJ}$$

5. Combine one mole of gaseous Li^+ and one mole of F^- to form one mole of solid LiF:

$$\text{Li}^+(g) + F^-(g) \longrightarrow \text{LiF}(s) \qquad \Delta H_5^\circ = ?$$

The reverse of step 5, as we saw earlier

$$\text{energy} + \text{LiF}(s) \longrightarrow \text{Li}^+(g) + F^-(g)$$

defines the lattice energy of LiF. Thus the lattice energy must have the same magnitude as ΔH_5° but an opposite sign. Although we cannot determine ΔH_5° directly, we can calculate its value by recognizing that the sum of steps 1 through 5 produces the overall reaction for the formation of LiF from its elements.

1.	$\text{Li}(s) \longrightarrow \text{Li}(g)$	$\Delta H_1^\circ = 155.2 \text{ kJ}$
2.	$\tfrac{1}{2}F_2(g) \longrightarrow F(g)$	$\Delta H_2^\circ = 75.3 \text{ kJ}$
3.	$\text{Li}(g) \longrightarrow \text{Li}^+(g) + e^-$	$\Delta H_3^\circ = 520 \text{ kJ}$
4.	$F(g) + e^- \longrightarrow F^-(g)$	$\Delta H_4^\circ = -323 \text{ kJ}$
5.	$\text{Li}^+(g) + F^-(g) \longrightarrow \text{LiF}(s)$	$\Delta H_5^\circ = ?$
	$\text{Li}(s) + \tfrac{1}{2}F_2(g) \longrightarrow \text{LiF}(s)$	$\Delta H_{\text{overall}}^\circ = -594.1 \text{ kJ}$

According to Hess's law we can write

$$\Delta H_{\text{overall}}^\circ = \Delta H_1^\circ + \Delta H_2^\circ + \Delta H_3^\circ + \Delta H_4^\circ + \Delta H_5^\circ$$

or

$$-594.1 \text{ kJ} = 155.2 \text{ kJ} + 75.3 \text{ kJ} + 520 \text{ kJ} - 323 \text{ kJ} + \Delta H_5^\circ$$

Hence

$$\Delta H_5^\circ = -1022 \text{ kJ}$$

Thus the lattice energy of LiF is $+1022$ kJ/mol.

Figure 10.11 summarizes the entire Born-Haber cycle for LiF. In this cycle we saw that steps 1, 2, and 3 all require input of energy. On the other hand, steps 4 and 5 release energy. Because ΔH_5° is a large negative quantity, the lattice energy of LiF is a large positive quantity, which accounts for the stability of LiF in the solid state. The greater the lattice energy, the more stable the ionic compound. Keep in mind that lattice energy is *always* a

The two F atoms in a F_2 molecule are held together by another type of chemical bond called the *covalent bond*. The energy required to break this bond is called the *bond dissociation energy*. More on this in Chapter 11.

The Born-Haber cycle provides a systematic and convenient way to analyze the energetics of ionic compound formation, but note that the actual mechanism by which LiF is formed need not follow these five simplified steps. This pathway just allows us, with the assistance of Hess's law, to examine the quantities of energy involved with this class of chemical reactions.

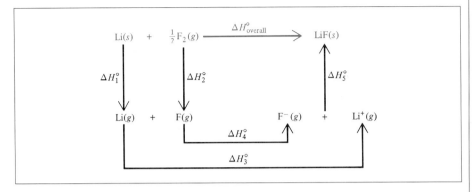

FIGURE   10.11   *The Born-Haber cycle for the formation of solid LiF.*

positive quantity because the separation of ions in a solid into ions in the gas phase is necessarily an endothermic process.

Table 10.2 lists the lattice energies and the melting points of several common ionic compounds. Note the unusually high lattice energy values for $MgCl_2$, Na_2O, and MgO. The first two of these ionic compounds involve a double-charged cation (Mg^{2+}) and a double-charged anion (O^{2-}), while the third involves the interaction of two double-charged species (Mg^{2+} and O^{2-}). The coulombic attractions among such double-charged species, or between them and single-charged ions, are much stronger than between single-charged anions and cations.

Note the unusually high melting point of MgO in Table 10.2.

TABLE 10.2 Lattice Energies and Melting Points of Some Alkali Metal and Alkaline Earth Metal Halides

Compound	Lattice Energy (kJ/mol)	Melting Point (°C)
LiF	1022	845
LiCl	828	610
LiBr	787	550
LiI	732	450
NaCl	788	801
NaBr	736	750
NaI	686	662
KCl	699	772
KBr	674	735
KI	632	680
$MgCl_2$	2527	714
Na_2O	2570	Sub*
MgO	3890	2800

*Na_2O sublimes at 1275°C.

SUMMARY

1. The Lewis dot symbol shows the number of valence electrons possessed by an atom of a given element. Lewis dot symbols are useful mainly for the representative elements.

2. Elements most likely to form ionic compounds are those with low ionization energies (such as the alkali metals and the alkaline earth metals, which form the cations) and those with high negative electron affinities (such as the halogens and oxygen, which form the anions).

3. An ionic bond consists of the electrostatic forces of attraction between positive and negative ions. An ionic crystal is a large collection of ions in which the positive and negative charges are balanced. The structure of a solid ionic compound maximizes the attractive forces among the ions and keeps the repulsions low.

4. Electron configurations of ions are written according to the Pauli exclusion principle and Hund's rule.

5. Ions with the noble gas electron configuration and configurations of the type $ns^2np^6nd^{10}$ and $ns^2(n - 1)d^{10}$ are usually stable. A number of transition metal ions whose configurations do not fit these patterns are also stable.

6. In general, anions are larger than cations. For isoelectronic ions, the ionic radius decreases from unipositive ion to dipositive ion to tripositive ion. Anions with two negative charges are larger than those with one negative charge.

7. Ionic compounds have high melting points and boiling points, and high heats of fusion and heats of vaporization. They are poor conductors of heat and electricity in the solid state. Ionic compounds are good conductors of electricity in the molten state. Ionic compounds containing only representative elements are usually colorless; those containing transition metals usually have distinctive colors.

8. Many ionic compounds can undergo displacement and precipitation reactions.

9. Lattice energy is a measure of the stability of an ionic compound. It can be calculated by employing the Born-Haber cycle, which is based on Hess's law.

KEY WORDS

Born-Haber cycle, p. 293
Ionic bond, p. 280

Ionic radius, p. 286
Lewis dot symbol, p. 279

PROBLEMS

More challenging problems are marked with an asterisk.

Lewis Dot Symbols

10.1 Explain the meaning of a Lewis dot symbol.

10.2 Without referring to Figure 10.1, write Lewis dot symbols for the atoms of the following elements: (a) Be, (b) Na, (c) Mg, (d) Al, (e) Si, (f) S, (g) Cl, (h) Sr, (i) Cs, (j) As, (k) I.

10.3 Write Lewis dot symbols for the following ions: (a) Li^+, (b) Cl^-, (c) S^{2-}, (d) Mg^{2+}, (e) N^{3-}.

10.4 Write Lewis dot symbols for the following species: (a) Br, (b) Br^-, (c) Pb, (d) Pb^{2+}.

10.5 Write Lewis dot symbols for the following atoms and ions: (a) Ga, (b) Ga^{3+}, (c) Rb, (d) Rb^+, (e) P, (f) P^{3-}, (g) Xe, (h) Se^{2-}.

Ionic Bond and Ionic Compounds

10.6 Explain how ionization energy and electron affinity determine whether elements will combine to form ionic compounds.

10.7 Name five metals and five nonmetals that are most likely to form ionic compounds.

10.8 Give an example of an ionic compound that contains only nonmetallic elements.

10.9 The term "molar mass" was introduced in Chapter 2 to replace "gram molecular weight." What is the advantage in using the term "molar mass" when we discuss ionic compounds?

10.10 Explain what an ionic bond is.

*10.11 An ionic bond is formed between a cation A^+ and an anion B^-. How would the energy of the ionic bond [see Equation (10.2)] be affected by the following changes? (a) Doubling the radius of A^+, (b) tripling the charge on A^+, (c) doubling the charges on A^+ and B^-, (d) decreasing the radii of A^+ and B^- to half of their original values.

10.12 Use Lewis dot symbols to represent the transfer of electrons between the following atoms to form cations and anions: (a) Na and F, (b) K and S, (c) Ba and O, (d) Al and N.

*10.13 Describe the following reactions by means of Lewis dot formulas of the reactants and products. (You need to balance the equations.)

(a) $Sr + Se \longrightarrow SrSe$
(b) $Ca + H \longrightarrow CaH_2$
(c) $Li + N \longrightarrow Li_3N$
(d) $Al + S \longrightarrow Al_2S_3$

10.14 Write the empirical formulas of and name the ionic compounds that are formed from the following pairs of ions: (a) Rb^+ and I^-, (b) Cs^+ and SO_4^{2-} (c) Sr^{2+} and N^{3-}, (d) Al^{3+} and S^{2-}.

10.15 For each pair of the following elements, state whether the binary compound formed is likely to be ionic or molecular. Write the empirical formula and name the compound. (a) I and F, (b) K and Br, (c) Mg and F, (d) Al and I.

Electron Configurations of Ions

10.16 Write ground-state electron configurations for the following ions: (a) Li^+, (b) H^-, (c) N^{3-}, (d) F^-, (e) S^{2-}, (f) Al^{3+}, (g) Se^{2-}, (h) Br^-, (i) Rb^+, (j) Sr^{2+}, (k) Sn^{2+}, (l) Te^{2-}, (m) Ba^{2+}, (n) Pb^{2+}, (o) In^{3+}, (p) Tl^+, (q) Tl^{3+}.

10.17 Write the ground-state electron configurations of the following ions, which play important roles in biochemical processes in our bodies: (a) Na^+, (b) Mg^{2+}, (c) Cl^-, (d) K^+, (e) Ca^{2+}, (f) Fe^{2+}, (g) Cu^{2+}, (h) Zn^{2+}.

10.18 Write ground-state electron configurations of the following transition metal ions: (a) Sc^{3+}, (b) Ti^{4+}, (c) V^{5+}, (d) Cr^{3+}, (e) Mn^{2+}, (f) Fe^{2+}, (g) Fe^{3+}, (h) Co^{2+}, (i) Ni^{2+}, (j) Cu^+, (k) Cu^{2+}, (l) Ag^+, (m) Au^+, (n) Au^{3+}, (o) Pt^{2+}.

10.19 Name the ions with $+3$ charges that have the following electron configurations: (a) $[Ar]3d^3$, (b) $[Ar]$, (c) $[Kr]4d^6$, (d) $[Xe]4f^{14}5d^6$.

10.20 Which of the following species are isoelectronic with each other? C, Cl^-, Mn^{2+}, B^-, Ar, Zn, Fe^{3+}, Ge^{2+}.

Ionic Radii

10.21 In each of the following pairs indicate which one of the species is smaller in size: (a) Cl or Cl^-; (b) Na or Na^+; (c) O^{2-} or S^{2-}; (d) Mg^{2+} or Al^{3+}; (e) Au^+ or Au^{3+}.

10.22 List the following ions in order of increasing ionic radius: N^{3-}, Na^+, F^-, Mg^{2+}, O^{2-}.

10.23 Which of these ions is larger, Fe^{2+} or Fe^{3+}? Explain.

10.24 Which of these anions is larger, N^{3-} or P^{3-}? Explain.

10.25 Arrange the following anions in order of increasing size: O^{2-}, Te^{2-}, Se^{2-}, S^{2-}. Explain your sequence.

*10.26 The H^- ion and the He atom have two $1s$ electrons each. Which of the two species is larger? Explain.

Physical and Chemical Properties

10.27 Briefly describe some general physical properties of ionic compounds.

10.28 Beryllium forms a compound with chlorine that has the empirical formula $BeCl_2$. How would you determine whether it is an ionic compound? (The compound is not soluble in water.)

10.29 Two solutions contain K_2SO_4 and $CuSO_4$, respectively. One solution is colorless, and the other solution is blue in color. Identify these two solutions.

10.30 Which substance can conduct electricity in the melt state, ICl or KCl?

10.31 Use the solubility rules in Section 3.2 to predict the outcomes of the following reactions (and balance the equations):

(a) $Ca(NO_3)_2(aq) + Na_2CO_3(aq) \longrightarrow$
(b) $BaCl_2(aq) + K_2SO_4(aq) \longrightarrow$
(c) $NH_4I(aq) + Pb(NO_2)_2(aq) \longrightarrow$
(d) $CH_3COOAg(aq) + SrBr_2(aq) \longrightarrow$

Lattice Energy

10.32 What is lattice energy? What role does it play in the study of the stability of ionic compounds?

10.33 For each of the following pairs of ionic compounds, specify which compound has the higher lattice energy: (a) KCl or MgO; (b) LiF or LiBr; (c) Mg_3N_2 or NaCl.

10.34 Which of the following do not represent stable ionic compounds? Explain. (a) RbI, (b) Mg_3N_2, (c) LiF_2, (d) $CaBr_3$, (e) CsO.

*10.35 Use the Born-Haber cycle outlined in Section 10.6 for LiF to calculate the lattice energy of NaCl. (Heat of sublimation of Na $= 108$ kJ/mol and $\Delta H_f^\circ(NaCl) = -411$ kJ/mol. Energy needed to dissociate one-half mole of Cl_2 into Cl atoms $= 121.4$ kJ.)

*10.36 Calculate the lattice energy of calcium chloride given that the heat of sublimation of Ca is 121 kJ/mol and $\Delta H_f^\circ(CaCl_2) = -795$ kJ/mol.

Miscellaneous Problems

10.37 What are polyatomic ions? Give some examples of polyatomic cations and polyatomic anions.

10.38 Arrange the following isoelectronic species in order of (a) increasing ionic radius, and (b) increasing ionization energy: O^{2-}, F^-, Na^+, Mg^{2+}.

*10.39 The atomic radius of K is 216 pm and that of K^+ is 133 pm. Calculate the percent decrease in volume that occurs when K(g) is converted to $K^+(g)$. (The volume of a sphere is $\frac{4}{3}\pi r^3$, where r is the radius of the sphere.)

*10.40 The atomic radius of F is 72 pm and that of F^- is 136 pm. Calculate the percent increase in volume that occurs when F(g) is converted to $F^-(g)$.

*10.41 Use the ionization energy (see Figure 9.6) and electron affinity (see Figure 9.9) values to calculate the energy change (in kJ) for the following reactions:

(a) $Li(g) + I(g) \longrightarrow Li^+(g) + I^-(g)$
(b) $Na(g) + F(g) \longrightarrow Na^+(g) + F^-(g)$
(c) $K(g) + Cl(g) \longrightarrow K^+(g) + Cl^-(g)$

11

CHEMICAL BONDING II: COVALENT COMPOUNDS

In this chapter we continue our exploration of the microscopic and representational worlds of chemical bonds. Despite the importance of ionic bonds and ionic substances, if you look around the room, chances are good that most of the matter you see is held together by another kind of chemical bond—the covalent bond. If your visual inventory of your current surroundings includes plastics, natural or synthetic fabrics, or wood products, covalent bonds are largely responsible for such materials. And although they are invisible to the eye, the gases that compose the atmosphere and the air you breathe are also made up of covalently bonded substances. In fact *you* are largely composed of covalent substances. We will explore this fundamentally important type of chemical bond in the following pages.

11.1 THE COVALENT BOND

Although the existence of molecules was hypothesized as early as the seventeenth century, chemists had little knowledge of how and why molecules formed until early in this century. The first major breakthrough was Gilbert Lewis's discovery of the role of electrons in chemical bond formation. In particular, Lewis developed the concept that a chemical bond involves two atoms in a molecule sharing a pair of electrons. For instance, according to Lewis, the formation of a chemical bond in H_2 can be described as

$$H\cdot \; + \; \cdot H \longrightarrow H\!:\!H$$

This type of electron pairing is an example of the ***covalent bond,*** which is *a bond in which two electrons are shared by two atoms.* For the sake of simplicity, the shared pair of electrons is often represented by a single line. Thus the covalent bond in the hydrogen molecule can be written as H—H.

When dealing with covalent bonding among atoms other than hydrogen, we need be concerned only with the valence electrons. Consider the fluorine molecule (F_2). The electron configuration of F is $1s^2 2s^2 2p^5$; thus each F atom has seven valence electrons, and the F_2 molecule is represented as

$$:\!\ddot{F}\!:\!\ddot{F}\!: \qquad \text{or} \qquad :\!\ddot{F}\!-\!\ddot{F}\!:$$

Note that there are some *valence electrons that are not involved in covalent bond formation;* these are called ***nonbonding electrons,*** or ***lone pairs.*** Thus each F in F_2 has three lone pairs of electrons:

$$\text{lone pairs} \longrightarrow :\!\ddot{F}\!-\!\ddot{F}\!: \longleftarrow \text{lone pairs}$$

In the water molecule, the electron configuration of O is $1s^2 2s^2 2p^4$; thus it has six valence electrons. The O atom forms two covalent bonds with the two H atoms in H_2O:

$$H\!:\!\ddot{O}\!:\!H \qquad \text{or} \qquad H\!-\!\ddot{O}\!-\!H$$

Here we see that the O atom has two lone pairs. The hydrogen atom has no lone pairs because its only electron is used to form a covalent bond.

The structures we have shown for H_2, F_2, and H_2O are called Lewis formulas. A ***Lewis formula*** is *a representation of covalent bonding using Lewis symbols, in which shared electron pairs are shown either as lines or as pairs of dots between two atoms, and lone pairs are shown as pairs of dots on individual atoms.* Only valence electrons are shown in a Lewis formula.

Compounds containing only covalent bonds are called ***covalent compounds.*** There are two types of covalent compounds. One contains discrete molecular units and are thus called *molecular covalent compounds;* examples are water, carbon dioxide, and hydrogen chloride. The other type exists in an extensive three-dimensional structure and does not contain discrete molecular units; these compounds include solid beryllium chloride $(BeCl_2)$ and silicon dioxide (SiO_2). Such substances are sometimes called *network covalent compounds,* to distinguish them from molecular covalent

This discussion applies only to representative elements.

Covalent compounds are the same as molecular compounds, which we discussed in previous chapters.

compounds. Unless otherwise stated, we will mean the molecular covalent type when referring to covalent compounds in this chapter.

Two atoms held together by one electron pair are said to be joined by a *single bond*. In many compounds, however, *two atoms share two or more pairs of electrons*. Such compounds contain one or more **multiple bonds.** If two atoms share two pairs of electrons the covalent bond is called a *double bond*. Double bonds are found in molecules like carbon dioxide (CO_2):

$$\ddot{O}::C::\ddot{O} \quad\text{ or }\quad \ddot{O}=C=\ddot{O}$$

Another example of a molecule with a double bond is ethylene (C_2H_4):

A *triple bond* arises when two atoms share three pairs of electrons, as for example in the nitrogen molecule (N_2):

$$:N::N: \quad\text{ or }\quad :N\equiv N:$$

The acetylene molecule (C_2H_2) also contains a triple bond, in this case between two carbon atoms:

$$H:C::C:H \quad\text{ or }\quad H-C\equiv C-H$$

Note that in ethylene and acetylene all the valence electrons are used in bonding and that there are no lone pairs on the carbon atoms.

You have now seen two major classes of bonding: ionic bonding and covalent bonding. The compounds resulting from these kinds of bonds differ in their properties. There are two types of forces of attraction in covalent compounds. The first type is the force that holds the atoms together in a molecule. A quantitative measure of this attraction is given by bond energy, discussed in Section 11.7. The second type of force of attraction exists *between* molecules, called *intermolecular forces* ("inter-" means between). Because intermolecular forces are usually quite weak (compared to forces holding atoms together in a molecule), molecules of a covalent compound are not held together tightly. Consequently covalent compounds are usually gases, liquids, or low-melting solids. On the other hand the electrostatic forces holding the ions together in an ionic compound are usually very strong so that ionic compounds are *always* solids with high melting points. Many ionic compounds are soluble in water, and the resulting aqueous solutions conduct electricity because the compounds are strong electrolytes. Most covalent compounds are either insoluble in water or their aqueous solutions generally do not conduct electricity because the compounds are nonelectrolytes. Molten ionic compounds conduct electricity because they contain mobile cations and anions, whereas liquid or molten covalent compounds do not conduct electricity because no ions are present. Table 11.1 compares some of the general properties of a typical ionic compound, sodium chloride, with those of a covalent compound, carbon tetrachloride (CCl_4).

You will be introduced to the skills involved in writing proper Lewis formulas at the end of Section 11.4. For the present we are simply examining typical Lewis formulas to become familiar with the language associated with them.

With the exception of carbon monoxide, stable molecules containing carbon do not have lone pairs on the carbon atoms.

If intermolecular forces are weak, it will be relatively easy to break up aggregates of molecules to form liquids (from solids) and gases (from liquids).

TABLE 11.1   Comparison of Some General Properties of an Ionic Compound and a Covalent Compound

Property	NaCl	CCl₄
Appearance	White solid	Colorless liquid
Melting point (°C)	801	−23
Molar heat of fusion (kJ/mol)	30.2	2.5
Boiling point (°C)	1413	76.5
Molar heat of vaporization (kJ/mol)	600	30
Density (g/cm³)	2.17	1.59
Solubility in water	High	Very low
Electrical conductivity		
Solid	Poor	Poor
Liquid	Good	Poor

11.2   ELECTRONEGATIVITY

In a sense the covalent bond and the ionic bond represent the extreme situations in bonding. Covalent bonding is the sharing of an electron pair by two atoms, whereas ionic bonding involves a complete transfer of one or more electrons from one atom to another. However, in the majority of cases we find that chemical bonds are neither purely covalent nor purely ionic. Rather, there is only a partial transfer of an electron from one atom to the other. Thus the bond holding the two atoms together is appropriately described as partly covalent and partly ionic.

Consider the hydrogen fluoride (HF) molecule. Experimental evidence indicates that the partial transfer of an electron (or the shift in electron density as it is more commonly described) is from the hydrogen atom to the fluorine atom (Figure 11.1). We can represent the electron density shift by adding an arrow (⊢→) above the Lewis formula for HF:

$$\overset{\longleftarrow}{\text{H—F}}$$

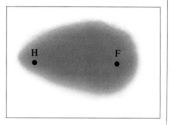

FIGURE 11.1 *Electron density distribution in the HF molecule. The dots represent the positions of the nuclei.*

This "unequal sharing" of the bonding electron pair results in a relatively greater electron density near the fluorine atom and a correspondingly lower electron density near the hydrogen atom. The charge separation in the molecule can be represented as follows:

$$\overset{\delta^+ \;\; \delta^-}{\text{H—F}}$$

where δ (delta) denotes the fractional electron charge separation. Since the electrons are not shared equally between the H and F atoms, the H—F bond is not purely covalent, as in the H₂ or F₂ molecule. Such a bond is more correctly described as a *polar covalent bond*, or simply a *polar bond*.

A property that helps us distinguish a purely covalent bond from a polar bond is the **electronegativity** of the elements, that is, *the ability of an atom to attract electrons toward itself in a chemical bond*. As we might expect, the electronegativity of an element is related to its electron affinity

1A	2A	3B	4B	5B	6B	7B	8B			1B	2B	3A	4A	5A	6A	7A
H 2.1																
Li 1.0	Be 1.5											B 2.0	C 2.5	N 3.0	O 3.5	F 4.0
Na 0.9	Mg 1.2											Al 1.5	Si 1.8	P 2.1	S 2.5	Cl 3.0
K 0.8	Ca 1.0	Sc 1.3	Ti 1.5	V 1.6	Cr 1.6	Mn 1.5	Fe 1.8	Co 1.9	Ni 1.9	Cu 1.9	Zn 1.6	Ga 1.6	Ge 1.8	As 2.0	Se 2.4	Br 2.8
Rb 0.8	Sr 1.0	Y 1.2	Zr 1.4	Nb 1.6	Mo 1.8	Tc 1.9	Ru 2.2	Rh 2.2	Pd 2.2	Ag 1.9	Cd 1.7	In 1.7	Sn 1.8	Sb 1.9	Te 2.1	I 2.5
Cs 0.7	Ba 0.9	La–Lu 1.0–1.2	Hf 1.3	Ta 1.5	W 1.7	Re 1.9	Os 2.2	Ir 2.2	Pt 2.2	Au 2.4	Hg 1.9	Tl 1.8	Pb 1.9	Bi 1.9	Po 2.0	At 2.2
Fr 0.7	Ra 0.9															

and ionization energy. Thus an atom such as fluorine, which has a high electron affinity (tends to pick up electrons easily) and a high ionization energy (does not lose electrons easily), has a high electronegativity. On the other hand, sodium, which has a low electron affinity and a low ionization energy, has a low electronegativity.

Linus Pauling (1901–   ) devised a method for calculating *relative* electronegativities of most elements. These values are shown in Figure 11.2. This chart is important because it shows trends and relationships among electronegativity values. Figure 11.3 shows the periodic trends in electro-

FIGURE 11.2 *The electronegativities of the more common elements. Group 8A elements (the noble gases) are not shown because most of them do not form compounds.*

Electronegativity values do not have units associated with them.

FIGURE 11.3 *Variation of electronegativity with atomic number. The halogens have the highest electronegativities and the alkali metals the lowest.*

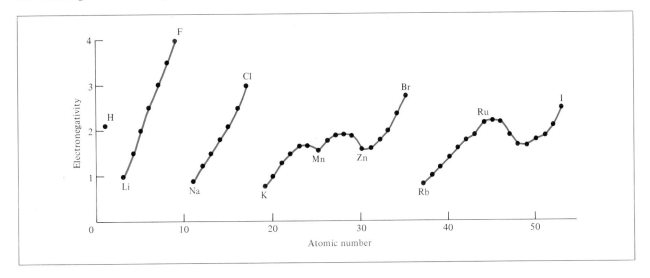

negativity. In general, electronegativity increases from left to right across a period in the periodic table, reflecting the decreasing metallic character of the elements. Within each group, electronegativity decreases with increasing atomic number, indicating increasing metallic character (Figure 11.2). Note that the transition metals do not follow these trends. The most electronegative elements (the halogens, oxygen, nitrogen, and sulfur) are found near the upper-right-hand corner of the periodic table, and the least electronegative elements (the alkali and alkaline earth metals) are clustered near the lower-left-hand corner.

Atoms of elements with widely different electronegativities tend to form ionic bonds with each other, since the atom of the less electronegative element gives up its electron(s) to the atom of the more electronegative element. Atoms of elements with more similar electronegativities tend to form covalent bonds or polar covalent bonds with each other because usually only a small shift in electron density takes place. These trends and characteristics are what we would expect from our knowledge of the elements' ionization energies and electron affinities.

A word about the difference between electronegativity and electron affinity is appropriate here. Both properties express the tendency of an atom to attract electrons. However, electron affinity refers to an isolated atom's attraction for an additional electron, whereas electronegativity expresses the attraction of an atom in a chemical bond (with another atom) for the shared electrons.

Although there is no sharp distinction between an extreme polar covalent bond and an ionic bond, the following rule is helpful in distinguishing between them. An ionic bond forms when the electronegativity difference between the two bonding atoms is 1.6 or more. This rule applies to most but not all ionic compounds. Sometimes chemists use the quantity *percent ionic character* to describe the nature of a bond. A purely ionic bond has 100 percent ionic character, whereas a purely covalent bond has 0 percent ionic character. Figure 11.4 shows the relation between electronegativity difference and percent ionic character. As you can see, the greater the

These tendencies are often summarized by the observation that an ionic bond generally involves an atom of a metallic element and an atom of a nonmetallic element, whereas a covalent bond generally involves atoms of nonmetallic elements.

No known compound has 100 percent ionic character.

FIGURE 11.4 *Relation between electronegativity difference and percent ionic character. In reality, the largest difference there can be in a bond is 3.3, which is the difference between the most electronegative element F, (4.0), and the least electronegative element, Cs (0.7).*

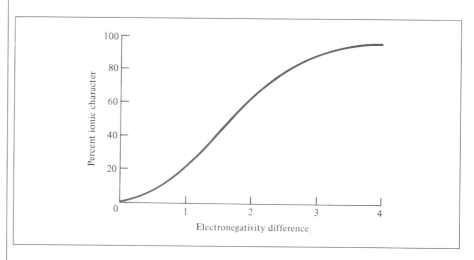

difference in electronegativities between the atoms in a bond, the greater the ionic character of the bond.

EXAMPLE 11.1

Classify the following bonds as ionic, polar covalent, or covalent: (a) the bond in HCl, (b) the bond in KF, and (c) the CC bond in H_3CCH_3.

Answer

(a) In Figure 11.2 we see that the electronegativity difference between H and Cl is 0.9, which is appreciable but not large enough (by the 1.6 rule) to qualify HCl as an ionic compound. Therefore, the bond between H and Cl is polar covalent.
(b) The electronegativity difference between K and F is 3.2, which is well above the 1.6 mark; therefore, the bond between K and F is ionic.
(c) The two C atoms are identical in every respect—they are bonded to each other and each is bonded to three other H atoms. Therefore, the bond between them is purely covalent.

Similar examples: Problems 11.7, 11.9, 11.11.

11.3 THE OCTET RULE

A useful pattern emerges as we examine the total number of valence electrons around each atom in a covalent compound. For example, consider the hydrogen molecule. The two electrons in the single bond spend part of their time in the region of each H atom. In that sense each atom has two electrons around itself as in the case of a helium atom. This fact can be illustrated by drawing a circle around each atom as follows:

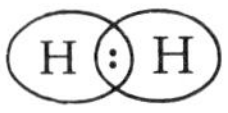

$$2e^- \quad 2e^-$$

By applying the same electron-counting procedure to F in HF and to Cl in Cl_2, we can write

$$2e^- \quad 8e^- \qquad 8e^- \quad 8e^-$$

We see that both F and Cl have eight electrons around each atom; F in HF has the neon configuration, and Cl in Cl_2 has the argon configuration.

We saw in earlier chapters that a complete shell of two or eight electrons constitutes a very stable configuration. From this knowledge, Lewis formulated the **octet rule,** which may be stated as follows: *An atom other than hydrogen tends to form bonds until it is surrounded by eight valence electrons.* Therefore, we can say that a covalent bond forms when there are not enough electrons for each individual atom to have a complete octet. By sharing electrons in a covalent bond, the individual atoms can complete their octets.

The octet rule applies also to molecules with multiple bonds. As the following diagrams show, the number of electrons around each atom except hydrogen is eight:

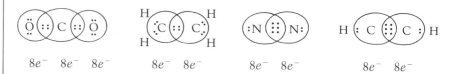

$8e^-$ $8e^-$ $8e^-$ $8e^-$ $8e^-$ $8e^-$ $8e^-$ $8e^-$ $8e^-$

The octet rule helps us predict the stability of molecules containing the second-period elements, but it breaks down when applied to elements in the third period and beyond (to be discussed later). Although the octet rule and the Lewis formula do not present a complete picture of bonding, they are very useful in understanding the bonding scheme in many compounds and are still frequently employed by chemists in discussing the properties and reactions of molecules. For this reason, you should become familiar with writing Lewis formulas of compounds. The basic steps are as follows:

1. Write the skeletal structure of the compound; that is, determine what atoms are bonded to what other atoms. For simple compounds this task is fairly easy. For more complex compounds, we must either be given the information or make an intelligent guess about it. The following facts are useful in determining skeletal structures. Hydrogen and fluorine *always* occupy the terminal (end) position in the Lewis formula. Nitrogen, oxygen, sulfur, chlorine, bromine, and iodine can occupy either the terminal or a central position. Phosphorus *always* occupies a central position, and carbon seldom occupies the terminal position. In many cases, the central atom is *less* electronegative than the surrounding atoms.
2. Count the total number of valence electrons present. (If necessary, refer to Figure 9.1.) For polyatomic anions, add the number of negative charges to that total. For example, for the CO_3^{2-} ion we add 2 because the $2-$ charge indicates that there are 2 more electrons than are provided by the neutral atoms. For polyatomic cations, we subtract the number of positive charges from this total. Thus, for NH_4^+ we subtract 1 because the $1+$ charge indicates a loss of one electron from the group of neutral atoms.
3. Draw a single covalent bond between the central atom and each of the surrounding atoms. Complete the octets of the atoms bonded to the central atom. (Remember that the valence shell of a hydrogen atom is complete with only two electrons.) Electrons belonging to the central or surrounding atoms must be shown as lone pairs if they are not involved in bonding. The maximum number of electrons that can be used is that determined in step 2.
4. If the octet rule is not met for the central atom, then we must write double or triple bonds between the surrounding atoms and the central atom, using the lone pairs on the surrounding atoms.

EXAMPLE 11.2

Write the Lewis formula for nitrogen trichloride (NCl_3), in which all three Cl atoms are bonded to the N atom.

Answer

Step 1

The skeletal structure of NCl_3 is

$$Cl$$
$$N$$
$$Cl \quad Cl$$

Step 2

The outer-shell electron configurations of N and Cl are $2s^2 2p^3$ and $3s^2 3p^5$, respectively. Thus there are $5 + (3 \times 7)$, or 26 valence electrons to account for in NCl_3.

Step 3

We draw a single covalent bond between N and each Cl, and complete the octets:

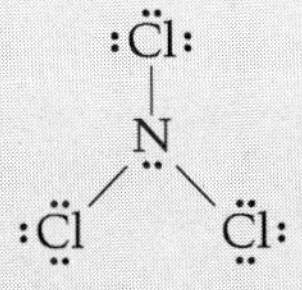

$$:\ddot{C}l:$$
$$|$$
$$N$$
$$:\ddot{C}l \quad \ddot{C}l:$$

Since this structure satisfies the octet rule for all the atoms, step 4 is not required. To check, we count the valence electrons in NCl_3 (in chemical bonds and in lone pairs). The result is 26, the same as the number of valence electrons on three Cl atoms and one N atom.

Similar example: Problem 11.15.

EXAMPLE 11.3

What is the Lewis formula for carbon tetrabromide (CBr_4)?

Answer

Step 1

We are not given the skeletal structure of the compound. However, we know that carbon does not usually occupy a terminal position. Furthermore, carbon is less electronegative than bromine (see Figure 11.2), so that it is most likely to occupy a central position:

$$Br$$
$$Br \quad C \quad Br$$
$$Br$$

(Continued)

Step 2

The outer-shell electron configurations of C and Br are $2s^2 2p^2$ and $4s^2 4p^5$, respectively. Thus there are $4 + (4 \times 7)$, or 32 valence electrons to account for in CBr_4.

Step 3

We draw a single covalent bond between the central atom and each of the end atoms and complete the octets of the end atoms:

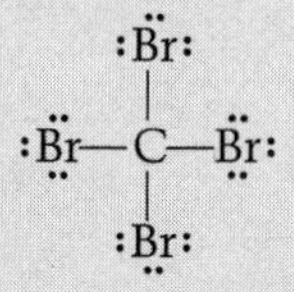

$$
\begin{array}{c}
:\ddot{Br}: \\
| \\
:\ddot{Br} - C - \ddot{Br}: \\
| \\
:\ddot{Br}:
\end{array}
$$

Since this structure satisfies the octet rule for all the atoms, step 4 is not required. Checking the total number of valence electrons in CBr_4 (in chemical bonds and in lone pairs), we find 32, the number we counted in step 2. Note that there are no lone pairs on the C atom.

Similar example: Problem 11.15.

EXAMPLE 11.4

The octet rule also applies to polyatomic anions and cations.

Write the Lewis formula for the nitrate ion (NO_3^-), in which the oxygen atoms are bonded to the nitrogen atom.

Answer

Step 1

The skeletal structure of NO_3^- is

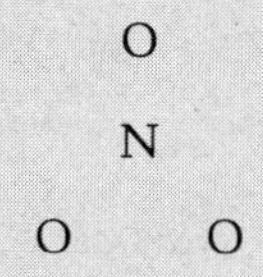

$$
\begin{array}{c}
O \\
\\
N \\
\\
O \qquad O
\end{array}
$$

Step 2

The outer-shell electron configurations of N and O are $2s^2 2p^3$ and $2s^2 2p^4$, respectively, and the ion itself has one negative charge. Thus the total number of electrons is $5 + (3 \times 6) + 1$, or 24.

Step 3

We draw a single covalent bond between N and each O and comply with the octet rule for the O atoms:

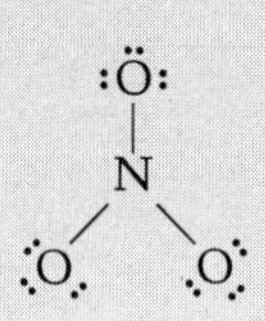

$$
\begin{array}{c}
:\ddot{O}: \\
| \\
N \\
/ \quad \backslash \\
:\ddot{O}. \qquad .\ddot{O}:
\end{array}
$$

Step 4

We see that the octet rule is satisfied for the O atoms but not for the N atom. Therefore we move a lone pair from one of the O atoms to form another bond with N. Now the octet rule is also satisfied for the N atom:

$$
\left[
\begin{array}{c}
:\ddot{O}: \\
\parallel \\
N \\
:\ddot{O}. \quad .\ddot{O}:
\end{array}
\right]^{-}
$$

As a final check, we make sure that there are twenty-four valence electrons in the Lewis formula of the nitrate ion.

Similar example: Problem 11.16.

11.4 FORMAL CHARGE AND LEWIS FORMULA

We have seen that a covalent bond appears to be the result of electron sharing by two atoms, each of which contributes one electron. In many cases, however, *one of the atoms donates both electrons to form a covalent bond.* Such a bond is called a ***coordinate covalent bond.*** Although the properties of a coordinate covalent bond do not differ from those of a normal covalent bond (because all electrons are alike no matter what their source), the distinction is useful for keeping track of valence electrons in Lewis formulas. Consider the ozone (O_3) molecule as an example. In O_3, the central O atom is bonded to two end O atoms. Proceeding by steps as we did in Examples 11.3 and 11.4 we draw the skeletal structure of O_3 and then we add bonds and electrons so that the octet rule is satisfied for the two end atoms. Each O atom has six valence electrons but the central atom donates four valence electrons to form two coordinate covalent bonds with the end atoms:

$$
\begin{array}{c}
\ddot{O} \\
:\ddot{O}. \quad .\ddot{O}:
\end{array}
$$

You can see that the octet rule is not satisfied for the central atom. To remedy this, we convert a lone pair on one of the end atoms into an additional bond between that end atom and the central atom, as follows:

$$
\begin{array}{c}
\ddot{O} \\
:\ddot{O}. \quad .\ddot{O}:
\end{array}
$$

Now there is only one coordinate covalent bond—the single bond between the two O atoms. In the O=O bond, each atom donates two electrons, so that no coordinate covalent bond is present there.

The existence of a coordinate covalent bond can be determined by the *formal charges* on the atoms. The formal charge on an atom in a Lewis formula is defined as follows:

Another name for a coordinate covalent bond is *dative bond.*

Ozone is a toxic, light blue gas (boiling point: $-111.3°C$) with a pungent odor.

$$\begin{array}{c}\text{formal charge on}\\ \text{an atom in a}\\ \text{Lewis formula}\end{array} = \begin{array}{c}\text{total number of}\\ \text{valence electrons}\\ \text{in the free atom}\end{array} - \begin{array}{c}\text{total number}\\ \text{of nonbonding}\\ \text{electrons}\end{array} - \frac{1}{2}\left(\begin{array}{c}\text{total number}\\ \text{of bonding}\\ \text{electrons}\end{array}\right) \quad (11.1)$$

For the ozone molecule, we can calculate the formal charges on the atoms using Equation (11.1) as follows:

- *The central O atom.* The central atom (see the preceding Lewis formula) has six valence electrons, one lone pair (or two nonbinding electrons), and three bonds (or six bonding electrons). Substituting in Equation (11.1) we write

$$\text{formal charge} = 6 - 2 - \tfrac{1}{2}(6) = +1$$

- *The end O atom in O=O.* This atom has six valence electrons, two lone pairs (or four nonbonding electrons), and two bonds (or four bonding electrons). Thus we write

$$\text{formal charge} = 6 - 4 - \tfrac{1}{2}(4) = 0$$

- *The end O atom in O—O.* This atom has six valence electrons, three lone pairs (or six nonbonding electrons), and one bond (or two bonding electrons). Thus we write

$$\text{formal charge} = 6 - 6 - \tfrac{1}{2}(2) = -1$$

We can now write the Lewis formula for ozone including the formal charges as

When you write formal charges, the following rules will be helpful:

- For neutral molecules, the sum of the formal charges must add up to zero. (This rule applies, for example, to the O_3 molecule.)
- For cations, the sum of the formal charges must equal the positive charge.
- For anions, the sum of formal charges must equal the negative charge.

Keep in mind that formal charges do not indicate actual charge separations within the molecule. In the O_3 molecule, for example, there is no evidence that the central atom bears a net +1 charge or that one of the end atoms bears a −1 charge. Writing these charges on the atoms in the Lewis formula merely helps us keep track of the valence electrons in the molecule. As we will see shortly, formal charges help us decide which Lewis formula is more plausible for a given molecule.

EXAMPLE 11.5

Write formal charges for the nitrate ion.

Answer

The Lewis formula for the nitrate ion was developed in Example 11.4. The formal charges on the atoms can be calculated as follows:

The N atom: formal charge $= 5 - 0 - \frac{1}{2}(8) = +1$
The O atom in N=O: formal charge $= 6 - 4 - \frac{1}{2}(4) = 0$
The O atom in N—O: formal charge $= 6 - 6 - \frac{1}{2}(2) = -1$

Thus the Lewis formula for NO_3^- with formal charges is

$$
\begin{array}{c}
: \ddot{O} : \\
\| \\
\overset{+}{N} \\
\diagup \quad \diagdown \\
_- : \ddot{O} \cdot \qquad \cdot \ddot{O} : _-
\end{array}
$$

Note that the sum of the formal charges is $-2 + 1 = -1$, the same as the charge on the nitrate ion.

Formal charges are often useful in selecting a plausible Lewis formula for a given compound. The guidelines we follow are

- A Lewis formula in which there are no formal charges is preferable to one in which formal charges are present.
- Lewis formulas with large formal charges $(+2, +3,$ and/or $-2, -3,$ and so on) are less plausible than those with small formal charges.
- In choosing among Lewis formulas having similar distributions of formal charges, the most plausible Lewis formula is that in which negative formal charges are placed on the more electronegative atoms.

EXAMPLE   11.6

Formaldehyde, a liquid with a disagreeable odor, has been used traditionally as a preservative for dead animals. Its molecular formula is CH_2O. Draw the most likely Lewis formula for the compound.

Answer

The two possible skeletal structures are

$$
\begin{array}{cc}
\text{H} \quad \text{C} \quad \text{O} \quad \text{H} & \begin{array}{c} \text{H} \\ \quad \text{C} \quad \text{O} \\ \text{H} \end{array} \\
\text{(a)} & \text{(b)}
\end{array}
$$

Following the procedures in previous examples we can draw the Lewis formulas as

$$
\begin{array}{cc}
\text{H}-\overset{-}{\ddot{\text{C}}}=\overset{+}{\ddot{\text{O}}}-\text{H} & \begin{array}{c} \text{H} \\ \diagdown \\ \text{C}=\ddot{\text{O}} \\ \diagup \\ \text{H} \end{array} \\
\text{(a)} & \text{(b)}
\end{array}
$$

Since (b) carries no formal charges, it is the more likely structure. This conclusion is confirmed by experimental evidence.

Can you suggest two other reasons why (a) is less plausible?

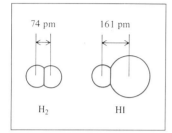

FIGURE 11.5 *The bond length between two atoms is the distance between the two nuclei of the atoms. Here the bond lengths are defined for diatomic molecules H_2 and HI.*

Bond length provides information about the strength of the bond and the size and shape of the molecule.

11.5 THE CONCEPT OF RESONANCE

In our earlier discussion of the O_3 molecule, you saw that the octet rule for the central atom can be satisfied by drawing a double bond between it and an end O atom. However, we can add the double bond in either of two places, as shown by these two equivalent Lewis formulas:

Arbitrarily choosing one of these two Lewis formulas to represent ozone presents some problems. For one thing, it makes the task of accounting for the known bond lengths in O_3 difficult.

Bond length is *the distance between the centers of two bonded atoms in a molecule* (Figure 11.5). Chemists know from experience that bond length depends not only on the nature of the bonded atoms but also on whether the bond joining the atoms is a single bond, a double bond, or a triple bond. Table 11.2 shows some typical bond lengths. As you can see, for a given pair of atoms, triple bonds are shorter than double bonds, which, in turn, are shorter than single bonds. On the basis of these data and the proposed Lewis formula for O_3, we would expect that the O—O bond would be longer than the O=O bond. However, experimental evidence shows that both oxygen-to-oxygen bonds are equal in length (128 pm); therefore, neither of the two Lewis formulas shown accurately represents the molecule.

The way we resolve this problem is to have *both* Lewis formulas represent the ozone molecule:

Each of the two Lewis formulas is called a **resonance structure** or **resonance form**. A resonance structure, then, is *one of two or more alternative Lewis formulas for a single molecule that cannot be described fully with only one Lewis formula*. The $\longleftrightarrow$ symbol indicates that the structures shown are resonance forms. The term **resonance** itself means *the use of two or more Lewis formulas to represent a particular molecule*. There is an interesting analogy for resonance. A medieval European traveler returned home from Africa and described a rhinoceros as a cross between a griffin and a unicorn, two familiar but imaginary animals. Similarly, we describe ozone, a real molecule, in terms of the two familiar but nonexistent molecules shown.

A common misconception is the notion that a molecule such as ozone somehow shifts quickly back and forth from one resonance structure to the other. Keep in mind that *neither* resonance structure adequately represents the actual molecule, which has its own unique, stable structure. "Resonance" is a human invention, designed to compensate partially for limitations in these simple bonding models. To extend the analogy, a rhinoceros is its "own" self, not some oscillation between mythical griffin and unicorn!

Other examples of resonance structures are those for the nitrate ion:

TABLE 11.2
Average Bond Lengths of Some Common Single, Double, and Triple Bonds

Bond Type	Bond Length (pm)
C—H	107
C—O	143
C=O	121
C—C	154
C=C	133
C≡C	120
C—N	143
C=N	138
C≡N	116
N—O	136
N=O	122
O—H	96

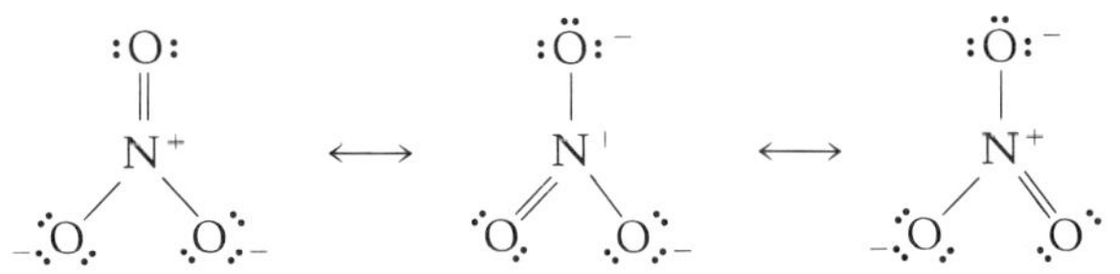

All nitrogen-to-oxygen bonds are equivalent. Therefore, the properties of the nitrate ion are best described by considering its resonance structures together, not separately.

The concept of resonance applies equally well to organic systems. A well-known example is the benzene molecule (C_6H_6):

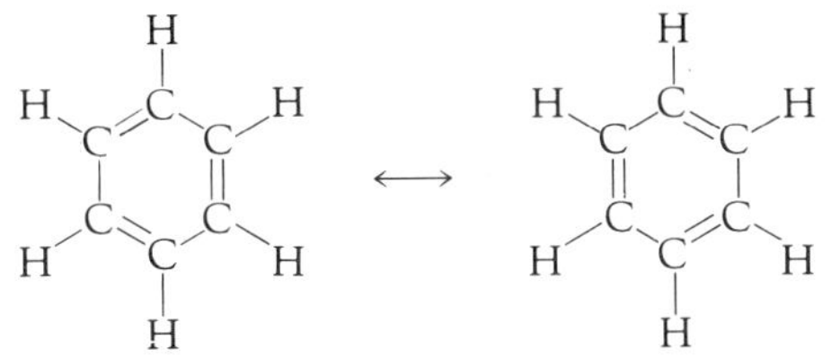

A simpler way of drawing the structure of the benzene molecule or of other compounds containing the "benzene ring" is to show only the skeleton and not the carbon and hydrogen atoms. In this way the resonance structures are represented by

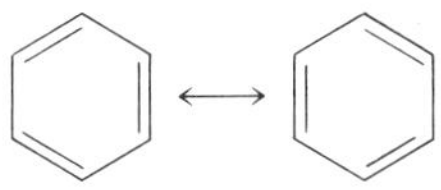

Note that the C atoms at the corners of the hexagon and the H atoms are all omitted. Only the bonds between the C atoms are shown.

If benzene actually existed as one of its resonance forms, there would be two different bond lengths between adjacent C atoms, one characteristic of the single bond and the other of the double bond. In fact, the distances between adjacent C atoms in benzene are *all* 140 pm, which is between the length of a C—C bond (154 pm) and that of a C=C bond (133 pm), shown in Table 11.2.

The following rules will help you in drawing resonance structures:

- The positions of electrons, but not those of atoms, can be rearranged in different resonance structures. In other words, the same atoms must be bonded to one another in all the resonance structures for a given species.
- The most reasonable resonance structures are those in which most or all atoms obey the octet rule.

EXAMPLE 11.7

Draw resonance structures (including formal charges) for dinitrogen tetroxide (N_2O_4), which has the following skeletal arrangement:

(Continued)

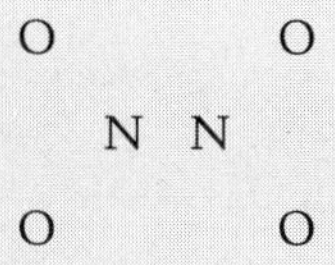

$$O \qquad O$$
$$N \quad N$$
$$O \qquad O$$

Answer

Since each nitrogen atom has five valence electrons and each oxygen atom has six valence electrons, the total number of valence electrons is $(2 \times 5) + (4 \times 6) = 34$. The following Lewis formula satisfies the octet rule for both the N and O atoms:

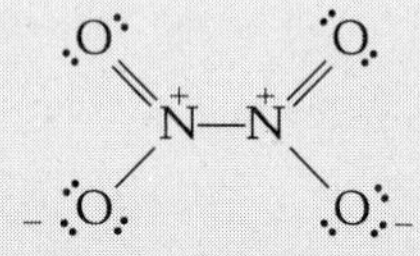

However, since there is no preference as to where we should draw the N—O and N=O bonds, we must include the following resonance structures:

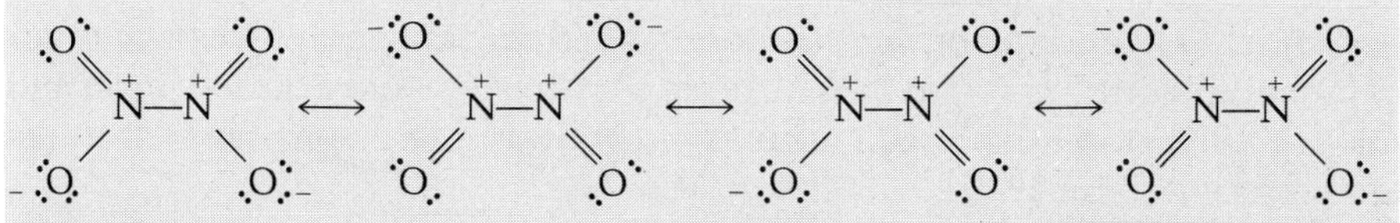

Similar examples: Problems 11.30, 11.35, 11.38.

A final note on resonance. Although resonance structures provide a more accurate description of the properties of a molecule, we will often use only one Lewis formula to represent a molecule, for simplicity.

11.6 EXCEPTIONS TO THE OCTET RULE

Now that we know the general features of the octet rule, we can ask the question, Why does the octet rule work? Elements in the second period have only $2s$ and $2p$ subshells, which can hold a total of eight electrons. When an atom of one of these elements forms a chemical compound, it can acquire the noble gas electron configuration [Ne] by sharing electrons with other atoms in the same molecule. That is why the octet rule works so well for many molecules containing second-period elements. However, there are a number of important exceptions to the octet rule that give us further insight into the nature of chemical bonding. Following are three types of these exceptions to the octet rule for representative elements.

The Incomplete Octet

In some compounds we find that the number of electrons surrounding the central atom is fewer than eight. Consider, for example, beryllium, which is a Group 2A element. The electron configuration of beryllium is $1s^2 2s^2$; thus it has two valence electrons in the $2s$ orbital. In the solid state, beryllium chloride consists of extensive long chains in which each Be atom is bonded to four Cl atoms and each Cl atom is bonded to two Be atoms. Heating the solid produces a vapor that consists of discrete $BeCl_2$ units. The Lewis formula of the $BeCl_2$ molecule is

$$:\ddot{Cl}—Be—\ddot{Cl}:$$

We see that while both Cl atoms achieve the noble gas electron configuration [Ar], the Be atom has only four electrons surrounding itself. Thus the octet rule is not satisfied for beryllium.

You might wonder whether it is possible to complete the octet for Be by using the two lone pairs on the Cl atoms to form additional bonds as follows:

$$\overset{+}{\ddot{Cl}}=\overset{2-}{Be}=\overset{+}{\ddot{Cl}}$$

Although the octet rule is now satisfied for Be, this resonance structure is not too plausible because it involves large formal charge separations and places the negative charges on the less electronegative beryllium atom. Many covalent compounds of beryllium do not obey the octet rule.

Elements in Group 3A, particularly boron and aluminum, also tend to form compounds in which fewer than eight electrons surround each atom. Take boron as an example. Since its electron configuration is $1s^2 2s^2 2p^1$, it has a total of three valence electrons. Boron forms with the halogens a class of compounds of the general formula BX_3, where X denotes a halogen atom. Thus, in boron trifluoride there are only six electrons around the boron atom:

$$\begin{array}{c} :\ddot{F}: \\ | \\ :\ddot{F}—B \\ | \\ :\ddot{F}: \end{array}$$

As in the case of $BeCl_2$, we can satisfy the octet rule for B by drawing the following resonance structure:

$$\begin{array}{c} \overset{+}{:F:} \\ \| \\ :\ddot{F}—B^- \\ | \\ :\ddot{F}: \end{array}$$

However, this resonance structure is less plausible because it involves formal charge separations and places the negative charge on the less electronegative boron atom.

Although boron trifluoride is stable, it has a tendency to pick up an unshared electron pair from an atom in another compound, as shown by its reaction with ammonia:

$$\begin{array}{ccccc} :\ddot{F}: & H & & :\ddot{F}: & H \\ | & | & & | & | \\ :\ddot{F}—B & + & :N—H & \longrightarrow & :\ddot{F}—B^-—N^+—H \\ | & | & & | & | \\ :\ddot{F}: & H & & :\ddot{F}: & H \end{array}$$

This structure satisfies the octet rule for all of the B, N, and F atoms.

Unlike beryllium, which forms mostly covalent compounds, the rest of the Group 2A elements form mostly ionic compounds.

Boron trifluoride is a colorless gas (boiling point: $-100°C$) with a pungent odor.

The B—N bond is a coordinate covalent bond because the N atom is donating both of the electrons.

Odd-Electron Molecules

Some molecules contain an *odd* number of electrons. Among these are chlorine dioxide (ClO_2), nitric oxide (NO), and nitrogen dioxide (NO_2):

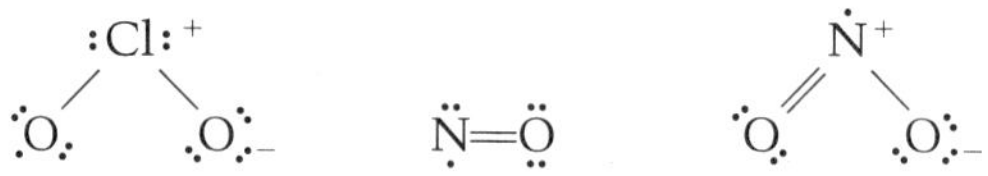

Since we need an even number of electrons for complete pairing (to reach eight), the octet rule clearly can *never* be satisfied for all the atoms in any of these molecules.

The Expanded Octet

Atoms of the second-period elements can *never* have more than eight electrons around themselves in a compound.

In a number of compounds there are more than eight valence electrons around an atom; this is called an *expanded octet*. Expanded octets occur only around atoms of elements in and beyond the third period of the periodic table. In addition to the $3s$ and $3p$ orbitals, elements in the third period also have $3d$ orbitals that can be used in bonding. Consider sulfur hexafluoride, which is a very stable compound. The electron configuration of sulfur is $[Ne]3s^23p^4$. In SF_6, each of sulfur's six valence electrons forms a covalent bond with a fluorine atom, giving rise to twelve electrons around the central sulfur atom:

Sulfur hexafluoride is a colorless, odorless gas (boiling point: $-64°C$).

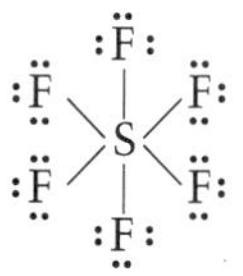

In the next chapter we will see that these twelve electrons, or six bonding pairs, are accommodated in six orbitals that originate from one $3s$, three $3p$, and two of the five $3d$ orbitals. However, sulfur also forms many compounds in which it does not violate the octet rule. Thus in sulfur dioxide S has only eight electrons around itself and therefore obeys the octet rule:

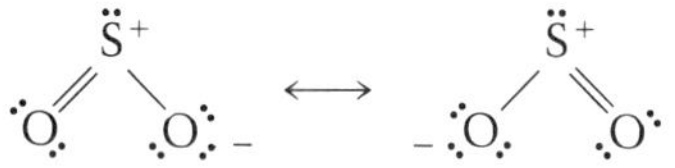

EXAMPLE 11.8

Draw three resonance structures for the odd-electron molecule ClO_2.

Answer

We can draw three resonance structures for ClO_2 as follows:

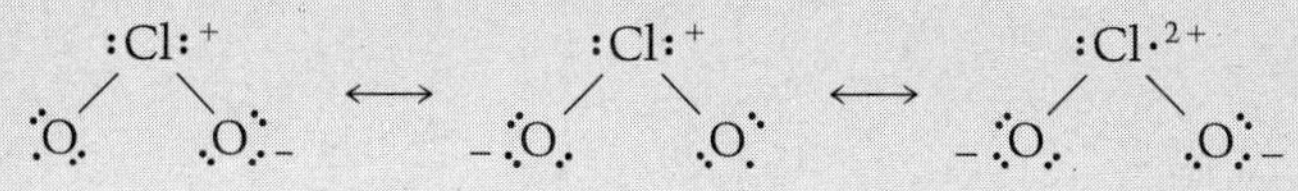

The first two resonance structures are equivalent but the third is different. The third structure is less plausible because there is a greater separation of formal charges. Therefore, we can say that the third resonance structure makes less of a contribution to the overall properties of ClO_2 than the first two resonance structures.

EXAMPLE 11.9

Draw the Lewis formula for aluminum triiodide (AlI_3).

Answer

The outer-shell electron configuration of Al is $3s^23p^1$. The Al atom forms three covalent bonds with the I atoms as follows:

$$:\ddot{\overset{..}{I}}:$$
$$|$$
$$:\ddot{\overset{..}{I}}—Al$$
$$|$$
$$:\ddot{I}:$$

Like boron, aluminum is a Group 3A element.

Although the octet rule is satisfied for the I atoms, there are only six valence electrons around the Al atom. This molecule is an example of the incomplete octet.

Similar example: Problem 11.39.

EXAMPLE 11.10

Draw the Lewis formula for phosphorus pentafluoride (PF_5), in which all five F atoms are bonded directly to the P atom.

Answer

The outer-shell electron configurations for P and F are $3s^23p^3$ and $2s^22p^5$, respectively, and so the total number of valence electrons is $5 + (5 \times 7)$, or 40. The Lewis formula of PF_5 is

$$:\ddot{\overset{..}{F}}:$$
$$| \quad \ddot{\overset{..}{F}}:$$
$$:\ddot{\overset{..}{F}}—P \big\langle$$
$$| \quad \ddot{F}:$$
$$:\ddot{F}:$$

Like sulfur, phosphorus is a third-period element.

Although the octet rule is satisfied for the F atoms, there are ten valence electrons around the P atom. This is an example of the expanded octet.

Similar example: Problem 11.52.

11.7   STRENGTH OF THE COVALENT BOND

Bond Dissociation Energy and Bond Energy

We have seen that the Lewis theory of chemical bonding depicts a covalent bond as the sharing of two electrons between the atoms. However, the Lewis approach does not indicate the relative strengths of covalent bonds. For example, the bonds in H_2 and Cl_2 are represented by identical single lines:

$$H—H \qquad :\ddot{C}l—\ddot{C}l:$$

yet we know from experience that it takes more energy to break the bond in H_2 than in Cl_2. In this sense we conclude that H_2 is the more stable molecule of the two. A quantitative measure of the stability of a molecule is **bond dissociation energy,** which is the *enthalpy change required to break a particular bond in a mole of gaseous diatomic molecules.* For the hydrogen molecule

$$H_2(g) \longrightarrow H(g) + H(g) \qquad \Delta H° = 436.4 \text{ kJ}$$

This equation tells us that to break the covalent bonds in one mole of gaseous H_2 molecules, we must supply 436.4 kJ of energy.

Similarly for the less stable chlorine molecule

$$Cl_2(g) \longrightarrow Cl(g) + Cl(g) \qquad \Delta H° = 242.7 \text{ kJ}$$

Bond dissociation energy can also be applied to diatomic molecules containing unlike elements, such as HCl

$$HCl(g) \longrightarrow H(g) + Cl(g) \qquad \Delta H° = 431.9 \text{ kJ}$$

as well as to molecules containing double and triple bonds

$$O_2(g) \longrightarrow O(g) + O(g) \qquad \Delta H° = 498.7 \text{ kJ}$$

$$N_2(g) \longrightarrow N(g) + N(g) \qquad \Delta H° = 941.4 \text{ kJ}$$

Measuring the strength of covalent bonds becomes more complicated when we consider polyatomic molecules. For example, measurements show that the energy needed to break the O—H bond in H_2O is not a constant:

$$H_2O(g) \longrightarrow H(g) + OH(g) \qquad \Delta H° = 502 \text{ kJ}$$

$$OH(g) \longrightarrow H(g) + O(g) \qquad \Delta H° = 427 \text{ kJ}$$

In each case, an O—H bond is broken, but the first step is more endothermic than the second. The difference in the two $\Delta H°$ values suggests that the O—H bond itself undergoes change, presumably brought about by the changed chemical environment. But the variation is usually not very great, as indicated by the two $\Delta H°$ values for the O—H bond, which are roughly comparable in H_2O and many other molecules.

The example of the O—H bonds in H_2O shows that when dealing with polyatomic molecules, we can speak only of approximate or *average* bond energies. Put another way, except for diatomic molecules, the energy of a particular bond type is not constant but varies somewhat from molecule

We specify the gaseous state here because bond dissociation energies in solids and liquids are affected by neighboring molecules.

It is important to understand that bond dissociation energy is a precisely measured quantity that applies to the strength of a bond in a diatomic molecule. On the other hand, *bond energy* is an average quantity used to describe the strength of a particular bond in a polyatomic molecule.

TABLE 11.3 Some Bond Dissociation Energies of Diatomic Molecules* and Average Bond Energies

Bond	Bond Energy (kJ/mol)	Bond	Bond Energy (kJ/mol)
H—H	436.4	C—S	255
H—N	393	C=S	477
H—O	460	N—N	393
H—S	368	N=N	418
H—P	326	N≡N	941.4
H—F	568.2	N—O	176
H—Cl	431.9	N—P	209
H—Br	366.1	O—O	142
H—I	298.3	O=O	498.7
C—H	414	O—P	502
C—C	347	O=S	469
C=C	620	P—P	197
C≡C	812	P=P	489
C—N	276	S—S	268
C=N	615	S=S	352
C≡N	891	F—F	150.6
C—O	351	Cl—Cl	242.7
C=O	781	Br—Br	192.5
C—P	263	I—I	151.0

* Bond dissociation energies for diatomic molecules (in color) have more significant figures than bond energies for polyatomic molecules. This is because, for diatomic molecules, the bond dissociation energies are directly measurable quantities and not averaged over many compounds.

to molecule. Table 11.3 lists the bond dissociation energies and average bond energies for a number of common covalent bonds.

Use of Bond Energies in Thermochemistry

Chapter 5 considered the energy changes that accompany chemical reactions. In comparing the thermochemical changes of a large number of reactions, we are struck by the wide variation in the enthalpies of different reactions. For example, the combustion of hydrogen gas in oxygen gas is fairly exothermic:

$$H_2(g) + \tfrac{1}{2}O_2(g) \longrightarrow H_2O(l) \qquad \Delta H° = -285.8 \text{ kJ}$$

On the other hand, the formation of glucose ($C_6H_{12}O_6$) from water and carbon dioxide, best achieved by photosynthesis, is highly endothermic:

$$6CO_2(g) + 6H_2O(l) \longrightarrow C_6H_{12}O_6(s) + 6O_2(g) \qquad \Delta H° = 2801 \text{ kJ}$$

To account for such variations, we need to examine the stability of individual reactant and product molecules. After all, most chemical reactions involve the making and breaking of bonds. Therefore, knowing the bond energies and hence the stability of molecules should tell us something about the thermochemical nature of reactions that molecules undergo.

In many cases it is possible to estimate the enthalpy of reaction by using the average bond energies. Recall that energy is always required to break chemical bonds and that chemical bond formation is always accompanied by a release of energy. To estimate the enthalpy of a reaction, then, all we need to do is count the total number of bonds broken and formed in the reaction and record all of the corresponding energy changes. The enthalpy of reaction in the *gas phase* is given by

$$\Delta H° = \Sigma BE(\text{reactants}) - \Sigma BE(\text{products})$$
$$= \text{total energy input} - \text{total energy released} \qquad (11.2)$$

where *BE* stands for average bond energy and Σ is the summation sign. Equation (11.2) as written takes care of the sign convention for $\Delta H°$. Thus if the total energy input is greater than the total energy released, $\Delta H°$ is positive and the reaction is endothermic. On the other hand, if more energy is released than absorbed, $\Delta H°$ is negative and the reaction is exothermic. Note that if reactants and products are all diatomic molecules, then Equation (11.2) should yield accurate results, since the bond dissociation energies of diatomic molecules are accurately known. If some or all of the reactants and products are polyatomic molecules, Equation (11.2) will yield only approximate results because average bond energies will be used in the calculation.

For diatomic molecules, Equation (11.2) becomes equivalent to Equation (5.4), so that the results obtained from these two equations should be in excellent agreement.

EXAMPLE 11.11

Use Equation (11.2) to calculate the enthalpy of reaction for the process

$$H_2(g) + Cl_2(g) \longrightarrow 2HCl(g)$$

Compare your result with that obtained using Equation (5.4).

Answer

The first step is to count the number of bonds broken and the number of bonds formed. This is best done by using a table:

Refer to Table 11.3 for bond dissociation energy values for these diatomic molecules.

Type of Bonds Broken	Number of Bonds Broken	Bond Energy (kJ/mol)	Energy Change (kJ)
H—H (H_2)	1	436.4	436.4
Cl—Cl (Cl_2)	1	242.7	242.7

Type of Bonds Formed	Number of Bonds Formed	Bond Energy (kJ/mol)	Energy Change (kJ)
H—Cl (HCl)	2	431.9	863.8

Next, we obtain the total energy input and total energy released:

$$\text{total energy input} = 436.4 \text{ kJ} + 242.7 \text{ kJ} = 679.1 \text{ kJ}$$

$$\text{total energy released} = 863.8 \text{ kJ}$$

Using Equation (11.2), we write

$$\Delta H° = 679.1 \text{ kJ} - 863.8 \text{ kJ} = -184.7 \text{ kJ}$$

Alternatively, we can use Equation (5.4) and the data in Appendix 1 to calculate the enthalpy of reaction:

$$\Delta H° = 2\Delta H_f°(\text{HCl}) - [\Delta H_f°(\text{H}_2) + \Delta H_f°(\text{Cl}_2)]$$
$$= (2 \text{ mol})(-92.3 \text{ kJ/mol}) - 0 - 0$$
$$= -184.6 \text{ kJ}$$

As stated earlier, the agreement between Equations (11.2) and (5.4) is excellent if the reactants and products are diatomic molecules. If a reaction has one or more polyatomic molecules, either as reactants or products, then Equation (11.2) can be used only to *estimate* the enthalpy of the reaction, because in that case we have to use average bond energies.

EXAMPLE 11.12

Estimate the enthalpy change for the combustion of hydrogen gas:

$$2\text{H}_2(g) + \text{O}_2(g) \longrightarrow 2\text{H}_2\text{O}(g)$$

Answer

As in Example 11.11, we construct a table:

Type of Bonds Broken	Number of Bonds Broken	Bond Energy (kJ/mol)	Energy Change (kJ)
H—H (H$_2$)	2	436.4	872.8
O=O (O$_2$)	1	498.7	498.7

Type of Bonds Formed	Number of Bonds Formed	Bond Energy (kJ/mol)	Energy Change (kJ)
O—H (H$_2$O)	4	460	1840

Next, we obtain the total energy input and total energy released:

$$\text{total energy input} = 872.8 \text{ kJ} + 498.7 \text{ kJ} = 1372 \text{ kJ}$$

$$\text{total energy released} = 1840 \text{ kJ}$$

Using Equation (11.2) we write

$$\Delta H° = 1372 \text{ kJ} - 1840 \text{ kJ} = -468 \text{ kJ}$$

This result is only an estimate because we used the bond energy of O—H, which is an average quantity. Alternatively, we can use Equation (5.4) and the data in Appendix 1 to calculate the enthalpy of reaction:

$$\Delta H° = 2\Delta H_f°(\text{H}_2\text{O}) - [2\Delta H_f°(\text{H}_2) + \Delta H_f°(\text{O}_2)]$$
$$= (2 \text{ mol})(-241.8 \text{ kJ/mol}) - 0 - 0$$
$$= -483.6 \text{ kJ}$$

Similar example: Problem 11.46.

Note that the estimated value based on bond energies is quite close to the value calculated using ΔH_f° data. In general, the bond energy method works best for reactions that are either quite endothermic or quite exothermic, that is, for $\Delta H^\circ > 100$ kJ or for $\Delta H^\circ < -100$ kJ.

AN ASIDE ON THE PERIODIC TABLE
Representative Elements, Noble Gases, and the Octet Rule

We saw earlier in the chapter that the octet rule applies only to the representative elements. Even for these elements, there are several important exceptions (see Section 11.6). Figure 11.6 divides many of the representative elements into five categories. The first category consists only of hydrogen, which never has more than two valence electrons in any of its compounds. The second category consists of beryllium and the Group 3A elements. Compounds containing these elements either obey the octet rule or frequently have incomplete octets. The third category consists of carbon, nitrogen, oxygen, and fluorine. These elements usually obey the octet rule in their compounds. The fourth category consists of elements in Group 4A through Group 7A (starting with the third period). These elements either obey the octet rule or form expanded octets in their compounds. The last category consists of the noble gases krypton and xenon. The atoms of these elements *always* have ten or more valence electrons in their compounds and therefore they never obey the octet rule.

The other representative elements—the alkali metals and the alkaline earth metals (except beryllium)—usually form ionic compounds and therefore the octet rule does not apply to their compounds.

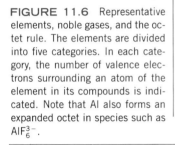

FIGURE 11.6 Representative elements, noble gases, and the octet rule. The elements are divided into five categories. In each category, the number of valence electrons surrounding an atom of the element in its compounds is indicated. Note that Al also forms an expanded octet in species such as AlF_6^{3-}.

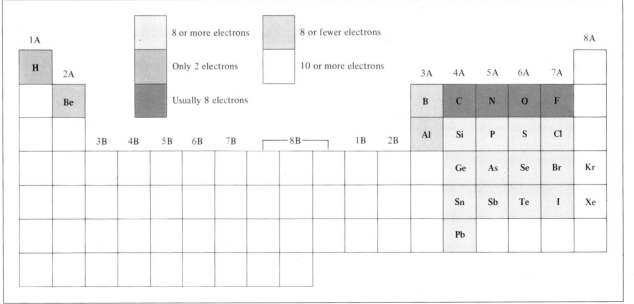

PLATE 1

THE GROUP 1A ELEMENTS

1A																	
Li																	
Na																	
K																	
Rb																	
Cs																	
Fr																	

(The Alkali Metals)

ELEMENTS

Lithium (Li)

Sodium (Na)

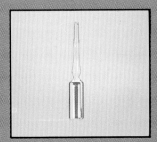

Potassium (K)

Rubidium (Rb)

Cesium (Cs)

MINERALS

Halite (NaCl)

Spodumene (LiAlSi$_2$O$_6$)

PLATE 2

THE GROUP 2A ELEMENTS

(The Alkaline Earth Metals)

2A
Be
Mg
Ca
Sr
Ba
Ra

ELEMENTS

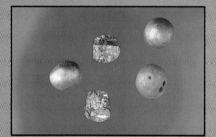

Beryllium (Be)

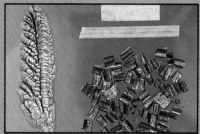

Magnesium (Mg)

Calcium (Ca)

Strontium (Sr)

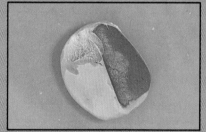

Barium (Ba)

Radium (Ra)

MINERALS

Beryl ($Be_3Al_2Si_6O_{18}$)

Dolomite ($CaCO_3 \cdot MgCO_3$)

Fluorite (CaF_2)

Celestite ($SrSO_4$)

MINING AND PRODUCTION

Magnesium bromide and magnesium chloride are obtained by solar evaporation of salt water.

"Pigs" of magnesium ready for use.

Left: Copper sheets await stamping into pennies.

Right: Molten manganese.

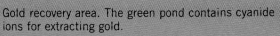

Steel making.

Gold recovery area. The green pond contains cyanide ions for extracting gold.

SUMMARY

1. In a covalent bond, two electrons are shared between two atoms. In multiple covalent bonds, two or three electron pairs are shared between two atoms. Some bonded atoms possess lone pairs, that is, pairs of valence electrons not involved in bonding. The arrangement of bonding electrons and lone pairs around each atom in a molecule is shown by the Lewis formula.
2. Electronegativity is a measure of the ability of an atom to attract electrons to itself in a chemical bond.
3. The octet rule predicts that atoms form covalent bonds in order to surround themselves with eight electrons each. When one atom in a covalently bonded pair donates two electrons to the bond, the Lewis formula can include the formal charge on each atom as a means of keeping track of the valence electrons. There are exceptions to the octet rule, particularly for covalent beryllium compounds, elements in Group 3A, and elements in the third period and beyond in the periodic table.
4. For some molecules or polyatomic ions, two or more Lewis formulas based on the same skeletal structure may satisfy the octet rule and appear chemically reasonable. Such resonance structures taken together represent the molecule or ion.
5. The strength of a covalent bond is measured in terms of its bond dissociation energy (for diatomic molecules) or its bond energy (for polyatomic molecules). Bond energies can be used to estimate the enthalpy of reactions.

KEY WORDS

Bond dissociation energy, p. 318
Bond length, p. 312
Coordinate covalent bond, p. 309
Covalent bond, p. 300
Covalent compound, p. 300
Electronegativity, p. 302
Lewis formula, p. 300

Lone pair, p. 300
Multiple bond, p. 301
Nonbonding electron, p. 300
Octet rule, p. 305
Resonance, p. 312
Resonance form, p. 312
Resonance structure, p. 312

PROBLEMS

More challenging problems are marked with an asterisk.

The Covalent Bond

11.1 What is Lewis's definition of a covalent bond?
11.2 Distinguish among single, double, and triple bonds in a molecule, and give examples.
11.3 Fluorine has seven valence electrons $(2s^2 2p^5)$, so seven covalent bonds could in principle be formed around the atom. Such a compound might be FH_7 or FCl_7 yet these compounds have never been prepared. Why?

Electronegativity and Bond Type

11.4 What is the difference between electronegativity and electron affinity?
11.5 Describe how the electronegativities of the elements vary according to their positions in the periodic table.
11.6 Distinguish between the covalent bond and the ionic bond. What is a polar covalent bond?
11.7 List the following bonds in order of increasing ionic character: the lithium to fluorine bond in LiF, the potassium to oxygen bond in K_2O, the

nitrogen to nitrogen bond in N_2, the sulfur to oxygen bond in SO_2, the chlorine to fluorine bond in ClF_3.

11.8 Arrange the following bonds in order of increasing ionic character: carbon to hydrogen, fluorine to hydrogen, bromine to hydrogen, sodium to iodine, potassium to fluorine, lithium to chlorine.

11.9 Four atoms are arbitrarily labeled D, E, F, and G. Their electronegativities are as follows: D = 3.8, E = 3.3, F = 2.8, and G = 1.3. If the atoms of these elements form the molecules DE, DG, EG, and DF, how would you arrange these molecules in order of increasing covalent bond character?

11.10 List the following bonds in order of increasing ionic character: potassium to fluorine, chlorine to chlorine, bromine to chlorine, silicon to carbon.

11.11 Classify the following bonds as ionic, polar covalent, or covalent, and give your reasons: (a) the CC bond in H_3CCH_3, (b) the KI bond in KI, (c) the NB bond in H_3NBCl_3, (d) the ClO bond in ClO_2, (e) the SiSi bond in $Cl_3SiSiCl_3$, (f) the SiCl bond in $Cl_3SiSiCl_3$, (g) the CaF bond in CaF_2.

Lewis Formulas and the Octet Rule

11.12 What is the difference between a Lewis dot symbol and a Lewis formula?

11.13 Explain the essential features of the Lewis octet rule.

11.14 Explain the concept of formal charge. Do formal charges on a molecule represent actual separation of charges?

11.15 Write Lewis formulas for the following molecules: (a) ICl, (b) PH_3, (c) CS_2, (d) P_4 (each P is bonded to three other P atoms), (e) H_2S, (f) HNO_2, (g) N_2H_4, (h) $HClO_3$, (i) $COBr_2$ (C is bonded to O and Br atoms).

11.16 Write Lewis formulas for the following ions: (a) O_2^{2-}, (b) C_2^{2-}, (c) NO^+, (d) SO_3^{2-}, (e) PO_4^{3-}, (f) NO_2^-, (g) NH_4^+.

11.17 The skeletal structure of acetic acid in the following formula is correct, but some of the bonds are wrong. (a) Explain what is wrong with some of the bonds. (b) Write the correct Lewis formula for the molecule.

$$\begin{array}{c} \text{H} \quad :\text{O}: \\ | \quad\;\; | \\ \text{H}{=}\text{C}{-}\text{C}{-}\ddot{\text{O}}{-}\text{H} \\ | \\ \text{H} \end{array}$$

*11.18 The following Lewis formulas are incorrect. Explain what is wrong with each one and give a correct Lewis formula for the molecule. (Relative positions of atoms are shown correctly.)

(a) $\text{H}{-}\ddot{\text{C}}{=}\ddot{\text{N}}$

(b) $\text{H}{=}\text{C}{=}\text{C}{=}\text{H}$

(c) $\ddot{\ddot{\text{O}}}{=}\text{Se}{=}\ddot{\ddot{\text{O}}}$

(d) $$\begin{array}{c} :\ddot{\text{F}} \qquad \ddot{\text{F}}: \\ \diagdown \quad\; \diagup \\ \ddot{\text{B}} \\ | \\ :\ddot{\text{F}} \end{array}$$

(e) $\text{H}{-}\ddot{\text{O}}{=}\ddot{\text{F}}:$

(f) $$\begin{array}{c} \text{H} \\ \diagdown \\ \quad \text{C}{-}\ddot{\text{F}}: \\ \diagup \\ \ddot{\text{O}} \end{array}$$

(g) $$\begin{array}{c} :\ddot{\text{F}} \qquad \ddot{\text{F}}: \\ \diagdown \quad\; \diagup \\ \text{N} \\ | \\ :\ddot{\text{F}}: \end{array}$$

(h) $$\begin{array}{c} \text{H}{-}\ddot{\text{O}} \\ \diagdown \\ \quad\;\; \text{S}{=}\ddot{\text{O}} \\ \diagup \\ \text{H}{-}\ddot{\text{O}} \end{array}$$

11.19 Write Lewis formulas for the following four isoelectronic species: (a) CO, (b) NO^+, (c) CN^-, (d) N_2.

11.20 Write Lewis formulas for the reaction

$$AlCl_3 + Cl^- \longrightarrow AlCl_4^-$$

What is the bond between Al and Cl in the product called?

11.21 Give three examples of compounds that do not satisfy the octet rule. Write a Lewis formula for each example you give.

11.22 Why is the octet rule not obeyed for many compounds containing elements in the third period of the periodic table and beyond?

*11.23 Write Lewis formulas for the following molecules: (a) XeF_2, (b) XeF_4, (c) XeF_6, (d) $XeOF_4$, (e) XeO_2F_2. In each case Xe is the central atom. Indicate those molecules in which the octet rule is not obeyed.

11.24 Write a Lewis formula for $SbCl_5$. Is the octet rule obeyed in this molecule?

11.25 Write a Lewis formula for the BeH_2 molecule. Is the octet rule satisfied for the Be atom?

11.26 Write Lewis formulas for SeF_4 and SeF_6. Is the octet rule satisfied for Se?

Resonance

11.27 Explain the concept of resonance.

11.28 Is it possible to "trap" a resonance structure of a compound for study? Explain.

11.29 The resonance concept is sometimes described in terms of a mule being a cross between a horse and a donkey. Compare this analogy with that used in this chapter, that is, the description of a rhinoceros as a cross between a griffin and a unicorn. Which description is more appropriate? Explain your choice.

*11.30 Write Lewis formulas for the species, include all resonance forms, and show formal charges: (a) HCO_2^-, (b) $CH_2NO_2^-$, (c) SO_3^-. Relative positions of the atoms are as follows:

11.31 Draw three resonance structures for the carbonate ion, CO_3^{2-}. Show formal charges.

11.32 Draw three resonance structures for the phosphite ion, HPO_3^{2-}. Show formal charges. (*Hint:* The H atom is bonded to the central P atom.)

11.33 Draw three reasonable resonance structures for the OCN^- ion. Show formal charges.

11.34 Draw four reasonable resonance structures for the PO_3F^{2-} ion. The central P atom is bonded to the three O atoms and the F atom. Show formal charges.

*11.35 Write Lewis formulas for hydrazoic acid, HN_3, and diazomethane, CH_2N_2. Are there resonance forms of these molecules? If so, draw their Lewis formulas. The skeletal structures of the molecules are

$$H$$

$$H\ N\ N\ N \qquad\qquad C\ N\ N$$

$$H$$

11.36 Draw four resonance structures for the chlorate ion, ClO_3^-, and five resonance structures for the perchlorate ion, ClO_4^-.

*11.37 Some resonance forms of the molecule CO_2 are given here. Explain why some of them are likely to be of little importance in describing the bonding in this molecule.

(a) $\ddot{O}{=}C{=}\ddot{O}$

(b) $:\overset{+}{O}{\equiv}C{-}\overset{-}{\underset{..}{O}}:$

(c) $:O{\equiv}\overset{+}{C}\ \ \overset{-}{\underset{..}{O}}:$

(d) $:\overset{-}{\underset{..}{O}}{-}\overset{2+}{C}{-}\overset{-}{\underset{..}{O}}:$

11.38 Draw three resonance structures for the molecule N_2O in which the atoms are arranged in the order NNO. Indicate formal charges.

11.39 The AlI_3 molecule (see Example 11.9) is an example of the incomplete octet. Draw three resonance structures of the molecule in which the octet rule is satisfied for both the Al and the I atoms. Show formal charges.

11.40 Draw reasonable resonance structures for the following ions: (a) HSO_4^-, (b) SO_4^{2-}, (c) HSO_3^-, (d) SO_3^{2-}.

11.41 Draw four reasonable resonance structures each for NO and NO_2. Show formal charges.

Bond Energy

11.42 Explain why the bond energy of a molecule is usually defined in terms of a gas-phase reaction.

11.43 Why are bond energy values always positive?

11.44 From the following data calculate the average bond energy for the N—H bond:

$$NH_3(g) \longrightarrow NH_2(g) + H(g) \qquad \Delta H° = 435\ kJ$$
$$NH_2(g) \longrightarrow NH(g) + H(g) \qquad \Delta H° = 381\ kJ$$
$$NH(g) \longrightarrow N(g) + H(g) \qquad \Delta H° = 360\ kJ$$

*11.45 The bond energy of $F_2(g)$ is 150.6 kJ/mol. Calculate $\Delta H_f°$ for $F(g)$.

11.46 For the reaction

$$H_2(g) + C_2H_4(g) \longrightarrow C_2H_6(g)$$

(a) Estimate the enthalpy of reaction, using the bond energy values in Table 11.3. (b) Calculate the enthalpy of reaction, using standard enthalpies of formation. ($\Delta H_f°$ for H_2, C_2H_4, and C_2H_6 are 0, 52.3 kJ/mol, and -84.7 kJ/mol, respectively.)

11.47 For the reaction

$$2C_2H_6(g) + 7O_2(g) \longrightarrow 4CO_2(g) + 6H_2O(g)$$

(a) Predict the enthalpy of reaction from the average bond energies in Table 11.3. (b) Calculate the enthalpy of reaction from the standard enthalpies of formation (see Appendix 1) of the reactant and product molecules, and compare the result with your answer for part (b).

Miscellaneous Problems

11.48 Define the following terms: (a) lone pair, (b) the octet rule, (c) electronegativity, (d) coordinate covalent bond, (e) bond length, (f) resonance, (g) resonance structure.

*11.49 Describe some characteristics of an ionic compound such as KF that would distinguish it from a covalent compound such as benzene (C_6H_6).

11.50 Is a coordinate covalent bond different from a normal covalent bond? Explain.

11.51 Discuss the atomic properties that are important in determining whether a particular compound will have ionic or covalent bonds.

11.52 Write Lewis formulas for BrF_3, ClF_5, and IF_7. Indicate those in which the octet rule is not obeyed.

11.53 Write three reasonable resonance structures of the azide ion N_3^- in which the atoms are arranged as NNN. Show formal charges.

12

CHEMICAL BONDING III: THE GEOMETRY OF MOLECULES

I n our study of chemical bonding we have been concerned mainly with how chemically bonded atoms are "connected," without considering that these interconnected atoms exist in a three-dimensional world. Lewis formulas, for example, are easily drawn on flat, two-dimensional paper. (This is, in fact, one of the conveniences of that model.) Unfortunately, Lewis formulas do not imply any particular geometry or dimensionality for the molecules or polyatomic ions being considered.

By contrast, atomic orbitals have associated with them clearly defined shapes, such as a subshell of three *p* orbitals positioned at right angles to one another. However, knowledge of the geometry of the orbitals in an individual, isolated atom is not the same as knowledge of the geometry of clusters of bonded atoms.

Atomic orbitals and Lewis formulas will be revisited as we consider the characteristic shapes of common molecules. You will soon see how both approaches can be extended to allow reliable predictions and explanations of a wide variety of molecular shapes. Macroscopic, observable properties of substances are profoundly influenced by such geometric considerations.

326

12.1   MOLECULAR GEOMETRY

Molecular geometry is concerned with the shape of molecules. Many physical and chemical properties—such as melting point, boiling point, and the types of reactions that molecules undergo—are affected by a molecule's shape.

We can make some simple predictions about the possible shapes that various molecules can assume purely on the basis of geometry. For instance, if we imagine atoms as spheres and the covalent bonds joining them as lines, we can generate a number of basic geometric shapes. In a diatomic molecule, then, the two bonded atoms lie side by side and the geometry of the molecule is linear and one-dimensional. A triatomic molecule (with three atoms) may also be linear and one-dimensional. If a triatomic molecule is not linear, the three atoms must lie in a common plane no matter how they are positioned relative to one another. This arrangement results in a planar molecule, which is two-dimensional. Molecules containing more than three atoms may be linear, planar, or three-dimensional.

In Chapter 11 we studied chemical bond formation in terms of Lewis formulas. However, Lewis formulas show only how atoms in a molecule share electrons; they do not show the geometry of the molecule. In order to predict and/or understand the shape of a molecule, we must extend the Lewis description of chemical bonding.

12.2   THE VALENCE-SHELL ELECTRON-PAIR REPULSION (VSEPR) MODEL

The geometric arrangement of atoms in a molecule can often be predicted reliably if we know the number of electrons surrounding a central atom. The basis of this approach is the idea that electrons in the valence shell of an atom repel one another. The **valence shell** is *the outermost electron-occupied shell of an atom, which holds the electrons that are usually involved in bonding.* In a covalent bond, a pair of electrons (often called the *bonding pair*) is responsible for holding two atoms together. However, in a polyatomic molecule, where there is more than one bond between the central atom and the surrounding atoms, the forces of repulsion between the electrons in the bonding pairs cause them to remain as far apart as possible. The shape that the molecule ultimately assumes (as defined by the positions of all the atoms) is such that this repulsion is at a minimum. Therefore this approach to the study of molecular geometry is called the **valence-shell electron-pair repulsion (VSEPR) model**, because *it accounts for the geometrical arrangements of shared and unshared electron pairs around a central atom in terms of the repulsions between electron pairs.*

The notion of "electron-pair repulsion" can be demonstrated simply by tying two or more inflated round balloons together, so that the balloons are in close mutual contact. Each balloon in such a cluster represents a valence-shell electron pair. Just as the balloons assume characteristic arrangements that minimize their mutual "repulsions" (that is, the balloons all try to avoid being crowded into the same space), VSEPR postulates similar elec-

The term "geometry" does not ordinarily apply to ionic solids. In an ionic solid, each cation is surrounded by a number of anions, and vice versa. We will discuss how these ions are packed together in Chapter 14.

The observation that the geometry of diatomic molecules is linear is simply an extension of a much more fundamental notion: Two points define a straight line!

VSEPR is pronounced "vesper."

tron-pair repulsions and predicts resulting geometries for many simple molecules.

With this model in mind, we can predict the geometry of molecules in a systematic way. In studying molecular geometry, it is convenient to divide molecules into two categories, based on whether or not the central atom has lone pairs. Before we discuss the two categories, make note of the following rules that are useful in considering any type of molecule:

- As far as electron-pair repulsion is concerned, double bonds and triple bonds can be treated as though they are single bonds between neighboring atoms.
- If two or more resonance structures can be drawn for a molecule, we may choose any one of them for our geometry study. Furthermore, formal charges need not be shown.

Molecules in Which the Central Atom Has No Lone Pairs

For simplicity we will consider molecules that contain atoms of only two elements, A and B, of which A is the central atom. These molecules have the general formula AB_x, where x is an integer 2, 3, ... (If $x = 1$, it is a diatomic molecule AB, which is linear by definition.) In the vast majority of cases, x is between 2 and 6.

Table 12.1 shows five possible arrangements of electron pairs around the central atom A (which in this category has no lone pairs). As a result of mutual repulsion, the electron pairs stay as far apart from one another as possible. Note that the table shows arrangements of the electron pairs but not the positions of the surrounding atoms. Molecules in this category that are of interest to us will have one of these five arrangements of electron pairs. We will therefore take a close look now at the geometry of molecules with the formulas AB_2, AB_3, AB_4, AB_5, and AB_6.

AB_2: Beryllium Chloride ($BeCl_2$).   The Lewis formula of beryllium chloride in the gaseous state (solid beryllium chloride does not exist as discrete molecular units) is

$$\overset{\displaystyle 180°}{Cl\overset{\frown}{}Be\overset{\frown}{}Cl}$$

Because the bonding pairs repel each other, they must be at opposite ends of a straight line in order for them to be as far apart as possible. Thus, the ClBeCl angle is 180°, and the molecule is linear (see Table 12.1).

AB_3: Boron Trifluoride (BF_3).   Boron trifluoride contains three covalent bonds, or bonding pairs:

$$\underset{\displaystyle F\diagup \diagdown F}{\overset{\displaystyle F}{\overset{\displaystyle |}{B}}} \quad 120°$$

TABLE 12.1 Arrangement of Electron Pairs About a Central Atom (A) in a Molecule

Number of Electron Pairs	Arrangement of Electron Pairs	Shape
2		Linear
3		Trigonal planar
4		Tetrahedral
5		Trigonal bipyramidal
6		Octahedral

In the most stable arrangement, the three B—F bonds point to the corners of an equilateral triangle. Thus, each of the three FBF angles is equal to 120° and all four atoms lie in the same plane. According to Table 12.1, the geometry of BF_3 is described as trigonal planar, so called because the three F atoms are at the corners of an equilateral triangle.

AB$_4$: Methane (CH$_4$). The Lewis formula of methane is

$$\begin{array}{c} H \\ | \\ H—C—H \\ | \\ H \end{array}$$

Since there are four bonding pairs, the geometry of CH$_4$ is tetrahedral. This is a three-dimensional shape, as shown in Table 12.1. Each HCH angle is 109.5°.

AB$_5$: Phosphorus Pentachloride (PCl$_5$). The Lewis formula of phosphorus pentachloride (in the gas phase) is

$$\begin{array}{c} Cl \\ Cl\backslash \ | \\ P—Cl \\ Cl \diagup\ | \\ Cl \end{array}$$

The only way to minimize the repulsive forces among the five bonding pairs is to arrange the P—Cl bonds in the form of a trigonal bipyramid (see Table 12.1). Note that there are two different ClPCl angles in this arrangement (90° and 120°). Any attempt to increase one of the ClPCl angles would, by necessity, decrease other ClPCl angles, and the net result would be a decrease in stability. By convention, we define the P—Cl bonds above and below the central triangular plane to be along the *axial* direction and the P—Cl bonds in the triangular plane to be along the *equatorial* direction.

AB$_6$: Sulfur Hexafluoride (SF$_6$). The Lewis formula of sulfur hexafluoride is

$$\begin{array}{c} F \\ F\backslash | \diagup F \\ >S< \\ F \diagup | \backslash F \\ F \end{array}$$

The octahedral arrangement of the six bonding pairs, shown in Table 12.1, provides the most stable form for the SF$_6$ molecule, in which each of the FSF angles is equal to 90°.

Table 12.2 shows the geometries of some simple molecules as predicted by the VSEPR model. Note that Table 12.2 differs from Table 12.1 in that the former shows the positions of *all* the atoms (and therefore the geometries of the molecules) while the latter shows only the central atom and the arrangement of all the electron pairs. As we will see shortly, Table 12.1 can also be used to study the geometry of molecules in which the central atom possesses one or more lone pairs.

A tetrahedron has four sides or four faces, all of which are triangles.

The structure of phosphorus pentachloride in the solid state is quite complex; it involves cations and anions.

The trigonal bipyramid shape can be visualized by imagining two three-sided pyramids joined by a common base. The five electron pairs would reside at the corners of this structure.

The use of the term "octahedral" (where the prefix "octa" means eight) for a six electron-pair structure sometimes leads to confusion. The regular structure defined by the six pairs of electrons has eight sides—an octahedron.

TABLE 12.2 Geometry of Some Simple Molecules and Ions in Which the Central Atoms Do Not Possess Lone Pairs

Molecule	Geometry		Examples
AB_2	Linear	B—A—B	$BeCl_2$, $HgCl_2$
AB_3	Trigonal planar		BF_3
AB_4	Tetrahedral		CH_4, NH_4^+
AB_5	Trigonal bipyramidal		PCl_5
AB_6	Octahedral		SF_6

Molecules in Which the Central Atom Possesses One or More Lone Pairs

Determining the geometry of a molecule is more complicated if the central atom contains lone pairs in addition to bonding pairs. In such molecules there are three types of repulsive forces—between bonding pairs, between lone pairs, and between a bonding pair and a lone pair. In general, the repulsive forces decrease in the following order:

lone pair vs. lone pair repulsion > lone pair vs. bonding pair repulsion > bonding pair vs. bonding pair repulsion

Electrons in a bonding pair are held by the attractive forces of the nuclei in the two bonded atoms. These electrons have less "spatial distribution," that is, they take up less space than lone pair electrons, which are associated with only one particular atom. Consequently, in a molecule that contains lone pair electrons, they occupy more space and they experience a greater repulsion from neighboring lone pairs and bonding pairs. To keep track of the total number of bonding pairs and lone pairs, we designate these molecules as $AB_x E_y$; A is the central atom, B is a surrounding atom, and E is a lone pair on A. Both x and y are integers; $x = 2, 3, \ldots$, and $y = 1, 2,$ $\ldots$. Thus the values of x and y indicate the number of surrounding atoms and number of lone pairs on the central atom, respectively. In our scheme here the simplest molecule would be a triatomic molecule with one lone pair on the central atom; the formula of this molecule is AB_2E.

For molecules in which the central atom has one or more lone pairs, we need to distinguish between the overall shape of the electron pairs and the geometry of the molecule. The overall shape of the electron pairs refers to the arrangement of *all* electron pairs on the central atom, both bonding pairs and lone pairs. On the other hand, the geometry of a molecule is described only in terms of the arrangement of its atoms, and hence only the bonding pairs. We will see shortly that if lone pairs are present on the central atom, the overall shape of the electron pairs is *not* the same as the geometry of the molecule.

We will now consider some specific examples.

AB_2E: Sulfur Dioxide (SO_2).   The Lewis formula of sulfur dioxide is

$$\ddot{S} \diagup\!\!\!\diagup \quad \diagdown \atop \ddot{O}. \qquad .\ddot{O}:$$

From the discussion in Section 11.5 we know that this represents only one of the two equivalent resonance structures; the other one is

$$\ddot{S} \diagup \quad \diagdown\!\!\!\diagdown \atop :\ddot{O}. \qquad .\ddot{O}.$$

However, as we mentioned earlier, we need use only one resonance structure to apply the VSEPR model. Furthermore, we do not need to show formal charges in order to predict geometry. The double bond in either SO_2 resonance structure can be treated as if it were a single bond. In our simplified scheme, then, the SO_2 molecule can be viewed as consisting of three electron pairs (on the central S atom), of which two are bonding pairs and one is a lone pair. In Table 12.1 we see that the overall arrangement of three electron pairs is trigonal planar. But because one of the electron pairs is a lone pair, the SO_2 molecule has a "bent" shape. Since the lone pair vs. bonding pair repulsion is greater than the bonding pair vs. bonding pair repulsion, the two sulfur-to-oxygen bonds are pushed closer together and the OSO angle is less than 120°. By experiment the OSO angle is found to be 119.5°.

AB_3E: Ammonia (NH_3).   The ammonia molecule contains three bonding pairs and one lone pair:

$$H—\ddot{N}—H \atop | \atop H$$

As Table 12.1 shows, the overall shape of four electron pairs is tetrahedral. But in NH_3 one of the electron pairs is a lone pair, so the geometry of NH_3

is pyramidal (so called because it looks like a pyramid, with the N atom at the apex). Because the lone pair repels the bonding pairs more strongly, the three N—H bonding pairs are pushed closer together; thus the HNH angle in ammonia is smaller than the ideal tetrahedral angle of 109.5° (Figure 12.1).

AB$_2$E$_2$: Water (H$_2$O).   A water molecule contains two bonding pairs and two lone pairs:

$$\overset{\displaystyle \cdot \overset{\cdot \cdot}{O} \cdot}{\underset{H \qquad H}{\diagup \diagdown}}$$

Since water is a triatomic molecule, by definition it has a planar structure. The overall shape of the four electron pairs in water is tetrahedral, the same electron-pair shape that is found in ammonia. However, unlike ammonia, there are two lone pairs on the central O atom. These lone pairs tend to be as far apart from each other as possible. Consequently, the two O—H bonding pairs are pushed toward each other, so that we can predict an even greater deviation from the tetrahedral angle than in NH$_3$. This prediction is correct, as Figure 12.1 shows.

AB$_4$E: Sulfur Tetrafluoride (SF$_4$).   The Lewis formula of SF$_4$ is

$$\overset{F \diagdown \quad \diagup F}{\underset{F \diagup \quad \diagdown F}{\overset{\cdot \cdot}{S}}}$$

The central sulfur atom has five electron pairs whose arrangement, according to Table 12.1, is trigonal bipyramidal. In the SF$_4$ molecule, however, one of the electron pairs is a lone pair, so that the molecule must have one of the following geometries:

(a)   (b)

In (a), the lone pair is located at the corner of the equilateral triangle (called the *equatorial position*), and in (b) it is located at a position perpendicular to the plane containing the three F atoms (called the *axial position*). It turns out that the total repulsions are less in (a), which is described as a distorted tetrahedron, than in (b). Therefore, SF$_4$ has the (a) geometry. In

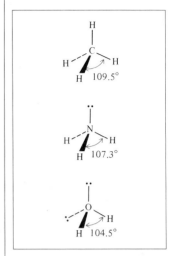

FIGURE 12.1   *The bond angles in CH$_4$, NH$_3$, and H$_2$O. The dotted line in each diagram represents a bond axis behind the plane of the paper, the wedged line represents a bond axis in front of the plane of the paper, and the thin solid lines represent bonds in the plane of the paper.*

By experiment we find that the angle between the axial F atoms and S is 186°, whereas that between the equatorial F atoms and S is 116°.

fact, for trigonal bipyramidal arrangement of electron pairs, the lone pairs usually "prefer" to be in the equatorial positions rather than the axial positions.

AB_4E_2: Xenon Tetrafluoride (XeF_4). The Lewis formula for xenon tetrafluoride is

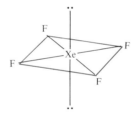

The Xe atom has four bonding pairs and two lone pairs. In Table 12.1 we can see that the overall arrangement of the six electron pairs is octahedral. In octahedral arrangements, the two lone pairs prefer to be in axial positions because they are farthest away from each other. Thus XeF_4 has a square planar molecular geometry (the four F atoms lie in the corners of a square, which is two-dimensional or planar). The two lone pairs are placed above and below the plane as follows:

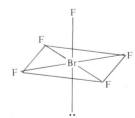

AB_5E: Bromine Pentafluoride (BrF_5). The Lewis formula of bromine pentafluoride is

The Br atom has five bonding pairs and one lone pair. The overall arrangement of six electron pairs is octahedral and the BrF_5 molecule has a square pyramidal geometry:

Table 12.3 shows the geometries of simple molecules in which the central atom contains one or more lone pairs.

The terms "axial" and "equatorial" apply also to octahedral geometry.

The overall shape of the molecule is that of a pyramid with a square base.

TABLE 12.3 Geometry of Simple Molecules in Which the Central Atoms Contain One or More Lone Pairs

Class of Molecule	Total Number of Electron Pairs	Number of Bonding Pairs	Number of Lone Pairs	Geometry	Examples
AB_2E	3	2	1	 Bent	SO_2
AB_3E	4	3	1	 Trigonal pyramidal	NH_3
AB_2E_2	4	2	2	 Bent	H_2O
AB_4E	5	4	1	 Distorted tetrahedron	IF_4^+, SF_4, XeO_2F_2
AB_3E_2	5	3	2	 T–shaped	ClF_3
AB_2E_3	5	2	3	 Linear	XeF_2, I_3^-
AB_5E	6	5	1	 Square pyramidal	BrF_5, $XeOF_4$
AB_4E_2	6	4	2	 Square planar	XeF_4, ICl_4^-

Guidelines for Applying the VSEPR Model

Having studied the geometries of molecules in the two categories (central atoms with and without lone pairs) we summarize some rules that will help you in applying the VSEPR model to all types of molecules:

- Write the Lewis formula of the molecule, considering only the electron pairs around the central atom (that is, the atom that is bonded to more than one other atom).
- Count the total number of electron pairs around the central atom (i.e., bonding pairs and lone pairs). As a good approximation, treat double and triple bonds as though they were single bonds. Refer to Table 12.1 to predict the overall shape of the electron pairs.
- Use Tables 12.2 and 12.3 to predict the geometry of the molecule.
- In predicting bond angles, note that a lone pair repels another lone pair or a bonding pair more strongly than a bonding pair repels another bonding pair. Remember that there is usually no easy way to predict bond angles accurately when the central atom possesses one or more lone pairs.

The VSEPR model, as you have already observed, generates reliable predictions of the geometries of a variety of molecular structures. Chemists accept and use the VSEPR approach because of its simplicity and utility. Although some theoretical concerns have been voiced as to whether "electron-pair repulsion" actually determines molecular shapes, it is safe to say that such an assumption leads to useful (and generally reliable) predictions. We need not ask more of any model at this stage in the study of chemistry.

EXAMPLE 12.1

Use the VSEPR model to predict the geometry of the following molecules and ions: (a) AsH_3, (b) OF_2, (c) $AlCl_4^-$, (d) I_3^-, (e) C_2H_4.

Answer

(a) The Lewis formula of AsH_3 is

$$H—\overset{\cdot\cdot}{As}—H$$
$$|$$
$$H$$

This molecule has three bonding pairs and one lone pair, which is similar to the combination of electron pairs in ammonia. Therefore, the geometry of AsH_3 is trigonal pyramidal. We cannot predict the HAsH angle accurately, but we know that it must be less than 109.5°.

(b) The Lewis formula of OF_2 is

$$\overset{\cdot\cdot}{\underset{F \quad\quad F}{O}}$$

Since there are two lone pairs on the O atom, the OF_2 molecule must have a bent shape, like that of H_2O. Again, all we can say about angle size is that the FOF angle should be less than 109.5°.

(c) The Lewis formula of $AlCl_4^-$ is

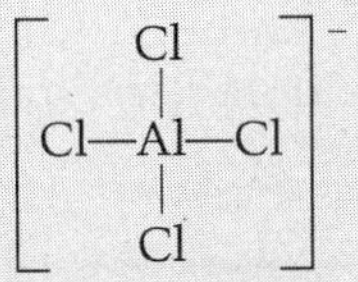

$$\begin{bmatrix} & Cl & \\ Cl-Al-Cl \\ & Cl & \end{bmatrix}^-$$

Since the central Al atom has no lone pairs and all four Al—Cl bonds are equivalent, the $AlCl_4^-$ ion should be tetrahedral, and the ClAlCl angles should all be 109.5°.

(d) The Lewis formula of I_3^- is

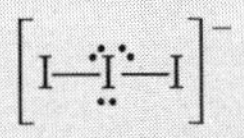

$$\begin{bmatrix} I-\ddot{I}-I \end{bmatrix}^-$$

The central I atom has two bonding pairs and three lone pairs. From Table 12.3 we see that the three lone pairs all lie in the triangular plane, and the I_3^- ion should be linear, like the XeF_2 molecule.

(e) The Lewis formula of C_2H_4 is

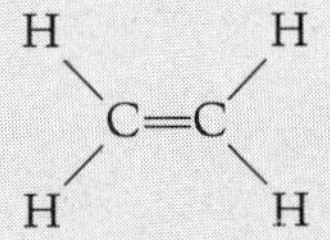

The C=C bond is treated as though it were a single bond. Because there are no lone pairs present, the arrangement around each C atom has a trigonal planar shape like BF_3, discussed earlier. Thus the predicted bond angles in C_2H_4 are:

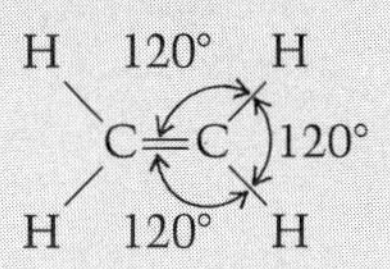

Similar examples: Problems 12.4, 12.5, 12.7, 12.8.

12.3 DIPOLE MOMENTS

Now that you know how to predict the geometries of molecules, you may wonder how the actual geometry of a molecule is determined. A number of spectroscopic techniques and X-ray measurements give us bond lengths and bond angles. But a simpler method—dipole moment measurement—is useful in studying molecular geometry. Although a dipole moment measurement does not provide bond lengths or bond angles, it can tell us about the overall geometry of a molecule.

In Section 11.2 we studied the hydrogen fluoride molecule and learned that, because the F atom is more electronegative than the H atom, there is a shift in electron density from H to F:

$$\overset{\longmapsto}{H-F}$$

and there are charge separations in the molecule as follows:

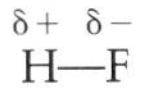

$$\overset{\delta+\ \ \delta-}{H-F}$$

The I_3^- ion is one of the few structures for which the bond angle (180°) can be predicted accurately even though the central atom contains lone pairs.

In C_2H_4, all six atoms lie in the same plane.

To review the concepts of electronegativity and bond polarity, see Section 11.2.

Consequently, the H—F bond in hydrogen fluoride is a polar bond. A quantitative measure of the polarity of the bond is the ***dipole moment*** (μ), which is *the product of the charge (Q) and the distance r between charges:*

$$\mu = Q \times r \qquad (12.1)$$

In a diatomic molecule like HF, the charge Q is equal in magnitude to $\delta+$ and $\delta-$.

By definition, the charges on both ends of an electrically neutral diatomic molecule must be equal in magnitude and opposite in sign. However, the quantity Q in Equation (12.1) refers only to the magnitude of the charge and not its sign, so μ is always positive. *A molecule that possesses a dipole moment* is called a ***polar molecule;*** *a molecule that does not possess a dipole moment is called a* ***nonpolar molecule.*** Thus all ***homonuclear diatomic molecules,*** that is, *diatomic molecules containing atoms of the same element* (for example, H_2, O_2, F_2, and so on), lack a dipole moment because there is no charge separation in these molecules. They are all nonpolar molecules. On the other hand, ***heteronuclear diatomic molecules,*** or *diatomic molecules containing atoms of different elements* (for example, HCl, CO, NO, and so on), generally possess a dipole moment. They are all polar molecules.

Dipole moments can be measured by studying the behavior of molecules between two oppositely charged plates (Figure 12.2). In the presence of an electric field (due to the $+$ and $-$ charges on the plates), polar molecules will align themselves with the more negative end oriented toward the positive plate and the more positive end toward the negative plate. Nonpolar molecules are unaffected by an electric field. Dipole moments are usually expressed in *debye* units (D) (after Peter Debye, 1884–1966). The conversion factor is

FIGURE 12.2 *Behavior of polar molecules in (a) the absence and (b) the presence of an external electric field. The arrangement shown in (b) is an idealized presentation; polar molecules do not line up perfectly as shown. Nonpolar molecules are not affected by the electric field.*

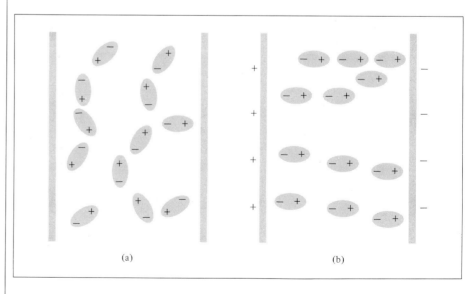

(a)   (b)

$$1 D = 3.33 \times 10^{-30} \text{ C m}$$

where C is coulomb and m is meter.

The dipole moment of a molecule containing three or more atoms depends on both polarity *and* molecular geometry. However, even if polar bonds are present, the molecule itself may not have a dipole moment. Consider the carbon dioxide (CO_2) molecule. Since CO_2 is a triatomic molecule, its geometry is either linear or bent:

linear molecule
(no dipole moment)

resultant
dipole moment

bent molecule
(has a dipole moment)

The arrows denote the net shift of electron density from the less electronegative atom (carbon) to the more electronegative atom (oxygen). In each case, the dipole moment of the entire molecule is made up of two *bond moments*, that is, individual dipole moments in the polar C=O bonds. The measured dipole moment is equal to the sum of these bond moments. The dipole moment, like the bond moment, is a *vector quantity*, that is, it has both magnitude and direction. It is clear that the two bond moments in CO_2 are equal in magnitude. Since they point in opposite directions in a linear CO_2 molecule, the sum or resultant dipole moment is zero. On the other hand, if the CO_2 molecule is bent, the two bond moments would partially reinforce each other, so that the molecule would have to possess a dipole moment. Experimental evidence shows that carbon dioxide has no dipole moment. Therefore we conclude that the carbon dioxide molecule has a linear shape, consistent with its nonpolar character. The linear nature of carbon dioxide has, in fact, been confirmed through other experimental measurements.

Table 12.4 lists the dipole moments of several polar diatomic and polyatomic molecules.

The **VSEPR** model predicts that CO_2 is a linear molecule.

TABLE 12.4 Dipole Moments of Some Polar Molecules

Molecule	Geometry	Dipole Moment (D)
HF	Linear	1.92
HCl	Linear	1.08
HBr	Linear	0.78
HI	Linear	0.38
H_2O	Bent	1.87
H_2S	Bent	1.10
NH_3	Pyramidal	1.46
SO_2	Bent	1.60

EXAMPLE 12.2

Predict whether each of the following molecules has a dipole moment: (a) IBr, (b) BF_3 (trigonal planar), (c) CH_2Cl_2 (tetrahedral).

Answer

(a) Since IBr (iodine bromide) is a diatomic molecule, it has a linear geometry. Bromine is more electronegative than iodine (see Figure 11.2), so IBr is polar, with bromine at the negative end.

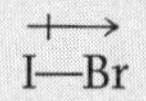

$$\overset{\longmapsto}{I-Br}$$

Thus the molecule possesses a dipole moment.
(b) Since fluorine is more electronegative than boron, each B—F bond in BF_3 (boron trifluoride) is polar and the three bond moments are equal. However, the symmetrical geometry of a trigonal planar shape means that the three bond moments exactly cancel one another:

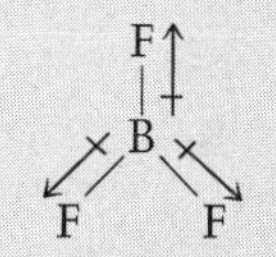

Consequently BF_3 does not have a dipole moment; it is a nonpolar molecule.
(c) The Lewis formula of CH_2Cl_2 (methylene chloride) is

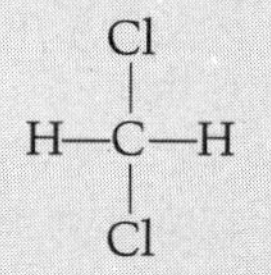

$$\begin{array}{c} Cl \\ | \\ H-C-H \\ | \\ Cl \end{array}$$

This molecule is similar to CH_4 in that it has an overall tetrahedral shape. However, because not all the bonds are identical, there are three different bond angles: HCH, HCCl, and ClCCl. These bond angles are close to, but not equal to, 109.5°. Since chlorine is more electronegative than carbon, which is more electronegative than hydrogen, the bond moments do not cancel and the molecule possesses a dipole moment:

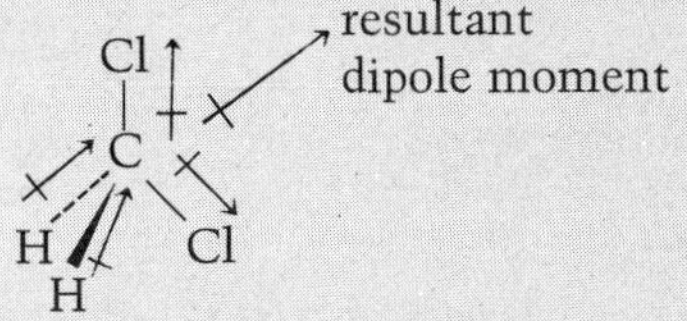

Thus CH_2Cl_2 is a polar molecule.

Similar example: Problem 12.16.

12.4 VALENCE BOND THEORY

The VSEPR model provides a relatively simple and straightforward method for predicting the geometry of molecules, and it is largely based on the Lewis formulas of molecules. As we said earlier, however, the Lewis theory of chemical bonding does not explain *why* chemical bonds exist. We need a better understanding of this topic to account for many observed molecular properties.

Lewis's idea of relating the formation of a covalent bond to the pairing of electrons was not only novel but highly provocative. Why would two negatively charged particles pair together to form a bond? Surely the electrostatic repulsion between them would lead to instability. Furthermore, the Lewis theory describes the single bond between the H atoms in H_2 and that between the F atoms in F_2 in essentially the same way, that is, the pairing of two electrons. Yet these two molecules have quite different bond dissociation energies (436.4 kJ/mol for H_2 and 150.6 kJ/mol for F_2). Such a discrepancy (and many others) cannot be explained by the Lewis theory. For a proper explanation of chemical bond formation we must therefore look to quantum mechanics. In fact, the quantum mechanical study of chemical bonding also provides a means for understanding molecular geometry.

At present, two quantum mechanical theories are used to describe the covalent bond and electronic structure of molecules. *Valence bond (VB) theory* assumes that the electrons in a molecule occupy atomic orbitals of the individual atoms. It permits us to retain our picture of individual atoms taking part in the bond formation. The second theory, called *molecular orbital (MO) theory*, assumes the formation of molecular orbitals from the atomic orbitals. The molecular orbital theory will be discussed in Chapter 13; here we will focus on the valence bond theory.

Let us start by considering the formation of an H_2 molecule from two H atoms. Lewis theory describes the H—H bond in terms of the pairing of the two electrons on the H atoms. In the framework of the valence bond theory, the covalent H—H bond is formed by the *overlap* of the two $1s$ orbitals in the H atoms. By overlap, we mean that the two orbitals share some common region in space.

Chapter 8 discussed electron configuration of isolated atoms. Here we must examine what happens to these atoms when they take part in bond formation. Initially, when the two atoms are far apart, there is no interaction. We say that the potential energy of this system (that is, the two H atoms) is zero. As the atoms approach each other, each electron is attracted by the nucleus in the other atom, and at the same time, the electrons repel each other, as do the nuclei. While the atoms are still separated, attraction is stronger than repulsion, so that the potential energy of the system *decreases* (that is, it becomes more negative) as the atoms approach each other (Figure 12.3). This trend continues until the potential energy reaches a minimum value. At this point, where the system has the lowest potential energy, it is most stable. This condition corresponds to a substantial overlap of the $1s$ orbitals and the formation of a stable H_2 molecule. If the distance between the nuclei were to decrease further, the potential energy would rise steeply and finally become positive as a result of the increased electron–electron and nuclear–nuclear repulsions.

In accord with the law of conservation of energy, the decrease in potential energy as a result of H_2 formation must be accounted for by a release of energy. By experiment we find that as a H_2 molecule is formed by two H

Recall that an object possesses potential energy by virtue of its position.

FIGURE 12.3 *Variation of the potential energy of two H atoms with their distance of separation. At the minimum point, the H_2 molecule is in its most stable state and the bond length is 74 pm.*

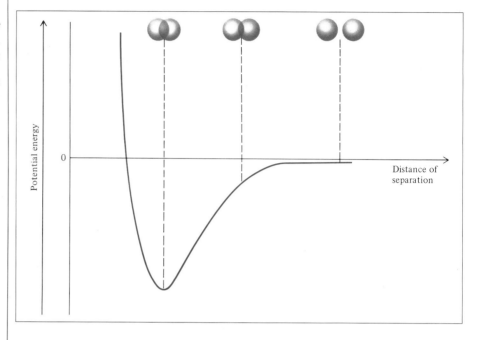

atoms, heat is given off. The converse is also true. To break a H—H bond, energy must be supplied to the molecule.

Thus we see that the valence bond theory gives a clearer picture of chemical bond formation than the Lewis theory does. The valence bond theory states that a stable molecule is formed from reacting atoms when the potential energy of the system has decreased to a minimum, whereas the Lewis theory ignores energy changes in chemical bond formation.

The orbital diagram of the F atom is shown on p. 237.

The concept of atomic orbital overlap applies equally well to diatomic molecules other than H_2. Thus a stable F_2 molecule is formed when the $2p$ orbitals (containing the unpaired electrons) in the two F atoms overlap to form a covalent bond. Similarly, the formation of the HF molecule can be explained by the overlap of the $1s$ orbital in H with the $2p$ orbital in F. In each case, we can account for the changes in potential energy as the distance between the reacting atoms changes. Because the orbitals involved are not the same kind in all cases, we can see why the bond energies and bond lengths in H_2, F_2, and HF might be different. As we stated earlier, the Lewis theory treats *all* the covalent bonds in the same manner, and it is not obvious from that theory why one covalent bond might differ from another.

12.5 HYBRIDIZATION OF ATOMIC ORBITALS

The orbital diagram of the N atom is shown on p. 237.

Theoretically, the concept of atomic orbital overlap should apply also to polyatomic molecules. However, in dealing with polyatomic molecules a satisfactory bonding scheme must also account for their geometry. Consider the NH_3 molecule. Because the nitrogen atom has three unpaired electrons (one in each of the three $2p$ orbitals), we might expect the three N—H covalent bonds to be formed by the overlap of the hydrogen $1s$ orbitals and the nitrogen $2p$ orbitals. If this were the case, the HNH bond angles in NH_3 would all be 90° because, as Figure 8.16 shows, the three $2p$ orbitals are mutually perpendicular to one another. But experimental evidence tells us that the angles are all 107.3° (see Figure 12.1). It might be argued that because nitrogen is more electronegative than hydrogen, the hydrogen atoms all bear a fractional positive charge and the forces of repulsion among these atoms increase the size of the HNH angles. Although such repulsive forces do exist, their magnitude is not great enough to increase the angles from 90° to 107.3°. Discrepancies of this type are noted also for other polyatomic molecules. In addition, bond formation often cannot be accounted for in terms of the ground-state electron configuration of the central atom.

One way to explain bonding in polyatomic molecules is to introduce the concept of **hybridization.** Hybridization is *the process of mixing the atomic orbitals in an atom (usually the central atom) to generate a set of new atomic orbitals,* called **hybrid orbitals.** Hybrid orbitals, which are *atomic orbitals obtained when two or more nonequivalent orbitals of the same atom combine,* are then used to form covalent bonds. It is important to understand that this "mixing" of atomic orbitals is a human contrivance—a set of mathematical procedures that extends the utility of our atomic orbital model. To put it another way, hybridization takes place in the calculations of chemists, not in actual molecules. We accept the concept of hybridization not because of its objective reality, but because it gives results that are consistent with our knowledge of molecular bonding and molecular geometry.

The following points are useful for an understanding of hybridization:

- The concept of hybridization does not apply to isolated atoms. It is used only to explain a bonding scheme in a molecule.
- Hybridization is the mixing of at least two nonequivalent atomic orbitals, for example, s and p orbitals. Therefore, a hybrid orbital is not a pure atomic orbital. Hybrid orbitals have very different shapes from pure atomic orbitals.
- The number of hybrid orbitals generated is equal to the number of atomic orbitals that participate in the hybridization process.
- For a given hybridization process, all the hybrid orbitals are equivalent in every respect, except in their relative orientations in space.
- Hybridization requires an input of energy; however, the system more than recovers this energy during bond formation.
- Covalent bonds in polyatomic molecules are formed by the overlap of a hybrid orbital and a pure atomic orbital, or of two hybrid orbitals. Therefore, the hybridization bonding scheme is still within the framework of the valence bond theory, that is, electrons in a molecule are assumed to occupy hybrid orbitals of the individual atoms.

We will now apply the hybridization concept to the study of bonding and geometry in some polyatomic molecules. The molecules are grouped according to the state of hybridization, that is, the mixing of a specific number of s and p orbitals of the central atom.

Be: $1s^2 2s^2$

sp Hybridization

Consider the $BeCl_2$ (beryllium chloride) molecule, which is linear. The orbital diagram for the valence electrons in Be is

We know that in its ground state Be will not form covalent bonds with Cl because its electrons are paired in the $2s$ orbital. So we turn to hybridization. The hybridization process can be visualized as follows. First a $2s$ electron is promoted to a $2p$ orbital, resulting in

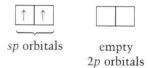

Now there are two different Be orbitals available for bonding (the $2s$ and $2p$ orbitals). However, we find this is a contradiction of known results from experiment. In the actual $BeCl_2$ molecule, the two Be—Cl bonds are identical in every respect. But if two Cl atoms were to combine with Be in its excited state, one Cl atom would share a $2s$ electron and the other Cl would share a $2p$ electron, making two nonequivalent Be—Cl bonds. Thus the $2s$ and $2p$ orbitals must be mixed, or hybridized, to form two equivalent sp hybrid orbitals:

Hybrid orbitals are named by specifying the type and number of atomic orbitals that participated in the hybridization. Since one *s* and one *p* atomic orbitals were involved here, the two hybrid orbitals are called *sp* hybrid orbitals.

sp orbitals   empty
$2p$ orbitals

These two hybrid orbitals lie along the same line, so that the angle between them is 180°. Each of the Be—Cl bonds is then formed by the overlap of a Be sp hybrid orbital and a Cl $3p$ orbital, and the resulting $BeCl_2$ molecule has a linear geometry (Figure 12.4).

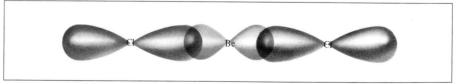

FIGURE 12.4 *The linear geometry of $BeCl_2$ can be explained by assuming the Be atom to be sp-hybridized. The two sp hybrid orbitals overlap with the two 2p orbitals of chlorine to form two covalent bonds.*

sp² Hybridization

Next we will look at the BF_3 (boron trifluoride) molecule, which has a planar geometry. Considering only the valence electrons, the orbital diagram of B is

B: $1s^2 2s^2 2p^1$

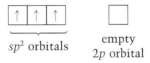

First, we promote a $2s$ electron to an empty $2p$ orbital:

Then, mixing the $2s$ orbital with the two $2p$ orbitals generates three sp^2 hybrid orbitals:

These three sp^2 orbitals lie in a plane and the angle between any two orbitals is 120°. Each of the B—F bonds is formed by the overlap of a B sp^2 hybrid orbital and a $2p$ orbital in a F atom (Figure 12.5). The BF_3 molecule is therefore planar and all the FBF angles are 120°.

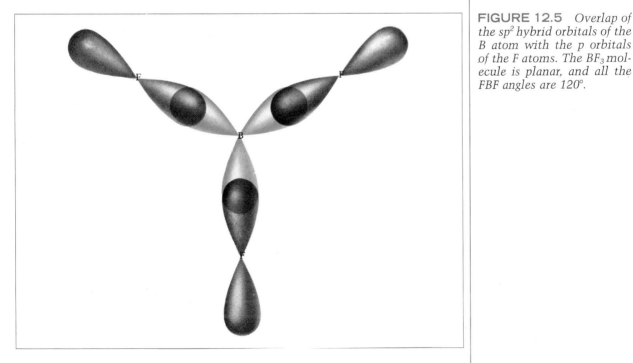

FIGURE 12.5 *Overlap of the sp^2 hybrid orbitals of the B atom with the p orbitals of the F atoms. The BF_3 molecule is planar, and all the FBF angles are 120°.*

C: $1s^2 2s^2 2p^2$

sp^3 Hybridization

An example of a molecule containing a sp^3-hybridized atom is CH_4 (methane), which has a tetrahedral geometry. Considering only the valence electrons, we can represent the orbital diagram of C as

$$2s \qquad 2p$$

First we promote a $2s$ electron to the empty $2p$ orbital:

$$2s \qquad 2p$$

Then we mix the $2s$ orbital with the three $2p$ orbitals and get four sp^3 orbitals:

$$sp^3 \text{ orbitals}$$

These four equivalent hybrid orbitals are directed toward the four corners of a regular tetrahedron. Figure 12.6 shows how the C sp^3 hybrid orbitals and the H $1s$ orbitals overlap to form four covalent C—H bonds. Thus CH_4 has a tetrahedral shape and all the HCH angles are 109.5°.

The NH_3 molecule provides us with another example of sp^3 hybridization. The ground-state electron configuration of N is $1s^2 2s^2 2p^3$, so that the orbital diagram for the sp^3-hybridized N atom is

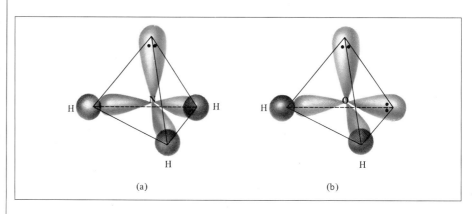

$$sp^3 \text{ orbitals}$$

Three of the four hybrid orbitals are used to form covalent N—H bonds, and the fourth hybrid orbital is used to accommodate the lone pair on nitrogen [Figure 12.7(a)]. Repulsion between the lone pair electrons and those in the bonding orbitals decreases the HNH bond angle from 109.5° to 107.3°.

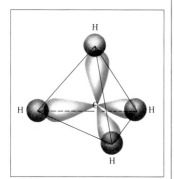

FIGURE 12.6 *The formation of four bonds between the carbon sp^3 hybrid orbitals and the hydrogen $1s$ orbitals in CH_4.*

FIGURE 12.7 *(a) The NH_3 molecule. The N atom is sp^3-hybridized. It forms three bonds with the H atoms. One of the sp^3 hybrid orbitals is occupied by the lone pair. (b) The H_2O molecule. The O atom is sp^3-hybridized. It forms two bonds with the H atoms. Two of the sp^3 hybrid orbitals are occupied by the two lone pairs.*

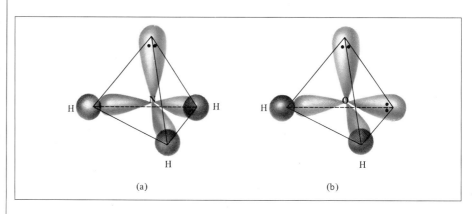

(a)   (b)

The geometry of the water molecule too can be explained by assuming that the O atom is sp^3-hybridized. The ground-state electron configuration of O is $1s^2 2s^2 2p^4$, so that the orbital diagram for the sp^3-hybridized O atom is

sp^3 orbitals

Two of the four hybrid orbitals are used to form O—H bonds, and the other two are used to accommodate the two lone pairs [Figure 12.7(b)]. The strong repulsion between the lone pairs and the electrons in the bonding orbitals further reduces the HOH angle to 104.5°.

You may have noticed an interesting connection between hybridization and the octet rule. Regardless of the type of hybridization, an atom starting with one s and three p orbitals will still possess four orbitals, which can accommodate a total of eight electrons in a compound. For elements in the second period of the periodic table, eight is the maximum number of electrons that an atom of any of these elements can accommodate. This is the reason that the octet rule is usually obeyed by the second-period elements. An atom of a third-period element may use only the $3s$ and $3p$ orbitals to form hybrid orbitals in a compound. If it does, the octet rule again is observed in the compound. However, as you will see shortly, such an atom may use $3d$ orbitals in addition to $3s$ and $3p$ orbitals to form hybrid orbitals. When this happens, the octet rule no longer applies.

The main second-period exceptions are Be and B, which normally associate themselves in bonding with four and six valence electrons, respectively.

Table 12.5 lists the sp, sp^2, and sp^3 hybridizations (as well as other types of hybridizations to be discussed shortly) together with the shapes of the hybrid orbitals. In order to assign a suitable state of hybridization of the central atom in a molecule, we must have some idea about the geometry of the molecule. We can start by drawing the Lewis formula of the molecule and predict the overall shape of the electron pairs (both bonding pairs and lone pairs) using the VSEPR model (see Table 12.1). We can then deduce the state of hybridization of the central atom by matching the shape of the electron pairs with that of the hybrid orbitals in Table 12.5. The following example illustrates this procedure.

This procedure works for molecules that do not contain multiple bonds. Hybridization in molecules containing multiple bonds will be discussed shortly.

EXAMPLE 12.3

Assign the hybridization state of the central (underlined) atom in each of the following molecules: (a) $\underline{Hg}Cl_2$, (b) $\underline{Al}I_3$, and (c) $\underline{P}F_3$. Describe the hybridization process and determine the molecular geometry in each case.

Answer

(a) The ground-state electron configuration of Hg is $[Xe]6s^2 4f^{14} 5d^{10}$; therefore, the Hg atom has two valence electrons (the $6s$ electrons). The Lewis formula of $HgCl_2$ is

Cl—Hg—Cl

(Continued)

TABLE 12.5 Important Hybrid Orbitals and Their Geometries

Pure Atomic Orbitals of the Central Atom	Hybridization of the Central Atom	Number of Hybrid Orbitals	Shape of Hybrid Orbitals	Examples
s, p	sp	2	180° Linear	$BeCl_2$
s, p, p	sp^2	3	120° Planar	BF_3
s, p, p, p	sp^3	4	109.5° Tetrahedral	CH_4, NH_4^+
s, p, p, p, d	sp^3d	5	90° 120° Trigonal bipyramidal	PCl_5
s, p, p, p, d, d	sp^3d^2	6	90° 90° Octahedral	SF_6

There are no lone pairs on the Hg atom, so that the shape of the two electron pairs is linear (see Table 12.1). From Table 12.5 we conclude that Hg is *sp*-hybridized because it has the geometry of the two *sp* hybrid orbitals. The hybridization process can be imagined to take place as follows. First we draw the orbital diagram for the ground state of Hg:

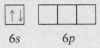

6*s*   6*p*

By promoting a 6*s* electron to the 6*p* orbital, we get the excited state:

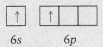

6*s*   6*p*

The 6*s* and 6*p* orbitals then mix to form two *sp* hybrid orbitals:

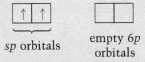

sp orbitals   empty 6*p*
                 orbitals

The two Hg—Cl bonds are formed by the overlap of the Hg *sp* hybrid orbitals with the 3*p* orbitals of the Cl atoms. Thus HgCl₂ is a linear molecule.

(b) The ground-state electron configuration of Al is [Ne]$3s^2 3p^1$. Therefore, Al has three valence electrons. The Lewis formula of AlI₃ is

$$
\begin{array}{c}
I \\
| \\
I—Al \\
| \\
I
\end{array}
$$

There are three bonding pairs and no lone pairs on the Al atom. From Table 12.1 we see that the shape of three electron pairs is trigonal planar, and from Table 12.5 we conclude that Al must be *sp²*-hybridized in AlI₃. The orbital diagram of the ground-state Al atom is

3*s*   3*p*

By promoting a 3*s* electron into the 3*p* orbital we obtain the following excited state:

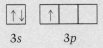

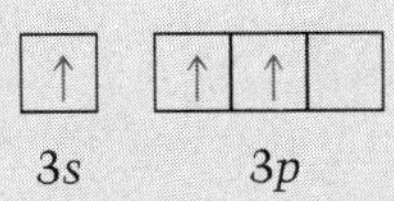

3*s*   3*p*

The 3*s* and two 3*p* orbitals then mix to form three *sp²* hybrid orbitals:

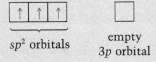

sp² orbitals   empty
                 3*p* orbital

The *sp²* hybrid orbitals overlap with the 5*p* orbitals of I to form three covalent Al—I bonds. We predict the AlI₃ molecule to be planar, and all the IAlI angles are 120°.

(Continued)

(c) The ground-state electron configuration of P is $[Ne]3s^23p^3$. Therefore, the P atom has five valence electrons. The Lewis formula of PF_3 is

$$F—\overset{\displaystyle ..}{P}—F$$
$$|$$
$$F$$

There are three bonding pairs and one lone pair on the P atom. In Table 12.1 we see that the overall shape of four electron pairs is tetrahedral, and from Table 12.5 we conclude that P must be sp^3-hybridized. The orbital diagram of the ground-state P atom is

$$\boxed{\uparrow\downarrow} \quad \boxed{\uparrow}\,\boxed{\uparrow}\,\boxed{\uparrow}$$
$$3s \qquad\quad 3p$$

By promoting a $3s$ electron into the $3p$ orbital we obtain the following excited state:

$$\boxed{\uparrow} \quad \boxed{\uparrow}\,\boxed{\uparrow}\,\boxed{\uparrow\downarrow}$$
$$3s \qquad\quad 3p$$

The $3s$ and $3p$ orbitals then mix to form four sp^3 hybrid orbitals:

$$\boxed{\uparrow}\,\boxed{\uparrow}\,\boxed{\uparrow}\,\boxed{\uparrow\downarrow}$$
$$\underbrace{}$$
$$sp^3 \text{ orbitals}$$

As in the case of NH_3, which we discussed earlier, one of the sp^3 hybrid orbitals is used to accommodate the lone pair on P. The other three sp^3 hybrid orbitals are used to form covalent P—F bonds with the $2p$ orbitals of F. We predict the geometry of the molecule to be pyramidal; the FPF angle should be somewhat less than 109.5°.

Similar examples: Problems 12.24, 12.25.

Hybridization of s, p, and d Orbitals

We have seen that hybridization neatly explains bonding that involves atoms of the second-period elements, which have $2s$ and $2p$ orbitals but no $2d$ orbitals. For elements in the third period and beyond, however, we cannot always account for molecular geometry by assuming the hybridization of only s and p orbitals. To understand the formation of molecules with trigonal bipyramid and octahedral geometries, for instance, we must include d orbitals in the hybridization concept.

Consider the SF_6 molecule as an example. In Section 12.2 we saw that this molecule has octahedral geometry, which is also the shape of the six electron pairs. From Table 12.5 we see that the S atom must be sp^3d^2-hybridized in SF_6. The ground-state electron configuration of S is $[Ne]3s^23p^4$:

$$\boxed{\uparrow\downarrow} \quad \boxed{\uparrow\downarrow}\,\boxed{\uparrow}\,\boxed{\uparrow} \quad \boxed{}\,\boxed{}\,\boxed{}\,\boxed{}\,\boxed{}$$
$$3s \qquad\quad 3p \qquad\qquad\qquad 3d$$

Since the $3d$ level is relatively low in energy, we can promote both $3s$ and $3p$ electrons to two of the $3d$ orbitals:

3s   3p   3d

Mixing the $3s$, three $3p$, and two $3d$ orbitals generates six sp^3d^2 hybrid orbitals:

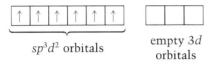

sp^3d^2 orbitals   empty $3d$
                          orbitals

The six S—F bonds are formed by the overlap of the S hybrid orbitals and the $2p$ orbitals of the F atoms. Since there are twelve electrons around the S atom, the octet rule is not satisfied. *The use of d orbitals in addition to s and p orbitals to form covalent bonds* is an example of **valence-shell expansion,** which corresponds to the expanded octet case discussed in Section 11.6. Returning for a moment to the second-period elements, note that because these elements do not have $2d$ levels, they can never expand their valence shells and hence their atoms can never be surrounded by more than eight electrons in any of their compounds.

EXAMPLE 12.4

Describe the hybridization state of phosphorus in phosphorus pentabromide (PBr_5).

Answer

The Lewis formula of PBr_5 is

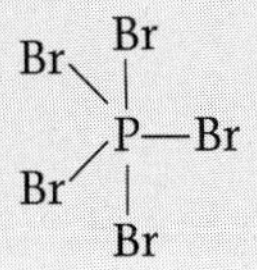

$$
\begin{array}{c}
Br \quad\; Br \\
\quad\searrow \;|\; \\
\quad P{-}Br \\
\quad\nearrow \;|\; \\
Br \quad\; Br
\end{array}
$$

In Table 12.1 we see that the shape of five electron pairs is trigonal bipyramidal. Referring to Table 12.5 we find that this is the shape of five sp^3d hybrid orbitals. Thus P must be sp^3d-hybridized in PBr_5. The ground-state electron configuration of P is $[Ne]3s^23p^3$. To describe the hybridization process, we start with the orbital diagram for the ground state of P

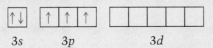

3s   3p   3d

Promoting a $3s$ electron into a $3d$ orbital results in the excited state:

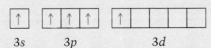

3s   3p   3d

(Continued)

Mixing the one $3s$, three $3p$, and one $3d$ orbitals generates five equivalent sp^3d hybrid orbitals:

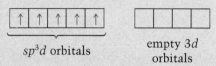

sp^3d orbitals   empty $3d$ orbitals

These hybrid orbitals overlap with the $4p$ orbitals of Br to form five covalent P—Br bonds. Since there are no lone pairs on the P atom, the geometry of PBr_5 is trigonal bipyramidal, the same as the electron pairs.

Similar example: Problem 12.45.

12.6 HYBRIDIZATION IN MOLECULES CONTAINING DOUBLE AND TRIPLE BONDS

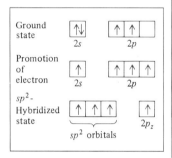

FIGURE 12.8 *Orbital diagrams showing the changes involved in the sp^2 hybridization of a carbon atom. In this case, the 2s orbital is mixed with only two 2p orbitals to form three equivalent sp^2 hybrid orbitals. This process leaves an electron in the unhybridized orbital, the $2p_z$ orbital.*

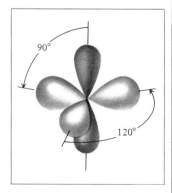

FIGURE 12.9 *Each carbon atom in the C_2H_4 molecule has three sp^2 hybrid orbitals and one unhybridized $2p_z$ orbital that is perpendicular to the plane of the hybrid orbitals.*

The concept of hybridization can be applied also to molecules containing double and triple bonds. Consider the ethylene molecule (C_2H_4) as an example. In Example 12.1 we saw that C_2H_4 contains a carbon–carbon double bond and has planar geometry. Both the geometry and the bonding can be understood if we assume that each carbon atom is sp^2-hybridized. Figure 12.8 shows orbital diagrams of this hybridization process. We assume that only the $2p_x$ and $2p_y$ orbitals combine with the 2s orbital, and that the $2p_z$ orbital remains unchanged. Figure 12.9 shows that the $2p_z$ orbital is perpendicular to the plane of the hybrid orbitals. Now how do we account for the bonding of the C atoms? As Figure 12.10(a) shows, each carbon atom uses the three sp^2 hybrid orbitals to form two bonds with the two hydrogen 1s orbitals and one bond with the sp^2 hybrid orbital of the adjacent C atom. In addition, the two unhybridized $2p_z$ orbitals of the C atoms form another bond by overlapping sideways [Figure 12.10(b)].

A distinction should be made between the two types of covalent bonds found in C_2H_4. *A covalent bond formed by orbitals overlapping end-to-end*, as shown in Figure 12.10(a), *has its electron density concentrated between the nuclei of the bonding atoms*, and is called a **sigma bond** (**σ bond**). Thus, the three bonds formed by each C atom in Figure 12.10(a) are all sigma bonds. On the other hand, *a covalent bond formed by sideways overlapping orbitals has its electron density concentrated above and below the plane of the nuclei of the bonding atoms* and is called a **pi bond** (**π bond**). The two C atoms form a pi bond as shown in Figure 12.10(b). Figure 12.10(c) shows the orientation of both sigma and pi bonds, and Figure 12.11 is another way of looking at the planar C_2H_4 molecule and the formation of the pi bond. Although we normally represent the carbon–carbon double bond as C=C (as in a Lewis formula), it is important to keep in mind that the two bonds are different types: One is a sigma bond and the other is a pi bond.

The acetylene molecule (C_2H_2) contains a carbon–carbon triple bond. Since the molecule is linear, we can explain the geometry and bonding by assuming that each C atom is sp-hybridized (Figure 12.12). As Figure 12.13

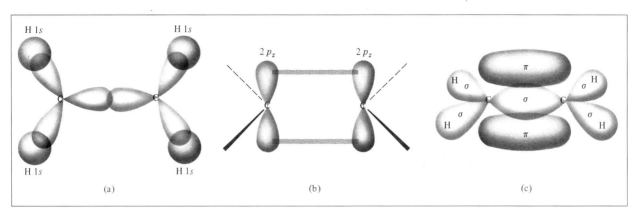

(a)   (b)   (c)

FIGURE 12.10   *Bonding in C_2H_4. (a) Top view showing the sigma bonds between carbon atoms and between carbon and hydrogen atoms. All the atoms lie in the same plane; therefore C_2H_4 is a planar molecule. (b) Side view, the two bonds showing how overlap of the two $2p_z$ orbitals on the two carbon atoms will take place, leading to formation of a pi bond. (c) Diagram showing the sigma bonds and the pi bond. Note that the pi bond lies above and below the plane of the molecule.*

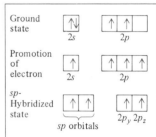

FIGURE 12.12   *Orbital diagrams showing the changes involved in the sp hybridization of a carbon atom. In this case, the 2s orbital is mixed with only one 2p orbital to form two sp hybrid orbitals. This process leaves an electron in each of the two unhybridized 2p orbitals, namely the $2p_y$ and $2p_z$ orbitals.*

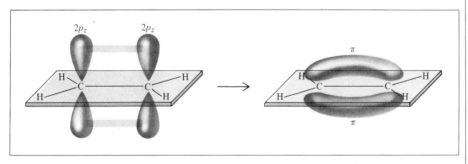

FIGURE 12.11   *Another view of the pi bond formation in the C_2H_4 molecule. Note that all six atoms lie in the same plane.*

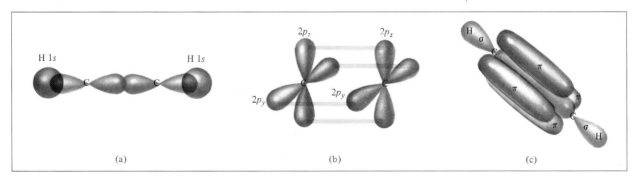

(a)   (b)   (c)

FIGURE 12.13   *Bonding in C_2H_2. (a) Top view showing the sigma bonds between carbon atoms and between carbon and hydrogen atoms. All the atoms lie along a straight line; therefore acetylene is a linear molecule. (b) Side view showing the overlap of the two $2p_y$ orbitals and of the two $2p_z$ orbitals of the two carbon atoms, which leads to formation of two pi bonds. (c) Diagram showing both the sigma and pi bonds.*

shows, the two sp hybrid orbitals of each C atom form one sigma bond with a hydrogen $1s$ orbital and another sigma bond with the other C atom. In addition, two pi bonds are formed by the sideways overlap of the unhybridized $2p_y$ and $2p_z$ orbitals. Thus the C≡C bond is made up of one sigma bond and two pi bonds.

A useful rule that helps us predict the state of hybridization in molecules containing multiple bonds is the following. If the central atom forms a double bond, it is sp^2-hybridized; if it forms two double bonds or a triple bond, it is sp-hybridized. Note that this rule applies only to atoms of the second-period elements. As regards atoms of third-period elements and beyond that form multiple bonds, the problem of predicting is more involved and will not be dealt with here.

EXAMPLE 12.5

Describe the bonding scheme in the formaldehyde molecule whose Lewis formula is

$$\text{H} \diagdown \atop \text{H} \diagup \text{C}=\overset{\displaystyle ..}{\underset{\displaystyle ..}{\text{O}}}$$

Assume that the O atom is unhybridized.

Answer

We note that the C atom (a second-period element) has a double bond; therefore, it is sp^2-hybridized. The orbital diagram of the O atom is

↑↓		↑	↑↓	↑

$$2s \qquad\qquad 2p$$

One of its orbitals $(2p_x)$ forms a sigma bond with C, and the other orbital $(2p_z)$ forms a pi bond with the unhybridized $2p_z$ orbital of C (Figure 12.14). The two lone pairs on the O atom then are the electrons in the $2s$ and $2p_y$ orbitals. The

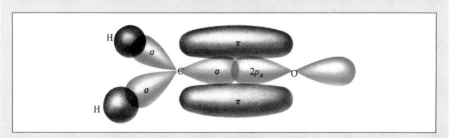

FIGURE 12.14 *The bonding scheme in the formaldehyde molecule. A sigma bond is formed by the overlap of the sp^2 hybrid orbital of carbon and the $2p_x$ orbital of oxygen; a pi bond is formed by the overlap of the $2p_z$ orbitals of the carbon and oxygen atoms. The two lone pairs on oxygen are placed in the $2s$ and $2p_y$ orbitals (not shown).*

The choice of the $2p$ orbitals for bonding and for holding the lone pair is completely arbitrary.

two remaining sp^2 orbitals of the C atom form two sigma bonds with the H atoms, as in ethylene.

Similar examples: Problems 12.23, 12.34.

In summary, the hybridization approach to molecular geometry is quantum mechanical, based on the concept of atomic orbitals. To make use of it we must have a rough idea about the shape of the electron pairs and then try to determine the state of hybridization of the central atom, working with Tables 12.1 and 12.5. For molecules containing multiple bonds, we can deduce the state of hybridization of the central atom (if it belongs to a second-period element), depending on whether it forms a double bond, two double bonds, or a triple bond. The hybridization procedure is more involved than the VSEPR model, but it also provides a more detailed description of chemical bonding.

AN ASIDE ON THE PERIODIC TABLE
Structures of the Nonmetallic Elements

Our discussion of molecular geometry and bonding in this chapter has provided a background for us to examine the structures of some common nonmetals. The nature of metallic bonding will be dealt with in a later chapter.

Figure 12.15 shows the most common nonmetallic elements in their positions in the

FIGURE 12.15 The most common nonmetallic elements, shown in their positions in the periodic table. The stable elemental form of hydrogen is diatomic hydrogen molecule, H_2.

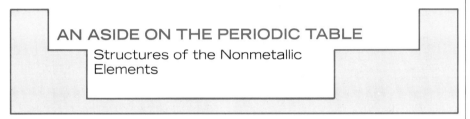

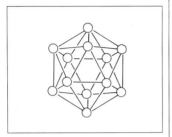

FIGURE 12.16 The basic unit in elemental boron has an icosahedral shape. The boron-to-boron bond is not the conventional two-electron pair discussed in Chapters 11 and 12.

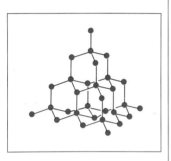

FIGURE 12.17 The structure of diamond. Each carbon is tetrahedrally bonded to four other carbon atoms.

periodic table. We will study the structure of these elements in the order of their increasing group numbers.

Group 3A: Boron (B)

Boron is unique among the elements in the structural complexity of its elemental forms. The structure unit that is basic to all its allotropic forms (see p. 35 for definition of allotrope) is an *icosahedron*, shown in Figure 12.16. An icosahedron is a complex three-dimensional structure having twenty equal faces. The B atoms occupy the twelve corners of the icosahedron. Each B atom is bonded to five other B atoms within the structure and to a sixth B atom outside of the icosahedron. However, both the structure and the nature of bonding in elemental boron are beyond the scope of this textbook.

Group 4A: Carbon (C), Silicon (Si), Germanium (Ge)

Carbon. Carbon exists in two allotropic forms—diamond and graphite. Each C atom in diamond is sp^3-hybridized; it uses the hybrid orbitals to form four covalent bonds with adjacent C atoms (Figure 12.17). The resulting structure is a giant three-dimensional network that is extremely hard. In graphite, the C atoms are arranged in six-membered rings (Figure 12.18). Each atom is covalently bonded to three other atoms. The planar geometry of the rings and the bonding suggest that the C atoms are sp^2-hybridized. The layers of rings are held together by weak forces. The covalent bonds in graphite account for its hardness; however, because the layers can slide over one another, graphite is slippery to the touch and useful as a lubricant.

Silicon and Germanium. The elemental forms of Si and Ge have the diamond structure. Thus we can assume that Si and Ge atoms are sp^3-hybridized in giant three-dimensional networks.

Group 5A: Nitrogen (N), Phosphorus (P), Arsenic (As)

Nitrogen. Nitrogen exists in the diatomic form, N_2, and is therefore a linear molecule.

Phosphorus. Phosphorus exists in several allotropic forms. One common form is white phosphorus, a waxy solid (see Color Plate 5) consisting of P_4 molecules (Figure 12.19). The four P atoms in P_4 occupy the corners of a tetrahedron. Each P atom is bonded to three other P atoms by single bonds. However, this arrangement requires that each PPP angle be only 60°, which is much smaller than the angle in any other molecule that we have discussed so far. Even without invoking hybridization, the three p orbitals on each P atom are at 90° from each other! The 60° bond angle requires that the bonding electron pairs in the molecule be crowded rather closely together. This "strain" in the P_4 molecule is believed to be responsible for its high chemical reactivity.

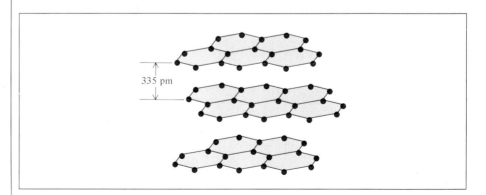

335 pm

FIGURE 12.18 The structure of graphite. The distance between successive layers is 335 pm.

Arsenic. Arsenic in its most stable form is a gray solid (see Color Plate 5) consisting of sheets of covalently bonded As atoms. In the vapor phase As is known to exist as tetrahedral As_4 molecules that resemble P_4 molecules.

Group 6A: Oxygen (O), Sulfur (S), Selenium (Se)

Oxygen. Oxygen exists in two allotropic forms, O_2 and ozone (O_3). The more stable form, O_2, is a diatomic molecule and therefore has a linear structure. The ozone molecule, as we saw earlier, has a bent shape. The central atom in O_3 is sp^2-hybridized.

Sulfur. Sulfur exists in several allotropic forms. The stable form at ordinary temperatures is orthorhombic sulfur, which is a yellow solid (see Color Plate 6). Orthorhombic sulfur consists of discrete S_8 molecules whose shape is that of a "puckered" ring (Figure 12.20). Because the SSS angles in S_8 are all 108.0°, which is quite close to the ideal tetrahedral angle (109.5°), the S atom is assumed to be sp^3-hybridized.

Selenium. At room temperature the stable form of Se consists of Se_8 molecules. The bonding and geometry of Se_8 are similar to those in S_8.

Group 7A: The Halogens

The elemental forms of the halogens (see Color Plate 7) all consist of diatomic molecules (F_2, Cl_2, Br_2, and I_2) which have a linear geometry.

Group 8A: The Noble Gases

The noble gases (He, Ne, Ar, Kr, Xe, Rn) all exist as monatomic gases (see Color Plate 8).

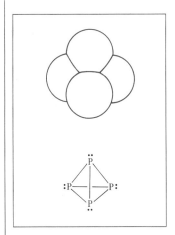

FIGURE 12.19 The structure of the P_4 molecule.

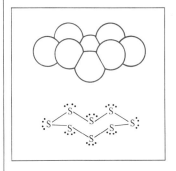

FIGURE 12.20 The structure of the S_8 molecule.

SUMMARY

1. The VSEPR model for predicting molecular geometry is based on the assumption that valence-shell electron pairs repel one another and tend to stay as far apart as possible.

2. According to the VSEPR model, molecular geometry can be predicted from the number of bonding electron pairs and lone pairs. Lone pairs repel other pairs more strongly than bonding pairs do and thus distort bond angles from those of the ideal geometry.

3. The dipole moment is a measure of the charge separation in molecules containing atoms of different electronegativity. The dipole moment of a molecule is the resultant of whatever bond moments are present in a molecule. Information about molecular geometry can be obtained from dipole moment mesurements.

4. In valence bond theory, hybridized atomic orbitals are formed by the combination and rearrangement of (for example, *s* and *p*) orbitals of the same atom. The hybridized orbitals are equal to each other in energy and electron density.

5. Valence-shell expansion can be explained by invoking hybridization of *s*, *p*, and *d* orbitals.

6. In *sp* hybridization, the two hybrid orbitals lie in a straight line; in sp^2 hybridization, the three hybrid orbitals are directed toward the corners of a triangle; in sp^3 hybridization, the four hybrid orbitals are directed

toward the corners of a tetrahedron; in sp^3d hybridization, the five hybrid orbitals are directed toward the corners of a trigonal bipyramid; in sp^3d^2 hybridization, the six hybrid orbitals are directed toward the corners of an octahedron.

7. In an sp^2-hybridized atom (for example, carbon), the one unhybridized p orbital can form a pi bond with another p orbital. A carbon–carbon double bond consists of a sigma bond and a pi bond. In an sp-hybridized carbon atom, the two unhybridized p orbitals can form two pi bonds with two p orbitals on another atom (or atoms). A carbon–carbon triple bond consists of one sigma bond and two pi bonds.

KEY WORDS

Dipole moment, p. 337
Heteronuclear diatomic molecule,
   p. 338
Homonuclear diatomic molecule,
   p. 338
Hybrid orbital, p. 343
Hybridization, p. 343

Nonpolar molecule, p. 338
Pi bond, p. 352
Polar molecule, p. 338
Sigma bond, p. 352
Valence shell, p. 327
Valence-shell expansion, p. 351
VSEPR model, p. 327

PROBLEMS

More challenging problems are marked with an asterisk.

VSEPR

12.1 Discuss the salient features of the VSEPR method.

12.2 Within the framework of the VSEPR method, what are the arrangements of two, three, four, five, and six valence-shell electron pairs about a central atom?

12.3 In the trigonal bipyramidal arrangement, why does a lone pair occupy an equatorial position rather than an axial position?

12.4 Predict the geometries of the following species using the VSEPR method: (a) PCl_3, (b) $CHCl_3$, (c) SiH_4, (d) $TeCl_4$.

12.5 What are the geometries of the following ions? (a) NH_4^+, (b) NH_2^-, (c) CO_3^{2-}, (d) ICl_2^-, (e) ICl_4^-, (f) AlH_4^-, (g) $SnCl_5^-$, (h) H_3O^+, (i) BeF_4^{2-}, (j)ClO_2^-.

*12.6 Predict the geometry of the odd-electron molecules ClO_2 and NO_2.

12.7 What are the geometries of the following species? (a) $AlCl_3$, (b) $ZnCl_2$, (c) $ZnCl_4^-$, (d) PF_3.

12.8 Predict the geometry of the following molecules using the VSEPR method: (a) $HgBr_2$, (b) N_2O (ar-rangement of atoms is NNO), (c) SCN^- (arrange-ment of atoms is SCN).

*12.9 Describe the geometry around each of the three central atoms in the CH_3COOH molecule.

Dipole Moment and Molecular Geometry

12.10 What is the necessary condition for a molecule to possess a dipole moment?

12.11 How is it possible for a molecule to have bond dipole moments yet be nonpolar?

12.12 The bonds in beryllium hydride (BeH_2) molecules are polar, yet the dipole moment of the molecule is zero. Explain.

12.13 Explain why an atom cannot have a dipole mo-ment.

12.14 Does the molecule OCS have a higher or lower dipole moment than CS_2?

12.15 Arrange the following molecules in order of in-creasing dipole moment: H_2O, H_2S, H_2Te, H_2Se. (See Table 12.4.)

12.16 List the following molecules in order of increasing dipole moment: H_2O, CBr_4, H_2S, HF, NH_3, CO_2. (See Table 12.4.)

*12.17 Arrange the following compounds in order of increasing dipole moment:

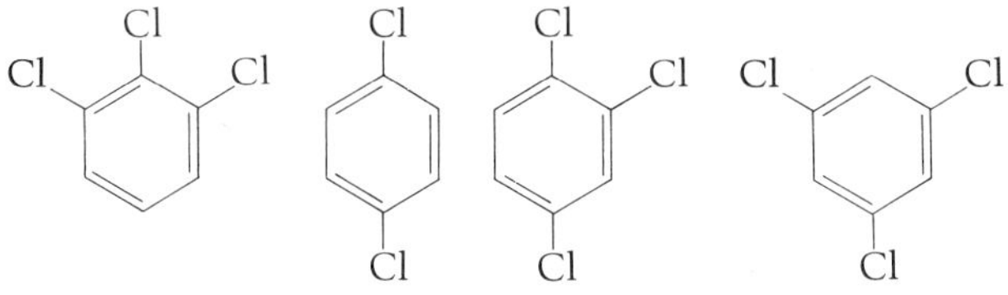

(a) (b) (c) (d)

12.18 Which of the following molecules has a higher dipole moment?

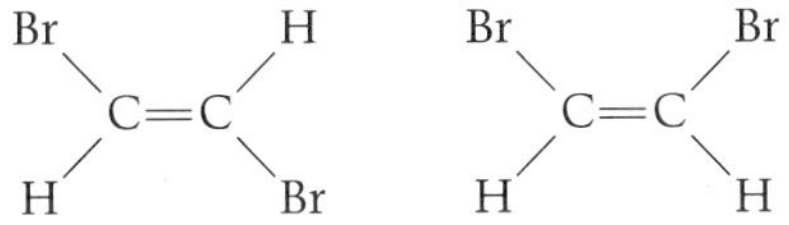

(a) (b)

12.19 The dipole moments of the hydrogen halides decrease from HF to HI (see Table 12.4). Explain this trend.

Hybridization

12.20 What is the hybridization of atomic orbitals? Why is it impossible for an isolated atom to exist in the hybridized state?

12.21 What is the size of the angle between two hybrid orbitals on the same atom: (a) sp and sp hybrid orbitals, (b) sp^2 and sp^2 hybrid orbitals, (c) sp^3 and sp^3 hybrid orbitals.

12.22 Which of the following pairs of atomic orbitals of adjacent nuclei can overlap to form a sigma bond? Which overlap to form a pi bond? Which cannot overlap (no bond)? Consider the x axis to be the internuclear axis (that is, the line joining the nuclei of the two atoms. (a) $1s$ and $1s$, (b) $1s$ and $2p_x$, (c) $2p_x$ and $2p_y$, (d) $3p_y$ and $3p_y$, (e) $2p_x$ and $2p_x$, (f) $1s$ and $2s$.

*12.23 What is the total number of sigma bonds and pi bonds in each of the following molecules?

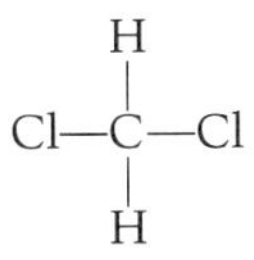

(a) 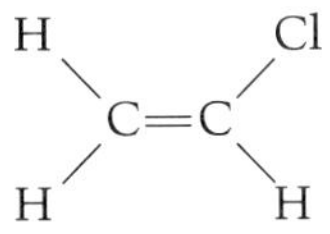 (b)

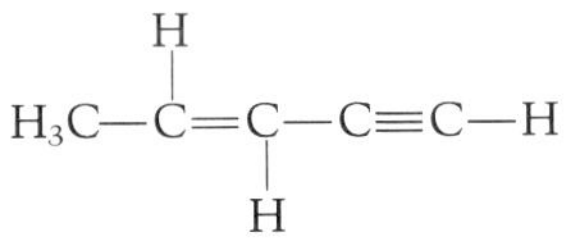

(c)

12.24 Describe the bonding scheme in the molecule AsH_3 in terms of hybridization.

12.25 What is the hybridization of Si in SiH_4 and in $H_3Si—SiH_3$?

12.26 Draw a diagram representing the formation of a double bond between two carbon atoms in ethylene (C_2H_4).

12.27 What are the hybrid orbitals in the carbon atoms of the following molecules?

(a) $H_3C—CH_3$
(b) $H_3C—CH=CH_2$
(c) $CH_3—CH_2—OH$
(d) $CH_3CH=O$
(e) CH_3COOH

12.28 Indicate the hybrid orbitals used by nitrogen atoms in the following species: (a) NH_3, (b) $H_2N—NH_2$, (c) NO_3^-.

12.29 Indicate the hybrid orbitals used by carbon atoms in the following species: (a) CO, (b) CO_2, (c) CN^-.

*12.30 Describe the change in hybridization (if any) of the Al atom in the following reaction:

$$AlCl_3 + Cl^- \longrightarrow AlCl_4^-$$

*12.31 Consider the reaction

$$BF_3 + NH_3 \longrightarrow F_3B—NH_3$$

Describe the changes in hybridization (if any) of the B and N atoms as a result of this reaction.

*12.32 The allene molecule $H_2C=C=CH_2$ has a linear shape (that is, the three C atoms lie on a straight line). What are the hybridization states of the carbon atoms? Draw diagrams to show the formation of sigma bonds and pi bonds in this molecule.

*12.33 What is the state of hybridization of the central N atom in the azide ion, N_3^-? (Arrangement of atoms: NNN)

*12.34 What are the hybridization states of the C, N, and O atoms in this molecule?

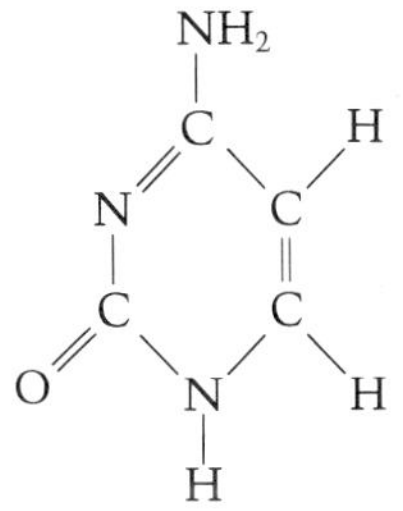

Miscellaneous Problems

12.35 How would you distinguish between a sigma bond and a pi bond?

12.36 Predict the bond angles for the following molecules: (a) $BeCl_2$, (b) BCl_3, (c) CCl_4, (d) CH_3Cl, (e) Hg_2Cl_2 (arrangement of atoms: ClHgHgCl), (f) $SnCl_2$, (g) SCl_2, (h) H_2O_2, (i) SnH_4.

12.37 Write Lewis formulas for the following, together with other information requested: (a) BF_3. Shape: planar or nonplanar? (b) ClO_3^-. Shape: planar or nonplanar? (c) H_2O. Show the direction of the resultant dipole moment. (d) OF_2. Polar or nonpolar molecule? (e) NO_2. Estimate the ONO bond angle.

12.38 Draw the Lewis formula for mercury(II) bromide ($HgBr_2$). Is this molecule linear or bent? How would you prove its geometry?

12.39 Briefly compare the approaches employed by the VSEPR and hybridization methods in the study of molecular geometry.

12.40 Give Lewis formulas for the following, together with the other information requested: (a) SO_3. Polar or nonpolar molecule? (b) PF_3. Polar or nonpolar molecule? (c) F_3SiH. Show the direction of the resultant dipole moment. (d) SiH_3^-. Planar or pyramidal shape? (e) Cl_2CH_2. Polar or nonpolar molecule?

12.41 Which of the following species are tetrahedral? $SiCl_4$, SeF_4, XeF_4, CI_4, $CdCl_4^{2-}$.

12.42 Which of the following molecules are linear? ICl_2^-, IF_2^+, OF_2, SnI_2, $CdBr_2$.

12.43 Draw the Lewis formula for the $BeCl_4^{2-}$ ion. Predict its geometry and describe the state of hybridization of the Be atom.

*12.44 The following molecules (AX_4Y_2) all have octahedral geometry. Group the molecules that are equivalent to each other.

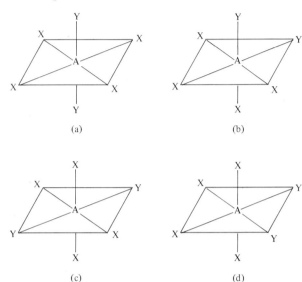

(a)   (b)

(c)   (d)

12.45 Describe the state of hybridization of phosphorus in phosphorus pentafluoride (PF_5).

13
CHEMICAL BONDING IV: MOLECULAR ORBITALS

This chapter introduces and develops a second approach to chemical bonding, an approach that complements the valence bond theory and extends our ability to account for and predict the bonding characteristics of molecules.

Regardless of the number of bonding models considered, each must eventually account for the same known characteristics of various substances. Thus the question is not which bonding model is "correct," but rather which bonding model gives us the most satisfactory explanation of specific molecular properties. As you might suspect (simply due to the presence here of this chapter), chemists have found value in aspects of at least two different approaches to chemical bonding. This chapter will give you the opportunity to compare the valence bond approach, which you have already studied, with the molecular orbital model—another useful way to think about the structure of matter.

13.1 MOLECULAR ORBITAL THEORY

As we saw in Chapter 12, valence bond theory is one of the two quantum mechanical approaches that explain bonding in molecules. It accounts, at least qualitatively, for the stability of the covalent bond in terms of over-lapping atomic orbitals. Through the use of hybridization, valence bond theory can explain molecular geometries predicted by the VSEPR model. However, the assumption that electrons in a molecule occupy atomic or-bitals of the individual atoms can only be an approximation, since each bonding electron in a molecule must be in an orbital that is characteristic of the molecule as a whole.

In some cases, valence bond theory cannot satisfactorily account for observed properties of molecules. Consider the oxygen molecule, whose Lewis formula is

$$\ddot{O}=\ddot{O}$$

According to valence bond theory, any molecule that has an even number of electrons should be diamagnetic, since all the electrons are assumed to be paired. By experiment we find that the oxygen molecule is paramagnetic, with two unpaired electrons. This finding suggests a fundamental defi-ciency in valence bond theory, one that justifies searching for an alternative bonding approach capable of accounting for the properties of molecules, including such a common molecule as O_2.

The magnetic and other properties of molecules are sometimes better explained by another quantum mechanical approach called *molecular or-bital (MO) theory.* Molecular orbital theory describes covalent bonds in terms of **molecular orbitals,** which *result from interaction of the atomic orbitals of the bonding atoms and are associated with the entire molecule.* That is how a molecular orbital differs from an atomic orbital—an atomic orbital is associated with only one atom. We must realize that neither theory is perfect in dealing with all aspects of bonding; each has its strengths and weaknesses. This book will use both theories, emphasizing one or the other as the situation warrants.

MO theory also allows us to predict geometry, but we will consider here only its applica-tion to bond formation.

13.2 BONDING AND ANTIBONDING MOLECULAR ORBITALS

The overlap of the $1s$ orbitals of two hydrogen atoms can lead to the formation of two types of molecular orbitals: the bonding molecular orbital and the antibonding molecular orbital. A **bonding molecular orbital** is *of lower energy and greater stability than the atomic orbitals from which it was formed.* An **antibonding molecular orbital** is *of higher energy and lower stability than the atomic orbitals from which it was formed.* As the names "bonding" and "antibonding" suggest, placing electrons in a bonding molecular orbital yields a stable covalent bond, whereas placing electrons in an antibonding molecular orbital results in an unstable bond.

In the bonding molecular orbital the electron density is greatest between the nuclei of the bonding atoms. In the antibonding molecular orbital, on the other hand, the electron density decreases to zero between the nuclei.

Although the bonding and an-tibonding molecular orbitals have distinctly different ener-gies, the sum of their ener-gies equals the energies of the two atomic orbitals from which they were formed—an example of conservation of energy at the molecular level.

We can understand this distinction if we recall that electrons in orbitals have wave characteristics (see Section 8.1). A unique wave property allows waves of the same type to interact with each other in such a way that the resultant wave has either an enhanced amplitude or a diminished amplitude. In the former case, we call the interaction *constructive interference;* in the latter case, it is *destructive interference* (Figure 13.1).

The formation of bonding molecular orbitals corresponds to constructive interference (the buildup in amplitude is analogous to the buildup of electron density between the two nuclei). The formation of antibonding molecular orbitals corresponds to destructive interference (the decrease in amplitude is analogous to the decrease in electron density between the two nuclei). The constructive and destructive interactions between the two $1s$ orbitals in the H_2 molecule, then, lead to the formation of a sigma bonding molecular orbital (σ_{1s}) and a sigma antibonding molecular orbital ($\sigma_{1s}^{\star}$):

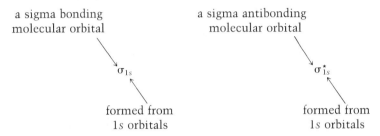

where the asterisk denotes an antibonding molecular orbital.

In a **sigma molecular orbital** (bonding or antibonding) *the electron density is concentrated around a line between the two nuclei of the bonding atoms.* Two electrons in a sigma molecular orbital form a sigma bond (see Section 12.6). Remember that a single covalent bond (such as H—H or F—F) is almost always a sigma bond.

Figure 13.2 shows the *molecular orbital energy level diagram*—that is, the relative energy levels of the orbitals produced in the formation of the H_2 molecule, and the constructive and destructive interactions between the two $1s$ orbitals. Notice that in the antibonding molecular orbital there is a *node,* or zero electron density, between the nuclei. The nuclei are repelled by each other's positive charges, rather than held together. Electrons in the bonding molecular orbital have less energy (and hence greater stability) than they would have in the isolated atoms. On the other hand, electrons in the antibonding molecular orbital have more energy (and less stability) than they would have in the isolated atoms.

So far we have used the hydrogen molecule to illustrate molecular orbital formation, but the concept is equally applicable to other molecules. In the H_2 molecule we consider only the interaction between $1s$ orbitals; with more complex molecules we need to consider additional atomic orbitals as well. Nevertheless, for all s orbitals, the process is the same as for $1s$ orbitals. Thus, the interaction between two $2s$ or $3s$ orbitals can be understood in terms of the molecular orbital energy level diagram and the formation of bonding and antibonding molecular orbitals shown in Figure 13.2.

For p orbitals the process is more complex because they can interact with each other in two different ways. For example, two $2p$ orbitals can approach

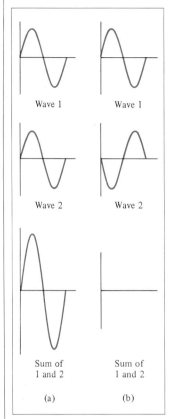

Wave 1

Wave 1

Wave 2

Wave 2

Sum of 1 and 2

Sum of 1 and 2

(a)

(b)

FIGURE 13.1 *The constructive interference (a) and destructive interference (b) of two waves of the same wavelength and amplitude.*

It is the attraction between the electron density in the bonding molecular orbital and the nuclei that holds the two atoms together.

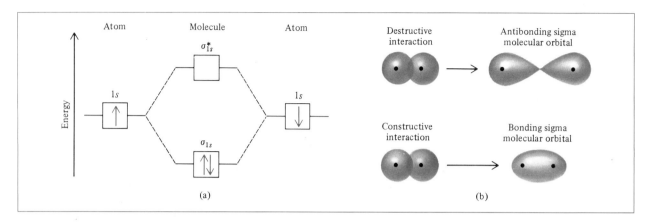

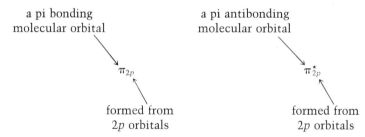

FIGURE 13.2 *(a) Energy levels of bonding and antibonding molecular orbitals in the H_2 molecule. Note that the two electrons in the σ_{1s} must have opposite spins in accord with the Pauli exclusion principle. Keep in mind that the higher the energy of the molecular orbital, the less stable the electrons in that molecular orbital. (b) Constructive and destructive interactions between the two hydrogen 1s orbitals that lead to the formation of a bonding and an antibonding molecular orbital. In the bonding molecular orbital, there is a buildup of the electron density between the nuclei, which acts as a negatively charged "glue" holding the positively charged nuclei together.*

each other end-to-end to produce a sigma bonding and a sigma antibonding molecular orbital, as shown in Figure 13.3(a). Alternatively, the two *p* orbitals can overlap sideways to generate a bonding or an antibonding pi molecular orbital depending on whether the interaction is constructive or destructive [Figure 13.3(b)]:

a pi bonding molecular orbital   a pi antibonding molecular orbital

π_{2p} $\pi^{\star}_{2p}$

formed from 2p orbitals   formed from 2p orbitals

In a ***pi molecular orbital*** (bonding or antibonding), *the electron density is concentrated above and below the line joining the two nuclei of the bonding atoms.* Two electrons in a pi molecular orbital form a pi bond (see Section 12.6). A double bond almost always comprises a sigma bond and a pi bond; a triple bond is always a sigma bond plus two pi bonds.

13.3 MOLECULAR ORBITAL CONFIGURATIONS

To understand properties of molecules, we must know their electron configurations, that is, the distribution of electrons among various molecular orbitals. The procedure for determining the distribution is analogous to that used to determine the electron configurations of atoms (see Section 8.9).

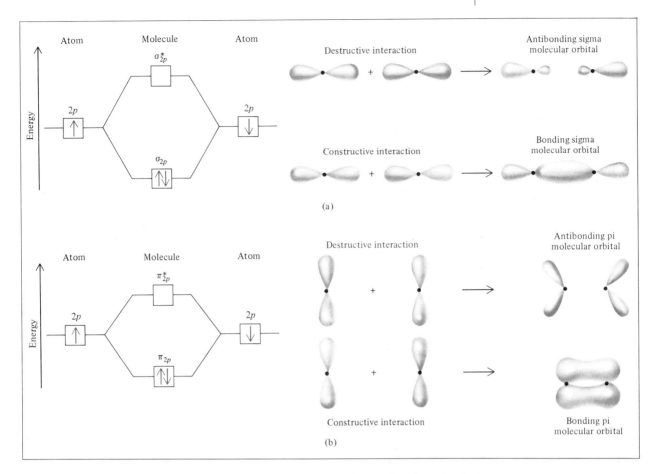

FIGURE 13.3 *Two possible interactions and the corresponding molecular orbitals for two equivalent orbitals. (a) When the p orbitals overlap each other end-to-end, a sigma bonding and a sigma antibonding molecular orbital are formed. (b) When the p orbitals overlap each other side-to-side, a pi bonding and a pi antibonding molecular orbital are formed. Normally, a sigma bonding molecular orbital is more stable than a pi bonding molecular orbital, since the side-to-side interaction leads to a smaller overlap of the p orbitals than the end-to-end interaction. We assume here that the $2p_x$ orbitals take part in the sigma molecular orbital formation. The $2p_y$ and $2p_z$ orbitals can interact to form only π molecular orbitals. The behavior shown in (b) represents the interaction between the $2p_y$ orbitals or the $2p_z$ orbitals.*

Rules Governing Molecular Electron Configuration and Stability

In order to write the electron configuration of any molecule, we must first know the order of increasing energy of the molecular orbitals that are formed in the molecule. Once the order is known, we can use the following rules to guide us in filling these molecular orbitals with electrons. The rules also help us understand the stabilities of the molecular orbitals:

• The number of molecular orbitals formed is always equal to the number of atomic orbitals combined.

- The more stable the bonding molecular orbital, the less stable the corresponding antibonding molecular orbital.
- In a stable molecule, the number of electrons in bonding molecular orbitals is always greater than that in antibonding molecular orbitals.
- Like an atomic orbital, each molecular orbital can accommodate up to two electrons, with opposite spins in accordance with the Pauli exclusion principle.
- When electrons are added to molecular orbitals having the same energy, the most stable arrangement is that predicted by Hund's rule, that is, electrons enter these molecular orbitals with parallel spins.
- The number of electrons in the molecular orbitals is equal to the sum of all the electrons on the bonding atoms.

Hydrogen and Helium

In Section 13.4 we will study molecules formed by atoms of the second-period elements. Before we do, it will be instructive to predict the relative stabilities of the following simple species H_2^+, H_2, He_2^+, and He_2, using the molecular orbital energy level diagrams shown in Figure 13.4. The σ_{1s} and σ_{1s}^* orbitals can accommodate a maximum of four electrons. The total number of electrons increases from one for H_2^+ to four for He_2. Pauli's exclusion principle must be obeyed, so each molecular orbital can accommodate a maximum of two electrons with opposite spins. We are concerned only with the ground-state electron configurations in these cases.

In comparing the stabilities of these species—and, in fact, in general—it is useful to introduce a quantity called ***bond order,*** defined as

$$\text{bond order} = \tfrac{1}{2}\left(\begin{array}{c}\text{number of electrons} \\ \text{in bonding MOs}\end{array} - \begin{array}{c}\text{number of electrons} \\ \text{in antibonding MOs}\end{array}\right) \quad (13.1)$$

where MO is the common abbreviation for "molecular orbital." The bond order gives us a measure of the strength of the bond. For a single covalent bond, the bond order is one. For example, if there are two electrons in the bonding molecular orbital and none in the antibonding molecular orbital the bond order is one. A bond order of zero (or a negative value) means the bonds have no stability, and the molecule cannot exist. It should be understood that bond order can only be used qualitatively for purposes of comparison. For example, two electrons in a bonding sigma molecular orbital

The quantitative measure of the strength of a bond is bond dissociation energy (or bond energy), discussed in Section 11.7.

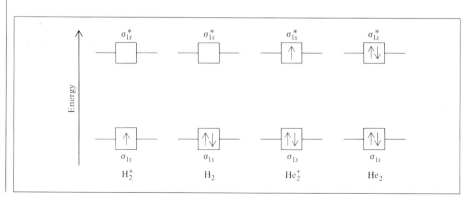

FIGURE 13.4 *Energy levels of the bonding and antibonding molecular orbitals in H_2^+, H_2, He_2^+, and He_2. In all these species, the molecular orbitals are formed by the interaction of two 1s orbitals.*

or two electrons in a bonding pi molecular orbital would have a bond order of one. Yet, these two bonds must differ in bond strength (and bond length) because of the differences in extent of overlap of atomic orbitals.

We are ready now to make predictions about the stabilities of H_2^+, H_2, He_2^+, and He_2 (see Figure 13.4). The H_2^+ molecule ion has only one electron, in the σ_{1s} orbital. Since a covalent bond consists of two electrons in a bonding molecular orbital, H_2^+ has only half of one bond, or a bond order of $\frac{1}{2}$. Thus, we predict that the H_2^+ molecule may be a stable species. The electron configuration of H_2^+ is given by $(\sigma_{1s})^1$.

Note that the superscript in $(\sigma_{1s})^1$ indicates that there is one electron in the sigma bonding molecular orbital.

The H_2 molecule has two electrons, both of which are in the σ_{1s} orbital. According to our scheme, two electrons equal one full bond; therefore, the H_2 molecule has a bond order of one, or one full covalent bond. The electron configuration of H_2 is given by $(\sigma_{1s})^2$.

As for the He_2^+ molecule ion, we find the first two electrons in the σ_{1s} orbital and the third electron in the $\sigma_{1s}^\star$ orbital. Because the antibonding molecular orbital is destabilizing, we expect He_2^+ to be less stable than H_2. Roughly speaking, the instability resulting from the electron in the $\sigma_{1s}^\star$ orbital is canceled by the stability gained by having an electron in the σ_{1s} orbital. Thus, the bond order is $\frac{1}{2}$ and the overall stability of He_2^+ is similar to that of the H_2^+ molecule. The electron configuration of He_2^+ is given by $(\sigma_{1s})^2(\sigma_{1s}^\star)^1$.

In He_2 there would be two electrons in the σ_{1s} orbital and two electrons in the $\sigma_{1s}^\star$ orbital, and the molecule would have no net stability, since the bond order is zero. The electron configuration of He_2 would be given by $(\sigma_{1s})^2(\sigma_{1s}^\star)^2$.

To summarize, we can arrange these four species in the order of decreasing stability:

$$H_2 > H_2^+, He_2^+ > He_2$$

The hydrogen molecule is, of course, a stable species. Our simple molecular orbital method predicts that H_2^+ and He_2^+ also possess some stability, since they both have bond orders of $\frac{1}{2}$. Indeed, their existence has been demonstrated by experiments. It turns out that H_2^+ is somewhat more stable than He_2^+, since there is only one electron in the hydrogen molecule ion and therefore it has less electron–electron repulsion. Furthermore, H_2^+ also has less nuclear repulsion than He_2^+. As for He_2, our prediction that it would have no stability is correct; no one has ever found any evidence for the existence of He_2 molecules.

13.4 HOMONUCLEAR DIATOMIC MOLECULES OF SECOND-PERIOD ELEMENTS

We are now ready to study the ground-state electron configuration in molecules containing second-period elements. We will consider only homonuclear diatomic molecules because they are simpler to deal with.

Figure 13.5 shows the molecular orbital energy level diagram for the Li_2 molecule, which involves only s orbitals. The situation is more complex when the bonding also involves p orbitals. It was mentioned earlier that two p orbitals can form either a sigma bond or a pi bond. Since there are

FIGURE 13.5 *Molecular orbital energy level diagram for the Li_2 molecule. The six electrons in Li_2 (Li is $1s^2 2s^1$) are in the σ_{1s}, σ_{1s}^*, and σ_{2s} orbitals. Since there are two electrons each in σ_{1s} and σ_{1s}^* (just as in He_2), there is no net bonding or antibonding effect. Therefore, the single covalent bond in Li_2 is the result of the two electrons in the bonding molecular orbital σ_{2s}.*

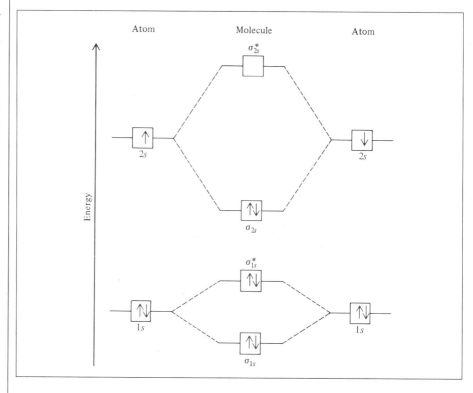

Figure 13.5 also provides a useful insight regarding antibonding molecular orbitals. Note that although the antibonding orbital (σ_{1s}^*) has higher energy and is thus less stable than the bonding orbital (σ_{1s}), this antibonding orbital has *greater* stability than the σ_{2s} bonding orbital. This simply reminds us that the concepts of stability and energy applied here are relative terms rather than absolutes.

three p orbitals for each atom of a second-period element, we know that one sigma and two pi molecular orbitals will result from the constructive interaction. The molecular orbitals are called σ_{2p_x}, π_{2p_y}, and π_{2p_z} orbitals, where the subscripts $2p_x$, $2p_y$, and $2p_z$ denote the origin of the atomic orbitals that take part in forming the molecular orbitals. As Figure 13.3 shows, a σ molecular orbital normally has a greater overlap of the two p orbitals than a π molecular orbital does, so that we would expect the former to be lower in energy. However, we must also consider the influence of other filled orbitals. In reality the energies of molecular orbitals actually increase as follows:

$$\sigma_{1s} < \sigma_{1s}^* < \sigma_{2s} < \sigma_{2s}^* < \pi_{2p_y} = \pi_{2p_z} < \sigma_{2p_x} < \pi_{2p_y}^* = \pi_{2p_z}^* < \sigma_{2p_x}^*$$

The order of reversal of the σ_{2p_x} and the π_{2p_y} and π_{2p_z} energies may be qualitatively understood as follows. The electrons in the σ_{1s}, σ_{1s}^*, σ_{2s}, and σ_{2s}^* orbitals tend to concentrate along the line between the two nuclei. Since the σ_{2p_x} orbital has a geometry similar to that of the σ_s orbitals, its electrons will concentrate in the same region. Consequently, electrons in a σ_{2p_x} orbital will experience a greater repulsion as a result of the filled σ_s and σ_s^* orbitals than when they are in the π_{2p_y} and π_{2p_z} orbitals.

With these concepts and Figure 13.6, which shows the order of increasing molecular orbital energies, we can write the electron configurations and predict the magnetic properties and bond orders of second-period homo-

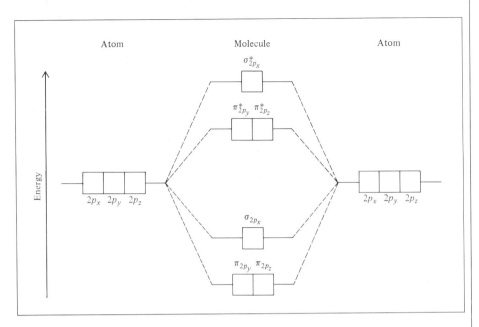

FIGURE 13.6 *General molecular orbital energy level diagram for the second-period homonuclear diatomic molecules. For simplicity, the σ_{1s} and σ_{2s} orbitals have been omitted. Note that in these molecules the σ_{2p_x} orbital is higher in energy than either the π_{2p_y} or the π_{2p_z} orbitals. This means that electrons in the σ_{2p_x} orbitals are more unstable than those in π_{2p_y} and π_{2p_z}. This reversal in relative stability arises as a result of the different interactions between the electrons in the σ_{2p_x} orbital on the one hand and the π_{2p_y} and π_{2p_z} orbitals on the other hand, with the electrons in the lower σ_s orbitals.*

nuclear diatomic molecules. We will start with lithium and move across the period.

The Lithium Molecule (Li₂)

The electron configuration of Li is $1s^2 2s^1$, so Li_2 has a total of six electrons. According to Figure 13.5, these electrons (two each) are positioned in the σ_{1s}, $\sigma_{1s}^{\star}$, and σ_{2s} molecular orbitals. Since there are two more electrons in the bonding molecular orbitals than in antibonding orbitals, the bond order is one [see Equation (13.1)]. We conclude that the Li_2 molecule is stable and should be diamagnetic. Indeed, diamagnetic Li_2 molecules are known to exist in the vapor phase at high temperatures.

The Beryllium Molecule (Be₂)

The Be atom has the electron configuration of $1s^2 2s^2$, and so there are eight electrons in Be_2. The electrons are placed in the σ_{1s}, $\sigma_{1s}^{\star}$, σ_{2s}, and $\sigma_{2s}^{\star}$ orbitals (see Figure 13.5). Because there are equal numbers of electrons in the bond-

Figure 13.5 applies to all the elements in the second period.

ing and antibonding molecular orbitals, the bond order of Be_2 is zero [Equation (13.1)]. As we would expect, the Be_2 molecule does not exist.

The Boron Molecule (B_2)

The electron configuration of B is $1s^2 2s^2 2p^1$. Boron is the first element in the second period with p electrons that participate in bonding. There are a total of ten electrons in B_2; the first eight are placed in the first four molecular orbitals (that is, σ_{1s}, $\sigma_{1s}^\star$, σ_{2s}, and $\sigma_{2s}^\star$). According to Hund's rule, the last two electrons are placed in the π_{2p_y} and π_{2p_z} orbitals with parallel spins (see Figure 13.6). Consequently, the B_2 molecule has a bond order of one and is predicted to be paramagnetic with two unpaired electrons. These properties are confirmed by studying B_2 in the vapor phase.

    An interesting feature about B_2 is that the single bond in the molecule is a pi bond (because of the two electrons in the two pi molecular orbitals). We had said earlier that in the vast majority of cases single bonds are sigma bonds. The boron molecule is one of the few exceptions.

The Carbon Molecule (C_2)

The carbon atom has the electron configuration $1s^2 2s^2 2p^2$; thus there are twelve electrons in the C_2 molecule. From the preceding discussion, we see that as we go from B_2 to C_2 the two additional electrons are also placed in the π_{2p_y} and π_{2p_z} orbitals. Therefore, C_2 has a bond order of two and is diamagnetic. Again, diamagnetic C_2 molecules have been detected in the vapor state. Note that in the double bond in C_2 both bonds are pi bonds (because of the four electrons in the two pi molecular orbitals). In most other molecules, a double bond is made up of a sigma bond and a pi bond.

The Nitrogen Molecule (N_2)

The electron configuration of N is $1s^2 2s^2 2p^3$, and there are fourteen electrons in N_2. Continuing our buildup process, we place the last two electrons in the σ_{2p_x} orbital (see Figure 13.6). Thus N_2 has a bond order of three and should be diamagnetic. As we know, the nitrogen molecule is a very stable species that exists as a gas at room temperature. The triple bond in N_2 is made up of one sigma bond and two pi bonds. This is the pattern for *all* triple bonds.

The Oxygen Molecule (O_2)

The oxygen atom has the electron configuration $1s^2 2s^2 2p^4$; thus there are sixteen electrons in O_2. According to Hund's rule, the two electrons gained in going from N_2 to O_2 are placed in the $\pi_{2p_y}^\star$ and $\pi_{2p_z}^\star$ orbitals with parallel spins (see Figure 13.6). Consequently, the O_2 molecule has a bond order of two and is expected to be paramagnetic. These predictions are confirmed by experiments.

As we stated in Section 13.1, valence bond theory does not account for the magnetic properties of the oxygen molecule. To show the two unpaired electrons on O_2, we need to draw another resonance structure (in addition to the one on p. 362) that shows two unpaired electrons:

$$\cdot \ddot{O} - \ddot{O} \cdot$$

This structure, however, is unsatisfactory on at least two counts. First, it implies the presence of a single covalent bond, but experimental evidence provides strong support for believing that there is a double bond in this molecule. Second, it places seven valence electrons around each oxygen atom, a "violation" of the octet rule.

The correct prediction that O_2 is paramagnetic was one of the earlier triumphs of the molecular orbital theory.

The Fluorine Molecule (F₂)

The electron configuration of F is $1s^2 2s^2 2p^5$; thus F_2 has eighteen electrons. The two electrons gained in going from O_2 to F_2 are also placed in the antibonding pi molecular orbitals (see Figure 13.6). We predict that the fluorine molecule has a bond order of one and is diamagnetic; this is borne out by experiments.

The Neon Molecule (Ne₂)

With neon ($1s^2 2s^2 2p^6$), we come to the last element of the second period. Because the two electrons gained in going from F_2 to Ne_2 are placed in the antibonding sigma molecular orbital ($\sigma^*_{2p_x}$), the neon molecule has a bond order of zero and does not exist.

Table 13.1 summarizes the general properties of Li_2, B_2, C_2, N_2, O_2, and F_2.

EXAMPLE 13.1

Write the ground-state electron configuration for the O_2 molecule.

Answer

The O_2 molecule has a total of sixteen electrons. Using the order of increasing energies of the molecular orbitals on p. 368 and referring to Table 13.1, we write the ground-state electron configuration of O_2 as

$$(\sigma_{1s})^2 (\sigma^*_{1s})^2 (\sigma_{2s})^2 (\sigma^*_{2s})^2 (\pi_{2p_y})^2 (\pi_{2p_z})^2 (\sigma_{2p_x})^2 (\pi^*_{2p_y})^1 (\pi^*_{2p_z})^1$$

The magnetic properties of O_2 had been known for some time; however, the origin of its paramagnetism was not understood.

TABLE 13.1 Properties of Homonuclear Diatomic Molecules of the Second-Period Elements†

	Li_2	B_2	C_2	N_2	O_2	F_2
$\sigma^\star_{2p_x}$	☐	☐	☐	☐	☐	☐
$\pi^\star_{2p_y},\ \pi^\star_{2p_z}$	☐ ☐	☐ ☐	☐ ☐	☐ ☐	↑ | ↑	↑↓ | ↑↓
σ_{2p_x}	☐	☐	☐	↑↓	↑↓	↑↓
$\pi_{2p_y},\ \pi_{2p_z}$	☐ ☐	↑ | ↑	↑↓ | ↑↓	↑↓ | ↑↓	↑↓ | ↑↓	↑↓ | ↑↓
$\sigma^\star_{2s}$	☐	↑↓	↑↓	↑↓	↑↓	↑↓
σ_{2s}	↑↓	↑↓	↑↓	↑↓	↑↓	↑↓
Bond order	1	1	2	3	2	1
Bond length (pm)	267	159	131	110	121	142
Bond dissociation energy (kJ/mol)	104.6	288.7	627.6	941.4	498.7	150.6
Magnetic properties	Diamagnetic	Paramagnetic	Diamagnetic	Diamagnetic	Paramagnetic	Diamagnetic

† For simplicity the σ_{1s} and $\sigma^\star_{1s}$ orbitals are omitted. These two orbitals hold a total of four electrons.

EXAMPLE 13.2

The N_2^+ ion can be prepared by bombarding the N_2 molecule with fast-moving electrons. Predict the following properties of N_2^+: (a) electron configuration, (b) bond order, (c) magnetic character, and (d) bond length relative to the bond length of N_2 (is it longer or shorter?).

Answer

(a) Since N_2^+ has one fewer electron than N_2, its electron configuration is (see Table 13.1)

$$(\sigma_{1s})^2(\sigma^\star_{1s})^2(\sigma_{2s})^2(\sigma^\star_{2s})^2(\pi_{2p_y})^2(\pi_{2p_z})^2(\sigma_{2p_x})^1$$

(b) The bond order of N_2^+ is found by using Equation (13.1):

$$\text{bond order} = \tfrac{1}{2}(9 - 4) = 2.5$$

(c) N_2^+ has one unpaired electron, so it is paramagnetic.
(d) Since the electrons in the bonding molecular orbitals are responsible for holding the atoms together, N_2^+ should have a weaker and, therefore, longer bond than N_2. (In fact, the bond length of N_2^+ is 112 pm, compared to 110 pm for N_2.)

Similar examples: Problems 13.7, 13.8, 13.9, 13.10.

13.5 DELOCALIZED MOLECULAR ORBITALS

We have so far discussed chemical bonding in terms of electron pairs. However, the properties of a molecule often cannot be explained accurately by a single structure. A case in point is the O_3 molecule, which was discussed in Section 11.5. There we overcame the dilemma by introducing the concept of resonance. In this section we will tackle the problem in another way—by applying the molecular orbital approach. As in Section 11.5, the benzene molecule and the nitrate ion will be used as examples.

The Benzene Molecule

The benzene molecule (C_6H_6) is a planar hexagonal molecule in which the carbon atoms are situated at the corners. All the carbon–carbon bonds (and the carbon–hydrogen bonds) are equal in length and strength, and the CCC and HCC angles are equal to 120°. Therefore, each carbon atom is sp^2-hybridized; it forms three sigma bonds with two adjacent carbon atoms and a hydrogen atom (Figure 13.7). This arrangement leaves an unhybridized $2p_z$ orbital on each carbon atom, perpendicular to the plane of the benzene molecule, or the *benzene ring*, as it is often called. So far the description resembles the configuration in ethylene (C_2H_4) as discussed in Section 12.6, except that there are now six unhybridized $2p_z$ orbitals in a cyclic arrangement.

Because of their similar shape and orientation, each $2p_z$ orbital overlaps with two others, one on each adjacent carbon atom. According to the rules listed in Section 13.3, the interaction of six $2p_z$ orbitals leads to the formation of six pi molecular orbitals, of which three are bonding and three are antibonding. A benzene molecule in the ground state therefore has six electrons in the three pi bonding molecular orbitals (two electrons with paired spins in each orbital).

Figure 13.8 shows a bonding pi molecular orbital in benzene. A characteristic of these pi molecular orbitals is that *they are not confined between two adjacent bonding atoms,* as they are in the case of ethylene, *but actually extend over three or more atoms.* Therefore, electrons residing in any of these orbitals are free to move around the benzene ring. Molecular orbitals of this type are called ***delocalized molecular orbitals.*** For this reason, the structure of benzene is sometimes represented as

in which the circle indicates that the pi bonds between carbon atoms are not confined to individual pairs of atoms; rather, the pi electron densities are evenly distributed throughout the benzene molecule. The carbon and hydrogen atoms are not shown in the simplified diagram.

We can now describe each carbon-to-carbon linkage as containing a sigma bond and a "partial" pi bond. The bond order between any two adjacent carbon atoms is therefore somewhere between one and two. Thus molecular orbital theory offers an alternative description of the benzene molecule as opposed to the resonance approach, which is based on valence bond theory.

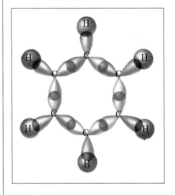

FIGURE 13.7 *Top view showing the sigma bond framework in the benzene molecule. Each carbon atom is sp^2-hybridized and forms sigma bonds with two adjacent carbon atoms and another sigma bond with a hydrogen atom.*

The resonance structures of benzene are shown on p. 313.

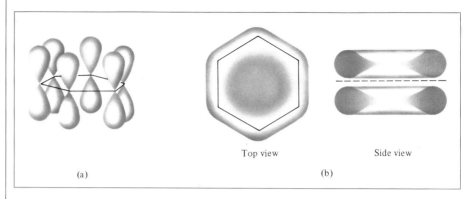

Top view   Side view

(a)   (b)

FIGURE 13.8 *(a) The six $2p_z$ orbitals on the carbon atoms in benzene. (b) The top view and side view of a delocalized molecular orbital formed by the overlap of the $2p_z$ orbitals. The delocalized molecular orbital possesses pi symmetry and lies above and below the plane of the benzene ring. Actually, these $2p_z$ orbitals can combine in six different ways to yield three bonding molecular orbitals and three antibonding molecular orbitals. The one shown in (b) is energetically the most stable.*

The Nitrate Ion

Cyclic compounds like benzene are not the only ones with delocalized molecular orbitals. Let's look at bonding in the nitrate ion (NO_3^-). The planar structure of the nitrate ion can be understood by assuming the nitrogen atom to be sp^2-hybridized. The N atom forms sigma bonds with three O atoms. Thus the unhybridized $2p_z$ orbital of the N atom can overlap simultaneously with the $2p_z$ orbitals of the three O atoms (Figure 13.9). The result is a delocalized molecular orbital that extends over all four nuclei in such a way that the electron densities (and hence the bond orders) in the nitrogen-to-oxygen bonds are all the same. Molecular orbital theory therefore provides an alternative explanation of the properties of the nitrate ion as compared to the resonance structures of the ion shown on p. 313.

We should note that molecules with delocalized molecular orbitals are

FIGURE 13.9 *The bonding scheme in the nitrate ion. The nitrogen atom forms three sigma bonds with the three oxygen atoms. In addition, the $2p_z$ orbitals of the nitrogen and oxygen atoms overlap to form delocalized molecular orbitals, so that there is also a partial pi bond between the nitrogen atom and each of the three oxygen atoms.*

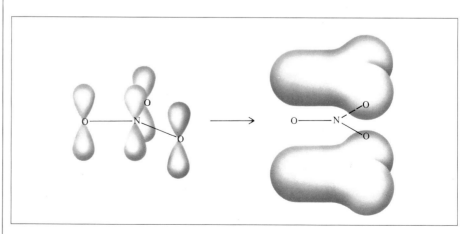

generally more stable than those containing molecular orbitals extending over only two atoms. For example, the benzene molecule, which contains delocalized molecular orbitals, is chemically less reactive (and hence more stable) than molecules containing "localized" C=C bonds, such as the ethylene molecule.

SUMMARY

1. Molecular orbital theory describes bonding in terms of the combination and rearrangement of atomic orbitals to form orbitals that are associated with the molecule as a whole.
2. Bonding molecular orbitals increase electron density between the nuclei and are lower in energy than individual atomic orbitals. In antibonding molecular orbitals the electron density decreases to zero between the nuclei, and the energy level is higher than that of the individual atomic orbitals.
3. We write electron configurations for molecular orbitals as we do for atomic orbitals, filling in electrons in the order of increasing energy levels. The number of molecular orbitals always equals the number of atomic orbitals that were combined. In filling the molecular orbitals, the Pauli exclusion principle and Hund's rule are followed.
4. Molecules are stable if the number of electrons in bonding molecular orbitals is greater than that in antibonding molecular orbitals.
5. Delocalized molecular orbitals, in which electrons are free to move around a whole molecule or group of atoms, are formed by electrons in p orbitals of adjacent atoms. Delocalized molecular orbitals are an alternative to resonance structures in explaining observed molecular properties.

KEY WORDS

Antibonding molecular orbital, p. 362
Bond order, p. 366
Bonding molecular orbital, p. 362
Delocalized molecular orbital, p. 373

Molecular orbital, p. 362
Pi molecular orbital, p. 364
Sigma molecular orbital, p. 363

PROBLEMS

More challenging problems are marked with an asterisk.

Molecular Orbital Theory

13.1 Explain in molecular orbital terms the changes in H—H internuclear distance that occur as the molecule H_2 is ionized first to H_2^+ and then to H_2^{2+}.

13.2 The formation of the H_2 molecule from two H atoms is an energetically favorable process. Yet statistically there is less than a 100 percent chance that any two H atoms will undergo the reaction. Apart from energy consideration, how would you account for this observation simply from the electron spins in the two H atoms?

13.3 Explain the significance of bond order. Can bond order be used to quantitatively compare the strengths of chemical bonds?

13.4 Draw a molecular orbital energy level diagram for each of the following species: He_2, HHe, He_2^+. Compare their relative stabilities in terms of bond orders. (You can treat HHe as a diatomic molecule with three electrons.)

13.5 Arrange the following species in order of increasing stability: Li_2, Li_2^+, Li_2^-. Justify your choice with a molecular orbital energy level diagram.

13.6 Use molecular orbital theory to explain why the Be_2 molecule does not exist.

13.7 Explain why the bond order of N_2 is greater than that of N_2^+, but the bond order of O_2 is less than that of O_2^+.

13.8 Use molecular orbital theory to compare the relative stabilities of F_2 and F_2^+.

*13.9 Compare the relative stability of the following species and indicate their magnetic properties (that is, diamagnetic or paramagnetic): O_2, O_2^+, O_2^- (superoxide ion), O_2^{2-} (peroxide ion).

13.10 Which of these species has a longer bond, B_2 or B_2^+? Explain.

13.11 Acetylene (C_2H_2) has a tendency to lose two protons (H^+) and form the carbide ion (C_2^{2-}), which is present in a number of ionic compounds, such as CaC_2 and MgC_2. Describe the bonding scheme in the C_2^{2-} ion in terms of molecular orbital theory. Compare the bond order in C_2^{2-} with that in C_2.

13.12 Compare the Lewis and molecular orbital theory treatments of the oxygen molecule.

*13.13 A single bond is almost always a sigma bond and a double bond is almost always made up of a sigma bond and a pi bond. There are very few exceptions to this rule. Use the B_2 and C_2 molecules to show that they are examples of the exceptions.

13.14 Define the following terms: bonding molecular orbital, antibonding molecular orbital, pi molecular orbital, sigma molecular orbital, bond order.

Delocalized Molecular Orbitals

*13.15 Both ethylene (C_2H_4) and benzene (C_6H_6) contain the C=C bond. The reactivity of ethylene is greater than that of benzene. For example, ethylene readily reacts with molecular bromine, whereas benzene is normally quite inert toward molecular bromine and many other compounds. Explain this difference in reactivity.

13.16 Explain why the symbol on the left is a better representation for benzene molecules than that on the right:

*13.17 Determine which of these molecules has a more delocalized orbital and justify your choice:

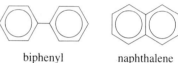

biphenyl   naphthalene

(*Hint:* Both molecules contain two benzene rings. In naphthalene, the two rings are fused together, whereas in biphenyl the two rings are joined by a single bond. In the latter case, the two rings can rotate about the single bond.)

*13.18 Nitryl fluoride (FNO_2) is very reactive chemically. The fluorine and oxygen atoms are bonded to the nitrogen atom. (a) Write a Lewis formula for FNO_2. (b) Indicate the hybrid orbitals used by the nitrogen atom. (c) Describe the bonding in terms of the molecular orbital theory. Where would you expect delocalized molecular orbitals?

*13.19 Describe the bonding in the carbonate ion (CO_3^{2-}) in terms of delocalized molecular orbitals.

*13.20 What is the state of hybridization of the central O atom in O_3? Describe the bonding in O_3 in terms of delocalized molecular orbitals.

14
LIQUIDS AND SOLIDS

All gases can be transformed, under suitable conditions, to the liquid or solid state. Thus it would seem plausible that an extension of the kinetic molecular theory of gases would help account for the properties of these two condensed states. In fact, this is the case—with one major qualification.

Ideal gas behavior was explained assuming no intermolecular forces were present. Gas molecules interacted with each other only through random molecular collisions; at other times a given gas molecule was assumed to be uninfluenced by any other molecules in the system. By contrast, the units that compose liquids and solids must experience attractive forces to account for their close proximity.

In this chapter we will examine the nature of these between-unit forces, and see how these microscopic interactions help account for the large-scale, macroscopic properties of liquids and solids.

14.1   THE KINETIC MOLECULAR THEORY OF LIQUIDS AND SOLIDS

The kinetic molecular theory was used in Chapter 6 to explain the behavior of gases. You will recall that a gaseous system is a collection of molecules in constant, random motion. The distances between molecules are so great (compared to their diameters) that at ordinary temperatures and pressures (say, 25°C and 1 atm), there is no appreciable interaction among the molecules. This rather simple description explains several characteristic properties of gases. Because there is a great deal of empty space in a gas, that is, space not occupied by molecules, gases can be readily compressed. The lack of strong forces *among* molecules allows a gas to expand to the volume of its container. The large amount of empty space also explains why gases have very low densities under normal conditions.

Liquids and solids are quite a different story. The major difference between the condensed state (liquids or solids) and the gaseous state is in the distance between molecules. In a liquid the molecules are held together more closely, so that there is very little empty space. Thus liquids are much more difficult to compress than gases. Molecules in the liquid are held by one or more types of attractive forces, which will be discussed in the next section. A liquid also has a definite volume, since molecules in a liquid do not break away from the attractive forces. The molecules do, however, move past one another, and so a liquid can flow, can be poured, and assumes the shape of its container.

In a solid, molecules are held rigidly in position with virtually no freedom of motion. Many solids are characterized by long-range order, that is, the molecules are arranged in regular configurations in three dimensions. The amount of empty space in a solid is even less than that in a liquid. Thus solids are almost incompressible and possess definite shape and volume. With very few exceptions (water being the most important), the density of the solid is higher than that of the liquid for a given substance. Table 14.1 summarizes some of the characteristic properties of the three states of matter, and Figure 14.1 is a molecular view of the differences among a solid, a liquid, and a gas.

TABLE 14.1   Characteristic Properties of Gases, Liquids, and Solids

State of Matter	Volume/Shape	Density	Compressibility	Motion of Molecules
Gas	Assumes the volume and shape of its container	Low	Very compressible	Very free motion
Liquid	Has a definite volume but assumes the shape of its container	High	Only slightly compressible	Slide past one another freely
Solid	Has a definite volume and shape	High	Virtually incompressible	Vibrate about fixed positions

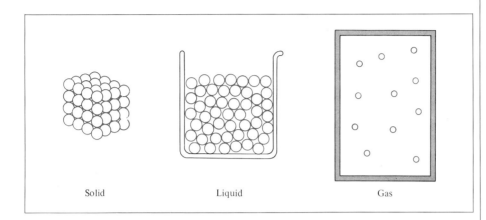

Solid   Liquid   Gas

14.2 INTERMOLECULAR FORCES

As you know, *attractive forces exist among molecules.* These forces, called **intermolecular forces,** are responsible for the nonideal behavior of gases, discussed in Chapter 6. They are also responsible for the existence of the condensed states of matter—liquids and solids. When the temperature of a gas is lowered, the average kinetic energy of its molecules decreases. Eventually, at a sufficiently low temperature, the molecules no longer have enough energy to break away from one another's attraction. At this point, the molecules aggregate to form small drops of liquid. This phenomenon of going from the gaseous to liquid state is known as *condensation.*

In contrast to intermolecular forces are **intramolecular forces,** or *forces that hold atoms together in a molecule.* (Chemical bonding, discussed in Chapters 10 to 13, involves intramolecular forces.) Intramolecular forces are responsible for the stability of individual molecules, whereas intermolecular forces are primarily responsible for the bulk properties of matter (for example, melting point and boiling point).

Generally, intermolecular forces are much weaker than intramolecular forces. Thus it usually requires much less energy to evaporate a liquid than to break the bonds in the molecules. For cxample, the molar heat of vaporization of water at its boiling point is 40.79 kJ/mol, whereas it takes about 930 kJ of energy to break the two O—H bonds in one mole of water molecules. The boiling points of substances often reflect the strength of the intermolecular forces operating among the molecules. At the boiling point, the attraction among molecules must be overcome so that they can enter the vapor phase. If more energy must be supplied to substance A to separate molecules that are held together by stronger intermolecular forces than in substance B, then the boiling point of substance A is higher than that of B.

In order to understand the properties of condensed matter, we must have a good understanding of intermolecular forces, of which there are many different types. Some of them can be explained by Coulomb's law. Quantum mechanics provides an explanation for others. Depending on the physical state of a substance (that is, gas, liquid, or solid) and the nature of the molecules, more than one type of interaction may play a role in the total attraction between molecules, as we will see below.

Dipole—Dipole Forces

Dipole–dipole forces are *forces that act between polar molecules,* that is, between molecules that possess dipole moments (see Section 12.3). Their origin is electrostatic, and they can be understood in terms of Coulomb's law. The larger the dipole moments, the greater the force. Figure 14.2 shows the orientation of polar molecules in a solid. In liquids the molecules are not held as rigidly as in a solid, but they tend to align themselves in such a way that, on the average, the attractive interaction is at a maximum.

Ion—Dipole Forces

Coulomb's law also explains **ion–dipole forces,** which *occur between an ion (either a cation or an anion) and a dipole* (Figure 14.3). The strength of this interaction depends on the charge and size of the ion and the magnitude of the dipole. The charges on cations are generally more concentrated, since cations are usually smaller than anions. Therefore, for equal charges, a cation interacts more strongly with dipoles than an anion does.

Hydration, discussed in Section 5.6, is an example of ion–dipole interaction. In an aqueous NaCl solution, the Na^+ and Cl^- ions are surrounded by water molecules, which have a large dipole moment (1.87 D). In this way the water molecules act as an electrical insulator that keeps the ions apart. This process explains what happens when an ionic compound dissolves in water. Carbon tetrachloride (CCl_4), on the other hand, is a nonpolar molecule, and therefore it lacks the ability to participate in ion–dipole interaction. What we find in practice is that carbon tetrachloride (and other nonpolar liquids) is a poor solvent for ionic compounds.

Dispersion Forces

So far only ionic species and molecules with dipole moments have been discussed. You may be wondering what kind of attractive interaction exists among neutral and nonpolar molecules. To answer this question, we will first consider the arrangement shown in Figure 14.4. If we place an ion or a polar molecule near a neutral atom (or a nonpolar molecule), the electron density of the atom (or molecule) is distorted by the force exerted by the ion or the polar molecule. The resulting dipole in the atom (or molecule) is said to be an **induced dipole** because *the separation of positive and negative charges in the neutral atom (or a nonpolar molecule) is due to the proximity of an ion or a polar molecule.* The attractive interaction between an ion and the induced dipole is called *ion-induced dipole interaction,* and the attractive interaction between a polar molecule and the induced dipole is called *dipole-induced dipole interaction.*

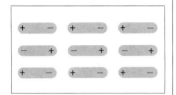

FIGURE 14.2 *A possible arrangement of molecules with a permanent dipole moment in the solid state. The dipoles are aligned for maximum attractive interaction.*

FIGURE 14.3 *Two types of ion–dipole interactions.*

Ionic compounds generally have very low solubilities in nonpolar solvents.

The ease with which a dipole moment can be induced depends not only on the charge on the ion or the strength of the dipole but also on the polarizability of the neutral atom or molecule. ***Polarizability*** is *the ease with which the electron density in the neutral atom (or molecule) can be distorted.* Generally, the larger the number of electrons and the more diffuse the electron cloud in the atom or molecule, the greater its polarizability. By *diffuse cloud* we mean an electron cloud that is spread over an appreciable volume, so that the electrons are not held tightly by the nucleus. In general, *p* orbitals have more diffuse electron clouds than *s* orbitals, and pi molecular orbitals have more diffuse electron clouds than sigma molecular orbitals.

Polarizability gives us the clue to why gases containing neutral atoms or nonpolar molecules (for example, He and N_2) are able to condense. When we say that the helium atom has no dipole moment, we are speaking of an *isolated* helium atom. But in a collection of helium atoms moving around in a container, each atom collides with other atoms many times per second. Since a collision almost always distorts the symmetrical shape of the charge cloud, a *temporary* dipole is created within the atom. This temporary dipole can instantaneously induce dipoles in its nearest neighbors (Figure 14.5). Many such dipoles are created and destroyed every moment, but because their orientations are completely random, the dipole moment of helium averaged over a given period of time is zero. The important point is that at any instant the temporary dipoles attract one another—at low temperatures (and reduced atomic speeds) this attraction will hold the atoms together, causing the helium gas to condense.

A quantum mechanical interpretation of temporary dipoles was provided by Fritz London (1900–1954) in 1930. London showed that the magnitude of this attractive interaction is directly proportional to the polarizability of the atom or molecule. As we might expect, *the attractive forces that arise as a result of temporary dipoles induced in the atoms or molecules,* called ***dispersion forces,*** may be quite weak. This is certainly true of helium, which has a boiling point of only 4.2 K or −269°C. (Note that helium has only two electrons that are tightly held in the 1s orbital. Therefore, the helium atom has a small polarizability.)

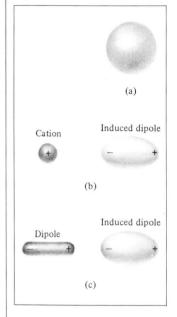

FIGURE 14.4 *(a) Spherical charge distribution in a helium atom. Distortion caused by (b) the approach of a cation, and (c) by the approach of a dipole.*

Strictly speaking, the forces between two nonbonded helium atoms should be called inter*atomic* forces and not inter*molecular* forces. However, for simplicity we use the term "intermolecular forces" to describe forces between two nonbonded species, regardless of whether they are atoms or molecules.

FIGURE 14.5 *Induced dipoles interacting with each other. Such patterns exist only momentarily; new arrangements are formed in the next instant. This type of interaction is responsible for the condensation of nonpolar gases.*

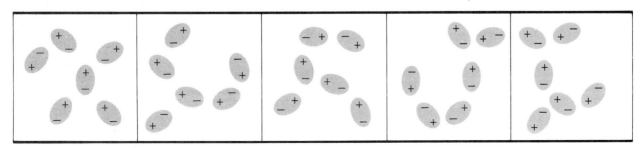

TABLE 14.2 Melting Points of Similar Nonpolar Compounds

Compound	Melting Point (°C)
CH_4	-182.5
C_2H_6	-183.3
CF_4	-183.7
CCl_4	-23.0
CBr_4	90.0
CI_4	171.0

As mentioned earlier, atoms and molecules containing large numbers of electrons, especially those that have p electrons in their valence shells, have large polarizabilities. Consequently, dispersion forces between atoms and between nonpolar molecules can sometimes be stronger than the dipole–dipole forces between polar molecules.

The melting point of a substance, like the boiling point, can be used as an indication of the strength of intermolecular forces. In general, the stronger the intermolecular forces, the more tightly the molecules are held in the solid, and the higher the melting point. Table 14.2 compares the melting points of some substances that consist of nonpolar molecules. Note that the melting point increases also as the number of electrons in the molecule increases. The only attractive forces present in these systems are the dispersion forces.

For a dramatic illustration, let us compare the melting points of CH_3F $(-141.8°C)$ and CCl_4 $(-23°C)$. Although CH_3F has a dipole moment of 1.8 D, it melts at a much lower temperature than CCl_4, a nonpolar molecule. CCl_4 melts at a higher temperature simply because it contains more electrons. As a result, the dispersion forces among CCl_4 molecules are stronger than the dispersion forces plus the dipole–dipole forces among CH_3F molecules. (Keep in mind that dispersion forces exist among species of any type, whether they are neutral or bear a net charge, and whether they are polar or nonpolar.)

EXAMPLE 14.1

What type(s) of intermolecular forces exist between the following pairs: (a) HBr and H_2S, (b) Cl_2 and CBr_4, (c) I_2 and NO_3^-, (d) NH_3 and C_6H_6?

Answer

(a) Both HBr and H_2S are polar molecules so the forces between them are dipole–dipole forces. In addition, there are also dispersion forces between the molecules. (b) Both Cl_2 and CBr_4 are nonpolar so there are only dispersion forces between these molecules. (c) I_2 is nonpolar so the forces between it and the ion NO_3^- are ion-induced dipole forces and dispersion forces. (d) NH_3 is polar and C_6H_6 is nonpolar. The forces are dipole-induced dipole forces and dispersion forces.

Similar example: Problem 14.16.

van der Waals Forces and van der Waals Radii

The *dipole–dipole, dipole-induced dipole, and dispersion forces* make up what chemists have commonly referred to as **van der Waals forces**, after the Dutch physicist Johannes van der Waals (see Section 6.9). These forces are all attractive in nature and play an important role in determining the physical properties of substances, such as melting points, boiling points, and solubilities.

The distance between molecules (or monatomic species) in a solid or liquid is determined by a balance between the van der Waals forces of attraction and the forces of repulsion between electrons and between nuclei. The **van der Waals radius** is *one-half the distance between two equivalent nonbonded atoms in their most stable arrangement*, that is, when the net attractive forces are at a maximum.

The main difference between the atomic radius and ionic radius (see Sections 9.3 and 10.4), on the one hand, and the van der Waals radius, on the other hand, is that the latter applies only to atoms that are not bonded to each other. For instance, two He atoms will not combine to form a stable He_2 molecule, yet He atoms do attract one another through dispersion forces. In liquid helium, we can imagine that the atoms are drawn to one another until the net attractive forces are at their maximum. If any two neighboring atoms are pushed together more closely, repulsive forces (between the electrons and between the nuclei) cause the system to lose stability. Under the condition of maximum attraction, half the distance measured between any two adjacent He atoms in liquid helium is the van der Waals radius of He atom.

As another example, consider iodine molecules (I_2) in the solid state (Figure 14.6). The I_2 molecules are packed in such a way that the total potential energy of the system is at a minimum (which corresponds to maximum net attraction). In this case, half the distance between two adjacent iodine atoms in two *different* I_2 molecules is the van der Waals radius of an iodine atom. Figure 14.7 shows the van der Waals radii of selected atoms.

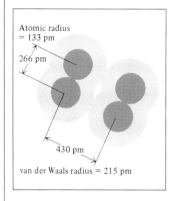

Atomic radius = 133 pm

266 pm

430 pm

van der Waals radius = 215 pm

FIGURE 14.6 *The relationship between van der Waals radius and atomic radius for I_2. In solid I_2 the molecules pack together so that the shortest distance between iodine nuclei in adjacent I_2 molecules is 430 pm. The van der Waals radius of iodine is defined as half this distance, or 215 pm. On the other hand, the atomic radius of iodine is half the distance between two iodine nuclei in the same molecule, which is half of 266 pm, or 133 pm.*

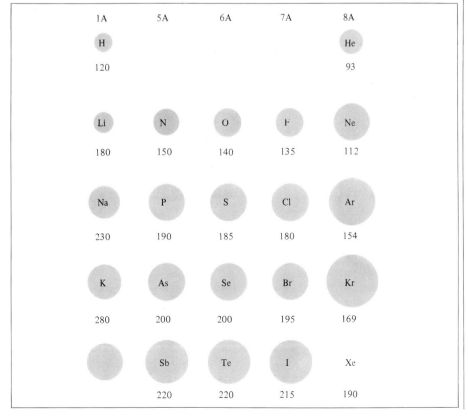

1A	5A	6A	7A	8A
H				He
120				93
Li	N	O	F	Ne
180	150	140	135	112
Na	P	S	Cl	Ar
230	190	185	180	154
K	As	Se	Br	Kr
280	200	200	195	169
	Sb	Te	I	Xe
	220	220	215	190

FIGURE 14.7 *Van der Waals radii (in pm) of a number of familiar atoms.*

The Hydrogen Bond

Note that these atoms possess at least one lone pair that can interact with the hydrogen atom in hydrogen bonding.

The ***hydrogen bond*** is *a special type of dipole–dipole interaction between the hydrogen atom in a polar bond such as O—H or N—H, and the electronegative atoms O, N, or F.* This interaction is written

A—H---B or A—H---A

A and B represent O, N, or F; A—H is one molecule and B is a part of another molecule, and the dashed line represents the hydrogen bond. The three atoms usually lie along a straight line, but the angle AHB (or AHA) can deviate as much as 30° from linearity.

The average energy of a hydrogen bond is quite large for a dipole–dipole interaction (up to 40 kJ/mol). Thus, hydrogen bonds play a very important role in determining the structures and properties of many compounds. Figure 14.8 shows several examples of hydrogen bonding.

Early evidence for hydrogen bonding was seen in the study of boiling points of compounds. Normally, the boiling points of a series of similar compounds containing elements in the same group increase with increasing molar mass. But, as Figure 14.9 shows, some exceptions were noticed for the hydrogen compounds of Groups 5A, 6A, and 7A elements. In each of these series, the lightest compound (NH_3, H_2O, HF) has the *highest* boiling point, contrary to our expectations based on molar mass. This study and other related observations led chemists to postulate the existence of the hydrogen bond. In solid HF, for example, the molecules do not exist as individual units; instead, they form long zig-zag chains:

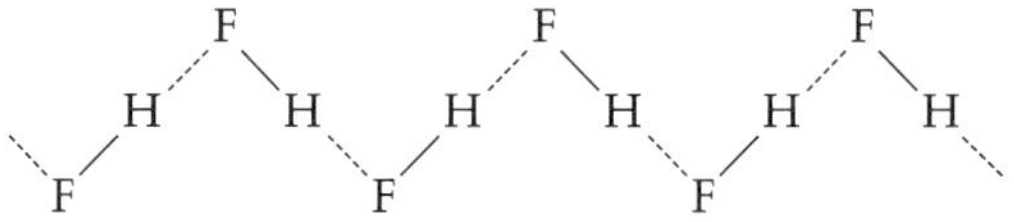

This is an example of intermolecular hydrogen bonding. In the liquid state the zig-zag chains are broken, but the molecules are still extensively hy-

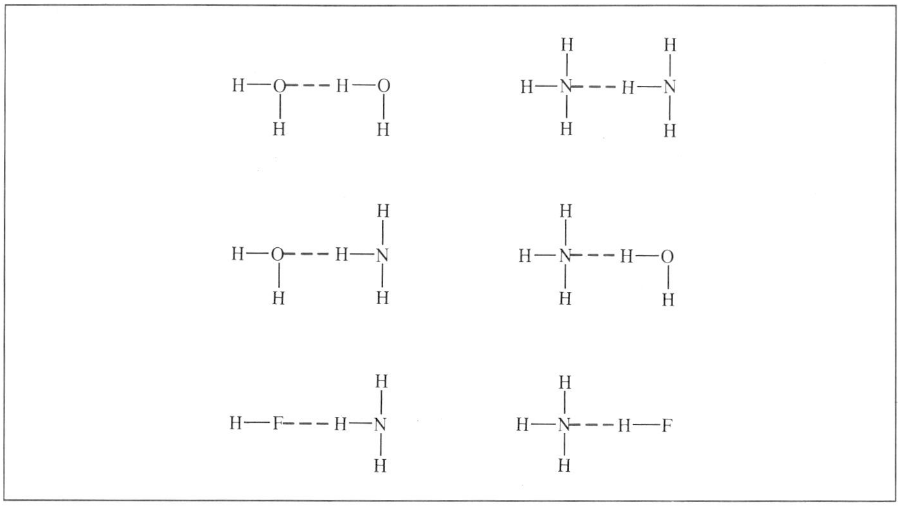

FIGURE 14.8 *Examples of hydrogen bonding. Solid lines represent covalent bonds and dotted lines represent hydrogen bonds.*

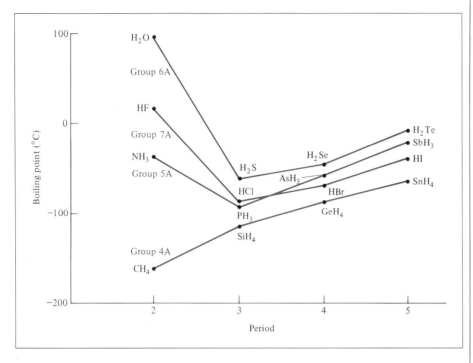

FIGURE 14.9 *Boiling points of the hydrogen compounds of Group 4A, 5A, 6A, and 7A elements. Although normally the boiling point should increase as we move down a group, we see that three compounds (NH₃, H₂O, and HF) behave differently. The anomaly can now be explained in terms of intermolecular hydrogen bonding.*

drogen-bonded to one another. Molecules held together by hydrogen bonding are difficult to break apart, so that liquid HF has an unusually high boiling point.

The strength of the hydrogen bond is determined by the coulombic interaction between the lone pair electrons of the electronegative atom and the hydrogen nucleus. It may seem odd at first that the boiling point of HF is lower than that of water. Fluorine is more electronegative than oxygen, and so we would expect a stronger hydrogen bond to exist in liquid HF than in H_2O. But H_2O is unique in that each molecule is capable of forming *four* intermolecular hydrogen bonds, and the H_2O molecules are therefore held together more strongly. We will return to this very important property of water in the next section.

EXAMPLE 14.2

Which of the following can form hydrogen bonds with water: CH_3OCH_3, CH_4, F^-, HCOOH, and Na^+?

Answer

There are no electronegative elements in either CH_4 or Na^+. Therefore, only CH_3OCH_3, F^-, and HCOOH can form hydrogen bonds with water (Figure 14.10).

Similar example: Problem 14.11.

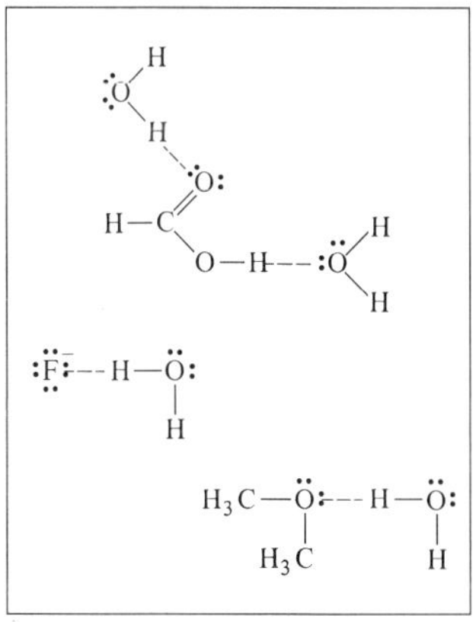

The various types of intermolecular forces discussed so far are all attractive in nature. Keep in mind, though, that molecules also exert repulsive

FIGURE 14.10 *Hydrogen bond formation between various solutes and water molecules.*

forces on one another. Thus when two molecules are brought into contact with each other the repulsion between the electrons and between the nuclei in the molecules will come into play. The magnitude of the repulsive force rises very steeply as the distance separating the molecules in a condensed state is shortened. This is the reason that liquids and solids are so hard to compress. In these states, the molecules are already in close contact with one another, so that they greatly resist being compressed.

14.3   THE LIQUID STATE

Now that you are familiar with intermolecular forces, we can look at the properties of substances in condensed states, which are largely determined by those forces. We will start with the liquid state. A great many interesting and important chemical reactions occur in water and other liquid solvents, as we will discover in subsequent chapters. In this section we will look at two phenomena associated with liquids: surface tension and viscosity. We will also discuss the structure and properties of water.

Surface Tension

We have seen that one property of liquids is their tendency to assume the shapes of their containers. Why, then, does water bead up on a newly waxed car instead of forming a sheet over it? The answer to this question lies in intermolecular forces.

Molecules within a liquid are pulled in all directions; there is no tendency for them to be pulled in any one way. However, molecules at the surface are pulled only downward and sideways but not upward (Figure 14.11). These attractions thus tend to pull the molecules into the liquid and cause the surface to behave as if it were tightened like an elastic film. Therefore, since there is little or no attraction between polar water molecules and the wax molecules (which are essentially nonpolar) on a freshly waxed car, a drop of water will assume the shape of a small round bead.

For any given quantity of liquid, the sphere offers the minimum surface area.

FIGURE 14.11 *Intermolecular forces acting on a molecule in the surface layer of a liquid and in the interior region of the liquid.*

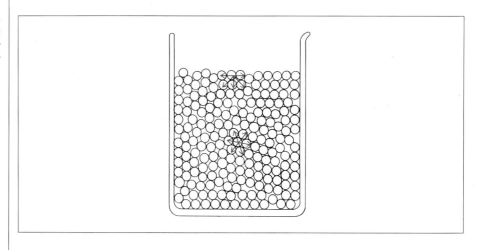

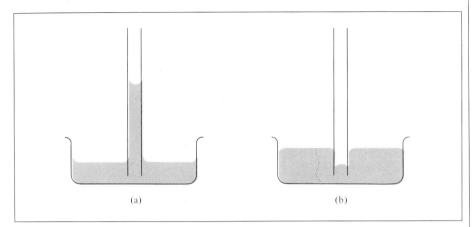

FIGURE 14.12 *(a) When adhesion is greater than cohesion, the liquid (for example, water) rises in the capillary tube. (b) When cohesion is greater than adhesion, a depression of the liquid (for example, mercury) in the capillary tube results. Note that the meniscus in the tube of water is concave, or rounded downward, whereas that in the tube of mercury is convex, or rounded upward.*

A measure of the elasticlike force existing in the surface of a liquid is surface tension. The **surface tension** of a liquid is *the amount of energy required to stretch or increase the surface by unit area.* As expected, liquids containing molecules that possess strong intermolecular forces also have high surface tensions. For example, because of hydrogen bonding, water has a considerably greater surface tension than most common liquids.

Another way that surface tension manifests itself is in *capillary action.* Figure 14.12(a) shows water rising spontaneously in a capillary tube. A thin film of water adheres to the wall of the glass tube. The surface tension of water causes this film to contract and, as it does, it pulls the water up the tube. Two types of forces bring about capillary action. One is *the intermolecular attraction between like molecules* (in this case, the water molecules), called **cohesion.** The other, which is called **adhesion,** is *an attraction between unlike molecules,* such as those in water and in the walls of a glass tube. If adhesion is stronger than cohesion, as it is in Figure 14.12(a), the contents of the tube will be pulled upward along the walls. This process continues until the adhesive force is balanced by the weight of the water in the tube. This action is by no means universal among liquids, as Figure 14.12(b) shows. Cohesion in mercury is greater than the adhesion between mercury and glass, so that a depression in the liquid level actually occurs when a capillary tube is dipped into mercury.

Viscosity

The expression "slow as molasses in January" owes its truth to another physical property of liquids, called viscosity. **Viscosity** is *a measure of a fluid's resistance to flow.* The greater the viscosity, the more slowly the liquid flows. The viscosity of a liquid always decreases as temperature increases; thus hot molasses flows much faster than cold molasses. Lubricating motor oils are rated according to their viscosities, and motorists are usually advised to use a high viscosity oil in summer and a low viscosity oil in winter. Because the average operating temperature of a car's engine can be quite high in summer, starting with a more viscous lubricant ensures

TABLE 14.3   Viscosity of Some Common Liquids at 20°C

Liquid	Viscosity (N s/m²)*
Acetone (C_3H_6O)	0.000316
Benzene (C_6H_6)	0.000625
Carbon tetrachloride (CCl_4)	0.000969
Ethanol (C_2H_5OH)	0.001200
Ethyl ether ($C_2H_5OC_2H_5$)	0.000233
Glycerol ($C_3H_8O_3$)	1.49
Mercury (Hg)	0.001554
Water (H_2O)	0.00101
Blood	0.004

* In SI units, viscosities are expressed as newton-second per meter squared.

that the oil will not become too thin. In cold weather, the temperature of the engine will not get quite as high, and a more viscous oil would be less effective.

Liquids that have strong intermolecular forces have higher viscosities than those that have weak intermolecular forces (Table 14.3). Water has a higher viscosity than many other liquids because of its ability to form hydrogen bonds. Interestingly, the viscosity of glycerol is significantly higher than that of all other liquids. Glycerol has the structure

$$CH_2-OH$$
$$|$$
$$CH-OH$$
$$|$$
$$CH_2-OH$$

Like water, glycerol can form hydrogen bonds. We see that each glycerol molecule has three —OH groups that can participate in hydrogen bonding with other glycerol molecules. Furthermore, because of their shape the molecules have a great tendency to become entangled rather than to slip past one another as the molecules in less viscous liquids do. These interactions contribute to its high viscosity.

Glycerol is a clear, odorless, syrupy liquid used in explosives, ink, and lubricants, among other things.

The Structure and Properties of Water

Water is so common a substance that we often overlook its unique nature. All life processes involve water. Water is an excellent solvent for many ionic compounds, as well as for other substances capable of forming hydrogen bonds with water. Because of strong intermolecular hydrogen bonding, water has a high specific heat (see Table 5.3). Thus, water can absorb a substantial amount of heat while its temperature rises only slightly. The converse is also true: Water can give off much heat with only a slight decrease in its temperature. For this reason, the huge quantities of water that are present in our lakes and oceans can effectively moderate the climate of adjacent land areas by absorbing heat in the summer and giving off heat in the winter, with only small changes in the temperature of the body of water.

If water did not possess the ability to form hydrogen bonds, it would be a gas at room temperature.

The most striking property of water is that the solid form is less dense than the liquid form: An ice cube floats at the surface of water in a glass. This is a virtually unique property. The density of almost all other substances is greater in the solid state than in the liquid state. The fact that ice is less dense than water has a profound ecological significance. Consider, for example, the temperature changes in a lake that is in a country with a cold climate. As the temperature of the water near the surface is lowered, its density increases. The colder, denser water sinks toward the bottom, while the warmer water, which is less dense, rises to the top. This normal convection motion continues until the temperature throughout the body of water reaches 4°C. Below this temperature, the density of water decreases with decreasing temperature (Figure 14.13), so that it no longer sinks. On further cooling, the water begins to freeze at the surface. The ice layer formed does not sink; it even acts as a thermal insulator for the water below it. If ice were denser than water, it would sink to the bottom of the lake and the water would freeze upward; eventually it would freeze solid. Most living organisms in the lake could not survive. Fortunately, because of the unique properties of water, this does not happen.

To understand why water is different, we have to examine the electronic structure in the H_2O molecule. A water molecule is made up of one oxygen atom joined by two covalent bonds to two hydrogen atoms. As we saw in Chapter 11, there are two pairs of nonbonding electrons, or two lone pairs, on the oxygen atom:

Although water plays an important role in chemical, biological, and ecological systems, the structure of its liquid form is not completely understood.

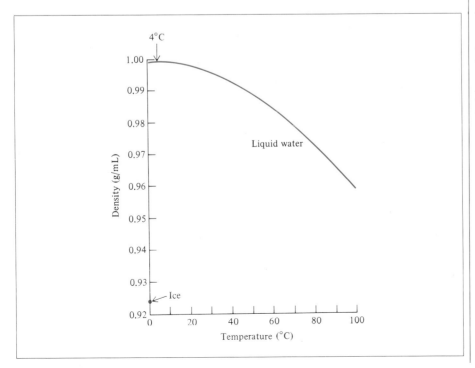

FIGURE 14.13 *Plot of density versus temperature for liquid water. Note the large difference between the density of ice and that of liquid water at 0°C. The maximum density of water is reached at 4°C.*

The O atom in water is *sp*³-hybridized. The four H atoms are tetrahedrally arranged about the O atom.

Although many compounds are capable of forming intermolecular hydrogen bonds, there is a significant difference between H_2O and other polar molecules such as NH_3 and HF. In water, the number of hydrogen bonds around each oxygen atom is equal to two, the same as the number of lone electron pairs on the oxygen atom. Thus, water molecules are joined together in an extensive three-dimensional network in which each oxygen atom is bonded to four hydrogen atoms by two covalent bonds and two hydrogen bonds. This equality in the number of hydrogen atoms and lone pairs is *not a* characteristic of NH_3 or HF or, for that matter, any other molecule capable of forming hydrogen bonds. Consequently, these other molecules can form rings or chains, but not three-dimensional structures.

The highly ordered three-dimensional structure of ice (Figure 14.14) prevents the molecules from getting too close to one another. But consider what happens when ice is heated and melts. There is evidence that the three-dimensional structure remains largely intact, although the bonds may become bent and distorted. At the melting point, a small number of water molecules possess sufficient kinetic energy to pull free from the intermolecular hydrogen bonds. These molecules become trapped in the cavities of the three-dimensional structure. In other words, there are more molecules per unit volume in liquid water than in ice. Thus, since density = mass/volume, the density of water is greater than that of ice. With further heating, more water molecules are released from intermolecular hydrogen bonding, so the density tends to increase with temperature. Of course, at the same time, water expands as it is heated, and thus its density begins to decrease. These two processes—the trapping of free water molecules in cavities and

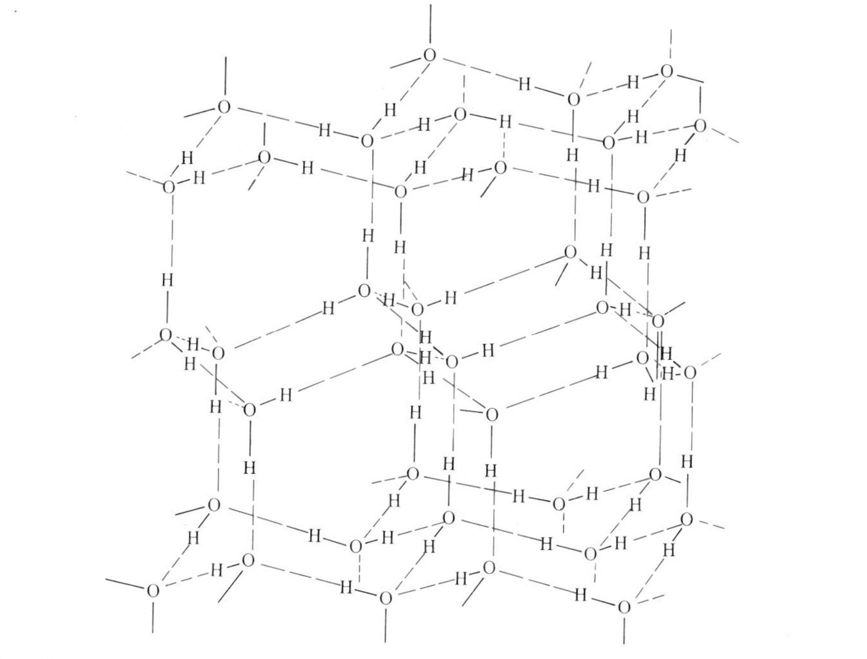

FIGURE 14.14 *The three-dimensional structure of ice. Each O atom is bonded to four H atoms. The covalent bonds are shown by short solid lines and the weaker hydrogen bonds by long dashed lines between O and H. The empty space in the structure accounts for ice's low density.*

thermal expansion—act in opposite directions. From 0°C to 4°C, the trapping prevails and water becomes progressively denser. Beyond 4°C, however, thermal expansion predominates and the density of water decreases with increasing temperature (see Figure 14.13).

Having discussed the liquid state we can now take a look at solids. The next two sections will cover the properties of solids and the final section of the chapter will consider interconversion of the three states of matter.

14.4 CRYSTALLINE SOLIDS

Solids can be divided into two categories: crystalline solids and amorphous solids. A **crystalline solid,** such as ice or sodium chloride, *possesses rigid and long-range order; its atoms, molecules, or ions occupy specific positions. The center of each of the positions is called a* **lattice point,** *and the geometrical order of these lattice points is called the* **crystal structure.** The arrangement of atoms, molecules, or ions in a crystalline solid is such that the net attractive intermolecular forces are at their maximum.

The basic repeating unit of the arrangement of atoms, molecules, or ions in a crystalline solid is a **unit cell.** Figure 14.15 shows a unit cell and its extension, and Figure 14.16 shows the fourteen different types of unit cells from which all crystals can be generated. Table 14.4 gives the angles and shows how the cell edge lengths relate to each other, in the seven simple units (cubic, orthorhombic, triclinic, rhombohedral, hexagonal, monoclinic, and tetragonal). Any of these unit cells, when repeated in space, forms the lattice structure characteristic of a crystalline solid.

Consider, for example, the cubic unit cell. For simplicity, we can assume that each lattice point is occupied by an atom. The geometry of the cubic unit cell is particularly simple, since all sides and all angles are equal. As Figure 14.17 shows, the location of the atoms determines whether we call a cubic unit cell a *simple cubic cell* (scc), a *body-centered cubic cell* (bcc), or a *face-centered cubic cell* (fcc). Because every unit cell in a crystalline solid is adjacent to other unit cells, many of the atoms are shared by neighboring unit cells. For example, in all types of cubic cells, each corner

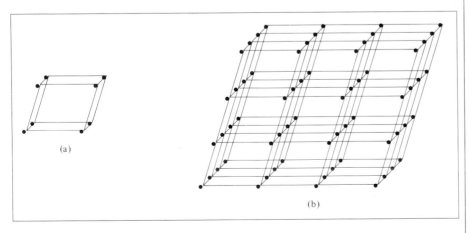

(a)

(b)

FIGURE 14.15 *(a) A unit cell and (b) its extension into three dimensions. The black spheres represent either atoms or molecules.*

FIGURE 14.16 *The four-teen different types of unit cells from which all crystals can be generated. The sides and angles in the simple cubic unit cell are named (a, b, c, and α, β, γ) as in Table 14.4.*

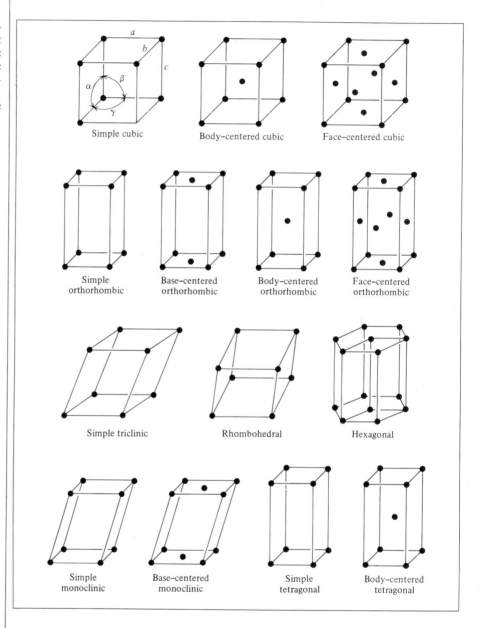

TABLE 14.4   The Seven Simple Unit Cells

Type	Edge Length*	Angle*
Cubic	$a = b = c$	$\alpha = \beta = \gamma = 90°$
Orthorhombic	$a \neq b \neq c$	$\alpha = \beta = \gamma = 90°$
Triclinic	$a \neq b \neq c$	$\alpha \neq \beta \neq \gamma \neq 90°$
Rhombohedral	$a = b = c$	$\alpha = \beta = \gamma \neq 90°$
Hexagonal	$a = b \neq c$	$\alpha = \beta = 90°, \gamma = 120°$
Monoclinic	$a \neq b \neq c$	$\alpha = \gamma = 90°, \beta \neq 90°$
Tetragonal	$a = b \neq c$	$\alpha = \beta = \gamma = 90°$

* Defined in Figure 14.16 (see simple cubic).

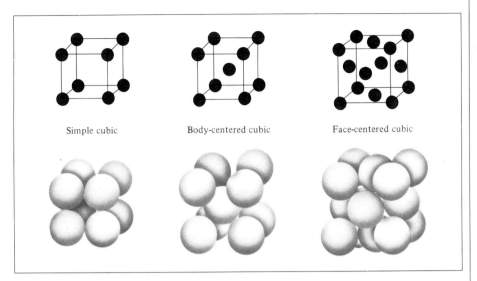

FIGURE 14.17 *Three types of cubic cells. In reality, the spheres representing atoms, molecules, or ions are in contact with one another in these cubic cells.*

Simple cubic   Body-centered cubic   Face-centered cubic

atom is shared by eight unit cells [Figure 14.18(a)]; a face-centered atom is shared by two unit cells [Figure 14.18(b)].

Figure 14.16 may seem to suggest that atoms in a crystalline solid are confined within small volumes. Actually, each atom occupies appreciable volume and is in contact with its neighbors (see Figure 14.17). The forces responsible for the stability of any crystal can be attractive ionic forces, covalent bonds, van der Waals forces, hydrogen bonds, or a combination of some of these forces.

Packing Spheres

We can understand the general geometric requirements for crystal formation by considering the different ways of packing a number of identical spheres (Ping-Pong balls, for example) to form an ordered three-dimensional

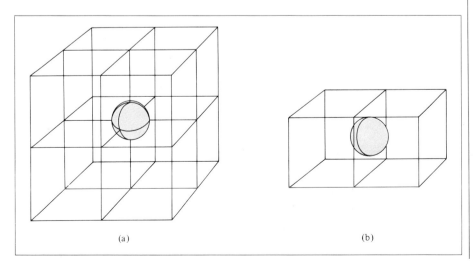

(a)   (b)

FIGURE 14.18 *(a) A corner atom in any cubic cell is shared among eight unit cells; (b) a face-centered atom in a cubic cell is shared between two unit cells.*

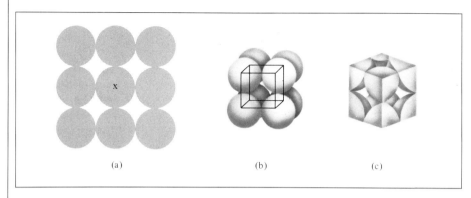

FIGURE 14.19 *Arrangement of identical spheres to give a simple cubic cell. (a) Top view of one layer of spheres. (b) Definition of a simple cubic cell. (c) Since each sphere is shared by eight unit cells and there are eight corners in a cube, there will be the equivalent of one complete sphere inside a simple cubic unit cell.*

structure. In the simplest case, a layer of spheres can be arranged as shown in Figure 14.19(a). The three-dimensional structure can be generated by placing a layer above and below this layer in such a way that spheres in one layer are directly over the spheres in the layer below it. This procedure can be extended to generate many, many layers, as in the case of a crystal. Focusing on the sphere marked with "x" we see that it is in contact with four spheres in its own layer, one sphere in the layer above, and one sphere in the layer below. Thus each sphere in this arrangement is said to have a **coordination number** of 6 because it has six immediate neighbors. The coordination number is therefore *the number of atoms (or ions) surrounding an atom (or ion) in a crystal lattice.* Figure 14.19(b) shows that the basic, repeating unit in this array of spheres is a simple cubic unit cell. Within each cell there is the equivalent of one sphere [Figure 14.19(c)].

Packing Efficiency. The **packing efficiency,** or *percentage of the cell space occupied by the spheres,* is an important crystal property. For one thing, it determines the density of the crystal. To find the packing efficiency of a simple cubic cell, let a be the length of the edge of the cubic unit cell and r the radius of the spheres, so that $a = 2r$ (Figure 14.20). Then

$$\text{volume of one sphere} = \frac{4\pi}{3}r^3 = \frac{4\pi}{3}\left(\frac{a}{2}\right)^3$$
$$\text{volume of the unit cell} = a^3$$

Since each corner sphere is shared by eight unit cells and there are eight corners in a cube, there will be the equivalent of only one complete sphere *inside* a simple cubic unit cell [see Figure 14.19(c)]. We can now write

$$\text{packing efficiency} = \frac{\text{volume of spheres inside the cell}}{\text{volume of the cell}} \times 100\%$$

FIGURE 14.20 *The relation between atomic radius, r, and edge length, a, of a simple cubic cell (a = 2r).*

$$= \frac{\frac{4\pi}{3}\left(\frac{a}{2}\right)^3}{a^3} \times 100\%$$

$$= \frac{\frac{4\pi a^3}{24}}{a^3} \times 100\% = \frac{\pi}{6} \times 100\%$$

$$= 52\%$$

A body-centered cubic arrangement is shown in Figure 14.21. The second layer fits into the depressions of the first layer and the third layer into the depressions of the second layer. The coordination number of each sphere in this structure is 8 (each sphere is in contact with four spheres in the layer above and four spheres in the layer below), and the packing efficiency is 68 percent.

Closest Packing. The efficiency of packing can be further increased. We start with the structure shown in Figure 14.22(a). The sphere marked "a" is in contact with six neighbors marked "x." Around each first-layer sphere there are six depressions marked "b" and "c." In the second layer, spheres are packed as closely as possible to the first layer by resting in the depressions between the spheres in the first layer. If the depressions marked "b" are occupied by second-layer spheres, then those marked "c" cannot be, and vice versa. Figure 14.22(b) shows the second-layer spheres resting over the "b" depressions. Note that the "a" sphere touches three spheres in the second layer.

There are two ways that third-layer spheres may cover the second layer to achieve what is called ***closest packing,*** which refers to *the most efficient arrangements for spheres.* The spheres may fit into the depressions marked so that each third-layer sphere is directly over a first-layer sphere. We call

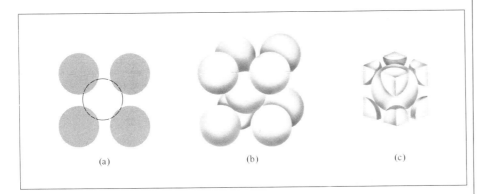

(a) (b) (c)

FIGURE 14.21 *Arrangement of identical spheres forming a body-centered cube. (a) Top view. (b) Definition of a body-centered cubic unit cell. (c) There is the equivalent of two complete spheres inside a body-centered cubic unit cell. Note that the spheres do not touch along the edges, but they do touch along the cube diagonal. Consequently, the edge length of the unit cell, a, is equal to $4r/\sqrt{3}$, where r is the radius of the sphere.*

FIGURE 14.22 *(a) In a close-packed layer, each sphere (see sphere marked "a") is in contact with six other spheres (marked "x"). There are six depressions marked "b" and "c" around "a." (b) In the second layer, spheres are fitted into the depressions marked "b."*

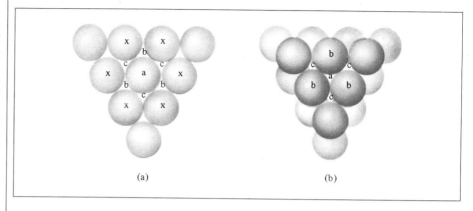

(a)   (b)

this arrangement the *hexagonal close-packed (hcp) structure* [Figure 14.23(a)]. Alternatively, the third-layer spheres may fit into the depressions that lie directly over the depressions in the first layer. In this case, we obtain the *cubic close-packed (ccp) structure* [Figure 14.23(b)]. As you can see, the ccp structure has a face-centered cubic unit cell. In both the hcp and ccp structures each sphere has a coordination number of 12 (each sphere is in contact with six spheres in its own layer, three spheres in the layer

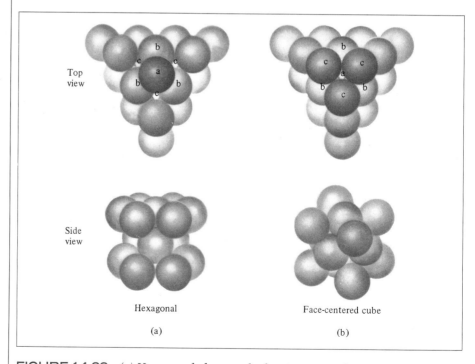

Top view

Side view

Hexagonal   Face-centered cube

(a)   (b)

FIGURE 14.23 *(a) Hexagonal close-packed structure. In this arrangement, each third-layer sphere is directly over a first-layer sphere. (b) Cubic close-packed structure. In this arrangement, each third-layer sphere fits into a depression that is directly over a depression in the first layer. The spheres touch along the face diagonals of the face-centered cube. The unit cell edge length, a, is equal to $4r/\sqrt{2}$, where r is the radius of the sphere.*

above, and three spheres in the layer below). The packing efficiency is the same for both: 74 percent.

EXAMPLE 14.3

The metal zinc has a hexagonal close-packed structure with a density of 7.14 g/cm^3. Calculate the radius of a Zn atom in picometers.

Answer

First we calculate the volume of the hcp structure occupied by one mole of Zn. Let V_{hcp} be the molar volume, so that we can write

$$V_{hcp} = \frac{\text{molar mass of Zn}}{\text{density of Zn}}$$
$$= \frac{65.37 \text{ g Zn/mol Zn}}{7.14 \text{ g/cm}^3}$$
$$= 9.16 \text{ cm}^3/\text{mol Zn}$$

This is a molar volume, so it contains 6.02×10^{23} Zn atoms. From the packing efficiency of hcp structure, we know that 74 percent of the volume is occupied by the Zn atoms (the other 26 percent is empty space); that is

volume of Zn atoms = $0.74 \times 9.16 \text{ cm}^3/\text{mol Zn} = 6.78 \text{ cm}^3/\text{mol Zn}$

The volume of a single Zn atom is given by

$$\text{volume of one Zn atom} = \frac{6.78 \text{ cm}^3}{1 \text{ mol Zn}} \times \frac{1 \text{ mol Zn}}{6.02 \times 10^{23} \text{ Zn atoms}}$$
$$= 1.13 \times 10^{-23} \text{ cm}^3$$

The volume of a sphere of radius r is

$$V = \tfrac{4}{3}\pi r^3$$

so for one spherical Zn atom

$$1.13 \times 10^{-23} \text{ cm}^3 = \tfrac{4}{3}\pi r^3$$
$$r^3 = 2.70 \times 10^{-24} \text{ cm}^3$$
$$r = 1.39 \times 10^{-8} \text{ cm}$$
$$= 139 \text{ pm}$$

This value is in excellent agreement with the known value of 138 pm for the radius of a Zn atom.

Similar example: Problem 14.31.

Remember that density is an intensive property so that it is the same for one gram or one mole of a substance.

X-Ray Diffraction of Crystals

Virtually all we know about crystal structure has been learned from X-ray diffraction studies. ***X-ray diffraction*** refers to *the scattering of X rays by the units of a regular crystalline solid.* The scattering (or diffraction) patterns obtained are used to deduce the arrangements of particles in the solid lattice.

In Section 13.2 we discussed the interference phenomenon to illustrate wave properties (see Figure 13.1). Since X rays are one form of electromagnetic radiation, we would expect them to exhibit similar behavior under suitable conditions. In 1912 the German physicist Max von Laue (1879–1960) correctly suggested that, because the wavelength of X rays is comparable in magnitude to the distances between lattice points in a crystal, the lattice should be able to *diffract* X rays. An X-ray diffraction pattern is a result of interference in the waves associated with X rays.

Figure 14.24 shows a typical arrangement for producing an X-ray diffraction pattern of a crystal. A beam of X rays is directed at a mounted crystal. Atoms in the crystal absorb some of the incoming radiation and then reemit it; the process is called the *scattering of X rays.*

To understand how a diffraction pattern may be generated, consider the scattering of X rays by atoms in two parallel planes (Figure 14.25). Initially, the two incident rays are *in phase* with each other (that is, their maxima and minima occur at the same positions). The upper wave is scattered or reflected by an atom in the first layer, while the lower wave is scattered by an atom in the second layer. In order for these two scattered waves to be in phase again, the extra distance traveled by the lower wave must be an integral multiple of the wavelength (λ) of the X ray, that is

$$\text{BC} + \text{CD} = 2d \sin \theta = n\lambda \qquad n = 1, 2, 3, \ldots \qquad (14.1)$$

where θ is the angle between the X rays and the plane of the crystal and d is the distance between adjacent planes. Equation (14.1) is known as the Bragg equation [after William H. Bragg (1862–1942) and Sir William L. Bragg (1890–1972)]. The reinforced waves produce a dark spot on a photographic film for each value of θ that satisfies the Bragg equation. It might appear that only one spacing (that is, d) could be measured for a cubic crystal. Actually, several crystal spacings are measurable, since various diagonal planes besides those shown in Figure 14.25 can diffract X rays.

Reinforced waves are waves that have interacted constructively (see Figure 13.1).

EXAMPLE 14.4

X rays of wavelength 0.154 nm strike an aluminum crystal; the rays are reflected at an angle of 19.3°. Assuming that $n = 1$, calculate the spacing between the planes of aluminum atoms (in pm) that is responsible for this angle of reflection. The conversion factor is obtained from 1 nm = 1000 pm.

Answer

From Equation (14.1)

$$d = \frac{n\lambda}{2 \sin \theta} = \frac{\lambda}{2 \sin \theta}$$

$$= \frac{0.154 \text{ nm} \times \dfrac{1000 \text{ pm}}{1 \text{ nm}}}{2 \sin 19.3°}$$

$$= 233 \text{ pm}$$

Similar examples: Problems 14.42, 14.43, 14.44.

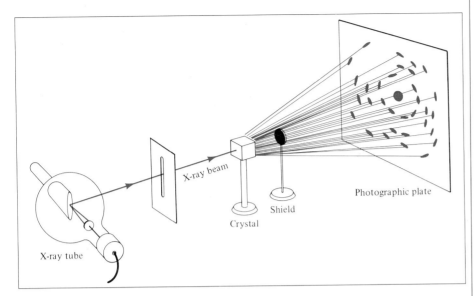

FIGURE 14.24 *An arrangement for obtaining the X-ray diffraction pattern of a crystal. The shield prevents the strong undiffracted X rays from damaging the photographic plate.*

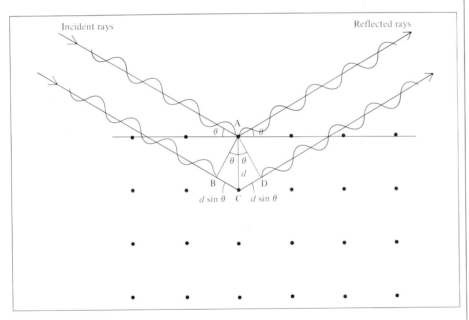

FIGURE 14.25 *Reflection of X rays from two layers of atoms. The lower wave travels a distance 2d sin θ longer than the upper wave does. For the two waves to be in phase again after reflection, it must be true that 2d sin θ = nλ, where λ is the wavelength of the X ray and n = 1, 2, 3, ...*

The X-ray diffraction technique offers the most accurate method for determining bond lengths and bond angles in molecules. Because X rays are scattered by electrons, chemists can construct an electron density contour map from the diffraction patterns by using a complex mathematical procedure. Basically, an *electron density contour map* tells us the relative values of electron densities at various locations in a molecule. The densities reach a maximum near the center of each atom. In this manner, we can determine the positions of the nuclei and hence the geometric parameters of the molecule. Figure 14.26 shows the bond lengths of the anthracene molecule ($C_{14}H_{10}$) that were determined from its electron density contour map.

FIGURE 14.26 *(a) Electron density contour map of the anthracene molecule. The positions of the carbon nuclei are at the centers of the contour maps. Normally the positions of hydrogen atoms cannot be determined by the X-ray diffraction technique. (b) Carbon-to-carbon bond lengths (in pm) in the anthracene molecule as deduced from the X-ray study. For simplicity, the bond angles are omitted from this diagram.*

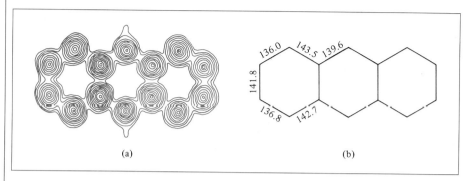

(a)   (b)

Types of Crystals

The structure and properties of crystals are determined by the kinds of forces that hold the particles together. We can classify any crystal as one of four types: ionic, covalent, molecular, or metallic (Table 14.5).

Ionic Crystals. There are two important characteristics to note about ionic crystals: (1) Charged species are present, and (2) anions and cations are generally quite different in size. Consider the alkali metal halides, for example. In these crystals, the number of anions surrounding a cation must equal the number of cations surrounding an anion in order to maximize net attraction.

Knowing the radii of the ions is helpful in understanding the structure and stability of these compounds. There is no way to measure the radius

TABLE 14.5 Types of Crystals and General Properties

Type of Crystal	Units at Lattice Points	Force(s) Holding the Units Together	General Properties	Examples
Ionic	Positive and negative ions	Electrostatic attraction	Hard, brittle, high melting point, poor conductor of heat and electricity	NaCl, LiF, MgO
Covalent	Atoms	Covalent bond	Hard, high melting point, poor conductor of heat and electricity	C (diamond), SiO_2 (quartz)
Molecular*	Molecules or atoms	Dispersion forces, dipole–dipole forces, hydrogen bonds	Soft, low melting point, poor conductor of heat and electricity	Ar, CO_2, I_2, H_2O, $C_{12}H_{22}O_{11}$ (sucrose)
Metallic	Atoms	Metallic bond	Soft to hard, low to high melting point, good conductor of heat and electricity	All metallic elements; for example, Na, Mg, Fe, Cu

* Included in this category are crystals made up of individual atoms.

of an individual ion, but under favorable circumstances it is possible to estimate the radii of a number of ions. For example, if we know the radius of I^- in KI is about 216 pm we can proceed to determine the radius of K^+ ion in KI, the radius of Cl^- ion in KCl, and so on.

Figure 10.8 (p. 288) gives the radii of a number of ions. Keep in mind that these values are averaged over many different compounds. Let us consider the NaCl crystal. Figure 14.27 shows that the edge length of the unit cell of NaCl is twice the sum of the two ionic radii. Using the values in Figure 10.4, we find the same edge length to be 2(95 + 181) pm or 552 pm. But the edge length shown in Figure 14.27 is 564 pm. The discrepancy between these two values tells us that the radius of an ion actually varies slightly from one compound to another.

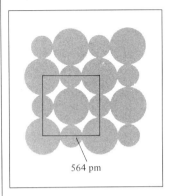

564 pm

FIGURE 14.27 *Relation between the radii of Na^+ and Cl^- ions and the unit cell dimension. Here the cell edge length is equal to twice the sum of the two ionic radii.*

EXAMPLE 14.5

How many Na^+ and Cl^- ions are in each NaCl unit cell?

Answer

NaCl has a face-centered cubic lattice. As Figure 2.5 shows, one whole Na^+ ion is at the center of the unit cell and there are twelve Na^+ ions at the edges. Since each edge Na^+ ion is shared by four unit cells, the total number of Na^+ ions is $1 + (12 \times \frac{1}{4}) = 4$. Similarly, there are six Cl^- ions at the face centers and eight Cl^- ions at the corners. Each face-centered ion is shared by two unit cells and each corner ion is shared by eight unit cells (see Figure 14.18), so the total number of Cl^- ions is $(6 \times \frac{1}{2}) + (8 \times \frac{1}{8}) = 4$. Thus there are four Na^+ ions and four Cl^- ions in each NaCl unit cell. Figure 14.28 shows the portions of the Na^+ and Cl^- ions *within* a unit cell.

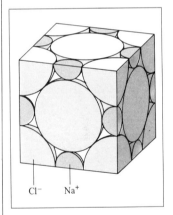

Cl^- Na^+

FIGURE 14.28 *Portions of Na^+ and Cl^- ions within a face-centered cubic cell.*

EXAMPLE 14.6

The edge length of the NaCl unit cell is 564 pm. What is the density of NaCl in g/cm^3?

Answer

From Example 14.5 we see that there are four Na^+ ions and four Cl^- ions in each unit cell. The total mass (in amu) in a unit cell is therefore

$$\text{mass} = 4(22.99 \text{ amu} + 35.45 \text{ amu}) = 233.8 \text{ amu}$$

The volume of the unit cell is $(564 \text{ pm})^3$. The density of the unit cell is

$$\text{density} = \text{mass/volume}$$

$$= \frac{(233.8 \text{ amu})\left(\dfrac{1 \text{ g}}{6.02 \times 10^{23} \text{ amu}}\right)}{(564 \text{ pm})^3 \left(\dfrac{1 \text{ cm}}{1 \times 10^{10} \text{ pm}}\right)^3}$$

$$= 2.16 \text{ g/cm}^3$$

Most ionic crystals have high melting points, an indication of the strong cohesive force holding the ions together. A measure of the stability of ionic crystals is the lattice energy (see Section 10.6); the higher the lattice energy, the more stable the compound. Table 10.2 lists the lattice energies of some alkali and alkaline earth metal halides together with their melting points.

Covalent Crystals. In covalent crystals, atoms are held together entirely by covalent bonds. Well-known examples are the two allotropes of carbon: graphite and diamond (see p. 356).

A diamond melts at about 3550°C. Since this process involves breaking strong covalent bonds, such a high melting point is not surprising.

In diamond each carbon atom is bonded to four other atoms. The strong covalent bonds in three dimensions contribute to diamond's unusual hardness (it is the hardest material known). In graphite, carbon atoms are arranged in six-membered rings (see Figure 12.18). The atoms are all sp^2-hybridized; each atom is covalently bonded to three other atoms. The remaining unhybridized $2p$ orbital is used in pi bonding. In fact, in each layer of graphite there is the kind of delocalized molecular orbital that is present in benzene (see Section 13.5). Because electrons are free to move around in this extensively delocalized molecular orbital, graphite is a good conductor of electricity in directions along the planes of carbon atoms. The

The central electrode in flashlight batteries is made of graphite.

layers are held together by the weak van der Waals forces. The covalent bonds in graphite account for its hardness; however, because the layers can slide over one another, graphite is slippery to the touch and is used as a lubricant. It is also used in pencils and typewriter ribbons.

Another type of covalent crystal is quartz (SiO_2). The arrangement of Si atoms in quartz is similar to that of carbon in diamond, but an oxygen atom is located between each pair of Si atoms. Since Si and O have different electronegativities (see Figure 11.2), the Si—O bond is polar. Nevertheless, SiO_2 is similar to diamond in many respects, such as hardness and high melting point.

The melting point of quartz is 1610°C.

Molecular Crystals. In a molecular crystal, molecules occupy lattice points and the attractive forces between them are van der Waals forces and/or hydrogen bonding. An example of a molecular crystal is solid sulfur dioxide (SO_2), in which the predominant attractive force is dipole–dipole interaction. Intermolecular hydrogen bonding is mainly responsible for maintaining the three-dimensional ice lattice (see Figure 14.14). Figure 14.29 shows the unit cell of molecular iodine (I_2), in which the molecules are held together only by dispersion forces.

"Molecular crystals" include crystals containing only individual atoms, such as crystalline argon.

With the exception of ice, molecules in molecular crystals are packed together as closely as their size and shape allow. This explains why the densities of most solids are greater than those of the corresponding liquids. Because van der Waals forces and hydrogen bonding are generally quite weak (compared to the covalent bond), molecular crystals are considerably less stable than ionic and covalent crystals. Indeed, most molecular crystals melt below 100°C.

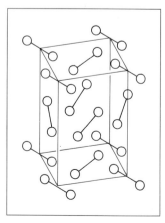

FIGURE 14.29 *The unit cell of I_2. Each sphere represents an iodine atom. The size of molecules has been reduced to show their positions. The unit cell is face-centered orthorhombic (see Figure 14.16).*

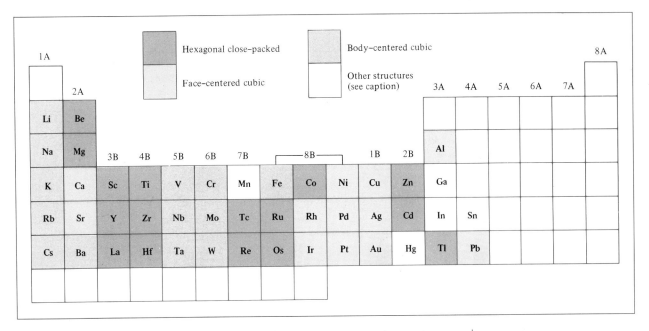

FIGURE 14.30 *Crystal structures of various metals. The metals are shown in their positions in the periodic table. Mn has a cubic structure, Ga an orthorhombic structure, In and Sn the tetragonal structure, and Hg the rhombohedral structure.*

Metallic Crystals. In a sense, the structure of metallic crystals is the simplest to deal with, since every lattice point in a crystal is occupied by an atom of the same metal. Metallic crystals are generally body-centered cubic, face-centered cubic, or hexagonal close-packed (Figure 14.30). As we saw earlier, the packing efficiency of these crystal structures is quite high, and thus metallic elements are usually very dense.

The bonding in metals is quite different from that in other types of crystals. The bonding electrons in a metal are highly delocalized over the entire crystal. In fact, metal atoms in a crystal can be imagined as an array of positive ions immersed in a sea of delocalized valence electrons (Figure 14.31). The great cohesive force resulting from delocalization is responsible for the strength of metals. The mobility of the delocalized electrons accounts for metals being good conductors of heat and electricity.

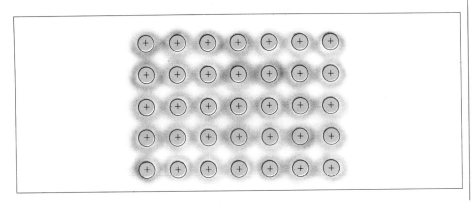

FIGURE 14.31 *A cross-section of the crystal structure of a metal. Each circled positive charge represents the nucleus and inner electrons of a metal atom. The gray area surrounding the positive metal ions indicates the mobile sea of electrons.*

14.5 AMORPHOUS SOLIDS

We stated earlier that not all solids are crystalline in structure. ***Amorphous solids,*** such as glass, *lack a regular three-dimensional arrangement of atoms.* In this section, we will briefly discuss some interesting and important properties of glass.

Glass is one of mankind's most valuable and versatile materials. It is also one of the oldest—glass articles date as far back as 1000 B.C. The term ***glass*** is commonly used to mean *the optically transparent fusion product of inorganic materials that has cooled to a rigid state without crystallizing.* In some respects glass behaves more like a liquid than a solid. X-ray diffraction studies show that glass lacks long-range periodic order. Furthermore, like liquids, it is capable of flowing. The rate of flow at room temperature is very slow and becomes noticeable only after a number of years. If you visit an ancient cathedral or an old New England building, you may see the increased thickness of glass near the bottoms of windows.

There are about 800 different types of glass in common use today. Table 14.6 shows the composition and properties of three of them. Figure 14.32 shows two-dimensional schematic representations of crystalline quartz, quartz glass, and a multicomponent glass.

The color of glass is due largely to the presence of metal ions (as oxides). For example, green glass contains iron(III) oxide, Fe_2O_3, or copper(II) oxide, CuO; yellow glass contains uranium(IV) oxide, UO_2; blue glass contains cobalt(II) and copper(II) oxides, CoO and CuO; and red glass contains small particles of gold and copper.

> Note that most of the ions are derived from the transition metals.

TABLE 14.6 Composition and Properties of Three Types of Glass

Name	Composition	Properties and Uses
Pure quartz glass	100% SiO_2	Low thermal expansion, transparent to wide range of wavelengths. Used in optical research.
Pyrex glass	SiO_2, 60–80% B_2O_3, 10–25% Al_2O_3, small amount	Low thermal expansion; transparent to visible and infrared, but not to UV radiation. Used mainly in laboratory and household cooking glassware.
Soda-lime glass	SiO_2, 75% Na_2O, 15% CaO, 10%	Easily attacked by chemicals and sensitive to thermal shocks. Transmits visible light, but absorbs UV radiation. Used mainly in windows and bottles.

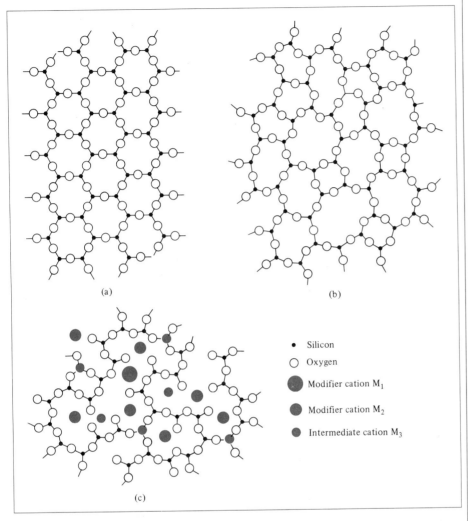

(a)

(b)

- Silicon
○ Oxygen
● Modifier cation M_1
● Modifier cation M_2
● Intermediate cation M_3

(c)

FIGURE 14.32 *Two-dimensional representation of (a) crystalline quartz, (b) noncrystalline quartz glass, and (c) a multicomponent glass.*

14.6 PHASE CHANGES

The discussions in Chapter 6 and in this chapter have given us an overview of the properties of the three states of matter. However, solids, liquids, and gases do not always remain in one state. As temperature changes they undergo what is known as **phase changes**, or *transformation from one phase to another.* As Figure 14.1 shows, molecules in the solid state have the most order, whereas molecules in the gas phase have the greatest randomness. We know from experience that energy (usually in the form of heat) is required to bring about solid-to-liquid-to-gas changes, which are accompanied by increases in randomness or disorder. As we study phase transformations, keep in mind the relationship between energy change and the increase or decrease in molecular order. These parameters will help us understand the nature of these physical changes.

Liquid–Vapor Equilibrium

Vapor Pressure. Molecules in a liquid are not held in fixed positions; rather, they are in constant motion, although they lack the total freedom of gaseous molecules. Because liquids are denser than gases, the collision rate among molecules is much higher in the liquid phase than in the gas phase.

Figure 14.33 shows the kinetic energy distribution of molecules in a liquid. *At any given temperature, a certain number of the molecules in a liquid possess sufficient kinetic energy to escape from the surface.* This process is called *evaporation,* or *vaporization.*

When a liquid evaporates, its molecules in the gas phase exert a vapor pressure. Consider the apparatus shown in Figure 14.34. Before the evaporation process starts, the mercury levels in the manometer are equal. As soon as a few molecules leave the liquid, a vapor phase is established. The vapor pressure is measurable only when a fair amount of vapor is present. The process of evaporation does not continue indefinitely, however. Eventually, the mercury levels stabilize and no further changes are observable.

What happened at the molecular level? In the beginning, there is only one-way traffic: Molecules are moving from the liquid to the empty space. Soon the molecules in the space above the liquid establish a vapor phase. *As the concentration of the molecules in the vapor phase increases, some*

> The difference between a gas and a vapor is explained in Section 6.1.

FIGURE 14.33 *Kinetic energy distribution curves of molecules in a liquid at (a) a temperature T_1 and (b) a higher temperature T_2. Note that at the higher temperature the curve flattens out. The shaded areas represent the number of molecules possessing kinetic energy equal to or greater than a certain kinetic energy E_1. The higher the temperature, the greater the number of molecules with high kinetic energy.*

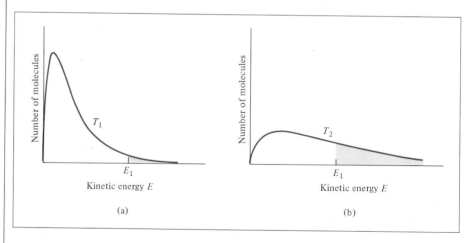

(a)   (b)

FIGURE 14.34 *Measurement of the vapor pressure of a liquid (a) before the evaporation begins and (b) at equilibrium. In (b) the number of molecules leaving the liquid is equal to the number of molecules returning to the liquid. The difference in the mercury levels (h) gives the equilibrium vapor pressure of the liquid at the specified temperature.*

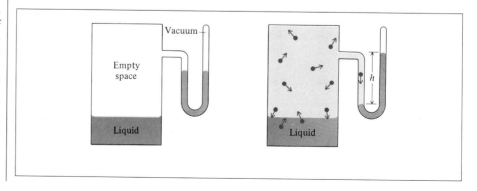

FIGURE 14.35 *Comparison of the rates of evaporation and condensation at constant temperature. Initially, there are relatively few molecules in the vapor phase, and so the rate of condensation is low. As the concentration of molecules in the vapor phase builds up, the rate of condensation also increases. The rate of evaporation remains constant with time. At dynamic equilibrium, the rate of evaporation is equal to the rate of condensation and no net change takes place.*

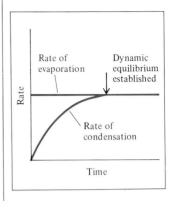

molecules return to the liquid phase, a process called **condensation.** Condensation occurs because a molecule striking the liquid surface loses some of its kinetic energy and becomes trapped by intermolecular forces in the liquid.

The rate of evaporation is constant at any given temperature, and the rate of condensation increases with increasing concentration of molecules in the vapor phase. A state of **dynamic equilibrium** *(when the rate of a forward process is exactly balanced by the rate of the reverse process)* is reached when the rates of condensation and evaporation become equal (Figure 14.35). *The vapor pressure measured under dynamic equilibrium of condensation and evaporation* is called the **equilibrium vapor pressure.** We often use the simpler term "vapor pressure" when we talk about the equilibrium vapor pressure of a liquid. This is acceptable as long as we know the meaning of the abbreviated term.

It is important to remember that the equilibrium vapor pressure is the *maximum* vapor pressure that a liquid exerts at a given temperature and that it is a constant at constant temperature.

Equilibrium vapor pressure is independent of the amount of liquid as long as there is some liquid present.

Plots of vapor pressure versus temperature for three different liquids are shown in Figure 14.36. We see in Figure 14.33 that the number of molecules with higher kinetic energies is greater at the higher temperature, and therefore so is the evaporation rate. For this reason, the vapor pressure of a liquid always increases with temperature (see Figure 14.36). For example, the vapor pressure of water is 17.5 mmHg at 20°C, but it rises to 760 mmHg at 100°C (see Table 6.3).

Heat of Vaporization and Boiling Point. A measure of how strongly molecules are held in a liquid is its **molar heat of vaporization (ΔH_{vap}),**

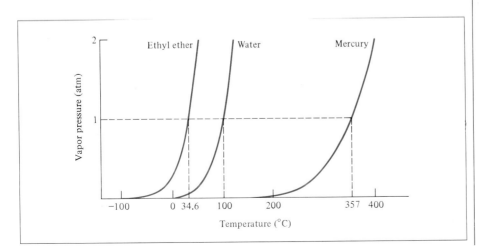

FIGURE 14.36 *The increase in vapor pressure as temperature is raised, for three liquids. The normal boiling points of the liquids (at 1 atm) are shown on the horizontal axis.*

TABLE 14.7 Molar Heats of Vaporization for Selected Liquids

Substance	Boiling Point* (°C)	ΔH_{vap} (kJ/mol)
Argon (Ar)	−186	6.3
Methane (CH_4)	164	9.2
Ethyl ether ($C_2H_5OC_2H_5$)	34.6	26.0
Ethanol (C_2H_5OH)	78.3	39.3
Benzene (C_6H_6)	80.1	31.0
Water (H_2O)	100	40.79
Mercury (Hg)	357	59.0

* Measured at 1 atm.

usually defined as *the energy (in kilojoules) required to vaporize one mole of a liquid* (Table 14.7). The value of ΔH_{vap} is directly related to the strength of intermolecular forces that exist in the liquid. If intermolecular attraction is strong, then molecules in a liquid cannot escape easily into the vapor phase. Consequently, the liquid has a low vapor pressure. It also has a high molar heat of vaporization because much energy is needed to increase the kinetic energy of individual molecules to free them from intermolecular attraction.

A practical way to demonstrate the molar heat of vaporization is by rubbing alcohol on your hands. The heat from your hands increases the kinetic energy of the alcohol molecules. The alcohol evaporates rapidly, thus extracting heat from your hands and cooling them. The process is similar to perspiration, which is one of the effective means by which the human body maintains a constant temperature. Because of the strong intermolecular hydrogen bonding that exists in water, a considerable amount of energy is needed to vaporize the water of perspiration from the body's surface. This energy is supplied by the heat generated in various metabolic processes.

Metabolism refers to the chemical processes in an organism by which the stored energy in food is used for growth and function.

You have already seen that the vapor pressure of a liquid increases with temperature. For every liquid there exists a temperature at which it begins to boil. The **boiling point** is *the temperature at which the vapor pressure of a liquid is equal to the external atmospheric pressure.* The *normal* boiling point of a liquid is the boiling point when the external pressure is 1 atm.

The boiling action is accompanied by the formation of bubbles within the liquid. When a bubble forms, the liquid originally occupying that space is pushed aside, and the level of the liquid in the container is forced to rise upward. The pressure exerted *on* the bubble is largely atmospheric pressure, plus some *hydrostatic pressure* (that is, pressure due to the presence of liquid). The pressure *inside* the bubble is due solely to the vapor pressure of the liquid. When the vapor pressure becomes equal to the external pressure, the bubble rises to the surface of the liquid and bursts. If the vapor pressure in the bubble were lower than the external pressure, the bubble would collapse before it could rise. We can thus conclude that the boiling point of a liquid depends on the external atmospheric pressure. (We usually ignore the small contribution due to the hydrostatic pressure.) For example, at 1 atm water boils at 100°C, but if the pressure is reduced to 0.5 atm, water boils at only 82°C.

Since the boiling point is defined in terms of the vapor pressure of the liquid, we expect the boiling point to be related to the molar heat of vaporization: The higher the ΔH_{vap}, the higher the boiling point. The data in Table 14.7 roughly confirm our prediction. Ultimately, the magnitudes of boiling point and ΔH_{vap} are determined by the strength of intermolecular forces. For example, argon (Ar) and methane (CH_4), which have weak dispersion forces, have low boiling points and small molar heats of vaporization (see Table 14.7). Ethyl ether ($C_2H_5OC_2H_5$) has a dipole moment, and the dipole–dipole forces account for its moderately high boiling point and ΔH_{vap}. Both ethanol (C_2H_5OH) and water have strong hydrogen bonding, which accounts for their high boiling points and large ΔH_{vap} values. Strong metallic bonding causes mercury to have the highest boiling point and ΔH_{vap} of this group of liquids. Interestingly, the boiling point of benzene, which is nonpolar, is comparable to that of ethanol. Benzene has a large polarizability due to its electrons in the delocalized pi molecular orbitals, and the dispersion forces among benzene molecules can become as strong as or even stronger than dipole–dipole forces and/or hydrogen bonds.

Critical Temperature and Pressure. The opposite of evaporation is condensation. In principle, a gas can be made to liquefy by either one of two techniques. By cooling a sample of gas we decrease the kinetic energy of its molecules, and eventually molecules aggregate to form small drops of liquid. Alternatively, we may apply pressure to the gas. Under compression, the average distance between molecules is reduced so that they are, in this way too, held together by mutual attraction. Industrial liquefaction processes use a combination of these two methods.

Every substance has *a temperature,* called the **critical temperature** (T_c), *above which its gas form cannot be made to liquefy, no matter how great the applied pressure.* This is also *the highest temperature at which a substance can exist as a liquid. The minimum pressure that must be applied to bring about liquefaction at the critical temperature* is called the **critical pressure** (P_c). The existence of the critical temperature can be qualitatively explained as follows. The intermolecular attraction is a finite quantity for any given substance. At temperatures below T_c, this force is sufficiently strong to hold the molecules together (under some appropriate pressure) in a liquid. Above T_c, molecular motion becomes so energetic that the molecules can always break away from this attraction.

Table 14.8 lists the critical temperatures and critical pressures of a number of common substances. The critical temperature of a substance is determined by the strength of its intermolecular force. Substances such as benzene, ethanol, mercury, and water, which have strong intermolecular forces, also have high critical temperatures compared to the other substances listed in the table.

Putting it another way, above the critical temperature there is no fundamental distinction between a liquid and a gas— we simply have a fluid.

Keep in mind that intermolecular forces are independent of temperature, whereas the kinetic energy of molecules increases with temperature.

TABLE 14.8 Critical Temperatures and Critical Pressures of Selected Substances

Substance	T_c (°C)	P_c (atm)
Ammonia (NH_3)	132.4	111.5
Argon (Ar)	−186	6.3
Benzene (C_6H_6)	288.9	47.9
Carbon dioxide (CO_2)	31.0	73.0
Ethanol (C_2H_5OH)	243	63.0
Ethyl ether ($C_2H_5OC_2H_5$)	192.6	35.6
Molecular hydrogen (H_2)	−239.9	12.8
Mercury (Hg)	1462	1036
Methane (CH_4)	−83.0	45.6
Molecular nitrogen (N_2)	−147.1	33.5
Molecular oxygen (O_2)	−118.8	49.7
Water (H_2O)	374.4	219.5

Liquid–Solid Equilibrium

The transformation of liquid to solid is called *freezing*, and the reverse process is called *melting* or *fusion*. The **melting point** of a solid (or the freezing point of a liquid) is *the temperature at which solid and liquid phases coexist in equilibrium.* The *normal* melting point (or the *normal* freezing point) of a substance is the melting point (or freezing point) measured at 1 atm pressure. We generally omit the word "normal" in referring to the melting point of a substance at 1 atm.

The most familiar liquid–solid equilibrium occurs between water and ice. At 0°C and 1 atm, the dynamic equilibrium is represented by

$$\text{ice} \rightleftharpoons \text{water}$$

A practical illustration of this dynamic equilibrium is provided by a glass of ice water. As the ice cubes melt to form water, some of the water between ice cubes may freeze, thus joining the cubes together. This is not a true dynamic equilibrium; since the glass is not kept at 0°C, all the ice cubes will eventually melt away.

Because molecules are more strongly held in the solid state than in the liquid state, heating is required to bring about the solid–liquid phase transition. *The energy (in kilojoules) required to melt one mole of a solid* is called the **molar heat of fusion (ΔH_{fus})**. Table 14.9 shows the molar heats of fusion for the substances listed in Table 14.7. A comparison of the data in the two tables shows that ΔH_{fus} is generally smaller than ΔH_{vap} for the same substance. This is consistent with the fact that molecules in a liquid are still fairly closely packed together, so that some energy is needed to bring about the rearrangement from solid to liquid. On the other hand, when a liquid evaporates, its molecules become completely separated from one another and considerably more energy is required to overcome the attractive force.

"Fusion" refers to the process of melting. Thus a "fuse" breaks an electrical circuit when a metallic strip melts due to the heat generated by excessively high electrical current.

TABLE 14.9 Molar Heats of Fusion for Selected Substances

Substance	Melting Point* (°C)	ΔH_{fus} (kJ/mol)
Argon (Ar)	−190	1.3
Methane (CH_4)	−183	0.84
Ethyl ether ($C_2H_5OC_2H_5$)	−116.2	6.90
Ethanol (C_2H_5OH)	−117.3	7.61
Benzene (C_6H_6)	5.5	10.9
Water (H_2O)	0	6.01
Mercury (Hg)	−39	23.4

* Measured at 1 atm.

Heating and Cooling Curves. *Heating curves* such as the one shown in Figure 14.37 are helpful in the study of phase transitions. Figure 14.37 shows that when a solid sample is heated, its temperature increases gradually until point A is reached. At this point, the solid begins to melt. During the melting period (A → B), heat is being absorbed by the system, yet its temperature remains constant. The heat helps the molecules overcome the cohesive force in the solid. Once the sample has melted completely (point B), the heat absorbed increases the average kinetic energy of the liquid molecules, and the liquid temperature rises (B → C). The vaporization process (C → D) can be explained similarly. The temperature remains constant during the period when the increased kinetic energy is overcoming the cohesive forces in the liquid. When all molecules are in the gas phase, the temperature rises again.

As we would expect, the *cooling curve* of a substance is the reverse of its heating curve. If we remove heat from a gas sample at a steady rate, its temperature will decrease. As the liquid is being formed, heat is given off by the system, since its potential energy is decreasing. For this reason, the

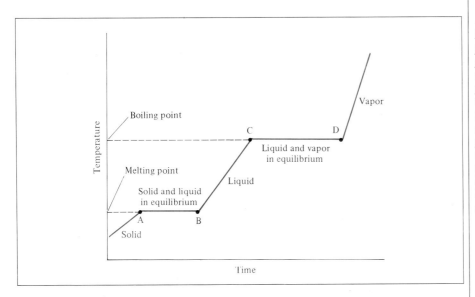

FIGURE 14.37 *A typical heating curve, from the solid phase through the liquid phase to the gas phase of a substance. Because heat of fusion is usually smaller than heat of vaporization, a substance melts in less time than it takes to boil. This explains why AB is shorter than CD. The steepness of the solid, liquid, and vapor heating lines is determined by the specific heat of the substance in each state.*

temperature of the system remains constant over the condensation period (D → C). After all the vapor has been condensed, the temperature of the liquid begins to drop. Continued cooling of the liquid finally leads to freezing (B → A).

A liquid can be temporarily cooled to below its freezing point. This phenomenon, called **supercooling,** may occur when heat is removed from a liquid so rapidly that the molecules literally have no time to assume the ordered structure of a solid. A supercooled liquid is unstable; gentle stirring or the addition to it of a small "seed" crystal of the same substance will quickly convert it to a solid.

Solid–Vapor Equilibrium

Solids, too, may undergo evaporation and, therefore, possess a vapor pressure. Consider the following dynamic equilibrium:

$$\text{solid} \rightleftharpoons \text{vapor}$$

The process in which molecules go directly from the solid into the vapor phase is called **sublimation,** and the reverse process (that is, *from vapor directly to solid*) is called **deposition.** Naphthalene (the substance used to make moth balls) has a fairly high (equilibrium) vapor pressure for a solid (1 mmHg at 53°C); thus its pungent odor quickly permeates an enclosed space. Because molecules are more tightly held in a solid, its vapor pressure is generally much smaller than that of the corresponding liquid. *The energy (in kilojoules) required for subliming one mole of a solid,* called the **molar heat of sublimation (ΔH_{sub})** of a substance, is given by the sum of the molar heats of fusion and vaporization:

$$\Delta H_{sub} = \Delta H_{fus} + \Delta H_{vap} \tag{14.2}$$

Equation (14.2) is an illustration of Hess's law (see Section 5.5). The enthalpy, or heat change, for the overall process is the same whether the substance changes directly from the solid to the vapor form or goes from solid to liquid and then to vapor.

Figure 14.38 summarizes the types of phase changes discussed in this section.

This equation holds only if all the phase changes occur at the same temperature. If that is not the case, the equation can be used only as an approximation.

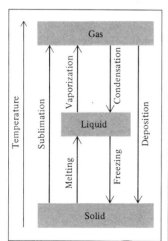

FIGURE 14.38   *The various phase changes that a substance may undergo.*

EXAMPLE 14.7

How much heat (in kilojoules) is required to melt 86.0 g of ice at 0°C?

Answer

We need to convert 86.0 g into moles. Then, with the molar heat of fusion from Table 14.9, we can calculate the amount of heat required for the melting process:

$$\text{heat required} = 86.0 \text{ g H}_2\text{O} \times \frac{1 \text{ mol H}_2\text{O}}{18.0 \text{ g H}_2\text{O}} \times \frac{6.01 \text{ kJ}}{1 \text{ mol H}_2\text{O}}$$

$$= 28.7 \text{ kJ}$$

Similar example: Problem 14.58.

EXAMPLE 14.8

Calculate the amount of energy (in kilojoules) needed to heat 346 g of water from 0°C to 182°C. Assume that the specific heat of water is 4.184 J/g°C over the entire liquid range and the specific heat of steam is 1.99 J/g°C.

Answer

The calculation can be broken down into three steps.

Step 1: Heating water from 0°C to 100°C

Using Equation (5.5) we write

$$
\begin{aligned}
q_1 &= mp\Delta t \\
&= (346 \text{ g})(4.184 \text{ J/g°C})(100°C - 0°C) \\
&= 1.45 \times 10^5 \text{ J} \\
&= 145 \text{ kJ}
\end{aligned}
$$

Step 2: Evaporating 346 g of water at 100°C

In Table 14.7 we see $\Delta H_{vap} = 40.79$ kJ/mol for water, so

$$
\begin{aligned}
q_2 &= 346 \text{ g H}_2\text{O} \times \frac{1 \text{ mol H}_2\text{O}}{18.0 \text{ g H}_2\text{O}} \times \frac{40.79 \text{ kJ}}{1 \text{ mol H}_2\text{O}} \\
&= 784 \text{ kJ}
\end{aligned}
$$

Step 3: Heating steam from 100°C to 182°C

$$
\begin{aligned}
q_3 &= mp\Delta t \\
&= (346 \text{ g})(1.99 \text{ J/g°C})(182°C - 100°C) \\
&= 5.65 \times 10^4 \text{ J} \\
&= 56.5 \text{ kJ}
\end{aligned}
$$

The overall energy required is given by

$$
\begin{aligned}
q_{overall} &= q_1 + q_2 + q_3 \\
&= 145 \text{ kJ} + 784 \text{ kJ} + 56.5 \text{ kJ} \\
&= 986 \text{ kJ}
\end{aligned}
$$

Similar example: Problem 14.59.

14.7 PHASE DIAGRAMS

The overall relationships among the solid, liquid, and vapor phases are best represented in a single graph known as a phase diagram. A *phase diagram shows the conditions at which a substance exists as a solid, liquid, or gas.* In this section we will briefly discuss the phase diagrams of water and carbon dioxide.

Water

Figure 14.39(a) shows the phase diagram of water. The graph is divided into three regions, each of which represents a pure phase. The line separating any two regions indicates conditions under which these two phases can

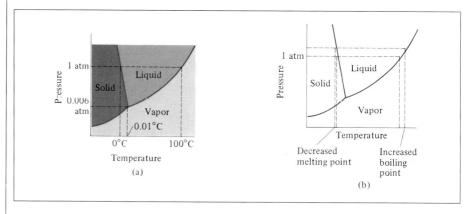

FIGURE 14.39 *(a) The phase diagram of water. Each solid line between two phases specifies the conditions (pressure and temperature) under which the two phases can exist in equilibrium. The point at which all three phases can exist in equilibrium (0.006 atm and 0.01°C) is called the* triple point. *Note that the liquid–vapor line extends only to the critical temperature of water. (b) This phase diagram tells us that increasing the pressure on ice lowers its melting point and that increasing the pressure on liquid water raises its boiling point.*

exist in equilibrium. For example, the curve between the liquid and vapor phases shows the variation of vapor pressure with temperature (compare this curve with Figure 14.36). The other two curves similarly indicate conditions for equilibrium between ice and liquid water and between ice and water vapor. The point at which all three curves meet is called the **triple point.** For water, this point is at 0.01°C and 0.006 atm. This is the only *condition under which all three phases can be in equilibrium with one another.*

Phase diagrams enable us to predict changes in the melting point and boiling point of a substance as a result of changes in the external pressure, as well as directions of phase transitions brought about by changes in temperature and pressure. The normal melting point and boiling point of water, measured at 1 atm, are 0°C and 100°C, respectively. What would happen if the melting and boiling were carried out at some other pressure? Figure 14.39(b) shows clearly that increasing the pressure above 1 atm will raise the boiling point and lower the melting point. A decrease in pressure will lower the boiling point and raise the melting point.

Aspects of the phase diagram of water have some practical applications. For example, the effect of increased external pressure on boiling point explains why pressure cookers save time in the kitchen. A pressure cooker is a sealed container that allows steam to escape only when it exceeds a predetermined pressure. The pressure above the water in the cooker is the sum of atmospheric pressure and steam pressure. Consequently, the water in the pressure cooker will boil above 100°C and the food in it will be hotter and cook faster. We can also use Figure 14.39(b) to explain ice skating. Because skates have very thin blades or runners, an average adult can exert

Water in a pressurized automobile radiator remains liquid at temperatures higher than 100°C. If the radiator cap is removed from an extremely hot car, however, the lowered external pressure may allow water to boil out with considerable vigor!

a pressure equivalent to several hundred atmospheres on the ice. (Remember that pressure is defined as force per unit area.) Thus the ice under the skate runners melts below 0°C, and the film of water formed under each runner facilitates the movement of the skater over ice. Friction between the blades and ice also helps melt the ice.

Carbon Dioxide

The phase diagram of carbon dioxide (Figure 14.40) is generally similar to that of water, with one important exception—the slope of the curve between solid and liquid is positive. In fact, this holds true for almost all other substances. Water behaves differently because ice is *less* dense than liquid water. The triple point of carbon dioxide is at 5.2 atm and −57°C.

An interesting observation can be made about Figure 14.40. As you can see, the entire liquid phase lies well above atmospheric pressure; therefore, it is impossible for solid carbon dioxide to melt at 1 atm. Instead, when solid CO_2 is heated to −78°C at 1 atm, it sublimes. In fact, solid carbon dioxide is called Dry Ice because it looks like ice and *does not melt* (Figure 14.41).

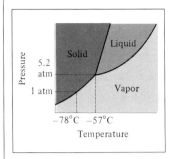

FIGURE 14.40 *The phase diagram of carbon dioxide. This diagram differs from Figure 14.39(a) in that the solid–liquid boundary line has a positive slope. The triple point is at 5.2 atm and −57°C. The liquid–vapor line does not extend beyond the critical temperature of CO_2. As you can see, the liquid phase is not stable below 5.2 atm, so that only the solid and vapor phases of carbon dioxide can exist under atmospheric conditions.*

FIGURE 14.41 *At room temperature, Dry Ice does not melt; it can only sublime.*

AN ASIDE ON THE PERIODIC TABLE
The Melting Points, Boiling Points, and Densities of the Elements

In this chapter we discussed various properties of liquids and solids. In particular, we saw that intermolecular forces play a central role in determining the bulk properties of substances. Three properties that indicate the strength of intermolecular forces and how atoms of the elements are packed together in condensed states are melting point, boiling point, and density. Figure 14.42 shows these quantities for a number of elements, located in their positions in the periodic table. The following points are worth noting:

- As we move across a period, the variations in boiling points and melting points reflect the changes in bonding and structure of the elements. Take the third-period elements as an example. Sodium, magnesium, and aluminum are metals; atoms in these substances are held together by bonding that is characteristic of the metallic state. Except for sodium, the melting points of these metals are fairly high. Silicon is a covalent crystal; each Si atom is covalently bonded to four other Si atoms (silicon crystallizes in the diamond lattice). Consequently, the solid is extremely hard and stable. The strong bonding explains the very high melting point of silicon. The first four elements all exist in extensive three-dimensional networks. There is an abrupt change in the melting point from silicon to phosphorus. As we saw in Chapter 12 (p. 356), phosphorus exists in discrete P_4 units. The weak dispersion forces among the nonpolar P_4 molecules account for the low melting point. Elemental sulfur exists as S_8 molecules. The stronger dispersion forces among the nonpolar S_8 molecules resulting from the increased number of electrons account for the rise in melting point from phosphorus to sulfur. Because chlorine exists as a nonpolar diatomic molecule and argon is a monatomic species, their melting points are quite low. Note that the last four elements all form molecular crystals, which, as we know, usually have lower melting points than other types of crystals.
- Explanations of the trends in boiling points are similar to those noted for the melting points.
- All the densest elements are metals. Interestingly, some of the least dense elements are also metals. The alkali metals and some of the alkaline earth metals (Be, Mg, and Ca) are among the lightest and softest (that is, easiest to compress) solids known. Aluminum is another light metal; it finds many industrial applications partly because of its low density.

SUMMARY

1. All substances exist in one of three states: gas, liquid, or solid. The major difference between the condensed state and the gaseous state is the distance of separation between molecules.
2. Intermolecular forces act between molecules or between molecules and ions. Generally, these forces are much weaker than bonding forces.
3. Dipole–dipole forces and ion–dipole forces attract molecules with dipole moments.

Key:

Au
2807 — Boiling point (°C)
1064 — Melting point (°C)
19.3 — Density (g/cm³)

1A	2A	3B	4B	5B	6B	7B	8B	8B	8B	1B	2B	3A	4A	5A	6A	7A	8A
H_2 −252.9 −259.1 0.071																	He −268.9 −272.2 0.126
Li 1347 180.5 0.534	Be 2970* 1278 1.85											B ? 2300 2.34	C 4827 3680* 2.25	N_2 −195.8 −209.9 0.81	O_2 −182.96 −218.4 1.14	F_2 −188.1 −219.6 1.51	Ne −246.0 −248.7 1.20
Na 882.9 97.8 0.97	Mg 1090 648.8 1.74											Al 2467 660 2.7	Si 2355 1410 2.33	P_4 280 44.2 1.82	S_8 444.6 112.8 2.07	Cl_2 −34.6 −101.0 1.56	Ar −185.7 −189.2 1.40
K 774 63.65 0.86	Ca 1484 839 1.54	Sc 2831 1541 2.99	Ti 3287 1660 4.5	V 3380 1890 5.96	Cr 2672 1857 7.20	Mn 1962 1244 7.20	Fe 2750 1535 7.86	Co 2870 1495 8.9	Ni 2732 1453 8.90	Cu 2567 1083 8.92	Zn 907 419.6 7.14	Ga 2403 29.8 5.90	Ge 2830 937.4 5.35	As 613* ? 5.72	Se_8 685 217 4.81	Br_2 58.8 −7.2 3.12	Kr −152.3 −156.6 2.6
Rb 688 38.89 1.53	Sr 1384 769 2.6	Y 3338 1522 4.47	Zr 4377 1852 6.49	Nb 4742 2468 8.57	Mo 4612 2617 10.2	Tc ? 2140 11.5	Ru 3900 2310 12.3	Rh 3727 1966 12.4	Pd 3140 1552 12.0	Ag 2212 961.9 10.5	Cd 765 320.9 8.64	In 2080 156.6 7.30	Sn 2270 231.9 7.28	Sb 1750 631 6.68	Te 989.8 449.5 6.0	I_2 184 113.5 4.93	Xe −107.1 −111.9 3.06
Cs 678.4 28.4 1.88	Ba 1640 725 3.51	La 3457 921 6.17	Hf 4602 2227 13.3	Ta 5425 2996 16.6	W 5660 3410 19.35	Re 5627 3180 20.53	Os 5027 3045 22.48	Ir 4130 2410 22.4	Pt 3827 1772 21.45	Au 2807 1064 19.3	Hg 356.6 −38.9 13.59	Tl 1457 303.5 11.85	Pb 1740 327.5 11.4	Bi 1560 271.3 9.80			

FIGURE 14.42 The melting points, boiling points, and densities of the most stable allotropic forms (if there is more than one) of the elements. Molecular formulas are shown for elements that exist in discrete molecular units. For elements that exist as gases under atmospheric conditions (H_2, N_2, O_2, F_2, Cl_2, and the noble gases), the densities refer to their liquid state at the boiling point. The asterisk indicates that the element undergoes sublimation at the temperature.

4. Dispersion forces are the result of temporary dipole moments induced in ordinarily nonpolar molecules. The extent to which a dipole moment can be induced in a molecule is called its polarizability. The term "van der Waals forces" refers to the total effect of dipole–dipole, dipole-induced dipole, and dispersion forces. The van der Waals radius is one-half the distance at which these attractive forces are at their maximum between nonbonded atoms.

5. Hydrogen bonding is a relatively strong dipole–dipole force that acts between a polar bond containing a hydrogen atom and the bonded electronegative atoms, O, N, or F. Hydrogen bonds between water molecules are particularly strong.

6. Liquids tend to assume a geometry that ensures the minimum surface area. Surface tension is the energy needed to expand a liquid surface area; strong intermolecular forces lead to greater surface tension.

7. Viscosity is a measure of the resistance of a liquid to flow; it decreases with increasing temperature.

8. Water molecules in both the solid and liquid states form a three-dimensional network in which each oxygen atom is covalently bonded to two hydrogen atoms and is hydrogen-bonded to two hydrogen atoms. This unique structure accounts for the fact that ice is less dense than liquid water, a property that allows life to survive under the ice in ponds and lakes in cold climates.

9. Water is also ideally suited for its ecological role by its high specific heat, another property imparted by its strong hydrogen bonding. Large bodies of water are able to moderate the climate by giving off and absorbing substantial amounts of heat with only small changes in the water temperature.

10. All solids are either crystalline (with a regular structure of atoms, ions, or molecules) or amorphous (without this regular structure). Glass is an example of an amorphous solid.

11. Any crystal structure can be classified as one of seven basic types, built up of unit cells that are repeated throughout the crystal. X-ray diffraction has provided much of our knowledge about crystal structure.

12. The four types of crystals and the forces that hold their atoms, ions, or molecules together are ionic crystals, held together by ionic bonding; covalent crystals, covalent bonding; molecular crystals, van der Waals forces and/or hydrogen bonding; and metallic crystals, metallic bonding.

13. For a liquid in a closed vessel, dynamic equilibrium is established between evaporation and condensation. The vapor pressure over the liquid under these conditions is the equilibrium vapor pressure, which is often referred to more simply as "vapor pressure."

14. At the boiling point, the vapor pressure of a liquid equals the external pressure. The molar heat of vaporization of a liquid is the energy required to vaporize one mole of the liquid. The molar heat of fusion of a solid is the energy required to melt one mole of the solid.

15. The relationships among the phases of a single substance are represented by a phase diagram, in which each region represents a pure phase and the boundaries between the regions show the temperatures

and pressures at which the two phases are in equilibrium. At the triple point, all three phases are in equilibrium.

KEY WORDS

PROBLEMS

More challenging problems are marked with an asterisk.

Intermolecular Forces

14.1 Define the following terms: (a) dipole–dipole interaction, (b) dipole-induced dipole interaction, (c) ion–dipole interaction, (d) dispersion forces, (e) van der Waals forces.

14.2 Explain clearly the difference between the temporary dipole moment discussed in this chapter and the permanent dipole moment in a polar molecule.

14.3 Give some evidence that all neutral atoms and molecules exert attractive forces on one another.

14.4 Carbon tetrachloride (CCl_4) has no permanent dipole moment, yet its boiling point ($76.7°C$) is considerably higher than that of sulfur dioxide ($-10°C$), which has a dipole moment of 1.60 D. Explain these observations.

14.5 Which of the following substances are most likely to exist as gases at room temperature? (a) LiBr, (b) Ne, (c) CH_4, (d) CO_2, (e) $CaCO_3$. Explain.

14.6 If you lived in Alaska, which of the following natural gases would you keep in an outdoor storage tank in winter: methane (CH_4), propane (C_3H_8), or butane (C_4H_{10})? Explain.

14.7 The binary hydrogen compounds of the Group 4A elements are CH_4 ($-162°C$), SiH_4 ($-112°C$), GeH_4 ($-88°C$), and SnH_4 ($-52°C$). The temperatures in parentheses denote the boiling points. Explain the increase in boiling points from CH_4 to SnH_4.

14.8 Explain what van der Waals radii represent. How do they differ from atomic radii and ionic radii?

14.9 What is a hydrogen bond? Name the elements that can take part in hydrogen bond formation.

14.10 Ammonia is both a donor and an acceptor of hydrogen in hydrogen bond formation. Draw a diagram to show the hydrogen bonding of an ammonia molecule with two other ammonia molecules.

14.11 Which of the following species are capable of hydrogen bonding among themselves? (a) C_2H_6, (b) HI, (c) KF, (d) BeH_2, (e) HCOOH.

*__14.12__ Arrange the following in order of increasing boiling point: RbF, CO_2, CH_3OH, CH_3Br. Explain your choice.

14.13 Diethyl ether has a boiling point of 34.5°C and 1-butanol has a boiling point of 117°C:

$$\underset{\text{diethyl ether}}{H-\overset{\overset{\displaystyle H}{|}}{\underset{\underset{\displaystyle H}{|}}{C}}-\overset{\overset{\displaystyle H}{|}}{\underset{\underset{\displaystyle H}{|}}{C}}-O-\overset{\overset{\displaystyle H}{|}}{\underset{\underset{\displaystyle H}{|}}{C}}-\overset{\overset{\displaystyle H}{|}}{\underset{\underset{\displaystyle H}{|}}{C}}-H}$$

$$\underset{\text{1-butanol}}{H-\overset{\overset{\displaystyle H}{|}}{\underset{\underset{\displaystyle H}{|}}{C}}-\overset{\overset{\displaystyle H}{|}}{\underset{\underset{\displaystyle H}{|}}{C}}-\overset{\overset{\displaystyle H}{|}}{\underset{\underset{\displaystyle H}{|}}{C}}-\overset{\overset{\displaystyle H}{|}}{\underset{\underset{\displaystyle H}{|}}{C}}-OH}$$

These two compounds have the same numbers and types of atoms. Explain the difference in their boiling points. (*Hint:* Which of the two can take part in hydrogen bonding?)

*__14.14__ Explain the difference in the melting points of the following compounds:

m.p. 45°C　　m.p. 115°C

(*Hint:* Only one of the two can form intramolecular hydrogen bonds.)

14.15 People living in the northeastern United States who have swimming pools usually drain the water before winter every year. Why?

14.16 List the types of intermolecular forces that exist among molecules (or basic units) in each of the following species: (a) benzene (C_6H_6), (b) H_2O_2, (c) PF_3, (d) NaCl, (e) CS_2.

14.17 Which member of each of the following pairs of substances would you expect to have a higher boiling point? (a) O_2 and N_2, (b) SO_2 and CO_2, (c) HF and HI.

14.18 Explain in terms of intermolecular forces why (a) NH_3 has a higher boiling point than CH_4, (b) KCl has a higher melting point than I_2, and (c) naphthalene ($C_{10}H_8$) is more soluble in benzene than is LiBr.

14.19 Which of the following statements are false? (a) Dipole–dipole interactions between molecules are greatest if the molecules possess only temporary dipole moments (b) All compounds containing hydrogen atoms can participate in hydrogen bond formation. (c) Dispersion forces exist between all atoms, molecules, and ions. (d) The extent of ion-induced dipole interaction depends only on the charge on the ion.

14.20 What kind of attractive forces must be overcome in order to (a) melt ice, (b) boil molecular bromine, (c) melt solid iodine, and (d) dissociate F_2 into F atoms?

14.21 Which substance in each of the following pairs would you expect to have the higher boiling point? (a) Ne or Xe, (b) CO_2 or CS_2, (c) CH_4 or Cl_2, (d) F_2 or LiF, (e) NH_3 or PH_3. Explain.

*__14.22__ The following compounds have the same number and type of atoms. Which one would you expect to have a higher boiling point?

$$CH_3-CH_2-CH_2-CH_3 \qquad CH_3-\overset{\overset{\displaystyle CH_3}{|}}{CH}-CH_3$$

(*Hint:* Molecules that can be stacked together more easily would have a greater intermolecular attraction.)

The Liquid State

14.23 Explain why liquids, unlike gases, are virtually incompressible.

14.24 Predict which liquid has a greater surface tension, ethanol (C_2H_5OH) or dimethyl ether (CH_3OCH_3).

*__14.25__ Draw diagrams showing the capillary action for both (a) water and (b) mercury in three tubes of different radii.

14.26 Despite the fact that stainless steel is much denser than water, a razor blade can be made to float on water. Explain.

14.27 Predict the viscosity of ethylene glycol

$$\overset{\displaystyle CH_2-OH}{\underset{\displaystyle CH_2-OH}{|}}$$

relative to that of ethanol and glycerol (see Table 14.3).

14.28 Which liquid would you expect to have a greater viscosity, water or diethyl ether? The structure of diethyl ether is shown in Problem 14.13.

The Solid State

14.29 What is the coordination number of each sphere in (a) a simple cubic lattice, (b) a body-centered cubic lattice, and (c) a face-centered cubic lattice? Assume the spheres to be of equal size.

*14.30 Calculate the number of spheres in the simple cubic, body-centered cubic, and face-centered cubic cells. Assume the spheres to be of equal size and that they are only at the lattice points. Also calculate the packing efficiency for each type of cell.

*14.31 The metal polonium crystallizes in a primitive cubic lattice (the Po atoms occupy only the lattice points). If the unit cell length is 336 pm, what is the atomic radius of Po?

14.32 Tungsten crystallizes in a body-centered cubic lattice (the W atoms occupy only the lattice points). How many W atoms are present in a unit cell?

*14.33 The atomic radius of gold is 144 pm and gold crystallizes in a cubic close-packed structure (the face-centered cube). Calculate the density of the metal. (*Hint:* See Figure 14.23 for the relationship between the unit cell edge length and the radius of the atom.)

*14.34 Metallic iron crystallizes in a cubic lattice. The unit cell edge length is 287 pm. The density of iron is 7.87 g/cm^3. How many iron atoms are there within a unit cell?

*14.35 Silver crystallizes in a cubic close-packed structure (the face-centered cube). The unit cell edge length is 408.7 pm. Calculate the density of silver.

*14.36 Barium metal crystallizes in a body-centered cubic lattice (the Ba atoms are at the lattice points only). The unit cell edge length is 502 pm, and the density of the metal is 3.50 g/cm^3. Using this information, calculate Avogadro's number. (*Hint:* First calculate the volume occupied by one mole of Ba atoms in the unit cells. Next calculate the volume occupied by one Ba atom in the unit cell.)

14.37 Vanadium crystallizes in a body-centered cubic lattice (the V atoms occupy only the lattice points). How many V atoms are present in a unit cell?

*14.38 Europium crystallizes in a body-centered cubic lattice (the Eu atoms occupy only the lattice points). The density of Eu is 5.26 g/cm^3. Calculate the unit cell edge length in pm.

*14.39 Copper crystallizes in a face-centered cubic lattice (the Cu atoms are at the lattice points only). If the density of metallic copper is 8.96 g/cm^3, what is the unit cell edge length in pm?

*14.40 Crystalline silicon has a cubic structure. The unit cell edge length is 543 pm. The density of the solid is 2.33 g/cm^3. Calculate the number of Si atoms in one unit cell.

14.41 A face-centered cubic cell contains 8 X atoms at the corners of the cell and 6 Y atoms at the faces. What is the empirical formula of the solid?

14.42 When X rays of wavelength 0.090 nm are diffracted by a metallic crystal, the angle of first-order diffraction (n = 1) is measured to be 15.2°. What is the distance (in pm) between the layers of atoms responsible for the diffraction?

14.43 The distance between layers in a NaCl crystal is 282 pm. X rays are diffracted from these layers at an angle of 23.0°. Assuming that n = 1, calculate the wavelength of the X rays in nm.

14.44 X rays of wavelength 0.065 nm are diffracted from a crystal at an angle of 46°. Assuming that n = 1, calculate the distance (in pm) between layers in the crystal.

14.45 Describe the general properties of (a) covalent crystals, (b) molecular crystals, (c) ionic crystals, (d) metallic crystals. Give two examples of each type of crystal.

14.46 Classify the crystalline state of the following substances as ionic crystals, covalent crystals, molecular crystals, or metallic crystals: (a) CO_2, (b) B, (c) S_8, (d) KBr, (e) Mg, (f) SiO_2, (g) LiCl, (h) Cr. (The structure of B is discussed on p. 356.)

14.47 Explain why ice is less dense than liquid water.

14.48 Explain why diamond is harder than graphite.

14.49 Carbon and silicon belong to Group 4A of the periodic table and have the same valence electron configuration (ns^2np^2). Why does silicon dioxide (SiO_2) have a much higher melting point than carbon dioxide (CO_2)?

*14.50 The interionic distances of several alkali halides are

NaCl	NaBr	NaI	KCl	KBr	KI
282 pm	299 pm	324 pm	315 pm	330 pm	353 pm

(a) Plot lattice energy vs. interionic distance for these compounds. (b) Plot lattice energy vs. the reciprocal of the interionic distance. Which graph is more linear? Why? (For lattice energies see Table 10.2.)

14.51 Which has a greater density, crystalline SiO_2 or amorphous SiO_2?

Phase Changes

14.52 The vapor pressure of a liquid in a closed container depends on which of the following? (a) the volume above the liquid, (b) the amount of liquid present, (c) temperature.

14.53 Referring to Figure 14.36, estimate the boiling points of ethyl ether, water, and mercury at 0.5 atm.

14.54 Wet clothes dry more quickly on a hot and dry day than on a hot, humid day. Explain.

14.55 Explain why the molar heat of vaporization of a substance is invariably greater than its molar heat of fusion.

14.56 Which of the following phase transitions gives off more heat? (a) one mole of steam to one mole of water at 100°C, or (b) one mole of water to one mole of ice at 0°C.

14.57 A beaker of water is heated to boiling by a Bunsen burner. Would adding another burner raise the boiling point of water? Explain.

14.58 Calculate the amount of heat (in kJ) required to convert 74.6 g of water to steam at 100°C.

14.59 How much heat (in kJ) is needed to convert 866 g of ice at −10°C to steam at 126°C? (The specific heats of ice and steam are 2.03 J/g°C and 1.99 J/g°C, respectively.)

14.60 The molar heats of fusion and sublimation of molecular iodine are 15.27 kJ/mol and 62.30 kJ/mol, respectively. Use Equation (14.2) to estimate the molar heat of vaporization of liquid iodine.

14.61 How is the rate of evaporation of a liquid affected by (a) temperature, (b) the surface area of a liquid exposed to air, (c) intermolecular forces?

14.62 The following are liquid compounds and their boiling points: butane, −0.5°C; ethanol, 78.3°C; toluene, 110.6°C. At −10°C, which of these liquids would you expect to have the highest vapor pressure? Which the lowest? Explain.

14.63 Why is the critical temperature of water greater than that of most other substances?

14.64 Explain why no substance can exist in a liquid state at a temperature above its critical temperature.

14.65 Describe one practical consequence of the existence of a critical temperature for a gas.

14.66 If one substance has a higher critical temperature than another, what can you deduce about their relative intermolecular forces?

14.67 Freeze-dried coffee is prepared by freezing a sample of brewed coffee and then removing the ice component by vacuum pumping the sample. Describe the phase changes taking place during these processes.

14.68 A student hangs wet clothes outdoors on a winter day when the temperature is −15°C. After a few hours, the clothes are found to be fairly dry. Describe the phase changes in this drying process.

14.69 Use one of the phase changes as an example and explain what is meant by a dynamic equilibrium.

14.70 Steam at 100°C causes greater damage to the skin than water at 100°C. Why?

14.71 The greater the molar heat of vaporization of a liquid, the greater its vapor pressure. True or false?

14.72 A pressure cooker is a sealed container that allows steam to escape when it exceeds a predetermined pressure. How does this device reduce the time needed for cooking?

14.73 The blades of ice skates are quite thin, so that the pressure exerted on ice by a skater can be rather substantial. Explain how this fact enables a person to skate on ice.

*14.74 A length of wire is placed on top of a block of ice. The ends of the wire extend over the edges of the ice. A heavy weight is attached to each end. It is found that the ice under the wire gradually melts, so that the wire slowly moves through the ice block. At the same time, the water above the wire refreezes. Explain the phase changes that accompany this phenomenon.

14.75 Why is solid carbon dioxide called Dry Ice?

*14.76 The boiling point and freezing point of sulfur dioxide are −10°C and −72.7°C (at 1 atm), respectively. The triple point is −75.5°C and 1.65 × 10⁻³ atm, and its critical point is at 157°C and 78 atm. On the basis of this information, draw a rough sketch of the phase diagram of SO_2.

14.77 Consider the phase diagram of water shown below:

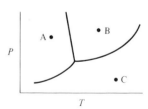

Label the regions. Predict what would happen if we did the following: (a) starting at A, we raise the temperature at constant pressure; (b) starting at C, we lower the temperature at constant pressure; (c) starting at B, we lower the pressure at constant temperature.

Miscellaneous Problems

*14.78 The properties of gases, liquids, and solids differ in a number of respects. How would you use the kinetic molecular theory (see Section 6.7) to explain the following observations? (a) Ease of compressibility decreases from gas to liquid to solid. (b) Solids retain a definite shape, but gases and liquids do not. (c) For most substances, the volume of a given amount of material increases as it changes from solid to liquid to gas.

14.79 A small drop of oil in water assumes a spherical shape. Explain. (*Hint:* Oil is made up of nonpolar molecules, which tend to avoid contact with water.)

14.80 Define the following terms: (a) evaporation, (b) condensation, (c) vapor pressure, (d) molar heat of vaporization, (e) molar heat of fusion, (f) melting point, (g) boiling point, (h) sublimation.

14.81 Define and/or explain the following terms: (a) crystalline solid, (b) amorphous solid, (c) surface tension, (d) viscosity, (e) cooling curve, (f) heating curve, (g) supercooled liquid, (h) phase diagram.

14.82 Name the kinds of attractive forces that must be overcome in order to (a) boil liquid ammonia, (b) melt solid phosphorus (P_4), (c) dissolve CsI in liquid HF, (d) melt potassium metal.

14.83 Select the substance in each pair that should have the higher boiling point. In each case identify the principal intermolecular forces involved and account briefly for your choice. (a) K_2S or $(CH_3)_3N$, (b) Br_2 or $CH_3CH_2CH_2CH_3$.

14.84 At $-35°C$, liquid HI has a higher vapor pressure than liquid HF. Explain.

14.85 Explain the critical temperature phenomenon in terms of the kinetic molecular theory.

*14.86 The melting points of the fluorides of the second-period elements are, approximately, LiF, 845°C; BeF_2, 800°C; BF_3, $-127°C$; CF_4, $-150°C$; NF_3, $-207°C$; OF_2, $-223.8°C$. Account for the variations in melting points in terms of the structures of the species and the strength of intermolecular forces that operate.

14.87 The standard enthalpy of formation of gaseous molecular bromine is 30.7 kJ/mol. From these data calculate the molar heat of vaporization of molecular bromine at 25°C. (*Hint:* The standard enthalpy of formation of liquid molecular bromine is zero.)

15
PHYSICAL PROPERTIES OF SOLUTIONS

In previous chapters we studied several macroscopic properties of solutions in which water was the solvent and the solute was either a molecular or an ionic substance. Such aqueous solutions are important in a variety of ways. Many chemical and nearly all biological processes take place in a water environment. And "wet chemistry"—chemical reactions occurring in water solution—is most likely a major focus of the laboratory program you are following in this course. In this chapter we will explore a number of macroscopic properties of solutions in which the solvent is a liquid and the solute is a solid, liquid, or gas. We will see that molecular-level interactions between units of the solvent and solute (as well as solvent–solvent and solute–solute interactions) help account for the characteristic behavior of a wide range of common solutions.

15.1   TYPES OF SOLUTIONS

We previously noted (see Section 3.2) that a solution represents a homogeneous mixture of two or more substances. Since this definition places no restrictions on the nature of the substances involved, we can consider six basic types of solutions, depending on the original states (solid, liquid, or gas) of the solution components. Table 15.1 gives examples of each of these types.

Our focus here will be on solutions involving at least one liquid component—that is, gas–liquid, liquid–liquid, and solid–liquid solutions. And, perhaps not too surprisingly, the liquid we will most often consider as the solvent in our study of solutions is water. But let's start our discussion of solutions at the molecular level, where all "dissolving" ultimately must take place.

TABLE 15.1   Types of Solutions

Component 1	Component 2	State of Resulting Solution	Examples
Gas	Gas	Gas	Air
Gas	Liquid	Liquid	Soda water (CO_2 in water)
Gas	Solid	Solid	H_2 gas in palladium
Liquid	Liquid	Liquid	Ethanol in water
Solid	Liquid	Liquid	NaCl in water
Solid	Solid	Solid	Brass (Cu/Zn), solder (Sn/Pb)

15.2   A MOLECULAR VIEW OF THE SOLUTION PROCESS

Molecules in liquid and solid states are held together by attractive intermolecular forces (see Section 14.2). These forces also play a central role in the formation of solutions. The saying "like dissolves like" is based on the action of intermolecular forces.

When one substance (the solute) dissolves in another (the solvent)—for example, a solid dissolving in a liquid—particles of the solute disperse uniformly throughout the solvent. The solute particles occupy positions that are normally taken by solvent molecules. The ease with which a solute particle may replace a solvent molecule depends on three types of interactions:

- solvent–solvent interaction
- solvent–solute interaction
- solute–solute interaction

We use the term "solute particle" rather than "solute molecule" because the dissolved species may be in the molecular or ionic form.

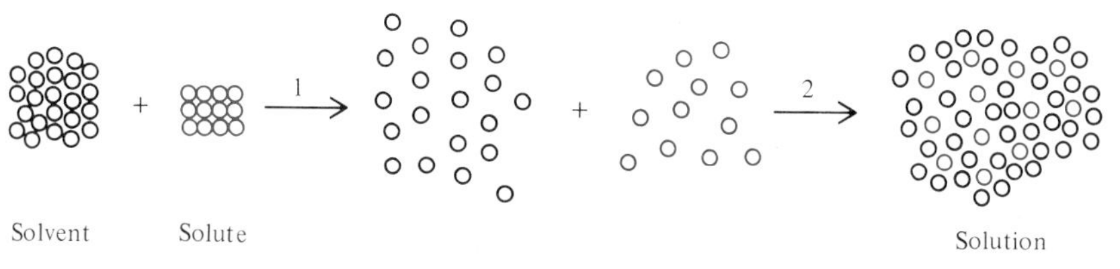

Solvent   Solute   Solution

We can imagine the solution process as taking place in two separate steps (Figure 15.1). Step 1 involves the separation of solvent molecules and of solute molecules, and step 2 involves the mixing of solvent and solute molecules to form a solution. Step 1 requires energy input to break up attractive intermolecular forces, whereas step 2 gives off energy to the surroundings. This analysis helps us understand the thermochemical nature of the solution process (see Section 5.6). If the solute–solvent attraction is stronger than the solvent–solvent attraction and solute–solute attraction, the solution process will be favorable, that is, the solute will be soluble in the solvent. Such a solution process is exothermic. If the solute–solvent interaction is weaker than the solvent–solvent and solute–solute interactions, then only a relatively small amount of the solute will be dissolved in the solvent and the solution process will be endothermic.

You may wonder why a solute dissolves in a solvent if the attraction among its own molecules is stronger than that between its molecules and the solvent molecules. The dissolution process, and indeed all physical and chemical processes, is governed by two factors. One is the energy factor, which in our discussion here determines whether a solution process will be exothermic or endothermic. As we mentioned, only exothermic solution processes are related to favorable solute–solvent interactions.

The other factor that needs to be considered is the disorder or randomness that results when solute and solvent molecules mix to form a solution. In the pure state, the solvent and solute possess a fair degree of order. By order we mean the more or less regular arrangement of atoms, molecules, or ions in three-dimensional space. Much of this order is destroyed when the solute dissolves in the solvent (see Figure 15.1). Therefore, the solution process is *always* accompanied by an increase in disorder or randomness. It is the increase in disorder of the system that favors the solubility of any substance. This explains why a substance can be soluble in a solvent even though the solution process is endothermic.

A good illustration of the importance of the increase in disorder as a driving force in the solution process is provided by the mixing of two nonreacting gases, such as N_2 and O_2. Under atmospheric conditions these gases behave almost like ideal gases, so that the intermolecular interaction between them is negligible. Consequently, the mixing process is neither exothermic nor endothermic. Yet we know that these gases can be mixed in any proportion, which means that the process is highly favorable. Indeed, it is the disorder created when these two pure gases are mixed to form a mixture (or a solution) of gases that is responsible for the process taking place so readily.

This gaseous solution—assuming a certain proportion of oxygen gas and nitrogen gas and mixed with small amounts of carbon dioxide and argon—is known, of course, as "air"!

15.3 SOLUTIONS OF LIQUIDS IN LIQUIDS

Both carbon tetrachloride [CCl_4, b.p. (boiling point) = 76.5°C] and benzene (C_6H_6, b.p. = 80.1°C) are nonpolar liquids. The only intermolecular forces present in these substances are dispersion forces (see Section 14.2). When these two liquids are mixed, they readily dissolve in each other. This is because the attraction between CCl_4 and C_6H_6 molecules is comparable in magnitude to that between CCl_4 molecules and that between C_6H_6 molecules. When *two liquids are completely soluble in each other in all proportions*, as in this case, they are said to be **miscible.** Ethanol (C_2H_5OH) and water also are miscible liquids. In a solution of ethanol and water the solute–solvent interaction takes the form of hydrogen bonding that is comparable in magnitude to the hydrogen bonding between water molecules and between ethanol molecules.

What would happen if we tried to dissolve carbon tetrachloride in water? To form a solution, the CCl_4 molecules would have to replace some of the H_2O molecules. However, the attractive forces between CCl_4 and H_2O molecules are dipole-induced dipole and dispersion forces, which in this case are much weaker than the hydrogen bonds in water (and the dispersion forces in CCl_4). Consequently, the two liquids do not mix; they are said to be *immiscible.* What we obtain, then, are two distinct liquid phases: one consisting of water with a very small amount of CCl_4 in it, and one consisting of CCl_4 with a very small amount of water in it (Figure 15.2).

These examples illustrate the "like dissolves like" saying. Other examples are sugar and ethylene glycol (a substance used as an antifreeze) dissolving in water (sugar and ethylene glycol molecules can form hydrogen bonds with water) and elemental sulfur (containing the nonpolar S_8 molecules) dissolving in the nonpolar carbon disulfide (CS_2) solvent. The forces of attraction between the S_8 molecules and the CS_2 molecules are solely dispersion forces.

The "like dissolves like" saying, while useful, should be applied with caution. For example, acetic acid (CH_3COOH) is totally miscible with water. We can understand this fact because CH_3COOH molecules can form hydrogen bonds with H_2O molecules. Yet acetic acid is also soluble in nonpolar solvents, such as CCl_4 and C_6H_6! Experimental evidence shows that in nonpolar solvents acetic acid forms **dimers,** which are *molecules made up of two identical molecules.* The acetic acid dimer has the structure

$$H_3C-C\diagup\substack{O\text{----}H-O \\ O-H\text{----}O}\diagdown C-CH_3$$

in which two (*di-*) molecules of the acid are bonded together. As the formula shows, the dimer is held together by two hydrogen bonds. Although acetic acid is a polar molecule, the dimeric species is much less polar because the symmetric structure created by hydrogen bonding reduces the polarity of the molecules. The attractive forces between the dimers and CCl_4 or C_6H_6 molecules are mainly dispersion forces.

In either gas–gas or liquid–liquid solutions, distinctions between solutes and solvents become more a matter of definition than physical reality. In these cases the solvent is usually assumed to be the component present in the greatest amount.

A sugar molecule is like a water molecule only in the sense that both molecules exert similar types of intermolecular forces, that is, hydrogen bonding.

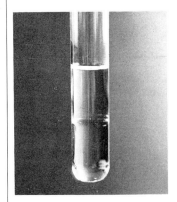

FIGURE 15.2 *A two-phase mixture of water and carbon tetrachloride. The more dense CCl_4 forms the bottom layer.*

15.4 SOLUTIONS OF SOLIDS IN LIQUIDS

To facilitate the discussion of the solubility of solids in liquids, we can divide solids into four categories, as outlined in Section 14.6: ionic, covalent, molecular, and metallic.

Ionic Crystals

> The solubility of a solid in a liquid does not depend on whether it is in crystalline or noncrystalline form.

Section 5.6 described the solution process of ionic compounds in some detail. Using sodium chloride as an example, we saw that ions are stabilized in solution by hydration, which involves ion–dipole interaction. We also analyzed the heat of solution in terms of the lattice energy of the compound and the heat of hydration. In general, we can predict that ionic compounds should be much more soluble in polar solvents, such as water, liquid ammonia, and liquid hydrogen fluoride, than in nonpolar solvents, such as benzene and carbon tetrachloride. Since the molecules of nonpolar solvents lack a dipole moment, they cannot effectively solvate the Na^+ and Cl^- ions. The predominant intermolecular interaction is ion-induced dipole interaction, which is much weaker than ion–dipole interaction. Consequently, the solubility of ionic compounds in nonpolar solvents is in general extremely low. A set of guidelines for predicting the solubility of many inorganic ionic compounds in water is given in Section 3.2.

Covalent Crystals

Covalent crystals, such as graphite and quartz (SiO_2), generally do not dissolve in any solvent, polar or nonpolar.

Molecular Crystals

The attractive forces in a molecular crystal are relatively weak dipole–dipole, dispersion type, and/or hydrogen bonding. Consider naphthalene ($C_{10}H_8$) as an example. In a naphthalene crystal the $C_{10}H_8$ molecules are held together solely by dispersion forces. Thus we can predict that naphthalene should dissolve readily in nonpolar solvents, such as benzene, but much less so in water. This is indeed the case. On the other hand, we find that urea (H_2NCONH_2) is much more soluble in water and ethanol

> Urea is an organic compound that is used as a fertilizer, in plastics, and in medicine, among other things.

$$H_2N—C—NH_2$$
$$\overset{\|}{O}$$

than in CCl_4 or C_6H_6 because it has the capacity to form hydrogen bonds with water. (The H atoms and the O and N atoms in urea can form hydrogen bonds with H_2O and C_2H_5OH molecules.)

Metallic Crystals

In general, metals are not soluble in any solvent, whether it is polar or nonpolar. A number of metals instead react chemically with the solvent.

For example, the alkali metals and some alkaline earth metals (Ca, Sr, and Ba) react with water to produce hydrogen gas and the corresponding metal hydroxide. The alkali metals also "dissolve" in liquid ammonia to produce a beautiful blue solution containing solvated electrons:

$$Na(s) \xrightarrow{NH_3} Na^+ + e^-$$

The Na^+ ions and the electrons are stabilized by ion–dipole interaction with the polar NH_3 molecules.

> Although we use the term "dissolve," this is really a chemical rather than a physical process.

EXAMPLE 15.1

Predict the relative solubilities in the following cases: (a) Br_2 in carbon tetrachloride (μ = 0 D) or water (μ = 1.87 D), (b) KCl in benzene (μ = 0 D) or liquid ammonia (μ = 1.46 D).

Answer

(a) Br_2 is a nonpolar molecule and therefore should be more soluble in CCl_4, which is also nonpolar, than in water. The only intermolecular forces between Br_2 and CCl_4 are dispersion forces.
(b) KCl is an ionic compound. For it to dissolve, the individual K^+ and Cl^- ions must be stabilized by ion–dipole interaction. Since benzene does not have a dipole moment, potassium chloride should be more soluble in liquid ammonia, a polar molecule with a large dipole moment.

Similar example: Problem 15.9.

15.5 CONCENTRATION UNITS

A quantitative study of a solution requires that we know its *concentration*, that is, the amount of solute dissolved in a given amount of solvent. Chemists use several different concentration units in their work, each of which has advantages as well as limitations. The particular concentration unit used is generally determined by the kind of measurement made of the solution, to be discussed later in this section. First, though, let us examine four kinds of concentration units: percent by mass, mole fraction, molarity, and molality.

Types of Concentration Units

Percent by Mass. The *percent by mass* (also called the *percent by weight* or the *weight percent*) is defined as

$$\text{percent by mass of solute} = \frac{\text{mass of solute}}{\text{mass of solute + mass of solvent}} \times 100\%$$
$$= \frac{\text{mass of solute}}{\text{mass of soln}} \times 100\%$$

The percent by mass has no units because it is a ratio of two similar quantities.

EXAMPLE 15.2

A sample of 0.892 g of potassium chloride (KCl) is dissolved in 54.6 g of water. What is the percent by mass of KCl in this solution?

Answer

$$\text{percent by mass of KCl} = \frac{\text{mass of solute}}{\text{mass of soln}} \times 100\%$$

$$= \frac{0.892 \text{ g}}{0.892 \text{ g} + 54.6 \text{ g}} \times 100\%$$

$$= 1.61\%$$

Similar example: Problem 15.12.

Mole Fraction (X). The concept of mole fraction was first introduced in Section 6.6. The mole fraction of a component of a solution, say, component A, is written X_A and is defined as

$$\text{mole fraction of component A} = X_A = \frac{\text{moles of A}}{\text{sum of moles of all components}}$$

The mole fraction has no units, since it is a ratio of two similar quantities.

EXAMPLE 15.3

A chemist prepared a solution by adding 200.4 g of pure ethanol (C_2H_5OH) to 143.9 g of water. Calculate the mole fractions of these two components. The molar masses of ethanol and water are 46.02 g and 18.02 g, respectively.

Answer

The number of moles of C_2H_5OH and H_2O present are

$$\text{moles of } C_2H_5OH = 200.4 \text{ g } C_2H_5OH \times \frac{1 \text{ mol } C_2H_5OH}{46.02 \text{ g } C_2H_5OH}$$

$$= 4.355 \text{ mol } C_2H_5OH$$

$$\text{moles of } H_2O = 143.9 \text{ g } H_2O \times \frac{1 \text{ mol } H_2O}{18.02 \text{ g } H_2O}$$

$$= 7.986 \text{ mol } H_2O$$

In a two-component system made up of A and B molecules, the mole fraction of A is given by

$$\text{mole fraction of A} = \frac{\text{moles of A}}{\text{sum of moles of A and B}}$$

Using this equation we can write the mole fractions of ethanol and water as

$$X_{C_2H_5OH} = \frac{4.355 \text{ mol}}{(4.355 + 7.986) \text{ mol}} = 0.3529$$

$$X_{H_2O} = \frac{7.986 \text{ mol}}{(4.355 + 7.986) \text{ mol}} = 0.6471$$

By definition, the sum of the mole fractions of all components in a solution must be 1. Thus

$$X_{C_2H_5OH} + X_{H_2O} = 0.3529 + 0.6471 = 1.0000$$

Molarity (M). The molarity unit was first introduced in Section 4.4; it is defined as the number of moles of solute in one liter of solution, that is

$$\text{molarity} = \frac{\text{moles of solute}}{\text{liters of soln}}$$

Thus, molarity has the units of mol/L. For example, the molar mass of acetic acid (CH_3COOH) is 60.05 g, so one liter of solution containing 60.05 g of the substance is one molar, or 1.000 M acetic acid. Similarly, if 6.005 g of acetic acid is dissolved in enough water to make up 100 mL of solution, the concentration is again 1.000 M. As long as we know the amount of solute present in a given volume of solution, we can always calculate its molarity. By convention, we use square brackets [ ] to represent mol/L; for example, in 2.0 M CH_3COOH, the concentration of CH_3COOH is written

$$[CH_3COOH] = 2.0 \ M = 2.0 \text{ mol/L}$$

and in 0.34 M CH_3COOH

$$[CH_3COOH] = 0.34 \ M = 0.34 \text{ mol/L}$$

It is important to keep in mind that molarity refers only to the amount of solute originally dissolved in water or other solvent and does not deal with subsequent processes (such as the ionization of an acid or the dissociation of an ionic compound).

EXAMPLE 15.4

What is the molarity of 153.2 mL of an aqueous solution containing 24.42 g of lithium chlorate $(LiClO_3)$? The molar mass of $LiClO_3$ is 90.39 g.

Answer

The first step is to calculate the number of moles of $LiClO_3$ in the solution:

$$\text{moles of } LiClO_3 = 24.42 \text{ g } LiClO_3 \times \frac{1 \text{ mol } LiClO_3}{90.39 \text{ g } LiClO_3}$$

$$= 0.2702 \text{ mol } LiClO_3$$

(Continued)

For additional examples of
molarity calculations, refer to
Examples 4.8 to 4.10.

This is the number of moles in 153.2 mL of solution. Now we can calculate the moles in one liter of solution, or the molarity of the solution:

$$\text{molarity} = \frac{\text{moles of solute}}{\text{liters of soln}}$$

$$[\text{LiClO}_3] = \frac{0.2702 \text{ mol LiClO}_3}{153.2 \text{ mL}} \times \frac{1000 \text{ mL}}{1 \text{ L}}$$

$$= 1.764 \text{ mol LiClO}_3/\text{L}$$

$$= 1.764 \text{ } M$$

Molality (m). *Molality* is *the number of moles of solute dissolved in 1 kg (1000 g) of solvent*—that is

$$\text{molality} = \frac{\text{moles of solute}}{\text{mass of solvent (kg)}}$$

For example, to prepare a one *molal*, or 1 *m*, sodium sulfate (Na_2SO_4) aqueous solution, we need to dissolve 1 mole (142.0 g) of the substance in 1000 g (1 kg) of water. Depending on the nature of the solute–solvent interaction, the final volume of the solution will be either greater or less than 1000 mL. It is also possible, though very unlikely, that the final volume could be equal to 1000 mL.

EXAMPLE 15.5

Calculate the molality of a sulfuric acid solution containing 24.4 g of sulfuric acid in 198 g of water. The molar mass of sulfuric acid is 98.08 g.

Answer

From the known molar mass of sulfuric acid, we can calculate the molality in two steps. First we need to find the number of grams of sulfuric acid dissolved in 1000 g (1 kg) of water. Next we must convert the number of grams into the number of moles. Combining these two steps we write

$$\text{molality} = \frac{\text{moles of solute}}{\text{mass of solvent (kg)}}$$

$$= \frac{24.4 \text{ g H}_2\text{SO}_4}{198 \text{ g H}_2\text{O}} \times \frac{1000 \text{ g H}_2\text{O}}{1 \text{ kg H}_2\text{O}} \times \frac{1 \text{ mol H}_2\text{SO}_4}{98.08 \text{ g H}_2\text{SO}_4}$$

$$= 1.26 \text{ mol H}_2\text{SO}_4/\text{kg H}_2\text{O}$$

$$= 1.26 \text{ } m$$

Similar example: Problem 15.14.

EXAMPLE 15.6

How many grams of water are needed to dissolve 18.7 g of ammonium nitrate (NH_4NO_3) in order to prepare a 0.542 m solution?

Answer

The number of moles of NH_4NO_3 in 18.7 g of NH_4NO_3 is

$$\text{moles} = 18.7 \text{ g NH}_4\text{NO}_3 \times \frac{1 \text{ mol NH}_4\text{NO}_3}{80.06 \text{ g NH}_4\text{NO}_3} = 0.234 \text{ mol NH}_4\text{NO}_3$$

Since

$$\text{molality} = \frac{\text{moles of solute}}{\text{mass of solvent (kg)}}$$

we write

$$\text{mass of water} = \frac{\text{mol NH}_4\text{NO}_3}{\text{molality}}$$

$$= 0.234 \text{ mol NH}_4\text{NO}_3 \times \frac{1 \text{ kg H}_2\text{O}}{0.542 \text{ mol NH}_4\text{NO}_3}$$

$$= 0.432 \text{ kg H}_2\text{O}$$
$$= 432 \text{ g H}_2\text{O}$$

Comparison of Concentration Units

As mentioned, the four concentration terms have various uses. For instance, the mole fraction unit is not generally used to express the concentrations of solutions for titrations and gravimetric analyses, but it is useful for calculating partial pressures of gases (see Section 6.6) and in dealing with vapor pressures of solutions (to be discussed later in this chapter).

We saw in Section 4.4 that the advantage of molarity is that it is generally easier to measure the volume of a solution, using precisely calibrated volumetric flasks, than to weigh the solvent. For this reason, molarity is often preferred over molality. On the other hand, molality is independent of temperature, since the concentration is expressed in number of moles of solute and mass of solvent, whereas the volume of a solution generally increases with increasing temperature. A 1.00 M solution at 25°C may become 0.97 M at 45°C because of volume expansion. This concentration dependence on temperature can significantly affect the accuracy of an experiment.

Percent by mass is similar to molarity in that it too is independent of temperature, since it is defined in terms of ratio of mass of solute to mass of solution. Furthermore, we do not need to know the molar mass of the solute to calculate the percent by mass.

The following examples will show that these terms are interconvertible if the density of the solution is known.

EXAMPLE 15.7

Calculate the molarity of a 0.396 molal glucose ($C_6H_{12}O_6$) solution. The molar mass of glucose is 180.2 g and the density of the solution is 1.16 g/mL.

Answer

The mass of the solution must be converted to volume in working problems of this type. The density of the solution is used as a conversion factor. Since a 0.396 m solution contains 0.396 mole of glucose in 1 kg of water, the total mass of the solution is

$$\left(0.396 \text{ mol } C_6H_{12}O_6 \times \frac{180.2 \text{ g}}{1 \text{ mol } C_6H_{12}O_6}\right) + 1000 \text{ g } H_2O \text{ solution} = 1071 \text{ g}$$

From the known density of the solution (1.16 g/mL), we can calculate the molarity as follows:

$$
\begin{aligned}
\text{molarity} &= \frac{0.396 \text{ mol } C_6H_{12}O_6}{1071 \text{ g soln}} \times \frac{1.16 \text{ g soln}}{1 \text{ mL soln}} \times \frac{1000 \text{ mL soln}}{1 \text{ L soln}} \\
&= \frac{0.429 \text{ mol } C_6H_{12}O_6}{1 \text{ L soln}} \\
&= 0.429 \text{ } M
\end{aligned}
$$

EXAMPLE 15.8

The density of a 2.45 M aqueous methanol (CH_3OH) solution is 0.976 g/mL. What is the molality of the solution? The molar mass of methanol is 32.04 g.

Answer

The total mass of 1 liter of a 2.45 M solution of methanol is

$$1 \text{ L soln} \times \frac{1000 \text{ mL soln}}{1 \text{ L soln}} \times \frac{0.976 \text{ g soln}}{1 \text{ mL soln}} = 976 \text{ g soln}$$

Since this solution contains 2.45 moles of methanol, the amount of water in the solution is

$$976 \text{ g soln} - \left(2.45 \text{ mol } CH_3OH \times \frac{32.04 \text{ g } CH_3OH}{1 \text{ mol } CH_3OH}\right) = 898 \text{ g } H_2O$$

Now the molality of the solution can be calculated:

$$
\begin{aligned}
\text{molality} &= \frac{2.45 \text{ mol } CH_3OH}{898 \text{ g } H_2O} \times \frac{1000 \text{ g } H_2O}{1 \text{ kg } H_2O} \\
&= \frac{2.73 \text{ mol } CH_3OH}{1 \text{ kg } H_2O} \\
&= 2.73 \text{ } m
\end{aligned}
$$

Similar examples: Problems 15.15, 15.16.

EXAMPLE 15.9

Calculate the molality of a 35.4 percent (by mass) aqueous solution of phosphoric acid (H_3PO_4). The molar mass of phosphoric acid is 98.00 g.

Answer

100% − 35.4% = 64.6%, so the ratio of phosphoric acid to water in this solution is 35.4 g H_3PO_4 to 64.6 g H_2O. From the known molar mass of phosphoric acid, we can calculate the molality in two steps. First we need to find the number of grams of phosphoric acid dissolved in 1000 g (1 kg) of water. Next, we must convert the number of grams into the number of moles. Combining these two steps we write

$$\text{molality} = \frac{35.4 \text{ g } H_3PO_4}{64.6 \text{ g } H_2O} \times \frac{1000 \text{ g } H_2O}{1 \text{ kg } H_2O} \times \frac{1 \text{ mol } H_3PO_4}{98.00 \text{ g } H_3PO_4}$$

$$= 5.59 \text{ mol } H_3PO_4/\text{kg } H_2O$$

$$= 5.59 \; m$$

Similar example: Problem 15.20.

15.6 TEMPERATURE AND SOLUBILITY

Solid Solubility and Temperature

Temperature usually affects solubility. *At a given temperature, when the maximum amount of a substance has dissolved in a solvent,* we have a **saturated solution.** Before this point is reached, *when a solution contains less solute than it has the capacity to dissolve,* we have an **unsaturated solution.** A dynamic equilibrium process exists in a saturated solution. Take a saturated LiBr solution as an example. The solution can be prepared by adding an *excess* of solid LiBr to a beaker of water at 25°C. At every instant some of the Li^+ and Br^- ions in the solid dissolve in water, while an *equal* number of Li^+ and Br^- ions in solution aggregate to re-form solid LiBr. *The process in which dissolved solute comes out of solution and forms crystals* is called **crystallization.**

Figure 15.3 shows the temperature dependence of the solubility of some ionic compounds in water. In most but certainly not all cases the solubility of a solid substance increases with temperature. However, there is no clear correlation between the sign of ΔH_{soln} and the variation of solubility with temperature. For example, the solution process of $CaCl_2$ is exothermic, whereas that of NH_4NO_3 is endothermic. But the solubility of both compounds increases with increasing temperature. In general it is difficult to predict the change in solubility as temperature changes; such a dependence is normally determined by experiments.

Many substances can also form **supersaturated solutions,** which *contain more of the solute than is present in saturated solutions.* For instance, a supersaturated solution of sodium acetate (CH_3COONa) can be prepared by

FIGURE 15.3 *Tempera-ture dependence of the sol-ubility of some ionic com-pounds in water.*

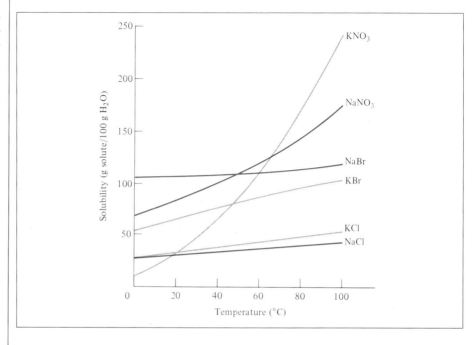

cooling a saturated solution of it. If the process is carried out slowly and carefully, the excess sodium acetate will not crystallize out as the temper-ature decreases. However, supersaturated solutions are unstable. Once the saturated solution has been prepared, addition of a very small amount of sodium acetate (called a *seed crystal*) will cause the excess sodium acetate to crystallize out of the solution immediately (Figure 15.4).

Note that both precipitation and crystallization describe the separation of excess solid substance from a supersaturated solution. However, the solids formed in the two processes differ in appearance. We normally think of precipitates as being made up of small particles, whereas crystals may be large and well-formed.

Precipitation reactions are first introduced in Section 3.6.

FIGURE 15.4 *From left to right: A supersaturated sodium acetate solution and the rapid appearance of sodium acetate crystals after the addition of a small seed crystal.*

Fractional Crystallization

As Figure 15.3 shows, the degree to which the solubility of a substance depends on temperature varies considerably. *Fractional crystallization,* or *the separation of a mixture of substances into pure components on the basis of their differing solubilities,* makes use of this fact. For example, suppose we have a sample of 90 g of KNO_3, which is contaminated with 10 g of NaCl. To purify the KNO_3 sample, we dissolve the mixture in 100 mL of water at 60°C and then gradually cool the solution to 0°C. At this temperature the solubilities of KNO_3 and NaCl are 12.1 g/100 g H_2O and 34.2 g/100 g H_2O, respectively. Thus (90 − 12) g, or 78 g, of KNO_3 will separate from the solution, but all of the NaCl will remain dissolved (Figure 15.5). In this manner, about 90 percent of the original KNO_3 can be obtained in pure form. The KNO_3 crystals can be separated from the solution by filtration.

Many of the commercially available solid inorganic and organic compounds that are used in the laboratory are purified by fractional crystallization. Generally the method works best if the compound to be purified has a steep solubility curve, that is, if it is considerably more soluble at high temperatures than at low temperatures. Otherwise, much of it will remain dissolved as the solution is cooled. Fractional crystallization works well also if the amount of impurity or impurities in the solution is relatively small.

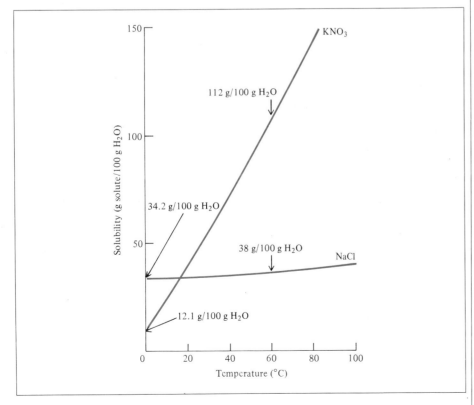

FIGURE 15.5 *The solubilities of KNO_3 and NaCl at 0°C and 60°C. The difference enables us to isolate one of these salts from a solution containing both of them, through fractional crystallization.*

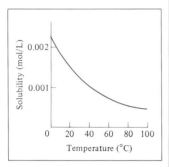

FIGURE 15.6 *Dependence on temperature of the solubility of O_2 gas in water. Note that the solubility decreases as temperature increases. The pressure of the gas over the solution is 1 atm.*

It is not clear whether fish would find this discussion to be on the lighter side.

Gas Solubility and Temperature

In contrast to the solubility of solids, the solubility of gases in water *always* decreases with increasing temperature (Figure 15.6). The next time you heat water in a beaker, notice the bubbles of air forming on the side of the glass before the water boils. As the temperature rises, the dissolved air molecules begin to "boil out" of the solution.

The reduced solubility of molecular oxygen in hot water has a direct bearing on **thermal pollution,** that is, *the heating of the environment— usually waterways—to temperatures that are harmful to its living inhabitants.* It is estimated that every year in the United States some 100,000 billion gallons of water are used for industrial cooling, mostly in electric power and nuclear power production. This process heats up the water, which is then released into the rivers and lakes from which it was taken. Ecologists have become increasingly concerned about the effect of thermal pollution on aquatic life. Fish, like all other cold-blooded animals, have much more difficulty coping with rapid temperature fluctuation in the environment than humans do. An increase in water temperature accelerates their rate of metabolism, which generally doubles with each increase of 10°C. The speedup of metabolism increases the fishes' need for oxygen gas at the same time that the supply of oxygen gas becomes more limited because of its lower solubility in heated water. Studies are in progress to devise effective ways to cool power plants that will minimize damage to the biological environment.

On the lighter side, knowledge of the variation of gas solubility with temperature can improve one's performance in a popular recreational sport—fishing. On a hot summer day, an experienced fisherman usually picks a deep spot in the river or lake to cast the bait. The oxygen gas content is greater in the deeper, cooler region, and most fish will be found there.

15.7 EFFECT OF PRESSURE ON THE SOLUBILITY OF GASES

For all practical purposes, external pressure does not influence the solubilities of liquids and solids, but it does have a major effect on the solubility of gases. The quantitative relationship between gas solubility and pressure is given by **Henry's law** (William Henry, 1775–1836), which states that *the solubility of a gas in a liquid (c) is proportional to the pressure of the gas over the solution:*

$$c \propto P$$

$$c = kP \tag{15.1}$$

If several gases are present, *P* becomes the partial pressure.

Note that c is the concentration (in mol/L) of the dissolved gas; P is the pressure (in atm) of the gas over the solution; and, for a given gas, k is a constant that depends only on temperature. The constant k has the units mol/L·atm. You can see that when the pressure of the gas is 1 atm, c is *numerically* equal to k.

Henry's law can be understood qualitatively in terms of kinetic molecular theory. The amount of gas that will dissolve in a solvent depends on how

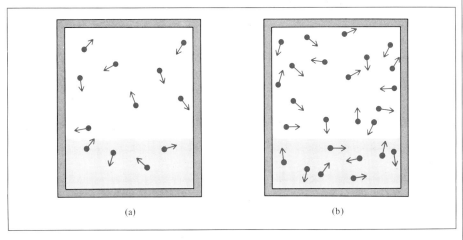

(a)   (b)

frequently the molecules in the gas phase collide with the liquid surface and subsequently become trapped by the condensed phase. Suppose we have a gas in dynamic equilibrium with a solution [Figure 15.7(a)]. At every instant, the number of gas molecules entering the solution is equal to the number of dissolved molecules moving into the gas phase. When the partial pressure is increased, more molecules will dissolve in the liquid because the number of molecules striking the surface of the liquid is greater. This process continues until the concentration of the solution is again such that the number of molecules leaving the solution per second is equal to that entering the solution [Figure 15.7(b)]. Because of the increased concentration of molecules in both the gas and solution phases, this number is greater in (b) than when the partial pressure was lower [Figure 15.7(a)].

EXAMPLE 15.10

The solubility of pure nitrogen gas at 25°C and 1 atm is 6.8×10^{-4} mol/L. What is the concentration of nitrogen gas dissolved in water under atmospheric conditions? Assume the partial pressure of nitrogen gas in atmosphere to be 0.78 atm.

Answer

The first step is to calculate the quantity k in Equation (15.1):

$$c = kP$$
$$6.8 \times 10^{-4} \text{ mol/L} = k(1 \text{ atm})$$
$$k = 6.8 \times 10^{-4} \text{ mol/L·atm}$$

Therefore, the solubility of nitrogen gas in water is

$$c = (6.8 \times 10^{-4} \text{ mol/L} \cdot \text{atm})(0.78 \text{ atm})$$
$$= 5.3 \times 10^{-4} \text{ mol/L}$$
$$= 5.3 \times 10^{-4} \text{ } M$$

The decrease in solubility is the result of lowering the pressure from 1 atm to 0.78 atm.

Similar example: Problem 15.28.

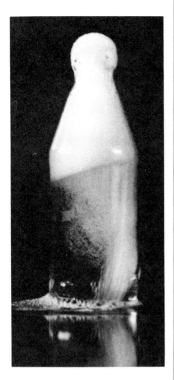

FIGURE 15.8 *The effervescence of a soft drink. The bottle was shaken slightly before opening to dramatize the escape of CO_2.*

Most gases obey Henry's law, although there are some important exceptions. For example, if the dissolved gas *reacts* with water, higher solubilities can result. The solubility of ammonia is higher than expected because of the reaction

$$NH_3 + H_2O \rightleftharpoons NH_4^+ + OH^-$$

Carbon dioxide also reacts with water, as follows:

$$CO_2 + H_2O \rightleftharpoons H_2CO_3$$

Another interesting example concerns the dissolution of molecular oxygen in blood. Normally, oxygen gas is only sparingly soluble in water (2.9×10^{-4} mol/L at 24°C and partial pressure of 0.2 atm). However, its solubility in blood increases dramatically because of the high content of hemoglobin (Hb) molecules in blood. As oxygen carriers, each hemoglobin molecule can bind up to four oxygen molecules, which are eventually delivered to the tissues for use in metabolism:

$$Hb + 4O_2 \rightleftharpoons Hb(O_2)_4$$

It is this process that accounts for the high solubility of molecular oxygen in blood.

A practical demonstration of Henry's law is provided every time we open a bottle of soda. Before the bottle is sealed, it is pressurized with a mixture of air and CO_2 saturated with water vapor. Because of carbon dioxide's high partial pressure, the amount of the gas dissolved in the soft drink is many times greater than the amount that could be dissolved under atmospheric conditions. Once the cap is removed, the pressurized gases escape, and the solubility of CO_2 is then determined only by its normal atmospheric partial pressure of 0.03 atm. Consequently, the excess dissolved CO_2 comes out of solution, causing the fizzing (Figure 15.8).

15.8 COLLIGATIVE PROPERTIES OF NONELECTROLYTE SOLUTIONS

Several important properties of solutions depend on the number of solute particles in solution and not on the nature of the solute particles. These properties are called **colligative properties** (or collective properties) because they are bound together through their common origin, that is, the number of solute particles present. They are vapor pressure lowering, boiling point elevation, freezing point depression, and osmotic pressure. For our discussion of colligative properties of nonelectrolyte solutions it is important to keep in mind that we are talking about relatively dilute solutions, that is, solutions whose concentrations are ≤0.2 *M*.

Vapor Pressure Lowering

To review the concept of equilibrium vapor pressure as it applies to pure liquids, see Section 14.6.

If a solute is **nonvolatile** (that is, it *does not have a measurable vapor pressure*) the vapor pressure of the solution is always less than that of the pure solvent. Thus, the relationship between solution vapor pressure and solvent vapor pressure depends on the concentration of the solute in the solution. This relationship is given by **Raoult's law** (Francois Marie Raoult, 1830–1901), which states that *the partial pressure of a solvent over a solution, P_1, is given by the vapor pressure of the pure solvent, P_1°, times the mole fraction of the solvent in the solution, X_1*:

$$P_1 = X_1 P_1^\circ \tag{15.2}$$

In a solution containing only one solute, $X_1 = 1 - X_2$, where X_2 is the mole fraction of the solute, so that Equation (15.2) can be rewritten as

$$P_1 = (1 - X_2)P_1^\circ$$
$$P_1^\circ - P_1 = \Delta P = X_2 P_1^\circ \tag{15.3}$$

We see that the decrease in vapor pressure, ΔP, is directly proportional to the concentration (measured in mole fraction) of the solute present.

EXAMPLE 15.11

At 25°C, the vapor pressure of pure water is 23.76 mmHg and that of a dilute aqueous urea solution is 22.98 mmHg. Estimate the molality of the solution.

Answer

From Equation (15.3) we write

$$\Delta P = (23.76 - 22.98) \text{ mmHg} = X_2(23.76 \text{ mmHg})$$
$$X_2 = 0.033$$

By definition

$$X_2 = \frac{n_2}{n_1 + n_2}$$

where n_1 and n_2 are the numbers of moles of solvent and solute, respectively. Since the solution is dilute, we can assume that n_1 is much larger than n_2, and we can write

$$X_2 = \frac{n_2}{n_1 + n_2} \simeq \frac{n_2}{n_1} \quad (n_1 \gg n_2)$$
$$n_2 = n_1 X_2$$

The number of moles of water in 1 kg of water is

$$1000 \text{ g H}_2\text{O} \times \frac{1 \text{ mol H}_2\text{O}}{18.02 \text{ g H}_2\text{O}} = 55.5 \text{ mol H}_2\text{O}$$

and the number of moles of urea present in 1 kg of water is

$$n_2 = n_1 X_2 = (55.5 \text{ mol})(0.033)$$
$$= 1.8 \text{ mol}$$

Thus, the concentration of the urea solution is 1.8 m.

Similar examples: Problems 15.33, 15.34.

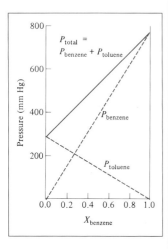

FIGURE 15.9 *The dependence of the partial pressures of benzene and toluene on their mole fraction in a benzene–toluene solution ($X_{toluene} = 1 - X_{benzene}$) at 80°C. This solution is said to be ideal because the vapor pressures obey Raoult's law [Equation (15.2)].*

An ideal solution has a zero heat of solution; that is, the solution process is neither exothermic nor endothermic. In practice there are very few known ideal solutions.

The longer the fractionating column, the more complete the separation of the volatile liquids.

If both components of a solution are **volatile** (that is, *have measurable vapor pressure*), the vapor pressure of the solution will be the sum of the individual partial pressures. Raoult's law holds equally well in this case:

$$P_A = X_A P_A^\circ$$
$$P_B = X_B P_B^\circ$$

where P_A and P_B are the partial pressures over the solution for components A and B, and P_A° and P_B° are the vapor pressures of the pure substances. The total vapor pressure is given by Dalton's law of partial pressure [Equation (6.11)]:

$$P_{total} = P_A + P_B$$

Figure 15.9 shows the dependence of the total vapor pressure (P_{total}) in a benzene–toluene solution on the composition of the solution. [Note that we need only express the composition of the solution in terms of the mole fraction of one component, which is the mole fraction of benzene in our discussion. For every value of $X_{benzene}$, the mole fraction of toluene, $X_{toluene}$, is given by $(1 - X_{benzene})$.] The benzene–toluene solution is one of the few examples of an **ideal solution,** which is *any solution that obeys Raoult's law.*

Fractional Distillation.   The discussion of vapor pressure of solutions has a direct bearing on **fractional distillation,** *a procedure for separating liquid components of a solution that is based on their different boiling points.* Fractional distillation is somewhat analogous to fractional crystallization (see Section 15.5). Suppose we want to separate a *binary system* (that is, a system with two components), say, benzene–toluene, one of the few ideal solutions known. Both benzene and toluene are relatively volatile, yet their boiling points are appreciably different (80.1°C and 110.6°C, respectively). When we boil a solution containing these two substances, the vapor formed will be somewhat richer in the more volatile component, benzene. If the vapor is condensed in a separate container and that liquid is boiled again, a still higher concentration of benzene will be obtained. By repeating this process many times, it is possible to separate benzene completely from toluene.

In practice, chemists use an apparatus such as that shown in Figure 15.10 to separate volatile liquids. The round-bottomed flask containing the benzene–toluene solution is fitted with a long column packed with small glass beads. When the solution boils, the vapor condenses on the lower portion of the beads and the liquid falls back into the distilling flask. As time goes on, the beads gradually become heated, allowing the vapor to move upward slowly. In essence, the packing material causes the benzene–toluene mixture to be subjected continuously to numerous vaporization–condensation steps. At each step the composition of the vapor within the column becomes progressively richer in the more volatile or lower boiling point component (in this case, benzene). Finally, essentially pure benzene vapor rises to the top of the column and is then condensed and collected in a receiving flask.

Fractional distillation is as important in industry as it is in the laboratory.

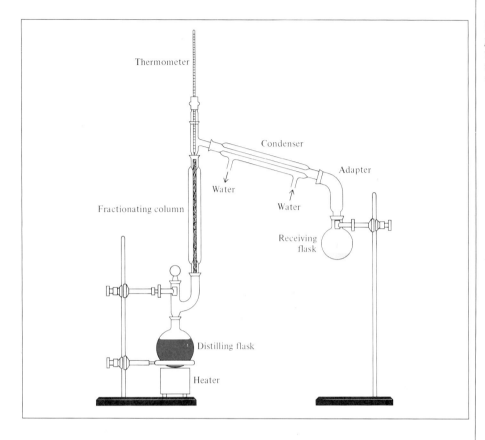

The petroleum industry employs fractional distillation on a large scale to separate from crude oil its various components.

Boiling Point Elevation

We have seen that the presence of a nonvolatile solute lowers the vapor pressure of a solution. It must therefore also affect the boiling point of the solution. The boiling point of a solution is the temperature at which its vapor pressure equals the external atmospheric pressure (see Section 14.6). Figure 15.11 shows the phase diagram of water and the changes that occur in an aqueous solution. Because at any temperature the vapor pressure of the solution is lower than that of the pure solvent, the liquid–vapor curve (the dotted line) for the solution lies below that for the pure solvent (the solid line). Consequently, the intersection between the dotted line and the horizontal line (starting at $P = 1$ atm) falls on a *higher* temperature value than the boiling point of the pure solvent. This graphical analysis shows that the boiling point of the solution will be higher than that of water. The boiling point elevation, ΔT_b, is defined as

$$\Delta T_b = T_b - T_b^\circ$$

Since $T_b > T_b^\circ$, ΔT_b is a positive quantity.

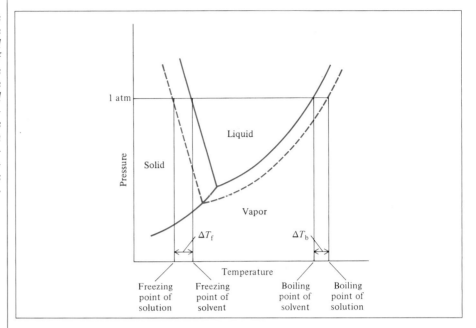

where T_b is the boiling point of the solution and T_b° the boiling point of the solvent. Because ΔT_b is proportional to the vapor pressure lowering, it is also proportional to the concentration (molality) of the solution, that is

$$\Delta T_b \propto m$$

$$\Delta T_b = K_b m \qquad (15.4)$$

where m is the concentration of the solute in molality units and K_b is the proportionality constant. The latter is called the *molal boiling point elevation constant*; it has the units of °C/m.

It is important to understand the choice of concentration unit here. We are dealing with a system (the solution) whose temperature is *not* kept constant, so we cannot express the concentration units in molarity because molarity changes with temperature.

Table 15.2 lists the K_b values of several common solvents. Using the K_b value for water and Equation (15.4), you can see that if the molality m of an aqueous solution is 1.00, the boiling point will be 100.52°C.

Freezing Point Depression

A nonscientist may be forever unaware of the boiling point elevation phenomenon, but a careful observer living in a cold climate is familiar with freezing point depression. Ice on frozen roads and sidewalks will melt when sprinkled with salts such as NaCl or CaCl$_2$. This method of thawing succeeds because it depresses the freezing point of water.

Figure 15.11 shows that the lowering of the vapor pressure of the solution causes the solid–liquid curve to shift to the left (see the dotted line). Consequently, this dotted line intersects the horizontal line at a temperature

TABLE 15.2 Boiling Point Elevation and Freezing Point Depression Constants of Several Common Liquids

Solvent	Normal Freezing Point (°C)†	K_f^* (°C/m)	Normal Boiling Point (°C)†	K_b (°C/m)
Water	0	1.86	100	0.52
Benzene	5.5	5.12	80.1	2.53
Ethanol	−114.6	1.99	78.4	1.22
Acetic acid	16.6	3.90	117.9	2.93
Chloroform	−63.5	—	61.7	3.63
Carbon tetrachloride	−23	—	76.5	5.03

*The K_f values of chloroform and carbon tetrachloride cannot be determined accurately due to experimental difficulties.
†Measured at 1 atm.

lower than the freezing point of water. The depression in freezing point, ΔT_f, is defined as

$$\Delta T_f = T_f^\circ - T_f$$

where T_f° is the freezing point of the pure solvent and T_f the freezing point of the solution. Again, ΔT_f is proportional to the concentration of the solution:

$$\Delta T_f \propto m$$

$$\Delta T_f = K_f m \qquad (15.5)$$

where m is the concentration of the solute in molality units, and K_f is the *molal freezing point depression constant* (see Table 15.2). Like K_b, K_f has the units of °C/m.

Note that whereas the solute must be nonvolatile in the case of boiling point elevation, no such restriction applies to freezing point depression. For example, methanol (CH_3OH), a fairly volatile liquid that boils at only 65°C, has sometimes been used as an antifreeze in automobile radiators.

> Since $T_f^\circ > T_f$, ΔT_f is a positive quantity.

> When an aqueous solution freezes, the solid first formed is almost always ice.

EXAMPLE 15.12

Ethylene glycol (EG), $CH_2(OH)CH_2(OH)$, is a common automobile antifreeze. It is cheap, water-soluble, and fairly nonvolatile (b.p. 197°C). Calculate the freezing point of a solution containing 651 g of this substance in 2505 g of water. Would you keep this substance in your car radiator during the summer? The molar mass of ethylene glycol is 62.01 g.

Answer

The number of moles of ethylene glycol in 1000 g of water is

$$651 \text{ g EG} \times \frac{1 \text{ mol EG}}{62.01 \text{ g EG}} \times \frac{1000 \text{ g}}{2505 \text{ g}} = 4.19 \text{ mol EG}$$

(Continued)

Thus, the molality of the solution is 4.19 m. From Equation (15.5) and Table 15.2 we have

$$\Delta T_f = (1.86°C/m)(4.19\ m)$$
$$= 7.79°C$$

Thus, the solution will freeze at $-7.79°C$. We can calculate boiling point elevation in the same way as follows:

$$\Delta T_b = (0.52°C/m)(4.19\ m)$$
$$= 2.2°C$$

Because the solution will boil at 102.2°C, it would be preferable to leave the antifreeze in your car radiator in summer to prevent the solution from boiling.

Similar examples: Problems 15.40, 15.42.

Osmotic Pressure

The phenomenon of osmotic pressure is illustrated in Figure 15.12. The left compartment of the apparatus contains pure solvent; the right compartment contains a solution. The two compartments are separated by a **semipermeable membrane,** which *allows solvent molecules to pass through but blocks the movement of solute molecules.* At the start, the water levels in the two tubes are equal [see Figure 15.12(a)]. After some time, the level in the right tube begins to rise; this continues until equilibrium is reached. *The net movement of solvent molecules through a semipermeable membrane from a pure solvent or from a dilute solution to a more concentrated solution* is called **osmosis.** The **osmotic pressure** (π) of

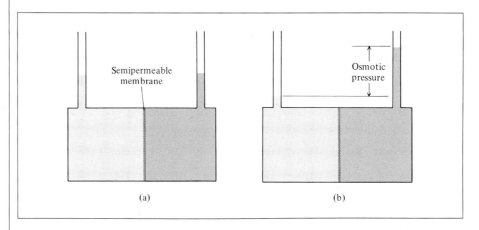

FIGURE 15.12 *The osmotic pressure phenomenon. (a) The levels of the pure solvent (left) and solution (right) are equal at the start. (b) During osmosis, the level on the solution side rises as a result of the net flow of the solvent from left to right. The osmotic pressure is equal to the hydrostatic pressure exerted by the column of fluid in the right tube at equilibrium. Basically the same effect is observed when the pure solvent is replaced by a more dilute solution than that on the right.*

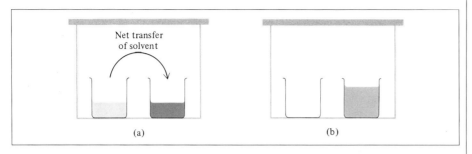

FIGURE 15.13 *In the
container unequal vapor
pressures lead to a transfer
of water from the left beaker
to the solution in the right
beaker. (a) At the beginning,
and (b) at equilibrium. This
driving force for solvent
transfer is analogous to the
osmotic phenomenon shown
in Figure 15.12.*

a solution is *the pressure required to stop osmosis from pure solvent into the solution.* As shown in Figure 15.12(b), this pressure can be measured directly from the difference in the fluid levels.

What causes water to move spontaneously from left to right in this case? Consider the vapor pressure of water and that of a solution (Figure 15.13). Because the vapor pressure of pure water is higher, there will be a tendency for net transfer of water to take place from the left beaker to the right one, and if given enough time, the transfer will be carried out to completion. A similar driving force causes water to move into the solution during osmosis.

Although osmosis is a common and well-studied phenomenon, relatively little is known as to how the semipermeable membrane stops some molecules yet allows others to pass through. In some cases, it is simply a matter of size. A semipermeable membrane may have a number of pores small enough to let only the solvent molecules pass through. In other cases, a different mechanism may be responsible for the membrane's selectivity—for example, the solvent's greater "solubility" in the membrane.

The osmotic pressure of a solution is given by

$$\pi = MRT \tag{15.6}$$

where M is the molarity of solution, R the gas constant (0.0821 L · atm/K · mol), and T the absolute temperature. The osmotic pressure, π, is expressed in atm. Since osmotic pressure measurements are carried out at constant temperature, we express the concentration here in terms of the more convenient units of molarity rather than molality.

As in the boiling point elevation and freezing point depression equations, we see that osmotic pressure too is directly proportional to the concentration of solution. This is what we would expect, bearing in mind that all colligative properties depend only on the number of solute particles in solution. If two solutions are of equal concentration and, hence, of the same osmotic pressure, they are said to be *isotonic.* If two solutions are of unequal osmotic pressures, the more concentrated solution is said to be *hypertonic* and the more dilute solution is described as *hypotonic* (Figure 15.14).

The osmotic pressure phenomenon manifests itself in many different ways. To study the contents of red blood cells, which are protected from the external environment by a semipermeable membrane, biochemists use a technique called *hemolysis.* The red blood cells are placed in a hypotonic solution, which causes water to move into the cells, as shown in Figure 15.14(b). The cells swell and eventually burst, releasing hemoglobin and other protein molecules.

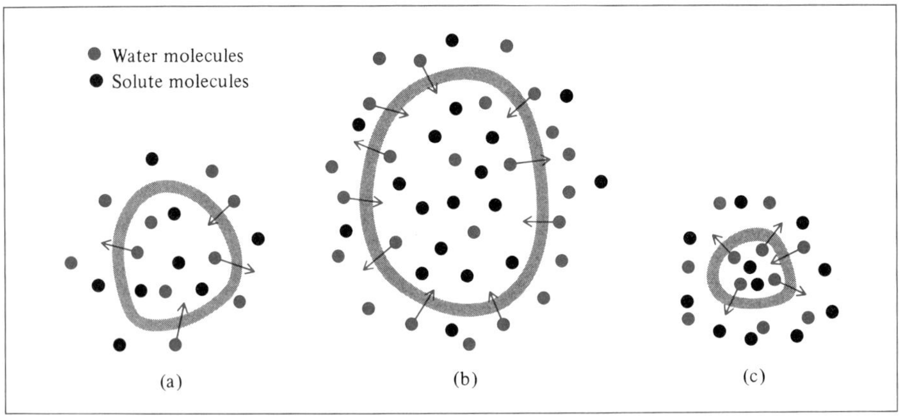

FIGURE 15.14 *A cell in (a) an isotonic solution, (b) a hypotonic solution, and (c) a hypertonic solution. The cell remains unchanged in (a), swells in (b), and shrinks in (c).*

Home preserving of jam and jelly has recently been revived as a popular (and money-saving) hobby in the United States. The large quantity of sugar used in this process is partly responsible for killing bacteria that may cause botulism. As Figure 15.14(c) shows, when a bacterial cell is in a hypertonic sugar solution, the intracellular water tends to move out of the bacterial cell to the more concentrated solution by osmosis. This process, known as *crenation*, causes the cell to shrink and, eventually, cease to function. The acidic medium due to the presence of citric acid and other acids in fruits also helps kill the bacteria.

Osmotic pressure is the major mechanism that forces water upward in plants. The leaves of trees constantly lose water to the air, in a process called *transpiration*, so the solute concentrations in leaf fluids increase. Water is pushed up through the trunk, branches, and stems by osmotic pressure, which may have to be as high as 10 to 15 atm in order to reach leaves at the tops of the tallest trees. (The tallest trees are California's redwoods, which reach about 120 m in height.)

The capillary action discussed in Section 14.3 is only responsible for the rise of water up to a few centimeters.

When we say that seawater has an osmotic pressure of 30 atm, we mean that when seawater is placed in the apparatus shown in Figure 15.12, we obtain a measurement of 30 atm.

EXAMPLE 15.13

The average osmotic pressure of seawater is about 30.0 atm at 25°C. Calculate the concentration (in molarity) of an aqueous solution of urea (NH_2CONH_2) that is isotonic with seawater.

Answer

A solution of urea that is isotonic with seawater must have the same osmotic pressure, 30.0 atm. Using Equation (15.6)

$$\pi = MRT$$

$$M = \frac{\pi}{RT} = \frac{30.0 \text{ atm}}{(0.0821 \text{ L·atm/K·mol})(298 \text{ K})}$$
$$= 1.23 \text{ mol/L}$$
$$= 1.23 \ M$$

Similar example: Problem 15.60.

Determining Molar Mass with Colligative Properties

The colligative properties of nonelectrolyte solutions provide us with a means of determining the molar mass of a solute. Theoretically, any of the four colligative properties can be used for this purpose. In practice, however, only freezing point depression and osmotic pressure are used to determine molar mass, because in those cases the changes are more pronounced.

EXAMPLE 15.14

A quantity of 7.85 g of a compound having the empirical formula C_5H_4 is dissolved in 301 g of benzene. The freezing point of the solution is 1.05°C below that of pure benzene. What are the molar mass and molecular formula of this compound?

Answer

From Equation (15.5) and Table 15.2 we can write

$$\text{molality} = \frac{\Delta T_f}{K_f} = \frac{1.05°C}{5.12°C/m} = 0.205 \ m$$

Thus, the solution contains 0.205 mole of solute per kilogram of benzene. Since the solution was prepared by dissolving 7.85 g of solute in 301 g of benzene, the number of grams of solute in a kilogram of solvent is therefore

$$\frac{\text{g solute}}{\text{kg benzene}} = \frac{7.85 \text{ g solute}}{301 \text{ g benzene}} \times \frac{1000 \text{ g benzene}}{1 \text{ kg benzene}}$$

$$= \frac{26.1 \text{ g solute}}{1 \text{ kg benzene}}$$

Now that we know 0.205 mole of solute is 26.1 g, the molar mass of solute must be

$$\text{molar mass} = 1 \text{ mol} \times \frac{26.1 \text{ g}}{0.205 \text{ mol}} = 127 \text{ g}$$

Since the formula mass of C_5H_4 is 64 g and the molar mass is found to be 127 g, the molecular formula of the compound is $C_{10}H_8$.

Similar examples: Problems 15.41, 15.43.

EXAMPLE 15.15

A solution is prepared by dissolving 35.0 g of hemoglobin (Hb) in enough water to make up one liter in volume. If the osmotic pressure of the solution is found to be 10.0 mmHg at 25°C, calculate the molar mass of hemoglobin.

Answer

First we need to calculate the concentration of the solution:

(Continued)

$$\pi = MRT$$

$$M = \frac{\pi}{RT}$$

$$= \frac{10.0 \text{ mmHg} \times \dfrac{1 \text{ atm}}{760 \text{ mmHg}}}{(0.0821 \text{ L} \cdot \text{atm/K} \cdot \text{mol})(298 \text{ K})}$$

$$= 5.38 \times 10^{-4} \, M$$

The volume of the solution is one liter, so it must contain 5.38×10^{-4} mol of Hb. We use this quantity to calculate the molar mass:

$$\text{moles of Hb} = \frac{\text{mass of Hb}}{\text{molar mass of Hb}}$$

$$\text{molar mass of Hb} = \frac{\text{mass of Hb}}{\text{moles of Hb}}$$

$$= \frac{35.0 \text{ g}}{5.38 \times 10^{-4} \text{ mol}}$$

$$= 6.51 \times 10^{4} \text{ g/mol}$$

Similar examples: Problems 15.50, 15.51.

A pressure such as 10.0 mmHg in the preceding example can be measured readily. For this reason, osmotic pressure measurements are among the most useful techniques for determining the molar masses of large molecules such as proteins. For contrast, let us estimate the depression in freezing point of the same hemoglobin solution. If a solution is quite dilute, we can assume that molarity is roughly equal to molality. (Molarity would be equal to molality if the density of the solution were 1 g/mL.) Hence, from Equation (15.5) we write

$$\Delta T_f = (1.86°C/m)(5.38 \times 10^{-4} \, m)$$
$$= 1.00 \times 10^{-3} \text{ °C}$$

Because of low sensitivity and experimental difficulties, vapor pressure lowering and boiling point elevation are not normally used for molar mass determination.

A thousandth of a degree is too small a temperature change to measure accurately. For this reason, the freezing point depression technique is more suitable for measuring the molar mass of smaller and more soluble molecules having molar masses of 500 g or less, since the freezing point depressions of their solutions are much greater.

15.9 COLLIGATIVE PROPERTIES OF ELECTROLYTE SOLUTIONS

The colligative properties of nonelectrolyte solutions are better understood than those of electrolyte solutions. The study of the "structure" of electrolyte solutions, that is, how each ion is arranged relative to its neighbors and how it interacts with the solvent molecules, has occupied some of the most prominent chemists of the past hundred years. Yet no completely satisfactory theory of electrolyte solution has been formulated.

Qualitatively, at least, we know the following facts. In a dilute solution, say, 0.01 M or less, each ion is surrounded by a number of water molecules. The degree of hydration depends on the nature of the ion: It is greater for small ions with high charges. The movement of cations and anions toward the negative and positive electrodes in a solution is influenced by the size of this "hydration sphere" (see Figure 5.7).

The interionic attraction in an electrolyte solution also affects its properties. It is reasonable to assume that each ion is surrounded, on the average, by ions bearing the opposite charge. Thus, we can think of each ion as being the center of an ionic atmosphere (Figure 15.15). Indeed, experimental evidence strongly supports this notion.

A fully hydrated (or solvated) ion is called a *free ion.* At higher concentrations, cations and anions tend to pair up to form **ion pairs.** An ion pair consists of *a cation and an anion held closely together by attractive forces with few or no water molecules between them* (Figure 15.16). The presence of ion pairs in a solution decreases the electrical conductivity; since the cation and anion in a neutral ion pair cannot move freely as individual units there can be no net migration in solution. Electrolytes containing multicharged ions such as Mg^{2+}, Al^{3+}, SO_4^{2-}, CO_3^{2-}, and PO_4^{3-} have a greater tendency to form ion pairs than 1:1 salts such as NaCl or KNO_3.

Dissociation of electrolytes into ions has a direct bearing on the colligative properties of solutions, which are determined only by the number of particles present. For example, the depression in freezing point of a 0.1 m NaCl solution would be about twice as great as that of a nonelectrolyte solution of 0.1 m containing cane sugar or urea as solute. The same holds true for boiling point elevation and osmotic pressure; thus Equations (15.4), (15.5), and (15.6) should be modified as follows:

$$\Delta T_b = iK_b m \qquad (15.7)$$

$$\Delta T_f = iK_f m \qquad (15.8)$$

$$\pi = iMRT \qquad (15.9)$$

where i is called the *van't Hoff factor* (Jacobus van't Hoff, 1852–1911) and is defined as follows:

$$i = \frac{\text{actual number of particles in solution after dissociation}}{\text{number of formula units initially dissolved in solution}}$$

Ion pairs effect an apparent decrease in the number of independently moving particles.

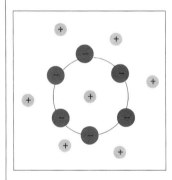

FIGURE 15.15   *An ionic atmosphere surrounding a cation. If the center ion is an anion, then the ionic atmosphere is made up mainly of cations.*

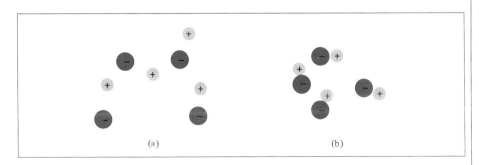

FIGURE 15.16   *(a) Free ions and (b) ion pairs in solution. Such an ion pair bears no net charge and therefore cannot conduct electricity in solution.*

(a)   (b)

For every unit of NaCl or KNO₃ that dissociates, we get two ions ($i = 2$); for every unit of Na₂SO₄ or MgCl₂ that dissociates, we get three ions ($i = 3$).

Thus i should be 1 for all nonelectrolytes. For strong electrolytes such as NaCl and KNO₃, i should be 2, and for strong electrolytes such as Na_2SO_4 and $MgCl_2$, i should be 3.

Table 15.3 shows the van't Hoff factor for several strong electrolytes. For 1:1 electrolytes such as HCl and NaCl, the agreement between measured and calculated i values is quite good. Significant deviations are observed for electrolytes containing Mg^{2+}, Fe^{3+}, and SO_4^{2-} ions, suggesting the formation of ion pairs.

TABLE 15.3 The van't Hoff Factor of 0.05 M Electrolyte Solutions at 25°C

Electrolyte	i (measured)	i (calculated)
Sucrose*	1.0	1.0
HCl	1.9	2.0
NaCl	1.9	2.0
MgSO₄	1.3	2.0
MgCl₂	2.7	3.0
FeCl₃	3.4	4.0

*Sucrose is a nonelectrolyte. It is listed here for comparison purposes only.

EXAMPLE 15.16

The osmotic pressures of 0.010 M solutions of potassium iodide (KI) and of sucrose at 25°C are 0.465 atm and 0.245 atm, respectively. Calculate the van't Hoff factor for KI.

Answer

This comparison works if the concentrations of the original solutions are the same.

Since osmotic pressure is directly proportional to the number of solute particles present in solution and since there are more particles in the KI solution than in the sucrose solution, we can express the van't Hoff factor for KI as follows:

$$i = \frac{\text{number of particles in KI soln}}{\text{number of particles in sucrose soln}}$$
$$= \frac{\text{osmotic pressure of KI soln}}{\text{osmotic pressure of sucrose soln}}$$
$$= \frac{0.465 \text{ atm}}{0.245 \text{ atm}}$$
$$= 1.90$$

Similar example: Problem 15.59.

15.10 WATER IN THE ENVIRONMENT

In this section we will turn our attention to the properties of natural waters. In particular, we will focus on the causes of water pollution and on ways to purify water that are based on some of the principles introduced earlier.

One gallon of drinking water is an ample daily supply for one person, but water use in our modern society far exceeds this essential minimum. For example, a bath requires about 40 gallons of water. Estimates show that average daily use of water in the United States amounts to 1500 gallons of water per person!

The oceans cover about 70 percent of Earth's surface and contain nearly 98 percent of Earth's total water supply (Figure 15.17). Because seawater contains NaCl and many other salts, it is unfit for human consumption or for agriculture. Most of the fresh water we use comes from lakes, rivers, and water stored underground. Before we can use this water for bathing and drinking, it must be purified to prevent disease and poisoning. For our protection, and for the protection of plants, animals, and the environment, water that has been used in homes or in industrial processes must also be purified before it is returned to lakes, rivers, and oceans.

Industry and agriculture use large amounts of water for cooling (at nuclear and conventional power plants) and for irrigation.

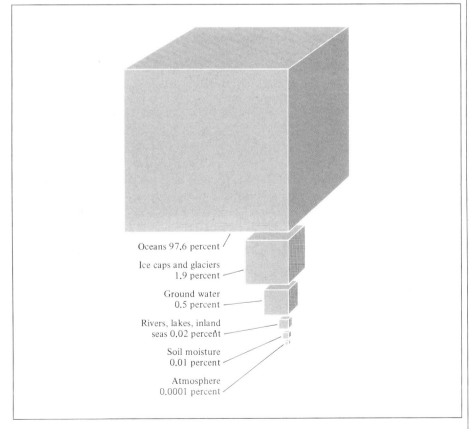

Oceans 97.6 percent

Ice caps and glaciers 1.9 percent

Ground water 0.5 percent

Rivers, lakes, inland seas 0.02 percent

Soil moisture 0.01 percent

Atmosphere 0.0001 percent

FIGURE 15.17 *The cubes show the relative amounts of water in storage on Earth. Nearly 98 percent of the water is in the oceans, which contain an estimated volume of 1.5×10^{21} liters of water. Most of the readily available fresh water is stored in porous rock beds as ground water. The quantity of water present in the atmosphere is relatively small but because it is actively transported, it plays a key role in the water cycle.*

Water Pollution

A **pollutant** is *a substance that is harmful to the biological environment.* A water pollutant might be thought of as anything in water that was not present when the water was in its natural state. Because water's "natural" state varies according to where it is found, this definition really means that the addition of *any* solute to water leads to water pollution. The only truly pure water would be 100 percent distilled water, which is fine for automobile batteries, steam irons, and the like, but is not pleasant to drink. Delicious-tasting spring water is not pure in this sense; rather, it has just the right combination of minerals to taste just right. Yet these same minerals can eventually clog a steam iron or shorten the life of a car battery. Thus, any substance can be a pollutant if too much of it is present. How much is too much depends on how the water is to be used.

A mineral is a naturally occurring substance with a characteristic range of chemical composition.

Water pollutants can be divided into two types: those that are natural and those introduced by human activities. Water coming into contact naturally with limestone ($CaCO_3$) or dolomite ($CaCO_3 \cdot MgCO_3$) dissolves Ca^{2+} and Mg^{2+} ions. These ions make the water unsuitable for some household and industrial uses, as we will see. Another undesirable ion in household water is Fe^{3+}, which reacts with hot water to form solid $Fe(OH)_3$, or rust, often seen in bathtubs. Particles of soil, sand, and minerals suspended in water as silt are another type of natural pollutant. These particles settle to the bottom of rivers and harbors, which then require expensive dredging to keep shipping open. They also get into reservoirs, thus reducing the reservoirs' capacities.

Man-made pollutants include human sewage, toxic substances, radioactive substances, and heat. We mentioned some consequences of thermal pollution in Section 15.6. Compared to natural pollutants, human-introduced pollutants usually cause more damage to the environment and are much more difficult to remove. Human and industrial wastes discharged into rivers and lakes are broken down by bacteria and other microorganisms in a manner very much like digestion of food in the human body. These microorganisms are *aerobic* in nature, that is, they need oxygen to function. One example of biological degradation is the conversion of carbohydrates, for which we can write the general formula $C_6H_{12}O_6$, into carbon dioxide and water:

$$C_6H_{12}O_6 + 6O_2 \longrightarrow 6CO_2 + 6H_2O$$

Other complex organic compounds are similarly broken down to smaller and less objectionable molecules. The measure of the oxygen required to sustain such a process is called *biological oxygen demand* (*BOD*). The oxygen consumed is replenished by natural aeration and from the photosynthesis of aquatic green plants. If BOD exceeds the rate of oxygen replenishment, the amount of dissolved oxygen gas will gradually diminish, threatening the survival of fish and other aquatic life. Eventually, *anaerobic* microorganisms, those that can function in the absence of dissolved O_2, take over. The decomposition of organic debris by anaerobic microorganisms produces the poisonous and rotten-smelling pollutant hydrogen sulfide

(H_2S), as well as ammonia, methane, and other substances. When a newspaper reports that a lake is "dying," it is describing this kind of slow, and sometimes irreversible, type of degradation of water quality. Human and animal wastes, and those that originate from meat packing or tanning operations, can also cause typhoid, dysentery, cholera, and other diseases.

Toxic substances found in waters polluted by human activity frequently include synthetic organic compounds such as solvents (carbon tetrachloride, chloroform, benzene, and the like) and insecticides, the best-known of which is dichlorodiphenyltrichloroethane, or DDT:

$$
\begin{array}{c}
\text{Cl} \\
| \\
\text{Cl}-\text{C}-\text{Cl} \\
| \\
\text{C} \\
| \\
\text{H}
\end{array}
$$

(DDT structure: two para-chlorophenyl rings attached to a central CH, which is bonded to a CCl₃ group)

Because DDT is extremely toxic to insects (it acts mainly as a respiratory poison), it was widely used all over the world to protect crops and forests from insect ravages and in tropical countries to prevent epidemics of malaria. DDT is an inert substance and is much more slowly biodegraded (broken down by natural biological processes) than normal organic wastes. Estimates show that about 30 percent of all the DDT ever produced is still present in lakes and oceans.

DDT also affects the life cycles of certain animals, notably the golden eagle, the peregrine falcon, and the California brown pelican. It moves up the food chain as follows: Over the years, a number of aquatic organisms have become resistant to DDT; they can acquire the substance by absorbing water or other organic matter contaminated with it and survive. Fish feed on these organisms and tend to accumulate the DDT residues in their fatty tissues. The molecules making up the fatty tissues and DDT molecules are both nonpolar. This is another example of "like dissolves like." Eventually, the pesticide finds its way into the bodies of the fish-eating birds mentioned above. Although we do not yet understand the precise mechanism by which DDT acts on these birds, the results are well-documented: The eggshells of poisoned birds become too thin to withstand normal nest activity. The reproductive capacities of some species have declined so drastically that the species may become extinct in our lifetime.

DDT is not the only toxic substance magnified in food chains. Another series of compounds, known as the *polychlorinated biphenyls* (*PCB*), have joined DDT as truly global pollutants. A typical compound in the PCB category is 2,4,6,2',4'-pentachlorobiphenyl:

(2,4,6,2',4'-pentachlorobiphenyl structure)

DDT is also harmful to humans.

Until recently, these compounds had been used as coolants in transformers, in packaging materials for food, in the manufacture of paint, and in many other areas. Like DDT, PCBs are insoluble in water but soluble in fats and oils. They are even more resistant to chemical and biological degradation than DDT. The toxicity levels of PCBs are believed to be comparable to those of DDT.

Many polluted bodies of water also contain certain extremely toxic elements, such as mercury, lead, and cadmium in trace amounts. Lead has been used for glazing pottery, tin plumbing, cooking vessels, and paints, and it is present in gasoline in the form of the antiknock additive tetraethyl lead, $Pb(C_2H_5)_4$. Humans have used mercury in various forms for almost three thousand years. Industrially, it functions as an electrode in the preparation of chlorine gas from NaCl solutions. Mercury, lead, and cadmium are harmful to humans because of their ability to attack enzymes, the biological catalysts essential for the proper functioning of our bodies. A severe case of lead or mercury poisoning will result in brain damage, convulsions, and sometimes death. Some historians have theorized that the fall of the Roman empire was the result of lead poisoning. The ruling class of Rome supposedly suffered from sterility and brain damage through contamination from lead vessels and lead pipes, which only the rich could afford.

Both lead and mercury compounds concentrate in food chains in much the same way that DDT and PCB do. What is worse, however, is the fact that the toxic effects of these metals on human tissues and organs are cumulative—our bodies cannot metabolize them. Once mercury enters water, it can be converted by anaerobic microorganisms into methylmercury cation, CH_3Hg^+, and dimethylmercury, $(CH_3)_2Hg$, both of which are extremely poisonous. Concentrations as low as ten parts per million (ppm) by mass in drinking water can be fatal.

Figure 15.18 shows the various pollutant elements that are found in water

Mercury vapor is highly toxic.

FIGURE 15.18 *Various pollutant elements in water supplies.*

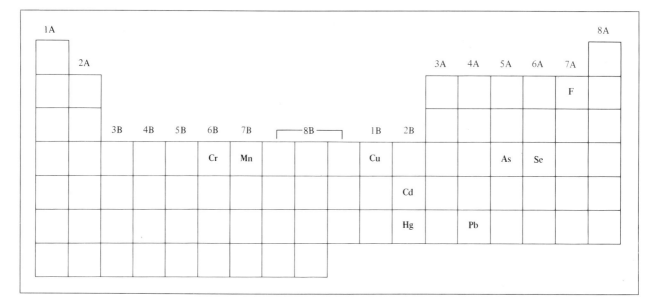

supplies. The metals are present as cations; fluorine is present as the F^- ion, arsenic as the AsO_4^{3-} ion, and selenium as the SeO_3^{2-} and/or SeO_4^{2-} ion. These elements vary a great deal in their toxicities and in the variety of toxic effects they bring about.

The Detergent Problem

Water containing Ca^{2+} and Mg^{2+} ions is called **hard water**, *and water that is mostly free of these ions* is called **soft water.** Although the presence of these ions in water is seldom harmful to the human body, Ca^{2+} and Mg^{2+} can considerably reduce the effectiveness of soap. Efforts to eliminate these ions by the addition of certain chemicals have created a new type of pollution in our natural water.

Soaps are sodium salts of organic molecules that have a long hydrocarbon chain at one end and a —COO⁻ group at the other end (Figure 15.19). The hydrocarbon chain is nonpolar and therefore tends to avoid interacting with the polar water molecules. For this reason, the chain is described as *hydrophobic,* meaning *water-fearing.* On the other hand, the —COO⁻ group interacts favorably with water molecules through ion–dipole interaction; it is called *hydrophilic,* meaning *water-liking.* The cleansing action of soap is the result of the dual nature of the hydrophobic hydrocarbon chain and the hydrophilic end group of its molecules. The hydrocarbon tail is readily soluble in oily substances that are made up of nonpolar groups (another example of "like dissolves like"), whereas the ionic —COO⁻ group prefers to be exposed to water. When enough soap molecules have surrounded an oil droplet, as shown in Figure 15.20, the entire group becomes soluble in water because the exterior portion is now largely hydrophilic. This is how greasy substances are removed by the action of soap.

A hydrocarbon chain is made up of only carbon and hydrogen atoms. It is usually nonpolar.

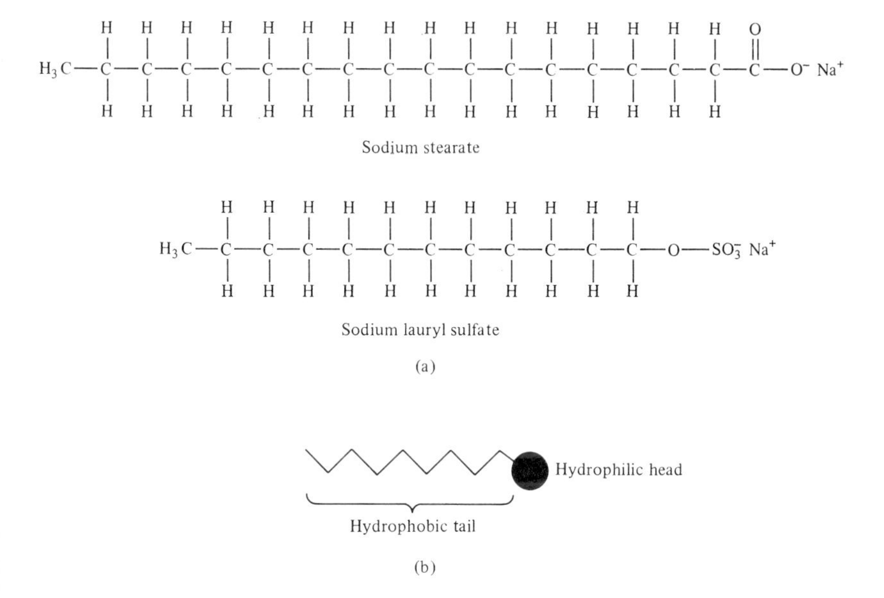

Sodium stearate

Sodium lauryl sulfate

(a)

Hydrophilic head

Hydrophobic tail

(b)

FIGURE 15.19 *(a) Structures of some common soap molecules. (b) A simplified representation of a soap molecule. The black circle denotes the hydrophilic head and the zig-zag lines are the hydrophobic tail.*

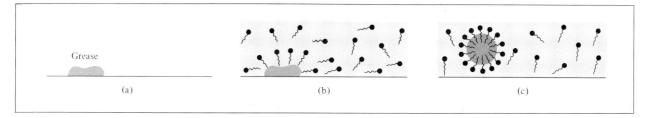

FIGURE 15.20 *The cleansing action of soap. (a) A greasy spot (b) can be removed by soap (c) because the hydrophobic tails of soap molecules interact with grease, and the whole system then becomes soluble in water.*

In hard water the Ca^{2+} and Mg^{2+} ions react with soap to form insoluble salts or curds, which often appear in the form of "bathtub rings":

$$Ca^{2+}(aq) + 2C_{17}H_{35}CO_2^-Na^+(aq) \longrightarrow Ca(C_{17}H_{35}CO_2)_2(s) + 2Na^+(aq)$$

This process continues until most of the Ca^{2+} and Mg^{2+} ions are removed. Thus, more soap is needed in hard water than in soft water. In regions with very hard water, the cost of extra soap can become significant. The removal of curds from clothes is also a considerable problem.

Detergents are an alternative to soap. Because they are synthetic cleaning compounds, detergents are often referred to as *syndets.* One of the earliest detergents developed was sodium lauryl sulfate, $C_{12}H_{25}OSO_3^-Na^+$ (see Figure 15.19). Like sodium stearate, it contains a long hydrocarbon chain and a polar end group (the sulfate group). Unlike soap, it does not form precipitates with Ca^{2+} or Mg^{2+} ions, so that it can be used in both soft water and hard water.

Commercial detergents are not pure compounds. In addition to sodium lauryl sulfate, a detergent may contain 20 to 40 percent of a "builder," sodium tripolyphosphate, $Na_5P_3O_{10}$, as well as corrosion inhibitors, perfumes, optical brighteners, and other substances. Sodium tripolyphosphate (Figure 15.21) is a **chelating agent** (from the Greek word "chele," meaning claw); it *forms soluble compounds with ions* (in this case Ca^{2+} and Mg^{2+}) *in solution.* For this reason the detergent can be used in hard water. Unfortunately, phosphates are also plant nutrients, so the dumping of waste washwater into lakes and rivers promotes the growth of algae. Once dead, algae are decomposed by aerobic microorganisms. Eventually, the dissolved oxygen content of the water is depleted by these microorganisms to the point that no aquatic life can exist.

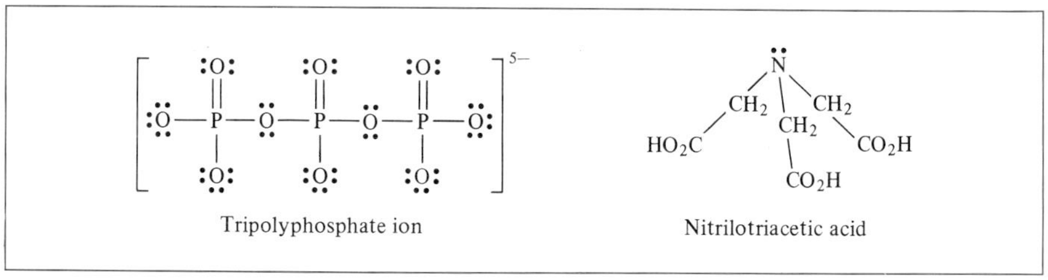

FIGURE 15.21 *Structures of the tripolyphosphate ion and nitrilotriacetic acid. These substances have the ability to tie up undesirable Ca^{2+} and Mg^{2+} ions in solution.*

Tripolyphosphate ion

Nitrilotriacetic acid

Several substitutes for "phosphate" detergents are now available. Sodium silicates produce a chelating effect that is similar to that of the phosphates. Another compound, nitrilotriacetic acid (NTA), has also been tested for some detergents. NTA, like sodium tripolyphosphate, can form soluble compounds with dipositive ions (see Figure 15.21). Since NTA is not a plant nutrient, its presence in a lake does not lead to a buildup of organic debris, as the phosphate detergents do. However, it is suspected of being mutagenic. A *mutagen* is a substance capable of causing permanent changes in the hereditary material of living organisms, resulting in new and usually undesirable inheritable characteristics. Sometimes the cure can be worse than the problem.

Water Purification

Municipal Water Treatment. What can be done to purify water once it is polluted? There are basically three steps involved in the treatment of sewage and waste water. *Primary treatment* is a physical separation process—sedimentation, for example—that removes much of the undissolved solid materials. The next step, *secondary treatment*, removes much of the organic matter by making use of aerobic microorganisms present in water. This is a carefully controlled process, since the quantity and type of organic matter vary from place to place, so the proper amount of oxygen and biological sludge must be provided for effective degradation. In the final step, *tertiary treatment*, nitrogen, phosphorus, and other nutrients (or other objectionable chemicals) are removed by specific chemical and physical techniques. At this point, the water is ready to be discharged back into natural bodies of water (Figure 15.22).

In the purification of water for household use, chemicals such as molecular chlorine and ozone are added to degrade dissolved organic substances, thus removing unpleasant odors and improving the taste. Municipal drinking water is usually treated with these chemicals, which also act as disinfectants. Since chlorine gas itself is a harmful substance, it must be used sparingly. Normally, a concentration of 1 ppm of chlorine by mass in water is sufficient to kill bacteria without causing any ill effects in humans. Organic materials that resist biodegradation can be removed at this stage by passing the water through a filter bed containing activated charcoal. Activated charcoal is obtained by the distillation of wood (that is, heating wood in a vacuum) and other carbonaceous materials. The carbon residue is activated by heating to 800°C with steam and the final product is a fine powder. The enormous surface area of the substance (about 1000 m^2 per gram!) enables it to *adsorb* gases and small solid particles suspended in water.

Water Softening. Several methods are available for softening hard water. The addition of *washing soda*, or sodium carbonate decahydrate (Na$_2$CO$_3$ · 10H$_2$O), removes Ca^{2+} and Mg^{2+} ions by forming insoluble salts. The equation is

$$Ca^{2+}(aq) + CO_3^{2-}(aq) \longrightarrow CaCO_3(s)$$

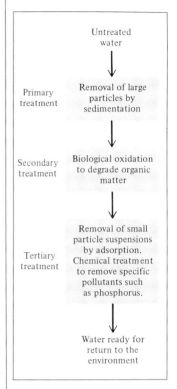

FIGURE 15.22 *The major steps in municipal sewage treatment.*

Adsorption is the process of drawing and holding gas, liquid, or dissolved substances onto a solid surface.

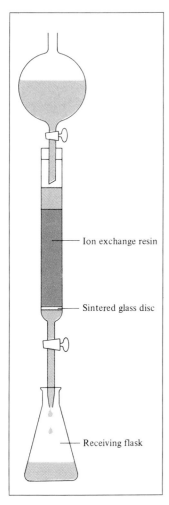

FIGURE 15.23 *Arrangement for an ion-exchange process. The solution containing undesirable ions is slowly poured down a column packed with certain ion-exchange resins, the sites of ion exchanges.*

FIGURE 15.24 *Structure of the ion-exchange resin zeolite, showing the large cavity in which cations may be trapped. Every Si is bonded to four O atoms and every Al is bonded to three O atoms.*

In the *lime-soda* method, both Na_2CO_3 and $Ca(OH)_2$ are added to *temporary hard water*. (Temporary hard water is water that contains bicarbonate ions, HCO_3^-, in addition to Ca^{2+} and Mg^{2+} ions.) The choice of Na_2CO_3 is obvious—carbonate ions are needed to precipitate Ca^{2+} and Mg^{2+} ions. It may seem strange, at first, that calcium hydroxide is used to remove calcium ions. However, the following equation shows the chemical reasoning for this choice:

$$Ca(HCO_3)_2(aq) + Ca(OH)_2(aq) \longrightarrow 2CaCO_3(s) + 2H_2O(l)$$

Again, most of the Ca^{2+} ions are removed from the solution as the insoluble $CaCO_3$.

Ion exchange is a particularly effective method for water softening. As the name suggests, this is an exchange process; *the undesirable ions are replaced with another type of ion* that does not affect the cleaning action of soaps. It is possible to have both cation-exchange and anion-exchange units. The former are used in household water softeners. Figure 15.23 shows the basic arrangement for an ion-exchange experiment. The solution under study is allowed to flow slowly down a column packed with the ion-exchange resin *zeolite* (Figure 15.24), a claylike material containing a silicon-oxygen-aluminum network. The negatively charged zeolite contains pores large enough to hold ions. Initially, the zeolite holds Na^+ ions. When hard water comes in contact with the zeolite, some of the Na^+ ions are replaced by the Ca^{2+} ions, as follows:

$$Ca^{2+}(aq) + 2Na^+R^-(s) \longrightarrow 2Na^+(aq) + Ca^{2+}R_2^{2-}(s)$$

where R denotes the zeolite resin. A used column can be regenerated by passing through it a saturated NaCl solution to "flush out" the Ca^{2+} ions occupying most of the holes. An ion-exchange unit can thus be used repeatedly, with little deterioration.

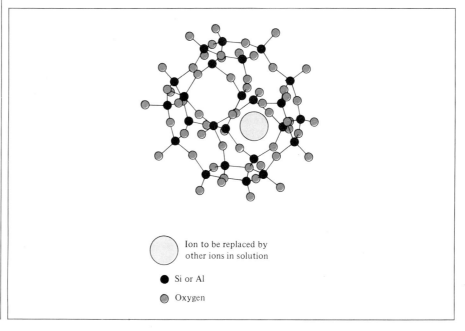

Ion to be replaced by other ions in solution

● Si or Al

◉ Oxygen

Desalination

Over the centuries, scientists have tried to find *ways of removing salts from seawater*, a process called ***desalination***, to augment the supply of fresh water. The ocean is an enormous and extremely complex aqueous solution. There are about 1.5×10^{21} liters of water in the ocean, 3.5 percent of which (by mass) consists of dissolved substances. Table 15.4 shows the concentrations of seven substances that, together, comprise more than 99 percent of the dissolved constituents of ocean water.

In an age that has seen astronauts land on the moon and great advances in science and medicine, desalination may seem to be a simple objective. But although the technology exists, the cost is prohibitive. Herein lies an interesting paradox of our technological society: It is often just as difficult to do something simple like desalination at a socially acceptable cost as it is to do something complex like sending an astronaut to the moon.

There are several methods of desalination in use. We will discuss four of them here.

Distillation. The oldest method of desalination, distillation accounts for about 90 percent of the approximately 500 million gallons per day capacity of the desalination systems currently in operation. The process involves vaporizing seawater and then condensing the pure water vapor. Most distillation systems use heat energy. In order to reduce the cost of distillation, attempts have been made to use solar radiation (Figure 15.25). This approach is attractive because sunshine is normally most intense in arid lands, where the need for water is also the greatest. Despite intensive research and development efforts, a number of engineering problems persist, and "solar stills" do not yet operate on a large scale.

Freezing. This method too has been under development for a number of years, but it has not been made commercially effective. It is based on the fact that when an aqueous solution freezes, the solid that separates from solution is almost all pure H_2O. The main advantage of freezing is its low energy consumption, compared with distillation. The heat of vaporization for water is 40.79 kJ/mol, while that of fusion is only 6.01 kJ/mol. The

TABLE 15.4 Dissolved Substances in Seawater

Ions	g/kg of Seawater
Cl^-	19.35
Na^+	10.76
SO_4^{2-}	2.71
Mg^{2+}	1.29
Ca^{2+}	0.41
K^+	0.39
HCO_3^-	0.14

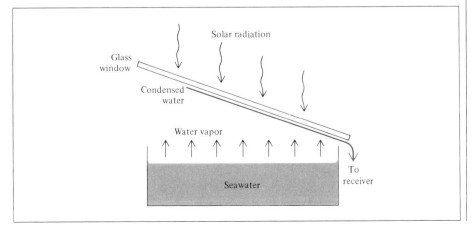

FIGURE 15.25 *A solar still for desalinating sea water.*

FIGURE 15.26 *Reverse osmosis. By applying enough pressure on the solution side, fresh water can be made to flow from right to left. The semipermeable membrane allows the flow of water molecules but not of dissolved ions.*

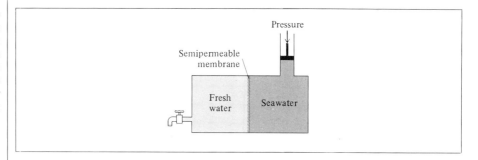

major disadvantages of freezing are associated with the growing, handling, and washing of ice crystals.

Reverse Osmosis. Both distillation and freezing involve phase changes that require considerable energy input. On the other hand, desalination by *reverse osmosis* does not involve a phase change and is economically more desirable. Reverse osmosis *uses high pressure to force water through a semipermeable membrane from a more concentrated solution into a less concentrated one.* Consider the apparatus shown in Figure 15.26. The osmotic pressure of seawater is about 30 atm—this is the pressure that must be applied to the saline solution in order to stop the flow of water from left to right. If the pressure on the salt solution were to be increased beyond 30 atm, the osmotic flow would be reversed, and fresh water would actually pass from the solution through the membrane into the left compartment. Desalination by reverse osmosis is considerably cheaper than by distillation, and it does not have the technical difficulties associated with the freezing method. The main practical obstacle to this method is the development of a membrane that is permeable to water but not to other dissolved substances and that can be used for prolonged periods under high-pressure conditions. Once this problem is solved, and present signs are encouraging, reverse osmosis could become a major technique for desalination.

Ion Exchange. Dissolved ions in seawater can be removed by passing the salt solution first through a cation-exchange column, followed by an anion-exchange column (Figure 15.27). An anion-exchange column consists of resins in which replaceable OH^- ions are embedded in a cationic network, similar to the manner in which Na^+ ions are embedded in a cation-exchange resin (shown in Figure 15.24). The exchange reactions are

Cation exchange: $Na^+(aq) + H^+R^-(s) \longrightarrow Na^+R^-(s) + H^+(aq)$

$$Ca^{2+}(aq) + 2H^+R^-(s) \longrightarrow Ca^{2+}R_2^{2-}(s) + 2H^+(aq)$$

Anion exchange: $Cl^-(aq) + R^+OH^-(s) \longrightarrow R^+Cl^-(s) + OH^-(aq)$

$$SO_4^{2-}(aq) + 2R^+OH^-(s) \longrightarrow R_2^{2+}SO_4^{2-}(s) + 2OH^-(aq)$$

where R^- represents the negatively charged resin in the cation exchange and R^+ the positively charged resin in the anion exchange. The H^+ and OH^- ions then combine according to the neutralization reaction

$$H^+(aq) + OH^-(aq) \longrightarrow H_2O(l)$$

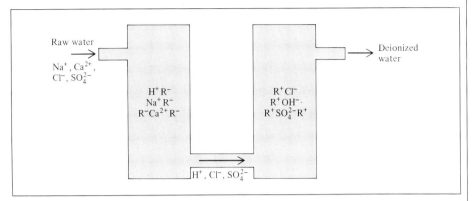

FIGURE 15.27 *A cation-exchange column and an anion-exchange column working in combination. In the cation-exchange column, the negatively-charged resin is represented by R^-, while in the anion-exchange column, the positively-charged resin is represented by R^+.*

The major disadvantage of the ion-exchange method is the cost of the cationic and anionic resins. For this reason, this technique has so far been used mainly in conjunction with distillation or reverse osmosis, as a final refinement step in desalination.

SUMMARY

1. Solutions are homogeneous mixtures of two or more substances, which may be solids, liquids, or gases.

2. The ease of dissolution of a solute in a solvent is governed by intermolecular forces. In addition to energy consideration, the other driving force for the solution process is the disorder created when molecules of the solute and solvent mix to form a solution.

3. The concentration of a solution can be expressed as percent by mass, mole fraction, molarity, and molality. These differing units are useful in differing circumstances.

4. Increasing the temperature usually increases the solubility of solid and liquid substances and always decreases the solubility of gases.

5. According to Henry's law, the solubility of a gas in a liquid is directly proportional to the partial pressure of the gas over the solution.

6. Raoult's law states that the partial pressure of a substance, say, A, is related to the mole fraction (X_A) of A and to the vapor pressure (P_A°) of pure A as follows: $P_A = X_A P_A^\circ$. An ideal solution obeys Raoult's law over the entire range of concentration. In practice, very few solutions behave ideally in this manner.

7. Vapor pressure lowering, boiling point elevation, freezing point depression, and osmotic pressure are colligative properties of solutions, that is, properties that depend only on the number of solute particles that are present and not on their nature.

8. In electrolyte solutions, the interaction between ions leads to the formation of ion pairs. The van't Hoff factor provides a measure of the extent of dissociation in solution.

9. Water carries both natural pollutants (such as metal ions) and those introduced by human activities. Among the latter are sewage, DDT

and other insecticides, synthetic organic compounds, lead, mercury, and other metal compounds, soaps and detergents, and phosphorus-containing compounds.

10. Water purification for various purposes includes softening, or removal of metal ions; disinfection and removal of unpleasant odors; and the primary, secondary, and tertiary treatments of waste water to remove particles, organic matter, and other pollutants. Desalination is the removal of salts from seawater.

KEY WORDS

Chelating agent, p. 458
Colligative properties, p. 440
Crystallization, p. 435
Desalination, p. 461
Dimer, p. 427
Fractional crystallization,
   p. 437
Fractional distillation, p. 442
Hard water, p. 457
Henry's law, p. 438
Hydrophilic, p. 457
Hydrophobic, p. 457
Ideal solution, p. 442
Ion exchange, p. 460
Ion pairs, p. 451
Miscible, p. 427

Molality, p. 432
Nonvolatile, p. 440
Osmosis, p. 446
Osmotic pressure, p. 446
Pollutant, p. 454
Raoult's law, p. 441
Reverse osmosis, p. 462
Saturated solution, p. 435
Semipermeable membrane, p. 446
Soft water, p. 457
Supersaturated solution,
   p. 435
Thermal pollution, p. 438
Unsaturated solution, p. 435
Volatile, p. 442

PROBLEMS

More challenging problems are marked with an asterisk.

Intermolecular Forces and Solubility

15.1 Give a brief description of the solution process at the molecular level. Use the dissolution of solid in liquid as an example.

15.2 Based on intermolecular force consideration, explain what "like dissolves like" means.

15.3 What is solvation? What are the factors that influence the extent to which solvation occurs?

* 15.4 As you know, some solution processes are endothermic and others are exothermic. Provide a molecular interpretation of the difference.

15.5 Explain why the solution process invariably leads to an increase in randomness.

15.6 Describe the factors that affect the solubility of a substance in a liquid.

15.7 Why is naphthalene ($C_{10}H_8$) more soluble than CsF in benzene?

15.8 Explain why ethanol (C_2H_5OH) is not soluble in cyclohexane (C_6H_{12}).

15.9 Arrange the following compounds in order of increasing solubility in water: O_2, LiCl, Br_2, methanol (CH_3OH).

*15.10 Explain the variations in solubility in water of the alcohols listed below:

Compound	Solubility in Water g/100 g, 20°C
CH_3OH	∞
CH_3CH_2OH	∞
$CH_3CH_2CH_2OH$	∞
$CH_3CH_2CH_2CH_2OH$	9
$CH_3CH_2CH_2CH_2CH_2OH$	2.7

(*Note:* ∞ means the alcohol and water are completely miscible in all proportions.)

15.11 Which of the alcohols listed in Problem 15.10 would you expect to be the best solvent for each of the following? (a) I_2, (b) KBr, (c) $CH_3CH_2CH_2CH_2CH_3$. Explain your answer.

Concentration Units

15.12 Calculate the percent by mass of the solute in each of the following aqueous solutions: (a) 5.50 g of NaBr in 78.2 g of solution, (b) 31.0 g of KCl in 152 g of water, (c) 4.5 g of toluene in 29 g of benzene.

15.13 Calculate the amount of water (in grams) that must be added to (a) 5.00 g of urea (H_2NCONH_2) in the preparation of a 16.2 percent by mass solution, (b) 26.2 g of $MgCl_2$ in the preparation of a 1.50 percent by mass solution.

15.14 Calculate the molality of each of the following solutions: (a) 14.3 g of sucrose ($C_{12}H_{22}O_{11}$) in 676 g of water, (b) 7.20 mole of ethylene glycol ($C_2H_6O_2$) in 3546 g of water.

15.15 Calculate the molality of each of the following aqueous solutions: (a) 2.50 M NaCl solution (density of solution = 1.08 g/mL), (b) 5.86 M ethanol solution (density of solution = 0.927 g/mL), (c) 48.2 percent by mass KBr solution.

15.16 Calculate the molality of each of the following aqueous solutions: (a) 1.22 M sugar ($C_{12}H_{22}O_{11}$) solution (density of solution = 1.12 g/mL), (b) 0.87 M NaOH solution (density of solution = 1.04 g/mL), (c) 5.24 M $NaHCO_3$ solution (density of solution = 1.19 g/mL).

15.17 The alcohol content of hard liquor is normally given in terms of the "proof," which is defined as twice the percentage by volume of ethanol (C_2H_5OH) present. Calculate the number of grams of alcohol present in 1.00 L of 75 proof gin. The density of ethanol is 0.798 g/mL.

15.18 The concentrated sulfuric acid we use in the laboratory is 98.0 percent H_2SO_4 by mass. Calculate the molality and molarity of the acid solution. The density of the solution is 1.83 g/mL.

*15.19 Calculate the molarity, molality, and the mole fraction of NH_3 for a solution of 30.0 g of NH_3 in 70.0 g of water. The density of the solution is 0.982 g/mL.

*15.20 The density of an aqueous solution containing 10.0 percent of ethanol (C_2H_5OH) by mass is 0.984 g/mL. (a) Calculate the molality of this solution. (b) Calculate its molarity. (c) What volume of the solution would contain 0.125 mole of ethanol? (d) Calculate the mole fraction of water in the solution.

15.21 It is estimated that 1.0 mL of seawater contains about 4.0×10^{-12} g of gold. The total volume of ocean water is 1.5×10^{21} L. Calculate the total amount of gold present in seawater. With this much gold out there, why hasn't someone become rich by mining gold from the ocean?

Solubility and Fractional Crystallization

15.22 A quantity of 3.20 g of a salt dissolves in 9.10 g of water to give a saturated solution at 25°C. What is the solubility (in g salt/100 g H_2O) of the salt?

15.23 The solubility of KNO_3 is 155 g per 100 g of water at 75°C and 38.0 g at 25°C. What mass (in grams) of KNO_3 will crystallize out of solution if exactly 100 g of its saturated solution at 75°C are cooled to 25°C?

15.24 A 50 g sample of impure $KClO_3$ (solubility = 7.1 g per 100 g H_2O at 20°C) is contaminated with 10 percent of KCl (solubility = 25.5 g per 100 g H_2O at 20°C). Calculate the minimum quantity of 20°C water needed to dissolve all the KCl from the sample. How much $KClO_3$ will be left after this treatment? (Assume that the solubilities are unaffected by the presence of the other compound.)

Gas Solubility

15.25 Discuss the factors that influence the solubility of a gas in a liquid.

15.26 A student is making observations of two beakers of water. One beaker is heated to 30°C and the other is heated to 100°C. In each case, bubbles form in the water. Are these bubbles of the same origin? Explain.

15.27 A man bought a goldfish in a pet shop. After returning home, he put the goldfish in a bowl of recently boiled water (the water had been allowed to cool quickly). A few minutes later the fish was found dead. Explain what happened to the fish.

15.28 The solubility of CO_2 in water at 25°C and 1 atm is 0.034 mol/L. What is its solubility under atmospheric conditions? (The partial pressure of CO_2 is 0.030 atm in air.)

15.29 A beaker of water is initially saturated with dissolved air. Explain what happens when He gas at 1 atm is bubbled through the solution for a long time.

15.30 Neither HCl nor NH_3 gas obeys Henry's law. Explain.

15.31 A miner working 260 m below sea level opened a carbonated soft drink during a lunch break. To his surprise, the soft drink tasted very "flat." Shortly afterward, the miner took an elevator to the surface. During the trip up, he could not stop belching. Why?

*15.32 The solubility of N_2 in blood at 37°C and at a partial pressure of 0.80 atm is 5.6×10^{-4} mol/L. A deep-sea diver breathes compressed air with the partial pressure of N_2 equal to 4.0 atm. Assume that the total volume of blood in the body is 5.0 liters. Calculate the amount of N_2 gas released (in liters) when the diver returns to the surface of the water, where the partial pressure of N_2 is 0.80 atm.

Colligative Properties

*15.33 How many grams of sucrose $(C_{12}H_{22}O_{11})$ must be added to 550 g of water to give a solution with a vapor pressure 3.0 mmHg less than that of pure water at 20°C? (The vapor pressure of water at 20°C is 17.5 mmHg.)

15.34 The vapor pressure of benzene is 100.0 mmHg at 26.1°C. Calculate the vapor pressure of a solution containing 24.6 g of camphor $(C_{10}H_{16}O)$ dissolved in 98.5 g of benzene. (Camphor is a solid of low volatility.)

15.35 The vapor pressures of ethanol (CH_3CH_2OH) and 1-propanol $(CH_3CH_2CH_2OH)$ at 35°C are 100 mmHg and 37.6 mmHg, respectively. Assume ideal behavior and calculate the partial pressures of ethanol and 1-propanol at 35°C over a solution of ethanol in 1-propanol, in which the mole fraction of ethanol is 0.300.

*15.36 The vapor pressure of ethanol (C_2H_5OH) at 20°C is 44 mmHg and the vapor pressure of methanol (CH_3OH) at the same temperature is 94 mmHg. A mixture of 30.0 g of methanol and 45.0 g of ethanol is prepared (and may be assumed to behave as an ideal solution). (a) Calculate the vapor pressure of methanol and ethanol above this solution at 20°C. (b) Calculate the mole fraction of methanol and ethanol in the vapor above this solution at 20°C. (c) Suggest a method for separating the two components of the solution.

15.37 Two beakers, one containing an 8.0 M LiCl aqueous solution and the other containing pure water, are placed under a tightly sealed bell jar at room temperature. After a few months one beaker is completely dry, whereas the other has increased in water by an amount equal to that originally present in the other beaker. Account for this phenomenon.

15.38 How many grams of urea (H_2NCONH_2) must be added to 450 g of water to give a solution with a vapor pressure 2.50 mmHg less than that of pure water at 30°C? (The vapor pressure of water at 30°C is 31.8 mmHg.)

*15.39 Which of the following two aqueous solutions has (a) a higher boiling point, (b) a higher freezing point, (c) a lower vapor pressure, and (d) a higher osmotic pressure: 0.35 m $CaCl_2$ or 0.90 m urea? State your reasons.

15.40 An aqueous solution contains the amino acid glycine (NH_2CH_2COOH). Assuming no ionization of the acid, calculate the molality of the solution if it freezes at $-1.1°C$.

15.41 Pheromones are compounds secreted by the females of many insect species to attract males. One of these compounds contains 80.78 percent C, 13.56 percent H, and 5.66 percent O. A solution of 1.00 g of this pheromone in 8.50 g of benzene freezes at 3.37°C. What are the molecular formula and molar mass of the compound? (The normal freezing point of pure benzene is 5.50°C.)

15.42 How many liters of the antifreeze ethylene glycol $[CH_2(OH)CH_2(OH)]$ would you add to a car radiator containing 6.50 L of water if the coldest winter temperature in your area is $-20°C$? Calculate the boiling point of this water–ethylene glycol mixture. The density of ethylene glycol is 1.11 g/mL.

15.43 A solution of 0.85 g of the organic compound mesitol in 100.0 g of benzene is observed to have a freezing point of 5.16°C. What are the molality of the mesitol solution and the molar mass of mesitol?

15.44 A solution of 2.50 g of a compound of empirical formula C_6H_5P in 25.0 g of benzene is observed to freeze at 4.3°C. Calculate the molar mass of the solute and its molecular formula.

15.45 A solution is prepared by condensing 4.00 L of a gas, measured at 27°C and 748 mmHg pressure, into 58.0 g of benzene. Calculate the freezing point of this solution.

15.46 An organic solid was extracted from gum arabic, and elemental analysis showed that it contained 40.0 percent C, 6.7 percent H, and 53.3 percent O. A solution of 0.650 g of the solid in 27.8 g of the solvent diphenyl gave a freezing point depression of 1.56°C. Calculate the molar mass and molecular formula of the solid. (K_f for the solvent diphenyl is 8.00°C/m.)

*15.47 Consider two aqueous solutions, one of sucrose $(C_{12}H_{22}O_{11})$ and the other of nitric acid (HNO_3), both of which freeze at $-1.5°C$. What other properties do these solutions have in common?

*15.48 The molar mass of benzoic acid (C_6H_5COOH) determined by the freezing point depression method, using benzene as the solvent, is found to be twice that expected for the molecular formula, $C_7H_6O_2$. Explain this apparent anomaly.

15.49 A quantity of 7.480 g of an organic compound is dissolved in water to make 300.0 mL of solution. The solution has an osmotic pressure of 1.43 atm at 27°C. When analyzed this compound is found to contain 41.8 percent C, 4.7 percent H, 37.3 percent O, and 16.3 percent N. Calculate the molecular formula of the compound.

15.50 A solution containing 0.8330 g of a protein of unknown structure in 170.0 mL of aqueous solution was found to have an osmotic pressure of 5.20 mmHg at 25° C. Determine the molar mass of the protein.

15.51 A solution of 6.85 g of a carbohydrate in 100.0 g of water has a density of 1.024 g/mL and an osmotic pressure of 4.61 atm at 20.0°C. Calculate the molar mass of the carbohydrate.

15.52 Explain why it is extremely important that fluids used in intravenous injections have approximately the same osmotic pressure as blood.

15.53 What are the normal freezing points and boiling points of the following solutions? (a) 21.2 g NaCl in 135 mL of water, (b) 15.4 g of urea in 66.7 mL of water.

*15.54 At 25°C the vapor pressure of pure water is 23.76 mmHg and that of seawater is 22.98 mmHg. Assuming that seawater contains only NaCl, estimate its concentration in molality units.

*15.55 Consider the following initial arrangement:

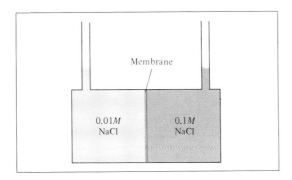

Membrane

0.01M NaCl   0.1M NaCl

What would happen if (a) the membrane is permeable to both water and the Na^+ and Cl^- ions, (b) permeable to water and Na^+ ions but not to Cl^- ions, (c) permeable to water but not to Na^+ and Cl^- ions?

15.56 Arrrange the following solutions in decreasing order of their expected freezing points: 0.10 m Na_3PO_4, 0.35 m NaCl, 0.20 m $MgCl_2$, 0.15 m $C_6H_{12}O_6$, 0.15 m CH_3COOH.

15.57 Both NaCl and $CaCl_2$ are used to melt ice on roads in winter. What advantages do these substances have over sucrose or urea in lowering the freezing point of water?

15.58 Arrange the following aqueous solutions in order of decreasing freezing point and explain your reasons: 0.50 m HCl, 0.50 m glucose, 0.50 m acetic acid.

15.59 The osmotic pressure of 0.010 M solutions of $CaCl_2$ and urea at 25°C are 0.605 atm and 0.245 atm, respectively. Calculate the van't Hoff factor for the $CaCl_2$ solution.

15.60 A 0.86 percent by mass solution of NaCl is called "physiological saline" because its osmotic pressure is equal to that of the solution in blood cells. Calculate the osmotic pressure of this solution at normal body temperature (37°C). Note that the density of the saline solution is 1.005 g/mL.

Miscellaneous Problems

15.61 Define the following terms: (a) solute, (b) solvent, (c) electrolyte, (d) nonelectrolyte, (e) solubility, (f) unsaturated solution, (g) saturated solution, (h) supersaturated solution, (i) crystallization, (j) precipitation.

15.62 Discuss the processes of fractional crystallization and fractional distillation and their applications.

15.63 Explain each of the following statements: (a) The boiling point of seawater is greater than that of pure water. (b) Carbon dioxide escapes from the solution when the cap is removed from a soft-drink bottle. (c) Concentrations of dilute solutions expressed in molalities are approximately equal to those expressed in molarities. (d) In discussing the colligative properties of a solution (with the exception of osmotic pressure), it is preferable to express the concentration units in molalities rather than in molarities. (e) Methanol (b.p. 65°C) is useful as an antifreeze, but it should be removed from the car radiator during the summer season

15.64 How does the ease of ion pair formation depend on (a) charges on the ions, (b) sizes of the ions, (c) nature of the solvent (polar vs. nonpolar)?

*15.65 Lysozyme is an enzyme that cleaves bacterial cell walls. A sample of lysozyme extracted from chicken egg white has a molar mass of 13,930 g. A quantity of 0.100 g of this enzyme is dissolved in 150 g of water at 25°C. Calculate the vapor pressure lowering, the depression in freezing point, the elevation in boiling point, and the osmotic pressure of this solution. (The vapor pressure of water at 25°C = 23.76 mmHg.)

*15.66 A solution of 1.00 g of anhydrous aluminum chloride, $AlCl_3$, in 50.0 g of water is found to have a

freezing point of $-1.11°C$. Explain this observation.

15.67 What minimum pressure must be applied to seawater at $25°C$ in order to carry out the reverse osmosis process? (Treat seawater as a 0.70 M NaCl solution.)

15.68 A cucumber placed in concentrated brine (salt water) shrivels into a pickle. Explain.

15.69 Explain why reverse osmosis is (theoretically) more desirable as a method for desalination than distillation or freezing.

*15.70 Solutions A and B have osmotic pressures of 2.4 atm and 4.6 atm, respectively, at a certain temperature. What is the osmotic pressure of a solution prepared by mixing equal volumes of A and B at the same temperature?

16
OXIDATION–REDUCTION REACTIONS

Reactions involving the transfer of electrons represent a diverse and important family of chemical changes. In this chapter we will consider such oxidation–reduction reactions in detail, focusing both on the molecular-level similarities in such processes and on the range of macroscopic applications we encounter in studying these changes. Our study of oxidation–reduction reactions is simplified by classifying the diversity of reactions in this family into several major categories of chemical change.

To a large extent, understanding the material presented in this chapter depends on a series of definitions that chemists have established to simplify their discussion of oxidation–reduction reactions. Thus you will encounter the vocabulary associated with these reactions, as well as a wide range of chemical examples all sharing the common trait that they involve electron transfer.

469

16.1 OXIDATION–REDUCTION REACTIONS: SOME DEFINITIONS

We discussed the formation of ionic compounds from metallic and non-metallic elements in Section 10.2. Let us return to the reaction between atoms of lithium and fluorine. As we saw, this reaction can be considered (for convenience) as two separate steps involving the loss of an electron by a Li atom and the gain of an electron by a F atom:

$$\cdot Li \longrightarrow Li^+ + e^-$$

$$:\ddot{F}\cdot + e^- \longrightarrow :\ddot{F}:^-$$

This representation of the fluorine–lithium reaction assumes isolated atoms of each reactant, for convenience. (See p. 280.)

Each of these steps is called a **half-reaction,** which *explicitly shows electrons involved in either oxidation or reduction.* The sum of the half-reactions gives the overall reaction

$$\cdot Li + :\ddot{F}\cdot + e^- \longrightarrow Li^+ + :\ddot{F}:^- + e^-$$

or

$$\cdot Li + :\ddot{F}\cdot \longrightarrow Li^+:\ddot{F}:^-$$

The *half-reaction that involves loss of electrons* is called an **oxidation reaction;** the *half-reaction that involves gain of electrons* is called a **reduction reaction.** In the above example, lithium is oxidized and acts as a **reducing agent** because it *donates an electron* to fluorine. Fluorine is reduced and acts as an **oxidizing agent** because it *accepts an electron* from lithium.

As another example, consider the formation of MgO. The two half-reactions are

$$\cdot Mg\cdot \longrightarrow Mg^{2+} + 2e^-$$

$$\cdot\ddot{O}\cdot + 2e^- \longrightarrow :\ddot{O}:^{2-}$$

and the overall reaction is

$$\cdot Mg\cdot + \cdot\ddot{O}\cdot + 2e^- \longrightarrow Mg^{2+} + :\ddot{O}:^{2-} + 2e^-$$

or

$$\cdot Mg\cdot + \cdot\ddot{O}\cdot \longrightarrow Mg^{2+}:\ddot{O}:^{2-}$$

In this case, Mg is oxidized and is the reducing agent because it loses two electrons, and oxygen is reduced and is the oxidizing agent because it accepts two electrons.

Oxidizing agents are *always* reduced and reducing agents are *always* oxidized. This statement, which sounds somewhat confusing, is simply a consequence of the definitions of the two processes.

Oxidation and reduction always occur together because in any reaction the total number of electrons lost by the reducing agent must equal the number of electrons gained by the oxidizing agent. **Redox reaction,** a term that combines "reduction" and "oxidation," is often used when discussing *oxidation–reduction reactions.*

16.2 OXIDATION NUMBERS

The definitions of oxidation and reduction in terms of the loss and gain of electrons apply to the formation of ionic compounds such as LiF and MgO, which we used as examples in Section 16.1. But what about compounds that are partly covalent and are partly ionic, such as HF? Actually, the difference between HF and LiF with regard to oxidation and reduction is only a matter of degree. Strictly speaking, no compounds are known to be 100 percent ionic.

For this reason, chemists have introduced the oxidation number to help keep account of electrons in chemical reactions. ***Oxidation number*** refers to the *number of charges an atom would have in a molecule if electrons were transferred completely in the direction indicated by the difference in electronegativity* (see Section 11.2). For example, since F is more electronegative than H, the oxidation number of fluorine in HF is -1, whereas that of hydrogen is $+1$. Thus, in a formal way, we treat electrons as if they were transferred completely from H to F.

We can now redefine redox reactions more generally in terms of oxidation number, as follows: *An element is said to be oxidized if its oxidation number is increased in a reaction; if the oxidation number of the element is decreased in a reaction, it is said to be reduced.* When molecular hydrogen reacts with molecular fluorine to form hydrogen fluoride

$$H_2(g) + F_2(g) \longrightarrow 2HF(g)$$

the oxidation number of hydrogen increases from 0 (in H_2) to $+1$ (in HF) and that of fluorine decreases from 0 (in F_2) to -1 (in HF). Thus hydrogen is the element oxidized and fluorine is the element reduced in this reaction.

It is important to remember that oxidation number has no physical reality. Assigning an oxidation number of $+3$, for instance, to an element in a compound does not necessarily mean that the atom of that element actually bears three net positive charges. We should think of oxidation number only as a useful way of looking at redox reactions.

Assigning Oxidation Numbers

There are several general rules for assigning oxidation numbers to elements in compounds.

1. The oxidation number of an atom of any element in its elemental state (that is, uncombined form) is zero, no matter how complex the molecule. Thus, all the atoms in H_2, F_2, Be, Li, Na, O_2, P_4, and S_8 have the same oxidation number: zero.
2. For an ion composed of only one atom, the oxidation number is equal to the charge on the ion. Thus K^+ ion has the oxidation number $+1$; Mg^{2+} ion, $+2$; Al^{3+} ion, $+3$; F^- ion, -1; O^{2-} ion, -2; and so on. All alkali metals have the oxidation number $+1$ and alkaline earth metals have the oxidation number $+2$ in their compounds.
3. The oxidation number of oxygen in most compounds (for example, H_2O and CaO) is -2, but it is different in the following cases. In OF_2, it has the oxidation number $+2$, because fluorine is more electronegative than oxygen. In hydrogen peroxide (H_2O_2) and in peroxide ion (O_2^{2-}), its oxidation number is -1. The Lewis formula for hydrogen peroxide is as follows:

The term *oxidation state* is used interchangeably with the term *oxidation number.*

Note that the oxidation numbers are always expressed on a "per atom" basis. Do not confuse the total charge on the peroxide anion $(2-)$ with the oxidation number of oxygen within the anion (-1).

$$H—\ddot{O}—\ddot{O}—H$$

A bond between identical atoms in a molecule makes no contribution to the oxidation number of those atoms because the electron pair of that bond is *equally* shared. Since H has the oxidation number $+1$, each O atom in H_2O_2 has the oxidation number -1. In the superoxide ion, O_2^-, each O atom has the oxidation number $-\frac{1}{2}$.

4. Fluorine has the oxidation number -1 in all of its compounds. This is a consequence of the fact that fluorine has the highest electronegativity of all the elements.

5. The oxidation number of hydrogen is $+1$, except when it is bonded to a metal as in LiH, NaH, and BaH_2, where its oxidation number is -1 because most metals are less electronegative than hydrogen.

6. In a neutral molecule, the sum of the oxidation numbers of all the atoms must be zero.

7. In a polyatomic ion, the sum of the oxidation numbers of all the elements in the ion must be equal to the net charge of the ion. For example, in the ammonium ion, NH_4^+, the oxidation number of nitrogen is -3 and that of hydrogen is $+1$. Thus, the sum of oxidation numbers is $-3 + (4 \times 1) = +1$, which is equal to the net charge of the ion.

EXAMPLE 16.1

Assign oxidation numbers to all the elements in the following compounds and ion: (a) Rb_2O, (b) $LiAlH_4$, (c) $Na_2Cr_2O_7$, and (d) NO_2^-.

Answer

(a) From rule 2 we see that rubidium has an oxidation number of $+1$ (Rb^+) and oxygen an oxidation number of -2 (O^{2-}).

(b) In lithium aluminum hydride ($LiAlH_4$), there are two different metals present. For the purpose of evaluating oxidation numbers only, we can write the formula as if it contained two units: (LiH) and (AlH_3). From rule 2 and rule 5 we see that the oxidation number of Li is $+1$ (Li^+), that of Al is $+3$ (Al^{3+}), and that of H is -1 (H^-).

(c) From rule 2, it is clear that the oxidation number of Na is $+1$ (Na^+). From rule 7, we see that the sum of the oxidation numbers in $Cr_2O_7^{2-}$ must be -2. We know that the oxidation number of O is -2 (rule 3), so all that remains is to determine the oxidation number of Cr, which we call x. The sum of the oxidation numbers in $Cr_2O_7^{2-}$ is

$$2(x) + 7(-2) = -2$$
$$x = +6$$

(d) Since oxygen is more electronegative than nitrogen (see Figure 11.2), it is assigned an oxidation number of -2 (rule 3). In order to have the overall charge equal to -1 (rule 7) the oxidation number of nitrogen must be $+3$.

Similar examples: Problems 16.6, 16.7, 16.8.

FIGURE 16.1 The oxidation numbers of elements in their compounds. The more common oxidation numbers are in color. The oxidation number +3 is common to all lanthanides and actinides in their compounds.

1A	2A	3B	4B	5B	6B	7B	8B	8B	8B	1B	2B	3A	4A	5A	6A	7A	8A
1 H +1, −1																	2 He
3 Li +1	4 Be +2											5 B +3	6 C +4, +2, −4	7 N +5, +4, +3, +2, +1, −3	8 O +2, $-\frac{1}{2}$, $-\frac{1}{2}$	9 F −1	10 Ne
11 Na +1	12 Mg +2											13 Al +3	14 Si +4, −4	15 P +5, +3, −3	16 S +6, +4, +2, −2	17 Cl +7, +6, +5, +4, +3, +1, −1	18 Ar
19 K +1	20 Ca +2	21 Sc +3	22 Ti +4, +3, +2	23 V +5, +4, +3, +2	24 Cr +6, +5, +4, +3, +2	25 Mn +7, +6, +4, +3, +2	26 Fe +3, +2	27 Co +3, +2	28 Ni +2	29 Cu +2, +1	30 Zn +2	31 Ga +3	32 Ge +4, −4	33 As +5, +3, −3	34 Se +6, +4, −2	35 Br +5, +3, +1, −1	36 Kr +4, +2
37 Rb +1	38 Sr +2	39 Y +3	40 Zr +4	41 Nb +5, +4	42 Mo +6, +4, +3	43 Tc +7, +6, +4	44 Ru +8, +6, +4, +3	45 Rh +4, +3, +2	46 Pd +4, +2	47 Ag +1	48 Cd +2	49 In +3	50 Sn +4, +2	51 Sb +5, +3, −3	52 Te +6, +4, −2	53 I +7, +5, +1, −1	54 Xe +6, +4, +2
55 Cs +1	56 Ba +2	57 La +3	72 Hf +4	73 Ta +5	74 W +6, +4	75 Re +7, +6, +4	76 Os +8, +4	77 Ir +4, +3	78 Pt +4, +2	79 Au +3, +1	80 Hg +2, +1	81 Tl +3, +1	82 Pb +4, +2	83 Bi +5, +3	84 Po +2	85 At −1	86 Rn

Periodic Variations of Oxidation Numbers

Figure 16.1 shows the known oxidation numbers of the more familiar elements, arranged according to their positions in the periodic table. The following features are especially important:

- Metallic elements have only positive oxidation numbers, whereas nonmetallic elements can have either positive or negative oxidation numbers.
- The highest oxidation number a representative element can have is its group number in the periodic table. For example, the halogens are in Group 7A, so their highest possible oxidation number is +7, which Cl, Br, and I exhibit in some of their compounds.

- The transition metals, unlike metals of the representative elements, usually have multiple oxidation numbers. Take the first row of the transition metals (Sc to Cu), for example. We note that the maximum oxidation number increases from $+3$ for Sc to $+7$ for Mn. Then it falls from Fe to Cu. Color Plate 20 shows the colors of vanadium ions in four different oxidation states ($+5$, $+4$, $+3$, and $+2$).

Knowledge of the oxidation numbers of the elements helps us write correct formulas for compounds. Table 16.1 shows the formulas of the chlorides, oxides, and hydrides of the elements in the second and third periods of the periodic table. In Section 9.6 we discussed the changes from metallic to nonmetallic properties in some of these compounds as we move from left to right.

The Stock system for inorganic nomenclature, introduced in Section 2.9, is based on the oxidation number concept. Note that the Stock system applies mainly to compounds containing metallic elements. Applied to compounds containing only nonmetallic elements the Stock system can sometimes be ambiguous. For example, according to it we would designate both NO_2 (nitrogen dioxide) and N_2O_4 (dinitrogen tetroxide) as nitrogen(IV) oxide, because the oxidation number of N in both of these compounds is $+4$.

We can also predict some properties of compounds from their oxidation numbers. In general, compounds whose central atoms have a high (positive) oxidation number tend to be covalent, and those whose central atoms have a low oxidation number tend to be ionic. Thus in Group 4A we find that lead(II) compounds are largely ionic, whereas lead(IV) compounds are

TABLE 16.1 Formulas of the Chlorides, Oxides, and Hydrides of Elements in the Second and Third Periods

Second Period	Li	Be	B	C	N	O	F
Chloride	$LiCl^*$	$BeCl_2^*$	BCl_3	CCl_4	NCl_3	Cl_2O	ClF
Oxide	Li_2O^*	BeO^*	$B_2O_3^*$	CO	N_2O		OF_2
				CO_2	NO		
					NO_2		
					N_2O_3		
					N_2O_4		
					N_2O_5		
Hydride	LiH^*	BeH_2^*	B_2H_6	CH_4	NH_3	H_2O	HF

Third Period	Na	Mg	Al	Si	P	S	Cl
Chloride	$NaCl^*$	$MgCl_2^*$	Al_2Cl_6	$SiCl_4$	PCl_3	S_2Cl_2	
					PCl_5	SCl_2	
Oxide	Na_2O^*	MgO^*	$Al_2O_3^*$	SiO_2^*	P_4O_6	SO_2	Cl_2O
	$Na_2O_2^*$				P_4O_{10}	SO_3	Cl_2O_7
Hydride	NaH^*	MgH_2^*	AlH_3^*	SiH_4	PH_3	H_2S	HCl

* Empirical formula.

mainly covalent. For example, $PbCl_2$ is a high-melting solid (m.p. 501°C) and an electrolyte, but $PbCl_4$ is a low-melting liquid (m.p. -15°C) and a nonelectrolyte. Their chemical properties differ too. The reason for the differences in their chemical properties is that a metal ion with high charges is unstable (and energetically unfavorable)—in this example the hypothetical Pb^{4+} ion—so the metal preferentially forms covalent bonds with other atoms in its high oxidation states.

16.3 TYPES OF REDOX REACTIONS

In Chapter 3 we classified chemical reactions into five types: combination, decomposition, displacement, metathesis, and acid–base neutralization. It is appropriate at this point to see how redox reactions fit into these categories. Neither metathesis nor acid–base neutralization reactions involve changes in oxidation numbers. Therefore, they are nonredox reactions. Some kinds of combination, decomposition, and displacement reactions, however, are redox reactions. In addition, there are some redox reactions that do not fit any of these categories.

Combination Reactions

Combination reactions involving one or more free elements are redox reactions. Some examples are

$$\overset{0}{Zn}(s) + \overset{0}{S}(s) \longrightarrow \overset{+2\ -2}{ZnS}(s)$$

$$\overset{0}{Cl_2}(g) + \overset{+3\ -1}{PCl_3}(l) \longrightarrow \overset{+5\ -1}{PCl_5}(s)$$

$$\overset{0}{N_2}(g) + 3\overset{0}{H_2}(g) \longrightarrow 2\overset{-3\ +1}{NH_3}(g)$$

where the number above each element denotes the oxidation number.

We show only oxidation numbers that undergo changes.

Decomposition Reactions

Decomposition reactions that produce one or more free elements are redox reactions. Some examples are

$$2\overset{+2\ -2}{HgO}(s) \longrightarrow 2\overset{0}{Hg}(l) + \overset{0}{O_2}(g)$$

$$2\overset{+1\ -2}{H_2O}(l) \longrightarrow 2\overset{0}{H_2}(g) + \overset{0}{O_2}(g)$$

Remember that this process is best carried out by electrolysis.

Displacement Reactions

Reactions in which an atom or an ion of one element is displaced from a compound by an atom of another element are normally redox reactions. The activity series, introduced in Section 3.5, helps us decide which substance acts as the oxidizing agent and which substance is the reducing agent. Referring to Figure 3.9 we see that elements listed at the top of the series are the strongest reducing agents (or weakest oxidizing agents), whereas those near the bottom are the weakest reducing agents (or strongest oxidizing agents). Some examples of redox displacement reactions are

$$\overset{0}{2Na}(s) + \overset{+1}{2H_2O}(l) \longrightarrow \overset{+1\ +1}{2NaOH}(aq) + \overset{0}{H_2}(g)$$

$$\overset{0}{Mg}(s) + \overset{+1}{2HCl}(aq) \longrightarrow \overset{+2}{MgCl_2}(aq) + \overset{0}{H_2}(g)$$

$$\overset{0}{Zn}(s) + \overset{+2}{CuSO_4}(aq) \longrightarrow \overset{+2}{ZnSO_4}(aq) + \overset{0}{Cu}(s)$$

$$\overset{0}{Cl_2}(g) + \overset{-1}{2KBr}(aq) \longrightarrow \overset{-1}{2KCl}(aq) + \overset{0}{Br_2}(l)$$

Disproportionation Reactions

In a **disproportionation reaction** an *element in one oxidation state is both oxidized and reduced.* One reactant in a disproportionation reaction contains an element that can have at least three oxidation states. One of the states is the reactant element itself, another is a higher state, and the third is lower. The thermal decomposition of hydrogen peroxide is an example of a disproportionation reaction:

$$\overset{-1}{2H_2O_2}(aq) \longrightarrow \overset{-2}{2H_2O}(l) + \overset{0}{O_2}(g)$$

Note that the oxidation number of oxygen (-1) in the reactant has both increased to zero in O_2 and decreased to -2 in H_2O. Another example is the thermal decomposition of nitrous acid:

$$\overset{+3}{3HNO_2}(aq) \longrightarrow \overset{+5}{HNO_3}(aq) + \overset{+2}{2NO}(g) + H_2O(l)$$

A third example involves the copper(I) ion. This ion is unstable in water and readily disproportionates to yield elemental Cu and the copper(II) ion:

$$\overset{+1}{2Cu^+}(aq) \longrightarrow \overset{0}{Cu}(s) + \overset{+2}{Cu^{2+}}(aq)$$

The preceding disproportionation reactions can be described as decomposition reactions. The two disproportionation reactions that follow, on the other hand, are not decomposition reactions:

$$\overset{0}{Cl_2}(g) + H_2O(l) \longrightarrow \overset{-1}{HCl}(aq) + \overset{+1}{HOCl}(aq)$$

The Mn^{3+} ion is unstable in aqueous solution and it disproportionates as follows:

$$\overset{+3}{2Mn^{3+}}(aq) + 2H_2O(l) \longrightarrow \overset{+2}{Mn^{2+}}(aq) + \overset{+4}{MnO_2}(s) + 4H^+(aq)$$

Elements that are most likely to undergo disproportionation are N, P, O, S, Cl, Br, I, Mn, Cu, Au, and Hg.

Miscellaneous Redox Reactions

There are a number of redox reactions that do not fit into any of the classes so far described. Two examples are

$$\overset{0}{3\text{Cu}}(s) + \overset{+5}{8\text{HNO}_3}(aq) \longrightarrow \overset{+2}{3\text{Cu}}\overset{+5}{(\text{NO}_3)_2}(aq) + \overset{+2}{2\text{NO}}(g) + 4\text{H}_2\text{O}(l)$$

$$\overset{-2}{2\text{ZnS}}(s) + \overset{0}{3\text{O}_2}(g) \longrightarrow \overset{-2}{2\text{ZnO}}(s) + \overset{+4\ -2}{2\text{SO}_2}(g)$$

We refer to such reactions as *miscellaneous redox reactions.*

EXAMPLE 16.2

Which of the following reactions involves oxidation–reduction? For the oxidation–reduction reactions, indicate the changes in oxidation numbers of the elements, and identify the oxidizing and reducing agents.

(a) $\quad (\text{NH}_4)_2\text{Cr}_2\text{O}_7(s) \longrightarrow \text{Cr}_2\text{O}_3(s) + \text{N}_2(g) + 4\text{H}_2\text{O}(g)$

(b) $\quad 3\text{AuCl}(s) \longrightarrow 2\text{Au}(s) + \text{AuCl}_3(aq)$

(c) $\quad \text{CH}_4(g) + 2\text{O}_2(g) \longrightarrow \text{CO}_2(g) + 2\text{H}_2\text{O}(l)$

(d) $\quad \text{CaCO}_3(s) \longrightarrow \text{CaO}(s) + \text{CO}_2(g)$

Answer

To identify oxidation–reduction reactions, we compare the oxidation numbers of each element on the left- and right-hand sides.

(a)

Left-Hand Side	*Right-Hand Side*
N = −3	N = 0
H = +1	H = +1
Cr = +6	Cr = +3
O = −2	O = −2

The oxidation number of N increases and that of Cr decreases, so this is a redox decomposition reaction. Since there is only one reactant, both the oxidized and reduced reagents are present in the compound $(\text{NH}_4)_2\text{Cr}_2\text{O}_7$.

(b)

Left-Hand Side	*Right-Hand Side*
Au = +1	Au = 0 and +3
Cl = −1	Cl = −1

Since the oxidation number of Au increases from +1 to +3 and decreases from +1 to 0, this is a disproportionation reaction.

(c)

Left-Hand Side	*Right-Hand Side*
C = −4	C = +4
H = +1	H = +1
O = 0	O = −2

(Continued)

The oxidation number of carbon decreases while that of oxygen increases; this is a redox reaction of the miscellaneous type. The oxidizing agent is molecular oxygen and the reducing agent is CH_4.

(d)

Left-Hand Side	Right-Hand Side
Ca = +2	Ca = +2
C = +4	C = +4
O = −2	O = −2

As you can see, there is no change in the oxidation number of any of the elements. Therefore, this is a decomposition reaction that does not involve oxidation–reduction.

Similar example: Problem 16.3.

16.4 THE NATURE OF REDOX REACTIONS

Many reactions of practical interest are redox reactions. For example, all combustions are redox processes. The action of household bleaching agents is due to the presence of the hypochlorite ion, OCl^-. It is this ion that oxidizes the color-bearing substances in stains, converting them to colorless compounds. We will briefly discuss four important processes in which redox reactions play central roles.

Explosives

Two well-known explosives are nitroglycerin and trinitrotoluene (TNT):

Nitroglycerin
$(C_3H_5N_3O_9)$

Trinitrotoluene (TNT)
$(C_7H_5N_3O_6)$

TNT is a solid. For every two moles of TNT that undergo decomposition, fifteen moles of gases are formed:

$$2C_7H_5N_3O_6(s) \longrightarrow 3N_2(g) + 7CO(g) + 5H_2O(g) + 7C(s)$$

It is the sudden formation of hot expanding gases that makes TNT a powerful explosive (Figure 16.2).

Nitroglycerin is a liquid. It detonates on shock and explodes violently:

$$4C_3H_5N_3O_9(l) \longrightarrow 6N_2(g) + 12CO_2(g) + O_2(g) + 10H_2O(g)$$

FIGURE 16.2 *TNT explosive at work.*

Note that for every four moles of nitroglycerin that undergo decomposition, twenty-nine moles of gases are formed. In 1867 the Swedish inventor Alfred Nobel (1833–1896) discovered that if nitroglycerin is absorbed in porous silica, it becomes an easily handled but still very effective explosive, which he named dynamite. Its manufacture and further development were the principal sources of his great wealth, with which he established the Nobel Prizes.

The Thermite Reaction

Like many redox combination and decomposition reactions, certain redox displacement reactions are highly exothermic. Of these, one of the best known is the *thermite reaction,* which involves the reaction of aluminum with a metal oxide:

$$2Al(s) + Fe_2O_3(s) \longrightarrow Al_2O_3(l) + 2Fe(l)$$

Because Al_2O_3 has a large negative enthalpy of formation ($\Delta H_f^\circ = -1670$ kJ/mol), this reaction releases a large quantity of thermal energy. With suitable amounts of reactants, the temperature of the reaction can reach 3000°C! Thermite reactions have been used to weld large masses of iron and steel (Figure 16.3) and the military has used thermite bombs as incendiary devices.

FIGURE 16.3 *The thermite reaction being applied to weld reinforcing rods for concrete at a construction site.*

Black and White Photography

Black and white photographic film contains small grains of silver bromide, evenly spread over a thin gelatin coating on a transparent surface. Exposure of the film to light activates the silver bromide:

$$AgBr + light \longrightarrow AgBr^*$$

The asterisk denotes AgBr affected by light. The exposed film is then treated with a developer, a solution containing a mild reducing agent. In the redox process that follows, the Ag^+ ions in the activated AgBr are preferentially reduced to metallic silver. The number of black metallic silver particles formed on the film is directly proportional to the amount of light that originally fell on the film.

This brief description summarizes the way in which a black and white negative is prepared. A positive print can be obtained by shining light through the negative onto another piece of photographic paper and repeating the development procedure. Because white regions of the subject appear black in a negative, and vice versa, this process inverts the light and dark areas of the negative to produce the desired picture (Figure 16.4).

Breath Analyzer

Redox reactions find use also in law enforcement, for example, in testing drivers suspected of being drunk. A sample of the driver's breath is drawn

Other silver halides too are light-sensitive. For example, a silver chloride precipitate, originally white in color, gradually turns pink-purple in the presence of light.

The unactivated AgBr is subsequently removed from the film. Otherwise, the entire film would gradually turn black because of the slow reduction of Ag^+ to Ag.

a   b

FIGURE 16.4   *The negative (a) and positive (b) of a black and white photograph.*

into an instrument, shown in Figure 16.5, in which it is treated with an oxidizing agent such as an acidic solution of potassium dichromate. The equation for the oxidation of alcohol or ethanol is

$$3CH_3CH_2OH + \underset{\substack{\text{potassium}\\ \text{dichromate}\\ \text{(orange-yellow)}}}{2K_2Cr_2O_7} + \underset{\substack{\text{sulfuric}\\ \text{acid}}}{8H_2SO_4} \longrightarrow$$

$$\underset{\substack{\text{acetic}\\ \text{acid}}}{3CH_3COOH} + \underset{\substack{\text{chromic}\\ \text{sulfate}\\ \text{(green)}}}{2Cr_2(SO_4)_3} + \underset{\substack{\text{potassium}\\ \text{sulfate}}}{2K_2SO_4} + 11H_2O$$

ethanol

As you can see, the dichromate ion is reduced to the chromic (Cr^{3+}) ion (Color Plate 21 shows the colors of $Cr_2O_7^{2-}$ and Cr^{3+} ions). On the other hand, ethanol is oxidized to acetic acid. The blood alcohol level in the driver can be readily determined by measuring the color change from orange-yellow to green.

For organic compounds, oxidation can be defined as the removal of H atoms. Note that each ethanol molecule has six H atoms whereas each acetic acid molecule has four H atoms.

FIGURE 16.5 *A breath analyzer used in police departments to test drunken drivers.*

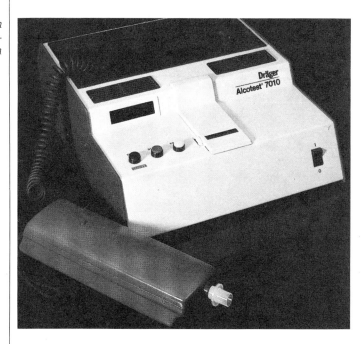

FIGURE 16.5 *A breath analyzer used in police de-partments to test drunken drivers.*

16.5   BALANCING REDOX EQUATIONS

Some redox processes, such as the reaction of sodium with molecular chlorine, or the combustion of sulfur in air and carbon in air, are so simple that the equations representing them can be balanced easily:

$$2Na(s) + Cl_2(g) \longrightarrow 2NaCl(s)$$

$$C(s) + O_2(g) \longrightarrow CO_2(g)$$

$$S(s) + O_2(g) \longrightarrow SO_2(g)$$

But many of the redox reactions we encounter are more complex than these. In principle, we can balance any redox reaction using the steps outlined in Section 3.1. However, many oxidation–reduction reactions are complicated enough to justify special balancing methods. Two redox balancing methods have been particularly useful to chemists—the oxidation number method and the ion–electron method. As you might expect, the oxidation number method is based on identifying and following changes in oxidation numbers in the redox equation. By contrast, the ion–electron method requires no explicit use of oxidation numbers; it focuses first on balancing the two half-reactions involved in the overall change. We will illustrate how both balancing methods can be usefully applied to specific redox equations.

Oxidation Number Method

Suppose we are asked to balance the equation representing the oxidation of elemental sulfur by nitric acid to yield sulfur dioxide and nitric oxide. We can balance the equation with the aid of the following steps.

Step 1. *Write the skeletal equation containing the oxidizing and reducing agents and products.*

Our skeletal equation is

$$S + HNO_3 \longrightarrow SO_2 + NO$$

Step 2. *Assign oxidation numbers to the atoms of each element on both sides of the equation and determine which elements have been oxidized and which were reduced. Determine the number of units of increase and decrease in oxidation number for each of these elements.*

A careful inspection of the oxidation number of each element on both sides of the equation shows that sulfur is the element oxidized, and nitrogen is the element reduced. We summarize as follows.

Oxidation Number		Change in
Left-Hand Side	*Right-Hand Side*	*Oxidation Number*
$S = 0$	$S = +4$	$+4$
$N = +5$	$N = +2$	-3

Step 3. *Equalize the increase and decrease in oxidation number by multiplying each substance oxidized or reduced by the appropriate coefficients. Use the same coefficients in the equation for the elements oxidized and reduced.*

From the above table we multiply S by 3 and N by 4 in order to equalize the change in oxidation numbers:

$$4(-3) = -12$$
$$3(+4) = +12$$
$$3S + 4HNO_3 \longrightarrow 3SO_2 + 4NO$$
$$0 \quad\quad +5 \quad\quad\quad +4 \quad\quad +2$$

Consequently, we place the coefficient 3 in front of S and the coefficient 4 in front of N.

Step 4. *Balance the remaining atoms by inspection. For reactions in acidic solution, add H^+ or H_2O or both to the equation as needed. For reactions in basic solution, add OH^- or H_2O or both to the equation as needed.*

Note that four H atoms appear on the left-hand side, so we need two molecules of H_2O on the right:

$$3S + 4HNO_3 \longrightarrow 3SO_2 + 4NO + 2H_2O$$

Step 5. *Verify that the equation contains the same numbers and types of atoms on both sides of the equation. For ionic equations, the charges too must balance on both sides of the equation.*

Following are two more examples illustrating the oxidation number method.

EXAMPLE 16.3

Write a balanced equation showing the oxidation of hydrochloric acid by potassium permanganate to yield manganese(II) chloride, molecular chlorine, and potassium chloride.

Answer

Step 1

First we write the skeletal equation as

$$KMnO_4 + HCl \longrightarrow MnCl_2 + Cl_2 + KCl$$

Step 2

We construct the following table:

Oxidation Number		Change in
Left-Hand Side	Right-Hand Side	Oxidation Number
$Cl = -1$	$Cl = 0$	$+1$
$Mn = +7$	$Mn = +2$	-5

We see that the HCl in this reaction has a dual function—it acts as a reducing agent, and it also provides chloride ions.

Step 3

Note that we cannot multiply Mn by 1 and Cl by 5 because of the subscript in Cl_2. Instead, we must make the change in oxidation number of both Mn and Cl equal to 10:

$$(2)(5)(+1) = +10$$
$$(2)(1)(-5) = -10$$
$$2KMnO_4 + 10HCl \longrightarrow 2MnCl_2 + 5Cl_2 + 2KCl$$
$$\quad\; +7 \qquad\quad -1 \qquad\quad +2 \qquad\quad 0$$

Step 4

To balance the oxygen atoms, we must write in eight H_2O molecules on the right-hand side. Then the number of HCl molecules must be increased to sixteen to balance the H atoms:

$$2KMnO_4 + 16HCl \longrightarrow 2MnCl_2 + 5Cl_2 + 2KCl + 8H_2O$$

Step 5

A final check shows that the number of each type of atom is the same on both sides of the equation.

Similar example: Problem 16.12.

Note that only 10 out of 16 Cl^- ions from HCl are oxidized. The other 6 Cl^- ions are unchanged.

EXAMPLE 16.4

Write a balanced equation showing the oxidation of iodide ion by permanganate ion to yield manganese(IV) oxide and iodate ion in neutral or slightly basic solution.

Answer

Step 1

The skeletal equation is

$$MnO_4^- + I^- \longrightarrow MnO_2 + IO_3^-$$

Step 2

We construct the following table.

Oxidation Number		Change in
Left-Hand Side	Right-Hand Side	Oxidation Number
I = −1	I = +5	+6
Mn = +7	Mn = +4	−3

Step 3

To equalize the changes in oxidation numbers, we multiply I by 3 and Mn by 6:

$$3(+6) = +18$$
$$6(-3) = -18$$

$$\underset{+7 \quad\quad -1 \quad\quad\quad +4 \quad\quad +5}{6MnO_4^- + 3I^- \longrightarrow 6MnO_2 + 3IO_3^-}$$

Dividing the coefficients throughout by 3, we obtain

$$2MnO_4^- + I^- \longrightarrow 2MnO_2 + IO_3^-$$

Step 4

Since the reaction occurs in basic solution, we can balance both mass and charge by adding H_2O and OH^-. There are three negative charges on the left and only one negative charge on the right. We balance the negative charges by adding two OH^- ions on the right:

$$2MnO_4^- + I^- \longrightarrow 2MnO_2 + IO_3^- + 2OH^-$$

To balance O and H atoms, we add one H_2O on the left:

$$2MnO_4^- + I^- + H_2O \longrightarrow 2MnO_2 + IO_3^- + 2OH^-$$

Step 5

A final check shows that both charge and mass are balanced.

Similar example: Problem 16.12.

Ion—Electron Method

The ion—electron method takes an approach that is rather different from the oxidation number method, but it produces the same result—a properly balanced redox equation. The overall reaction is divided into two half-reactions, one for oxidation and one for reduction. The equations for the two half-reactions are balanced separately and then added to give the overall balanced equation.

We pretend that the reaction takes place in two separate steps.

Suppose we are asked to balance the equation showing the oxidation of Fe(II) ion to Fe(III) ion by the dichromate ion in an acidic medium. The following steps will help us accomplish this task.

Step 1. *Write the skeletal equation for the reaction in ionic form. Separate the equation into two equations for the half-reactions*:

$$Fe^{2+} + Cr_2O_7^{2-} \longrightarrow Fe^{3+} + Cr^{3+}$$

The equations for the two half-reactions are

$$\text{oxidation:} \quad Fe^{2+} \longrightarrow Fe^{3+}$$

$$\text{reduction:} \quad Cr_2O_7^{2-} \longrightarrow Cr^{3+}$$

Step 2. *Balance the atoms in each half-reaction separately. For reactions in acidic medium, add H_2O to balance O atoms and H^+ to balance H atoms. For reactions in basic medium, first balance the atoms as you would in acidic medium. Then, for every H^+ ion, add an equal number of OH^- ions to both sides of the equation. Where H^+ and OH^- appear on the same side of the equation, combine the ions to give H_2O.*

We balance the atoms in each half-reaction. The oxidation reaction is already balanced for Fe. For the reduction step, we multiply the Cr^{3+} ion by 2 to balance Cr. Since the reaction takes place in acidic medium, we add seven H_2O molecules on the right-hand side to balance O atoms:

$$Cr_2O_7^{2-} \longrightarrow 2Cr^{3+} + 7H_2O$$

To balance H atoms, we add fourteen H^+ ions on the left-hand side:

$$14H^+ + Cr_2O_7^{2-} \longrightarrow 2Cr^{3+} + 7H_2O$$

Step 3. *Add electrons to one side of each half-reaction to balance the charges. If necessary, equalize the number of electrons in the two half-reactions by multiplying one or both half-reactions by appropriate coefficients.*

For the oxidation half-reaction we write

$$Fe^{2+} \longrightarrow Fe^{3+} + e^-$$

Since there are net twelve positive charges on the left-hand side of the reduction half-reaction and only six positive charges on the right-hand side, we must add six electrons on the left:

$$14H^+ + Cr_2O_7^{2-} + 6e^- \longrightarrow 2Cr^{3+} + 7H_2O$$

To equalize the number of electrons in both half-reactions, we multiply the oxidation half-reaction by 6:

$$6Fe^{2+} \longrightarrow 6Fe^{3+} + 6e^-$$

Step 4. *Add the two half-reactions together and balance the final equation by inspection. The electrons on both sides should cancel.*

The two half-reactions are added to give

$$14H^+ + Cr_2O_7^{2-} + 6Fe^{2+} + 6e^- \longrightarrow 2Cr^{3+} + 6Fe^{3+} + 7H_2O + 6e^-$$

The electrons on both sides cancel and we are left with the balanced net ionic equation:

$$14H^+ + Cr_2O_7^{2-} + 6Fe^{2+} \longrightarrow 2Cr^{3+} + 6Fe^{3+} + 7H_2O$$

Step 5. *Verify that the equation contains the same types and numbers of atoms and the same charges on both sides of the equation.*

A final check shows that the final equation is both "atomically" and "electrically" balanced.

Let's consider another example of balancing redox equation using the ion–electron method.

EXAMPLE 16.5

Write a balanced molecular equation showing the oxidation of sodium sulfide by potassium permanganate in slightly basic solution to yield elemental sulfur and manganese(IV) oxide.

Answer

Step 1

We start by writing the skeletal equation in ionic form:

$$S^{2-} + MnO_4^- \longrightarrow S + MnO_2$$

The two half-reactions are

$$\text{oxidation:} \quad S^{2-} \longrightarrow S$$
$$\text{reduction:} \quad MnO_4^- \longrightarrow MnO_2$$

Step 2

The oxidation half-reaction is already balanced for S. In the reduction half-reaction, to balance O atoms we add two H_2O on the right:

$$MnO_4^- \longrightarrow MnO_2 + 2H_2O$$

To balance H atoms, we add four H^+ ions on the left:

$$MnO_4^- + 4H^+ \longrightarrow MnO_2 + 2H_2O$$

Since the reaction occurs in a basic medium and there are four H^+ ions, we add four OH^- ions to both sides of the equation:

$$MnO_4^- + 4H^+ + 4OH^- \longrightarrow MnO_2 + 2H_2O + 4OH^-$$

Combining the H^+ and OH^- ions to form H_2O we write

$$MnO_4^- + 4H_2O \longrightarrow MnO_2 + 2H_2O + 4OH^-$$

or

$$MnO_4^- + 2H_2O \longrightarrow MnO_2 + 4OH^-$$

(Continued)

Step 3

Next, we balance the charges of the two half-reactions as follows:

$$S^{2-} \longrightarrow S + 2e^-$$

$$MnO_4^- + 2H_2O + 3e^- \longrightarrow MnO_2 + 4OH^-$$

To equalize the number of electrons, we multiply the oxidation half-reaction by 3 and the reduction step by 2:

$$3S^{2-} \longrightarrow 3S + 6e^-$$

$$2MnO_4^- + 4H_2O + 6e^- \longrightarrow 2MnO_2 + 8OH^-$$

Step 4

These two half-reactions are added to give

$$3S^{2-} + 2MnO_4^- + 4H_2O + 6e^- \longrightarrow 3S + 2MnO_2 + 8OH^- + 6e^-$$

or, simply

$$3S^{2-} + 2MnO_4^- + 4H_2O \longrightarrow 3S + 2MnO_2 + 8OH^-$$

To convert this equation into molecular form, we must add the cations given in the original statement of this example; we add six Na^+ and two K^+ ions on both sides:

$$3Na_2S + 2KMnO_4 + 4H_2O \longrightarrow 3S + 2MnO_2 + 6NaOH + 2KOH$$

Step 5

A final check shows that the net ionic equation is balanced for the numbers and types of atoms and for charges, and the molecular equation is balanced for the numbers and types of atoms.

Similar example: Problem 16.13.

16.6 QUANTITATIVE ASPECTS OF REDOX REACTIONS

In some respects, redox reactions are similar to acid–base reactions: Redox reactions involve the transfer of electrons, and acid–base reactions involve the transfer of protons. Just as an acid can be titrated against a base, we can titrate an oxidizing agent against a reducing agent, using the procedure outlined in Section 4.6. We can, for example, carefully add a solution containing an oxidizing agent to a solution containing a reducing agent. The *equivalence point* is reached when the reducing agent is completely oxidized by the oxidizing agent.

The equivalence point for acid–base titration is defined in Section 4.6.

Redox titrations normally involve an indicator that has distinctly different colors in its oxidized and reduced forms. In the presence of a large amount of the reducing agent, the color of the indicator is characteristic of its reduced form. The indicator assumes the color of its oxidized form when it is present in an oxidizing medium. At or near the equivalence point, there is a *sharp* change in the indicator's color as it changes from one form to the other, so the equivalence point can be readily determined.

Two common oxidizing agents you may encounter in the laboratory are potassium dichromate $(K_2Cr_2O_7)$ and potassium permanganate $(KMnO_4)$. These substances are available in very pure form, so that solutions of accurately known concentration can be prepared from them. Furthermore, the colors of the anions are distinctly different from those of the reduced species; for example

$$MnO_4^- \longrightarrow Mn^{2+}$$
$$\text{purple} \qquad \text{light pink}$$

$$Cr_2O_7^{2-} \longrightarrow Cr^{3+}$$
$$\text{orange-} \qquad \text{green}$$
$$\text{yellow}$$

The colors of these ions are shown in Color Plate 21.

In many cases, the oxidizing agent itself can be used as an *internal* indicator in a redox titration. This is particularly true of permanganate ion because the change in color from purple to light pink (if MnO_4^- is reduced to Mn^{2+}) is quite easy to follow.

Although the acidic solutions of both dichromate and permanganate are powerful oxidizing agents, permanganate is the stronger of the two. Both substances will oxidize iron(II) salts:

$$Fe^{2+} \longrightarrow Fe^{3+} + e^-$$

But only permanganate is powerful enough to oxidize chloride ions to chlorine gas:

This reaction should be avoided because chlorine gas is poisonous.

$$2Cl^- \longrightarrow Cl_2 + 2e^-$$

Therefore dichromate is preferred over permanganate as the oxidizing agent in titrating iron if Cl^- ions are also present. Occasionally a basic permanganate solution is used in a redox reaction, as Example 16.5 shows.

Redox titrations require the same type of calculations (based on the mole method) as acid–base neutralizations. The difference is that the equations and stoichiometry tend to be more complex for redox reactions.

EXAMPLE 16.6

A quantity of 17.00 mL of 0.1000 M $KMnO_4$ solution is needed to oxidize 25.00 mL of a $FeSO_4$ solution in an acidic medium. What is the concentration of the $FeSO_4$ solution? The net ionic reaction is

$$5Fe^{2+} + MnO_4^- + 8H^+ \longrightarrow Mn^{2+} + 5Fe^{3+} + 4H_2O$$

Answer

The number of moles of $KMnO_4$ in 17.00 mL of the solution is

$$\text{moles of } KMnO_4 = 17.00 \text{ mL soln} \times \frac{0.1000 \text{ mol } KMnO_4}{1 \text{ L soln}} \times \frac{1 \text{ L soln}}{1000 \text{ mL soln}}$$
$$= 1.700 \times 10^{-3} \text{ mol } KMnO_4$$

(Continued)

and the number of moles of $FeSO_4$ oxidized is

$$\text{moles of } FeSO_4 = 1.700 \times 10^{-3} \text{ mol } KMnO_4 \times \frac{5 \text{ mol } FeSO_4}{1 \text{ mol } KMnO_4}$$
$$= 8.500 \times 10^{-3} \text{ mol } FeSO_4$$

The concentration of the $FeSO_4$ solution is equal to moles of $FeSO_4$ per liter of solution:

$$[FeSO_4] = \frac{8.500 \times 10^{-3} \text{ mol } FeSO_4}{25.00 \text{ mL soln}} \times \frac{1000 \text{ mL soln}}{1 \text{ L soln}}$$
$$= 0.3400 \text{ mol/L}$$
$$= 0.3400 \ M$$

Similar examples: Problems 16.15, 16.16, 16.18.

Equivalent Mass

We can solve Example 16.6 another way by using the concept of an ***equivalent.*** *One equivalent of an oxidizing agent* is the *amount that gains one mole of electrons. One equivalent of a reducing agent* is the *amount that loses one mole of electrons.*

Equivalents are so defined that in *any* redox process, at the equivalence point

equivalents of oxidizing agent = equivalents of reducing agent

In Example 16.6

oxidation: $\quad Fe^{2+} \longrightarrow Fe^{3+} + e^-$
$\qquad\qquad\quad {\scriptstyle +2} \qquad\quad {\scriptstyle +3}$

reduction: $\quad 5e^- + MnO_4^- \longrightarrow Mn^{2+}$
$\qquad\qquad\qquad\quad {\scriptstyle +7} \qquad\qquad {\scriptstyle +2}$

Since one mole of Fe^{2+} loses one mole of electrons in the oxidation, it follows, therefore, that one mole of Fe^{2+} is equal to one equivalent of Fe^{2+}. In the reduction, however, one mole of MnO_4^- is equal to *five* equivalents of MnO_4^- because one mole of MnO_4^- gains *five* moles of electrons.

Just as we speak of one mole of a substance and the molar mass of the substance, we can also speak of one equivalent of a substance and the equivalent mass of the substance. The ***equivalent mass*** of a substance is the *mass of the substance in grams that gains or loses one mole of electrons in a redox reaction.* Returning to the preceding example it can be seen that the relationship between the molar mass and equivalent mass of each of the two substances is as follows, where n is the number of electrons transferred per formula unit:

	Molar Mass	Equivalent Mass	
$FeSO_4$	151.9 g	$\dfrac{151.9 \text{ g}}{n} = \dfrac{151.9 \text{ g}}{1}$	$= 151.9$ g
$KMnO_4$	158.0 g	$\dfrac{158.0 \text{ g}}{n} = \dfrac{158.0 \text{ g}}{5}$	$= 31.60$ g

Normality

For redox titrations, chemists often use the concentration unit **normality** (**N**), which is the *number of equivalents of an oxidizing agent or a reducing agent per liter of solution,* that is

$$\text{normality} = \frac{\text{equivalents of solute}}{\text{liters of soln}}$$

Normality is numerically related to molarity by the simple relationship

$$N = nM$$

Since the lowest number of electrons transferred per formula unit (n) is 1, we see that for a given solution normality is *never* less than molarity. For the reaction in Example 16.6, we can see the relationship between molarity and normality by writing

$$\text{KMnO}_4: \quad \text{normality} = 5(0.1000) \, N$$
$$= 0.5000 \, N$$

$$\text{FeSO}_4: \quad \text{normality} = 1(0.3400) \, N$$
$$= 0.3400 \, N$$

EXAMPLE 16.7

A quantity of 42.60 g of $K_2Cr_2O_7$ is dissolved in enough dilute HCl to make up exactly one liter of solution. Calculate the molarity and normality of the solution when $Cr_2O_7^{2-}$ reacts in a certain redox process according to the following half-reaction:

$$14H^+ + Cr_2O_7^{2-} + 6e^- \longrightarrow 2Cr^{3+} + 7H_2O$$

Answer

The molarity of the solution is

$$\text{molarity} = 42.60 \text{ g } K_2Cr_2O_7 \times \frac{1 \text{ mol } K_2Cr_2O_7}{294.2 \text{ g } K_2Cr_2O_7} \times \frac{1}{1 \text{ L soln}}$$

$$= 0.1448 \text{ mol/L}$$
$$= 0.1448 \, M$$

Each Cr accepts three electrons and there are two Cr per unit of $Cr_2O_7^{2-}$ (a total of six electrons), so the normality of the solution is

$$6(0.1448) \, N = 0.8688 \, N$$

Similar example: Problem 16.25.

We should keep in mind that the normality of a solution depends not only on its molarity but also on the reaction that is undergone by the solute, that is, the oxidizing or reducing agent. For example, the MnO_4^- ion can be reduced to Mn^{2+}, MnO_2, or MnO_4^{2-}, when the medium is acidic, neutral, or basic, respectively. The normality of the solution must be determined by the actual number of electrons transferred:

$$
\begin{aligned}
\text{acidic medium:} && MnO_4^- &\longrightarrow Mn^{2+} & 1\,M &= 5\,N \\
&& {\scriptstyle +7} && {\scriptstyle +2} \\[4pt]
\text{neutral medium:} && MnO_4^- &\longrightarrow MnO_2 & 1\,M &= 3\,N \\
&& {\scriptstyle +7} && {\scriptstyle +4} \\[4pt]
\text{basic medium:} && MnO_4^- &\longrightarrow MnO_4^{2-} & 1\,M &= 1\,N \\
&& {\scriptstyle +7} && {\scriptstyle +6}
\end{aligned}
$$

Having discussed how the normality of a solution actually depends on the reaction the solute undergoes, our next step is to see how normality can be used in redox titration calculations. From the definition of normality we write

$$
\text{normality} \times \text{volume} = \frac{\text{equivalents of solute}}{\text{liters of soln}} \times \text{liters of soln}
$$
$$
= \text{equivalents of solute}
$$

At the equivalence point, the product of normality and volume of the oxidizing agent must be equal to the product of normality and volume of the reducing agent:

$$
N_{ox}V_{ox} = N_{red}V_{red}
$$

or

$$
\text{equivalents of oxidizing agent} = \text{equivalents of reducing agent}
$$

where the subscripts *ox* and *red* denote oxidizing agent and reducing agent, respectively. The volumes of the oxidizing and reducing agents must be expressed in the same units (L or mL).

The use of normality would seem to greatly simplify routine solution stoichiometry problems, as Example 16.8 suggests.

EXAMPLE 16.8

How many milliliters of a 0.500 N $KMnO_4$ solution are needed to oxidize 25.0 mL of 0.638 N $FeSO_4$ in acidic solution?

Answer

At the equivalence point

$$
N_{ox}V_{ox} = N_{red}V_{red}
$$
$$
0.500\,N \times V_{KMnO_4} = 0.638\,N \times 25.0\ \text{mL}
$$
$$
V_{KMnO_4} = 31.9\ \text{mL}
$$

You may be struck by the apparent ease with which Example 16.8 was completed. Not even a balanced equation was given! In fact, the normality of the $KMnO_4$ solution in the problem had the stoichiometry of the reaction already "built in." This is a clear advantage for repeated calculations involving a particular redox system, such as this one. However, you should realize that if this $KMnO_4$ solution is used in other redox reactions involving different numbers of electrons, the normality of this oxidizing agent would change. Thus, normality is a reaction-dependent concentration unit. This represents both an advantage and a disadvantage, when it is compared to molarity.

16.7 SOME COMMON OXIDIZING AND REDUCING AGENTS

In this last section we will briefly examine the properties of several oxidizing and reducing agents that are commonly encountered in the laboratory.

Oxidizing Agents

Hydrogen Peroxide (H_2O_2). Hydrogen peroxide, a colorless, syrupy liquid (m.p. $-0.9°C$), is a strong oxidizing agent (Figure 16.6). It can oxidize Fe^{2+} ions to Fe^{3+} ions in an acidic solution. The half-reactions are

$$H_2O_2(aq) + 2H^+(aq) + 2e^- \longrightarrow 2H_2O(l)$$

$$2Fe^{2+}(aq) \longrightarrow 2Fe^{3+}(aq) + 2e^-$$

and the overall reaction is

$$H_2O_2(aq) + 2Fe^{2+}(aq) + 2H^+(aq) \longrightarrow 2Fe^{3+}(aq) + 2H_2O(l)$$

Hydrogen peroxide can also oxidize sulfite ions (SO_3^{2-}) to sulfate ions (SO_4^{2-})

$$H_2O_2(aq) + SO_3^{2-}(aq) \longrightarrow SO_4^{2-}(aq) + H_2O(l)$$

and sulfides (S^{2-}) to sulfates

$$4H_2O_2(aq) + S^{2-}(aq) \longrightarrow SO_4^{2-}(aq) + 4H_2O(l)$$

Hydrogen peroxide is regarded as a "clean" oxidizing agent, since it does not directly contribute any new reaction products to the aqueous system involved. Its half-reaction simply produces water molecules, as the equation suggests.

FIGURE 16.6 *A label of the reagent grade hydrogen peroxide solution used in chemical laboratories.*

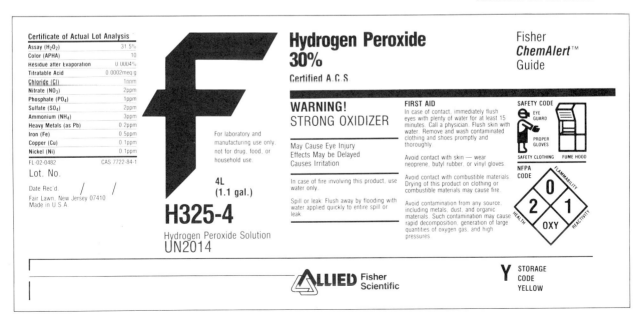

Sulfuric Acid (H_2SO_4). Sulfuric acid is a colorless, viscous liquid (m.p. 10.4°C). The concentrated sulfuric acid we use in the laboratory is 98 percent by mass, which corresponds to a concentration of 18 M (Figure 16.7). The oxidizing strength of sulfuric acid depends on its temperature and concentration. A cold, dilute sulfuric acid solution reacts with metals above hydrogen in the activity series (see Figure 3.9), liberating molecular hydrogen in a displacement reaction:

$$Mg(s) + H_2SO_4(aq) \longrightarrow MgSO_4(aq) + H_2(g)$$

This is a typical reaction of an active metal with an acid. The strength of sulfuric acid as an oxidizing agent is greatly enhanced when it is both hot and concentrated. In such a solution, the oxidizing agent is actually S in H_2SO_4 rather than the hydrated proton, $H^+(aq)$. The principal oxidizing action of sulfuric acid under these conditions is given by the half-reaction

$$2H_2SO_4(aq) + 2e^- \longrightarrow SO_4^{2-}(aq) + SO_2(g) + 2H_2O(l)$$

Note that only one of the S atoms is reduced to SO_2 (oxidation number of S changes from $+6$ to $+4$); the other S atom remains unchanged in the SO_4^{2-} ion. Thus copper reacts with concentrated sulfuric acid as follows:

$$Cu(s) + 2H_2SO_4(aq) \longrightarrow CuSO_4(aq) + SO_2(g) + 2H_2O(l)$$

Note that $H^+(aq)$ cannot oxidize Cu, but SO_4^{2-} in H_2SO_4 can.

Depending on the nature of the reducing agents, the sulfate ion may be further reduced to elemental sulfur or the sulfide ion:

$$H_2SO_4(aq) + 6H^+(aq) + 6e^- \longrightarrow S(s) + 4H_2O(l)$$

$$H_2SO_4(aq) + 6H^+(aq) + 8e^- \longrightarrow S^{2-}(aq) + 4H_2O(l)$$

FIGURE 16.7 *A label of the reagent grade concentrated sulfuric acid used in chemical laboratories.*

For example, reduction of H_2SO_4 by HI yields H_2S and I_2:

$$8HI(aq) + H_2SO_4(aq) \longrightarrow H_2S(aq) + 4I_2(s) + 4H_2O(l)$$

Concentrated sulfuric acid also oxidizes nonmetals. For example, it oxidizes carbon to carbon dioxide and sulfur to sulfur dioxide:

$$C(s) + 2H_2SO_4(aq) \longrightarrow CO_2(g) + 2SO_2(g) + 2H_2O(l)$$

$$S(s) + 2H_2SO_4(aq) \longrightarrow 3SO_2(g) + 2H_2O(l)$$

Nitric Acid (HNO_3). Nitric acid, a powerful oxidizing agent, is a liquid that boils at 82.6°C. It does not exist in the pure form in the liquid state since it decomposes spontaneously to some extent as follows:

$$4HNO_3(l) \longrightarrow 4NO_2(g) + 2H_2O(l) + O_2(g)$$

Figure 16.8 shows a label for the reagent grade aqueous nitric acid solution.

The oxidation number of N in HNO_3 is $+5$. The most common reduction products of nitric acid are NO_2 (N = $+4$), NO (N = $+2$), and NH_4^+ (N = -3). The half-reactions yielding these products are

$$NO_3^-(aq) + 2H^+(aq) + e^- \longrightarrow NO_2(g) + H_2O(l)$$

$$NO_3^-(aq) + 4H^+(aq) + 3e^- \longrightarrow NO(g) + 2H_2O(l)$$

$$NO_3^-(aq) + 10H^+(aq) + 8e^- \longrightarrow NH_4^+(aq) + 3H_2O(l)$$

It is rare for one of these reactions to occur with the total exclusion of the other two. The brown color often noticeable in reactions involving nitric acid results from the formation of NO_2.

On standing, a concentrated nitric acid solution turns slightly yellow as a result of NO_2 formation.

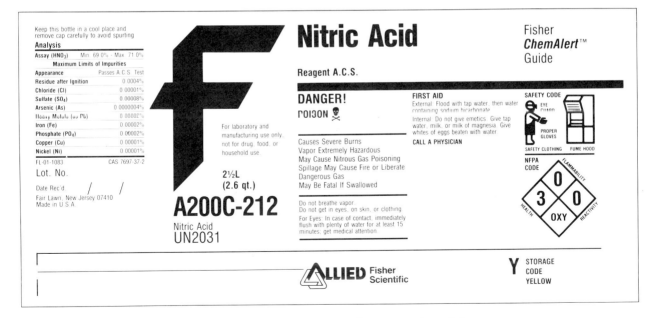

FIGURE 16.8 *A label of the reagent grade concentrated nitric acid used in chemical laboratories.*

Nitric acid will oxidize most metals to their corresponding cations. However, the oxidizing agent is the nitrate ion, NO_3^-, and not the hydrated proton. The nitrate ion is strong enough to oxidize metals both below and above hydrogen in the activity series (see Figure 3.9). For example, copper reacts with concentrated nitric acid as follows (see Color Plate 22):

$$Cu(s) + 4H^+(aq) + 2NO_3^-(aq) \longrightarrow Cu^{2+}(aq) + 2NO_2(g) + 2H_2O(l)$$

In the presence of a strong reducing agent, such as zinc metal, nitric acid can be reduced all the way to the ammonium ion:

$$4Zn(s) + 10H^+(aq) + NO_3^-(aq) \longrightarrow 4Zn^{2+}(aq) + NH_4^+(aq) + 3H_2O(l)$$

Aqua regia means "regal water."

Concentrated nitric acid does not oxidize gold. However, when the acid is added to concentrated hydrochloric acid in a 1:3 ratio by volume (one part HNO_3 to three parts HCl) the resulting solution, called *aqua regia*, can oxidize gold, as follows:

$$Au(s) + 3HNO_3(aq) + 4HCl(aq) \longrightarrow HAuCl_4(aq) + 3H_2O(l) + 3NO_2(g)$$

Oxyacids are defined in Section 2.9.

Concentrated nitric acid also oxidizes a number of nonmetals to their corresponding oxyacids:

$$P_4(s) + 20HNO_3(aq) \longrightarrow 4H_3PO_4(aq) + 20NO_2(g) + 4H_2O(l)$$

$$S(s) + 6HNO_3(aq) \longrightarrow H_2SO_4(aq) + 6NO_2(g) + 2H_2O(l)$$

Other Oxidizing Agents. Two other commonly encountered oxidizing agents are potassium dichromate ($K_2Cr_2O_7$) and potassium permanganate ($KMnO_4$). Some of their reactions were discussed in Section 16.6. As we saw there, the ions that undergo redox reactions are the dichromate ion, $Cr_2O_7^{2-}$, and the permanganate ion, MnO_4^-. These ions are powerful oxidizing agents that function best in acidic solutions.

Reducing Agents

Hydrogen Peroxide (H_2O_2). Earlier we described hydrogen peroxide as an oxidizing agent. However, in the presence of a stronger oxidizing agent than itself, H_2O_2 can act as a reducing agent. For example, note its reaction with an acidic permanganate solution:

$$5H_2O_2(aq) + 2MnO_4^-(aq) + 6H^+(aq) \longrightarrow 2Mn^{2+}(aq) + 5O_2(g) + 8H_2O(l)$$

In this reaction the oxidation number of ten oxygen atoms changes from -1 to 0 while that of the other eight oxygen atoms remains unchanged.

The Sulfite Ion (SO_3^{2-}). The sulfite ion is a fairly potent reducing agent. It can, for example, reduce nitric acid, as follows:

$$SO_3^{2-}(aq) + 2H^+(aq) + 2NO_3^-(aq) \longrightarrow SO_4^{2-}(aq) + 2NO_2(g) + H_2O(l)$$

As we saw earlier, the sulfite ion can also reduce hydrogen peroxide.

Metals. Metals above hydrogen in the activity series (see Figure 3.9) are all powerful reducing agents. Many of them can reduce the hydrogen in water and in hydrochloric acid to yield molecular hydrogen. Metals below hydrogen in the series can reduce only nitric acid and concentrated sulfuric acid.

SUMMARY

1. In oxidation–reduction, or redox, reactions, oxidation and reduction always occur simultaneously. Oxidation is characterized by the loss of electrons; reduction is characterized by the gain of electrons.
2. Oxidation numbers help us keep track of charge distribution and are assigned to all atoms in a compound or ion according to a specific set of rules. Oxidation can be defined as an increase in oxidation number; reduction can be defined as a decrease in oxidation number. Oxidation numbers are essentially bookkeeping devices; they have no fundamental physical meaning.
3. Many redox reactions can be classified as combination, decomposition, displacement, or disproportionation.
4. Redox equations can be balanced by the oxidation number method, in which changes in oxidation numbers are balanced; or by the ion–electron method, in which the oxidation and reduction half-reaction equations are balanced separately and then added together.
5. Redox titrations can be carried out in a manner similar to acid–base titrations. At the equivalence point, the number of equivalents of the oxidizing agent that reacted is equal to the number of equivalents of the reducing agent used.
6. Among the common oxidizing agents in the laboratory are hydrogen peroxide, concentrated sulfuric acid, and concentrated nitric acid.

KEY WORDS

Disproportionation reaction, p. 476
Equivalent, p. 490
Equivalent mass, p. 490
Half-reaction, p. 470
Normality, p. 491
Oxidation number, p. 471

Oxidation reaction, p. 470
Oxidizing agent, p. 470
Redox reaction, p. 470
Reducing agent, p. 470
Reduction reaction, p. 470

PROBLEMS

More challenging problems are marked with an asterisk.

Oxidation–Reduction Reactions

16.1 Is it possible to have a reaction in which oxidation occurs and reduction does not?

16.2 For the complete redox reactions given below (i) break down each reaction into its half-reactions;

(ii) identify the oxidizing agent; (iii) identify the reducing agent.

(a) $2Sr + O_2 \longrightarrow 2SrO$
(b) $16Na + S_8 \longrightarrow 8Na_2S$
(c) $2Li + H_2 \longrightarrow 2LiH$
(d) $2Cs + Br_2 \longrightarrow 2CsBr$
(e) $3Mg + N_2 \longrightarrow Mg_3N_2$

(f) $Zn + I_2 \longrightarrow ZnI_2$

(g) $2C + O_2 \longrightarrow 2CO$

16.3 Which of the following may be regarded as redox reactions?

(a) $Cl_2 + 2OH^- \longrightarrow Cl^- + ClO^- + H_2O$

(b) $Ca^{2+} + 2F^- \longrightarrow CaF_2$

(c) $NH_3 + H^+ \longrightarrow NH_4^+$

(d) $2CCl_4 + CrO_4^{2-} \longrightarrow$
$\qquad 2COCl_2 + CrO_2Cl_2 + 2Cl^-$

(e) $Ag^+ + 2NH_3 \longrightarrow Ag(NH_3)_2^+$

(f) $Ca + F_2 \longrightarrow CaF_2$

(g) $2Li + H_2 \longrightarrow 2LiH$

(h) $Ba(NO_3)_2 + Na_2SO_4 \longrightarrow 2NaNO_3 + BaSO_4$

(i) $CuO + H_2 \longrightarrow Cu + H_2O$

(j) $Zn + 2HCl \longrightarrow ZnCl_2 + H_2$

(k) $2FeCl_2 + Cl_2 \longrightarrow 2FeCl_3$

16.4 For the complete redox reactions given below, break down each reaction into half-reactions and identify the oxidizing agent and the reducing agent:

(a) $4Fe + 3O_2 \longrightarrow 2Fe_2O_3$

(b) $Cl_2 + 2NaBr \longrightarrow 2NaCl + Br_2$

(c) $Si + 2F_2 \longrightarrow SiF_4$

(d) $H_2 + Cl_2 \longrightarrow 2HCl$

Oxidation Number

16.5 Is the oxidation number an experimentally measurable quantity? Explain.

16.6 Give the oxidation numbers of all the elements in the following molecules and ions:

(a) SO, SO_2, SO_3, SO_3^{2-}, SO_4^{2-}

(b) ClO_2, ClO^-, ClO_2^-, ClO_3^-, ClO_4^-

(c) N_2O, NO, NO_2, N_2O_4, N_2O_5, NO_2^-, NO_3^-

16.7 Give the oxidation number of each atom in the following molecules and ions: (a) NaCN, (b) ClF, (c) IF_7, (d) ClO_2, (e) ClO^-, (f) ClO_2^-, (g) ClO_3^-, (h) ClO_4^-, (i) NaSCN, (j) CH_4, (k) C_2H_2, (l) C_2H_4, (m) K_2CrO_4, (n) $K_2Cr_2O_7$, (o) $KMnO_4$, (p) $NaHCO_3$, (q) Li_2, (r) $NaIO_3$, (s) KO_2, (t) PF_6^-, (u) H_2O_2, (v) Na_2O, (w) Na_2O_2, (x) $K_4Fe(CN)_6$, (y) $KAuCl_4$, (z) $K_3Fe(CN)_6$.

16.8 Give oxidation numbers for the underlined atoms in the following molecules and ions: (a) $\underline{Cs}_2O$, (b) $\underline{Ca}I_2$, (c) $\underline{Al}_2O_3$, (d) $H_3\underline{As}O_3$, (e) $\underline{Ti}O_2$, (f) $\underline{Mo}O_4^{2-}$, (g) $\underline{Pt}Cl_4^{2-}$, (h) $\underline{Pt}Cl_6^{2-}$, (i) $\underline{Sn}F_2$, (j) $\underline{Cl}F_3$, (k) $\underline{Sb}F_6^-$, (l) $\underline{P}_4$, (m) $\underline{Mn}O_4^-$, (n) $\underline{O}_2$, (o) $\underline{O}_3$.

16.9 Give the oxidation numbers of all the elements in the following molecules and ions: (a) Mg_3N_2, (b) CsO_2, (c) CaC_2, (d) CO_3^{2-}, (e) $C_2O_4^{2-}$, (f) ZnO_2^{2-}, (g) $NaBH_4$, (h) WO_4^{2-}.

16.10 Arrange the following species in order of increasing oxidation number of the sulfur atom: (a) H_2S, (b) S_8, (c) H_2SO_4, (d) S^{2-}, (e) HS^-, (f) SO_2, (g) SO_3.

16.11 Indicate the oxidation number of phosphorus in each of the following compounds: (a) HPO_3, (b) H_3PO_2, (c) H_3PO_3, (d) H_3PO_4, (e) $H_4P_2O_7$, (f) $H_5P_3O_{10}$.

Balancing Redox Equations

*16.12 Balance the following redox equations by the oxidation method:

(a) $C + H_2SO_4 \longrightarrow CO_2 + SO_2$

(b) $I_2O_5 + CO \longrightarrow I_2 + CO_2$

(c) $HI + HNO_3 \longrightarrow I_2 + NO$

(d) $Ag + H_2SO_4 \longrightarrow Ag_2SO_4 + SO_2$

(e) $Sb + HNO_3 \longrightarrow Sb_2O_5 + NO$

(f) $H_2S + HNO_3 \longrightarrow S + NO$

(g) $S + H_2SO_4 \longrightarrow SO_2 + H_2O$

(h) $PbO_2 + H_2SO_4 + Mn(NO_3)_2 \longrightarrow$
$\qquad PbSO_4 + HNO_3 + HMnO_4$

(i) $KOH + Cl_2 \longrightarrow KCl + KClO$

(j) $H_3AsO_4 + Zn + HNO_3 \longrightarrow$
$\qquad AsH_3 + Zn(NO_3)_2$

*16.13 Balance the following redox equations by the ion–electron method:

(a) $Mn^{2+} + H_2O_2 \longrightarrow MnO_2$ in basic solution

(b) $Cr_2O_7^{2-} + C_2O_4^{2-} \longrightarrow Cr^{3+} + CO_2$
$\qquad$ in acidic solution

(c) $ClO_3^- + Cl^- \longrightarrow Cl_2 + ClO_2$
$\qquad$ in acidic solution

(d) $S_2O_3^{2-} + I_2 \longrightarrow I^- + S_4O_6^{2-}$
$\qquad$ in acidic solution

(e) $H_2O_2 + Fe^{2+} \longrightarrow Fe^{3+}$ in acidic solution

(f) $Cu + HNO_3 \longrightarrow Cu^{2+} + NO$
$\qquad$ in acidic solution

(g) $Bi(OH)_3 + SnO_2^{2-} \longrightarrow SnO_3^{2-} + Bi$
$\qquad$ in basic solution

(h) $CN^- + MnO_4^- \longrightarrow CNO^- + MnO_2$
$\qquad$ in basic solution

(i) $Fe^{2+} + MnO_4^- \longrightarrow Fe^{3+} + Mn^{2+}$
$\qquad$ in acidic solution

(j) $Br_2 \longrightarrow BrO_3^- + Br^-$ in basic solution

16.14 Balance the following redox equation either by the oxidation method or by the ion–electron method. The reaction occurs in acidic solution.

$$MnO_4^- + SO_2 \longrightarrow Mn^{2+} + HSO_4^-$$

Redox Titrations

16.15 Oxidation of 25.0 mL of a solution containing Fe^{2+} requires 26.0 mL of 0.0250 M $K_2Cr_2O_7$ in acidic solution. Balance the following equation and calculate the molar concentration of Fe^{2+}:

$$Cr_2O_7^{2-} + Fe^{2+} + H^+ \longrightarrow Cr^{3+} + Fe^{3+}$$

16.16 The SO_2 often present as an air pollutant dissolves in H_2O to form H_2SO_3. The H_2SO_3 can then be titrated with $KMnO_4$ solution:

$$H_2SO_3 + MnO_4^- \longrightarrow SO_4^{2-} + Mn^{2+}$$

Calculate the number of grams of SO_2 in a 50.0 mL water sample if 7.37 mL of 0.00800 M $KMnO_4$ solution is required for titration. Note that the reaction is in acidic solution and the equation is not balanced.

16.17 A sample of iron ore weighing 0.2792 g was dissolved in dilute acid solution and all of the iron was converted to Fe(II) ions. The solution required 23.30 mL of 0.0971 N $KMnO_4$ for titration according to the following unbalanced equation:

$$Fe^{2+} + MnO_4^- \longrightarrow Fe^{3+} + Mn^{2+}$$

Calculate the percentage by mass of iron in the ore.

16.18 Oxidation of 25.0 mL of a solution containing H_2SO_3 requires 22.2 mL of 0.0862 M $K_2Cr_2O_7$ in acidic solution. Calculate the molar concentration of H_2SO_3.

16.19 From the following incomplete and unbalanced equation

$$Cr^{3+} + ClO_4^- \longrightarrow Cr_2O_7^{2-} + Cl^-$$

determine the equivalent mass of the ClO_4^- ion, expressed as a fraction of the molar mass.

16.20 Calculate the molarity of $K_2Cr_2O_7$ if 60.00 mL is consumed in a titration with 20.00 mL of 0.1800 N $FeSO_4$ according to the following reaction in acid solution:

$$Fe^{2+} + K_2Cr_2O_7 \longrightarrow Fe^{3+} + Cr^{3+}$$

*16.21 Iodate ion, IO_3^-, will oxidize SO_3^- to SO_4^{2-} in acidic solution. A 100.0 mL sample of solution containing 1.390 g of KIO_3 reacts with 32.5 mL of 0.500 M Na_2SO_3. What is the final oxidation number of the iodine after the reaction has occurred?

*16.22 A quantity of 25.0 mL of a solution containing both Fe^{2+} and Fe^{3+} ions is reacted with 23.0 mL of 0.0200 M $KMnO_4$. As a result, all of the Fe^{2+} ions are oxidized to Fe^{3+}. The Fe^{3+} ions are then all reduced to the Fe^{2+} ions by zinc metal (in dilute sulfuric acid). Finally, 25.0 mL of the solution containing only the Fe^{2+} ions required 40.0 mL of the same $KMnO_4$ solution for oxidation to Fe^{3+}. Calculate the molar concentrations of Fe^{2+} and Fe^{3+} ions in the original solution.

16.23 A quantity of 0.9768 g of Fe(II) salt consumes 32.33 mL of 0.1037 N $K_2Cr_2O_7$. Calculate the percent by mass of Fe^{2+} ion in the salt.

16.24 Oxalic acid $(H_2C_2O_4)$ can be oxidized by $KMnO_4$. (a) Balance the following equation by the ion–electron method in acid solution:

$$MnO_4^- + C_2O_4^{2-} \longrightarrow Mn^{2+} + CO_2$$

(b) Referring to (a), complete the following:

$$1 \ M \ KMnO_4 = ? \ N \ KMnO_4$$
$$1 \ M \ H_2C_2O_4 = ? \ N \ H_2C_2O_4$$

(c) If a 1.00 g sample of $H_2C_2O_4$ requires 24.0 mL of 0.0500 N $KMnO_4$ solution to reach an equivalence point, what is the percent by mass of $H_2C_2O_4$ in the sample?

16.25 Sulfite ions can be oxidized to sulfates by dichromate ions in acidic solution:

$$8H^+ + Cr_2O_7^{2-} + 3SO_3^{2-} \longrightarrow$$
$$2Cr^{3+} + 3SO_4^{2-} + 4H_2O$$

Calculate the normality and molarity of the $K_2Cr_2O_7$ solution if 28.42 mL of the solution reacts completely with a 25.00 mL solution of 0.6286 N Na_2SO_3.

Miscellaneous Problems

16.26 Define the following terms: (a) oxidation number, (b) oxidizing agent, (c) reducing agent, (d) disproportionation reaction, (e) equivalent, (f) equivalent mass, (g) normality.

16.27 Use the Stock system to name the following compounds: (a) $PbCl_2$, (b) $PbCl_4$, (c) Hg_2Cl_2, (d) $HgCl_2$, (e) $CuBr$, (f) $CuBr_2$, (g) Cr_2O_3, (h) CrO_3.

16.28 Write the chemical formulas for each of the following substances: (a) vanadium(V) oxide, (b) iron(II) oxide, (c) iron(III) oxide, (d) titanium(III) oxide, (e) nickel(II) nitride, (f) ruthenium(IV) chloride.

16.29 Ozone can be synthesized from molecular oxygen (O_2) by an electric discharge:

$$3O_2(g) \longrightarrow 2O_3(g)$$

Is this a redox reaction? Explain.

16.30 Indicate the most ionic compound in each group of compounds: (a) $SnCl_2$ and $SnCl_4$, (b) $TiCl_2$, $TiCl_3$, and $TiCl_4$.

16.31 Hydrochloric acid is not an oxidizing agent in the sense that sulfuric acid and nitric acid are. Explain why the chloride ion is not a strong oxidizing agent like SO_4^{2-} and NO_3^-. (*Hint:* Examine the oxidation numbers of Cl, S, and N.)

16.32 Complete and balance the following ionic equation, which represents a reaction that takes place in acid solution. Identify the oxidizing and reducing agents and determine the equivalent mass of each agent.

$$Ce^{4+} + As_2O_3 \longrightarrow H_3AsO_4 + Ce^{3+}$$

16.33 Calculate the mass of Cu^+ in grams in a sample that requires 36.0 mL of 0.146 M $KMnO_4$ solution to reach the equivalence point. The products are Mn^{2+} and Cu^{2+} ions. The reaction occurs in an acid solution.

17
CHEMICAL EQUILIBRIUM

Two complementary views of dynamic equilibrium will be developed in this chapter. At the macroscopic level we will explore the observable properties of chemical systems at equilibrium, examining, for example, the concentrations of reactants and products, and the influence of temperature, pressure, volume, and concentrations on the equilibrium state. At the molecular level, we will view the state of dynamic equilibrium as a natural consequence of reversible reactions—that is, of molecular changes that are in direct opposition to each other.

To this point in the book we have solved stoichiometry problems under the assumption that the reaction proceeds until the limiting reactant is fully consumed. However, as we will learn in this chapter, all reactants and products are present in chemical systems at equilibrium. Thus another focus of this chapter concerns the arithmetic associated with equilibrium reactions—the stoichiometry of reactions that do not go to completion.

17.1 THE CONCEPT OF EQUILIBRIUM

The concept of dynamic equilibrium was defined in Section 14.6, where we discussed the evaporation–condensation equilibrium of a liquid.

Equilibrium is a *state in which there are no observable changes as time goes by*. There are two kinds of equilibrium. In a system that is in **static equilibrium**—for example, a book resting on your desk—*no changes are taking place either at the macroscopic, observable level or at the molecular level*. In a *dynamic equilibrium,* on the other hand, there is much activity at the molecular level, although no observable change occurs. A situation in which two people hit a large number of tennis balls across the net to each other provides an example of dynamic equilibrium if the number of tennis balls observed on each side of the net remains constant as time goes by.

Chemists are generally more interested in the dynamic kind of equilibrium. In previous chapters we saw several examples of dynamic equilibrium that involve physical processes such as phase changes and dissolution of slightly soluble substances. If a mixture of ice and water is kept at 0°C and 1 atm in a closed container, no observable changes occur. The amount of each substance remains the same as long as these conditions are kept constant. Yet there is much activity at the molecular level. At any given moment, many H_2O molecules break away from the ice lattice and enter the liquid state, while an equal number of H_2O molecules join the solid ice network. The dynamic equilibrium of the system is shown by

$$H_2O(s) \rightleftharpoons H_2O(l)$$

where the $\rightleftharpoons$ sign, as you know, indicates that the process is reversible.

The vaporization of water in a closed container at a given temperature is another example of dynamic equilibrium, in which the numbers of H_2O molecules leaving and returning to the liquid phase per second are equal:

$$H_2O(l) \rightleftharpoons H_2O(g)$$

The pressure of the water vapor measured at this temperature is called the *equilibrium vapor pressure* (see Section 14.8).

Another example of dynamic equilibrium is found in a saturated solution. For example, in a saturated AgCl solution in contact with some solid AgCl we find

$$AgCl(s) \rightleftharpoons Ag^+(aq) + Cl^-(aq)$$

At every instant, a portion of solid AgCl breaks up into individual Ag^+ and Cl^- ions and enters the aqueous phase as hydrated species while the same number of Ag^+ and Cl^- ions combine in solution to form part of the solid AgCl network.

17.2 CHEMICAL EQUILIBRIUM

Equilibrium between two phases of the same substance, such as the liquid water–ice and the liquid water–water vapor systems, and solid and dissolved species, is called physical equilibrium because the changes that occur are physical processes. The study of physical equilibria leads to useful infor-

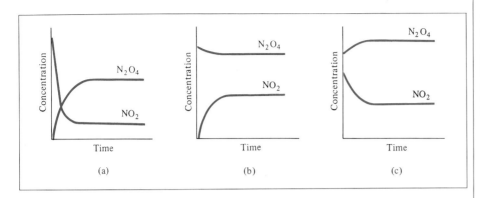

FIGURE 17.1 *Variations of the concentrations of NO$_2$ and N$_2$O$_4$ with time, in three situations. (a) Initially only NO$_2$ is present. (b) Initially only N$_2$O$_4$ is present. (c) Initially a mixture of NO$_2$ and N$_2$O$_4$ is present.*

mation such as the equilibrium vapor pressure and the solubility of sub-stances. However, of particular interest to chemists is the dynamic equilibrium that takes the form of a chemical reaction.

Let us begin by considering the reversible reaction involving nitrogen dioxide (NO$_2$) and dinitrogen tetroxide (N$_2$O$_4$). The progress of the reaction

$$N_2O_4(g) \rightleftharpoons 2NO_2(g)$$

can be monitored easily because N$_2$O$_4$ is a colorless gas, whereas NO$_2$ has a dark brown color sometimes visible in polluted air. Suppose that a known amount of N$_2$O$_4$ is injected into an evacuated flask. The brown color appears immediately, indicating the formation of NO$_2$ molecules. The color intensifies as the dissociation of N$_2$O$_4$ continues, until equilibrium is reached. Beyond that point, no further change in color is observed. By experiment we find that we can also reach the equilibrium state by starting with pure NO$_2$ or a mixture of NO$_2$ and N$_2$O$_4$. In each case, we observe an initial change in color, due to either the formation of NO$_2$ (if the color intensifies) or the depletion of NO$_2$ (if the color fades), and then the final state in which the color of NO$_2$ no longer changes. Depending on the temperature of the reacting system and the initial amounts of NO$_2$ and N$_2$O$_4$, the concentrations of NO$_2$ and N$_2$O$_4$ at equilibrium will differ from system to system (Figure 17.1).

Table 17.1 shows some experimental data for this reaction at 25°C. The gas concentrations are expressed in molarity units, which can be calculated

TABLE 17.1 The NO$_2$–N$_2$O$_4$ System at 25°C

Initial Concentrations (M)		Equilibrium Concentrations (M)		Ratio of Concentrations at Equilibrium	
				$\dfrac{[NO_2]}{[N_2O_4]}$	$\dfrac{[NO_2]^2}{[N_2O_4]}$
NO$_2$	N$_2$O$_4$	NO$_2$	N$_2$O$_4$		
0.000	0.670	0.0558	0.669	0.0834	4.65×10^{-3}
0.0500	0.450	0.0457	0.452	0.101	4.62×10^{-3}
0.0300	0.500	0.0480	0.491	0.0978	4.69×10^{-3}
0.0400	0.600	0.0524	0.592	0.0885	4.64×10^{-3}
0.200	0.000	0.0204	0.0898	0.227	4.63×10^{-3}

Recall that [ ] means mol/L.

A more general definition of the equilibrium constant is given below.

by knowing the number of moles of the gases present initially and at equilibrium and the volume of the flask in liters. Analysis of the data at equilibrium shows that although the ratio $[NO_2]/[N_2O_4]$ gives scattered values, the ratio $[NO_2]^2/[N_2O_4]$ gives a nearly constant value whose average is 4.65×10^{-3}. Because this constant value corresponds to the equilibrium situation, we say that the "equilibrium constant" for the reaction at 25°C is given by

$$K = \frac{[NO_2]^2}{[N_2O_4]} = 4.65 \times 10^{-3} \qquad (17.1)$$

Note that the exponent 2 for $[NO_2]$ in this expression is the same as the stoichiometric coefficient for NO_2 in the reversible reaction.

We can generalize this discussion by considering the following reaction:

$$aA + bB \rightleftharpoons cC + dD$$

where a, b, c, and d are the stoichiometric coefficients for the reacting species A, B, C, and D. The equilibrium constant for the reaction at a particular temperature is

$$K = \frac{[C]^c[D]^d}{[A]^a[B]^b} \qquad (17.2)$$

Equation (17.2) is the mathematical form of the *law of mass action*, first proposed by Cato Guldberg (1836–1902) and Peter Waage (1833–1900) in 1864. This law cannot be easily stated in words. What it means is that there is a condition, given by the **equilibrium constant,** that *relates the concentrations of reactants and products at equilibrium.* The equilibrium constant is defined by a quotient. The numerator is obtained by multiplying together the equilibrium concentrations of the products, each raised to a power equal to its stoichiometric coefficient in the balanced equation. The same procedure is applied to the equilibrium concentrations of reactants to obtain the denominator. The formulation of this law originally was based purely on empirical evidence, such as the study of reactions like NO_2–N_2O_4. However, the equilibrium constant is a quantity that has its origin in a scientific discipline called thermodynamics, discussed in Chapter 23.

The terms "reactants" and "products" may seem confusing in this context, since any substance serving as a reactant for the forward reaction serves also as a product for the reverse reaction. To avoid semantic difficulties, we will simply call substances to the right of the equilibrium arrows "products" and substances to the left of the arrows "reactants."

The Magnitude of the Equilibrium Constant

The value of the equilibrium constant depends on the nature of the reaction and the temperature. At a constant temperature we find that the magnitude of K varies greatly from one reaction to another. Let's consider three examples relating the magnitude of K to the relative quantities of the reactants and products formed at equilibrium.

Any number equal to or larger than 10 is said to be much greater than 1.

Example 1. If K is very large compared to 1, as in the following system at 2300°C

$$2O_3(g) \rightleftharpoons 3O_2(g) \qquad K = \frac{[O_2]^3}{[O_3]^2} = 2.54 \times 10^{12}$$

then an equilibrium mixture of O_2 and O_3 at that temperature will contain very little of O_3 as compared to O_2. Thus, if $[O_2] = 0.500\ M$ at equilibrium, then

$$[O_3]^2 = \frac{(0.500)^3}{2.54 \times 10^{12}}$$
$$= 4.92 \times 10^{-14}$$
$$[O_3] = 2.22 \times 10^{-7}\ M$$

Example 2. The opposite of Example 1 holds true if K is very small compared to 1. Consider the following equilibrium system, at 25°C:

$$Cl_2(g) \rightleftharpoons Cl(g) + Cl(g) \qquad K = \frac{[Cl]^2}{[Cl_2]} = 1.4 \times 10^{-38}$$

An equilibrium mixture at this temperature will contain mostly Cl_2 molecules and very few Cl atoms. Suppose, for example, the equilibrium concentration of Cl_2 is 0.76 M. Then

$$[Cl]^2 = (0.76)(1.4 \times 10^{-38})$$
$$= 1.1 \times 10^{-38}$$
$$[Cl] = 1.0 \times 10^{-19}\ M$$

As you can see, the concentration of Cl is very low compared to that of Cl_2.

Example 3. If K is neither very large nor very small compared to 1, then the quantities of reactants and products present at equilibrium will be comparable. Consider the following equilibrium system. At 830°C

$$CO(g) + H_2O(g) \rightleftharpoons H_2(g) + CO_2(g) \qquad K = \frac{[H_2][CO_2]}{[CO][H_2O]} = 5.10$$

If $[CO] = 0.200\ M$, $[H_2O] = 0.400\ M$, and $[H_2] = 0.300\ M$ at equilibrium, then

$$[CO_2] = \frac{(0.200)(0.400)}{(0.300)}(5.10) = 1.36\ M$$

17.3 VARIOUS WAYS OF EXPRESSING EQUILIBRIUM CONSTANTS

The concept of equilibrium constants is extremely important in chemistry. As you will soon see, they serve as the key to solving a wide variety of stoichiometry problems involving equilibrium systems. To use equilibrium constants, we must know how to express them in terms of the reactant and product concentrations. Our only guidance is the law of mass action [Equation (17.2)]. However, because the concentrations of the reactants and products can be expressed in several types of units and because the reacting species are not always in the same phase, there may be more than one way

Any number equal to or smaller than 0.1 is said to be much smaller than 1.

to express the equilibrium constant for the *same* reaction. To begin with, we will consider reactions in which the reactants and products are in the same phase. Then we will look at another type of equilibrium reaction.

Homogeneous Equilibria

The term **homogeneous equilibrium** applies to reactions in which *all reacting species are in the same phase.* The ionization of acetic acid in aqueous solution is an example:

$$CH_3COOH(aq) \rightleftharpoons CH_3COO^-(aq) + H^+(aq)$$

The equilibrium constant for this process, K_c, is given by

$$K_c = \frac{[CH_3COO^-][H^+]}{[CH_3COOH]} \tag{17.3}$$

An example of homogeneous gas-phase equilibrium is the dissociation of N_2O_4, which we discussed [Equation (17.1)]. The equilibrium constant was given by

$$K_c = \frac{[NO_2]^2}{[N_2O_4]}$$

Note that the subscript c in K_c denotes that the concentrations of the reacting species are expressed in moles per liter. The concentrations of reactants and products in gaseous reactions can also be expressed in terms of their partial pressures. From Equation (6.7) we see that at constant temperature the pressure P of a gas is directly related to the concentration in mol/L of the gas, that is, $P = (n/V)RT$. Thus, for the equilibrium process

$$N_2O_4(g) \rightleftharpoons 2NO_2(g)$$

we can write

$$K_P = \frac{P_{NO_2}^2}{P_{N_2O_4}} \tag{17.4}$$

where P_{NO_2} and $P_{N_2O_4}$ are the equilibrium partial pressures of NO_2 and N_2O_4, respectively. The subscript p in K_P tells us that equilibrium concentrations are expressed in terms of pressure.

    In general, K_c is not equal to K_P, since the partial pressures of reactants and products are not equal to their concentrations expressed in moles per liter. A simple relationship between K_P and K_c can be derived as follows. Let us consider the equilibrium

$$aA(g) \rightleftharpoons bB(g)$$

where a and b are stoichiometric coefficients. The equilibrium constant K_c is given by

$$K_c = \frac{[B]^b}{[A]^a}$$

and the expression for K_P is

$$K_P = \frac{P_B^b}{P_A^a}$$

where P_A and P_B are the partial pressures of A and B. Assuming ideal gas behavior

$$P_A V = n_A RT$$
$$P_A = \frac{n_A RT}{V}$$

where V is the volume of the container in liters. Also

$$P_B V = n_B RT$$
$$P_B = \frac{n_B RT}{V}$$

Substituting these relations into the expression for K_P, we obtain

$$K_P = \frac{\left(\dfrac{n_B RT}{V}\right)^b}{\left(\dfrac{n_A RT}{V}\right)^a}$$

$$= \frac{\left(\dfrac{n_B}{V}\right)^b}{\left(\dfrac{n_A}{V}\right)^a}(RT)^{b-a}$$

Now both n_A/V and n_B/V have the units of mol/L and can be replaced by [A] and [B], so that

$$K_P = \frac{[B]^b}{[A]^a}(RT)^{\Delta n}$$
$$= K_c(RT)^{\Delta n}$$

where

$$\Delta n = b - a$$
$$= \text{moles of gaseous products} - \text{moles of gaseous reactants}$$

Since pressures are usually expressed in atm, the gas constant R is given by $0.0821 \text{ L} \cdot \text{atm/K} \cdot \text{mol}$ (see Appendix 2), and we can write the relationship between K_P and K_c as

$$K_P = K_c(0.0821\ T)^{\Delta n} \tag{17.5}$$

To use this equation, the pressures in K_P must be in atm.

At this point we must say something about units and the equilibrium constant. It would seem that the equilibrium constants in Equations (17.1) and (17.3) should have the units of mol/L or M and the one in Equation (17.4) should have the unit of atm. It turns out the equilibrium constant (K_P or K_c) should be treated as a dimensionless quantity. Keep in mind that the equilibrium constant is simply a number without units.

EXAMPLE 17.1

Write K_c and K_P (if applicable) expressions for the following reversible reactions at equilibrium:

(a) $HF(aq) \rightleftharpoons H^+(aq) + F^-(aq)$

(b) $2NO(g) + O_2(g) \rightleftharpoons 2NO_2(g)$

(c) $2H_2S(g) + 3O_2(g) \rightleftharpoons 2H_2O(g) + 2SO_2(g)$

Answer

(a) Since there are no gases present, K_P does not apply and we have only K_c:

$$K_c = \frac{[H^+][F^-]}{[HF]}$$

(b)
$$K_c = \frac{[NO_2]^2}{[NO]^2[O_2]} \qquad K_P = \frac{P_{NO_2}^2}{P_{NO}^2 P_{O_2}}$$

(c)
$$K_c = \frac{[H_2O]^2[SO_2]^2}{[H_2S]^2[O_2]^3} \qquad K_P = \frac{P_{H_2O}^2 P_{SO_2}^2}{P_{H_2S}^2 P_{O_2}^3}$$

Similar examples: Problems 17.4, 17.6.

EXAMPLE 17.2

The following equilibrium process has been studied at 230°C:

$$2NO(g) + O_2(g) \rightleftharpoons 2NO_2(g)$$

The concentrations of the reacting species at equilibrium in one experiment are found to be $[NO] = 0.0542\ M$, $[O_2] = 0.127\ M$, and $[NO_2] = 15.5\ M$. Calculate the equilibrium constant (K_c) of the reaction at this temperature.

Answer

The equilibrium constant is given by

$$K_c = \frac{[NO_2]^2}{[NO]^2[O_2]}$$

Substituting the concentrations, we find that

$$K_c = \frac{(15.5)^2}{(0.0542)^2(0.127)} = 6.44 \times 10^5$$

Note that K_c is given without units.

Similar example: Problem 17.8.

EXAMPLE 17.3

The equilibrium constant K_P for the reaction

$$PCl_5(g) \rightleftharpoons PCl_3(g) + Cl_2(g)$$

is found to be 1.05 at 250°C. If the equilibrium partial pressures of PCl_5 and PCl_3 are 0.875 atm and 0.463 atm, respectively, what is the equilibrium partial pressure of Cl_2 at 250°C?

Answer

First, we write K_P in terms of the partial pressures of the reacting species.

$$K_P = \frac{P_{PCl_3} P_{Cl_2}}{P_{PCl_5}}$$

Knowing the partial pressures, we write

$$1.05 = \frac{(0.463)(P_{Cl_2})}{(0.875)}$$

or

$$P_{Cl_2} = \frac{(1.05)(0.875)}{(0.463)} = 1.98 \text{ atm}$$

Note that we have added atm as the unit for P_{Cl_2}.

Similar example: Problem 17.11.

EXAMPLE 17.4

The equilibrium constant (K_c) for the reaction

$$N_2O_4(g) \rightleftharpoons 2NO_2(g)$$

is 4.65×10^{-3} at 25°C. What is the value of K_P at this temperature?

Answer

From Equation (17.5) we write

$$K_P = K_c(0.0821 \ T)^{\Delta n}$$

Since $T = 298$ K and $\Delta n = 1$, we have

$$K_P = (4.65 \times 10^{-3})(0.0821 \times 298)$$
$$= 0.114$$

Note that K_P, like K_c, is treated as a dimensionless quantity. This example shows that we can get quite a different value for the equilibrium constant for the same reaction, depending on whether we are expressing the concentrations in moles per liter or atmospheres. In the special case of $\Delta n = 0$, we have $K_c = K_P$.

Similar examples: Problems 17.9, 17.12.

Any number raised to 0 power is equal to 1. Thus, $(0.0821 \ T)^0 = 1$.

Heterogeneous Equilibria

A reversible reaction involving reactants and products that are in different phases leads to a **heterogeneous equilibrium.** For example, when calcium carbonate is heated in a closed vessel, the following equilibrium is attained:

$$CaCO_3(s) \rightleftharpoons CaO(s) + CO_2(g)$$

The two solids and one gas constitute three separate phases. At equilibrium, the equilibrium constant is

$$K'_c = \frac{[CaO][CO_2]}{[CaCO_3]} \tag{17.6}$$

We use the prime here (K'_c) to distinguish the equilibrium constant from the modifiied one shown in Equation (17.7).

But how do we express the concentration of a solid substance? Because both $CaCO_3$ and CaO are pure solids, their "concentrations" do not change as the reaction proceeds. This follows from the fact that the molar concentration of a pure solid (or a pure liquid) is a constant at a given temperature; it does not depend on the quantity of the substance present. For example, the molar concentration at 20°C of copper (density: 8.96 g/cm^3) is the same, whether we have one gram or one ton of the metal:

$$\frac{8.96 \text{ g}}{\text{cm}^3} \times \frac{1 \text{ mol}}{63.55 \text{ g}} = 0.141 \text{ mol/cm}^3 = 141 \text{ mol/L}$$

From the above discussion we can express the equilibrium constant for the decomposition of $CaCO_3$ in a different way. Rearranging Equation (17.6), we obtain

$$\frac{[CaCO_3]}{[CaO]}K'_c = [CO_2] \tag{17.7}$$

Since both $[CaCO_3]$ and $[CaO]$ are constants and K'_c is an equilibrium constant, all the terms on the left-hand side of the equation are constants. We can simplify the equation by writing

$$\frac{[CaCO_3]}{[CaO]}K'_c = K_c = [CO_2]$$

where K_c, the "new" equilibrium constant, is now conveniently expressed in terms of a single concentration, that of CO_2. Keep in mind that the value of K_c does not depend on how much $CaCO_3$ and CaO are present, as long as some quantity of each is present at equilibrium. (Of course, without some $CaCO_3$ or CaO present we could not establish an equilibrium condition.)

Alternatively, we can express the equilibrium constant as

$$K_P = P_{CO_2} \tag{17.8}$$

The equilibrium constant in this case is numerically equal to the pressure of the CO_2 gas, an easily measurable quantity.

EXAMPLE 17.5

Write the equilibrium constant expression K_c, and K_P if applicable, for each of the following heterogeneous systems:

(a) $(NH_4)_2Se(s) \rightleftharpoons 2NH_3(g) + H_2Se(g)$

(b) $AgCl(s) \rightleftharpoons Ag^+(aq) + Cl^-(aq)$

(c) $Ca(HCO_3)_2(aq) \rightleftharpoons CaCO_3(s) + H_2O(l) + CO_2(g)$

Answer

(a) The equilibrium constant is given by

$$K_c' = \frac{[NH_3]^2[H_2Se]}{[(NH_4)_2Se]}$$

However, since $(NH_4)_2Se$ is a solid, we write the new equilibrium constant as

$$K_c = [NH_3]^2[H_2Se]$$

where $K_c = K_c'[(NH_4)_2Se]$. Alternatively, we can express the equilibrium constant K_P in terms of the partial pressures of NH_3 and H_2Se:

$$K_P = P_{NH_3}^2 P_{H_2Se}$$

(b)
$$K_c' = \frac{[Ag^+][Cl^-]}{[AgCl]}$$

$$K_c = [Ag^+][Cl^-]$$

Again, we have incorporated $[AgCl]$ into K_c because $AgCl$ is a solid.

(c)
$$K_c' = \frac{[CaCO_3][H_2O][CO_2]}{[Ca(HCO_3)_2]}$$

Because the water produced in the reaction is negligible compared to the water solvent, the concentration of water is kept fairly constant. Thus both $[CaCO_3]$ and $[H_2O]$ may be treated as constants, so that

$$K_c = \frac{[CO_2]}{[Ca(HCO_3)_2]}$$

CO_2 is a gas, but $Ca(HCO_3)_2$ is a species dissolved in water. Therefore we cannot write a K_P for this equilibrium.

Similar example: Problem 17.5.

In one liter or 1000 g of water there are 1000 g/(18.02 g/mol) or 55.5 moles of water. Therefore, the "concentration" of water, or $[H_2O]$, is 55.5 mol/L or 55.5 M. This value can also be used for water in most solutions.

EXAMPLE 17.6

Consider the following heterogeneous equilibrium:

$$CaCO_3(s) \rightleftharpoons CaO(s) + CO_2(g)$$

At 800°C, the pressure of CO_2 is 0.236 atm. Calculate (a) K_P and (b) K_c for the reaction at this temperature.

Continued)

Answer

(a) From Equation (17.8) we write

$$K_P = P_{CO_2}$$
$$= 0.236$$

(b) From Equation (17.5) we write

$$K_P = K_c(0.0821 \ T)^{\Delta n}$$

$T = 800 + 273 = 1073 \ K$ and $\Delta n = 1$, so we substitute these in the equation and obtain

$$0.236 = K_c(0.0821 \times 1073)$$
$$K_c = 2.68 \times 10^{-3}$$

Similar example: Problem 17.13.

Multiple Equilibria

The reactions we have considered so far are all relatively simple. Let us now consider a more complicated situation, in which the product molecules in one equilibrium system are involved in a second equilibrium process:

$$A + B \rightleftharpoons C + D$$
$$C + D \rightleftharpoons E + F$$

The products formed in the first reaction, C and D, react further to form products E and F. At equilibrium we can write two separate equilibrium constants:

$$K'_c = \frac{[C][D]}{[A][B]}$$

and

$$K''_c = \frac{[E][F]}{[C][D]}$$

The overall reaction is given by the sum of the two reactions

$$
\begin{array}{ll}
A + B \rightleftharpoons C + D & K'_c \\
C + D \rightleftharpoons E + F & K''_c \\
\hline
\text{overall reaction:} \quad A + B \rightleftharpoons E + F & K_c
\end{array}
$$

and the equilibrium constant K_c for the overall reaction is

$$K_c = \frac{[E][F]}{[A][B]}$$

We obtain this same expression if we take the product of the expressions for K'_c and K''_c:

$$K'_c K''_c = \frac{[C][D]}{[A][B]} \times \frac{[E][F]}{[C][D]} = \frac{[E][F]}{[A][B]}$$

Therefore we arrive at

$$K_c = K_c'K_c'' \tag{17.9}$$

We can now make an important statement about multiple equilibria: *If a reaction can be expressed as the sum of two or more reactions, the equilibrium constant for the overall reaction is given by the product of the equilibrium constants of the individual reactions.*

Among the many known examples of multiple equilibria is the ionization of diprotic and polyprotic acids in aqueous solution. The following equilibrium constants have been determined for carbonic acid (H_2CO_3) at 25°C:

$$H_2CO_3(aq) \rightleftharpoons H^+(aq) + HCO_3^-(aq) \qquad K_c' = \frac{[H^+][HCO_3^-]}{[H_2CO_3]} = 4.2 \times 10^{-7}$$

$$HCO_3^-(aq) \rightleftharpoons H^+(aq) + CO_3^{2-}(aq) \qquad K_c'' = \frac{[H^+][CO_3^{2-}]}{[HCO_3^-]} = 4.8 \times 10^{-11}$$

The overall reaction is the sum of these two reactions

$$H_2CO_3(aq) \rightleftharpoons 2H^+(aq) + CO_3^{2-}(aq)$$

and the corresponding equilibrium constant is given by

$$K_c = \frac{[H^+]^2[CO_3^{2-}]}{[H_2CO_3]}$$

Using Equation (17.9) we arrive at

$$\begin{aligned} K_c &= K_c'K_c'' \\ &= (4.2 \times 10^{-7})(4.8 \times 10^{-11}) \\ &= 2.0 \times 10^{-17} \end{aligned}$$

The Form of *K* and the Equilibrium Equation

Before closing this section, we should note the two following important rules about writing equilibrium constants:

- When we reverse an equilibrium equation, the equilibrium constant becomes the reciprocal of the original equilibrium constant. Thus if we write the NO_2–N_2O_4 equilibrium as

$$N_2O_4(g) \rightleftharpoons 2NO_2(g)$$

then

$$K_c = \frac{[NO_2]^2}{[N_2O_4]} = 4.65 \times 10^{-3}$$

However, we can represent the equilibrium equally well as

$$2NO_2(g) \rightleftharpoons N_2O_4(g)$$

and the equilibrium constant is now given by

$$K_c' = \frac{[N_2O_4]}{[NO_2]^2} = \frac{1}{4.65 \times 10^{-3}} = 215$$

The reciprocal of *x* is $\frac{1}{x}$.

You can see that $K_c = 1/K_c'$ or $K_c K_c' = 1.00$ Either K_c or K_c' is a valid equilibrium constant, but it is meaningless to say that the equilibrium constant for the NO_2–N_2O_4 system is 4.65×10^{-3} or 215, unless we also specify how the equilibrium equation is written.

- The value of K depends on how the equilibrium equation is balanced. Consider the following:

$$N_2O_4(g) \rightleftharpoons 2NO_2(g) \qquad K_c = \frac{[NO_2]^2}{[N_2O_4]}$$

$$\tfrac{1}{2}N_2O_4(g) \rightleftharpoons NO_2(g) \qquad K_c' = \frac{[NO_2]}{[N_2O_4]^{\frac{1}{2}}}$$

Looking at the exponents we see that $K_c' = \sqrt{K_c}$. In Table 17.1 we find $K_c = 4.65 \times 10^{-3}$; therefore $K_c' = 0.0682$.

This example illustrates once again the need to write the particular chemical equation when quoting the numerical value of an equilibrium constant. In general, remember that if you double a chemical equation throughout, the corresponding equilibrium constant will be the square of the original value; if you triple the equation, the equilibrium constant will be the cube of the original value, and so on.

EXAMPLE 17.7

The reaction for the production of ammonia can be written in a number of ways:

(a) $N_2(g) + 3H_2(g) \rightleftharpoons 2NH_3(g)$

(b) $\tfrac{1}{2}N_2(g) + \tfrac{3}{2}H_2(g) \rightleftharpoons NH_3(g)$

(c) $\tfrac{1}{3}N_2(g) + H_2(g) \rightleftharpoons \tfrac{2}{3}NH_3(g)$

Write the equilibrium constant expression for each formulation. (Express the concentrations of the reacting species in mol/L.)
(d) How are the equilibrium constants related to one another?

Answer

(a) $$K_a = \frac{[NH_3]^2}{[N_2][H_2]^3}$$

(b) $$K_b = \frac{[NH_3]}{[N_2]^{\frac{1}{2}}[H_2]^{\frac{3}{2}}}$$

(c) $$K_c = \frac{[NH_3]^{\frac{2}{3}}}{[N_2]^{\frac{1}{3}}[H_2]}$$

(d) $$K_a = K_b^2$$
$$K_a = K_c^3$$
$$K_b^2 = K_c^3 \qquad \text{or} \qquad K_b = K_c^{\frac{3}{2}}$$

Similar examples: Problems 17.10, 17.17.

Summary of the Rules for Writing Equilibrium Constant Expressions

- The concentrations of the reacting species in the condensed phase are expressed in mol/L; in the gaseous phase, the concentrations can be expressed in mol/L or in atm. K_c is related to K_P by a simple equation [Equation (17.5)].
- The concentrations of pure solids and pure liquids do not appear in the equilibrium constant expressions.
- The equilibrium constant (K_c or K_P) is treated as a dimensionless quantity.
- In quoting a value for the equilibrium constant, we must specify the balanced equation and the temperature.
- If a reaction can be expressed as the sum of two or more reactions, the equilibrium constant for the overall reaction is given by the product of the equilibrium constants of the individual reactions.

17.4   WHAT DOES THE EQUILIBRIUM CONSTANT TELL US?

We have seen that the equilibrium constant for a given reaction can be calculated from known equilibrium concentrations; remember, of course, that the equilibrium constant is constant only if the temperature does not change. Once we know the value of the equilibrium constant, we can use Equation (17.2) to calculate unknown equilibrium concentrations. In general, the equilibrium constant helps us predict the direction in which a reaction mixture will proceed to achieve equilibrium, and to calculate the concentrations of reactants and products once equilibrium has been reached. These uses of the equilibrium constant will be explored in this section.

Predicting the Direction of a Reaction

Suppose we start with a reaction mixture consisting of both reactants and products. The two key questions are: (1) Is the system at equilibrium? (2) If not, in which direction will the net reaction proceed to achieve equilibrium?

Let us consider the reaction between hydrogen and iodine molecules to produce hydrogen iodide molecules:

$$H_2(g) + I_2(g) \rightleftharpoons 2HI(g)$$

The equilibrium constant (K_c) for the reaction is 54.3 at 430°C. We can answer the two questions by using the ***reaction quotient*** (Q_c), defined as a *number equal to the ratio of product concentrations to reactant concentrations, each raised to the power of its stoichiometric coefficient at some point other than equilibrium.* For the reaction we are considering, the reaction quotient is given by

$$Q_c = \frac{(\text{initial concentration of HI})^2}{(\text{initial concentration of H}_2)(\text{initial concentration of I}_2)}$$
$$= \frac{(\text{HI})^2}{(\text{H}_2)(\text{I}_2)}$$

Note that we use parentheses ( ) for the concentrations in mol/L to remind us that these concentrations are *initial* concentrations and not necessarily equilibrium concentrations. If Q_c is greater than K_c, the equilibrium constant, then there must be too much HI (compared to H_2 and I_2) for equilibrium to exist. Consequently, the net reaction will proceed from right to left to decrease (HI) and to increase (H_2) and (I_2) until Q_c equals K_c. On the other hand, if Q_c is smaller than K_c, then the net reaction will progress from left to right until Q_c is equal to K_c. If Q_c is equal to K_c at the start, then the system is already at equilibrium and no net changes will occur. We summarize these three cases as follows:

- $Q_c > K_c$ The ratio of initial concentrations of products to reactants is too large. To reach equilibrium, products must be converted to reactants. The system proceeds from right to left (consuming products, forming reactants) to reach equilibrium.
- $Q_c = K_c$ The initial concentrations are equilibrium concentrations. The system is at equilibrium.
- $Q_c < K_c$ The ratio of initial concentrations of products to reactants is too small. To reach equilibrium, reactants must be converted to products. The system proceeds from left to right (consuming reactants, forming products) to reach equilibrium.

EXAMPLE 17.8

At the start of a reaction, there are 0.0218 mol H_2, 0.0145 mol I_2, and 0.0783 mol HI in a 3.50 L stainless steel reaction vessel at 430°C ($K_c = 54.3$). Decide whether the system is at equilibrium. If not, predict which way the reaction will proceed.

Answer

The equilibrium reaction is

$$H_2(g) + I_2(g) \rightleftharpoons 2HI(g)$$

The initial concentrations of the reacting species are

$$(H_2) = \frac{0.0218 \text{ mol}}{3.50 \text{ L}} = 0.00623 \ M$$

$$(I_2) = \frac{0.0145 \text{ mol}}{3.50 \text{ L}} = 0.00414 \ M$$

$$(HI) = \frac{0.0783 \text{ mol}}{3.50 \text{ L}} = 0.0224 \ M$$

The reaction quotient is

$$Q_c = \frac{[HI]^2}{[H_2][I_2]} = \frac{(0.0224)^2}{(0.00623)(0.00414)}$$
$$= 19.5$$

Since this value is smaller than the equilibrium constant (54.3), the system is not at equilibrium. The net result will be an increase in the concentration of HI and a decrease in the concentrations of H_2 and I_2; that is, the net reaction will proceed from left to right until equilibrium is reached.

Similar examples: Problems 17.26, 17.29.

Calculating Equilibrium Concentrations

If we know the equilibrium constant for a particular reaction, we can calculate the concentrations in the equilibrium mixture from a knowledge of the initial concentrations. Depending on the information given, the calculation may be straightforward or it may be more involved. The situations we normally encounter are (1) only the initial reactants' concentrations or only the initial products' concentrations are given and (2) both the initial reactants' and products' concentrations are given. In solving equilibrium constant problems, we follow three basic steps:

1. Express the equilibrium concentrations of all species in terms of a single unknown, which we call x.
2. Write the equilibrium constant expression in terms of the equilibrium concentrations. Knowing the value of the equilibrium constant, we can solve for x.
3. Having solved for x, calculate the equilibrium concentrations of all species.

The following examples illustrate the application of this approach.

EXAMPLE 17.9

The equilibrium constant (K_c) is 24.0 for the reaction $A \rightleftharpoons B$ at a certain temperature. Suppose that A is initially present at a concentration of 0.850 mol/L. Calculate the concentrations of both A and B at equilibrium at this temperature.

Answer

Step 1

At equilibrium, some of the A molecules will be converted to the B molecules. From the stoichiometry of the reaction we see that 1 mol A $\approx$ 1 mol B. Let x be the equilibrium concentration of B in mol/L; therefore the equilibrium concentration of A must be $(0.850 - x)$ mol/L. We can now summarize the changes in concentrations as follows:

(Continued)

	A	$\rightleftharpoons$	B
Initial	0.850 M		0 M
Change	$-x$ M		$+x$ M
Equilibrium	$(0.850 - x)$ M		x M

A positive $(+)$ change represents an increase and a negative $(-)$ change indicates a decrease in concentration at equilibrium.

Step 2

$$K_c = \frac{[B]}{[A]}$$

$$24.0 = \frac{x}{0.850 - x}$$

$$x = 0.816 \ M$$

Step 3

At equilibrium

$$[A] = (0.850 - 0.816) \ M = 0.034 \ M$$

$$[B] = 0.816 \ M$$

Similar examples: Problems 17.28, 17.32.

EXAMPLE 17.10

A mixture of 0.500 mol H_2 and 0.500 mol I_2 was placed in a 1.00 L stainless steel flask at 430°C. Calculate the concentrations of H_2, I_2, and HI at equilibrium. The equilibrium constant (K_c) for the reaction $H_2(g) + I_2(g) \rightleftharpoons 2HI(g)$ is 54.3 at this temperature.

Answer

Step 1

The stoichiometry of the reaction is 1 mol H_2 reacting with 1 mol I_2 to yield 2 mol HI. Let x be the depletion in concentration (mol/L) of either H_2 or I_2 at equilibrium. It follows that the equilibrium concentration of HI must be $2x$. We summarize the changes in concentrations as follows:

	H_2	+	I_2	$\rightleftharpoons$	2HI
Initial	0.500 M		0.500 M		0.000 M
Change	$-x$ M		$-x$ M		$+2x$ M
Equilibrium	$(0.500 - x)$ M		$(0.500 - x)$ M		$2x$ M

Step 2

The equilibrium constant is given by

$$K_c = \frac{[HI]^2}{[H_2][I_2]}$$

Substituting

$$54.3 = \frac{(2x)^2}{(0.500 - x)(0.500 - x)}$$

Taking the square root of both sides, we get

$$7.37 = \frac{2x}{0.500 - x}$$
$$x = 0.393 \ M$$

Step 3

At equilibrium, the concentrations are

$$[H_2] = (0.500 - 0.393) \ M = 0.107 \ M$$

$$[I_2] = (0.500 - 0.393) \ M = 0.107 \ M$$

$$[HI] = 2 \times 0.393 \ M = 0.786 \ M$$

Similar example: Problem 17.37.

EXAMPLE 17.11

Starting with 4.20 mol HI in a 9.60 L stainless steel reaction vessel, for the reaction $H_2(g) + I_2(g) \rightleftharpoons 2HI(g)$ calculate the concentrations of H_2, I_2, and HI at equilibrium at 430°C ($K_c = 54.3$).

Answer

The initial concentration of HI is

$$\frac{4.20 \ \text{mol}}{9.60 \ \text{L}} = 0.438 \ M$$

Step 1

The stoichiometry of the reaction is such that for every two moles of HI reacted, one mole of H_2 and one mole of I_2 are formed. Let $2x$ be the depletion in concentration (mol/L) of HI at equilibrium. It follows that the equilibrium concentrations of both H_2 and I_2 must be $x \ M$. Next we construct the following:

	H_2	+	I_2	$\rightleftharpoons$	2HI
Initial	0.000 M		0.000 M		0.438 M
Change	+x M		+x M		−2x M
Equilibrium	x M		x M		(0.438 − 2x) M

Step 2

The equilibrium constant for the reaction is given by

$$K_c = \frac{[HI]^2}{[H_2][I_2]}$$

(Continued)

Substituting:

$$54.3 = \frac{(0.438 - 2x)^2}{x^2}$$

Taking the square root of both sides, we get

$$7.37 = \frac{(0.438 - 2x)}{x}$$
$$x = 0.0467 \; M$$

Step 3

At equilibrium, the concentrations are

$$[H_2] = 0.0467 \; M$$

$$[I_2] = 0.0467 \; M$$

$$[HI] = (0.438 - 2 \times 0.0467) \; M = 0.345 \; M$$

Similar example: Problem 17.36.

EXAMPLE 17.12

Referring to Example 17.8, calculate the concentrations of H_2, I_2, and HI at equilibrium.

Answer

Step 1

Let x be the depletion in concentration (mol/L) for H_2 and I_2 at equilibrium. From the stoichiometry of the reaction it follows that the increase in concentration for HI must be $2x$. Next we write

	H_2	+	I_2	$\rightleftharpoons$	2HI
Initial	0.00623 M		0.00414 M		0.0224 M
Change	$-x \; M$		$-x \; M$		$+2x \; M$
Equilibrium	$(0.00623 - x) \; M$		$(0.00414 - x) \; M$		$(0.0224 + 2x) \; M$

Step 2

The equilibrium constant is

$$K_c = \frac{[HI]^2}{[H_2][I_2]}$$

Substituting

$$54.3 = \frac{(0.0224 + 2x)^2}{(0.00623 - x)(0.00414 - x)}$$

It is not possible to solve the above equation by the square root shortcut, as the starting concentrations $[H_2]$ and $[I_2]$ are unequal. Instead, we must first reduce the equation to a quadratic expression (of the form $ax^2 + bx + c = 0$):

$$50.3x^2 - 0.653x + 8.98 \times 10^{-4} = 0$$

The solution for a quadratic equation (see Appendix 4) is

$$x = \frac{-b \pm \sqrt{b^2 - 4ac}}{2a}$$

Here we have $a = 50.3$, $b = 0.653$, and $c = 8.98 \times 10^{-4}$, so that

$$x = \frac{0.653 \pm \sqrt{(0.653)^2 - 4(50.3)(8.98 \times 10^{-4})}}{2 \times 50.3}$$

$$x = 0.0114 \ M \quad \text{or} \quad x = 0.00156 \ M$$

The first solution is physically impossible, since the amounts of H_2 and I_2 reacted would be more than those originally present. The second solution gives the correct answer. Note that in solving quadratic equations of this type, one answer is always physically impossible, so the choice of which value to use for x is easy to make.

Step 3

At equilibrium, the concentrations are

$$[H_2] = (0.00623 - 0.00156) \ M = 0.00467 \ M$$

$$[I_2] = (0.00414 - 0.00156) \ M = 0.00258 \ M$$

$$[HI] = (0.0224 + 2 \times 0.00156) \ M = 0.0255 \ M$$

Similar example: Problem 17.31.

Examples 17.9–17.12 show that we can calculate the concentrations of all the reacting species at equilibrium if we know the equilibrium constant and the initial concentrations. This information is of great value for estimating the yield of a reaction. For example, if the reaction between H_2 and I_2 to form HI were to go to completion, the number of moles of HI formed in Example 17.10 would be 2×0.500 mol, or 1.00 mol. However, because of the equilibrium process, the actual amount of HI formed can be no more than 2×0.393 mol, or 0.786 mol.

17.5 FACTORS THAT AFFECT CHEMICAL EQUILIBRIUM

Chemical equilibrium represents a balance between forward and reverse reactions. In most cases, this balance is quite delicate. Changes in experimental conditions may disturb the balance and shift the equilibrium position so that more or less of the desired product will be formed. At our disposal are the following experimentally controllable variables: concentration, pressure, volume, and temperature. Here we will examine how each of these variables affects a reacting system at equilibrium. In addition, we will examine the effect of a catalyst on equilibrium. (A *catalyst* is a substance that increases the speed of a reaction without itself being used up.)

When we say that the equilibrium *shifts* to the right, for example, we mean that reaction proceeds in the forward direction to a greater extent than in the reverse direction.

Le Chatelier's Principle

There is a general rule that helps us predict the direction in which an equilibrium reaction will move when a change in concentration, pressure, volume, or temperature occurs. The rule, known as **Le Chatelier's principle** (Henry Le Chatelier, 1850–1936), states that *if an external stress is applied to a system at equilibrium, the system will adjust itself in such a way that the stress is partially offset.* The word "stress" here means a change in concentration, pressure, volume, or temperature. We will use Le Chatelier's principle to assess the effects of such changes.

Unfortunately, human stress cannot be reduced in ways described by this principle.

Changes in Concentrations

Iron(III) thiocyanate [$Fe(SCN)_3$] dissolves readily in water to give a deep red solution. The equilibrium between dissolved but undissociated $Fe(SCN)_3$ and the Fe^{3+} and SCN^- ions is given by

$$Fe(SCN)_3(aq) \rightleftharpoons \underset{\text{pale yellow}}{Fe^{3+}(aq)} + \underset{\text{colorless}}{3SCN^-(aq)}$$
$$\underset{\text{red}}{}$$

The dissociation of $Fe(SCN)_3$ actually takes place stepwise, so ions such as $Fe(SCN)_2^+$ and $FeSCN^{2+}$ are also present in solution. However, this simplified equation is sufficient for our discussion.

What would happen if we were to add some sodium thiocyanate (NaSCN) to this solution? In this case, the stress applied to the equilibrium system is an increase in the concentration of SCN^- (from the dissociation of NaSCN). To offset this stress, some Fe^{3+} ions will react with the SCN^- ions and the equilibrium is shifted from right to left:

$$Fe(SCN)_3(aq) \longleftarrow Fe^{3+}(aq) + 3SCN^-(aq)$$

Both Na^+ and NO_3^- are colorless spectator ions.

Consequently, the red color of the solution deepens (Figure 17.2). Similarly, if we add iron(III) nitrate [$Fe(NO_3)_3$] to the original solution, the red color will also deepen because the additional Fe^{3+} ions [from $Fe(NO_3)_3$] will shift the equilibrium from right to left.

Now suppose we add some oxalic acid ($H_2C_2O_4$) to the original solution. Oxalic acid ionizes in water to form the oxalate ion, $C_2O_4^{2-}$, which binds strongly to the Fe^{3+} ions. The formation of the stable colorless ion

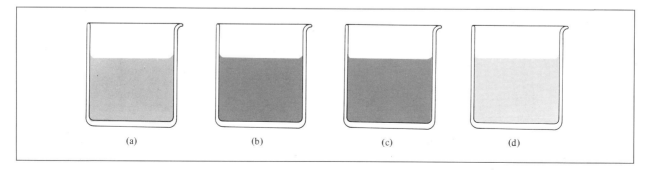

(a)   (b)   (c)   (d)

FIGURE 17.2 *Effect of change in concentration on the position of equilibrium. (a) An aqueous $Fe(SCN)_3$ solution. (b) After the addition of some NaSCN to the solution in (a). (c) After the addition of some $Fe(NO_3)_3$ to the solution in (a). (d) After the addition of some $H_2C_2O_4$ to the solution in (a).*

$Fe(C_2O_4)_3^{3-}$ removes free Fe^{3+} ions in solution. Consequently, more $Fe(SCN)_3$ units will dissociate and the equilibrium is shifted from left to right:

$$Fe(SCN)_3(aq) \longrightarrow Fe^{3+}(aq) + 3SCN^-(aq)$$

The red color of the solution will become lighter.

This experiment demonstrates that at equilibrium all reactants and products are present in the reacting system. Second, increasing the concentrations of the products (Fe^{3+} or SCN^-) shifts the equilibrium to the left, and decreasing the concentration of the product Fe^{3+} shifts the equilibrium to the right. These results are predicted by Le Chatelier's principle.

Oxalic acid is sometimes used to remove bathtub rings that consist of rust, or Fe_2O_3.

Since Le Chatelier's principle simply summarizes the observed behavior of equilibrium systems, it is incorrect to say that a given equilibrium shift occurs "because of" Le Chatelier's principle.

EXAMPLE 17.13

At 350°C, the equilibrium constant (K_c) for the reaction

$$N_2(g) + 3H_2(g) \rightleftharpoons 2NH_3(g)$$

is 2.37×10^{-3}. In a certain experiment, the equilibrium concentrations are $[N_2] = 0.683\ M$, $[H_2] = 8.80\ M$, and $[NH_3] = 1.05\ M$. Suppose some of the NH_3 is quickly removed from the mixture so that its concentration is reduced to $0.774\ M$. (a) Use Le Chatelier's principle to predict the direction that the net reaction will shift to reach a new equilibrium. (b) Confirm your prediction in (a) by calculating the reaction quotient (Q_c) and compare its value with K_c.

Answer

(a) The stress applied to the system is the removal of NH_3. To offset this stress, more N_2 and H_2 gases will react to produce NH_3 until a new equilibrium is established. The net reaction will therefore shift from left to right, that is

$$N_2(g) + 3H_2(g) \longrightarrow 2NH_3(g)$$

(b) At the instant when some of the NH_3 is removed, the system is no longer at equilibrium. The reaction quotient is given by

$$Q_c = \frac{(NH_3)^2}{(N_2)(H_2)^3}$$

$$= \frac{(0.774)^2}{(0.683)(8.80)^3}$$

$$= 1.29 \times 10^{-3}$$

Since this value is smaller than 2.37×10^{-3}, the net reaction will shift from left to right until Q_c is equal to K_c.

Changes in Volume and Pressure

Changes in pressure ordinarily do not affect the concentrations of reacting species in condensed phases (say, in an aqueous solution) because liquids and solids are virtually incompressible. On the other hand, concentrations of gases are greatly affected by changes in pressure. Let us look again at Equation (6.7):

$$PV = nRT$$

$$P = \left(\frac{n}{V}\right) RT$$

The term (n/V) is the concentration of the gas in mol/L. Thus P and V are related to each other inversely: The greater the pressure, the smaller is the volume, and vice versa.

Suppose that the equilibrium system

$$N_2O_4(g) \rightleftharpoons 2NO_2(g)$$

is in a cylinder fitted with a movable piston, at some constant temperature. What happens if we increase the pressure on the gases by pushing down on the piston? Since the volume decreases, the concentration of both NO_2 and N_2O_4 will increase (because the number of moles of NO_2 and N_2O_4 per liter is now greater). Since the system is no longer at equilibrium, we write

$$Q_c = \frac{(NO_2)^2}{(N_2O_4)}$$

Because the concentration of NO_2 is squared, the numerator will increase to a greater extent than the denominator. Thus $Q_c > K_c$ and the net reaction will shift to the left until $Q_c = K_c$. Conversely, a decrease in pressure (increase in volume) would result in $Q_c < K_c$; the net reaction would shift to the right until $Q_c = K_c$.

In general, an increase in pressure (decrease in volume) favors the net reaction that decreases the total number of moles of gases, and a decrease in pressure (increase in volume) favors the net reaction that increases the total number of moles of gases. For systems in which there would be no change in the number of moles of gases, a pressure (or volume) change has no effect on the position of equilibrium.

Note that it is possible to change the pressure of a system without changing its volume. Suppose the NO_2–N_2O_4 system is contained in a stainless steel vessel whose volume is constant. We can increase the total pressure in the vessel by adding an inert gas (helium, for example) to the equilibrium system. Adding helium to the equilibrium mixture at constant volume will increase the total gas pressure and decrease the mole fractions of both NO_2 and N_2O_4; but the partial pressure of each gas, given by the product of its mole fraction and total pressure (see Section 6.6), will not change. Thus the presence of an inert gas in such a case does not affect the equilibrium.

EXAMPLE 17.14

Consider the following equilibrium systems:
(a) $2PbS(s) + 3O_2(g) \rightleftharpoons 2PbO(s) + 2SO_2(g)$

(b) $PCl_5(g) \rightleftharpoons PCl_3(g) + Cl_2(g)$

(c) $H_2(g) + CO_2(g) \rightleftharpoons H_2O(g) + CO(g)$

Predict the direction of the net reaction in each case as a result of increasing the pressure (decreasing the volume) on the system at constant temperature.

Answer

(a) We consider only the gaseous molecules. In the balanced equation the number of moles of gaseous products is 2 and that of reactants is 3; therefore the net reaction will shift to the right.
(b) The number of moles of products is 2 and that of reactants is 1; therefore the net reaction will shift to the left.
(c) The number of moles of products is equal to the number of moles of reactants, so a change in pressure has no effect on the equilibrium.

Similar example: Problem 17.41.

Changes in Temperature

The formation of NO_2 from N_2O_4 is an endothermic process:

$$N_2O_4(g) \longrightarrow 2NO_2(g) \qquad \Delta H° = 58.0 \text{ kJ}$$

Under equilibrium conditions, the quantity of heat absorbed is exactly the same as that liberated in the exothermic process:

$$2NO_2(g) \longrightarrow N_2O_4(g) \qquad \Delta H° = -58.0 \text{ kJ}$$

Thus, the net heat effect at equilibrium is zero. What happens if the gases are heated at constant volume? Here the stress applied to the system is heat. To offset the stress, an endothermic reaction is needed, that is, the decomposition of N_2O_4 to NO_2, to absorb the additional thermal energy. Consequently, the equilibrium constant for the reversible reaction $N_2O_4(g) \rightleftharpoons 2NO_2(g)$, given by

$$K_c = \frac{[NO_2]^2}{[N_2O_4]}$$

will increase with increasing temperature (see Color Plate 23). In general, it is helpful to remember that an increase in temperature favors an endothermic reaction, and a decrease in temperature favors an exothermic reaction.

The Effect of a Catalyst

We saw earlier that a catalyst enhances the rate of a chemical reaction. In Section 6.6, for example, we learned that the decomposition of $KClO_3$ is facilitated by the catalyst MnO_2 (see Figure 6.13). For a reversible reaction, a catalyst increases the rates in *both* directions. Since equilibrium is attained when the forward rate of a reaction equals the reverse rate, the presence of a catalyst does not shift the position of an equilibrium, as do changes in concentration, pressure, volume, and temperature. In fact, adding a catalyst to a reaction mixture that is not at equilibrium will speed up both the forward and reverse rates to achieve an equilibrium mixture much faster. The same equilibrium mixture could be obtained without the catalyst, but we might have to wait much longer for it to happen.

Summary of Factors that May Affect the Equilibrium Position

We have considered four ways to affect a reacting system at equilibrium. It is important to remember that of the four *only a change in temperature will change the equilibrium constant.* Changes in concentration, pressure, and volume can alter the equilibrium concentrations of the reacting mixture, but they cannot change the equilibrium constant as long as the temperature does not change. A catalyst can help establish equilibrium faster, but it too has no effect on the equilibrium constant, nor on the equilibrium concentrations of the reacting species.

EXAMPLE 17.15

Consider the following equilibrium process:

$$N_2F_4(g) \rightleftharpoons 2NF_2(g) \qquad \Delta H° = 38.5 \text{ kJ}$$

Predict the changes in the equilibrium if (a) the reacting mixture is heated, (b) NF_2 gas is removed from the reacting mixture at constant temperature and volume, (c) the pressure on the reacting mixture is decreased at constant temperature, and (d) an inert gas, such as helium, is added to the reacting mixture at constant volume and temperature.

Answer

(a) Since the forward reaction is endothermic, an increase in temperature will favor the formation of NF_2. The equilibrium constant

$$K_c = \frac{[NF_2]^2}{[N_2F_4]}$$

will therefore increase with increasing temperature.

(b) The stress here is the removal of NF_2 gas. To offset it, more N_2F_4 will decompose to form NF_2. The equilibrium constant K_c remains unchanged, however.

(c) A decrease in pressure (which is accompanied by an increase in gas volume) will favor the formation of more gas molecules, that is, the forward reaction. Thus, more NF_2 gas will be formed. The equilibrium constant will remain unchanged.

(d) Adding helium to the equilibrium mixture at constant volume will not shift the equilibrium.

Similar examples: Problems 17.42, 17.44, 17.45.

17.6 APPLYING LE CHATELIER'S PRINCIPLE TO AN INDUSTRIAL PROCESS—THE HABER SYNTHESIS OF AMMONIA

NH$_3$ ranked 3rd in quantity of all chemicals produced in 1984 in the United States.

Ammonia is a valuable chemical used in fertilizer, the manufacture of nitric acid, and many other applications. About 14 million tons of ammonia were produced in the United States in 1984. Industrially ammonia is prepared

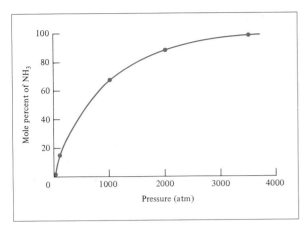

FIGURE 17.3 *Mole percent of NH_3 versus total pressure of the gases at 425°C. Mole percent of NH_3 is the ratio of number of moles of NH_3 to the total number of moles of reactants and product, multiplied by 100%.*

from the molecular nitrogen in the atmosphere and molecular hydrogen obtained as a by-product in the thermal breakdown of petroleum. Since huge quantities of ammonia are produced at various plants, even a slight improvement in the yield can significantly reduce the cost of production. Let us apply Le Chatelier's principle to help us maximize the yield in the production of ammonia.

The equation for the production of ammonia is

$$N_2(g) + 3H_2(g) \rightleftharpoons 2NH_3(g) \qquad \Delta H° = -92.6 \text{ kJ}$$

Looking at this equation we can form two conclusions. First, since one mole of N_2 reacts with three moles of H_2 to produce two moles of NH_3, a higher yield of NH_3 can be obtained at equilibrium if the reaction is carried out at high pressures. This is indeed the case, as shown by the plot of the mole percent of NH_3 versus the total pressure of the reacting system (Figure 17.3). Second, the exothermic nature of the forward reaction tells us that the equilibrium constant K_c, given by

$$K_c = \frac{[NH_3]^2}{[N_2][H_2]^3}$$

will decrease with increasing temperature (Table 17.2). Thus, for maximum yield of NH_3, the reaction should be run at the lowest possible temperature. The graph in Figure 17.4 shows that the yield of ammonia increases with decreasing temperature. A low-temperature operation is desirable in other

TABLE 17.2 Variation with Temperature of the Equilibrium Constant for the Synthesis of Ammonia

$t(°C)$	K_c
25	6.0×10^5
200	0.65
300	0.011
400	6.2×10^{-4}
500	7.4×10^{-5}

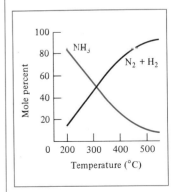

FIGURE 17.4 *The composition (mole percent) of ($H_2 + N_2$) and of NH_3 at equilibrium (for a certain starting mixture) as a function of temperature.*

respects too. If the operating temperature were below $-33.5°C$, which is the boiling point of NH_3, then the NH_3 formed could be quickly liquefied and removed from the reacting system. (Both N_2 and H_2 are still gases at this temperature.) Consequently, the net reaction would shift from left to right, just as desired.

How do our conclusions match the conditions that industrial chemists use in the manufacture of ammonia? We find that the operating pressures are indeed very high (between 500 atm and 1000 atm). Furthermore, in the industrial process the NH_3 never reaches its equilibrium value but is constantly removed from the reaction mixture in a continuous process. The only contradiction is that the industrial operation is usually carried out at about 500°C! It would seem that such a high-temperature operation would be more costly and would lower the yield of NH_3, which is counter to the desired result that could be obtained from what we have learned from Le Chatelier's principle. The justification for the choice of high temperature is that the *rate* of NH_3 production at room temperature is much too slow to be of any practical value. Even at 500°C the rate is not fast enough, so chemists also add a catalyst consisting of iron plus oxides of potassium and aluminum to further enhance the rate. These were the combinations first successfully tested by Fritz Haber (1868–1934) after whom the industrial synthesis of ammonia is named.

An ammonia plant is shown in Figure 17.5, and Figure 17.6 is a schematic diagram of the Haber synthesis of NH_3.

With very few exceptions, the rate of a reaction increases with increasing temperature.

Haber's original interest in ammonia synthesis was to further the production of military explosives in World War I. This process, originally dedicated to destructive uses, is now responsible for supporting a sizable proportion of the world's population through fertilizer manufacture for food production.

FIGURE 17.5   *An ammonia synthesis plant.*

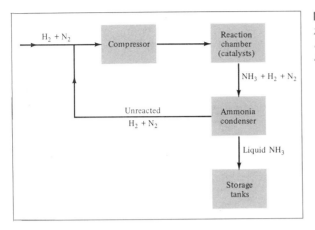

SUMMARY

1. Chemists are concerned with dynamic equilibria between phases (physical equilibria) and between reacting substances (chemical equilibria).
2. For the general chemical reaction

$$a\text{A} + b\text{B} \rightleftharpoons c\text{C} + d\text{D}$$

   the concentrations of reactants and products at equilibrium (in moles per liter) are related by the equilibrium constant expression

$$K_c = \frac{[\text{C}]^c[\text{D}]^d}{[\text{A}]^a[\text{B}]^b}$$

3. The equilibrium constant may also be expressed in terms of the equilibrium partial pressures of gases, as K_P.
4. An equilibrium in which all reactants and products are in the same phase is a homogeneous equilibrium. If the reactants and products are not all in the same phase, the equilibrium is called heterogeneous. The concentrations of pure solids and pure liquids are constant and need not appear in the equilibrium constant expression of a reaction.
5. If a reaction can be expressed as the sum of two or more reactions, the equilibrium constant for the overall reaction is given by the product of the equilibrium constants of the individual reactions.
6. The value of K depends on how the chemical equation is balanced, and the equilibrium constant for the reverse of a particular reaction is the reciprocal of the equilibrium constant of that reaction.
7. The reaction quotient (Q) is an expression in the same form as the equilibrium constant expression, but for a reaction that is not at equilibrium. If $Q > K$, the reaction will proceed from right to left to achieve equilibrium. If $Q < K$, the reaction will proceed from left to right to achieve equilibrium.
8. Le Chatelier's principle states that if an external stress is applied to a system at chemical equilibrium, the system will adjust to partially offset the stress.

9. Only a change in temperature will change the value of the equilibrium constant for a particular reaction. Changes in concentration, pressure, or volume will change the equilibrium concentrations of reactants and products. The addition of a catalyst will hasten the attainment of equilibrium.

KEY WORDS

Equilibrium, p. 502
Equilibrium constant, p. 504
Heterogeneous equilibrium, p. 510
Homogeneous equilibrium, p. 506

Le Chatelier's principle, p. 522
Reaction quotient, p. 515
Static equilibrium, p. 502

PROBLEMS

More challenging problems are marked with an asterisk.

Concept of Equilibrium

17.1 Explain the term "equilibrium." What is the difference between static equilibrium and dynamic equilibrium?

17.2 Give two examples of physical equilibrium and two examples of chemical equilibrium.

17.3 Briefly describe the importance of equilibrium in the study of chemistry.

Equilibrium Constant Expressions

17.4 Write the expression for the equilibrium constants (K_c and K_P, if applicable) of each of the following reactions:

(a) $H_2O(l) \rightleftharpoons H_2O(g)$
(b) $H_2O(g) + CO(g) \rightleftharpoons H_2(g) + CO_2(g)$
(c) $2Mg(s) + O_2(g) \rightleftharpoons 2MgO(s)$
(d) $PCl_5(g) \rightleftharpoons PCl_3(g) + Cl_2(g)$
(e) $CaCO_3(s) + 2H^+(aq) \rightleftharpoons$
$\qquad Ca^{2+}(aq) + CO_2(g) + H_2O(l)$

17.5 Write the expressions for the equilibrium constants K_P of the following thermal decompositions:

(a) $2NaHCO_3(s) \rightleftharpoons$
$\qquad Na_2CO_3(s) + CO_2(g) + H_2O(g)$
(b) $2CaSO_4(s) \rightleftharpoons 2CaO(s) + 2SO_2(g) + O_2(g)$

17.6 Write equilibrium constant expressions for K_c and for K_P (if applicable) for the following processes:

(a) $2CO_2(g) \rightleftharpoons 2CO(g) + O_2(g)$
(b) $3O_2(g) \rightleftharpoons 2O_3(g)$

(c) $CO(g) + Cl_2(g) \rightleftharpoons COCl_2(g)$
(d) $H_2O(g) + C(s) \rightleftharpoons CO(g) + H_2(g)$
(e) $HCOOH(aq) \rightleftharpoons H^+(aq) + HCOO^-(aq)$
(f) $2HgO(s) \rightleftharpoons 2Hg(l) + O_2(g)$
(g) $Ni(s) + 4CO(g) \rightleftharpoons Ni(CO)_4(g)$

17.7 Write the equilibrium constant expressions for K_c and for K_P (if applicable) for the following reactions:

(a) $2NO_2(g) + 7H_2(g) \rightleftharpoons 2NH_3(g) + 4H_2O(l)$
(b) $2ZnS(s) + 3O_2(g) \rightleftharpoons 2ZnO(s) + 2SO_2(g)$
(c) $C(s) + CO_2(g) \rightleftharpoons 2CO(g)$
(d) $N_2O_5(g) \rightleftharpoons 2NO_2(g) + \frac{1}{2}O_2(g)$
(e) $C_6H_5COOH(aq) \rightleftharpoons$
$\qquad C_6H_5COO^-(aq) + H^+(aq)$

Calculation of Equilibrium Constants

17.8 Consider the following equilibrium process at 700°C:

$$2H_2(g) + S_2(g) \rightleftharpoons 2H_2S(g)$$

Analysis shows that there are 2.50 moles of H_2, 1.35×10^{-5} mole of S_2, and 8.70 moles of H_2S present in a 12.0 L flask. Calculate the equilibrium constant K_c for the reaction.

17.9 What is the K_P for the equilibrium (at 1273°C)

$$2CO(g) + O_2(g) \rightleftharpoons 2CO_2(g)$$

if K_c is 2.24×10^{22} at the same temperature?

17.10 A reaction vessel contains NH_3, N_2, and H_2 at equilibrium at a certain temperature. The equilibrium concentrations are $[NH_3] = 0.25\ M$, $[N_2]$

= 0.11 M, and [H_2] = 1.91 M. Calculate the equilibrium constant (K_c) for the synthesis of ammonia if the reaction is represented as

(a) $N_2(g) + 3H_2(g) \rightleftharpoons 2NH_3(g)$
(b) $\frac{1}{2}N_2(g) + \frac{3}{2}H_2(g) \rightleftharpoons NH_3(g)$

17.11 Consider the following reaction:

$$N_2(g) + O_2(g) \rightleftharpoons 2NO(g)$$

If the equilibrium partial pressures of N_2, O_2, and NO are 0.15 atm, 0.33 atm, and 0.050 atm, respectively, at 2200°C, what is the K_P?

17.12 The equilibrium constant K_c for the reaction

$$I_2(g) \rightleftharpoons 2I(g)$$

is 3.8×10^{-5} at 727°C. Calculate K_c and K_P for the equilibrium

$$2I(g) \rightleftharpoons I_2(g)$$

at the same temperature.

17.13 The pressure of the reacting mixture at equilibrium

$$CaCO_3(s) \rightleftharpoons CaO(s) + CO_2(g)$$

is 0.105 atm at 350°C. Calculate K_P and K_c for this reaction.

*17.14 A quantity of 2.50 moles of NOCl was initially in a 1.50 L reaction chamber at 400°C. After equilibrium was established, it was found that 28.0 percent of the NOCl had dissociated:

$$2NOCl(g) \rightleftharpoons 2NO(g) + Cl_2(g)$$

Calculate the equilibrium constant K_c for the reaction.

17.15 The equilibrium constant K_P for the reaction

$$PCl_5(g) \rightleftharpoons PCl_3(g) + Cl_2(g)$$

is 1.05 at 250°C. Start with a mixture of PCl_5, PCl_3, and Cl_2 at pressures 0.177 atm, 0.223 atm, and 0.111 atm, respectively, at 250°C. When the mixture comes to equilibrium at that temperature, which pressures will decrease and which will increase? Explain why.

17.16 Ammonium carbamate, $NH_4CO_2NH_2$, decomposes as follows:

$$NH_4CO_2NH_2(s) \rightleftharpoons 2NH_3(g) + CO_2(g)$$

At 40°C, the total gas pressure (NH_3 and CO_2) is 0.363 atm. Calculate the equilibrium constant, K_P.

17.17 The equilibrium constant (K_c) for the following reaction is 0.65 at 200°C.

$$N_2(g) + 3H_2(g) \rightleftharpoons 2NH_3(g)$$

(a) What is the value of K_P for this reaction?
(b) What is the value of K_c for $2NH_3(g) \rightleftharpoons N_2(g) + 3H_2(g)$?
(c) What is K_c for $\frac{1}{2}N_2(g) + \frac{3}{2}H_2(g) \rightleftharpoons NH_3(g)$?
(d) What are the values of K_P for the reactions described in (b) and (c)?

*17.18 At 20°C, the vapor pressure of water is 0.0231 atm. Calculate K_P and K_c for the "reaction"

$$H_2O(l) \rightleftharpoons H_2O(g)$$

17.19 Consider the following reaction at 1600°C:

$$Br_2(g) \rightleftharpoons 2Br(g)$$

It is found that after putting 1.05 mole of Br_2 in a 0.980 L flask, 1.20 percent of the Br_2 undergoes dissociation. Calculate the equilibrium constant (K_c) for the reaction.

*17.20 Consider the equilibrium

$$2NOBr(g) \rightleftharpoons 2NO(g) + Br_2(g)$$

If nitrosyl bromide, NOBr, is 34 percent dissociated at 25°C and the total pressure is 0.25 atm, calculate K_P and K_c for the dissociation at this temperature.

*17.21 Pure phosgene gas ($COCl_2$), 3.00×10^{-2} mole, was placed in a 1.50 L container. It was heated to 800 K and at equilibrium it was found that the pressure of CO was 0.497 atm. Calculate the equilibrium constant K_P for the reaction

$$CO(g) + Cl_2(g) \rightleftharpoons COCl_2(g)$$

Multiple Equilibria

17.22 The following equilibrium constants have been determined for hydrosulfuric acid at 25°C:

$$H_2S(aq) \rightleftharpoons H^+(aq) + HS^-(aq)$$
$$K_c' = 5.7 \times 10^{-8}$$

$$HS^-(aq) \rightleftharpoons H^+(aq) + S^{2-}(aq)$$
$$K_c'' = 1.2 \times 10^{-13}$$

Calculate the equilibrium constant for the following reaction at the same temperature:

$$H_2S(aq) \rightleftharpoons 2H^+(aq) + S^{2-}(aq)$$

17.23 Consider the following equilibria:

$$2SO_2(g) + O_2(g) \rightleftharpoons 2SO_3(g)$$
$$2NO_2(g) \rightleftharpoons 2NO(g) + O_2(g)$$

Express the equilibrium constant K_c for the process

$$SO_2(g) + NO_2(g) \rightleftharpoons SO_3(g) + NO(g)$$

in terms of the equilibrium constants for the first two processes.

17.24   Given the following equilibrium constant expressions at 1123 K:

$$C(s) + CO_2(g) \rightleftharpoons 2CO(g) \qquad K'_P = 1.3 \times 10^{14}$$
$$CO(g) + Cl_2(g) \rightleftharpoons COCl_2(g) \qquad K''_P = 6.0 \times 10^{-3}$$

Write the equilibrium constant (K_P) expression and calculate the equilibrium constant at 1123 K for

$$C(s) + CO_2(g) + 2Cl_2(g) \rightleftharpoons 2COCl_2(g)$$

17.25   At a certain temperature the following reactions have the constants shown:

$$S(s) + O_2(g) \rightleftharpoons SO_2(g) \qquad K'_c = 4.2 \times 10^{52}$$
$$2S(s) + 3O_2(g) \rightleftharpoons 2SO_3(g) \qquad K''_c = 9.8 \times 10^{128}$$

Calculate the equilibrium constant (K_c) for the reaction

$$2SO_2(g) + O_2(g) \rightleftharpoons 2SO_3(g)$$

Equilibrium Concentrations

17.26   The equilibrium constant K_P for the reaction

$$2SO_2(g) + O_2(g) \rightleftharpoons 2SO_3(g)$$

is 5.60×10^4 at 350°C. SO_2 and O_2 are mixed initially at 0.350 atm and 0.762 atm, respectively, at 350°C. When the mixture equilibrates, is the total pressure less than or greater than the sum of the initial pressures, 1.112 atm?

17.27   The equilibrium constant (K_P) for the reaction

$$2NO_2(g) \rightleftharpoons 2NO(g) + O_2(g)$$

is 158 at 1000 K. Calculate P_{O_2} if $P_{NO_2} = 0.400$ atm and $P_{NO} = 0.270$ atm.

17.28   The dissociation of molecular iodine into iodine atoms is represented as

$$I_2(g) \rightleftharpoons 2I(g)$$

At 1000 K, the equilibrium constant K_c for the reaction is 3.80×10^{-5}. Suppose you start with 0.0456 mole of I_2 in a 2.30 L flask at 1000 K. What are the concentrations of the gases at equilibrium?

17.29   The equilibrium constant K_c for the synthesis of ammonia at 200°C is 0.65. Starting with $[H_2] = 0.76\ M$, $[N_2] = 0.60\ M$, and $[NH_3] = 0.48\ M$, when this mixture comes to equilibrium, which gases will increase in concentration and which will decrease in concentration?

17.30   The equilibrium constant K_c for the decomposition of phosgene, $COCl_2$, is 4.63×10^{-3} at 527°C:

$$COCl_2(g) \rightleftharpoons CO(g) + Cl_2(g)$$

Calculate the equilibrium partial pressures of all the components starting with pure phosgene at 0.760 atm.

17.31   Consider the following equilibrium process at 686°C:

$$CO_2(g) + H_2(g) \rightleftharpoons CO(g) + H_2O(g)$$

The equilibrium concentrations of the reacting species are $[CO] = 0.050\ M$, $[H_2] = 0.045\ M$, $[CO_2] = 0.086\ M$, and $[H_2O] = 0.040\ M$. (a) Calculate K_c for the reaction at 686°C. (b) If the equilibrium concentration of CO_2 were raised to 0.50 mol/L, what would be the concentrations of all the gases at equilibrium?

17.32   The equilibrium constant K_c for the dissociation of N_2O_4 is 4.65×10^{-3} at 25°C

$$N_2O_4(g) \rightleftharpoons 2NO_2(g)$$

Starting with 0.100 mole of N_2O_4 in a 5.00 L reaction vessel, calculate the concentrations of N_2O_4 and NO_2 at equilibrium.

17.33   Consider the heterogeneous equilibrium process:

$$C(s) + CO_2(g) \rightleftharpoons 2CO(g)$$

At 700°C, the total pressure of the system is found to be 4.50 atm. If the equilibrium constant K_P is 1.52, calculate the equilibrium partial pressures of CO_2 and CO.

17.34   For the reaction

$$H_2(g) + CO_2(g) \rightleftharpoons H_2O(g) + CO(g)$$

at 700°C, $K_c = 0.534$. Calculate the number of moles of H_2 formed at equilibrium if a mixture of 0.300 mole of CO and 0.300 mole of H_2O is heated to 700°C in a 10.0 L container.

17.35   A sample of pure NO_2 gas heated to 1000 K decomposes:

$$2NO_2(g) \rightleftharpoons 2NO(g) + O_2(g)$$

The equilibrium constant K_P is 158. Analysis shows that the partial pressure of O_2 is 0.25 atm at equilibrium. Calculate the pressure of NO and NO_2 in the mixture.

17.36   The equilibrium constant K_c for the reaction

$$H_2(g) + Br_2(g) \rightleftharpoons 2HBr(g)$$

is 2.18×10^6 at 730°C. Starting with 3.20 mol HBr in a 12.0 L reaction vessel, calculate the concentrations of H_2, Br_2, and HBr at equilibrium.

17.37   The equilibrium constant K_c for the reaction

$$H_2(g) + CO_2(g) \rightleftharpoons H_2O(g) + CO(g)$$

is 4.2 at 1650°C. Initially 0.80 mol H_2 and 0.80 mol CO_2 are injected into a 5.0 L flask. Calculate the concentration of each species at equilibrium.

Le Chatelier's Principle

17.38 Explain Le Chatelier's principle. How can this principle help us obtain maximum yields from reactions?

17.39 Use Le Chatelier's principle to explain why the equilibrium vapor pressure of a liquid increases with increasing temperature.

17.40 Consider the following equilibrium systems:

(a) $A \rightleftharpoons 2B$ $\Delta H° = 20.0$ kJ
(b) $A + B \rightleftharpoons C$ $\Delta H° = -5.4$ kJ
(c) $A \rightleftharpoons B$ $\Delta H° = 0.0$ kJ

Predict the change in the equilibrium constant K_c that would occur in each case if the temperature of the reacting system were raised.

17.41 What effect does an increase in pressure have on each of the following systems at equilibrium?

(a) $A(s) \rightleftharpoons 2B(s)$
(b) $2A(l) \rightleftharpoons B(l)$
(c) $A(s) \rightleftharpoons B(g)$
(d) $A(g) \rightleftharpoons B(g)$
(e) $A(g) \rightleftharpoons 2B(g)$

(The temperature is kept constant in each case.)

17.42 Consider the equilibrium

$$3O_2(g) \rightleftharpoons 2O_3(g) \qquad \Delta H° = 284 \text{ kJ}$$

What would be the effect on the position of equilibrium of (a) increasing the total pressure on the system by decreasing its volume, (b) adding O_2 to the reaction mixture, (c) decreasing the temperature?

*17.43 Consider the gas-phase reaction

$$2CO(g) + O_2(g) \rightleftharpoons 2CO_2(g)$$

Predict the shift in the equilibrium position when helium gas is added to the equilibrium mixture (a) at constant pressure and (b) at constant volume.

17.44 Consider the following equilibrium process:

$$PCl_5(g) \rightleftharpoons PCl_3(g) + Cl_2(g) \qquad \Delta H° = 92.5 \text{ kJ}$$

Predict the direction of the shift in equilibrium when (a) the temperature is raised, (b) more chlorine gas is added to the reaction mixture, (c) some PCl_3 is removed from the mixture, (d) the pressure on the gases is increased, (e) a catalyst is added to the reaction mixture.

17.45 Consider the reaction

$$2SO_2(g) + O_2(g) \rightleftharpoons 2SO_3(g) \qquad \Delta H° = -198.2 \text{ kJ}$$

Comment on the changes in the concentrations of SO_2, O_2, and SO_3 at equilibrium if we were to (a) increase the temperature, (b) increase the pressure, (c) increase SO_2, (d) add a catalyst, (e) add helium at constant volume.

17.46 In the uncatalyzed reaction at 100°C

$$N_2O_4(g) \rightleftharpoons 2NO_2(g)$$

the pressure of the gases at equilibrium are $P_{N_2O_4} = 0.377$ atm and $P_{NO_2} = 1.56$ atm. What would happen to these pressures if a catalyst were present?

Miscellaneous Problems

17.47 Give two examples each of a homogeneous equilibrium and a heterogeneous equilibrium.

17.48 Consider the statement: "The equilibrium constant of a reacting mixture of solid NH_4Cl and gaseous NH_3 and HCl is 0.316." List two important pieces of information that are missing from this statement.

*17.49 Explain why the value of the equilibrium constant depends on temperature.

17.50 Initially pure NOCl gas was heated to 240°C in a 1.00 L container. At equilibrium it was found that the total pressure was 1.00 atm and the NOCl pressure was 0.64 atm.

$$2NOCl(g) \rightleftharpoons 2NO(g) + Cl_2(g)$$

(a) Calculate the partial pressures of NO and Cl_2 in the system. (b) Calculate the equilibrium constant K_P.

17.51 Consider the following reaction:

$$N_2(g) + O_2(g) \rightleftharpoons 2NO(g)$$

The equilibrium constant K_P for the reaction is 1.0×10^{-15} at 25°C and 0.050 at 2200°C. Is the formation of nitric oxide endothermic or exothermic? Explain your answer.

17.52 Baking soda (sodium bicarbonate) undergoes thermal decomposition as follows:

$$2NaHCO_3(s) \rightleftharpoons Na_2CO_3(s) + CO_2(g) + H_2O(g)$$

Would we obtain more CO_2 and H_2O by adding extra baking soda to the reaction mixture in (a) a closed vessel or (b) an open vessel?

17.53 Consider the following reaction at equilibrium:

$$A(g) \rightleftharpoons 2B(g)$$

From the following data, calculate the equilibrium constant (both K_P and K_c) at each temperature. Is the reaction endothermic or exothermic?

Temperature (°C)	[A]	[B]
200	0.0125	0.843
300	0.171	0.764
400	0.250	0.724

*17.54 The equilibrium constant K_P for the reaction

$$2H_2O(g) \rightleftharpoons 2H_2(g) + O_2(g)$$

is found to be 2×10^{-42} at 25°C. (a) What is K_c for the reaction at the same temperature? (b) The very small value of K_P (and K_c) indicates that the reaction overwhelmingly favors the formation of water molecules. Explain why, despite this fact, a mixture of hydrogen and oxygen gases can be kept at room temperature without any change.

17.55 At a certain temperature and a total pressure of 1.2 atm, the partial pressures of an equilibrium mixture

$$2A(g) \rightleftharpoons B(g)$$

are $P_A = 0.60$ atm and $P_B = 0.60$ atm. (a) Calculate the K_P for the reaction at this temperature. (b) If the total pressure were increased to 1.5 atm, what would be the partial pressures of A and B at equilibrium?

17.56 Consider the following equilibrium system:

$$C(graphite) \rightleftharpoons C(diamond) \qquad \Delta H° = 1.90 \text{ kJ}$$

The densities of diamond and graphite are 3.52 g/cm^3 and 2.25 g/cm^3, respectively. Is the formation of diamond from graphite favored by (a) high or low temperature and (b) high or low pressure?

18
ACIDS AND BASES I: GENERAL PROPERTIES

Acids and bases are among the most important and useful chemical substances, not only in industrial processes and research laboratories, but in homes as well. Our explanation of acids and bases in this chapter takes place at several related levels. We will survey the common macroscopic properties of these substances and examine three sets of acid-base definitions that have historically helped chemists see molecular-level similarities among a diversity of chemical changes.

The concept of acid and base strength will be examined in terms of the structure of molecules. This discussion will serve as an introduction to the quantitative treatment of acids and bases in the next chapter.

535

18.1 GENERAL PROPERTIES OF ACIDS AND BASES

Before we study the definitions of acids and bases and the reactions they undergo, let us summarize some of their general properties. These properties collectively provide operational definitions of acids and bases—definitions that permit us to identify them readily in the laboratory.

Acids

1. They have a sour taste; for example, vinegar owes its taste to acetic acid, and lemons and other citrus fruits contain citric acid.
2. They cause color changes in plant dyes; for example, they change the color of litmus from blue to red.
3. The nonoxidizing acids (for example, HCl) react with metals above hydrogen in the activity series (p. 84) to produce hydrogen gas.
4. They react with carbonates and bicarbonates (for example, Na_2CO_3, $CaCO_3$, $NaHCO_3$) to produce CO_2 gas.
5. They react with bases to form salts and water (see Section 3.7).
6. Aqueous acid solutions conduct electricity.

Bases

1. They have a bitter taste.
2. They feel slippery; for example, soaps, which contain bases, exhibit this property.
3. They cause color changes in plant dyes; for example, they change the color of litmus from red to blue.
4. They react with acids to form salts and water.
5. Aqueous base solutions conduct electricity.

18.2 DEFINITIONS OF ACIDS AND BASES

Arrhenius Acids and Bases

In the late nineteenth century Svante Arrhenius (1859–1927) formulated a theory of acids and bases that defines an acid as a substance that ionizes in water to produce H^+ ions and a base as a substance that ionizes in water to produce OH^- ions. Two Arrhenius acids are

$$HCl(aq) \longrightarrow H^+(aq) + Cl^-(aq)$$

$$HNO_3(aq) \longrightarrow H^+(aq) + NO_3^-(aq)$$

Arrhenius bases are ionic metal hydroxides such as NaOH and $Ba(OH)_2$

$$NaOH(aq) \longrightarrow Na^+(aq) + OH^-(aq)$$

$$Ba(OH)_2(aq) \longrightarrow Ba^{2+}(aq) + 2OH^-(aq)$$

The reaction between hydrochloric acid and sodium hydroxide is

$$HCl(aq) + NaOH(aq) \longrightarrow NaCl(aq) + H_2O(l)$$

Acids and bases were first introduced in Section 2.9.

Caution: It is not advisable to taste chemicals.

This is the basis of the "fizz" that is heard when certain brands of antacids are dissolved in water.

Recall that the H^+ ion is actually a proton.

Using Arrhenius's definitions, we can represent the acid-base neutralizations by the following net ionic equation:

$$H^+(aq) + OH^-(aq) \longrightarrow H_2O(l)$$

Arrhenius's interpretation of acid-base behavior satisfactorily explains reactions of protonic acids (that is, acids that contain ionizable H atoms such as HCl and H_2SO_4) with metal hydroxides [for example, NaOH and $Ba(OH)_2$]. The acid-base titrations discussed in Section 3.7 depend on this kind of acid-base behavior. However, although the use of Arrhenius's definitions is widespread, they do have some serious limitations. First, they apply only to acid-base reactions in aqueous solutions. Second, they do not explain the fact that substances such as NH_3, which do not contain the hydroxide group, increase the OH^- ion concentration when added to water.

All ions are hydrated in aqueous solution (see Section 5.6). We will discuss the structure of the hydrated proton, or $H^+(aq)$, later.

Brønsted–Lowry Acids and Bases

A broader and more general definition of acids and bases was provided independently by Johannes Brønsted (1879–1947) and Thomas Lowry (1879–1936) in 1932. According to their theory, an acid—referred to today as a **Brønsted–Lowry acid**—is a *substance capable of donating a proton*, and a base—or a **Brønsted–Lowry base**—is a *substance capable of accepting a proton*. When HCl gas dissolves in water, it acts as an acid, donating a proton to the solvent (H_2O):

$$HCl(aq) + H_2O(l) \longrightarrow H_3O^+(aq) + Cl^-(aq)$$

H_3O^+ is called the *hydronium ion.* The solvent acts as a base in accepting a proton. Note that water can play the role of either a base or an acid in Brønsted–Lowry acid-base reactions. (We will return shortly to this very important property of water.)

Note that the Brønsted–Lowry definitions, unlike the Arrhenius definitions, do not restrict acid and base behavior to water solutions.

An extension of the Brønsted–Lowry definition of acids and bases is the concept of the **conjugate acid-base pair,** which can be defined as *an acid and its conjugate base or a base and its conjugate acid.* The conjugate base of a Brønsted–Lowry acid is the species that remains after a proton has been removed from the acid. Conversely, a conjugate acid results from the addition of a proton to a Brønsted–Lowry base. Every Brønsted–Lowry acid has a conjugate base, and every Brønsted–Lowry base has a conjugate acid.

Conjugate means joined together.

For example, the chloride ion Cl^- is the conjugate base formed from the acid HCl, and H_2O is the conjugate base of the acid H_3O^+. Similarly, the ionization of acetic acid can be represented as follows:

$$\underset{\text{acid}_1}{CH_3COOH(aq)} + \underset{\text{base}_2}{H_2O(l)} \rightleftharpoons \underset{\text{acid}_2}{H_3O^+(aq)} + \underset{\text{base}_1}{CH_3COO^-(aq)}$$

The subscripts 1 and 2 designate the two conjugate acid-base pairs. Thus the acetate ion (CH_3COO^-) is the conjugate base of CH_3COOH.

The hydroxide ion (OH^-) is a Brønsted–Lowry base because it can accept a proton:

$$H_3O^+(aq) + OH^-(aq) \longrightarrow 2H_2O(l)$$

The formula of a conjugate base always has one fewer hydrogen atom and one more negative charge (or one fewer positive charge) than the formula of the corresponding acid.

TABLE 18.1 Some Common Acids and Their Conjugate Bases

Acid		Conjugate Base	
Name	Formula	Name	Formula
Hydrochloric acid	HCl	Chloride ion	Cl^-
Nitric acid	HNO_3	Nitrate ion	NO_3^-
Hydrocyanic acid	HCN	Cyanide ion	CN^-
Perchloric acid	$HClO_4$	Perchlorate ion	ClO_4^-
Sulfuric acid	H_2SO_4	Hydrogen sulfate ion	HSO_4^-
Hydrogen sulfate ion	HSO_4^-	Sulfate ion	SO_4^{2-}
Carbonic acid	H_2CO_3	Bicarbonate ion	HCO_3^-
Bicarbonate ion	HCO_3^-	Carbonate ion	CO_3^{2-}
Ammonium ion	NH_4^+	Ammonia	NH_3

The Brønsted–Lowry definition also allows us to classify ammonia as a base because of its ability to accept a proton:

$$NH_3(aq) \;+\; H_2O(l) \rightleftharpoons NH_4^+(aq) \;+\; OH^-(aq)$$
$$\text{base}_1 \qquad \text{acid}_2 \qquad\quad \text{acid}_1 \qquad\quad \text{base}_2$$

In this case, NH_4^+ is the conjugate acid of the base NH_3, and OH^- is the conjugate base of the acid H_2O.

Table 18.1 lists a number of common acids, together with their conjugate bases.

EXAMPLE 18.1

Identify the conjugate acid-base pairs in the following reaction:

$$NH_3(aq) \;+\; HF(aq) \rightleftharpoons NH_4^+(aq) \;+\; F^-(aq)$$

Answer

The conjugate acid-base pairs are (1) HF (acid) and F^- (base) and (2) NH_4^+ (acid) and NH_3 (base).

Similar example: Problem 18.5.

The Hydrated Proton

The H^+ ion is hydrated in aqueous solution. The reaction between H^+ and water is usually given by

$$H^+(aq) \;+\; H_2O(l) \longrightarrow H_3O^+(aq)$$

However, evidence suggests that the proton may be associated with more than one water molecule (Figure 18.1). In practice, a quantitative treatment of acid ionization is unaffected by whether we write H^+, H_3O^+, $H_7O_3^+$, or $H_9O_4^+$ for the proton in water. In this book we will write the formula of a

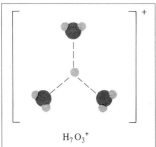

$$H_7O_3^+$$

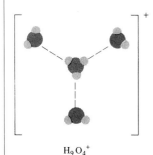

$$H_9O_4^+$$

FIGURE 18.1 *Probable structures for the hydrated H^+ ion in water: $H_7O_3^+$ and $H_9O_4^+$. The colored dashes indicate hydrogen bonds.*

proton in aqueous solution either as H^+ or H_3O^+. The formula H^+ is less cumbersome in calculations involving hydrogen ion concentrations (see next section) and in calculations involving equilibrium constants (see Chapter 19), whereas the formula H_3O^+ is more useful when discussing Brønsted–Lowry acid-base properties.

18.3 AUTOIONIZATION OF WATER AND THE pH SCALE

As we have learned in preceding chapters, water is a unique solvent. One of its special properties is the ability to act both as an acid and as a base. We have seen that water functions as a base in reactions with acids such as HCl and CH_3COOH and that it functions as an acid in reactions with bases such as NH_3. In fact, water itself undergoes ionization to a small extent as follows:

$$H_2O(l) \rightleftharpoons H^+(aq) + OH^-(aq) \tag{18.1}$$

This reaction is sometimes called the *autoionization* of water. To describe the acid-base properties of water in the Brønsted–Lowry framework, it is preferable to express the autoionization as

$$H-\ddot{O}: + H-\ddot{O}: \rightleftharpoons \left[H-\ddot{O}-H \right]^+ + H-\ddot{O}:^-$$
$$\quad | \qquad\quad | \qquad\qquad\quad | $$
$$\quad H \qquad\quad H \qquad\qquad\quad H$$

or

$$\underset{\text{acid}_1}{H_2O} + \underset{\text{base}_2}{H_2O} \rightleftharpoons \underset{\text{acid}_2}{H_3O^+} + \underset{\text{base}_1}{OH^-}$$

The acid-base conjugate pairs are (1) $acid_1$ and $base_1$ (H_2O and OH^-) and (2) $acid_2$ and $base_2$ (H_3O^+ and H_2O).

The Ion Product of Water

In the study of acid-base reactions in aqueous solutions, the important quantity is the hydrogen ion concentration. (As we will see later, the hydroxide ion concentration is related to the hydrogen ion concentration, so we need not consider it here.) Although water undergoes autoionization, it is a very weak electrolyte and, therefore, a poor electrical conductor.

Expressing the hydrated proton as H^+ rather than H_3O^+, we can write the equilibrium constant for the autoionization of water [see Equation (18.1)] as

$$K_c = \frac{[H^+][OH^-]}{[H_2O]}$$

Since very few water molecules are ionized, the concentration of water, that is, $[H_2O]$, remains virtually unchanged, so that

$$K_c[H_2O] = K_w = [H^+][OH^-] \tag{18.2}$$

Tap water and water from underground sources do conduct electricity, because they contain many dissolved ions.

Recall that in pure water, $[H_2O] = 55.5\ M$ (see p. 511).

K_w, the "new" equilibrium constant, is called the ***ion-product constant,*** which is the *product of the molar concentrations of H^+ and OH^- ions at a particular temperature.* In pure water at 25°C, the concentrations of H^+ and OH^- ions are equal and found to be $[H^+] = 1.0 \times 10^{-7}$ M and $[OH^-] = 1.0 \times 10^{-7}$ M. Thus, from Equation (18.2)

$$K_w = (1.0 \times 10^{-7})(1.0 \times 10^{-7}) = 1.0 \times 10^{-14} \qquad \text{at 25°C}$$

Whenever $[H^+] = [OH^-]$, the aqueous solution is said to be neutral. In an acidic solution there is an excess of H^+ ions, or $[H^+] > [OH^-]$. In a basic solution there is an excess of hydroxide ions, or $[H^+] < [OH^-]$. In practice we can adjust the concentration of either H^+ or OH^- ions in solution, but we cannot vary both of them independently. If we adjust the solution so that $[H^+] = 1.0 \times 10^{-6}$ M, the OH^- concentration *must* change to

$$[OH^-] = \frac{K_w}{[H^+]} = \frac{1.0 \times 10^{-14}}{1.0 \times 10^{-6}} = 1.0 \times 10^{-8}\ M$$

It is important to remember that, because K_w is an equilibrium constant, its value changes with temperature. Thus at 40°C, $K_w = 3.8 \times 10^{-14}$ and Equation (18.2) becomes

$$3.8 \times 10^{-14} = [H^+][OH^-] = K_w$$

Since the solution is still neutral, $[H^+] = [OH^-]$. Therefore, at this temperature the concentrations are

$$[H^+] = \sqrt{3.8 \times 10^{-14}} = 1.9 \times 10^{-7}\ M$$

and

$$[OH^-] = \sqrt{3.8 \times 10^{-14}} = 1.9 \times 10^{-7}\ M$$

A neutral solution at 40°C will have a higher H^+ ion concentration (1.9×10^{-7} M) than a neutral solution at 25°C (1.0×10^{-7} M). Unless otherwise stated, we will assume in all subsequent discussions that the temperature is 25°C.

To dramatize these concentrations, imagine that you could randomly draw out and examine 10 particles (H_2O molecules, H^+, or OH^- ions) per second from a beaker of water. On average, it would take you nearly two years, working nonstop, to extract one H^+ ion!

K_w increases with temperature because ionization of water is an endothermic process.

EXAMPLE 18.2

The concentration of OH^- ions in a certain household ammonia cleaning solution is 0.0025 M. Calculate the concentration of the H^+ ions.

Answer

From Equation (18.2) we write

$$[H^+] = \frac{K_w}{[OH^-]} = \frac{1.0 \times 10^{-14}}{0.0025} = 4.0 \times 10^{-12}\ M$$

Since $[H^+] < [OH^-]$, the solution is basic, as we would expect from the earlier discussion of the reaction of ammonia with water.

Similar example: Problem 18.8.

pH—A Measure of Acidity

Because the concentrations of H^+ and OH^- ions are frequently very small numbers and therefore inconvenient to work with, Soren Sorensen (1868–1939) in 1909 proposed a more practical measure called pH. The **pH** of a solution is defined as

$$pH = -\log [H^+] \qquad (18.3)$$

See Appendix 4 for a discussion of logarithm.

Thus we see that the *pH of a solution is given by the negative logarithm of the hydrogen ion concentration (in mol/L)*.

Keep in mind that Equation (18.3) is simply a definition designed to give us convenient numbers to work with. The definition of pH has no theoretical significance. Furthermore, the term $[H^+]$ in Equation (18.3) pertains only to the *numerical part* of the expression for hydrogen ion concentration, for we cannot take the logarithm of units. Thus, like the equilibrium constant, the pH of a solution is a dimensionless quantity.

It is appropriate to digress briefly at this point to comment on the rule for handling significant figures in logarithms. The decimal places in the log of a number equal the significant figures in the number. For example

$$\begin{aligned}
\log 6.7 \times 10^{-4} &= \log 6.7 + \log (1 \times 10^{-4}) \\
&= 0.83 - 4.00 \\
&= -3.17
\end{aligned}$$

The number "4" is exact (because it is an integer), whereas "0.83" is the correct two significant figure representation of 6.7 in 6.7×10^{-4}. Thus the "17" in the final answer (-3.17) tells us that the log has two significant figures.

Since pH is simply a way to express hydrogen ion concentration, acidic and basic solutions at 25°C can be identified by their pH values, as follows:

$$\text{acidic solutions:} \quad [H^+] > 1.0 \times 10^{-7}\,M, \text{pH} < 7.00$$

$$\text{basic solutions:} \quad [H^+] < 1.0 \times 10^{-7}\,M, \text{pH} > 7.00$$

$$\text{neutral solutions:} \quad [H^+] = 1.0 \times 10^{-7}\,M, \text{pH} = 7.00$$

In the laboratory, the pH of a solution is measured with a pH meter (Figure 18.2). Electrodes attached to the meter are dipped into the test

In each case the pH has only two significant figures. The "7" in 7.00 is used only to locate the decimal point in the number 1.0×10^{-7}.

FIGURE 18.2 *A pH meter commonly used in the laboratory to determine the pH of a solution. Although many pH meters have scales marked with values from 1.0 to 14.0, pH values can, in fact, be smaller than 1 and greater than 14.*

TABLE 18.2 The pHs of Some Common Fluids

Sample	pH Value
Gastric juice in the stomach	1.0–2.0
Lemon juice	2.4
Vinegar	3.0
Grapefruit juice	3.2
Orange juice	3.5
Urine	4.8–7.5
Water exposed to air*	5.5
Saliva	6.4–6.9
Milk	6.5
Pure water	7.0
Blood	7.35–7.45
Tears	7.4
Milk of magnesia	10.6
Household ammonia	11.5

*Water exposed to air for a long period of time absorbs atmospheric CO_2 to form carbonic acid, H_2CO_3.

solution and the pH is read directly on the scale. Table 18.2 lists the pHs of several common fluids.

A scale similar to the pH scale can be devised using the negative logarithm of the hydroxide ion concentration. Thus we define pOH as

$$pOH = -\log [OH^-] \tag{18.4}$$

Now consider again the ion-product constant for water:

$$[H^+][OH^-] = K_w = 1.0 \times 10^{-14}$$

Taking the negative logarithm of both sides, we obtain

$$-(\log [H^+] + \log [OH^-]) = -\log (1.0 \times 10^{-14})$$
$$-\log [H^+] - \log [OH^-] = 14.00$$

From the definitions of pH and pOH we obtain

$$pH + pOH = 14.00 \tag{18.5}$$

Equation (18.5) provides us with another way to express the relationship between the H^+ ion concentration and the OH^- ion concentration.

EXAMPLE 18.3

The concentration of H^+ ions in a bottle of table wine was 3.2×10^{-4} M right after the cork was removed. Only half of the wine was consumed. The other half, after it had been standing open to the air for a month, was found to have a hydrogen ion concentration equal to 1.0×10^{-3} M. Calculate the pH of the wine on these two occasions.

Answer

On the first occasion, $[H^+] = 3.2 \times 10^{-4}\ M$, which we substitute in Equation (18.3).

$$pH = -\log [H^+]$$
$$= -\log (3.2 \times 10^{-4}) = 3.49$$

On the second occasion, $[H^+] = 1.0 \times 10^{-3}\ M$, so that

$$pH = -\log (1.0 \times 10^{-3}) = 3.00$$

The increase in hydrogen ion concentration (or decrease in pH) is largely the result of the conversion of some of the alcohol (ethanol) to acetic acid, a reaction that takes place in the presence of molecular oxygen.

Similar examples: Problems 18.9, 18.10, 18.11.

EXAMPLE 18.4

The pH of rainwater in a certain region in the U.S. Northeast on a particular day was 4.82. Calculate the H^+ ion concentration of the rainwater.

Answer

From Equation (18.3)

$$4.82 = -\log [H^+]$$

Taking the antilog of both sides of this equation (see Appendix 4) gives

$$1.5 \times 10^{-5}\ M = [H^+]$$

Similar examples: Problems 18.9, 18.10, 18.11.

EXAMPLE 18.5

For an NaOH solution $[OH^-]$ is $2.9 \times 10^{-4}\ M$. Calculate the pH of the solution.

Answer

We use Equation (18.4):

$$pOH = -\log [OH^-]$$
$$= -\log (2.9 \times 10^{-4})$$
$$= 3.54$$

Now we use Equation (18.5):

$$pH + pOH = 14.00$$
$$pH = 14.00 - pOH$$
$$= 14.00 - 3.54 = 10.46$$

Similar examples: Problems 18.9, 18.10, 18.11.

18.4   STRENGTHS OF ACIDS AND BASES

The strength of an acid is determined by the extent of its ionization in solution. Suppose we have two aqueous solutions containing monoprotic acids HA and HB, respectively. In the framework of the Brønsted–Lowry theory, let us say that HA is a stronger acid than HB because it can transfer a proton to water (which acts as a Brønsted–Lowry base in this case) more readily than HB can.

> The initial concentrations of HA and HB are the same.

$$HA(aq) + H_2O(l) \rightleftharpoons H_3O^+(aq) + A^-(aq)$$

$$HB(aq) + H_2O(l) \rightleftharpoons H_3O^+(aq) + B^-(aq)$$

At equilibrium, the solution containing HA will have a higher concentration of H^+ ions and a lower pH than that containing HB.

Strong acids—hydrochloric acid (HCl), nitric acid (HNO_3), perchloric acid ($HClO_4$), and sulfuric acid (H_2SO_4), for example—are all strong electrolytes. For most practical purposes, they may be assumed to be completely ionized in water.

$$HCl(aq) + H_2O(l) \longrightarrow H_3O^+(aq) + Cl^-(aq)$$

$$HNO_3(aq) + H_2O(l) \longrightarrow H_3O^+(aq) + NO_3^-(aq)$$

$$HClO_4(aq) + H_2O(l) \longrightarrow H_3O^+(aq) + ClO_4^-(aq)$$

$$H_2SO_4(aq) + H_2O(l) \longrightarrow H_3O^+(aq) + HSO_4^-(aq)$$

Note that H_2SO_4 is a diprotic acid (we consider only the first stage of ionization here).

> The percent of ionization refers only to the equilibrium condition.

Most acids, however, are classified as weak or moderately weak because they ionize only partially in water. For example, in a 0.10 M CH_3COOH solution, the percent of ionization of the acid is 1.3; only 13 of every 1000 molecules of CH_3COOH dissolved in water exist in the ionized form.

In general, the extent of ionization can vary greatly from acid to acid. For example, both hydrofluoric acid (HF) and hydrocyanic acid (HCN) are classified as weak acids, yet in 0.10 M solutions the percents of ionization of HF and HCN at equilibrium are 8.4 and 0.0070 at 25°C, respectively. Thus HF is a stronger acid than both HCN and CH_3COOH. The limited ionization of weak acids is related to the equilibrium constant, a relationship we will deal with in the next chapter.

Much of what we have said so far about acids applies also to bases. The strength of a base refers to its ability to accept a proton from a reference acid, which is normally the solvent. Hydroxides of alkali metals and alkaline earth metals, such as NaOH, KOH, and $Ba(OH)_2$, are strong bases. These substances are all strong electrolytes that ionize completely in solution:

> A reference acid is an acid chosen to compare the strengths of several bases.

> All alkali metal hydroxides are soluble. Of the alkaline metal hydroxides, $Be(OH)_2$, $Mg(OH)_2$, $Ca(OH)_2$ are insoluble, $Sr(OH)_2$ is slightly soluble, and $Ba(OH)_2$ is soluble.

$$NaOH(aq) \longrightarrow Na^+(aq) + OH^-(aq)$$

$$KOH(aq) \longrightarrow K^+(aq) + OH^-(aq)$$

$$Ba(OH)_2(aq) \longrightarrow Ba^{2+}(aq) + 2OH^-(aq)$$

On the other hand, ammonia is a weak base. Measurements show that in a 0.10 M NH_3 solution at 25°C, only 13 of every 1000 dissolved NH_3 molecules undergo the reaction

$$NH_3(aq) + H_2O(l) \rightleftharpoons NH_4^+(aq) + OH^-(aq)$$

so its "percent ionization" is also 1.3 percent. The fact that the same percent ionization is found for NH_3 as for CH_3COOH in 0.10 M solutions is coincidental.

Table 18.3 lists some important conjugate acid-base pairs, according to their relative strengths. We can make several observations in examining this table:

- From the trends in acid and base strengths, we see that if an acid is strong, its conjugate base must be weak, and vice versa. For example, HI is a stronger acid than HNO_2; therefore, I^- is a weaker base than NO_2^-.

- H_3O^+ is the strongest acid that can exist in aqueous solution. Acids stronger than H_3O^+ react with water to produce H_3O^+ and their conjugate bases. Thus HBr, which is a stronger acid than H_3O^+, reacts with water completely to form H_3O^+ and Br^-:

$$HBr(aq) + H_2O(l) \longrightarrow H_3O^+(aq) + Br^-(aq)$$

Acids weaker than H_3O^+ react with water to a much smaller extent, producing H_3O^+ and their conjugate bases. For example, the following equilibrium lies primarily to the left:

$$HF(aq) + H_2O(l) \rightleftharpoons H_3O^+(aq) + F^-(aq)$$

- The OH^- ion is the strongest base that can exist in aqueous solution. Bases stronger than OH^- react with water to produce OH^- and their conjugate acids. For example, the amide ion (NH_2^-) is a stronger base than OH^-, so it reacts with water completely as follows:

$$NH_2^-(aq) + H_2O(l) \longrightarrow NH_3(aq) + OH^-(aq)$$

For this reason the amide ion does not exist in aqueous solutions.

Note that NH_3 does not ionize like an acid such as HCl because it does not split up to form ions.

A bottle labeled "0.10 M HCl" actually contains 0.10 M H_3O^+, with no molecular HCl present at all!

The amide ion does exist in liquid ammonia.

TABLE 18.3   Relative Strengths of Conjugate Acid-Base Pairs

Acid		Conjugate Base
$HClO_4$		ClO_4^-
HI	Strong acids.	I^-
HBr	Assumed to be 100	Br^-
HCl	percent ionized in	Cl^-
H_2SO_4	aqueous solution.	HSO_4^-
HNO_3		NO_3^-
H_3O^+		H_2O
HSO_4^-		SO_4^{2-}
HCOOH	Weak acids.	$HCOO^-$
HNO_2	At equilibrium, there	NO_2^-
HF	is a mixture of	F^-
CH_3COOH	nonionized acid	CH_3COO^-
NH_4^+	molecules, as well as	NH_3
HCN	the H^+ ions and the	CN^-
H_2O	conjugate base.	OH^-
NH_3		NH_2^-*

Acid strength increases ↑   Base strength increases ↓

*NH_2^- is called the amide ion.

EXAMPLE 18.6

Calculate the pH of (a) a 1.0×10^{-5} M HCl solution and (b) a 0.020 M Ba(OH)$_2$ solution.

Answer

(a) Since HCl is a strong acid, it is completely ionized in solution:

$$HCl(aq) \longrightarrow H^+(aq) + Cl^-(aq)$$

The concentrations of all the species (HCl, H$^+$, and Cl$^-$) before and after ionization can be represented as follows:

	HCl(aq) $\longrightarrow$	H$^+$(aq) +	Cl$^-$(aq)
Initial	1.0×10^{-5} M	0.0 M	0.0 M
Change	-1.0×10^{-5} M	$+1.0 \times 10^{-5}$ M	$+1.0 \times 10^{-5}$ M
Final	0.0 M	1.0×10^{-5} M	1.0×10^{-5} M

A positive (+) change represents an increase and a negative (−) change indicates a decrease in concentration. Thus

$$[H^+] = 1.0 \times 10^{-5} \ M$$
$$pH = -\log (1.0 \times 10^{-5})$$
$$= 5.00$$

(b) Ba(OH)$_2$ is a strong base; each Ba(OH)$_2$ produces two OH$^-$ ions:

$$Ba(OH)_2(aq) \longrightarrow Ba^{2+}(aq) + 2OH^-(aq)$$

The changes in the concentrations of all the species can be represented as follows:

	Ba(OH)$_2$(aq) $\longrightarrow$	Ba^{2+}(aq) +	2OH$^-$(aq)
Initial	0.020 M	0.00 M	0.00 M
Change	-0.020 M	$+0.020$ M	$+2(0.020)$ M
Final	0.00 M	0.020 M	0.040 M

Thus

$$[OH^-] = 0.040 \ M$$
$$pOH = -\log 0.040 = 1.40$$

Therefore

$$pH = 14.00 - pOH$$
$$= 14.00 - 1.40$$
$$= 12.60$$

Note that in both (a) and (b) we neglected the very small contribution to [H$^+$] and [OH$^-$] due to autoionization of water.

EXAMPLE 18.7

Predict the direction of the following reaction in aqueous solution:

$$HNO_2(aq) + CN^-(aq) \rightleftharpoons HCN(aq) + NO_2^-(aq)$$

Answer

From Table 18.3 we see that HNO_2 is a stronger acid than HCN. Thus CN^- is a stronger base than NO_2^-. The net reaction will proceed from left to right because HNO_2 is a better proton donor than HCN (and CN^- is a better proton acceptor than NO_2^-).

Similar example: Problem 18.26.

The Leveling Effect

We have seen that one way to compare the strengths of acids is to choose a reference base (which is usually the solvent) and measure the extent to which the acid donates a proton to the base. However, since the reaction between any acid stronger than H_3O^+ and water goes completely to the right, strong acids such as $HClO_4$, HCl, HNO_3, and H_2SO_4 will appear to be of equal strength in aqueous solution. Therefore we cannot compare the relative strengths of all acids on the basis of their readiness to lose a proton in aqueous solution. *The inability of a solvent to differentiate among the relative strengths of all acids stronger than the solvent's conjugate acid* is known as the **leveling effect** because the solvent is said to level the strengths of these acids, making them seem identical.

A simple analogy to the leveling effect is the following. Suppose an average adult and an Olympic weight lifter want to compare their strengths. First they attempt to lift a 15-kg weight. However, since they both can do this with ease, the result is inconclusive. But when they try to lift a 50-kg weight, the Olympic weight lifter lifts it readily but the average adult must struggle to do it or may even fail. The weights here correspond to the solvents in which we study the ionization of acids. The lighter weight corresponds to water, which exerts the leveling effect that renders comparison of acid strengths inconclusive.

To compensate for the leveling effect of water we can use a more weakly basic solvent like acetic acid. (This procedure corresponds to the use of a heavier weight to compare the relative strengths of the average adult and the Olympic weight lifter.) Acetic acid can function as a base by accepting a proton:

Pure acetic acid is also called glacial acetic acid.

$$CH_3COOH + H^+ \rightleftharpoons CH_3COOH_2^+$$

The structure of a protonated acetic acid molecule is

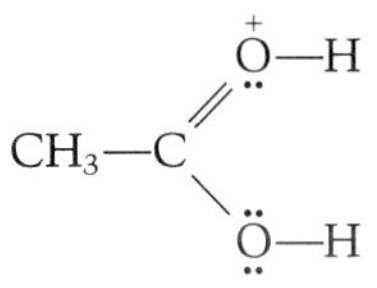

This is one of two equivalent resonance structures.

Since acetic acid is a much weaker base than water, it is not as easily protonated. Thus there are appreciable differences in the extent to which the following reactions proceed from left to right:

$$HNO_3 + CH_3COOH \rightleftharpoons CH_3COOH_2^+ + NO_3^-$$

$$HCl + CH_3COOH \rightleftharpoons CH_3COOH_2^+ + Cl^-$$

$$HClO_4 + CH_3COOH \rightleftharpoons CH_3COOH_2^+ + ClO_4^-$$

$$H_2SO_4 + CH_3COOH \rightleftharpoons CH_3COOH_2^+ + HSO_4^-$$

Measuring the extent of ionization of these and other acids in acetic acid solvent shows that their relative strengths increase as follows:

$$HNO_3 < H_2SO_4 < HCl < HBr < HI < HClO_4$$

18.5 MOLECULAR STRUCTURE AND STRENGTH OF ACIDS

What makes one acid stronger than another? How can we predict the relative strengths of acids in a series of similar compounds? To begin with, we must realize that the strength of an acid depends on a number of factors, such as the properties of the solvent, temperature, and, of course, the molecular structure of the acid. When we compare the strengths of two acids, we can eliminate some variables by considering their properties in the same solvent and at the same temperature. In this way we need consider only the structures of the acids.

Let us consider a certain acid HX. The factors that determine the strength of the acid are the polarity and bond energy of the H—X bond. The more polar the bond, the more readily the acid is ionized into H^+ and X^- ions. On the other hand, strong bonds (that is, bonds with high bond energies) are less easily ionized than weaker ones.

In this section we will concentrate on two types of acids: ***binary acids*** (or *acids that contain only two different elements*) and ***ternary acids*** (or *acids that contain three different elements*).

Binary Acids

The halogens form a particularly important series of binary acids called the *hydrohalic acids*. As we saw earlier, the strengths of the hydrohalic acids increase in the following order:

$$HF \ll HCl < HBr < HI$$

The ≪ sign means "much smaller than."

Because of the leveling effect of water, the relative strengths of the strong acids HCl, HBr, and HI must be evaluated in some other solvent, such as acetic acid. As Table 11.3 (p. 319) shows, HF has the strongest bond of the four hydrogen halides. Consequently, more energy is required to break the H—F bond, so HF is a weak acid. At the other extreme in the series, HI has the lowest bond energy, so HI is the strongest acid of the group. In this series of acids the polarity of the bond actually decreases from HF to HI because F is the most electronegative of the halogens (see Figure 11.2). This property should enhance the acidity of HF relative to the other acids, but its magnitude is not great enough to outweigh the trend in bond energies.

The relative importance of bond energy versus bond polarity in determining acid strength is reversed for binary compounds in a particular period of the periodic table. Consider the binary hydrogen compounds of the second-period elements: CH_4, NH_3, H_2O, and HF. Methane (CH_4) has no measurable acidic properties whatsoever, whereas hydrofluoric acid (HF) is an acid of measurable strength in water. The strengths of the compounds as acids increase as follows:

$$CH_4 < NH_3 < H_2O < HF$$

This trend is the reverse of what we would expect on the basis of bond energy considerations. (From Table 11.3 we see that the HF molecule has the highest bond energy of the four molecules.) We can explain the trend by noting the increase in electronegativity as we go from C to F. As electronegativity increases, the H—X bond becomes more polar (X denotes the C, N, O, or F atoms), so the compound has a greater trendency to ionize into H^+ and X^- ions.

Ternary Acids

Most of the ternary acids are oxyacids (see Section 2.9), which are characterized by the presence of one or more O—H bonds. Figure 18.3 shows the Lewis formulas of a number of oxyacids. The strength of the acid depends on the tendency of the O—H bond to become ionized:

$$\text{—Z—O—H} \longrightarrow \text{—Z—O}^- + H^+$$

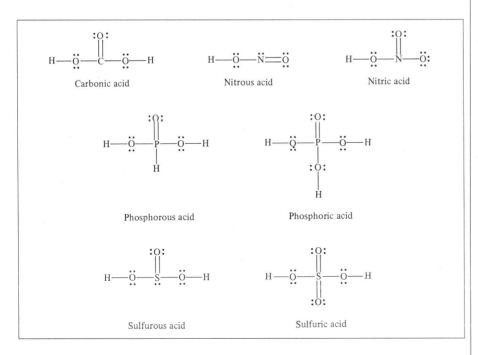

FIGURE 18.3 *Lewis formulas of some common oxyacids. For simplicity, the formal charges have been omitted.*

In most oxyacids Z is a non-metallic element.

Z denotes the central atom. Any factor that draws the electrons toward the Z atom strengthens the Z—O bond and weakens the O—H bond. The more readily the O—H bond is broken, the stronger is the oxyacid. Factors that favor a strong Z—O bond are high electronegativity and high oxidation state of the Z atom, because they increase the atom's ability to attract electrons. To compare their strengths, it is convenient to divide the oxyacids into two groups.

1. *Oxyacids that contain different central atoms whose elements are in the same group of the periodic table and have the same oxidation number.* In this group, the strengths of the acids increase with increasing electronegativity of the central atom, as the following example illustrates:

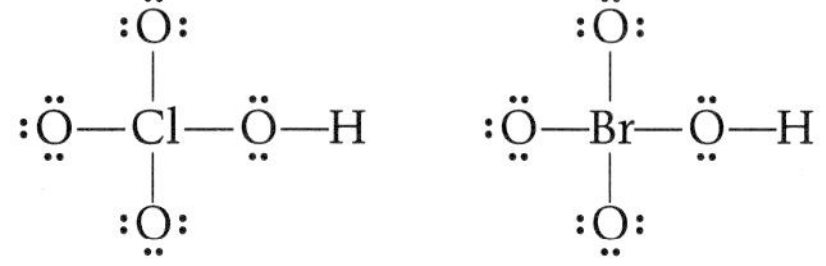

$$\ddot{\text{O}}:\qquad\qquad \ddot{\text{O}}:$$
$$:\ddot{\text{O}}—\text{Cl}—\ddot{\text{O}}—\text{H}\qquad :\ddot{\text{O}}—\text{Br}—\ddot{\text{O}}—\text{H}$$
$$:\ddot{\text{O}}:\qquad\qquad :\ddot{\text{O}}:$$

Cl and Br have the same oxidation number, +7. However, since Cl is more electronegative than Br (see Figure 11.2), the Cl—O bond (in the Cl—O—H group) is stronger than the Br—O bond (in the Br—O—H group). Consequently, perchloric acid has a weaker O—H bond and therefore the relative strengths are

$$\text{HClO}_4 > \text{HBrO}_4$$

2. *Oxyacids that have the same central atom, but different number of attached groups.* In this group the acid strength increases as the oxidation number of the central atom increases. Consider the oxyacids of chlorine shown in Figure 18.4. The higher the oxidation number, the stronger the Cl—O bond in the Cl—O—H group and, therefore, the weaker the O—H bond. Thus HClO_4 is the stronger acid, and the acid strength decreases as follows:

$$\text{HClO}_4 > \text{HClO}_3 > \text{HClO}_2 > \text{HOCl}$$

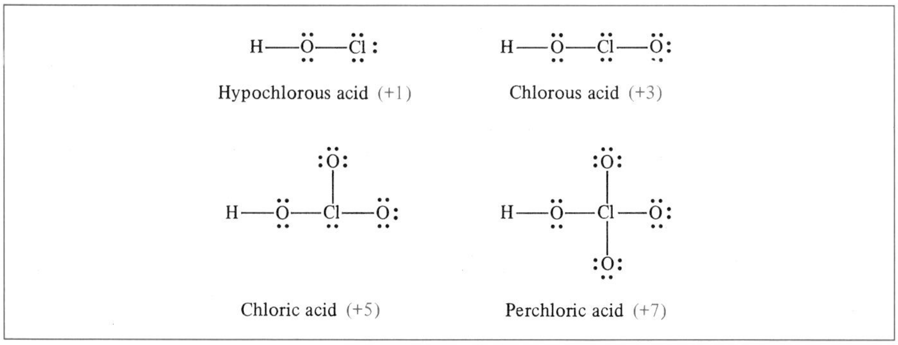

FIGURE 18.4 *Lewis formulas of oxyacids of chlorine. The numbers in color indicate the oxidation number of the Cl atom. For simplicity, the formal charges have been omitted.*

H—$\ddot{\text{O}}$—$\ddot{\text{Cl}}$:   H—$\ddot{\text{O}}$—$\ddot{\text{Cl}}$—$\ddot{\text{O}}$:

Hypochlorous acid (+1)   Chlorous acid (+3)

$:\ddot{\text{O}}:$ $:\ddot{\text{O}}:$
H—$\ddot{\text{O}}$—Cl—$\ddot{\text{O}}$:   H—$\ddot{\text{O}}$—Cl—$\ddot{\text{O}}$:
                                        $:\ddot{\text{O}}:$

Chloric acid (+5)   Perchloric acid (+7)

EXAMPLE 18.8

Predict the relative strengths of the oxyacids in each of the following groups:
(a) HOCl, HOBr, and HOI; (b) HNO_3 and HNO_2.

Answer

(a) These acids all have the same structure and the halogens all have the same oxidation number $(+1)$. Since the electronegativity decreases from Cl to I, the strength of the X—O bond (where X denotes a halogen atom) decreases from HOCl to HOI, and the strength of the O—H bond increases from HOCl to HOI. Thus the acid strength decreases as follows:

$$HOCl > HOBr > HOI$$

(b) The structures of HNO_3 and HNO_2 are shown in Figure 18.3. Since the oxidation number of N is $+5$ in HNO_3 and $+3$ in HNO_2, HNO_3 is a stronger acid than HNO_2.

Similar examples: Problems 18.33, 18.34.

18.6 SOME TYPICAL ACID-BASE REACTIONS

Having discussed the strengths of acids and bases, we will look now at some typical acid-base reactions in aqueous solution. Acid-base reactions are generally characterized by the following equation:

$$Acid + base \longrightarrow salt + water$$

The equilibrium constants for these reactions are very large; therefore, for all practical purposes, they are assumed to go to completion, as indicated by the single arrow. However, the kind of equation we write for an acid-base reaction depends on whether the acid and the base are strong or weak. We will therefore divide these reactions into four different categories. For simplicity, we will limit our discussion to monoprotic acids and to bases like ammonia and the alkali metal hydroxides.

Reactions of Strong Acids with Strong Bases

Strong acids and strong bases are completely ionized in solution. Therefore, the reaction of a strong acid with a strong base can be represented by the net ionic equation

$$H^+(aq) + OH^-(aq) \longrightarrow H_2O(l)$$

This is the equation for the neutralization reaction between HCl and KOH, for example. Because KOH is a strong base, the K^+ ion does not react with H_2O to form KOH and H^+. The Cl^- ion is an extremely weak Brønsted–Lowry base (it is the conjugate base of the strong acid HCl). Therefore, the Cl^- ion also does not react with H_2O. A solution prepared by mixing

The K^+ ion does not qualify as a Brønsted–Lowry acid because the ion (or its hydrated form) has no tendency to donate protons.

equimolar amounts of a strong monoprotic acid and an alkali metal hydroxide is neutral, with a pH of 7.

An illustration of strong acid–strong base neutralization is provided by the treatment of acid indigestion. An average adult's stomach produces between 2 and 3 liters of gastric juice (mostly hydrochloric acid) daily. The pH of the gastric juice is about 1.5, which corresponds to a hydrogen ion concentration of 0.03 M. This concentration is strong enough to dissolve zinc metal! The purpose of this highly acidic medium is to digest food. However, if the acid content exceeds this limit, the surplus H^+ ions can cause muscular contraction, pain, swelling, inflammation, and bleeding of the stomach.

One way to reduce the hydrogen ion concentration in the stomach is to take an antacid, which acts to reduce the hydrogen ion concentration. A common antacid is milk of magnesia, an aqueous suspension of small particles of $Mg(OH)_2$. The neutralization reaction is

$$2HCl(aq) + Mg(OH)_2(s) \longrightarrow MgCl_2(aq) + 2H_2O(l)$$

Reactions of Weak Acids with Strong Bases

Weak acids are largely nonionized in solution, so the ionic equation representing the reaction between a weak acid such as CH_3COOH and a strong base such as NaOH is

$$CH_3COOH(aq) + OH^-(aq) \longrightarrow CH_3COO^-(aq) + H_2O(l)$$

where the OH^- ions are supplied by the base. In our discussion of acid strengths we noted that a strong acid has a weak conjugate base. It therefore follows that a weak acid should have a strong conjugate base. While these statements are generally true, we must be careful in applying them since the strengths of acids and bases are used in a comparative and not in an absolute sense. $HClO_4$ is the strongest acid (of the group in Table 18.3) so its conjugate base ClO_4^- is the weakest base. At the other extreme, NH_3 is an extremely weak acid so its conjugate base NH_2^- is a very strong base. However, in dealing with acids that fall between these two extreme limits we often encounter situations in which a weak acid may have a weak conjugate base. A case in point is the weak acid CH_3COOH and its weak conjugate base (CH_3COO^-). Because CH_3COO^- is a weak base, its reaction with water is incomplete, as indicated by the double arrow:

$$\underset{\text{base}_1}{CH_3COO^-(aq)} + \underset{\text{acid}_2}{H_2O(l)} \rightleftharpoons \underset{\text{acid}_1}{CH_3COOH(aq)} + \underset{\text{base}_2}{OH^-(aq)}$$

Nevertheless, the resulting solution is basic because of the OH^- ions produced. Note that the Na^+ ion does not react with water because NaOH is a strong base.

Reactions of Strong Acids with Weak Bases

A typical reaction between a strong acid and a weak base is the reaction between HNO_3 and NH_3. The ionic equation for this reaction is

The quantitative relationship between weak acids and their conjugate bases will be given in the next chapter.

$$H^+(aq) + NH_3(aq) \longrightarrow NH_4^+(aq)$$

where the H^+ ions are supplied by the acid. Since NH_4^+ is the weak conjugate acid of the weak base NH_3, it will react with water as follows:

$$\underset{\text{acid}_1}{NH_4^+(aq)} + \underset{\text{base}_2}{H_2O(l)} \rightleftharpoons \underset{\text{base}_1}{NH_3(aq)} + \underset{\text{acid}_2}{H_3O^+(aq)}$$

Thus, a solution prepared by mixing equimolar amounts of HNO_3 and NH_3 will be acidic. (Note that this reaction produces an excess of H^+ ions.) The NO_3^- ion is an extremely weak conjugate base of the strong acid HNO_3 and therefore does not react with water.

Reactions of Weak Acids with Weak Bases

Neither weak acids nor weak bases ionize appreciably in solution. Thus the equation representing the reaction between a weak acid such as CH_3COOH and a weak base such as NH_3 must be written in molecular form:

$$CH_3COOH(aq) + NH_3(aq) \longrightarrow CH_3COO^-(aq) + NH_4^+(aq)$$

In this case, both the anion and the cation will react with water:

$$CH_3COO^-(aq) + H_2O(l) \rightleftharpoons CH_3COOH(aq) + OH^-(aq)$$
$$NH_4^+(aq) + H_2O(l) \rightleftharpoons NH_3(aq) + H_3O^+(aq)$$

Whether the resulting solution will be acidic, basic, or neutral depends on the relative extent to which the CH_3COO^- and NH_4^+ ions react with water. We will present a quantitative treatment of the reactions between the cation and anion of a salt (CH_3COONH_4 in our discussion) with water in the next chapter.

18.7 THE ACID AND BASE PROPERTIES OF OXIDES AND HYDROXIDES

Acidic, Basic, and Amphoteric Oxides

As we saw in Chapter 9, oxides can be classified as acidic, basic, or amphoteric. Therefore, a discussion of acid-base reactions would be incomplete without examining the properties of these compounds.

Figure 18.5 shows the formulas of a number of binary oxides of the representative elements in their highest oxidation states. Note that all alkali metal oxides and all alkaline earth metal oxides except BeO are basic. BeO and several metallic oxides in Groups 3A and 4A are amphoteric. Nonmetallic oxides in which the oxidation number of the representative element is high are acidic (for example, N_2O_5, SO_3, and Cl_2O_7), while those in which the oxidation number of the representative element is low (for example, CO and NO) do not show any measurable acidic properties. No nonmetallic oxides are known to have basic properties.

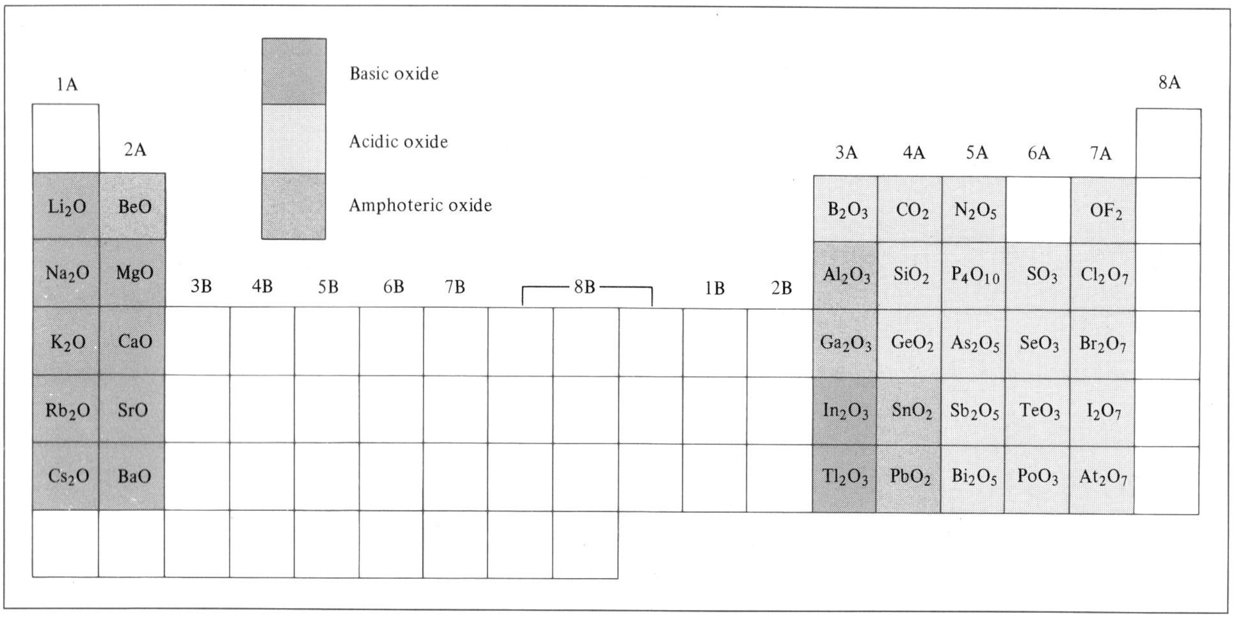

1A														3A	4A	5A	6A	7A	8A
Li_2O	BeO													B_2O_3	CO_2	N_2O_5		OF_2	
Na_2O	MgO	3B	4B	5B	6B	7B	—	8B	—	1B	2B			Al_2O_3	SiO_2	P_4O_{10}	SO_3	Cl_2O_7	
K_2O	CaO													Ga_2O_3	GeO_2	As_2O_5	SeO_3	Br_2O_7	
Rb_2O	SrO													In_2O_3	SnO_2	Sb_2O_5	TeO_3	I_2O_7	
Cs_2O	BaO													Tl_2O_3	PbO_2	Bi_2O_5	PoO_3	At_2O_7	

Basic oxide

Acidic oxide

Amphoteric oxide

FIGURE 18.5 *The formulas of a number of oxides of the representative elements in their highest oxidation states.*

The basic metallic oxides react with water to form the corresponding metal hydroxides:

$$Na_2O(s) + H_2O(l) \longrightarrow 2NaOH(aq)$$

$$BaO(s) + H_2O(l) \longrightarrow Ba(OH)_2(aq)$$

The reactions between acidic oxides and water are as follows:

$$CO_2(g) + H_2O(l) \rightleftharpoons H_2CO_3(aq)$$

$$SO_3(g) + H_2O(l) \rightleftharpoons H_2SO_4(aq)$$

$$N_2O_5(g) + H_2O(l) \rightleftharpoons 2HNO_3(aq)$$

$$P_4O_{10}(s) + 6H_2O(l) \rightleftharpoons 4H_3PO_4(aq)$$

$$Cl_2O_7(g) + H_2O(l) \rightleftharpoons 2HClO_4(aq)$$

Reactions between acidic oxides and bases and those between basic oxides and acids resemble normal acid-base reactions in that the products are a salt and water:

$$\underset{\substack{\text{acidic} \\ \text{oxide}}}{CO_2(g)} + \underset{\text{base}}{2NaOH(aq)} \longrightarrow \underset{\text{salt}}{Na_2CO_3(aq)} + \underset{\text{water}}{H_2O(l)}$$

$$\underset{\substack{\text{basic} \\ \text{oxide}}}{BaO(s)} + \underset{\text{acid}}{2HNO_3(aq)} \longrightarrow \underset{\text{salt}}{Ba(NO_3)_2(aq)} + \underset{\text{water}}{H_2O(l)}$$

Note that BeO is also amphoteric. Be resembles Al as a result of the diagonal relationship (p. 267).

As Figure 18.5 shows, aluminum oxide (Al_2O_3) is an amphoteric oxide. Depending on the reaction conditions, it can behave either as an acidic oxide or a basic oxide. For example, Al_2O_3 acts as a base with hydrochloric acid

$$Al_2O_3(s) + 6HCl(aq) \longrightarrow 2AlCl_3(aq) + 3H_2O(l)$$

and acts as an acid with sodium hydroxide

$$Al_2O_3(s) + 2NaOH(aq) + 3H_2O(l) \longrightarrow 2NaAl(OH)_4(aq)$$

Note that only a salt, $NaAl(OH)_4$ [containing the Na^+ and $Al(OH)_4^-$ ions], is formed in the latter reaction; no water is produced. Nevertheless, this reaction can still be classified as an acid-base reaction because Al_2O_3 neutralizes NaOH.

Some transition metal oxides in which the metal has a high oxidation number act as acidic oxides. Two examples are manganese(VII) oxide (Mn_2O_7) and chromium(VI) oxide (CrO_3), both of which react with water to produce acids:

$$Mn_2O_7(l) + H_2O(l) \longrightarrow 2HMnO_4(aq)$$
$$\text{permanganic acid}$$

$$CrO_3(s) + H_2O(l) \longrightarrow H_2CrO_4(aq)$$
$$\text{chromic acid}$$

Basic and Amphoteric Hydroxides

The alkali and alkaline earth metal hydroxides [except $Be(OH)_2$], as we have seen, are basic in properties. A number of hydroxides, especially those in Groups 3A and 4A, are amphoteric. For example, aluminum hydroxide reacts with both acids and bases. The net ionic equations are as follows:

$$Al(OH)_3(s) + 3H^+(aq) \longrightarrow Al^{3+}(aq) + 3H_2O(l)$$

$$Al(OH)_3(s) + OH^-(aq) \rightleftharpoons Al(OH)_4^-(aq)$$

All amphoteric hydroxides are insoluble compounds.

It is interesting to note that beryllium hydroxide, like aluminum hydroxide, also exhibits amphoterism:

$$Be(OH)_2(s) + 2H^+(aq) \longrightarrow Be^{2+}(aq) + 2H_2O(l)$$

$$Be(OH)_2(s) + 2OH^-(aq) \rightleftharpoons Be(OH)_4^-(aq)$$

This is another example of the diagonal relationship between Be and Al.

Figure 18.6 shows the classification of metal hydroxides according to the positions of the metals in the periodic table. While many metal hydroxides are basic and some are amphoteric, there are no metal hydroxides that exhibit only acidic properties.

18.8 LEWIS ACIDS AND BASES

Acid-base properties so far have been discussed in terms of the Brønsted–Lowry theory. To behave as a Brønsted–Lowry base, for example, a substance must be able to accept protons. Thus both the hydroxide ion and ammonia are bases:

$$H^+ + {}^-\!:\!\ddot{O}\!-\!H \longrightarrow H\!-\!\ddot{O}\!-\!H$$

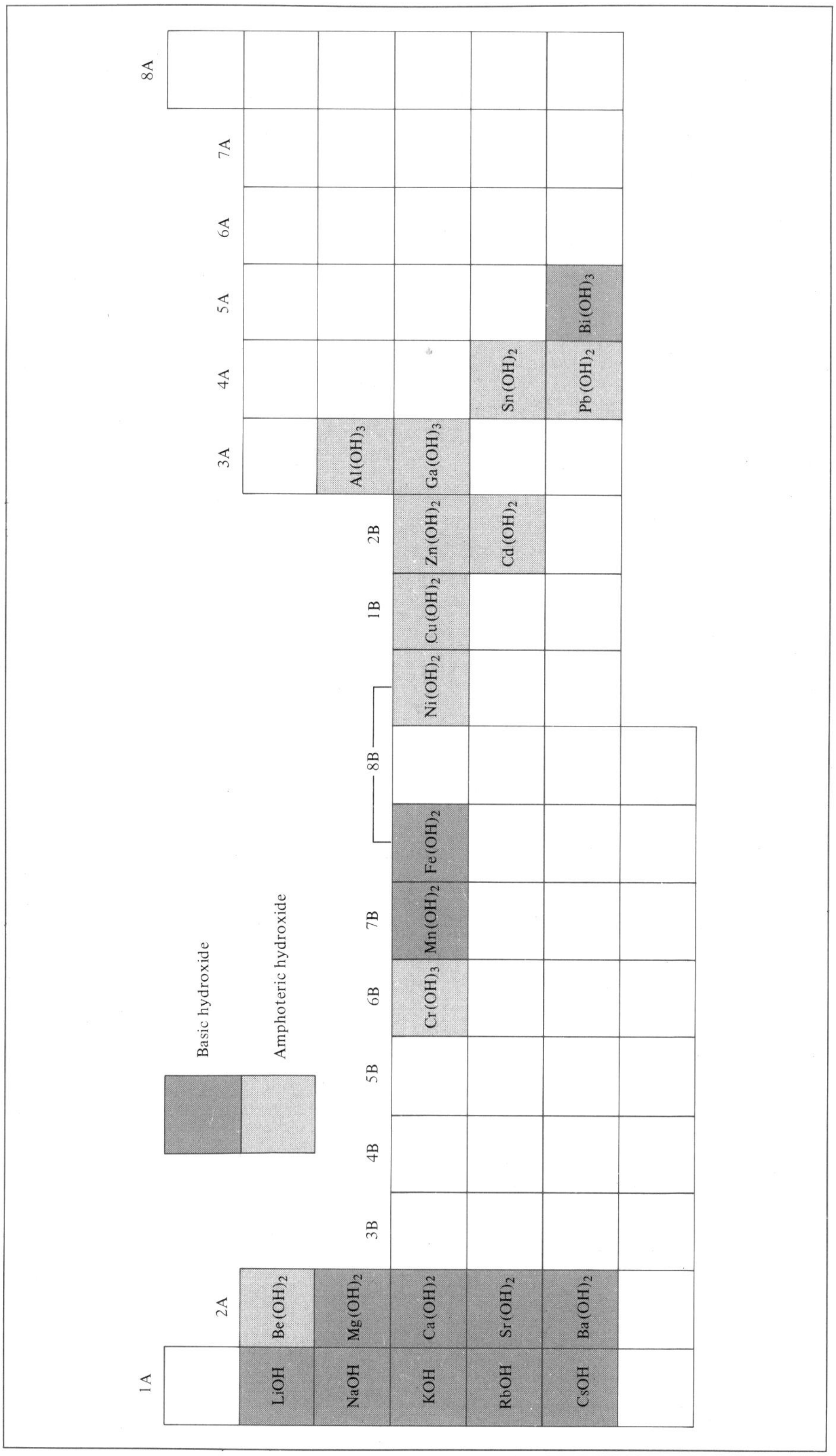

FIGURE 18.6 *Classification of metal hydroxides.*

In each case, the atom to which the proton becomes attached possesses at least one unshared pair of electrons. This characteristic property of the OH^- ion, of NH_3, and of other Brønsted–Lowry bases suggests a more general definition of acids and bases.

G. N. Lewis formulated such a definition. According to Lewis's definition, a base is *a substance that can donate a pair of electrons,* and an acid is *a substance that can accept a pair of electrons.* For example, in the protonation of NH_3 discussed above, NH_3 acts as a **Lewis base** because it donates a pair of electrons to the proton H^+, which acts as a **Lewis acid** by accepting the pair of electrons.

The Lewis concept significantly broadens the list of substances that can be classified as acids and bases. Consider, for example, the following reaction between boron trifluoride (BF_3) and ammonia:

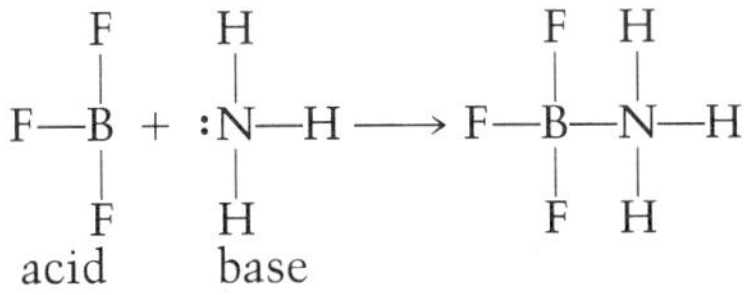

In Section 12.5 we saw that the B atom in BF_3 is sp^2-hybridized. The vacant, unhybridized $2p$ orbital accepts the pair of electrons from NH_3. So we see that BF_3 functions as an acid within the Lewis framework, even though it does not contain an ionizable proton.

Another Lewis acid containing boron is boric acid. Boric acid (a weak acid used in eyewash) is an oxyacid with the following structure:

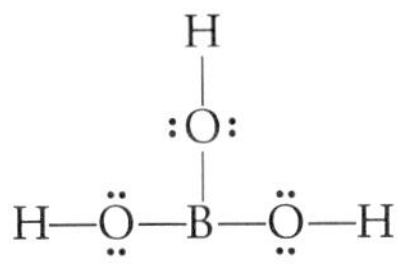

However, unlike other oxyacids shown in Figure 18.3, boric acid does not ionize in water to produce a H^+ ion. Instead, its reaction with water is

$$B(OH)_3(aq) + H_2O(l) \rightleftharpoons B(OH)_4^-(aq) + H^+(aq)$$

Thus it behaves as a Lewis acid by accepting a pair of electrons from the hydroxide ion that is derived from the H_2O molecule.

The hydration of carbon dioxide to produce carbonic acid

$$CO_2(g) + H_2O(l) \rightleftharpoons H_2CO_3(aq)$$

can be understood in Lewis's framework as follows. The first step involves the donation of a lone pair on the oxygen atom in H_2O to the carbon atom in CO_2. (An orbital is vacated on the C atom to accommodate the lone pair by removing the electron pair in the C—O pi bond.)

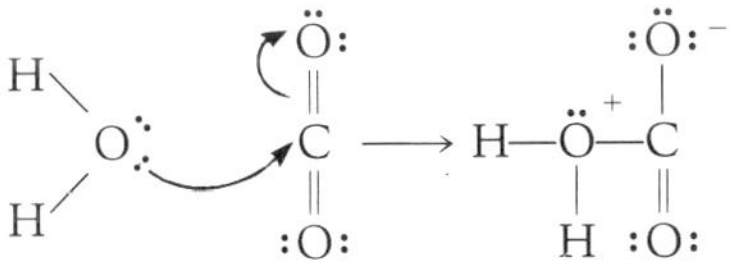

Lewis presented his theory of acids and bases in 1923, the same year in which Brønsted and Lowry introduced theirs.

All Brønsted–Lowry bases are Lewis bases.

Lewis acid-base reactions in which the acid does not contain an ionizable hydrogen atom do not lead to the formation of a salt and water.

Therefore, H_2O is a Lewis base and CO_2 is a Lewis acid. Next, a proton is transferred onto the O atom bearing a negative charge to form H_2CO_3.

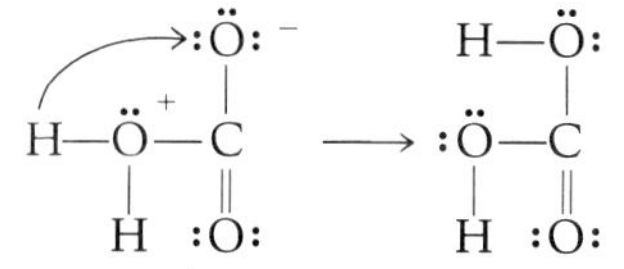

The shifts of the electrons and the proton are indicated by the curved arrows.

Other examples of Lewis acid-base reactions are

$$Ag^+(aq) + 2NH_3(aq) \rightleftharpoons Ag(NH_3)_2^+(aq)$$
$$\text{acid} \qquad\quad \text{base}$$

$$Cd^{2+}(aq) + 4I^-(aq) \rightleftharpoons CdI_4^{2-}(aq)$$
$$\text{acid} \qquad\quad \text{base}$$

$$Ni(s) + 4CO(g) \rightleftharpoons Ni(CO)_4(g)$$
$$\text{acid} \qquad \text{base}$$

It is important to note that the hydration of metal ions in solution is in itself a Lewis acid-base reaction. Thus, when copper(II) sulfate ($CuSO_4$) dissolves in water, each Cu^{2+} ion is associated with six water molecules as $Cu(H_2O)_6^{2+}$. In this case, the Cu^{2+} ion acts as the acid and the H_2O molecules act as the base.

Although the Lewis definition of acids and bases is of greater significance because of its generality, we normally speak of "an acid" and "a base" in terms of the Brønsted–Lowry definition. The term "Lewis acid" usually is reserved for substances that can accept a pair of electrons but do not contain ionizable hydrogen atoms.

EXAMPLE 18.9

Identify the Lewis acid and Lewis base in each of the following reactions:
(a) $SnCl_4(s) + 2Cl^-(aq) \rightleftharpoons SnCl_6^{2-}(aq)$

(b) $Hg^{2+}(aq) + 4CN^-(aq) \rightleftharpoons Hg(CN)_4^{2-}(aq)$

(c) $Co^{3+}(aq) + 6NH_3(aq) \rightleftharpoons Co(NH_3)_6^{3+}(aq)$

Answer

(a) In this reaction, $SnCl_4$ accepts two pairs of electrons from the Cl^- ions. Therefore $SnCl_4$ is the Lewis acid and Cl^- is the Lewis base.
(b) Here the Hg^{2+} ion accepts four pairs of electrons from the CN^- ions. Therefore Hg^{2+} is the Lewis acid and CN^- is the Lewis base.
(c) In this reaction, the Co^{3+} ion accepts six pairs of electrons from the NH_3 molecules. Therefore, Co^{3+} is the Lewis acid and NH_3 is the Lewis base.

Similar example: Problem 18.39.

AN ASIDE ON THE PERIODIC TABLE
Common Inorganic Binary Acids and Oxyacids

Having discussed the general properties of acids and bases in this chapter, we can survey the common inorganic acids, some of which you have probably used in the laboratory (for example, HCl, HNO_3, H_2SO_4, and H_3PO_4). Figure 18.7 shows these acids arranged according to the positions of the central elements in the periodic table. As you can see, in most cases the central atom belongs to an electronegative nonmetallic element (Groups 3A to 7A). Elements such as nitrogen, phosphorus, sulfur, and the halogens form the largest number of binary acids and oxyacids. The oxyacids that contain a central metal atom are chromic acid (H_2CrO_4) and permanganic acid ($HMnO_4$). In each case the metal atom is in its highest oxidation state ($+6$ for Cr and $+7$ for Mn). In studying these acids, you should pay particular attention to the following: (1) their nomenclature, (2) their relative strengths, and (3) their oxidizing ability. The nomenclature of a number of acids was introduced in Chapter 2. Of all the acids shown in Figure 18.7, only HNO_3, H_2SO_4 and $HClO_4$ are commonly available strong oxidizing agents. The oxidizing properties of HNO_3, H_2SO_4 and the MNO_4^- ion were discussed in Chapter 16.

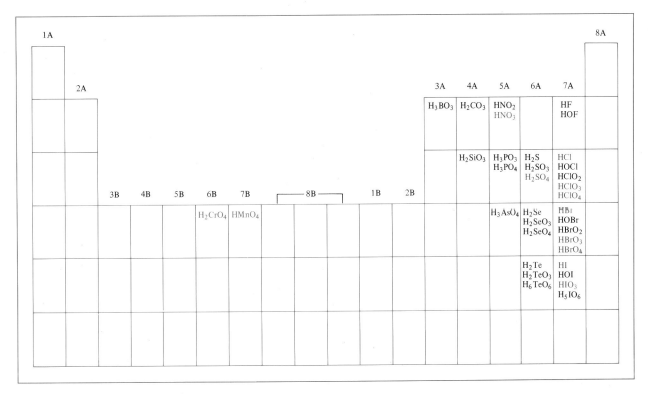

FIGURE 18.7   Some common binary acids and oxyacids. Strong acids are shown in color.

SUMMARY

1. Arrhenius acids increase hydrogen ion concentration in aqueous solution and Arrhenius bases increase hydroxide ion concentration in solution.
2. Brønsted–Lowry acids donate protons and Brønsted–Lowry bases accept protons. This is the definition normally involved when the terms "acid" and "base" are used.
3. The acidity of an aqueous solution is expressed as its pH, which is defined as the negative logarithm of the hydrogen ion concentration (in mol/L).
4. At 25°C, an acidic solution has pH < 7, a basic solution has pH > 7, and a neutral solution has pH = 7.
5. The strength of an acid or a base is measured by the extent of its ionization in solution.
6. The solvent plays an important role in determining the strength of an acid. The leveling effect is the inability of a solvent to differentiate among the relative strengths of all acids stronger than its conjugate acid.
7. In aqueous solution, the following are classified as strong acids: $HClO_4$, HI, HBr, HCl, H_2SO_4 (first stage of ionization), and HNO_3. The following are classified as strong bases in aqueous solution: hydroxides of alkali metals and alkaline earth metals (except beryllium).
8. The relative strengths of acids can be qualitatively explained in terms of their molecular structures.
9. Most oxides can be classified as acidic, basic, or amphoteric. Metal hydroxides are either basic or amphoteric.
10. Lewis acids accept pairs of electrons and Lewis bases donate pairs of electrons. The term "Lewis acid" is often reserved for substances that can accept electron pairs but do not contain ionizable hydrogen atoms.

KEY WORDS

Binary acid, p. 548
Brønsted–Lowry acid, p. 537
Brønsted–Lowry base, p. 537
Conjugate acid-base pair, p. 537
Ion-product constant, p. 540

Leveling effect, p. 547
Lewis acid, p. 557
Lewis base, p. 557
pH, p. 541
Ternary acid, p. 548

PROBLEMS

More challenging problems are marked with an asterisk.

Brønsted–Lowry Acids and Bases

18.1 Define Brønsted–Lowry acids and bases. How do Brønsted–Lowry's definitions differ from Arrhenius's definitions of acids and bases?

18.2 Classify each of the following species as a Brønsted–Lowry acid or base, or both: (a) H_2O, (b) OH^-, (c) H_3O^+, (d) NH_3, (e) NH_4^+, (f) NH_2^-, (g) NO_3^-, (h) CO_3^{2-}, (i) HBr, (j) HCN.

18.3 What are the names and formulas of the conjugate bases of the following acids? (a) HNO_2, (b) H_2SO_4, (c) H_2S, (d) HCN, (e) HCOOH (formic acid)

18.4 Write balanced ionic equations for the following: (a) NaOH solution reacts with NH_4Cl to form H_2O and the conjugate base of NH_4^+. (b) HCl solution reacts with CH_3COONa to form the conjugate acid of CH_3COO^-.

18.5 Identify the acid-base conjugate pairs in each of the following reactions:

(a) $CH_3COO^- + HCN \rightleftharpoons CH_3COOH + CN^-$
(b) $NH_2^- + NH_3 \rightleftharpoons NH_3 + NH_2^-$
(c) $HF + NH_3 \rightleftharpoons NH_4^+ + F^-$
(d) $HCO_3^- + HCO_3^- \rightleftharpoons H_2CO_3 + CO_3^{2-}$
(e) $H_2PO_4^- + NH_3 \rightleftharpoons HPO_4^{2-} + NH_4^+$
(f) $HClO + CH_3NH_2 \rightleftharpoons CH_3NH_3^+ + ClO^-$
(g) $CO_3^{2-} + H_2O \rightleftharpoons HCO_3^- + OH^-$
(h) $Zn(OH)_2 + 2OH^- \rightleftharpoons ZnO_2^{2-} + 2H_2O$
(i) $CH_3COO^- + H_2O \rightleftharpoons CH_3COOH + OH^-$

18.6 Give the conjugate acid of each of the following bases: (a) HS^-, (b) HCO_3^-, (c) CO_3^{2-}, (d) $H_2PO_4^-$, (e) HPO_4^{2-}, (f) PO_4^{3-}, (g) HSO_4^-, (h) SO_4^{2-}, (i) HSO_3^-, (j) SO_3^{2-}.

18.7 Give the conjugate base of each of the following acids: (a) $CH_2ClCOOH$, (b) HIO_4, (c) H_3PO_4, (d) $H_2PO_4^-$, (e) HPO_4^{2-}, (f) H_2SO_4, (g) HSO_4^-, (h) H_2SO_3, (i) HSO_3^-, (j) NH_4^+, (k) H_2S, (l) HS^-, (m) $HOCl$.

pH and pOH Calculations

18.8 Indicate whether the following solutions are acidic, basic, or neutral at 25°C: (a) 0.62 M NaOH, (b) 1.4×10^{-3} M HCl, (c) 2.5×10^{-11} M H^+, (d) 3.3×10^{-10} M OH^-, (e) 1.0×10^{-7} M OH^-.

18.9 Calculate the hydrogen ion concentration for solutions with the following pH values: (a) 2.42, (b) 11.21, (c) 6.96, (d) 15.00.

18.10 Calculate the hydrogen ion concentration in mol/L for each of the following solutions: (a) a solution whose pH is 5.20, (b) a solution whose pH is 16.00, (c) a solution whose hydroxide concentration is 3.7×10^{-9} M.

18.11 Calculate the pH of each of the following solutions: (a) 0.0010 M HCl, (b) 0.76 M KOH, (c) 2.8×10^{-4} M $Ba(OH)_2$, (d) 5.2×10^{-4} M HNO_3.

18.12 How much NaOH (in grams) is needed to prepare a 546 mL solution with a pH of 10.00?

18.13 Is it possible for a solution to have a negative pH?

18.14 A solution is made by dissolving 18.4 g of HCl in 662 mL of water. Calculate the pH of the solution. (Assume that the volume of the solution is also 662 mL.)

18.15 Consider the process

$$H_2O(l) \rightleftharpoons H^+(aq) + OH^-(aq) \qquad \Delta H° = 38.7 \text{ kJ}$$

Will $[H^+]$ be greater or smaller than 1.0×10^{-7} M at 30°C?

18.16 Calculate the pH of water at 40°C, given that K_w is 3.8×10^{-14} at this temperature.

18.17 Fill in the word acidic, basic, or neutral for the following cases:

(a) pOH > 7; solution is _____
(b) pOH = 7; solution is _____
(c) pOH < 7; solution is _____

18.18 The pOH of a solution is 9.40. Calculate the hydrogen ion concentration of the solution.

18.19 Calculate the number of moles of KOH in 5.50 mL of a 0.360 M KOH solution. What is the pOH of the solution?

Strengths of Acids and Bases

18.20 Explain what is meant by the strength of an acid.

18.21 Classify each of the following species as a weak or strong acid: (a) HNO_3, (b) HF, (c) H_2SO_4, (d) HSO_4^-, (e) H_2CO_3, (f) HCO_3^-, (g) HCl, (h) HCN, (i) HNO_2.

18.22 Classify each of the following species as a weak or strong base: (a) LiOH, (b) CN^-, (c) H_2O, (d) ClO_4^-, (e) NH_2^-.

18.23 You are given two aqueous solutions containing a strong acid (HA) and a weak acid (HB), respectively. Describe how you would compare the strengths of these two acids by (a) pH measurement, (b) electrical conductance measurement, (c) studying the rate of hydrogen gas evolution when these solutions are reacted with an active metal such as Mg or Zn.

18.24 Which of the following statements are true regarding a 0.10 M solution of a weak acid HA?

(a) The pH is 1.00.
(b) $[H^+] \gg [A^-]$
(c) $[H^+] = [A^-]$
(d) The pH is less than 1.

18.25 Which of the following statements are true regarding a 1.0 M solution of a strong acid HA?

(a) $[A^-] > [H^+]$
(b) The pH is 0.00.
(c) $[H^+] = 1.0$ M
(d) $[HA] = 1.0$ M

18.26 Predict the direction that predominates in this reaction:

$$CH_3COO^-(aq) + H_2O(l) \rightleftharpoons$$
$$CH_3COOH(aq) + OH^-(aq)$$

18.27 What is the strongest acid that can exist in (a) water, (b) glacial acetic acid, (c) liquid ammonia?

18.28 What is the strongest base that can exist in (a) water, (b) liquid ammonia?

18.29 Without referring to the book, write the formulas of all the strong acids and strong bases you have learned so far in this course.

18.30 Predict whether the following reaction will proceed from left to right to any measurable extent:

$$CH_3COOH(aq) + Cl^-(aq) \longrightarrow$$

18.31 H_2SO_4 is a strong acid and HSO_4^- is a weak acid. Account for this difference.

18.32 What is meant by the leveling effect?

18.33 Compare the strengths of the following pairs of oxyacids: (a) H_2SO_4 and H_2SeO_4, (b) H_2SO_3 and H_2SeO_3, (c) H_3PO_4 and H_3AsO_4, (d) $HBrO_4$ and HIO_4.

18.34 Predict the acid strengths of the following compounds: H_2O, H_2S, and H_2Se.

*18.35 Which of the following acids is the stronger: CH_3COOH or $CH_2ClCOOH$? Explain your choice.

*18.36 Consider the following compounds:

phenol   methanol

Experimentally it is found that phenol is a stronger acid than methanol. Explain this difference in terms of the structures of the conjugate bases. (*Hint:* A more stable conjugate base favors ionization. Only one of the conjugate bases can be stabilized by resonance.)

Lewis Acids and Bases

18.37 What are the Lewis definitions of an acid and a base? In what way are they more general than the Brønsted–Lowry definitions?

18.38 Classify each of the following species as a Lewis acid or a Lewis base: (a) CO_2, (b) H_2O, (c) I^-, (d) SO_2, (e) NH_3, (f) OH^-, (g) H^+, (h) BCl_3.

18.39 Describe the following reaction according to the Lewis theory of acids and bases:

$$AlCl_3(s) + Cl^-(aq) \longrightarrow AlCl_4^-(aq)$$

18.40 Which would be considered a stronger Lewis acid: (a) BF_3 or BCl_3, (b) Fe^{2+} or Fe^{3+}? Explain.

*18.41 Describe the hydration of SO_2 as a Lewis acid-base reaction. (*Hint:* Follow the procedure for the hydration of CO_2 in Section 18.8.)

Miscellaneous Problems

18.42 Write balanced equations for the following neutralization reactions: (a) $KOH + H_3PO_4$, (b) $NH_3 + H_2SO_4$, (c) $Mg(OH)_2 + HClO_4$, (d) $NaOH + H_2CO_3$.

18.43 Use the ionization of HCN in water as an example to illustrate the meaning of dynamic equilibrium.

18.44 Give an example of (a) a weak acid that contains oxygen atoms, (b) a weak acid that does not contain oxygen atoms, (c) a neutral molecule that acts as a Lewis acid, (d) a neutral molecule that acts as a Lewis base, (e) a weak acid that contains two ionizable H atoms, (f) a conjugate acid-base pair, both of which react with HCl to give carbon dioxide gas.

18.45 Is the concentration of H^+ ions from the autoionization of water in a 0.10 M HCl solution greater than, equal to, or smaller than 1.0×10^{-7} M? (*Hint:* Apply Le Chatelier's principle.)

18.46 A typical reaction between an antacid and gastric juice is

$$NaHCO_3(aq) + HCl(aq) \longrightarrow$$
$$NaCl(aq) + H_2O(l) + CO_2(g)$$

Calculate the volume (in L) of carbon dioxide generated from 0.350 g of sodium bicarbonate and excess gastric juice at 1.00 atm and 37.0°C.

18.47 Arrange the oxides in each of the following groups in order of increasing basicity: (a) K_2O, Al_2O_3, BaO, (b) CrO_3, CrO, Cr_2O_3.

*18.48 Explain why metal oxides tend to be basic if the oxidation number of the metal is low and acidic if the oxidation number of the metal is high. (*Hint:* Metallic compounds with low oxidation numbers of the metals are more ionic than those in which the oxidation numbers of the metals are high.)

18.49 $Zn(OH)_2$ is an amphoteric hydroxide. Write balanced ionic equations to show its reaction with (a) HCl and (b) NaOH [the product is $Zn(OH)_4^{2-}$].

18.50 $Al(OH)_3$ is an insoluble compound. It dissolves in excess NaOH in solution. Write a balanced ionic equation for this reaction and classify the reaction.

18.51 Explain why all Brønsted–Lowry bases are Lewis bases.

19

ACIDS AND BASES II: ACID AND BASE IONIZATIONS AND SALT HYDROLYSIS

We have already noted that the solvent water has the capacity to act either as a Brønsted–Lowry acid or a Brønsted–Lowry base, under certain conditions. In this chapter we will examine a variety of related acid base reactions involving water's direct participation either as a proton donor or proton acceptor. These reactions fall into two major categories of chemical change—acid-base ionization reactions and hydrolysis reactions.

In both cases, direct application of acid-base definitions and equilibrium constant expressions leads to useful insights that help us understand these important families of chemical reaction. This chapter will thus extend your ability to deal with the "arithmetic" of acid-base systems, as well as clarify some important molecular-level perspectives on acids and bases.

563

19.1 WEAK ACIDS AND ACID IONIZATION CONSTANTS

In Chapter 18 we discussed acid strength in terms of molecular structure. Quantitative measure of the strength of an acid is based on the equilibrium concept. Consider a weak monoprotic acid, HA. When it dissolves in water and reaches equilibrium, we have

$$HA(aq) + H_2O(l) \rightleftharpoons H_3O^+(aq) + A^-(aq)$$

or simply

$$HA(aq) \rightleftharpoons H^+(aq) + A^-(aq)$$

The equilibrium constant for this *acid ionization,* which we will call the **acid ionization constant, K_a,** is given by

$$K_a = \frac{[H^+][A^-]}{[HA]}$$

All concentrations in this equation are equilibrium concentrations.

At a given temperature, the strength of the acid HA is measured quantitatively by the magnitude of K_a. The larger K_a, the stronger the acid—that is, the greater the concentration of H^+ ions at equilibrium due to its ionization.

In strong acids, such as $HClO_4$ and HNO_3, the molecules are completely ionized, and calculation of the concentration of H^+ ions in solution is straightforward, if we know the initial acid concentration (see Example 18.6). Since there is no measurable concentration of nonionized acid molecules at equilibrium, K_a must be extremely large. Thus any acid stronger than H_3O^+ (see Table 18.3) is simply labeled a strong acid in aqueous solution; it does not have a K_a value.

Any acid weaker than H_3O^+ is a weak acid in aqueous solution. Because the ionization of weak acids is never complete, all species (the nonionized acid, H^+ ions, and A^- ions) are present at equilibrium (Figure 19.1). Table 19.1 lists values of K_a for a number of weak acids. Note that within this group there is actually great variation in acid strength. For example, even though both hydrofluoric acid (HF) and hydrocyanic acid (HCN) are classified as weak acids (see Table 18.3). HF ($K_a = 7.1 \times 10^{-4}$) is considerably stronger than HCN ($K_a = 4.9 \times 10^{-10}$).

We can calculate K_a from the initial concentration of the acid and the pH of the solution, and we can use K_a and the initial concentration of the acid to calculate equilibrium concentrations of all the species and pH of solution. In calculating the equilibrium concentrations in a weak acid, we follow essentially the same procedure that was outlined in Section 17.4; the systems may be different, but the calculations are based on the same principle, the law of mass action [Equation (17.2)]. The three basic steps are:

1. Express the equilibrium concentrations of all species in terms of a single unknown, which we call x.
2. Write the acid ionization constant in terms of the equilibrium concentrations. Knowing the value of K_a, we can solve for x.

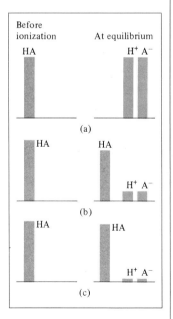

FIGURE 19.1 *Diagram illustrating the extent of ionization of (a) a strong acid that undergoes 100 percent ionization, (b) a weak acid, and (c) a very weak acid.*

TABLE 19.1 Ionization Constants of Some Weak Acids at 25°C

Name of Acid	Formula	Structure	K_a	Conjugate Base	K_b
Acetic acid	CH_3COOH	$CH_3-\overset{\displaystyle O}{\overset{\|}{C}}-O-H$	1.8×10^{-5}	CH_3COO^-	5.6×10^{-10}
Acetylsalicylic acid (aspirin)	$C_9H_8O_4$	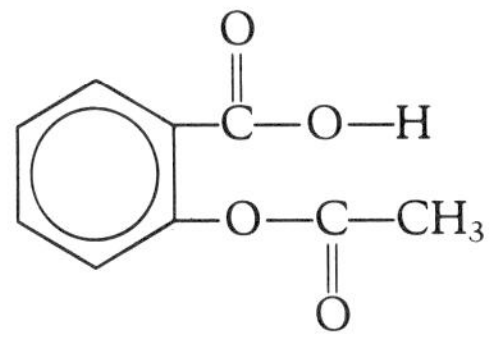	3.0×10^{-4}	$C_9H_7O_4^-$	3.3×10^{-11}
Ascorbic acid*	$C_6H_8O_6$	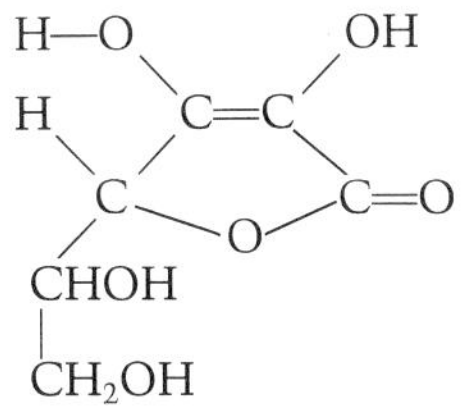	8.0×10^{-5}	$C_6H_7O_6^-$	1.3×10^{-10}
Benzoic acid	C_6H_5COOH	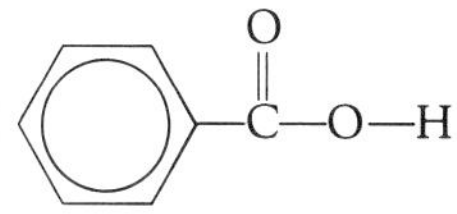	6.5×10^{-5}	$C_6H_5COO^-$	1.5×10^{-10}
Formic acid	$HCOOH$	$H-\overset{\displaystyle O}{\overset{\|}{C}}-O-H$	1.7×10^{-4}	$HCOO^-$	5.9×10^{-11}
Hydrocyanic acid	HCN	$H-C\equiv N$	4.9×10^{-10}	CN^-	2.0×10^{-5}
Hydrofluoric acid	HF	$H-F$	7.1×10^{-4}	F^-	1.4×10^{-11}
Nitrous acid	HNO_2	$O=N-O-H$	4.5×10^{-4}	NO_2^-	2.2×10^{-11}
Phenol	C_6H_5OH	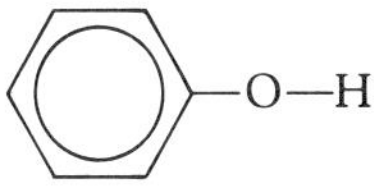	1.3×10^{-10}	$C_6H_5O^-$	7.7×10^{-5}

* For ascorbic acid it is the upper left hydroxyl group that is associated with the ionization constant.

3. Having solved for x, calculate the equilibrium concentrations of all species and/or the pH of the solution.

Unless otherwise stated, we will assume that the temperature is 25°C for all such calculations.

EXAMPLE 19.1

Calculate the concentrations of the nonionized acid and the ions of a 0.100 M formic acid (HCOOH) solution at equilibrium.

(Continued)

Answer

Step 1

Table 19.1 shows that HCOOH is a weak acid. Since it is monoprotic, one HCOOH molecule ionizes to give one H^+ ion and one $HCOO^-$ ion. Let x be the equilibrium concentration of H^+ and $HCOO^-$ ions in mol/L. Then the equilibrium concentration of HCOOH must be $(0.100 - x)$ mol/L or $(0.100 - x)$ M. We can now summarize the changes in concentrations as follows:

	$HCOOH(aq) \rightleftharpoons$	$H^+(aq) +$	$HCOO^-(aq)$
Initial	0.100 M	0.000 M	0.000 M
Change	$-x$ M	$+x$ M	$+x$ M
Equilibrium	$(0.100 - x)$ M	x M	x M

Step 2

Referring to Table 19.1

$$K_a = \frac{[H^+][HCOO^-]}{[HCOOH]} = 1.7 \times 10^{-4}$$

$$\frac{x^2}{0.100 - x} = 1.7 \times 10^{-4}$$

This equation can be rewritten as

$$x^2 + 1.7 \times 10^{-4}x - 1.7 \times 10^{-5} = 0$$

See Appendix 4 for a discussion of the quadratic equation.

which fits a quadratic equation of the form $ax^2 + bx + c = 0$.

To avoid solving a quadratic equation we can often apply a simplifying approximation to this type of problem. Since HCOOH is a weak acid, the extent of its ionization must be small. Therefore x is a small number compared to 0.100. As a general rule, if the quantity x, which is subtracted from the original concentration of the acid (0.100 M in this case), is equal to or less than 5 percent of the original concentration, we can assume $0.100 - x \simeq 0.100$. This removes x from the denominator of the equilibrium constant expression and we avoid a quadratic equation. If x is more than 5 percent of the original value, then we must solve the quadratic equation. Normally we can apply the approximation in cases where K_a is small (equal to or less than 1×10^{-4}) and the initial concentration of the acid is high (equal to or greater than 0.1 M). In case of doubt, we can always solve for x by the approximate method and then check the validity of our approximation. Assuming that $0.100 - x \simeq 0.100$, then

The $\simeq$ sign means "approximately equal to."

$$\frac{x^2}{0.100 - x} \simeq \frac{x^2}{0.100} = 1.7 \times 10^{-4}$$
$$x^2 = 1.7 \times 10^{-5}$$

Taking the square roots on both sides, we obtain

$$x = 4.1 \times 10^{-3} \, M$$

Step 3

At equilibrium, therefore

$$[H^+] = 4.1 \times 10^{-3} \, M$$
$$[HCOO^-] = 4.1 \times 10^{-3} \, M$$
$$[HCOOH] = (0.100 - 0.0041) \, M$$
$$= 0.096 \, M$$

To check the validity of our approximation

$$\frac{0.0041 \ M}{0.100 \ M} \times 100\% = 4.1\%$$

This shows that the quantity x is less than 5 percent of the original concentration of the acid. Thus our approximation was justified.

Note that we omitted the contribution to $[H^+]$ by water. Except in very dilute acid solutions, this omission is always valid, because the hydrogen ion concentration due to water is negligibly small compared to that due to the acid.

Similar example: Problem 19.1.

EXAMPLE 19.2

Calculate the pH of a 0.050 M nitrous acid (HNO_2) solution.

Answer

Step 1

From Table 19.1 we see that HNO_2 is a weak acid. Letting x be the equilibrium concentration of H^+ and NO_2^- ions in mol/L, we summarize:

	$HNO_2(aq)$	$\rightleftharpoons$	$H^+(aq)$	$+ \ NO_2^-(aq)$
Initial	0.050 M		0.00 M	0.00 M
Change	$-x \ M$		$+x \ M$	$+x \ M$
Equilibrium	$(0.050 - x) \ M$		$x \ M$	$x \ M$

Step 2

Referring to Table 19.1,

$$K_a = \frac{[H^+][NO_2^-]}{[HNO_2]} = 4.5 \times 10^{-4}$$

$$\frac{x^2}{0.050 - x} = 4.5 \times 10^{-4}$$

Applying the approximation $0.050 - x \simeq 0.050$, we obtain

$$\frac{x^2}{0.050 - x} \simeq \frac{x^2}{0.050} = 4.5 \times 10^{-4}$$

$$x^2 = 2.3 \times 10^{-5}$$

Taking the square roots on both sides gives

$$x = 4.8 \times 10^{-3} \ M$$

To test the approximation

$$\frac{0.0048 \ M}{0.050 \ M} \times 100\% = 9.6\%$$

(Continued)

This shows that x is more than 5 percent of the original concentration. Thus our approximation is *not* valid, and we must solve the quadratic equation in step 2, which simplifies to

$$x^2 + 4.5 \times 10^{-4}x - 2.3 \times 10^{-5} = 0$$

Using the quadratic formula,

$$x = \frac{-b \pm \sqrt{b^2 - 4ac}}{2a}$$

$$= \frac{-4.5 \times 10^{-4} \pm \sqrt{(4.5 \times 10^{-4})^2 - 4(1)(-2.3 \times 10^{-5})}}{2(1)}$$

Thus

$$x = 4.6 \times 10^{-3}\ M \quad \text{or} \quad x = -5.0 \times 10^{-3}\ M$$

For this kind of problem, one of the solutions is always physically impossible.

The second solution is physically impossible, since the concentrations of ions produced as a result of ionization cannot be negative. Therefore, the solution to the quadratic equation is given by the positive root—that is, $x = 4.6 \times 10^{-3}\ M$.

Step 3

At equilibrium

$$[H^+] = 4.6 \times 10^{-3}\ M$$

and

$$pH = -\log(4.6 \times 10^{-3})$$
$$= 2.34$$

For the rule for handling significant figures in logarithms, see p. 541.

Similar examples: Problems 19.2, 19.4.

EXAMPLE 19.3

The pH of a 0.100 M solution of a weak monoprotic acid (HA) is 2.85. What is the K_a of the acid?

Answer

In this case we are given the pH, from which we can obtain the equilibrium concentrations, and we are asked to calculate the acid ionization constant. We can follow the same basic steps.

Step 1

First we need to calculate the hydrogen ion concentration from the pH value.

$$pH = -\log[H^+]$$
$$2.85 = -\log[H^+]$$

Taking the antilog of both sides, we get

$$[H^+] = 1.4 \times 10^{-3}\ M$$

Next we summarize the changes:

	$HA(aq)$	$\rightleftharpoons$	$H^+(aq)$	+	$A^-(aq)$
Initial	0.100 M		0.000 M		0.000 M
Change	$-0.0014\ M$		$+0.0014\ M$		$+0.0014\ M$
Equilibrium	$(0.100 - 0.0014)\ M$		0.0014 M		0.0014 M

Step 2

The acid ionization constant is given by

$$K_a = \frac{[H^+][A^-]}{[HA]} = \frac{(0.0014)(0.0014)}{(0.100 - 0.0014)}$$
$$= 2.0 \times 10^{-5}$$

Similar examples: Problems 19.2, 19.4.

Percent Ionization

Another way to compare the strengths of monoprotic acids is to compare their percent ionization at a given concentration. The percent ionization of a monoprotic acid is defined as

$$\% \text{ ionization} = \frac{\text{hydrogen ion concentration at equilibrium}}{\text{initial concentration of acid}} \times 100\%$$

$$= \frac{\text{equilibrium concentration of conjugate base}}{\text{initial concentration of acid}} \times 100\%$$

Since we are dealing with a monoprotic acid, the concentration of hydrogen ion produced by the acid must be equal to the concentration of the conjugate base (A^-) produced.

EXAMPLE 19.4

Calculate the percent ionization of (a) a 0.60 M hydrofluoric acid (HF) solution and of (b) a 0.60 M hydrocyanic acid (HCN) solution.

Answer

(a) HF

Step 1

Let x be the concentrations of H^+ and F^- ions at equilibrium in mol/L. We summarize:

	$HF(aq)$	$\rightleftharpoons$	$H^+(aq)$	+	$F^-(aq)$
Initial	0.60 M		0.00 M		0.00 M
Change	$-x\ M$		$+x\ M$		$+x\ M$
Equilibrium	$(0.60 - x)\ M$		$x\ M$		$x\ M$

(Continued)

Step 2

Referring to Table 19.1

$$K_a = \frac{[H^+][F^-]}{[HF]} = 7.1 \times 10^{-4}$$

$$\frac{x^2}{0.60 - x} = 7.1 \times 10^{-4}$$

Assuming that $0.60 - x \simeq 0.60$, then

$$\frac{x^2}{0.60 - x} \simeq \frac{x^2}{0.60} = 7.1 \times 10^{-4}$$

$$x^2 = 4.3 \times 10^{-4}$$

$$x = 0.021 \ M$$

Because HF is monoprotic, the concentration of hydrogen ions produced by the acid at equilibrium is equal to that of the F^- ions. Thus

$$\% \text{ ionization} = \frac{0.021 \ M}{0.60 \ M} \times 100\%$$

$$= 3.5\%$$

(b) HCN

Step 1

Let x be the concentrations of H^+ and CN^- ions at equilibrium in mol/L. We summarize:

	HCN(aq)	$\rightleftharpoons$	$H^+(aq)$	+ $CN^-(aq)$
Initial	0.60 M		0.00 M	0.00 M
Change	−x M		+x M	+x M
Equilibrium	(0.60 − x) M		x M	x M

Step 2

Referring to Table 19.1, we obtain

$$K_a = \frac{[H^+][CN^-]}{[HCN]} = 4.9 \times 10^{-10}$$

$$\frac{x^2}{0.60 - x} = 4.9 \times 10^{-10}$$

We assume that $0.60 - x \simeq 0.60$:

$$\frac{x^2}{0.60 - x} \simeq \frac{x^2}{0.60} = 4.9 \times 10^{-10}$$

$$x^2 = 2.9 \times 10^{-10}$$

$$x = 1.7 \times 10^{-5} \ M$$

Like HF, HCN is monoprotic. Therefore

$$\% \text{ ionization} = \frac{1.7 \times 10^{-5} \ M}{0.60 \ M} \times 100\%$$

$$= 0.0028\%$$

Thus, at the same concentration, HF ionizes to a much greater extent than does HCN. Note that the form of the percent ionization calculation is exactly the same as the calculation used to check the approximation that avoids a quadratic equation solution. Since the simplifying assumption is regarded as valid for acids whose percent ionization is less than 5 percent, both calculations in Example 19.4 qualify.

Similar examples: Problems 19.5, 19.6.

The extent to which a weak acid ionizes depends on the initial concentration of the acid. The more dilute the solution, the greater the percent ionization (Figure 19.2). In qualitative terms, when an acid is diluted, initially the number of particles (nonionized acid molecules plus ions) per unit volume is reduced. According to Le Chatelier's principle (see Section 17.5), to counteract this "stress" (that is, the dilution), the equilibrium shifts from unionized acid to H^+ and its conjugate base to produce more particles (ions).

It is important to keep in mind that at a constant temperature the equilibrium constant K_a, unlike the percent ionization, is independent of the acid concentration.

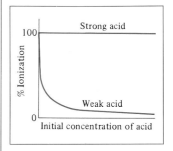

FIGURE 19.2 *Dependence of percent ionization of acids on initial concentration. Note that at very low concentrations all acids (weak and strong) are almost completely ionized.*

EXAMPLE 19.5

Compare the percent ionization of HF at 0.60 *M* and at 0.00060 *M*.

Answer

In Example 19.4 we found the percent ionization of 0.60 *M* HF to be 3.5 percent. In order to compare we must find the percent ionization of 0.00060 *M* HF.

Step 1

Let *x* be the concentration in mol/L of H^+ and F^- at equilibrium. We summarize the changes:

	HF(*aq*)	$\rightleftharpoons$	H^+(*aq*) +	F^-(*aq*)
Initial	0.00060 *M*		0.00 *M*	0.00 *M*
Change	$-x$ *M*		$+x$ *M*	$+x$ *M*
Equilibrium	(0.00060 $-$ *x*) *M*		*x M*	*x M*

Step 2

$$\frac{x^2}{0.00060 - x} = 7.1 \times 10^{-4}$$

Since the concentration of the acid is very low and the ionization constant is fairly large, the approximation method is not applicable. We express the equation in quadratic form, and then substitute in the quadratic formula:

(Continued)

$$x^2 + 7.1 \times 10^{-4}x - 4.3 \times 10^{-7} = 0$$

$$x = \frac{-7.1 \times 10^{-4} \pm \sqrt{(7.1 \times 10^{-4})^2 - 4(1)(-4.3 \times 10^{-7})}}{2(1)}$$

$$x = 3.9 \times 10^{-4} \, M$$

Therefore

$$\text{\% ionization} = \frac{3.9 \times 10^{-4} \, M}{0.00060 \, M} \times 100\%$$
$$= 65\%$$

The large percent ionization here shows clearly that the approximation $0.00060 - x \simeq 0.00060$ would not be valid. It also confirms our prediction that the extent of acid ionization increases with dilution.

Similar examples: Problems 19.5, 19.6.

19.2 WEAK BASES AND BASE IONIZATION CONSTANTS

Strong bases such as the alkali metal and alkaline earth metal (except beryllium) hydroxides are completely ionized in water:

$$\text{NaOH}(aq) \longrightarrow \text{Na}^+(aq) + \text{OH}^-(aq)$$

$$\text{KOH}(aq) \longrightarrow \text{K}^+(aq) + \text{OH}^-(aq)$$

$$\text{Ba(OH)}_2(aq) \longrightarrow \text{Ba}^{2+}(aq) + 2\text{OH}^-(aq)$$

pH calculations for these solutions are straightforward (see Example 18.6).
   Weak bases are treated like weak acids. When ammonia dissolves in water, it undergoes the reaction

$$\text{NH}_3(aq) + \text{H}_2\text{O}(l) \rightleftharpoons \text{NH}_4^+(aq) + \text{OH}^-(aq)$$

The production of hydroxide ions in this reaction (which we call *base ionization*) means that, in this solution at 25°C, $[\text{OH}^-] > [\text{H}^+]$, and therefore pH > 7.
   The equilibrium constant for the reaction is

$$K = \frac{[\text{NH}_4^+][\text{OH}^-]}{[\text{NH}_3][\text{H}_2\text{O}]}$$

Again, because very few water molecules are consumed by the reaction (compared to the total concentration of water), we can treat $[\text{H}_2\text{O}]$ as a constant. Thus

$$K[\text{H}_2\text{O}] = K_b = \frac{[\text{NH}_4^+][\text{OH}^-]}{[\text{NH}_3]}$$
$$= 1.8 \times 10^{-5}$$

where K_b, the *equilibrium constant for base ionization*, is called the **base ionization constant.** Table 19.2 lists several common weak bases and their

Strictly speaking, NH$_3$ does not ionize in solution the way an acid does. Nevertheless, we use the term "ionization constant" to describe the reaction between NH$_3$ and H$_2$O to produce OH$^-$ ions and its conjugate acid, NH$_4^+$.

TABLE 19.2 Ionization Constants of Some Common Weak Bases at 25°C

Name of Base	Formula	Structure	K_b*	Conjugate Acid	K_a
Ammonia	NH_3	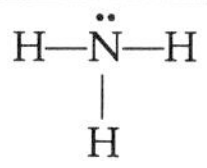 H—N̈—H with H below	1.8×10^{-5}	NH_4^+	5.6×10^{-10}
Aniline	$C_6H_5NH_2$	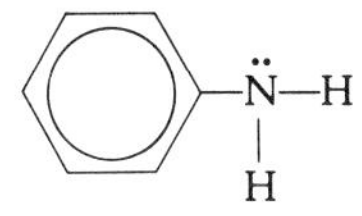 —N̈—H with H below (phenyl)	3.8×10^{-10}	$C_6H_5\overset{+}{N}H_3$	2.6×10^{-5}
Caffeine	$C_8H_{10}N_4O_2$	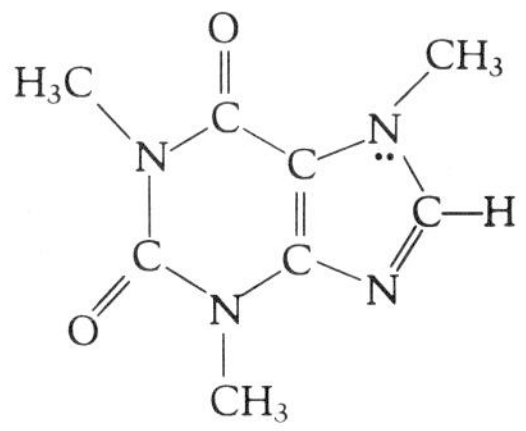	4.1×10^{-4}	$C_8H_{11}\overset{+}{N}_4O_2$	2.4×10^{-14}
Ethylamine	$C_2H_5NH_2$	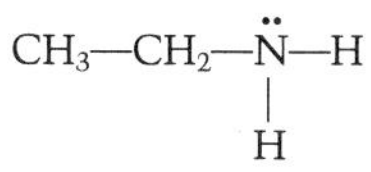 CH_3—CH_2—N̈—H with H below	5.6×10^{-4}	$C_2H_5\overset{+}{N}H_3$	1.8×10^{-11}
Methylamine	CH_3NH_2	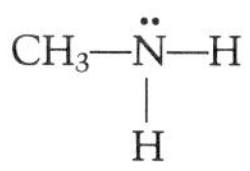 CH_3—N̈—H with H below	4.4×10^{-4}	$CH_3\overset{+}{N}H_3$	2.3×10^{-11}
Pyridine	C_5H_5N	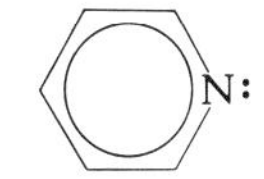 N: (ring)	1.7×10^{-9}	$C_5H_5\overset{+}{N}H$	5.9×10^{-6}
Urea	N_2H_4CO	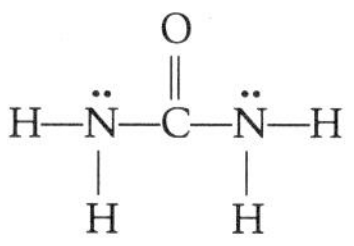 H—N̈—C(=O)—N̈—H with H below each N	1.5×10^{-14}	$H_2NCO\overset{+}{N}H_3$	0.67

* The nitrogen atom with the lone pair accounts for each compound's basicity. In the case of urea, K_b can be associated with either nitrogen atom.

ionization constants. Note that the basicity of all these compounds is due to the lone pair on the nitrogen atom.

EXAMPLE 19.6

What is the pH of a 0.400 M ammonia solution?

Answer

The procedure is essentially the same as the one we use for weak acids.

Step 1

Let x be the concentration in mol/L of NH_4^+ and OH^- ions at equilibrium. Next we summarize:

(Continued)

	$NH_3(aq)$	$+ H_2O(l) \rightleftharpoons$	$NH_4^+(aq)$	$+ OH^-(aq)$
Initial	0.400 M		0.000 M	0.000 M
Change	$-x$ M		$+x$ M	$+x$ M
Equilibrium	$(0.400 - x)$ M		x M	x M

Step 2

Using the base ionization constant listed in Table 19.2, we write

$$K_b = \frac{[NH_4^+][OH^-]}{[NH_3]} = 1.8 \times 10^{-5}$$

$$\frac{x^2}{0.400 - x} = 1.8 \times 10^{-5}$$

You should confirm the validity of this approximation.

Applying the approximation $0.400 - x \approx 0.400$ gives

$$\frac{x^2}{0.400 - x} \simeq \frac{x^2}{0.400} = 1.8 \times 10^{-5}$$

$$x^2 = 7.2 \times 10^{-6}$$

$$x = 2.7 \times 10^{-3} \, M$$

Step 3

At equilibrium, $[OH^-] = 2.7 \times 10^{-3} \, M$. Thus

$$pOH = -\log (2.7 \times 10^{-3})$$

$$= 2.57$$

$$pH = 14.00 - 2.57$$

$$= 11.43$$

Note that we omitted the contribution to $[OH^-]$ from water.

Similar examples: Problems 19.8, 19.9.

19.3 THE RELATIONSHIP BETWEEN CONJUGATE ACID-BASE IONIZATION CONSTANTS

An important relationship between the acid ionization constant and the ionization constant of its conjugate base can be derived as follows. Using acetic acid as an example, we write

$$CH_3COOH(aq) \rightleftharpoons H^+(aq) + CH_3COO^-(aq)$$

$$K_a = \frac{[H^+][CH_3COO^-]}{[CH_3COOH]}$$

The conjugate base, CH_3COO^-, reacts with water according to the equation

$$CH_3COO^-(aq) + H_2O(l) \rightleftharpoons CH_3COOH(aq) + OH^-(aq)$$

and we can write the base ionization constant as

$$K_b = \frac{[CH_3COOH][OH^-]}{[CH_3COO^-]}$$

The product of these two ionization constants is given by

$$K_a K_b = \frac{[H^+][CH_3COO^-]}{[CH_3COOH]} \times \frac{[CH_3COOH][OH^-]}{[CH_3COO^-]}$$
$$= [H^+][OH^-]$$
$$= K_w$$

This result may seem strange at first, but can be understood by realizing that the sum of reactions (1) and (2) below is simply the autoionization of water.

(1) $CH_3COOH(aq) \rightleftharpoons H^+(aq) + CH_3COO^-(aq)$ K_a
(2) $CH_3COO^-(aq) + H_2O(l) \rightleftharpoons CH_3COOH(aq) + OH^-(aq)$ K_b

(3) $H_2O(l) \rightleftharpoons H^+(aq) + OH^-(aq)$ K_w

This example illustrates one of the rules we learned about chemical equilibria: When two reactions are added to give a third reaction, the equilibrium constant for the third reaction is the product of the equilibrium constants for the two added reactions (see Section 17.3). Thus, for any conjugate acid-base pair it is always true that

$$K_a K_b = K_w \qquad (19.1)$$

Expressing Equation (19.1) in the following ways

$$K_a = \frac{K_w}{K_b} \qquad K_b = \frac{K_w}{K_a}$$

enables us to draw an important conclusion: The stronger the acid (the larger K_a), the weaker its conjugate base (the smaller K_b), and vice versa (see Tables 19.1 and 19.2).

We can use Equation (19.1) to calculate the K_b of the conjugate base (CH_3COO^-) of CH_3COOH as follows. We find the K_a value of CH_3COOH in Table 19.1 and write

$$K_b = \frac{K_w}{K_a}$$
$$= \frac{1.0 \times 10^{-14}}{1.8 \times 10^{-5}} = 5.6 \times 10^{-10}$$

You can see that both K_a and K_b are small numbers. This illustrates the statement made in Section 18.6 that we often encounter both a weak acid and a weak conjugate base of the same acid (p. 552).

19.4 DIPROTIC AND POLYPROTIC ACIDS

The treatment of diprotic and polyprotic acids is more involved, because these substances may yield more than one hydrogen ion per molecule. These acids ionize in a stepwise manner; that is, they lose one proton at a time. An ionization constant expression can be written for each step of ionization. As a result, two or more equilibrium constant expressions must often be used to calculate the concentrations of species present in the acid solution. Some examples follow.

For H_2S

$$H_2S(aq) \rightleftharpoons H^+(aq) + HS^-(aq) \qquad K_{a_1} = \frac{[H^+][HS^-]}{[H_2S]}$$

$$HS^-(aq) \rightleftharpoons H^+(aq) + S^{2-}(aq) \qquad K_{a_2} = \frac{[H^+][S^{2-}]}{[HS^-]}$$

For H_2CO_3

$$H_2CO_3(aq) \rightleftharpoons H^+(aq) + HCO_3^-(aq) \qquad K_{a_1} = \frac{[H^+][HCO_3^-]}{[H_2CO_3]}$$

$$HCO_3^-(aq) \rightleftharpoons H^+(aq) + CO_3^{2-}(aq) \qquad K_{a_2} = \frac{[H^+][CO_3^{2-}]}{[HCO_3^-]}$$

In each case, the conjugate base in the first step of ionization becomes the acid in the second step of ionization.

Table 19.3 shows the ionization constants of several diprotic acids and a polyprotic acid. You can see that for a given acid, the first ionization constant is much larger than the second ionization constant, and so on. This trend seems logical when we realize that it is easier to remove a H^+ ion from a neutral molecule than from a negatively charged ion derived from the molecule.

As the following examples show, calculation involving ionization constants of diprotic or polyprotic acids is more complex than the calculation for a monoprotic acid because it involves more than one stage of ionization.

EXAMPLE 19.7

Calculate the pH of a 0.010 M H_2SO_4 solution.

Answer

We note that H_2SO_4 is a strong acid (for the first stage of ionization) and that HSO_4^- is a weak acid. We summarize the changes in the first stage of ionization:

	$H_2SO_4(aq) \longrightarrow$	$H^+(aq)$	$+ HSO_4^-(aq)$
Initial	0.010 M	0.00 M	0.00 M
Change	-0.010 M	$+0.010$ M	$+0.010$ M
Final	0.00 M	0.010 M	0.010 M

For the second stage of ionization, we proceed as in the case of a weak monoprotic acid.

Step 1

Let x be the concentration in mol/L of H^+ and SO_4^{2-} produced by the ionization of HSO_4^-. The total concentration of the H^+ ion at equilibrium must be the sum of H^+ ion concentrations due to both stages of ionization, that is, (0.010 $+ x$) M.

	$HSO_4^-(aq) \rightleftharpoons$	$H^+(aq)$	$+ SO_4^{2-}(aq)$
Initial	0.010 M	0.010 M	0.00 M
Change	$-x$ M	$+x$ M	$+x$ M
Equilibrium	(0.010 $- x$) M	(0.010 $+ x$) M	x M

(Continued)

TABLE 19.3 Ionization Constants of Some Common Diprotic and Polyprotic Acids in Water at 25°C

Name of Acid	Formula	Structure	K_a	Conjugate Base	K_b
Carbonic acid	H_2CO_3	H—O—C(=O)—O—H	4.2×10^{-7}	HCO_3^-	2.4×10^{-8}
Bicarbonate ion	HCO_3^-	H—O—C(=O)—O$^-$	4.8×10^{-11}	CO_3^{2-}	2.1×10^{-4}
Hydrosulfuric acid	H_2S	H—S—H	5.7×10^{-8}	HS^-	1.8×10^{-7}
Bisulfide ion	HS^-	H—S$^-$	1.2×10^{-13}	S^{2-}	8.3×10^{-2}
Oxalic acid	$C_2H_2O_4$	H—O—C(=O)—C(=O)—O—H	6.5×10^{-2}	$C_2HO_4^-$	1.5×10^{-13}
Hydrogen oxalate ion	$C_2HO_4^-$	H—O—C(=O)—C(=O)—O$^-$	6.1×10^{-5}	$C_2O_4^{2-}$	1.6×10^{-10}
Phosphoric acid	H_3PO_4	H—O—P(=O)(—O—H)—O—H	7.5×10^{-3}	$H_2PO_4^-$	1.3×10^{-12}
Dihydrogen phosphate ion	$H_2PO_4^-$	H—O—P(=O)(—O—H)—O$^-$	6.2×10^{-8}	HPO_4^{2-}	1.6×10^{-7}
Hydrogen phosphate ion	HPO_4^{2-}	H—O—P(=O)(—O$^-$)—O$^-$	4.8×10^{-13}	PO_4^{3-}	2.1×10^{-2}
Sulfuric acid	H_2SO_4	H—O—S(=O)(=O)—O—H	very large	HSO_4^-	very small
Bisulfate ion	HSO_4^-	H—O—S(=O)(=O)—O$^-$	1.3×10^{-2}	SO_4^{2-}	7.7×10^{-13}
Sulfurous acid	H_2SO_3	H—O—S(=O)—O—H	1.3×10^{-2}	HSO_3^-	7.7×10^{-13}
Bisulfite ion	HSO_3^-	H—O—S(=O)—O$^-$	6.3×10^{-8}	SO_3^{2-}	1.6×10^{-7}

From Table 19.3 we write

$$K_{a_2} = \frac{[H^+][SO_4^{2-}]}{[HSO_4^-]} = 1.3 \times 10^{-2}$$

$$\frac{(0.010 + x)x}{0.010 - x} = 1.3 \times 10^{-2}$$

Since the K_{a_2} of HSO_4^- is quite large, we must solve the quadratic equation, which simplifies to

$$x^2 + 0.023x - 1.3 \times 10^{-4} = 0$$

$$x = \frac{-0.023 \pm \sqrt{(0.023)^2 - 4(1)(-1.3 \times 10^{-4})}}{2(1)}$$

$$= 4.7 \times 10^{-3} \, M$$

The total H^+ ion concentration at equilibrium is the sum of the concentrations due to both stages of ionization $(0.010 + 4.7 \times 10^{-3}) \, M$ or $0.015 \, M$. Finally

$$pH = -\log[H^+]$$
$$= -\log 0.015$$
$$= 1.82$$

Similar example: Problem 19.14.

EXAMPLE 19.8

Hydrogen sulfide (H_2S) is responsible for the odor of rotten eggs. When dissolved in water, it becomes a weak diprotic acid, called hydrosulfuric acid. Calculate the concentrations of H_2S, HS^-, H^+, and S^{2-} in a saturated aqueous solution of H_2S of concentration $0.10 \, M$.

Answer

We begin with the first stage of ionization.

Step 1

	$H_2S(aq)$	$\rightleftharpoons H^+(aq)$	$+ HS^-(aq)$
Initial	$0.10 \, M$	$0.00 \, M$	$0.00 \, M$
Change	$-x \, M$	$+x \, M$	$+x \, M$
Equilibrium	$(0.10 - x) \, M$	$x \, M$	$x \, M$

Step 2

With the ionization constant from Table 19.3, we write

$$K_{a_1} = \frac{[H^+][HS^-]}{[H_2S]} = 5.7 \times 10^{-8}$$

$$\frac{x^2}{0.10 - x} = 5.7 \times 10^{-8}$$

Since K_{a_1} is very small and the initial concentration is fairly large, we can apply the approximation $0.10 - x \simeq 0.10$.

$$\frac{x^2}{0.10 - x} \simeq \frac{x^2}{0.10} = 5.7 \times 10^{-8}$$

$$x = 7.5 \times 10^{-5} \, M$$

Step 3

After equilibrium is reached for the first stage of ionization the concentrations are

$$[H^+] = 7.5 \times 10^{-5} \, M$$
$$[HS^-] = 7.5 \times 10^{-5} \, M$$
$$[H_2S] = (0.10 - 7.5 \times 10^{-5}) \, M$$
$$= 0.10 \, M$$

Having followed the rules for significant figures, we find that the concentration of H_2S remained essentially unchanged at $0.10 \, M$.

Next we consider the second stage of ionization.

Step 1

Let y be the equilibrium concentration of S^{2-} in mol/L. Thus the equilibrium concentration of HS^- must be $(7.5 \times 10^{-5} - y) \, M$. We have

	$HS^-(aq)$	$\rightleftharpoons$	$H^+(aq)$	$+ \, S^{2-}(aq)$
Initial	$7.5 \times 10^{-5} \, M$		$7.5 \times 10^{-5} \, M$	$0.00 \, M$
Change	$-y \, M$		$+y \, M$	$+y \, M$
Equilibrium	$(7.5 \times 10^{-5} - y) \, M$		$(7.5 \times 10^{-5} + y) \, M$	$y \, M$

Step 2

Using the ionization constant in Table 19.3, we write

$$K_{a_2} = \frac{[H^+][S^{2-}]}{[HS^-]} = 1.2 \times 10^{-13}$$

$$\frac{(7.5 \times 10^{-5} + y)y}{(7.5 \times 10^{-5} - y)} = 1.2 \times 10^{-13}$$

Because the ionization constant is extremely small, the amount of S^{2-} ions produced at equilibrium must also be very small. Thus we can safely make the approximations

$$7.5 \times 10^{-5} + y \simeq 7.5 \times 10^{-5}$$
$$7.5 \times 10^{-5} - y \simeq 7.5 \times 10^{-5}$$

and substitute them in the equation, which reduces to

$$y = 1.2 \times 10^{-13} \, M$$

Step 3

At equilibrium, therefore

$$[H_2S] = 0.10 \, M$$
$$[HS^-] = 7.5 \times 10^{-5} \, M$$
$$[H^+] = 7.5 \times 10^{-5} \, M$$
$$[S^{2-}] = 1.2 \times 10^{-13} \, M$$

Similar examples: Problems 19.15, 19.16, 19.17.

Example 19.8 allows us to make some generalizations about the ionization of weak diprotic acids. If, for a solution containing the acid H_2A as the only solute, K_{a_1} is greater than K_{a_2} by at least a factor of 1,000, then

1. The concentration of H^+ ions at equilibrium may be assumed to result only from the first stage of ionization.
2. The concentration of A^{2-} ions is numerically equal to the second ionization constant (K_{a_2}).

Sometimes it is convenient to combine the first and second stages of the ionization of a diprotic acid in a single equation; for example, the two stages of ionization of H_2S can be combined:

$$\begin{array}{ll} H_2S(aq) \rightleftharpoons H^+(aq) + HS^-(aq) & K_{a_1} \\ HS^-(aq) \rightleftharpoons H^+(aq) + S^{2-}(aq) & K_{a_2} \\ \hline \text{overall:} \quad H_2S(aq) \rightleftharpoons 2H^+(aq) + S^{2-}(aq) & K \end{array}$$

And the equilibrium constant K is calculated as follows:

$$K = \frac{[H^+]^2[S^{2-}]}{[H_2S]} = K_{a_1}K_{a_2}$$

$$K_{a_1}K_{a_2} = (5.7 \times 10^{-8})(1.2 \times 10^{-13})$$
$$= 6.8 \times 10^{-21}$$

We must not conclude that the concentration of H^+ ions at equilibrium is only twice the S^{2-} concentration, as the above overall equation seems to suggest. As Example 19.8 shows, at equilibrium $[H^+]$ is much larger than $[S^{2-}]$ because the extent of ionization of H_2S is much greater than that of HS^-.

Phosphoric acid is responsible for much of the "tangy" flavor of popular cola drinks.

An important polyprotic acid is phosphoric acid (H_3PO_4), which has three ionizable hydrogen atoms:

$$H_3PO_4(aq) \rightleftharpoons H^+(aq) + H_2PO_4^-(aq) \qquad K_{a_1} = \frac{[H^+][H_2PO_4^-]}{[H_3PO_4]} = 7.5 \times 10^{-3}$$

$$H_2PO_4^-(aq) \rightleftharpoons H^+(aq) + HPO_4^{2-}(aq) \qquad K_{a_2} = \frac{[H^+][HPO_4^{2-}]}{[H_2PO_4^-]} = 6.2 \times 10^{-8}$$

$$HPO_4^{2-}(aq) \rightleftharpoons H^+(aq) + PO_4^{3-}(aq) \qquad K_{a_3} = \frac{[H^+][PO_4^{3-}]}{[HPO_4^{2-}]} = 4.8 \times 10^{-13}$$

We see that phosphoric acid is a weak polyprotic acid and that the ionization constants decrease rapidly. Thus we can predict that, in a solution containing phosphoric acid, the concentration of the nonionized acid is the highest and the only other species present in significant concentrations are H^+ and $H_2PO_4^-$ ions.

19.5 HYDROLYSIS

The term **hydrolysis** describes the *reaction of a substance with water*. (It derives from the Greek words "hydro," meaning "water," and "lysis," meaning "to split apart.") **Salt hydrolysis** is the *reaction of the anion or cation, or both, of a salt with water*.

Salts are formed by acid-base reactions. Because cations of salts are conjugate acids of the bases, and anions of salts are conjugate bases of the acids, there is really no fundamental difference between salt hydrolysis and acid-base reaction. In fact, as we will see, the equilibrium constants of salt hydrolysis are identical to the acid and base ionization constants.

In much of the discussion that follows, the cation involved in hydrolysis is the ammonium (NH_4^+) ion. However, a large class of metal ions also undergoes hydrolysis, and we will discuss their properties later in this section.

Salt Hydrolysis

In studying the reaction of a salt with water, we will consider four different types of salts: (1) salts of strong acids and strong bases, (2) salts of weak acids and strong bases, (3) salts of strong acids and weak bases, and (4) salts of weak acids and weak bases. The formation of these salts was discussed briefly in Section 18.6.

Salts of Strong Acids and Strong Bases.   An example of a salt derived from a strong acid and a strong base is sodium chloride:

$$NaOH(aq) + HCl(aq) \longrightarrow NaCl(aq) + H_2O(l)$$

When NaCl dissolves in water, it is completely dissociated into Na^+ and Cl^- ions:

$$NaCl(s) \xrightarrow{H_2O} Na^+(aq) + Cl^-(aq)$$

Because Na^+ and Cl^- ions do not react with water, no further changes will take place.

$$Na^+(aq) + H_2O(l) \longrightarrow \text{no reaction}$$

$$Cl^-(aq) + H_2O(l) \longrightarrow \text{no reaction}$$

Since NaOH is a strong base, Na^+ has no tendency to react with H_2O to form NaOH and H^+. Similarly, HCl is a strong acid, so Cl^- does not react with H_2O to form HCl and OH^-. Thus a solution containing NaCl remains neutral. The same behavior characterizes other salts of strong acids and strong bases, such as $NaClO_4$, LiI, KBr, and $BaCl_2$. Keep in mind that cations of alkali metals and alkaline earth metals (except beryllium) as well as anions of all strong acids (see Table 18.3) do not undergo hydrolysis.

The Cl^- ion is an extremely weak Brønsted–Lowry base.

Salts of Weak Acids and Strong Bases.   When sodium acetate (CH_3COONa), a salt derived from a weak acid (acetic acid) and a strong base (sodium hydroxide), dissolves in water, the solution becomes basic. Most salts are strong electrolytes, so we can always assume that they are completely dissociated in aqueous solution. Thus the first step in the reaction is the complete dissociation of sodium acetate:

$$CH_3COONa(s) \xrightarrow{H_2O} CH_3COO^-(aq) + Na^+(aq)$$

The Na^+ ion, as we noted earlier, does not react with water. On the other hand, the CH_3COO^- ion, being the conjugate base of a weak acid, undergoes hydrolysis as follows:

Refer to Table 18.3 (p. 545) for relative strengths of conjugate acid-base pairs.

$$CH_3COO^-(aq) + H_2O(l) \rightleftharpoons CH_3COOH(aq) + OH^-(aq)$$

It is this reaction that produces an excess of OH^- ions and makes the solution basic; the pH of a sodium acetate solution is greater than 7.

A measure of the extent of hydrolysis is given by the equilibrium constant for this reaction, which is identical to the base ionization constant for CH_3COO^-, the conjugate base (see Table 19.1):

$$K_b = \frac{[CH_3COOH][OH^-]}{[CH_3COO^-]} = 5.6 \times 10^{-10}$$

From Equation (19.1)

$$K_b = \frac{K_w}{K_a} = \frac{1.0 \times 10^{-14}}{1.8 \times 10^{-5}}$$
$$= 5.6 \times 10^{-10}$$

Since each CH_3COO^- ion that hydrolyzes produces one OH^- ion, the concentration of OH^- at equilibrium is the same as the concentration of CH_3COO^- that hydrolyzed. We can define the percent hydrolysis as

$$\% \text{ hydrolysis} = \frac{[CH_3COO^-]_{hydrolyzed}}{[CH_3COO^-]_{initial}} \times 100\%$$
$$= \frac{[OH^-]_{equilibrium}}{[CH_3COO^-]_{initial}} \times 100\%$$

EXAMPLE 19.9

Calculate the pH of a 0.15 M solution of sodium acetate (CH_3COONa). What is the percent hydrolysis?

Answer

We note that CH_3COONa is the salt formed from a weak acid (CH_3COOH) and a strong base (NaOH). Consequently, only the anion (CH_3COO^-) will hydrolyze. The initial dissociation of the salt is

$$CH_3COONa(s) \xrightarrow{H_2O} CH_3COO^-(aq) + Na^+(aq)$$
$$\qquad\qquad\qquad\quad 0.15\ M \qquad\quad 0.15\ M$$

We can now treat the hydrolysis of CH_3COO^-. The acetate ion acts as a weak Brønsted–Lowry base.

Step 1

Let x be the equilibrium concentration of CH_3COOH and OH^- ions in mol/L. We summarize the changes:

	$CH_3COO^-(aq)$ + $H_2O(l)$ $\rightleftharpoons$	$CH_3COOH(aq)$ +	$OH^-(aq)$
Initial	0.15 M	0.00 M	0.00 M
Change	$-x\ M$	$+x\ M$	$+x\ M$
Equilibrium	$(0.15 - x)\ M$	$x\ M$	$x\ M$

Remember that CH₃COONa is a strong electrolyte.

Step 2

$$K_b = \frac{[CH_3COOH][OH^-]}{[CH_3COO^-]} = 5.6 \times 10^{-10}$$

$$\frac{x^2}{0.15 - x} = 5.6 \times 10^{-10}$$

Since K_b is a very small number and the initial concentration of the base is large, we can apply the approximation $0.15 - x \simeq 0.15$:

$$\frac{x^2}{0.15 - x} \simeq \frac{x^2}{0.15} = 5.6 \times 10^{-10}$$

$$x = 9.2 \times 10^{-6} \, M$$

Step 3

At equilibrium

$$[OH^-] = 9.2 \times 10^{-6} \, M$$
$$pOH = -\log(9.2 \times 10^{-6})$$
$$= 5.04$$
$$pH = 14.00 - 5.04$$
$$= 8.96$$

Thus the solution is basic, as we would expect. The percent hydrolysis is given by

$$\% \text{ hydrolysis} = \frac{9.2 \times 10^{-6} \, M}{0.15 \, M} \times 100\%$$
$$= 0.0061\%$$

The result shows that only a small portion of the anion undergoes hydrolysis. The small percent hydrolysis also justifies our approximation $(0.15 - x \simeq 0.15)$.

Similar example: Problem 19.22.

Salts of Strong Acids and Weak Bases. When a salt that is derived from a strong acid and a weak base dissolves in water, the solution becomes acidic. Ammonium chloride (NH_4Cl), which is the product of the reaction between hydrochloric acid and ammonia, is an example. The first step is the complete dissociation of ammonium chloride:

$$NH_4Cl(s) \xrightarrow{H_2O} NH_4^+(aq) + Cl^-(aq)$$

Cl^- ions do not hydrolyze, but NH_4^+ ions act as a Brønsted–Lowry acid and react with water as follows:

$$NH_4^+(aq) + H_2O(l) \rightleftharpoons NH_3(aq) + H_3O^+(aq)$$

or simply

$$NH_4^+(aq) \rightleftharpoons NH_3(aq) + H^+(aq)$$

Thus the pH of the solution is less than 7 due to the presence of the excess H^+ ions. The equilibrium constant for the hydrolysis of the NH_4^+ ions is the same as the acid ionization constant:

By coincidence, K_a of NH_4^+ is the same as K_b of CH_3COO^-.

$$K_a = \frac{[NH_3][H^+]}{[NH_4^+]} = 5.6 \times 10^{-10}$$

EXAMPLE 19.10

What is the pH and percent hydrolysis of a 0.10 M NH_4Cl solution?

Answer

We note that NH_4Cl is the salt formed from a strong acid (HCl) and a weak base (NH_3). Consequently, only the cation (NH_4^+) of the salt will hydrolyze. The initial dissociation of the salt is

Remember that NH_4Cl is a strong electrolyte.

$$NH_4Cl(s) \xrightarrow{H_2O} \underset{0.10\ M}{NH_4^+(aq)} + \underset{0.10\ M}{Cl^-(aq)}$$

We can now treat the hydrolysis of the cation as the ionization of the acid.

Step 1

We represent the hydrolysis of the cation NH_4^+, and let x be the equilibrium concentration of NH_3 and H^+ ions in mol/L:

	$NH_4^+(aq)$	$\rightleftharpoons$	$NH_3(aq)$	+ $H^+(aq)$
Initial	0.10 M		0.00 M	0.00 M
Change	$-x$ M		$+x$ M	$+x$ M
Equilibrium	$(0.10 - x)$ M		x M	x M

Step 2

From Table 19.2 we obtain the K_a for NH_4^+:

$$K_a = \frac{[NH_3][H^+]}{[NH_4^+]} = 5.6 \times 10^{-10}$$

$$\frac{x^2}{0.10 - x} = 5.6 \times 10^{-10}$$

Applying the approximation $0.10 - x \simeq 0.10$, we get

$$\frac{x^2}{0.10 - x} \simeq \frac{x^2}{0.10} = 5.6 \times 10^{-10}$$

$$x = 7.5 \times 10^{-6}\ M$$

Thus the pH is given by

$$pH = -\log(7.5 \times 10^{-6})$$
$$= 5.12$$

The percent hydrolysis is

$$\% \text{ hydrolysis} = \frac{7.5 \times 10^{-6}\ M}{0.10\ M} \times 100\%$$
$$= 0.0075\%$$

We see that the extent of hydrolysis is very small in a 0.10 M ammonium chloride solution and that our approximation $(0.10 - x \simeq 0.10)$ was justified.

Similar example: Problem 19.23.

Salts of Weak Acids and Weak Bases.   From our discussion so far we can predict that both the cation and the anion of a salt derived from a weak acid and a weak base will undergo hydrolysis. However, whether a solution containing such a salt is acidic, basic, or neutral depends on the relative strengths of the weak acid and the weak base. Since the mathematics associated with this type of system is relatively involved, we will limit ourselves to making reliable qualitative predictions about these solutions. We consider three situations.

- $K_b > K_a$. If K_b for the conjugate base is greater than K_a for the conjugate acid, then the solution must be basic because the anion derived from the weak acid will hydrolyze to a greater extent than the cation derived from the weak base. Consider ammonium cyanide (NH_4CN) as an example. The NH_4^+ ion is the conjugate acid of the weak base NH_3, and CN^- is the conjugate base of the weak acid HCN. For the hydrolysis

$$NH_4^+(aq) \rightleftharpoons NH_3(aq) + H^+(aq)$$

we know from Table 19.2 that $K_a = 5.6 \times 10^{-10}$. On the other hand, K_b for the hydrolysis of CN^- ions

$$CN^-(aq) + H_2O(l) \rightleftharpoons HCN(aq) + OH^-(aq)$$

is obtained from Equation (19.1) and Table 19.1:

$$K_b = \frac{K_w}{K_a} = \frac{1.0 \times 10^{-14}}{4.9 \times 10^{-10}}$$
$$= 2.0 \times 10^{-5}$$

We see that $K_b > K_a$, so we conclude that the CN^- ions will hydrolyze to a greater extent than the NH_4^+ ions. Consequently, at equilibrium there will be more OH^- ions than H^+ ions and the solution will be basic.

- $K_b < K_a$. Conversely, if K_b is smaller than K_a, as in the case of ammonium nitrite (NH_4NO_2), the solution will be acidic because the NH_4^+ ions will hydrolyze to a greater extent than the NO_2^- ions:

$$NH_4^+ \rightleftharpoons NH_3(aq) + H^+(aq)$$
$$NO_2^-(aq) + H_2O(l) \rightleftharpoons HNO_2(aq) + OH^-(aq)$$

where

$$K_b = \frac{[HNO_2][OH^-]}{[NO_2^-]}$$

TABLE 19.4 Hydrolysis of Four Types of Salts

Salt Derived From	Examples	Ions that Undergo Hydrolysis	pH of Solution
Strong acid and strong base	NaCl, KI, RbBr, BaCl$_2$	None	$\simeq 7$
Weak acid and strong base	KNO$_2$ CH$_3$COONa	Anion	>7
Strong acid and weak base	NH$_4$Cl NH$_4$NO$_3$	Cation	<7
Weak acid and weak base	NH$_4$NO$_2$ CH$_3$COONH$_4$ NH$_4$CN	Anion and cation	<7 if $K_b < K_a$ $\simeq 7$ if $K_b \simeq K_a$ >7 if $K_b > K_a$

Again, using Equation (19.1) and Table 19.1, we obtain

$$K_b = \frac{K_w}{K_a} = \frac{1.0 \times 10^{-14}}{4.5 \times 10^{-4}}$$
$$= 2.2 \times 10^{-11}$$

- $K_a \simeq K_b$. If K_a is approximately equal to K_b, as in the case of ammonium acetate, the solution will be nearly neutral.

Table 19.4 summarizes the behavior in aqueous solution of the four types of salts that we have discussed.

Metal Ion Hydrolysis

Solutions containing small, highly charged metal cations are often acidic in aqueous solution. For example, when aluminum chloride (AlCl$_3$) dissolves in water, the Al^{3+} ions exist in the hydrated form as Al(H$_2$O)$_6^{3+}$ (Figure 19.3). Let us consider a particular metal ion bond to the oxygen atom of one of the six water molecules in Al(H$_2$O)$_6^{3+}$:

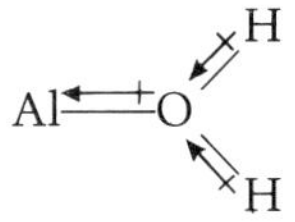

The positively charged Al^{3+} ion reduces the electron density about the oxygen atom, making the O—H bond more polar. Consequently, the H atoms have a greater tendency to become ionized than those in water molecules not involved in hydration. The resulting ionization process can be written as

$$Al(H_2O)_6^{3+}(aq) + H_2O(l) \rightleftharpoons Al(OH)(H_2O)_5^{2+}(aq) + H_3O^+(aq)$$

or simply

$$Al(H_2O)_6^{3+}(aq) \rightleftharpoons Al(OH)(H_2O)_5^{2+}(aq) + H^+(aq)$$

The extent of hydrolysis is greatest for the smallest and most highly charged ions, such as Al^{3+}, Cr^{3+}, Fe^{3+}, Be^{2+}, and Bi^{3+}, because a highly

The flow of electron density is toward the Al^{3+} ion.

Note that the hydrated aluminum ion qualifies as a proton donor and thus as a Brønsted–Lowry acid in this reaction.

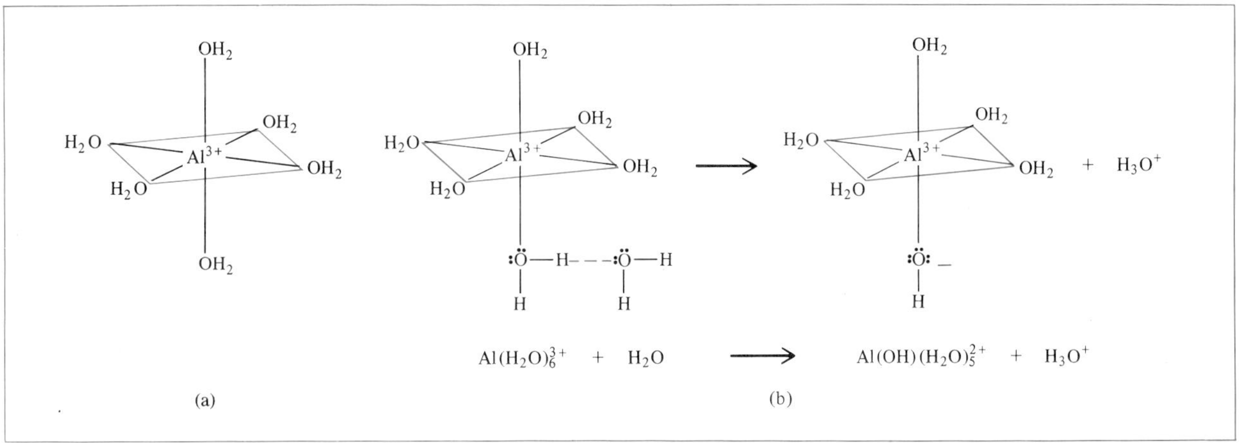

FIGURE 19.3 *(a) The $Al(H_2O)_6^{3+}$ ion. The six H_2O molecules surround the Al^{3+} ion octahedrally. (b) The attraction of the small Al^{3+} ion for the lone pairs on the oxygen atoms is so great that the O—H bonds (in a H_2O molecule attached to the metal cation) are weakened, making loss of a proton (or H^+) to an incoming H_2O molecule possible. This metal hydrolysis makes the solution acidic.*

charged small cation can more effectively polarize the O—H bond, facilitating ionization. The hydrolysis of such cations may involve more than one step. For example, the hydrolysis of the hydrated Al^{3+} ion may proceed according to the reaction

$$Al(OH)(H_2O)_5^{2+}(aq) \rightleftharpoons Al(OH)_2(H_2O)_4^+(aq) + H^+(aq)$$

and so on.

SUMMARY

1. The acid ionization constant K_a is larger for stronger acids and smaller for weaker acids. K_b similarly expresses the strengths of bases.
2. Percent ionization is another measure of the strength of acids. The more dilute a solution of a weak acid, the greater the percent ionization of the acid.
3. The product of the ionization constant of an acid and the ionization constant of its conjugate base is equal to the ion-product constant of water, that is, $K_aK_b = K_w$.
4. Most salts are strong electrolytes and dissociate completely into ions in solution. The reaction of these ions with water, called salt hydrolysis, can lead to acidic or basic solutions. In salt hydrolysis, the conjugate bases of weak acids yield basic solutions, and the conjugate acids of weak bases yield acidic solutions.
5. Small, highly charged metal ions such as Al^{3+} and Fe^{3+} hydrolyze to yield acidic solutions.

KEY WORDS

Acid ionization constant, p. 564

Base ionization constant, p. 572

Hydrolysis, p. 580

Salt hydrolysis, p. 580

PROBLEMS[†]

More challenging problems are marked with an asterisk.

Weak Acid Ionization Constants

19.1 Calculate the concentrations of all the species in a 0.10 M benzoic acid (C_6H_5COOH) solution at equilibrium ($K_a = 6.5 \times 10^{-5}$).

19.2 The pH of a 0.060 M weak monoprotic acid is 3.44. Calculate the K_a of the acid.

19.3 A quantity of 0.0560 g of acetic acid is dissolved in enough water to make 50.0 mL of solution. Calculate the concentrations of H^+, CH_3COO^-, and CH_3COOH at equilibrium (K_a for acetic acid $= 1.8 \times 10^{-5}$).

19.4 What is the original molarity of a solution of formic acid (HCOOH) whose pH is 3.26 at equilibrium?

19.5 Calculate the percent ionization of benzoic acid at the following concentrations: (a) 0.20 M, (b) 0.00020 M.

19.6 A 0.040 M solution of a monoprotic acid in water is 14 percent ionized. Calculate the ionization constant of the acid.

*19.7 (a) Calculate the percent ionization of a 0.20 M solution of the monoprotic acetylsalicylic acid (aspirin). (b) The pH of gastric juice in the stomach of a certain individual is 1.00. After a few aspirin tablets have been swallowed, the concentration of acetylsalicylic acid in the stomach is 0.20 M. Calculate the percent ionization of the acid under these conditions. What effect does the nonionized acid have on the membranes lining the stomach ($K_a = 3.0 \times 10^{-4}$)?

Weak Base Ionization Constant; K_a–K_b Relationship

19.8 Calculate the pH for each of the following solutions: (a) 0.10 M NH_3, (b) 0.050 M pyridine, (c) 0.260 M methylamine.

19.9 The pH of a 0.30 M solution of a weak base is 10.66. What is the K_b of the base?

19.10 What is the original molarity of a solution of ammonia whose pH is 11.22?

19.11 In a 0.080 M NH_3 solution, what percent of the NH_3 is present as NH_4^+?

19.12 Use NH_3 and its conjugate acid NH_4^+ to derive the relationship $K_aK_b = K_w$.

Diprotic and Polyprotic Acids

19.13 Malonic acid ($C_3H_4O_4$) is a diprotic acid. Explain what that means.

19.14 What are the concentrations of HSO_4^-, SO_4^{2-}, and H^+ in a 0.20 M $KHSO_4$ solution? (*Hint:* H_2SO_4 is a strong acid; K_a for $HSO_4^- = 1.3 \times 10^{-2}$.)

*19.15 Oxalic acid ($C_2H_2O_4$) is a diprotic acid (see Table 19.3). Calculate the concentrations of $C_2H_2O_4$, $C_2HO_4^-$, $C_2O_4^{2-}$, and H^+ in a 0.10 M oxalic acid solution.

*19.16 Calculate the concentrations of H^+, HCO_3^-, and CO_3^{2-} in a 0.025 M H_2CO_3 solution.

*19.17 Calculate the concentration of sulfide ions (S^{2-}) in a saturated solution of H_2S (0.10 M). The pH of the solution is 1.50 after the addition of HCl.

*19.18 Calculate the concentrations of all species in a 0.100 M H_3PO_4 solution.

Salt Hydrolysis; Metal Ion Hydrolysis

19.19 Specify which of the following salts will undergo hydrolysis: KF, $NaNO_3$, NH_4NO_2, $MgSO_4$, KCN, C_6H_5COONa, RbI, $NaHCO_3$, $CaCl_2$, HCOOK.

19.20 A certain salt, MX (containing the M^+ and X^- ions), is dissolved in water and the pH of the resulting solution is 7.0. Can you say anything about the strengths of the acid and the base from which the salt is derived?

19.21 Predict whether a solution containing the salt K_2HPO_4 will be acidic, neutral, or basic. (*Hint:* You need to consider both the ionization and hydrolysis of HPO_4^{2-}.)

19.22 Calculate the pH of a 0.36 M CH_3COONa solution.

19.23 Calculate the pH of a 0.42 M NH_4Cl solution.

*19.24 Calculate the pH of a 0.20 M ammonium acetate (CH_3COONH_4) solution.

† Unless otherwise stated, the temperature is assumed to be 25°C for all problems.

19.25 How many grams of NaCN would you need to dissolve in enough water to make up an exactly 250 mL solution whose pH is 10.00?

19.26 What is the pH of a 0.20 M K_2S solution?

19.27 In a certain experiment a student finds the pHs of 0.10 M solutions of three salts KX, KY, and KZ are 7.0, 9.0, and 11.0, respectively. Arrange the acids HX, HY, and HZ in the order of increasing acid strength.

19.28 Which ion of the alkaline earth metals is most likely to undergo hydrolysis?

19.29 Predict the pH of the aqueous solutions containing the following salts: (a) KBr, (b) $Al(NO_3)_3$, (c) $BaCl_2$, (d) $Bi(NO_3)_3$.

Miscellaneous Problems

19.30 Why is it necessary to specify the temperature when we discuss K_a and K_b values?

19.31 Why do we usually not quote K_a values for strong acids?

19.32 Compare the pH of a 0.040 M HCl solution with that of a 0.040 M H_2SO_4 solution.

19.33 The first and second ionization constants of a diprotic acid H_2A are K_{a_1} and K_{a_2} at a certain temperature. Under what conditions will $[A^{2-}] = K_{a_2}$?

19.34 The K_a of formic acid is 1.7×10^{-4} at 25°C. Will the acid become stronger or weaker at 40°C? Explain.

19.35 Use the data in Table 19.1 to calculate the equilibrium constant for the following reaction:

$$CH_3COOH(aq) + NO_2^-(aq) \Longrightarrow$$
$$CH_3COO^-(aq) + HNO_2(aq)$$

(*Hint:* You need to apply the rule of multiple equilibria. Break the above equation into two separate equations for the ionizations of CH_3COOH and HNO_2.)

20

ACIDS AND BASES III: BUFFER SOLUTIONS AND ACID-BASE TITRATIONS

This chapter provides a detailed survey of the major categories of acid-base titrations and also considers buffers—an interesting family of chemical solutions that are resistant to pH changes. Buffer solutions play important roles in many chemical and biological systems. For example, the pH of our blood is maintained at about 7.4 by a number of buffer systems. A deviation of only three or four tenths of a pH unit can lead to disease or death. The molecular basis for buffer behavior and for the major types of acid-base titrations will be explained.

590

20.1 THE COMMON ION EFFECT

Our discussion of acid-base ionization and salt hydrolysis in Chapter 19 was limited to solutions containing a single solute. What happens when a solution contains two different dissolved compounds? In the following example, both sodium acetate and acetic acid are in solution and they dissociate and ionize to produce CH_3COO^- ions:

$$CH_3COONa(s) \xrightarrow{H_2O} CH_3COO^-(aq) + Na^+(aq)$$

$$CH_3COOH(aq) \rightleftharpoons CH_3COO^-(aq) + H^+(aq)$$

CH_3COONa is a strong electrolyte, so it dissociates completely in solution, but CH_3COOH, a weak acid, ionizes only slightly. According to Le Chatelier's principle, the addition of CH_3COO^- ions from CH_3COONa to a solution of CH_3COOH will suppress the ionization of CH_3COOH (that is, shift the equilibrium from right to left), thereby decreasing the hydrogen ion concentration. Thus a solution containing both CH_3COOH and CH_3COONa will be *less* acidic than a solution containing only CH_3COOH at the same concentration. The shift in equilibrium of the acetic acid ionization is caused by the additional acetate ions (CH_3COO^-) which come from the salt.

The shift in equilibrium caused by the addition of a compound having an ion in common with the dissolved substances is called the **common ion effect.** The common ion effect plays an important role in determining the pH of a solution and the solubility of a slightly soluble salt. We will deal with the latter in Chapter 21. Here we will study the common ion effect as it relates to the pH of a solution. Keep in mind that despite its distinctive name, the common ion effect is simply a special case of Le Chatelier's principle.

Let us consider the pH of a solution containing a weak acid (HA) and a soluble salt of the weak acid (such as NaA). We start by writing

$$HA(aq) + H_2O(l) \rightleftharpoons H_3O^+(aq) + A^-(aq)$$

or simply

$$HA(aq) \rightleftharpoons H^+(aq) + A^-(aq)$$

The ionization constant K_a is given by

$$K_a = \frac{[H^+][A^-]}{[HA]} \qquad (20.1)$$

Remember that all concentration terms (mol/L) are equilibrium concentrations.

Rearranging Equation (20.1) gives

$$[H^+] = \frac{K_a[HA]}{[A^-]}$$

Taking the negative logarithm of both sides, we obtain

$$-\log[H^+] = -\log K_a - \log\frac{[HA]}{[A^-]}$$

$$-\log[H^+] = -\log K_a + \log\frac{[A^-]}{[HA]}$$

or

$$pH = pK_a + \log \frac{[A^-]}{[HA]} \qquad (20.2)$$

where

$$pK_a = -\log K_a \qquad (20.3)$$

Equation (20.2) is called the *Henderson–Hasselbalch equation.* In a more general form it can be expressed as

$$pH = pK_a + \log \frac{[\text{conjugate base}]}{[\text{acid}]} \qquad (20.4)$$

In our example, HA is the acid and A^- is the conjugate base. Thus, if we know K_a and the concentrations of the acid and the salt of the acid, we can calculate the pH of the solution.

It is important to remember that the Henderson–Hasselbalch equation is derived from the equilibrium constant expression. It is valid regardless of the source of the conjugate base (that is, whether it comes from the acid alone or is supplied by both the acid and its salt).

In solving problems involving the common ion effect we are usually given the starting concentrations of a weak acid HA and its salt, such as NaA. As long as the concentrations of these species are reasonably high ($\geq 0.1\ M$), we can neglect the ionization of the acid and the hydrolysis of the salt. Thus we can use the starting concentrations as the equilibrium concentrations in Equation (20.1) or Equation (20.4).

> pK_a is related to K_a as pH is related to $[H^+]$. Remember that the stronger the acid (that is, K_a is large), the smaller is pK_a.

EXAMPLE 20.1

(a) Calculate the pH of a solution containing 0.20 M CH_3COOH and 0.30 M CH_3COONa. (b) What would be the pH of a 0.20 M CH_3COOH solution with no salt present?

Answer

(a) Sodium acetate is a strong electrolyte, so it dissociates completely in solution

$$CH_3COONa(s) \xrightarrow{H_2O} \underset{0.30\ M}{CH_3COO^-(aq)} + \underset{0.30\ M}{Na^+(aq)}$$

The equilibrium concentrations of both the acid and the conjugate base are assumed to be the same as the starting concentrations; that is

$$[CH_3COOH] = 0.20\ M \quad \text{and} \quad [CH_3COO^-] = 0.30\ M$$

This is a valid assumption because (1) CH_3COOH is a weak acid and the extent of hydrolysis of the CH_3COO^- ion is very small (see Example 19.9), and (2) the presence of CH_3COO^- ions further suppresses the ionization of CH_3COOH, and the presence of CH_3COOH further suppresses the hydrolysis of the CH_3COO^- ions. From Equation (20.1) we get

$$[H^+] = \frac{K_a[HA]}{[A^-]}$$
$$= \frac{(1.8 \times 10^{-5})(0.20)}{0.30}$$
$$= 1.2 \times 10^{-5}\,M$$

Thus

$$pH = -\log[H^+]$$
$$= -\log(1.2 \times 10^{-5})$$
$$= 4.92$$

Alternatively, we can calculate the pH of the solution by using the Henderson–Hasselbalch equation. In this case we need to first calculate pK_a of the acid [see Equation (20.3)]:

$$pK_a = -\log K_a$$
$$= -\log(1.8 \times 10^{-5})$$
$$= 4.74$$

We can calculate the pH of the solution by substituting the value of pK_a and the concentrations of the acid and its conjugate base in Equation (20.4):

$$pH = pK_a + \log\frac{[CH_3COO^-]}{[CH_3COOH]}$$
$$= 4.74 + \log\frac{0.30\,M}{0.20\,M}$$
$$= 4.92$$

(b) To calculate the pH of a weak acid, we follow the procedure used in Section 19.1.

Step 1

Let x be the equilibrium concentration of H^+ and CH_3COO^- in mol/L. We summarize as follows:

	$CH_3COOH(aq) \rightleftharpoons$	$H^+(aq) +$	$CH_3COO^-(aq)$
Initial	0.20 M	0.00 M	0.00 M
Change	$-x\,M$	$+x\,M$	$+x\,M$
Equilibrium	$(0.20 - x)\,M$	$x\,M$	$x\,M$

Step 2

$$\frac{[H^+][CH_3COO^-]}{[CH_3COOH]} = 1.8 \times 10^{-5}$$
$$\frac{x^2}{0.20 - x} = 1.8 \times 10^{-5}$$

Assuming that $0.20 - x \simeq 0.20$, we obtain

$$\frac{x^2}{0.20 - x} \simeq \frac{x^2}{0.20} = 1.8 \times 10^{-5}$$
$$x = 1.9 \times 10^{-3}\,M$$

(Continued)

Step 3

At equilibrium, therefore

$$[H^+] = 1.9 \times 10^{-3} \, M$$

$$pH = -\log (1.9 \times 10^{-3})$$

$$= 2.72$$

Thus, without the common ion effect, the pH of a 0.20 M CH_3COOH solution is 2.72, considerably lower than 4.92, which is the pH in the presence of CH_3COONa, as we calculated in (a). The presence of the common ion (CH_3COO^-) in (a) suppresses the ionization of the acid (CH_3COOH).

Similar example: Problem 20.3.

The common ion effect also operates in a solution containing a weak base (for example, NH_3) and a salt of the base (for example, NH_4Cl). At equilibrium

$$NH_4^+(aq) \rightleftharpoons NH_3(aq) + H^+(aq)$$

This is also the equilibrium constant for the hydrolysis of NH_4^+.

so that

$$K_a = \frac{[NH_3][H^+]}{[NH_4^+]}$$

We can derive the Henderson–Hasselbalch equation for this system as follows. Rearranging the above equation we obtain

$$[H^+] = \frac{K_a[NH_4^+]}{[NH_3]}$$

Taking the negative logarithm of both sides gives

$$-\log [H^+] = -\log K_a - \log \frac{[NH_4^+]}{[NH_3]}$$

$$-\log [H^+] = -\log K_a + \log \frac{[NH_3]}{[NH_4^+]}$$

or

$$pH = pK_a + \log \frac{[NH_3]}{[NH_4^+]}$$

A solution containing both NH_3 and its salt NH_4Cl is *less* basic than a solution containing only NH_3. The common ion NH_4^+ suppresses the ionization of NH_3 in the solution containing both the base and the salt.

20.2   BUFFER SOLUTIONS

A **buffer solution** is a *solution of (a) a weak acid or base and (b) its salt; both components must be present. The solution has the ability to resist changes in pH upon the addition of small amounts of either acid or base.*

Buffers are very important to chemical and biological systems. The pH in the human body varies greatly from one fluid to another; for example, the pH of blood is about 7.4, whereas the gastric juice in our stomachs has a pH of about 1.5. These pH values, which are crucial for the proper functioning of enzymes and the balance of osmotic pressure, must be maintained by buffers in most cases.

A buffer solution must contain an acid to react with OH^- ions that may be added to it and a base to react with any added H^+ ions. Furthermore, the acid and the base components of the buffer must not consume each other in a neutralization reaction. These requirements are satisfied by an acid-base conjugate pair (a weak acid and its conjugate base or a weak base and its conjugate acid).

One of the simplest buffers is the acetic acid–sodium acetate system. A solution containing these two substances has the ability to neutralize either added acid or added base, as follows. If a base is added to the buffer system, the OH^- ions will be consumed by the acid in the buffer

$$CH_3COOH(aq) + OH^-(aq) \longrightarrow CH_3COO^-(aq) + H_2O(l)$$

in which the OH^- ions are supplied by the base. If an acid is added, the H^+ ions will be consumed by the conjugate base in the buffer, according to the equation

$$CH_3COO^-(aq) + H^+(aq) \longrightarrow CH_3COOH(aq)$$

in which the acetate ions are provided by the dissociation of the sodium acetate:

$$CH_3COONa(s) \xrightarrow{H_2O} CH_3COO^-(aq) + Na^+(aq)$$

As you can see, this buffer system is identical to the common ion effect situation described in Example 20.1.

EXAMPLE 20.2

(a) Calculate the pH of a buffer system containing 1.0 M CH_3COOH and 1.0 M CH_3COONa. (b) What is the pH of the buffer system after the addition of 0.10 mole of gaseous HCl to one liter of the solution? Assume that the volume of the solution does not change when the HCl is added.

Answer

(a) The pH of the buffer system before the addition of HCl can be calculated according to the procedure described in Example 20.1. Assuming negligible ionization of the acetic acid and hydrolysis of the acetate ions, we have, at equilibrium

$$[CH_3COOH] = 1.0\ M \quad \text{and} \quad [CH_3COO^-] = 1.0\ M$$

$$K_a = \frac{[H^+][CH_3COO^-]}{[CH_3COOH]} = 1.8 \times 10^{-5}$$

(Continued)

$$[H^+] = \frac{K_a[CH_3COOH]}{[CH_3COO^-]}$$

$$= \frac{(1.8 \times 10^{-5})[CH_3COOH]}{[CH_3COO^-]}$$

$$= \frac{(1.8 \times 10^{-5})(1.0)}{(1.0)}$$

$$= 1.8 \times 10^{-5}\ M$$

$$pH = -\log(1.8 \times 10^{-5})$$

$$= 4.74$$

Thus when the concentrations of the acid and the conjugate base are the same, the pH of the buffer is equal to the pK_a of the acid.

(b) After the addition of HCl, complete ionization of HCl acid occurs:

$$HCl(aq) \longrightarrow H^+(aq) + Cl^-(aq)$$
$$0.10\ mol \qquad 0.10\ mol \quad 0.10\ mol$$

Originally, there were 1.0 mol CH_3COOH and 1.0 mol CH_3COO^- present in one liter of the solution. After neutralization of the HCl acid by CH_3COO^-, which we write as

$$CH_3COO^-(aq) + H^+(aq) \longrightarrow CH_3COOH(aq)$$
$$0.10\ mol \qquad 0.10\ mol \qquad 0.10\ mol$$

the number of moles of acetic acid and acetate ions present are

$$CH_3COOH:\quad (1.0 + 0.1)\ mol = 1.1\ mol$$

$$CH_3COO^-:\quad (1.0 - 0.1)\ mol = 0.90\ mol$$

Next we calculate the hydrogen ion concentration:

$$[H^+] = \frac{K_a[CH_3COOH]}{[CH_3COO^-]}$$

$$= \frac{(1.8 \times 10^{-5})(1.1)}{0.90}$$

$$= 2.2 \times 10^{-5}\ M$$

The pH of the solution becomes

$$pH = -\log(2.2 \times 10^{-5})$$

$$= 4.66$$

Note that since the volume of the solution is the same for both species, we replaced the ratio of their molar concentrations by the ratio of the number of moles present, that is, (1.1 mol/L)/(0.90 mol/L) = (1.1 mol/0.90 mol).

Similar examples: Problems 20.9, 20.12.

Hydrochloric acid is a strong electrolyte.

We see that there is a decrease in pH of 0.08 (becoming more acidic) as a result of the addition of HCl. We can also compare the change in H^+ ion concentrations as follows:

before addition of HCl: $[H^+] = 1.8 \times 10^{-5} M$

after addition of HCl: $[H^+] = 2.2 \times 10^{-5} M$

Thus the H^+ ion concentration increases by a factor of

$$\frac{2.2 \times 10^{-5} M}{1.8 \times 10^{-5} M} = 1.2$$

To appreciate the effectiveness of the CH_3COONa/CH_3COOH buffer, let us find out what would happen if 0.10 mol HCl is added to one liter of water, and compare it to the result in the preceding example. In this case

before addition of HCl: $[H^+] = 1.0 \times 10^{-7} M$

after addition of HCl: $[H^+] = 0.10 M$

Thus, as a result of adding the HCl, the H^+ ion concentration increases by a factor of

$$\frac{0.10 M}{1.0 \times 10^{-7} M} = 1.0 \times 10^6$$

amounting to a millionfold increase! This comparison shows that a properly chosen buffer solution can maintain a fairly constant H^+ ion concentration, or pH.

The Henderson–Hasselbalch equation for the CH_3COONa/CH_3COOH buffer system is

$$pH = pK_a + \log \frac{[CH_3COO^-]}{[CH_3COOH]}$$

You can see that pK_a and the log of the ratio of the concentrations of CH_3COO^- and CH_3COOH together determine the pH of the buffer solution. Conversely, once we know the pH of the solution, we can readily calculate the relative amounts of the acid and its conjugate base. For example, if the pH of a certain CH_3COONa/CH_3COOH buffer system is 3.86,

$$pH = pK_a + \log \frac{[\text{conjugate base}]}{[\text{acid}]}$$

$$3.86 = 4.74 + \log \frac{[CH_3COO^-]}{[CH_3COOH]}$$

$$-0.88 = \log \frac{[CH_3COO^-]}{[CH_3COOH]}$$

The K_a of CH_3COOH is 1.8×10^{-5} so its pK_a is 4.74.

Taking the antilog of both sides of this equation, we obtain

$$\frac{[CH_3COO^-]}{[CH_3COOH]} = 0.13$$

This ratio tells us that the concentration of the conjugate base is only 0.13 times that of the acid.

The relationship between pH and the amount of either the acid or the conjugate base present is best understood by studying the *distribution curve* shown in Figure 20.1. In this case, the distribution curve gives the fraction of acetic acid and of acetate ion present in solution as a function of pH. At

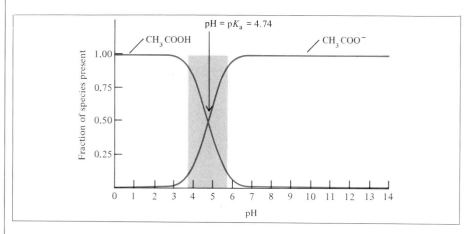

FIGURE 20.1 *Distribution curves for acetic acid and acetate ion as a function of pH. The fraction of species present is given by the ratio of the concentration of either CH_3COOH or CH_3COO^- to the total concentration of CH_3COOH plus CH_3COO^- in solution. Thus, at very low pHs (very acidic media), most of the species present in solution are CH_3COOH molecules, and the concentration of CH_3COO^- is very low. The reverse holds true for very high pHs (very basic media). The shaded region indicates the pH range over which the CH_3COO^-/CH_3COOH buffer system functions most effectively. When $[CH_3COO^-] = [CH_3COOH]$, or when the fraction is 0.5, the pH of the solution is numerically equal to the pK_a of the acid (4.74).*

low pHs the concentration of CH_3COOH is much greater than that of the CH_3COO^- ion, because the presence of H^+ ions drives the equilibrium from right to left (Le Chatelier's principle):

$$CH_3COOH(aq) \rightleftharpoons H^+(aq) + CH_3COO^-(aq)$$

Thus, the species present are mostly nonionized acetic acid molecules. The opposite effect occurs at high pHs. Here, the hydroxide ions deplete the concentration of the acid, so the concentration of the acetate ion increases:

$$CH_3COOH(aq) + OH^-(aq) \rightleftharpoons CH_3COO^-(aq) + H_2O(l)$$

Now there are many more CH_3COO^- ions present than there are CH_3COOH molecules. In order for the buffer to function properly, it must contain comparable amounts of the acid and its conjugate base. There is only a rather limited range (shown by the shaded region in Figure 20.1) over which the buffer is most effective. The range is often called the **buffer range,** that is, *the pH range in which a buffer is effective,* and is defined by the following expression:

$$\text{pH range} = pK_a \pm 1.00 \tag{20.5}$$

Since pK_a for acetic acid is 4.74, the buffer range for the CH_3COO^-/CH_3COOH system is given by pH = 3.74 to 5.74.

Note that the buffer functions best at pH = 4.74, that is, when $[CH_3COOH] = [CH_3COO^-]$. An additional requirement for a buffer system to function effectively is that the concentrations of the weak acid and its conjugate base should be sufficiently high so that the buffer can neutralize

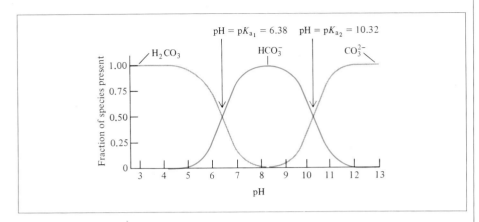

FIGURE 20.2   *Distribution curves for carbonic acid, bicarbonate ion, and carbonate ion, as a function of pH. The interpretation of these curves is similar to that in Figure 20.1. Note that at any given pH value, there are only two predominant species (H_2CO_3 and HCO_3^-, or HCO_3^- and CO_3^{2-}) present in solution. When $[H_2CO_3]$ = $[HCO_3^-]$, we find that pH = pK_{a_1} = 6.38. When $[HCO_3^-]$ = $[CO_3^{2-}]$, we find that pH = pK_{a_2} = 10.32.*

an appreciable amount of added H^+ or OH^- ions without significantly changing its pH.

The CH_3COO^-/CH_3COOH buffer system has no physiological importance. On the other hand, the bicarbonate–carbonic acid (HCO_3^-/H_2CO_3) buffer plays an important role in many biological systems. Because H_2CO_3 is diprotic, there are actually two different buffer systems, that is, HCO_3^-/H_2CO_3 and CO_3^{2-}/HCO_3^-, with HCO_3^- acting as the conjugate base of the first system and the acid of the second. From Table 19.3 we can calculate the buffer ranges as follows:

$$HCO_3^-/H_2CO_3 \qquad pH = pK_{a_1} \pm 1.00 = 6.38 \pm 1.00$$
$$\text{buffer range:} \qquad pH = 5.38 \text{ to } 7.38$$

$$CO_3^{2-}/HCO_3^- \qquad pH = pK_{a_2} \pm 1.00 = 10.32 \pm 1.00$$
$$\text{buffer range:} \qquad pH = 9.32 \text{ to } 11.32$$

Figure 20.2 shows the distribution curves for these two buffers. They clearly function over quite different pH regions.

Now a question arises: How do we prepare a buffer solution with a specific pH? Equation (20.4) indicates that if [acid] $\simeq$ [conjugate base], then pH $\simeq pK_a$. Thus, to prepare a buffer solution, we choose a weak acid whose pK_a is equal to or close to the desired pH. This choice not only gives the correct pH value of the buffer system, but it also ensures that we have comparable amounts of the acid and its conjugate base present; these are prerequisites for the buffer system to function effectively.

EXAMPLE 20.3

Describe how you would prepare a "phosphate buffer" at a pH of about 7.40.

Answer

We write three stages of ionization of phosphoric acid as follows. The K_a values are obtained from Table 19.3 and the pK_a values are found by employing Equation (20.4).

(Continued)

$$H_3PO_4(aq) \rightleftharpoons H^+(aq) + H_2PO_4^-(aq) \qquad K_{a_1} = 7.5 \times 10^{-3}; \; pK_{a_1} = 2.12$$

$$H_2PO_4^-(aq) \rightleftharpoons H^+(aq) + HPO_4^{2-}(aq) \qquad K_{a_2} = 6.2 \times 10^{-8}; \; pK_{a_2} = 7.21$$

$$HPO_4^{2-}(aq) \rightleftharpoons H^+(aq) + PO_4^{3-}(aq) \qquad K_{a_3} = 4.8 \times 10^{-13}; \; pK_{a_3} = 12.32$$

The most suitable of the three buffer systems is $HPO_4^{2-}/H_2PO_4^-$, because pK_a of the acid $H_2PO_4^-$ is closest to the desired pH. From the Henderson–Hasselbalch equation we write

$$pH = pK_a + \log \frac{[\text{conjugate base}]}{[\text{acid}]}$$

$$7.40 = 7.21 + \log \frac{[HPO_4^{2-}]}{[H_2PO_4^-]}$$

$$\log \frac{[HPO_4^{2-}]}{[H_2PO_4^-]} = 0.19$$

Taking the antilog, we obtain

$$\frac{[HPO_4^{2-}]}{[H_2PO_4^-]} = 1.5$$

Thus, one way to prepare a phosphate buffer with a pH of 7.40 is to dissolve disodium hydrogen phosphate (Na_2HPO_4) and sodium dihydrogen phosphate (NaH_2PO_4) in molar ratio of 1.5:1.0 in water. For example, we could dissolve 1.5 mole of Na_2HPO_4 and 1.0 mole of NaH_2PO_4 in enough water to make up a one liter solution.

20.3   A CLOSER LOOK AT ACID-BASE TITRATIONS

Acid-base titrations were first discussed in Section 4.6. We will now follow the changes in pH during the course of an acid-base titration. We will consider three categories: (1) titrations involving a strong acid and a strong base, (2) titrations involving a weak acid and a strong base, and (3) titrations involving a strong acid and a weak base. Titrations involving a weak acid and a weak base are more complicated, due to the hydrolysis of both the cation and the anion of the salt formed. These titrations will not be dealt with here. The reactions we will study are similar to those discussed in Section 18.6.

Titrations Involving a Strong Acid and a Strong Base

The reaction between a strong acid (say, HCl) and a strong base (say, NaOH) is represented by

$$HCl(aq) + NaOH(aq) \longrightarrow NaCl(aq) + H_2O(l)$$

or in terms of a net ionic equation by

$$H^+(aq) + OH^-(aq) \longrightarrow H_2O(l)$$

Consider the addition of a 0.100 M NaOH solution (from a buret) to an Erlenmeyer flask containing 25.0 mL of 0.100 M HCl. Figure 20.3 shows

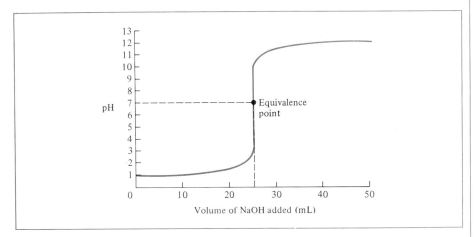

FIGURE 20.3   *pH profile of a strong acid–strong base titration. A 0.100 M NaOH solution is added from a buret to 25.0 mL of a 0.100 M HCl solution in an Erlenmeyer flask (see Figure 4.7). This curve is sometimes referred to as a titration curve.*

the pH profile of the titration (also known as the *titration curve*). Before the addition of NaOH, the pH of the acid is given by $-\log(0.100)$, or 1.00. When NaOH is added, the pH of the solution increases slowly at first. Near the equivalence point the pH begins to rise steeply and, at the equivalence point (that is, the point at which equimolar amounts of acid and base have reacted), the curve rises almost vertically. In a strong acid-strong base titration both the hydrogen ion and hydroxide ion concentrations are very small at the equivalence point; consequently, the addition of a drop of the base can cause a large increase in $[OH^-]$ and hence the pH of the solution. Beyond the equivalence point, the pH increases slowly with the addition of NaOH.

It is possible to calculate the pH of the solution at every stage of titration. Here are three sample calculations:

1. *After the addition of 10.0 mL of 0.100 M NaOH to 25.0 mL of 0.100 M HCl.*

The total volume of the solution is 35.0 mL. The number of moles of NaOH in 10.0 mL is

$$10.0 \text{ mL} \times \frac{0.100 \text{ mol NaOH}}{1 \text{ L NaOH}} \times \frac{1 \text{ L}}{1000 \text{ mL}} = 1.00 \times 10^{-3} \text{ mol}$$

The number of moles of HCl originally present in 25.0 mL of solution is

$$25.0 \text{ mL} \times \frac{0.100 \text{ mol HCl}}{1 \text{ L HCl}} \times \frac{1 \text{ L}}{1000 \text{ mL}} = 2.50 \times 10^{-3} \text{ mol}$$

Thus, the amount of HCl left after partial neutralization is (2.50×10^{-3}) $- (1.00 \times 10^{-3})$, or 1.50×10^{-3} mol. Next, the concentration of H^+ ions in 35.0 mL of solution is found as follows:

$$\frac{1.50 \times 10^{-3} \text{ mol HCl}}{35.0 \text{ mL}} \times \frac{1000 \text{ mL}}{1 \text{ L}} = 0.0429 \text{ mol HCl/L}$$
$$= 0.0429 \text{ M HCl}$$

For convenience we will use only three significant figures for titration calculations and two significant figures (the two decimal places) for pH values.

Keep in mind that 1 mol NaOH ⇌ 1 mol HCl.

Thus $[H^+] = 0.0429\ M$. Therefore the pH of the solution is

$$pH = -\log 0.0429 = 1.37$$

2. After the addition of 25.0 mL of 0.100 M NaOH to 25.0 mL of 0.100 M HCl.

Neither the Na^+ ion nor the Cl^- ion undergoes hydrolysis.

   This is a simple calculation, because it involves a complete neutralization reaction. At the equivalence point, $[H^+] = [OH^-]$ and the pH of the solution is 7.00.

3. After the addition of 35.0 mL of 0.100 M NaOH to 25.0 mL of 0.100 M HCl.

   The total volume of the solution is 60.0 mL. The number of moles of NaOH present in 35.0 mL is

$$35.0\ \text{mL} \times \frac{0.100\ \text{mol NaOH}}{1\ \text{L NaOH}} \times \frac{1\ \text{L}}{1000\ \text{mL}} = 3.50 \times 10^{-3}\ \text{mol}$$

The number of moles of HCl in 25.0 mL solution is 2.50×10^{-3}. After complete neutralization of HCl, the number of moles of NaOH left is $(3.50 \times 10^{-3}) - (2.50 \times 10^{-3})$, or 1.00×10^{-3} mol. The concentration of NaOH in 60.0 mL of solution is

$$\frac{1.00 \times 10^{-3}\ \text{mol NaOH}}{60.0\ \text{mL}} \times \frac{1000\ \text{mL}}{1\ \text{L}} = 0.0167\ \text{mol NaOH/L}$$
$$= 0.0167\ M\ \text{NaOH}$$

Thus $[OH^-] = 0.0167\ M$ and $pOH = -\log 0.0167 = 1.78$. Hence, the pH of the solution is

$$pH = 14.00 - pOH$$
$$= 14.00 - 1.78$$
$$= 12.22$$

Table 20.1 shows a more complete set of data for the titration.

TABLE 20.1 Titration Data for HCl versus NaOH*

Volume of NaOH Added (mL)	Total Volume (mL)	Mole Excess of Acid or Base	pH of Solution
0.0	25.0	$2.50 \times 10^{-3}\ H^+$	1.00
10.0	35.0	1.50×10^{-3}	1.37
20.0	45.0	5.00×10^{-4}	1.95
24.0	49.0	1.00×10^{-4}	2.69
24.5	49.5	5.00×10^{-5}	3.00
25.0	50.0	0.00	7.00
25.5	50.5	$5.00 \times 10^{-5}\ OH^-$	11.00
26.0	51.0	1.00×10^{-4}	11.29
30.0	55.0	5.00×10^{-4}	11.96
40.0	65.0	1.50×10^{-3}	12.36
50.0	75.0	2.50×10^{-3}	12.52

* The titration is carried out by slowly adding 0.100 M NaOH solution from a buret to 25.0 mL of a 0.100 M HCl solution in an Erlenmeyer flask.

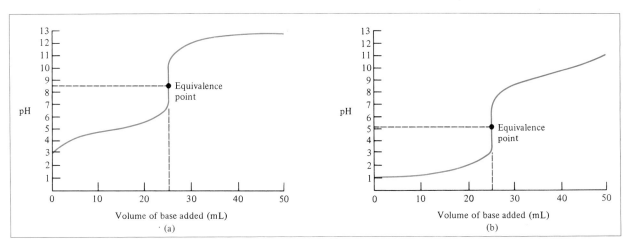

FIGURE 20.4 *Titration curves. (a) Weak acid versus strong base. (b) Strong acid versus weak base. In each case, a 0.100 M base solution is added from a buret to 25.0 mL of a 0.100 M acid solution in an Erlenmeyer flask. Note that as a result of salt hydrolysis, the pH at the equivalence point is greater than 7 for (a) and less than 7 for (b).*

Titrations Involving a Weak Acid and a Strong Base

Consider the neutralization reaction between acetic acid (a weak acid) and sodium hydroxide (a strong base):

$$CH_3COOH(aq) + NaOH(aq) \longrightarrow CH_3COONa(aq) + H_2O(l)$$

This equation can be simplified to

$$CH_3COOH(aq) + OH^-(aq) \longrightarrow CH_3COO^-(aq) + H_2O(l)$$

The acetate ion undergoes hydrolysis as follows:

$$CH_3COO^-(aq) + H_2O(l) \rightleftharpoons CH_3COOH(aq) + OH^-(aq)$$

Therefore, at the equivalence point the pH will be *greater than* 7 as a result of the excess OH^- ions formed [Figure 20.4(a)].

EXAMPLE 20.4

Calculate the pH in the titration of 25.0 mL of 0.100 *M* acetic acid by sodium hydroxide after the addition to the acid solution of (a) 10.0 mL of 0.100 *M* NaOH, (b) 25.0 mL of 0.100 *M* NaOH, (c) 35.0 mL of 0.100 *M* NaOH.

Answer

The neutralization reaction is

$$CH_3COOH(aq) + NaOH(aq) \longrightarrow CH_3COONa(aq) + H_2O(l)$$

In each of the following cases [(a), (b), and (c)] we first calculate the number of moles of NaOH added to the acetic acid solution. Next, we calculate the

(Continued)

number of moles of the acid (or the base) left over after neutralization. Then we determine the pH of the solution.

(a) The number of moles of NaOH in 10.0 mL is

$$10.0 \text{ mL} \times \frac{0.100 \text{ mol NaOH}}{1 \text{ L NaOH soln}} \times \frac{1 \text{ L}}{1000 \text{ mL}} = 1.00 \times 10^{-3} \text{ mol}$$

The number of moles of CH_3COOH originally present in 25.0 mL of solution is

$$25.0 \text{ mL} \times \frac{0.100 \text{ mol } CH_3COOH}{1 \text{ L } CH_3COOH \text{ soln}} \times \frac{1 \text{ L}}{1000 \text{ mL}} = 2.50 \times 10^{-3} \text{ mol}$$

Thus, the amount of CH_3COOH left after all added base has been neutralized is

$$(2.50 \times 10^{-3} - 1.00 \times 10^{-3}) \text{ mol} = 1.50 \times 10^{-3} \text{ mol}$$

The amount of CH_3COONa formed is 1.00×10^{-3} mole, as can be seen from the equation

$$\begin{array}{llll} CH_3COOH(aq) & + \text{ NaOH}(aq) & \longrightarrow CH_3COONa(aq) & + H_2O(l) \\ 1.00 \times 10^{-3} \text{ mol} & 1.00 \times 10^{-3} \text{ mol} & 1.00 \times 10^{-3} \text{ mol} \end{array}$$

At this stage we have a buffer system made up of CH_3COONa and CH_3COOH. To calculate the pH of this solution we write

$$\begin{aligned} [H^+] &= \frac{[CH_3COOH]K_a}{[CH_3COO^-]} \\ &= \frac{(1.50 \times 10^{-3})(1.8 \times 10^{-5})}{1.00 \times 10^{-3}} \\ &= 2.7 \times 10^{-5} \text{ } M \end{aligned}$$

$$\begin{aligned} pH &= -\log (2.7 \times 10^{-5}) \\ &= 4.57 \end{aligned}$$

Since the volume of the solution is the same for CH_3COOH and CH_3COO^-, the ratio of the number of moles present is equal to the ratio of their molar concentrations.

(b) These data (that is, 25.0 mL of 0.100 M NaOH reacting with 25.0 mL of 0.100 M CH_3COOH) correspond to the equivalence point. The number of moles of both NaOH and CH_3COOH in 25.0 ml are 2.50×10^{-3} mol, so the amount of the salt formed is

$$\begin{array}{llll} CH_3COOH(aq) & + \text{ NaOH}(aq) & \longrightarrow CH_3COONa(aq) & + H_2O(l) \\ 2.50 \times 10^{-3} \text{ mol} & 2.50 \times 10^{-3} \text{ mol} & 2.50 \times 10^{-3} \text{ mol} \end{array}$$

The total volume is 50.0 mL so the concentration of the salt is

$$\begin{aligned} [CH_3COONa] &= \frac{2.50 \times 10^{-3} \text{ mol}}{50.0 \text{ mL}} \times \frac{1000 \text{ mL}}{1 \text{ L}} \\ &= 0.0500 \text{ mol/L} = 0.0500 \text{ } M \end{aligned}$$

The next step is to calculate the pH of the solution that results from the hydrolysis of the CH_3COO^- ions. Following the procedure described in Example 19.9, we find that the pH of the solution at the equivalence point is 8.72.

(c) After the addition of 35.0 mL of NaOH the solution is well past the equivalence point. At this stage we have two types of species that are responsible for making the solution basic: CH_3COO^- and OH^-. However, since OH^- is a much stronger base than CH_3COO^-, we can safely neglect the CH_3COO^- ions and calculate the pH of the solution using only the concentration of the OH^- ions. Only 25.0 mL of NaOH is needed for complete neutralization, so the number of moles of NaOH left after neutralization is

$$(35.0 - 25.0) \text{ mL} \times \frac{0.100 \text{ mol NaOH}}{1 \text{ L NaOH soln}} \times \frac{1 \text{ L}}{1000 \text{ mL}} = 1.00 \times 10^{-3} \text{ mol}$$

The total volume of the combined solutions is now 60.0 mL, so we calculate OH^- concentration as follows:

$$[OH^-] = \frac{1.00 \times 10^{-3} \text{ mol}}{60.0 \text{ mL}} \times \frac{1000 \text{ mL}}{1 \text{ L}}$$
$$= 0.0167 \text{ mol/L} = 0.0167 \text{ } M$$

$$pOH = -\log 0.0167$$
$$= 1.78$$

$$pH = 14.00 - pOH$$
$$= 14.00 - 1.78$$
$$= 12.22$$

Similar example: Problem 20.21.

Titrations Involving a Strong Acid and a Weak Base

Consider the titration between hydrochloric acid (a strong acid) and ammonia (a weak base):

$$HCl(aq) + NH_3(aq) \longrightarrow NH_4Cl(aq)$$

or simply

$$H^+(aq) + NH_3(aq) \longrightarrow NH_4^+(aq)$$

The pH at the equivalence point is *less than* 7 as a result of salt hydrolysis [Figure 20.4(b)]:

$$NH_4^+(aq) + H_2O(l) \longrightarrow NH_3(aq) + H_3O^+(aq)$$

or simply

$$NH_4^+(aq) \longrightarrow NH_3(aq) + H^+(aq)$$

EXAMPLE 20.5

Calculate the pH in the titration of hydrochloric acid by ammonia after adding (a) 10.0 mL of 0.100 M NH_3, (b) 25.0 mL of 0.100 M NH_3, and (c) 35.0 mL of 0.100 M NH_3, to 25.0 mL of 0.100 M HCl.

(Continued)

Answer

The neutralization reaction is

$$HCl(aq) + NH_3(aq) \longrightarrow NH_4Cl(aq)$$

In each of the following cases [(a), (b), and (c)] we first calculate the number of moles of NH_3 added to the hydrochloric acid solution. Next, we calculate the number of moles of the acid (or the base) left over after neutralization. Then we determine the pH of the solution.

(a) The number of moles of NH_3 in 10.0 mL of 0.100 M NH_3 is

$$10.0 \text{ mL} \times \frac{0.100 \text{ mol } NH_3}{1 \text{ L } NH_3 \text{ soln}} \times \frac{1 \text{ L}}{1000 \text{ mL}} = 1.00 \times 10^{-3} \text{ mol}$$

The number of moles of HCl originally present in 25.0 mL of solution is

$$25.0 \text{ mL} \times \frac{0.100 \text{ mol HCl}}{1 \text{ L HCl soln}} \times \frac{1 \text{ L}}{1000 \text{ mL}} = 2.50 \times 10^{-3} \text{ mol}$$

Thus, the amount of HCl left after all added base has been neutralized is

$$(2.50 \times 10^{-3} - 1.00 \times 10^{-3}) \text{ mol} = 1.50 \times 10^{-3} \text{ mol}$$

At this stage, two types of species in solution act to decrease the pH of the solution: The NH_4^+ ions hydrolyze to produce H^+ ions and the HCl ionizes to yield H^+ ions. Since the percent hydrolysis of NH_4^+ is small (see Example 19.10) and HCl is a strong acid, the pH of the solution is largely determined by the remaining acid, whose concentration is

$$[HCl] = \frac{1.50 \times 10^{-3} \text{ mol}}{35.0 \text{ mL}} \times \frac{1000 \text{ mL}}{1 \text{ L}}$$
$$= 0.0429 \text{ mol/L} = 0.0429 \text{ } M$$

Thus the pH of the solution is

$$pH = -\log 0.0429$$
$$= 1.37$$

(b) These data (that is, 25.0 mL of 0.100 M NH_3 reacting with 25.0 mL of 0.100 M HCl) correspond to the equivalence point. The amount of the salt formed is

$$\begin{array}{ccc} HCl(aq) & + NH_3(aq) & \longrightarrow NH_4Cl(aq) \\ 2.50 \times 10^{-3} \text{ mol} & 2.50 \times 10^{-3} \text{ mol} & 2.50 \times 10^{-3} \text{ mol} \end{array}$$

The total volume is 50.0 mL so the concentration of NH_4Cl is

$$[NH_4Cl] = \frac{2.50 \times 10^{-3} \text{ mol}}{50.0 \text{ mL}} \times \frac{1000 \text{ mL}}{1 \text{ L}}$$
$$= 0.0500 \text{ mol/L} = 0.0500 \text{ } M$$

NH_4Cl is a strong electrolyte.

Following the procedure described in Example 19.10, we find that the pH of the solution due to the hydrolysis of NH_4^+ ions is 5.28.

(c) After adding 35.0 mL of NH_3 all of the HCl will be neutralized and the solution will now contain NH_3 and NH_4^+, which are the ingredients for a buffer system. Since 25.0 mL of NH_3 is used to react with HCl, the number of moles of NH_3 left over is that present in 10.0 mL of the solution, which is 1.00 $\times$

10^{-3} mol [see part (a)]. From part (b) we find that the number of moles of NH_4^+ present is 2.50×10^{-3}. Therefore the H^+ concentration is

$$[H^+] = \frac{[NH_4^+]K_a}{[NH_3]}$$
$$= \frac{(2.50 \times 10^{-3})(5.6 \times 10^{-10})}{1.00 \times 10^{-3}}$$
$$= 1.4 \times 10^{-9} \text{ mol/L} = 1.4 \times 10^{-9} \ M$$

Thus the pH of the solution is

$$pH = -\log (1.4 \times 10^{-9})$$
$$= 8.85$$

Similar example: Problem 20.22.

Since the volume of the solution is the same for NH_4^+ and NH_3, the ratio of the number of moles present is equal to the ratio of their molar concentrations.

20.4 ACID-BASE INDICATORS

In this section we will discuss the role of indicators in acid-base titrations. You will recall from Chapter 4 that an indicator is usually a weak organic acid or base that has distinctly different colors in its nonionized and ionized forms. These two forms are related to the pH of the solution in which the indicator is dissolved, as we will soon see. The color change of an indicator can be used to follow the progress of a titration.

Indicators do not all change color at the same pH, so the choice of indicator for a particular titration depends on the nature of acid and base used (that is, whether they are strong or weak). Let us consider a weak monoprotic acid, HIn, which acts as an indicator. In solution

$$HIn(aq) \rightleftharpoons H^+(aq) + In^-(aq)$$

If the indicator is in a sufficiently acidic medium, the equilibrium, according to Le Chatelier's principle, will shift to the left and the predominant color of the indicator will be that of the nonionized form (HIn). On the other hand, in a basic medium the equilibrium will shift to the right and the color of the conjugate base (In^-) will predominate. Roughly speaking, we can use the following ratios of concentrations to predict the perceived color of the indicator:

We assume that HIn and In^- have different colors.

$$\frac{[HIn]}{[In^-]} \geq 10 \qquad \text{color of acid (HIn) predominates}$$

$$\frac{[In^-]}{[HIn]} \geq 10 \qquad \text{color of conjugate base } (In^-) \text{ predominates}$$

If $[HIn] \simeq [In^-]$ then the indicator color is a combination of the colors of HIn and In^-.

In Section 4.6 we mentioned that phenolphthalein is a suitable indicator for the titration of NaOH and HCl. Phenolphthalein is colorless in acidic and neutral solutions, but reddish pink in basic solutions. Measurements show that at pH < 8.3 the indicator is colorless but it begins to turn reddish pink when the pH exceeds 8.3. The steepness of the pH curve in Figure 20.3 tells us that, as we approach the equivalence point, the addition of a very small quantity of NaOH (say 0.05 mL, which is about the volume of a drop from the buret) will bring about a large increase in the pH of the solution. What is important, however, is the fact that the steep portion of the pH profile covers the range over which phenolphthalein changes from colorless to reddish pink. Whenever such a matching occurs, the indicator can be used to locate the equivalence point of the titration.

Many acid-base indicators are plant pigments. For example, by boiling chopped red cabbage in water we can extract the pigments that exhibit many different colors at various pHs, as shown in Color Plate 24.

Table 20.2 lists a number of indicators commonly used in acid-base titrations. The choice of an indicator depends on the strengths of the acid and base in a particular titration, as discussed in Section 20.3. The criterion for choosing the proper indicator or indicators for a given titration, therefore, is whether the pH range over which the indicator changes color corresponds with the steep portion of the titration curves shown in Figures 20.3 and 20.4. If this criterion is not met, then the indicator will not accurately locate the position of the equivalence point.

TABLE 20.2 Some Common Acid-Base Indicators

Indicator	Color in acid	Color in base	pH Range*
Thymol blue	Red	Yellow	1.2–2.8
Bromophenol blue	Yellow	Blue	3.0–4.6
Methyl orange	Orange	Yellow	3.1–4.4
Methyl red	Red	Yellow	4.2–6.3
Chlorophenol blue	Yellow	Red	4.8–6.4
Bromothymol blue	Yellow	Blue	6.0–7.6
Cresol red	Yellow	Red	7.2–8.8
Phenolphthalein	Colorless	Reddish pink	8.3–10.0

* The pH range is defined as the range over which the indicator changes from the acid color to the base color.

EXAMPLE 20.7

Which indicator or indicators listed in Table 20.2 would you use for the acid-base titrations shown in (a) Figure 20.3, (b) Figure 20.4(a), and (c) Figure 20.4(b)?

Answer

(a) Near the equivalence point, the pH of the solution changes abruptly from 4 to 10. Therefore, except for thymol blue, bromophenol blue, and methyl orange, all of the other indicators are suitable for use in the titration. (b) Here the steep portion covers the pH range between 7 and 10; therefore, the suitable indicators are cresol red and phenolphthalein. (c) Here the steep portion of the pH curve covers the pH range between 3 and 7; therefore, the suitable indicators are bromophenol blue, methyl orange, methyl red, and chlorophenol blue.

Similar example: Problem 20.31.

SUMMARY

1. The common ion effect tends to suppress the ionization of a weak acid or a weak base. Its action can be explained by Le Chatelier's principle.
2. A buffer solution is a combination of a weak acid and its weak conjugate base; the solution reacts with small amounts of added acid or base in such a way that the pH of the solution remains nearly constant. Buffer systems play an important role in maintaining the pH of body fluids.
3. The pH at the equivalence point of an acid-base titration depends on the hydrolysis of the salt formed in the neutralization reaction. For strong acid–strong base titrations, the pH at the equivalence point is 7; for weak acid–strong base titrations, the pH at the equivalence point is greater than 7; for strong acid–weak base titrations, the pH at the equivalence point is less than 7.
4. Acid-base indicators are weak organic acids or bases that can be used to detect the equivalence point in an acid-base neutralization reaction.

KEY WORDS

Buffer range, p. 598
Buffer solution, p. 594

Common ion effect, p. 591

PROBLEMS

More challenging problems are marked with an asterisk.

Common Ion Effect

20.1 Define the term "common ion effect." Explain the common ion effect in terms of Le Chatelier's principle.

20.2 Describe the change in pH (increase, decrease, or no change) that results from each of the following additions: (a) potassium acetate to an acetic acid solution, (b) ammonium nitrate to an ammonia solution, (c) sodium formate (HCOONa) to a formic acid (HCOOH) solution, (d) potassium chloride to a hydrochloric acid solution, (e) barium iodide to a barium nitrite solution.

20.3 Determine the pH of (a) a 0.40 M CH_3COOH solution and of (b) a solution that is 0.40 M in CH_3COOH and 0.20 M in CH_3COONa.

20.4 Determine the pH of (a) a 0.20 M NH_3 solution and of (b) a solution that is 0.20 M in NH_3 and 0.30 M in NH_4Cl.

20.5 Determine the hydrogen ion and acetate ion concentrations in (a) a 0.100 M CH_3COOH solution and (b) the concentrations of the same ions in a solution that is 0.100 M in both CH_3COOH and HCl.

Buffers

20.6 Define the terms "buffer" and "buffer range." What constitutes a buffer solution?

20.7 Specify which of the following systems can be classified as a buffer system: (a) KCl/HCl, (b) NH_3/NH_4NO_3, (c) Na_2HPO_4/NaH_2PO_4, (d) KNO_2/HNO_2, (e) $KHSO_4$/H_2SO_4, (f) HCOOK/HCOOH.

20.8 A diprotic acid, H_2A, has the following ionization constants: $K_{a_1} = 1.1 \times 10^{-3}$ and $K_{a_2} = 2.5 \times 10^{-6}$. In order to make up a buffer solution of pH 5.80, which combination would you choose: $NaHA$/H_2A or Na_2A/$NaHA$?

20.9 Calculate the pH of 1.00 L of the buffer 1.00 M CH_3COONa/1.00 M CH_3COOH before and after the addition of (a) 0.080 mol NaOH and (b) 0.12 mol HCl. (Assume that there is no change in volume.)

20.10 What is the pH of the buffer 0.10 M Na_2HPO_4/0.15 M KH_2PO_4?

20.11 Calculate the pH of the buffer system 0.15 M NH_3/0.35 M NH_4Cl.

20.12 Calculate the pH of a buffer solution prepared by adding 20.5 g of CH_3COOH and 17.8 g of CH_3COONa to enough water to make 5.00×10^2 mL of solution.

20.13 The pH of a bicarbonate–carbonic acid buffer is 8.00. Calculate the ratio of the concentration of carbonic acid to that of the bicarbonate ion.

20.14 Calculate the pH of the 0.20 M NH_3/0.20 M NH_4Cl buffer. What is the pH of the buffer after the addition of 10.0 mL of 0.10 M HCl to 65.0 mL of the buffer?

20.15 The pH of a sodium acetate/acetic acid buffer is 4.50. Calculate the ratio $[CH_3COO^-]/[CH_3COOH]$.

20.16 The pH of blood plasma is 7.40. Assuming the principal buffer system is HCO_3^-/H_2CO_3, calculate the ratio $[HCO_3^-]/[H_2CO_3]$. Is this buffer more effective against an added acid or an added base?

20.17 What mass in grams of sodium acetate must be dissolved in 0.400 L of 0.200 M acetic acid to make a buffer solution with a pH of 4.56? (Assume that the volume of the solution remains constant.)

Acid-Base Titrations

20.18 A quantity of 12.5 mL of 0.500 M H_2SO_4 exactly neutralizes 50.0 mL of NaOH. What is the concentration of the NaOH solution?

20.19 A quantity of 5.00 g of a diprotic acid is dissolved in water and made up to exactly 250 mL. Calculate the molar mass of the acid if 25.0 mL of this solution required 11.1 mL of 1.00 M KOH for neutralization.

20.20 Calculate the pH at the equivalence point for the following titrations: (a) 0.10 M HCl vs. 0.10 M NH_3, (b) 0.10 M CH_3COOH vs. 0.10 M NaOH.

*20.21 Exactly 100 mL of 0.100 M nitrous acid solution is titrated against a 0.100 M NaOH solution. Compute the pH for (a) the initial solution, (b) the point when 80.0 mL of the base has been added, (c) the equivalence point, (d) the point at which 105 mL of the base has been added.

*20.22 Calculate the pH at the following points in the titration of 50.0 mL of 0.100 M methylamine (CH_3NH_2), $K_b = 4.4 \times 10^{-4}$, with a 0.200 M HCl solution: (a) at the start before any acid has been added, (b) after adding 15.0 mL of HCl solution, (c) at the equivalence point.

*20.23 A sample of 0.1276 g of an unknown monoprotic acid was dissolved in 25.0 mL of water and titrated with 0.0633 M NaOH solution. The volume of base required to reach the equivalence point was 18.4 mL. (a) Calculate the molar mass of the acid. (b) After 10.0 mL of base had been added in the titration, the pH was determined to be 5.87. What is the K_a of the unknown acid?

20.24 A solution is made up by mixing exactly 500 mL of 0.167 M NaOH with exactly 500 mL 0.100 M CH_3COOH. Calculate the equilibrium concentrations of H^+, CH_3COOH, CH_3COO^-, OH^-, and Na^+.

20.25 Sketch titration curves for the following acid-base titrations: (a) HCl vs. NaOH, (b) HCl vs. NH_3, (c) CH_3COOH vs. NaOH. In each case, the acid is added to the base in an Erlenmeyer flask. Your graphs should show pH as the y axis and volume of acid added as the x axis.

Indicators

20.26 Explain how an acid-base indicator works in a titration.

20.27 What are the criteria for choosing an indicator for a particular acid-base titration?

20.28 The amount of indicator used in an acid-base titration must be kept small. Why?

20.29 A student carried out an acid-base titration by adding NaOH solution from a buret to an Erlenmeyer flask containing HCl solution and using phenolphthalein as indicator. At the equivalence point, she observed a faint reddish pink color. However, after a few minutes, the solution gradually turned colorless. What do you suppose happened?

*20.30 The ionization constant K_a of an indicator HIn is 1.0×10^{-6}. The color of the nonionized form is red and that of the ionized form is yellow. What is the color of this indicator in a solution whose pH is 4.00? (Hint: The color of an indicator can be estimated by considering the ratio $[HIn]/[In^-]$. If the ratio is equal to or greater than 10, the color will be that of the nonionized form. If the ratio is equal to or smaller than 0.1, the color will be that of the ionized form.)

20.31 Referring to Table 20.2, specify which indicator or indicators you would use for the following titrations: (a) HCOOH vs. NaOH, (b) HCl vs. KOH, (c) HNO_3 vs. NH_3.

Miscellaneous Problems

20.32 Cacodylic acid is $(CH_3)_2AsO_2$. Its ionization constant is 6.4×10^{-7}. (a) Calculate the pH of 50.0 mL of a 0.10 M solution of the acid. (b) Calculate the pH of 25.0 mL of 0.15 M $(CH_3)_2AsO_2Na$. (c) Mix the solutions in part (a) and part (b). Calculate the pH of the resulting solution.

20.33 A quantity of 0.560 g of KOH is added to 25.0 mL of 1.00 M HCl. Excess Na_2CO_3 is then added to the solution. What mass (in grams) of CO_2 is formed?

20.34 Calculate the number of molecules of water in

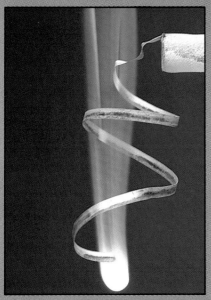

PLATE 12 The light blue flame seen when sulfur burns in air.

PLATE 11 (Left) A piece of magnesium ribbon burning in air.

PLATE 13 Top (left and right): Colors of $CuSO_4 \cdot 5H_2O$ and $CuSO_4$. Bottom (left and right): Colors of $CoCl_2 \cdot 6H_2O$ and $CoCl_2$.

PLATE 14 Left to right: When a piece of zinc metal is placed in an aqueous $CuSO_4$ solution, zinc atoms enter the solution as Zn^{2+} ions and Cu^{2+} ions are converted to metallic copper (first beaker). In time, the blue color of the solution disappears (second beaker). When a piece of copper wire is placed in an aqueous $AgNO_3$ solution, copper atoms enter the solution as Cu^{2+} ions and Ag^+ ions are converted to metallic silver (third beaker). Gradually, the solution acquires the characteristic blue color due to the presence of the Cu^{2+} ions in solution (fourth beaker).

PLATE 15 Left: An aqueous KBr solution. Right: After bubbling chlorine gas through the solution for a long time, most of the bromide ions are converted to liquid bromine.

PLATE 16 Hydrogen gas burning in air.

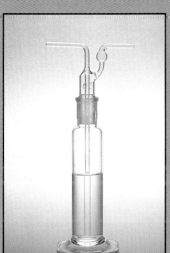

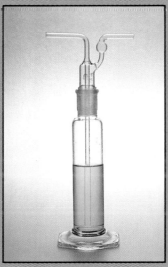

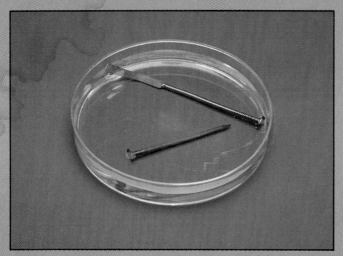

PLATE 26   An iron nail that is cathodically protected by a piece of zinc strip does not rust in water while an iron nail without such a protection rusts readily.

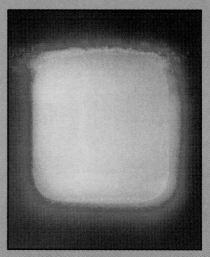

PLATE 27   The red glow of the radio-active plutonium-238 isotope. The orange color is due to the presence of its oxide.

PLATE 28   The explosion of a thermonuclear bomb.

PLATE 29   (Bottom left) Crystals of a salt composed of sodium anions and a complex of sodium cations with certain chelating agents.

PLATE 30   Left: Xenon gas (colorless) and PtF_6 (red gas) separated from each other. Right: When the two gases are mixed, a yellow-orange solid compound is formed.

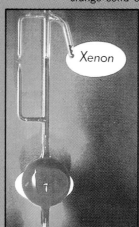

oxalic acid hydrate $H_2C_2O_4 \cdot xH_2O$, from the following data: 5.00 g of the compound is made up to exactly 250 mL solution and 25.0 mL of this solution requires 15.9 mL of 0.500 M sodium hydroxide solution for neutralization.

*20.35 A volume of 25.0 mL of 0.100 M HCl is titrated against a 0.100 M NH$_3$ solution added to it from a buret. Calculate the pH values of the solution (a) after 10.0 mL of NH$_3$ solution has been added, (b) after 25.0 mL of NH$_3$ solution has been added, (c) after 35.0 mL of NH$_3$ solution has been added.

20.36 The pK_a of butyric acid (HBut) is 4.7. Calculate K_b for the butyrate ion (But$^-$).

21
PRECIPITATION REACTIONS AND SOLUBILITY EQUILIBRIA

S olid deposits in tea kettles, the formation of limestone caves, and the identification of common ions in solution are all examples of applications of chemical equilibrium principles. Such macroscopic examples of solubility equilibria will be examined in this chapter.

The chemical systems considered here may seem unique, but you will find that molecular-level insights regarding dynamic equilibrium, Le Chatelier's principle, and calculations based on equilibrium constant expressions are basic to these systems. Thus this chapter is as much an application of knowledge you have already learned as it is an important extension of our understanding of chemical equilibrium.

21.1 SOLUBILITY AND SOLUBILITY PRODUCT

The general rules for predicting the solubility of ionic compounds were introduced earlier in Section 3.2. These solubility rules, although useful, do not allow us to make quantitative predictions regarding the extent to which given ionic compounds will dissolve in water. To develop such a quantitative treatment, all that is needed is the realization that solubility can be analyzed as a special case of chemical equilibrium.

Consider a saturated solution of silver chloride that is in contact with solid silver chloride. The solubility equilibrium can be represented as

$$AgCl(s) \rightleftharpoons Ag^+(aq) + Cl^-(aq)$$

Since AgCl is a strong electrolyte, all of the AgCl that dissolves in water dissociates completely into Ag^+ and Cl^- ions. We know from Chapter 17 that for heterogeneous reactions the concentration of the solid is a constant. Thus we can write the equilibrium constant for the dissolution of AgCl (see Example 17.5) as

$$K_{sp} = [Ag^+][Cl^-]$$

where K_{sp} is called the solubility product constant or simply the solubility product. In general, the **solubility product** of a compound is the *product of the molar concentrations of the constituent ions, each raised to the power of its stoichiometric coefficient in the equilibrium equation.* The value of K_{sp} indicates the solubility of a compound—the smaller the K_{sp}, the less soluble is the compound.

Because AgCl contains only two monovalent ions, its solubility product expression is particularly simple to write. The following are three cases that are more complex.

- MgF_2

$$MgF_2(s) \rightleftharpoons Mg^{2+}(aq) + 2F^-(aq)$$
$$K_{sp} = [Mg^{2+}][F^-]^2$$

- Ag_2S

$$Ag_2S(s) \rightleftharpoons 2Ag^+(aq) + S^{2-}(aq)$$
$$K_{sp} = [Ag^+]^2[S^{2-}]$$

- $Ca_3(PO_4)_2$

$$Ca_3(PO_4)_2(s) \rightleftharpoons 3Ca^{2+}(aq) + 2PO_4^{3-}(aq)$$
$$K_{sp} = [Ca^{2+}]^3[PO_4^{3-}]^2$$

Table 21.1 lists the solubility products of a number of salts of low solubilities. Soluble salts such as NaCl and KNO_3, which have very large K_{sp} values, are not listed in the table.

Consider an aqueous solution containing Ag^+ and Cl^- ions at 25°C. Any one of the following conditions may exist: (1) the solution is unsaturated, (2) the solution is saturated, or (3) the solution is supersaturated. Following the procedure in Section 17.4 we use Q, called the *ion product*, to represent the product of the molar concentrations of the ions raised to the power of their stoichiometric coefficients:

Unless otherwise stated, all discussion of solubility refers to water as solvent at a temperature of 25°C.

Not listing K_{sp} values for soluble salts is somewhat analogous to not listing K_a values for strong acids in Table 19.1.

$$Q = (Ag^+)(Cl^-)$$

The parentheses ( ) remind us that these concentrations do not necessarily correspond to those involved in an equilibrium state between the ions and the solid. The possible relationships between Q and K_{sp} are

$Q < K_{sp}$ unsaturated solution
$(Ag^+)(Cl^-) < 1.6 \times 10^{-10}$

$Q = K_{sp}$ saturated solution
$[Ag^+][Cl^-] = 1.6 \times 10^{-10}$

$Q > K_{sp}$ supersaturated solution; some AgCl
$(Ag^+)(Cl^-) > 1.6 \times 10^{-10}$ precipitate will form until the product of the ionic concentrations is equal to 1.6×10^{-10}

TABLE 21.1 Solubility Products of Some Slightly Soluble Salts at 25°C

Name	Formula	K_{sp}
Aluminum hydroxide	$Al(OH)_3$	1.8×10^{-33}
Barium carbonate	$BaCO_3$	8.1×10^{-9}
Barium fluoride	BaF_2	1.7×10^{-6}
Barium sulfate	$BaSO_4$	1.1×10^{-10}
Bismuth sulfide	Bi_2S_3	1.6×10^{-72}
Cadmium sulfide	CdS	1.4×10^{-28}
Calcium carbonate	$CaCO_3$	8.7×10^{-9}
Calcium fluoride	CaF_2	4.0×10^{-11}
Chromium(III) hydroxide	$Cr(OH)_3$	3.0×10^{-29}
Cobalt(II) sulfide	CoS	4.0×10^{-21}
Copper(II) sulfide	CuS	1.0×10^{-44}
Copper(I) bromide	$CuBr$	4.2×10^{-8}
Copper(I) iodide	CuI	5.1×10^{-12}
Copper(II) hydroxide	$Cu(OH)_2$	2.2×10^{-20}
Iron(II) hydroxide	$Fe(OH)_2$	1.6×10^{-14}
Iron(III) hydroxide	$Fe(OH)_3$	1.1×10^{-36}
Iron(II) sulfide	FeS	3.7×10^{-19}
Lead(II) carbonate	$PbCO_3$	3.3×10^{-14}
Lead(II) chloride	$PbCl_2$	2.4×10^{-4}
Lead(II) fluoride	PbF_2	4.1×10^{-8}
Lead(II) iodide	PbI_2	1.4×10^{-8}
Lead(II) sulfide	PbS	3.4×10^{-28}
Magnesium hydroxide	$Mg(OH)_2$	1.2×10^{-11}
Manganese(II) sulfide	MnS	1.4×10^{-15}
Mercury(I) chloride	Hg_2Cl_2	3.5×10^{-18}
Mercury(II) sulfide	HgS	4.0×10^{-54}
Nickel(II) sulfide	NiS	1.4×10^{-24}
Silver bromide	$AgBr$	7.7×10^{-13}
Silver chloride	$AgCl$	1.6×10^{-10}
Silver iodide	AgI	8.3×10^{-17}
Silver sulfide	Ag_2S	1.8×10^{-49}
Tin(II) sulfide	SnS	1.0×10^{-26}
Strontium carbonate	$SrCO_3$	1.6×10^{-9}
Zinc hydroxide	$Zn(OH)_2$	1.8×10^{-14}
Zinc sulfide	ZnS	1.2×10^{-23}

Calculations involving K_{sp} can be considered in three categories:

- Calculating the concentration of one of the ions in a saturated solution, given the concentrations of other ions (Example 21.1).
- Calculating K_{sp} from solubility data (Examples 21.2 and 21.3).
- Calculating solubility from K_{sp} data (Examples 21.4 and 21.5).

In these examples we assume that the temperature is at 25°C in all calculations. All of the K_{sp} values are obtained from Table 21.1.

EXAMPLE 21.1

(a) In a saturated solution of cadmium sulfide, the concentration of the cadmium ions is 2.6×10^{-8} M. What is $[S^{2-}]$? (b) In a saturated solution of magnesium hydroxide, the concentration of the hydroxide ions is 8.3×10^{-4} M. What is $[Mg^{2+}]$?

Answer

(a) The solubility equilibrium is given by

$$CdS(s) \rightleftharpoons Cd^{2+}(aq) + S^{2-}(aq)$$

Therefore

$$K_{sp} = [Cd^{2+}][S^{2-}]$$
$$1.4 \times 10^{-28} = (2.6 \times 10^{-8})[S^{2-}]$$

$$[S^{2-}] = \frac{1.4 \times 10^{-28}}{2.6 \times 10^{-8}}$$
$$= 5.4 \times 10^{-21} \ M$$

(b) At equilibrium

$$Mg(OH)_2(s) \rightleftharpoons Mg^{2+}(aq) + 2OH^-(aq)$$

Therefore

$$K_{sp} = [Mg^{2+}][OH^-]^2$$

$$[Mg^{2+}] = \frac{1.2 \times 10^{-11}}{(8.3 \times 10^{-4})^2}$$
$$= 1.7 \times 10^{-5} \ M$$

Similar example: Problem 21.4.

EXAMPLE 21.2

It is found experimentally that the solubility of calcium sulfate is 0.67 g/L. Calculate the value of K_{sp} for calcium sulfate.

(Continued)

In this type of calculation we define solubility as grams of solute per liter of solution (g/L).

Answer

First we calculate the number of moles of $CaSO_4$ that are dissolved in one liter of solution:

$$\frac{0.67 \text{ g } CaSO_4}{1 \text{ L soln}} \times \frac{1 \text{ mol } CaSO_4}{136.2 \text{ g } CaSO_4} = 4.9 \times 10^{-3} \text{ mol/L}$$

In the solubility equilibrium

$$CaSO_4(s) \rightleftharpoons Ca^{2+}(aq) + SO_4^{2-}(aq)$$

we see that for every mole of $CaSO_4$ that dissolves, one mole of Ca^{2+} and one mole of SO_4^{2-} are produced. Thus, at equilibrium

$$[Ca^{2+}] = 4.9 \times 10^{-3} \text{ } M \quad \text{and} \quad [SO_4^{2-}] = 4.9 \times 10^{-3} \text{ } M$$

Now we can calculate K_{sp}:

$$\begin{aligned} K_{sp} &= [Ca^{2+}][SO_4^{2-}] \\ &= (4.9 \times 10^{-3})(4.9 \times 10^{-3}) \\ &= 2.4 \times 10^{-5} \end{aligned}$$

Similar examples: Problems 21.6, 21.8, 21.9.

EXAMPLE 21.3

The molar solubility of silver sulfate is 1.5×10^{-2} mol/L. Calculate the solubility product of the salt.

Answer

First we write the solubility equilibrium equation:

$$Ag_2SO_4(s) \rightleftharpoons 2Ag^+(aq) + SO_4^{2-}(aq)$$

From the stoichiometry we see that one mole of Ag_2SO_4 produces two moles of Ag^+ and one mole of SO_4^{2-} in solution. Therefore, when 1.5×10^{-2} mol Ag_2SO_4 is dissolved in one liter of solution, the concentrations are

$$[Ag^+] = 2(1.5 \times 10^{-2}) \text{ } M = 3.0 \times 10^{-2} \text{ } M$$

$$[SO_4^{2-}] = 1.5 \times 10^{-2} \text{ } M$$

Now we can calculate the solubility product constant:

$$\begin{aligned} K_{sp} &= [Ag^+]^2[SO_4^{2-}] \\ &= (3.0 \times 10^{-2})^2(1.5 \times 10^{-2}) \\ &= 1.4 \times 10^{-5} \end{aligned}$$

Similar example: Problem 21.7.

The calculation of solubility from K_{sp} data is more involved. In solving problems of this type, we follow essentially the same procedure we used in Section 17.4 to calculate equilibrium concentration from the equilibrium constant.

EXAMPLE 21.4

From the data in Table 21.1, calculate the molar solubility of calcium carbonate ($CaCO_3$), that is, the number of moles of $CaCO_3$ dissolved in one liter of solution.

Answer

Step 1

Let s be the molar solubility (in mol/L) of $CaCO_3$. Since one unit of $CaCO_3$ yields one Ca^{2+} ion and one CO_3^{2-} ion, at equilibrium both $[Ca^{2+}]$ and $[CO_3^{2-}]$ are equal to s. We summarize the changes in concentrations as follows:

$$CaCO_3(s) \rightleftharpoons Ca^{2+}(aq) + CO_3^{2-}(aq)$$

	Ca^{2+}	CO_3^{2-}
Initial	0.00 M	0.00 M
Change	$+s$ M	$+s$ M
Equilibrium	s M	s M

Step 2

$$K_{sp} = [Ca^{2+}][CO_3^{2-}]$$
$$8.7 \times 10^{-9} = (s)(s)$$
$$s = (8.7 \times 10^{-9})^{\frac{1}{2}} = 9.3 \times 10^{-5} \ M$$

Step 3

At equilibrium, therefore

$$[Ca^{2+}] = 9.3 \times 10^{-5} \ M$$

$$[CO_3^{2-}] = 9.3 \times 10^{-5} \ M$$

Because one mole of $CaCO_3$ yields one mole of Ca^{2+} and one mole of CO_3^{2-}, the molar solubility of $CaCO_3$ also is $9.3 \times 10^{-5} \ M$.

Similar example: Problem 21.7.

EXAMPLE 21.5

From the data in Table 21.1, calculate the solubility of bismuth sulfide (Bi_2S_3) in g/L.

Answer

Step 1

Let s be the molar solubility of Bi_2S_3. Since one unit of Bi_2S_3 yields two Bi^{3+} ions and three S^{2-} ions, at equilibrium $[Bi^{3+}]$ is $2s$ and $[S^{2-}]$ is $3s$. We summarize the changes in concentrations as follows:

$$Bi_2S_3(s) \rightleftharpoons 2Bi^{3+}(aq) + 3S^{2-}(aq)$$

	Bi^{3+}	S^{2-}
Initial	0.00 M	0.00 M
Change	$+2s$ M	$+3s$ M
Equilibrium	$2s$ M	$3s$ M

(Continued)

Step 2

$$K_{sp} = [Bi^{3+}]^2[S^{2-}]^3$$
$$1.6 \times 10^{-72} = (2s)^2(3s)^3 = 108s^5$$
$$s^5 = \frac{1.6 \times 10^{-72}}{108} = 1.5 \times 10^{-74}$$

Solving for s, we get

$$s = 1.7 \times 10^{-15} \, M$$

Knowing the molar mass of Bi_2S_3 (514.2 g) and its molar solubility, we can calculate its solubility as follows:

$$\text{solubility of } Bi_2S_3 = \frac{1.7 \times 10^{-15} \text{ mol } Bi_2S_3}{1 \text{ L soln}} \times \frac{514.2 \text{ g } Bi_2S_3}{1 \text{ mol } Bi_2S_3}$$
$$= 8.7 \times 10^{-13} \text{ g/L}$$

Similar example: Problem 21.5.

In carrying out calculations involving solubility and/or solubility product, it is important to keep the following points in mind:

- The solubility is the quantity of substance that dissolves in a certain quantity of water. In solubility equilibria calculations, it is usually expressed as *grams* of solute per liter of solution. Molar solubility is the number of *moles* of solute per liter of solution.
- The solubility product is an equilibrium constant.
- Both solubility and solubility product refer to a *saturated solution*.

As Examples 21.1–21.5 show, solubility and solubility product are related. If we know one, we can calculate the other, but each quantity provides us

TABLE 21.2   Relationship Between K_{sp} and Molar Solubility (s)

Compound	K_{sp} Expression	Equilibrium Concentration (M) cation	anion	Relation Between K_{sp} and s
AgCl	$[Ag^+][Cl^-]$	s	s	$K_{sp} = s^2$; $s = (K_{sp})^{\frac{1}{2}}$
BaSO$_4$	$[Ba^{2+}][SO_4^{2-}]$	s	s	$K_{sp} = s^2$; $s = (K_{sp})^{\frac{1}{2}}$
Ag$_2$CO$_3$	$[Ag^+]^2[CO_3^{2-}]$	$2s$	s	$K_{sp} = 4s^3$; $s = \left(\dfrac{K_{sp}}{4}\right)^{\frac{1}{3}}$
PbF$_2$	$[Pb^{2+}][F^-]^2$	s	$2s$	$K_{sp} = 4s^3$; $s = \left(\dfrac{K_{sp}}{4}\right)^{\frac{1}{3}}$
Al(OH)$_3$	$[Al^{3+}][OH^-]^3$	s	$3s$	$K_{sp} = 27s^4$; $s = \left(\dfrac{K_{sp}}{27}\right)^{\frac{1}{4}}$
Bi$_2$S$_3$	$[Bi^{3+}]^2[S^{2-}]^3$	$2s$	$3s$	$K_{sp} = 108s^5$; $s = \left(\dfrac{K_{sp}}{108}\right)^{\frac{1}{5}}$

with different information. Table 21.2 shows the relationship between molar solubility and solubility product for a number of ionic compounds.

Predicting Precipitation Reactions

From a knowledge of the solubility rules (see Section 3.2) and the solubility products listed in Table 21.1, we can predict whether a precipitate will form when we mix two solutions. Example 21.6 illustrates the steps involved in the calculation.

EXAMPLE 21.6

Exactly 200 mL of 0.0040 M $BaCl_2$ is added to exactly 600 mL of 0.0080 M K_2SO_4. Will a precipitate form?

Answer

According to solubility rule 4 on p. 76 the only precipitate that might form is $BaSO_4$:

$$Ba^{2+}(aq) + SO_4^{2-}(aq) \longrightarrow BaSO_4(s)$$

The number of moles of Ba^{2+} present in the original 200 mL of solution is

$$200 \text{ mL} \times \frac{0.0040 \text{ mol Ba}^{2+}}{1 \text{ L soln}} \times \frac{1 \text{ L}}{1000 \text{ mL}} = 8.0 \times 10^{-4} \text{ mol Ba}^{2+}$$

The total volume after combining the two solutions is 800 mL. The concentration of Ba^{2+} in the 800 mL volume is

$$[Ba^{2+}] = \frac{8.0 \times 10^{-4} \text{ mol}}{1 \text{ L soln}} \times \frac{1000 \text{ mL}}{800 \text{ mL}}$$
$$= 1.0 \times 10^{-3} \text{ M}$$

The number of moles of SO_4^{2-} in the original 600 mL solution is

$$600 \text{ mL} \times \frac{0.0080 \text{ mol SO}_4^{2-}}{1 \text{ L soln}} \times \frac{1 \text{ L}}{1000 \text{ mL}} = 4.8 \times 10^{-3} \text{ mol SO}_4^{2-}$$

The concentration of SO_4^{2-} in the 800 mL of the combined solution is

$$[SO_4^{2-}] = \frac{4.8 \times 10^{-3} \text{ mol}}{1 \text{ L soln}} \times \frac{1000 \text{ mL}}{800 \text{ mL}}$$
$$= 6.0 \times 10^{-3} \text{ M}$$

Now we must compare Q and K_{sp}. From Table 21.1, the K_{sp} for $BaSO_4$ is 1.1×10^{-10}. As for Q

$$Q = (Ba^{2+})(SO_4^{2-}) = (1.0 \times 10^{-3})(6.0 \times 10^{-3})$$
$$= 6.0 \times 10^{-6}$$

Compare Q and K_{sp}

$$Q = 6.0 \times 10^{-6}$$

$$K_{sp} = 1.1 \times 10^{-10}$$

$$\therefore Q > K_{sp}$$

(Continued)

Thus the solution is supersaturated so that some of the $BaSO_4$ will precipitate out of solution until

$$[Ba^{2+}][SO_4^{2-}] = 1.1 \times 10^{-10}$$

Similar examples: Problems 21.10, 21.11, 21.12.

21.2 SEPARATION OF IONS BY FRACTIONAL PRECIPITATION

Sometimes it is desirable to remove one type of ion from solution by precipitation while leaving other ions in solution. For instance, the addition of sulfate ions to a solution containing both potassium and barium ions causes $BaSO_4$ to precipitate out, thereby removing most of the Ba^{2+} ions from the solution. The other "product" that we obtain, K_2SO_4, is soluble and will remain in solution. The $BaSO_4$ precipitate can be separated from the solution by filtration.

Even when *both* products are insoluble, we can still achieve some degree of separation by choosing the proper reagent to bring about precipitation. Consider a solution that contains Cl^-, Br^-, and I^- ions. One way to separate these ions is to convert them into insoluble silver halides. In Table 21.1 we find the solubility products:

Compound	K_{sp}
AgCl	1.6×10^{-10}
AgBr	7.7×10^{-13}
AgI	8.3×10^{-17}

As the K_{sp} values show, the solubility of the halides decreases from AgCl to AgI. Thus, when a soluble compound, such as silver nitrate, is slowly added to this solution, AgI begins to precipitate first, followed by AgBr and then AgCl.

The following example describes the separation of only two ions (Cl^- and Br^-), but the procedure can be applied to a solution containing more than two different types of ions, if precipitates of differing solubility can be formed.

EXAMPLE 21.7

Silver nitrate is slowly added to a solution that is 0.020 M in Cl^- ions and 0.020 M in Br^- ions. Calculate the concentration of Ag^+ ions (in mol/L) required to initiate (a) the precipitation of AgBr and (b) the precipitation of AgCl.

Answer

(a) From the K_{sp} values, we know that AgBr will precipitate before AgCl, so for AgBr we write

$$K_{sp} = [Ag^+][Br^-]$$

Since $[Br^-] = 0.020\ M$, the concentration of Ag^+ that must be exceeded to initiate the precipitation of AgBr is

$$[Ag^+] = \frac{K_{sp}}{[Br^-]} = \frac{7.7 \times 10^{-13}}{0.020}$$
$$= 3.9 \times 10^{-11}\ M$$

Thus, $[Ag^+] > 3.9 \times 10^{-11}\ M$ is required to start the precipitation of AgBr.

(b) For AgCl

$$K_{sp} = [Ag^+][Cl^-]$$

$$[Ag^+] = \frac{K_{sp}}{[Cl^-]} = \frac{1.6 \times 10^{-10}}{0.020}$$
$$= 8.0 \times 10^{-9}\ M$$

Therefore, $[Ag^+] > 8.0 \times 10^{-9}\ M$ is needed to initiate the precipitation of AgCl.

Similar examples: Problems 21.15, 21.16.

For the process described in part (b) of Example 21.7, what is the concentration of Br^- ions in solution just before AgCl begins to precipitate? To answer this question we let $[Ag^+] = 8.0 \times 10^{-9}\ M$. Then

$$[Br^-] = \frac{K_{sp}}{[Ag^+]}$$
$$= \frac{7.7 \times 10^{-13}}{8.0 \times 10^{-9}}$$
$$= 9.6 \times 10^{-5}\ M$$

The percent of Br^- remaining in solution (called *unprecipitated*) at the critical concentration of Ag^+ is

$$\%\ Br^- = \frac{[Br^-]_{unppt'd}}{[Br^-]_{original}} \times 100\%$$
$$= \frac{9.6 \times 10^{-5}\ M}{0.020\ M} \times 100\%$$
$$= 0.48\%\ unprecipitated$$

Thus, $(100 - 0.48)\%$ or 99.52% of Br^- will have precipitated just before AgCl begins to precipitate. By this procedure, the Br^- ions can be quantitatively separated from the Cl^- ions by fractional precipitation.

21.3   THE COMMON ION EFFECT AND SOLUBILITY

The common ion effect was introduced in Chapter 20 in our discussion of acid and base ionizations. Here we will examine the relationship between the common ion effect and solubility.

As we have noted, the solubility product is an equilibrium constant; precipitation of a salt from the solution occurs whenever the ion product exceeds K_{sp} for that salt. In a saturated solution of AgCl, for example, the ion product $[Ag^+][Cl^-]$ is, of course, equal to K_{sp}. Furthermore, simple stoichiometry tells us that $[Ag^+] = [Cl^-]$. But this equality does not hold in all situations.

Suppose we study a solution containing two dissolved substances that share a common ion, say, AgCl and $AgNO_3$. In addition to the dissociation of AgCl, the following process also contributes to the total concentration of the common silver ions in solution:

$$AgNO_3(s) \xrightarrow{H_2O} Ag^+(aq) + NO_3^-(aq)$$

If $AgNO_3$ is added to a saturated AgCl solution, the increase in $[Ag^+]$ will make the ion product greater than the solubility product:

$$Q = (Ag^+)(Cl^-) > K_{sp}$$

To reestablish equilibrium, some AgCl will precipitate out of the solution, in accordance with Le Chatelier's principle, until the ion product is once again equal to K_{sp}. The effect of adding a common ion, then, is a *decrease* in the solubility of the salt (AgCl) in solution. Note that in this case $[Ag^+]$ is no longer equal to $[Cl^-]$ at equilibrium; rather, $[Ag^+] > [Cl^-]$.

EXAMPLE 21.8

Calculate the solubility of silver chloride (in g/L) in a 6.5×10^{-3} M silver nitrate solution.

Answer

Step 1

Since $AgNO_3$ is a strong electrolyte, it dissociates completely:

$$AgNO_3(s) \xrightarrow{H_2O} \underset{6.5 \times 10^{-3}\ M}{Ag^+(aq)} + \underset{6.5 \times 10^{-3}\ M}{NO_3^-(aq)}$$

Let s be the molar solubility of AgCl in $AgNO_3$ solution. We summarize the changes in concentrations as follows:

	AgCl(s) $\rightleftharpoons$	Ag$^+$(aq)	+ Cl$^-$(aq)
Initial		6.5×10^{-3} M	0.0 M
Change		$+s$ M	$+s$ M
Equilibrium		$(6.5 \times 10^{-3} + s)$ M	s M

Step 2

$$K_{sp} = [Ag^+][Cl^-]$$
$$1.6 \times 10^{-10} = (6.5 \times 10^{-3} + s)(s)$$

Since AgCl is quite insoluble and the presence of Ag^+ ions (from $AgNO_3$) further lowers the solubility of AgCl, s must be a very small number compared to 6.5×10^{-3}. Therefore, applying the approximation $6.5 \times 10^{-3} + s \simeq 6.5 \times 10^{-3}$, we obtain

$$1.6 \times 10^{-10} = 6.5 \times 10^{-3}s$$
$$s = 2.5 \times 10^{-8}\ M$$

Step 3

At equilibrium

$$[Ag^+] = (6.5 \times 10^{-3} + 2.5 \times 10^{-8})\ M \simeq 6.5 \times 10^{-3}\ M$$

$$[Cl^-] = 2.5 \times 10^{-8}\ M$$

and so our approximation $6.5 \times 10^{-3} + 2.5 \times 10^{-8} \simeq 6.5 \times 10^{-3}$ was justified in step 2. Since all the Cl^- ions must come from AgCl, the amount of AgCl dissolved in $AgNO_3$ solution also is $2.5 \times 10^{-8}\ M$. Then, knowing the molar mass of AgCl (143.4 g), we can calculate the solubility of the AgCl as follows:

$$\text{solubility of AgCl in } AgNO_3 \text{ solution} = \frac{2.5 \times 10^{-8}\ \text{mol AgCl}}{1\ \text{L soln}} \times \frac{143.4\ \text{g AgCl}}{1\ \text{mol AgCl}}$$
$$= 3.6 \times 10^{-6}\ \text{g/L}$$

Similar examples: Problems 21.19, 21.20.

For comparison with the result obtained in Example 21.8, let us find the solubility of AgCl in pure water. We start with

$$K_{sp} = [Ag^+][Cl^-] = 1.6 \times 10^{-10}$$

Hence

$$[Ag^+] = [Cl^-] = 1.3 \times 10^{-5}\ M$$

The solubility of AgCl in g/L therefore is

$$\text{solubility of AgCl in water} = \frac{1.3 \times 10^{-5}\ \text{mol AgCl}}{1\ \text{L soln}} \times \frac{143.4\ \text{g AgCl}}{1\ \text{mol AgCl}}$$
$$= 1.9 \times 10^{-3}\ \text{g/L}$$

As we might have expected, the solubility of AgCl is lower in the presence of the common ion Ag^+ in $AgNO_3$ solution than it is in pure water.

EXAMPLE 21.9

What is the molar solubility of lead(II) iodide in a 0.050 M solution of sodium iodide?

(Continued)

Remember that, at a given temperature, only the solubility of a compound is affected (decreased) by the common ion effect. Its solubility product, which is an equilibrium constant, remains the same whether or not other substances are present in the solution.

Answer

Step 1

Sodium iodide is a strong electrolyte, so

$$NaI(s) \xrightarrow{H_2O} Na^+(aq) + I^-(aq)$$
$$\phantom{NaI(s) \xrightarrow{H_2O} Na^+(aq)}\; 0.050\ M \quad 0.050\ M$$

Let s be the molar solubility of PbI_2 in mol/L. We summarize the changes in concentrations as follows:

	$PbI_2(s) \rightleftharpoons$	$Pb^{2+}(aq)\ +$	$2I^-(aq)$
Initial		0.00 M	0.050 M
Change		+s M	+2s M
Equilibrium		s M	(0.050 + 2s) M

Step 2

$$K_{sp} = [Pb^{2+}][I^-]^2$$
$$1.4 \times 10^{-8} = (s)(0.050 + 2s)^2$$

The low solubility of PbI_2 and the common ion (I^-) effect allow us to make the approximation $0.050 + 2s \approx 0.050$ so that

$$1.4 \times 10^{-8} = (0.050)^2 s$$
$$s = 5.6 \times 10^{-6}\ M$$

Step 3

At equilibrium

$$[Pb^{2+}] = 5.6 \times 10^{-6}\ M$$

$$[I^-] = (0.050\ M) + 2(5.6 \times 10^{-6}\ M) \approx 0.050\ M$$

Since all the Pb^{2+} ions must come from PbI_2, the molar solubility of PbI_2 in the NaI solution also is $5.6 \times 10^{-6}\ M$.

As an exercise, show that the molar solubility of PbI_2 in pure water is $1.5 \times 10^{-3}\ M$.

Similar example: Problem 21.21.

21.4   pH AND SOLUBILITY

This can be regarded as another application of Le Chatelier's principle.

In addition to factors we have discussed, the solubility of many substances depends also on the pH of solution. Consider the solubility equilibrium of magnesium hydroxide at 25°C:

$$Mg(OH)_2(s) \rightleftharpoons Mg^{2+}(aq) + 2OH^-(aq)$$

for which

$$K_{sp} = [Mg^{2+}][OH^-]^2 = 1.2 \times 10^{-11}$$

To see the effect of pH on the solubility of $Mg(OH)_2$, let us first calculate the pH of a saturated $Mg(OH)_2$ solution. Let s be the solubility of $Mg(OH)_2$ in mol/L. Proceeding as in Example 21.4

$$K_{sp} = (s)(2s)^2 = 4s^3$$
$$4s^3 = 1.2 \times 10^{-11}$$
$$s^3 = 3.0 \times 10^{-12}$$
$$s = 1.4 \times 10^{-4}\ M$$

At equilibrium, therefore

$$[OH^-] = 2 \times 1.4 \times 10^{-4}\ M = 2.8 \times 10^{-4}\ M$$

$$pOH = -\log (2.8 \times 10^{-4}) = 3.55$$

$$pH = 14.00 - 3.55 = 10.45$$

Thus, in a medium whose pH is less than 10.45, the solubility of $Mg(OH)_2$ would increase. This follows from the fact that a lower pH indicates a higher $[H^+]$ and thus a lower $[OH^-]$, as we would expect from $K_w = [H^+][OH^-]$. Consequently, $[Mg^{2+}]$ can be higher to maintain the equilibrium condition, so more $Mg(OH)_2$ will dissolve. The dissolution process and the effect of extra H^+ ions can be summarized as follows:

$$Mg(OH)_2(s) \rightleftharpoons Mg^{2+}(aq) + 2OH^-(aq)$$
$$2H^+(aq) + 2OH^-(aq) \rightleftharpoons 2H_2O(l)$$

overall: $$Mg(OH)_2(s) + 2H^+(aq) \rightleftharpoons Mg^{2+}(aq) + 2H_2O(l)$$

If the pH of the medium were higher than 10.45, $[OH^-]$ would be higher and the solubility of $Mg(OH)_2$ would decrease because of the common ion (OH^-) effect.

The pH also influences the solubility of salts that contain a basic anion. For example, the solubility equilibrium for BaF_2 is

$$BaF_2(s) \rightleftharpoons Ba^{2+}(aq) + 2F^-(aq)$$

and

$$K_{sp} = [Ba^{2+}][F^-]^2$$

In an acidic medium, the high $[H^+]$ will shift the following equilibrium from left to right:

$$H^+(aq) + F^-(aq) \rightleftharpoons HF(aq)$$

Remember that HF is a weak acid.

As $[F^-]$ decreases, $[Ba^{2+}]$ must increase to maintain the equilibrium condition. Thus more BaF_2 will dissolve. The dissolution process and the effect of pH on the solubility of BaF_2 can be summarized as follows:

$$BaF_2(s) \rightleftharpoons Ba^{2+}(aq) + 2F^-(aq)$$
$$2H^+(aq) + 2F^-(aq) \rightleftharpoons 2HF(aq)$$

overall: $$BaF_2(s) + 2H^+(aq) \rightleftharpoons Ba^{2+}(aq) + 2HF(aq)$$

The solubilities of salts containing anions that do not hydrolyze (for example, Cl^-, Br^-, and I^-) are unaffected by pH.

The following example shows the effect of pH on the solubility of a slightly soluble substance.

EXAMPLE 21.10

At 25°C the molar solubility of $Mg(OH)_2$ in pure water is 1.4×10^{-4} M. Calculate its molar solubility in a buffer medium whose pH is (a) 12.00 and (b) 9.00.

Answer

(a) We write

$$Mg(OH)_2(s) \rightleftharpoons Mg^{2+}(aq) + 2OH^-(aq)$$

$$pH = 12.00$$

$$pOH = 14.00 - 12.00 = 2.00$$

$$[OH^-] = 1.0 \times 10^{-2} \ M$$

Therefore

$$K_{sp} = [Mg^{2+}][OH^-]^2 = 1.2 \times 10^{-11}$$

$$[Mg^{2+}] = \frac{1.2 \times 10^{-11}}{(1.0 \times 10^{-2})^2}$$

$$= 1.2 \times 10^{-7} \ M$$

Due to the common ion effect, the molar solubility of $Mg(OH)_2$ is markedly lower than its solubility in pure water (1.4×10^{-4} M).

(b) In this case we write

$$pH = 9.00$$

$$pOH = 14.00 - 9.00 = 5.00$$

$$[OH^-] = 1.0 \times 10^{-5} \ M$$

$$[Mg^{2+}] = \frac{1.2 \times 10^{-11}}{(1.0 \times 10^{-5})^2} = 0.12 \ M$$

Since 1 mol $Mg(OH)_2 \rightleftharpoons$ 1 mol Mg^{2+}, the molar solubility of $Mg(OH)_2$ is 0.12 M. The increase in molar solubility from (a) to (b) is the result of the removal of OH^- ions by the extra H^+ ions present.

Similar example: Problem 21.28.

EXAMPLE 21.11

Calculate the concentration of aqueous ammonia necessary to initiate the precipitation of zinc hydroxide from a 0.0030 M solution of $ZnCl_2$.

Answer

The equilibria of interest are

$$Zn(OH)_2(s) \rightleftharpoons Zn^{2+}(aq) + 2OH^-(aq)$$

$$NH_3(aq) + H_2O(l) \rightleftharpoons NH_4^+(aq) + OH^-(aq)$$

First we find the OH^- ion concentration above which $Zn(OH)_2$ begins to precipitate. We write

$$K_{sp} = [Zn^{2+}][OH^-]^2 = 1.8 \times 10^{-14}$$

Since $ZnCl_2$ is a strong electrolyte, $[Zn^{2+}] = 0.0030\ M$ and

$$[OH^-]^2 = \frac{1.8 \times 10^{-14}}{0.0030} = 6.0 \times 10^{-12}$$
$$[OH^-] = 2.4 \times 10^{-6}\ M$$

Next, we calculate the concentration of NH_3 that will supply $2.4 \times 10^{-6}\ M$ OH^- ions, which is the concentration in a saturated $Zn(OH)_2$ solution. Let x be the initial concentration of NH_3 in mol/L. We summarize the changes in concentrations as a result of the ionization of NH_3 as follows:

	$NH_3(aq) + H_2O(l) \rightleftharpoons$	$NH_4^+(aq)$	$+$	$OH^-(aq)$
Initial	$x\ M$	$0.00\ M$		$0.00\ M$
Change	$-2.4 \times 10^{-6}\ M$	$+2.4 \times 10^{-6}\ M$		$+2.4 \times 10^{-6}\ M$
Equilibrium	$(x - 2.4 \times 10^{-6})\ M$	$2.4 \times 10^{-6}\ M$		$2.4 \times 10^{-6}\ M$

Substituting the equilibrium concentrations in the expression for the ionization constant

$$K_b = \frac{[NH_4^+][OH^-]}{[NH_3]} = 1.8 \times 10^{-5}$$
$$\frac{(2.4 \times 10^{-6})(2.4 \times 10^{-6})}{(x - 2.4 \times 10^{-6})} = 1.8 \times 10^{-5}$$

Solving for x, we obtain

$$x = 2.7 \times 10^{-6}\ M$$

Therefore, the initial concentration of NH_3 must be slightly greater than $2.7 \times 10^{-6}\ M$ to initiate the precipitation of $Zn(OH)_2$.

Similar examples: Problems 21.29, 21.30.

21.5 COMPLEX ION EQUILIBRIA AND SOLUBILITY

Some metal ions, especially those of transition metals, have the ability to form complex ions in solution. A ***complex ion*** may be defined as an *ion containing a central metal cation bonded to one or more molecules or ions*. The formation of complex ions is of great importance in many chemical and biological systems. Here we will consider their effect on solubility.

Copper(II) sulfate ($CuSO_4$) dissolves in water to produce a blue solution. Hydrated copper(II) ions are responsible for this color. Many other sulfates (Na_2SO_4, for example) are colorless. According to our definition, the hydrated copper ion—$Cu(H_2O)_6^{2+}$—is a complex ion, and when we write $Cu^{2+}(aq)$ we mean the hydrated ion.

Adding a *few drops* of concentrated ammonia solution to a $CuSO_4$ solution causes a light blue precipitate, copper(II) hydroxide, to form:

$$Cu^{2+}(aq) + 2OH^-(aq) \longrightarrow Cu(OH)_2(s)$$

where the OH^- ions are supplied by the ammonia solution. If an *excess* of NH_3 is then added, the blue precipitate redissolves to produce a beautiful dark blue solution, due to the formation of the complex ion $Cu(NH_3)_4^{2+}$ (see Color Plate 25):

$$Cu(OH)_2(s) + 4NH_3(aq) \rightleftharpoons Cu(NH_3)_4^{2+}(aq) + 2OH^-(aq)$$

A measure of the tendency of a particular metal ion to form a particular complex ion is given by the **formation constant K_f** (also called the **stability constant**), which is the *equilibrium constant for the complex ion formation*. The larger the K_f, the more stable is the complex ion. Table 21.3 lists the formation constants of a number of complex ions. The formation of the $Cu(NH_3)_4^{2+}$ ion can be expressed as

$$Cu^{2+}(aq) + 4NH_3(aq) \rightleftharpoons Cu(NH_3)_4^{2+}(aq)$$

for which the formation constant is

$$K_f = \frac{[Cu(NH_3)_4^{2+}]}{[Cu^{2+}][NH_3]^4}$$
$$= 5.0 \times 10^{13}$$

The very large value of K_f in this case indicates the great stability of the complex ion in solution and accounts for the very low concentration of copper(II) ions at equilibrium.

TABLE 21.3 Formation Constants of Selected Complex Ions in Water at 25°C

Complex Ion	Equilibrium Expression	Formation Constant (K_f)
$Ag(NH_3)_2^+$	$Ag^+ + 2NH_3 \rightleftharpoons Ag(NH_3)_2^+$	1.5×10^7
$Ag(CN)_2^-$	$Ag^+ + 2CN^- \rightleftharpoons Ag(CN)_2^-$	1.0×10^{21}
$Cu(CN)_4^{2-}$	$Cu^{2+} + 4CN^- \rightleftharpoons Cu(CN)_4^{2-}$	1.0×10^{25}
$Cu(NH_3)_4^{2+}$	$Cu^{2+} + 4NH_3 \rightleftharpoons Cu(NH_3)_4^{2+}$	5.0×10^{13}
$Cd(CN)_4^{2-}$	$Cd^{2+} + 4CN^- \rightleftharpoons Cd(CN)_4^{2-}$	7.1×10^{16}
CdI_4^{2-}	$Cd^{2+} + 4I^- \rightleftharpoons CdI_4^{2-}$	2.0×10^6
$HgCl_4^{2-}$	$Hg^{2+} + 4Cl^- \rightleftharpoons HgCl_4^{2-}$	1.7×10^{16}
HgI_4^{2-}	$Hg^{2+} + 4I^- \rightleftharpoons HgI_4^{2-}$	2.0×10^{30}
$Hg(CN)_4^{2-}$	$Hg^{2+} + 4CN^- \rightleftharpoons Hg(CN)_4^{2-}$	2.5×10^{41}
$Co(NH_3)_6^{3+}$	$Co^{3+} + 6NH_3 \rightleftharpoons Co(NH_3)_6^{3+}$	5.0×10^{31}

Complex ion formations are examples of Lewis acid-base reactions in which the metal ion acts as the Lewis acid and the neutral molecule or anion acts as the Lewis base (see Section 18.8).

EXAMPLE 21.12

A quantity of 0.20 mole of $CuSO_4$ is added to a liter of 1.20 *M* NH_3 solution. What is the concentration of Cu^{2+} ions at equilibrium?

Answer

The addition of $CuSO_4$ to the NH_3 solution results in the reaction

$$Cu^{2+}(aq) + 4NH_3(aq) \rightleftharpoons Cu(NH_3)_4^{2+}(aq)$$

Since K_f is very large (5.0×10^{13}), the reaction lies mostly to the right. As a good approximation, we can assume that essentially all of the dissolved Cu^{2+} ions end up as $Cu(NH_3)_4^{2+}$ ions. Thus the amount of NH_3 consumed in forming the complex ions is 4×0.20 M, or 0.80 M. (Note that 0.20 mol Cu^{2+} is initially present in solution and four NH_3 molecules are needed to "complex" one Cu^{2+} ion.) The concentration of NH_3 at equilibrium therefore is $(1.20 - 0.80)$ M, or 0.40 M, and that of $Cu(NH_3)_4^{2+}$ is 0.20 M, the same as the initial concentration of Cu^{2+}. Since $Cu(NH_3)_4^{2+}$ does dissociate to a slight extent, we call the concentration of Cu^{2+} ions at equilibrium x and write

$$K_f = \frac{[Cu(NH_3)_4^{2+}]}{[Cu^{2+}][NH_3]^4} = 5.0 \times 10^{13}$$

$$\frac{0.20}{x(0.40)^4} = 5.0 \times 10^{13}$$

Solving for x, we obtain

$$x = 1.6 \times 10^{-13} \ M = [Cu^{2+}]$$

The small value of $[Cu^{2+}]$ at equilibrium, compared to 0.20 M, certainly justifies our approximation.

Similar example: Problem 21.34.

The effect of complex ion formation generally is to *increase* the solubility of a substance, as the following example shows.

EXAMPLE 21.13

Calculate the molar solubility of silver chloride in a 1.0 M NH_3 solution.

Answer

Step 1

The equilibria reactions are

$$AgCl(s) \rightleftharpoons Ag^+(aq) + Cl^-(aq)$$
$$K_{sp} = [Ag^+][Cl^-] = 1.6 \times 10^{-10}$$

$$Ag^+(aq) + 2NH_3(aq) \rightleftharpoons Ag(NH_3)_2^+(aq)$$
$$K_f = \frac{[Ag(NH_3)_2^+]}{[Ag^+][NH_3]^2} = 1.5 \times 10^7$$

overall: $\quad AgCl(s) + 2NH_3(aq) \rightleftharpoons Ag(NH_3)_2^+(aq) + Cl^-(aq)$

The equilibrium constant K for the overall reaction is the product of the equilibrium constants of the individual reactions (see Section 17.3):

(Continued)

$$K = K_{sp}K_f = \frac{[Ag(NH_3)_2^+][Cl^-]}{[NH_3]^2}$$
$$= (1.6 \times 10^{-10})(1.5 \times 10^7)$$
$$= 2.4 \times 10^{-3}$$

Let s be the molar solubility of AgCl (mol/L). We summarize the changes in concentrations as a result of the formation of the complex ion $[Ag(NH_3)_2^+]$ as follows:

	AgCl(s) +	2NH₃(aq)	⇌	Ag(NH₃)₂⁺(aq) +	Cl⁻(aq)
Initial		1.0 M		0.0 M	0.0 M
Change		$-2s$ M		$+s$ M	$+s$ M
Equilibrium		$(1.0 - 2s)$ M		s M	s M

The formation constant for $Ag(NH_3)_2^+$ is quite large so that most of the silver ions exist in the complexed form. In the absence of ammonia we have, at equilibrium, $[Ag^+] = [Cl^-]$. As a result of complex ion formation, however, we can write $[Ag(NH_3)_2^+] = [Cl^-]$.

Step 2

$$K = \frac{(s)(s)}{(1.0 - 2s)^2}$$
$$2.4 \times 10^{-3} = \frac{s^2}{(1.0 - 2s)^2}$$

Taking the square root of both sides, we obtain

$$0.049 = \frac{s}{1.0 - 2s}$$
$$s = 0.045 \ M$$

Step 3

At equilibrium, 0.045 mole of AgCl will dissolve in one liter of 1.0 M NH₃ solution.

FIGURE 21.1 *From left to right: Formation of AgCl precipitate when a AgNO₃ solution is added to a NaCl solution. Upon the addition of a NH₃ solution, the AgCl precipitate dissolves due to the formation of the soluble Ag(NH₃)₂⁺ ion.*

We saw earlier (p. 623) that the molar solubility of AgCl in pure water is 1.3 $\times 10^{-5}$ M. Thus, the formation of the complex ion $Ag(NH_3)_2^+$ enhances the solubility of AgCl. Figure 21.1 shows the formation of AgCl precipitate and its subsequent dissolution as a result of complex ion formation.

According to Le Chatelier's principle, the removal of Ag^+ ions from solution to form $Ag(NH_3)_2^+$ ions will cause more AgCl to dissolve.

21.6 APPLICATION OF THE SOLUBILITY PRODUCT PRINCIPLE IN QUALITATIVE ANALYSIS

In Section 4.5, we discussed the principle of gravimetric analysis, by which we measure the amount of an ion in an unknown sample. Here we will briefly discuss **qualitative analysis**, or *the determination of the types of ions present in a solution.* We will focus on the cations.

There are some twenty common cations that can be analyzed readily in aqueous solution. These cations can be divided into five groups, according to the solubility products of their insoluble salts (Table 21.4). Since an unknown solution may contain any one or all twenty ions, any analysis must be carried out systematically from group 1 through group 5. We will not give a comprehensive discussion of how each cation can be identified; instead, we will outline the procedure for separating these twenty ions by adding precipitating reagents to an unknown solution.

Do not confuse the groups in Table 21.4, which are based on solubility products, with those of the periodic table, which are based on the electron configurations of the elements.

- *Group 1 Cations.* Dilute HCl is added to the unknown solution. Only the Ag^+, Hg_2^{2+}, and Pb^{2+} ions precipitate as insoluble chlorides. All other ions form soluble chlorides and remain in solution.

TABLE 21.4 Separation of Cations into Groups According to Their Precipitation Reactions to Various Reagents

Group	Cation	Precipitating Reagents	Insoluble Salt	K_{sp}
1	Ag^+	HCl	AgCl	1.6×10^{-10}
	Hg_2^{2+}	↓	Hg_2Cl_2	3.5×10^{-18}
	Pb^{2+}		$PbCl_2$	2.4×10^{-4}
2	Bi^{3+}	H_2S	Bi_2S_3	1.6×10^{-72}
	Cd^{2+}	in acidic	CdS	1.4×10^{-28}
	Cu^{2+}	solutions	CuS	1.0×10^{-44}
	Sn^{2+}	↓	SnS	1.0×10^{-26}
3	Al^{3+}	H_2S	$Al(OH)_3$	1.8×10^{-33}
	Co^{2+}	in basic	CoS	4.0×10^{-21}
	Cr^{3+}	solutions	$Cr(OH)_3$	3.0×10^{-29}
	Fe^{2+}		FeS	3.7×10^{-19}
	Mn^{2+}		MnS	1.4×10^{-15}
	Ni^{2+}		NiS	1.4×10^{-24}
	Zn^{2+}	↓	ZnS	1.2×10^{-23}
4	Ba^{2+}	Na_2CO_3	$BaCO_3$	8.1×10^{-9}
	Ca^{2+}		$CaCO_3$	8.7×10^{-9}
	Sr^{2+}	↓	$SrCO_3$	1.6×10^{-9}
5	K^+	No precipitating	None	
	Na^+	reagent	None	
	NH_4^+		None	

- *Group 2 Cations.* After the chloride precipitates have been removed by filtration, hydrogen sulfide is reacted with the unknown solution, which now contains hydrochloric acid. H_2S is a diprotic acid; therefore

$$H_2S(aq) \rightleftharpoons H^+(aq) + HS^-(aq) \qquad K_{a_1} = 5.7 \times 10^{-8}$$

$$HS^-(aq) \rightleftharpoons H^+(aq) + S^{2-}(aq) \qquad K_{a_2} = 1.2 \times 10^{-13}$$

The small values of the acid ionization constants mean that the sulfide ion concentration in solution is very low. Moreover, the H^+ ions supplied by the HCl in solution produce a common ion effect, which further decreases the S^{2-} ion concentration. (According to Le Chatelier's principle, the presence of additional H^+ ions will drive the above equilibria from right to left.) The extremely low sulfide ion concentration in solution means that only those metal sulfides with the smallest K_{sp} values will precipitate out of solution, namely, Bi_2S_3, CdS, CuS, and SnS (see Table 21.4). These precipitates are removed from the solution by filtration.

- *Group 3 Cations.* At this stage, sodium hydroxide is added to the solution to make it alkaline. In the presence of excess OH^- ions, ionization of H_2S and HS^- shifts from left to right and, hence, the concentration of the S^{2-} ions increases. Consequently

$$[M^{2+}][S^{2-}] > K_{sp}$$

where M^{2+} represents the metal cations (Co^{2+}, Fe^{2+}, Mn^{2+}, Ni^{2+}, and Zn^{2+}). Therefore, the sulfides of these metals will precipitate out of solution. Note that the Al^{3+} and Cr^{3+} ions actually precipitate as the hydroxides $Al(OH)_3$ and $Cr(OH)_3$, rather than the sulfides, because the hydroxides are less soluble. The solution is then filtered to remove the insoluble sulfides and hydroxides.

- *Group 4 Cations.* Sodium carbonate is added to the basic solution. Normally, carbonate ions have a tendency to hydrolyze in pure water, according to the equation

$$CO_3^{2-}(aq) + H_2O(l) \rightleftharpoons HCO_3^-(aq) + OH^-(aq)$$

and since most bicarbonate salts are soluble, no metal ions precipitate out of the solution. However, in the basic solution, the excess OH^- ions drive the equilibrium from right to left (Le Chatelier's principle) and the increase in the CO_3^{2-} ion concentration results in the precipitation of the insoluble carbonates $BaCO_3$, $CaCO_3$, and $SrCO_3$. These precipitates are removed by filtration.

- *Group 5 Cations.* At this stage, the only cations possibly remaining in solution are Na^+, K^+, and NH_4^+. The presence of NH_4^+ can be determined by adding sodium hydroxide:

$$NaOH(aq) + NH_4^+(aq) \longrightarrow Na^+(aq) + H_2O(l) + NH_3(g)$$

The ammonia gas is detected either by its characteristic odor or by observing a piece of wet red litmus paper turning blue when placed above (not in contact with) the solution. To confirm the presence of Na^+ and K^+ ions, we usually use a flame test, as follows: A piece of platinum wire (chosen because platinum is inert) is moistened with the solution and is then held over a Bunsen burner flame. Each type of metal ion gives a characteristic color when heated in this manner. For example, the color emitted by Na^+ ions is yellow, that of K^+ ions is violet, and that of Sr^{2+} ions is crimson.

Figure 21.2 is a flow chart for the separation of metal ions.

Two points regarding qualitative analysis must be mentioned. First, the separation of the cations into groups is made as selective as possible; that is, the chosen anions added must be such that they will precipitate the fewest types of cations. For example, all the cations in group 1 form insoluble sulfides. Thus, if H_2S is reacted with the solution at the start, as many as seven different sulfides might precipitate out of solution (group 1 *and* group 2 sulfides), an undesirable outcome. Second, the separation of cations at each step must be carried out as completely as possible. For example, if we do not add enough HCl to the unknown solution to remove all the group 1 cations, they will precipitate with the group 2 cations as insoluble sulfides; this would interfere with further chemical analysis and lead to erroneous conclusions.

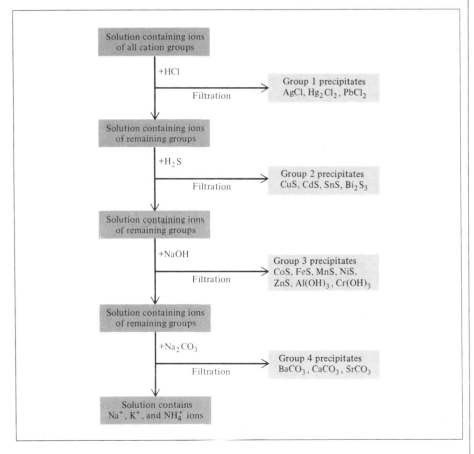

FIGURE 21.2 *A flow chart for the separation of cations in qualitative analysis.*

EXAMPLE 21.14

A solution contains the group 2 cation Sn^{2+} and the group 3 cation Fe^{2+}, both at 0.10 M. The solution is saturated with H_2S ($[H_2S] = 0.10\ M$). Calculate the range of hydrogen ion concentrations that will result only in the precipitation of Sn^{2+} ions as SnS.

Answer

Since the metal ion concentrations are fixed at 0.10 M, the precipitation of one ion and not the other must be controlled by the sulfide ion concentration. Therefore, we need to calculate the sulfide ion concentrations corresponding to the cation concentrations. For SnS

$$K_{sp} = [Sn^{2+}][S^{2-}] = 1.0 \times 10^{-26}$$

$$[S^{2-}] = \frac{1.0 \times 10^{-26}}{0.10} = 1.0 \times 10^{-25}\ M$$

Thus a sulfide ion concentration greater than $1.0 \times 10^{-25}\ M$ will result in the precipitation of SnS. For FeS

$$K_{sp} = [Fe^{2+}][S^{2-}] = 3.7 \times 10^{-19}$$

$$[S^{2-}] = \frac{3.7 \times 10^{-19}}{0.10} = 3.7 \times 10^{-18}\ M$$

Thus a sulfide ion concentration greater than $3.7 \times 10^{-18}\ M$ will cause FeS to precipitate.

In Section 19.4 we saw that for the equilibrium process

$$H_2S(aq) \rightleftharpoons 2H^+(aq) + S^{2-}(aq)$$

the equilibrium constant K is

$$K = \frac{[H^+]^2[S^{2-}]}{[H_2S]} = 6.8 \times 10^{-21}$$

We can now calculate $[H^+]$ for each of the $[S^{2-}]$ values. For SnS

$$\frac{[H^+]^2(1.0 \times 10^{-25})}{0.10} = 6.8 \times 10^{-21}$$

$$[H^+]^2 = 6.8 \times 10^3$$

$$[H^+] = 82\ M$$

This result shows that in order to prevent SnS from precipitating, $[H^+]$ must be equal to or greater than 82 M! Since no solution can have such a high acidity, SnS will precipitate no matter what the hydrogen ion concentration may be.

For FeS

$$\frac{[H^+]^2(3.7 \times 10^{-18})}{0.10} = 6.8 \times 10^{-21}$$

$$[H^+]^2 = 1.8 \times 10^{-4}$$

$$[H^+] = 0.013\ M$$

Thus, to prevent FeS from precipitating, the hydrogen ion concentration must be equal to or greater than 0.013 M.

In summary, for the selective precipitation of SnS we must have

$$[H^+] \leq 82 \, M$$

and

$$[H^+] \geq 0.013 \, M$$

or

$$0.013 \, M \leq [H^+] \leq 82 \, M$$

Similar examples: Problems 21.41, 21.42.

21.7 PRECIPITATION REACTIONS IN NATURE AND IN ENGINEERING PROBLEMS

Many examples can be described regarding the roles of precipitation reactions in nature and in our lives. Here we will consider two interesting cases.

Formation of Sinkholes, Stalagmites, and Stalactites

If you have visited the Carlsbad Caverns in New Mexico, you must have been impressed by the fantastic rock formations, called *stalactites*, the icicle-like forms that hang from the cavern ceiling, and *stalagmites*, the columns that rise upward from the cavern floor. How are these objects formed?

The principal nonsilicate mineral in rocks is limestone, $CaCO_3$. The solubility product for the process

$$CaCO_3(s) \rightleftharpoons Ca^{2+}(aq) + CO_3^{2-}(aq)$$

is

$$K_{sp} = [Ca^{2+}][CO_3^{2-}] = 8.7 \times 10^{-9}$$

Thus $CaCO_3$ is rather insoluble. However, $CaCO_3$ dissolves readily in acid solutions as a result of the following reaction:

$$CaCO_3(s) + 2HCl(aq) \longrightarrow Ca^{2+}(aq) + 2Cl^-(aq) + H_2O(l) + CO_2(g)$$

All carbonates are decomposed by acids.

Since soil moisture commonly contains humic acids derived from the decay of vegetation, most groundwater can dissolve limestone. Furthermore, water containing dissolved CO_2 is acidic and reacts with $CaCO_3$ as follows:

$$CaCO_3(s) + CO_2(aq) + H_2O(l) \rightleftharpoons Ca^{2+}(aq) + 2HCO_3^-(aq)$$

The extent to which this reaction takes place depends on the partial pressure of CO_2. At high partial pressures of CO_2 the amount of dissolved CO_2 is high and the above equilibrium is shifted to the right; at low partial pressures, the equilibrium is shifted to the left.

When acidic surface waters seep underground, they slowly dissolve away the limestone deposits. If the limestone deposits are close to the surface of

According to Henry's law (see Section 15.6), the solubility of a gas in a solution is directly proportional to its partial pressure over the solution.

FIGURE 21.3   *Down-ward-growing, icicle-like stalactites and upward-growing, columnar stalag-mites. It may take hundreds of years for these structures to form.*

the earth, their dissolution leads to the collapse of the thin layer of earth that lies above them. The resulting depressions in the landscape are called *sinkholes*. On the other hand, the dissolution of deep deposits of limestone produces underground caves. When a solution containing Ca^{2+} and HCO_3^- ions drips through the cracks of a cave's ceiling into the cave, where the partial pressure of carbon dioxide is lower than it is in the soil, the solution becomes supersaturated in carbon dioxide. The escape of the CO_2 gas from the solution shifts the equilibrium to the left:

$$CaCO_3(s) + CO_2(aq) + H_2O(l) \longleftarrow Ca^{2+}(aq) + 2HCO_3^-(aq)$$

Consequently, a precipitate of $CaCO_3$ forms and stalactites and stalagmites begin to develop (Figure 21.3). In time, stalactites and stalagmites may join to form columns that reach from the floor to the ceiling of the cave.

Precipitation Problems in Engineering

A precipitation problem frequently encountered in the boilers of steam generators and in hot water pipes also involves limestone and, to a lesser

FIGURE 21.4 *Boiler scale in a hot water pipe. The scale consists mostly of* $CaCO_3$ *and some* $MgCO_3$.

extent, dolomite (a mineral made up of $CaCO_3$ and $MgCO_3$ in 1:1 ratio). As mentioned earlier, water containing dissolved CO_2 helps to dissolve $CaCO_3$:

$$CaCO_3(s) + CO_2(aq) + H_2O(l) \longrightarrow Ca^{2+}(aq) + 2HCO_3^-(aq)$$

When solutions containing calcium and bicarbonate ions are boiled or heated, the reaction is reversed:

$$Ca^{2+}(aq) + 2HCO_3^-(aq) \longrightarrow CaCO_3(s) + CO_2(g) + H_2O(l)$$

Carbon dioxide gas is driven off from the solution, and calcium carbonate precipitates from the solution. Solid $CaCO_3$ formed in this way is the main component of the scale that accumulates in boilers, water heaters, pipes, and tea kettles. The thick layer formed reduces heat transfer and decreases the efficiency and durability of boilers, pipes, and appliances. In household hot water pipes $CaCO_3$ can restrict or totally block the flow of water (Figure 21.4). A simple way to remove these deposits is to introduce a small amount of hydrochloric acid:

> Recall that gas solubility in water decreases with increasing temperature.

$$CaCO_3(s) + 2HCl(aq) \longrightarrow Ca^{2+}(aq) + 2Cl^-(aq) + H_2O(l) + CO_2(g)$$

SUMMARY

1. The solubility product, K_{sp}, expresses the equilibrium between a solid and its ions in solution. Solubility can be found from K_{sp} and vice versa.
2. The presence of a common ion from another source decreases the solubility of a salt.
3. The solubility of slightly soluble salts containing basic anions increases as the hydrogen ion concentration increases. The solubility of salts with anions derived from strong acids are unaffected by pH.
4. Complex ions are formed in solution by the combination of a metal cation with molecules or anions. The formation constant, K_f, measures the tendency toward the formation of a specific complex ion.
5. Qualitative analysis is the identification of cations and anions in solution.

KEY WORDS

Complex ion, p. 627
Formation constant, p. 628
Qualitative analysis, p. 631

Solubility product, p. 613
Stability constant, p. 628

PROBLEMS†

More challenging problems are marked with an asterisk.

Solubility; Solubility Product

21.1 Explain the difference between solubility and solubility product of an insoluble substance such as $BaSO_4$.

21.2 Write the solubility product expression for the ionic compound A_xB_y.

21.3 Write the K_{sp} expressions for the following compounds: (a) CuBr, (b) ZnC_2O_4, (c) Ag_2CrO_4, (d) Hg_2Cl_2, (e) $AuCl_3$, (f) $Mn_3(PO_4)_2$.

21.4 Calculate the concentration of ions in the following saturated solutions:

(a) [I^-] in AgI solution with [Ag^+] = 1.2×10^{-8} M
(b) [Al^{3+}] in $Al(OH)_3$ with [OH^-] = 7.4×10^{-9} M
(c) [S^{2-}] in Bi_2S_3 solution with [Bi^{3+}] = 8.1×10^{-7} M

21.5 Calculate the solubility (g/L) of the following compounds: (a) CdS, (b) $Fe(OH)_3$, (c) Ag_2S.

21.6 From the solubility data given, calculate the solubility products for the following compounds:

(a) SrF_2, 7.3×10^{-2} g/L
(b) $PbCrO_4$, 4.5×10^{-5} g/L
(c) Ag_3PO_4, 6.7×10^{-3} g/L

21.7 The K_{sp} for manganese(II) carbonate is 1.8×10^{-11}. What is the molar solubility of this substance?

21.8 The solubility of an ionic compound MX (molar mass = 346 g) is 4.63×10^{-3} g/L. What is the K_{sp} for the compound?

21.9 The solubility of an ionic compound M_2X_3 (molar mass = 288 g) is 3.6×10^{-17} g/L. What is the K_{sp} for the compound?

21.10 A sample of 20.0 mL of 0.10 M $Ba(NO_3)_2$ is added to 50.0 mL of 0.10 M Na_2CO_3. Will $BaCO_3$ precipitate?

21.11 A volume of 75 mL of 0.060 M NaF is mixed with 25 mL of 0.15 M $Sr(NO_3)_2$. Calculate the concentrations in the final solution of NO_3^-, Na^+, Sr^{2+}, and F^-. (K_{sp} for $SrF_2 = 2.0 \times 10^{-10}$.)

21.12 If 2.00 mL of 0.200 M NaOH is added to 1.00 L of 0.100 M $CaCl_2$, will precipitation occur? [K_{sp} of $Ca(OH)_2 = 5.4 \times 10^{-6}$.]

21.13 What is the pH of a saturated solution of zinc hydroxide?

21.14 The pH of a saturated solution of a metal hydroxide MOH is 9.68. Calculate the K_{sp} for the compound.

Fractional Precipitation

21.15 Solid NaI is slowly added to a solution that is 0.010 M in Cu^+ and 0.010 M in Ag^+. (a) Which compound will begin to precipitate first? (b) Calculate [Ag^+] when CuI just begins to precipitate. (c) What percent of Ag^+ remains in solution at this point?

*21.16 The solubility products of AgCl, Ag_2S, and Ag_3PO_4 are, respectively, 1.6×10^{-10}, 1.8×10^{-49}, and 1.8×10^{-18}. (a) If Ag^+ is added (without changing the volume) to 1.00 L of solution containing 0.10 mol Cl^-, 0.10 mol S^{2-}, and 0.10 mol PO_4^{3-}, which compounds will precipitate first, second, and third? (b) When the salt of the second anion precipitates, what will be the concentration of the first anion, in mol/L?

*21.17 Find the approximate pH range suitable for the separation of Fe^{3+} and Zn^{2+} by precipitation of $Fe(OH)_3$ from a solution that is initially 0.010 M in Fe^{3+} and Zn^{2+}.

Common Ion Effect

21.18 Use $CaCO_3$ and $Ca(NO_3)_2$ to show how the presence of a common ion can reduce the solubility of an insoluble salt.

21.19 The molar solubility of AgCl in 6.5×10^{-3} M $AgNO_3$ is 2.5×10^{-8} M. In deriving the K_{sp} from these data, which of the following assumptions are reasonable?

† The temperature is assumed to be 25° C for all the problems.

(a) K_{sp} is the same as solubility.

(b) K_{sp} of AgCl is the same in 6.5×10^{-3} M $AgNO_3$ as in pure water.

(c) Solubility of AgCl is independent of $[AgNO_3]$.

(d) $[Ag^+]$ in solution does not change significantly upon the addition of AgCl to 6.5×10^{-3} M $AgNO_3$.

(e) $[Ag^+]$ in solution after the addition of AgCl to 6.5×10^{-3} M $AgNO_3$ is the same as it would be in pure water.

21.20 Calculate the solubility in g/L of AgBr (a) in pure water and (b) in 0.0010 M NaBr.

21.21 The solubility product of $PbBr_2$ is 8.9×10^{-6}. Determine the molar solubility (a) in pure water, (b) in 0.20 M KBr solution, (c) in 0.20 M $Pb(NO_3)_2$ solution.

21.22 How many grams of ZnS will dissolve in 3.0×10^2 mL of 0.050 M $Zn(NO_3)_2$?

21.23 Calculate the molar solubility of AgCl in a solution made by dissolving 10.0 g of $CaCl_2$ in 1.00 L of solution.

21.24 Calculate the molar solubility of $BaSO_4$ (a) in water and (b) in a solution containing 1.0 M SO_4^{2-} ions.

pH and Solubility

21.25 Which of the following ionic compounds will be more soluble in acid solution than in water? (a) $BaSO_4$, (b) $PbCl_2$, (c) $Fe(OH)_3$, (d) Bi_2S_3, (e) $CaCO_3$.

21.26 Which of the following substances will be more soluble in acid solution than in pure water? (a) CuI, (b) Ag_2SO_4, (c) MnS, (d) $Zn(OH)_2$, (e) BaC_2O_4, (f) $Ca_3(PO_4)_2$.

21.27 What is the pH of a saturated solution of aluminum hydroxide?

21.28 Calculate the molar solubility of $Fe(OH)_2$ at (a) pH 8.00 and (b) pH 10.00.

21.29 The solubility product of $Pb(OH)_2$ is 4.2×10^{-15}. What minimum OH^- concentration must be attained (e.g., by adding NaOH) to make the Pb^{2+} concentration in a solution of $Pb(NO_3)_2$ less than 1.0×10^{-10} M?

21.30 Calculate whether or not a precipitate will form if 2.00 mL of 0.60 M NH_3 is added to 1.0 L of 1.0×10^{-3} M of $FeSO_4$.

Complex Ions

21.31 Write the formation constant expressions for the following complex ions: (a) $Zn(OH)_4^{2-}$, (b) $Co(NH_3)_6^{3+}$, (c) HgI_4^{2-}.

21.32 Explain the formation of complexes in Table 21.3 in terms of Lewis acid-base theory.

21.33 Explain, with balanced ionic equations, why (a) AgI dissolves in ammonia solution, (b) AgBr dissolves in NaCN solution, (c) CdS dissolves in NaI solution, (d) HgS dissolves in KCl solution.

21.34 A quantity of 2.50 g of $CuSO_4$ is dissolved in 9.0×10^2 mL of 0.30 M NH_3. What are the concentrations of Cu^{2+}, $Cu(NH_3)_4^{2+}$, and NH_3 at equilibrium?

21.35 Calculate the concentrations of Cd^{2+}, $Cd(CN)_4^{2-}$, and CN^- at equilibrium when 0.50 g of $Cd(NO_3)_2$ dissolves in 5.0×10^2 mL of 0.50 M NaCN.

★21.36 If NaOH is added to 0.010 M Al^{3+}, which will be the predominant species at equilibrium: $Al(OH)_3$ or $Al(OH)_4^-$? The pH of the solution is 14.00. [K_f for $Al(OH)_4^- = 2.0 \times 10^{33}$.]

21.37 Both Ni^{2+} and Zn^{2+} form complex ions with NH_3. Write balanced equations for the reactions. However, $Zn(OH)_2$ is soluble in 6 M NaOH and $Ni(OH)_2$ is not. Explain.

Qualitative Analysis

21.38 In a group 1 analysis, a student obtained a precipitate containing both AgCl and $PbCl_2$. Suggest one reagent that would allow her to separate AgCl(s) from $PbCl_2$(s).

21.39 In a group 1 analysis, a student adds hydrochloric acid to the unknown solution to make $[Cl^-] = 0.15$ M. Some $PbCl_2$ precipitates. (a) Calculate the concentration of Pb^{2+} remaining in solution. (b) At this concentration of Pb^{2+}, will it precipitate as PbS in group 2, where $[S^{2-}] = 1.0 \times 10^{-21}$ M?

21.40 What is the maximum possible concentration of Cu^{2+} in a solution that is saturated with H_2S ($[H_2S] = 0.10$ M) at pH 3.00?

21.41 At pH = 1.00, will the compound MS precipitate from a solution saturated with H_2S (0.10 M) if $[M^{2+}] = 0.20$ M initially? (K_{sp} for MS = 6.0×10^{-20}.)

21.42 In adjusting the pH of an unknown solution before precipitating group 2 cations, a student fixes the pH at 1.50 instead of 0.50. How will this error affect the behavior of Zn^{2+} if it is present at 0.010 M concentration? Assume $[H_2S] = 0.10$ M. (*Hint:* Zn^{2+} normally precipitates as a group 3 sulfide.)

21.43 Both KCl and NH_4Cl are white solids. Suggest one reagent that would allow you to distinguish between these two compounds.

Miscellaneous Problems

21.44 Describe a simple test that would allow you to distinguish between $AgNO_3$(s) and $Cu(NO_3)_2$(s).

21.45 Write the electron configuration for Al^{3+} and predict the geometry of the $Al(OH)_4^-$ ion.

21.46 Suggest a reagent that would allow you to separate Al^{3+} from Fe^{3+}.

21.47 All metal hydroxides react with acids. However, a number of them also react with bases. Such compounds are said to be amphoteric. Identify the amphoteric hydroxides among the following compounds: NaOH, $Al(OH)_3$, $Cu(OH)_2$, $Mg(OH)_2$, $Zn(OH)_2$, $Fe(OH)_3$, $Ni(OH)_2$. Write ionic equations for the complex ion formation between the amphoteric hydroxide and the OH^- ions.

22
CHEMICAL KINETICS

Chemical changes differ greatly in the actual rate at which they take place. For example, the slow deterioration of marble statues due to the action of acid rain can be contrasted to the fast, violently exothermic reactions of explosives. Despite the apparent rapidity with which some reactions occur, no chemical reaction occurs instantaneously. The study of reaction rates provides us with important insights regarding chemical reactions.

In this chapter we will study the rate of chemical reactions at two related levels. On the one hand, we will explore macroscopic concerns such as how reaction rates are measured and how empirical equations are developed and used to describe how fast reactions occur. At the molecular level, we will see how reaction rate data can provide clues to the detailed nature by which given chemical changes occur.

641

22.1   THE RATE OF A REACTION

The area of chemistry concerned with the speeds, or rates, at which chemical reactions occur is called **chemical kinetics.** The word "kinetic" pertains to motion. (Recall that in Chapter 6 we studied the kinetic molecular theory of gases, which describes the random motion of gas molecules.) The rate of a reaction, or **reaction rate,** is the *change of the concentration of reactant or product with time.* (For simplicity, we will speak of "the rate" rather than "the reaction rate.")

We know that any reaction can be represented by the general equation

$$\text{reactants} \longrightarrow \text{products}$$

which tells us that during the course of a reaction, reactant molecules are consumed while product molecules are formed. Thus, we can follow the progress of a reaction by monitoring either the decrease in concentration of the reactants or the increase in concentration of the products.

Figure 22.1 shows the progress of a simple reaction in which reactant molecules A are converted to product molecules B:

$$A \longrightarrow B$$

The decrease in the number of A molecules and the increase in the number of B molecules with time are shown in Figure 22.2.

To see how the rate of a chemical reaction is actually measured, we will consider two specific examples: the reaction of molecular bromine with formic acid and the thermal decomposition of dinitrogen pentoxide. After we have gained an understanding of how rates are measured experimentally, we will learn to write rate expressions using the balanced equations.

Reaction Between Molecular Bromine and Formic Acid.   In aqueous solutions, molecular bromine reacts with formic acid (HCOOH) as follows:

$$Br_2(aq) + HCOOH(aq) \longrightarrow 2Br^-(aq) + 2H^+(aq) + CO_2(g)$$

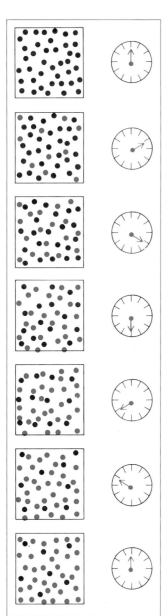

FIGURE 22.1   *Diagram of an A → B reaction taking place in the interval of 60 s. Initially, only A molecules (black dots) are present. As time progresses, B molecules (colored dots) are formed.*

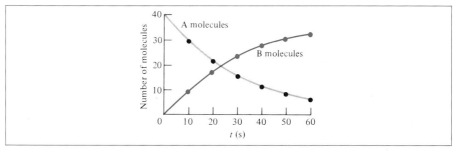

FIGURE 22.2   *The rate of the reaction A → B, represented as the decrease of A molecules with time and as the increase of B molecules with time.*

Molecular bromine has a characteristic red-brown color. As the reaction progresses, the concentration of Br_2 steadily decreases. This change can be monitored easily by measuring the decrease in intensity of molecular bromine's color with a spectrometer. Table 22.1 shows the molar concentration of Br_2 at different times; these data are plotted in Figure 22.3. Note that time zero is just after the mixing of the reactants.

The rate of the reaction can be defined as the change in the reactant concentration over a certain time interval, that is

$$\text{rate} = -\frac{[Br_2]_{final} - [Br_2]_{initial}}{t_{final} - t_{initial}}$$

$$= -\frac{\Delta[Br_2]}{\Delta t}$$

where $\Delta[Br_2] = [Br_2]_{final} - [Br_2]_{initial}$ and $\Delta t = t_{final} - t_{initial}$. Because the concentration of Br_2 *decreases* during the time interval, $\Delta[Br_2]$ is a negative quantity. But the rate of a reaction is a positive quantity, so a minus sign is needed in the rate expression to make the rate positive.

From Table 22.1 we can calculate the rate over the first 50-second time interval as follows:

$$\text{rate} = -\frac{\Delta[Br_2]}{\Delta t}$$

$$= -\frac{(0.0101 - 0.0120)\ \text{mol/L}}{50.0\ \text{s}}$$

$$= 3.80 \times 10^{-5}\ M/s$$

The rate during the next 50-second time interval (from $t = 50.0$ s to $t = 100.0$ s) is similarly given by

$$\text{rate} = -\frac{(0.00846 - 0.0101)\ \text{mol/L}}{50.0\ \text{s}} = 3.28 \times 10^{-5}\ M/s$$

Both the Br^- ion and formic acid are colorless.

Since [ ] is mol/L or *M*, and time is in s, the units of rate are *M*/s.

TABLE 22.1
Data for Bromine Concentration in the Reaction Between Molecular Bromine and Formic Acid at 25°C

Time (s)	$[Br_2]$ (mol/L)
0.0	0.0120
50.0	0.0101
100.0	0.00846
150.0	0.00710
200.0	0.00596
250.0	0.00500
300.0	0.00420
350.0	0.00353
400.0	0.00296

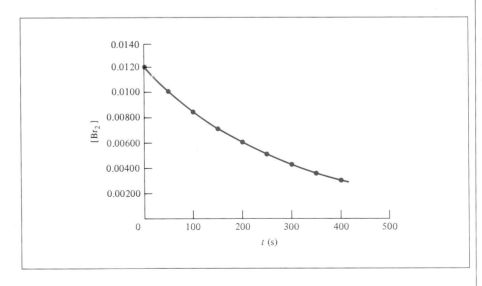

FIGURE 22.3 *Change in molecular bromine concentration as a function of time, in a reaction between molecular bromine and formic acid.*

These calculations demonstrate that the rate of the reaction is not a constant, but changes with the concentration, or number, of reactant molecules present. Initially, when the concentration of Br_2 is high, the rate is correspondingly high. As time goes on, the rate gradually decreases and eventually becomes zero when all of the molecular bromine has reacted.

The rate of the reaction also depends on the concentration of formic acid. However, by using a large excess of formic acid in the reaction mixture we can ensure that the concentration of formic acid will remain virtually constant throughout the course of the reaction. Consequently, under this condition the change in the amount of formic acid present in solution will have no effect on the rate.

Calculating rates of a reaction in the manner just described is unsatisfactory in the sense that it gives us only average rates. The rate of 3.80×10^{-5} M/s is not associated with any particular instant of time; rather, it represents the average value of the rates from time zero to 50.0 s. If we had chosen the first 100 seconds as the time interval, the rate would have been 3.54×10^{-5} M/s, and so on.

There is much advantage in speaking of the rate of a reaction at a specific time. For one thing, it makes the comparison of two reaction rates more meaningful. We have seen that the value of an average rate depends on the time interval we choose. However, we can gradually eliminate this arbitrary choice of time interval by calculating the rate over a smaller and smaller time interval. In fact, when the interval is made infinitesimally small, the rate becomes the *slope* of the concentration versus time curve at a particular time (Figure 22.4). In this way we can speak of the reaction rate at a particular time. Furthermore, the rate always has the same value at that instant for the same concentrations of reactants, as long as the temperature is kept constant. To distinguish this rate from the average rate, we call it the *instantaneous rate*. In the following discussion, we will frequently refer to the instantaneous rate merely as the rate.

Table 22.2 lists the rates of the reaction at specific times. These rates are obtained from the curve in Figure 22.4. The rate at $t = 0$ is called the

FIGURE 22.4 *The rates of the molecular bromine–formic acid reaction at $t = 0$, 150 s, and 300 s are given by the negative slopes of the curve at these times.*

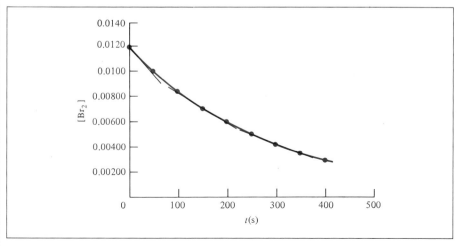

TABLE 22.2   Rates of the Reaction Between Molecular Bromine and Formic Acid at 25°C

Time (s)	$[Br_2]$	Rate (M/s)	$k = \dfrac{rate}{[Br_2]} \ (1/s)$
0.0	0.0120	4.20×10^{-5}	3.50×10^{-3}
50.0	0.0101	3.52×10^{-5}	3.49×10^{-3}
100.0	0.00846	2.96×10^{-5}	3.50×10^{-3}
150.0	0.00710	2.49×10^{-5}	3.51×10^{-3}
200.0	0.00596	2.09×10^{-5}	3.51×10^{-3}
250.0	0.00500	1.75×10^{-5}	3.50×10^{-3}
300.0	0.00420	1.48×10^{-5}	3.52×10^{-3}
350.0	0.00353	1.23×10^{-5}	3.48×10^{-3}
400.0	0.00296	1.04×10^{-5}	3.51×10^{-3}

initial rate, which is the rate of the reaction immediately after the mixing of the reactants.

Figure 22.5 is a plot of the rates versus Br_2 concentration. The linear relationship shows that the rate is directly proportional to the concentration; the higher the concentration, the higher is the rate:

$$\text{rate} \propto [Br_2]$$
$$= k[Br_2]$$

The term k is known as the **rate constant,** *a constant of proportionality between the reaction rate and the concentrations of reactants.* Rearrangement of the equation gives

$$k = \frac{\text{rate}}{[Br_2]}$$

It is important to understand that k is *not* affected by the concentration of Br_2. To be sure, the *rate* is greater at a higher concentration and smaller at

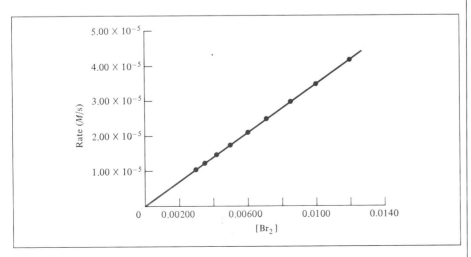

FIGURE 22.5   *Plot of rates versus molecular bromine concentrations, in the reaction between molecular bromine and formic acid. The straight line relationship shows that the rate of reaction is directly proportional to the molecular bromine concentration.*

a lower concentration of Br_2, but the *ratio*—rate/[Br_2]—remains the same. As we will see in Section 22.3, the value of any rate constant depends only on the temperature.

From Table 22.2 we can calculate the rate constant for the reaction. Taking the data for $t = 0$, we write

$$
\begin{aligned}
k &= \frac{\text{rate}}{[Br_2]} \\
&= \frac{4.20 \times 10^{-5}\ M/s}{0.0120\ M} \\
&= 3.50 \times 10^{-3}/s
\end{aligned}
$$

Similarly, we can use the data for $t = 50$ s to calculate k again, and so on. The slight variations in these values of k listed in Table 22.2 (see last column) are due to experimental deviations in rate measurements.

Thermal Decomposition of Dinitrogen Pentoxide. This reaction occurs in the gas phase as follows:

$$2N_2O_5(g) \xrightarrow{\Delta} 4NO_2(g) + O_2(g)$$

The reaction can also be studied in an organic solvent such as carbon tetrachloride (CCl_4):

$$2N_2O_5 \text{ (in } CCl_4) \xrightarrow{\Delta} 4NO_2 \text{ (in } CCl_4) + O_2(g)$$

Both N_2O_5 and NO_2 are soluble in CCl_4, but O_2 is not. Thus, as the reaction progresses, we can monitor the amount of O_2 evolved at various time intervals, using the apparatus shown in Figure 22.6. The gas volume, measured under laboratory conditions, can be converted to a common standard, that is, STP. Since one mole of an ideal gas occupies 22.4 L at STP, the number of moles of O_2 evolved can be readily calculated from the measured volume. Looking at the equation for the reaction, we see that 2 mol N_2O_5 $\backsimeq$ 1 mol O_2. Therefore, the decrease in N_2O_5 concentration can also be determined, as the following example shows.

FIGURE 22.6 *Apparatus for measuring the rate of thermal decomposition of* N_2O_5.

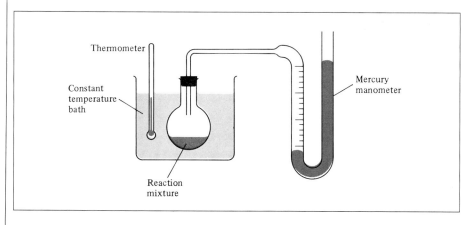

EXAMPLE 22.1

In a typical experiment, the decomposition of N_2O_5 in 100 mL of a 2.00 M solution in CCl_4 generated 0.258 L of O_2 (at STP) after 200 min at 45°C. Calculate the concentration of unreacted N_2O_5 at that time.

Answer

The number of moles of O_2 produced is

$$\text{moles of } O_2 = 0.258 \text{ L } O_2 \times \frac{1 \text{ mol } O_2}{22.4 \text{ L } O_2}$$
$$= 0.0115 \text{ mol } O_2$$

The number of moles of N_2O_5 decomposed is twice that of O_2, namely, $2 \times 0.0115 = 0.0230$ mol. Originally there was 0.200 mole of N_2O_5 in 100 mL of solution. Therefore, the concentration of unreacted N_2O_5 in solution after the given time interval is

$$[N_2O_5] = \frac{(0.200 - 0.0230) \text{ mol } N_2O_5}{100 \text{ mL soln}} \times \frac{1000 \text{ mL soln}}{1 \text{ L soln}}$$
$$= \frac{1.77 \text{ mol}}{1 \text{ L soln}}$$
$$= 1.77 \ M$$

Following the procedure outlined in Example 22.1, we can calculate the concentration of unreacted N_2O_5 after various time intervals. The results are given in Table 22.3 and plotted in Figure 22.7.

The rate of N_2O_5 decomposition at any time can be measured as described in the section about the reaction of molecular bromine with formic acid.

Reaction Rate and Stoichiometry

Having discussed the measurement of reaction rates, we will now see how rate expressions are written from chemical equations. For stoichiometri-

TABLE 22.3 Data of the Thermal Decomposition of N_2O_5 in CCl_4 at 45°C

Time (min)	$[N_2O_5]$	$\Delta[N_2O_5]$*	Volume (L) of O_2 at STP
0	2.00	0	0
200	1.77	0.23	0.258
400	1.56	0.44	0.493
600	1.38	0.62	0.694
800	1.22	0.78	0.874
1000	1.08	0.92	1.03
1200	0.95	1.05	1.18

*$\Delta[N_2O_5]$ is the change in N_2O_5 concentration between $t = 0$ and $t = t$. For example, the third entry in this column, 0.44, is the difference between 2.00 (at $t = 0$) and 1.56 (at $t = 400$ min).

FIGURE 22.7 *The decrease in the concentration of N_2O_5 and the increase in the volume of O_2 as a function of time, in the decomposition of N_2O_5.*

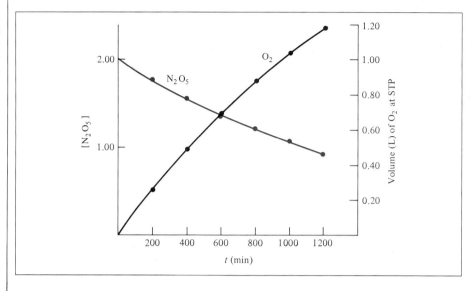

FIGURE 22.7 *The decrease in the concentration of N_2O_5 and the increase in the volume of O_2 as a function of time, in the decomposition of N_2O_5.*

cally simple reactions such as A → B, we can express the rate either as

$$\text{rate} = -\frac{\Delta[A]}{\Delta t} \quad \text{or} \quad \text{rate} = \frac{\Delta[B]}{\Delta t}$$

(The rate of product formation does not require the minus sign because $\Delta[B]$ is a positive quantity.) For more complex reactions, we must be careful in writing the rate expressions. Consider, for example, the reaction

$$2A \longrightarrow B$$

Two moles of A disappear for each mole of B that forms—that is, the rate of disappearance of A is twice as fast as the rate of appearance of B. We write the rate either as

$$\text{rate} = -\frac{1}{2}\frac{\Delta[A]}{\Delta t} \quad \text{or} \quad \text{rate} = \frac{\Delta[B]}{\Delta t}$$

In general, for the reaction

$$aA + bB \longrightarrow cC + dD$$

the rate is given by

$$\text{rate} = -\frac{1}{a}\frac{\Delta[A]}{\Delta t} = -\frac{1}{b}\frac{\Delta[B]}{\Delta t} = \frac{1}{c}\frac{\Delta[C]}{\Delta t} = \frac{1}{d}\frac{\Delta[D]}{\Delta t}$$

EXAMPLE 22.2

Write the rate expressions for the following reactions:

(a) $I^-(aq) + OCl^-(aq) \longrightarrow Cl^-(aq) + OI^-(aq)$

(b) $3O_2(g) \longrightarrow 2O_3(g)$

(c) $4NH_3(g) + 5O_2(g) \longrightarrow 4NO(g) + 6H_2O(g)$

Answer

(a) Since each of the stoichiometric coefficients equals 1

$$\text{rate} = -\frac{\Delta[I^-]}{\Delta t} = -\frac{\Delta[OCl^-]}{\Delta t} = \frac{\Delta[Cl^-]}{\Delta t} = \frac{\Delta[OI^-]}{\Delta t}$$

(b) Here the coefficients are 3 and 2, so

$$\text{rate} = -\frac{1}{3}\frac{\Delta[O_2]}{\Delta t} = \frac{1}{2}\frac{\Delta[O_3]}{\Delta t}$$

(c) In this reaction

$$\text{rate} = -\frac{1}{4}\frac{\Delta[NH_3]}{\Delta t} = -\frac{1}{5}\frac{\Delta[O_2]}{\Delta t} = \frac{1}{4}\frac{\Delta[NO]}{\Delta t} = \frac{1}{6}\frac{\Delta[H_2O]}{\Delta t}$$

Similar example: Problem 22.4.

22.2 THE RATE LAWS

We saw in the preceding section that the rate of a reaction is related to the concentrations of the reacting species. In fact, for a great many reactions the rate is proportional to the product of the concentrations of the reactants raised to some power. Consider the reaction

$$a\text{A} + b\text{B} \longrightarrow c\text{C} + d\text{D}$$

The forward rate (from left to right) is

$$\begin{aligned}\text{rate} &\propto [\text{A}]^x[\text{B}]^y \\ &= k[\text{A}]^x[\text{B}]^y\end{aligned} \tag{22.1}$$

where k is the rate constant and x and y are themselves constants for a given reaction. Equation (22.1) is known as the **rate law,** an *expression relating the rate of a reaction to the rate constant and the concentrations of the reactants.* If we know the values of k, x, and y, we can always calculate the rate of the reaction, given the concentrations of A and B. As we will see shortly, like k, x and y must be determined experimentally.

> Rate laws are always determined experimentally.

The sum of the powers to which all reactant concentrations appearing in the rate law are raised is called the overall **reaction order.** This value enables us to predict the effect on the reaction rate of a change in the concentration of one of the reactants. Suppose, for example, that for a certain reaction $x = 1$ and $y = 2$. The rate law for this reaction is

> Reaction order is always defined in terms of reactant (not product) concentrations.

$$\text{rate} = k[\text{A}][\text{B}]^2$$

Let us assume that initially $[\text{A}] = 1.0\ M$ and $[\text{B}] = 1.0\ M$. The equation tells us that if we double the concentration of A from 1.0 M to 2.0 M at constant [B], we also double the rate:

> This reaction is first order in A, second order in B, and third order overall ($1 + 2 = 3$).

$$[\text{A}] = 1.0\ M \quad \begin{aligned}\text{rate}_1 &= k(1.0\ M)(1.0\ M)^2 \\ &= k(1.0\ M^3)\end{aligned}$$

$$[\text{A}] = 2.0\ M \quad \begin{aligned}\text{rate}_2 &= k(2.0\ M)(1.0\ M)^2 \\ &= k(2.0\ M^3)\end{aligned}$$

Hence

$$rate_2 = 2(rate_1)$$

On the other hand, if we double the concentration of B from 1.0 M to 2.0 M at constant [A], the rate will be increased by a factor of 4 because of the power 2 in the exponent:

$$[B] = 1.0\ M \qquad rate_1 = k(1.0\ M)(1.0\ M)^2$$
$$= k(1.0\ M^3)$$

$$[B] = 2.0\ M \qquad rate_2 = k(1.0\ M)(2.0\ M)^2$$
$$= k(4.0\ M^3)$$

Hence

$$rate_2 = 4(rate_1)$$

If, for a certain reaction, $x = 0$ and $y = 1$, then the rate law is

$$rate = k[A]^0[B]$$
$$= k[B]$$

This reaction is zero order in A and first order in B (first order overall). Thus the rate of the reaction is *independent* of the concentration of A present.

It is important to keep in mind that the exponents x and y in Equation (22.1) are not necessarily related to the stoichiometric coefficients in the balanced equation; that is, in general it is *not* true that for $aA + bB \rightarrow cC + dD$, $a = x$ and $b = y$. Furthermore, depending on the nature of the chemical process, the reaction order can be an integer, zero, or even a noninteger. Consider the following examples.

- For the thermal decomposition of N_2O_5

$$2N_2O_5(g) \longrightarrow 4NO_2(g) + O_2(g)$$

  the rate law is

$$rate = k[N_2O_5]$$

  and not rate $= k[N_2O_5]^2$, as we might have inferred from the balanced equation.
- The rate law for the thermal decomposition of acetaldehyde (CH_3CHO)

$$CH_3CHO(g) \longrightarrow CH_4(g) + CO(g)$$

  has been determined experimentally to be

$$rate = k[CH_3CHO]^{3/2}$$

  and not rate $= k[CH_3CHO]$.
- The reaction of hydrogen peroxide with iodide ion in an acidic medium is

$$H_2O_2(aq) + 3I^-(aq) + 2H^+(aq) \longrightarrow I_3^-(aq) + 2H_2O(l)$$

  and the rate law is

$$rate = k[H_2O_2][I^-]$$

Thus the reaction is first order in H_2O_2, first order in I^-, and zero order in H^+. The concentration of H^+ ions has *no* effect on the rate of the reaction. Therefore the overall reaction order is 2.

Sometimes, by coincidence, the reaction orders are the same as the stoichiometric coefficients. For example, the reaction

$$H_2(g) + I_2(g) \longrightarrow 2HI(g)$$

has the rate law expression

$$\text{rate} = k[H_2][I_2]$$

which is first order in H_2 and first order in I_2. This is really the exception rather than the rule. In general, *the order of a reaction must be determined by experiment; it cannot be deduced from the overall balanced equation.*

Determination of Reaction Order

If a reaction involves only one reactant, the rate law can be readily determined by measuring the rate of the reaction as a function of the reactant's concentration. For example, if the rate doubles when the concentration of the reactant doubles, then the rate is first order in the reactant. If the rate quadruples when the concentration doubles, the reaction is second order in the reactant.

For a reaction involving more than one reactant, we can find the rate law by measuring the dependence of the reaction rate on the concentration of each reactant, one at a time. We fix the concentrations of all but one reactant and record the rate of the reaction as a function of the concentration of that reactant. Any changes in the rate must be due only to changes in that substance. The dependence thus observed gives us the order in that particular reactant. The same procedure is then applied to the next reactant, and so on. (This method is known as the *isolation method.*)

In practice, it is preferable to observe the dependence only of the *initial* rate on reactant concentrations, because as the reaction proceeds, the concentrations of the reactants decrease and it may become difficult to measure the changes accurately. Also, there may be a reverse reaction of the type

$$\text{products} \longrightarrow \text{reactants}$$

which would introduce error in the rate measurement. Both of these complications are virtually absent during the early stages of the reaction.

EXAMPLE 22.3

The rate of the reaction $A + 2B \rightarrow C$ has been observed at 25°C. From the following data, determine the rate law for the reaction and calculate the rate constant.

(Continued)

Experiment	Initial [A]	Initial [B]	Initial Rate (M/s)
1	0.100	0.100	5.50×10^{-6}
2	0.200	0.100	2.20×10^{-5}
3	0.400	0.100	8.80×10^{-5}
4	0.100	0.300	1.65×10^{-5}
5	0.100	0.600	3.30×10^{-5}

Answer

We assume that the rate law takes the form

$$\text{rate} = k[A]^x[B]^y$$

The data from experiments 1 and 2 show that, at a constant concentration (0.100 M) of B, the rate of the reaction increases by a factor of 4 when we double the concentration of A. Therefore, the reaction must be second order in A. Data from experiments 4 and 5 show that, at a constant concentration of A, the rate of the reaction is doubled when we double the concentration of B. Therefore, the reaction must be first order in B and the rate law is

$$\text{rate} = k[A]^2[B]$$

Thus the overall reaction order is (2 + 1), or 3.

The rate constant k can be calculated using the data from any one of the experiments. Since

$$k = \frac{\text{rate}}{[A]^2[B]}$$

data from experiment 1 give us

$$k = \frac{5.50 \times 10^{-6} \, M/s}{(0.100 \, M)^2(0.100 \, M)}$$
$$= 5.50 \times 10^{-3}/M^2 s$$

Similar examples: Problems 22.11, 22.56.

Rate law expressions enable us to calculate the rate of a reaction from a knowledge of the rate constant and reactant concentrations. They also can be converted into equations that allow us to determine the concentrations of reactants at any time during the course of a reaction. We will illustrate this application by considering two of the simplest kinds of rate laws—that is, those applying to reactions that are first order overall and to reactions that are second order overall.

First-Order Reactions

A **_first-order reaction_** is a _reaction whose rate depends on the reactant concentration raised to the first power._ In a first-order reaction of the type

$$A \longrightarrow \text{product}$$

the rate is

$$\text{rate} = -\frac{\Delta[A]}{\Delta t}$$

Also, from the rate law we know that

$$\text{rate} = k[A]$$

Thus

$$-\frac{\Delta[A]}{\Delta t} = k[A] \qquad (22.2)$$

We can determine the units of the first-order rate constant k by transposing:

$$k = -\frac{\Delta[A]}{[A]} \frac{1}{\Delta t} = \frac{M}{M\,\text{s}} = \frac{1}{\text{s}}$$

(The minus sign does not enter into the evaluation of units.) Using calculus, we can show that

$$\log \frac{[A]_0}{[A]_t} = \frac{kt}{2.303} \qquad (22.3)$$

where $[A]_0$ and $[A]_t$ are the concentrations of A at times $t = 0$ and $t = t$, respectively.

It should be understood that $t = 0$ need not correspond to the beginning of the experiment; it can be any arbitrarily chosen time when we start to monitor the change in the concentration of A.

Equation (22.3) can be rearranged as follows:

$$\log [A]_0 - \log [A]_t = \frac{kt}{2.303}$$

or

$$\log [A]_t = -\frac{kt}{2.303} + \log [A]_0 \qquad (22.4)$$

Equation (22.4) takes the form of the linear equation $y = mx + b$, in which m is the slope of the line that is the graph of the equation.

$$\log [A]_t = -\frac{k}{2.303}(t) + \log [A]_0$$
$$\begin{array}{ccccc} \updownarrow & & \updownarrow & \updownarrow & \updownarrow \\ y & = & m & x + & b \end{array}$$

Thus a plot of $\log [A]_t$ versus t (or y versus x) gives a straight line with a slope of $-k/2.303$ (or m). This allows us to calculate the rate constant k. Figure 22.8 shows the characteristics of a first-order reaction.

There are many known first-order reactions. The decomposition of N_2O_5, which we discussed earlier, is first order in N_2O_5. Another example is the conversion of *cis*-2-butene to *trans*-2-butene:

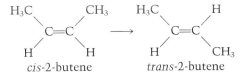

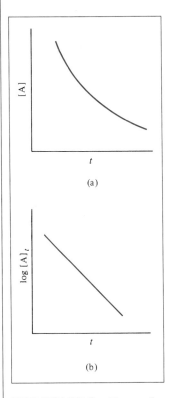

FIGURE 22.8 *First-order reaction characteristics. (a) Decrease of reactant concentration with time. (b) Plot of the straight-line relationship to obtain the rate constant. The slope of the line is equal to $-k/2.303$.*

To review logarithm and significant figures, see p. 541.

The following example shows how Equation (22.3) can be used in different ways to obtain useful information about a first-order reaction.

EXAMPLE 22.4

The conversion of cyclopropane to propene in the gas phase

$$CH_2\text{---}CH_2 \quad \longrightarrow \quad CH_3\text{---}CH\text{==}CH_2$$
$$\text{(with CH}_2\text{ above)}$$
$$\text{cyclopropane} \qquad\qquad \text{propene}$$

is a first-order reaction with a rate constant of 6.71×10^{-4}/s at 250°C. (a) If the initial concentration of cyclopropane was 0.25 M, what is the concentration after 4.5 min? (b) How long will it take for the concentration of cyclopropane to decrease from 0.25 M to 0.15 M? (c) How long will it take to convert 72 percent of the starting material?

Answer

(a) Applying Equation (22.3)

$$\log \frac{[A]_0}{[A]_t} = \frac{kt}{2.303}$$

$$\log \frac{0.25}{x} = \frac{(6.71 \times 10^{-4}\text{/s})(4.5 \times 60\text{ s})}{2.303}$$

where x is the desired concentration. Solving the equation, we obtain

$$\log \frac{0.25}{x} = 0.079$$

$$x = 0.21\ M$$

(b) Again using Equation (22.3)

$$\log \frac{0.25}{0.15} = \frac{(6.71 \times 10^{-4}\text{/s})t}{2.303}$$

$$t = 7.6 \times 10^2\text{ s}$$

$$= 13\text{ min}$$

(c) In a calculation of this type, we do not need to know the actual concentration of the starting material. We can arbitrarily assume that it is 1.00. Thus the concentration of cyclopropane after time t is $(1.00 - 0.72)$, or 0.28. From Equation (22.3), we write

$$t = \frac{2.303}{k} \log \frac{[A]_0}{[A]_t}$$

$$= \frac{2.303}{6.71 \times 10^{-4}\text{/s}} \log \frac{1.00}{0.28}$$

$$= 1.9 \times 10^3\text{ s} \quad \text{or} \quad 32\text{ min}$$

Note that it was not necessary to include the units of concentration. Whatever they were, they would have been canceled in log $([A]_0/[A]_t)$.

Similar examples: Problems 22.14, 22.15.

Because k is given in unit of 1/s, we need to convert 4.5 minutes to seconds.

The **half-life** of a reaction, $t_{\frac{1}{2}}$, is the *time required for the concentration of a reactant to decrease to half of its initial concentration*. We can obtain an expression for $t_{\frac{1}{2}}$ for a first-order reaction as follows. From Equation (22.3) we write

$$t = \frac{2.303}{k} \log \frac{[A]_0}{[A]_t}$$

By definition, when $t = t_{\frac{1}{2}}$, $[A]_t = [A]_0/2$, so

$$t_{\frac{1}{2}} = \frac{2.303}{k} \log \frac{[A]_0}{[A]_0/2} \tag{22.5}$$

$$t_{\frac{1}{2}} = \frac{2.303}{k} \log 2 = \frac{0.693}{k} \tag{22.6}$$

Equation (22.6) tells us that the half-life of a first-order reaction is independent of the initial concentration of the reactant. Thus, it takes the same time for the concentration of the reactant to decrease from 1.0 M to 0.50 M, say, as it does for a decrease in concentration from 0.10 M to 0.050 M. Measuring the half-life of a reaction is one way to determine the rate constant of a first-order reaction.

EXAMPLE 22.5

At 45°C, the first-order decomposition

$$2N_2O_5 \longrightarrow 4NO_2 + O_2$$

in CCl_4 has a rate constant of 6.2×10^{-4}/min. (a) What is the rate constant in terms of seconds? (b) Calculate the half-life of the reaction in seconds.

Answer

(a) Let the rate constant be k so that

$$k = \frac{6.2 \times 10^{-4}}{\text{min}} \times \frac{1 \text{ min}}{60 \text{ s}} = 1.0 \times 10^{-5}/\text{s}$$

(b) According to Equation (22.6)

$$t_{\frac{1}{2}} = \frac{0.693}{k}$$

$$= \frac{0.693}{1.0 \times 10^{-5}/\text{s}}$$

$$= 6.9 \times 10^4 \text{ s}$$

Thus it will take 6.9×10^4 seconds, or about 19 hours, for *any* initial concentration of N_2O_5 to decrease to half its value.

Similar example: Problem 22.24.

We have developed two sets of equations—Equations (22.3) and (22.4) and Equations (22.5) and (22.6)—that can help us evaluate a first-order rate constant. From experimental data of concentration versus time, we can determine k graphically (as shown in Figure 22.8) or use the half-life method (as the next example will show).

Example 22.6 involves gases, and in dealing with gases it is usually convenient to express the concentrations in units of pressure (atm or mmHg) rather than molarity. From the ideal gas equation [Equation (6.7)] we can write $P = (n/V)RT$. This equation shows that the pressure P (or the partial pressure) of a gas is directly proportional to its concentration (in mol/L) if both V and T remain constant.

EXAMPLE 22.6

The rate of decomposition of azomethane is studied by monitoring the partial pressure of the reactant as a function of time:

$$CH_3-N=N-CH_3(g) \longrightarrow N_2(g) + C_2H_6(g)$$

The data obtained at 300°C are shown in the following table:

Partial Pressure of Azomethane (mmHg)	284	220	193	170	150	132
Time (s)	0	100	150	200	250	300

(a) Are these values consistent with first-order kinetics? If so, determine the rate constant (b) by plotting the data as shown in Figure 22.8 and (c) by the half-life method.

Answer

(a) The partial pressure of azomethane at any time is directly proportional to its concentration (in mol/L), because it is the product of its mole fraction and the total pressure of the system (Section 6.6). The higher the concentration, the greater its mole fraction. Therefore Equation (22.4) can be written in terms of partial pressures as

$$\log P_t = -\frac{k}{2.303}(t) + \log P_0$$

where P_0 and P_t are the partial pressures of azomethane at time $t = 0$ and $t = t$. This equation has the form of the linear equation $y = mx + b$. Figure 22.9(a), which is based on the data given in the table below, shows that a plot of $\log P_t$ versus t yields a straight line, so the reaction is indeed first order.

$\log P_t$	2.453	2.342	2.286	2.230	2.176	2.121
t (s)	0	100	150	200	250	300

(b) From Equation (22.4) we see that the slope of the line for a first-order reaction is equal to $-k/2.303$. In Figure 22.9(a) the slope is -1.11×10^{-3}/s.

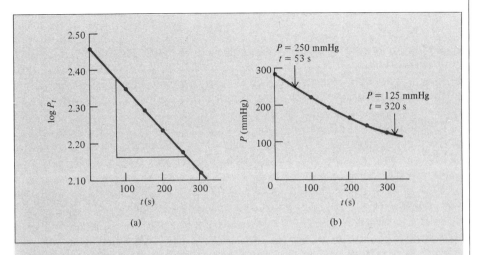

FIGURE 22.9 *(a) Plot of log P_t versus time. The slope of the line is calculated from two pairs of coordinates:*

$$slope = \frac{2.163 - 2.370}{(263 - 76)s} = -1.11 \times 10^{-3}/s$$

According to Equation (22.4), the slope is equal to $-k/2.303$. (b) The half-life method for determining the rate constant of the first-order reaction. The time it takes for the pressure to drop from 250 mmHg (an arbitrary choice) to 125 mmHg is the half-life of the reaction. Here $t_{\frac{1}{2}} = 320$ s $- 53$ s or 267 s.

Therefore

$$-\frac{k}{2.303} = slope$$
$$-k = 2.303(slope)$$
$$= 2.303(-1.11 \times 10^{-3}/s)$$
$$k = 2.56 \times 10^{-3}/s$$

(c) To calculate the rate constant by the half-life method we need to plot the partial pressure versus time, as shown in Figure 22.9(b), because the given data do not show the time it takes for the partial pressure to decrease to half of its value. From the curve we find that $t_{\frac{1}{2}} = 267$ s, so the rate constant can be calculated by using Equation (22.6):

$$t_{\frac{1}{2}} = \frac{0.693}{k}$$
$$k = \frac{0.693}{267 \text{ s}}$$
$$= 2.60 \times 10^{-3}/s$$

The small difference between the two rate constants calculated in (b) and (c) for the same reaction should not be surprising. Treating experimental data graphically always introduces some uncertainty, as evidenced by the slight disagreement of our results.

Similar examples: Problems 22.17, 22.18, 22.19.

Second-Order Reactions

A *second-order reaction* is a *reaction whose rate depends on reactant concentration raised to the second power or on the concentrations of two different reactants, each raised to the first power.* The simpler type (the former case) involves only one kind of reactant molecule:

$$A \longrightarrow \text{product}$$

where

$$\text{rate} = -\frac{\Delta[A]}{\Delta t}$$

From the rate law

$$\text{rate} = k[A]^2$$

Thus

$$-\frac{\Delta[A]}{\Delta t} = k[A]^2$$

We can determine the units of the second-order rate constant k by solving for k:

$$k = -\frac{\Delta[A]}{[A]^2}\frac{1}{\Delta t} = \frac{M}{M^2 \text{ s}} = \frac{1}{M \text{ s}}$$

(Remember, we do not use the minus sign for units.)

The other type of second-order reaction is represented as

$$A + B \longrightarrow \text{product}$$

where

$$\text{rate} = -\frac{\Delta[A]}{\Delta t} = -\frac{\Delta[B]}{\Delta t}$$

From the rate law

$$\text{rate} = k[A][B]$$

Thus

$$-\frac{\Delta[A]}{\Delta t} = -\frac{\Delta[B]}{\Delta t} = k[A][B]$$

In this case, the rate of decrease in A is the same as the rate of decrease in B. The reaction is first order in A and first order in B, so it has an overall order of 2. Again, the units of k are $1/M$ s.

By means of calculus we can obtain the following expressions for "A → product" reactions:

$$\frac{1}{[A]_t} = \frac{1}{[A]_0} + kt \tag{22.7}$$

(The corresponding equation "A + B → product" reactions is too complex to discuss here.) From Equation (22.7) we can obtain an equation for the half-life of a second-order reaction by setting $[A]_t = [A]_0/2$:

$$\frac{1}{[A]_0/2} = \frac{1}{[A]_0} + kt_{\frac{1}{2}}$$

Solving for $t_{\frac{1}{2}}$ we obtain

$$t_{\frac{1}{2}} = \frac{1}{k[A]_0} \qquad (22.8)$$

Note that the half-life of a second-order reaction is *not* independent of the initial concentration, as in the case of a first-order reaction [see Equation (22.6)]. This is one way to distinguish a first-order reaction from a second-order reaction.

EXAMPLE 22.7

The recombination of iodine atoms to form molecular iodine in the gas phase

$$I(g) + I(g) \longrightarrow I_2(g)$$

follows second-order kinetics and has the high rate constant $7.0 \times 10^9/M$ s at 23°C. (a) Assuming that the initial concentration of I was 0.086 M, calculate the concentration after 2.0 min. (b) What would the half-lives of the reaction be if the initial concentrations of I were 0.60 M and 0.42 M?

Answer

(a) We begin with Equation (22.7):

$$\frac{1}{[A]_t} = \frac{1}{[A]_0} + kt$$

$$\frac{1}{x} = \frac{1}{0.086\ M} + (7.0 \times 10^9/M\ s)(2.0 \times 60\ s)$$

where x is the desired concentration. Solving the equation, we get

$$x = 1.2 \times 10^{-12}\ M$$

This is such a low concentration that it is virtually undetectable. The very large rate constant for the reaction means that practically all the I atoms combine after only two minutes of reaction time.

(b) We need Equation (22.8) for this part.

$[I]_0 = 0.60\ M$

$$t_{\frac{1}{2}} = \frac{1}{k[A]_0}$$

$$= \frac{1}{(7.0 \times 10^9/M\ s)(0.60\ M)}$$
$$= 2.4 \times 10^{-10}\ s$$

$[I]_0 = 0.42\ M$

$$t_{\frac{1}{2}} = \frac{1}{(7.0 \times 10^9/M\ s)(0.42\ M)}$$
$$= 3.4 \times 10^{-10}\ s$$

(Continued)

These results show that, as mentioned earlier, the half-life of a second-order reaction is not a constant but depends on the initial concentration of the reactant(s).

Similar example: Problem 22.25.

We have now discussed first-order reactions and second-order reactions. Reactions whose order is zero are rare. Mathematically speaking, zero-order reactions are easy to deal with. The rate law is

$$\text{rate} = k[A]^0$$
$$= k$$

Thus the rate of a zero-order reaction is a *constant*, independent of its reactant concentrations. Third-order reactions are quite complex; they will not be studied in this book.

22.3 ACTIVATION ENERGY AND TEMPERATURE DEPENDENCE OF RATE CONSTANTS

The Collision Theory of Chemical Kinetics

With very few exceptions, reaction rates increase with increasing temperature. For example, the time required to hard-boil an egg is much shorter if the "reaction" is carried out at 100°C (about 10 minutes) than at 80°C (about 30 minutes). Conversely, an effective way to preserve foods is to store them at subzero temperatures, thereby slowing the rate of bacterial decay. Figure 22.10 shows a typical dependence of the rate constant of a reaction on temperature. In order to explain this behavior, we must ask how reactions get started in the first place.

It seems logical to assume—and it is generally the case—that chemical reactions occur as a result of collisions among reacting molecules. In terms of the *collision theory* of chemical kinetics, then, we expect the rate of a reaction to be directly proportional to the number of molecular collisions per second, or to the frequency of molecular collisions:

$$\text{rate} \propto \frac{\text{number of collisions}}{\text{s}}$$

This simple relationship explains the dependence of rate on concentration.

Consider the reaction of A molecules with B molecules to form some product. Suppose that each product molecule is formed by the direct combination of an A molecule and a B molecule. If we doubled the concentration of A, say, then the number of A–B collisions would also double, because there would be twice as many A molecules that could collide with B molecules (Figure 22.11). Consequently, the rate would increase by a factor

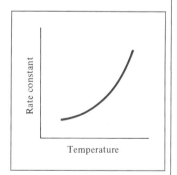

FIGURE 22.10 *Dependence of rate constant on temperature. The rate constants of most reactions increase with increasing temperature.*

of 2. Similarly, doubling the concentration of B molecules would increase the rate twofold. Thus, we can express the rate law as

$$\text{rate} = k[\text{A}][\text{B}]$$

that is, the reaction is first order in both A and B and obeys second-order kinetics.

The collision theory is intuitively appealing, but the relationship between rate and molecular collision is more complicated than you might expect. The preceding discussion seems to imply that a reaction always occurs when an A and a B molecule collide. However, this assumption is usually not borne out. Calculations based on the kinetic molecular theory show that, at ordinary pressures (say, 1 atm) and temperatures (say, 298 K), there are about 1×10^{27} binary collisions (collisions between two molecules) in 1 mL of volume every second, in the gas phase. An even larger number is obtained for processes occurring in liquids. If every binary collision led to a product, then most reactions would be complete almost instantaneously. In practice, we find that the rates of reactions differ greatly. This means that, in many cases, collisions alone do not guarantee that a reaction will take place.

Any molecule in motion possesses kinetic energy; the faster the motion, the greater the kinetic energy. But a fast-moving molecule will not break up into fragments on its own. To react, it must collide with another molecule. To use a simple analogy, a car traveling at 80 km/h (50 mph) will not begin to disintegrate on its own (assuming that it is in good running condition), but if it meets another car in a head-on collision, then quite a few pieces of both cars will fly apart. When molecules collide, part of their kinetic energy is converted to vibrational energy. If the initial kinetic energies are large, then the molecules will vibrate so strongly as to break some of the chemical bonds, which is the first step toward product formation. If the initial kinetic energies are small, the molecules will merely bounce off each other intact. (Note that this is an important difference between car collisions and molecular collisions.) Energetically speaking, there is a limit in collision energies below which no reaction occurs.

We postulate that, in order to react, the colliding molecules must have a total kinetic energy equal to or greater than the **activation energy** (E_a), which is the *minimum amount of energy required to initiate a chemical reaction.* Lacking this energy, the molecules remain intact and no change results from the collision. *The species temporarily formed by the reactant molecules as a result of the collision before they form the product* is called the **activated complex.**

Figure 22.12 shows two different energy profiles for the reaction

$$\text{A} + \text{B} \longrightarrow \text{C} + \text{D}$$

If the products are more stable than the reactants, then the reaction will be accompanied by a release of heat; that is, the reaction is exothermic [Figure 22.12(a)]. On the other hand, if the products are less stable than the reactants, then heat will be absorbed by the reacting mixture from the surroundings and we have an endothermic reaction [Figure 22.12(b)]. In both cases we plot the energy of the reacting system versus the reaction

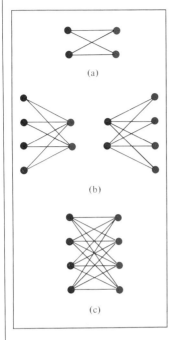

FIGURE 22.11 *Dependence of the rate of reaction on number of collisions. We concentrate here only on A–B collisions, which lead to formation of products. (a) There are 4 possible collisions among two A and two B molecules. (b) Doubling the number of either type of molecule (but not both) increases the number of collisions to 8. (c) Doubling both the A and B molecules increases the number of collisions to 16.*

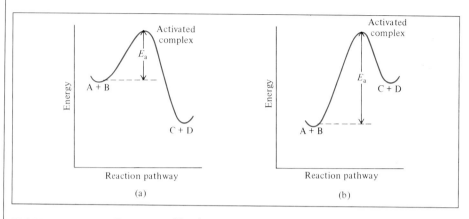

FIGURE 22.12 *Energy profiles for (a) exothermic and (b) endothermic reactions. These plots show the variation in energy of the reactant molecules (A and B) as a function of their geometries during the course of a reaction. The activated complex is a highly unstable species that possesses the maximum amount of energy. The difference between the energies of the reactants and that of the activated complex is equal to the activation energy. Note that the products (C and D) are more stable than the reactants in (a), and the reverse is true for (b).*

pathway. Qualitatively, these plots show the variation in the energies of molecules A and B as a function of their geometries (that is, the changes in bond lengths and bond angles as they are being converted to molecules C and D) during the course of the reaction.

In order for A and B to react, they must possess the necessary activation energy. We can think of this energy as the barrier that prevents less energetic molecules from reacting. Because the number of reacting molecules in an ordinary reaction is very large, there is a tremendous spread in the speed and, hence, in the kinetic energies of the molecules. Normally, only a small fraction of these molecules—the very fast-moving ones—can take part in the reaction. The increase in the rate (or the rate constant) with temperature can now be explained: The speeds of the molecules conform to the Maxwell distributions shown in Figure 6.15. Compare the speed distributions at two different temperatures. Since a larger number of energetic molecules is present at the higher temperature, the rate of product formation is also greater at the higher temperature.

The Arrhenius Equation

In 1889 Svante Arrhenius showed that the dependence of the rate constant of a reaction on temperature can be expressed by the following equation, now known as the *Arrhenius equation*:

$$k = Ae^{-E_a/RT} \qquad (22.9)$$

where E_a is the activation energy of the reaction (in kJ/mol), R is the gas constant (8.314 J/K·mol), T is the absolute temperature, and e is the base of the natural logarithm scale (see Appendix 4). The quantity A represents the collision frequency, and is called the *frequency factor*. It can be treated as a constant for a given reacting system over a fairly wide temperature range. Equation (22.9) shows that the rate constant is directly proportional to A and, therefore, to the collision frequency. Further, because of the minus sign associated with E_a/RT, the rate constant decreases with increasing activation energy and increases with increasing temperature. This equation can be expressed in a more useful form by taking the natural logarithm of both sides

$$\ln k = \ln A e^{-E_a/RT}$$
$$= \ln A - \frac{E_a}{RT}$$

or in terms of common logarithms (base 10):

$$\log k = \log A - \frac{E_a}{2.303RT} \tag{22.10}$$

Equation (22.10) can take the form of a linear equation:

$$\log k = -\frac{E_a}{2.303R}\left(\frac{1}{T}\right) + \log A$$

$$\begin{array}{ccccc}
\updownarrow & & \updownarrow & \updownarrow & \updownarrow \\
y & = & m & x \ + & b
\end{array}$$

Thus, a plot of $\log k$ versus $1/T$ gives a straight line whose slope m is equal to $-E_a/2.303R$ and whose intercept b with the ordinate (the y axis) is $\log A$.

EXAMPLE 22.8

The rate constants for the decomposition of acetaldehyde

$$CH_3CHO(g) \longrightarrow CH_4(g) + CO(g)$$

were measured at five different temperatures. The data are shown below. Plot $\log k$ versus $1/T$ and determine the activation energy (in kJ/mol) for the reaction.

T (K)	700	730	760	790	810
k (1/$M^{\frac{1}{2}}$s)	0.011	0.035	0.105	0.343	0.789

Answer

We need to plot $\log k$ (y) versus $1/T$ (x). From the given data we obtain

$1/T$ (1/K)	1.43×10^{-3}	1.37×10^{-3}	1.32×10^{-3}	1.27×10^{-3}	1.23×10^{-3}
$\log k$	-1.96	-1.46	-0.979	-0.465	-0.103

(Continued)

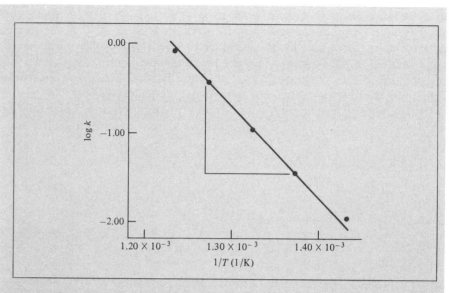

FIGURE 22.13 *Plot of log k versus 1/T. The slope of the line is calculated from two pairs of coordinates:*

$$slope = \frac{-1.80 - (-0.30)}{(1.41 - 1.25) \times 10^{-3}/K} = -9.38 \times 10^3 \, K$$

The slope is equal to $-E_a/2.303RT$. The intercept of the straight line on the log k axis is not shown here because it would make the horizontal axis too long.

These data, when plotted, yield the graph shown in Figure 22.13. The slope of the straight line is -9.38×10^3 K. From the linear form of Equation (22.10)

$$slope = -\frac{E_a}{2.303R} = -9.38 \times 10^3 \, K$$

$$E_a = (2.303)(8.314 \, J/K \cdot mol)(9.38 \times 10^3 \, K)$$
$$= 1.80 \times 10^5 \, J/mol$$
$$= 1.80 \times 10^2 \, kJ/mol$$

It is important to note that although the rate constant itself has the units $1/M^{\frac{1}{2}}$s, the quantity log k has no units (we cannot take the logarithm of any unit).

Similar examples: Problems 22.27, 22.28.

An equation relating the rate constants k_1 and k_2 at temperatures T_1 and T_2 can be used to calculate the activation energy or to find the rate constant at another temperature if the activation energy is known. To derive such an equation we start with Equation (22.10):

$$\log k_1 = \log A - \frac{E_a}{2.303RT_1} \tag{22.11}$$

$$\log k_2 = \log A - \frac{E_a}{2.303RT_2} \tag{22.12}$$

Subtracting Equation (22.12) from Equation (22.11) gives

$$\log k_1 - \log k_2 = -\frac{E_a}{2.303RT_1} + \frac{E_a}{2.303RT_2}$$

$$\log \frac{k_1}{k_2} = \frac{E_a}{2.303R}\left(\frac{1}{T_2} - \frac{1}{T_1}\right) \tag{22.13}$$

EXAMPLE 22.9

The rate constant of a first-order reaction is 3.46×10^{-2}/s at 298 K. What is the rate constant at 350 K if the activation energy for the reaction is 50.2 kJ/mol?

Answer

Substituting in Equation (22.13)

$$\log \frac{3.46 \times 10^{-2}}{k_2} = \frac{50.2 \times 10^3 \text{ J/mol}}{2.303(8.314 \text{ J/K} \cdot \text{mol})}\left(\frac{1}{350 \text{ K}} - \frac{1}{298 \text{ K}}\right)$$

where k_2 is the rate constant at 350 K. Solving the equation gives

$$\log \frac{3.46 \times 10^{-2}}{k_2} = -1.31$$

$$\frac{3.46 \times 10^{-2}}{k_2} = 0.049$$

$$k_2 = 0.71/\text{s}$$

Similar examples: Problems 22.33, 22.34, 22.36.

The Arrhenius equation is quite useful when studying reactions involving simple species (atoms or diatomic molecules). But for more complex reactions the quantity A in Equation (22.9) does not depend solely on the collision frequency. To understand why, let us consider the reaction between carbon monoxide and nitrogen dioxide that produces carbon dioxide and nitric oxide.

$$CO(g) + NO_2(g) \longrightarrow CO_2(g) + NO(g)$$

The reaction will take place if the reactants collide as shown in Figure 22.14(a). On the other hand, the collision shown in Figure 22.14(b) will not yield products, even though the reacting species may possess the necessary activation energy, because the molecules are not properly oriented.

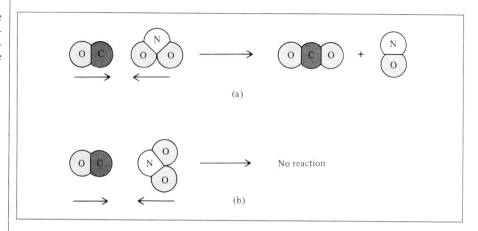

(a)

No reaction

(b)

To use a travel metaphor, an overall chemical equation specifies the origin and destination, but not the actual route followed during a trip. The reaction mechanism is comparable to the actual route taken.

22.4 REACTION MECHANISMS

Elementary Steps and Molecularity

As we mentioned earlier, an overall balanced chemical equation does not tell us much about how a reaction actually takes place. In many cases, it merely represents the sum of *a series of simple reactions* that are often called the **elementary steps** because they represent the *progress of the overall reaction at the molecular level. The sequence of elementary steps that leads to product formation is* called the **reaction mechanism.**

An elementary step in which only one reacting molecule participates, as in the conversion of cyclopropane to propene discussed in Example 22.4, is called a **unimolecular reaction.** *An elementary step that involves two molecules is called a* **bimolecular reaction.** An example is

$$CO(g) + O_2(g) \longrightarrow CO_2(g) + O(g)$$

Very few **termolecular reactions,** or *reactions that involve the participation of three molecules in one elementary step,* are known. A probable termolecular reaction is

$$2NO(g) + Br_2(g) \longrightarrow 2NOBr(g)$$

Keep in mind that all elementary steps involving uni-, bi-, and termolecular reactions occur exactly as shown by the equation.

Knowing the **molecularity of a reaction,** that is, the *number of molecules reacting in an elementary step,* enables us to deduce the rate law. Suppose we have the following unimolecular reaction:

$$A \longrightarrow product$$

Because this *is* the process that occurs at the molecular level, it follows that the rate of the reaction should be directly proportional to the number of A molecules present, or the concentration of A; that is

$$rate = k[A]$$

Thus the reaction is first order in A. Referring to the collision theory of chemical kinetics discussed earlier, we see that for a bimolecular reaction involving A and B molecules, the product is formed as a result of the collision between an A and a B molecule:

$$A + B \longrightarrow \text{product}$$

The rate of the reaction is given by

$$\text{rate} = k[A][B]$$

Thus the order for each reactant in a reaction that describes an elementary step is equal to the stoichiometric coefficient for the reactant in the equation representing that step. In a termolecular reaction, the product is formed as a result of the *simultaneous* encounter of three molecules, which is a less likely event than a bimolecular collision. For this reason, termolecular reactions are rare.

Reactions Involving More Than One Elementary Step

If all reactions occurred in a single elementary step, the study of reaction mechanisms would be very simple. But most reactions involve more than one elementary step. Our tasks are to determine the details of the steps and deduce the order of the reaction. Let us illustrate our approach with the decomposition of hydrogen peroxide in the presence of iodide ions. In an alkaline (basic) medium, the overall reaction is

$$2H_2O_2(aq) \longrightarrow 2H_2O(l) + O_2(g)$$

By experiment, the rate law is found to be

$$\text{rate} = k[H_2O_2][I^-]$$

Thus the reaction is first order with respect to both H_2O_2 and I^-. You can see that decomposition does not occur according to the overall balanced reaction. If it did the reaction would be second order in H_2O_2 (as a result of the collision of two H_2O_2 molecules). What's more, the I^- ion, which is not even in the overall equation, appears in the rate law expression. How can we reconcile these facts?

We can account for the observed rate law by assuming that the reaction takes place in two separate elementary steps, each of which is a bimolecular reaction:

$$H_2O_2 + I^- \xrightarrow{k_1} H_2O + OI^- \qquad (1)$$

$$H_2O_2 + OI^- \xrightarrow{k_2} H_2O + O_2 + I^- \qquad (2)$$

If we further assume that the second step is much faster than the first step (that is, if $k_2 \gg k_1$), then the overall rate is controlled by the rate of the first step (1), which is aptly called the **rate-determining step.** The rate-determining step is the *slowest step in the sequence of steps leading to the formation of products.* In our example, after the OI^- ion is formed in the

In studying reaction mechanisms, the first step invariably is the determination of the rate law of the reaction.

An analogy for the rate-determining step is the flow of traffic along a narrow road. Assuming that cars cannot pass one another on the road, the rate at which the cars travel is governed by the slowest-moving car.

first step it is immediately consumed in the second step (2). So the rate of the reaction can be determined from the first step alone:

$$\text{rate} = k[\text{H}_2\text{O}_2][\text{I}^-]$$

In postulating any reaction mechanism, the sum of all the elementary steps must give the overall equation. Adding (1) and (2) we obtain

$$2\text{H}_2\text{O}_2 \longrightarrow 2\text{H}_2\text{O} + \text{O}_2$$

because the hypoiodite (OI$^-$) ions cancel out. Species such as OI$^-$ are called **_intermediates_** because they *appear in the mechanism of the reaction (that is, the elementary steps) but not in the overall balanced equation.* Keep in mind that an intermediate is always formed in an initial elementary step and is consumed in a later elementary step. (Note that the I$^-$ ion also does not appear in the overall equation. However, I$^-$ differs from OI$^-$ in that the former is present at the start of the reaction. The function of I$^-$ is to speed up the reaction, that is, it is a catalyst. We will discuss the subject of catalysis in Section 22.5.)

How can we find out whether our proposed mechanism is correct? We could try to detect the presence of the OI$^-$ ions by spectroscopic means; evidence of their existence would support our scheme. However, you must realize that for complex reactions it is very difficult to prove the uniqueness of any particular mechanism.

Let us summarize the sequence involved in studying reaction mechanisms. First, collect data (measurement of rates). Second, formulate the rate law (determine the rate constant and order of reaction). Third, postulate the elementary steps and elucidate the mechanism of the reaction (Figure 22.15). The elementary steps must satisfy two requirements: (1) They should be consistent with the overall balanced equation and (2) the rate-determining step should predict the same rate law as that determined experimentally. Keep in mind that to support any particular reaction scheme, we need to detect the presence of any intermediate(s) formed in one or more elementary steps.

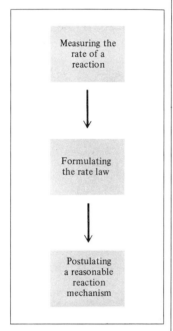

FIGURE 22.15 *Sequence of steps in the study of reaction mechanism.*

EXAMPLE 22.10

The gas-phase decomposition of nitrous oxide (N$_2$O) is believed to occur via two elementary steps:

$$\text{N}_2\text{O} \xrightarrow{k_1} \text{N}_2 + \text{O} \tag{1}$$

$$\text{N}_2\text{O} + \text{O} \xrightarrow{k_2} \text{N}_2 + \text{O}_2 \tag{2}$$

The rate law is rate $= k[\text{N}_2\text{O}]$. (a) Write the equation for the overall reaction. (b) Which of the species are intermediates? (c) What can you say about the relative rates of steps (1) and (2)?

Answer

(a) The sum of steps (1) and (2) gives the overall reaction

$$2\text{N}_2\text{O} \longrightarrow 2\text{N}_2 + \text{O}_2$$

(b) Since the O atom is produced in the first elementary step and it does not appear in the overall balanced equation, it is an intermediate.

(c) If we assume that step (1) is the rate-determining step (that is, if $k_2 \gg k_1$), then the rate of the overall reaction is given by

$$\text{rate} = k_1[N_2O]$$

and $k = k_1$.

Similar example: Problem 22.40.

The ultimate goal in chemical kinetics is to determine the details of each of the elementary steps in the course of an overall reaction. For example, we need to know the orientation of reactant molecules relative to each other in the activated complex, the sequence of the breaking and making of chemical bonds, and the energetics involved in these changes. This knowledge should allow us to predict reaction order and to calculate the activation energy and the rate constant. But chemists have had rather limited success in this endeavor. Except for the simplest of reactions, it is extremely difficult to obtain an accurate energy profile of a reaction. Thus, chemists must often use their own ingenuity to gather information about kinetics.

In one instance, chemists wanted to know which C—O bond is broken in the reaction between methyl acetate and water, in order to gain a better understanding of the reaction mechanism.

$$\underset{\text{methyl acetate}}{CH_3-\overset{\overset{\displaystyle O}{\|}}{C}-O-CH_3} + H_2O \longrightarrow \underset{\text{acetic acid}}{CH_3-\overset{\overset{\displaystyle O}{\|}}{C}-OH} + \underset{\text{methanol}}{CH_3OH}$$

The two possibilities are

$$\underset{(a)}{CH_3-\overset{\overset{\displaystyle O}{\|}}{C}\,\vdots\,O-CH_3} \qquad \underset{(b)}{CH_3-\overset{\overset{\displaystyle O}{\|}}{C}-O\,\vdots\,CH_3}$$

Although the final product is the same, regardless of which bond is broken, the activation energies required for the processes shown in (a) and (b) must somehow be different. Consequently, the rates of these two processes must also differ. To choose between the schemes (a) and (b), chemists use water containing the oxygen-18 isotope instead of ordinary water (which contains the oxygen-16 isotope). It is found that only the acetic acid formed contains the oxygen-18 isotope:

$$CH_3-\overset{\overset{\displaystyle O}{\|}}{C}-{}^{18}O-H$$

Thus, the reaction must have occurred via bond-breaking scheme (a), because the product formed via scheme (b) would retain both of its original

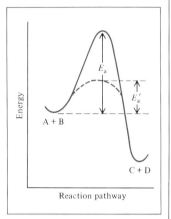

FIGURE 22.16 *Comparison of the activation energy barriers of an uncatalyzed reaction and the same reaction catalyzed. Note that the catalyst lowers the energy barrier but does not affect the actual energies of the reactants or products.*

A rise in temperature also increases the rate of a reaction. However, at elevated temperatures the products formed may undergo other reactions, thereby reducing the yield.

A catalyst lowers the activation energy for both the forward and reverse reactions.

oxygen atoms. This example gives some idea of how inventive chemists must be in studying reaction mechanisms.

22.5 CATALYSIS

A **catalyst** is a *substance that increases the rate of a chemical reaction without itself being consumed.* A catalyst can react to form an intermediate, but it is regenerated in a subsequent step of the reaction. For example, in the laboratory preparation of molecular oxygen, a sample of potassium chlorate is heated, as shown in Figure 3.6; the reaction is

$$2KClO_3(s) \xrightarrow{\Delta} 2KCl(s) + 3O_2(g)$$

However, this thermal decomposition is very slow in the absence of a catalyst. The rate of decomposition can be drastically increased by adding a small amount of the catalyst manganese dioxide (MnO_2), a black powdery substance. All of the MnO_2 can be recovered at the end of the reaction. Another example, as we saw in the last section, involves the I^- ions that act as a catalyst for the decomposition of hydrogen peroxide. Again, no I^- ions are consumed in the overall reaction.

Regardless of its nature, a catalyst speeds up a reaction by providing a different set of elementary steps with more favorable kinetics. According to Equation (22.9) we see that the rate constant k (and hence the rate) of a reaction depends on the frequency factor A and the activation energy E_a—the larger the A or the smaller the E_a, the greater the rate. In many cases, a catalyst increases the rate by lowering the activation energy for the reaction.

Let us assume that the following reaction has a certain rate constant k and an activation energy E_a.

$$A + B \xrightarrow{k} C + D$$

In the presence of a catalyst, however, the rate constant is k_c (called the *catalytic rate constant*):

$$A + B \xrightarrow{k_c} C + D$$

By the definition of a catalyst, $k_c > k$. Figure 22.16 shows the energy profiles for both reactions. Note that the total energies of the reactants (A and B) and those of the products (C and D) remain the same; the only difference is a lowering of activation energy from E_a to E_a'. Because a catalyst enhances the rate of the reverse reaction to the same extent as it does the forward reaction, its presence therefore cannot shift the position of equilibrium. (see p. 525).

We will now discuss three types of catalysis.

Heterogeneous Catalysis

Heterogeneous catalysis is by far the most important type of catalysis in industrial processes. In heterogeneous catalysis, the catalyst is a solid and the reactants and products are either gases or liquids.

The Haber Synthesis of Ammonia. In Section 17.6 we discussed the Haber synthesis of ammonia in terms of Le Chatelier's principle. In 1905, after testing literally thousands of compounds at various temperatures and pressures, Fritz Haber discovered that iron plus a few percent of oxides of potassium and aluminum (K_2O and Al_2O_3) served to catalyze the reaction of molecular hydrogen with molecular nitrogen, yielding ammonia at about 500°C.

$$N_2(g) + 3H_2(g) \rightleftharpoons 2NH_3(g)$$

The initial step in the Haber process involves the adsorption or binding of reactant molecules to the specific sites on the surface of the catalyst. The adsorption process weakens the covalent bonds in H_2 and in N_2 molecules so that the molecules eventually dissociate into atoms. These highly reactive atomic species then combine to form the desired ammonia molecules, which eventually leave the surface of the catalyst (Figure 22.17).

Catalytic Converters. Auto exhaust is a major source of air pollution. At high temperatures inside a running car's engine, nitrogen and oxygen gases react to form nitric oxide:

$$N_2(g) + O_2(g) \rightleftharpoons 2NO(g)$$

The forward reaction is endothermic so that the equilibrium constant increases with increasing temperature.

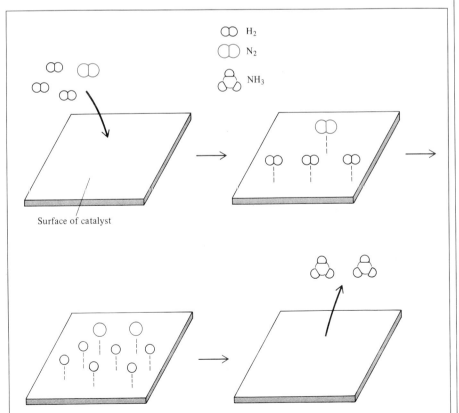

FIGURE 22.17 *The catalytic action in the synthesis of ammonia. First the H_2 and N_2 molecules bind to surface of the catalyst. This interaction weakens the covalent bonds in the molecules, which eventually causes the molecules to dissociate. The highly reactive H and N atoms combine to form NH_3 molecules, which then leave the surface.*

FIGURE 22.18 *A two-stage catalytic converter in an automobile.*

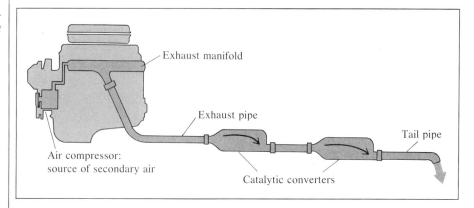

When released to the atmosphere, NO rapidly combines with O_2 to form NO_2, a poisonous dark brown gas with a choking odor:

$$2NO(g) + O_2(g) \longrightarrow 2NO_2(g)$$

Other undesirable gases emitted by an automobile are carbon monoxide (CO) and various unburned hydrocarbons.

A *hydrocarbon* is a compound containing only carbon and hydrogen atoms. Hydrocarbons are the main constituents of gasoline.

Most autos are now manufactured with catalytic converters (Figure 22.18). An efficient catalytic converter must serve two purposes: to oxidize CO and unburned hydrocarbons to CO_2 and H_2O and to reduce NO and NO_2 to N_2 and O_2. Hot exhaust gases into which air has been injected are passed through the first chamber of one converter to accelerate the complete burning of hydrocarbons and to decrease CO emission. (A cross section of the catalytic converter, containing Pt or Pd, or a transition metal oxide such as CuO or Cr_2O_3, is shown in Figure 22.19). However, since high temperatures increase NO production, a second chamber containing a different catalyst (again a transition metal or a transition metal oxide) operating at a lower temperature is required to dissociate NO into N_2 and O_2 prior to discharge through the tailpipe.

Homogeneous Catalysis

In *homogeneous catalysis* the reactants, products, and catalyst are all dispersed in a single phase, usually the liquid phase. For example, the reaction of ethyl acetate with water to form acetic acid and ethanol is normally too slow to be measured.

FIGURE 22.19 *A cross-sectional view of a catalytic converter used in automobiles. The beads contain Pt, Pd, and Rh, and are designed to catalyze the combustion of CO and hydrocarbons.*

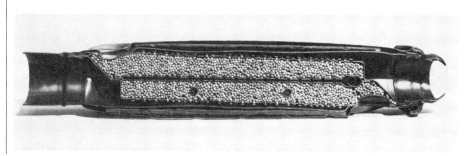

$$CH_3-\overset{\overset{\displaystyle O}{\|}}{C}-O-C_2H_5 \; + \; H_2O \longrightarrow CH_3-\overset{\overset{\displaystyle O}{\|}}{C}-OH + C_2H_5OH$$
ethyl acetate   acetic acid   ethanol

In the absence of the catalyst, the rate law is given by:

$$\text{rate} = k[CH_3COOC_2H_5]$$

However, the reaction can be catalyzed by acids. In the presence of hydro-chloric acid the rate is given by

$$\text{rate} = k_c[CH_3COOC_2H_5][H^+]$$

Acid and base catalyses are the most important types of homogeneous catalysis in liquid solution.

Homogeneous catalysis can also take place in the gas phase. A well-known example of catalyzed gas-phase reactions is the lead chamber process, which for many years was the major process for the manufacture of sulfuric acid. Starting with sulfur, we can expect the production of sulfuric acid to occur in the following steps:

$$S(s) \; + \; O_2(g) \longrightarrow SO_2(g)$$
$$2SO_2(g) \; + \; O_2(g) \longrightarrow 2SO_3(g)$$
$$H_2O(l) \; + \; SO_3(g) \longrightarrow H_2SO_4(aq)$$

In the actual process, however, sulfur dioxide is not converted directly to sulfur trioxide; rather, the oxidation is more efficiently carried out in the presence of the catalyst nitrogen dioxide:

$$2SO_2(g) \; + \; 2NO_2(g) \longrightarrow 2SO_3(g) \; + \; 2NO(g)$$
$$\underline{2NO(g) \; + \; O_2(g) \longrightarrow 2NO_2(g)}$$

overall reaction: $\quad\quad 2SO_2(g) \; + \; O_2(g) \longrightarrow 2SO_3(g)$

Note that there is no net loss of NO_2 in the overall reaction, which is a necessary condition for it to function as a catalyst.

In recent years chemists have devoted much effort to developing a class of metallic compounds that are homogeneous catalysts. These compounds are soluble in various organic solvents and can therefore catalyze reactions in the same phase as the dissolved reactants. Many of the processes they catalyze are organic reactions. For example, a red-violet compound of rhodium, $[(C_6H_5)_3P]_3RhCl$, catalyzes the conversion of C=C bonds to C—C bonds by molecular hydrogen as follows:

$$\overset{|\quad|}{\underset{|\quad|}{C=C}} + H_2 \longrightarrow \overset{|\quad\quad|}{\underset{\underset{\displaystyle H \;\; H}{|\quad\quad|}}{-C-C-}}$$

Such reactions are of importance in food, pharmaceutical, and other commercial processes. The advantages of homogeneous catalysis over heterogeneous catalysis are: (1) the reactions can often be carried out under atmospheric conditions, thus reducing costs of production and minimizing the decomposition of products at high temperatures, (2) catalysts can often be designed to function selectively for a particular type of reaction, and (3) costs of catalysts are lower compared to those of the many precious metals (for example, platinum and gold) that are used in heterogeneous catalysis.

We assume that $k_c \gg k$ so that the rate is determined solely by the catalyzed portion of the reaction.

Enzyme Catalysis

Of all the intricate processes that have evolved in living cells, none is more striking or more essential than enzyme-catalyzed reactions. **Enzymes** are *biological catalysts*. The amazing fact about enzymes is not only that they can increase the rate of biochemical reactions by factors ranging from 10^6 to 10^{12}, but that they are highly specific. They act only on certain molecules, called *substrates,* while leaving the rest of the system unaffected. It has been estimated that an average living cell may contain some 3000 different enzymes, each of them catalyzing a specific reaction in which substrates are converted into the appropriate products. But enzymes do not operate independently. There is, so to speak, a built-in control which turns on certain enzymes for action when the need arises and then turns them off when sufficient amounts of products have been formed.

One theory for the specificity of enzymes—the so-called "lock-and-key" theory—was developed by the German chemist Emil Fischer (1852–1919) in 1894. An enzyme usually contains one or more *active sites,* where reactions with substrates take place; the rest of the molecule maintains the three-dimensional integrity of the network. According to Fischer's hypothesis, the active site is assumed to have a rigid structure, similar to a lock. A substrate molecule has the complementary structure and functions like a key (Figure 22.20). Although it is appealing in many respects, this theory has some glaring discrepancies. For instance, substrates of quite different sizes and shapes are known to bind to the same type of enzyme. Therefore the enzyme's specificity cannot be explained in terms of structural rigidity.

Chemists now know that an enzyme molecule (or at least its active site) actually possesses a fair amount of flexibility. During a reaction it changes to accommodate the substrate molecule. Just how these changes are brought about is not entirely clear.

The mathematical treatment of enzyme kinetics is quite complex, even when we know the basic steps involved in the reaction. A simplified scheme is as follows:

$$E + S \rightleftharpoons ES$$
$$ES \xrightarrow{k} P + E$$

where E, S, and P represent enzyme, substrate, and product, respectively. It is often assumed that the formation of the enzyme–substrate intermediate

FIGURE 22.20 *The lock-and-key model of an enzyme's specificity for substrate molecules.*

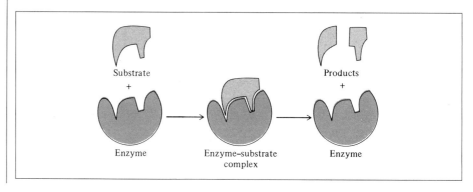

Substrate
+

Enzyme

Enzyme–substrate complex

Products
+

Enzyme

(ES) and its decomposition back to enzyme and substrate molecules occur rapidly and that the rate-determining step is the formation of product. In general, the rate of such an enzyme-catalyzed reaction is given by

$$\text{rate} = \frac{\Delta[\text{P}]}{\Delta t} = k[\text{ES}]$$

The concentration of the ES intermediate is itself proportional to the amount of the substrate present, and a plot of the rate versus the concentration of substrate typically yields a curve such as that shown in Figure 22.21. Initially the rate rises rapidly with increasing substrate concentration. However, above a certain concentration all the active sites are occupied, and the rate becomes zero order in the substrate, that is, the rate remains the same with increasing substrate concentration. At and beyond this point, the rate of formation of product depends only on how fast the ES intermediate breaks down, and not on the number of substrate molecules present.

All active sites are occupied at and beyond this substrate concentration
Rate of product formation
[S]

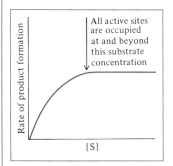

FIGURE 22.21 *Plot of the rate of product formation versus substrate concentration in an enzyme-catalyzed reaction. Note that after the substrate concentration has reached a certain value (that is, when all the enzymes are tied up by the substrate molecules to form the ES intermediate), the rate in the substrate concentration becomes zero order, that is, the rate no longer changes with increasing substrate concentration.*

AN ASIDE ON THE PERIODIC TABLE

Metals Used in Heterogeneous and Homogeneous Catalyses

Many metals and their compounds are used as catalysts for a variety of reactions. Figure 22.22 shows the most important metals in industrial and laboratory catalyses according to their positions in the periodic table. As you can see, most of the metals belong to the transition metal family. These metals possess incompletely filled *d* sub-

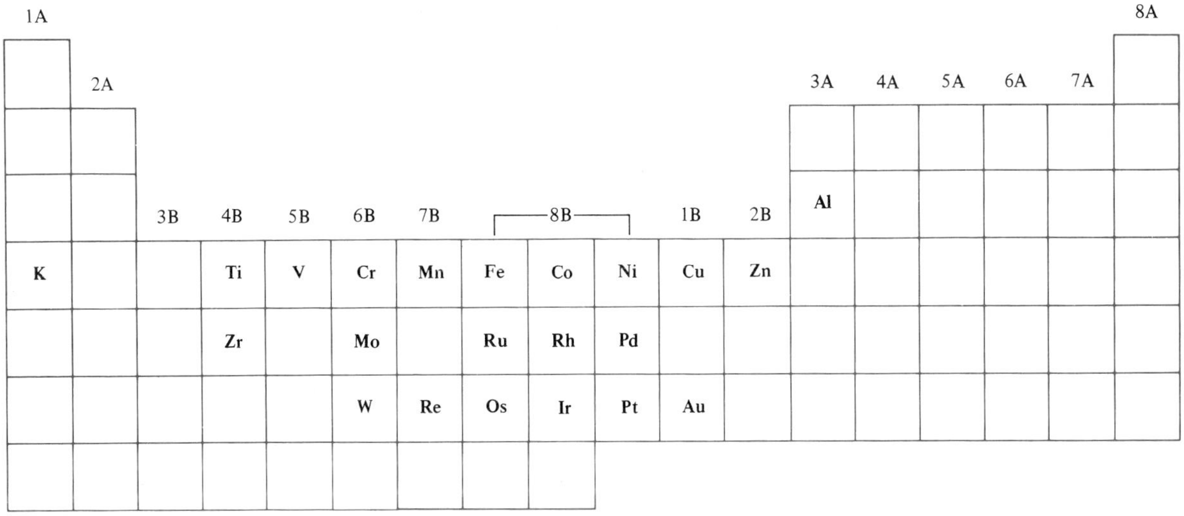

1A
8A
2A
3A
4A
5A
6A
7A
3B
4B
5B
6B
7B
8B
1B
2B
Al
K
Ti
V
Cr
Mn
Fe
Co
Ni
Cu
Zn
Zr
Mo
Ru
Rh
Pd
W
Re
Os
Ir
Pt
Au

FIGURE 22.22 Metals that are most frequently used in heterogeneous and homogeneous catalyses.

TABLE 22.4 Some Reactions Catalyzed by Metals and/or Metal Compounds*

Reaction	Catalysts	Comments
$N_2 + 3H_2 \longrightarrow 2NH_3$	Fe plus K_2O and Al_2O_3	Haber process for the synthesis of ammonia
$4NH_3 + 5O_2 \longrightarrow 4NO + 6H_2O$	Pt and Rh	First step in the manufacture of nitric acid
$2SO_2 + O_2 \longrightarrow 2SO_3$	Pt or V_2O_5	An intermediate step in the manufacture of sulfuric acid
$CO + H_2O \longrightarrow CO_2 + H_2$	ZnO and CuO	Industrial preparation of hydrogen gas
$CO + 3H_2 \longrightarrow CH_4 + H_2O$	Ni	Industrial preparation of methane
$CO + 2H_2 \longrightarrow CH_3OH$	Zn and Cr_2O_3	Industrial preparation of methanol
$\underset{\mid}{\overset{\mid}{C}}{=}\underset{\mid}{\overset{\mid}{C}} + H_2 \longrightarrow \underset{\underset{H}{\mid}}{\overset{\mid}{-C-}}\underset{\underset{H}{\mid}}{\overset{\mid}{C-}}$	Ni, or Pt, or Pd	Of importance to various industrial processes
Same as above	$[(C_6H_5)_3P]_3RhCl$	Homogeneous catalysis
Vulcanization	ZnO	A process to cross-link molecules in rubber for better performance in tires
$2KClO_3 \longrightarrow 2KCl + 3O_2$	MnO_2	Laboratory preparation of molecular oxygen

* Unless otherwise stated, all reactions involve heterogeneous catalysis.

shells, which allow them to act both as electron donors and electron acceptors. This versatility enables them to interact with several reactant molecules. Nontransition metals lack this ability and are not widely used as catalysts. In heterogeneous catalysis, the catalysts are either metals or simple compounds of the metals, such as the oxides and chlorides. On the other hand, more complex compounds of the metals are involved in homogeneous catalyses. Table 22.4 lists some of the reactions catalyzed by metals and/or their compounds.

SUMMARY

1. The rate of a chemical reaction is shown by the change in the concentration of reactant or product molecules over a period of time. The rate is not constant, but varies continuously as concentration changes.

2. The rate law is an expression relating the rate of a reaction to the rate constant and the concentrations of the reactants raised to appropriate powers. The rate constant k for a given reaction changes only with temperature.

3. Reaction order is the sum of the powers to which reactant concentrations are raised in the rate law. The rate law and the reaction order cannot be determined from the stoichiometry of the overall equation for a reaction; they must be determined by experiments.

4. The half-life of a reaction (the time it takes for the concentration of a reactant to decrease by one-half) can be used to determine the rate constant of a first-order reaction.

5. According to collision theory, a reaction occurs when molecules collide with sufficient energy, called the activation energy, to break the bonds and initiate the reaction.

6. The rate constant and the activation energy are related by the Arrhenius equation, $k = Ae^{-E_a/RT}$.

7. The overall balanced equation for a reaction may be the sum of a series of simple reactions, called elementary steps. The complete series of elementary steps for a reaction is the reaction mechanism.

8. The rate-determining step is the slowest step in a reaction mechanism.

9. A catalyst speeds up a reaction usually by lowering the value of E_a. A catalyst can be recovered unchanged at the end of a reaction.

10. In heterogeneous catalysis, which is of great industrial importance, the catalyst is a solid and the reactants are gases or liquids. In homogeneous catalysis, the catalyst and the reactants are in the same phase. Enzymes are catalysts in living systems.

KEY WORDS

Activated complex, p. 661
Activation energy, p. 661
Bimolecular reaction, p. 666
Catalyst, p. 670
Chemical kinetics, p. 642
Elementary step, p. 666
Enzyme, p. 674
First-order reaction, p. 652
Half-life, p. 655
Intermediate, p. 668

Molecularity of a reaction, p. 666
Rate constant, p. 645
Rate law, p. 649
Rate-determining step, p. 667
Reaction mechanism, p. 666
Reaction order, p. 649
Reaction rate, p. 642
Second-order reaction, p. 658
Termolecular reaction, p. 666
Unimolecular reaction, p. 666

PROBLEMS

More challenging problems are marked with an asterisk.

Reaction Rate

22.1 Distinguish among the following terms: average rate, instantaneous rate, initial rate.

22.2 What are the advantages of measuring the initial rate of a reaction?

22.3 What are the units of the rate of a reaction?

22.4 Write the reaction rate expressions for the following reactions:

(a) $H_2(g) + I_2(g) \longrightarrow 2HI(g)$
(b) $2H_2(g) + O_2(g) \longrightarrow 2H_2O(g)$
(c) $5Br^-(aq) + BrO_3^-(aq) + 6H^+(aq) \longrightarrow$
$3Br_2(aq) + 3H_2O(l)$

22.5 Consider the reaction

$$N_2(g) + 3H_2(g) \longrightarrow 2NH_3(g)$$

Suppose that at a particular moment during the reaction molecular hydrogen is reacting at the rate of 0.074 M/s. (a) What is the rate at which ammonia is being formed? (b) What is the rate at which molecular nitrogen is reacting?

*22.6 Consider the reaction

$$C_2H_5I(aq) + H_2O(l) \longrightarrow$$
$$C_2H_5OH(aq) + H^+(aq) + I^-(aq)$$

How could you follow the progress of the reaction by measuring the electrical conductance of the solution?

Rate Laws

22.7 Explain what is meant by the rate law of a reaction.

22.8 What are the units for the rate constants of first-order and second-order reactions?

22.9 The rate constant of a first-order reaction is 66/s. What is the rate constant in terms of minutes?

22.10 The rate law for the reaction

$$NH_4^+(aq) + NO_2^-(aq) \longrightarrow N_2(g) + 2H_2O(l)$$

is given by rate = $k[NH_4^+][NO_2^-]$. At 25°C, the rate constant is $3.0 \times 10^{-4}/M$ s. Calculate the rate of the reaction at this temperature if $[NH_4^+] =$ 0.26 M and $[NO_2^-]$ = 0.080 M.

22.11 Consider the reaction

$$A + B \longrightarrow products$$

From the following data, obtained at a certain temperature, determine the order of the reaction and calculate the rate constant:

[A]	[B]	Rate (M/s)
1.50	1.50	3.20×10^{-1}
1.50	2.50	3.20×10^{-1}
3.00	1.50	6.40×10^{-1}

22.12 Determine the overall orders of the reactions to which the following rate laws apply: (a) rate = $k[NO_2]^2$; (b) rate = k; (c) rate = $k[H_2][Br_2]^{\frac{1}{2}}$; (d) rate = $k[NO]^2[O_2]$.

22.13 What are the overall orders of the reactions to which the following rate laws apply?

(a) rate = $k[Cl_2][H_2]$
(b) rate = $k[NO_2]^2[O_2]$
(c) rate = $k[H_2]^{\frac{1}{2}}[C_2H_4]$

22.14 A certain first-order reaction is 35.5 percent complete in 4.90 min at 25°C. What is its rate constant?

22.15 Consider the reaction

$$A \longrightarrow B$$

The rate of the reaction is 1.6×10^{-2} M/s when the concentration of A is 0.35 M. Calculate the rate constant if the reaction is (a) first order in A and (b) second order in A.

22.16 The following rate data, obtained at 250 K, describe the reaction

$$F_2(g) + 2ClO_2(g) \longrightarrow 2FClO_2(g)$$

$[F_2]$	$[ClO_2]$	$\dfrac{-\Delta[F_2]}{\Delta t}$ (M/s)
0.10	0.010	1.2×10^{-3}
0.10	0.040	4.8×10^{-3}
0.20	0.010	2.4×10^{-3}

(a) Deduce the rate law for the reaction. (b) Calculate the rate constant. (c) Calculate the rate of the reaction at the time when $[F_2]$ = 0.010 M and $[ClO_2]$ = 0.020 M.

*22.17 At 500 K butadiene gas converts to cyclobutene gas:

$$CH_2{=}CH{-}CH{=}CH_2 \longrightarrow$$

butadiene (g)   cyclobutene (g)

From the following rate data determine graphically the order of the reaction in butadiene and the rate constant.

Time from Start (s)	Concentration of Butadiene (M)
195	0.0162
604	0.0147
1246	0.0129
2180	0.0110
4140	0.0084
8135	0.0057

22.18 The following gas-phase reaction was studied at 290°C by observing the change in pressure as a function of time in a constant volume vessel.

$$ClCO_2CCl_3(g) \longrightarrow 2COCl_2(g)$$

Time (s)	0	181	513	1164
P (mmHg)	15.76	18.88	22.79	27.08

Is the reaction first or second order in [$ClCO_2CCl_3$]?

22.19 Azomethane, $C_2H_6N_2$, decomposes at 297°C in the gas phase as follows:

$$C_2H_6N_2(g) \longrightarrow C_2H_6(g) + N_2(g)$$

From the data in the following table determine the order of the reaction in [$C_2H_6N_2$] and the rate constant.

Time (min)	0	15	30	48	75
[$C_2H_6N_2$]	0.36	0.30	0.25	0.19	0.13

22.20 Consider the reaction

$$X + Y \longrightarrow Z$$

The following data are obtained at 360 K:

Initial Rate of Disappearance of X (M/s)	[X]	[Y]
0.147	0.10	0.50
0.127	0.20	0.30
4.064	0.40	0.60
1.016	0.20	0.60
0.508	0.40	0.30

(a) Determine the order of the reaction. (b) Determine the initial rate of disappearance of X when the concentration of X is 0.20 M and that of Y is 0.30 M.

22.21 The rate constant for the second-order reaction

$$2NOBr(g) \longrightarrow 2NO(g) + Br_2(g)$$

is 0.80/M s at 10°C. Starting with a concentration of 0.086 M of NOBr, calculate its concentration after 22 s.

Half-Life

22.22 Define the half-life of a reaction. How does the half-life of a first-order reaction differ from that of a second-order reaction?

22.23 What is the half-life of a compound if 75 percent of a given amount of the compound decomposes in 60 min? Assume first-order kinetics.

22.24 The thermal decomposition of phosphine (PH$_3$) into phosphorus and molecular hydrogen is a first-order reaction:

$$4PH_3(g) \longrightarrow P_4(g) + 6H_2(g)$$

The half-life of the reaction is 35.0 s at 680°C. Calculate (a) the first-order rate constant for the

reaction and (b) the time required for 95 percent of the phosphine to decompose.

22.25 The following reaction is second order in A:

$$A \longrightarrow product$$

At a certain temperature the second-order rate constant is 1.46/M s. Calculate the half-life of the reaction if the initial concentration of A is 0.86 M.

Activation Energy

22.26 Define activation energy. How does the magnitude of activation energy affect the rate of a reaction?

22.27 Variation of the rate constant with temperature for the first-order reaction

$$2N_2O_5(g) \longrightarrow 2N_2O_4(g) + O_2(g)$$

is given in the following table. Determine graphically the activation energy for the reaction.

T(K)	k(1/s)
273	7.87×10^3
298	3.46×10^5
318	4.98×10^6
338	4.87×10^7

22.28 The first-order rate constant for the reaction A → B has been measured at a series of temperatures:

k(1/s)	t(°C)
10.6	10
47.4	20
162	30
577	40

Determine graphically the activation energy for the reaction.

22.29 The reaction of G_2 with E_2 to form 2EG is exothermic, and the reaction of G_2 with X_2 to form 2XG is endothermic. The activation energy of the exothermic reaction is greater than that of the endothermic reaction. Sketch the potential energy profile diagrams for these two reactions on the same graph.

22.30 Given the same concentrations, the reaction

$$CO(g) + Cl_2(g) \longrightarrow COCl_2(g)$$

at 250°C is 1.50×10^3 times faster than the same reaction at 150°C. Calculate the energy of activation for this reaction. Assume the frequency factor to remain constant.

22.31 The combustion of methane in oxygen is a highly exothermic reaction, yet a mixture of methane and oxygen gas can be kept indefinitely without any apparent change. Explain.

22.32 Draw a potential energy curve for the reaction

$$A + B \longrightarrow C$$

given the following conditions: (a) C is more stable than A and B combined and (b) C is less stable than A and B combined. Which reaction is exothermic and which reaction is endothermic?

22.33 The rate constants of a certain second-order reaction are $3.20 \times 10^{-2}/M$ s at 25°C and $9.45 \times 10^{-1}/M$ s at 150°C. Calculate the activation energy for this reaction.

22.34 For the reaction

$$NO(g) + O_3(g) \longrightarrow NO_2(g) + O_2(g)$$

the frequency factor A is 8.7×10^{12}/s and the activation energy is 63 kJ/mol. What is the rate constant for the reaction at 75°C?

22.35 The rate constant of a first-order reaction is 4.60×10^{-4}/s at 350°C. If the activation energy is 104 kJ/mol, calculate the temperature at which its rate constant is 8.80×10^{-4}/s.

22.36 The rate constant for the decomposition of the AB molecule at 25°C is 5.8×10^{-2}/s. Calculate the rate constant at 125°C if the activation energy is 52 kJ/mol.

22.37 The rate constant of a certain reaction triples when the temperature is raised from 26°C to 92°C. What is the activation energy of the reaction?

22.38 The rate constants of some reactions double with every 10-degree rise in temperature. Assume a reaction takes place at 295 K and 305 K. What must the activation energy be for the rate constant to double as described?

Reaction Mechanism

22.39 Explain what is meant by the rate-determining step of a reaction.

22.40 The rate law for the reaction

$$2NO(g) + Cl_2(g) \longrightarrow 2NOCl(g)$$

is given by rate = $k[NO][Cl_2]$. (a) What is the order of the reaction? (b) A mechanism involving the following steps has been proposed for the reaction:

$$NO(g) + Cl_2(g) \longrightarrow NOCl_2(g)$$
$$NOCl_2(g) + NO(g) \longrightarrow 2NOCl(g)$$

If this mechanism is correct, what does it imply about the relative rates of these two steps?

22.41 When methyl phosphate is heated in acid solution, it reacts with water:

$$CH_3OPO_3H_2 + H_2O \longrightarrow CH_3OH + H_3PO_4$$

If the reaction is carried out in water enriched with ^{18}O, the oxygen-18 isotope is found in the phosphoric acid product but not in the methanol. What does this tell us about the mechanism of the reaction?

*22.42 For the reaction $X_2 + Y + Z \rightarrow XY + XZ$ it is found that doubling the concentration of X_2 doubles the reaction rate, tripling the concentration of Y triples the rate, and doubling the concentration of Z has no effect. (a) What is the rate law for this reaction? (b) Why is it that the change in the concentration of Z has no effect on the rate? (c) Suggest a mechanism for the reaction that is consistent with the rate law.

22.43 The destruction of ozone in the stratosphere by Freons (halogen-containing hydrocarbons) is believed to take place as follows:

$$Cl + O_3 \longrightarrow ClO + O_2$$
$$O_3 \xrightarrow{\text{radiation}} O + O_2$$
$$ClO + O \longrightarrow Cl + O_2$$

where Cl comes from the Freons. (a) What is the role of Cl in the above reactions? (b) What is the species O called in this scheme? (c) What is the overall reaction?

22.44 Consider the following elementary step:

$$X + 2Y \longrightarrow XY_2$$

(a) Write a rate law for this reaction. (b) If the initial rate of formation of XY_2 is 3.8×10^{-3} M/s and the initial concentrations of X and Y are 0.26 M and 0.88 M, what is the rate constant of the reaction?

22.45 Explain why termolecular reactions are rare.

22.46 Reactions can be classified as unimolecular, bimolecular, and so on. Why are there no zero-molecular reactions?

Catalysis

22.47 What are the characteristics of a catalyst?

22.48 Distinguish between homogeneous catalysis and heterogeneous catalysis.

22.49 Consider the decomposition of hydrogen peroxide:

$$2H_2O_2 \longrightarrow 2H_2O + O_2$$

The reaction can occur on its own or it can be catalyzed by an enzyme. The activation energies for the uncatalyzed and enzyme-catalyzed reactions at 25°C are found to be 75 kJ/mol and 0.58 kJ/mol, respectively. Calculate the ratios of the rate constants for the reaction, assuming that the frequency factor is the same in both cases.

22.50 Most reactions, including enzyme-catalyzed reactions, proceed faster at higher temperatures. However, for a given enzyme, the rate drops off abruptly at a certain temperature. Account for this behavior. (*Hint:* The three-dimensional structure of an enzyme molecule is maintained by relatively weak intermolecular forces within the molecule.)

22.51 A certain reaction is known to have a small rate constant at room temperature. Is it possible to make the reaction proceed at a fast rate without changing the temperature?

22.52 The concentrations of enzymes in cells are usually quite small. Explain.

Miscellaneous Problems

22.53 Define the following terms: (a) reaction order, (b) reaction rate, (c) rate constant, (d) half-life, (e) reaction mechanism, (f) molecularity of a reaction.

22.54 List four important factors that can influence the rate of a chemical reaction.

22.55 "The rate constant for the reaction

$$NO_2(g) + CO(g) \longrightarrow NO(g) + CO_2(g)$$

is $1.64 \times 10^{-6}/M$ s." What is incomplete about this statement?

22.56 The bromination of acetone is acid-catalyzed:

$$CH_3COCH_3 + Br_2 \xrightarrow[\text{catalyst}]{H^+}$$
$$CH_3COCH_2Br + H^+ + Br^-$$

The rate of disappearance of bromine was measured for several different concentrations of acetone, bromine, and H^+ ions at a certain temperature:

	$[CH_3COCH_3]$	$[Br_2]$	$[H^+]$	Rate of Disappearance of Br_2 (M/s)
(a)	0.30	0.050	0.050	5.7×10^{-5}
(b)	0.30	0.10	0.050	5.7×10^{-5}
(c)	0.30	0.050	0.10	1.2×10^{-4}
(d)	0.40	0.050	0.20	3.1×10^{-4}
(e)	0.40	0.050	0.050	7.6×10^{-5}

(a) What is the rate law for the reaction? (b) Determine the rate constant.

*22.57 The rate of evolution of $O_2(g)$ from aqueous solutions of hypochlorite ion, OCl^-, depends on the hydrogen ion concentration. (a) From the rate data obtained from three experiments (see table below) conducted at 298 K, determine the rate law for the reaction

$$2OCl^-(aq) \longrightarrow O_2(g) + 2Cl^-(aq)$$

For each experiment the amount of HCl given was mixed with the amount of OCl^- listed.

Amount of HCl Solution	Amount of OCl^- Solution	Rate of O_2 Evolution
25.0 mL of 0.10 M	25.0 mL of 0.010 M	35.0 mL/min
25.0 mL of 0.20 M	25.0 mL of 0.010 M	70.0 mL/min
25.0 mL of 0.10 M	25.0 mL of 0.0050 M	9.0 mL/min

(b) What is the function of hydrogen ion in this reaction?

22.58 Iron reacts with oxygen to form Fe_3O_4. All other conditions being the same, predict which of the following iron objects will react faster with oxygen: iron bar, iron wire, or iron filings? Explain your choice.

22.59 Use the Arrhenius equation to show why the rate of a reaction (a) decreases with increasing activation energy and (b) increases with increasing temperature.

23
THERMO-DYNAMICS

A mixture of hydrogen and oxygen gases reacts explosively (if heated by a burning match) to form water, but a beaker of water treated the same way will not decompose to form hydrogen and oxygen gases. Chemical kinetics, the subject of Chapter 22, tells us how fast a reaction proceeds. To answer the question whether a reaction will start in the first place, we must turn to thermodynamics.

Thermodynamics is a scientific discipline that deals with the interconversion of heat and other forms of energy. When applied to chemistry, thermodynamics enables us to predict the feasibility of a reaction and the extent to which a reaction will proceed before it reaches equilibrium. The equilibrium constant was defined in terms of the law of mass action in Chapter 17. In this chapter we will see that the equilibrium constant also has a basis in thermodynamics.

23.1   SOME DEFINITIONS

Thermodynamics is the *scientific study of the interconversion of heat and other forms of energy.* In contrast to many areas of study in chemistry, thermodynamics always treats large samples of matter and deals only with macroscopic properties. Furthermore, unlike the kinetic molecular theory of gases (see Chapter 6) or quantum mechanics (see Chapter 8), the laws of thermodynamics are not based on specific models involving the structure of matter and therefore are unaffected by our changing concepts of atoms and molecules.

Before we study the laws of thermodynamics, we should establish several basic definitions. The word "system" is frequently used in thermodynamics. At this point, it will be helpful for you to review the definition of system and the different types of systems introduced in Section 5.1. Two other important terms, *state* and *state functions,* are discussed below.

State and State Function

In previous chapters we used the word "state" to denote the phase of a substance (solid state, liquid state, or gas state). But in thermodynamics, when we speak of the **state of a system,** we mean the *values of all pertinent macroscopic variables—for example, composition, volume, pressure, and temperature.* If we know the values of all these quantities, then we know everything (thermodynamically speaking) that there is to know about the system. In practice, a number of mathematical relationships help us calculate these variables. For instance, for a given amount of an ideal gas sample, the relationship among pressure, volume, amount of gas, and temperature is given by Equation (6.7):

$$PV = nRT$$

Thus, for a fixed amount of gas (*n* is a constant), we need to know any two of the three variables (*P, V,* and *T*) in order to calculate the other one. This equation is called an *equation of state.*

Pressure, volume, and temperature are said to be **state functions,** *properties that are determined by the state the system is in.* Once we specify the state of a system by fixing the values of a few of the state functions, the values of all other state functions are also fixed. Referring again to the ideal gas equation, we see that once the volume and temperature of one mole of an ideal gas are specified, the pressure must assume the value $P = RT/V.$

An important property of state functions is that when the state of a system changes, the magnitude of change in any state function depends only on the initial and final states of the system and not on how the change is accomplished. Let us assume that the change involves the expansion of a gas from an initial volume V_i (1 L) to a final volume V_f (2 L). The change or the increase in volume is

$$\begin{aligned} \Delta V &= V_f - V_i \\ &= 2\ L - 1\ L \\ &= 1\ L \end{aligned}$$

Note that Δ means "final − initial."

Recall that an object possesses potential energy by virtue of its position.

where ΔV (delta V) denotes the change in volume. The change can be brought about in an infinite number of ways. We can let the gas expand directly from 1 L to 2 L as described, or we can allow it to expand to 5 L and then compress it down to 2 L, and so on. No matter how we do it, the change in volume is always 1 L. The same applies to changes in pressure and temperature.

Another state function is energy. Using potential energy as an example, we find that the net increase in gravitational potential energy when we go from the second floor to the third floor of a building is always the same, no matter how we get there.

Not all changes can be described in terms of the initial and final states alone. For example, the locations of New York and San Francisco are fixed with relation to each other, but the distance traveled by a person going from one city to the other depends on the means of transportation. It may be 4986 km by air, 5062 km by car, and so on. Thus, although the initial and final states are the same in each case, the amount of change, or the distance traveled, depends on the path taken. Therefore the distance traveled between the two cities is *not* a state function. Nor, as we will see in the next section, are two important thermodynamic quantities, work and heat.

23.2   THE FIRST LAW OF THERMODYNAMICS

The ***first law of thermodynamics*** essentially describes conservation of energy; it states that *energy can be converted from one form to another, but cannot be created or destroyed.* In other words, the total energy in the universe is a constant.

It would be impossible to prove the validity of the first law of thermodynamics if we had to determine the total energy content of the universe. Even determining the total energy content of 1 mL of oxygen gas would be extremely difficult. Fortunately, in testing the validity of the first law of thermodynamics, we need to be concerned only with the *change* in the internal energy of a system that results from the difference between its *initial state* and its *final state*. The change in internal energy, ΔE, is given by

$$\Delta E = E_f - E_i \qquad (23.1)$$

where E_i and E_f are the internal energies of the system in its initial and final states, respectively.

Consider the reaction between two moles of hydrogen gas and one mole of oxygen gas to produce two moles of water:

$$2H_2(g) + O_2(g) \longrightarrow 2H_2O(l)$$

In this case our system is composed of the reactant molecules (H_2 and O_2) and product molecules (H_2O). We do not know the total internal energy content of either the reactant molecules or the product molecules, but we can accurately measure the *change* in energy content, ΔE, given by

ΔE = energy content of 2 mol $H_2O(l)$ − [energy content of 2 mol $H_2(g)$
$$+ 1 \text{ mol } O_2(g)]$$

= E(product) − E(reactants)

By experiment, we find that the total internal energy content of the system decreases as a result of the *exothermic* reaction [that is, E(product) < E(reactants), or ΔE is a negative quantity]. Energy (mainly in the form of heat) is lost by the system to the surroundings. Since the total energy of the universe is a constant, it follows that the sum of energy changes must be zero, that is

$$\Delta E_{sys} + \Delta E_{surr} = 0$$

or

$$\Delta E_{sys} = -\Delta E_{surr}$$

where the subscripts "sys" and "surr" denote system and surroundings, respectively. Thus, if one system undergoes an energy change, ΔE_{sys}, the rest of the universe, or the surroundings, must undergo a change in energy that is equal in magnitude but opposite in sign. It follows that energy gained in one place must have been lost somewhere else.

It is equally true that the energy lost by one system must show up elsewhere in the universe as an energy gain. Furthermore, because energy can be changed from one form to another, the energy lost by one system can be gained by another system in a different form. For example, the energy lost by burning oil in a power plant may ultimately turn up in our homes as electrical energy, heat, light, and so on.

In chemistry, we are normally interested in the changes associated with the system, not with its surroundings. Therefore, a more useful form of the first law is

$$\Delta E = q + w \qquad (23.2)$$

(We drop the subscript "sys" for simplicity.) Equation (23.2) says that the change in the internal energy ΔE of a system is the sum of the heat exchange q between system and surroundings and the work done w on (or by) the system. In keeping with the sign convention used for thermochemical processes (see Section 5.2), q is positive for an endothermic process and negative for an exothermic process. The sign convention for work is that w is positive for work done on the system by the surroundings and negative for work done by the system on the surroundings. Table 23.1 summarizes the sign conventions for q and w.

TABLE 23.1   Sign Conventions for Work and Heat

Process	Sign
Work done *by* the system *on* the surroundings	−
Work done *on* the system *by* the surroundings	+
Heat absorbed by the system from the surroundings (*endothermic* process)	+
Heat absorbed by the surroundings from the system (*exothermic* process)	−

For convenience, we can omit the word "internal" when discussing the energy of a system.

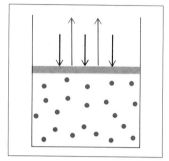

FIGURE 23.1 *The expansion of a gas against a constant external pressure (such as atmospheric pressure). The gas is in a cylinder fitted with a weightless movable piston.*

Work done (w) is equal to $-P\Delta V$. For gas expansion, $\Delta V > 0$, so $-P\Delta V$ is a negative quantity. For gas compression, $\Delta V < 0$ and $-P\Delta V$ is a positive quantity. Thus the signs for w are in accordance with our convention.

The internal energy of a system has two components: kinetic energy and potential energy. The kinetic energy component consists of the *translational motion* (that is, motion from one place to another) of individual molecules, vibration and rotation of molecules, and the movement of electrons. Potential energy is determined by repulsive forces between electrons and between nuclei, and attractive forces between electrons and nuclei; intermolecular interactions also contribute to the potential energy of the system. As mentioned earlier, it is impossible to measure all these contributions accurately, so we cannot calculate the total energy of a system with any certainty. We get around this by concentrating on the *changes* in energy, which *can* be determined experimentally.

Work and Heat

In Chapter 1 we defined *work* as force (F) multiplied by distance (d):

$$w = Fd$$

In thermodynamics, work has a broader range of meaning that includes mechanical work, electrical work, and so on. In this chapter we will concentrate on mechanical work; in Chapter 24 we will discuss the nature of electrical work.

A useful example of mechanical work is the expansion of a gas (Figure 23.1). Suppose that a gas is in a cylinder fitted with a movable piston, at a certain temperature, pressure, and volume. As it expands, the gas pushes the piston upward against a constant opposing external pressure P. The work done by the gas on the surroundings is

$$w = -P\Delta V$$

where ΔV, the change in volume, is given by ($V_f - V_i$). This follows from the fact that pressure × volume can be expressed as (force/area) × volume:

$$\underset{\text{pressure} \;\cdot\; \text{volume}}{\frac{F}{d^2} \quad \times \quad d^3} \quad = Fd = w$$

where F is the opposing force and d has the dimension of length, d^2 has the dimensions of area, and d^3 has the dimensions of volume. Thus the product of pressure and volume is equal to force times distance, or work. You can see that for a given increase in volume (that is, for a certain value of ΔV), the work done depends on the magnitude of the external, or opposing, pressure P. The value of P thus defines the "path" of the process. If P is zero (that is, if the gas is expanding against a vacuum), the work done must also be zero. If P is some positive, nonzero value, then the work done is given by $-P\Delta V$. Remember, however, that in order for the gas to expand, the pressure of the gas must be greater than the opposing pressure.

EXAMPLE 23.1

One mole of an ideal gas initially at 25°C undergoes an expansion in volume from 1.0 L to 4.0 L at constant temperature. Calculate the work done if the gas expands (a) against a vacuum, (b) against a constant external pressure of 0.50 atm, (c) against a constant external pressure of 3.0 atm.

Answer

(a) Since the external pressure is zero, the work w done during expansion is

$$w = -P\Delta V$$
$$= -0(4 - 1)\ L$$
$$= 0$$

(b) Here the external pressure is 0.50 atm, so

$$w = -P\Delta V$$
$$= -(0.50\ \text{atm})\ (4 - 1)\ L$$
$$= -1.5\ L\cdot atm$$

It would be more convenient to express w in units of energy, so from Appendix 2 we write

$$R = 0.08206\ L\cdot atm/K\cdot mol$$

and

$$R = 8.314\ J/K\cdot mol$$

Hence

$$0.08206\ L\cdot atm/K\cdot mol = 8.314\ J/K\cdot mol$$
$$1\ L\cdot atm = \frac{8.314}{0.08206}\ J$$
$$1\ L\cdot atm = 101.3\ J$$

This is the conversion factor we need. Now we can write

$$w = -1.5\ L\cdot atm \times \frac{101.3\ J}{1\ L\cdot atm}$$
$$= -1.5 \times 10^2\ J$$

(c) In this case

$$w = -P\Delta V = -(3.0\ \text{atm})\ (4 - 1)\ L$$
$$= -9.0\ L\cdot atm$$
$$= -9.0\ L\cdot atm \times \frac{101.3\ J}{1\ L\cdot atm}$$
$$= -9.1 \times 10^2\ J$$

Example 23.1 clearly shows that work is not a state function. We *cannot* write $\Delta w = w_f - w_i$ for a change, because such an equation implies that the work done depends only on the initial state (that is, when $V = 1.0$ L) and final state (when $V = 4.0$ L) and nothing else. The amount of work performed depends on how the process is carried out since it is determined by the value of the external, opposing pressure.

Having discussed the nature of work we now turn our attention to heat. Heat is associated with energy of motion at the molecular level. We are perhaps most familiar with linear, or translational, motion of molecules. But molecules can display two other forms of motion: rotation and vibra-

We use lowercase letters (such as "w") to represent thermodynamic quantities that are not state functions.

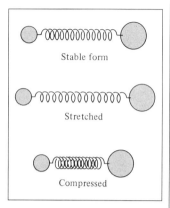

FIGURE 23.2 *The vibrational motion of a diatomic molecule such as HCl. The chemical bond between the atoms can be stretched and compressed like a spring.*

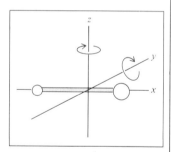

FIGURE 23.3 *Rotation of a diatomic molecule about the y and z axes. Rotating a diatomic or a linear polyatomic molecule about its internuclear axis (the x axis) does not constitute a true rotation because such a motion would not shift the positions of the atoms. During the rotation, the center of gravity of the molecule, which is also the origin of the x, y, and z axes, remains unchanged. A nonlinear polyatomic molecule such as H_2O can rotate about all three axes.*

tion. Let us take a diatomic molecule, HCl, as an example. This molecule can vibrate (Figure 23.2) and it can rotate in two different ways (Figure 23.3).

The freedom of molecules to execute translational, rotational, and vibrational motion depends very much on the state of the substance. In a solid, the molecules can vibrate about a mean position and may even rotate slightly. But translational motion in a solid normally is totally absent. In liquids, the molecules vibrate and rotate and, to some extent, translate. In gases, the molecules vibrate, rotate, and move rapidly from place to place. Table 23.2 summarizes these facts about molecular energy.

At a given temperature, the amount of ***thermal energy*** (that is, *energy associated with molecular motion*) contained in a substance depends on the amount of the substance present and the nature of the molecules that make up the substance. When we heat a substance with a Bunsen burner, the temperature of the substance rises because its thermal energy increases. The rise in temperature is the result of the increase in one, two, or all three types of molecular motion shown in Table 23.2. The higher the temperature, the more energetic are the various molecular motions. Conversely, if we cool the substance by placing it in contact with a colder object, the molecular motion will diminish, as indicated by a drop in temperature.

Heat is not a state function because the heat associated with a given change in state depends on the path taken; that is, we cannot write $\Delta q = q_f - q_i$. It is important to note that the quantity $(q + w)$ in Equation (23.2) does *not* depend on the path taken, because it is always equal to the change in energy ΔE. Energy, as we saw earlier, *is* a state function.

EXAMPLE 23.2

The work done when a gas is compressed in a cylinder like that shown in Figure 23.1 is found to be 299 J. During this process, there is a heat transfer of 70.3 J from the gas to the surroundings. Calculate the change in energy of the gas.

Answer

We write Equation (23.2):

$$\Delta E = q + w$$

Compression is work done on the gas, so the sign for w (from Table 23.1) is positive: $w = 299$ J. From the direction of heat transfer (system to surroundings), we know that q is negative; $q = -70.3$ J. Thus

$$\Delta E = (-70.3 \text{ J}) + (299 \text{ J})$$
$$= 229 \text{ J}$$

As a result of the compression and heat transfer, the energy of the gas increases by 229 J.

Similar example: Problem 23.2.

TABLE 23.2 Molecular Motion in Gases, Liquids, and Solids

Phase	Molecular Motion		
	Translation	*Rotation*	*Vibration*
Gas	Free	Free	Free
Liquid	Restricted	Somewhat restricted	Free
Solid	Absent	Very restricted	Free

A practical application of the first law of thermodynamics is the making of artificial snow. If you are an avid downhill skier, sooner or later you will ski on artificial snow. How is this stuff made? The secret of snowmaking is in the equation $\Delta E = q + w$. A snowmaking machine contains a mixture of compressed air and water vapor at about 20 atm. At the appropriate moment, the mixture is sprayed into the atmosphere (Figure 23.4). Because of the large difference in pressure, the outcoming air and water vapor undergo expansion so rapidly that, as a good approximation, no heat exchange occurs between the system (air and water) and its surroundings; that is, $q = 0$ and

$$\Delta E = q + w = w$$

Since the system does work on the surroundings, w is negative and there is a decrease in its energy. In Section 6.7 we saw that the average kinetic energy of a gas is directly proportional to the absolute temperature [Equation (6.13)]. Since kinetic energy is part of the total energy of the system,

FIGURE 23.4 *A snow-making machine.*

the change in energy can be equated to the change in kinetic energy as follows:

$$\Delta E = C\Delta T$$

where C is a positive constant and ΔT is the change in absolute temperature of the system. Because ΔE is negative, ΔT must also be negative, and it is this cooling effect (or the decrease in the kinetic energy of the molecules) that is responsible for the formation of snow. Although we need only water to form snow, the presence of air, which also cools on expansion, helps to lower the temperature of water vapor.

Enthalpy

Chemical reactions can be carried out under a variety of conditions. Two of the most important are the conditions of *constant volume* and *constant pressure*.

If a reaction is run at constant volume, $\Delta V = 0$ and no work will result from this change. From Equation (23.2) it follows that

The subscript $_V$ reminds us of the constant-volume process.

$$\Delta E = q + w = q_V$$

that is, the change in energy is equal to the heat change. Bomb calorimetry, discussed in Section 5.7, is such a constant-volume process.

Constant-volume conditions are often inconvenient and sometimes impossible to achieve. Most reactions occur under conditions of constant pressure (usually atmospheric pressure), also discussed in Section 5.7. If a reaction conducted at constant pressure results in a net increase in the number of moles of a gas, then the system will do work on the surroundings (expansion). Conversely, if more gas molecules are consumed than are produced, work will be done on the system by the surroundings (compression).

Caution: This reaction can be hazardous.

When a small piece of sodium metal is added to a beaker of water, the reaction that takes place is

$$2Na(s) + 2H_2O(l) \longrightarrow 2NaOH(aq) + H_2(g)$$

The energy released during the reaction is ΔE. One of the products is gaseous hydrogen, which must force its way into the atmosphere by pushing back the air. Consequently, some of the energy produced by the reaction will be used to do the work of pushing back a volume of air (ΔV) against atmospheric pressure (P). The heat of the reaction under the constant-pressure condition is ΔH, which is the enthalpy change; that is, $q_P = \Delta H$ (see Section 5.3). We sum up the situation at constant pressure by writing

The subscript $_P$ reminds us of the constant-pressure process.

$$\begin{aligned} \Delta E &= q_P + w \\ &= \Delta H - P\Delta V \end{aligned}$$

or

$$\Delta H = \Delta E + P\Delta V \qquad (23.3)$$

Equation (23.3) allows us to define enthalpy of a system as

$$H = E + PV \qquad (23.4)$$

where E is the energy of the system, and P and V are the pressure and volume of the system, respectively. According to Equation (23.4), enthalpy has the units of energy, and it is a state function because E, P, and V are all state functions.

Equation (23.4) allows us to explain Hess's law, which was introduced in Section 5.5, as follows: Because H is a state function, ΔH is independent of path. Hence the enthalpy change for any process is the same whether it takes one step or many steps.

23.3   ENTROPY AND THE SECOND LAW OF THERMODYNAMICS

Spontaneous Processes

One of the main objectives in studying thermodynamics, as far as chemists are concerned, is to be able to predict whether or not a reaction will occur when reactants are brought together under a special set of conditions (for example, at a certain temperature, pressure, and concentration). A reaction that does occur under the given set of conditions is called a *spontaneous reaction*; if it does not occur under that set of conditions it is said to be *nonspontaneous*. We have all witnessed many spontaneous chemical and physical processes both in and outside the laboratory: the rusting of an iron nail when exposed to moist air, the violent reaction of a piece of sodium metal with water, the dissolution of a lump of sugar in a cup of coffee, the melting of an ice cube in your hand, and so on.

After many observations we know that reactions that occur spontaneously in one direction cannot also take place spontaneously in the opposite direction. For example, a rusty nail cannot become shiny again if left alone. Nor can dissolved sugar suddenly reappear in its original form. However, an important question remains: Is there a thermodynamic quantity that can help us predict whether a process will occur spontaneously?

It seems logical to assume that spontaneous processes occur so as to decrease the energy of a system. This assumption helps us explain why things fall downward and why springs unwind. In chemical reactions, too, we find that a large number of exothermic reactions are spontaneous at room temperature. An example is the combustion of methane:

$$CH_4(g) \ + \ 2O_2(g) \longrightarrow CO_2(g) \ + \ 2H_2O(l) \qquad \Delta H^\circ \ = \ -890.4 \text{ kJ}$$

Another example is the acid-base neutralization reaction, which can be represented by

$$H^+(aq) \ + \ OH^-(aq) \longrightarrow H_2O(l) \qquad \Delta H^\circ \ = \ -56.2 \text{ kJ}$$

But the assumption that spontaneous processes decrease a system's energy fails in a number of cases. Consider a solid to liquid phase transition such as

$$H_2O(s) \longrightarrow H_2O(l) \qquad \Delta H^\circ \ = \ 6.01 \text{ kJ}$$

Experience tells us that ice melts spontaneously above 0°C even though the process is endothermic. As another example, consider the cooling that results when ammonium nitrate dissolves in water:

$$NH_4NO_3(s) \xrightarrow{H_2O} NH_4^+(aq) + NO_3^-(aq) \qquad \Delta H° = 25 \text{ kJ}$$

The dissolution process is spontaneous, yet it is also endothermic. The decomposition of mercury(II) oxide is an example of an endothermic reaction that is nonspontaneous at room temperature but becomes spontaneous when the temperature is raised:

$$2HgO(s) \xrightarrow{\Delta} 2Hg(l) + O_2(g) \qquad \Delta H° = 90.7 \text{ kJ}$$

From a study of the examples mentioned and many more cases we come to the following conclusion: Exothermicity favors the spontaneity of a reaction but does not guarantee it. It is possible for an exothermic reaction to be nonspontaneous, just as it is for an endothermic reaction to be spontaneous. Consideration of energy changes alone is not enough to predict spontaneity or nonspontaneity. Thus, it becomes necessary to look for another thermodynamic quantity (in addition to enthalpy) that can help us predict the direction of chemical reactions. This quantity turns out to be *entropy*, as we will see shortly.

It is important to keep in mind that just because a reaction is spontaneous does not necessarily mean that it will occur at an observable rate. A spontaneous reaction may be very fast, as in the case of acid-base neutralization, or extremely slow, as in the rusting of a nail. Thermodynamics can tell us whether a reaction will occur, but it does not say how fast it will occur. Reaction rates are the subject of chemical kinetics (see Chapter 22).

Entropy

In addition to knowing the enthalpy change, we need to know the entropy change in order to predict the spontaneity of a process. **Entropy (S)** is a *direct measure of the randomness or disorder of a system*. In other words, entropy describes the extent to which atoms, molecules, or ions are distributed in a disorderly fashion in a given region in space. The greater the disorder of a system, the greater its entropy. It is possible to determine the *absolute* entropy of a substance, something we cannot do for energy or enthalpy, but the details of entropy measurements need not concern us here. A few general points about entropy will serve our needs in this chapter:

Like energy and enthalpy, entropy is an extensive property.

- Listed entropy values are for substances at 1 atm and 25°C; these are called *standard* entropies $(S°)$.
- Entropies of both elements and compounds are positive (that is, $S° > 0$). In contrast, the standard enthalpy of formation $(\Delta H_f°)$ for elements in their stable forms is zero, but for compounds it may be positive or negative.

- The units of entropy are J/K and J/K·mol. The latter units refer to one mole of the substance. (Because entropy values are usually quite small, we use joules rather than kilojoules.)

For any substance, the particles in the solid state are more ordered than those in the liquid state, which in turn are more ordered than those in the gaseous state. Consider the entropies of the following substances at 25°C:

Substance	$S°$ (J/K·mol)
$H_2O(l)$	69.9
$H_2O(g)$	188.7
$Br_2(l)$	152.3
$Br_2(g)$	245.3
$I_2(s)$	116.7
$I_2(g)$	260.6
C(*diamond*)	2.44
C(*graphite*)	5.69

Appendix 1 lists standard entropy values of a variety of elements and compounds.

Water vapor has a greater entropy than liquid water, bromine vapor has a greater entropy than liquid bromine, and iodine vapor has a greater entropy than solid iodine. Both diamond and graphite are solids, but diamond has a more ordered structure (see Figures 12.17 and 12.18). Consequently, diamond has a smaller entropy than graphite. For any substance, the entropy always increases in the following order:

$$S_{solid} < S_{liquid} < S_{gas}$$

The comparison is valid only for the same molar amount of the substance.

Even for different substances, it is sometimes true that solids have lower entropies than liquids, which have lower entropies than gases.

Like energy and enthalpy, entropy is a state function. Consider a certain process in which a system changes from some initial state to some final state. The entropy change for the process, ΔS, is

$$\Delta S = S_f - S_i$$

where S_f and S_i are the entropies of the system in the final and initial states, respectively. If the change results in an increase in randomness, or disorder, then $S_f > S_i$, or $\Delta S > 0$. Figure 23.5 shows several processes that lead to an increase in entropy. You can see that in each case, the system changes from a more ordered state to a less ordered one. It is clear that both melting and vaporization processes have $\Delta S > 0$. When an ionic solid dissolves in water, the highly ordered crystalline structure of the solid and part of the ordered structure of water are destroyed. Consequently, the resulting solution possesses a greater disorder than the pure solute and pure solvent. Heating also increases the entropy of a system. The higher the temperature, the more energetic the various types of molecular motion (see p. 688). This means an increase in randomness at the molecular level, which results in an increase in entropy.

FIGURE 23.5 *Processes that lead to an increase in entropy of the system: (a) melting: $S_{liquid} > S_{solid}$; (b) vaporization: $S_{vapor} > S_{liquid}$; (c) dissolving: $S_{soln} > (S_{solvent} + S_{solute})$; (d) heating: $S_{T_2} > S_{T_1}$.*

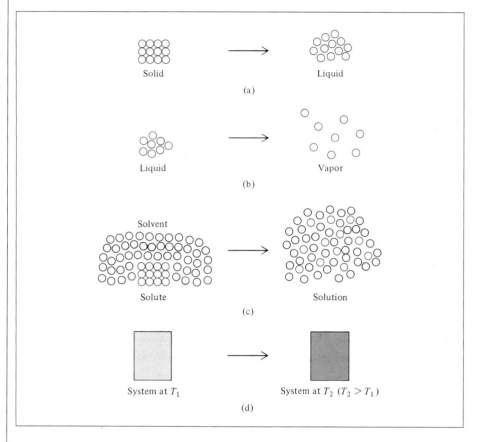

Solid → Liquid

(a)

Liquid → Vapor

(b)

Solvent Solute → Solution

(c)

System at T_1 → System at T_2 ($T_2 > T_1$)

(d)

EXAMPLE 23.3

Predict whether the entropy change is greater than or less than zero for each of the following processes: (a) freezing liquid bromine, (b) evaporating a beaker of ethanol at room temperature, (c) dissolving potassium iodide in water, (d) cooling nitrogen gas from 80°C to 20°C.

Answer

(a) This is a liquid to solid phase transition. The system becomes more ordered so that $\Delta S < 0$. (b). This is a liquid to vapor phase transition. The system becomes more disordered and $\Delta S > 0$. (c) A solution is invariably more disordered than its components (the solute and solvent). Therefore, $\Delta S > 0$. (d) Cooling decreases molecular motion; therefore, $\Delta S < 0$.

Similar example: Problem 23.15.

The Second Law of Thermodynamics

Having discussed the nature of entropy, we can now introduce the ***second law of thermodynamics:*** *The entropy of the universe increases in a spon-*

taneous process and remains unchanged in an equilibrium process. Since the universe is made up of the system and the surroundings, the entropy change in the universe (ΔS_{univ}) for any process is the *sum* of the entropy changes in the system (ΔS_{sys}) and in the surroundings (ΔS_{surr}). Mathematically, we can express the second law of thermodynamics as follows:

$$\text{a spontaneous process:} \quad \Delta S_{univ} = \Delta S_{sys} + \Delta S_{surr} > 0 \qquad (23.5)$$

$$\text{an equilibrium process:} \quad \Delta S_{univ} = \Delta S_{sys} + \Delta S_{surr} = 0 \qquad (23.6)$$

This law says that for a spontaneous process ΔS_{univ} must be greater than zero, but it does not place a restriction on either ΔS_{sys} or ΔS_{surr}. Thus, it is possible for either ΔS_{sys} or ΔS_{surr} to be negative, as long as the sum of these two quantities is greater than zero. For an equilibrium process, ΔS_{univ} is zero. In this case ΔS_{sys} and ΔS_{surr} must be equal in magnitude, but opposite in sign.

Entropy Changes in Chemical Reactions

Our next step is to calculate entropy changes in chemical reactions. The procedure for calculating such changes is quite similar to that for calculating the enthalpy of reaction, outlined in Section 5.4. For the reaction

$$a\text{A} + b\text{B} \longrightarrow c\text{C} + d\text{D}$$

the standard entropy change, $\Delta S°$, is given by

$$\Delta S° = [cS°(\text{C}) + dS°(\text{D})] - [aS°(\text{A}) + bS°(\text{B})]$$

or, in general

$$\Delta S° = \Sigma nS°(\text{products}) - \Sigma mS°(\text{reactants}) \qquad (23.7)$$

where m and n are stoichiometric coefficients.

The units of $S°$ in this equation are J/K·mol.

Equation (23.7) is analogous to Equation (5.4).

EXAMPLE 23.4

From the absolute entropy values in Appendix 1, calculate the standard entropy changes for the following reactions at 25°C:

(a) $\text{CaCO}_3(s) \longrightarrow \text{CaO}(s) + \text{CO}_2(g)$

(b) $\text{N}_2(g) + 3\text{H}_2(g) \longrightarrow 2\text{NH}_3(g)$

(c) $\text{H}_2(g) + \text{Cl}_2(g) \longrightarrow 2\text{HCl}(g)$

Answer

We can calculate $\Delta S°$ by using Equation (23.7).

(a) $\Delta S° = [S°(\text{CaO}) + S°(\text{CO}_2)] - [S°(\text{CaCO}_3)]$
   $= (1 \text{ mol})(39.8 \text{ J/K·mol}) + (1 \text{ mol})(213.6 \text{ J/K·mol}) - (1 \text{ mol})(92.9 \text{ J/K·mol})$
   $= 160.5 \text{ J/K}$

(Continued)

Thus, when one mole of $CaCO_3$ decomposes to form one mole of CaO and one mole of gaseous CO_2, there is an increase in entropy equal to 160.5 J/K.

(b) $\Delta S° = [2S°(NH_3)] - [S°(N_2) + 3S°(H_2)]$
$= (2 \text{ mol})(193 \text{ J/K·mol}) - [(1 \text{ mol})(192 \text{ J/K·mol}) + (3 \text{ mol})(131 \text{ J/K·mol})]$
$= -199 \text{ J/K}$

This result shows that when one mole of gaseous nitrogen reacts with three moles of gaseous hydrogen to form two moles of gaseous ammonia, there is a decrease in entropy equal to -199 J/K.

(c) $\Delta S° = [2S°(HCl)] - [S°(H_2) + S°(Cl_2)]$
$= (2 \text{ mol})(187 \text{ J/K·mol}) - [(1 \text{ mol})(131 \text{ J/K·mol}) + (1 \text{ mol})(223 \text{ J/K·mol})]$
$= 20 \text{ J/K}$

Thus, the formation of two moles of gaseous HCl from one mole of gaseous H_2 and one mole of gaseous Cl_2 results in a small increase in entropy equal to 20 J/K.

It is important to note that the entropy changes calculated in Example 23.4 pertain only to the system (that is, the reactant and product molecules); therefore, $\Delta S°$ can be either positive or negative for a spontaneous reaction. We can draw some general conclusions from the results.

- If a reaction produces more gas molecules than it consumes [Reaction (a) in Example 23.4], $\Delta S°$ is positive. If the total number of gas molecules diminishes [Reaction (b) in Example 23.4], $\Delta S°$ is negative.
- If there is no net change in the total number of gas molecules [Reaction (c) in Example 23.4], then $\Delta S°$ is either a small positive or small negative number.

These conclusions are based on the fact that gases invariably have greater entropy than liquids and solids. For reactions involving only liquids and solids prediction is more difficult, although the general rule still holds in many cases that an increase in the total number of molecules and/or ions is accompanied by an increase in entropy.

EXAMPLE 23.5

Predict whether the entropy change of the system in each of the following reactions is positive or negative.

(a) $Ag^+(aq) + Cl^-(aq) \longrightarrow AgCl(s)$

(b) $NH_4Cl(s) \longrightarrow NH_3(g) + HCl(g)$

(c) $H_2(g) + Br_2(g) \longrightarrow 2HBr(g)$

Answer

(a) The Ag^+ and Cl^- ions are free to move in solution, whereas AgCl is a solid. Furthermore, the number of particles decreases from left to right. Therefore, ΔS is negative.

(b) Since the solid is converted to two gaseous products, ΔS is positive.
(c) We see that the same number of moles of gas is involved in the reactants as in the product. Therefore we cannot predict the sign of ΔS but we know the change must be quite small.

Similar examples: Problems 23.17, 23.18, 23.19.

23.4 GIBBS FREE ENERGY

The second law of thermodynamics tells us that to be spontaneous a reaction must lead to an increase in the entropy of the universe, that is, $\Delta S_{univ} > 0$. In order to determine the sign of ΔS_{univ} for a reaction, however, we would need to calculate both ΔS_{sys} and ΔS_{surr}. Because we are usually concerned only with what happens in a particular system rather than what happens in the entire universe, and because the calculation of ΔS_{surr} often can be quite difficult, it is desirable to have another thermodynamic function that will help us determine whether a reaction will occur spontaneously, even though we consider only the system itself.

Earlier we mentioned that enthalpy change (of the system) is one of the two factors that help us predict the direction of a reaction. Entropy change (of the system) is the other factor. Together, enthalpy and entropy enable us to define a thermodynamic function, called the **Gibbs free energy** (Josiah Willard Gibbs, 1839–1903), or simply **free energy (G).** The free energy is the *energy available to do useful work.* As we will see, free energy change for a process carried out at constant pressure and temperature can be used to predict the spontaneity of a reaction. Mathematically, the free energy of a system, G, is given by

$$G = H - TS \tag{23.8}$$

where H is the enthalpy, T the temperature (in kelvins), and S the entropy, of the system, respectively. You can see that G has the units of energy (both H and TS have the units of energy) and, like H, T, and S, it is a state function.

The change in free energy (ΔG) of a system for a process *at constant temperature* is given by

$$\Delta G = \Delta H - T\Delta S \tag{23.9}$$

The relationship between ΔG and the spontaneity of a reaction at constant temperature and pressure may be summarized as follows:

$\Delta G < 0$ A spontaneous reaction.
$\Delta G > 0$ A nonspontaneous reaction. The reaction is spontaneous in the reverse direction.
$\Delta G = 0$ The system is at equilibrium. There is no net change.

One way to appreciate the significance of free energy is to realize that the adjective "free" describes the usable, or available, energy—energy in forms that could be used to do work. Thus, if a particular reaction is accompanied by a release of usable energy (that is, if ΔG is negative), this fact alone guarantees that it must be spontaneous.

Calculating Free-Energy Changes

Because free-energy change under conditions of constant pressure and temperature can be used to predict the outcome of a process, we will now look at ways to calculate its value. Consider the following reaction:

$$a\text{A} + b\text{B} \longrightarrow c\text{C} + d\text{D}$$

If the reaction is carried out under standard-state conditions, that is, reactants in their standard states are converted to products in their standard states, the free-energy change is called the standard free-energy change, $\Delta G°$. Table 23.3 summarizes the conventions used by chemists to define the standard states of pure substances as well as solutions. For the above reaction the standard free-energy change is given by

$$\Delta G° = [c\Delta G_f°(\text{C}) + d\Delta G_f°(\text{D})] - [a\Delta G_f°(\text{A}) + b\Delta G_f°(\text{B})]$$

or, in general

$$\Delta G° = \Sigma n\Delta G_f°(\text{products}) - \Sigma m\Delta G_f°(\text{reactants}) \qquad (23.10)$$

where m and n are stoichiometric coefficients. The standard free energy of formation of a compound $(\Delta G_f°)$ is the free-energy change that occurs when one mole of the compound is synthesized from its elements in their standard states. For the combustion of graphite

$$\text{C}(graphite) + \text{O}_2(g) \longrightarrow \text{CO}_2(g)$$

the standard free-energy change [from Equation (23.10)] is

$$\Delta G° = \Delta G_f°(\text{CO}_2) - [\Delta G_f°(\text{C}) + \Delta G_f°(\text{O}_2)]$$

As in the case of the standard enthalpy of formation (p. 130), we define the standard free energy of formation of any element in its stable form as zero. Thus

$$\Delta G_f°(\text{C}) = 0 \qquad \text{and} \qquad \Delta G_f°(\text{O}_2) = 0$$

The standard free-energy change for the reaction is numerically equal to the standard free energy of formation of CO_2:

$$\Delta G° = \Delta G_f°(\text{CO}_2)$$

Appendix 1 lists the values of $\Delta G_f°$ for a number of compounds.

Note that $\Delta G°$ is in kJ, but $\Delta G_f°$ is in kJ/mol. The equation holds because the coefficient in front of $\Delta G_f°$ (one in this case) has the unit mol.

TABLE 23.3 Conventions for Standard States

State of Matter	Standard State
Gas	1 atm pressure
Liquid	Pure liquid
Solid	Pure solid
Elements*	$\Delta G_f°$ (element) $= 0$
Solution	1 molar concentration

* The most stable allotropic form at 25°C.

EXAMPLE 23.6

Calculate the standard free-energy changes for the following reactions at 25°C:

(a) $CH_4(g) + 2O_2(g) \longrightarrow CO_2(g) + 2H_2O(l)$

(b) $2MgO(s) \longrightarrow 2Mg(s) + O_2(g)$

Answer

(a) According to Equation (23.10), we write

$$\Delta G° = [\Delta G_f°(CO_2) + 2\Delta G_f°(H_2O)] - [\Delta G_f°(CH_4) + 2\Delta G_f°(O_2)]$$

We insert the values from Appendix 1:

$$\Delta G° = [(1 \text{ mol})(-394.4 \text{ kJ/mol}) + (2 \text{ mol})(-237.2 \text{ kJ/mol})]$$
$$- [(1 \text{ mol})(-50.8 \text{ kJ/mol}) + (2 \text{ mol})(0 \text{ kJ/mol})]$$
$$= -818.0 \text{ kJ}$$

(b) The equation is

$$\Delta G° = [2\Delta G_f°(Mg) + \Delta G_f°(O_2)] - [2\Delta G_f°(MgO)]$$

From data in Appendix 1 we write

$$\Delta G° = [(2 \text{ mol})(0 \text{ kJ/mol}) + (1 \text{ mol})(0 \text{ kJ/mol})]$$
$$- [(2 \text{ mol})(-569.6 \text{ kJ/mol})]$$
$$= 1139 \text{ kJ}$$

Similar examples: Problems 23.22, 23.27.

In the preceding example the large negative value of $\Delta G°$ in (a) means that the combustion of methane is a highly spontaneous process under standard-state conditions, whereas the decomposition of MgO in (b) is nonspontaneous because $\Delta G°$ is a large, positive quantity. We stress that a large, negative $\Delta G°$ does not tell us anything about the actual *rate* of the spontaneous process.

Applications of Equation (23.9)

Free energy is useful because it allows us to predict whether a given process is spontaneous. In order to predict the sign of ΔG, we need to determine both ΔH and ΔS, as shown by Equation (23.9). The factors that tend to make ΔG negative are a negative ΔH (an exothermic reaction) and a positive ΔS (a reaction that results in an increase in disorder of the system). In addition, temperature may also influence the direction of a spontaneous reaction. We will discuss the four possibilities below.

- If both ΔH and ΔS are positive, then ΔG will be negative only when the $T\Delta S$ term is greater in magnitude than ΔH. This condition is met when T is large.
- If ΔH is positive and ΔS is negative, ΔG will always be positive, regardless of temperature (because the $-T\Delta S$ term will always be positive).

TABLE 23.4   Factors Affecting the Sign of ΔG in the Relationship $\Delta G = \Delta H - T\Delta S$

ΔH	ΔS	ΔG	Example
+	+	Reaction proceeds spontaneously at high temperatures. At low temperatures, reaction is spontaneous in the reverse direction.	$H_2(g) + I_2(g) \longrightarrow 2HI(g)$
+	−	ΔG is always positive. Reaction is spontaneous in the reverse direction at all temperatures.	$3O_2(g) \longrightarrow 2O_3(g)$
−	+	ΔG is always negative. Reaction proceeds spontaneously at all temperatures.	$2H_2O_2(l) \longrightarrow 2H_2O(l) + O_2(g)$
−	−	Reaction proceeds spontaneously at low temperatures. At high temperatures, the reverse reaction becomes spontaneous.	$NH_3(g) + HCl(g) \longrightarrow NH_4Cl(s)$

- If ΔH is negative and ΔS is positive, then ΔG will always be negative regardless of temperature.
- If ΔH is negative and ΔS is negative, then ΔG will be negative only when $T\Delta S$ is smaller in magnitude than ΔH. This condition is met when T is small.

What temperature will cause ΔG to be negative for the first and last cases depends on the actual values of ΔH and ΔS of the system. Table 23.4 summarizes the effects of the four possibilities just described.

We will now consider two specific applications of Equation (23.9).

Temperature and Chemical Reactions.   Calcium oxide (CaO), also called quicklime, is an important inorganic compound used as a starting material in a number of industrial processes (for example, in water treatment, manufacture of glass and cement, and food processing). It is prepared by the thermal decomposition of calcium carbonate ($CaCO_3$), which is found in limestone, chalk, and marble:

$$CaCO_3(s) \longrightarrow CaO(s) + CO_2(g)$$

This reaction is usually carried out in a kiln. The important information for the practical chemist is the temperature at which the decomposition becomes appreciable (spontaneous). We can make a reliable estimate of that temperature as follows. First we calculate $\Delta H°$ [using Equation (5.4)] and $\Delta S°$ [using Equation (23.7)] for the reaction at 25°C, using the data in Appendix 1.

$$
\begin{aligned}
\Delta H° &= [\Delta H_f°(CaO) + \Delta H_f°(CO_2)] - [\Delta H_f°(CaCO_3)] \\
&= [(1\ \text{mol})(-635.6\ \text{kJ/mol}) + (1\ \text{mol})(-393.5\ \text{kJ/mol})] \\
&\qquad\qquad\qquad\qquad\qquad - [(1\ \text{mol})(-1206.9\ \text{kJ/mol})] \\
&= 177.8\ \text{kJ}
\end{aligned}
$$

$$
\begin{aligned}
\Delta S° &= [S°(CaO) + S°(CO_2)] - [S°(CaCO_3)] \\
&= [(1\ \text{mol})(39.8\ \text{J/K·mol}) + (1\ \text{mol})(213.6\ \text{J/K·mol})] \\
&\qquad\qquad\qquad\qquad\qquad - [(1\ \text{mol})(92.9\ \text{J/K·mol})] \\
&= 160.5\ \text{J/K}
\end{aligned}
$$

For reactions carried out under standard-state conditions, Equation (23.9) takes the form

$$\Delta G° = \Delta H° - T\Delta S°$$

so we obtain

$$\Delta G° = 177.8 \text{ kJ} - (298 \text{ K})(160.5 \text{ J/K})\left(\frac{1 \text{ kJ}}{1000 \text{ J}}\right)$$

$$= 130.0 \text{ kJ}$$

Since $\Delta G°$ is a large positive quantity, we conclude that the reaction is not favored at 25°C (or 298 K). In order to make $\Delta G°$ negative, we first have to find the temperature at which $\Delta G°$ is zero, that is

$$0 = \Delta H° - T\Delta S°$$

or

$$T = \frac{\Delta H°}{\Delta S°}$$

$$= \frac{(177.8 \text{ kJ})(1000 \text{ J/1 kJ})}{160.5 \text{ J/K}}$$

$$= 1108 \text{ K}$$

$$= 835°C$$

At a temperature higher than 835°C, $\Delta G°$ will become negative, indicating that the decomposition is favorable. For example, at 840°C or 1113 K:

$$\Delta G° = \Delta H° - T\Delta S°$$

$$= 177.8 \text{ kJ} - (1113 \text{ K})(160.5 \text{ J/K})\left(\frac{1 \text{ kJ}}{1000 \text{ J}}\right)$$

$$= -0.8 \text{ kJ}$$

Two points are worth making about such a calculation. First, we used the $\Delta H°$ and $\Delta S°$ values at 25°C to calculate changes that occur at a much higher temperature. Since both $\Delta H°$ and $\Delta S°$ change with temperature, this approach will not give us an accurate value of $\Delta G°$ although it is good enough for "ball park" estimates. Second, we should not be misled into thinking that nothing happens below 835°C and that at 835°C $CaCO_3$ suddenly begins to decompose. Far from it. The fact that $\Delta G°$ is a positive value at some temperature below 835°C does not mean that no CO_2 is produced. It only says that the pressure of the CO_2 gas formed at that temperature will be below 1 atm (its standard state; see Table 23.3). As Figure 23.6 shows, the pressure of CO_2 at first increases very slowly with temperature; it becomes easily measurable above 700°C. The significance of 835°C is that this is the temperature at which the equilibrium pressure of CO_2 reaches 1 atm. Above 835°C, the equilibrium pressure of CO_2 exceeds 1 atm. (In Section 23.5 we will see how $\Delta G°$ is related to the equilibrium constant of a given reaction.)

The equilibrium constant of this reaction is $K_P = P_{CO_2}$.

EXAMPLE 23.7

Water gas is a fuel made by exposing steam to red-hot coke (coke is the product from the distillation of coal; it contains carbon):

$$H_2O(g) + C(s) \longrightarrow CO(g) + H_2(g)$$

(Continued)

FIGURE 23.6 *Equilibrium pressure of CO_2 from the decomposition of $CaCO_3$, as a function of temperature. This curve is calculated assuming that $\Delta H°$ and $\Delta S°$ of the reaction do not change with temperature.*

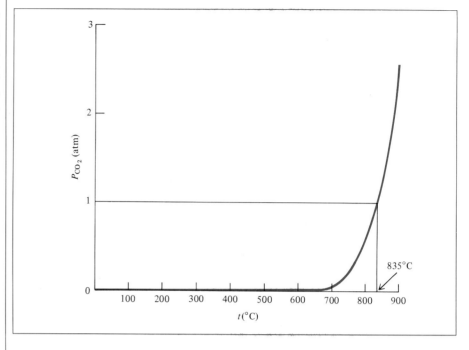

From the data in Appendix 1, calculate the temperature at which the reaction becomes favorable. Assume that both $\Delta H°$ and $\Delta S°$ are independent of temperature.

Answer

$$\Delta H° = [\Delta H_f°(CO) + \Delta H_f°(H_2)] - [\Delta H_f°(H_2O) + \Delta H_f°(C)]$$
$$= [(1 \text{ mol})(-110.5 \text{ kJ/mol}) + (1 \text{ mol})(0 \text{ kJ/mol})]$$
$$- [(1 \text{ mol})(-241.8 \text{ kJ/mol}) + (1 \text{ mol})(0 \text{ kJ/mol})]$$
$$= 131.3 \text{ kJ}$$

$$\Delta S° = [S°(CO) + S°(H_2)] - [S°(H_2O) + S°(C)]$$
$$= [(1 \text{ mol})(197.9 \text{ J/K·mol}) + (1 \text{ mol})(131.0 \text{ J/K·mol})]$$
$$- [(1 \text{ mol})(188.7 \text{ J/K·mol}) + (1 \text{ mol})(5.69 \text{ J/K·mol})]$$
$$= 134.5 \text{ J/K}$$

It is obvious from the given conditions that the reaction must take place at a fairly high temperature (in order to have red-hot coke). Setting $\Delta G° = 0$

$$0 = \Delta H° - T\Delta S°$$
$$T = \frac{\Delta H°}{\Delta S°}$$
$$= \frac{(131.3 \text{ kJ})(1000 \text{ J/1 kJ})}{134.5 \text{ J/K}}$$
$$= 976 \text{ K}$$
$$= 703°C$$

Because the temperature is an estimate, it is sufficient to express it in only three significant figures.

Thus the reaction is favored above 703°C.

Similar example: Problem 23.23.

Phase Transitions. Equation (23.9) can be applied also to phase transitions. At the transition temperature (that is, at the melting point or the boiling point) the system is at equilibrium ($\Delta G = 0$), so Equation (23.9) becomes

$$0 = \Delta H - T\Delta S$$

$$\Delta S = \frac{\Delta H}{T}$$

Let us first consider the ice–water equilibrium. For the ice $\rightarrow$ water transition, ΔH is the heat of fusion (see Table 14.9) and T is the melting point, so the entropy change is

$$\Delta S_{\text{ice}\rightarrow\text{water}} = \frac{6010 \text{ J/mol}}{273 \text{ K}}$$
$$= 22.0 \text{ J/K·mol}$$

Thus when one mole of ice melts at 0°C, there is an increase in entropy of 22.0 J/K. The increase in entropy is consistent with the increase in disorder from solid to liquid. Conversely, for the water $\rightarrow$ ice transition, the decrease in entropy is given by

$$\Delta S_{\text{water}\rightarrow\text{ice}} = \frac{-6010 \text{ J/mol}}{273 \text{ K}}$$
$$= -22.0 \text{ J/K·mol}$$

In the laboratory we normally carry out unidirectional changes, that is, either ice to water or water to ice. We can calculate entropy changes using the $\Delta S = \Delta H/T$ equation as long as the temperature remains at 0°C. The same procedure can be applied to the water–steam transition. In this case ΔH is the heat of vaporization and T is the boiling point of water.

The melting of ice is an endothermic process (ΔH is positive), whereas the freezing of water is exothermic (ΔH is negative).

EXAMPLE 23.8

The heats of fusion and vaporization of benzene are 10.9 kJ/mol and 31.0 kJ/mol, respectively. Calculate the entropy changes for the solid $\rightarrow$ liquid and liquid $\rightarrow$ vapor transitions for benzene. At 1 atm pressure, benzene melts at 5.5°C and boils at 80.1°C.

Answer

The entropy of fusion is given by

$$\Delta S_{\text{fus}} = \frac{\Delta H_{\text{fus}}}{T_{\text{mp}}}$$
$$= \frac{(10.9 \text{ kJ/mol})(1000 \text{ J/1 kJ})}{(5.5 + 273) \text{ K}}$$
$$= 39.1 \text{ J/K·mol}$$

(Continued)

and the entropy of vaporization is given by

$$\Delta S_{vap} = \frac{\Delta H_{vap}}{T_{bp}}$$
$$= \frac{(31.0 \text{ kJ/mol})(1000 \text{ J/1 kJ})}{(80.1 + 273) \text{ K}}$$
$$= 87.8 \text{ J/K·mol}$$

Since vaporization creates more disorder than the melting process, $\Delta S_{vap} > \Delta S_{fus}$.

Similar example: Problem 23.25.

23.5 FREE ENERGY AND CHEMICAL EQUILIBRIUM

In this section we will take a closer look at the difference between ΔG and $\Delta G°$ and at the relationship between free-energy changes and equilibrium constants.

The equations for free-energy change and standard free-energy change are

$$\Delta G = \Delta H - T\Delta S$$
$$\Delta G° = \Delta H° - T\Delta S°$$

It is important that we understand the conditions under which these equations should be used and what kind of information can be obtained from ΔG and $\Delta G°$. Let us illustrate by considering the following reaction:

$$\text{reactants} \longrightarrow \text{products}$$

The standard free-energy change for the reaction ($\Delta G°$) is given by

$$\Delta G° = G°(\text{products}) - G°(\text{reactants})$$

The quantity $\Delta G°$ represents the change when the reactants in their standard states are converted to products in their standard states (see Table 23.3 for definitions of standard states). We have already seen an example of calculating $\Delta G°$ for reactions from the known standard free energies of formation of reactants and products (see Example 23.6). Suppose we start the reaction in solution with all the reactants in their standard states (that is, they are all at 1 M concentration). As soon as the reaction starts, the standard-state condition no longer exists for the reactants or the products since neither remains at 1 M concentration. Under non-standard-state conditions, we must use ΔG rather than $\Delta G°$ to predict the direction of the reaction. The relationship between ΔG and $\Delta G°$ is

$$\Delta G = \Delta G° + 2.303RT \log Q \tag{23.11}$$

where R is the gas constant (8.314 J/K·mol), T the absolute temperature of the reaction, and Q is the reaction quotient defined on p. 515. We see that ΔG depends on two quantities: $\Delta G°$ and $2.303RT \log Q$. For a given reaction at temperature T, the value of $\Delta G°$ is fixed but that of $2.303RT \times \log Q$

is not, because Q varies according to the composition of the reacting mixture. Let us consider the two cases:

Case I: If $\Delta G°$ is a large negative value, the $2.303RT \log Q$ term will not become positive enough to match the $\Delta G°$ term until significant product formation has occurred.

Case II: If $\Delta G°$ is a large positive value, the $2.303RT \log Q$ term will be more negative than $\Delta G°$ is positive only as long as very little product formation has occurred and the concentration of the reactant is high relative to that of the product.

At equilibrium, by definition, $\Delta G = 0$ and $Q = K$, where K is the equilibrium constant. Thus

$$0 = \Delta G° + 2.303RT \log K$$

or

$$\Delta G° = -2.303RT \log K \tag{23.12}$$

For chemists, Equation (23.12) is one of the most important equations in thermodynamics because it relates the equilibrium constant of a reaction to the change in standard free energy. Thus, from a knowledge of K we can calculate $\Delta G°$, and vice versa.

For reactions having very large or very small equilibrium constants, it is generally very difficult, if not impossible, to measure the K values by monitoring the concentrations of the reacting species. Consider, for example, the formation of nitric oxide from molecular nitrogen and molecular oxygen:

$$N_2(g) + O_2(g) \rightleftharpoons 2NO(g)$$

At 25°C, the equilibrium constant K_c is

$$K_c = \frac{[NO]^2}{[N_2][O_2]} = 4.0 \times 10^{-31}$$

The very small K_c value means that the concentration of NO at equilibrium will be exceedingly low. In such a case the equilibrium constant is more conveniently obtained from the $\Delta G°$ of the reaction. (As we have seen, $\Delta G°$ can be calculated from the $\Delta H°$ and $\Delta S°$ values.) On the other hand, the equilibrium constant for the formation of hydrogen iodide from molecular hydrogen and molecular iodine is near unity at room temperature:

$$H_2(g) + I_2(g) \rightleftharpoons 2HI(g)$$

Thus it is easier to measure K for this reaction and then calculate $\Delta G°$ using Equation (23.12). This procedure involves less effort than measuring $\Delta H°$ and $\Delta S°$ of the reaction.

Figure 23.7 shows plots of free energy (G) of a reacting system versus the extent of the reaction, for two types of reactions. As you can see, if $\Delta G° < 0$, the products are favored over reactants at equilibrium. Conversely, if $\Delta G° > 0$, there will be more reactants than products at equilibrium.

Table 23.5 summarizes the three possible relations between $\Delta G°$ and K, as predicted by Equation (23.12).

Sooner or later a reversible reaction will reach equilibrium.

In Equation (23.12), K_P is used for reactions involving gases and K_c is used for reactions in solution.

K_c increases with temperature. This is the reason that a substantial amount of NO is formed in the engine of a running auto (see p. 671).

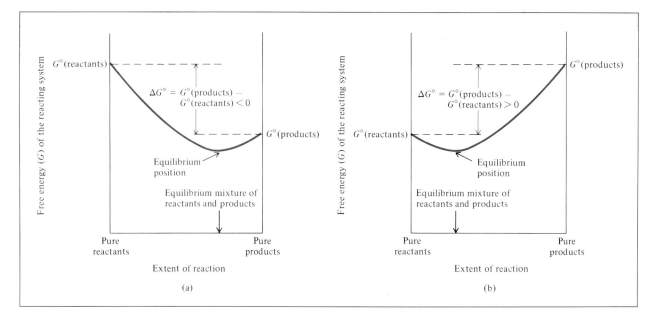

FIGURE 23.7 *(a) G°(products) < G°(reactants), so ΔG° < 0. At equilibrium, there is a significant conversion of reactants to products. (b) G°(products) > G°(reactants) so ΔG° > 0. At equilibrium, reactants are favored over products.*

TABLE 23.5 Relation Between $\Delta G°$ and K as Predicted by the Equation $\Delta G° = -2.303RT \log K$

K	$\log K$	$\Delta G°$	Comments
>1	Positive	Negative	Products are favored over reactants at equilibrium.
= 1	0	0	Products and reactants are equally favored at equilibrium.
<1	Negative	Positive	Reactants are favored over products at equilibrium.

EXAMPLE 23.9

Using data listed in Appendix 1, calculate the equilibrium constant for the following reaction at 25°C:

$$2H_2(g) + O_2(g) \rightleftharpoons 2H_2O(l)$$

Answer

According to Equation (23.10)

$$\Delta G° = [2\Delta G_f°(H_2O)] - [2\Delta G_f°(H_2) + \Delta G_f°(O_2)]$$
$$= [(2 \text{ mol})(-237.2 \text{ kJ/mol})] - [(2 \text{ mol})(0 \text{ kJ/mol}) + (1 \text{ mol})(0 \text{ kJ/mol})]$$
$$= -474.4 \text{ kJ}$$

Utilizing Equation (23.12)

$$\Delta G° = -2.303RT \log K$$

$$-474.4\text{kJ} \times \frac{1000 \text{ J}}{1 \text{ kJ}} = -(2.303)(8.314 \text{ J/K·mol})(298 \text{ K}) \log K$$

$$\log K = 83.1$$

Taking the antilog of both sides, we get

$$K = 1 \times 10^{83}$$

Similar examples: Problems 23.29, 23.30, 23.31.

As stated earlier, it is difficult to measure very large (and very small) K values directly because the concentrations of some species will be very small at equilibrium and therefore difficult to determine accurately.

EXAMPLE 23.10

The standard free-energy change for the reaction

$$N_2(g) + 3H_2(g) \rightleftharpoons 2NH_3(g)$$

is -33.2 kJ and the equilibrium constant (K_P) is 6.59×10^5 at 25°C. In a certain experiment, the initial pressures are $P_{H_2} = 0.250$ atm, $P_{N_2} = 0.870$ atm, and $P_{NH_3} = 12.9$ atm. Calculate ΔG for the reaction at these pressures and predict the direction of reaction.

Answer

Equation (23.11) can be written as

$$\Delta G = \Delta G° + 2.303RT \log Q_P$$

$$= \Delta G° + 2.303RT \log \frac{P_{NH_3}^2}{P_{H_2}^3 P_{N_2}}$$

$$= -33.2 \times 1000 \text{ J} + (2.303)(8.314 \text{ J/K·mol})(298 \text{ K}) \times \log \frac{(12.9)^2}{(0.250)^3(0.870)}$$

$$= -33.2 \times 10^3 \text{ J} + 23.3 \times 10^3 \text{ J}$$

$$= -9.9 \times 10^3 \text{ J} = -9.9 \text{ kJ}$$

Since ΔG is negative, the net reaction proceeds from left to right. As an exercise, confirm this prediction by calculating the reaction quotient Q_P and comparing it with the equilibrium constant K_P.

Similar examples: Problems 23.28, 23.35.

EXAMPLE 23.11

In Chapter 21 we discussed the solubility product of slightly soluble substances. With the solubility product of silver chloride at 25°C (1.6×10^{-10}), calculate $\Delta G°$ for the process

$$AgCl(s) \rightleftharpoons Ag^+(aq) + Cl^-(aq)$$

(Continued)

Answer

Because this is a heterogeneous equilibrium, the solubility product is the equilibrium constant.

$$K_{sp} = [Ag^+][Cl^-] = 1.6 \times 10^{-10}$$

Using Equation (23.12) we obtain

$$\Delta G° = -(2.303)(8.314 \text{ J/K·mol})(298 \text{ K}) \log 1.6 \times 10^{-10}$$
$$= 5.6 \times 10^4 \text{ J}$$
$$= 56 \text{ kJ}$$

The large, positive $\Delta G°$ is consistent with the fact that the equilibrium lies mostly to the left.

Similar example: Problem 23.38.

AN ASIDE ON THE PERIODIC TABLE
Standard Entropies of Elements at 25°C

In this chapter we saw that entropy is a measure of the randomness of a system. Its value depends on parameters such as the type of substance present, its structure, and physical state. In previous chapters we discussed the physical states (p. 416) and structure of the elements (nonmetallic elements on p. 355 and metallic elements on p. 403). It is therefore appropriate for us to correlate the standard entropies of the elements with what we already know about their properties.

Figure 23.8 shows the standard entropies (in J/K·mol) of the more familiar elements. The following points are of particular interest:

- As expected, the standard entropies are greatest for gases and lowest for solids. Liquids have intermediate values.
- For elements in the same group and possessing the same structure, the standard entropies increase as we move down the group; the electronic structure becomes more complex with increasing atomic number, contributing to greater entropy values. Consider the Group 1A elements, the alkali metals. These metals (Li to Cs) all have a body-centered cubic lattice (see Figure 14.30) and are all solids at 25°C. Their standard entropies increase steadily from Li to Cs. The same trends, that is, the increase in standard entropies as we move down a group, are observed for the Group 2A--4A elements and the transition metals, even though elements in the same group may not possess the same crystal lattice. In Groups 5A—7A we see rather abrupt changes in standard entropies because of the variation in physical states from one element to the next. Take the halogens as an example. Both elemental fluorine and chlorine exist as gases, so $S°$ increases from F_2 to Cl_2. However, since elemental bromine exists as a liquid and elemental iodine is a solid, $S°$ decreases from Cl_2 to Br_2 to I_2 even though iodine has the highest atomic number of the group. For the noble gases, $S°$ increases from He to Rn, as we would expect.

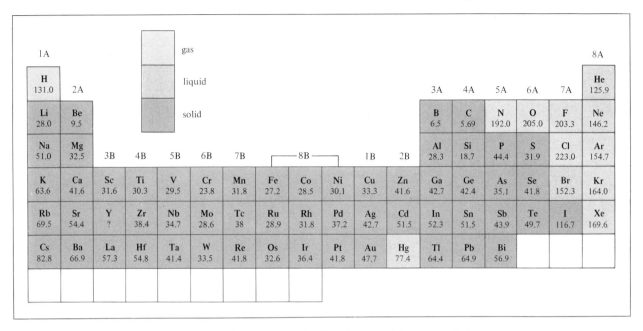

FIGURE 23.8 Standard entropies (J/K·mol) of the more familiar elements. Wherever applicable, the value refers to the most stable allotropic form of the element at 25°C.

- Unlike the trends within a group, there are no clear trends across a period because of the changes in physical state as we go from element to element.
- Both mercury and bromine are liquids. You might expect the standard entropy of mercury to be greater because it has a higher atomic number. However, because mercury exists in the metallic state and therefore has more order at the atomic level, it has a smaller $S°$ value than elemental bromine.

SUMMARY

1. The state of a system is defined by variables such as composition, volume, temperature, and pressure.
2. The value of a change in a state function depends only on the initial and final states of the system, and not on the path. Energy is a state function; work and heat are not.
3. Energy can be converted from one form to another, but it cannot be created or destroyed (first law of thermodynamics). In chemistry we are concerned mainly with thermal energy, electrical energy, and mechanical energy which is usually associated with pressure–volume work.
4. Change in enthalpy (ΔH) is equal to the heat of a process that occurs at constant pressure; ΔH is also equal to $\Delta E + P\Delta V$ for a constant-pressure process.

5. Entropy is a measure of the disorder of a system. Any spontaneous process in the universe must lead to an increase in entropy (second law of thermodynamics) or, stated mathematically, $\Delta S_{univ} = \Delta S_{sys} + \Delta S_{surr} > 0$.

6. The standard entropy change of a chemical reaction can be calculated from the absolute entropies of reactants and products.

7. Under constant temperature and pressure conditions, the free-energy change (ΔG) is less than zero for a spontaneous process and greater than zero for a nonspontaneous process. For an equilibrium process, $\Delta G = 0$.

8. For a chemical or physical process at constant temperature and pressure, $\Delta G = \Delta H - T\Delta S$. This equation can be used to predict the spontaneity of a process.

9. The standard free-energy change for a reaction ($\Delta G°$) can be found from the standard free energies of formation of reactants and products.

10. The equilibrium constant of a reaction and the standard free-energy change of the reaction are related by the equation $\Delta G° = -2.303RT \times \log K$.

KEY WORDS

Entropy, p. 692
First law of thermodynamics, p. 684
Free energy, p. 697
Gibbs free energy, p. 697
Second law of thermodynamics, p. 694

State function, p. 683
State of a system, p. 683
Thermal energy, p. 688
Thermodynamics, p. 683

PROBLEMS

More challenging problems are marked with an asterisk.

First Law of Thermodynamics

23.1 Explain the sign conventions in the equation $\Delta E = q + w$.

23.2 A gas expands at constant temperature and does P–V work on the surroundings equal to 325 J. At the same time, it absorbs 127 J of heat from the surroundings. Calculate the change in energy of the gas.

*23.3 Define the term "enthalpy change" and outline how you would go about measuring the enthalpy change occurring (a) when a sample of metallic magnesium dissolves in sulfuric acid and (b) when a sample of molten naphthalene freezes.

23.4 A sample of 2.10 moles of crystalline acetic acid, initially at 17.0°C, is allowed to melt at 17.0°C, and is then heated to 118.1°C (its normal boiling point) at 1.00 atm. The sample is allowed to va-

porize at 118.1°C and is then rapidly quenched to 17.0°C, re-forming the crystalline solid. Calculate ΔH for the total process as described.

23.5 The standard enthalpy change $\Delta H°$ for the thermal decomposition of silver nitrate according to the following equation is +78.67 kJ:

$$AgNO_3(s) \longrightarrow AgNO_2(s) + \tfrac{1}{2}O_2(g)$$

The standard enthalpy of formation of $AgNO_3(s)$ is −123.02 kJ/mol. Calculate the standard enthalpy of formation of $AgNO_2(s)$.

23.6 (a) Calculate $\Delta H°$ for the following reaction:

$$3N_2H_4(l) \longrightarrow 4NH_3(g) + N_2(g)$$
hydrazine

The standard enthalpy of formation of hydrazine, $N_2H_4(l)$, is 50.42 kJ/mol. (b) Both hydrazine and ammonia will burn in oxygen to produce $H_2O(l)$ and $N_2(g)$. (i) Write balanced equations for each of these processes and calculate $\Delta H°$ for each of them. (ii) On a mass basis (per kg) would hydra-zine or ammonia be the better fuel?

*23.7 Calculate the work done when 50.0 g of tin dis-solves in excess acid at 1.00 atm and 25.0°C:

$$Sn(s) + 2H^+(aq) \longrightarrow Sn^{2+}(aq) + H_2(g)$$

Assume ideal gas behavior.

*23.8 In pumping a bicycle tire, a student notices that the valve stem warms up. Use the first law of thermodynamics to account for this warming ef-fect.

*23.9 Calculate the work done in joules by the reaction

$$2Na(s) + 2H_2O(l) \longrightarrow 2NaOH(aq) + H_2(g)$$

when 0.34 g of Na reacts with water to form hydrogen gas at 0°C and 1.0 atm.

*23.10 Calculate the work done in joules when 1.0 mole of water vaporizes at 1.0 atm and 100°C. Assume that the volume of liquid water is negligible com-pared to that of steam at 100°C. Assume ideal gas behavior.

Spontaneous Processes

23.11 Indicate which of the following processes are spontaneous and which are nonspontaneous: (a) dissolving table salt (NaCl) in hot soup, (b) climb-ing Mt. Everest, (c) spreading fragrance in a room by removing the cap from a perfume bottle, (d) separating $^{235}UF_6$ from $^{238}UF_6$ by gaseous effusion.

23.12 Which of the following processes are spontaneous and which are nonspontaneous at a given tem-perature?

(a) $NaNO_3(s) \xrightarrow{H_2O} NaNO_3(aq)$ saturated soln

(b) $NaNO_3(s) \xrightarrow{H_2O} NaNO_3(aq)$ unsaturated soln

(c) $NaNO_3(s) \xrightarrow{H_2O} NaNO_3(aq)$ supersaturated soln

23.13 The reaction $NH_3(g) + HCl(g) \rightarrow NH_4Cl(s)$ pro-ceeds spontaneously at 25°C even though there is a decrease in disorder in the system (gases are converted to a solid). Explain.

Entropy

23.14 In each pair of substances listed here, choose the one having the larger standard entropy at 25°C. Explain the basis for your choice. (a) Li(s) or Li(l), (b) $C_2H_5OH(l)$ or $CH_3OCH_3(l)$, (c) He(g) or Ne(g), (d) CO(g) or $CO_2(g)$, (e) graphite or diamond, (f) $NO_2(g)$ or $N_2O_4(g)$. The same molar amount is used in the comparison.

23.15 How does the entropy change when (a) a solid is melted and (b) a liquid is vaporized? Why is the increase in entropy greater for vaporization than for fusion for the same substance?

23.16 Arrange the following substances (1 mole each) in the order of increasing entropy at 25°C: (a) Ne(g), (b) $SO_2(g)$, (c) Na(s), (d) NaCl(s), (e) $NH_3(g)$. Give the reasons for your arrangement.

23.17 Predict (without consulting Appendix 1), and give reasons for your predictions, whether the entropy change is positive or negative for each of the fol-lowing reactions:

(a) $2KClO_4(s) \longrightarrow 2KClO_3(s) + O_2(g)$
(b) $H_2O(g) \longrightarrow H_2O(l)$
(c) $S(s) + O_2(g) \longrightarrow SO_2(g)$
(d) $2Na(s) + 2H_2O(l) \longrightarrow 2NaOH(aq) + H_2(g)$
(e) $CH_4(g) + 2O_2(g) \longrightarrow CO_2(g) + 2H_2O(l)$
(f) $N_2(g) \longrightarrow 2N(g)$
(g) $2LiOH(aq) + CO_2(g) \longrightarrow$
$$Li_2CO_3(aq) + H_2O(l)$$
(h) $NH_4Cl(s) \longrightarrow NH_3(g) + HCl(g)$

23.18 Discuss qualitatively the sign of the entropy change expected for each of the following pro-cesses:

(a) $PCl_3(l) + Cl_2(g) \longrightarrow PCl_5(s)$
(b) $2HgO(s) \longrightarrow 2Hg(l) + O_2(g)$
(c) $H_2(g) \longrightarrow 2H(g)$
(d) $H_2(g) + Cl_2(g) \longrightarrow 2HCl(g)$
(e) $U(s) + 3F_2(g) \longrightarrow UF_6(s)$

23.19 Predict whether the entropy change is positive or negative for each of the following reactions:

(a) $Zn(s) + 2HCl(aq) \longrightarrow ZnCl_2(aq) + H_2(g)$
(b) $O(g) + O(g) \longrightarrow O_2(g)$
(c) $NH_4NO_3(s) \longrightarrow N_2O(g) + 2H_2O(g)$
(d) $Ba^{2+}(aq) + SO_4^{2-}(aq) \longrightarrow BaSO_4(s)$
(e) $2H_2O_2(l) \longrightarrow 2H_2O(l) + O_2(g)$

23.20 The molar heat of vaporization of ethanol is 39.3 kJ/mol and the boiling point of ethanol is 78.3°C. Calculate ΔS for the vaporization of 0.50 mol ethanol.

Free Energy

23.21 Why is it more convenient to predict the direction of a reaction in terms of ΔG_{sys} rather than ΔS_{univ}?

23.22 Calculate $\Delta G°$ for the following reactions at 25°C:

(a) $N_2(g) + O_2(g) \longrightarrow 2NO(g)$
(b) $H_2O(l) \longrightarrow H_2O(g)$
(c) $2C_2H_2(g) + 5O_2(g) \longrightarrow 4CO_2(g) + 2H_2O(l)$
(d) $2Mg(s) + O_2(g) \longrightarrow 2MgO(s)$
(e) $2SO_2(g) + O_2(g) \longrightarrow 2SO_3(g)$

(*Hint:* Look up the standard free energies of formation of the reactants and products in Appendix 1.)

23.23 From the following ΔH and ΔS values, predict whether each of the reactions would be spontaneous at 25°C. If not, at what temperature might the reaction become spontaneous? Reaction A: $\Delta H = 10.5$ kJ, $\Delta S = 30$ J/K; Reaction B: $\Delta H = 1.8$ kJ, $\Delta S = -113$ J/K; Reaction C: $\Delta H = -126$ kJ, $\Delta S = 84$ J/K; Reaction D: $\Delta H = -11.7$ kJ, $\Delta S = -105$ J/K.

*23.24 Calculate $\Delta G°$ for the following reaction at 25°C:

$$2C(graphite) + H_2(g) \longrightarrow C_2H_2(g)$$

Is this reaction spontaneous under standard-state conditions? If not, suggest a way to bring about the synthesis of acetylene (C_2H_2) from graphite and hydrogen.

*23.25 Use the following data to determine the normal boiling point, in K, of mercury. What assumptions must you make in order to do the calculation?

$$Hg(l) \qquad \Delta H_f° = 0 \text{ (by definition)}$$
$$S° = 77.4 \text{ J/K·mol}$$

$$Hg(g) \qquad \Delta H_f° = 60.78 \text{ kJ/mol}$$
$$S° = 174.7 \text{ J/K·mol}$$

23.26 A certain reaction is known to have a $\Delta G°$ value of -122 kJ. Will the reaction necessarily occur if the reagents are mixed together?

23.27 Calculate $\Delta G°$ for the combustion of ethane (C_2H_6):

$$2C_2H_6(g) + 7O_2(g) \longrightarrow 4CO_2(g) + 6H_2O(l)$$

(See Appendix 1 for thermodynamic data.)

Free Energy and Chemical Equilibrium

23.28 Consider the reaction

$$2NO_2(g) \rightleftharpoons N_2O_4(g)$$

Using values listed in Appendix 1, calculate (a) $\Delta G°$ for the reaction, (b) K_P for the reaction, and (c) ΔG for the reaction at 25°C, if the partial pressures of NO_2 and N_2O_4 are 1.50 atm and 2.40 atm, respectively.

23.29 For the reaction

$$H_2(g) + I_2(g) \rightleftharpoons 2HI(g)$$

$\Delta G° = 2.60$ kJ. Calculate K_P for the reaction at 25°C.

23.30 For the equilibrium reaction

$$2H_2O(g) \rightleftharpoons 2H_2(g) + O_2(g)$$

at 25°C, calculate $\Delta G°$ and K_P for the reaction.

23.31 Calculate $\Delta G°$ and K_P for the following equilibrium reaction at 25°C. The $\Delta G_f°$ values are, for $Cl_2(g)$, 0; for $PCl_3(g)$, -286 kJ/mol; for $PCl_5(g)$, -325 kJ/mol.

$$PCl_5(g) \rightleftharpoons PCl_3(g) + Cl_2(g)$$

23.32 Referring to the preceding problem calculate ΔG for the reaction if the partial pressures of the initial mixture are $P_{PCl_5} = 0.0029$ atm, $P_{PCl_3} = 0.27$ atm, and $P_{Cl_2} = 0.40$ atm.

23.33 Calculate $\Delta G°$ and K_P for the following processes at 25°C:

(a) $H_2(g) + Br_2(l) \rightleftharpoons 2HBr(g)$
(b) $\frac{1}{2}H_2(g) + \frac{1}{2}Br_2(l) \rightleftharpoons HBr(g)$

Account for the differences in $\Delta G°$ and K_P obtained for (a) and (b).

23.34 Consider the decomposition of calcium carbonate:

$$CaCO_3(s) \rightleftharpoons CaO(s) + CO_2(g)$$

Calculate the pressure in atm of CO_2 in an equilibrium process at (a) 25°C and (b) 800°C. Given that $\Delta H° = 177.8$ kJ and $\Delta S° = 160.5$ J/K.

23.35 The equilibrium constant (K_P) for the reaction

$$H_2(g) + CO_2(g) \rightleftharpoons H_2O(g) + CO(g)$$

is 4.40 at 2000 K. (a) Calculate $\Delta G°$ for the reaction. (b) Calculate ΔG for the reaction if the partial pressures are $P_{H_2} = 0.25$ atm, $P_{CO_2} = 0.78$ atm, $P_{H_2O} = 0.66$ atm, and $P_{CO} = 1.20$ atm.

23.36 At 25°C, $\Delta G°$ for the process

$$H_2O(l) \rightleftharpoons H_2O(g)$$

is 8.6 kJ. Calculate the "equilibrium constant" for the process.

23.37 Referring to Figure 23.7, draw a diagram for reactions for which $\Delta G° = 0$.

23.38 Consider the following reaction at 25°C:

$$Fe(OH)_2(s) \rightleftharpoons Fe^{2+}(aq) + 2OH^-(aq)$$

Calculate $\Delta G°$ for the reaction. [K_{sp} for $Fe(OH)_2$ is 1.6×10^{-14}.]

23.39 Calculate K_P for the following reaction at 25°C:

$$N_2(g) + O_2(g) \rightleftharpoons 2NO(g)$$

From your result, what can you conclude about the stability of a mixture of O_2 and N_2 gases in the atmosphere?

23.40 The equilibrium constant (K_P) for the reaction

$$CO(g) + Cl_2(g) \rightleftharpoons COCl_2(g)$$

is 5.62×10^{35} at 25°C. Calculate $\Delta G_f°$ for $COCl_2$ at 25°C.

23.41 Calculate $\Delta G°$ for the process

$$C(diamond) \longrightarrow C(graphite)$$

Is the reaction spontaneous at 25°C? If so, why is it that diamonds do not become graphite on standing?

*23.42 Calculate the pressure of O_2 (in atm) over a sample of NiO at 25°C if $\Delta G° = 212$ kJ for the reaction

$$NiO(s) \rightleftharpoons Ni(s) + \tfrac{1}{2}O_2(g)$$

Miscellaneous Problems

23.43 Explain what is meant by a state function.

23.44 What is an equation of state?

23.45 Which of the following thermodynamic functions are associated only with the first law of thermodynamics: S, E, G, and H?

*23.46 In discussion of the second law of thermodynamics, the phrase "arrow of time" is sometimes used in connection with entropy. Discuss the meaning of this phrase.

23.47 A student placed 1 g of each of three compounds A, B, and C in a container and found that after one week no change had occurred. Offer some possible explanations for the fact that no reactions took place. Assume that A, B, and C are totally miscible liquids.

23.48 Predict the signs of ΔH, ΔS, and ΔG of the system for the following processes at 1 atm: (a) ammonia melts at $-60°C$, (b) ammonia melts at $-77.7°C$, (c) ammonia melts at $-100°C$. (The normal melting point of ammonia is $-77.7°C$.)

*23.49 Give a detailed example of each of the following, with an explanation: (a) a thermodynamically spontaneous process; (b) a process that would violate the first law of thermodynamics; (c) a process that would violate the second law of thermodynamics; (d) an irreversible process; (e) an equilibrium process.

*23.50 Gaseous diffusion is a spontaneous process. In most cases, the enthalpy change is quite small. Use Equation (23.9) to predict the entropy change of the system (that is, whether $\Delta S > 0$ or $\Delta S < 0$).

*23.51 Ammonium nitrate dissolves spontaneously and endothermically in water. What can you deduce about the sign of ΔS for the solution process?

*23.52 When a stretched sheet of rubber (e.g., from a balloon) is allowed to contract rapidly, it cools down (try it!). Use Equation (23.9) to deduce the sign of ΔS when a sheet of rubber is stretched. Are the molecules in rubber more ordered when the rubber is stretched?

24
ELECTRO–CHEMICAL REACTIONS AND ELECTROLYTIC REACTIONS

The corrosion of iron, the operation of a flashlight battery, and the electrolysis of water to form hydrogen and oxygen gases all share a key characteristic: Each represents a chemical change involving electron transfer or oxidation-reduction. Commonly encountered oxidation-reduction reactions include both spontaneous and nonspontaneous changes. (The corrosion of iron is, unfortunately, highly spontaneous, whereas the electrolysis of water is nonspontaneous, for example.) The notion of free energy introduced in Chapter 23 will help clarify our study of oxidation-reduction reactions.

The observable, macroscopic changes associated with various oxidation-reduction reactions often have little in common (compare, for example, explosives with a battery). At the molecular level, only two fundamental processes are at work—the loss of one or more electrons (oxidation) by a species and the gain of one or more electrons (reduction) by another species.

24.1 GALVANIC CELLS

In Chapter 16 we discussed redox reactions in terms of electron transfer from a reducing agent to an oxidizing agent. This kind of transfer of electrons can be used to generate electricity if the reactants (that is, the reducing and oxidizing agents) are kept apart during the course of the reaction. An example illustrating such a conversion of chemical energy to electrical energy is the reaction between zinc metal and copper(II) sulfate.

When a piece of zinc metal is put into an aqueous solution of copper(II) sulfate ($CuSO_4$), two things happen. Zinc atoms enter the solution as Zn^{2+} ions and, more noticeably, Cu^{2+} ions leave the solution and deposit on the Zn metal as metallic copper. As a result of the conversion of Cu^{2+} ions to copper metal, the blue color of the solution (due to the hydrated Cu^{2+} ions) gradually fades (see Color Plate 14). When most or all of the Cu^{2+} ions have been reduced to Cu metal, the solution becomes colorless. The chemical changes that occur can be represented by

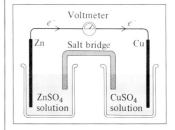

FIGURE 24.1 *Diagram of a galvanic cell. Zn and Cu electrodes are immersed in $ZnSO_4$ and $CuSO_4$ solutions, respectively. The salt bridge (an inverted U tube) provides an electrically conducting medium between the two solutions. It contains an inert electrolyte solution (such as KCl solution). The openings of the U tube are loosely plugged with cotton balls (not shown) to prevent the KCl solution from running into the containers, while allowing the anions and cations to move across. Electrons flow externally from the Zn electrode (anode) to the Cu electrode (cathode).*

$$\text{oxidation:} \qquad Zn(s) \longrightarrow Zn^{2+}(aq) + 2e^-$$
$$\text{reduction:} \qquad \underline{Cu^{2+}(aq) + 2e^- \longrightarrow Cu(s)}$$
$$\text{overall:} \qquad Zn(s) + Cu^{2+}(aq) \longrightarrow Zn^{2+}(aq) + Cu(s)$$

Reversing the role of the metals—that is, putting a piece of copper into a zinc sulfate ($ZnSO_4$) solution—does not produce a change.

We can draw at least two conclusions from this simple experiment. First, zinc has a greater tendency to be oxidized than does copper. Second, in this redox reaction there is a net transfer of two electrons from Zn to Cu^{2+}.

We can apply this knowledge to generate electricity by means of an *electrochemical cell,* designed so that electrons will flow along an external circuit. Figure 24.1 shows the essential components of a *galvanic cell* (Luigi Galvani, 1737–1798), also called a *voltaic cell,* in which the production of electricity by a spontaneous redox reaction is accomplished. A bar of zinc metal is dipped into a $ZnSO_4$ solution, and a bar of copper metal is dipped into a $CuSO_4$ solution. The galvanic cell operates on the principle that the oxidation of Zn to Zn^{2+} and the reduction of Cu^{2+} to Cu can be made to take place in separate locations (and simultaneously) with the transfer of electrons through the external wire. The zinc bar and the copper bar are called *electrodes.* By definition, the *electrode at which oxidation occurs* is called the **anode;** the *electrode at which reduction occurs* is called the **cathode.**

For this system, the *oxidation and reduction reactions at the electrodes,* which are called **half-cell reactions,** are

$$\text{Zn electrode (anode):} \qquad Zn(s) \longrightarrow Zn^{2+}(aq) + 2e^-$$
$$\text{Cu electrode (cathode):} \qquad Cu^{2+}(aq) + 2e^- \longrightarrow Cu(s)$$

Note that unless the two solutions are separated from each other, the Cu^{2+} ions will react directly with the zinc bar

$$Cu^{2+}(aq) + Zn(s) \longrightarrow Cu(s) + Zn^{2+}(aq)$$

and there will be no flow of electrons.

Half-cell reactions are similar to the half-reactions used in the ion-electron method of balancing redox reactions (see Section 16.5).

To complete the electric circuit, the solutions must be connected by a conducting medium through which the cations and anions can move. This requirement is satisfied by a *salt bridge,* which, in its simplest form, is an inverted U tube containing an inert electrolyte such as KCl or NH_4NO_3 whose ions will not react with other ions in solution or with the electrodes. During the course of the overall redox reaction, electrons flow externally from the anode (Zn electrode) to the cathode (Cu electrode). In the solution, the cations (Zn^{2+}, Cu^{2+}, and K^+) move toward the cathode, while the anions (SO_4^{2-} and Cl^-) move in the opposite direction, toward the anode.

The fact that electrons flow from one electrode to the other indicates that there is a voltage difference between the two electrodes. This *voltage difference between the electrodes,* called the **electromotive force,** or **emf** ($\mathscr{E}$), can be measured by connecting a voltmeter between the two electrodes as shown in Figure 24.1. The emf of a galvanic cell is usually measured in volts; it is also referred to as *cell voltage* or *cell potential.* We will see later that the emf of a cell depends not only on the nature of the electrodes and the ions, but also on the concentrations of the ions and the temperature at which the cell is operated.

The conventional notation for representing galvanic cells is the *cell diagram.* For the galvanic cell just described, using KCl as the electrolyte in the salt bridge and assuming 1 *M* concentrations, the cell diagram is

$$Zn(s)\big|Zn^{2+}(aq, 1\ M)\big|KCl(sat'd)\big|Cu^{2+}(aq, 1\ M)\big|Cu(s)$$

where the vertical lines represent phase boundaries. For example, the zinc electrode is a solid and the Zn^{2+} ions (from $ZnSO_4$) are in solution. Thus we draw a line between Zn and Zn^{2+} to show the phase boundaries. Note that there is also a line between the $ZnSO_4$ solution and the KCl solution in the salt bridge because these two solutions are not mixed and therefore constitute two separate phases. The other lines can be similarly explained. By convention, the anode is written first at the left and the other components appear in the order in which we would encounter them in moving from the anode to the cathode.

24.2 STANDARD REDUCTION POTENTIALS

When the concentrations of the Cu^{2+} and Zn^{2+} ions are both 1.0 *M*, we find that the emf of the cell shown in Figure 24.1 is 1.10 V at 25°C. This voltage must be related directly to the redox reactions, but how? Just as the overall cell reaction can be thought of as the sum of two half-cell reactions, the measured emf of the cell can be treated as the difference of the electrical potentials at the Zn and Cu electrodes. Knowing one of these electrode potentials, we could obtain the other by subtraction (from 1.10 V). It is impossible to measure the potential of just a single electrode, but if we arbitrarily set the potential value of a particular electrode at zero we can use it to determine the relative potentials of other electrodes. The hydrogen electrode, shown in Figure 24.2, is appropriate for this purpose. Hydrogen gas is bubbled into a hydrochloric acid solution at 25°C. The platinum electrode serves two purposes. First, it provides a surface on which

The emf of a galvanic cell is independent of the volume of the solutions and the sizes of the electrodes.

This method is analogous to choosing the surface of the ocean as our reference for altitude, arbitrarily assigning it to be zero meters, and then referring to any terrestrial altitude as being a certain number of meters above or below sea level.

the dissociation of hydrogen molecules can take place:

$$H_2 \longrightarrow 2H^+ + 2e^-$$

Second, it serves as an electrical conductor to the external circuit. Under standard-state conditions (that is, when the pressure of H_2 is 1 atm and the concentration of the HCl solution is 1 M) and at 25°C, the potential for the following reduction is defined to be *exactly* zero:

For definitions of standard states, see Table 23.3 on p. 698.

$$2H^+(aq, 1\ M) + 2e^- \longrightarrow H_2(g, 1\ atm) \qquad \mathscr{E}° = 0\ V$$

The superscript ° denotes standard-state conditions. *For a reduction reaction at an electrode when all solutes are 1 M and all gases are at 1 atm, the voltage* is called the **standard reduction potential.** Thus, the standard reduction potential of the hydrogen electrode is zero. The electrode itself is known as the *standard hydrogen electrode (SHE).*

We can use the standard hydrogen electrode to measure the potentials of other kinds of electrodes. Figure 24.3(a) shows a galvanic cell with an SHE and a zinc electrode. The cell diagram is

$$Zn(s)|Zn^{2+}(aq, 1\ M)\|KCl(sat'd)\|H^+(aq, 1\ M)|H_2(g, 1\ atm)|Pt(s)$$

When all the reactants are in their standard states (that is, H_2 at 1 atm, H^+ and Zn^{2+} ions at 1 M), the emf of the cell is 0.76 V. From the direction of electron flow we can write the half-cell reactions as follows:

anode (oxidation): $\qquad\qquad Zn(s) \longrightarrow Zn^{2+}(aq, 1\ M) + 2e^-$
cathode (reduction): $\quad 2H^+(aq, 1\ M) + 2e^- \longrightarrow H_2(g, 1\ atm)$

overall: $\quad Zn(s) + 2H^+(aq, 1\ M) \longrightarrow Zn^{2+}(aq, 1\ M) + H_2(g, 1\ atm)$

Note that even though H_2 is formed by the reduction of H^+ ions in solution, hydrogen gas still has to be bubbled in at the platinum electrode to satisfy the standard-state condition of 1 atm.

The **standard emf** of the cell ($\mathscr{E}°_{cell}$) is defined as

$$\mathscr{E}°_{cell} = \mathscr{E}°_{substance\ reduced} - \mathscr{E}°_{substance\ oxidized}$$

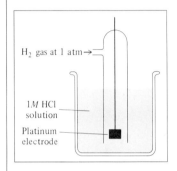

FIGURE 24.2 *A hydrogen electrode operating under standard-state conditions. Hydrogen gas at 1 atm is bubbled through a 1 M HCl solution. The platinum electrode is part of the hydrogen electrode.*

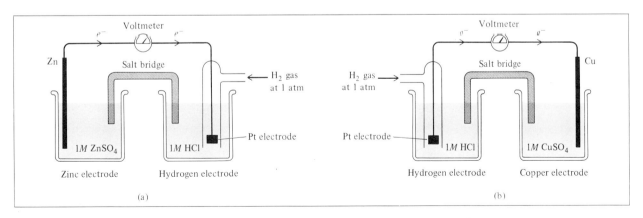

FIGURE 24.3 *(a) A cell consisting of a zinc electrode and a hydrogen electrode. (b) A cell consisting of a copper electrode and a hydrogen electrode. Both cells are operating under standard state conditions. Note that in (a) the SHE acts as the cathode, but in (b) the SHE acts as the anode.*

that is, the *difference between the standard reduction potential of the substance that undergoes reduction and the standard reduction potential of the substance that undergoes oxidation.* We can also express the standard emf of the cell as follows:

$$\mathscr{E}^\circ_{cell} = \mathscr{E}^\circ_{cathode} - \mathscr{E}^\circ_{anode} \tag{24.1}$$

Because $\mathscr{E}^\circ_{SHE} = 0$ by definition, we can write

$$0.76 \text{ V} = \mathscr{E}^\circ_{SHE} - \mathscr{E}^\circ_{Zn}$$
$$= 0 - \mathscr{E}^\circ_{Zn}$$
$$\mathscr{E}^\circ_{Zn} = -0.76 \text{ V}$$

Thus the standard reduction potential of Zn is -0.76 V; that is

$$Zn^{2+}(aq, 1 \text{ } M) + 2e^- \longrightarrow Zn(s) \qquad \mathscr{E}^\circ_{Zn} = -0.76 \text{ V}$$

Similarly, the standard reduction potential of Cu can be determined by using the arrangement shown in Figure 24.3(b). The cell diagram is

$$Pt(s)|H_2(g, 1 \text{ atm})|H^+(aq, 1 \text{ } M)\|KCl(sat'd)\|Cu^{2+}(aq, 1 \text{ } M)|Cu(s)$$

Because the electrons in this cell flow from the SHE to the Cu electrode, the half-cell reactions are

anode (oxidation): $\qquad\qquad\qquad H_2(g, 1 \text{ atm}) \longrightarrow 2H^+(aq, 1 \text{ } M) + 2e^-$
cathode (reduction): $\qquad\qquad Cu^{2+}(aq, 1 \text{ } M) + 2e^- \longrightarrow Cu(s)$

overall: $\qquad H_2(g, 1 \text{ atm}) + Cu^{2+}(aq, 1 \text{ } M) \longrightarrow 2H^+(aq, 1 \text{ } M) + Cu(s)$

In this case the potentials are

$$\mathscr{E}^\circ_{cell} = \mathscr{E}^\circ_{Cu} - \mathscr{E}^\circ_{SHE}$$
$$0.34 \text{ V} = \mathscr{E}^\circ_{Cu} - 0$$
$$\mathscr{E}^\circ_{Cu} = 0.34 \text{ V}$$

Thus the standard reduction potential for Cu is 0.34 V; that is

$$Cu^{2+}(aq, 1 \text{ } M) + 2e^- \longrightarrow Cu(s) \qquad \mathscr{E}^\circ_{Cu} = 0.34 \text{ V}$$

Referring to the cell shown in Figure 24.1, we can now write

anode (oxidation): $\qquad\qquad\qquad Zn(s) \longrightarrow Zn^{2+}(aq, 1 \text{ } M) + 2e^-$
cathode (reduction): $\qquad Cu^{2+}(aq, 1 \text{ } M) + 2e^- \longrightarrow Cu(s)$

overall: $\qquad Zn(s) + Cu^{2+}(aq, 1 \text{ } M) \longrightarrow Zn^{2+}(aq, 1 \text{ } M) + Cu(s)$

and the emf of the cell is

$$\mathscr{E}^\circ_{cell} = \mathscr{E}^\circ_{Cu} - \mathscr{E}^\circ_{Zn}$$
$$= 0.34 \text{ V} - (-0.76 \text{ V})$$
$$= 1.10 \text{ V}$$

This example illustrates how we can use the *sign* of the emf of the cell to predict the spontaneity of a redox reaction. Under standard-state conditions for reactants and products, the redox reaction is spontaneous in the forward direction if the standard emf of the cell is positive. If it is negative, the reaction is spontaneous in the opposite direction. It is important to keep in mind that a negative $\mathscr{E}^\circ_{cell}$ does *not* mean that a reaction will not occur if the reactants are mixed at 1 M concentrations. It merely means that the equilibrium, when reached, will lie to the left. We will examine the relationship between $\mathscr{E}°_{cell}$ and $\Delta G°$ later.

Table 24.1 lists the standard reduction potentials of a number of electrodes. Note that by definition the SHE has an $\mathscr{E}°$ value of 0.00 V. The negative standard reduction potentials increase above it and the positive standard reduction potentials increase below it. It is important to understand the following points about the table.

The activity series on p. 84 is based on data given in Table 24.1.

- The $\mathscr{E}°$ values apply to the half-cell reactions as read in the forward (left to right) direction.
- The more positive $\mathscr{E}°$, the greater the tendency for the substance to be reduced. For example, the half-cell reaction

$$F_2(g, 1 \text{ atm}) + 2e^- \longrightarrow 2F^-(aq, 1 M) \qquad \mathscr{E}° = 2.87 \text{ V}$$

has the highest $\mathscr{E}°$ value among all the half-cell reactions. Thus F_2 is the *strongest* oxidizing agent because it has the greatest tendency to be reduced. At the other extreme is the reaction

$$Li^+(aq, 1 M) + e^- \longrightarrow Li(s) \qquad \mathscr{E}° = -3.05 \text{ V}$$

which has the most negative $\mathscr{E}°$ value. Thus the Li^+ ion is the *weakest* oxidizing agent because it is the most difficult species to reduce. Conversely, we say that the F^- ion is the weakest reducing agent and that the Li metal is the strongest reducing agent. Under standard-state conditions, the oxidizing agents (the species on the left-hand side of the half-reactions in Table 24.1) increase in strength from top to bottom and the reducing agents (the species on the right-hand side of the half-reactions) increase in strength from bottom to top.
- The half-cell reactions are reversible. Depending on the conditions, any electrode can act either as an anode or as a cathode. Earlier we saw that the SHE is the cathode (H^+ is reduced to H_2) when placed in a cell with zinc and that it becomes the anode (H_2 is oxidized to H^+) when placed in a cell with copper.
- Under standard-state conditions, any species on the left of a given half-cell reaction will react spontaneously with a species that appears on the right of any half-cell reaction located *above it* in Table 24.1. This is sometimes called the *diagonal rule*. In the case of the Cu/Zn galvanic cell

$$Zn^{2+}(aq, 1 M) + 2e^- \longrightarrow Zn(s) \qquad \mathscr{E}° = -0.76 \text{ V}$$

$$Cu^{2+}(aq, 1 M) + 2e^- \longrightarrow Cu(s) \qquad \mathscr{E}° = +0.34 \text{ V}$$

We see that the substance on the left of the second half-cell reaction is Cu^{2+} and the substance on the right in the first half-cell reaction is Zn. Therefore, as we saw earlier, Zn will reduce Cu^{2+} spontaneously to form Zn^{2+} and Cu.
- $\mathscr{E}°$ changes sign whenever we reverse a half-cell reaction. Thus, if

$$Mn^{2+}(aq, 1 M) + 2e^- \longrightarrow Mn(s) \qquad \mathscr{E}° = -1.18 \text{ V}$$

then

$$Mn(s) \longrightarrow Mn^{2+}(aq, 1 M) + 2e^- \qquad \mathscr{E}° = 1.18 \text{ V}$$

TABLE 24.1 Standard Reduction Potentials at 25°C*

Half-reaction	$\mathscr{E}°(V)$
$Li^+(aq) + e^- \longrightarrow Li(s)$	−3.05
$K^+(aq) + e^- \longrightarrow K(s)$	−2.93
$Ba^{2+}(aq) + 2e^- \longrightarrow Ba(s)$	−2.90
$Sr^{2+}(aq) + 2e^- \longrightarrow Sr(s)$	−2.89
$Ca^{2+}(aq) + 2e^- \longrightarrow Ca(s)$	−2.87
$Na^+(aq) + e^- \longrightarrow Na(s)$	−2.71
$Mg^{2+}(aq) + 2e^- \longrightarrow Mg(s)$	−2.37
$Be^{2+}(aq) + 2e^- \longrightarrow Be(s)$	−1.85
$Al^{3+}(aq) + 3e^- \longrightarrow Al(s)$	−1.66
$Mn^{2+}(aq) + 2e^- \longrightarrow Mn(s)$	−1.18
$2H_2O + 2e^- \longrightarrow H_2(g) + 2OH^-(aq)$	−0.83
$Zn^{2+}(aq) + 2e^- \longrightarrow Zn(s)$	−0.76
$Cr^{3+}(aq) + 3e^- \longrightarrow Cr(s)$	−0.74
$Fe^{2+}(aq) + 2e^- \longrightarrow Fe(s)$	−0.44
$Cd^{2+}(aq) + 2e^- \longrightarrow Cd(s)$	−0.40
$PbSO_4(s) + 2e^- \longrightarrow Pb(s) + SO_4^{2-}(aq)$	−0.31
$Co^{2+}(aq) + 2e^- \longrightarrow Co(s)$	−0.28
$Ni^{2+}(aq) + 2e^- \longrightarrow Ni(s)$	−0.25
$Sn^{2+}(aq) + 2e^- \longrightarrow Sn(s)$	−0.14
$Pb^{2+}(aq) + 2e^- \longrightarrow Pb(s)$	−0.13
$2H^+(aq) + 2e^- \longrightarrow H_2(g)$	0.00
$Sn^{4+}(aq) + 2e^- \longrightarrow Sn^{2+}(aq)$	+0.13
$Cu^{2+}(aq) + e^- \longrightarrow Cu^+(aq)$	+0.15
$SO_4^{2-}(aq) + 4H^+(aq) + 2e^- \longrightarrow SO_2(g) + 2H_2O$	+0.20
$AgCl(s) + e^- \longrightarrow Ag(s) + Cl^-(aq)$	+0.22
$Cu^{2+}(aq) + 2e^- \longrightarrow Cu(s)$	+0.34
$O_2(g) + 2H_2O + 4e^- \longrightarrow 4OH^-(aq)$	+0.40
$I_2(s) + 2e^- \longrightarrow 2I^-(aq)$	+0.53
$MnO_4^-(aq) + 2H_2O + 3e^- \longrightarrow MnO_2(s) + 4OH^-(aq)$	+0.59
$O_2(g) + 2H^+(aq) + 2e^- \longrightarrow H_2O_2(aq)$	+0.68
$Fe^{3+}(aq) + e^- \longrightarrow Fe^{2+}(aq)$	+0.77
$Ag^+(aq) + e^- \longrightarrow Ag(s)$	+0.80
$Hg_2^{2+}(aq) + 2e^- \longrightarrow 2Hg(l)$	+0.85
$2Hg^{2+}(aq) + 2e^- \longrightarrow Hg_2^{2+}(aq)$	+0.92
$NO_3^-(aq) + 4H^+(aq) + 3e^- \longrightarrow NO(g) + 2H_2O$	+0.96
$Br_2(l) + 2e^- \longrightarrow 2Br^-(aq)$	+1.07
$O_2(g) + 4H^+(aq) + 4e^- \longrightarrow 2H_2O$	+1.23
$MnO_2(s) + 4H^+(aq) + 2e^- \longrightarrow Mn^{2+}(aq) + 2H_2O$	+1.23
$Cr_2O_7^{2-}(aq) + 14H^+(aq) + 6e^- \longrightarrow 2Cr^{3+}(aq) + 7H_2O$	+1.33
$Cl_2(g) + 2e^- \longrightarrow 2Cl^-(aq)$	+1.36
$Au^{3+}(aq) + 3e^- \longrightarrow Au(s)$	+1.50
$MnO_4^-(aq) + 8H^+(aq) + 5e^- \longrightarrow Mn^{2+}(aq) + 4H_2O$	+1.51
$Ce^{4+}(aq) + e^- \longrightarrow Ce^{3+}(aq)$	+1.61
$PbO_2(s) + 4H^+(aq) + SO_4^{2-}(aq) + 2e^- \longrightarrow PbSO_4(s) + 2H_2O$	+1.70
$H_2O_2(aq) + 2H^+(aq) + 2e^- \longrightarrow 2H_2O$	+1.77
$Co^{3+}(aq) + e^- \longrightarrow Co^{2+}(aq)$	+1.82
$O_3(g) + 2H^+(aq) + 2e^- \longrightarrow O_2(g) + H_2O(l)$	+2.07
$F_2(g) + 2e^- \longrightarrow 2F^-(aq)$	+2.87

Increasing strength as oxidizing agent (↓)

Increasing strength as reducing agent (↑)

* For all half-reactions the concentration is 1 *M* for dissolved species and the pressure is 1 atm for gases.

- Changing the stoichiometric coefficients of a half-cell reaction *does not* affect the value of $\mathscr{E}°$ because electrode potentials are intensive properties. For example, from Table 24.1

$$I_2(s) + 2e^- \longrightarrow 2I^-(aq, 1\ M) \qquad \mathscr{E}° = 0.53\ \text{V}$$

but $\mathscr{E}°$ does not change if we write the reaction

$$2I_2(s) + 4e^- \longrightarrow 4I^-(aq, 1\ M) \qquad \mathscr{E}° = 0.53\ \text{V}$$

EXAMPLE 24.1

Arrange the following species in order of increasing strength as oxidizing agents: MnO_4^- (in acidic solution), Sn^{2+}, Al^{3+}, Co^{3+}, and Ag^+. Assume all species are in their standard states.

Answer

Consulting Table 24.1, we write the half-reactions in the order in which they appear there:

$$Al^{3+}(aq, 1\ M) + 3e^- \longrightarrow Al(s) \qquad \mathscr{E}° = -1.66\ \text{V}$$

$$Sn^{2+}(aq, 1\ M) + 2e^- \longrightarrow Sn(s) \qquad \mathscr{E}° = -0.14\ \text{V}$$

$$Ag^+(aq, 1\ M) + e^- \longrightarrow Ag(s) \qquad \mathscr{E}° = 0.80\ \text{V}$$

$$MnO_4^-(aq, 1\ M) + 8H^+(aq, 1\ M) + 5e^- \longrightarrow Mn^{2+}(aq, 1\ M) + 4H_2O(l)$$
$$\mathscr{E}° = 1.51\ \text{V}$$

$$Co^{3+}(aq, 1\ M) + e^- \longrightarrow Co^{2+}(aq, 1\ M)$$
$$\mathscr{E}° = 1.82\ \text{V}$$

Remember that the more positive the standard reduction potential, the greater the tendency of the species to be reduced, or the stronger the species as an oxidizing agent. We find that the oxidizing agents increase in strength as follows:

$$Al^{3+} < Sn^{2+} < Ag^+ < MnO_4^- < Co^{3+}$$

Similar example: Problem 24.9.

EXAMPLE 24.2

Predict what will happen if molecular bromine (Br_2) is added to a solution containing NaCl and NaI at 25°C. Assume all species are in their standard states.

Answer

To predict what redox reaction(s) will take place, we need to compare the standard reduction potentials for the following half-reactions:

$$I_2(s) + 2e^- \longrightarrow 2I^-(aq, 1\ M) \qquad \mathscr{E}° = 0.53\ \text{V}$$

$$Br_2(l) + 2e^- \longrightarrow 2Br^-(aq, 1\ M) \qquad \mathscr{E}° = 1.07\ \text{V}$$

$$Cl_2(g, 1\ \text{atm}) + 2e^- \longrightarrow 2Cl^-(aq, 1\ M) \qquad \mathscr{E}° = 1.36\ \text{V}$$

(Continued)

Applying the diagonal rule we see that Br_2 will oxidize I^- but will not oxidize Cl^-. Therefore, the only redox reaction that will occur appreciably under standard-state conditions is

oxidation:	$2I^-(aq, 1\ M) \longrightarrow I_2(s) + 2e^-$	
reduction:	$Br_2(l) + 2e^- \longrightarrow 2Br^-(aq, 1\ M)$	
overall:	$2I^-(aq, 1\ M) + Br_2(l) \longrightarrow I_2(s) + 2Br^-(aq, 1\ M)$	

Note that the Na^+ ions are inert and do not enter into the redox reaction.

Similar examples: Problems 24.6, 24.10.

EXAMPLE 24.3

A galvanic cell consists of an Mg electrode in a 1.0 M $Mg(NO_3)_2$ solution and an Ag electrode in a 1.0 M $AgNO_3$ solution. Calculate the standard emf of this electrochemical cell at 25°C.

Answer

Table 24.1 gives the standard reduction potentials of the two electrodes:

$$Mg^{2+}(aq, 1\ M) + 2e^- \longrightarrow Mg(s) \qquad \mathscr{E}° = -2.37\ V$$

$$Ag^+(aq, 1\ M) + e^- \longrightarrow Ag(s) \qquad \mathscr{E}° = 0.80\ V$$

Applying the diagonal rule, we see that Ag^+ will oxidize Mg:

anode (oxidation):	$Mg(s) \longrightarrow Mg^{2+}(aq, 1\ M) + 2e^-$
cathode (reduction):	$2Ag^+(aq, 1\ M) + 2e^- \longrightarrow 2Ag(s)$
overall:	$Mg(s) + 2Ag^+(aq, 1\ M) \longrightarrow Mg^{2+}(aq, 1\ M) + 2Ag(s)$

Note that in order to balance the overall equation we multiplied the reduction of Ag^+ by 2. We can do so because $\mathscr{E}°$ is not affected by this procedure. We find the emf of the cell by using Equation (24.1):

$$\begin{aligned} \mathscr{E}°_{cell} &= \mathscr{E}°_{Ag} - \mathscr{E}°_{Mg} \\ &= 0.80\ V - (-2.37\ V) \\ &= 3.17\ V \end{aligned}$$

Similar examples: Problems 24.3, 24.4, 24.5.

24.3 SPONTANEITY OF REDOX REACTIONS

Table 24.1 enables us to predict the outcome of redox reactions under standard-state conditions, whether they take place in an electrochemical cell (where the reducing agent and oxidizing agent are physically separated from each other) or in a beaker (where the reactants are all mixed together). Our next step is to see how $\mathscr{E}°_{cell}$ is related to other thermodynamic quantities.

In Chapter 23 we saw that the free-energy change (decrease) in a spontaneous process is the energy available to do work (p. 697). In fact, for a process carried out at constant temperature and pressure

$$\Delta G = w_{max}$$

where w_{max} is the maximum amount of work that can be done.

In a galvanic cell, chemical energy is converted into electrical energy. Electrical energy in this case is the product of the emf of the cell and the total electrical charge (in coulombs) that passes through the cell:

$$\text{electrical energy} = \text{volts} \times \text{coulombs}$$
$$= \text{joules}$$

The total charge is determined by the number of moles of electrons (n) that pass through the circuit. By definition

$$\text{total charge} = nF$$

where F, the Faraday constant (Michael Faraday, 1791–1867), is the electrical charge contained in one mole of electrons. Experimentally, it has been found that one ***faraday*** is *equivalent to 96,487 coulombs*, or 96,500 coulombs when rounded off to three significant figures. Thus

$$1\ F = 96,500\ \text{C/mol}$$

Since

$$1\ J = 1\ C \times 1\ V$$

we can also express the units of faraday as

$$1\ F = 96,500\ \text{J/V·mol}$$

A common device for measuring the cell's emf is the *potentiometer,* which can precisely match the voltage of the cell without actually draining any current from the cell. Thus the emf that is measured is the *maximum* voltage that the cell can achieve. This value is used to calculate the maximum amount of electrical energy that can be obtained from the chemical reaction. This energy is used to do electrical work (w_{ele}), so

$$w_{max} = w_{ele}$$
$$= -nF\mathscr{E}_{cell}$$

The negative sign on the right-hand side indicates that the electrical work is done by the system on the surroundings. Now, since

$$\Delta G = w_{max}$$

we obtain

$$\Delta G = -nF\mathscr{E}_{cell} \qquad (24.2)$$

Both n and F are positive quantities and ΔG is negative for a spontaneous process, so $\mathscr{E}_{cell}$ is positive. For reactions in which reactants and products are in their standard states, Equation (24.2) becomes

$$\Delta G° = -nF\mathscr{E}°_{cell} \qquad (24.3)$$

n is equal to the number of electrons exchanged between the reducing agent and oxidizing agent in the overall redox equation.

The sign convention for electrical work is the same as that for *P–V* work discussed in Section 23.2.

Here again, $\mathscr{E}^\circ_{\text{cell}}$ is positive for a spontaneous process.

In Section 23.5 we saw that the standard free-energy change ΔG° for a reaction is related to its equilibrium constant as follows:

$$\Delta G^\circ = -2.303RT \log K \qquad (23.12)$$

Therefore, from Equations (24.3) and (23.12) we obtain

$$-nF\mathscr{E}^\circ_{\text{cell}} = -2.303RT \log K$$

Solving for $\mathscr{E}^\circ_{\text{cell}}$

$$\mathscr{E}^\circ_{\text{cell}} = \frac{2.303RT}{nF} \log K \qquad (24.4)$$

When $T = 298$ K, Equation (24.4) can be simplified by substituting for R and F:

$$\mathscr{E}^\circ_{\text{cell}} = \frac{2.303(8.314 \text{ J/K·mol})(298 \text{ K})}{n(96,500 \text{ J/V·mol})} \log K$$

$$= \frac{0.0591 \text{ V}}{n} \log K \qquad (24.5)$$

Thus, if any one of the three quantities ΔG°, K, or $\mathscr{E}^\circ_{\text{cell}}$ is known, each of the other two can be calculated using Equation (23.12), Equation (24.3), or Equation (24.4). We can summarize the relationships among ΔG°, K, and $\mathscr{E}^\circ_{\text{cell}}$ and characterize the spontaneity of a redox reaction as follows:

ΔG°	K	$\mathscr{E}^\circ_{\text{cell}}$	Reaction under Standard-State Conditions
Negative	> 1	Positive	Spontaneous
0	$= 1$	0	At equilibrium
Positive	< 1	Negative	Nonspontaneous. Reaction is spontaneous in the reverse direction.

EXAMPLE 24.4

Using data in Table 24.1, calculate $\mathscr{E}^\circ$ for the reactions of mercury with (a) 1 M HCl and (b) 1 M HNO$_3$. Which acid will oxidize Hg to Hg$_2^{2+}$ under standard-state conditions?

Answer

(a) HCl: First we write the half-reactions:

oxidation: $\qquad\qquad\qquad 2\text{Hg}(l) \longrightarrow \text{Hg}_2^{2+}(aq, M) + 2e^-$

reduction: $\qquad 2\text{H}^+(aq, 1\ M) + 2e^- \longrightarrow \text{H}_2(g, 1 \text{ atm})$

overall: $\qquad 2\text{Hg}(l) + 2\text{H}^+(aq, 1\ M) \longrightarrow \text{Hg}_2^{2+}(aq, 1\ M) + \text{H}_2(g, 1 \text{ atm})$

The standard emf, $\mathscr{E}°$, is given by (we omit the subscript "cell" because this reaction is not carried out in an electrochemical cell):

$$\mathscr{E}° = \mathscr{E}°_{SHE} - \mathscr{E}°_{Hg}$$
$$= 0.00 - (0.85\ V)$$
$$= -0.85\ V$$

Since $\mathscr{E}°$ is negative, we conclude that mercury is not oxidized by hydrochloric acid under standard-state conditions.

(b) HNO_3: The reactions are

oxidation:
$$3[2Hg(l) \longrightarrow Hg_2^{2+}(aq,\ 1\ M) + 2e^-]$$

reduction:
$$2[NO_3^-(aq,\ 1\ M) + 4H^+(aq,\ 1\ M) + 3e^- \longrightarrow NO(g,\ 1\ atm) + 2H_2O(l)]$$

overall:
$$6Hg(l) + 2NO_3^-(aq,\ 1\ M) + 8H^+(aq,\ 1\ M) \longrightarrow 3Hg_2^{2+}(aq,\ 1\ M) + 2NO(g,\ 1\ atm) + 4H_2O(l)$$

Thus

$$\mathscr{E}° = \mathscr{E}°_{NO_3^-} - \mathscr{E}°_{Hg}$$
$$= 0.96\ V - (0.85\ V)$$
$$= 0.11\ V$$

Since $\mathscr{E}°$ is positive, the reaction is spontaneous under standard-state conditions.

EXAMPLE 24.5

Calculate the equilibrium constant for the following reaction at 25°C:

$$Sn(s) + 2Cu^{2+}(aq) \rightleftharpoons Sn^{2+}(aq) + 2Cu^+(aq)$$

Answer

The two half-reactions for the overall process are

oxidation:
$$Sn(s) \longrightarrow Sn^{2+}(aq) + 2e^-$$
reduction:
$$2Cu^{2+}(aq) + 2e^- \longrightarrow 2Cu^+(aq)$$

In Table 24.1 we find that $\mathscr{E}°_{Sn} = -0.14\ V$ and $\mathscr{E}°_{Cu^{2+}/Cu^+} = 0.15\ V$. Thus

$$\mathscr{E}° = \mathscr{E}°_{Cu^{2+}/Cu^+} - \mathscr{E}°_{Sn}$$
$$= 0.15\ V - (-0.14\ V)$$
$$= 0.29\ V$$

Equation (24.5) can be written

$$\log K = \frac{n\mathscr{E}°}{0.0591\ V}$$

In the half-reactions we find $n = 2$. Therefore

$$\log K = \frac{(2)(0.29\ V)}{0.0591\ V} = 9.8$$
$$K = 6 \times 10^9$$

Similar examples: Problems 24.18, 24.19.

EXAMPLE 24.6

Calculate the standard free-energy change for the following reaction at 25°C:

$$2Au(s) + 3Ca^{2+}(aq, 1\ M) \longrightarrow 2Au^{3+}(aq, 1\ M) + 3Ca(s)$$

Answer

First we break up the overall reaction into half-reactions:

oxidation: $2Au(s) \longrightarrow 2Au^{3+}(aq, 1\ M) + 6e^-$
reduction: $3Ca^{2+}(aq, 1\ M) + 6e^- \longrightarrow 3Ca(s)$

In Table 24.1 we find that $\mathscr{E}^\circ_{Au} = 1.50$ V and $\mathscr{E}^\circ_{Ca} = -2.87$ V. Therefore

$$\begin{aligned}
\mathscr{E}^\circ &= \mathscr{E}^\circ_{Ca} - \mathscr{E}^\circ_{Au} \\
&= -2.87\ V - 1.50\ V \\
&= -4.37\ V
\end{aligned}$$

Now we use Equation (24.3):

$$\Delta G^\circ = -nF\mathscr{E}^\circ$$

The half-reactions show that $n = 6$, so

$$\begin{aligned}
\Delta G^\circ &= -(6\ mol)(96{,}500\ J/V \cdot mol)(-4.37\ V) \\
&= 2.53 \times 10^6\ J \\
&= 2.53 \times 10^3\ kJ
\end{aligned}$$

The large positive value of ΔG° tells us that the reaction is not spontaneous under standard-state conditions at 25°C.

Similar example: Problem 24.16.

24.4 EFFECT OF CONCENTRATION ON CELL EMF

The Nernst Equation

We have so far concentrated on redox reactions in which reactants and products are in their standard states. However, standard-state conditions are often difficult and sometimes impossible to maintain. The relationship between the emf of a cell and the concentrations of reactants and products in a redox reaction of the type

$$a\mathrm{A} + b\mathrm{B} \longrightarrow c\mathrm{C} + d\mathrm{D}$$

can be derived as follows. From Equation (23.11) we write

$$\Delta G = \Delta G^\circ + 2.303RT \log Q$$

Since $\Delta G = -nF\mathscr{E}$ and $\Delta G^\circ = -nF\mathscr{E}^\circ$, the equation can be expressed as

$$-nF\mathscr{E} = -nF\mathscr{E}^\circ + 2.303RT \log Q$$

Dividing the equation through by $-nF$, we get

$$\mathscr{E} = \mathscr{E}^\circ - \frac{2.303RT}{nF} \log Q \tag{24.6}$$

We omit the subscript "cell" in $\mathscr{E}$ and $\mathscr{E}^\circ$ for convenience.

Equation (24.6) is known as the Nernst equation (Walter Hermann Nernst, 1864–1941) in which Q, the reaction quotient (see Section 17.4), is given by

$$Q = \frac{[C]^c[D]^d}{[A]^a[B]^b}$$

At 298 K, Equation (24.6) can be rewritten as

$$\mathscr{E} = \mathscr{E}° - \frac{0.0591\ V}{n} \log Q \qquad (24.7)$$

At equilibrium, there is no net transfer of electrons, so $\mathscr{E} = 0$ and $Q = K$, where K is the equilibrium constant. In this case, Equation (24.6) takes the form

$$0 = \mathscr{E}° - \frac{2.303RT}{nF} \log K$$

or

$$\mathscr{E}° = \frac{2.303RT}{nF} \log K$$

which is the same as Equation (24.4).

The Nernst equation enables us to calculate $\mathscr{E}$ as a function of reactant and product concentrations in a redox reaction. Referring to the galvanic cell in Figure 24.1

$$Zn(s) + Cu^{2+}(aq) \longrightarrow Zn^{2+}(aq) + Cu(s)$$

The Nernst equation for this cell at 25°C can be written as

$$\mathscr{E} = 1.10\ V - \frac{0.0591\ V}{2} \log \frac{[Zn^{2+}]}{[Cu^{2+}]}$$

If the ratio $[Zn^{2+}]/[Cu^{2+}]$ is less than 1, $\log [Zn^{2+}]/[Cu^{2+}]$ is a negative number so that the second term on the right-hand side of the above equation is positive. Under this condition $\mathscr{E}$ is greater than the standard emf $\mathscr{E}°$. If the ratio is greater than 1, $\mathscr{E}$ is smaller than $\mathscr{E}°$. Table 24.2 shows the variation of $\mathscr{E}$ with ion concentrations for the cell reaction that we

Remember that concentrations of pure solids (and pure liquids) do not appear in **Q**.

TABLE 24.2 Variation of $\mathscr{E}$ with Ion Concentrations (M) for the Cell Reaction Shown in Figure 24.1 ($t = 25°C$)

$[Cu^{2+}]$	$[Zn^{2+}]$	$\dfrac{[Zn^{2+}]}{[Cu^{2+}]}$	$\log \dfrac{[Zn^{2+}]}{[Cu^{2+}]}$	$\mathscr{E}\ (V)$
1.0	1.0×10^{-4}	1.0×10^{-4}	-4.00	1.22
1.0	1.0×10^{-3}	1.0×10^{-3}	-3.00	1.19
1.0	1.0×10^{-2}	1.0×10^{-2}	-2.00	1.16
1.0	1.0×10^{-1}	1.0×10^{-1}	-1.00	1.13
1.0	1.0	1.0	0.00	1.10
1.0×10^{-1}	1.0	1.0×10^{1}	1.00	1.07
1.0×10^{-2}	1.0	1.0×10^{2}	2.00	1.04
1.0×10^{-3}	1.0	1.0×10^{3}	3.00	1.01
1.0×10^{-4}	1.0	1.0×10^{4}	4.00	0.98

FIGURE 24.4 *Plot of $\mathscr{E}$ versus $\log([Zn^{2+}]/[Cu^{2+}])$ at 25°C. When $[Zn^{2+}]/[Cu^{2+}] = 1$, $\log [Zn^{2+}]/[Cu^{2+}] = 0$ and $\mathscr{E} = \mathscr{E}° = 1.10$ V.*

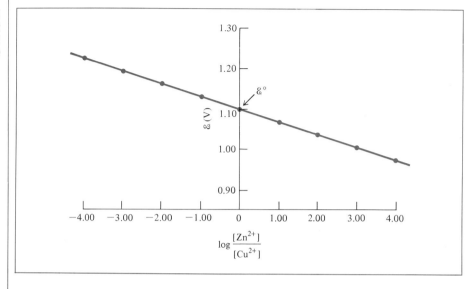

have been discussing. The above equation predicts that a plot of $\mathscr{E}$ versus $\log [Zn^{2+}]/[Cu^{2+}]$ should yield a straight line with a negative slope. This prediction is confirmed by Figure 24.4.

EXAMPLE 24.7

Predict whether the following reaction would proceed spontaneously as written at 298 K

$$Co(s) + Fe^{2+}(aq) \longrightarrow Co^{2+}(aq) + Fe(s)$$

given that $[Co^{2+}] = 0.15\ M$ and $[Fe^{2+}] = 0.68\ M$.

Answer

The half-reactions are

$$\text{oxidation:} \quad Co(s) \longrightarrow Co^{2+}(aq) + 2e^-$$
$$\text{reduction:} \quad Fe^{2+}(aq) + 2e^- \longrightarrow Fe(s)$$

In Table 24.1 we find that $\mathscr{E}°_{Co} = -0.28$ V and $\mathscr{E}°_{Fe} = -0.44$ V. Therefore, the standard emf is

$$\mathscr{E}° = \mathscr{E}°_{Fe} - \mathscr{E}°_{Co}$$
$$= -0.44\ V - (-0.28\ V)$$
$$= -0.16\ V$$

From Equation (24.7) we write

$$\mathscr{E} = \mathscr{E}° - \frac{0.0591\ V}{n} \log \frac{[Co^{2+}]}{[Fe^{2+}]}$$
$$= -0.16\ V - \frac{0.0591\ V}{2} \log \frac{0.15}{0.68}$$
$$= -0.16\ V + 0.019\ V$$
$$= -0.14\ V$$

Since $\mathscr{E}$ is negative (or ΔG is positive), the reaction is *not* spontaneous in the direction written.

Similar examples: Problems 24.22, 24.25.

It is interesting to determine at what ratio of $[Co^{2+}]$ to $[Fe^{2+}]$ the reaction in Example 24.7 will become spontaneous. We start with the Nernst equation as rewritten for 298 K [Equation (24.7)]:

$$\mathscr{E} = \mathscr{E}° - \frac{0.0591 \text{ V}}{n} \log Q$$

We set $\mathscr{E}$ equal to zero, since this corresponds to the equilibrium situation.

$$0 = -0.16 \text{ V} - \frac{0.0591 \text{ V}}{2} \log \frac{[Co^{2+}]}{[Fe^{2+}]}$$

$$\log \frac{[Co^{2+}]}{[Fe^{2+}]} = -5.4$$

Taking the antilog of both sides we obtain

$$\frac{[Co^{2+}]}{[Fe^{2+}]} = 4.0 \times 10^{-6}$$

Thus for the reaction to be spontaneous, the ratio $[Co^{2+}]/[Fe^{2+}]$ must be smaller than 4.0×10^{-6}.

EXAMPLE 24.8

Consider the galvanic cell shown in Figure 24.3(a). In a certain experiment, the emf $(\mathscr{E})$ of the cell is found to be 0.54 V at 25°C. Suppose that $[Zn^{2+}] = 1.0 \text{ } M$ and $P_{H_2} = 1.0$ atm. Calculate the molar concentration of H^+.

Answer

The overall cell reaction is

$$Zn(s) + 2H^+(aq, \text{ ? } M) \longrightarrow Zn^{2+}(aq, 1 \text{ } M) + H_2(g, 1 \text{ atm})$$

As we saw earlier (p. 717), the standard emf for the cell is 0.76 V. From Equation (24.7), we write

$$\mathscr{E} = \mathscr{E}° - \frac{0.0591 \text{ V}}{n} \log \frac{[Zn^{2+}] P_{H_2}}{[H^+]^2}$$

$$0.54 \text{ V} = 0.76 \text{ V} - \frac{0.0591 \text{ V}}{2} \log \frac{(1.0)(1.0)}{[H^+]^2}$$

$$-0.22 \text{ V} = -\frac{0.0591 \text{ V}}{2} \log \frac{1}{[H^+]^2}$$

(Continued)

$$7.4 = \log \frac{1}{[H^+]^2}$$
$$7.4 = -2 \log [H^+]$$
$$\log [H^+] = -3.7$$
$$[H^+] = 2 \times 10^{-4} \, M$$

Similar examples: Problems 24.23, 24.24.

The preceding example shows that a galvanic cell whose cell reaction involves H^+ ions can be used to measure $[H^+]$ or pH. The pH meter described in Section 18.3 is based on this principle. For practical considerations the electrodes employed in a pH meter are quite different from the SHE and zinc electrode in the galvanic cell (Figure 24.5).

Concentration Cells

We have seen that an electrode potential depends on the concentrations of the ions used. In practice, a cell may be constructed from two half-cells composed of the *same* material but differing in ion concentrations. Such a cell is called a *concentration cell.*

Consider a situation in which zinc electrodes are put into two aqueous solutions of zinc sulfate at 0.10 *M* and 1.0 *M* concentrations. The two solutions are connected by a salt bridge, and the electrodes are joined by a piece of wire in an arrangement like that shown in Figure 24.1. According to Le Chatelier's principle, the tendency for the reduction

$$Zn^{2+}(aq) + 2e^- \longrightarrow Zn(s)$$

increases with increasing concentration of Zn^{2+} ions. Therefore, reduction should occur in the more concentrated compartment and oxidation should take place on the more dilute side. The cell diagram is

$$Zn(s)|Zn^{2+}(aq, 0.10 \, M)\|KCl(sat'd)\|Zn^{2+}(aq, 1.0 \, M)|Zn(s)$$

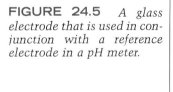

FIGURE 24.5 *A glass electrode that is used in conjunction with a reference electrode in a pH meter.*

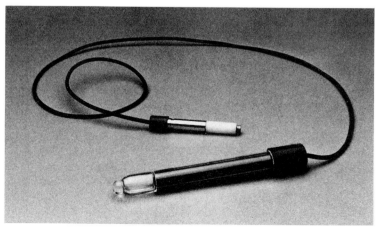

and the half-reactions are

oxidation: $\qquad$ $Zn(s) \longrightarrow Zn^{2+}(aq, 0.10\ M) + 2e^-$
reduction: $\underline{\quad Zn^{2+}(aq, 1.0\ M) + 2e^- \longrightarrow Zn(s) \qquad\qquad}$
overall: $\qquad Zn^{2+}(aq, 1.0\ M) \longrightarrow Zn^{2+}(aq, 0.10\ M)$

The emf of the cell is

$$\mathscr{E} = \mathscr{E}° - \frac{0.0591\ V}{n} \log \frac{[Zn]_{dil}}{[Zn]_{conc}}$$

where the subscripts "dil" and "conc" refer to the 0.10 M and 1.0 M concentrations, respectively. $\mathscr{E}°$ for this cell is zero (the *same* electrode and type of ions are involved), so

$$\mathscr{E} = -\frac{0.0591\ V}{2} \log \frac{0.10}{1.0}$$
$$= 0.0296\ V$$

The emf of concentration cells is usually small and decreases continually during the operation of the cell as the concentrations in the two compartments approach each other. When the concentrations of the ions in the two compartments are the same, $\mathscr{E}$ becomes zero.

24.5 BATTERIES

A **battery** is an electrochemical cell, or often, *several electrochemical cells connected in series, that can be used as a source of direct electric current at a constant voltage.* Although the operation of a battery is similar in principle to that of the galvanic cells described in Section 24.1, a battery has the advantage of being completely self-contained and requiring no auxiliary components such as a salt bridge. We will describe several types of batteries that are in widespread use.

The Dry Cell Battery

The most common dry cell, that is, a cell without fluid component, is the *Leclanché cell*, used in flashlights and transistor radios. The anode of the cell consists of a zinc can or container that is in contact with manganese dioxide (MnO_2) and an electrolyte. The electrolyte consists of ammonium chloride and zinc chloride in water, to which starch is added to thicken the solution to a pastelike consistency so that it is less likely to leak (Figure 24.6). A carbon rod serves as the cathode, which is immersed in the electrolyte in the center of the cell. The cell reactions are

anode: $\qquad$ $Zn(s) \longrightarrow Zn^{2+}(aq) + 2e^-$
cathode: $\underline{\ 2NH_4^+(aq) + 2MnO_2(s) + 2e^- \longrightarrow Mn_2O_3(s) + 2NH_3(aq) + H_2O(l)\ }$
overall: $\ Zn(s) + 2NH_4^+(aq) + 2MnO_2(s) \longrightarrow$
$\qquad\qquad\qquad Zn^{2+}(aq) + 2NH_3(aq) + H_2O(l) + Mn_2O_3(s)$

Actually, this equation is an oversimplification, because the reactions that occur in the cell are quite complex. The voltage produced by a dry cell is

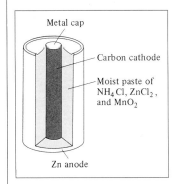

FIGURE 24.6 *Interior section of a dry cell, the kind used in flashlights and transistor radios. Actually, the cell is not completely dry, as it contains a moist electrolyte paste.*

about 1.5 V, but it drops slightly during use because of the accumulation of Zn^{2+} ions near the zinc anode.

The Mercury Battery

The mercury battery is used extensively in medicine and electronic industries and is more expensive than the common dry cell. Contained in a stainless steel cylinder, the mercury battery consists of a zinc anode (amalgamated with mercury) in contact with a strongly alkaline electrolyte containing zinc oxide and mercury(II) oxide. The cell reactions are

$$
\begin{aligned}
\text{anode:} \quad & Zn(Hg) + 2OH^-(aq) \longrightarrow ZnO(s) + H_2O(l) + 2e^- \\
\text{cathode:} \quad & HgO(s) + H_2O(l) + 2e^- \longrightarrow Hg(l) + 2OH^-(aq) \\
\hline
\text{overall:} \quad & Zn(Hg) + HgO(s) \longrightarrow ZnO(s) + Hg(l)
\end{aligned}
$$

Because there is no change in electrolyte composition during operation—the overall cell reaction involves only solid substances—the mercury battery provides a more constant voltage (1.35 V) than the Leclanché cell. It also has a considerably higher capacity and longer life. These qualities make the mercury battery ideal for use in pacemakers, hearing aids, electric watches, and light meters.

The Lead Storage Battery

The lead storage battery commonly used in automobiles consists of six identical cells joined together in series. Each cell has a lead anode and a cathode made of lead dioxide (PbO_2) packed on a metal plate (Figure 24.7). Both the cathode and the anode are immersed in an aqueous solution of sulfuric acid, which acts as the electrolyte. The cell reactions are

$$
\begin{aligned}
\text{anode:} \quad & Pb(s) + SO_4^{2-}(aq) \longrightarrow PbSO_4(s) + 2e^- \\
\text{cathode:} \quad & PbO_2(s) + 4H^+(aq) + SO_4^{2-}(aq) + 2e^- \longrightarrow PbSO_4(s) + 2H_2O(l) \\
\hline
\text{overall:} \quad & Pb(s) + PbO_2(s) + 4H^+(aq) + 2SO_4^{2-}(aq) \longrightarrow 2PbSO_4(s) + 2H_2O(l)
\end{aligned}
$$

Under normal operating conditions, each cell produces 2 V; a total of 12 V from the six cells is used to power the ignition circuit of the automobile and its other electrical systems. The lead storage battery can deliver large amounts of current for a short time, such as the time it takes to start up the engine.

Unlike the Leclanché cell and the mercury battery, the lead storage battery is rechargeable. Recharging the battery means reversing the normal electrochemical reaction by applying an external voltage at the cathode and the anode. (This kind of process is called *electrolysis*, which we will discuss in Section 24.7.) The reactions that replenish the original materials are

$$
\begin{aligned}
\text{anode:} \quad & PbSO_4(s) + 2e^- \longrightarrow Pb(s) + SO_4^{2-}(aq) \\
\text{cathode:} \quad & PbSO_4(s) + 2H_2O(l) \longrightarrow PbO_2(s) + SO_4^{2-}(aq) + 4H^+(aq) + 2e^- \\
\hline
\text{overall:} \quad & 2PbSO_4(s) + 2H_2O(l) \longrightarrow Pb(s) + PbO_2(s) + 4H^+(aq) + 2SO_4^{2-}(aq)
\end{aligned}
$$

The overall reaction is exactly the opposite of the normal cell reaction.

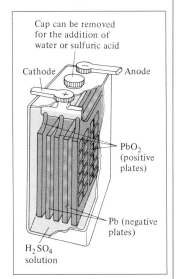

FIGURE 24.7 *Interior section of a lead storage battery. Under normal operating conditions, the concentration of the sulfuric acid solution is about 38 percent by mass.*

Two aspects of the operation of a lead storage battery are worth noting. First, because the electrochemical reaction consumes sulfuric acid, the degree to which the battery has been discharged can be checked by measuring the density of the electrolyte with a hydrometer, as is usually done at gas stations. The density of the fluid in a "healthy," fully charged battery should be equal to or greater than 1.2 g/cm^3. Second, people living in cold climates sometimes have trouble starting their cars because the battery has "gone dead." Thermodynamic calculations show that the emf of many electrochemical cells decreases with decreasing temperature. However, for a lead storage battery, the temperature coefficient is about 1.5×10^{-4} V/°C; that is, there is a decrease in voltage of 1.5×10^{-4} V for every degree drop in temperature. Thus, even allowing for a 40°C change in temperature, the decrease in voltage amounts to only 6×10^{-3} V, which is about

$$\frac{6 \times 10^{-3} \text{ V}}{12 \text{ V}} \times 100\% = 0.05\%$$

of the operating voltage, an insignificant change. The real cause of a battery's apparent breakdown is an increase in the viscosity of the electrolyte as the temperature decreases. For the battery to function properly, the electrolyte must be fully conducting. However, the ions move much more slowly in a viscous medium, so the resistance of the fluid increases, leading to a decrease in the power output of the battery. If an apparently "dead battery" is warmed to near room temperature, it recovers its ability to deliver normal power.

Fuel Cells

Fossil fuels are a major source of energy, but the conversion of fossil fuel into electrical energy is a highly inefficient process. Consider the combustion of methane:

$$CH_4(g) + 2O_2(g) \longrightarrow CO_2(g) + 2H_2O(l) + \text{energy}$$

To generate electricity, heat produced by the reaction is first used to convert water to steam, which then drives a turbine that drives a generator. An appreciable amount of energy in the form of heat is lost to the surroundings at each step; the most efficient power plant now in existence converts only about 40 percent of the original chemical energy into electricity. Because combustion reactions are redox reactions, it would be more desirable to carry them out directly by electrochemical means, thereby greatly increasing the efficiency of power production. This objective has been accomplished by devices known as fuel cells.

In its simplest form, a hydrogen–oxygen fuel cell consists of an electrolyte solution, such as sulfuric acid or potassium hydroxide solution, and two inert electrodes. Hydrogen and oxygen gases are bubbled through the anode and cathode compartments (Figure 24.8), where the following reactions take place:

anode:	$2H_2(g) + 4OH^-(aq) \longrightarrow 4H_2O(l) + 4e^-$
cathode:	$O_2(g) + 2H_2O(l) + 4e^- \longrightarrow 4OH^-(aq)$
overall:	$2H_2(g) + O_2(g) \longrightarrow 2H_2O(l)$

FIGURE 24.8 *A hydrogen-oxygen fuel cell. The temperature of the electrolyte solution is kept at about 200°C. The Ni and NiO embedded in the porous carbon electrodes are electrocatalysts. When the cell is in operation, the electrodes are connected to an electrical appliance by conducting wires.*

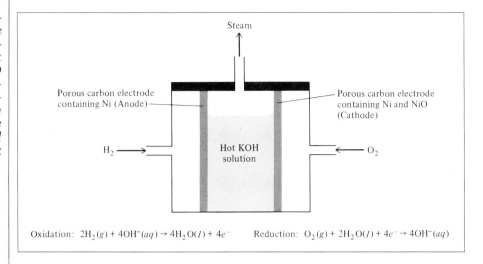

Oxidation: $2H_2(g) + 4OH^-(aq) \rightarrow 4H_2O(l) + 4e^-$ Reduction: $O_2(g) + 2H_2O(l) + 4e^- \rightarrow 4OH^-(aq)$

The standard emf of the cell can be calculated as follows, with data from Table 24.1:

$$
\begin{aligned}
\mathscr{E}^\circ &= \mathscr{E}^\circ_{\text{cathode}} - \mathscr{E}^\circ_{\text{anode}} \\
&= 0.40 \text{ V} - (-0.83 \text{ V}) \\
&= 1.23 \text{ V}
\end{aligned}
$$

Thus the cell reaction is spontaneous under standard-state conditions. Note that the reaction is the same as the hydrogen combustion reaction, but the oxidation and reduction are carried out separately at the anode and the cathode. Like platinum in the standard hydrogen electrode discussed earlier, the electrodes have a twofold function. First, they serve as electrical conductors. Second, the electrodes provide the necessary surfaces for the initial decomposition of the molecules into atomic species, prior to electron transfer. They are *electrocatalysts*. Metals such as platinum, nickel, and rhodium are good electrocatalysts.

In addition to the H_2–O_2 system, a number of other fuel cells have been developed. Among these is the propane–oxygen fuel cell. The half-cell reactions are

anode:	$C_3H_8(g) + 6H_2O(l) \longrightarrow 3CO_2(g) + 20H^+(aq) + 20e^-$
cathode:	$5O_2(g) + 20H^+(aq) + 20e^- \longrightarrow 10H_2O(l)$
overall:	$C_3H_8(g) + 5O_2(g) \longrightarrow 3CO_2(g) + 4H_2O(l)$

The overall reaction is identical to the burning of propane in oxygen.

Unlike batteries, fuel cells do not store chemical energy. Reactants must be constantly resupplied and products must be constantly removed from a fuel cell. In this respect, a fuel cell resembles an engine more than it does a battery. However, the fuel cell does not operate like a heat engine and therefore is not subject to the same kind of thermodynamic limitations in energy conversion. (A heat engine is any device that converts heat into work. The second law of thermodynamics imposes the restriction that it is impossible to convert heat *completely* into work. In practice, any such conversion must result in the loss of part of the energy, as heat, to the surroundings.)

Properly designed fuel cells may be as much as 70 percent efficient, about twice as efficient as an internal combustion engine. In addition, fuel-cell generators are free of the noise, vibration, heat transfer, thermal pollution, and other problems normally associated with conventional power plants. Despite all these features, fuel cells are not yet in large-scale operation. A major problem lies in the choice and availability of suitable electrocatalysts able to function efficiently for long periods of time without contamination. The most successful application of fuel cells to date has been in space vehicles.

24.6 CORROSION

Corrosion is the term usually applied to the *deterioration of metals by an electrochemical process.* We see many examples of corrosion around us—iron rust, silver tarnish, the green patina formed on copper and brass. Corrosion causes enormous damage to buildings, bridges, ships, and cars. One estimate put the cost of metallic corrosion to the U.S. economy in 1984 at about 80 billion dollars, or 3 percent of the gross national product for that year! This section will discuss some of the fundamental processes that occur in corrosion and methods used to protect metals against corrosion.

By far the most familiar example of corrosion is the formation of rust on iron. Oxygen gas and water must be present for iron to rust. Although the reactions involved are quite complex and not completely understood, the main steps are believed to be as follows. A region of the metal's surface serves as the anode, where oxidation occurs:

$$Fe(s) \longrightarrow Fe^{2+}(aq) + 2e^-$$

The electrons given up by iron reduce atmospheric oxygen to water at the cathode, which is another region of the same metal's surface:

$$O_2(g) + 4H^+(aq) + 4e^- \longrightarrow 2H_2O(l)$$

The overall redox reaction is

$$2Fe(s) + O_2(g) + 4H^+(aq) \longrightarrow 2Fe^{2+}(aq) + 2H_2O(l)$$

With data from Table 24.1, we find the standard emf for this process:

$$\begin{aligned} \mathscr{E}° &= \mathscr{E}°_{cathode} - \mathscr{E}°_{anode} \\ &= 1.23 \text{ V} - (-0.44 \text{ V}) \\ &= 1.67 \text{ V} \end{aligned}$$

Note that this reaction occurs in an acidic medium; the H^+ ions are supplied in part by the reaction of atmospheric carbon dioxide with water to form H_2CO_3.

The Fe^{2+} ions formed at the anode are further oxidized by oxygen:

$$4Fe^{2+}(aq) + O_2(g) + (4 + 2x)H_2O(l) \longrightarrow 2Fe_2O_3 \cdot xH_2O(s) + 8H^+(aq)$$

This hydrated form of iron(III) oxide is known as rust. The amount of water associated with the iron oxide varies, so we represent the formula as $Fe_2O_3 \cdot xH_2O$.

FIGURE 24.9 *The electrochemical process involved in rust formation. The H^+ ions are supplied by H_2CO_3, which forms when CO_2 dissolves in water.*

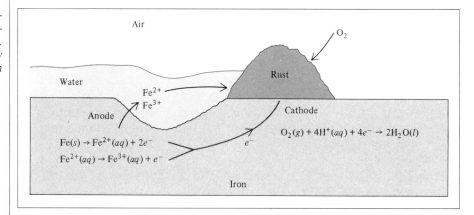

Figure 24.9 shows the mechanism of rust formation. The electric circuit is completed by the migration of electrons and ions; this is the reason that rusting occurs so rapidly in salt water. In cold climates, salts (NaCl or $CaCl_2$) spread on roadways to melt ice and snow are a major cause of rust formation on automobiles.

We stated earlier that metallic corrosion is not limited to iron. Consider aluminum, a metal used to make many useful things, including airplanes and beverage cans. Aluminum has a much greater tendency to oxidize than does iron; in Table 24.1 we see that Al has a more negative standard reduction potential than Fe. Based on this fact alone, we might expect to see airplanes slowly corrode away in rainstorms, and soda cans transformed into piles of corroded aluminum. These processes do not occur because a protective layer of aluminum oxide (Al_2O_3) forms on its surface when the metal is exposed to air; the rust that forms on the surface of iron is too porous to protect the underlying metal from further corrosion.

Coinage metals such as copper and silver also corrode, but much more slowly.

$$Cu(s) \longrightarrow Cu^{2+}(aq) + 2e^-$$

$$Ag(s) \longrightarrow Ag^+(aq) + e^-$$

In normal atmospheric exposure, copper forms a layer of copper carbonate ($CuCO_3$), a green substance also called *patina*, that protects the metal underneath the layer from further corrosion. Likewise, silverware that comes into contact with foodstuffs develops a layer of silver sulfide (Ag_2S).

A number of methods have been devised to protect metals from corrosion. Most of these methods are aimed at preventing rust formation. The most obvious approach is to coat the metal surface with paint. However, if the paint is scratched, pitted, or dented to expose even the smallest area of bare metal, rust will form under the paint layer. The surface of iron metal can be made inactive by a process called *passivation*. A thin oxide layer is formed when the metal is treated with a strong oxidizing agent such as concentrated nitric acid or a solution of potassium dichromate ($K_2Cr_2O_7$). In fact, potassium dichromate is often added to cooling systems and radiators to prevent rust formation.

The tendency for iron to oxidize is greatly reduced when it forms an alloy with other metals. For example, when iron is alloyed with chromium and nickel to become stainless steel, the layer of chromium oxide that is formed protects the iron from corrosion.

An iron container can be covered with a layer of another metal such as tin or zinc. A "tin can" is made by applying a thin layer of tin over iron. Rust formation is prevented as long as the tin layer remains intact. However, once the surface has been scratched, rusting occurs rapidly. If we look up the standard reduction potentials we find that iron acts as the anode and tin as the cathode in the corrosion process:

$$Fe^{2+}(aq) + 2e^- \longrightarrow Fe(s) \qquad \mathscr{E}° = -0.44 \text{ V}$$

$$Sn^{2+}(aq) + 2e^- \longrightarrow Sn(s) \qquad \mathscr{E}° = -0.14 \text{ V}$$

The protective process is different for zinc-plated, or *galvanized,* iron. Zinc is more easily oxidized than iron (see Table 24.1)

$$Zn^{2+}(aq) + 2e^- \longrightarrow Zn(s) \qquad \mathscr{E}° = -0.76 \text{ V}$$

so if a scratch exposes the iron, the zinc is still attacked. In this case, the zinc metal serves as the anode and the iron is the cathode. The term *cathodic protection* describes the process in which the metal that is to be protected from corrosion is made the cathode in what amounts to an electrochemical cell. Color Plate 26 shows how an iron nail can be protected from rusting by connecting the nail to a piece of zinc. Without such protection, an iron nail quickly rusts in water. Rusting of underground iron pipes and iron storage tanks can be prevented or greatly reduced by connecting them to metals such as zinc and magnesium, which are more easily oxidized than iron (Figure 24.10).

24.7 ELECTROLYSIS

So far we have concentrated on spontaneous redox reactions, which result in the conversion of chemical energy into electrical energy. In this section

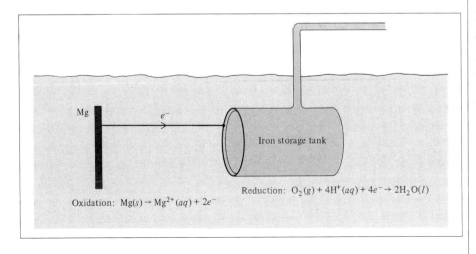

FIGURE 24.10 *Cathodic protection of an iron storage tank (cathode) by magnesium, a more electropositive metal (anode). Since only the magnesium is depleted in the electrochemical process, it is sometimes called the sacrificial anode.*

Oxidation: $Mg(s) \rightarrow Mg^{2+}(aq) + 2e^-$

Reduction: $O_2(g) + 4H^+(aq) + 4e^- \rightarrow 2H_2O(l)$

Mg

Iron storage tank

e^-

we will study the reverse process, ***electrolysis,*** in which *electrical energy is used to cause a nonspontaneous chemical reaction to occur.* As the following examples show, electrolysis is based on the same principles as those that apply to electrochemical processes carried out in galvanic cells.

Electrolysis of Molten Sodium Chloride

Sodium chloride, an ionic compound, can be electrolyzed when molten. Figure 24.11(a) is a diagram of an arrangement for large-scale electrolysis of NaCl. In molten NaCl, the cations and anions are the Na^+ and Cl^- ions, respectively. Figure 24.11(b) shows the reactions that occur at the electrodes. The *electrolytic cell* contains a pair of electrodes connected to the battery. The battery serves as an "electron pump," driving electrons to the cathode, where reduction occurs, and withdrawing electrons from the anode, where oxidation occurs. The reactions at the electrodes are

$$\text{anode (oxidation):} \quad 2Cl^- \longrightarrow Cl_2(g) + 2e^-$$
$$\text{cathode (reduction):} \quad \underline{2Na^+ + 2e^- \longrightarrow 2Na(l)}$$
$$\text{overall:} \quad 2Na^+ + 2Cl^- \longrightarrow 2Na(l) + Cl_2(g)$$

This process is one of the major sources of pure sodium metal and chlorine gas.

Electrolysis of Water

Water in a beaker under normal atmospheric conditions (1 atm and 25°C) will not spontaneously decompose to form hydrogen and oxygen gas

$$2H_2O(l) \longrightarrow 2H_2(g) + O_2(g) \qquad \Delta G° = 474.4 \text{ kJ}$$

FIGURE 24.11 *(a) A practical arrangement for the electrolysis of molten NaCl (m. pt. = 801°C). The sodium metal formed at the cathodes is in the liquid state. Since liquid sodium metal is lighter than molten NaCl, the sodium floats to the surface as shown, and is collected. Chlorine gas forms at the anode and is collected at the top. (b) A simplified diagram showing the electrode reactions during the electrolysis of molten NaCl.*

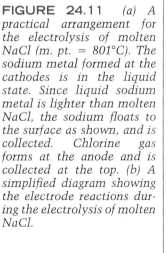

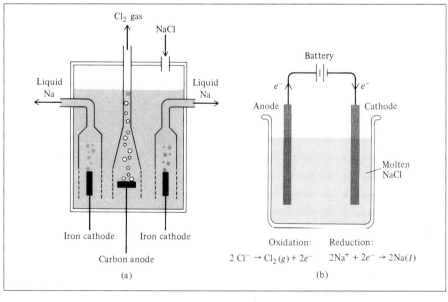

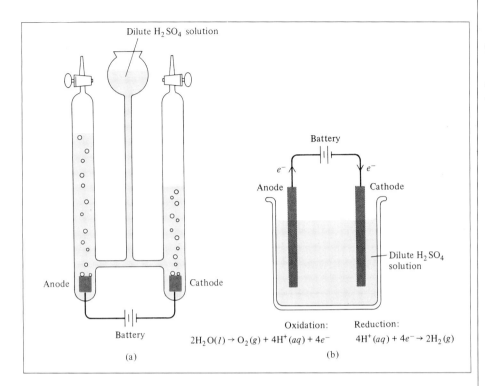

FIGURE 24.12 *(a) Apparatus for small-scale electrolysis of water. The volume of hydrogen gas generated is twice that of oxygen gas. (b) A simplified diagram showing the electrode reactions during electrolysis of water.*

because the standard free-energy change for the reaction is a large positive quantity. However, this reaction may be induced by electrolyzing water in a cell like that shown in Figure 24.12(a). This electrolytic cell consists of a pair of electrodes made of an inert (that is, nonreactive) metal, such as platinum, immersed in water. When the electrodes are connected to the battery, nothing happens because there are not enough ions in pure water to carry much of an electric current. (Remember that at 25°C, pure water has only 1×10^{-7} M H^+ ions and 1×10^{-7} M OH^- ions.)

The cell reaction starts when a few drops of sulfuric acid are added to the water. Almost immediately, gas bubbles are observed at both electrodes. We can predict the outcome of the electrolysis by looking up the standard reduction potential values in Table 24.1. The oxidation that occurs at the anode is

$$2H_2O(l) \longrightarrow O_2(g) + 4H^+(aq) + 4e^-$$

(The sulfate ions are much more difficult to oxidize than water; consequently, they act as spectator ions in solution.) Possible reductions that might occur at the cathode are

$$2H_2O(l) + 2e^- \longrightarrow H_2(g) + 2OH^-(aq) \qquad \mathscr{E}° = -0.83 \text{ V}$$

$$4H^+(aq) + 4e^- \longrightarrow 2H_2(g) \qquad \mathscr{E}° = 0.00 \text{ V}$$

Since the second reduction has a more positive $\mathscr{E}°$ value, it is the preferred reaction at the cathode. We can now represent the processes at the electrodes as

Remember that the more positive the $\mathscr{E}°$ value, the greater the tendency for the reduction to occur.

anode (oxidation): $\qquad 2H_2O(l) \longrightarrow O_2(g) + 4H^+(aq) + 4e^-$

cathode (reduction): $4H^+(aq) + 4e^- \longrightarrow 2H_2(g)$

overall: $\qquad 2H_2O(l) \longrightarrow 2H_2(g) + O_2(g)$

The predictions represented by these equations are indeed confirmed by experiment.

Electrolysis of an Aqueous Sodium Chloride Solution

This is the most complicated of the three examples of electrolysis considered here, because aqueous sodium chloride solution contains several species that can be oxidized and reduced. The possible oxidations that can occur at the anode are

(1) $\quad 2H_2O(l) \longrightarrow O_2(g) + 4H^+(aq) + 4e^-$

(2) $\quad 2Cl^-(aq) \longrightarrow Cl_2(g) + 2e^-$

In Table 24.1 we find

$$O_2(g) + 4H^+(aq) + 4e^- \longrightarrow 2H_2O(l) \qquad \mathscr{E}° = 1.23 \text{ V}$$

$$Cl_2(g) + 2e^- \longrightarrow 2Cl^-(aq) \qquad \mathscr{E}° = 1.36 \text{ V}$$

The standard reduction potentials of (1) and (2) are not greatly different, but the values do suggest that H_2O should be preferentially oxidized at the anode. However, by experiment we find that the gas liberated at the anode is Cl_2, not O_2! In studying electrolytic processes we sometimes find that the voltage required for a reaction is considerably higher than the electrode potential indicates. *The additional voltage required to cause electrolysis is called the **overvoltage.*** The overvoltage for O_2 formation is quite high. Therefore, under normal operating conditions Cl_2 instead of O_2 gas is actually formed at the anode.

The possible reductions that can occur at the cathode are

(3) $\quad Na^+(aq) + e^- \longrightarrow Na(s) \qquad \mathscr{E}° = -2.71 \text{ V}$

(4) $\quad 2H_2O(l) + 2e^- \longrightarrow H_2(g) + 2OH^-(aq) \qquad \mathscr{E}° = -0.83 \text{ V}$

(5) $\quad 2H^+(aq) + 2e^- \longrightarrow H_2(g) \qquad \mathscr{E}° = 0.00 \text{ V}$

Thus reduction becomes more difficult from (5) to (4) to (3); that is, hydrogen ions should accept electrons from the cathode more readily than do water molecules or sodium ions. However, the pH of an aqueous NaCl solution is near 7 (remember that NaCl does not hydrolyze), so the concentration of H^+ ions is about 1×10^{-7} M, which is much too low to make (5) a reasonable choice to represent the net change at the cathode. For this reason, we generally use (4) to describe the cathode reaction.

Thus, the reactions in the electrolysis of aqueous sodium chloride are

anode (oxidation): $\qquad 2Cl^-(aq) \longrightarrow Cl_2(g) + 2e^-$

cathode (reduction): $2H_2O(l) + 2e^- \longrightarrow H_2(g) + 2OH^-(aq)$

overall: $\quad 2H_2O(l) + 2Cl^-(aq) \longrightarrow H_2(g) + Cl_2(g) + 2OH^-(aq)$

Because Cl_2 is more easily reduced than O_2, it follows that it would be more difficult to oxidize Cl^- than H_2O at the anode.

As the overall reaction shows, the concentration of the Cl^- ions decreases during electrolysis and that of the OH^- ions increases. Therefore, in addition to H_2 and Cl_2, the useful by-product NaOH can be obtained by evaporating the aqueous solution at the end of the electrolysis.

EXAMPLE 24.9

An aqueous Na_2SO_4 solution is electrolyzed, using the apparatus shown in Figure 24.12(a). If the products formed at the anode and cathode are oxygen gas and hydrogen gas, describe the electrolysis in terms of the reactions at the electrodes.

Answer

Before we consider the electrode reactions, we should note that the Na^+ ions are not reduced at the cathode and SO_4^{2-} ions are not oxidized at the anode. We draw these conclusions from the electrolyses of aqueous sodium chloride solution and of water in the presence of sulfuric acid (H_2SO_4). Thus the electrode reactions are

$$\begin{array}{rl} \text{anode:} & 2H_2O(l) \longrightarrow O_2(g) + 4H^+(aq) + 4e^- \\ \text{cathode:} & 4H_2O(l) + 4e^- \longrightarrow 2H_2(g) + 4OH^-(aq) \\ \hline \text{overall:} & 2H_2O(l) \longrightarrow 2H_2(g) + O_2(g) \end{array}$$

Similar example: Problem 24.38.

> We assume that the H^+ and OH^- ions at the anode and cathode are allowed to mix to form H_2O.

Electrolysis has many important applications in industry, mainly in the extraction and purification of metals. We will discuss some of these applications in Chapter 27.

Quantitative Aspects of Electrolysis

The quantitative treatment of electrolysis was developed primarily by Faraday. He observed that the mass of product formed (or reactant consumed) at an electrode is proportional to both the amount of electricity transferred at the electrode and the molar mass of the substance in question. For example, in the electrolysis of molten NaCl, the cathode reaction tells us that one Na atom is produced when one Na^+ ion accepts an electron from the electrode. To reduce one mole of Na^+ ions, we must supply Avogadro's number (6.02×10^{23}) of electrons to the cathode. On the other hand, the stoichiometry of the anode reaction shows that oxidation of two Cl^- ions yields one chlorine molecule. Therefore, the formation of one mole of Cl_2 results in the transfer of two moles of electrons from the Cl^- ions to the anode. Similarly, it takes two moles of electrons to reduce one mole of Mg^{2+} ions and three moles of electrons to reduce one mole of Al^{3+} ions:

$$Mg^{2+} + 2e^- \longrightarrow Mg$$

$$Al^{3+} + 3e^- \longrightarrow Al$$

Therefore

$$2 F \approx 1 \text{ mol Mg}^{2+}$$

$$3 F \approx 1 \text{ mol Al}^{3+}$$

where F is the faraday.

In an electrolysis experiment, we generally measure the current (in amperes, A) that passes through an electrolytic cell in a given period of time. The relationship between charge (in coulombs, C) and current is

$$1 \text{ C} = 1 \text{ A} \times 1 \text{ s}$$

that is, a coulomb is the quantity of electrical charge passing any point in the circuit in one second when the current is one ampere.

EXAMPLE 24.10

A current of 0.452 A is passed for 1.50 hours through an electrolytic cell containing molten $MgCl_2$. Write the electrode reactions and calculate the quantity of products (in grams) formed at the electrodes.

Answer

Since the only ions present in molten $MgCl_2$ are Mg^{2+} and Cl^-, the reactions are

$$\text{anode:} \quad 2Cl^- \longrightarrow Cl_2(g) + 2e^-$$
$$\text{cathode:} \quad \underline{Mg^{2+} + 2e^- \longrightarrow Mg}$$
$$\text{overall:} \quad Mg^{2+} + 2Cl^- \longrightarrow Mg + Cl_2(g)$$

The quantities of Mg metal and chlorine gas formed depend on the number of electrons that pass through the electrolytic cell, which in turn depends on the current flow and time, or charge.

$$? \text{ C} = 0.452 \text{ A} \times 1.50 \text{ h} \times \frac{3600 \text{ s}}{1 \text{ h}} \times \frac{1 \text{ C}}{1 \text{ A} \cdot \text{s}} = 2.44 \times 10^3 \text{ C}$$

Since $1 F = 96,500$ C and $2 F$ are required to reduce one mole of Mg^{2+} ions, the mass of Mg metal formed at the cathode is calculated as follows:

$$? \text{ g Mg} = 2.44 \times 10^3 \text{ C} \times \frac{1 F}{96,500 \text{ C}} \times \frac{1 \text{ mol Mg}}{2 F} \times \frac{24.31 \text{ g Mg}}{1 \text{ mol Mg}} = 0.307 \text{ g Mg}$$

The anode reaction indicates that one mole of chlorine is produced per two faradays of electricity. Hence the mass of chlorine gas formed is

$$? \text{ g Cl}_2 = 2.44 \times 10^3 \text{ C} \times \frac{1 F}{96,500 \text{ C}} \times \frac{1 \text{ mol Cl}_2}{2 F} \times \frac{70.90 \text{ g Cl}_2}{1 \text{ mol Cl}_2} = 0.896 \text{ g Cl}_2$$

Similar examples: Problems 24.35, 24.36, 24.47.

Quantitative electrolysis also enables us to determine the charge of an ion. The charge is calculated by electrolyzing a solution containing the ion and relating the quantity of the element produced (either oxidized or reduced) to the number of coulombs of charge passing through the cell.

EXAMPLE 24.11

An aqueous solution of a platinum salt is electrolyzed by passing a current of 2.50 A for 2.00 h. As a result, 9.09 g of metallic platinum are formed at the cathode. Calculate the charge on the platinum ions in this solution.

Answer

Our first step is to calculate the quantity of charge (in coulombs) passing through the cell:

$$? \text{ C} = 2.00 \text{ h} \times \frac{3600 \text{ s}}{1 \text{ h}} \times \frac{2.50 \text{ C}}{1 \text{ s}} = 1.80 \times 10^4 \text{ C}$$

The number of faradays corresponding to this quantity of charge is

$$? F = 1.80 \times 10^4 \text{ C} \times \frac{1 F}{96,500 \text{ C}} = 0.187 F$$

The half-cell reaction for the reduction is

$$\text{Pt}^{n+} + ne^- \longrightarrow \text{Pt}$$

1 mol	$n(6.02 \times 10^{23})$	1 mol
195.1 g	$n(96,500 \text{ C})$	195.1 g
195.1 g	nF	195.1 g

where n is the number of charges on the platinum ion. The mass of Pt produced is

$$? \text{ g Pt} = 0.187 F \times \frac{1 \text{ mol Pt}}{nF} \times \frac{195.1 \text{ g Pt}}{1 \text{ mol Pt}} = 9.09 \text{ g Pt}$$

Solving for n, we get

$$n = \frac{36.5}{9.09} = 4$$

Thus the platinum ion is Pt^{4+}.

Similar example: Problem 24.48.

AN ASIDE ON THE PERIODIC TABLE:

Standard Reduction Potentials and the Reactivity of Selected Metals

The reactivity of metals was discussed previously in terms of their tendencies to form oxides, in Chapter 3 (p. 91), and heats of atomization, in Chapter 5 (p. 144). The listing of standard reduction potentials in Table 24.1 allows us to compare the metals with respect to their reactivities toward water and nonoxidizing acids such as hydrochloric acid and oxidizing acids such as nitric acid. Figure 24.13 shows metals in Groups 1A–4A, Groups 1B–2B, and the first-row transition metals, classified into four categories. In

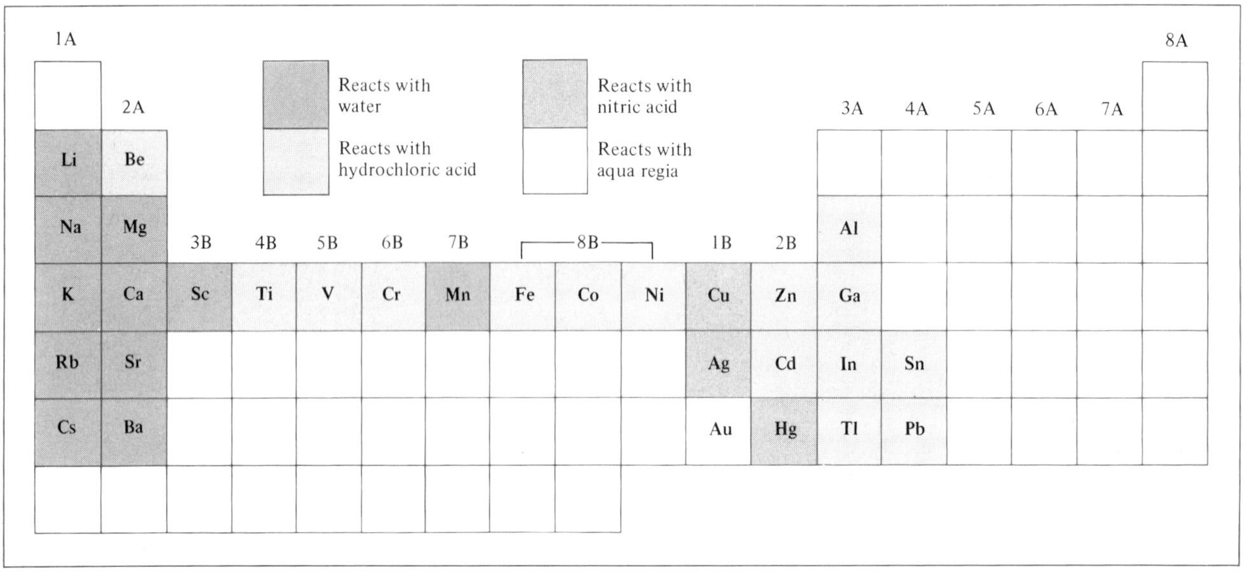

FIGURE 24.13 Classification of the reactivity of some common metals.

studying this classification, bear in mind that there is not always a clear correlation between the metals' standard reduction potentials and their reactivities toward water and acids. The reason is that many of the metals form a protective oxide layer which makes them appear less reactive than their standard reduction potentials indicate.

The first category consists of the alkali and alkaline earth metals which, as you know, are the most reactive of the metallic elements. The standard reduction potentials of these metals are all more negative than that of water; for example

$$Na^+(aq) + e^- \longrightarrow Na(s) \qquad \mathscr{E}° = -2.71 \text{ V}$$

$$2H_2O(l) + 2e^- \longrightarrow H_2(g) + 2OH^-(aq) \qquad \mathscr{E}° = -0.83 \text{ V}$$

Therefore, they can decompose water to form the corresponding metal hydroxides and hydrogen gas. However, there is much variation in reactivity within this group. For example, while the alkali metals all react with cold water, beryllium, the first member of the alkaline earth metals, does not react with water or steam. Beryllium's inactivity in this respect is due to the formation of a protective oxide (BeO) layer. Magnesium reacts only with steam. The rest of the alkaline earth metals all react with water.

The second category consists of metals whose standard reduction potentials are more positive than water but more negative than the following half-reaction

$$2H^+(aq) + 2e^- \longrightarrow H_2(g) \qquad \mathscr{E}° = 0.00 \text{ V}$$

Thus these metals cannot decompose water but do react with hydrochloric acid to form the corresponding metal chlorides and hydrogen gas. Note that aluminum is included in this category even though its standard reduction is -1.66 V which is strong enough to reduce water. Like Be, Al forms a protective oxide (Al_2O_3) layer that renders the metal less reactive.

The third category consists of metals such as copper, mercury, and silver whose standard reduction potentials are all positive. Therefore, they react neither with water nor with hydrochloric acid. These metals can be oxidized by a strong oxidizing agent such as concentrated nitric acid, however, because their standard reduction potentials are more negative than that of the following half-reaction:

$$NO_3^-(aq) + 4H^+(aq) + 3e^- \longrightarrow NO(g) + 2H_2O(l) \qquad \mathscr{E}° = +0.96 \text{ V}$$

In Section 16.7 we discussed the reactions of concentrated nitric acid with metals such as copper, and Example 24.4 was involved with the reaction between mercury and nitric acid.

Finally, metals such as gold and platinum are so unreactive that they are even unaffected by concentrated nitric acid. They can be oxidized, however, by treating the metals with a mixture of nitric acid and hydrochloric acid in a 1:3 ratio by volume (such a mixture is called *aqua regia*):

$$Au(s) + 4HCl(aq) + 3HNO_3(aq) \longrightarrow HAuCl_4(aq) + 3H_2O(l) + 3NO_2(g)$$

SUMMARY

1. All electrochemical reactions involve the transfer of electrons and are therefore redox reactions.
2. In a galvanic cell, electricity is produced by a spontaneous chemical reaction. The oxidation at the anode and the reduction at the cathode take place separately, and the electrons flow through an external circuit.
3. The two parts of a galvanic cell are the half-cells, and the reactions at the electrodes are the half-cell reactions. A salt bridge allows ions to flow between the half-cells.
4. The electromotive force (emf) of a cell is the voltage difference between the two electrodes. In the external circuit, electrons flow from the anode to the cathode in a galvanic cell. In solution, the anions move toward the anode and the cations move toward the cathode.
5. The quantity of electricity carried by one mole of electrons is called a faraday, which is equal to 96,500 coulombs.
6. Standard reduction potentials show the relative likelihood of half-cell reduction reactions and can be used to predict the products, direction, and spontaneity of redox reactions between various substances.
7. The decrease in free energy of the system in a spontaneous redox reaction is equal to the electrical work done by the system on the surroundings, or $\Delta G = -nF\mathscr{E}$.
8. The equilibrium constant for a redox reaction can be found from the standard electromotive force of a cell.
9. The Nernst equation gives the relationship between the cell emf and the concentrations of the reactants and products under nonstandard-state conditions.
10. Batteries, which consist of one or more electrochemical cells, are used widely as self-contained power sources. Some of the better-known batteries are the dry cell, such as the Leclanché cell, the mercury battery, and the lead storage battery used in automobiles. Fuel cells produce electrical energy from a continuous supply of reactants.
11. The corrosion of metals, the most familiar example of which is the rusting of iron, is an electrochemical phenomenon.
12. Electric current from an external source is used to drive a nonspontaneous chemical reaction in an electrolytic cell. In the solution, anions move to the anode, where oxidation occurs, and cations move to the cathode, where reduction occurs.

KEY WORDS

Anode, p. 715

Battery, p. 731

Cathode, p. 715

Corrosion, p. 735

Electrolysis, p. 738

Electromotive force (emf), p. 716

Faraday, p. 723

Half-cell reaction, p. 715

Overvoltage, p. 740

Standard emf, p. 717

Standard reduction potential, p. 717

PROBLEMS

More challenging problems are marked with an asterisk.

Galvanic Cells

24.1 Explain the basic features of a galvanic cell. Why do the two components in a galvanic cell have to be separated from each other?

24.2 Explain the function of a salt bridge in a galvanic cell.

24.3 Calculate the standard emf of a cell that uses the Mg/Mg^{2+} and Cu/Cu^{2+} half-cell reactions at 25°C. Write the equation for the cell reaction occurring under standard-state conditions.

24.4 Consider a galvanic cell consisting of a magnesium electrode in contact with 1.0 M $Mg(NO_3)_2$ and a cadmium electrode in contact with 1.0 M $Cd(NO_3)_2$. Calculate $\mathscr{E}°$ for the cell and draw a diagram showing the cathode, the anode, and the direction of electron flow.

24.5 Calculate the standard emf of a cell that uses Ag/Ag^+ and Al/Al^{3+} half-cell reactions. Write the cell reaction occurring under standard-state conditions.

Spontaneity of Redox Reactions

24.6 Predict whether Fe^{3+} can oxidize I^- to I_2 under standard-state conditions.

24.7 Predict whether the following reaction would occur spontaneously under standard-state conditions.

$$Mg(s) + 2Ag^+(aq) \longrightarrow Mg^{2+}(aq) + 2Ag(s)$$

24.8 Predict whether the following reactions would occur spontaneously in aqueous solution at 25°C. Assume that the initial concentrations of dissolved species are all 1.0 M.

(a) $Ca(s) + Cd^{2+}(aq) \longrightarrow Ca^{2+}(aq) + Cd(s)$

(b) $2Br^-(aq) + Sn^{2+}(aq) \longrightarrow Br_2(l) + Sn(s)$

(c) $2Ag(s) + Ni^{2+}(aq) \longrightarrow 2Ag^+(aq) + Ni(s)$

(d) $Cu^+(aq) + Fe^{3+}(aq) \longrightarrow Cu^{2+}(aq) + Fe^{2+}(aq)$

24.9 Specify which of the following reagents can oxidize H_2O to $O_2(g)$ under standard-state conditions: $H^+(aq)$, $Cl^-(aq)$, $Cl_2(g)$, $Cu^{2+}(aq)$, $Pb^{2+}(aq)$, $MnO_4^-(aq)$ (in acid).

24.10 Predict whether a spontaneous reaction will occur when (a) a piece of silver wire is dipped in a $ZnSO_4$ solution; (b) iodine is added to a $NaBr$ solution; (c) a piece of zinc metal is dipped in a $NiSO_4$ solution; (d) chlorine gas is bubbled through a KI solution. Assume all species to be in their standard states.

24.11 For each of the following redox reactions (i) write the half-reactions; (ii) write a balanced equation for the whole reaction; (iii) determine in which direction the reaction will proceed spontaneously under standard-state conditions.

(a) $H_2(g) + Ni^{2+}(aq) \longrightarrow H^+(aq) + Ni(s)$

(b) $Ni^{2+}(aq) + Cd(s) \longrightarrow Ni(s) + Cd^{2+}(aq)$

(c) $MnO_4^-(aq) + Cl^-(aq) \longrightarrow$
$\qquad Mn^{2+}(aq) + Cl_2(g)$ (in acid solution)

(d) $Ce^{3+}(aq) + H^+(aq) \longrightarrow Ce^{4+}(aq) + H_2(g)$

(e) $Cr(s) + Zn^{2+}(aq) \longrightarrow Cr^{3+}(aq) + Zn(s)$

*24.12 Consider the half-cell reactions involving the following oxidized and reduced forms of a number of elements: $Sn^{2+}(aq)$ and $Sn(s)$; $Cl_2(g)$ and $Cl^-(aq)$; $Ca^{2+}(aq)$ and $Ca(s)$; $Fe^{2+}(aq)$ and $Fe(s)$; $Ag^+(aq)$ and $Ag(s)$. (a) Of the species listed, which is the strongest oxidizing agent under standard-state conditions? (b) Which is the strongest reducing agent under standard-state conditions? (c) What are the standard cell potential and the overall spontaneous cell reaction for the galvanic cell constructed from the $Ag(s)$, $Ag^+(aq)$ and the $Fe(s)$, $Fe^{2+}(aq)$ half-cells? (d) Draw a diagram of this cell as you would construct it in order to measure its potential and indicate the direction of electron flow.

24.13 Consider the following half-reactions:

$$MnO_4^-(aq) + 8H^+(aq) + 5e^- \longrightarrow$$
$$Mn^{2+}(aq) + 4H_2O(l)$$

$$NO_3^-(aq) + 4H^+(aq) + 3e^- \longrightarrow NO(g) + 2H_2O(l)$$

Predict whether NO_3^- ions will oxidize Mn^{2+} to MnO_4^- under standard-state conditions.

24.14 Specify which species in each pair is a better oxidizing agent under standard-state conditions: (a) Br_2 or Au^{3+}, (b) H_2 or Ag^+, (c) Cd^{2+} or Cr^{3+}, (d) O_2 in acidic media or O_2 in basic media.

24.15 Specify which species in each pair is a better reducing agent under standard-state conditions: (a) Na or Li, (b) H_2 or I_2, (c) Fe^{2+} or Ag, (d) Br^- or Co^{2+}.

Relationships among $\mathscr{E}°$, $\Delta G°$, and K

24.16 Calculate $\Delta G°$ and K_c for the following reactions at 25°C.

(a) $Mg(s) + Pb^{2+}(aq) \rightleftharpoons Mg^{2+}(aq) + Pb(s)$
(b) $Br_2(l) + 2I^-(aq) \rightleftharpoons 2Br^-(aq) + I_2(s)$
(c) $O_2(g) + 4H^+(aq) + 4Fe^{2+}(aq) \rightleftharpoons$
$\qquad 2H_2O(l) + 4Fe^{3+}(aq)$
(d) $2Al(s) + 3I_2(s) \longrightarrow 2Al^{3+}(aq) + 6I^-(aq)$

*__24.17__ Given that $\mathscr{E}° = 0.52$ V for the reduction $Cu^+(aq) + e^- \rightarrow Cu(s)$, calculate $\mathscr{E}°$, $\Delta G°$, and K for the following reaction at 25°C: $2Cu^+(aq) \rightarrow Cu^{2+}(aq) + Cu(s)$.

24.18 Use the standard reduction potentials to find the equilibrium constant for each of the following reactions at 25°C.

(a) $Br_2(l) + 2I^-(aq) \rightleftharpoons 2Br^-(aq) + I_2(s)$
(b) $2Ce^{4+}(aq) + 2Cl^-(aq) \rightleftharpoons$
$\qquad Cl_2(g) + 2Ce^{3+}(aq)$
(c) $5Fe^{2+}(aq) + MnO_4^-(aq) + 8H^+(aq) \longrightarrow$
$\qquad Mn^{2+}(aq) + 4H_2O + 5Fe^{3+}(aq)$

24.19 The equilibrium constant for the reaction

$$Sr(s) + Mg^{2+}(aq) \rightleftharpoons Sr^{2+}(aq) + Mg(s)$$

is 2.69×10^{12} at 25°C. Calculate $\mathscr{E}°$ for a cell made up of the Sr/Sr^{2+} and Mg/Mg^{2+} half-cells.

24.20 What is the equilibrium constant for the following reaction at 25°C?

$$Mg(s) + Zn^{2+}(aq) \rightleftharpoons Mg^{2+}(aq) + Zn(s)$$

24.21 What spontaneous reaction will occur in aqueous solution, under standard-state conditions, among the ions Ce^{4+}, Ce^{3+}, Fe^{3+}, and Fe^{2+}? Calculate $\Delta G°$ and K_c for the reaction.

The Nernst Equation

24.22 What is the potential of a cell made up of Zn/Zn^{2+} and Cu/Cu^{2+} half-cells at 25°C if $[Zn^{2+}] = 0.25$ M and $[Cu^{2+}] = 0.15$ M?

24.23 Calculate the standard potential of the cell consisting of the Zn/Zn^{2+} half-cell and the SHE. What will be the emf of the cell if $[Zn^{2+}] = 0.45$ M, $P_{H_2} = 2.0$ atm, and $[H^+] = 1.8$ M?

24.24 What is the emf of a cell consisting of a Pb/Pb^{2+} half-cell and a Pt/H_2/H^+ half-cell if $[Pb^{2+}] = 0.10$ M, $[H^+] = 0.050$ M, and $P_{H_2} = 1.0$ atm?

24.25 Calculate $\mathscr{E}°$, $\mathscr{E}$, and ΔG from the following cell reactions.

(a) $Mg(s) + Sn^{2+}(aq) \longrightarrow Mg^{2+}(aq) + Sn(s)$
$\qquad [Mg^{2+}] = 0.045$ M, $[Sn^{2+}] = 0.035$ M
(b) $3Zn(s) + 2Cr^{3+}(aq) \longrightarrow 3Zn^{2+}(aq) + 2Cr(s)$
$\qquad [Cr^{3+}] = 0.010$ M, $[Zn^{2+}] = 0.0085$ M

24.26 Referring to Figure 24.1 calculate the ratio $[Cu^{2+}]/[Zn^{2+}]$ at which the following reaction will become spontaneous at 25°C:

$$Cu(s) + Zn^{2+}(aq) \longrightarrow Cu^{2+}(aq) + Zn(s)$$

Batteries and Fuel Cells

24.27 Explain clearly the differences between a primary galvanic cell—one that is not rechargeable—and a storage cell (for example, the lead storage battery), which is rechargeable.

24.28 Discuss the advantages and disadvantages of fuel cells over conventional power plants in producing electricity.

*__24.29__ The hydrogen/oxygen fuel cell is described in Section 24.5. (a) What volume of $H_2(g)$, stored at 25°C at a pressure of 155 atm, would be needed to run an electric motor drawing a current of 8.5 amperes for 3.0 h? (b) What volume (L) of air at 25°C and 1.00 atm will have to pass into the cell per minute to run the motor? Assume that air is 20 percent O_2 by volume, and that all the O_2 is consumed in the cell. The other components of air do not affect the fuel-cell reactions. Assume ideal gas behavior.

*__24.30__ Calculate the standard emf of the propane fuel cell discussed on p. 734 at 25°C, given that $\Delta G_f°$ for propane is -23.5 kJ/mol.

Corrosion

24.31 Steel hardware, including nuts and bolts, is often coated with a thin cadmium plating. Explain the function of the cadmium layer.

24.32 "Galvanized iron" is steel sheet coated with zinc; "tin" cans are made of steel sheet coated with tin. Discuss the functions of these coatings and the electrochemistry of the corrosion reactions that occur if an electrolyte contacts the scratched surface of a galvanized iron sheet or a tin can.

*24.33 Tarnished silver contains Ag_2S. The tarnish can be removed from silverware by placing the silver in an aluminum pan containing an inert electrolyte solution, such as NaCl. Explain the electrochemical principle for this procedure.

24.34 How does the tendency of iron to rust depend on the pH of solution?

Electrolysis

24.35 The half-reaction at an electrode is

$$Mg^{2+}(molten) + 2e^- \longrightarrow Mg(s)$$

Calculate the number of grams of magnesium that can be produced by passing 1.00 faraday through the electrode.

24.36 Consider the electrolysis of molten calcium chloride, $CaCl_2$. (a) Write the electrode reactions. (b) How many grams of calcium metal can be produced by passing 0.50 A for 30 min?

24.37 In the electrolysis of water one of the half-reactions is

$$2H_2O(l) \longrightarrow O_2(g) + 4H^+(aq) + 4e^-$$

Suppose 0.076 L of O_2 is collected at 25°C and 755 mmHg. How many faradays of electricity must have passed through the solution?

24.38 Explain why different products are obtained in the electrolysis of molten $ZnCl_2$ and the electrolysis of an aqueous solution of $ZnCl_2$.

24.39 Would it be cheaper to produce a ton of sodium or a ton of aluminum by electrolysis? (Consider only the cost of electricity.)

24.40 How many faradays of electricity are required to produce (a) 0.84 L of O_2 at exactly 1 atm and 25°C from aqueous H_2SO_4 solution; (b) 1.50 L of Cl_2 at 750 mmHg and 20°C from molten NaCl; (c) 6.0 g of Sn from molten $SnCl_2$?

24.41 Calculate the amounts of Cu and Br_2 produced at inert electrodes by passing a current of 4.50 A through a solution of $CuBr_2$ for 1 h.

24.42 In the electrolysis of an aqueous $AgNO_3$ solution, 0.67 g of Ag is deposited after a certain period of time. (a) Write the half-reaction for the reduction of Ag^+. (b) What is the probable oxidation half-reaction? (c) Calculate the quantity of electricity used, in coulombs.

24.43 A steady current was passed through molten $CoSO_4$ until 2.35 g of metallic cobalt was produced. Calculate the number of coulombs of electricity used.

*24.44 A constant electric current flows for 3.75 h through two electrolytic cells connected in series. One contains a solution of $AgNO_3$ and the second a solution of $CuCl_2$, and 2.00 g of silver is deposited in the first cell. (a) How many grams of copper are deposited in the second cell? (b) What is the current flowing, in amperes?

24.45 What is the hourly production rate of chlorine gas (in kg) from an electrolytic cell using aqueous NaCl electrolyte and carrying a current of 1.500×10^3 amperes? The anode efficiency for the oxidation of Cl^- is 93.0 percent.

*24.46 Chromium plating is applied by electrolysis to objects suspended in a dichromate solution, according to the following (unbalanced) half-reaction:

$$Cr_2O_7^{2-}(aq) + e^- + H^+(aq) \longrightarrow Cr(s) + H_2O(l)$$

How long (in h) would it take to apply a chromium plating of thickness 1.0×10^{-2} mm to a car bumper of surface area 0.25 m^2 in an electrolysis cell carrying a current of 25.0 amperes? (The density of chromium = 7.19 g/cm^3.)

24.47 The passage of a current of 0.750 A for 25.0 min deposited 0.369 g of copper from a $CuSO_4$ solution. From this information, calculate the molar mass of copper.

24.48 A quantity of 0.300 g of copper was deposited from a $CuSO_4$ solution by passing a current of 3.00 A for 304 s. Calculate the value of the faraday constant.

24.49 In a certain electrolysis experiment, 1.44 g of Ag were deposited in one cell (containing an aqueous $AgNO_3$ solution), while 0.120 g of an unknown metal X was deposited in another cell (containing an aqueous XCl_3 solution) in series. Calculate the molar mass of X, given that the ion derived from X bears a +3 charge.

24.50 If the cost of electricity to produce magnesium by the electrolysis of molten magnesium chloride is $155 per ton of metal, what is the cost (in dollars) of electricity to produce (a) 10.0 tons of aluminum, (b) 30.0 tons of sodium, (c) 50.0 tons of calcium?

24.51 In the electrolysis of water one of the half-reactions is

$$4H^+(aq) + 4e^- \longrightarrow 2H_2(g)$$

Suppose 0.845 L of H_2 is collected at 25°C and 782 mmHg. How many faradays of electricity must have passed through the solution?

24.52 A current of 6.00 A passes for 3.40 h through an electrolytic cell containing dilute sulfuric acid. If the volume of O_2 gas generated at the anode is

4.26 L (at STP), calculate the charge (in coulombs) on an electron.

Miscellaneous Problems

24.53 Define the following terms: (a) anode, (b) cathode, (c) electrolysis, (d) galvanic cell, (e) faraday, (f) Nernst equation.

24.54 Complete the following table.

$\mathcal{E}$	ΔG	Cell Reaction
> 0	> 0	
$= 0$		

State whether the cell reaction is spontaneous, nonspontaneous, or at equilibrium.

24.55 Distinguish between processes carried out in an electrolytic cell and those carried out in a galvanic cell.

24.56 Most useful galvanic cells give voltages of no more than 1.5 to 2.5 V. Explain why this is so. What are the prospects for developing practical galvanic cells with voltages of 5 V or more?

*24.57 Consider the cell composed of the SHE and a half-cell using the reaction $Ag^+(aq) + e^- \rightarrow Ag(s)$. (a) Calculate the standard cell potential. (b) What is the spontaneous cell reaction under standard-state conditions? (c) Calculate the cell potential when $[H^+]$ in the hydrogen electrode is changed to (i) 1.0×10^{-2} M, and (ii) 1.0×10^{-5} M, all other reagents being held at standard-state conditions. (d) Based on this cell arrangement, suggest a design for a pH meter.

*24.58 From the following information, calculate the solubility product of AgBr.

$$AgBr(s) + e^- \longrightarrow Ag(s) + Br^-(aq) \quad \mathcal{E}° = 0.07 \text{ V}$$
$$Ag^+(aq) + e^- \longrightarrow Ag(s) \quad \quad \quad \mathcal{E}° = 0.80 \text{ V}$$

*24.59 A silver rod and a SHE are dipped into a saturated aqueous solution of silver oxalate, $Ag_2C_2O_4$, at 25°C. The measured potential difference between the rod and the SHE is 0.589 V, the rod being positive. Calculate the solubility product constant for silver oxalate.

*24.60 A galvanic cell consists of a silver electrode in contact with 346 mL of 0.100 M AgNO₃ solution and a magnesium electrode in contact with 288 mL of 0.100 M Mg(NO₃)₂ solution. (a) Calculate

$\mathcal{E}$ for the cell at 25°C. (b) A current is drawn from the cell until 1.20 g of silver has been deposited at the silver electrode. Calculate $\mathcal{E}$ for the cell at this stage of operation.

24.61 Use data in Table 24.1 to explain why nitric acid is able to oxidize Cu to Cu^{2+} but hydrochloric acid cannot.

24.62 Consider a galvanic cell for which the anode reaction is

$$Pb(s) \longrightarrow Pb^{2+}(aq, 0.10 \ M) + 2e^-$$

and the cathode reaction is

$$VO^{2+}(aq, 0.10 \ M) + 2H^+(aq, 0.10 \ M) + e^- \longrightarrow \\ V^{3+}(aq, 1.0 \times 10^{-5} \ M) + H_2O(l)$$

At 25°C the measured emf of the cell is 0.67 V. Calculate (a) the standard reduction potential for the VO^{2+}/V^{3+} couple; (b) the equilibrium constant for the reaction

$$Pb(s) + 2VO^{2+}(aq) + 4H^+(aq) \rightleftharpoons \\ Pb^{2+}(aq) + 2V^{3+}(aq) + 2H_2O(l)$$

*24.63 Explain why chlorine gas can be prepared by electrolyzing an aqueous solution of NaCl but fluorine gas cannot be prepared by electrolyzing an aqueous solution of NaF.

24.64 Suppose you are asked to verify experimentally the electrode reactions shown in Example 24.9. In addition to the apparatus and the solution, you are also given two pieces of litmus paper, one blue and the other red. Describe what steps you would take in this experiment.

24.65 Calculate the emf of the following concentration cell at 25°C:

$$Cu(s)|Cu^{2+}(aq, 0.080 \ M)\|KCl(sat'd)\| \\ Cu^{2+}(aq, 1.2 \ M)|Cu(s)$$

24.66 The cathode reaction in the Leclanché cell is given by

$$2MnO_2(s) + Zn^{2+} + 2e^- \longrightarrow ZnMn_2O_4(s)$$

If a Leclanché cell produces a current of 0.0050 A, calculate how long (in h) this current supply can last if there is initially 4.0 g of MnO₂ present in the cell. Assume that there is an excess of Zn^{2+} ions.

25
NUCLEAR CHEMISTRY

To this point in our development of chemistry principles we have focused attention primarily on the arrangement and behavior of electrons surrounding the nuclei of atoms. However, it is atomic number—the number of protons held in the nucleus of an atom—that is the fundamental feature that distinguishes the 109 known elements from one another. Thus the synthesis of new elements necessarily involves altering the number of protons (and neutrons) held in the nuclei of atoms.

Such alterations in the composition of atomic nuclei are the focus of this chapter. Progress in nuclear science has led to major changes in modern life, both in research laboratories and in everyday affairs. The topics that have contributed to these changes include radioactivity, half-life, and the destructive and constructive applications of nuclear reactions. These and the potential benefits and burdens associated with nuclear changes will be explored here.

750

25.1 THE NATURE OF NUCLEAR REACTIONS

With the exception of hydrogen ($^1_1$H), all nuclei contain two kinds of fundamental particles, called *protons* and *neutrons*. Some nuclei are unstable; they emit particles and/or electromagnetic radiation spontaneously (see Section 7.2). This phenomenon is known as *radioactivity. Nuclei can also undergo change as a result of bombardment by neutrons or by other nuclei,* and nuclear changes that are induced in this way are known as **nuclear transmutation.** Both radioactive decay and nuclear transmutation are examples of *nuclear reactions.* As Table 25.1 shows, nuclear reactions differ from ordinary chemical reactions in several important ways.

Before we discuss specific nuclear reactions, let us see how to write and balance the equations. Writing an equation for a nuclear reaction is more involved than writing an equation for an ordinary chemical reaction. In addition to writing the symbols for various chemical elements, we must also explicitly indicate protons, neutrons, and electrons. In fact, we must show the numbers of protons and neutrons present in *every* species in such an equation.

The symbols for elementary particles are as follows:

$$^1_1\text{p or }^1_1\text{H} \qquad ^1_0\text{n} \qquad ^{\ 0}_{-1}e \text{ or } ^{\ 0}_{-1}\beta \qquad ^{\ 0}_{+1}e \text{ or } ^{\ 0}_{+1}\beta \qquad ^4_2\text{He or } ^4_2\alpha$$

$$\text{proton} \qquad \text{neutron} \qquad \text{electron} \qquad \text{positron} \qquad \text{alpha particle}$$

Following the notations used in Section 7.5, the superscript in each case denotes the mass number and the subscript is the atomic number. Thus, the "atomic number" of a proton is 1, since there is one proton present, and the "mass number" is also 1, because there is one proton but no neutrons present. On the other hand, the "mass number" of a neutron is 1, but its "atomic number" is zero, because there are no protons present. For the electron, the "mass number" is zero (there are neither protons nor neutrons present), but the "atomic number" is −1, because the electron possesses a unit negative charge.

The symbol $^{\ 0}_{-1}e$ represents an electron in or from an atomic orbital. The

The structure of the atomic nucleus was discussed in Chapter 7.

TABLE 25.1 Comparison of Chemical Reactions and Nuclear Reactions

Chemical Reactions	*Nuclear Reactions*
1. Atoms are rearranged by the breaking and forming of chemical bonds.	1. Elements (or isotopes of the same elements) are converted from one to another.
2. Only electrons in atomic orbitals are involved in the breaking and forming of bonds.	2. Protons, neutrons, electrons, and other elementary particles may be involved.
3. Reactions are accompanied by absorption or release of relatively small amounts of energy.	3. Reactions are accompanied by absorption or release of tremendous amounts of energy.
4. Rates of reaction are influenced by temperature, pressure, concentration, and catalysts.	4. Rates of reaction normally are not affected by external parameters.

symbol $_{-1}^{0}\beta$ represents an electron that, although physically identical to any other electron, comes from a nucleus and not from an atomic orbital.

The positron has the same mass as the electron, but bears a $+1$ charge. The α particle has two protons and two neutrons, so its atomic number is 2 and its mass number is 4.

In balancing any nuclear equation, we observe the following rules:

- The total number of protons and neutrons in the products and in the reactants must be the same (conservation of mass number).
- The total number of nuclear charges in the products and in the reactants must be the same (conservation of atomic number).

If the atomic numbers and mass numbers of all the species but one in a nuclear equation are known, the unknown species can be identified by applying these rules, as shown in the following example.

EXAMPLE 25.1

Balance the following nuclear equations (that is, identify the product X):

(a) $_{84}^{212}Po \longrightarrow _{82}^{208}Pb + X$

(b) $_{55}^{137}Cs \longrightarrow _{56}^{137}Ba + X$

(c) $_{11}^{20}Na \longrightarrow _{10}^{20}Ne + X$

Answer

(a) The mass number and atomic number are 212 and 84 respectively on the left-hand side and 208 and 82 respectively on the right-hand side. Thus, X must have a mass number of 4 and an atomic number of 2, which means that it is an α particle. The balanced equation is

$$_{84}^{212}Po \longrightarrow _{82}^{208}Pb + _{2}^{4}He$$

(b) In this case the mass number is the same on both sides of the equation, but the atomic number of the product is one more than that of the reactant. The only way this change can come about is to have a neutron in the Cs nucleus transformed into a proton and an electron, that is, $_{0}^{1}n \rightarrow _{1}^{1}p + _{-1}^{0}\beta$ (note that this process does not alter the mass number). Thus, the balanced equation is

$$_{55}^{137}Cs \longrightarrow _{56}^{137}Ba + _{-1}^{0}\beta$$

(c) As in (b), the mass number is the same on both sides of the equation, but the atomic number of the product is one *fewer* than that of the reactant. The only way this change can come about is to have one of the protons in the Na nucleus transformed into a neutron and a positron, that is, $_{1}^{1}p \rightarrow _{0}^{1}n + _{+1}^{0}\beta$ (note that this process does not alter the mass number). Thus the balanced equation is

$$_{11}^{20}Na \longrightarrow _{10}^{20}Ne + _{+1}^{0}\beta$$

Similar example: Problem 25.3.

We use the $_{-1}^{0}\beta$ notation here because the electron came from the nucleus.

25.2 NUCLEAR STABILITY

The volume of a nucleus is very small compared to the total volume of the atom, but the nucleus has a very high density because the protons and neutrons are tightly packed in the small space (see Section 7.6). Since all protons are positively charged, Coulomb's law would lead us to expect strong repulsive forces that would cause the components of the nucleus to fly apart. However, in addition to the repulsion, there also exist short-range attractive forces between proton and proton, proton and neutron, and neutron and neutron. The stability of any nucleus is therefore determined by the difference between coulombic repulsion and these short-range nuclear forces. We will not discuss the nature of these forces here. Instead, let us consider the empirical rules for predicting nuclear stability.

An unstable nucleus results in radioactive decay.

The principal factor for determining whether a nucleus is stable is the *neutron-to-proton ratio* (n:p). For stable atoms of elements of low atomic number, the n:p value is close to 1. As the atomic number increases, the neutron-to-proton ratios of the stable nuclei become greater than 1. This deviation arises at higher atomic numbers because a larger number of neutrons is needed to stabilize the nucleus by counteracting the strong repulsion among the protons. The following rules are useful in predicting nuclear stability.

- Nuclei that contain 2, 8, 20, 50, 82, or 126 protons or neutrons are generally more stable than nuclei that do not possess these numbers. For example, there are ten stable isotopes of tin (Sn) with the atomic number 50, and only two stable isotopes of antimony (Sb) with the atomic number 51. The numbers 2, 8, 20, 50, 82, and 126 are called *magic numbers*. These numbers have a significance in nuclear stability similar to the numbers of electrons associated with the very stable noble gases (that is, 2, 10, 18, 36, 54, and 86 electrons).
- Nuclei with even numbers of both protons and neutrons are generally more stable than those with odd number of nucleons (Table 25.2).
- All isotopes of the elements starting with polonium (Po, $Z = 84$) are radioactive. All isotopes of technetium (Tc, $Z = 43$) and promethium (Pm, $Z = 61$) are also radioactive.

Figure 25.1 shows a plot of the number of neutrons versus the number of protons in various isotopes. The stable nuclei are located in an area of the graph known as the *belt of stability*. Most of the radioactive nuclei lie outside this belt. Above the stability belt, the nuclei have higher neutron-

TABLE 25.2 Number of Stable Isotopes with Even and Odd Numbers of Protons and Neutrons

Protons	Neutrons	Number of Stable Isotopes
odd	odd	8
odd	even	50
even	odd	52
even	even	157

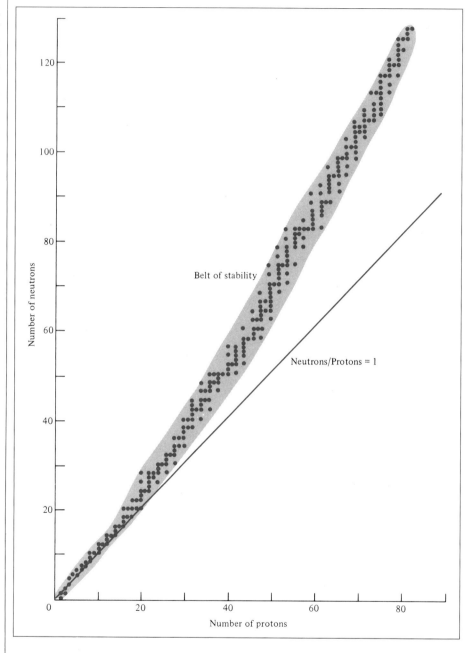

to-proton ratios than those within the belt (for the same number of protons). To lower this ratio (and hence move down toward the belt of stability), these nuclei undergo the following process, called *beta particle emission*:

$$_0^1n \longrightarrow \, _1^1p + \, _{-1}^0\beta$$

Beta particle emission leads to an increase in the number of protons and a simultaneous decrease in the number of neutrons. Some examples are

$$^{14}_{6}C \longrightarrow {}^{14}_{7}N + {}^{0}_{-1}\beta$$

$$^{40}_{19}K \longrightarrow {}^{40}_{20}Ca + {}^{0}_{-1}\beta$$

$$^{97}_{40}Zr \longrightarrow {}^{97}_{41}Nb + {}^{0}_{-1}\beta$$

Below the stability belt the nuclei have lower neutron-to-proton ratios than those in the belt (for the same number of protons). To increase this ratio (and hence move up toward the belt of stability), these nuclei either emit a positron

$$^{1}_{1}p \longrightarrow {}^{1}_{0}n + {}^{0}_{+1}\beta$$

or undergo electron capture. An example of positron emission is

$$^{38}_{19}K \longrightarrow {}^{38}_{18}Ar + {}^{0}_{+1}\beta$$

Electron capture is the capture of an electron by the nucleus in an atom, usually a $1s$ electron. This process has the same net effect as positron emission:

$$^{37}_{18}Ar + {}^{0}_{-1}e \longrightarrow {}^{37}_{17}Cl$$

$$^{55}_{26}Fe + {}^{0}_{-1}e \longrightarrow {}^{55}_{25}Mn$$

A quantitative measure of nuclear stability is the nuclear binding energy, discussed in Section 7.6.

> Note that we use ${}_{-1}^{0}e$ rather than ${}_{-1}^{0}\beta$ for the electron here because it came from an atomic orbital and not from the nucleus.

25.3   NATURAL RADIOACTIVITY

Nuclei outside the belt of stability, as well as nuclei with more than 83 protons, tend to be unstable. Radioactivity is the spontaneous emission, by unstable nuclei, of particles or electromagnetic radiation, or both. The three types of radiation are α particles (or doubly charged helium nuclei, He^{2+}), β particles (or electrons), and γ rays, which are very short wavelength (0.1 nm to 10^{-4} nm) electromagnetic waves, as discussed in Section 7.2.

When a radioactive nucleus disintegrates, the products formed may also be unstable and therefore will undergo further disintegration. This process is repeated until a stable product finally is formed. Starting with the original radioactive nucleus, the sequence of disintegration steps is called a *decay series*. Table 25.3 shows the decay series of naturally occurring uranium-238, which involves 16 steps. This decay scheme is known as the *uranium series*.

It is important that we be able to balance the nuclear reaction for any one of the steps. For example, the first step is the decay of uranium-238 to thorium-234, with the emission of an α particle. Hence, the reaction is

$$^{238}_{92}U \longrightarrow {}^{234}_{90}Th + {}^{4}_{2}He$$

The next step is represented by

$$^{234}_{90}Th \longrightarrow {}^{234}_{91}Pa + {}^{0}_{-1}\beta$$

and so on. In a discussion of radioactive decay steps, the beginning radioactive isotope is sometimes called the *parent* and the product the *daughter*.

TABLE 25.3   The Uranium Decay Series

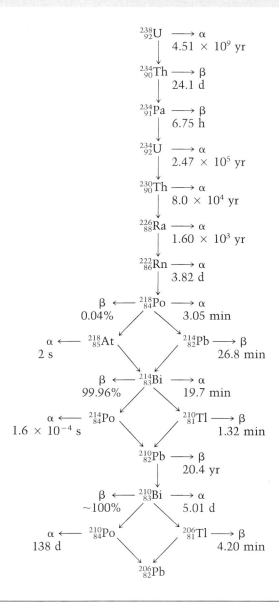

Kinetics of Radioactive Decay

An interesting feature about radioactive decay is that they all obey first-order kinetics. This means that the rate of radioactive decay at any time t is given by

$$\text{rate of decay at time } t = kN$$

where k is the first-order rate constant and N is the number of radioactive nuclei present at time t. The half-life for such a reaction is given by [see Equation (22.6)]

$$t_{\frac{1}{2}} = \frac{0.693}{k}$$

EXAMPLE 25.2

Strontium-90 is a β emitter with a half-life of 28.8 years. How many years will it take for 99.0 percent of a given sample of $^{90}_{38}Sr$ released in an atmospheric test of an atomic bomb to disintegrate?

Answer

The first step is to calculate the first-order constant for the decay. From Equation (22.6)

$$k = \frac{0.693}{t_{\frac{1}{2}}} = \frac{0.693}{28.8 \text{ yr}} = 2.41 \times 10^{-2}/\text{yr}$$

Next we write [see Equation (22.3)]

$$k = \frac{2.303}{t} \log \frac{N_0}{N_t}$$

where N_0 and N_t are the fractions of strontium-90 nuclei present at $t = 0$ and $t = t$. From the data given, $N_0 = 1.00$ (100 percent) and $N_t = 1.00 - 0.99 = 0.0100$; therefore

$$t = \frac{2.303}{2.41 \times 10^{-2}/\text{yr}} \log \frac{1.00}{0.0100}$$
$$= 191 \text{ yr}$$

Similar examples: Problems 25.14, 25.18.

The half-lives (and hence the rate constants) of radioactive decay vary greatly from nucleus to nucleus. For example, using Table 25.3, we find two extreme cases:

$$^{238}_{92}U \longrightarrow ^{234}_{90}Th + ^{4}_{2}He \qquad t_{\frac{1}{2}} = 4.51 \times 10^{9} \text{ yr}$$

$$^{214}_{84}Po \longrightarrow ^{210}_{82}Pb + ^{4}_{2}He \qquad t_{\frac{1}{2}} = 1.6 \times 10^{-4} \text{ s}$$

The ratio of these two rate constants after conversion to the same time unit is about 1×10^{21}, an enormously large number. Further, the rate constants are unaffected by changes in the environment, such as temperature and pressure. These highly unusual features are not encountered in ordinary chemical reactions (see Table 25.1).

We do not have to spend 4.51×10^9 years to make a half-life measurement of uranium-238. Its value can be calculated from the rate constant using Equation (22.6).

Dating Based on Radioactive Decay

The half-lives of radioactive decay have been used as "atomic clocks" to determine the ages of various objects. Two examples of dating by radioactive decay measurements will be described here.

Radiocarbon Dating. Earth's atmosphere is constantly bombarded by cosmic rays from outer space. They consist of electrons, neutrons, and atomic nuclei. These rays have extremely high penetrating power. One of the important reactions between the atmosphere and cosmic rays is the capture of neutrons by nitrogen atoms to produce carbon and hydrogen:

$$^{14}_{7}N + ^{1}_{0}n \longrightarrow ^{14}_{6}C + ^{1}_{1}H$$

The unstable carbon atoms are eventually converted into $^{14}CO_2$, which mixes with the ordinary carbon dioxide ($^{12}CO_2$) in the air. The radioactive carbon-14 isotope decays according to the equation

$$^{14}_{6}C \longrightarrow ^{14}_{7}N + ^{0}_{-1}\beta$$

As we have just discussed, the rate of decay (measured by the number of electrons emitted per second) obeys first-order kinetics—that is

$$rate = kN$$

In this case, N is the number of ^{14}C nuclei present. If we measure the rate of radioactive decay at a given time ($t = 0$) and then at a time interval t later, we can determine the first-order rate constant using Equation (22.3) as follows:

$$\log \frac{N_0}{N_t} = \frac{kt}{2.303}$$

$$k = \frac{2.303}{t} \log \frac{N_0}{N_t}$$

$$= \frac{2.303}{t} \log \frac{(\text{decay rate at } t = 0)}{(\text{decay rate at } t = t)}$$

With the measured first-order rate constant k (1.21×10^{-4}/yr) we can calculate the half-life for the decay using Equation (22.6):

$$t_{\frac{1}{2}} = \frac{0.693}{1.21 \times 10^{-4}/\text{yr}} = 5.73 \times 10^3 \text{ yr}$$

Thus, half of any given amount of ^{14}C isotope will disappear by decay over a period of 5730 years.

The carbon-14 isotopes enter all living organisms via plant photosynthesis, as well as through air and water. Plants are eaten by animals, which exhale carbon-14 in CO_2. Eventually, the carbon-14 isotopes participate in many aspects of the carbon cycle. The carbon lost by decay is constantly replenished by the production of new carbon-14 atoms in the atmosphere, and a dynamic equilibrium is established, whereby the ratio of ^{14}C to ^{12}C remains constant in living matter. When a plant or an animal dies, though, its supply of carbon-14 isotopes is no longer replenished, so the ratio in the organism decreases as ^{14}C decays. We find this happening also in carbon atoms trapped in Egyptian mummies and in coal, petroleum, and wood preserved underground. After a number of years, there are proportionately fewer carbon-14 isotopes in, say, a mummy than in a living person.

In 1955, Willard F. Libby (1908–1980) suggested that this fact could be used to estimate the length of time the carbon-14 isotope in a particular specimen has been decaying without replenishment. For this purpose we rearrange Equation (22.3) as follows:

$$t = \frac{2.303}{k} \log \frac{\text{decay rate of fresh sample}}{\text{decay rate of old sample}}$$

where t is the age of the sample. This ingenious technique is based on a remarkably simple idea involving rate measurements; its success depends on how accurately we can measure the rate of decay. In fresh samples, the ratio $^{14}C/^{12}C$ is about $1/10^{12}$, that is, there is one carbon-14 isotope for every 1×10^{12} carbon-12 isotopes present. Very sensitive equipment is needed to monitor the radioactive decay. Precision is more difficult with older samples because there are even fewer carbon-14 nuclei present. But despite these difficulties, radiocarbon dating has become an extremely valuable tool for estimating the age of archeological artifacts, paintings, and other objects dating back 1000 to 50,000 years.

Dating Using Uranium-238 Isotopes. Because some members of the uranium series have very long half-lives (see Table 25.3), it is particularly suitable for estimating the age of rocks in Earth and of extraterrestrial objects. The half-life for the first step ($^{238}_{92}U$ to $^{234}_{90}Th$) is 4.51×10^9 yr. This is about 20,000 times greater than the second largest value (that is, 2.47×10^5 yr, which is the half-life for $^{234}_{92}U$ to $^{230}_{90}Th$). Therefore, as a good approximation we can assume the half-life for the overall process (that is, from $^{238}_{92}U$ to $^{206}_{82}Pb$) to be governed solely by the first step:

We can think of the first step as the rate-determining step in the overall process.

$$^{238}_{92}U \longrightarrow \, ^{206}_{82}Pb + 8\,^4_2He + 6\,_{-1}^{0}\beta \qquad t_{\frac{1}{2}} = 4.51 \times 10^9 \text{ yr}$$

In naturally occurring uranium minerals, we should and do find some lead-206 isotopes formed by radioactive decay. Assuming that no lead was present when the mineral was formed and that the mineral has not undergone chemical changes that would allow the lead-206 isotope to be separated from the parent uranium-238, it is possible to estimate the age of the rocks from the mass ratio of $^{206}_{82}Pb$ to $^{238}_{92}U$. The preceding equation tells us that for every mole or 238 g of uranium that undergo complete decay, one mole or 206 g of lead are formed. If only half a mole of uranium-238 has undergone decay, the mass ratio Pb-206/U-238 becomes

$$\frac{206 \text{ g}/2}{238 \text{ g}/2} = 0.866$$

and the process would have taken a half-life or 4.51×10^9 years to complete. Ratios lower than 0.866 mean that the rocks are less than 4.51×10^9 years old, and higher ratios suggest a greater age. Interestingly, studies based on the uranium series as well as other decay series put the age of the oldest rocks, and, therefore, probably the age of Earth itself, at 4.5×10^9 or 4.5 billion years.

25.4 ARTIFICIAL RADIOACTIVITY

Nuclear Transmutation

The scope of nuclear chemistry would have been rather narrow if study had been limited to natural radioactive elements. An experiment performed by

Rutherford in 1919, however, suggested the possibility of observing artificial radioactivity. When he bombarded a sample of nitrogen with α particles, the following reaction took place:

$$^{14}_{7}\text{N} + ^{4}_{2}\text{He} \longrightarrow ^{17}_{8}\text{O} + ^{1}_{1}\text{p}$$

This reaction demonstrated for the first time the feasibility of converting one element into another—that is, of *nuclear transmutation.*

The above reaction can be abbreviated as $^{14}_{7}\text{N}(\alpha,\text{p})^{17}_{8}\text{O}$. Note that in the parentheses the bombarding particle is written first, followed by the ejected particle. Nuclear transmutations differ from radioactive decay in that the latter are spontaneous processes; consequently, in decay equations only *one* reactant appears on the left side of the equation.

EXAMPLE 25.3

Write the balanced equation for the nuclear reaction $^{56}_{26}\text{Fe}(\text{d},\alpha)^{54}_{25}\text{Mn}$, where d represents the deuterium nucleus (that is, $^{2}_{1}\text{H}$).

Answer

The abbreviation tells us that when iron-56 is bombarded with a deuterium nucleus, it produces the manganese-54 nucleus plus an α particle. Thus, the equation for this reaction is

$$^{56}_{26}\text{Fe} + ^{2}_{1}\text{H} \longrightarrow ^{4}_{2}\text{He} + ^{54}_{25}\text{Mn}$$

Similar example: Problem 25.4.

Although light elements are generally not radioactive, they can be made so by bombarding their nuclei with appropriate particles. As we saw earlier, the radioactive carbon-14 isotope can be prepared by bombarding nitrogen-14 with neutrons:

$$^{14}_{7}\text{N} + ^{1}_{0}\text{n} \longrightarrow ^{14}_{6}\text{C} + ^{1}_{1}\text{H}$$

Tritium $(^{3}_{1}\text{H})$ is prepared according to the following reaction:

$$^{6}_{3}\text{Li} + ^{1}_{0}\text{n} \longrightarrow ^{3}_{1}\text{H} + ^{4}_{2}\text{He}$$

Tritium decays with the emission of β particles:

$$^{3}_{1}\text{H} \longrightarrow ^{3}_{2}\text{He} + ^{0}_{-1}\beta \qquad t_{\frac{1}{2}} = 12.5 \text{ yr}$$

Many synthetic isotopes are prepared by using neutrons as projectiles. This is particularly convenient because neutrons carry no charges and are therefore not repelled by the targets—the nuclei. The situation is different when the projectiles are positively charged particles—for example, when protons or α particles are used, as in

$$^{27}_{13}\text{Al} + ^{4}_{2}\text{He} \longrightarrow ^{30}_{15}\text{P} + ^{1}_{0}\text{n}$$

To react with the aluminum nuclei, the α particles must possess very large kinetic energies in order to overcome the electrostatic repulsion between them and the target atoms. Charged species are forced to acquire great speed in *particle accelerators,* which make use of the influence of external electric and magnetic fields on charged particles. Alternating the polarity on specially constructed plates causes the particles to accelerate along a spiral path. When they have the energy necessary to carry out the desired nuclear reaction, they are guided out of the accelerator to strike a target substance (Figure 25.2).

Various designs have been developed for particle accelerators, one of which accelerates particles along a linear path of about 3 km (2 miles). It is now possible to accelerate particles to a speed well above 90 percent of the speed of light. (According to Einstein's theory of relativity, it is impossible for a particle to move *at* the speed of light. The only exception is the photon, which has a zero rest mass.) The extremely energetic particles produced in accelerators are employed by physicists to smash atomic nuclei to bits. Studying the debris from such disintegrations provides valuable information about nuclear structure and binding forces (Figure 25.3).

The Transuranium Elements

The advent of particle accelerators has also made possible the synthesis of *elements with atomic numbers greater than 92,* called **transuranium elements.** Since neptunium ($Z = 93$) was first prepared in 1940, sixteen other transuranium elements have been synthesized. All isotopes of these elements are radioactive. Table 25.4 shows their preparations.

Naming of Elements 104 and Beyond.   Until recently, the honor of naming a newly synthesized element went to its discoverers. However, there

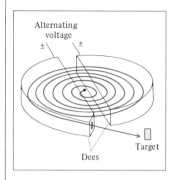

FIGURE 25.2 *Schematic diagram of a cyclotron accelerator. The particle (an ion) to be accelerated starts at the center (the black spot) and is forced to move in a spiral path by the influence of electrical and magnetic fields until it emerges at a high velocity. The magnetic fields are perpendicular to the plane of the dees, which are hollow and serve as electrodes.*

FIGURE 25.3 *A particle accelerator.*

TABLE 25.4 The Transuranium Elements

Atomic Number	Name	Symbol	Preparation
93	Neptunium	Np	$^{238}_{92}U + ^1_0n \longrightarrow ^{239}_{93}Np + ^0_{-1}\beta$
94	Plutonium	Pu	$^{239}_{93}Np \longrightarrow ^{239}_{94}Pu + ^0_{-1}\beta$
95	Americium	Am	$^{239}_{94}Pu + ^1_0n \longrightarrow ^{240}_{95}Am + ^0_{-1}\beta$
96	Curium	Cm	$^{239}_{94}Pu + ^4_2He \longrightarrow ^{242}_{96}Cm + ^1_0n$
97	Berkelium	Bk	$^{241}_{95}Am + ^4_2He \longrightarrow ^{243}_{97}Bk + 2\,^1_0n$
98	Californium	Cf	$^{242}_{96}Cm + ^4_2He \longrightarrow ^{245}_{98}Cf + ^1_0n$
99	Einsteinium	Es	$^{238}_{92}U + 15\,^1_0n \longrightarrow ^{253}_{99}Es + 7\,^0_{-1}\beta$
100	Fermium	Fm	$^{238}_{92}U + 17\,^1_0n \longrightarrow ^{255}_{100}Fm + 8\,^0_{-1}\beta$
101	Mendelevium	Md	$^{253}_{99}Es + ^4_2He \longrightarrow ^{256}_{101}Md + ^1_0n$
102	Nobelium	No	$^{246}_{96}Cm + ^{12}_6C \longrightarrow ^{254}_{102}No + 4\,^1_0n$
103	Lawrencium	Lr	$^{252}_{98}Cf + ^{10}_5B \longrightarrow ^{257}_{103}Lr + 5\,^1_0n$
104	Unnilquadium	Unq	$^{249}_{98}Cf + ^{12}_6C \longrightarrow ^{257}_{104}Unq + 4\,^1_0n$
105	Unnilpentium	Unp	$^{249}_{98}Cf + ^{15}_7N \longrightarrow ^{260}_{105}Unp + 4\,^1_0n$
106	Unnilhexium	Unh	$^{249}_{98}Cf + ^{18}_8O \longrightarrow ^{263}_{106}Unh + 4\,^1_0n$
107	Unnilseptium	Uns	$^{209}_{83}Bi + ^{54}_{24}Cr \longrightarrow ^{262}_{107}Uns + ^1_0n$
108	Unniloctium	Uno	$^{208}_{82}Pb + ^{58}_{26}Fe \longrightarrow ^{265}_{108}Uno + ^1_0n$
109	Unnilennium	Une	$^{209}_{83}Bi + ^{58}_{26}Fe \longrightarrow ^{266}_{109}Une + ^1_0n$

has been considerable controversy over the naming of some of the transuranium elements that have been independently synthesized by different groups of workers at about the same time. To resolve the controversies, and to systematize the naming of these elements, the International Union of Pure and Applied Chemistry (IUPAC) has recommended the use of three-letter symbols for elements having atomic numbers greater than 103. All the names end in *-ium*, for metals and nonmetals alike. The following roots are used to derive the names of the elements:

0 = nil		5 = pent	
1 = un		6 = hex	
2 = bi		7 = sept	
3 = tri		8 = oct	
4 = quad		9 = enn	

To form the name of an element, the roots are put together in the order of the digits that make up the atomic number, ending with *-ium*. For example, element 104 is unnilquadium (Unq), element 116 would be ununhexium (Uuh), element 202 would be binilbium (Bnb), and so on. (The *i* in *bi* is omitted when the root occurs before the *ium*).

25.5 NUCLEAR FISSION

In a ***nuclear fission*** event, a *heavy nucleus (mass number > 200) divides to form smaller nuclei of intermediate mass and one or more neutrons.* Because the heavy nucleus is less stable than its products (see Figure 7.5), this process releases a large amount of energy.

The first nuclear fission reaction studied was that of uranium-235 bom-

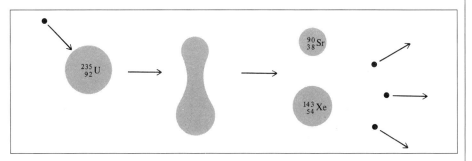

FIGURE 25.4 *Upon capture of a neutron (black sphere), a U-235 nucleus undergoes fission to yield two smaller nuclei. On the average, three neutrons are emitted for every U-235 nucleus that divides.*

barded with slow neutrons, whose speed is comparable to that of air molecules at room temperature. Uranium-235 in such bombardment undergoes fission, as shown in Figure 25.4. Actually, the reaction is very complex: More than thirty different elements have been found among the fission products (Figure 25.5). A representative reaction is

Slow neutrons are also called *thermal neutrons.*

$$^{235}_{92}\text{U} + ^{1}_{0}\text{n} \longrightarrow ^{90}_{38}\text{Sr} + ^{143}_{54}\text{Xe} + 3\,^{1}_{0}\text{n}$$

Although many heavy nuclei can be made to undergo fission, only the fission of naturally occurring uranium-235 and the artificial plutonium-239 has any practical importance. As Figure 7.5 shows, the binding energy per nucleon for uranium-235 is less than the binding energies for elements with mass numbers near 90 and 150. Therefore, when a uranium-235 nucleus is split into two smaller nuclei, a certain amount of energy is released. Let us estimate the magnitude of this energy, using the preceding reaction as an example. Table 25.5 shows the nuclear binding energies of uranium-235 and its fission products. The difference between the binding energies of the reactants and products is $[(1.23 \times 10^{-10} + 1.92 \times 10^{-10})\,\text{J} - 2.82 \times 10^{-10}\,\text{J}]$, or 3.3×10^{-11} J per nucleus. For one mole of uranium-235, the energy released would be $(3.3 \times 10^{-11})(6.02 \times 10^{23})$, or 2.0×10^{13} J. This is an extremely exothermic reaction, considering that the heat of combustion of one ton of coal is only about 8×10^{7} J.

The significant feature of uranium-235 fission is not just the enormous amount of energy released, but the fact that more neutrons are produced than are originally captured in the process. This property makes possible a **nuclear chain reaction,** which is a *self-sustaining sequence of nuclear fission reactions.* The neutrons generated during the initial stages of fission can induce fission in other uranium-235 nuclei, which in turn produce more neutrons, and so on. In less than a second, the reaction can become uncontrollable, liberating a tremendous amount of heat to the surroundings.

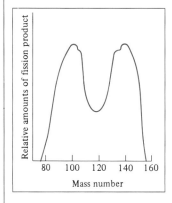

FIGURE 25.5 *The relative yields of the products resulting from the fission of U-235, as a function of mass number.*

TABLE 25.5 Nuclear Binding Energies of Uranium-235 and Its Fission Products

Isotope	Nuclear Binding Energy per Nucleus
Uranium-235	2.82×10^{-10} J
Strontium-90	1.23×10^{-10} J
Xenon-143	1.92×10^{-10} J

FIGURE 25.6 *Two types of nuclear fission. (a) If the mass of U-235 is subcritical, no chain reaction will result. Many of the neutrons produced will escape to the surroundings. (b) In a critical mass, most of the neutrons will be captured by U-235 nuclei and an uncontrollable chain reaction will occur.*

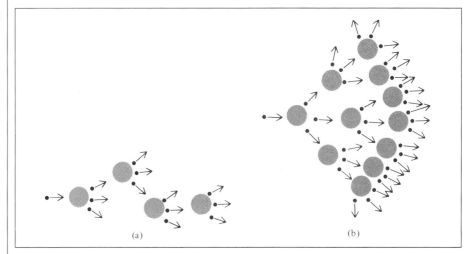

(a)   (b)

Figure 25.6 shows two types of fission reactions. In order for a chain reaction to occur, enough uranium-235 must be present in the sample to capture the neutrons. Otherwise, many of the neutrons will escape from the sample and the chain reaction will not occur [Figure 25.6(a)]. In this situation, the mass of the sample is said to be *subcritical*. Figure 25.6(b) shows what happens when the amount of the fissionable material is equal to or greater than the **critical mass,** the *minimum mass of fissionable material required to generate a self-sustaining nuclear chain reaction.* In this case, most of the neutrons will be captured by uranium-235 and a chain reaction will occur.

The Atomic Bomb

The first application of nuclear fission was in the development of the atomic bomb. How is such a bomb made and detonated? The crucial factor in the bomb's design is the determination of the critical mass for the bomb. A small atomic bomb is equivalent to 20,000 tons of TNT. Since one ton of TNT releases about 4×10^9 J of energy, 20,000 tons would produce 8×10^{13} J. Earlier we saw that one mole, or 235 g, of uranium-235 liberates 2.0×10^{13} J of energy; thus the mass of the isotope present in a small bomb must be at least

TNT stands for trinitrotoluene and is a powerful explosive (see p. 478).

$$235 \text{ g} \times \frac{8 \times 10^{13} \text{ J}}{2.0 \times 10^{13} \text{ J}} \simeq 1 \times 10^3 \text{ g} = 1 \text{ kg}$$

For obvious reasons, an atomic bomb is never assembled with the critical mass already present. Instead, the critical mass is formed by using a conventional explosive (for example, TNT), as shown in Figure 25.7. Uranium-235 was the fissionable material in the bomb dropped on Hiroshima, Japan, on August 6, 1945. Plutonium-239 was used in the bomb exploded over Nagasaki three days later. The fission reactions were similar in these two cases, as was the extent of the destruction.

Nuclear Reactors

A peaceful but controversial application of nuclear fission is the generation of electricity using heat from a controlled chain reaction in a nuclear reactor. Currently, nuclear reactors provide about 8 percent of the electrical energy in the United States. This is a small but by no means negligible contribution to the nation's energy production. Several different types of nuclear reactors are in existence; we will briefly discuss the main features of three of them: light water reactors, heavy water reactors, and breeder reactors.

Light Water Reactors. Most of the nuclear reactors in the United States are *light water reactors.* Figure 25.8 is a schematic diagram of such a reactor and Figure 25.9 shows the refueling process in the core of a nuclear reactor.

An important aspect of the fission process is the speed of the neutrons. Slow neutrons split uranium-235 nuclei more efficiently than do fast ones. Because fission reactions are so exothermic, the neutrons produced usually move at high velocities. Hence, for greater efficiency they must be slowed down before they can be used to induce nuclear disintegration. To accomplish this goal, scientists use **moderators,** which are *substances that can reduce the kinetic energy of neutrons.* A good moderator must satisfy several requirements: It must be a fluid so it can be used also as a coolant; it should possess a high specific heat; it should be nontoxic and inexpensive (as very large quantities of it are necessary); and it should resist conversion into a radioactive substance by neutron bombardment. There is no substance that fits all of these requirements, although water comes closer than many others that have been considered.

The nuclear fuel consists of uranium, usually in the form of its oxide, U_3O_8. Naturally occurring uranium contains about 0.7 percent of the uranium-235 isotope, which is too low a concentration to sustain a small-scale

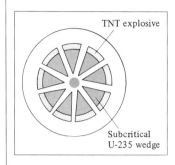

FIGURE 25.7 *Schematic cross-section of an atomic bomb. The TNT explosives are set off first. The explosion forces the sections of fissionable material together to form an amount considerably larger than the critical mass. The nuclear chain reaction is triggered by neutrons from the neutron source at the center.*

Nuclear reactors using H_2O as moderator are called light water reactors because H is the lightest isotope of the element hydrogen.

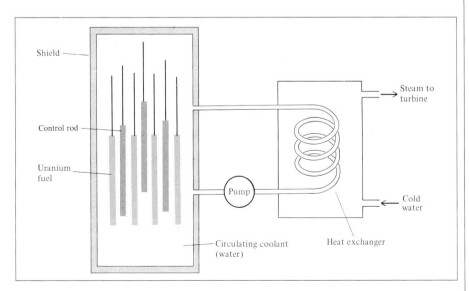

FIGURE 25.8 *Schematic diagram of a nuclear fission reactor. The fission process is controlled by Cd or B rods. The heat generated by the process is used to produce steam for the generation of electricity via a heat-exchange system.*

FIGURE 25.9 *Refueling in the core of a nuclear reactor.*

chain reaction; therefore, it must be enriched to 3 or 4 percent for the effective operation of a light water reactor. In principle, the only difference between an atomic bomb and a nuclear reactor is that the chain reaction that takes place in a nuclear reactor is kept under control at all times. The factor limiting the rate of the reaction is the number of neutrons present. This can be controlled by lowering cadmium or boron rods between the fuel elements. These rods capture neutrons according to the equations:

$$^{113}_{48}\text{Cd} + ^{1}_{0}\text{n} \longrightarrow ^{114}_{48}\text{Cd} + \gamma$$

$$^{10}_{5}\text{B} + ^{1}_{0}\text{n} \longrightarrow ^{7}_{3}\text{Li} + ^{4}_{2}\text{He}$$

where γ denotes gamma rays. Without the control rods, the heat generated would melt down the reactor core, releasing radioactive materials into the environment.

Nuclear reactors have rather elaborate cooling systems that absorb the heat generated by the nuclear reaction and transfer it outside the reactor core, where it is used to produce enough steam to drive an electric generator.

In this respect, a nuclear power plant is similar to a conventional power plant that burns fossil fuel. In both cases, large quantities of cooling water are needed to condense steam for reuse. Thus, most nuclear power plants are built near a river or a lake. Unfortunately, this method of cooling causes thermal pollution (see Section 15.5.).

Heavy Water Reactors. Another type of nuclear reactor uses D_2O, or heavy water, as the moderator, rather than H_2O. Heavy water slows down the neutrons emerging from the fission reaction less efficiently than does light water. Consequently, there is no need to use enriched uranium for fission in a *heavy water reactor*. The faster moving neutrons travel greater distances and hence the probability that they will strike the proper targets— uranium-235 isotopes—is correspondingly high. Eventually, most of the uranium-235 isotopes will take part in the fission process.

The main advantage of a heavy water reactor is that it eliminates the need for building expensive uranium enrichment facilities. However, D_2O must be prepared either by fractional distillation or electrolysis of ordinary water, which can be very expensive considering the quantity of water used in a nuclear reactor. In countries such as Canada and Norway, where hydroelectric power is abundant, the cost of producing D_2O by electrolysis can be kept reasonably low. At present, Canada is the only nation successfully using heavy water nuclear reactors. The fact that no enriched uranium is required in a heavy water reactor allows a country to enjoy the benefits of nuclear power without undertaking work that is closely associated with weapons technology.

D_2O costs about $150 per pound.

Breeder Reactors. Both light water and heavy water nuclear reactors depend on a constant supply of uranium ores. However, the world's supply of uranium is quite limited. Estimates show that at the present rate of consumption, we will exhaust the total supply of uranium around the year 2020. A new type of nuclear reactor, called the ***breeder reactor,*** is being developed in anticipation of this shortage. A breeder reactor uses uranium fuel, but it differs from conventional nuclear reactors in that *it produces more fissionable materials than it uses.*

It is known that when uranium-238 is bombarded with fast neutrons, the following reactions take place:

$$^{238}_{92}U + ^{1}_{0}n \longrightarrow ^{239}_{92}U$$
$$^{239}_{92}U \longrightarrow ^{239}_{93}Np + ^{0}_{-1}\beta \qquad t_{\frac{1}{2}} = 23.4 \text{ min}$$
$$^{239}_{93}Np \longrightarrow ^{239}_{94}Pu + ^{0}_{-1}\beta \qquad t_{\frac{1}{2}} = 2.35 \text{ days}$$

In this manner, the nonfissionable uranium-238 is transmuted into the fissionable isotope plutonium-239, which has a half-life of 24,400 years (see Color Plate 27).

In a typical breeder reactor, nuclear fuel is mixed with uranium-238 so that breeding takes place within the core. For every uranium-235 (or plutonium-239) nucleus undergoing fission, more than one neutron is captured by uranium-238 to generate plutonium-239. Thus, the stockpile of fissionable material can be steadily increased as the starting nuclear fuels are consumed. An important factor is the *doubling time,* the time required to produce as much net additional nuclear fuel as was originally present in the reactor. About 7 to 10 years are needed to regenerate the sizable amount of material needed to refuel the original reactor and to fuel another reactor of comparable size.

A *fertile isotope* is one that can be transmuted to a fissionable isotope.

Another fertile isotope is $^{232}_{90}$Th. Upon capturing slow neutrons, thorium is transmuted to uranium-233, which, like uranium-235, is a fissionable isotope:

$$^{232}_{90}\text{Th} + ^{1}_{0}\text{n} \longrightarrow {}^{233}_{90}\text{Th}$$

$$^{233}_{90}\text{Th} \longrightarrow {}^{233}_{91}\text{Pa} + {}^{0}_{-1}\beta \qquad t_{\frac{1}{2}} = 22 \text{ min}$$

$$^{233}_{91}\text{Pa} \longrightarrow {}^{233}_{92}\text{U} + {}^{0}_{-1}\beta \qquad t_{\frac{1}{2}} = 27.4 \text{ days}$$

Uranium-233 is stable enough for long-term storage. The amounts of uranium-238 and thorium-232 in Earth's crust are relatively plentiful (4 ppm and 12 ppm by mass, respectively). It therefore appears that breeder reactors offer the most promising prospect if nuclear fission is to become a major source of energy.

There are many people, however, including environmentalists, who regard nuclear fission as a highly undesirable method of energy production. Many fission products such as strontium-90 are dangerous radioactive isotopes with long half-lives. Plutonium-239, used as a nuclear fuel and produced in breeder reactors, is one of the most toxic substances known. It is an α emitter with a half-life of 24,400 years. Accidents, too, present many dangers, as the event in 1979 at the nuclear power plant at Three Mile Island in Pennsylvania clearly demonstrated. Furthermore, the problem of radioactive waste disposal has not yet been satisfactorily resolved. There have been many suggestions as to where to store or dispose of nuclear wastes, including burial deep underground, burial beneath the ocean floor, and storage in deep geological formations. The ideal disposal site would seem to be the sun, where a bit more radiation would make little difference. But that kind of operation would require 100 percent reliability in space technology.

Because of the potential hazards, the future of nuclear reactors is uncertain. What was once hailed as the ultimate solution to our energy needs in the twenty-first century is now being debated and questioned by both the scientific community and by laypeople. It seems likely that the controversy regarding the pros and cons of nuclear reactors will continue for some time.

25.6 NUCLEAR FUSION

In contrast to the nuclear fission process, **nuclear fusion,** the *combining of small nuclei into larger ones*, is largely exempt from the waste disposal problem.

Figure 7.5 shows that for the lightest elements, nuclear stability increases with increasing mass number. This behavior suggests that if two light nuclei combine or fuse together to form a larger, more stable nucleus, an appreciable amount of energy will be released in the process. This is the basis for ongoing research into the harnessing of nuclear fusion for production of energy.

Nuclear fusion occurs constantly in the sun. The sun is made up mostly of hydrogen and helium. In its interior, where temperatures reach about 15 million degrees Celsius, the following fusion reactions are believed to take place:

$$^1_1H + ^2_1H \longrightarrow ^3_2He$$
$$^3_2He + ^3_2He \longrightarrow ^4_2He + 2\,^1_1H$$
$$^1_1H + ^1_1H \longrightarrow ^2_1H + ^0_{+1}\beta$$

Because *fusion reactions take place only at very high temperatures*, they are often called **thermonuclear reactions.**

A major concern in choosing the proper nuclear fusion process for energy production is the temperature necessary to carry out the process. Some of the more promising reactions are shown in the table.

Reaction	Energy Released
$^2_1H + ^2_1H \longrightarrow ^3_1H + ^1_1H$	6.3×10^{-13} J
$^2_1H + ^3_1H \longrightarrow ^4_2He + ^1_0n$	2.8×10^{-12} J
$^6_3Li + ^2_1H \longrightarrow 2\,^4_2He$	3.6×10^{-12} J

These reactions take place at extremely high temperatures, of the order of 100 million degrees Celsius. The first reaction is particularly attractive because the world's supply of deuterium is virtually inexhaustible. The total volume of water on Earth is about 1.5×10^{21} L. Since the natural abundance of deuterium is 1.5×10^{-2} percent, the total amount of deuterium present is roughly 4.5×10^{21} g, or 5.0×10^{15} tons. The cost of preparing deuterium is minimal compared to the value of the energy released by the reaction.

All this "looks good on paper," but no large-scale controlled fusion reactions have yet been achieved. Although the necessary scientific knowledge exists, the technical difficulties have not been solved. The basic problem is finding a way to hold the nuclei together long enough at the appropriate temperature for fusion to occur. At temperatures of about 1×10^{7}°C, molecules cannot exist, and most or all atoms are stripped of all electrons. *This state of matter, in which a gaseous system consists of positive ions and electrons*, is called **plasma.** The problem of containing this plasma is a formidable one. What solid container can exist at such temperatures? None, unless the amount of plasma present is small, but then the solid surface would immediately cool the sample and quench the fusion reaction. Magnetic confinement, which employs carefully shaped magnets to keep the ions trapped in a dense, free-floating plasma, is the most promising answer to the container problem.

The Hydrogen Bomb

The technical problems inherent in the design of a nuclear fusion reactor are absent in a hydrogen bomb, also called a thermonuclear bomb. Here the objective is all power and no control. Hydrogen bombs do not contain gaseous hydrogen or gaseous deuterium; they contain solid lithium deuteride (LiD), which can be packed very tightly. The detonation of a hydrogen bomb involves two states—first a fission reaction and then a fusion reaction. The required temperature for fusion is derived from an atomic bomb. Immediately after the atomic bomb explodes, the following fusions occur, releasing vast amounts of energy (see Color Plate 28).

$$^6_3\text{Li} + {}^2_1\text{H} \longrightarrow 2\ {}^4_2\text{He}$$

$$^2_1\text{H} + {}^2_1\text{H} \longrightarrow {}^3_1\text{H} + {}^1_1\text{H}$$

There is no critical mass in a fusion bomb, and the force of explosion is limited only by the quantity of reactants present. Thermonuclear bombs are described as being "cleaner" than atomic bombs because they do not produce radioactive isotopes except for tritium, which is a weak β-particle emitter ($t_{\frac{1}{2}} = 12.5$ yr), and the products from the fission starter. Their damaging effects on the environment can be aggravated, however, by incorporating in the construction some nonfissionable material such as cobalt. Upon bombardment by neutrons, cobalt-59 is converted to cobalt-60, which is a very strong γ-ray emitter with a half-life of 5.2 years. The presence of these radioactive cobalt isotopes in the debris or fallout from a thermonuclear explosion would cause the deaths of those who survived the initial blast.

25.7   APPLICATIONS OF ISOTOPES

Both radioactive and stable isotopes have many applications in science and medicine. This section will discuss a few representative examples.

Structural Determination

One of the most important applications of radioisotopes is in determining the ages of rocks and other objects. In Section 25.3 we discussed techniques of carbon-14 and uranium-238 dating. Isotopes are also widely used in the study of chemical reaction mechanisms. We saw an example of the use of oxygen-18 isotope in the study of ester hydrolysis in Section 22.4. Now we will consider the use of a radioactive isotope in determining structure.

The formula of the thiosulfate ion is $S_2O_3^{2-}$. For some years chemists were uncertain as to whether the two sulfur atoms occupied equivalent positions in the ion. The thiosulfate ion is prepared by treatment of the sulfite ion with elemental sulfur:

$$SO_3^{2-}(aq) + S(s) \longrightarrow S_2O_3^{2-}(aq) \tag{25.1}$$

When thiosulfate is treated with dilute acid, the sulfite ion is re-formed and elemental sulfur precipitates:

$$S_2O_3^{2-}(aq) \xrightarrow{\text{H}^+} SO_3^{2-}(aq) + S(s) \tag{25.2}$$

If this sequence is carried out starting with the elemental sulfur enriched in radioactive sulfur-35 isotope, the isotope acts as a "label" for S atoms. All the labels are found in the sulfur precipitate in reaction (25.2); none of them appears in the final sulfite ions. Thus it is clear that the two atoms of sulfur in $S_2O_3^{2-}$ are not structurally equivalent, such as

$$[:\ddot{O}-\ddot{S}-\ddot{O}-\ddot{S}-\ddot{O}:]^{2-}$$

otherwise, the radioactive isotope would be present in both the elemental sulfur precipitate and the sulfite ion. Based on other studies, we know that the structure of the thiosulfate ion is

$$\left[\begin{array}{c} :\ddot{O}: \\ | \\ :\ddot{O}-S-\ddot{S}: \\ | \\ :\ddot{O}: \end{array}\right]^{2-}$$

Study of Photosynthesis

The study of photosynthesis is also rich with examples of isotope applications. The overall photosynthesis reaction can be represented as

$$6CO_2 + 6H_2O \longrightarrow C_6H_{12}O_6 + 6O_2$$

A question that arose early in studies of photosynthesis was whether the molecular oxygen was derived from water, from carbon dioxide, or from both. By the use of $H_2^{18}O$, it was demonstrated that the evolved oxygen came from water, and none came from carbon dioxide, because the O_2 formed contained only the ^{18}O isotopes. This result established the mechanism in which water molecules are "split" by light:

$$2H_2O + h\nu \longrightarrow O_2 + 4H^+ + 4e^-$$

The protons and electrons can then be used to drive energetically unfavorable reactions which are necessary for a plant's growth and function.

The radioactive carbon-14 isotope helped to determine the path of carbon in photosynthesis. Starting with $^{14}CO_2$, it was possible to isolate the intermediate products during photosynthesis and measure the amount of radioactivity of each carbon-containing compound. In this manner, the path from CO_2 through various intermediate compounds to carbohydrate could be clearly charted. *Isotopes, especially radioactive isotopes that are used to trace the path of the atoms of an element in a chemical or biological process*, are called ***tracers.***

Isotopes in Medicine

Tracers are used also for diagnosis in medicine. Sodium-24 (a β emitter with a half-life of 14.8 h) injected into the bloodstream as a salt solution can be monitored to trace the flow of blood and detect possible constrictions or obstructions in the circulatory system. Iodine-131 (a β emitter with a half-life of 8 days) has been used to test the activity of the thyroid gland. A malfunctioning thyroid can be detected by giving the patient a drink of a solution containing a known amount of $Na^{131}I$ and measuring the radioactivity just above the thyroid to see if the iodine is absorbed at the normal

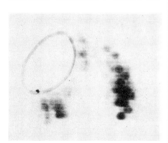

FIGURE 25.10 *(a) A normal thyroid, (b) an enlarged thyroid, and (c) a cancerous thyroid. These photos were taken by a linear photo scanner, using radioactive iodine-131.*

rate (Figure 25.10). Of course, the amounts of radioisotope used in the human body must always be kept small; otherwise, the patient might suffer permanent damage from the high-energy radiation.

A major advantage of using radioactive isotopes as tracers is that they are easy to detect. Their presence even in very small amounts can be detected by photographic techniques or by devices known as counters. Figure 25.11 is a diagram of a Geiger counter, an instrument widely used in scientific work and medical laboratories to detect radiation.

25.8   BIOLOGICAL EFFECTS OF RADIATION

In this section we will examine briefly the effects of radiation on biological systems. We will include both high-energy radiation and low-energy radiation, such as microwaves (see Figure 8.4).

Before we start this investigation, let us define quantitative ways of measuring radiation. The fundamental unit of radioactivity is the *curie* (Ci); one curie corresponds to exactly 3.70×10^{10} nuclear disintegrations per second. This decay rate is the equivalent of 1 g of radium. A millicurie (mCi) is one-thousandth of a curie. Thus 10 millicuries of a carbon-14 sample is the quantity that undergoes

$$(10 \times 10^{-3})(3.70 \times 10^{10}) = 3.70 \times 10^{8}$$

disintegrations per second. The intensity of radiation depends on the number of disintegrations, as well as on the energy and type of radiation emitted. One common unit for the absorbed dose of radiation is the *rad* (*r*adiation *a*bsorbed *d*ose), which is the amount of radiation that results in the absorption of 1×10^{-5} J per gram of irradiated material. The biological effect of radiation depends on the part of the body irradiated and the type of radiation. For this reason, the rad is often multiplied by a factor called RBE

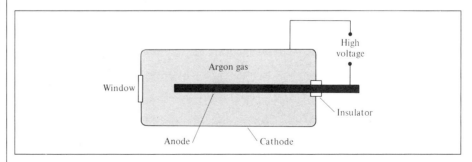

FIGURE 25.11   *Schematic diagram of a Geiger counter. Radiation (α, β, or γ rays) entering through the window causes ionization of the argon gas, which allows a small current to flow between the electrodes. Not shown: this current is amplified and is used to flash a light or operate a counter with a clicking sound.*

(*r*elative *b*iological *e*ffectiveness). The product is called a rem (*r*oentgen *e*quivalent for *m*an)

$$1 \text{ rem} = 1 \text{ rad} \times 1 \text{ RBE}$$

Of the three types of radiation, α particles usually have the least penetrating power and are therefore least harmful. β particles are more penetrating than α particles, but less so than γ rays, which are the most dangerous. γ rays have very short wavelengths and high energies. Furthermore, since they carry no charge, they cannot be stopped by shielding materials as easily as α and β particles. However, if α or β emitters are ingested, their damaging effects are greatly aggravated because the organs will be constantly subject to damaging radiation at close range.

The chemical basis of radiation damage, especially that produced by γ rays and X rays, is that of ionizing radiation. A γ ray has enough energy to remove electrons from atoms and molecules in its path, leading to the formation of ions and free radicals. Free radicals are molecular fragments having one or more unpaired electrons; they are usually short-lived and highly reactive. For example, when water is irradiated with γ rays or X rays, the following reactions take place:

$$H_2O \xrightarrow{\text{radiation}} H_2O^+ + e^-$$
$$H_2O^+ + H_2O \longrightarrow H_3O^+ + \cdot OH$$

The electron (in the hydrated form) can subsequently react with water or with a hydrogen ion to form atomic hydrogen, and with oxygen to produce the superoxide ion:

$$e^- + H_2O \longrightarrow H\cdot + OH^-$$
$$e^- + H^+ \longrightarrow H\cdot$$
$$e^- + O_2 \longrightarrow \cdot O_2^-$$
$$\text{superoxide ion}$$

In the tissues, these free radicals attack a host of organic compounds, such as enzymes and other protein molecules, to produce new radicals ($R\cdot$):

$$RH + \cdot OH \longrightarrow R\cdot + H_2O$$
$$RH + H\cdot \longrightarrow R\cdot + H_2$$

where R represents some organic group. The free radicals can also react with deoxyribonucleic acid (DNA), the genetic materials.

Organic compounds can, of course, be ionized directly by high-energy radiation. The reactions are similar to those of water:

$$RH \xrightarrow{\text{radiation}} RH^+ + e^-$$
$$RH^+ \longrightarrow R\cdot + H^+$$

In the presence of oxygen, the organic radicals form highly reactive peroxide radicals

$$R\cdot + O_2 \longrightarrow RO_2\cdot$$
$$\text{peroxide radical}$$

Current safety standards in the United States state that nuclear workers may be exposed to no more than 5 rem per year and members of the general public should be exposed to no more than 0.5 rem of manmade radiation per year.

which prevent the restoration of the initial structure of the organic molecule. Peroxides cannot form in the absence of oxygen, however, and many organic radicals may be converted to their original structure; for example

$$R\cdot \ + \ H\!-\!R' \longrightarrow RH \ + \ \cdot R'$$

It has long been known that exposure to high-energy radiation can induce cancer in humans and other animals. Cancer is characterized by uncontrolled cellular growth. However, it is also well established that cancer cells can be destroyed by proper radiation treatment. In radiation therapy, a compromise is sought. The radiation to which the patient is exposed must be sufficient to destroy cancer cells without killing too many normal cells and, hopefully, without inducing another form of cancer.

Radiation damage to living systems is generally classified as *somatic* and *genetic*. Somatic damage describes injuries that affect the organism during its own lifetime. Sunburn, skin rash, cancer, and cataracts are examples of somatic damage. Genetic damage means inheritable changes or mutations in the reproduction cells. For example, a person whose chromosomes have been damaged or altered by radiation may have deformed offspring.

Finally, we note that radiation damage to biological systems is caused not only by high-energy particles and short-wavelength radiation. Microwaves have wavelengths between 10 and 30 cm and definitely belong in the nonionizing radiation category. Yet exposure to microwave radiation can cause dizziness, skin burns, eye injuries, and cataracts. This is because microwaves are capable of speeding up the rotation of molecules. In a condensed medium, this motion leads eventually to the production of heat, which is essentially the way food is cooked in a microwave oven. The eyes and other organs that cannot dissipate heat easily are most vulnerable to microwave radiation damage.

Chromosomes are the part of the cell structure that contain the genetic materials.

AN ASIDE ON THE PERIODIC TABLE
Stable and Radioactive Isotopes of the Elements

From the point of view of nuclear chemistry, we can divide elements into three categories: (1) elements having both stable and radioactive isotopes; (2) elements having only radioactive isotopes; and (3) synthetic elements having only radioactive isotopes. This classification is shown in Figure 25.12. With the exception of technetium ($Z = 43$) and promethium ($Z = 61$), elements up to and including atomic number 83 (bismuth) have both stable and radioactive isotopes. As discussed earlier in this chapter, the transuranium elements are all synthetic. The difference between elements in categories (2) and (3) is that the former are radioactive decay products of some parent isotopes. Furthermore, some of these elements (for example, radon, thorium, and uranium) are found in nature. On the other hand, synthetic elements are created only in the

FIGURE 25.12 Numbers of known stable and radioactive isotopes of all the elements.

Legend:
- Both radioactive and stable isotopes
- Only radioactive isotopes
- Synthetic element; only radioactive isotopes
- Atomic number (top of box)
- Number of stable isotopes (bottom left)
- Number of radioactive isotopes (bottom right)

Example: C, atomic number 6, with 2 stable isotopes and 5 radioactive isotopes.

Each entry below is given as: Atomic number — Symbol — (stable isotopes | radioactive isotopes)

Group 1A / 8A and main groups:

Element	Z	Stable	Radioactive
H	1	2	1
He	2	2	2
Li	3	2	2
Be	4	1	3
B	5	2	4
C	6	2	5
N	7	2	4
O	8	3	5
F	9	1	5
Ne	10	3	5
Na	11	1	6
Mg	12	3	5
Al	13	1	7
Si	14	3	5
P	15	1	6
S	16	4	6
Cl	17	2	9
Ar	18	3	5
K	19	3	8
Ca	20	6	8
Sc	21	1	14
Ti	22	5	4
V	23	2	5
Cr	24	4	5
Mn	25	1	10
Fe	26	4	6
Co	27	1	13
Ni	28	5	7
Cu	29	2	9
Zn	30	5	10
Ga	31	2	11
Ge	32	5	12
As	33	1	13
Se	34	6	14
Br	35	2	16
Kr	36	6	17
Rb	37	2	19
Sr	38	4	13
Y	39	1	20
Zr	40	5	16
Nb	41	1	21
Mo	42	7	13
Tc	43	0	22
Ru	44	7	9
Rh	45	1	19
Pd	46	6	15
Ag	47	2	25
Cd	48	8	14
In	49	2	32
Sn	50	10	18
Sb	51	2	27
Te	52	8	22
I	53	1	23
Xe	54	9	22
Cs	55	1	20
Ba	56	7	17
La	57	2	16
Hf	72	6	15
Ta	73	2	18
W	74	5	16
Re	75	2	20
Os	76	7	12
Ir	77	2	23
Pt	78	6	28
Au	79	1	20
Hg	80	7	19
Tl	81	2	26
Pb	82	4	24
Bi	83	1	18
Po	84	0	33
At	85	0	21
Rn	86	0	20
Fr	87	0	21
Ra	88	0	16
Ac	89	0	11
Unq	104	0	1
Unp	105	0	3
Unh	106	0	1
Uns	107	0	1
Uno	108	0	1
Une	109	0	1

Lanthanides and Actinides:

Element	Z	Stable	Radioactive
Ce	58	4	15
Pr	59	1	14
Nd	60	7	10
Pm	61	0	14
Sm	62	7	13
Eu	63	2	19
Gd	64	7	11
Tb	65	1	23
Dy	66	7	14
Ho	67	1	28
Er	68	6	10
Tm	69	1	17
Yb	70	7	9
Lu	71	2	21
Th	90	0	12
Pa	91	0	13
U	92	0	15
Np	93	0	15
Pu	94	0	16
Am	95	0	13
Cm	96	0	13
Bk	97	0	8
Cf	98	0	12
Es	99	0	12
Fm	100	0	10
Md	101	0	3
No	102	0	7
Lr	103	0	2

laboratory; they are not the decay products of other isotopes and they are not found in nature. As you study Figure 25.12 keep in mind that new isotopes of elements are still frequently discovered or created in the laboratory.

SUMMARY

1. For stable nuclei of low atomic number, the neutron-to-proton ratio is close to 1. For heavier stable nuclei, the ratio becomes greater than 1.

2. All nuclei with 84 or more protons are unstable and radioactive. Nuclei with even atomic numbers are more stable than those with odd atomic numbers.

3. Radioactive nuclei emit α particles, β particles, or γ rays. The equation for a nuclear reaction includes the particles emitted, and both the mass numbers and the atomic numbers must balance.

4. Uranium-238 is the parent of a natural radioactive decay series which can be used to determine the ages of rocks.

5. Artificially radioactive elements are created by the bombardment of other elements by accelerated neutrons, protons, or α particles.

6. Nuclear fission is the splitting of a large nucleus into two smaller nuclei, plus neutrons. When these neutrons are captured efficiently by other nuclei, an uncontrollable chain reaction can occur.

7. Nuclear reactors use the heat from a controlled nuclear fission reaction to produce power. The three important types of reactors are light water reactors, heavy water reactors, and breeder reactors.

8. Nuclear fusion, the type of reaction that occurs in the sun, is the combination of two light nuclei to form one heavy nucleus. Fusion takes place only at very high temperatures—so high that controlled large-scale nuclear fusion has so far not been achieved.

9. Radioactive isotopes are easy to detect, and thus make excellent tracers in chemical reactions and in medical practice.

10. High-energy radiation damages living systems by causing ionization and the formation of free radicals.

KEY WORDS

Breeder reactor, p. 767
Critical mass, p. 764
Moderators, p. 765
Nuclear chain reaction, p. 763
Nuclear fission, p. 762
Nuclear fusion, p. 768

Nuclear transmutation, p. 751
Plasma, p. 769
Thermonuclear reaction, p. 769
Tracer, p. 771
Transuranium elements, p. 761

PROBLEMS

More challenging problems are marked with an asterisk.

Nuclear Reactions

25.1 How do nuclear reactions differ from ordinary chemical reactions?

25.2 Write complete nuclear equations for the following processes: (a) tritium, ^3_1H, undergoes β decay; (b) ^{242}Pu undergoes α particle emission; (c) ^{131}I undergoes β decay; (d) ^{251}Cf emits an α particle.

25.3 Complete the following nuclear equations and identify X in each case:

(a) $^{26}_{12}Mg + ^1_1p \longrightarrow ^4_2He + X$
(b) $^{59}_{27}Co + ^2_1H \longrightarrow ^{60}_{27}Co + X$
(c) $^{235}_{92}U + ^1_0n \longrightarrow ^{94}_{36}Kr + ^{139}_{56}Ba + 3X$
(d) $^{235}_{92}U + ^1_0n \longrightarrow ^{99}_{40}Sr + ^{135}_{52}Te + 2X$
(e) $^{53}_{24}Cr + ^4_2He \longrightarrow ^1_0n + X$
(f) $^{20}_8O \longrightarrow ^{20}_9F + X$
(g) $^{135}_{53}I \longrightarrow ^{135}_{54}Xe + X$
(h) $^{40}_{19}K \longrightarrow ^{0}_{-1}\beta + X$
(i) $^{59}_{27}Co + ^1_0n \longrightarrow ^{56}_{25}Mn + X$
(j) $^{78}_{33}As \longrightarrow ^{0}_{-1}\beta + X$

25.4 Write balanced nuclear equations for the following reactions and identify X.

(a) $^{15}_7N$ (p, α) $^{12}_6C$
(b) $^{27}_{13}Al$ (d, α) $^{25}_{12}Mg$
(c) $^{55}_{25}Mn$ (n, γ) X
(d) $^{80}_{34}Se$ (d, p) X
(e) ^9_4Be (d, 2p) ^9_3Li
(f) $^{106}_{46}Pd$ (α, p) $^{109}_{47}Ag$

25.5 Describe how you would prepare astatine-211, starting with bismuth-209.

25.6 A long-cherished dream of alchemists was to produce gold from cheaper and more abundant elements. This dream was finally realized when $^{198}_{80}Hg$ was converted into gold upon neutron bombardment. Write a balanced equation for this reaction.

Nuclear Stability

25.7 State the general rules for predicting nuclear stability.

25.8 For each pair of isotopes listed, predict which one is less stable: (a) ^6_3Li or ^9_3Li, (b) $^{22}_{11}Na$ or $^{25}_{11}Na$, (c) $^{48}_{20}Ca$ or $^{48}_{21}Sc$.

25.9 For each pair of elements listed, predict which one has more stable isotopes: (a) Co or Ni, (b) F or Se, (c) Ag or Cd.

25.10 The nucleus of nitrogen-18 lies above the stability belt. Write an equation for a nuclear reaction by which nitrogen-18 can achieve stability.

25.11 In each pair of isotopes shown, indicate the one you would expect to be radioactive: (a) $^{20}_{10}$Ne and $^{17}_{10}$Ne, (b) $^{40}_{20}$Ca and $^{45}_{20}$Ca, (c) $^{95}_{42}$Mo and $^{92}_{43}$Tc, (d) $^{195}_{80}$Hg and $^{196}_{80}$Hg, (e) $^{209}_{83}$Bi and $^{242}_{96}$Cm.

Radioactive Decay

25.12 Fill in the blanks in the following radioactive decay series:

(a) ^{232}Th $\xrightarrow{\alpha}$ _____ $\xrightarrow{\beta}$ _____ $\xrightarrow{\beta}$ ^{228}Th

(b) ^{235}U $\xrightarrow{\alpha}$ _____ $\xrightarrow{\beta}$ _____ $\xrightarrow{\alpha}$ ^{227}Ac

(c) _____ $\xrightarrow{\alpha}$ ^{233}Pa $\xrightarrow{\beta}$ _____ $\xrightarrow{\alpha}$ _____

25.13 A radioactive substance decays as follows:

Time (days)	Mass (g)
0	500
1	389
2	303
3	236
4	184
5	143
6	112

Calculate the first-order decay constant and the half-life of the reaction.

25.14 The radioactive decay of Tl-206 to Pb-206 has a half-life of 4.20 min. Starting with 5.00×10^{22} atoms of Tl-206, calculate the number of such atoms left after 42.0 min.

25.15 A freshly isolated sample of ^{90}Y was found to have an activity of 9.8×10^5 disintegrations per minute at 1.00 PM on December 3, 1982. At 2.15 PM on December 17, 1982, its activity was redetermined and found to be 2.6×10^4 disintegrations per minute. Calculate the half-life of ^{90}Y.

25.16 In the thorium decay series, thorium-232 loses a total of 6 alpha particles and 4 beta particles in a 10-stage process. What is the final isotope produced?

25.17 Why do radioactive decay series obey first-order kinetics?

25.18 Strontium-90 is one of the products of the fission of uranium-235. This isotope of strontium is radioactive, with a half-life of 28.8 yr. Calculate how long (in yr) it will take for 1.00 g of the isotope to be reduced to 0.200 g by decay.

25.19 Consider the following decay series:

$$A \longrightarrow B \longrightarrow C \longrightarrow D$$

where A, B, and C are radioactive with half-lives of 4.50 s, 15.0 days, and 1.00 s, respectively, and D is nonradioactive. Starting with 1.00 mole of A, and none of B, C, or D, calculate the number of moles of A, B, C, and D left after 30 days.

25.20 The radioactive potassium-40 isotope decays to argon-40 with a half-life of 1.2×10^9 yr. (a) Write a balanced equation for the reaction. (b) A sample of moon rock is found to contain 18 percent potassium-40 and 82 percent argon by mass. Calculate the age of the rock in years.

Nuclear Reactors

25.21 Discuss the differences between a light water and a heavy water nuclear fission reactor.

25.22 What are the advantages of a breeder reactor over a conventional nuclear fission reactor?

25.23 No form of energy production is without risk. Make a list of the risks to society involved in fueling and operating a conventional coal-fired electric power plant, and compare them with the risks of fueling and operating a nuclear fission powered electric plant.

25.24 Compare the advantages and disadvantages of nuclear fission and nuclear fusion as energy sources.

Application of Isotopes

25.25 Describe how you would use a radioactive iodine isotope to demonstrate that the following process is in dynamic equilibrium:

$$PbI_2(s) \rightleftharpoons Pb^{2+}(aq) + 2I^-(aq)$$

25.26 Consider the following redox reaction:

$$IO_4^-(aq) + 2I^-(aq) + H_2O(l) \longrightarrow I_2(s) + IO_3^-(aq) + 2OH^-(aq)$$

When KIO_4 is added to a solution containing iodide ions labeled with radioactive iodine-128, all the radioactivity appears in I_2 and none in the IO_3^- ion. What can you deduce about the mechanism for the redox process?

25.27 Explain how you might use a radioactive tracer to show that ions are not completely motionless in crystals.

25.28 Each hemoglobin molecule, the oxygen-carrier in blood, contains four Fe atoms. Explain how you would use the radioactive $^{59}_{26}$Fe ($t_{\frac{1}{2}}$ = 46 days) to show that the iron in a certain diet is converted into hemoglobin.

Miscellaneous Problems

25.29 Distinguish between the terms in each of the following pairs: (a) fission and fusion, (b) electron and positron, (c) γ rays and β rays, (d) natural radioactivity and artificial radioactivity.

25.30 Define each of the following terms: (a) nuclear binding energy, (b) cyclotron, (c) transuranium element, (d) mass defect, (e) isotope tracer, (f) breeder reactor.

25.31 Chlorine-39 is radioactive and undergoes β decay. What differences would you expect in the chemical behavior of HCl molecules containing chlorine-39, compared to those containing chlorine-35? Cl-35 is nonradioactive.

25.32 Why does it require such a high temperature to initiate a nuclear fusion reaction, whereas a nuclear fission reaction occurs readily at room temperature?

25.33 Explain why light elements, such as H and Li, are suitable for fusion processes, and heavy elements, such as U and Th, are suitable for fission processes.

25.34 Name the elements with atomic numbers (a) 110, (b) 116, (c) 125, (d) 218.

25.35 Write atomic numbers of the following elements: (a) ununbium, (b) untrioctium, (c) unsepthexium, (d) binilquadium.

25.36 Nuclei with an even number of protons and an even number of neutrons are more stable than those with an odd number of protons and/or odd number of neutrons. What is the significance of the even number of protons and of neutrons in this case?

*25.37 Tritium, ^3_1H, is radioactive and decays by electron emission. Its half-life is 12.5 years. In ordinary water the ratio of ^1H to ^3H atoms is 1.0×10^{17} to 1. (a) Write a balanced nuclear equation for tritium decay. (b) How many disintegrations will be observed per minute in a 1.00 kg sample of water?

25.38 Explain why the critical mass of a fissionable isotope is necessary for initiation of a nuclear explosion.

25.39 Balance the following equations, which are for nuclear reactions that are known to occur in the explosion of an atomic bomb:

(a) $^{235}_{92}U + ^1_0n \longrightarrow ^{140}_{56}Ba + 3^1_0n + X$
(b) $^{235}_{92}U + ^1_0n \longrightarrow ^{144}_{55}Cs + ^{90}_{37}Rb + 2X$
(c) $^{235}_{92}U + ^1_0n \longrightarrow ^{87}_{35}Br + 3^1_0n + X$
(d) $^{235}_{92}U + ^1_0n \longrightarrow ^{160}_{62}Sm + ^{72}_{30}Zn + 4X$

25.40 (a) What is the activity, in millicuries, of a 0.500 g sample of $^{237}_{93}Np$? (This isotope decays by α particle emission and has a half-life of 2.20×10^6 yr.) (b) Write a balanced nuclear equation for the decay of $^{237}_{93}Np$.

26

CHEMISTRY OF SELECTED NON-METALLIC ELEMENTS

U p to this point, we have focused mainly on fundamental principles—theories of chemical bonding, intermolecular forces, rates and mechanisms of chemical reactions, equilibrium, the laws of thermodynamics, electrochemistry, and nuclear chemistry. These topics provide both a macroscopic and a microscopic view of the structure and properties of matter. In this and the next three chapters we will take a more systematic approach to the study of the properties of elements and their compounds. This aspect of chemistry is commonly referred to as *descriptive chemistry*.

To study the descriptive chemistry of all the known elements comprehensively would require a book 50 times the length of this one, so we will consider only the more familiar elements. We will emphasize (1) their occurrence and preparation, (2) the physical and chemical properties of some of their compounds, and (3) their uses in modern society and their roles in biological systems. As we have done throughout the book, we will look for periodic trends as a way to help us understand the facts.

779

26.1   METALS, NONMETALS, AND METALLOIDS

Most elements can be classified as metallic or nonmetallic by a number of easily identifiable properties. Metals are lustrous in appearance, solid at room temperature (with the exception of mercury), good conductors of heat and electricity, malleable (can be hammered flat), and ductile (can be drawn into wire). Most of them have low electronegativities (see Figure 11.2). The oxidation numbers of metals in compounds are almost always positive.

The properties of nonmetallic elements are quite varied and therefore are more difficult to characterize. A number of nonmetals are gases in their elemental state: hydrogen, oxygen, nitrogen, fluorine, chlorine, and the noble gases. Only one nonmetallic element, bromine, is a liquid. All the remaining nonmetals are solids at room temperature. Unlike metals, non-metallic elements are poor conductors of heat and electricity, and many of them exhibit both positive and negative oxidation numbers in their compounds.

There is a small group of elements with properties that lie between those of metals and nonmetals. These elements are sometimes called metalloids (see Figure 1.4).

With the exception of hydrogen, all the nonmetals are concentrated in the upper right-hand corner of the periodic table. In Figure 11.2 we see that the electronegativities of the elements increase from left to right across any period and decrease from top to bottom in any group. These trends are consistent with the fact that compounds formed between metals and non-metals tend to be ionic, with metals as the cations and nonmetals as the anions.

In this chapter we will discuss the chemistry of a number of common and important nonmetallic elements: H, B, C, and Si; three members of the nitrogen family (N, P, and As); two members of the oxygen family (O and S); the halogens (F, Cl, Br, and I); and the noble gases (He, Ne, Ar, Kr, and Xe).

26.2   HYDROGEN

Hydrogen is the simplest element known—its most common atomic form contains only one proton and one electron. In the atomic state hydrogen tends to combine in an exothermic reaction that results in molecular hydrogen:

$$H(g) \; + \; H(g) \longrightarrow H_2(g) \qquad \Delta H° \; = \; -436.4 \text{ kJ}$$

Molecular hydrogen is a colorless, odorless, and nonpoisonous gas that boils at $-252.9°C$ (20.3 K).

Hydrogen is the most abundant element in the universe, accounting for about 70 percent of the universe's total mass. It is the third most abundant element in Earth's crust (see Figure 1.9). Unlike Jupiter and Saturn, Earth does not have a strong enough gravitational pull to retain the lightweight H_2 molecules, so hydrogen is not found in our atmosphere.

We saw in Chapter 9 that there is no totally suitable position for hydrogen in the periodic table. The electron configuration of H is $1s^1$. It resembles the alkali metals in one respect—it can be oxidized to the H^+ ion, which exists in aqueous solutions in the hydrated form. On the other hand, hydrogen can be reduced to the hydride ion (H^-), which is isoelectronic with helium ($1s^2$). In this respect, hydrogen resembles the halogens in that it forms the uninegative ion. Hydrogen forms a large number of covalent compounds. It also takes part in hydrogen bond formation (see Section 14.2). We will discuss the properties of some of hydrogen's compounds shortly.

Hydrogen gas plays an important role in industrial processes, and thus a number of methods have been developed for preparing it commercially. One method passes steam over heated iron:

$$3Fe(s) \,+\, 4H_2O(g) \xrightarrow{\Delta} Fe_3O_4(s) \,+\, 4H_2(g)$$

Another passes steam over a bed of red-hot coke:

$$C(s) \,+\, H_2O(g) \xrightarrow{\Delta} CO(g) \,+\, H_2(g)$$

The mixture of carbon monoxide and hydrogen gas produced in the latter reaction is commonly known as *water gas*. Because both CO and H_2 burn in air, water gas was used as a fuel for many years. But since CO is poisonous and an H_2–O_2 mixture can be explosive, water gas has been replaced by natural gases, such as methane and propane.

Two other major industrial sources of hydrogen gas are the reaction between propane and steam in the presence of a catalyst:

$$C_3H_8(g) \,+\, 3H_2O(g) \xrightarrow{\Delta} 3CO(g) \,+\, 7H_2(g)$$

and the thermal decomposition of some hydrocarbons, for example

$$C_6H_{14}(g) \xrightarrow{\Delta} C_6H_6(g) \,+\, 4H_2(g)$$

Small quantities of hydrogen gas can be prepared conveniently in the laboratory by the reaction of zinc with dilute hydrochloric acid (Figure 26.1):

$$Zn(s) \,+\, 2HCl(aq) \longrightarrow ZnCl_2(aq) \,+\, H_2(g)$$

Both of these processes involve the reduction of H_2O to H_2.

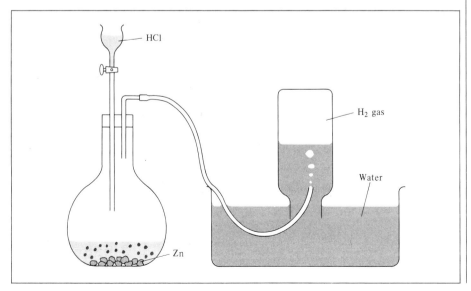

FIGURE 26.1 *Apparatus for the laboratory preparation of hydrogen gas. The gas is collected over water as in the case of oxygen gas (see Figure 6.13).*

These reactions are too violent to be suitable for laboratory preparation of hydrogen gas.

Hydrogen gas is also produced by the reaction between an alkali metal or an alkaline earth metal (Ca or Ba) and water:

$$2Na(s) + 2H_2O(l) \longrightarrow 2NaOH(aq) + H_2(g)$$

$$Ca(s) + 2H_2O(l) \longrightarrow Ca(OH)_2(aq) + H_2(g)$$

Very pure hydrogen gas can be obtained by the electrolysis of water (see p. 738).

Binary Hydrides

Binary hydrides are compounds containing hydrogen and another element, either a metal or a nonmetal. Depending on structure and properties, these hydrides can be broadly divided into three types: (1) ionic hydrides, (2) covalent hydrides, and (3) interstitial hydrides.

Ionic Hydrides. *Ionic hydrides* are formed when molecular hydrogen combines directly with any alkali metal or with some of the alkaline earth metals (Ca, Sr, and Ba):

$$2Li(s) + H_2(g) \longrightarrow 2LiH(s)$$

$$Ca(s) + H_2(g) \longrightarrow CaH_2(s)$$

Ionic hydrides are all solids that have the high melting points which are characteristic of ionic compounds. The anion in these compounds is the hydride ion, H^-, which is a very strong Brønsted–Lowry base. It readily accepts a proton from a proton donor such as water:

$$H^-(aq) + H_2O(l) \longrightarrow OH^-(aq) + H_2(g)$$

As a result of their high reactivity with water, ionic hydrides are frequently used to remove traces of water from organic solvents.

This is an example of the diagonal relationship between Be and Al (see p. 267).

Covalent Hydrides. In a *covalent hydride* the hydrogen atom is covalently bonded to the atom of another element. There are two types of covalent hydrides—those containing discrete molecular units, such as CH_4 and NH_3, and those having complex polymeric structure, such as $(BeH_2)_x$ and $(AlH_3)_x$, where x is a very large number. (A *polymer* is a large molecule made up of hundreds or thousands of atoms that are linked together covalently.)

Figure 26.2 shows the binary ionic and covalent hydrides of the representative elements. Across a given period the physical and chemical properties of these compounds change from ionic to covalent. Consider, for example, the hydrides of the second-period elements:

$$LiH \quad BeH_2 \quad B_2H_6 \quad CH_4 \quad NH_3 \quad H_2O \quad HF$$

LiH is an ionic compound with a high melting point (680°C). The structure of BeH_2 is that of a polymer; it is a covalent compound. The molecules B_2H_6 and CH_4 are nonpolar. On the other hand, NH_3, H_2O, and HF are all polar molecules in which the hydrogen atom is the *positive* end of the polar

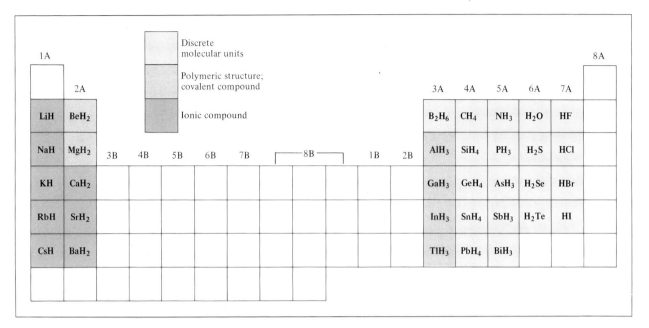

FIGURE 26.2 *Binary hydrides of the representative elements. In cases where hydrogen forms more than one compound with the same element, only the formula of the simplest hydride is shown. The properties of many of the transition metal hydrides are not well characterized.*

bond. Of this group of hydrides (NH_3, H_2O, and HF) only HF is acidic in water.

As we move down any group in Figure 26.2, for example, Group 2A, the compounds change from covalent (BeH_2 and MgH_2) to ionic (CaH_2, SrH_2, and BaH_2).

Interstitial Hydrides. Molecular hydrogen forms a number of hydrides with transition metals. In some of these compounds, the ratio of hydrogen atoms to metal atoms is *not* a constant. Such compounds are called *interstitial hydrides*. For example, depending on conditions the formula for titanium hydride can vary between $TiH_{1.8}$ and TiH_2.

Many of the interstitial hydrides retain metallic properties, such as electrical conductivity. Yet it is known that hydrogen is definitely bonded to the metal in these compounds, although the exact nature of bonding is often not clear.

A particularly interesting reaction is that between H_2 and palladium (Pd). Hydrogen gas is readily adsorbed onto the surface of the palladium metal where it dissociates into atomic hydrogen. The H atoms then "dissolve" into the metal. On heating and under the pressure of H_2 gas on one side of the metal, these atoms diffuse through the metal and recombine to form molecular hydrogen, which emerges as the gas from the other side. Because no other gas behaves in this way with palladium, this process has been used to separate hydrogen gas from other gases.

These compounds are sometimes called *nonstoichiometric compounds*. **Note that they do not obey the law of definite proportion discussed in Section 2.8.**

TABLE 26.1 Properties of H_2O and D_2O

Property	H_2O	D_2O
Molar mass (g/mol)	18.0	20.0
Melting point (°C)	0	3.8
Boiling point (°C)	100	101.4
Density at 4°C (g/cm³)	1.000	1.108

Isotopes of Hydrogen

The $_1^1H$ isotope is also called *protium*.

Hydrogen has three isotopes: $_1^1H$ (hydrogen), $_1^2H$ (deuterium, symbol D), and $_1^3H$ (tritium, symbol T). The natural abundances of the stable hydrogen isotopes are: hydrogen, 99.985 percent, and deuterium, 0.015 percent. Tritium is a radioactive isotope with a half-life of about 12.5 years.

Table 26.1 compares some of the common properties of H_2O with those of D_2O. As we have noted, deuterium oxide, or heavy water as it is commonly called, is used in some nuclear reactors as a coolant and a moderator of nuclear reactions. D_2O can be separated from H_2O by fractional distillation because H_2O boils at a lower temperature, as Table 26.1 shows. Another technique for its separation is electrolysis of water. Since H_2 gas is formed about eight times as fast as D_2 during electrolysis, the water remaining in the electrolytic cell becomes progressively enriched with D_2O. Interestingly, the Dead Sea, which for thousands of years has entrapped water that has no outlets other than through evaporation, has a higher $[D_2O]/[H_2O]$ ratio than water found elsewhere.

Although D_2O chemically resembles H_2O in all respects, it is a toxic substance. This is so because deuterium is heavier than hydrogen; thus its compounds often react more slowly than those of the lighter isotope. Regular drinking of D_2O instead of H_2O could prove fatal because of the slower rate of transfer of D^+ compared to that of H^+ in the acid-base reactions involved in enzyme catalysis. This *kinetic isotope effect* is also manifest in acid ionization constants. For example, the ionization constant of acetic acid

$$CH_3COOH(aq) \rightleftharpoons CH_3COO^-(aq) + H^+(aq) \qquad K_a = 1.8 \times 10^{-5}$$

is about three times as large as that of deuterated acetic acid:

$$CH_3COOD(aq) \rightleftharpoons CH_3COO^-(aq) + D^+(aq) \qquad K_a = 6 \times 10^{-6}$$

Hydrogenation

In addition to being used in the synthesis of ammonia (see Section 22.5), hydrogen is also used in the hydrogenation of organic compounds. **Hydrogenation** is the *addition of hydrogen, especially to compounds with double or triple carbon-carbon bonds.* For example, in the presence of a catalyst such as platinum, hydrogen can be added to the triple bond in acetylene as follows:

$$H-C\equiv C-H \xrightarrow[\text{catalyst}]{H_2} \quad \begin{array}{c} H \\ \diagdown \\ H \end{array} C=C \begin{array}{c} H \\ \diagup \\ H \end{array}$$

acetylene   ethylene

If there is enough hydrogen present, the hydrogenation process will proceed further:

$$\begin{array}{c} H \\ \diagdown \\ H \end{array} C=C \begin{array}{c} H \\ \diagup \\ H \end{array} \xrightarrow[\text{catalyst}]{H_2} \quad H-\overset{\displaystyle H}{\underset{\displaystyle H}{C}}-\overset{\displaystyle H}{\underset{\displaystyle H}{C}}-H$$

ethylene   ethane

Hydrogenation is an important process in the food industry. For example, although vegetable oils have considerable nutritional value, some of them must be hydrogenated because of their unsavory flavor due to their inappropriate molecular structures (that is, there are too many C=C bonds present). When they are exposed to air, the *polyunsaturated* molecules in various oils undergo oxidation to yield products with unpleasant tastes (oil that has oxidized is said to have gone *rancid*). Hydrogenation reduces the number of double bonds in the molecule but does not completely eliminate them. If all the double bonds are eliminated, the oil becomes hard and brittle. Under controlled conditions, suitable cooking oils and margarine may be prepared by hydrogenation of vegetable oils extracted from cottonseed, corn, and soybeans.

The Hydrogen Economy

Because the world's fossil fuel reserve is being depleted at an alarmingly fast rate, there have been intensive efforts in recent years to develop a method of obtaining hydrogen gas to use as an alternate energy source. Hydrogen gas could replace gasoline to power automobiles (after considerable modification of the engine, of course) or be used with oxygen gas in fuel cells to generate electricity (see p. 733). One major advantage of using hydrogen gas in these ways is that the reactions are essentially free of pollutants; the end product formed in a hydrogen-powered engine or in a fuel cell would be water, just as in the burning of hydrogen gas in air:

$$2H_2(g) + O_2(g) \longrightarrow 2H_2O(l)$$

Of course, the potential success of a so-called hydrogen economy would depend on how cheaply we could produce hydrogen gas and how easily we could store it.

As you know, hydrogen gas can be obtained from water by electrolysis, but this method consumes too much energy. A more attractive approach, which is currently in early stages of development, uses solar energy to "split" water molecules. In this scheme, a *catalyst* (a complex molecule containing one or more transition metal atoms, such as ruthenium) absorbs a photon from solar radiation and becomes energetically excited. In its excited state, the catalyst is capable of reducing water to molecular hydrogen.

The total volume of ocean water is about 1×10^{21} liters. Thus the ocean contains an almost inexhaustible supply of hydrogen.

In another scheme, which is based on chemical rather than photon-induced reaction, both hydrogen and oxygen gas can be produced in a thermochemical water-splitting cycle. Consider the following sequence of reactions:

$$CaBr_2(s) + H_2O(g) \xrightarrow{\Delta} CaO(s) + 2HBr(g)$$
$$Hg(l) + 2HBr(g) \xrightarrow{\Delta} HgBr_2(s) + H_2(g)$$
$$HgBr_2(s) + CaO(s) \longrightarrow HgO(s) + CaBr_2(s)$$
$$HgO(s) \xrightarrow{\Delta} Hg(l) + \tfrac{1}{2}O_2(g)$$

As you can see, in this "cycle" of reactions one mole of H_2O molecules is converted to one mole of H_2 and half a mole of O_2 molecules; the other starting compounds ($CaBr_2$, Hg, and $HgBr_2$) are regenerated at the end of the cycle so that they can be used again.

Some of the interstitial hydrides we have discussed are suitable candidates for storing hydrogen gas. The reactions that form these hydrides are usually reversible, so hydrogen gas can be obtained simply by reducing the pressure of the hydrogen gas above the metal. The advantages of using interstitial hydrides are (1) many metals have a high capacity to take up hydrogen gas—sometimes up to three times as many hydrogen atoms as there are metal atoms, and (2) because these hydrides are solids, they can be stored and transported more easily than gases or liquids.

26.3 BORON

Since boron has semimetallic properties, it is classified as a metalloid. It is a rare element, constituting less than 0.0003 percent of Earth's crust by mass. Boron does not occur pure in nature; it exists in the form of oxygen compounds. Some of these compounds are borax, $Na_2B_4O_7 \cdot 10H_2O$; ortho-boric acid, H_3BO_3; and kernite, $Na_2B_4O_7 \cdot 4H_2O$. Elemental boron, a transparent, crystalline substance that is almost as hard as diamond (see Color Plate 3), can be prepared by reducing boron oxide with magnesium powder:

$$B_2O_3(s) + 3Mg(s) \xrightarrow{\Delta} 2B(s) + 3MgO(s)$$

The structure of elemental boron is quite complex (see Figure 12.16).

A binary compound formed between boron and hydrogen is called a **borane.** There are a number of boranes, the simplest and most important of which is diborane, B_2H_6. There has been a great deal of interest in the bonding scheme of this molecule. Because the electron configuration of boron is $1s^2 2s^2 2p^1$, there is a total of twelve valence electrons in diborane (three for each boron atom and six for the hydrogen atoms). Thus it is impossible to represent diborane as

```
  H   H
  |   |
H—B—B—H
  |   |
  H   H
```

FIGURE 26.3 *Three-center molecular orbitals in diborane. Each molecular orbital is formed by the overlap of the sp³ hybrid orbitals of boron and the 1s orbital of hydrogen. No more than two electrons can be placed in each molecular orbital.*

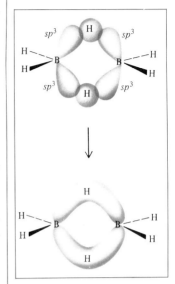

because this structure would require fourteen valence electrons (each of the seven covalent bonds would be made up of two electrons). Measurements have shown that there are actually two types of H atoms in diborane, in the ratio of 4:2. In the simple molecular orbital description, each B atom is assumed to be sp^3-hybridized. Four of the six H atoms are bonded to borons by sigma bonds. The other two are held together by *three-center molecular orbitals* (Figure 26.3). Each of the three-center bonds extends over three atoms: the two boron atoms and a bridging hydrogen atom.

The three-center bond is not restricted to diborane. Bridging hydrogen atoms are present also in higher boranes, such as B_5H_9 and B_6H_{10} (Figure 26.4). Because of their high heats of combustion

$$B_2H_6(g) + 3O_2(g) \longrightarrow B_2O_3(s) + 3H_2O(l) \qquad \Delta H° = -2137.7 \text{ kJ}$$

the boranes were once considered for use as high-energy fuels for jet engines and rockets. However, these compounds are expensive to prepare and difficult to store, so the idea was abandoned.

The boron halides (BF_3, BCl_3, BBr_3, and BI_3) are another series of interesting compounds. Since they have planar structures (see Section 12.2) the boron atom in each is assumed to be sp^2-hybridized (Figure 26.5). The unhybridized $2p_z$ orbital is vacant and can accept a pair of electrons from another molecule to form an addition compound:

$$BF_3 + NH_3 \longrightarrow F_3B{-}NH_3$$

As a result, the boron atom is sp^3-hybridized and the octet rule is satisfied for boron. This is a Lewis-type acid-base reaction as discussed in Section 18.8.

The only important oxide of boron is boron oxide (B_2O_3). It is prepared by heating boric acid. Boron oxide is used in the production of boron, in heat-resistant glassware, in electronic components, and in herbicides. Like

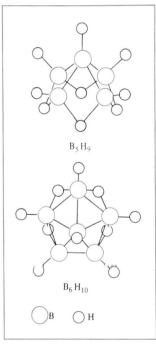

FIGURE 26.4 *Structures of B_5H_9 and B_6H_{10}.*

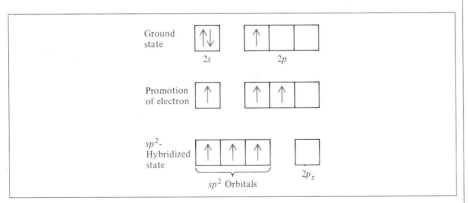

FIGURE 26.5 *The sp^2 hybridization of a boron atom. The vacant $2p_z$ orbital can accept a pair of electrons from a donor atom.*

most of the nonmetallic oxides, it is acidic (see Figure 18.5). When dissolved in water, it yields boric acid

$$B_2O_3(s) + 3H_2O(l) \longrightarrow 2B(OH)_3(aq)$$

which is moderately soluble in water. Boric acid, which is often written as H_3BO_3, is a very weak, exclusively monoprotic acid (Figure 26.6). It behaves as a Lewis acid by accepting a pair of electrons from the hydroxide ion, which originates from the H_2O molecule:

$$B(OH)_3(aq) + H_2O(l) \rightleftharpoons B(OH)_4^-(aq) + H^+(aq) \qquad K_a = 7.3 \times 10^{-10}$$

Its weak acid strength and antiseptic properties make boric acid suitable as an eyewash. Industrially, boric acid is used to make heat-resistant glass, called *borosilicate glass.*

26.4 CARBON AND SILICON

Carbon

Although it constitutes only about 0.09 percent by mass of Earth's crust, carbon is an essential element of living matter. It is found in the free state in the form of diamond and graphite (see Color Plate 4), and occurs combined in natural gas, petroleum, and coal. (*Coal* is a natural dark-brown to black solid used as a fuel; it is formed from fossilized plants and consists of amorphous carbon with various organic and some inorganic compounds.) Carbon combines with oxygen to form carbon dioxide in the atmosphere and occurs as carbonate in limestone and chalk.

Carbon exists in two allotropic forms as diamond and graphite (see Figures 12.17 and 12.18). Figure 26.7 shows the phase diagram of carbon. As you can see, graphite is the stable form of carbon at 1 atm and 25°C. Owners of diamond jewelry need not be alarmed, however, for the rate of the spontaneous process

$$C(diamond) \longrightarrow C(graphite) \qquad \Delta G° = -2.87 \text{ kJ}$$

is extremely slow; millions of years may pass before a diamond turns to graphite.

Synthetic diamond can be prepared from graphite by applying very high pressures and temperatures. Figure 26.8 shows a synthetic diamond and its starting material, graphite. Synthetic diamonds generally lack the optical properties of natural diamonds. They are useful, however, as abrasives and in cutting concrete and many other hard substances, including metals and alloys. The uses of graphite are described on p. 402.

Carbides and Cyanides.   *Carbon combines with metals to form a number of ionic compounds* called **carbides,** in which carbon is in the form of C_2^{2-} or C^{4-} ions. Two examples are CaC_2 and Be_2C. These ions are strong Brønsted–Lowry bases and react with water as follows:

$$C_2^{2-}(aq) + 2H_2O(l) \longrightarrow 2OH^-(aq) + C_2H_2(g)$$

$$C^{4-}(aq) + 4H_2O(l) \longrightarrow 4OH^-(aq) + CH_4(g)$$

Pyrex glassware is made of borosilicate glass.

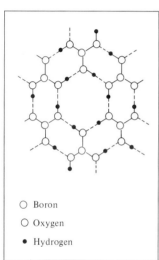

○ Boron

○ Oxygen

● Hydrogen

FIGURE 26.6 *Layer structure of boric acid. The dotted lines denote the hydrogen bonds.*

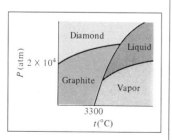

FIGURE 26.7 *Phase diagram of carbon. Note that under atmospheric conditions, graphite is the stable form of carbon.*

FIGURE 26.8 *Synthetic diamond and the starting material—graphite.*

Carbon also forms a covalent compound with silicon, SiC. Silicon carbide is called *carborundum* and is prepared as follows:

$$SiO_2(s) + 3C(s) \xrightarrow{\Delta} SiC(s) + 2CO(g)$$

The compound is also formed by heating silicon with carbon at 1500°C. Carborundum is almost as hard as diamond and has the diamond structure; each carbon atom is tetrahedrally bonded to four Si atoms, and vice versa. It is used mainly for cutting, grinding, and polishing metals and glasses.

Another important class of carbon compounds consists of the **cyanides**, which *contain the anion group* $:C≡N:^-$. Cyanide ions are extremely toxic because they bind almost irreversibly to the Fe(III) ion in cytochrome oxidase, a key enzyme in metabolic processes. Hydrogen cyanide is even more dangerous because of its volatility (b.p. 26°C). It has the odor of bitter almonds. A few tenths of 1 percent by volume of HCN in air can cause death within minutes. Hydrogen cyanide can be prepared by treating sodium cyanide or potassium cyanide with acid:

$$NaCN(s) + HCl(aq) \longrightarrow NaCl(aq) + HCN(aq)$$

Because HCN (in solution, called hydrocyanic acid) is a very weak acid ($K_a = 4.9 \times 10^{-10}$), most of the HCN produced in this reaction is in the non-ionized form and leaves the solution as hydrogen cyanide gas. For this reason, acids should never be mixed with metal cyanides in the laboratory without proper ventilation.

HCN is the gas used for execution in gas chambers.

Oxides of Carbon. Of the several oxides of carbon, the most important are carbon monoxide (CO) and carbon dioxide (CO_2). Carbon monoxide is a colorless, odorless gas formed by the incomplete combustion of carbon or carbon-containing compounds:

$$2C(s) + O_2(g) \longrightarrow 2CO(g)$$

Industrially, CO is prepared by passing steam over heated coke. Carbon monoxide burns readily in oxygen to form carbon dioxide:

$$2CO(g) + O_2(g) \longrightarrow 2CO_2(g) \qquad \Delta H° = -566 \text{ kJ}$$

Carbon monoxide is not an acidic oxide (it differs from carbon dioxide in that regard) and is only slightly soluble in water.

Although CO is relatively unreactive, it is a very poisonous gas because it has the unusual ability to bind very strongly to hemoglobin, the oxygen carrier in blood. Both molecular oxygen and carbon monoxide bind to the Fe(II) ion in hemoglobin. Unfortunately, the affinity of hemoglobin for CO is about 200 times greater than that for O_2. Hemoglobin molecules with tightly bound CO (called carboxyhemoglobin) cannot carry the oxygen needed for metabolic processes. A small amount of carbon monoxide intake can cause drowsiness and headache; death may result when about half the hemoglobin molecules are complexed with CO. The best first-aid response to carbon monoxide poisoning is to remove the victim immediately to an atmosphere with a plentiful oxygen supply or to give mouth-to-mouth resuscitation.

ppm means "parts per million."

Carbon monoxide is a major air pollutant. In cities with heavy automobile traffic, the concentration of CO in air is about 40 ppm by volume, approximately 95 percent of which is due to automobile exhaust. Note also that the concentration of carboxyhemoglobin in the blood of people who smoke is two to five times higher than that in nonsmokers.

Carbon dioxide is produced when any form of carbon or any carbon-containing compounds are burned in an excess of oxygen. Many carbonates give off CO_2 when heated or when treated with acid:

$$CaCO_3(s) \xrightarrow{\Delta} CaO(s) + CO_2(g)$$

$$CaCO_3(s) + 2HCl(aq) \longrightarrow CaCl_2(aq) + H_2O(l) + CO_2(g)$$

CO_2 is also a by-product of the fermentation of sugar:

$$\underset{\text{glucose}}{C_6H_{12}O_6(aq)} \xrightarrow{\text{yeast}} \underset{\text{ethanol}}{2C_2H_5OH(aq)} + 2CO_2(g)$$

and an end product of metabolism in animals:

$$C_6H_{12}O_6(aq) + 6O_2(g) \longrightarrow 6CO_2(g) + 6H_2O(l)$$

Carbon dioxide is suffocating but not toxic.

Carbon dioxide is a colorless and odorless gas. Unlike carbon monoxide, CO_2 is nontoxic. It is an acidic oxide:

$$CO_2(g) + H_2O(l) \rightleftharpoons H_2CO_3(aq)$$

(The reaction between CO_2 and H_2O is an example of Lewis acid-base reaction discussed on p. 557.)

Carbon dioxide is used in beverages, in fire extinguishers, and in the manufacture of baking soda, $NaHCO_3$, and soda ash, Na_2CO_3. Solid carbon dioxide (called Dry Ice) is used as a refrigerant. Dry Ice is also used in cloud seeding to induce rain. Clouds consist of tiny water droplets. In order for these droplets to aggregate into raindrops, nuclei—small particles onto which water molecules can cluster—must be present in the clouds. Normally, ice crystals that can act as nuclei for raindrop formation should form at 0°C. However, because of supercooling, they seldom develop unless the

FIGURE 26.9 *Effects produced by seeding a cloud deck with Dry Ice. Within an hour a hole (showing the disappearance of cloud) developed in the seeded area.*

temperature is below $-10°C$. To promote ice crystal formation, clouds are sometimes seeded with granulated Dry Ice, which is dispersed into the clouds from an airplane. When solid carbon dioxide sublimes (see Figure 14.41), it absorbs heat from the surrounding clouds and reduces the temperature below that required for ice crystal formation. Consequently, the presence of Dry Ice can often induce a small-scale rainfall (Figure 26.9).

Although carbon dioxide is only a trace gas in Earth's atmosphere, with a concentration of about 0.03 percent by volume, it plays a critical role in controlling climate. Carbon dioxide molecules in the atmosphere absorb some of the infrared radiation (from the sun) reflected from the surface of the Earth and become energetically excited:

$$CO_2 + h\nu \longrightarrow CO_2^*$$

where $h\nu$ denotes the energy of a photon in the infrared region and the asterisk a molecule in the excited state. Energetically excited molecules are unstable, and they quickly lose the excess energy through spontaneous emission of radiation. Part of this radiation is emitted to outer space and part of it back to the surface of Earth. Consequently, the temperature at and near the surface of Earth gradually rises.

Figure 26.10 shows the variation of carbon dioxide concentration over a period of years, as measured in Hawaii. The seasonal oscillations are caused by the removal of CO_2 by photosynthesis during the growing season in the northern hemisphere and its buildup during the fall and winter months. Clearly, the trend points to an increase in CO_2. The current rate of increase is about 1 ppm by volume per year, equivalent to 9×10^9 tons of CO_2! Scientists have estimated that by the year 2000 the CO_2 concentration will exceed preindustrial levels by about 25 percent. Some meteorologists pre-

CO_2's influence on Earth's temperature is often called the *greenhouse effect*. The glass roof of a greenhouse, like CO_2, transmits visible sunlight and absorbs some of the outgoing infrared radiation, thereby trapping the heat. This analogy is not quite exact, however, because the temperature rise in the greenhouse is due also to the restriction of air circulation inside.

FIGURE 26.10 *Yearly variation of carbon dioxide concentration at Mauna Loa, Hawaii. The general trend clearly points to increase of carbon dioxide in the atmosphere.*

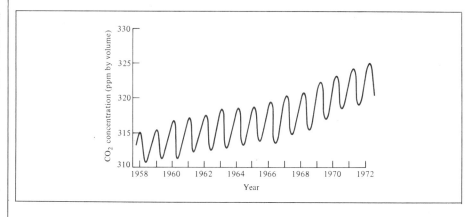

dict that this increase will raise Earth's average temperature by about 3°C. Although this seems like a small temperature change, it is actually large enough to affect the delicate thermal balance on Earth, and it could cause glaciers and ice caps to melt. This, in turn, would raise the sea level, perhaps resulting in flooded coastal areas. However, the significance of the long-term effect is not yet fully understood.

The Carbon Cycle.   Figure 26.11 shows the carbon cycle in our global ecosystem. The transfer of carbon dioxide to and from the atmosphere is an essential part of the carbon cycle, which begins with the process of photosynthesis conducted by plants and certain microorganisms:

$$6CO_2 + 6H_2O \longrightarrow C_6H_{12}O_6 + 6O_2$$

Carbohydrates and other complex carbon-containing molecules are consumed by animals, which respire and release carbon dioxide. After plants and animals die, they are decomposed by microorganisms in the soil; the carbon in their tissues is oxidized to carbon dioxide and returns to the atmosphere. In addition, there is a dynamic equilibrium between atmospheric CO_2 and carbonates in the oceans and lakes. Two other major sources of CO_2 production are volcanoes and combustion in homes and factories.

> This reaction is not spontaneous and must be promoted by radiant energy (visible light).

Silicon

Although both carbon and silicon belong to Group 4A of the periodic table, and therefore have similar outer electron configurations—ns^2np^2—they have quite different chemical properties. The differences between the chemistry of carbon and the chemistry of silicon not only are interesting and useful, but also provide us with some insight into the evolution of life.

Silicon is the second most abundant element, constituting about 26 percent by mass of Earth's crust. It does not occur in the free form and is most commonly combined with oxygen as silicates. Some silicate minerals are asbestos ($H_4Mg_3Si_2O_9$), zircon ($ZrSiO_4$), and beryl ($Be_3Al_2Si_6O_{18}$). As silica (SiO_2), it accounts for most beach sands.

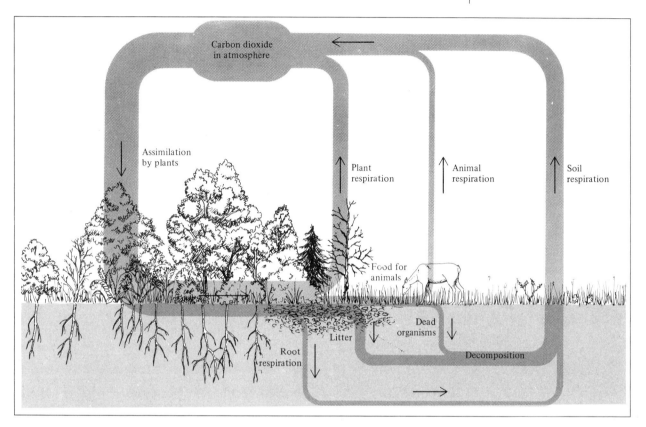

FIGURE 26.11 *The carbon cycle begins with the fixation of atmospheric carbon dioxide by the process of photosynthesis, conducted by plants and microorganisms. In this process, carbon dioxide and water react to form carbohydrates; the reaction simultaneously releases molecular oxygen. Some of the carbohydrate is consumed directly by the plant for energy; the carbon dioxide so generated is released either through the plant's leaves or through its roots. Part of the carbon fixed by plants is consumed by animals, which respire carbon dioxide. Plants and animals die and are ultimately decomposed by microorganisms in the soil; the carbon in their tissues is oxidized to carbon dioxide and returns to the atmosphere. The widths of the pathways shown are roughly proportional to the quantities involved. A similar carbon cycle takes place within the sea.*

Silicon can be prepared by heating a mixture of silica sand and coke to about 3000°C:

$$SiO_2(s) + 2C(s) \xrightarrow{\Delta} Si(l) + 2CO(g)$$

Pure silicon has the diamond structure (see Figure 12.17); each Si atom is tetrahedrally bonded to four other Si atoms. No silicon analog of graphite is known. Silicon combines with carbon to form silicon carbide, or carborundum, as discussed earlier.

Silicon forms a series of covalent hydrides called **silanes,** which have the general formula Si_nH_{2n+2}, such as SiH_4, Si_2H_6, Si_3H_8, Si_4H_{10}, etc. Like the hydrocarbons, the lower silanes are gases at room temperature. Chemically,

Ultrapure silicon is used in electronics (for example, the chips in your calculator).

they are much more reactive than their carbon counterparts. For example, when exposed to air, monosilane ignites spontaneously:

$$SiH_4(g) + 2O_2(g) \longrightarrow SiO_2(s) + 2H_2O(l)$$

In many respects, the silanes behave more like metal hydrides than like hydrocarbons. In the presence of an alkaline catalyst, they will reduce water to hydrogen:

$$SiH_4(aq) + 2H_2O(l) \longrightarrow SiO_2(s) + 4H_2(g)$$

Monosilane (SiH_4) is a strong reducing agent. It will reduce Fe(III) to Fe(II) and will precipitate silver from solutions of its salts:

$$SiH_4(g) + 4Fe^{3+}(aq) \longrightarrow Si(s) + 4H^+(aq) + 4Fe^{2+}(aq)$$

$$SiH_4(g) + 4Ag^+(aq) \longrightarrow Si(s) + 4H^+(aq) + 4Ag(s)$$

Methane (CH_4), the carbon counterpart of SiH_4, does not undergo such reactions.

No silicon analogs of ethylene and acetylene are known and only a very few compounds containing Si=Si and Si=C bonds have been prepared.

Silicon forms several covalent compounds with the halogens. For example, the action of hydrofluoric acid on silica produces the gaseous compound silicon tetrafluoride:

$$SiO_2(s) + 4HF(aq) \longrightarrow 2H_2O(l) + SiF_4(g)$$

Silicon burns vigorously in an atmosphere of chlorine to produce silicon tetrachloride:

$$Si(s) + 2Cl_2(g) \longrightarrow SiCl_4(g)$$

As we saw in Section 14.5, SiO_2 is a chief component of glass. However, because silica dissolves slowly in a strong alkaline solution

$$SiO_2(s) + 2OH^-(aq) \longrightarrow SiO_3^{2-}(aq) + H_2O(l)$$

concentrated alkaline solutions such as NaOH and KOH must be stored in polyethylene containers rather than in glass bottles.

About 90 percent of Earth's crust consists of **silicates,** which are *anion groups consisting of Si and O atoms.* This is the reason that oxygen and silicon are the first and second most abundant elements in nature (see Figure 1.9). Although silicates vary in the ratio of Si to O atoms, the basic unit in all silicates consists of a Si atom tetrahedrally bonded to four O atoms (Figure 26.12). In higher silicates, these units are joined together by sharing an O atom between two Si atoms (Figure 26.13).

The fibrous silicate mineral known as asbestos has been used for thousands of years. Because of its high chemical and thermal stability, asbestos has until recently been used widely as a thermal insulator in buildings. It is now well established that prolonged exposure to airborne suspensions of asbestos fiber dust can be very dangerous to the lung tissues and the digestive tract, and often results in cancer. The symptoms, however, may take years to develop.

Silicon has no known role in the human body, but *diatoms*, which are microscopic unicellular marine or freshwater algae, make their skeletons

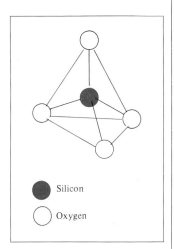

Silicon

Oxygen

FIGURE 26.12 *The tetrahedral arrangement of the SiO_4^{2-} ion. This is the basic unit for all the higher silicates.*

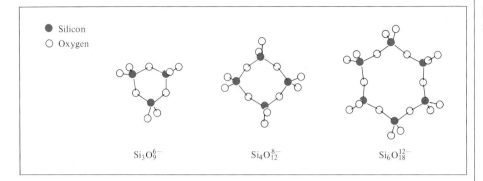

FIGURE 26.13 *Structures of several silicates.*

from SiO_2 and have an active metabolism involving Si. It has also been shown that silicon is an essential element in baby chicks, but its precise function is not known. In view of the fact that silicon is 146 times more plentiful than carbon and exhibits some of the same properties, scientists have wondered why it does not play a more significant part in directing the chemistry of living forms.

There are at least two reasons that favor the role of carbon as the basis for life on Earth. First, carbon dioxide is a monomeric (that is, a single molecular unit), stable molecule that is readily soluble in water. In contrast, SiO_2 does not exist in discrete molecular form, but comprises a giant Si—O network that forms a macromolecule of silicon dioxide (see Figure 14.32). Further, since silicon dioxide is not soluble in water, it cannot take part in acid-base reactions. The second reason concerns **catenation,** which is *the linking of like atoms.* Carbon has the unique ability to form long chains (consisting of more than fifty C atoms) and stable rings with five or six members. This versatility is responsible for the millions of organic compounds found on Earth. Silicon atoms, on the other hand, can form only relatively short chains. Therefore, the number of different silicon compounds that exist is severely limited.

The low stability of many silicon compounds is the result of the weak Si—Si bond (about 226 kJ/mol) as compared to the comparatively strong C—C bond (about 347 kJ/mol).

The Silicones. We turn now to a particularly important class of silicon compounds—the *silicones,* which consist of long chains of Si and O atoms:

$$-Si-O-Si-O-Si-O-$$

Silicones can be prepared by the hydrolysis of chlorotrimethylsilane and dichlorodimethylsilane:

$$2(CH_3)_3SiCl \quad + \quad n(CH_3)_2SiCl_2 \quad \xrightarrow{H_2O} \quad (CH_3)_3SiO \left(\begin{array}{c} CH_3 \\ | \\ Si-O \\ | \\ CH_3 \end{array} \right)_n Si(CH_3)_3$$

chlorotrimethylsilane   dichlorodimethylsilane

By varying the length of the polymeric chain (that is, the value of n) and the number and nature of the organic groups (—CH_3 or —C_6H_5), the silicone polymers formed can be volatile fluids, viscous oils, rubberlike substances, resins, or plastic materials.

The usefulness of silicones lies in the fact that they possess a combination of desirable properties not found in other polymers. Because the Si—O bonds are quite strong, they can withstand heat. The viscosity of low molar mass silicones does not change appreciably with temperature, and they are therefore suitable as grease and lubricating oils at low temperatures. Most silicones are inert in the presence of sunlight, water, and many other chemicals. Because silicones do not interact with the human body, silicone rubber is becoming important in surgery. It can be implanted in human tissues with no adverse effects. Films and tubing of silicone rubber can replace biological membranes and damaged arteries.

26.5 NITROGEN, PHOSPHORUS, AND ARSENIC

Nitrogen

About 78 percent of air by volume is nitrogen. The most important mineral sources of nitrogen are saltpeter (KNO_3) and Chile saltpeter ($NaNO_3$). Nitrogen is an essential element of life; it is in proteins and nucleic acids.

Molecular nitrogen is obtained by fractional distillation of air (the boiling points of liquid nitrogen and liquid oxygen are $-196°C$ and $-183°C$, respectively). In the laboratory, very pure nitrogen gas can be prepared by the thermal decomposition of ammonium nitrite:

> Molecular nitrogen will boil off before molecular oxygen during the fractional distillation of liquid air.

$$NH_4NO_2(s) \xrightarrow{\Delta} 2H_2O(g) + N_2(g)$$

The N_2 molecule contains a triple bond and is therefore very stable with respect to dissociation into atomic species. However, nitrogen forms a large number of compounds with hydrogen and oxygen in which the oxidation number of nitrogen varies from -3 to $+5$ (Table 26.2). Most of the nitrogen compounds are covalent; however, when heated with certain metals, nitrogen forms ionic nitrides containing N^{3-} ions:

$$6Li(s) + N_2(g) \xrightarrow{\Delta} 2Li_3N(s)$$

The nitride ion is a strong Brønsted–Lowry base and reacts with water to produce ammonia and hydroxide ion:

$$N^{3-}(aq) + 3H_2O(l) \longrightarrow NH_3(g) + 3OH^-(aq)$$

Figure 26.14 shows the binary nitrogen compounds of some of the representative elements.

Ammonia. Ammonia is one of the best-known nitrogen compounds. It is prepared industrially from nitrogen and hydrogen by the Haber process (see Sections 17.6 and 22.5). It can be prepared in the laboratory by treating ammonium chloride with sodium hydroxide:

$$NH_4Cl(aq) + NaOH(aq) \longrightarrow NaCl(aq) + H_2O(l) + NH_3(g)$$

TABLE 26.2 Common Compounds of Nitrogen

Oxidation Number	Compound	Formula	Structure

-3	Ammonia	NH_3	H—N̈—H
			    |
			    H

-2	Hydrazine	N_2H_4	H—N̈—N̈—H
			    | |
			    H H

-1	Hydroxylamine	NH_2OH	H—N̈—Ö—H
			    |
			    H

0	Nitrogen* (dinitrogen)	N_2	:N≡N:
$+1$	Nitrous oxide	N_2O	:N≡N—Ö:
$+2$	Nitric oxide	NO	:Ṅ═Ö
$+3$	Nitrous acid	HNO_2	Ö═N̈—Ö—H
$+4$	Nitrogen dioxide	NO_2	:Ö—Ṅ═Ö

$+5$	Nitric acid	HNO_3	Ö═N—Ö—H
			    |
			   :Ö:

* We list the element here as a reference.

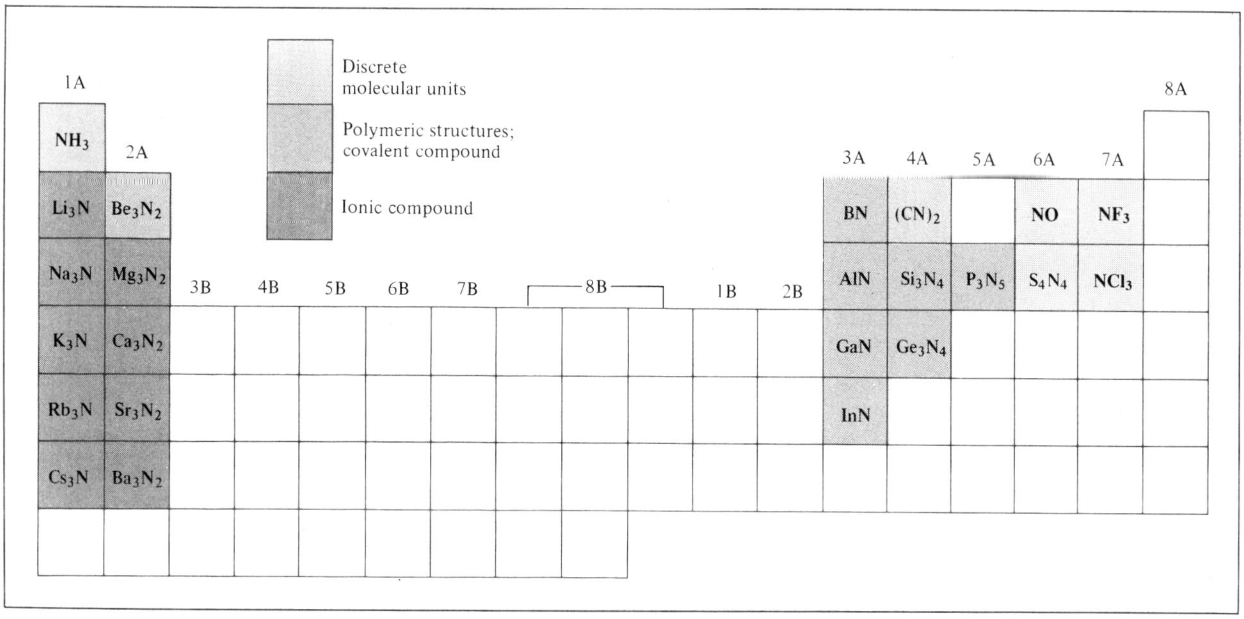

FIGURE 26.14 *Binary nitrogen compounds of some of the representative elements.*

The evolution of ammonia gas is facilitated by heating the solution. In terms of Brønsted–Lowry definition of acids and bases, the above reaction can be written in ionic form as

$$\underset{\text{acid}_1}{\text{NH}_4^+(aq)} + \underset{\text{base}_2}{\text{OH}^-(aq)} \longrightarrow \underset{\text{base}_1}{\text{NH}_3(aq)} + \underset{\text{acid}_2}{\text{H}_2\text{O}(l)}$$

Ammonia can also be prepared by the hydrolysis of a metal nitride:

$$\text{Mg}_3\text{N}_2(s) + 6\text{H}_2\text{O}(l) \longrightarrow 3\text{Mg(OH)}_2(s) + 2\text{NH}_3(g)$$

Ammonia is a colorless gas (b.p. $-33.4°C$) with an irritating odor. About three-quarters of the ammonia produced annually in the United States (19 million tons in 1984) is used in fertilizers (Appendix 5 lists the "top 25" chemicals produced in the United States). In aqueous solution ammonia behaves like a weak base:

$$\text{NH}_3(aq) + \text{H}_2\text{O}(l) \rightleftharpoons \text{NH}_4^+(aq) + \text{OH}^-(aq) \qquad K_b = 1.8 \times 10^{-5}$$

Liquid ammonia resembles water in that it undergoes autoionization

$$2\text{NH}_3(l) \rightleftharpoons \text{NH}_4^+ + \text{NH}_2^-$$

or simply

$$\text{NH}_3(l) \rightleftharpoons \text{H}^+ + \text{NH}_2^-$$

The amide ion is a strong Brønsted–Lowry base and does not exist in water (see Section 18.4).

where NH_2^- is called the *amide ion*. Note that both H^+ and NH_2^- are solvated with the NH_3 molecules. (Here is an example of ion-dipole interaction.) At $-50°C$, the ion product $([\text{H}^+][\text{NH}_2^-])$ is about 1×10^{-33}, considerably smaller than 1×10^{-14} for water at 25°C. Nevertheless liquid ammonia is a suitable solvent for many electrolytes, especially in cases where a more basic medium is required or if the solutes react with water.

One of the most unusual properties of liquid ammonia is its ability to dissolve many of the alkali and alkaline earth metals. In liquid ammonia, such a metal's atoms are converted to the cations by losing valence electrons:

$$\text{Na} \longrightarrow \text{Na}^+ + e^-$$

$$\text{Ca} \longrightarrow \text{Ca}^{2+} + 2e^-$$

Both the cation and the electron exist in the solvated form; the solvated electrons are responsible for the characteristic blue color of such solutions. Metal-ammonia solutions are powerful reducing agents (because they contain free electrons); they are useful in synthesizing both organic and inorganic compounds. It was discovered recently that the hitherto unknown alkali metal anions, M^-, are also formed in such solutions. These anions are stabilized by complex ion formation with certain large organic molecules. This means that an ammonia solution of an alkali metal contains ion pairs such as Na^+Na^-, K^+K^-, and Cs^+Cs^-! (Keep in mind that in each case the metal cation exists as a complex ion.) In fact, these "salts" are so stable that they can be isolated in crystalline form (see Color Plate 29). This finding is of considerable theoretical interest, for it shows clearly that the alkali metals can have a -1 oxidation number, although it is not found in ordinary compounds.

Oxides and Oxyacids of Nitrogen. There are many nitrogen oxides, but we will discuss only three important ones here: nitrous oxide, nitric oxide, and nitrogen dioxide.

Nitrous oxide (N_2O) is a colorless gas with a pleasing odor and sweet taste. It is prepared by heating ammonium nitrate to about 270°C:

$$NH_4NO_3(s) \xrightarrow{\Delta} N_2O(g) + 2H_2O(g)$$

Nitrous oxide resembles molecular oxygen in that it supports combustion. It does so because it decomposes when heated to form molecular nitrogen and molecular oxygen:

$$2N_2O(g) \xrightarrow{\Delta} 2N_2(g) + O_2(g)$$

It is chiefly used as an anesthetic in dental and other minor surgery. Nitrous oxide is also called "laughing gas" because a person inhaling the gas becomes somewhat giddy. No satisfactory explanation has yet been proposed for this unusual physiological response.

Nitric oxide (NO) is a colorless and slightly toxic gas. The reaction of N_2 and O_2 in the atmosphere

$$N_2(g) + O_2(g) \rightleftharpoons 2NO(g) \qquad \Delta H° = 180.8 \text{ kJ}$$

might be considered a way of fixing nitrogen. ***Nitrogen fixation*** *is the conversion of molecular nitrogen into nitrogen compounds.* The equilibrium constant for the above reaction is very small at room temperature but increases rapidly with temperature (for example, in a running auto engine).

The equilibrium constant (K_P) for this reaction is only 6.4×10^{-16} at 25°C, so very little NO will form at that temperature. An appreciable amount of nitric oxide is formed in the atmosphere, however, by the action of lightning. In the laboratory, the gas can be prepared by the reduction of dilute nitric acid with copper:

$$3Cu(s) + 8HNO_3(aq) \longrightarrow 3Cu(NO_3)_2(aq) + 4H_2O(l) + 2NO(g)$$

The nitric oxide molecule is paramagnetic, containing one unpaired electron. It can be represented by the following resonance structures:

$$\ddot{N}\!=\!\ddot{O} \longleftrightarrow \ddot{N}\!=\!\dot{O}^{\,+}$$

As we noted in Chapter 11, this molecule does not obey the octet rule. Nitric oxide dimerizes in the solid state to form N_2O_2, which is a diamagnetic molecule. Nitric oxide forms brown fumes of nitrogen dioxide instantaneously in the presence of air:

$$2NO(g) + O_2(g) \longrightarrow 2NO_2(g)$$

Consequently, we do not know what it would smell like. It is only slightly soluble in water, and the resulting solution is not acidic.

Unlike nitrous oxide and nitric oxide, nitrogen dioxide is a highly toxic yellow-brown gas with a choking odor. In the laboratory nitrogen dioxide is prepared by the action of concentrated nitric acid on copper (see Color Plate 22):

According to Le Chatelier's principle, the forward endothermic reaction is favored by heating.

The formation of NO in the atmosphere is promoted by electrical energy.

NO_2 is a major component of smog.

$$Cu(s) + 4HNO_3(aq) \longrightarrow Cu(NO_3)_2(aq) + 2H_2O(l) + 2NO_2(g)$$

Nitrogen dioxide is paramagnetic. It has a strong tendency to dimerize to dinitrogen tetroxide, which is a diamagnetic molecule:

$$2NO_2 \rightleftharpoons N_2O_4$$

This reaction occurs in both the gas phase and the liquid phase.

Nitrogen dioxide is an acidic oxide; it reacts rapidly with cold water to form both nitrous acid and nitric acid:

$$2NO_2(g) + H_2O(l) \longrightarrow HNO_2(aq) + HNO_3(aq)$$

Neither N_2O nor NO reacts with water.

This is a disproportionation reaction (see p. 476) in which the oxidation number of nitrogen changes from $+4$ (in NO_2) to $+3$ (in HNO_2) and $+5$ (in HNO_3). Note that this reaction is quite different from that between CO_2 and H_2O, in which only one acid (carbonic acid) is formed.

Nitric acid is one of the most important inorganic acids. The major industrial method of producing nitric acid is the *Ostwald process*. The starting materials, ammonia and molecular oxygen, are heated in the presence of a platinum-rhodium alloy catalyst to about 1000°C:

$$4NH_3(g) + 5O_2(g) \longrightarrow 4NO(g) + 6H_2O(g)$$

The nitric oxide is subsequently oxidized to nitrogen dioxide:

$$2NO(g) + O_2(g) \longrightarrow 2NO_2(g)$$

When dissolved in water, NO_2 forms both nitrous acid and nitric acid, as noted previously. On heating, nitrous acid decomposes to give nitric acid and nitric oxide:

This is an example of disproportionation. The oxidation number of N changes from $+3$ to $+5$ and $+2$.

$$3HNO_2(aq) \xrightarrow{\Delta} HNO_3(aq) + H_2O(l) + 2NO(g)$$

The concentrated nitric acid used in the laboratory is 68 percent HNO_3 by mass (density: 1.42 g/cm³), which corresponds to 15.7 *M*. The oxidizing properties of nitric acid are discussed in Section 16.7.

Nitric acid is used in the manufacture of fertilizers, dyes, drugs, and explosives (see discussion of nitroglycerin and TNT in Section 16.4).

The Nitrogen Cycle. Figure 26.15 summarizes the major processes involved in the cycle of nitrogen in nature. Through *biological* and *industrial* fixation, atmospheric nitrogen gas is converted into compounds suitable for assimilation (for example, nitrates) by higher plants. As with nitric oxide, a major mechanism for producing nitrates from nitrogen gas is lightning—about 30 million tons of HNO_3 are produced this way annually. The nitric acid is converted to nitrate salts in the soil. These nutrients are taken up by plants that, in turn are ingested by animals to make proteins and other essential biomolecules.

The term *denitrification* applies to processes that reverse nitrogen fixation. For example, certain anaerobic organisms decompose animal wastes as well as dead plants and animals to produce free molecular nitrogen from nitrates, thereby completing the cycle.

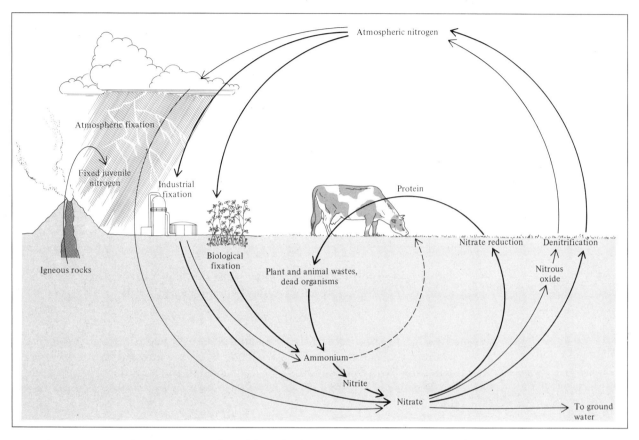

FIGURE 26.15 *The nitrogen cycle. Although the supply of nitrogen in the atmosphere is virtually inexhaustible, it must be combined with hydrogen or oxygen before it can be assimilated by higher plants, which in turn are consumed by animals. Juvenile nitrogen is nitrogen that has not previously participated in the nitrogen cycle.*

Phosphorus

Like nitrogen, phosphorus is a member of the Group 5A family; in many respects the chemistry of phosphorus resembles that of nitrogen. Phosphorus occurs most commonly in nature as *phosphate rocks*, which are mostly $Ca_3(PO_4)_2$ and $Ca_5(PO_4)_3F$. Free phosphorus can be obtained by heating calcium phosphate with coke and silica sand:

$$2Ca_3(PO_4)_2(s) + 10C(s) + 6SiO_2(s) \xrightarrow{\Delta} 6CaSiO_3(s) + 10CO(g) + P_4(g)$$

There are several allotropic forms of phosphorus, but only two—white phosphorus and red phosphorus (see Color Plate 5)—are of importance. White phosphorus consists of discrete tetrahedral P_4 molecules (Figure 26.16; also see Figure 12.19). A white solid (m.p. 44.2°C), it is insoluble in water but quite soluble in carbon disulfide (CS_2) and in organic solvents

FIGURE 26.16 *Structures of white and red phosphorus. Red phosphorus is believed to consist of a chain structure, as shown.*

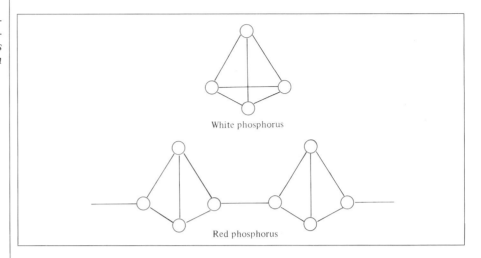

White phosphorus

Red phosphorus

such as chloroform ($CHCl_3$). White phosphorus is a highly toxic substance. It bursts into flames spontaneously when exposed to air:

$$P_4(s) + 5O_2(g) \longrightarrow P_4O_{10}(s)$$

The high reactivity of white phosphorus is believed to be the result of the strain in the P_4 molecule: The P—P bonds are compressed in the tetrahedral structure. White phosphorus was once used in the manufacture of matches, but because of its toxicity, it has been replaced by tetraphosphorus trisulfide, P_4S_3.

When heated in the absence of air, white phosphorus is slowly converted to red phosphorus at about 300°C:

$$nP_4 \text{ (white phosphorus)} \xrightarrow{\Delta} (P_4)_n \text{ (red phosphorus)}$$

Red phosphorus has a polymeric structure (see Figure 26.16) and is more stable and less volatile than white phosphorus.

Unlike N_2, elemental phosphorus contains no P=P or P≡P bonds.

Hydride of Phosphorus. The most important hydride of phosphorus is phosphine (PH_3), a colorless, very poisonous gas formed by heating white phosphorus in concentrated sodium hydroxide:

$$P_4(s) + 3NaOH(aq) + 3H_2O(l) \xrightarrow{\Delta} 3NaH_2PO_2(aq) + PH_3(g)$$
$$\text{sodium} \qquad \text{phosphine}$$
$$\text{hypophosphite}$$

Phosphine is moderately soluble in water and more soluble in carbon disulfide and organic solvents. Unlike those of ammonia, its aqueous solutions are neutral. In liquid ammonia, phosphine dissolves to give $NH_4^+PH_2^-$. Phosphine is a strong reducing agent; it reduces many metal salts to the metal. The gas burns in air at about 150°C:

$$PH_3(g) + 2O_2(g) \longrightarrow H_3PO_4(s)$$

Halides of Phosphorus. Phosphorus forms two types of binary compounds with halogens: the trihalides (PX_3) and the pentahalides (PX_5), where

X denotes a halogen atom. In contrast, nitrogen can form only trihalides (NX_3). Unlike nitrogen, phosphorus has a close-lying $3d$ subshell that can be used for valence-shell expansion. To account for the bonding in PCl_5, for example, it is necessary to invoke a hybridization process involving $3s$, $3p$, and $3d$ orbitals of phosphorus, or an sp^3d hybridization (see Example 12.4). The five sp^3d hybrid orbitals also explain satisfactorily the trigonal bipyramid geometry of the PCl_5 molecule (see Table 12.5). Nitrogen cannot form a pentahalide because five orbitals would be required in such a compound and only four are available in nitrogen (the one $2s$ and three $2p$ orbitals).

Phosphorus trichloride is prepared by heating white phosphorus in an atmosphere of chlorine:

$$P_4(l) + 6Cl_2(g) \xrightarrow{\Delta} 4PCl_3(g)$$

A colorless liquid (b.p. 76°C), PCl_3 has a pyramidal structure:

$$\ddot{P}$$
$$Cl \quad | \quad Cl$$
$$Cl$$

It is hydrolyzed according to the equation

$$PCl_3(l) + 3H_2O(l) \longrightarrow \underset{\text{phosphorous acid}}{H_3PO_3(aq)} + 3HCl(g)$$

In the presence of an excess of chlorine gas, PCl_3 is converted to phosphorus pentachloride, which is a light yellow solid:

$$PCl_3(l) + Cl_2(g) \longrightarrow PCl_5(s)$$

X-ray studies have shown that solid phosphorus pentachloride exists as $[PCl_4^+][PCl_6^-]$, in which the PCl_4^+ ion has the tetrahedral geometry and the PCl_6^- ion has the octahedral geometry. In the gas phase PCl_5 (which has the trigonal bipyramidal geometry) is in equilibrium with PCl_3 and Cl_2:

$$PCl_5(g) \rightleftharpoons PCl_3(g) + Cl_2(g)$$

Phosphorus pentachloride reacts with water as follows:

$$PCl_5(s) + 4H_2O(l) \longrightarrow H_3PO_4(aq) + 5HCl(aq)$$

Oxides and Oxyacids of Phosphorus. The two important oxides of phosphorus are tetraphosphorus hexaoxide, P_4O_6, and tetraphosphorus decaoxide, P_4O_{10}. The oxides are obtained by burning white phosphorus in limited and excess amounts of oxygen gas, respectively:

$$P_4(s) + 3O_2(g) \longrightarrow P_4O_6(s)$$

$$P_4(s) + 5O_2(g) \longrightarrow P_4O_{10}(s)$$

Figure 26.17 shows their structures. Both oxides are acidic, that is, they are converted to acids in water. The compound P_4O_{10} is a white flocculent powder (m.p. 420°C) that has a great affinity for water:

$$P_4O_{10}(s) + 6H_2O(l) \longrightarrow 4H_3PO_4(aq)$$

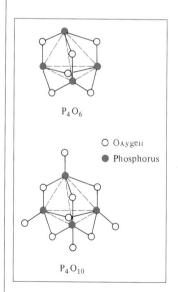

P_4O_6

○ Oxygen
● Phosphorus

P_4O_{10}

FIGURE 26.17 *The structures of P_4O_6 and P_4O_{10}. Note the tetrahedral arrangement of the P atoms in both molecules.*

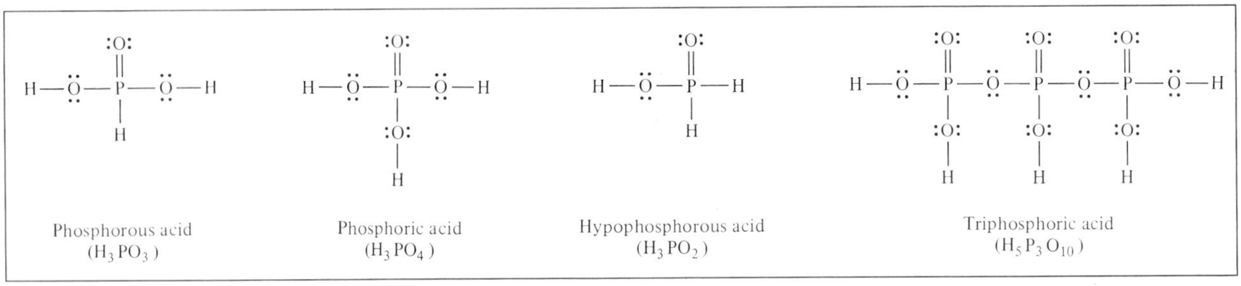

FIGURE 26.18 *Structures of some common phosphorus-containing acids.*

Phosphorous acid
(H_3PO_3)

Phosphoric acid
(H_3PO_4)

Hypophosphorous acid
(H_3PO_2)

Triphosphoric acid
($H_5P_3O_{10}$)

Phosphoric acid is the most important of the phosphorus-containing oxyacids.

For this reason, it is often used for drying gases and for removing water from solvents.

There are several acids containing phosphorus; some examples are phosphorous acid, H_3PO_3; phosphoric acid, H_3PO_4; hypophosphorous acid, H_3PO_2; and triphosphoric acid, $H_5P_3O_{10}$ (Figure 26.18). Phosphoric acid is also called orthophosphoric acid; it is a weak triprotic acid (see Section 19.4). It is prepared industrially by the reaction of calcium phosphate with sulfuric acid:

$$Ca_3(PO_4)_2(s) + 3H_2SO_4(aq) \longrightarrow 2H_3PO_4(aq) + 3CaSO_4(aq)$$

In the pure form phosphoric acid is a colorless solid (m.p. 42.2°C). The phosphoric acid we use in the laboratory is usually an 82 percent H_3PO_4 solution (by mass). Phosphoric acid has a tendency to undergo a **condensation reaction.** In a condensation reaction, *two small molecules combine to form a larger molecule; water is invariably one of the products.* When heated, two molecules of H_3PO_4 combine to form pyrophosphoric acid and water (Figure 26.19).

Phosphoric acid and phosphates have many commercial applications in detergents, fertilizers, flame retardants, and toothpastes, and as buffers in carbonated beverages to maintain a constant pH. The action of phosphorus

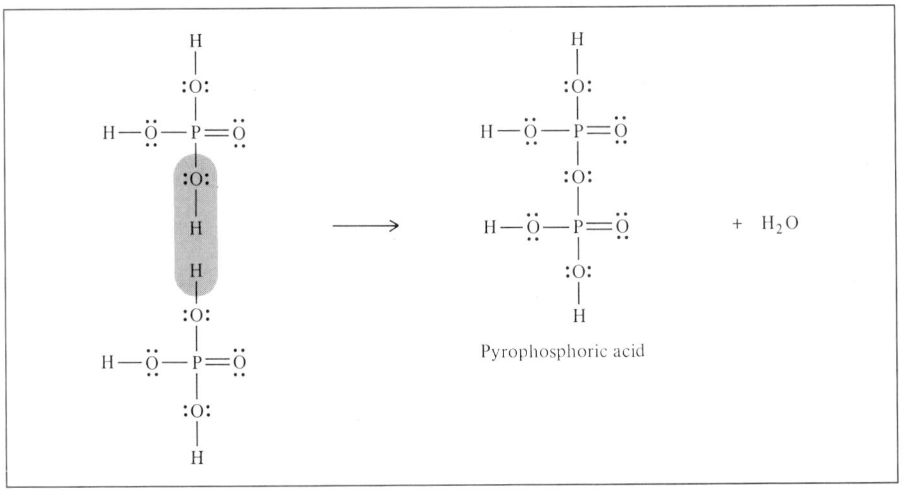

FIGURE 26.19 *Formation of pyrophosphoric acid by the combination of two phosphoric acid molecules.*

Pyrophosphoric acid

as a flame retardant depends on its ability to increase significantly the conversion of organic matter to char (elemental carbon) during burning, and thus to decrease the formation of flammable carbon-containing gases. Combustion is thus inhibited because char does not burn readily and because the amount of combustible gases, such as hydrocarbons, is greatly reduced. The use of the flame retardant *tris*(2,3-dibromopropyl) phosphate (called Tris)

$$
\begin{array}{c}
\text{O} \\
\|\\
\text{P} \\
\diagup\;|\;\diagdown \\
\text{O}\quad|\quad\text{O} \\
\end{array}
$$

CH₂	O	CH₂
CHBr	CH₂	CHBr
CH₂Br	CHBr	CH₂Br
	CH₂Br	

was banned in 1977 when it was discovered that mice and rats treated with the chemical developed kidney cancer. The discovery attracted much attention because this substance has been used to treat about half of the millions of children's sleeping garments sold annually in the United States.

Like nitrogen, phosphorus is an element that is essential to life. It constitutes only about 1 percent by mass of the human body, but it is a very important 1 percent. About 23 percent of the human skeleton is mineral matter. The phosphorus content of this mineral matter, calcium phosphate $Ca_3(PO_4)_2$, is 20 percent. Our teeth are basically $Ca_3(PO_4)_2$ and $Ca_5(PO_4)_3OH$. The phosphates are important components of the genetic materials deoxyribonucleic acid (DNA) and ribonucleic acid (RNA).

Arsenic

In Group 5A arsenic is the last element that is not a metal. As you know, the metallic character of the elements increases as we move down any group. Nitrogen and phosphorus are nonmetals, arsenic is a metalloid, and the last two members of the group, antimony (Sb) and bismuth (Bi), are metals. Note that this trend from nonmetallic to metallic properties closely parallels that observed for Group 4A elements.

The common minerals of arsenic are the sulfides As_4S_4 and As_2S_3. Arsenic is also found combined with a number of metals, such as Fe, Co, and Ni. Elemental arsenic can be obtained by heating FeAsS:

$$FeAsS(s) \xrightarrow{\Delta} FeS(s) + As(g)$$

As discussed in Chapter 12, arsenic in its most stable form is a gray solid consisting of sheets of covalently bonded As atoms (p. 357). In the vapor phase As exists as tetrahedral As_4 molecules that resemble P_4 molecules. In dry air, arsenic is stable but it is slowly converted to the oxide As_4O_6. It does not react with water but is oxidized by hot concentrated nitric acid to arsenic acid:

$$As(s) + 5HNO_3(aq) \longrightarrow H_3AsO_4(aq) + 5NO_2(g) + H_2O(l)$$

The hydride of arsenic, arsine (AsH_3), is a colorless poisonous gas. In fact, most compounds of arsenic are toxic, although the element itself is not.

Arsenic compounds are mainly used in agriculture as herbicides and pesticides. The oxide As_4O_6 is used to decolorize bottle glass.

26.6   OXYGEN AND SULFUR

Oxygen

Oxygen is by far the most abundant element in Earth's crust, constituting about 46 percent of its mass. The atmosphere contains about 21 percent of molecular oxygen by volume (23 percent by mass). Like nitrogen, oxygen occurs in the free state as a diatomic molecule (O_2). In the laboratory, oxygen gas can be obtained by heating potassium chlorate (see Figure 3.6):

$$2KClO_3(s) \xrightarrow{\Delta} 2KCl(s) + 3O_2(g)$$

The reaction is usually catalyzed by manganese(IV) dioxide (MnO_2). Pure oxygen gas can be prepared by electrolyzing water (p. 738). Industrially, oxygen gas is prepared by the fractional distillation of liquefied air. The boiling points of O_2 and N_2 are $-183°C$ and $-196°C$, respectively. Therefore, liquid nitrogen will boil off first and liquid oxygen will remain. Oxygen gas is colorless and odorless.

Oxygen is a fundamental building block of practically all vital biomolecules, accounting for about a fourth of the atoms in living matter. Molecular oxygen is the essential oxidant in the metabolic breakdown of food molecules. Without it, a human being cannot survive for more than five minutes.

Properties of Diatomic Oxygen.   Oxygen has two allotropes, O_2 and O_3. When we speak of molecular oxygen, we normally mean O_2. The O_3 molecule is called *ozone*, and will be discussed shortly.

The O_2 molecule is paramagnetic, containing two unpaired electrons (see Section 13.4). Molecular oxygen is a strong oxidizing agent. The half-reactions for the reductions in neutral (or basic) and acidic media are

$$O_2(g) + 2H_2O(l) + 4e^- \longrightarrow 4OH^-(aq) \qquad \mathscr{E}° = 0.40 \text{ V}$$

$$O_2(g) + 4H^+(aq) + 4e^- \longrightarrow 2H_2O(l) \qquad \mathscr{E}° = 1.23 \text{ V}$$

Oxides, Peroxides, and Superoxides.   Oxygen forms three types of oxides: the normal oxide (or simply the oxide), which contains the O^{2-} ion; the peroxide, which contains the O_2^{2-} ion; and the superoxide, which contains the O_2^- ion:

$\ddot{\text{O}}::\ddot{\text{O}}$	$:\ddot{\text{O}}:^{2-}$	$:\ddot{\text{O}}:\ddot{\text{O}}:^{2-}$	$:\ddot{\text{O}}:\dot{\text{O}}:^-$
oxygen	oxide	peroxide	superoxide

In Chapter 9 we discussed the formation of these oxides with alkali metals. These ions are strong Brønsted–Lowry bases and react with water as follows:

oxide: $\qquad O^{2-}(aq) + H_2O(l) \longrightarrow 2OH^-(aq)$

peroxide: $\qquad 2O_2^{2-}(aq) + 2H_2O(l) \longrightarrow O_2(g) + 4OH^-(aq)$

superoxide: $\qquad 4O_2^-(aq) + 2H_2O(l) \longrightarrow 3O_2(g) + 4OH^-(aq)$

Note that the reaction of O^{2-} with water is a hydrolysis reaction, but those involving O_2^{2-} and O_2^- are redox processes.

The nature of bonding in oxides changes across any period in the periodic table (see Figure 18.5). Oxides of elements on the left side of the periodic table, such as those of the alkali metals and alkaline earth metals, are generally ionic solids and have high melting points. Oxides of the metalloids and of the metallic elements toward the middle of the periodic table are also solids, but they have much less ionic character. Oxides of nonmetals are covalent compounds that generally exist as liquids or gases at room temperature. The acidic character of the oxides increases from left to right. Consider the oxides of the third-period elements:

$$\underbrace{Na_2O \quad MgO}_{basic} \quad \underbrace{Al_2O_3}_{amphoteric} \quad \underbrace{SiO_2 \quad P_4O_{10} \quad SO_3 \quad Cl_2O_7}_{acidic}$$

The basic character of the oxides increases as we move down a particular group. For example, MgO is only slightly soluble in water (0.006 g/L H_2O) but dissolves readily in acidic solutions:

$$MgO(s) + 2H^+(aq) \longrightarrow Mg^{2+}(aq) + H_2O(l)$$

whereas BaO is more soluble in water (35 g/L H_2O):

$$BaO(s) + H_2O(l) \longrightarrow Ba(OH)_2(aq)$$

The best known peroxide is hydrogen peroxide (H_2O_2). It is a colorless, syrupy liquid (m.p. $-0.9°C$), prepared in the laboratory by the action of cold dilute sulfuric acid on barium peroxide octahydrate:

$$BaO_2 \cdot 8H_2O(s) + H_2SO_4(aq) \longrightarrow BaSO_4(s) + H_2O_2(aq) + 8H_2O(l)$$

The structure of hydrogen peroxide is shown in Figure 26.20. Using the VSEPR method we see that the H—O and O—O bonds are bent around each oxygen atom, similar to the structure of water. The lone pair–bonding pair repulsion is greater in H_2O_2 than in H_2O, so that the HOO angle is only 97° (compared to 104.5° for HOH in H_2O).

The standard enthalpy of formation ($\Delta H_f°$) of hydrogen peroxide is -187.6 kJ/mol; thus we might expect the compound to be stable. In fact, hydrogen peroxide readily decomposes on heating or even in the presence of dust particles or certain metals, including iron and copper:

$$2H_2O_2(l) \longrightarrow 2H_2O(l) + O_2(g) \qquad \Delta H° = -196.4 \text{ kJ}$$

The reason for its apparent instability is that the products are more stable than hydrogen peroxide, as indicated by the exothermic nature of the reaction.

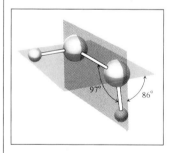

FIGURE 26.20 *Structure of H_2O_2.*

This is a disproportionation reaction. The oxidation number of O changes from -1 to -2 and 0.

Hydrogen peroxide is miscible with water in all proportions, due to its ability to hydrogen-bond with water. The dilute hydrogen peroxide solutions (3 percent by mass) that are available in drugstores are used as mild antiseptics; more concentrated H_2O_2 solutions are employed as bleaching agents for textiles, fur, and hair. The high heat of decomposition of hydrogen peroxide also makes it a suitable component in rocket fuel.

Hydrogen peroxide is a strong oxidizing agent; the half-reaction is

$$H_2O_2(aq) + 2H^+(aq) + 2e^- \longrightarrow 2H_2O(l) \qquad \mathscr{E}° = 1.77 \text{ V}$$

Some of its oxidizing actions are discussed in Section 16.7.

Hydrogen peroxide can act also as a reducing agent toward substances that are stronger oxidizing agents than itself. The half-reaction is

$$H_2O_2(aq) \longrightarrow 2H^+(aq) + O_2(g) + 2e^-$$

For example, hydrogen peroxide reduces silver oxide to metallic silver:

$$H_2O_2(aq) + Ag_2O(s) \longrightarrow 2Ag(s) + H_2O(l) + O_2(g)$$

and permanganate (MnO_4^-) to manganese(II) in an acidic solution:

$$5H_2O_2(aq) + 2MnO_4^-(aq) + 6H^+(aq) \longrightarrow 2Mn^{2+}(aq) + 5O_2(g) + 8H_2O(l)$$

If we want to determine hydrogen peroxide concentration, this reaction can be carried out as a redox titration, using a standard permanganate solution.

There are relatively few known superoxides, or compounds containing the O_2^- ion. In general, only the most reactive alkali metals (K, Rb, and Cs) form superoxides. We have seen that when a superoxide ion reacts with water, oxygen gas is evolved. This property makes potassium superoxide useful as an oxygen source in masks worn by firefighters and other rescue workers in places where the air supply is limited. The reaction is

$$2KO_2(s) + 2H_2O(l) \longrightarrow 2KOH(aq) + O_2(g) + H_2O_2(aq)$$

Because superoxides produce hydrogen peroxide when dissolved in water, they can also be used as oxidizing agents.

We should take note of the fact that both the peroxide ion and superoxide ion are by-products of metabolism. Because these ions are highly reactive, they can inflict great damage on various cellular components. Fortunately, through evolution our bodies have been equipped with the enzymes needed to convert these toxic substances to water and molecular oxygen.

Ozone.   Ozone is a rather toxic, light blue gas (b.p. $-111.3°C$) with a pungent odor. The odor of ozone is evident wherever significant electrical discharges are occurring (for example, near a subway train). Ozone can be prepared from molecular oxygen, either photochemically or by subjecting O_2 to an electrical discharge (Figure 26.21):

$$3O_2(g) \longrightarrow 2O_3(g) \qquad \Delta G° = 326.8 \text{ kJ}$$

Since the standard free energy of formation of ozone is a large positive quantity—$\Delta G_f° = (326.8/2)$ kJ/mol or 163.4 kJ/mol—it is not surprising that ozone is less stable than molecular oxygen. The ozone molecule has a bent structure in which the OOO angle is $116.5°$:

Note that the oxidation number of the O atom changes from -1 in H_2O_2 to 0 in O_2, whereas in both Ag_2O and H_2O it is -2.

The source of the water is moisture in the breath.

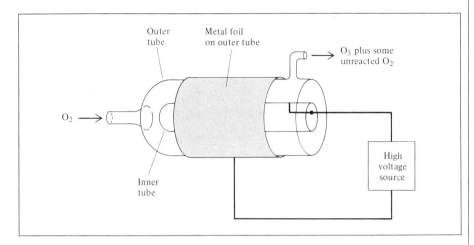

FIGURE 26.21 *Prepara-tion of O_3 from O_2 by elec-trical discharge. The outside of the outer tube and the in-side of the inner tube are coated with metal foils that are connected to a high volt-age source. (The metal foil on the inside of the inner tube is not shown.) During electrical discharge, O_2 gas is passed through the tube. The O_3 gas formed plus some unreacted O_2 gas exit at the upper-right-hand tube.*

Ozone is mainly used to purify drinking water, to deodorize air and sewage gases, and to bleach waxes, oils, and textiles.

Ozone is a very powerful oxidizing agent—its oxidizing power is exceeded only by that of molecular fluorine (see Table 24.1). Its oxidizing actions in basic and acidic solutions are represented by the following half-reactions:

$$O_3(g) + H_2O(l) + 2e^- \longrightarrow O_2(g) + 2OH^-(aq) \qquad \mathscr{E}° = +1.24 \text{ V}$$

$$O_3(g) + 2H^+(aq) + 2e^- \longrightarrow O_2(g) + H_2O(l) \qquad \mathscr{E}° = +2.07 \text{ V}$$

For example, when ozone is passed through an acidic solution of potassium iodide, the I^- ions are oxidized to molecular iodine, as indicated by its brown color:

$$O_3(g) + 2I^-(aq) + 2H^+(aq) \longrightarrow I_2(s) + O_2(g) + H_2O(l)$$

Ozone can oxidize Fe^{2+} ions to Fe^{3+} ions in acidic solution:

$$O_3(g) + 2Fe^{2+}(aq) + 2H^+(aq) \longrightarrow 2Fe^{3+}(aq) + O_2(g) + H_2O(l)$$

It can also oxidize sulfides of many metals to the corresponding sulfates:

$$4O_3(g) + PbS(s) \longrightarrow PbSO_4(s) + 4O_2(g)$$

In all these reactions, molecular oxygen is one of the products. But ozone also takes part in redox processes in which no oxygen gas is liberated:

$$O_3(g) + 3SO_2(g) \longrightarrow 3SO_3(g)$$

$$O_3(g) + 3Sn^{2+}(aq) + 6H^+(aq) \longrightarrow 3Sn^{4+}(aq) + 3H_2O(l)$$

Ozone oxidizes all the common metals except gold and platinum. In fact, a convenient test for ozone is based on its action on mercury. When exposed to ozone, the metal becomes dull looking and sticks to glass tubing (instead of flowing freely through it). This behavior is attributed to the change in surface tension caused by the formation of mercury(II) oxide:

$$O_3(g) + 3Hg(l) \longrightarrow 3HgO(s)$$

The amount of ozone in the atmosphere is generally very small, yet its presence has important effects on living matter. Through a very complex sequence of reactions the presence of ozone can promote smog formation, which is harmful to our health. Conversely, the presence of ozone in the stratosphere (about 40 km from the surface of the Earth) has a beneficial effect. Although the concentration of ozone in the stratosphere is quite low (about 10 ppm by volume), it is sufficient to absorb most of the sun's harmful wavelength radiation (between 200 nm and 300 nm):

$$O_3(g) + h\nu \longrightarrow O_2(g) + O(g)$$

This falls in the UV region (see Figure 8.4).

where $h\nu$ denotes the energy of a photon of that wavelength. The atomic oxygen produced reacts with molecular oxygen to regenerate ozone:

$$O(g) + O_2(g) \longrightarrow O_3(g)$$

so that no net depletion of ozone occurs over a period of time. Without this "ozone shield," the damaging effects of UV radiation would gradually destroy life on Earth.

Some scientists have become alarmed about the environmental effects of certain chlorofluoromethanes. These compounds, which bear the commercial name Freon, have the following formulas: CCl_3F (Freon 11), CCl_2F_2 (Freon 12), $CHClF_2$ (Freon 22), $CClF_2CClF_2$ (Freon 14), and $CClF_2CF_3$ (Freon 115). Because they are readily liquefied, relatively inert, and exhibit high vapor pressure, the Freons are used mainly as aerosol propellants in spray cans and as refrigerants. Once released into the atmosphere, they slowly diffuse up to the stratosphere, where they are decomposed by UV radiation of wavelengths between 175 nm and 220 nm, in the following gas-phase reactions:

$$CFCl_3 \longrightarrow CFCl_2 + Cl$$

$$CCl_2F_2 \longrightarrow CClF_2 + Cl$$

The free chlorine atoms can then undergo the following reactions:

$$Cl + O_3 \longrightarrow ClO + O_2$$
$$ClO + O \longrightarrow Cl + O_2$$

The net result (adding the two equations) shows the depletion of ozone that takes place:

$$O_3 + O \longrightarrow 2O_2$$

where the O atoms are supplied by the photochemical decomposition of ozone ($O_3 \rightarrow O_2 + O$). Note that the Cl atoms play the role of catalyst in these reactions, since Cl is not used up itself; ClO is an intermediate because it is produced in the first step and consumed in the second step.

How bad is "Freon pollution"? Because of uncertainties in measuring the concentrations and due to the complexities of the reactions involved, there is currently much debate on this point. Estimates of this effect range from the view that no appreciable damage has been done to the dire prediction that so much Freon has already been released to the atmosphere that irreparable depletion of ozone will occur during the next twenty to thirty years. Only time will tell.

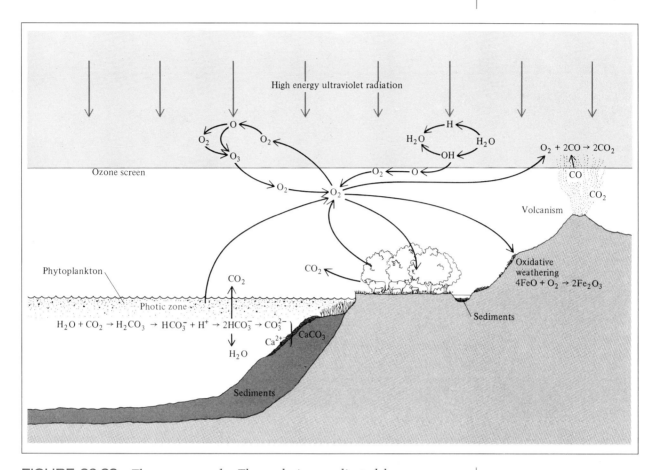

FIGURE 26.22 *The oxygen cycle. The cycle is complicated because oxygen appears in so many chemical forms and combinations, primarily as molecular oxygen, in water, and in organic and inorganic compounds.*

The Oxygen Cycle. Figure 26.22 shows the main processes in the global oxygen cycle. The cycle is complicated by the fact that oxygen appears in so many different chemical forms. Atmospheric oxygen is removed through respiration and various industrial processes. About 65 percent of the total U.S. oxygen production in 1984 (420 billion cubic feet) was used in the steel industry. The depletion of oxygen is intimately related to the production of carbon dioxide and water. Photosynthesis is the major mechanism by which molecular oxygen is regenerated from carbon dioxide. As in the carbon and nitrogen cycles, the overall balance of oxygen on Earth is ultimately tied to the energy supplied by the sun.

Sulfur

Although sulfur is not a very abundant element (it constitutes only about 0.06 percent of Earth's crust by mass), it is readily available because it occurs commonly in nature in the elemental form. The largest known

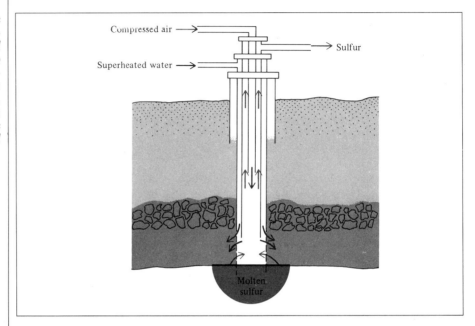

reserves of sulfur are in sedimentary deposits. In addition, sulfur occurs widely in gypsum ($CaSO_4 \cdot 2H_2O$) and various sulfide minerals.

Sulfur is extracted from underground deposits by the *Frasch process*, shown in Figure 26.23 (Herman Frasch, 1851–1914). In this process, superheated water (liquid water heated to about 160°C under high pressure to prevent it from boiling) is pumped down the outermost pipe to melt the sulfur. Next, compressed air is forced down the innermost pipe. Liquid sulfur mixed with air forms an emulsion that is less dense than water and therefore rises to the surface as it is forced up the middle pipe. Sulfur produced in this manner, which amounts to about 10 million tons per year, has a purity of about 99.5 percent.

There are several allotropic forms of sulfur, the most important being the rhombic and monoclinic forms. Rhombic sulfur is thermodynamically the most stable form; it has a puckered S_8 ring structure:

$$S_{\diagdown S \diagdown} \, ^{S \diagdown S \diagdown} \, ^{S}$$

It is a yellow, tasteless, and odorless solid (m.p. 112°C) that is insoluble in water but soluble in carbon disulfide (see Color Plate 6). On heating, it is slowly converted to monoclinic sulfur (m.p. 119°C), which also consists of the S_8 units. When liquid sulfur is heated above 150°C, the rings begin to break up and the entangling of the sulfur chains results in a sharp increase in the liquid's viscosity. Further heating tends to rupture the chains, and the viscosity begins to decrease.

Like nitrogen, sulfur shows a wide variation of oxidation numbers in its compounds (Table 26.3). The best known hydrogen compound of sulfur is

TABLE 26.3 Common Compounds of Sulfur

Oxidation Number	Compound	Formula	Structure
-2	Hydrogen sulfide	H_2S	
0	Sulfur*	S_8	
$+1$	Disulfur dichloride	S_2Cl_2	
$+2$	Sulfur dichloride	SCl_2	
$+4$	Sulfur dioxide	SO_2	
$+6$	Sulfur trioxide	SO_3	

* We list the element here as a reference.

hydrogen sulfide, which is prepared by the action of an acid on a sulfide; for example

$$FeS(s) + H_2SO_4(aq) \longrightarrow FeSO_4(aq) + H_2S(g)$$

Nowadays hydrogen sulfide used in qualitative analysis (see Section 21.6) is prepared by the hydrolysis of thioacetamide:

thioacetamide   acetic acid

Hydrogen sulfide is a colorless gas (b.p. $-60.2°C$) with an offensive odor resembling that of rotten eggs. (The smell of rotten eggs actually does come from hydrogen sulfide, which is formed by the bacterial decomposition of sulfur-containing proteins.) Hydrogen sulfide is a highly toxic substance that, like hydrogen cyanide, attacks respiratory enzymes. It is a very weak diprotic acid (see Table 19.3). In acidic solution, H_2S is a relatively weak reducing agent. In basic solution, the S^{2-} ion is a stronger reducing agent. For example, it is oxidized by permanganate to elemental sulfur:

$$3S^{2-}(aq) + 2MnO_4^-(aq) + 4H_2O(l) \longrightarrow 3S(s) + 2MnO_2(s) + 8OH^-(aq)$$

Oxides of Sulfur.   Sulfur has two important oxides: sulfur dioxide (SO_2) and sulfur trioxide (SO_3). Sulfur dioxide is formed when sulfur burns in air:

$$S(s) + O_2(g) \longrightarrow SO_2(g)$$

In the laboratory, it can be prepared by the action of an acid on a sulfite; for example

$$2HCl(aq) + Na_2SO_3(aq) \longrightarrow 2NaCl(aq) + H_2O(l) + SO_2(g)$$

or by the action of concentrated sulfuric acid on copper:

$$Cu(s) + 2H_2SO_4(aq) \longrightarrow CuSO_4(aq) + 2H_2O(l) + SO_2(g)$$

Sulfur dioxide (b.p. $-10°C$) is a colorless gas with a pungent odor, and it is quite toxic. An acidic oxide, it dissolves in water to form sulfurous acid:

$$SO_2(g) + H_2O(l) \longrightarrow H_2SO_3(aq)$$

> **Sulfurous acid has never been isolated in pure form.**

Sulfurous acid is a weak diprotic acid (see Table 19.3). Sulfur dioxide is slowly oxidized to sulfur trioxide, but the reaction rate can be greatly enhanced by a platinum or vanadium oxide catalyst:

$$2SO_2(g) + O_2(g) \longrightarrow 2SO_3(g)$$

Sulfur trioxide dissolves in water to form sulfuric acid:

$$SO_3(g) + H_2O(l) \longrightarrow H_2SO_4(aq)$$

Nature is responsible for about two-thirds of the sulfur discharged into the atmosphere. Volcanoes are the major source of sulfur (Figure 26.24).

FIGURE 26.24   *The eruption of Mt. St. Helens in southwest Washington on May 18, 1980. Volcanoes are a major source of sulfur (in the form of SO_2) and other air pollutants.*

Industrial input is mostly in the form of SO_2, with fair amounts of SO_3 resulting from combustion of sulfur-containing fuels—coal and oil—by power plants. Smelting of sulfide ores and some chemical operations, such as sulfuric acid production (to be discussed later), also produce substantial quantities of sulfur oxides.

Sulfur dioxide is a major component of air pollution. However, because SO_2 is not easily dissociated by the sun's radiation, it is generally not involved in photochemical formation of smog. Nevertheless, SO_2 molecules may be excited by solar radiation without dissociating, the result being the formation of sulfur trioxide:

$$SO_2 + h\nu \longrightarrow SO_2^\star$$
$$SO_2^\star + O_2 \longrightarrow SO_3 + O$$

where the asterisk denotes the excited molecules. Another way that sulfur trioxide may be formed from sulfur dioxide is by oxidation with ozone:

$$SO_2 + O_3 \longrightarrow SO_3 + O_2$$

In addition to their toxic effects on humans and plants, the presence of SO_2 and SO_3 in the atmosphere poses another serious environmental problem: the "acid rain" phenomenon. Precipitation in the northeastern United States, for example, has a mean pH of about 4. Atmospheric CO_2 in equilibrium with rainwater would not be expected to result in a pH less than 5.5, so the low pH is due to the presence of SO_2 and nitrogen oxides from combustion. The problem is not unique to the United States; any area that has a high density of industrial installations will receive precipitation that has a pH of 4 or lower.

Both SO_2 and SO_3 are converted to their acids (H_2SO_3 and H_2SO_4) by rainwater, and the resulting acids can corrode buildings and statues made of limestone and marble ($CaCO_3$). A typical reaction is

$$CaCO_3(s) + H_2SO_4(aq) \longrightarrow CaSO_4(aq) + H_2O(l) + CO_2(g)$$

Sulfur dioxide too can attack calcium carbonate:

$$2CaCO_3(s) + 2SO_2(g) + O_2(g) \longrightarrow 2CaSO_4(s) + 2CO_2(g)$$

Every year, acid rain causes hundreds of millions of dollars worth of damage. The term "stone leprosy" is used by some environmental chemists to describe the corrosion of stone by sulfur dioxide (Figure 26.25).

In addition to its damaging effects to buildings, acid rain is also extremely harmful to vegetation and aquatic life. Many well-documented cases show dramatically how acid rain has destroyed agricultural and forest lands and killed fish in lakes.

There are two ways to minimize the effects of SO_2 pollution. The most direct approach is to remove sulfur from coal and oil, but this is technologically difficult to accomplish. A cheaper (although less efficient) way is to remove SO_2 as it is being formed. For example, powdered limestone is injected into the power plant boiler along with coal. At a high temperature the following decomposition occurs:

$$\underset{\text{limestone}}{CaCO_3(s)} \longrightarrow \underset{\text{lime}}{CaO(s)} + CO_2(g)$$

FIGURE 26.25 *Photos of a statue, taken about sixty years apart (in 1908 and 1969), show the damaging effects of air pollutants such as sulfur dioxide.*

The lime then reacts with SO_2 produced by combustion to form calcium sulfite and calcium sulfate:

$$CaO(s) + SO_2(g) \longrightarrow CaSO_3(s)$$

$$2CaO(s) + 2SO_2(g) + O_2(g) \longrightarrow 2CaSO_4(s)$$

It is interesting to compare these reactions with the corrosion of stone buildings by SO_2. The treatment has certain drawbacks; for one, the injected limestone has a tendency to coat boiler tubes and partially plug the boiler. Test operations are underway with a wet system in which flue gas from the boiler is cleaned with water containing limestone particles or lime to remove most of its SO_2 content before it exits through the stack.

Sulfuric Acid. Sulfuric acid has often been described as the cornerstone of the chemical industry because of its many applications in the production of key chemicals. About half the sulfuric acid produced in the world is used to manufacture phosphate and ammonium sulfate fertilizers; the remaining half is used in the production of paints, fibers, plastics, detergents, soaps, and dyestuffs, as well as in metallurgy and petroleum refining. Production of sulfuric acid in the United States amounted to about 41 million tons in 1984, surpassing all other chemicals (see Appendix 5).

Sulfuric acid is a diprotic acid; it undergoes two stages of ionization in water, as follows:

$$H_2SO_4(aq) \longrightarrow H^+(aq) + HSO_4^-(aq)$$
$$HSO_4^-(aq) \rightleftharpoons H^+(aq) + SO_4^{2-}(aq) \qquad K_a = 1.3 \times 10^{-2}$$

Industrially, sulfuric acid is prepared by the *contact process*. Elemental sulfur is first burned in air to form sulfur dioxide:

$$S(s) + O_2(g) \longrightarrow SO_2(g)$$

Then the sulfur dioxide is oxidized to sulfur trioxide:

$$2SO_2(g) + O_2(g) \longrightarrow 2SO_3(g)$$

This reaction is catalyzed by V_2O_5.

Next, sulfur trioxide is dissolved in concentrated sulfuric acid to form *oleum*, or fuming sulfuric acid, $H_2S_2O_7$:

$$SO_3(g) + H_2SO_4(aq) \longrightarrow H_2S_2O_7(aq)$$

Finally, oleum is treated with water to liberate sulfuric acid:

$$H_2S_2O_7(aq) + H_2O(l) \longrightarrow 2H_2SO_4(aq)$$

Since two molecules of H_2SO_4 are formed for every $H_2S_2O_7$ molecule decomposed, there is a net increase in sulfuric acid. This roundabout procedure is necessary because a direct reaction between SO_3 and H_2O produces a fine fog of acid that is difficult to condense.

Sulfuric acid is a colorless, viscous liquid (m.p. 10.4°C). The concentrated sulfuric acid we use in the laboratory is 98 percent H_2SO_4 by mass (density: 1.84 g/cm³), which corresponds to a concentration of 18 *M*. The oxidizing strength of sulfuric acid is discussed in Section 16.7.

The great affinity of concentrated sulfuric acid for water makes it a good drying agent. Care must be taken, however, for its reaction with water generates large amounts of heat, sometimes enough to boil off the solution and cause it to splatter. Organic compounds containing hydrogen and oxygen in the ratio of 2:1 are sometimes charred by concentrated sulfuric acid (Figure 26.26):

$$C_{12}H_{22}O_{11}(s) \xrightarrow{H_2SO_4} 12C(s) + 11H_2O(l)$$
$$\text{sucrose}$$

FIGURE 26.26 *Sucrose (front) before and (back) after treatment with concentrated sulfuric acid.*

Other Compounds of Sulfur.   Carbon disulfide, a colorless, flammable liquid (b.p. 46°C) is formed by heating carbon and sulfur at high temperatures:

$$C(s) \ + \ 2S(l) \xrightarrow{\Delta} CS_2(l)$$

It is only slightly soluble in water. Carbon disulfide is a good solvent for sulfur, phosphorus, iodine, and other nonpolar substances, such as waxes and rubber.

Another interesting compound of sulfur is sulfur hexafluoride (SF_6), which is prepared by heating sulfur in an atmosphere of fluorine:

$$S(l) \ + \ 3F_2(g) \xrightarrow{\Delta} SF_6(g)$$

Sulfur hexafluoride is a nontoxic, colorless gas (b.p. $-63.8°C$). It is the most inert of all sulfur compounds; it resists attack even by molten KOH. The structure and bonding of SF_6 were discussed in Chapter 11 and Chapter 12.

Finally we note that many elements occur mainly in sulfide minerals (Figure 26.27). Of these minerals, pyrite (FeS_2) is one of the most abundant and is therefore a major source for both sulfur and iron (see Color Plate 10).

Pyrite is commonly called "fool's gold." It looks like gold and has certainly fooled many fortune hunters over the years.

26.7   THE HALOGENS

The halogens (fluorine, chlorine, bromine, and iodine) are the most reactive nonmetals (see Color Plate 7). Table 26.4 lists some of the properties of these elements. The element astatine also belongs to the Group 7A family. However, all isotopes of astatine are radioactive; the longest-lived isotope is astatine-210, which has a half-life of 8.3 h. Therefore it is both difficult and expensive to study astatine in the laboratory.

FIGURE 26.27 *Elements that occur in nature as sulfide minerals.*

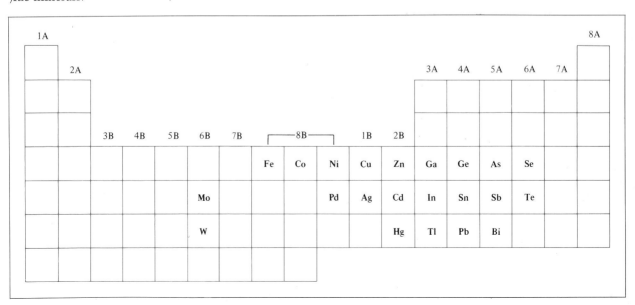

TABLE 26.4 Properties of the Halogens

Property	F	Cl	Br	I
Valence electron configuration	$2s^2 2p^5$	$3s^2 3p^5$	$4s^2 4p^5$	$5s^2 5p^5$
Melting point (°C)*	−223	−102	−7	114
Boiling point (°C)*	−187	−35	59	183
Appearance*	Pale yellow gas	Yellow-green gas	Red-brown liquid	Dark violet vapor Dark metallic-looking solid
Atomic radius (pm)	72	99	114	133
Ionic radius (pm)†	136	181	195	216
Ionization energy (kJ/mol)	1680	1251	1139	1003
Electronegativity	4.0	3.0	2.8	2.5
Standard reduction potential (V)*	2.87	1.36	1.07	0.53
Bond energy (kJ/mol)*	150.6	242.7	192.5	151.0

* These values and descriptions apply to the diatomic species X_2 where X represents a halogen atom. The half-reaction is: $X_2(g) + 2e^- \rightarrow 2X^-(aq)$.
† Refers to the anion X^-.

The halogens form a very large number of compounds. In the elemental state, they form diatomic molecules, X_2. In nature, however, because of their high reactivity, halogens are always found combined with other elements. Chlorine, bromine, and iodine occur as halides in seawater, and fluorine occurs in the minerals fluorspar (CaF_2) and cryolite (Na_3AlF_6).

Preparations and General Properties of the Halogens

Because fluorine and chlorine are strong oxidizing agents, they must be prepared by electrolytic rather than chemical oxidation of the fluoride and chloride ions. Electrolysis does not work for aqueous solutions of fluorides, however, because fluorine is a stronger oxidizing agent than oxygen. From Table 24.1 we find that

$$O_2(g) + 4H^+(aq) + 4e^- \longrightarrow 2H_2O(l) \qquad \mathscr{E}° = 1.23 \text{ V}$$

$$F_2(g) + 2e^- \longrightarrow 2F^-(aq) \qquad \mathscr{E}° = 2.87 \text{ V}$$

If F_2 were formed by the electrolysis of an aqueous fluoride solution, it would immediately oxidize water to oxygen. For this reason, fluorine is prepared by electrolyzing liquid hydrogen fluoride containing potassium fluoride to increase its conductivity, at about 70°C (Figure 26.28):

anode: $\qquad 2F^- \longrightarrow F_2(g) + 2e^-$
cathode: $\qquad 2H^+ + 2e^- \longrightarrow H_2(g)$

overall reaction: $\qquad 2HF(l) \longrightarrow H_2(g) + F_2(g)$

Chlorine gas is prepared industrially by the electrolysis of molten NaCl (see p. 738) or by the *electrolysis of a concentrated aqueous NaCl solution* (called *brine*). The latter process is called the **chlor-alkali process.** Figure

FIGURE 26.28 *Electrolytic cell for the preparation of fluorine gas. Note that because H_2 and F_2 tend to form an explosive mixture, these gases must be separated from each other.*

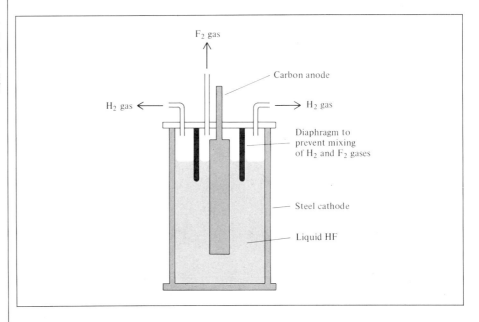

26.29 shows the arrangement for the electrolysis. The cathode is a liquid mercury pool at the bottom of the cell, and the anode is made of either graphite or titanium coated with platinum. Brine is continuously passed through the cell as shown. The electrode reactions are

$$\text{anode:} \quad 2Cl^-(aq) \longrightarrow Cl_2(g) + 2e^-$$
$$\text{cathode:} \quad 2Na^+(aq) + 2e^- \xrightarrow{\text{Hg}(l)} 2Na/Hg$$
$$\text{overall reaction:} \quad 2NaCl(aq) \longrightarrow 2Na/Hg + Cl_2(g)$$

where Na/Hg denotes the formation of sodium amalgam. (An *amalgam* is a substance made by combining mercury with another metal or metals.) The chlorine gas generated this way is very pure. The sodium amalgam does not react with the brine solution but decomposes when treated with pure water outside the cell:

$$2Na/Hg + 2H_2O(l) \longrightarrow 2NaOH(aq) + H_2(g) + 2Hg(l)$$

The useful by-products of the chlor-alkali process are therefore sodium hydroxide and hydrogen gas, and the mercury can be cycled back into the cell for reuse.

FIGURE 26.29 *The chlor-alkali process. The cathode contains mercury. The sodium-mercury amalgam is treated with water outside the cell to produce sodium hydroxide and hydrogen gas.*

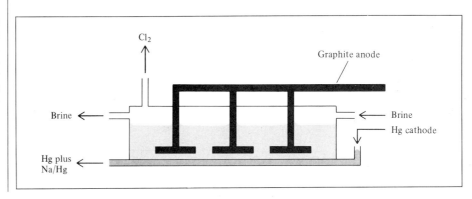

Molecular bromine and iodine are much more easily prepared from their halides than molecular chlorine. From Table 24.1, we write

$$Br_2(l) + 2e^- \longrightarrow 2Br^-(aq) \qquad \mathscr{E}° = 1.07 \text{ V}$$

$$I_2(s) + 2e^- \longrightarrow 2I^-(aq) \qquad \mathscr{E}° = 0.53 \text{ V}$$

Since chlorine is a stronger oxidizing agent than either bromine or iodine

$$Cl_2(g) + 2e^- \longrightarrow 2Cl^-(aq) \qquad \mathscr{E}° = 1.36 \text{ V}$$

free Br_2 or I_2 can be generated by bubbling chlorine gas through solutions containing bromide or iodide ions:

$$Cl_2(g) + 2Br^-(aq) \longrightarrow 2Cl^-(aq) + Br_2(l)$$

$$Cl_2(g) + 2I^-(aq) \longrightarrow 2Cl^-(aq) + I_2(s)$$

Bromine is prepared commercially from seawater by this process.

In the laboratory, chlorine, bromine, and iodine can be prepared by heating the alkali halides (NaCl, KBr, or KI) in concentrated sulfuric acid in the presence of manganese(IV) oxide. A representative reaction is

$$MnO_2(s) + 2H_2SO_4(aq) + 2NaCl(aq) \longrightarrow$$
$$MnSO_4(aq) + Na_2SO_4(aq) + 2H_2O(l) + Cl_2(g)$$

Although all halogens are highly reactive and toxic, the magnitude of reactivity and toxicity generally decreases from fluorine to iodine. Most of the halides can be classified into two categories. The fluorides and chlorides of many metallic elements, especially those belonging to the alkali metal and alkaline earth metal (except beryllium) families, are ionic compounds. Most of the halides of nonmetals such as sulfur and phosphorus are covalent compounds. Figure 26.30 shows the fluorides of the representative elements.

FIGURE 26.30 *Binary fluorides of some representative elements.*

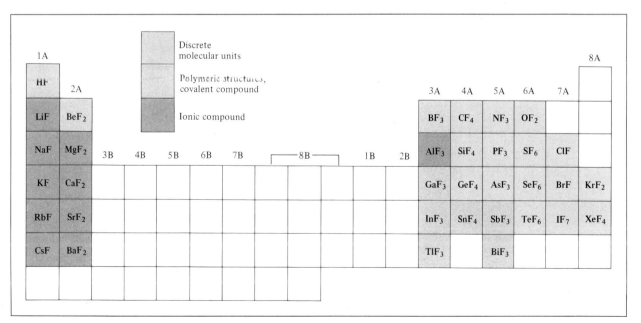

In Table 26.4 we see that the general valence electron configuration of the halogens is ns^2np^5, where $n = 2, 3, 4,$ and 5. The oxidation number is the number of valence electrons a halogen atom shares with the atom of another element. For example, when a halogen atom shares all seven of its valence electrons with an atom of a more electronegative element (as in Cl_2O_7), it is assigned an oxidation number of $+7$. If a halogen atom achieves a noble gas electron configuration by accepting an electron (as in NaCl), it has an oxidation number of -1. Thus the oxidation number of the halogens can vary between -1 and $+7$. The only exception is fluorine. Because it is the most electronegative element, it can have only two oxidation numbers: 0 (as in F_2) and -1 (as in LiF).

Recall that the first member of a group usually differs in properties from the rest of the members in the same group (p. 267).

The chemistry of fluorine differs in many ways from that of the rest of the halogens. The following are some of the important differences:

- Fluorine is the most reactive of the halogens. The difference in reactivity between fluorine and chlorine is greater than that between chlorine and bromine. (The F—F bond energy does not follow the trend of other halogens; see Table 26.4.)
- Hydrogen fluoride (HF) is a liquid (b.p. 19.5°C) as a result of strong intermolecular hydrogen bonding, whereas all other hydrogen halides are gases at room temperature.
- Hydrofluoric acid is a weak acid, whereas all other hydrohalic acids (HCl, HBr, and HI) are strong acids.
- Fluorine reacts with cold sodium hydroxide solution to produce oxygen difluoride:

$$2F_2(g) + 2NaOH(aq) \longrightarrow 2NaF(aq) + H_2O(l) + OF_2(g)$$

The same reaction with chlorine or bromine, on the other hand, produces a halide and a hypohalite:

This is a disproportionation reaction. The oxidation number of X changes from 0 to -1 and $+1$.

$$X_2(g) + 2NaOH(aq) \longrightarrow NaX(aq) + NaOX(aq) + H_2O(l)$$

where X stands for Cl or Br. Iodine does not react under the same conditions.

- Silver fluoride (AgF) is soluble. All other silver halides (AgCl, AgBr, and AgI) are insoluble (see solubility rules on p. 76).

The Hydrogen Halides

The hydrogen halides, an important class of halogen compounds, can be formed by the direct combination of the elements:

$$H_2(g) + X_2(g) \rightleftharpoons 2HX(g)$$

where X denotes a halogen atom. These reactions (especially those involving F_2 and Cl_2) can occur with explosive violence. In the laboratory, HF and HCl can be prepared by reacting the salts of the halides with concentrated sulfuric acid:

$$CaF_2(s) + H_2SO_4(aq) \longrightarrow 2HF(g) + CaSO_4(s)$$

$$2NaCl(s) + H_2SO_4(aq) \longrightarrow 2HCl(g) + Na_2SO_4(aq)$$

Hydrogen bromide and hydrogen iodide cannot be prepared this way because they are oxidized further to elemental bromine and iodine. For example, the reaction between NaBr and H_2SO_4 is

$$2NaBr(s) + 2H_2SO_4(aq) \longrightarrow Br_2(l) + SO_2(g) + Na_2SO_4(aq) + 2H_2O(l)$$

Instead, hydrogen bromide is prepared by first reacting bromine with phosphorus to form phosphorus tribromide:

$$P_4(s) + 6Br_2(l) \longrightarrow 4PBr_3(l)$$

Next, PBr_3 is treated with water to yield HBr:

$$PBr_3(l) + 3H_2O(l) \longrightarrow 3HBr(g) + H_3PO_3(aq)$$

Hydrogen iodide can be prepared in the same manner.

The high reactivity of HF is demonstrated by the fact that it attacks silica and silicates:

$$6HF(aq) + SiO_2(s) \longrightarrow H_2SiF_6(aq) + 2H_2O(l)$$

This property makes HF suitable for etching glass and is the reason that hydrogen fluoride must be kept in plastic or inert metal (for example, Pt) containers.

Aqueous solutions of hydrogen halides are acidic. As mentioned earlier, the strength of the acids increases from HF to HI.

Oxyacids of the Halogens

The halogens also form a series of oxyacids with the following general formulas:

HOX	HXO_2	HXO_3	HXO_4
hypohalous acid	halous acid	halic acid	perhalic acid

Chlorous acid ($HClO_2$) is the only known halous acid. All the halogens except fluorine form halic and perhalic acids. The Lewis formulas of the chlorine oxyacids are

H:Ö:Cl: H:Ö:Cl:Ö: H:Ö:Cl:Ö: H:Ö:Cl:Ö:

hypochlorous acid   chlorous acid   chloric acid   perchloric acid

For a given halogen, the acid strength decreases from perhalic acid to hypohalous acid; the origin of this trend is discussed in Section 18.5.

Table 26.5 lists some of the halogen compounds. Note that periodic acid (HIO_4) does not appear. This is because it cannot be isolated in the pure form. Instead, a substance with the formula H_5IO_6 is often used to represent periodic acid.

The -1 oxidation state of all the halogens is the most stable; therefore, all compounds in which a halogen has an oxidation number greater than -1 are oxidizing agents.

TABLE 26.5 Common Compounds of Halogens*

Compound	F	Cl	Br	I
Hydrogen halide	HF (-1)	HCl (-1)	HBr (-1)	HI (-1)
Oxides	OF_2 (-1)	Cl_2O $(+1)$	Br_2O $(+1)$	I_2O_5 $(+5)$
		ClO_2 $(+4)$	BrO_2 $(+4)$	
		Cl_2O_7 $(+7)$		
Oxyacids	HOF (-1)	HOCl $(+1)$	HOBr $(+1)$	HOI $(+1)$
		$HClO_2$ $(+3)$		
		$HClO_3$ $(+5)$	$HBrO_3$ $(+5)$	HIO_3 $(+5)$
		$HClO_4$ $(+7)$		H_5IO_6 $(+7)$

* The number in parentheses indicates the oxidation number of the halogen.

Interhalogen Compounds

Because the halogens are very reactive elements, it is not surprising that they also form binary compounds among themselves. **Interhalogen compounds,** or *compounds formed between two different halogen elements,* have the formulas

$$XX' \quad XX_3' \quad XX_5' \quad XX_7'$$

where X and X' are two different halogens and X is the larger atom of the two. Many of these compounds can be prepared by direct combination:

$$Cl_2(g) + F_2(g) \longrightarrow 2ClF(g)$$

$$Cl_2(g) + 3F_2(g) \longrightarrow 2ClF_3(g)$$

$$Br_2(l) + 3F_2(g) \longrightarrow 2BrF_3(l)$$

Others require more indirect routes:

$$KCl(s) + 3F_2(g) \longrightarrow KF(s) + ClF_5(g)$$

$$KI(s) + 4F_2(g) \longrightarrow KF(s) + IF_7(g)$$

Many of the interhalogen compounds are unstable and react violently with water. Note that all of the interhalogen molecules with the exception of the XX' type violate the octet rule. The geometry of these compounds can be predicted by using the VSEPR model (Figure 26.31).

FIGURE 26.31 *Structures of some interhalogen compounds.*

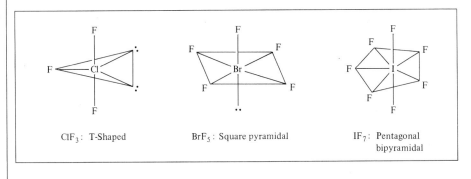

ClF_3: T-Shaped BrF_5: Square pyramidal IF_7: Pentagonal bipyramidal

Uses of the Halogens

Fluorine. The halogens and their compounds find many applications in industry, health care, and other areas. One is fluoridation, the practice of adding small quantities of fluorides (about 1 ppm by mass) such as NaF to drinking water to reduce dental caries. Hydroxyapatite, $Ca_5(PO_4)_3OH$, is the major constituent of bones and teeth. The presence of the hydroxide ion makes the substance susceptible to attack by acid (bacterial secretions, beverages with low pH, for example). Replacement of the hydroxide with fluoride ion makes teeth more resistant to decay. Fluorides are also used in toothpaste (as stannous fluoride, SnF_2) and other dentrifices.

One of the most important inorganic fluorides is uranium hexafluoride, UF_6, which is essential to the gaseous effusion process for separating isotopes of uranium (U-235 and U-238). Industrially, fluorine is used to produce Freons (discussed in Section 26.6) and polytetrafluoroethylene, a polymer better known as Teflon:

$$\left(\begin{array}{c} CF_2 \\ \quad CF_2 \quad \end{array} CF_2 \right)_n$$

Teflon is used in electrical insulators, high-temperature plastics, cooking utensils, and so on.

Another type of fluorine-containing compounds deserves some mention because of their potential biological importance. These are the *perfluorinated compounds*—organic compounds in which all of the hydrogen atoms are replaced by fluorine atoms. Two typical examples are

perfluorodecalin ($C_{10}F_{18}$)   perfluorotri-*n*-butylamine ($C_{12}NF_{27}$)

Inconceivable as it may seem, it has been demonstrated repeatedly that animals (including humans) can survive after having had their blood largely, or even totally, replaced by such compounds dissolved in a salt solution. For example, "bloodless" rats, with their blood replaced by perfluorotri-*n*-butylamine, can live up to five hours in an atmosphere containing 50 percent oxygen. Even more astounding is the fact that some animals can survive when totally immersed in a beaker filled with perfluoro compounds saturated with oxygen gas. As Figure 26.32 shows, the mouse is actually breathing the liquid!

The physiological effects of perfluorinated compounds are not yet clearly understood. However, it has been demonstrated that these compounds can be used as blood substitutes in humans in some cases. Among the many advantages of such an "artificial blood" would be (1) it could be prepared in large quantities at relatively low cost, (2) it could be kept (in frozen form)

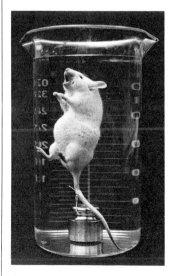

FIGURE 26.32 *A mouse breathing a liquid perfluoro compound saturated with oxygen gas.*

for a very long period of time, (3) its use would eliminate the transmission of diseases such as hepatitis and AIDS through blood transfusions, (4) it would be particularly valuable to patients with rare blood types, and (5) it might be used for patients who, for religious reasons, refuse to accept natural blood from a donor.

Chlorine. Chlorine plays an important biological role in the human body, because the chloride ion is the principal anion in intracellular and extracellular fluids. Chlorine is widely used as an industrial bleaching agent for paper and textiles. Ordinary household laundry bleach contains the active ingredient sodium hypochlorite (about 5 percent by mass), which is prepared by reacting chlorine gas with a cold solution of sodium hydroxide:

$$Cl_2(g) + 2NaOH(aq) \longrightarrow NaCl(aq) + NaOCl(aq) + H_2O(l)$$

Chlorine is also used to purify water and disinfect swimming pools. When chlorine dissolves in water, it undergoes the following reaction:

$$Cl_2(g) + H_2O(l) \longrightarrow HCl(aq) + HOCl(aq)$$

It is thought that the OCl^- ions destroy bacteria by oxidizing life-sustaining compounds within them.

Chlorinated methanes such as carbon tetrachloride and chloroform are useful organic solvents. Large quantities of chlorine are used to produce insecticides, such as DDT. However, in view of their damage to the environment, the use of many of these compounds is either totally banned or greatly restricted in the United States.

Bromine. So far as we know, bromine compounds occur naturally only in some marine organisms. Seawater is about 1×10^{-3} M in Br^-; therefore, it is the main source of bromine. Bromine is used to prepare ethylene dibromide ($BrCH_2CH_2Br$), which is used as an insecticide and as a scavenger for lead (that is, to combine with lead) in gasoline to keep lead from depositing in engines. Recent studies have shown that ethylene dibromide is a very potent carcinogen. Bromine combines directly with silver to form silver bromide ($AgBr$), which is used in photographic films (see p. 480).

Iodine. Iodine is not used as widely as the other halogens. A 50 percent (by mass) alcohol solution of iodine, known as tincture of iodine, is used medicinally as an antiseptic. Iodine is an essential constituent of the thyroid hormone thyroxine:

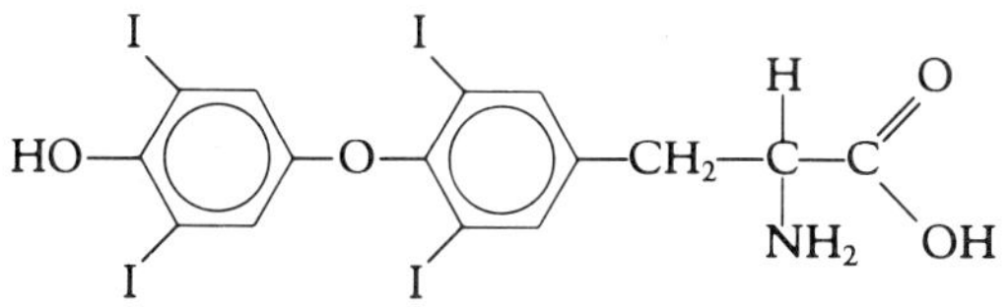

Iodine deficiency in the diet may result in enlargement of the thyroid gland (known as goiter). Iodized table salt sold in the United States usually contains 0.01 percent KI or NaI, which is more than sufficient to satisfy the 1

mg of iodine per week required for the formation of thyroxine in the human body.

A compound of iodine that deserves mention is silver iodide (AgI). It is a pale yellow solid that darkens when exposed to light. In this respect, it is similar to silver bromide. In Section 26.4 we saw that Dry Ice is used in cloud seeding. Silver iodide too can be used in this process. The advantage of using silver iodide is that enormous numbers of nuclei (that is, small particles on which ice crystals can form) become available. About 10^{15} nuclei are produced from 1 g of AgI by vaporizing an acetone solution of silver iodide in a hot flame, as it is dropped from an airplane.

26.8   THE NOBLE GASES

We come now to the Group 8A elements (sometimes known as the Group 0 elements): helium, neon, argon, krypton, xenon, and radon (see Color Plate 8). Because their outer s and p subshells are completely filled, these elements are chemically unreactive. Unlike the other nonmetallic gases, they occur naturally in the atomic state. As we will see later, not all these elements are totally inert. Table 26.6 lists some of their common properties. All isotopes of radon are radioactive; therefore, it is both difficult and expensive to study radon in the laboratory and it will not be discussed here.

Helium

Helium was detected in the sun by spectroscopic measurements of the emission spectrum before it was definitely identified on Earth. In the interior of the sun, helium is produced by the fusion of hydrogen atoms at temperatures exceeding 15 million degrees Celsius. Terrestrial helium results from alpha particles emitted by the radioactive decay of elements such as uranium and thorium in Earth's crust (such nuclear reactions are discussed in Chapter 25).

$$^{238}_{92}\text{U} \longrightarrow {}^{234}_{90}\text{Th} + {}^{4}_{2}\text{He}$$

$$^{230}_{90}\text{Th} \longrightarrow {}^{226}_{88}\text{Ra} + {}^{4}_{2}\text{He}$$

Helium nuclei (He^{2+}) readily pick up two electrons from the surroundings and become neutral helium atoms. The helium gas so formed mixes with natural gas and is found in concentrations ranging from 0.5 to 2.4 percent (by volume). It can be separated from other gases by applying enough pressure at a low temperature to liquefy all other gases, such as nitrogen, oxygen, and hydrocarbons. In Table 26.6 we see that helium has a very low boiling point of $-268.9°C$ (4.2 K). In fact, it is the most difficult gas to condense because the dispersion forces between the He atoms are so weak.

No stable compounds of helium have ever been prepared, hence there is no chemistry of helium to speak of. Helium itself, however, is a very interesting and important substance. The earliest practical use of helium was to replace hydrogen as a lifting gas in balloons. It has about 90 percent of the buoyancy of hydrogen, but it is not flammable and is nontoxic.

Note that the mechanism for cloud seeding is different for AgI. It does not lower the temperature as Dry Ice does.

TABLE 26.6 Properties of Noble Gases

Property	He	Ne	Ar	Kr	Xe
Valence electron configuration	$1s^2$	$2s^22p^6$	$3s^23p^6$	$4s^24p^6$	$5s^25p^6$
Atmospheric abundance (% by volume)	5×10^{-4}	0.0012	0.94	1.1×10^{-4}	9×10^{-6}
Melting point (°C)	−272.3	−248.7	−189.2	−157.1	−111.9
Boiling point (°C)	−268.9	−245.9	−185.7	−152.9	−106.9
First ionization energy (kJ/mol)	2373	2080	1521	1241	1167
Examples of compounds	None	None	None	KrF_2	XeF_4, XeO_4

Most helium produced today is used in the space industry to pressurize rocket fuels, as a coolant, and as a substance to dilute oxygen gas in spacecraft atmospheres. In this role, it has two advantages over nitrogen: It is lighter, and because it is less soluble in blood, it does not induce any unusual physiological effects in humans. The latter property makes it suitable to add to oxygen as a breathing mixture for divers.

Liquid helium has some unique thermodynamic properties. Figure 26.33 shows the specific heat of liquid helium as a function of temperature. When the liquid is cooled to −271.0°C (2.2 K), there is a very sharp transition, and the normal liquid helium (called helium I) becomes a superfluid (called helium II). As a superfluid, helium II is almost a perfect heat conductor. It also has nearly zero viscosity. This property is dramatically demonstrated by partly immersing a precooled beaker in helium II and observing the spontaneous climb of the liquid (as a thin layer) up the outer surface of the beaker and down into the beaker, until the levels inside and outside are equal.

An important application of liquid helium is in **superconductivity.** Superconductivity is the *loss of electrical resistance in certain metals, alloys, and compounds at low temperatures,* which a liquid helium medium provides. A superconducting substance has a particular transition temperature, below which it becomes a superconductor. For lead, for example, the transition temperature is 7.3 K; for niobium, it is 9.0 K. Besides its theoretical significance, superconductivity is of considerable interest in problems of energy consumption. Electricity generated in power plants is sometimes transmitted hundreds of miles. Although the cables (usually made up of copper) are good electrical conductors, they do have some resistance, which, over long distances, contributes significantly to the energy loss. (The energy loss is given by I^2R, where I is the electric current and R the resistance of the transmission cable.) Electric current along a superconducting medium would result in little or no energy loss because the resistance of the medium is very small.

Much research is now underway to develop materials that would become superconducting at temperatures higher than 4.2 K, so that liquid hydrogen (b.p. 20 K) or even liquid nitrogen (b.p. 77 K) could be used instead of the more expensive liquid helium.

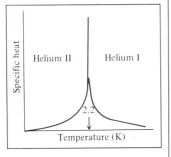

FIGURE 26.33 *The λ point of liquid helium. The choice of the name is based on the similarity between the curves and the Greek letter lambda.*

Neon and Argon

Both neon and argon have important practical uses. Neon is used in luminescent electric lighting, or "neon lights." Liquid neon is used in low-temperature research. Argon is used mainly in electric light bulbs to provide an inert atmosphere. Because of its inert nature, it is also employed to flush dissolved oxygen gas from molten metals in metallurgical processes.

Krypton and Xenon

The Group 8A elements were called the inert gases until 1963, and rightly so. No one had ever prepared a compound containing any of these elements. But an experiment carried out by Neil Bartlett (1932–   ) in that year shattered chemists' long-held view of these elements. Bartlett at that time was investigating the great oxidizing power of platinum hexafluoride, PtF_6. He noticed that when this compound came into contact with oxygen, it changed color and yielded an ionic substance with the formula $O_2^+PtF_6^-$. Since the ionization energy of O_2 (1176 kJ/mol) is very close to that of xenon (1167 kJ/mol), Bartlett reasoned that xenon should also be oxidized by PtF_6. He was correct. The reaction is rapid and visibly dramatic (see Color Plate 30):

$$Xe(g) + PtF_6(g) \longrightarrow Xe^+PtF_6^-(s)$$
$$\text{red} \qquad\qquad\quad \text{yellow-orange}$$

The report of this reaction triggered tremendous interest and excitement. Only a few months later, scientists at the Argonne National Laboratories succeeded in preparing binary xenon fluorides. By heating xenon and fluorine (in 1:5 ratio) in a nickel container at 400°C and 6 atm for one hour, they obtained xenon tetrafluoride (XeF_4) (Figure 26.34). Two other fluorides (XeF_2 and XeF_6) were also synthesized. A number of xenon–oxygen compounds (XeO_3, XeO_4) and ternary compounds containing Xe, F, and O ($XeOF_4$, XeO_3F_2) have also been prepared. The success with xenon suggested that other noble gases might undergo similar reactions. Indeed, a few compounds of krypton (KrF_2, for example) have since been prepared, but helium, argon, and neon remain "confirmed bachelors" to date. That this is so should not be surprising to us. As Table 26.6 shows, the ionization energies of the noble gases decrease from helium to xenon; therefore, the two elements on the far right, Kr and Xe, can be oxidized more easily than the others.

The structure and properties of noble gas compounds have been thoroughly investigated by many physical and chemical techniques. Today we know more about the chemistry of xenon than about some of the more abundant elements in spite of the fact that its compounds do not appear to have any commercial applications, and they are certainly *not* involved in natural biological processes.

At first it was thought that a unique type of bonding might be involved in the noble gas compounds. This turned out to be incorrect. For the most part, the geometry of these compounds can be accounted for satisfactorily

It has since been found that the product of this reaction is more complex than $Xe^+PtF_6^-$. However, this does not deter from Bartlett's achievement of making the first noble gas compound.

FIGURE 26.34 *Crystals of xenon tetrafluoride.*

by the VSEPR method outlined in Chapter 12. Figure 26.35 shows the structure of some Xe compounds as predicted by the VSEPR method. Unlike most other hexafluorides, such as SF_6, PtF_6, and UF_6, the XeF_6 molecule does not have an octahedral structure. In xenon hexafluoride, the Xe atom has fourteen valence electrons (eight from xenon plus one from each of the

FIGURE 26.35 *Structures of xenon compounds as predicted by the VSEPR method. The exact structure of XeF_6 (not shown here) has not been determined, although it is thought to be a distorted octahedron. Note that none of the compounds obey the octet rule.*

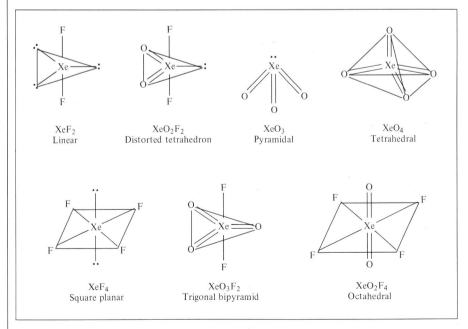

F atoms). Thus, the structure of XeF_6 must be based on the preferred arrangement of a total of seven electron pairs, of which one is a nonbonding pair.

Finally, we note two interesting properties of xenon. First, xenon has found some limited application in medicine as an anesthetic. Second, a 1978 study showed that when xenon is subjected to high pressure of about 320,000 atm, it becomes metallic; that is, it conducts electricity as metals do. While the applications of this phenomenon are not immediately obvious, it is nevertheless fascinating to see such a dramatic change in the properties of the substance—from a relatively unreactive gas to a metal—brought about by external influence.

SUMMARY

1. Hydrogen atoms contain one proton and one electron. They are the simplest atoms. Hydrogen combines with many metals and nonmetals to form hydrides; some hydrides are ionic and some are covalent.
2. There are three isotopes of hydrogen: ^1_1H, ^2_1H (deuterium), and ^3_1H (tritium). Heavy water contains deuterium.
3. Boron, a metalloid, forms unusual three-center bonds in its binary compounds with hydrogen, the boranes.
4. The important inorganic compounds of carbon are the carbides; the cyanides, most of which are extremely toxic; carbon monoxide, also toxic and a major air pollutant; and carbon dioxide, an end product of metabolism and a component in the global carbon cycle.
5. Silicon combines with oxygen in silicate minerals, in silica (SiO_2), and in silicones, which contain long silicon–oxygen chains. Silicone polymers with widely varying properties can be designed for a variety of uses.
6. Elemental nitrogen, N_2, contains a triple bond and is very stable. Compounds in which nitrogen has oxidation numbers from -3 to $+5$ are formed between nitrogen and hydrogen and/or oxygen atoms. Ammonia (NH_3) is one of the top 25 industrial chemicals.
7. White phosphorus, P_4, is highly toxic, very reactive, and flammable; the polymeric red phosphorus, $(P_4)_n$, is more stable. Phosphorus forms oxides and halides with oxidation numbers of $+3$ and $+5$ and several oxygen-containing acids.
8. Elemental oxygen, O_2, is paramagnetic and contains two unpaired electrons. Oxygen forms ozone (O_3), oxides (O^{2-}), peroxides (O_2^{2-}), and superoxides (O_2^-). Oxygen, the most abundant element in Earth's crust, is essential for life on Earth.
9. Sulfur is taken from Earth's crust by the Frasch process as a molten liquid. Sulfur exists in a number of allotropic forms and has a variety of oxidation numbers in its compounds.
10. Sulfuric acid is the cornerstone of the chemical industry. It is produced from sulfur via sulfur dioxide and sulfur trioxide in what is called the contact process.

11. The halogens are toxic and reactive, and are found only in compounds with other elements. Fluorine and chlorine are strong oxidizing agents and are prepared by electrolysis.

12. The reactivity, toxicity, and oxidizing ability of the halogens decrease from fluorine to iodine. The halogens all form binary acids (HX) and series of oxyacids.

13. The noble gases are quite unreactive and are present in the atmosphere in atomic form. Compounds of xenon and krypton have, quite unexpectedly, been prepared in recent years.

KEY WORDS

Borane, p. 786
Carbide, p. 788
Catenation, p. 795
Chlor-alkali process, p. 819
Condensation reaction, p. 804
Cyanide, p. 789

Hydrogenation, p. 784
Interhalogen compounds, p. 824
Nitrogen fixation, p. 799
Silane, p. 793
Silicate, p. 794
Superconductivity, p. 828

PROBLEMS

More challenging problems are marked with an asterisk.

Metals and Nonmetals

26.1 Without referring to Figure 1.4, determine for each of the following elements whether they are metals, metalloids, or nonmetals: (a) Cs, (b) Ge, (c) I, (d) Kr, (e) W, (f) Ga, (g) Te, (h) Bi.

26.2 List two chemical and two physical properties that distinguish a metal from a nonmetal.

26.3 Make a list of physical and chemical properties of chlorine (Cl_2) and magnesium. Comment on their differences with reference to the fact that one is a metal and the other is a nonmetal.

26.4 Carbon is usually classified as a nonmetal. However, the graphite used in "lead" pencils conducts electricity. Look at a pencil and list two nonmetallic properties of graphite.

Hydrogen

26.5 Explain why hydrogen has no unique position in the periodic table.

26.6 Describe two laboratory and two industrial preparations for hydrogen.

26.7 Hydrogen exhibits three types of bonding in its compounds. Describe each type of bonding with an example.

26.8 Elements number 17 and 20 form compounds with hydrogen. Look up the formulas for these two compounds and compare their chemical behavior in water.

26.9 Give an example of hydrogen as (a) an oxidizing agent and (b) a reducing agent.

26.10 Give the name of (a) an ionic hydride and (b) a covalent hydride. In each case describe the preparation and give the structure of the compound.

26.11 Compare the physical and chemical properties of the hydrides of each of the following elements: Na, Ca, C, N, O, Cl.

26.12 What are interstitial hydrides?

26.13 Suggest a physical method other than effusion that would allow you to separate hydrogen gas from neon gas.

26.14 Write a balanced equation to show the reaction between CaH_2 and H_2O. How many g of CaH_2 are needed to produce 26.4 L of H_2 gas at 20°C and 746 mmHg?

26.15 How many kg of water must be processed to obtain 2.0 L of D_2 at 25°C and 0.90 atm pressure? Assume that deuterium abundance is 0.015 percent and that recovery is 80 percent.

26.16 Predict the outcome of the following reactions:

(a) $CuO(s) + H_2(g) \longrightarrow$

(b) $Na_2O(s) + H_2(g) \longrightarrow$

*26.17 Lubricants used for watches usually consist of long-chain hydrocarbons. Oxidation by air forms solid polymers that eventually destroy the effectiveness of the lubricants. It is believed that one of the initial steps in the oxidation is removal of a hydrogen atom (hydrogen abstraction). By replacing the hydrogen atoms at reactive sites with deuterium atoms, it is possible to substantially slow down the overall oxidation rate. Why? (*Hint:* Consider the kinetic isotope effect.)

26.18 Describe briefly what is meant by the "hydrogen economy."

Boron

26.19 Describe an industrial preparation of boron.

26.20 The standard enthalpy of formation of $B_2H_6(g)$ is 31.4 kJ/mol. Find in Appendix 1 the ΔH_f° of C_2H_6 and C_2H_4. How do these compounds compare in thermodynamic stability?

26.21 Equal volumes of NH_3 gas and BF_3 gas at the same temperature and pressure react to form a solid product. What is its structure?

26.22 A gaseous compound contains only B and H. The density of the gas is 1.24 g/L at STP. Calculate the molar mass of the compound and suggest a molecular formula for it.

26.23 A solid compound of boron and hydrogen has the empirical formula B_5H_7. A solution consisting of 0.103 g of this solid dissolved in 5.00 g of benzene freezes 0.87°C below the freezing point of pure benzene. Determine the molecular formula of the compound of boron and hydrogen. The molal freezing point depression constant for benzene is 5.12°C/m.

*26.24 Write a balanced equation for the oxidation of boron to boric acid by nitric acid.

26.25 Consider the borohydride ion BH_4^-. (a) Is the ion isoelectronic with CH_4? (b) Describe the bonding in BH_4^- in terms of hybridization.

Carbon and Silicon

26.26 Describe the reaction between CO_2 and OH^- in terms of a Lewis acid-base reaction such as that shown on p. 557.

26.27 Draw a Lewis structure for the C_2^{2-} ion.

26.28 Balance the following equations:

(a) $Be_2C(s) + H_2O(l) \longrightarrow$
(b) $CaC_2(s) + H_2O(l) \longrightarrow$

26.29 Unlike $CaCO_3$, Na_2CO_3 does not yield CO_2 on heating. On the other hand, $NaHCO_3$ undergoes thermal decomposition to produce CO_2 and Na_2CO_3. (a) Write a balanced equation for the reaction. (b) How would you test for the CO_2 evolved?

26.30 Two solutions are labeled A and B. Solution A contains Na_2CO_3 and solution B contains $NaHCO_3$. Describe how you would distinguish between the two solutions if you were provided with a $MgCl_2$ solution. (*Hint:* You need to know the solubilities of $MgCO_3$ and $MgHCO_3$.)

*26.31 Magnesium chloride is dissolved in a solution containing sodium bicarbonate. On heating, a white precipitate is formed. Explain what causes the precipitation.

*26.32 A few drops of concentrated ammonia solution added to a calcium bicarbonate solution causes a white precipitate to form. Write a balanced equation for the reaction.

*26.33 Sodium hydroxide is hydroscopic; that is, it absorbs moisture when exposed to the atmosphere. A student placed a pellet of NaOH on a watch glass. A few days later, she noticed that the pellet was covered with a white solid. What is the identity of this solid? (*Hint:* Air contains CO_2.)

26.34 A piece of red-hot magnesium ribbon will continue to burn in an atmosphere of carbon dioxide even though CO_2 does not support combustion. Explain.

26.35 Is carbon monoxide isoelectronic with nitrogen (N_2)?

26.36 A concentration of 8.00×10^2 ppm by volume of CO is considered lethal to humans. Calculate the minimum mass of CO in grams that would become a lethal concentration in a closed room 17.6 m long, 8.80 m wide, and 2.64 m high. The temperature and pressure are 20.0°C and 756 mmHg, respectively.

26.37 Describe one way of preparing silicon.

26.38 What is the major component of sand?

26.39 What is the basic unit in silicates?

26.40 When heated strongly a silane decomposes into solid silicon and H_2. A sample yielded 0.465 g of Si and 0.0220 mole of H_2. Determine the empirical formula of the compound.

*26.41 Explain why it is that although silicon hydrides cannot form long-chain molecules, long chains of alternating silicon–oxygen atoms *can* make up large molecules, such as those found in the silicones.

26.42 Under certain conditions the reaction of Cl_2 with Mg_2Si forms a compound that has the empirical formula $SiCl_3$. Its molar mass is 269 g. What is the structure of the compound?

26.43 Suggest some reasons why carbon is more suitable as an essential element of terrestrial life than silicon would be.

*26.44 Explain why $SiCl_4$ undergoes hydrolysis, whereas CCl_4 does not.

*26.45 Glass containers are not suitable for storing very concentrated alkaline solutions. Why?

Nitrogen and Phosphorus

26.46 Describe a laboratory and an industrial preparation of nitrogen gas.

26.47 Nitrogen can be prepared by (a) passing ammonia over red-hot copper(II) oxide and by (b) heating ammonium dichromate [one of the products is Cr(III) oxide]. Write a balanced equation for each of the preparations.

26.48 Write balanced equations for the preparation of sodium nitrite by (a) heating sodium nitrate and (b) heating sodium nitrate with carbon.

26.49 Sodium amide ($NaNH_2$) reacts with water to produce sodium hydroxide and ammonia. Describe this reaction as a Brønsted–Lowry acid-base reaction.

26.50 Write a balanced equation for the formation of urea, $CO(NH_2)_2$, from carbon dioxide and ammonia. Should the reaction be run at a high or low pressure to maximize the yield?

26.51 Some farmers feel that lightning helps produce a better crop. What is the scientific basis for this belief?

*26.52 At 620 K, the vapor density of ammonium chloride relative to hydrogen (H_2) under the same conditions of temperature and pressure is 14.5, although, according to its formula mass, it should have a vapor density of 26.8. How would you account for this discrepancy?

26.53 What is meant by nitrogen fixation? Describe a process for fixation of nitrogen on an industrial scale.

26.54 Explain why nitric acid can be reduced but not oxidized.

26.55 Explain, giving one example in each case, why nitrous acid can act both as a reducing agent and as an oxidizing agent.

26.56 Write a balanced equation for each of the following processes: (a) On heating, ammonium nitrate produces nitric oxide, (b) on heating, potassium nitrate produces potassium nitrite and oxygen gas, (c) on heating, lead nitrate produces lead(II) oxide, nitrogen dioxide (NO_2), and oxygen gas.

26.57 Explain why, under normal conditions, the reaction between zinc and nitric acid does not produce hydrogen.

26.58 Potassium nitrite may be produced by heating a mixture of potassium nitrate and carbon. Write a balanced equation for this reaction. Calculate the theoretical yield of KNO_2 produced by heating 57.0 g of KNO_3 with an excess of carbon.

26.59 Consider the reaction

$$N_2(g) + O_2(g) \rightleftharpoons 2NO(g)$$

Given that the $\Delta G°$ for the reaction at 298 K is 173.4 kJ, calculate (a) the standard free energy of formation of NO, (b) K_P for the reaction, and (c) K_c for the reaction.

26.60 Predict the geometry of nitrous oxide (N_2O) by the VSEPR method and draw resonance structures for the molecule.

26.61 From the data in Appendix 1, calculate $\Delta H°$ for the synthesis of NO at 25°C:

$$4NH_3(g) + 5O_2(g) \longrightarrow 4NO(g) + 6H_2O(l)$$

26.62 Describe an industrial preparation of phosphorus.

*26.63 Explain why two N atoms can form a double bond or a triple bond, whereas two P atoms can form only a single bond. (Hint: Consider the effect of increasing atomic size on the sideway overlap in the formation of a pi bond.)

26.64 Why is the P_4 molecule unstable?

*26.65 Starting with elemental phosphorus (P_4), show how you would prepare phosphoric acid.

26.66 When 1.645 g of white phosphorus is dissolved in 75.5 g of CS_2, the solution boils at 46.709°C, whereas pure CS_2 boils at 46.300°C. The molal boiling point elevation constant for CS_2 is 2.34°C/m. Calculate the molar mass of white phosphorus and give the molecular formula.

26.67 Dinitrogen pentoxide can be prepared by the reaction between P_4O_{10} and HNO_3. Write a balanced equation for this reaction. Calculate the theoretical yield of N_2O_5 if 79.4 g of P_4O_{10} is used in the reaction and HNO_3 is the excess reagent. (Hint: One of the products is HPO_3.)

26.68 What is the hybridization of phosphorus in the phosphonium ion, PH_4^+?

*26.69 Explain why (a) NH_3 is more basic than PH_3, (b) NH_3 has a higher boiling point than PH_3, (c) PCl_5 exists but NCl_5 does not, (d) N_2 is more inert than P_4.

26.70 Solid PCl_5 exists as $[PCl_4^+][PCl_6^-]$. Draw Lewis formulas for these ions. Describe the state of hybridization of the P atoms.

Oxygen and Sulfur

26.71 Describe one industrial and one laboratory preparation of O_2.

26.72 Give an account of the various kinds of oxides that exist and illustrate each type by two examples.

26.73 Compare the structures of the following pairs of oxides: (a) CO_2 and SiO_2, (b) H_2O and H_2O_2, (c) MgO and NO.

26.74 Describe some unusual properties (both physical and chemical) of water.

26.75 What is the difference between hydrolysis and hydration?

26.76 Draw molecular orbital energy level diagrams for O_2, O_2^-, and O_2^{2-}.

26.77 Hydrogen peroxide can be prepared by treating barium peroxide with an aqueous solution of sulfuric acid. Write a balanced equation for this reaction.

26.78 One of the steps involved in the depletion of ozone in the stratosphere by nitric oxide may be represented as

$$NO(g) + O_3(g) \rightleftharpoons NO_2(g) + O_2(g)$$

From the data in Appendix 1, calculate $\Delta G°$, K_P, and K_c for the reaction at 25°C.

26.79 Hydrogen peroxide is unstable and decomposes readily:

$$2H_2O_2(aq) \longrightarrow 2H_2O(l) + O_2(g)$$

This reaction is accelerated by light, heat, or a catalyst. (a) Explain why hydrogen peroxide sold in drugstores is usually kept in dark bottles. (b) The concentrations of aqueous hydrogen peroxide solutions are normally expressed as percent by mass. In the decomposition of hydrogen peroxide, how many liters of oxygen gas can be produced at STP from 15.0 g of a 7.50 percent hydrogen peroxide solution?

26.80 SF_6 exists but OF_6 does not. Explain.

26.81 The total production of sulfuric acid in the United States in 1981 was 41 million tons. Calculate the amount of sulfur (in grams and moles) used in the production.

26.82 Sulfuric acid is a dehydrating agent. Write balanced equations for the reactions between sulfuric acid and the following substances: (a) HCOOH, (b) H_3PO_4, (c) HNO_3, (d) $HClO_3$. (*Hint:* Sulfuric acid is not decomposed by the dehydrating action.)

26.83 Explain why SCl_6, SBr_6, and SI_6 cannot be prepared.

26.84 Compare the physical and chemical properties of H_2O and H_2S.

26.85 Calculate the amount of $CaCO_3$ (in grams) that would be required to react with 50.6 g of SO_2 emitted by a power plant.

*26.86 The bad smell in water caused by hydrogen sulfide can be removed by the action of chlorine. The reaction is

$$H_2S(aq) + Cl_2(aq) \longrightarrow 2HCl(aq) + S(s)$$

If the hydrogen sulfide content of contaminated water is 22 ppm by mass, calculate the amount of Cl_2 (in grams) required to remove all the H_2S from 1.5×10^3 gallons of water, which is the average amount of water used daily per person in the United States (1 gallon = 3.785 L).

26.87 What are the oxidation numbers of O and F in HOF?

26.88 Concentrated sulfuric acid reacts with sodium iodide to produce molecular iodine, hydrogen sulfide, and sodium hydrogen sulfate. Write a balanced equation for the reaction.

26.89 Briefly describe the cause of acid rain and its damaging effects.

The Halogens

26.90 Describe an industrial preparation for each of the halogens.

26.91 Metal chlorides can be prepared in a number of ways: (a) direct combination of metal and molecular chlorine, (b) reaction between metal and hydrochloric acid, (c) acid-base neutralization, (d) metal carbonate treated with hydrochloric acid, (e) precipitation reaction. Give an example for each type of preparation.

*26.92 Show that chlorine, bromine, and iodine are very much alike by giving an account of their behavior (a) with hydrogen, (b) in producing silver salts, (c) as oxidizing agents, and (d) with sodium hydroxide. (e) In what respects is fluorine not a typical halogen element?

*26.93 Aqueous copper(II) sulfate solution is blue in color. When aqueous potassium fluoride is added to the $CuSO_4$ solution, a green precipitate is formed. If aqueous potassium chloride is added instead, a bright green solution is formed. Explain what happens in each case.

26.94 Draw structures for (a) $(HF)_2$ and (b) HF_2^-.

26.95 Sulfuric acid is a weaker acid than hydrochloric acid. Yet hydrogen chloride is evolved when concentrated sulfuric acid is added to sodium chloride. Explain.

26.96 A 375 gallon tank is filled with water containing 167 g of bromine in the form of Br^- ion. How many liters of Cl_2 gas at 1.00 atm and 20°C will be required to oxidize all of the bromide to molecular bromine?

*26.97 Hydrogen fluoride can be prepared by the action of sulfuric acid on sodium fluoride. Explain why hydrogen bromide cannot be prepared by the action of the same acid on sodium bromide.

26.98 What volume of bromine (Br_2) measured at 100°C and 700 mmHg pressure would be obtained if 2.00 L of dry chlorine (Cl_2), measured at 15.0°C and 760 mmHg, were absorbed by a potassium bromide solution?

26.99 Use the VSEPR method to predict the geometries of the following species: (a) IF_7, (b) I_3^-, (c) $SiCl_4$, (d) PF_5, (e) SF_4, (f) ClO_2^-.

26.100 Iodine pentoxide, I_2O_5, is sometimes used to remove carbon monoxide from the air by forming carbon dioxide and iodine. Write a balanced equation for this reaction and identify species that are oxidized and reduced.

The Noble Gases

26.101 As a group, the noble gases are more stable than other elements. Why?

26.102 Helium gas is used to dilute oxygen gas for deep-sea divers. If a diver has to work at a depth where the pressure is 4.0 atm, what must be the content of He (percent by volume) in a He/O_2 mixture so that the partial pressure of O_2 is 0.20 atm?

26.103 Why do all noble gas compounds violate the octet rule?

26.104 Balance the following equations:

(a) $XeF_4 + H_2O \longrightarrow XeO_3 + Xe + HF$
(b) $XeF_6 + H_2O \longrightarrow XeO_3 + HF$

26.105 Heated xenon(VI) fluoride undergoes decomposition:

$$XeF_6(s) \longrightarrow XeF_4(g) + F_2(g)$$

Starting with 8.96 g of XeF_6, calculate the combined volume of the product gases at 107°C and 374 mmHg. Assume complete decomposition.

26.106 Give oxidation numbers of Xe in the following compounds: XeF_2, XeF_4, XeF_6, $XeOF_4$, XeO_3, XeO_2F_2, XeO_4.

Miscellaneous Problems

26.107 Write a balanced equation for each of the following reactions: (a) Heating phosphorous acid yields phosphoric acid and phosphine (PH_3). (b) Lithium carbide reacts with hydrochloric acid to give lithium chloride and methane. (c) Bubbling HI gas through an aqueous solution of HNO_2 yields molecular iodine and nitric oxide. (d) Hydrogen sulfide is oxidized by chlorine to give HCl and SCl_2.

26.108 (a) Which of the following compounds has the greatest ionic character? PCl_5, $SiCl_4$, CCl_4, BCl_3. (b) Which of the following ions has the smallest ionic radius? F^-, C^{4-}, N^{3-}, O^{2-}. (c) Which of the following atoms has the highest ionization energy? F, Cl, Br, I. (d) Which of the following oxides is most acidic? H_2O, SiO_2, CO_2, XeO_4.

26.109 Starting with deuterium oxide, describe how you would prepare (a) ND_3 and (b) C_2D_2.

26.110 Both N_2O and O_2 support combustion. Suggest one physical and one chemical test to distinguish between the two gases.

26.111 What is the change in oxidation number for the following reaction?

$$3O_2 \longrightarrow 2O_3$$

27
METALS AND METALLURGY

W e have seen that elements can be classified as metals, nonmetals, or metalloids. Chapter 26 focused on familiar nonmetallic elements. In this chapter we will take a close look at the occurrence of metals, and the physical and chemical methods for extracting, refining, and purifying them. The chemistry of some familiar metals is the subject of Chapter 28.

TABLE 27.1 Principal Types of Minerals

Type	Minerals
Native metals	Ag, Au, Bi, Cu, Pd, Pt
Carbonates	$BaCO_3$ (witherite), $CaCO_3$ (calcite), $MgCO_3$ (magnesite), $CaCO_3 \cdot MgCO_3$ (dolomite), $PbCO_3$ (cerussite), $ZnCO_3$ (smithsonite)
Chromate	$PbCrO_4$ (crocoite)
Halides	CaF_2 (fluorite), $NaCl$ (halite), KCl (sylvite), Na_3AlF_6 (cryolite)
Hydroxides	$Mg(OH)_2$ (brucite), $Mg_3(Si_4O_{10})(OH)_2$ (talc)
Oxides	$Al_2O_3 \cdot 2H_2O$ (bauxite), Al_2O_3 (corundum), Fe_2O_3 (hematite), Fe_3O_4 (magnetite), Cu_2O (cuprite), MnO_2 (pyrolusite), SnO_2 (cassiterite), TiO_2 (rutile), ZnO (zincite)
Phosphates	$Ca_3(PO_4)_2$, $Ca_{10}(PO_4)_6(OH)_2$ (hydroxyapatite)
Silicates	$Be_3Al_2Si_6O_{18}$ (beryl), $ZrSiO_4$ (zircon), $NaAlSi_3O_8$ (albite)
Sulfides	Ag_2S (argentite), CdS (greenockite), Cu_2S (chalcocite), FeS_2 (pyrite), HgS (cinnabar), PbS (galena), ZnS (sphalerite)
Sulfates	$BaSO_4$ (barite), $CaSO_4$ (anhydrite), $PbSO_4$ (anglesite), $SrSO_4$ (celestite), $MgSO_4 \cdot 7H_2O$ (epsomite)

27.1 OCCURRENCE OF METALS

Most metals occur in nature in a chemically combined state as minerals. A **mineral** is a *naturally occurring substance with a characteristic range of chemical composition. The material of a mineral deposit concentrated enough to allow economical recovery of a desired metal is known as* **ore.** Table 27.1 lists the principal types of minerals, and Figure 27.1 is a classification of metals according to their minerals. Color Plates 1–10 show many of these minerals. In addition to the minerals found in Earth's crust, seawater is a rich source of some metal ions, such as Mg^{2+} and Ca^{2+}. Furthermore, vast areas of the ocean floor are covered with *manganese nodules,* which are made up mostly of manganese, along with iron, nickel, copper, and cobalt (Figure 27.2).

27.2 METALLURGY PROCESSES

Metallurgy is the *science and technology of separating metals from their ores and of compounding alloys.* The three principal steps in the recovery of a metal from its ore are preparation of the ore, production of the metal, and purification of the metal.

Preparation of the Ore

In the preliminary treatment of an ore, the desired mineral is separated from waste material—usually clay and silicate minerals—which are collec-

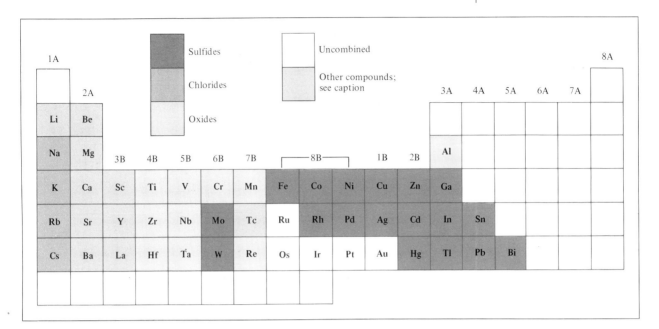

FIGURE 27.1 *Some of the best known minerals of the metals. The lithium-containing mineral is spodumene ($LiAlSi_2O_6$), the beryllium-containing mineral is beryl (see Table 27.1), and the minerals containing the rest of the alkaline earth metals are the carbonates and sulfates. The minerals for Sc, Y, and La are the phosphates. Some metals have more than one type of important mineral. For example, in addition to the sulfide, iron is found in hematite (Fe_2O_3) and magnetite (Fe_3O_4), and in addition to the oxide, aluminum is found in beryl.*

tively called the *gangue*. One very useful method for carrying out such a separation is called *flotation*. The ore is finely ground and added to water containing oil and detergent. The liquid mixture is then beaten or blown to form a froth. The oil preferentially wets the mineral particles, which are then carried to the top, while the gangue settles to the bottom. The froth

FIGURE 27.2 *Manganese nodules on the ocean floor.*

is skimmed off, allowed to collapse, and dried to recover the mineral particles.

Another physical separation process makes use of the magnetic properties of certain minerals. The mineral magnetite (Fe_3O_4) in particular can be separated from the gangue by using a strong electromagnet. Substances like iron and cobalt that are strongly attracted to magnets are called *ferromagnetic*.

Mercury can be used to treat ore for metal extraction. It forms alloys, or *amalgams*, with a number of metals. Mercury dissolves the silver and gold in an ore to form a liquid amalgam. The liquid is easily separated from the remaining ore. The gold or silver is recovered by distilling off the mercury.

Production of Metals

Because metals in their combined forms always have positive oxidation numbers, the production of a free metal is usually a reduction process. Preliminary operations may be necessary to convert the ore to a chemical state more suitable for reduction. For example, an ore may be *roasted* to drive off volatile impurities and at the same time to convert the carbonates and sulfides to the corresponding oxides, which can be reduced more conveniently to yield the pure metals.

$$CaCO_3(s) \xrightarrow{\Delta} CaO(s) + CO_2(g)$$

$$2PbS(s) + 3O_2(g) \xrightarrow{\Delta} 2PbO(s) + 2SO_2(g)$$

This last equation points up the fact that the conversion of sulfides to oxides is a major source of sulfur dioxide, a notorious air pollutant (p. 815). The development in recent years of processes for converting this by-product to sulfuric acid instead of releasing it into the air has helped to reduce SO_2 emissions in some parts of the country.

How a pure metal is obtained by reduction from its combined form depends on the standard reduction potential of the metal (see Table 24.1). Table 27.2 outlines the reduction processes for several metals. We will briefly describe the chemical and electrolytic reduction processes. In Section 27.3 we will discuss specific examples.

A more electropositive metal has a more negative standard reduction potential (see Table 24.1).

Chemical Reduction.   Theoretically we can use a more electropositive metal as a reducing agent to separate a less electropositive metal from its compound at high temperatures:

$$V_2O_5(s) + 5Ca(l) \longrightarrow 2V(l) + 5CaO(s)$$

$$TiCl_4(g) + 2Mg(l) \longrightarrow Ti(s) + 2MgCl_2(l)$$

$$Cr_2O_3(s) + 2Al(s) \longrightarrow 2Cr(l) + Al_2O_3(s)$$

$$3Mn_3O_4(s) + 8Al(s) \longrightarrow 9Mn(l) + 4Al_2O_3(s)$$

In some cases even molecular hydrogen can be used as a reducing agent, as in the preparation of tungsten (for making filaments in light bulbs) from tungsten(VI) oxide:

TABLE 27.2 Reduction Processes for Some Common Metals

	Metal	*Reduction Process*
Decreasing activity of metals ↓	Lithium, sodium, magnesium, calcium	Electrolytic reduction of the molten chloride
	Aluminum	Electrolytic reduction of anhydrous oxide (in molten cryolite)
	Chromium, manganese, titanium, vanadium, iron, zinc	Reduction of the metal oxide with a more electropositive metal, or reduction with coke and carbon monoxide
	Mercury, silver, platinum, copper, gold	These metals occur in the free (uncombined) state or can be obtained by roasting their sulfides

$$WO_3(s) + 3H_2(g) \longrightarrow W(s) + 3H_2O(g)$$

Reduction by carbon (in the form of coke) is the process that is responsible for the largest tonnage of metals (such as iron) produced. Some typical reductions are

$$2MO + C \longrightarrow 2M + CO_2$$

$$MO + C \longrightarrow M + CO$$

$$MO + CO \longrightarrow M + CO_2$$

where M represents a metal.

Electrolytic Reduction. Electrolytic reduction is suitable for very electropositive metals, such as sodium, magnesium, and aluminum. The process is usually carried out on the anhydrous molten oxide or halide of the metal:

$$2MO(l) \longrightarrow 2M \text{ (at cathode)} + O_2 \text{ (at anode)}$$

$$2MCl(l) \longrightarrow 2M \text{ (at cathode)} + Cl_2 \text{ (at anode)}$$

Purification of Metals

Metals prepared by reduction usually need further treatment to remove various impurities. The extent of purification, of course, depends on the use to be made of the metal. We will describe three common purification procedures.

Distillation. Metals that have low boiling points, such as mercury, magnesium, and zinc, may be separated from other metals by fractional distillation. One well-known method of fractional distillation is the *Mond process* (Ludwig Mond, 1839–1909) for the purification of nickel. Carbon monoxide gas is passed over the impure nickel metal at about 70°C to form the volatile tetracarbonylnickel (b.p. 43°C), a highly toxic substance, which is separated from the less volatile impurities by distillation:

$$Ni(s) + 4CO(g) \longrightarrow Ni(CO)_4(g)$$

Pure metallic nickel is then recovered from $Ni(CO)_4$ by heating the gas at 200°C:

$$Ni(CO)_4(g) \xrightarrow{\Delta} Ni(s) + 4CO(g)$$

The carbon monoxide that is released is recycled back into the process.

Electrolysis. Electrolysis is another important purification technique. The copper metal obtained by roasting copper sulfide usually contains a number of impurities, such as zinc, iron, silver, and gold. The more electropositive metals are removed by an electrolysis process in which the impure copper acts as the anode and *pure* copper acts as the cathode in a sulfuric acid solution containing Cu^{2+} ions (Figure 27.3). The reactions are

$$\text{anode (oxidation):} \qquad Cu(s) \longrightarrow Cu^{2+}(aq) + 2e^-$$
$$\text{cathode (reduction):} \qquad Cu^{2+}(aq) + 2e^- \longrightarrow Cu(s)$$

Iron and zinc, like the other more reactive metals in the copper anode, are also oxidized at the anode and enter the solution as Fe^{2+} and Zn^{2+}. They are not reduced at the cathode, however. The less electropositive metals, such as gold and silver, are not oxidized at the anode. Eventually, as the copper anode dissolves, these metals fall to the bottom of the cell. Thus, the net result of this electrolysis process is the transfer of copper from the anode to the cathode. Copper prepared this way has a purity greater than 99.5 percent (Figure 27.4).

The metal impurities separated from the copper anode are valuable by-products.

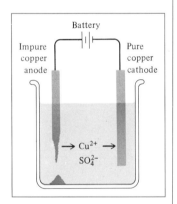

FIGURE 27.3 *The electrolytic purification of copper. The more electropositive metals, such as zinc and iron, are oxidized at the anode and enter the solution as Fe^{2+} and Zn^{2+}, while copper is transferred from the anode to the cathode as the Cu^{2+} ions in solution. The less electropositive metals, such as gold and silver, are not oxidized at the anode.*

FIGURE 27.4 *Cathodes of copper used in the electrorefining process.*

Zone Refining. Zone refining is often used to extract extremely pure metals. A portion of a metal rod containing the few impurities is heated electrically until it melts (Figure 27.5). Most impurities dissolve in the molten metal while pure metal crystallizes out from the bottom of the molten zone. When the molten zone carrying the impurities reaches the end of the rod, it is allowed to cool and is then cut off. Repetition of this procedure a number of times results in metal with a purity greater than 99.99 percent.

27.3 THE METALLURGIES OF FOUR METALS

Having discussed the general procedures for obtaining metals, we will now consider in some detail the metallurgies of four industrially important metals: sodium, aluminum, iron, and copper.

Sodium

Sodium occurs in nature in silicate minerals such as albite ($NaAlSi_3O_8$). Over a long period of time (on a geologic scale), silicate minerals are decomposed by wind and rain, and the sodium ions present are converted to more soluble compounds. Eventually, rain leaches these compounds from the soil and carries them to the sea. Sodium chloride accounts for about 3 percent (by mass) of seawater. Other sources of sodium are Chile saltpeter ($NaNO_3$) and cryolite (Na_3AlF_6).

Metallic sodium is most conveniently obtained from molten sodium chloride by electrolysis (see Section 24.7). The melting point of sodium chloride is rather high (801°C) and much energy is needed to keep large amounts of the substance molten. Adding a suitable substance, such as $CaCl_2$, lowers the melting point to about 600°C—a more convenient temperature for the electrolysis process.

Alternatively, sodium can be obtained by the electrolysis of molten sodium hydroxide (Figure 27.6). The half-cell reactions are

$$\text{anode (oxidation):} \quad 4OH^- \longrightarrow 2H_2O(g) + O_2(g) + 4e^-$$
$$\text{cathode (reduction):} \quad 4Na^+ + 4e^- \longrightarrow 4Na(l)$$
$$\text{overall:} \quad 4Na^+ + 4OH^- \longrightarrow 4Na(l) + 2H_2O(g) + O_2(g)$$

The design of the cell should be such that the steam liberated at the anode will not come in contact with sodium; otherwise, the following reaction would occur:

$$2Na(l) + 2H_2O(g) \longrightarrow 2NaOH(l) + H_2(g)$$

The advantage of using sodium hydroxide is its lower melting point (318.4°C). However, sodium hydroxide itself has to be made by electrolysis of an aqueous sodium chloride solution (see the chlor-alkali process discussed in Section 26.6).

The electrolysis method is applicable to the separation of many other alkali metals and to alkaline earth metals such as Li, Mg, and Ca.

Zone refining can be used also to purify nonmetals, such as silicon.

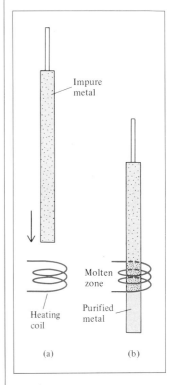

(a)   (b)

FIGURE 27.5 *The zone-refining technique for purifying metals. (a) The impure metal rod packed in a glass tube is lowered slowly through a heating coil. (For simplicity the glass tube is not shown.) (b) As the metal rod moves downward, the impurities dissolve in the molten portion of the metal while pure metal crystallizes out from the bottom of the molten zone.*

The technique for lowering the melting point of sodium chloride discussed above is based on freezing point depression. From Table 24.1 we see that Ca^{2+} ions are more difficult to reduce than Na^+ ions.

FIGURE 27.6 *An elec-trolytic cell for the prepara-tion of sodium from molten sodium hydroxide.*

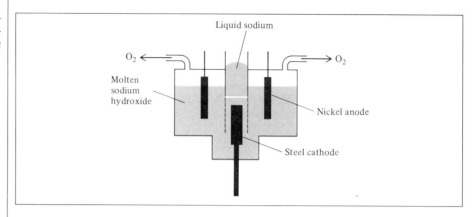

FIGURE 27.6 *An elec-trolytic cell for the prepara-tion of sodium from molten sodium hydroxide.*

Aluminum

The principal ore of aluminum is bauxite ($Al_2O_3 \cdot 2H_2O$), which is usually contaminated with silica (SiO_2), iron oxide, and titanium oxide. The ore is first heated in sodium hydroxide solution to convert the silica into soluble silicates:

$$SiO_2(s) + 2OH^-(aq) \longrightarrow SiO_3^{2-}(aq) + H_2O(l)$$

At the same time, aluminum oxide is converted to the aluminate ion (AlO_2^-):

$$Al_2O_3(s) + 2OH^-(aq) \longrightarrow 2AlO_2^-(aq) + H_2O(l)$$

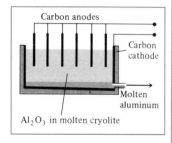

FIGURE 27.7 *Electro-lytic production of alumi-num based on the Hall pro-cess.*

Molten cryolite provides a good conducting medium for electrolysis.

Iron oxide and titanium oxide are unaffected by this treatment and are filtered off. Next, the solution is acidified to precipitate the insoluble aluminum hydroxide:

$$AlO_2^-(aq) + H_3O^+(aq) \longrightarrow Al(OH)_3(s)$$

After filtration, the aluminum hydroxide is heated to obtain aluminum oxide:

$$2Al(OH)_3(s) \longrightarrow Al_2O_3(s) + 3H_2O(g)$$

Anhydrous aluminum oxide, or *corundum*, is reduced to aluminum by the *Hall process* (Charles Hall, 1863–1914). Figure 27.7 shows a Hall electro-lytic cell, which contains a series of carbon anodes. The cathode is also made of carbon and constitutes the lining inside the cell. The key to the Hall process is the use of cryolite or Na_3AlF_6 (m.p. 1000°C) as solvent for aluminum oxide (m.p. 2045°C). The mixture is electrolyzed to produce aluminum and oxygen gas:

$$\begin{aligned}
\text{anode (oxidation):} &\quad 3[2O^{2-} \longrightarrow O_2(g) + 4e^-] \\
\text{cathode (reduction):} &\quad 4[Al^{3+} + 3e^- \longrightarrow Al(l)] \\
\text{overall:} &\quad 2Al_2O_3 \longrightarrow 4Al(l) + 3O_2(g)
\end{aligned}$$

Oxygen gas reacts with the carbon anodes (at elevated temperatures) to form carbon monoxide, which escapes as a gas. The liquid aluminum metal (m.p. 660.2°C) sinks to the bottom of the vessel, from which it can be drained from time to time during the procedure.

Iron

The principal ores of iron are hematite, Fe_2O_3, and magnetite, Fe_3O_4 (which may be thought of as consisting of FeO and Fe_2O_3). The cheapest way to recover iron is by reducing its ore with coke. Before we describe this process, let us examine the limitations imposed by thermodynamics, using the principles developed in Chapter 23. For simplicity, we will consider only the reduction from iron(II) oxide to iron.

The reduction can be carried out either with coke

$$(1) \quad C(s) + FeO(s) \longrightarrow CO(g) + Fe(l)$$

or with carbon monoxide gas

$$(2) \quad CO(g) + FeO(s) \longrightarrow CO_2(g) + Fe(l)$$

The temperatures at which these reactions become thermodynamically favorable can be understood by considering Figure 27.8. For a given reaction, the more negative $\Delta G°$ is at a certain temperature, the more favorable the reaction at that temperature. For the reverse reaction, $\Delta G°$ has the same magnitude but the opposite sign. Reaction (1) is given by (A) − (B) of Figure 27.8, that is

For convenience, we assume that the reduction is carried out under standard-state conditions.

$$
\begin{array}{lr}
\quad\ C + \tfrac{1}{2}O_2 \longrightarrow CO & \Delta G°(A) \\
+ \quad\quad\ FeO \longrightarrow Fe + \tfrac{1}{2}O_2 & -\Delta G°(B) \\
\hline
C + FeO + \tfrac{1}{2}O_2 \longrightarrow CO + Fe + \tfrac{1}{2}O_2 &
\end{array}
$$

or

$$(1) \quad C + FeO \longrightarrow CO + Fe$$

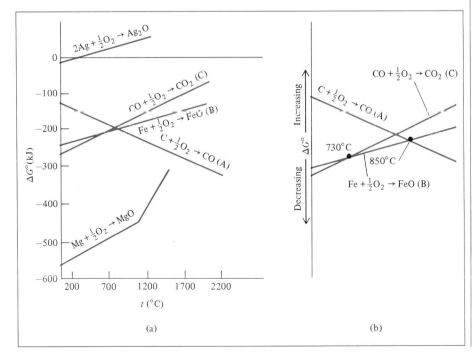

FIGURE 27.8   (a) Standard free-energy changes as a function of temperature, for the formation of certain oxides. (b) Expanded plots of $\Delta G°$ versus temperature for three of the reactions shown in (a)—the formation of CO, CO_2, and FeO.

At a given temperature

$$\Delta G°(1) = \Delta G°(A) - \Delta G°(B)$$

and the reaction will be spontaneous if $\Delta G°(B) > \Delta G°(A)$.

Similarly, reaction (2) is given by (C) − (B):

$$
\begin{array}{ll}
CO + \tfrac{1}{2}O_2 \longrightarrow CO_2 & \Delta G°(C) \\
+ \quad\quad FeO \longrightarrow Fe + \tfrac{1}{2}O_2 & -\Delta G°(B) \\
\hline
CO + FeO + \tfrac{1}{2}O_2 \longrightarrow CO_2 + Fe + \tfrac{1}{2}O_2 &
\end{array}
$$

or

$$(2) \quad CO + FeO \longrightarrow CO_2 + Fe$$

At a given temperature

$$\Delta G°(2) = \Delta G°(C) - \Delta G°(B)$$

and the reaction will be spontaneous if $\Delta G°(B) > \Delta G°(C)$.

Figure 27.8(b) shows that $\Delta G°$ for reaction (1) becomes favorable at temperatures greater than 850°C. On the other hand, a temperature of 730°C or lower favors reaction (2). These considerations play an important role in the design and operation of a blast furnace, shown in Figure 27.9. The iron ore, limestone (CaCO$_3$), and coke are introduced into the furnace at the top. Air or oxygen gas preheated to 1500°C is forced up through the furnace, where it combines with coke to form carbon monoxide:

$$2C(s) + O_2(g) \longrightarrow 2CO(g)$$

As the gas rises it cools sufficiently for reaction (2) to take place. By the time the ore works its way down to the bottom part of the furnace, most of it has been reduced to iron. The remaining portion is reduced by coke in the high-temperature region by reaction (1). The molten iron is run off to a receiver. The slag contains silicon dioxide and some manganese and phosphorus oxides. It forms the layer above the molten iron (because it is lighter) and it is run off, as shown in Figure 27.9.

Iron extracted in a blast furnace contains many impurities and is called *pig iron*; it may contain up to 4.5 percent carbon and some silicon, phosphorus, manganese, and sulfur. Pig iron is granular and brittle. It has a relatively low melting point (about 1180°C), so it can be cast; it is thus also called *cast iron*. The quality of iron can be vastly improved by treating it with oxygen gas at a high temperature to remove the small amounts of impurities and part of the carbon, in the form of the corresponding oxides. The end product is steel.

Figure 27.10 shows the process of steel making. A cylindrical vessel that can be rotated about the horizontal and vertical axes is charged with molten iron from the blast furnace and tilted to a vertical position. Pressurized oxygen gas is delivered via a water-cooled tube above the molten metal. Carbon becomes oxidized and is removed as carbon dioxide. During the blow of the oxygen gas, lime (CaO) is added to convert the silica into a molten slag of calcium silicate:

$$CaO(s) + SiO_2(s) \longrightarrow CaSiO_3(l)$$

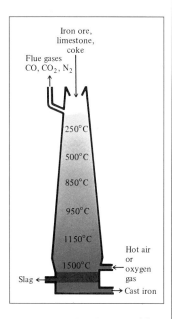

FIGURE 27.9 *A blast furnace. Iron ore, limestone, and coke are introduced at the top of the furnace. Iron is obtained from the ore by reduction with carbon.*

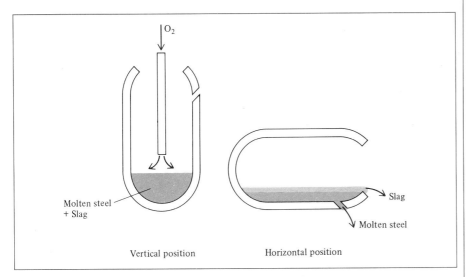

FIGURE 27.10 *Arrangement for steelmaking. The capacity of a typical vessel is 100 tons of cast iron.*

The molten metal is sampled at intervals. When the desired concentration of carbon has been reached, the vessel is rotated in a horizontal position so that the molten steel can be tapped off (see Color Plate 10). The properties and uses of steel depend on its composition (Table 27.3).

Copper

Copper ores contain mainly the sulfides, oxides, and carbonates of copper. We will describe the steps involved in extracting copper from its sulfide ore, chalcopyrite ($CuFeS_2$). The overall process is rather complex and difficult because the ore is so impure; it contains much clay and sand and, atom for atom, as much iron as copper.

The first step is the concentration of the ore by flotation. Next, the ore is roasted. By suitable adjustment of oxygen supply and temperature, most of the iron is converted to iron(II) oxide, while copper remains combined with sulfur:

$$2CuFeS_2(s) + 4O_2(g) \longrightarrow Cu_2S(s) + 2FeO(s) + 3SO_2(g)$$

TABLE 27.3 Types of Steel

Type	Composition (percent by mass)*								Uses
	C	Mn	P	S	Si	Ni	Cr	Others	
Plain	1.35	1.65	0.04	0.05	0.06	—	—	Cu (0.2–0.6)	Sheet products, tools
High-strength	0.25	1.65	0.04	0.05	0.15–0.9	0.4–1.0	0.3–1.3	Cu (0.01–0.08)	Construction
Stainless	0.03–1.2	1.0–10	0.04–0.06	0.03	1–3	1–22	4.0–27	—	Steam turbines, kitchen utensils, razor blades

* A single number indicates the maximum amount of the substance present.

Note that the oxidation number of copper is $+2$ in $CuFeS_2$. However, since CuS is less stable than Cu_2S at high temperatures, copper(II) sulfide is converted to copper(I) sulfide as shown. The product of the roasted ore is heated in a furnace with sand (SiO_2) to convert the iron oxide into a slag:

$$FeO(l) + SiO_2(l) \longrightarrow FeSiO_3(l)$$

The slag is poured off. Then the molten copper(I) sulfide is reacted with air to form sulfur dioxide and metallic copper:

$$Cu_2S(l) + O_2(g) \longrightarrow 2Cu(l) + SO_2(g)$$

Metallic copper prepared this way contains a number of impurities, which are removed by the electrolytic process described in Section 27.2.

27.4 HYDROMETALLURGY

The processes used for extracting metals all involve high-temperature operations. This approach is called *pyrometallurgy* and it derives from the ancient art of metal production. Pyrometallurgy has the disadvantage that high-temperature operations (up to about 1500°C) are expensive. Furthermore, roasting sulfide ores is an appreciable source of air pollution because of SO_2 emission. Metallurgists are seeking cheaper and cleaner ways to produce metals. A promising development, called *hydrometallurgy*, employs aqueous solutions at relatively low temperatures. In general, hydrometallurgy involves two separate steps: selective dissolution, or *leaching*, of a metal from an ore, and selective recovery of the metal from the solution. Leaching may involve physical, chemical, or electrolytic processes, as follows.

- *Physical method*
  In the simplest case, water is used to leach soluble chlorides and sulfates. For example

$$NaCl(s) \xrightarrow{H_2O} NaCl(aq)$$

- *Acid-base reactions*
  In most cases, aqueous acid or base solutions are used, rather than water, to dissolve ores. For example, aluminum hydroxide, which is amphoteric, dissolves in acids and bases:

$$Al(OH)_3(s) + 3H^+(aq) \longrightarrow Al^{3+}(aq) + 3H_2O(l)$$
$$Al(OH)_3(s) + OH^-(aq) \longrightarrow Al(OH)_4^-(aq)$$

- *Complex ion formation*
  Cyanide ions are used to extract gold and silver from other substances by forming soluble complex ions:

$$4Au(s) + 8CN^-(aq) + O_2(g) + 2H_2O(l) \longrightarrow 4Au(CN)_2^-(aq) + 4OH^-(aq)$$

The complex ion $Au(CN)_2^-$ is separated from the mixture (accompanied by a suitable cation such as Na^+) and treated with an electropositive metal such as zinc to recover the gold:

$$Zn(s) + 2Au(CN)_2^-(aq) \longrightarrow Zn(CN)_4^{2-}(aq) + 2Au(s)$$

- *Oxidation*
  Zinc sulfide, an insoluble substance, can be converted to the soluble zinc sulfate by bubbling oxygen through an aqueous suspension of zinc sulfide:

$$ZnS(s) + 2O_2(aq) \longrightarrow ZnSO_4(aq)$$

This reaction is slow at room temperature but the rate is greatly increased at a temperature of 200°C and under oxygen partial pressure of about 20 atm.

- *Electrolytic process*
  This process is used, for example, to recover nickel from its sulfide ore. With nickel sulfide as the anode of an electrolytic cell the following oxidation reactions occur:

$$Ni_3S_2(s) \longrightarrow Ni^{2+}(aq) + 2NiS(s) + 2e^-$$
$$NiS(s) \longrightarrow Ni^{2+}(aq) + S(s) + 2e^-$$

Although leaching is mainly an oxidation process, recovery of metals from the leaching solution is most often a reduction process. The dissolved metals may be reduced electrolytically

$$M^{n+}(aq) + ne^- \longrightarrow M(s)$$

or by a more active metal

$$M_1^{n+}(aq) + M_2(s) \longrightarrow M_1(s) + M_2^{n+}(aq)$$

where metal M_2 is more electropositive than metal M_1.

Let us now illustrate the principles of hydrometallurgy by considering two specific examples.

Extraction of Copper from Chalcopyrite

Earlier we discussed the extraction of copper from chalcopyrite, in which pyrometallurgical procedures are used. In hydrometallurgy the first step is the leaching process, which is carried out in an aqueous sulfuric acid solution in the presence of oxygen gas at room temperature:

$$2CuFeS_2(s) + H_2SO_4(aq) + 4O_2(g) \longrightarrow 2CuSO_4(aq) + Fe_2O_3(s) + 3S(s) + H_2O(l)$$

After iron(III) oxide and sulfur have been filtered out, the solution is electrolyzed to obtain metallic copper:

anode (oxidation): $\qquad 2H_2O(l) \longrightarrow O_2(g) + 4H^+(aq) + 4e^-$

cathode (reduction): $\quad 2Cu^{2+}(aq) + 4e^- \longrightarrow 2Cu(s)$

overall: $\qquad 2Cu^{2+}(aq) + 2H_2O(l) \longrightarrow 2Cu(s) + O_2(g) + 4H^+(aq)$

Converting the sulfide in the ore into elemental sulfur eliminates air pollution. Because it is performed at room temperature, the operation also reduces the cost significantly.

Extraction of Nickel from Nickel Sulfide

Nickel sulfide in the form of flotation concentrate (see Section 27.2) is leached by ammonia in the presence of air at 80°C to form diamminenickel(II) sulfate:

$$NiS(s) + 2NH_3(aq) + 2O_2(aq) \longrightarrow [Ni(NH_3)_2]SO_4(aq)$$

The solution is then heated to 180°C (under pressure) and reduced with hydrogen gas at 35 atm:

$$[Ni(NH_3)_2]SO_4(aq) + H_2(g) \longrightarrow Ni(s) + (NH_4)_2SO_4(aq)$$

Metallic nickel is recovered from the solution by filtration.

Hydrometallurgy is already in use on a commercial scale and in all probability it will eventually replace many of the pyrometallurgical methods.

27.5 BONDING IN METALS, NONMETALS, AND SEMICONDUCTING ELEMENTS

The physical and chemical differences between metals and nonmetals have been discussed in Section 26.1. Here we will take a close look at the nature of the metallic bond. A complete treatment of bonding in metals requires a thorough knowledge of quantum mechanics, which is beyond our needs here. However we can gain a qualitative understanding of the bonding in metals, metalloids, and nonmetals through a discussion of the "band theory."

Conductors

Consider the metal sodium. The electron configuration of Na is [Ne]$3s^1$, so each atom has one valence electron in the $3s$ orbital. In a metal, the atoms are packed closely together, so the energy levels of each sodium atom are affected by the immediate neighbors of the atom as a result of orbital overlaps. In Chapter 13 we saw that the interaction between two atomic orbitals leads to the formation of a bonding and an antibonding molecular orbital. Since the number of atoms in even a small piece of sodium metal is enormously large (on the order of 10^{20} atoms), the corresponding number of molecular orbitals formed is also very large. These molecular orbitals are so closely spaced on the energy scale that they are more appropriately described as forming a "band" (Figure 27.11). Note that because each level can accommodate two electrons, but each atom has only one electron to contribute, only the lower half of the energy levels (the bonding molecular

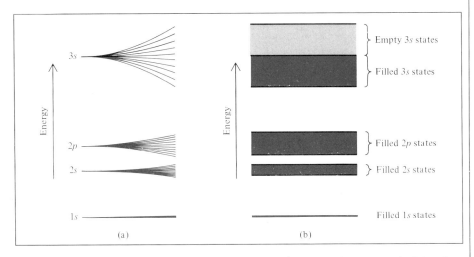

FIGURE 27.11 *Formation of conduction band in sodium metal. (a) When
sodium atoms are far apart there is no interaction among the corresponding atomic
orbitals (see the left portion of the diagram where the atomic orbitals 1s, 2s, etc.
retain their identities). When they are close to one another, as in the metal, the
orbitals overlap to form bonding and antibonding molecular orbitals. Because of
the very large number of atoms present, the molecular orbitals formed are closely
spaced, giving rise to bands. (b) The band structure in sodium metal. Note that
the 1s, 2s, and 2p bands are filled with electrons, whereas the band formed by the
overlap of 3s orbitals is only half-filled, since each sodium atom possesses only
one 3s electron.*

orbitals) are filled. This set of closely spaced filled levels is called the
valence band. The upper half of the energy levels corresponds to the empty
antibonding molecular orbitals. This set of closely spaced *empty* levels is
called the *conduction band.*

We can imagine sodium metal as an array of positive ions immersed in
a sea of delocalized valence electrons. The great cohesive force resulting
from the delocalization is partly responsible for the strength noted in most
metals. Because the valence band and the conduction band are adjacent to
each other, only a negligible amount of energy is needed to promote a
valence electron to the conduction band, where it is free to move through
the entire metal, since the conduction band is largely void of electrons.
This freedom of movement accounts for the fact that metals are *capable
of conducting electric current,* that is, they are good **conductors.**

Why don't substances like plastics and glass conduct electricity as metals
do? Figure 27.12 provides an answer to this question. Basically, the electrical
conductivity of a solid depends on the spacing and the state of occupancy
of the energy bands. Many metals resemble sodium in that their valence
bands are adjacent to their conduction bands and, therefore, these metals
readily act as conductors. In an **insulator,** the *gap between the valence
band and the conduction band is considerably greater than that in a metal;*
consequently, much more energy is needed to excite an electron into the
conduction band. Lacking this energy, free motion of the electrons is not
possible. Typical insulators are glass, wood, rubber, and plastics.

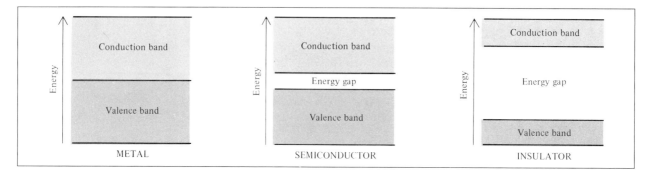

FIGURE 27.12 *Comparison of the energy gaps between valence band and conduction band in a metal, a semiconductor, and an insulator. In a metal the energy gap is virtually nonexistent; in a semiconductor the energy gap is small; and in an insulator the energy gap is very large—thus making the promotion of an electron from the valence band to the conduction band difficult.*

Semiconductors

A number of elements, especially Si and Ge in Group 4A, *have properties that are intermediate between those of metals and nonmetals* and are, therefore called **semiconducting elements.** The energy gap between the filled and empty bands for these solids is much smaller than that for insulators (see Figure 27.12). If the energy needed to excite electrons from the valence band into the conduction band is provided, the solid becomes a conductor. Note that this behavior is opposite that of the metals. A metal's ability to conduct electricity *decreases* with increasing temperature, because the enhanced vibration of atoms at higher temperatures tends to disrupt the flow of electrons.

The ability of a semiconductor to conduct electricity can also be enhanced by adding small amounts of certain impurities to the element (a process called *doping*). Let us consider what happens when a trace amount of boron or phosphorus is added to solid silicon. (In this doping, about five out of every million Si atoms are replaced by B or P atoms.) The structure of solid silicon is similar to that of diamond; each Si atom is covalently bonded to four Si atoms. Phosphorus ($[Ne]3s^2 3p^3$) has one more valence electron than silicon ($[Ne]3s^2 3p^2$), so there is a valence electron left over after four of them are used to form covalent bonds with silicon (Figure 27.13). This extra electron can be removed from the phosphorus atom by

FIGURE 27.13 *(a) Silicon crystal doped with phosphorus. (b) Silicon crystal doped with boron. Note the formation of a negative center in (a) and a positive center in (b).*

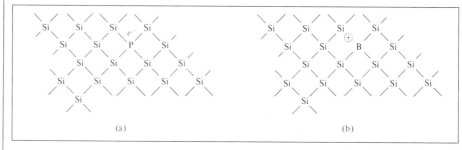

applying a voltage across the solid. The free electron can then move through the structure and function as a conduction electron. Impurities of this type are known as **donor impurities,** since they *provide conduction electrons. Solids containing donor impurities* are called **n-type semiconductors,** where n stands for negative (the charge of the "extra" electron).

The opposite effect occurs if boron is added to silicon. Since a boron atom has three valence electrons $(1s^2 2s^2 2p^1)$, it can form only three covalent bonds with silicon. Thus, for every boron atom in the silicon crystal, there is a single *vacancy* in a bonding orbital. It is possible to excite a valence electron from a nearby Si into this vacant orbital. A vacancy created at that Si atom can then be filled by an electron from a neighboring Si atom, and so on. In this manner, electrons can move through the crystal in one direction while the vacancies, or "positive holes," move in the opposite direction, and the solid becomes an electrical conductor. Semiconductors that *contain acceptor impurities* are called **p-type semiconductors,** where p stands for positive. *Impurities that are electron-deficient* are called **acceptor impurities.**

In both the p-type and n-type semiconductors the energy gap between the valence band and the conduction band is effectively reduced, so that only a small amount of energy is needed to excite the electrons. Typically, the conductivity of a semiconductor is increased by a factor of 100,000 or so by the presence of impurity atoms.

The growth of the semiconductor industry since the early 1960s has been truly remarkable. Today semiconductors are essential components of nearly all electronic equipment, ranging from radio and television sets to pocket calculators and computer facilities. One of the main advantages of solid-state devices over vacuum tube electronics is that the former can be made on a single "chip" of silicon no larger than the cross section of a pencil eraser. In this manner, much more equipment can be packed into a small volume—a point of particular importance in space travel, as well as in hand-held calculators and microprocessors (computers-on-a-chip).

AN ASIDE ON THE PERIODIC TABLE
A Portrayal of the Extraction of the Elements

In this chapter we covered various ways of extracting metals from their minerals. It is appropriate here to summarize our discussion, along with what we learned in Chapter 26 about the ways of obtaining pure nonmetallic elements (Figure 27.14). The methods used to obtain an element depend on the element's chemical reactivity. Thus electrolytic reduction and oxidation are used to prepare the most reactive metals (the alkali metals and most of the alkaline earth metals) and the most reactive nonmetals (molecular fluorine and molecular chlorine). Chemical reductions are used to obtain less reactive metals

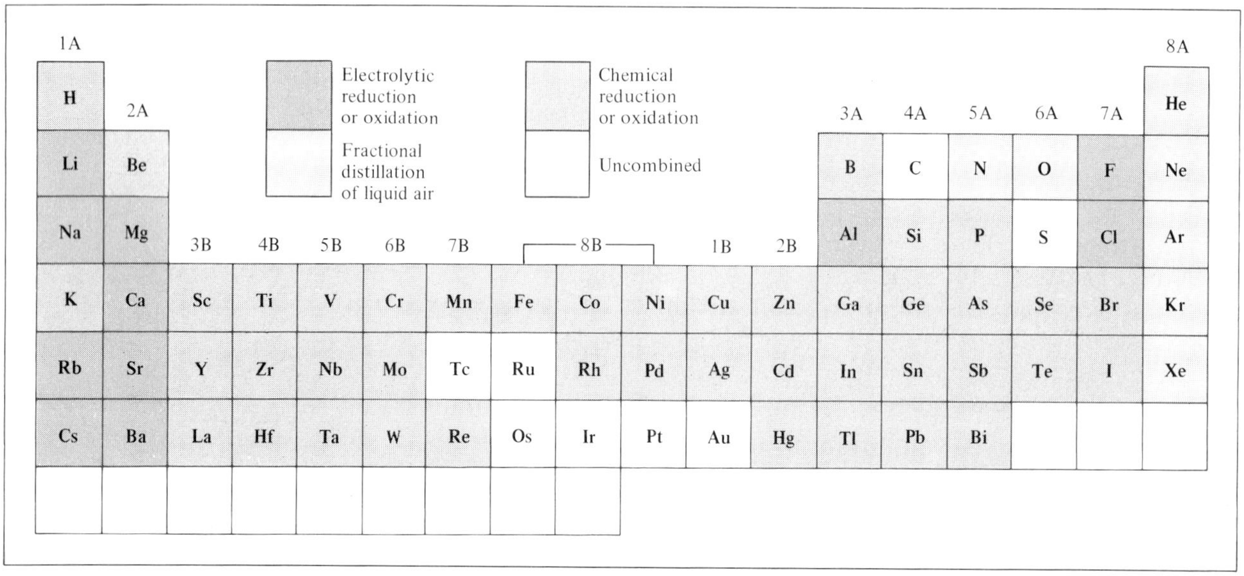

FIGURE 27.14 Various methods for obtaining pure elements. Although potassium is a very reactive metal, it is usually obtained by chemical rather than electrolytic reduction because potassium metal attacks carbon electrodes (see Section 28.1 for details). Note that oxygen and nitrogen gases, as well as the noble gases, are prepared by fractional distillation of liquid air. Technetium (Tc) is a synthetic element that is prepared by a nuclear reaction in the laboratory.

such as iron and zinc and nonmetals such as hydrogen gas and molecular bromine. The least reactive metals—for example, gold and platinum—are found in uncombined state in nature so no oxidation or reduction process is necessary.

SUMMARY

1. Depending on their reactivities, metals exist in nature either in the free or combined state.
2. To recover a metal from its ore, the ore must first be prepared. The metal is then separated (usually by a reduction process) and purified.
3. The common procedures for purifying metals are distillation, electrolysis, and zone refining.
4. Hydrometallurgy employs aqueous solutions at relatively low temperatures. It is cheaper than high-temperature operations (pyrometallurgy) and is largely nonpolluting.
5. Metallic bonding can be pictured as the force between positive ions immersed in a sea of electrons. The atomic orbitals merge to form energy bands, and a substance is a conductor when electrons can be readily promoted to the conduction band, where they are free to move through the substance.
6. In insulators, the energy gap between the valence band and the conduction band is so large that electrons cannot be promoted into the con-

duction band. In semiconductors, electrons can cross the energy gap at higher temperatures, and conductivity increases with increasing temperatures as more electrons are able to reach the conduction band.

7. *n*-type semiconductors contain donor impurities and extra electrons. *p*-type semiconductors contain acceptor impurities and positive holes.

KEY WORDS

Acceptor impurity, p. 853
Conductor, p. 851
Donor impurity, p. 853
Insulator, p. 851
Metallurgy, p. 838

Mineral, p. 838
n-type semiconductor, p. 853
Ore, p. 838
p-type semiconductor, p. 853
Semiconducting elements, p. 852

PROBLEMS

More challenging problems are marked with an asterisk.

Minerals

27.1 Why is pyrite (FeS_2) called fool's gold? (See Color Plate 10.)

27.2 Copper exists in nature in both the free (elemental) and combined states, whereas aluminum is never found in the free state. Explain.

27.3 Write chemical formulas for the following minerals: (a) calcite, (b) dolomite, (c) fluorite, (d) halite, (e) corundum, (f) magnetite, (g) beryl, (h) galena, (i) epsomite, (j) anhydrite.

27.4 Name the following minerals: (a) $MgCO_3$, (b) Na_3AlF_6, (c) Al_2O_3, (d) Ag_2S, (e) HgS, (f) ZnS, (g) $SrSO_4$, (h) $PbCO_3$, (i) MnO_2, (j) TiO_2.

Purification of Metals

27.5 Describe how the zone-refining process works.

27.6 In the Mond process for the purification of nickel, CO is passed over metallic nickel to give $Ni(CO)_4$:

$$Ni(s) + 4CO(g) \rightleftharpoons Ni(CO)_4(g)$$

Given that the standard free energies of formation of $CO(g)$ and $Ni(CO)_4(g)$ are -137.3 kJ/mol and -587.4 kJ/mol, calculate the equilibrium constant of the reaction at 80°C. (Assume ΔG_f° to be independent of temperature.)

27.7 Copper is purified by electrolysis (see Figure 27.3). A 5.00 kg anode is used in a cell where the current is 37.8 amp. How long (in h) must the current run to dissolve this anode and electroplate it on the cathode?

27.8 Consider the electrolytic procedure for purifying copper described in Figure 27.3. Suppose that a sample of copper contains the following impurities: Fe, Ag, Zn, Au, Co, Pt, and Pb. Which of the metals will be oxidized and dissolved in solution and which will be unaffected and simply form the sludge that accumulates at the bottom of the cell?

Metallurgical Processes

27.9 How would you obtain zinc from sphalerite (ZnS)?

*27.10 Starting with rutile (TiO_2), explain how you would obtain pure titanium metal. (*Hint:* First convert TiO_2 to $TiCl_4$. Next, reduce $TiCl_4$ with Mg. Look up physical properties of $TiCl_4$, Mg, and $MgCl_2$ in a chemistry handbook.)

27.11 A certain mine produces 2.0×10^8 kg of copper from chalcopyrite ($CuFeS_2$) each year. The ore contains only 0.80 percent Cu by mass. (a) If the density of the ore is 2.8 g/cm^3, calculate the volume (in cm^3) of ore removed each year. (b) Calculate the mass (in kg) of SO_2 produced by roasting (assume chalcopyrite to be the only source of sulfur).

27.12 Which of the following compounds would require electrolysis to yield the free metals? Ag_2S, $CaCl_2$, KCl, Fe_2O_3, Al_2O_3, $TiCl_4$.

27.13 Outline the procedures and reactions in the production of iron in a blast furnace.

27.14 Although iron is only about two-thirds as abundant as aluminum in the Earth's crust, mass for mass it costs only about one-quarter as much to produce. Why?

27.15 Can magnesium be used to obtain aluminum according to the following equation?

$$Al_2O_3(s) + 3Mg(s) \longrightarrow 2Al(s) + 3MgO(s)$$

(*Hint:* Calculate $\Delta G°$ for the reaction at 25°C and predict its value at a higher temperature.)

27.16 Before Hall invented his electrolytic process, aluminum was produced by reduction of its chloride with an active metal. Which metals would you use for the production of aluminum in that way?

27.17 The overall reaction for the electrolytic production of aluminum in the Hall process may be represented as

$$Al_2O_3(s) + 3C(s) \longrightarrow 2Al(l) + 3CO(g)$$

At 1000°C, the standard free-energy change for this process is 1160 kJ. (a) Calculate the minimum voltage required to produce one mole of aluminum at this temperature. (b) If the actual voltage applied is exactly three times the ideal value, calculate the energy required to produce 1.00 kg of the metal.

27.18 In steel making, nonmetallic impurities such as P, S, and Si are removed as the corresponding oxides. The inside of the furnace is usually lined with $CaCO_3$ and $MgCO_3$, which decompose at high temperatures to yield CaO and MgO. How do CaO and MgO help in the removal of the non-metallic oxides?

27.19 How long (in hours) will it take to deposit 664 g of Al in the Hall process with a current of 32.6 amp?

27.20 The principal silver ore is argentite (Ag_2S). Describe, with appropriate equations, how you would obtain metallic silver from the ore.

27.21 What is the advantage of hydrometallurgy over pyrometallurgy?

Conductors, Semiconductors, and Insulators

27.22 Discuss briefly the differences in bonding among metals, semiconductors, and insulators.

27.23 Describe the general characteristics of *n*-type and *p*-type semiconductors.

27.24 Which of the following elements would form *n*-type or *p*-type semiconductors with silicon? Ga, Sb, Al, As.

27.25 An early view of metallic bonding assumed that bonding in metals consisted of localized, shared electron-pair bonds between metal atoms. List the evidence that would help you to argue against this viewpoint.

Miscellaneous Problems

27.26 Define the following terms: (a) mineral, (b) ore, (c) metallurgy, (d) roasting, (e) leaching.

27.27 The quantity of 1.164 g of a certain metal sulfide was roasted in air. As a result, 0.972 g of the metal oxide was formed. If the oxidation number of the metal is +2, calculate the molar mass of the metal.

27.28 Referring to Figure 27.3, would you expect H_2O and H^+ to be reduced at the cathode and H_2O oxidized at the anode?

27.29 Referring to Figure 27.8, answer the following questions:
(a) Estimate the temperature at which the reaction

$$Ag_2O(s) \longrightarrow 2Ag(s) + \tfrac{1}{2}O_2(g)$$

becomes spontaneous.
(b) Account for the angle in the graph for MgO formation. (*Hint:* Look up the boiling point of magnesium in a handbook of chemistry.)
(c) Why does $\Delta G°$ for CO decrease with temperature and $\Delta G°$ for CO_2 increase with temperature?

27.30 A 0.450 g sample of steel contains manganese as an impurity. The sample is dissolved in acidic solution and the manganese is oxidized to the permanganate ion (MnO_4^-). The MnO_4^- ion is reduced to Mn^{2+} by reacting with 50.0 mL of 0.0800 *N* $FeSO_4$ solution. The excess Fe^{2+} ions are then oxidized to Fe^{3+} by 22.4 mL 0.0600 *N* $K_2Cr_2O_7$. Calculate the percent by mass of manganese in the sample.

27.31 Given that $\Delta G_f°(Fe_2O_3) = -741.0$ kJ/mol and that $\Delta G_f°(HgS) = -48.8$ kJ/mol, calculate $\Delta G°$ for the following reactions at 25°C:

(a) $HgS(s) \longrightarrow Hg(l) + S(s)$
(b) $2Fe_2O_3(s) \longrightarrow 4Fe(s) + 3O_2(g)$
(c) $2Al_2O_3(s) \longrightarrow 4Al(s) + 3O_2(g)$

27.32 Calculate the temperature (in °C) at which the equilibrium constant (K_P) for the following reaction is 1.00:

$$2CuO(s) \longrightarrow 2Cu(s) + O_2(g)$$

(Assume that $\Delta H°$ and $\Delta S°$ are temperature independent.)

27.33 When an inert atmosphere is needed in a metallurgical process, nitrogen is frequently used. However, in the reduction of $TiCl_4$ by Mg, helium is used. Explain why nitrogen is not suitable for this process.

28
CHEMISTRY OF SELECTED METALS

Chapter 26 focused on the more familiar non-metallic elements. Chapter 27 dealt with some of the methods used to extract and purify metals. Here we will turn our attention to some of the familiar metals. Once again, the emphasis will be on important chemical properties and the roles of metals and their compounds in industrial and biological processes. The transition metals will be discussed in the next chapter.

857

28.1　THE ALKALI METALS

As a group, the alkali metals (the Group 1A elements) are the most electropositive (or the least electronegative) elements known. They exhibit many similar properties. Table 28.1 lists some common properties of the alkali metals. From their electron configurations we expect the oxidation number of these elements in their compounds to be $+1$ since the cations would be isoelectronic with the noble gases. This is indeed the case.

These metals have low melting points and are soft enough to be sliced with a knife (see Color Plate 1). The alkali metals all possess a body-centered crystal structure (see Figure 14.30) with low packing efficiency. This accounts for their low densities. In fact, lithium is the lightest metal known (see Figure 14.42).

Because of their great chemical reactivity, the alkali metals never occur in elemental form; they are found combined with halide, sulfate, carbonate, and silicate ions. In this section we will describe the chemistry of the first three members of Group 1A: lithium, sodium, and potassium. The chemistry of rubidium and cesium is less important; all isotopes of francium, the last member of the group, are radioactive.

Lithium

Earth's crust is about 0.006 percent lithium by mass. The element is also present in seawater to the extent of about 0.1 ppm by mass. The most important lithium-containing mineral is spodumene ($LiAlSi_2O_6$). Lithium metal is obtained by the electrolysis of molten LiCl, to which some inert salts have been added to lower the melting point to about 500°C.

Like all alkali metals, lithium reacts with cold water to produce hydrogen gas:

$$2Li(s) + 2H_2O(l) \longrightarrow 2LiOH(aq) + H_2(g)$$

But, as we mentioned in Chapter 9 (p. 267), in some ways lithium resembles magnesium more than it does the other alkali metals. The following examples illustrate these similarities:

TABLE 28.1　Properties of Alkali Metals

	Li	Na	K	Rb	Cs
Valence electron configuration	$2s^1$	$3s^1$	$4s^1$	$5s^1$	$6s^1$
Density (g/cm³)	0.534	0.97	0.86	1.53	1.87
Melting point (°C)	179	97.6	63	39	28
Boiling point (°C)	1317	892	770	688	678
Atomic radius (pm)	155	190	235	248	267
Ionic radius (pm)*	60	95	133	148	169
Ionization energy (kJ/mol)	520	496	419	403	375
Electronegativity	1.0	0.9	0.8	0.8	0.7
Standard reduction potential (V)†	−3.05	−2.93	−2.71	−2.93	−2.92

* Refers to the cation M^+, where M denotes an alkali metal atom.
† The half-reaction is $M^+(aq) + e^- \rightarrow M(s)$

- On combustion, lithium forms the oxide (containing the O^{2-} ion); sodium and potassium form the peroxide (containing the O_2^{2-} ion).
- Lithium nitride is formed from the direct combination of the metal and molecular nitrogen. The nitrides of other alkali metals are formed more indirectly.
- Lithium carbonate and lithium phosphate are much less soluble than the carbonates and phosphates of other alkali metals.

Like the other alkali metal oxides, Li_2O is basic (see Figure 18.5) and reacts with water to yield the corresponding hydroxide:

$$Li_2O(s) + H_2O(l) \longrightarrow 2LiOH(aq)$$

The ability of lithium hydroxide to react with carbon dioxide to form lithium carbonate makes it a useful air purifier on space vehicles and submarines.

$$2LiOH(aq) + CO_2(g) \longrightarrow Li_2CO_3(s) + H_2O(l)$$

Lithium combines with molecular hydrogen at high temperatures to yield lithium hydride:

$$2Li(s) + H_2(g) \xrightarrow{\Delta} 2LiH(s)$$

Lithium hydride reacts readily with water, as follows:

$$2LiH(s) + 2H_2O(l) \longrightarrow 2LiOH(aq) + 2H_2(g)$$

This property makes LiH useful for drying organic solvents. Lithium aluminum hydride, $LiAlH_4$, is a powerful reducing agent that has many uses in organic synthesis. It can be prepared by reacting lithium hydride with aluminum chloride:

$$4LiH(s) + AlCl_3(s) \longrightarrow LiAlH_4(s) + 3LiCl(s)$$

Lithium chloride and bromide are highly **hydroscopic;** that is, they *have a great tendency to absorb water.* For this reason, they are sometimes used in dehumidifiers and air conditioners. Certain lithium salts, particularly lithium carbonate, are valuable drugs in the treatment of manic-depressive patients.

Sodium and Potassium

Sodium and potassium are about equally abundant in nature. They occur in silicate minerals such as albite ($NaAlSi_3O_8$) and orthoclase ($KAlSi_3O_8$). Over long periods of time (on a geologic scale), silicate minerals are slowly decomposed by wind and rain, and their sodium and potassium ions are converted to more soluble compounds. Eventually rain leaches these compounds out of the soil and carries them to the sea. Yet when we look at the composition of seawater, we find that the concentration ratio of sodium to potassium is about 28 to 1 (see Table 15.4). The reason for this uneven distribution is that potassium is essential to plant growth, while sodium is not. Plants take up many of the potassium ions along the way, while sodium ions are free to move on to the sea. Other minerals that contain sodium or

These properties parallel those of magnesium closely.

NaOH and KOH react with CO_2 similarly. However, because of its lower molar mass, LiOH is more suitable for use in space vehicles.

This reaction is less violent than that between the metal and water.

The stability of alkali halides can be studied by the Born-Haber cycle (see Section 10.6).

potassium are halite (NaCl), Chile saltpeter ($NaNO_3$), and sylvite (KCl).

Sodium metal is prepared by the electrolysis of molten NaCl in the Downs cell (see Figure 24.11). Metallic potassium cannot be easily prepared by the electrolysis of molten KCl because potassium metal reacts with the electrodes. Instead, it is usually prepared by the distillation of molten KCl in the presence of sodium vapor at 892°C. The reaction that takes place at this temperature is

$$Na(g) + KCl(l) \longrightarrow NaCl(l) + K(g)$$

Note that this is a chemical rather than electrolytic reduction.

This reaction may seem strange because we see in Table 24.1 that potassium is a stronger reducing agent than sodium. Thus we would not expect K^+ ions to be reduced by metallic sodium. However, potassium has a lower boiling point than sodium (770°C vs. 892°C), so it is more volatile at 892°C and distills off more easily. By continuously removing potassium as it is formed, the reaction can be made to go from left to right.

Sodium and potassium are both extremely reactive, but potassium is the more reactive of the two. They react with water to form the corresponding hydroxides, but sodium burns in air to form the peroxide

$$2Na(s) + O_2(g) \longrightarrow Na_2O_2(s)$$

whereas potassium burns in air to form superoxide:

$$K(s) + O_2(g) \longrightarrow KO_2(s)$$

Potassium oxide can be prepared by reacting metallic potassium with potassium nitrate at high temperatures:

$$10K(l) + 2KNO_3(s) \xrightarrow{\Delta} 6K_2O(s) + N_2(g)$$

We noted in Section 26.5 that alkali metals dissolve in liquid ammonia to produce beautiful blue solutions whose color is due to the solvated electrons. In the presence of certain organic compounds it is possible to form alkali metal anions, such as Na^- and K^-.

Sodium forms an alloy with potassium (25 percent Na and 75 percent K by mass) that is a liquid at room temperature. It is even more reactive and a stronger reducing agent than either sodium or potassium individually. Because of its high specific heat, this alloy is used as a heat exchanger.

Sodium and potassium are essential elements of living matter. Sodium ions and potassium ions are present in intracellular and extracellular fluids and they are essential for osmotic balance and enzyme functions. We will now describe the preparations and uses of several of the important compounds of sodium and potassium.

Sodium Chloride. Sodium chloride occurs in nature in rock salt, which was formed by the evaporation of prehistoric land-locked seas. It is also obtained by evaporating seawater (Figure 28.1). Sodium chloride is a white solid that is soluble in water. It has a high melting point (801°C) and is an ionic compound. Sodium chloride is used in food processing and in making soap. It is also used to melt ice on roads and sidewalks. Industrially, sodium chloride is used to prepare metallic sodium as described previously and in

FIGURE 28.1 *An evapo-
ration pool for the isolation
of sodium chloride.*

the production of sodium hydroxide and chlorine gas in the chlor-alkali process (see Section 26.7). It is also used in the production of sodium carbonate.

Sodium Carbonate.   Sodium carbonate, or *soda ash* (Na_2CO_3), is a white hydroscopic solid that melts at 851°C. Soda ash is an important and basic material for all kinds of industrial processes: in manufacturing—of soap and detergents, glass, drugs, and food additives—in water treatment, and so on. As Appendix 5 shows, it ranks tenth among chemicals produced in the United States.

The commercial process for producing soda ash, called the *Solvay process* (Ernest Solvay, 1838–1922), uses calcium carbonate and ammonia as starting materials. The first step in the Solvay process involves the thermal decomposition of calcium carbonate at about 1200°C:

$$CaCO_3(s) \xrightarrow{\Delta} CaO(s) + CO_2(g) \qquad (28.1)$$

Next, the CO_2 gas from reaction (28.1) and ammonia are bubbled through a cold saturated solution of sodium chloride to generate bicarbonate and ammonium ions:

$$CO_2(aq) + H_2O(l) \rightleftharpoons H^+(aq) + HCO_3^-(aq)$$

$$NH_3(aq) + H_2O(l) \rightleftharpoons NH_4^+(aq) + OH^-(aq)$$

or simply

$$NH_3(aq) + H_2CO_3(aq) \rightleftharpoons NH_4^+(aq) + HCO_3^-(aq) \qquad (28.2)$$

The Solvay process makes use of acid-base equilibria, solubility equilibria, and Le Chatelier's principle.

Recall that the ( ) symbol represents concentration that differs from the equilibrium concentration.

The production of bicarbonate ions is a crucial step in the Solvay process. Sodium bicarbonate is fairly soluble in cold water (about 70 g/L), but in the presence of excess Na^+ ions supplied by the saturated NaCl solution

$$(Na^+)(HCO_3^-) > K_{sp}$$

and precipitation occurs:

$$Na^+(aq) + HCO_3^-(aq) \longrightarrow NaHCO_3(s) \qquad (28.3)$$

Solid sodium bicarbonate is removed from the solution and is then heated to obtain the desired final product:

$$2NaHCO_3(s) \xrightarrow{\Delta} Na_2CO_3(s) + H_2O(g) + CO_2(g) \qquad (28.4)$$

The CO_2 gas is fed back into the vessel to react with NH_3 as shown by reaction (28.2).

This procedure is the quickest way to obtain Na_2CO_3 but not the cheapest, because NH_3 is quite expensive and must be recovered for reuse. Whenever there is a need to cut costs of production, chemists can usually find a way to do it. Recalling that one of the products in reaction (28.1) is CaO (or quicklime), we can convert it into calcium hydroxide (commonly called slaked lime) by treating it with water:

$$CaO(s) + H_2O(l) \longrightarrow Ca(OH)_2(s) \qquad (28.5)$$

Calcium hydroxide is only slightly soluble in water, so the concentration of OH^- ions due to its ionization is normally quite small:

$$Ca(OH)_2(s) \rightleftharpoons Ca^{2+}(aq) + 2OH^-(aq) \qquad (28.6)$$

However, in the presence of NH_4^+ ions, the equilibrium is shifted to the right as a result of the uptake of OH^- ions by

$$NH_4^+(aq) + OH^-(aq) \rightleftharpoons NH_3(aq) + H_2O(l) \qquad (28.7)$$

as ammonia gas is driven out of the solution by heat, some of which is provided by reaction (28.5). Then the ammonia gas is returned to the apparatus to generate more bicarbonate ions [see reaction (28.2)].

Figure 28.2 summarizes the steps in the Solvay process. The impressive feature of this process is that practically all the secondary products are used in the recovery of NH_3 or in the production of HCO_3^- ions. The exception is $CaCl_2$, which is discharged as a by-product. The world's soda ash manufacturers face the problem of having to dispose of millions of tons of $CaCl_2$ annually. Some of the $CaCl_2$ is used in making fast-setting cement and for melting ice on roads, but much of it is discharged into streams and rivers, adding to environmental pollution.

Sodium Hydroxide and Potassium Hydroxide. The properties of sodium hydroxide and potassium hydroxide are very similar. Both hydroxides are prepared by the electrolysis of aqueous NaCl and KCl solutions (see Figure 26.29); both hydroxides are strong bases and very soluble in water. Sodium hydroxide is used in the manufacture of soap and the purification of bauxite (see p. 844). Potassium hydroxide is used as an electrolyte in some storage

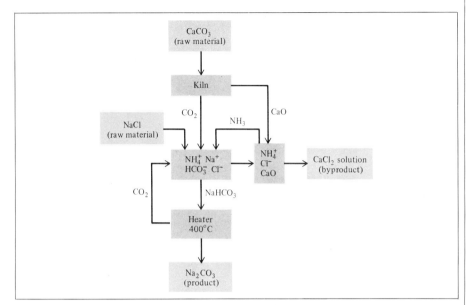

FIGURE 28.2 *Key steps in the Solvay process.*

batteries, and aqueous potassium hydroxide is used to remove carbon dioxide and sulfur dioxide from air.

Potassium Nitrate. Potassium nitrate is prepared beginning with the "reaction"

$$KCl(aq) + NaNO_3(aq) \longrightarrow KNO_3(aq) + NaCl(aq)$$

This process is carried out just below 100°C. Because KNO_3 is the least soluble salt at room temperature, it is separated from the solution by fractional crystallization. Like $NaNO_3$, KNO_3 decomposes when heated:

$$2KNO_3(s) \xrightarrow{\Delta} 2KNO_2(s) + O_2(g)$$

Gunpowder consists of potassium nitrate, wood charcoal, and sulfur in the approximate proportions of 6:1:1 by mass. When heated, the reaction is

$$2KNO_3(s) + S(s) + 3C(s) \xrightarrow{\Delta} K_2S(s) + N_2(g) + 3CO_2(g)$$

The sudden formation of hot expanding gases causes the explosion.

There is no net change in this process because all compounds are completely dissociated in solution.

28.2 THE ALKALINE EARTH METALS

The alkaline earth metals are somewhat less electropositive and less reactive than the alkali metals. Except for the first member of the family, beryllium, which resembles aluminum in some respects, the alkaline earth metals have similar chemical properties. Because their M^{2+} ions attain the stable noble gas electron configuration, the oxidation number of alkaline earth metals in the combined form is almost always +2. Table 28.2 lists some common properties of these metals.

All isotopes of radium are radioactive. Therefore it is both difficult and expensive to study radium in the laboratory.

TABLE 28.2 Properties of Alkaline Earth Metals

	Be	Mg	Ca	Sr	Ba
Valence electron configuration	$2s^2$	$3s^2$	$4s^2$	$5s^2$	$6s^2$
Density (g/cm^3)	1.86	1.74	1.55	2.6	3.5
Melting point (°C)	1280	650	838	770	714
Boiling point (°C)	2770	1107	1484	1380	1640
Atomic radius (pm)	112	160	197	215	222
Ionic radius (pm)*	31	65	99	113	135
First and second ionization energies (kJ/mol)	899 1757	738 1450	590 1145	548 1058	502 958
Electronegativity	1.5	1.2	1.0	1.0	0.9
Standard reduction potential (V)†	−1.85	−2.37	−2.87	−2.89	−2.90

* Refers to the cation M^{2+}, where M denotes an alkaline earth metal atom.
† The half-reaction is $M^{2+}(aq) + 2e^- \rightarrow M(s)$.

Beryllium

Beryllium is relatively rare, ranking 32 in order of elemental abundance in Earth's crust. The only important beryllium ore is beryl (beryllium aluminum silicate, $Be_3Al_2Si_6O_{18}$) (see Color Plate 2). Pure beryllium is obtained by first converting the ore to the oxide (BeO). The oxide is then converted to the chloride or fluoride. Beryllium fluoride is heated with magnesium in a furnace at about 1000°C to produce metallic beryllium:

$$BeF_2(s) + Mg(l) \longrightarrow Be(s) + MgF_2(s)$$

Beryllium is highly toxic, especially if inhaled as a dust. Its toxicity is the result of Be^{2+} ion's ability to compete with Mg^{2+} at many enzyme sites. Beryllium is used mainly in alloys with copper to enhance the hardness of copper. Be atoms do not absorb high-energy radiation readily, so beryllium is used also as a "window" for X-ray tubes. It is used as a moderator for neutrons produced in nuclear reactions (see Chapter 25).

Because only two electrons (the $1s$ electrons) shield the outer $2s$ electrons, the first and second ionization energies of Be are higher than those of other alkaline earth metals. On the other hand, it takes much less energy to promote one of the $2s$ electrons to the $2p$ orbital, followed by sp hybridization (see p. 344). Thus, in the gas phase beryllium forms a number of linear binary compounds with hydrogen and the halogens (BeH_2, $BeCl_2$, $BeBr_2$) that possess covalent character. The structures of beryllium hydride and beryllium chloride depend on their physical states. As a solid, beryllium chloride has the polymeric form

The chlorine atoms serve as bridges between adjacent Be atoms by making their lone pairs of electrons available for coordinate covalent bonding (see Section 11.4), which is indicated by arrows. Of course, once these coordinate covalent bonds have formed, they cannot be distinguished from the normal covalent bonds formed by the sharing of electrons from both Be and Cl. Each Be atom is roughly tetrahedrally bonded to four Cl atoms; that is, each Be atom is at the center of a tetrahedron (Figure 28.3). Figure 28.4 shows the sp^3 hybridization process of beryllium. Two of the four Cl atoms donate two pairs of electrons to two empty sp^3 hybrid orbitals on each Be atom.

Like solid $BeCl_2$, solid BeH_2 is a polymeric compound containing, in this case, bridging hydrogen atoms:

$$\begin{array}{c} \diagdown \\ Be \\ \diagup \end{array} \begin{array}{c} H \\ \diagup \quad \diagdown \\ \quad \\ \diagdown \quad \diagup \\ H \end{array} Be \begin{array}{c} H \\ \diagup \quad \diagdown \\ \quad \\ \diagdown \quad \diagup \\ H \end{array} \begin{array}{c} \diagup \\ Be \\ \diagdown \end{array}$$

In many of its compounds, beryllium forms four tetrahedral bonds, with Be^{2+} at the center, for example, in complexes like BeF_4^{2-}, $BeCl_4^{2-}$, and $BeBr_4^{2-}$. Beryllium forms complexes also with a number of organic molecules.

The chemistry of beryllium differs from that of the other Group 2A members (especially the heavier elements such as Ca, Sr, and Ba) in a number of ways.

- Beryllium is quite unreactive toward oxygen and water. Beryllium oxide, BeO, is formed only at elevated temperatures. Other alkaline earth metals are much more reactive toward oxygen and water.
- Beryllium oxide does not exhibit basic properties, which are characteristic of other alkaline earth metal oxides; that is, it does not undergo the following reaction:

$$MO(s) + H_2O(l) \longrightarrow M(OH)_2(s)$$

where M denotes an alkaline earth metal other than beryllium.

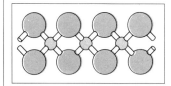

FIGURE 28.3 *Structure of $BeCl_2$ in the solid state. Each Be atom (small sphere) is roughly tetrahedrally bonded to four chlorine atoms (large spheres).*

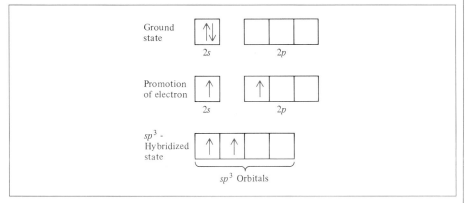

FIGURE 28.4 *The sp^3 hybridization of a Be atom in solid $BeCl_2$. Each Be atom has two vacant sp^3 hybrid orbitals, which can accept lone pairs from the bridging Cl atoms.*

- Beryllium fluoride and beryllium chloride are covalent compounds. These compounds do not conduct electricity in the molten state, in sharp contrast to the other molten alkaline earth metal halides.

As an example of the diagonal relationship (see Section 9.6), beryllium resembles aluminum in a number of respects. Thus, both Be^{2+} and Al^{3+} hydrolyze to yield acidic solutions:

$$Be(H_2O)_4^{2+}(aq) + H_2O(l) \rightleftharpoons Be(OH)(H_2O)_3^+(aq) + H_3O^+(aq)$$

$$Al(H_2O)_6^{3+}(aq) + H_2O(l) \rightleftharpoons Al(OH)(H_2O)_5^{2+}(aq) + H_3O^+(aq)$$

Both BeO and Al_2O_3 are amphoteric; that is, they react with acids and with bases:

$$BeO(s) + 2H^+(aq) + 3H_2O(l) \longrightarrow Be(H_2O)_4^{2+}(aq)$$

$$Al_2O_3(s) + 6H^+(aq) + 9H_2O(l) \longrightarrow 2Al(H_2O)_6^{3+}(aq)$$

and

$$BeO(s) + 2OH^-(aq) + H_2O(l) \longrightarrow Be(OH)_4^{2-}(aq)$$

$$Al_2O_3(s) + 6OH^-(aq) + 3H_2O(l) \longrightarrow 2Al(OH)_6^{3-}(aq)$$

In contrast, other alkaline earth metal oxides (MgO, CaO, SrO, and BaO) are all basic (see Figure 18.5).

Similarly, like $Al(OH)_3$, $Be(OH)_2$ is amphoteric (see Section 18.7). All other alkaline earth metal hydroxides are basic.

Magnesium

Magnesium is the eighth most plentiful element in Earth's crust (about 2.5 percent by mass). Among the principal magnesium ores are brucite, $Mg(OH)_2$; dolomite, $MgCO_3 \cdot CaCO_3$ (see Color Plate 2); and epsomite, $MgSO_4 \cdot 7H_2O$. Seawater is a good source of magnesium; there is about 1.3 g of magnesium in each kilogram of it. As is the case with most alkali and alkaline earth metals, metallic magnesium is obtained by the electrolysis of its molten chloride, $MgCl_2$ (obtained from seawater).

The chemistry of magnesium is intermediate between that of beryllium and the heavier Group 2A elements. Magnesium does not react with cold water but does react slowly with steam:

$$Mg(s) + H_2O(g) \longrightarrow MgO(s) + H_2(g)$$

It burns brilliantly in air to produce magnesium oxide (see Color Plate 11):

$$2Mg(s) + O_2(g) \longrightarrow 2MgO(s)$$

This property makes magnesium (in the form of thin ribbons or fibers) useful in flash photography (Figure 28.5).

Magnesium oxide reacts very slowly with water to form magnesium hydroxide, a solid suspension called *milk of magnesia* (Figure 28.6):

$$MgO(s) + H_2O(l) \longrightarrow Mg(OH)_2(s)$$

Milk of magnesia is used to relieve acid indigestion.

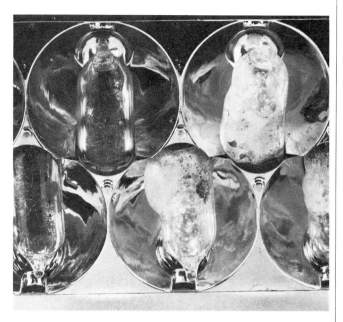

FIGURE 28.5 *The flash bulbs on the left contain thin magnesium ribbons. After use, the magnesium metal in the flash bulbs on the right is converted to magnesium oxide and magnesium nitride.*

Magnesium is a typical alkaline earth metal in that its hydroxide is a strong base. [The only alkaline earth hydroxide that is not a strong base is Be(OH)$_2$, which is amphoteric.] As Table 28.3 shows, all alkaline earth hydroxides except Ba(OH)$_2$ have low solubilities.

Magnesium reacts with dilute hydrochloric acid and with sulfuric acid to produce hydrogen gas:

$$Mg(s) + 2H^+(aq) \longrightarrow Mg^{2+}(aq) + H_2(g)$$

It reacts with concentrated nitric acid as follows:

$$3Mg(s) + 8HNO_3(aq) \longrightarrow 3Mg(NO_3)_2(aq) + 4H_2O(l) + 2NO(g)$$

FIGURE 28.6 *A commercially available milk of magnesia.*

TABLE 28.3 Solubilities of Alkaline Earth Hydroxides at 25°C

	Solubility Product K_{sp}	Molar Solubility (M)
Be(OH)$_2$	1×10^{-19}	3×10^{-7}
Mg(OH)$_2$	1.2×10^{-11}	1.4×10^{-4}
Ca(OH)$_2$	5.4×10^{-6}	0.011
Sr(OH)$_2$	3.1×10^{-4}	0.043
Ba(OH)$_2$	5.0×10^{-3}	0.11

On heating, magnesium reacts with a number of nonmetals, sometimes violently:

$$Mg(s) + Cl_2(g) \longrightarrow MgCl_2(s)$$

$$Mg(s) + S(l) \longrightarrow MgS(s)$$

$$3Mg(s) + N_2(g) \longrightarrow Mg_3N_2(s)$$

Magnesium nitride reacts with water to produce ammonia:

$$Mg_3N_2(s) + 6H_2O(l) \longrightarrow 3Mg(OH)_2(s) + 2NH_3(g)$$

The major uses of magnesium are in alloys, as a lightweight structural metal, for cathodic protection (see Section 24.6), in organic synthesis, and in batteries. Magnesium is essential to life, and Mg^{2+} ions are not toxic. It is estimated that the average adult ingests about 0.3 g of magnesium ions daily. Magnesium plays several important biological roles. It is present in intracellular and extracellular fluids. Magnesium ions are essential for the proper functioning of a number of enzymes. Magnesium is also present in the green plant pigment chlorophyll, which plays an important part in photosynthesis.

Calcium

Earth's crust contains about 3.4 percent calcium by mass. Calcium occurs in limestone, calcite, chalk, and marble as $CaCO_3$; in dolomite as $MgCO_3 \cdot CaCO_3$ (see Color Plate 2); in gypsum as $CaSO_4 \cdot 2H_2O$; and in fluorite as CaF_2. Metallic calcium is best prepared by the electrolysis of molten calcium chloride ($CaCl_2$).

As we read down Group 2A from beryllium to barium, we note an increase in metallic properties. Unlike beryllium and magnesium, calcium (like strontium and barium) reacts with cold water to yield the corresponding hydroxide, although the rate of reaction is much slower than the rates of reactions involving the alkali metals (see Figure 3.8):

$$Ca(s) + 2H_2O(l) \longrightarrow Ca(OH)_2(aq) + H_2(g)$$

The roles of calcium carbonate in hard water (p. 457) and in the preparation of sodium carbonate (p. 861) have been discussed.

Metallic calcium has rather limited uses. It is used mainly as an alloying agent for metals like aluminum and copper and in the preparation of beryllium metal from its compounds. It is also used as a dehydrating agent for organic solvents.

Calcium is an essential element in living matter. It is the major component of bones; the calcium ion is present in a complex phosphate salt, hydroxyapatite, $Ca_{10}(PO_4)_6(OH)_2$. A characteristic function of Ca^{2+} ions in living systems is the activation of a variety of metabolic processes. Calcium plays an important role in heart action, blood clotting, muscle contraction, and nerve transmission.

Strontium and Barium

Strontium and barium have properties similar to those of calcium except that, in general, they are more reactive. Strontium occurs as the carbonate $SrCO_3$ (strontianite) and as the sulfate $SrSO_4$ (celestite) (see Color Plate 2). There is no large-scale use of strontium. It is extracted by the electrolysis of its fused chloride ($SrCl_2$).

Barium occurs as the carbonate $BaCO_3$ (witherite) and as the sulfate $BaSO_4$ (barytes). Metallic barium can be prepared by the electrolysis of the fused chloride ($BaCl_2$) or by reduction of its oxide with aluminum:

$$3BaO(s) + 2Al(s) \longrightarrow 3Ba(s) + Al_2O_3(s)$$

Neither strontium nor barium is essential to living matter. But as environmental pollutants, both can have profound effects. Strontium-90 is a major nuclear fission product of uranium and plutonium. When an atomic bomb is exploded in the atmosphere, strontium-90 may be carried around the globe by winds. It settles on land and water alike, and reaches our bodies via a relatively short food chain. For example, cows eat contaminated grass and drink contaminated water, then pass along strontium-90 in their milk. Because Ca^{2+} and Sr^{2+} are quite similar chemically, strontium ions replace calcium ions in our bodies. A radioactive isotope with a half-life of about 28 years, strontium-90 emits β particles (electrons). Depending on the amount of $^{90}_{38}Sr$ present, constant exposure of the body to this radiation can seriously damage bone marrow, leading to anemia, leukemia, and other ailments. There is no cure for this type of radiation damage; it is impossible to remove the Sr^{2+} ions once they have replaced Ca^{2+} ions in the bones.

Strontium salts such as $Sr(NO_3)_2$ and $SrCO_3$ are used in red fireworks as well as in the familiar red warning flares on highways.

Unlike strontium, barium's toxicity to the human body is chemical rather than radioactive in nature. If taken into the body as a soluble salt (for example, $BaCl_2$), barium causes serious deterioration of the heart's function—a phenomenon known as *ventricular fibrillation*. Interestingly, the insoluble barium sulfate, $BaSO_4$, is nontoxic to humans, presumably because of the low Ba^{2+} ion concentration in solution. This dense barium salt absorbs X rays and acts as an opaque barrier to them. Thus, X-ray examination of a patient who has swallowed a solution containing finely divided $BaSO_4$ particles allows the radiologist to diagnose an ailment of the patient's digestive tract (Figure 28.7). Barium sulfate is quite insoluble

FIGURE 28.7 *Barium sulfate is opaque to X rays. An aqueous suspension of the substance is used to examine the path of digestive tracts.*

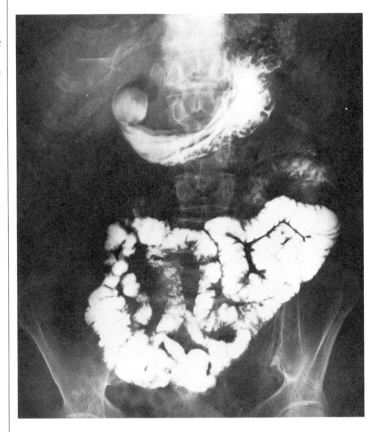

($K_{sp} = 1.1 \times 10^{-10}$), so it passes through the digestive system and no appreciable amounts of Ba^{2+} ions are taken up by the body.

28.3 ALUMINUM

Aluminum is the most abundant metal and the third most plentiful element in Earth's crust (7.5 percent by mass). The pure form does not occur in nature; its principal ore is bauxite ($Al_2O_3 \cdot 2H_2O$). Other minerals containing aluminum are orthoclase ($KAlSi_3O_8$), beryl ($Be_3Al_2Si_6O_{18}$) (see Color Plate 2), and cryolite (Na_3AlF_6). The major method for preparing metallic aluminum is the Hall process, described in Section 27.3.

Aluminum is one of the most versatile metals known. It has a low density (2.7 g/cm³) and high tensile strength (that is, it can be stretched or drawn out). Aluminum is malleable, it can be rolled into thin foils, and it is an excellent electrical conductor. Its conductivity is about 65 percent of copper's. However, because Al is cheaper and lighter than Cu, it is widely used in high-voltage transmission lines. Although aluminum's chief use is in aircraft construction, the pure metal itself is too soft and weak to withstand much strain. Its mechanical properties are greatly improved by alloying it with small amounts of metals such as copper, magnesium, and manganese, as well as silicon. Aluminum is not involved in living systems and is generally considered to be nontoxic.

As we read across the periodic table from left to right in a given period, we note a gradual decrease in metallic properties. Thus, although aluminum is considered an active metal, it does not react with water as do Na and Mg. Aluminum is an amphoteric element; it reacts with hydrochloric acid and with strong bases as follows:

$$2Al(s) + 6HCl(aq) \longrightarrow 2AlCl_3(aq) + 3H_2(g)$$

$$2Al(s) + 6NaOH(aq) \longrightarrow 2Na_3AlO_3(aq) + 3H_2(g)$$

Aluminum readily forms the oxide Al_2O_3 when exposed to air:

$$4Al(s) + 3O_2(g) \longrightarrow 2Al_2O_3(s)$$

A tenacious film of this oxide protects metallic aluminum from corrosion (see Section 24.6) and accounts for some of the inertness of aluminum.

Aluminum oxide has a very large exothermic enthalpy of formation ($\Delta H_f^\circ = -1670$ kJ/mol). This property makes aluminum suitable for use in solid propellants for rockets such as those used for the space shuttle *Columbia*. When a mixture of aluminum and ammonium perchlorate (NH_4ClO_4) is ignited, Al is oxidized to Al_2O_3 and the heat liberated in the reaction causes the gases that are formed to expand with great force. This action is responsible for lifting the rocket. The great affinity of aluminum for oxygen is illustrated nicely by the reaction of Al powder with a variety of metal oxides, particularly the transition metal oxides, to produce the corresponding metals. A typical reaction is

$$2Al(s) + Fe_2O_3(s) \longrightarrow Al_2O_3(l) + 2Fe(l) \qquad \Delta H^\circ = -852 \text{ kJ}$$

which can result in temperatures approaching 3000°C. This reaction is an example of the *thermite reaction*, which is used in the welding of steel and iron (see Figure 16.3).

Anhydrous aluminum chloride is prepared by heating the metal in dry chlorine or dry hydrogen chloride:

$$2Al(s) + 3Cl_2(g) \longrightarrow 2AlCl_3(s)$$

$$2Al(s) + 6HCl(g) \longrightarrow 2AlCl_3(s) + 3H_2(g)$$

Aluminum chloride exists as a dimer:

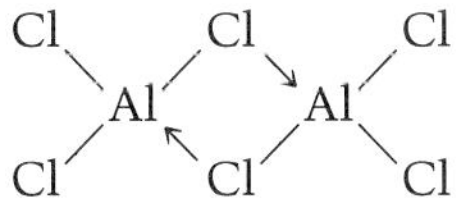

Each of the bridging chlorine atoms forms a normal covalent bond and a coordinate covalent bond with two aluminum atoms (compare with the structure of solid $BeCl_2$ on p. 864). Each aluminum atom is assumed to be sp^3-hybridized, so the vacant sp^3 hybrid orbital can accept a lone pair from the chlorine atom (Figure 28.8). Aluminum chloride undergoes hydrolysis as follows:

$$AlCl_3(s) + 3H_2O(l) \longrightarrow Al(OH)_3(s) + 3HCl(aq)$$

Aluminum hydroxide can be prepared by adding ammonia to a solution containing a soluble aluminum salt:

$$Al^{3+}(aq) + 3OH^-(aq) \longrightarrow Al(OH)_3(s)$$

FIGURE 28.8 *The sp^3 hybridization of an Al atom in $AlCl_3$ (or Al_2Cl_6). Each Al atom has one vacant sp^3 hybrid orbital that can accept a lone pair from the bridging Cl atom.*

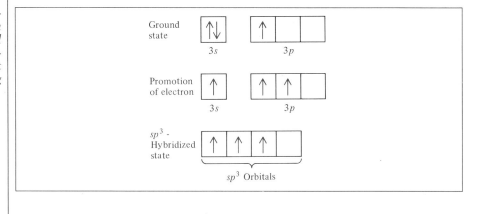

It is an amphoteric hydroxide:

$$Al(OH)_3(s) + 3H^+(aq) \longrightarrow Al^{3+}(aq) + 3H_2O(l)$$

$$Al(OH)_3(s) + OH^-(aq) \longrightarrow Al(OH)_4^-(aq)$$

In contrast to the boron hydrides, which are a well-defined series of compounds, aluminum hydride is a polymeric substance in which each aluminum atom is surrounded octahedrally by bridging hydrogen atoms (Figure 28.9). Thus the structure of aluminum hydride resembles that of the polymeric beryllium hydride discussed in Section 28.2.

When an aqueous mixture of aluminum sulfate and potassium sulfate is evaporated slowly, crystals of $KAl(SO_4)_2 \cdot 12H_2O$ are formed. Similar crystals can be formed by substituting Na^+ or NH_4^+ for K^+, and Cr^{3+} or Fe^{3+} for Al^{3+}. These compounds are called *alums*, and they have the general formula

$$M^+M^{3+}(SO_4)_2 \cdot 12H_2O \qquad \begin{array}{l} M^+: \;\; K^+, \, Na^+, \, NH_4^+ \\ M^{3+}: \;\; Al^{3+}, \, Cr^{3+}, \, Fe^{3+} \end{array}$$

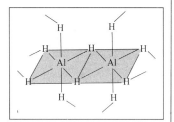

FIGURE 28.9 *Structure of aluminum hydride. Note that this compound is a polymer. Each Al atom is surrounded octahedrally by six bridging H atoms.*

28.4 TIN AND LEAD

Tin

Tin and lead are the only metallic elements in Group 4A. They were among the first metals used by humans. The principal ore of tin Sn(IV) oxide, or cassiterite (see Color Plate 4). Metallic tin is prepared by reducing SnO_2 with carbon at elevated temperatures:

$$SnO_2(s) + 2C(s) \longrightarrow Sn(l) + 2CO(g)$$

There are three allotropic forms of tin with the following transition temperatures:

$$\text{grey tin} \xrightarrow{\;13°C\;} \text{white tin} \xrightarrow{\;161°C\;} \gamma\text{-tin}$$

Metallic tin (that is, white tin) is soft and malleable. Its valence electron configuration is $5s^25p^2$; the metal forms compounds with oxidation numbers of $+2$ and $+4$. Tin(II) compounds (called *stannous compounds*) are generally more ionic and reducing, while tin(IV) compounds (called *stannic compounds*) are more covalent and oxidizing. Tin reacts with hydrochloric acid to give tin(II) chloride

$$Sn(s) + 2HCl(aq) \longrightarrow SnCl_2(aq) + H_2(g)$$

and with oxidizing acids such as nitric acid to give Sn(IV) compounds

$$Sn(s) + 4HNO_3(aq) \longrightarrow SnO_2(s) + 4NO_2(g) + 2H_2O(l)$$

Tin also reacts with hot concentrated aqueous solutions of sodium hydroxide and potassium hydroxide to form the stannate ion (SnO_3^{2-}):

$$Sn(s) + 2OH^-(aq) + H_2O(l) \longrightarrow SnO_3^{2-}(aq) + 2H_2(g)$$

Thus tin exhibits amphoteric properties. Tin combines directly with chlorine to form tin(IV) chloride, which is a liquid:

$$Sn(s) + 2Cl_2(g) \longrightarrow SnCl_4(l)$$

Unlike carbon and silicon, tin forms only two hydrides: SnH_4 and Sn_2H_6, neither of which is very stable. In both compounds, the tin atom is sp^3-hybridized.

Tin is used mainly to form alloys. For example, bronze is 20 percent tin and 80 percent copper; soft solder is 33 percent tin and 67 percent lead; and pewter is 85 percent tin, 6.8 percent copper, 6 percent bismuth, and 1.7 percent antimony. Another important tin alloy is Nb_3Sn, a superconducting material (see Section 26.8). Of course, tin is also used in the manufacture of "tin" cans (see Section 24.6).

Lead

The chief ore of lead is galena, PbS. Metallic lead is obtained by first roasting the sulfide in air:

$$2PbS(s) + 3O_2(g) \longrightarrow 2PbO(s) + 2SO_2(g)$$

The oxide is then reduced in a blast furnace with coke:

$$PbO(s) + C(s) \longrightarrow Pb(l) + CO(g)$$
$$PbO(s) + CO(g) \longrightarrow Pb(l) + CO_2(g)$$

Because both tin and lead are much less electropositive than the alkali and alkaline earth metals, they can be prepared more readily by chemical reduction than by the expensive process of electrolysis.

Like tin, lead forms compounds with oxidation numbers of +2 (called *plumbous compounds*) and +4 (called *plumbic compounds*). Lead(II) oxide, PbO, known as *litharge*, is used to glaze ceramic vessels. Practically all common lead compounds contain lead in the +2 oxidation state (PbO, $PbCl_2$, PbS, $PbSO_4$, $PbCO_3$, and $PbCrO_4$). Except for lead acetate and lead nitrate, most lead compounds are insoluble.

White lead, a hydroxy carbonate composed of lead, is prepared as follows:

$$2Pb(s) + O_2(g) + 2CH_3COOH(aq) \longrightarrow 2Pb(OH)(OOCCH_3)(aq)$$
$$6Pb(OH)(OOCCH_3)(aq) + 2CO_2(g) \longrightarrow$$
$$Pb_3(OH)_2(CO_3)_2(s) + 3Pb(OOCCH_3)_2(aq) + 2H_2O(l)$$
$$\text{white lead}$$

For a number of years, white lead was used as a pigment for paints, but because of its toxicity, its use in the United States has been banned.

Like tin compounds, the $+2$ oxidation state lead compounds are more ionic and more stable than the $+4$ oxidation state lead compounds. Lead(IV) oxide (PbO_2) is a covalent compound and a strong oxidizing agent. It can oxidize hydrochloric acid to molecular chlorine:

$$PbO_2(s) + 4HCl(aq) \longrightarrow PbCl_2(s) + Cl_2(g) + 2H_2O(l)$$

The two major uses of lead are in lead storage batteries (see Section 24.5) and in gasoline as the antiknocking agent tetraethyllead [$(C_2H_5)_4Pb$]. This compound is formed by heating ethyl chloride (C_2H_5Cl) with an alloy containing about 90 percent Pb and 10 percent Na by mass. It is a colorless, oily liquid (b.p. 200°C) that is soluble in most organic solvents but insoluble in water. Lead is relatively impenetrable to high-energy radiation, so it is used in protective shields for nuclear chemists, X-ray operators, and radiologists.

Lead is extremely toxic, and its effects in humans are cumulative, that is, they slowly build up. It enters the body either as inorganic lead (Pb^{2+}) ion or as tetraethyllead. Inhaled or ingested lead concentrates in the blood, the tissues, and the bones. It affects the central nervous system and impairs kidney functions. It can also inhibit the synthesis of hemoglobin molecules, which carry molecular oxygen through the bloodstream.

28.5 ZINC, CADMIUM, AND MERCURY

The Group 2B elements—zinc, cadmium, and mercury—resemble the alkaline earth metals in that they all have the same valence electron configuration, ns^2. Yet their chemical properties are quite different (Table 28.4). The difference arises as a result of the number of electrons in the next-to-outermost shell; Ca, Sr, and Ba have 8 electrons (s^2p^6), whereas Zn, Cd, and Hg have 18 electrons ($s^2p^6d^{10}$). Because the d orbitals do not shield outer electrons effectively from the nucleus, the increase in nuclear charge from

TABLE 28.4 Properties of the Group 2B Elements

	Zn	Cd	Hg
Valence electron configuration	$4s^23d^{10}$	$5s^24d^{10}$	$6s^25d^{10}$
Density (g/cm³)	7.14	8.64	13.59
Melting point (°C)	419.5	320.9	−38.9
Boiling point (°C)	907	767	357
Atomic radius (pm)	138	154	157
Ionic radius (pm)*	74	97	110
First and second ionization energies (kJ/mol)	906; 1733	867; 1631	1006; 1809
Electronegativity	1.6	1.7	1.9
Standard reduction potential (V)†	−0.76	−0.40	0.85

* Refers to the cation M^{2+}, where M denotes Zn, Cd, or Hg.
† The half-reaction is $M^{2+}(aq) + 2e^- \rightarrow M(s)$ for Zn and Cd. For Hg, it is $M_2^{2+}(aq) + 2e^- \rightarrow 2M(l)$.

calcium to zinc results in an increase in the ionization energies of the valence electrons. Further, we expect the zinc atom to be smaller than the calcium atom, and this is indeed the case (Ca: 197 pm; Zn: 138 pm). For similar reasons, cadmium and mercury lose electrons less readily than strontium and barium. Thus, the Group 2B elements are less reactive, less electropositive, and have a greater tendency to form covalent compounds than the alkaline earth metals.

Zinc and Cadmium

The properties of zinc and cadmium are so similar that we will describe their chemistries together. Not an abundant element (about 0.007 percent of Earth's crust by mass), zinc occurs principally as the mineral sphalerite (ZnS) (see Color Plate 9), also called zincblende. Metallic zinc is obtained by roasting the sulfide in air to convert it to the oxide and then reducing the oxide with finely divided carbon:

$$2ZnS(s) + 3O_2(g) \longrightarrow 2ZnO(s) + 2SO_2(g)$$

$$ZnO(s) + C(s) \longrightarrow Zn(s) + CO(g)$$

Cadmium is only about one-thousandth as abundant as zinc, and is present in small amounts in most zinc ores as CdS. Cadmium is obtained from flue dust emitted during the purification of zinc by distillation. Because cadmium is more volatile (b.p. 767°C) than zinc (b.p. 907°C), it evaporates first and concentrates in the first distillates.

Both zinc and cadmium are silvery metals in the pure state. Zinc is hard and brittle, but cadmium is soft enough to be cut with a knife.

The only known oxidation number of these metals is $+2$. Both metals react with strong acids to produce hydrogen gas:

$$Zn(s) + 2H^+(aq) \longrightarrow Zn^{2+}(aq) + H_2(g)$$

$$Cd(s) + 2H^+(aq) \longrightarrow Cd^{2+}(aq) + H_2(g)$$

Zinc and cadmium oxides are formed through direct combination of these metals with oxygen. Zinc is amphoteric, but cadmium is not. In addition to its reaction with strong acids, zinc also reacts with strong bases:

$$Zn(s) + 2OH^-(aq) + 2H_2O(l) \longrightarrow Zn(OH)_4^{2-}(aq) + H_2(g)$$

The inertness of cadmium toward basic solutions makes it a useful metal for plating other metals used in alkaline solutions.

Both Zn^{2+} and Cd^{2+} ions form a variety of similar complex ions in solution. Some examples are

$$Zn^{2+}(aq) + 4NH_3(aq) \rightleftharpoons Zn(NH_3)_4^{2+}(aq)$$

$$Zn^{2+}(aq) + 4CN^-(aq) \rightleftharpoons Zn(CN)_4^{2-}(aq)$$

$$Cd^{2+}(aq) + 4Cl^-(aq) \rightleftharpoons CdCl_4^{2-}(aq)$$

These are Lewis acid-base reactions.

These complexes are generally tetrahedral in structure, indicating sp^3 hybridization for the central metal atom.

Metallic zinc is used to form alloys. For example, brass is an alloy containing about 20 percent zinc and 80 percent copper by mass. Because zinc forms a protective oxide layer that withstands corrosion well, it is used to provide cathodic protection (see Section 24.6) for less electropositive metals. Iron protected this way, called *galvanized iron*, is prepared by dipping iron into molten zinc or by plating zinc on iron by electrolysis (using iron as the cathode in a solution containing Zn^{2+} ions).

Zinc sulfide is used in the white pigment lithopone, which contains a roughly equimolar mixture of ZnS and $BaSO_4$. Unlike the white lead discussed earlier, this substance is nontoxic. Zinc sulfide emits light when struck by X rays or an electron beam. Therefore it is used in screens of television sets, oscilloscopes, and X-ray fluoroscopes.

Zinc oxide, an important zinc compound, is prepared by burning the metal in air

$$2Zn(s) + O_2(g) \longrightarrow 2ZnO(s)$$

and by the thermal decomposition of zinc nitrate or zinc carbonate,

$$2Zn(NO_3)_2(s) \longrightarrow 2ZnO(s) + 4NO_2(g) + O_2(g)$$

$$ZnCO_3(s) \longrightarrow ZnO(s) + CO_2(g)$$

Zinc oxide is amphoteric:

$$ZnO(s) + 2HCl(aq) \longrightarrow ZnCl_2(aq) + H_2O(l)$$

$$ZnO(s) + 2NaOH(aq) \longrightarrow Na_2ZnO_2(aq) + H_2O(l)$$

The compound is used in the treatment of rubber. It is also used as a white pigment in paint, an ointment base, and in cement. An interesting property of zinc oxide is its photoconductivity. A *photoconductor* is a substance that conducts electric current when illuminated; when the light is turned off, it becomes an insulator. This property finds important commercial application in a process called *xerography* (from the Greek words "xeros" meaning dry and "graphein" meaning to write). Xerography is a dry process for photocopying. Figure 28.10 shows the basic steps involved in making a xerographic copy.

Zinc is an essential element in living matter; its presence in the human body is necessary for the activity of a number of enzymes such as carbonic anhydrase, which catalyzes the hydration and dehydration of carbon dioxide.

The major use of cadmium is to plate other metals. In the electroplating of steel, for example, the anode (Cd) and cathode (steel) are dipped into a solution containing $Cd(CN)_4^{2-}$ ions. Normally, a very thin film of metallic cadmium (about 0.0005 cm thick) is sufficient for protecting the metal.

Although cadmium has no biological significance, there is growing concern over its effect as an environmental pollutant. Like compounds of most heavy metals, cadmium compounds are exceedingly toxic. The common symptoms of cadmium poisoning are hypertension (high blood pressure), anemia, and kidney failure. Cadmium's action on our bodies is not as well understood as that of lead; presumably the Cd^{2+} ions also react with proteins. Electroplating industries are the main source of cadmium pollution, by discharging waste solutions into lakes and rivers, but there may be enough cadmium in the air alone to cause health problems.

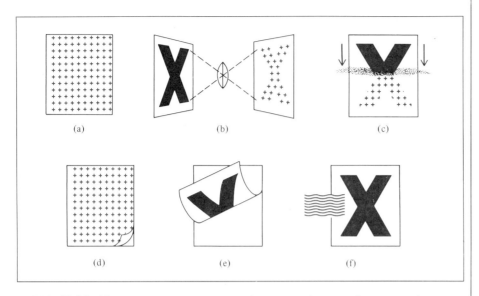

FIGURE 28.10 *Basic steps in xerography. (a) A photoconductive surface containing ZnO is given a positive electric charge (+). (b) The image of a document is exposed on the (a) surface. The charge drains away from the surface in all but the image area, which remains unexposed and charged. (c) Negatively charged powder is cascaded over the surface. The powder adheres electrostatically to the positively charged image area, making a visible image. (d) A sheet of plain paper is placed over the surface and given a positive charge. (e) The negatively charged powder image on the surface is electrostatically attracted to the positively charged paper. (f) The powder image is fused to the paper by heat.*

As mentioned earlier, zinc oxide is used to treat rubber. Because zinc and cadmium have similar properties, the oxide is always contaminated with a small amount of CdO. As an automobile tire wears, both ZnO and CdO are liberated to the atmosphere as fine dust, which may not settle for quite some time. Tobacco, too, contains a small amount of cadmium, so smokers probably have a higher body content of cadmium than nonsmokers.

Mercury

Mercury has been known since ancient times. It is the only metal that exists as a liquid at room temperature. (Two other metals—cesium, m.p. 28.7°C, and gallium, m.p. 29.8°C—come quite close to being liquids at room temperature.) Although mercury is rarer than gold and platinum (it constitutes about 8×10^{-6} percent of Earth's crust by mass), its sources are so much more concentrated that the metal can be obtained readily. The principal ore of mercury is mercury(II) sulfide, HgS, called *cinnabar* (see Color Plate 9). The ore is first concentrated by flotation and then roasted in air to yield mercury(II) oxide

$$2HgS(s) + 3O_2(g) \longrightarrow 2HgO(s) + 2SO_2(g)$$

which decomposes to yield mercury vapor:

$$2HgO(s) \longrightarrow 2Hg(g) + O_2(g)$$

Metallic mercury is a bright, silvery, dense liquid that freezes at $-38.9°C$ and boils at $357°C$. Liquid mercury dissolves many metals, such as copper, silver, gold, and the alkali metals, to form amalgams that may be either solids or liquids. The reactivity of a metal is generally lowered when it is amalgamated with mercury. For example, the sodium-mercury amalgam is so much less reactive than sodium that it decomposes water at a considerably slower rate. Such amalgams are useful as mild reducing agents in organic synthesis.

Unlike cadmium and zinc, mercury has both $+1$ (*mercurous*) and $+2$ (*mercuric*) oxidation numbers in its compounds. Because the standard reduction potential of mercury is positive, the metal does not react with water and does not decompose acids. Many of the mercury(II) salts (HgS and HgI_2, for example) are insoluble. This is particularly significant, for if HgS were soluble, rain and weathering might have distributed mercury throughout the planet in sufficient concentration to poison the environment. Life as we know it might have been impossible. (On the other hand, mercury might have become an essential element to other kinds of life systems.)

Mercury(II) oxide is formed by heating the metal in air slowly:

$$2Hg(l) + O_2(g) \longrightarrow 2HgO(s)$$

At higher temperatures, the oxide decomposes to yield elemental mercury and molecular oxygen again:

$$2HgO(s) \longrightarrow 2Hg(l) + O_2(g)$$

In fact, this was the famous experiment conducted by Joseph Priestley (1733–1804) in 1774 when he discovered oxygen.

Mercury with $+2$ oxidation number has a greater tendency to form covalent bonds than do zinc and cadmium. Mercury(II) chloride hydrolyzes only slightly because it is largely a covalent compound (as are $HgBr_2$ and HgI_2). The molecule has a linear structure, as predicted by the VSEPR model:

$$:\ddot{C}l\!-\!Hg\!-\!\ddot{C}l:$$

Mercury(II) fluoride, HgF_2, on the other hand, is essentially ionic. The Hg^{2+} ion also has a tendency to form very stable complex ions, such as $HgCl_4^{2-}$, $Hg(NH_3)_4^{2+}$, and $Hg(CN)_4^{2-}$.

For a number of years the true identity of mercurous ion was not clear; the two possibilities are Hg^+ and Hg_2^{2+}. If the mercury(I) ion existed as Hg^+ (electron configuration $6s^1$), it would be paramagnetic, but there is no experimental evidence to support this structure. Instead, the ion consists of two mercury atoms held together by a sigma bond:

$$[Hg\!-\!Hg]^{2+}$$

This was one of the first metal–metal bonds known in a compound.

The most common mercury(I) compound is probably mercury(I) chloride, Hg_2Cl_2, an insoluble white solid called *calomel*. At one time, calomel was used in medicine as a purgative. However, it has a tendency to disproportionate to Hg and $HgCl_2$, which is quite poisonous, so calomel is no longer used in internal medicine.

It has been estimated that mercury and its compounds have over two thousand uses. Only the more familiar and important applications will be mentioned here. Mercury has a large and uniform coefficient of volume expansion (with temperature) and is therefore suitable for thermometers. The liquid has a very high density (13.6 g/cm^3) and is employed in barometers. Although the conductivity of mercury is only about 2 percent that of copper, the advantages of its fluidity are so great that the metal is used in making electrical contacts in household light switches and thermostats. Mercury is also used as the cathode in the chlor-alkali process (p. 819).

Mercury is a metal poison that is cumulative, with notorious effects on neurological behavior. However, the toxicity of mercury very much depends on the physical and chemical states of the element. Mercury vapor is highly dangerous because it causes irritation and destruction of lung tissues. Figure 28.11 shows the temperature dependence of mercury vapor pressure. The toxicity of inorganic mercury compounds depends on their solubilities. For example, the insoluble Hg(I) chloride is not considered very toxic. Because Hg(II) salts are more soluble, they are considerably more toxic.

It was once believed that elemental mercury is sufficiently inert that it would stay at the bottom of a lake or river, and be slowly converted to the rather harmless HgS. This, unfortunately, is not the case. We now know that certain bacteria and microorganisms can metabolize mercury first to Hg^{2+} ions and eventually to the highly toxic CH_3Hg^+ and $(CH_3)_2Hg$ species. Fish take in bacteria, and these mercury species slowly concentrate in the fatty tissues of their bodies. Small fish are eaten by large fish and, at the end of this food chain, we find humans. Because of the cumulative effect, by the time we eat a walleye caught from a mercury-contaminated lake, the concentration of the metal in its body may be 50,000 times greater than that in the lake water! *The buildup of any poison along a food chain* is sometimes called **bioamplification.**

The harmful effects of mercury poisoning are usually slow to develop. Mercury poisoning can lead to brain damage and *erethism*, a condition that is characterized by mental and emotional disturbances.

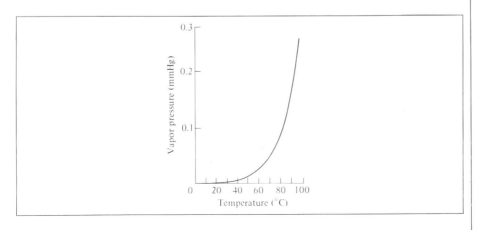

FIGURE 28.11 *Variation of mercury vapor pressure with temperature.*

SUMMARY

1. The alkali metals are the most reactive of all the metallic elements. They have the oxidation number $+1$ in their compounds. Under special conditions, some of them also form uninegative anions.
2. The chemistry of lithium resembles that of magnesium in some ways. This is an example of the diagonal relationship.
3. The alkaline earth metals are somewhat less reactive than the alkali metals. They almost always have the oxidation number $+2$ in their compounds. The properties of the alkaline earth elements become more metallic in going down their periodic table group.
4. The formation of bridged polymeric complexes (BeH_2, for example) and tetrahedral complexes is characteristic of beryllium, a highly toxic element.
5. The chemistry of beryllium resembles that of aluminum. This is another example of the diagonal relationship.
6. Strontium-90, a radioactive isotope, replaces calcium in bones and causes radiation damage.
7. Aluminum does not react with water and its oxide is amphoteric.
8. Tin and lead both form compounds with the oxidation numbers $+2$ and $+4$.
9. Zinc, cadmium, and mercury are less reactive, less electropositive, and have a greater tendency to form covalent compounds than the alkaline earth metals.
10. Zinc and cadmium have very similar properties, although zinc is amphoteric and cadmium is not. Both of these elements always have the oxidation number $+2$ in their compounds, and both form a variety of complex ions in solution.
11. Mercury forms Hg_2^{2+} and Hg^{2+} ions, and also forms a number of stable complex ions in solution. Mercury(II) compounds tend to be covalent.
12. Lead and its compounds, cadmium and its compounds, mercury vapor, soluble inorganic mercury, and organic mercury compounds are all very toxic substances.

KEY WORDS

Bioamplification, p. 879   Hydroscopic, p. 859

PROBLEMS

More challenging problems are marked with an asterisk.

Alkali and Alkaline Earth Metals

28.1 How is sodium metal prepared commercially?
28.2 Why is potassium usually not prepared electro-lytically from one of its salts?

28.3 Complete and balance the following equations:

(a) $K(s) + H_2O(l) \longrightarrow$
(b) $Li(s) + N_2(g) \longrightarrow$
(c) $CsH(s) + H_2O(l) \longrightarrow$
(d) $Li(s) + O_2(g) \longrightarrow$
(e) $Na(s) + O_2(g) \longrightarrow$

(f) $K(s) + O_2(g) \longrightarrow$

(g) $Rb(s) + O_2(g) \longrightarrow$

28.4 Write a balanced equation for each of the following reactions: (a) potassium reacts with water; (b) an aqueous solution of NaOH reacts with CO_2; (c) solid Na_2CO_3 reacts with an HCl solution; (d) solid $NaHCO_3$ reacts with an HCl solution; (e) solid $NaHCO_3$ is heated; (f) solid Na_2CO_3 is heated.

28.5 Describe how you would prepare NaOH, starting with NaCl.

28.6 Calculate the volume of CO_2 at 10.0°C and 746 mmHg pressure that is obtained by treating 25.0 g Na_2CO_3 with an excess of hydrochloric acid.

28.7 Write balanced equations for the reactions between metallic beryllium and (a) an acid, (b) a base.

28.8 Starting with Mg^{2+} ions in seawater, show how you would obtain pure magnesium.

28.9 From the thermodynamic data in Appendix 1, calculate the $\Delta H°$ values for the following decompositions.

(a) $MgCO_3(s) \longrightarrow MgO(s) + CO_2(g)$

(b) $CaCO_3(s) \longrightarrow CaO(s) + CO_2(g)$

Which of the two compounds is more easily decomposed by heat?

28.10 Starting with magnesium and concentrated nitric acid, describe how you would prepare magnesium oxide.

28.11 Describe two ways of preparing magnesium chloride.

28.12 The second ionization energy of magnesium is only about twice as great as the first, but the third ionization energy is ten times as great. Why does it take so much more energy to remove the third electron?

28.13 Helium contains the same number of electrons in its outer shell as the alkaline earth metals. Explain why helium is inert whereas the Group 2A metals are not.

28.14 List the sulfates of the Group 2A metals in order of increasing solubility in water. Explain the trend. (*Hint:* You need to consult a chemistry handbook.)

28.15 When exposed to air, calcium first forms calcium oxide, which is then converted to calcium hydroxide, and finally to calcium carbonate. Write a balanced equation for each step.

28.16 Write chemical formulas for (a) quicklime, (b) slaked lime, (c) lime water.

28.17 How would you prepare the following compounds? (a) BaO, (b) $Ba(OH)_2$, (c) $BaCO_3$, (d) $BaSO_4$.

28.18 Barium ions are toxic. Which of the following barium salts is more harmful if ingested: $BaCl_2$ or $BaSO_4$? Explain.

Aluminum

28.19 Describe the Hall process for preparing aluminum.

28.20 Aluminum does not rust as iron does. Why?

28.21 Aluminum forms the complex ions $AlCl_4^-$ and AlF_6^{3-}. Describe the shapes of these ions. $AlCl_6^{3-}$ does not form. Why?

*28.22** In basic solution aluminum metal is a strong reducing agent, being oxidized to AlO_2^-. Give balanced equations for the reaction of Al in basic solution with the following: (a) $NaNO_3$, to give ammonia; (b) water, to give hydrogen; (c) Na_2SnO_3, to give metallic tin.

28.23 Write a balanced equation for the thermal decomposition of aluminum nitrate to form aluminum oxide, nitrogen dioxide, and oxygen gas.

28.24 Describe some important properties of aluminum that make it one of the most versatile metals known.

28.25 Starting with aluminum, describe with balanced equations how you would prepare (a) Al_2Cl_6, (b) Al_2O_3, (c) $Al_2(SO_4)_3$, (d) $NH_4Al(SO_4)_2 \cdot 12H_2O$.

28.26 The pressure of gaseous Al_2Cl_6 increases more rapidly with temperature than that predicted by the ideal gas equation ($PV = nRT$) even though Al_2Cl_6 behaves like an ideal gas. Explain.

28.27 Explain the change in bonding when Al_2Cl_6 dissociates to form $AlCl_3$ in the gas phase.

Tin and Lead

28.28 Tin(II) chloride is a powerful reducing agent. Write balanced ionic equations for the reaction between Sn^{2+} and (a) acidified permanganate solution, (b) acidified dichromate solution, (c) Fe^{3+} ions, (d) Hg^{2+} ions, (e) Hg_2^{2+} ions.

28.29 A 0.753 g sample of bronze dissolves in nitric acid and gives 0.171 g of SnO_2. Calculate the percent by mass of tin in the bronze.

*28.30** Near room temperature tin exists in two allotropic forms called white tin and grey tin. White tin has a close-packed structure and is stable above 13°C, and grey tin, which has a diamond structure, is stable below this temperature. (a) Predict which of the two allotropic forms has a greater density. (b) Predict which of the two allotropic forms is a better conductor of electricity. (c) The absolute entropies of grey tin and white tin at 13°C are 44.4 J/K·mol and 53.6 J/K·mol, respectively. Calculate the enthalpy change for the reaction Sn(grey) → Sn(white) at this temperature.

28.31 Explain why Pb(II) compounds are usually ionic and Pb(IV) compounds are covalent.

28.32 Oil paintings containing lead(II) compounds as constituents of their pigments darken over the years. Suggest a chemical reason for the color change.

28.33 Despite the known hazards of lead, lead pipes are still in use in many places. In such places it is safer to drink water that contains carbonate ions than water without carbonate ions. Why?

28.34 Briefly describe the change in chemical properties as we move down in Group 4A from carbon to lead.

Zinc, Cadmium, and Mercury

28.35 Although zinc is not a toxic substance, water running through galvanized pipes often picks up substantial amounts of a toxic metal ion. Suggest what this metal ion might be.

28.36 Starting with zinc, describe how you would prepare (a) ZnO, (b) $Zn(OH)_2$, (c) $ZnCO_3$, (d) ZnS, (e) $ZnCl_2$.

28.37 Describe a method, giving equations, for the removal of mercury from a solution of $Cd(NO_3)_2$ and $Hg_2(NO_3)_2$.

28.38 Look up the melting points of the following compounds in a handbook: HgF_2, $HgCl_2$, $HgBr_2$, and HgI_2. What can you deduce about the bonding in these compounds?

28.39 Predict the pH of an aqueous HgF_2 solution, that is, whether it is higher or lower than 7.

28.40 The ingestion of a very small quantity of mercury is not considered too harmful. Would this statement still hold if the gastric juice in your stomach were mostly nitric acid instead of hydrochloric acid?

28.41 Given that

$$2Hg^{2+}(aq) + 2e^- \longrightarrow Hg_2^{2+}(aq) \qquad \mathscr{E}° = 0.92 \text{ V}$$
$$Hg_2^{2+}(aq) + 2e^- \longrightarrow 2Hg(l) \qquad \mathscr{E}° = 0.85 \text{ V}$$

Calculate $\Delta G°$ and K for the disproportionation at 25°C

$$Hg_2^{2+}(aq) \longrightarrow Hg^{2+}(aq) + Hg(l)$$

Miscellaneous Problems

28.42 Illustrate the diagonal relationship for (a) lithium and magnesium, (b) beryllium and aluminum.

28.43 Explain what is meant by amphoterism. Use compounds of beryllium and aluminum as examples.

28.44 Explain each of the following statements: (a) An aqueous solution of $AlCl_3$ is acidic. (b) $Al(OH)_3$ is soluble in NaOH solutions but not in NH_3 solution. (c) The melting point of Sn is lower than that of carbon. (d) Pb^{2+} is a weaker reducing agent than Sn^{2+}.

28.45 It has been shown that Na_2 species are formed in the vapor phase. Describe the formation of the "disodium molecule" in terms of a molecular orbital energy level diagram. Would you expect the alkaline earth metals to exhibit a similar property?

28.46 Write balanced equations for the following reactions: (a) The heating of aluminum carbonate. (b) The hydrolysis of the Be^{2+} ion. (The complex ion formed is $[Be(H_2O)_3(OH)]^+$.) (c) The reaction between lead(II) oxide and molecular hydrogen. (d) The reaction between $AlCl_3$ and K. (e) The reaction between Sn and hot concentrated H_2SO_4. (f) The reaction between Na_2CO_3 and $Ca(OH)_2$.

28.47 What reagents would you employ to separate the following pairs of ions in solution? (a) Na^+ and Ba^{2+}, (b) Mg^{2+} and Pb^{2+}, (c) Zn^{2+} and Hg^{2+}.

28.48 Write balanced equations for the following reactions: (a) Calcium oxide and dilute HCl solution. (b) A solution containing ammonium ions with $Ba(OH)_2$. (c) A solution containing Pb^{2+} ion and H_2S gas.

*28.49 Explain the following statements: (a) An aqueous solution of aluminum chloride has a pH less than 7; (b) molten beryllium chloride is a poor conductor of electricity; (c) the reaction between tin and concentrated nitric acid is different from those for most other metals such as zinc and magnesium; (d) mercury(II) iodide, an insoluble compound, is soluble in a solution containing iodide ions.

28.50 Compare the properties of the following Group 4A oxides: CO_2, SiO_2, SnO_2, PbO_2.

29

COORDINA- TION COMPOUNDS OF THE TRANSITION METALS

The elements in the series in the periodic table in which the d orbitals are gradually being filled are called the transition metals. The transition metals have widely varying and fascinating properties. In this chapter we will limit our discussion to the first-row transition metals from scandium to copper. One of the characteristics of these metals is their ability to form complex ions. We will examine the general properties of and bonding in compounds containing complex ions and some of their applications in industry, medicine, and other areas.

29.1 PROPERTIES OF THE TRANSITION METALS

You will recall that transition metals are those that have incompletely filled d subshells or readily give rise to ions that have incompletely filled d subshells (see Section 9.2). In this chapter we will concentrate on the first-row transition metals from scandium to copper, which are typical of all the transition metals. Table 29.1 lists some of their properties.

As we read across any period from left to right, atomic numbers increase, electrons are being added to the outer shell, and the nuclear charge is increasing by the addition of protons. In the third-period elements—sodium to argon—the outer electrons shield one another weakly from the extra nuclear charge. Consequently, atomic radii decrease rapidly from sodium to argon and the electronegativities and ionization energies increase steadily (see Figures 9.4, 9.6, and 11.2).

The properties are different for the transition metals. As we read across from scandium to copper, the nuclear charge is, of course, increasing, but electrons are being added to the inner $3d$ subshell. These $3d$ electrons shield the $4s$ electrons from the increasing nuclear charge somewhat more effectively than outer-shell electrons can shield one another, so the atomic radii decrease less rapidly. For the same reason, electronegativities and ionization energies increase only slightly compared to the increase from sodium to argon.

Although the transition metals are less electropositive (or more electronegative) than the alkali and alkaline earth metals, their standard reduction potentials seem to indicate that all of them except copper should react with strong acids such as hydrochloric acid to produce hydrogen gas. In fact, however, most transition metals are inert toward acids or react slowly because of a protective layer of oxide. A case in point is chromium: Despite its rather negative standard reduction potential (see Table 24.1), it is quite inert chemically because of the formation on its surfaces of chromium(III)

TABLE 29.1   Electron Configuration and Other Properties of the First-Row Transition Metals

	Sc	Ti	V	Cr	Mn	Fe	Co	Ni	Cu
Electron configuration									
M	$3d^14s^2$	$3d^24s^2$	$3d^34s^2$	$3d^54s^1$	$3d^54s^2$	$3d^64s^2$	$3d^74s^2$	$3d^84s^2$	$3d^{10}4s^1$
M^{2+}	—	$3d^2$	$3d^3$	$3d^4$	$3d^5$	$3d^6$	$3d^7$	$3d^8$	$3d^9$
M^{3+}	[Ne]	$3d^1$	$3d^2$	$3d^3$	$3d^4$	$3d^5$	$3d^6$	$3d^7$	$3d^8$
Electronegativity	1.3	1.5	1.6	1.6	1.5	1.8	1.9	1.9	1.9
Ionization energy (kJ/mol)									
First	631	658	650	652	717	759	760	736	745
Second	1235	1309	1413	1591	1509	1561	1645	1751	1958
Third	2389	2650	2828	2986	3250	2956	3231	3393	3578
Radius (pm)									
M	162	147	134	130	135	126	125	124	128
M^{2+}	—	90	88	85	80	77	75	69	72
M^{3+}	81	77	74	64	66	60	64	—	—

TABLE 29.2 Physical Properties of Elements K to Zn

	1A	2A	Transition Metals									2B
	K	Ca	Sc	Ti	V	Cr	Mn	Fe	Co	Ni	Cu	Zn
Atomic radius (pm)	235	197	162	147	134	130	135	126	125	124	128	138
Melting point (°C)	63.7	838	1539	1668	1900	1875	1245	1536	1495	1453	1083	419.5
Boiling point (°C)	760	1440	2730	3260	3450	2665	2150	3000	2900	2730	2595	906
Density (g/cm³)	0.86	1.54	3.0	4.51	6.1	7.19	7.43	7.86	8.9	8.9	8.96	7.14

oxide, Cr_2O_3. Consequently, chromium is commonly used as a protective and noncorrosive plating on other metals.

The following are some characteristics of the transition metals.

- *General physical properties*
  Most of the transition metals have a close-packed structure (see Figure 14.30) in which each atom has a coordination number of 12. Furthermore, these elements have relatively small atomic radii. The combined effect of closest packing and small atomic size results in strong metallic bonds. Therefore the transition metals have higher densities, higher melting points and boiling points, and higher heats of fusion and vaporization than the Group 1A and 2A metals, as well as the Group 2B metals (Table 29.2).

- *Variable oxidation states*
  The first-row transition elements possess electrons in the $3d$ and $4s$ subshells, whose energy levels are quite similar. Consequently, a particular element can form ions of roughly unchanged stability by losing various numbers of electrons. Figure 29.1 shows the oxidation states of the elements from scandium to copper. Note that the common oxidation states for each element include $+2$, $+3$, or both. The $+3$ oxidation states are more stable at the beginning of the series, whereas the $+2$ oxidation states are more stable toward the end. The highest

Sc	Ti	V	Cr	Mn	Fe	Co	Ni	Cu
				+7				
			+6	+6	+6			
		+5	+5	+5	+5			
	+4	+4	+4	+4	+4	+4		
+3	+3	+3	+3	+3	+3	+3	+3	+3
	+2	+2	+2	+2	+2	+2	+2	+2
								+1

FIGURE 29.1 *Oxidation states of the first-row transition metals. The most stable oxidation numbers are shown in color. The zero oxidation state is encountered in some compounds, such as $Ni(CO)_4$ and $Fe(CO)_5$.*

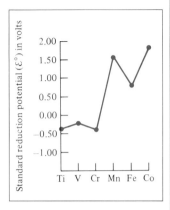

FIGURE 29.2 *Standard reduction potentials for the half reaction* $M^{3+} + e^- \rightarrow M^{2+}$ *for several transition metals.*

See Figure 22.21 for a listing of transition metals as catalysts.

oxidation state is $+7$, for manganese. Transition metals usually exhibit their highest oxidation states in compounds with very electronegative elements such as oxygen, fluorine, and chlorine; for example, V_2O_5, CrO_3, and Mn_2O_7.

The relative stability of the $+2$ and $+3$ oxidation states is indicated by the standard reduction potential $\mathscr{E}°$ for the half-cell reaction

$$M^{3+}(aq) + e^- \longrightarrow M^{2+}(aq)$$

in which M stands for a transition metal. The more positive the $\mathscr{E}°$ value, the more likely the reduction will proceed to the right, as shown, and the more stable is the $+2$ oxidation state compared to the $+3$ state. Figure 29.2 shows how $\mathscr{E}°$ varies for several metals. Manganese has a large positive $\mathscr{E}°$, which means that Mn^{2+} is more stable than Mn^{3+}. The stability of Mn^{2+} is the result of its half-filled subshell $3d^5$. Likewise the electron configuration of Fe^{3+} is $3d^5$, so Fe^{3+} should be more stable than Fe^{2+} (which has the electron configuration of $3d^6$). This is so, except that the difference in stability between Fe^{2+} and Fe^{3+} is not nearly as marked as that between Mn^{2+} and Mn^{3+}. The plot in Figure 29.2 shows also that Co^{2+} is more stable than Co^{3+}.

- *Color*
  Many solutions of transition metal ions, complex ions, and anions containing transition metals are distinctively colored (see Color Plates 19 and 21).
- *Magnetic properties*
  Many transition metal compounds are paramagnetic.
- *Complex ion formation*
  Transition metals have a pronounced tendency to form complex ions (see Section 21.5), such as $[Cu(NH_3)_4]^{2+}$, $[Co(NH_3)_6]^{3+}$, $[Ni(CN)_4]^{2-}$, and $[Fe(CN)_6]^{3-}$.
- *Catalytic properties*
  Many of the transition metals and their compounds are good catalysts for both inorganic and organic reactions. We have already described the use of iron as a catalyst in the Haber synthesis of ammonia (p. 671), the use of vanadium oxide (V_2O_5) in the production of sulfuric acid (p. 817), and the use of platinum to catalyze hydrogenation reactions (p. 784). Metals such as Cu and Ni as well as some of their compounds are also used as catalysts in many reactions.

29.2 COORDINATION COMPOUNDS

Recall that a complex ion contains a central metal bonded to one or more ions or molecules.

We have noted that transition metals have a distinct tendency to form complex ions. *A neutral species containing a complex ion* is called a **coordination compound.** Coordination compounds usually have rather complicated formulas in which we enclose the complex ion in brackets. Thus $[Ni(NH_3)_4]^{2+}$ is a complex ion and $[Ni(NH_3)_4]Cl_2$ is a coordination compound. Most, but not all, of the metals in coordination compounds are transition metals.

The molecules or ions that surround the metal in a complex ion are called **ligands.** The interactions between a metal atom and the ligands can be thought of as Lewis acid-base reactions. As we saw in Section 18.8, a Lewis base is a substance capable of donating one or more electron pairs. Every ligand has at least one unshared pair of valence electrons, and here are four examples:

$$\overset{\cdot\cdot}{\underset{H}{O}}\overset{\cdot\cdot}{\underset{H}{}} \qquad H\overset{\overset{\cdot\cdot}{N}}{\underset{H}{}}H \qquad :\overset{\cdot\cdot}{\underset{\cdot\cdot}{Cl}}:^{-} \qquad :C{\equiv}O:$$

Therefore, ligands play the role of Lewis bases. On the other hand, a transition metal atom (either in its neutral or positively charged state) acts as a Lewis acid, accepting (and sharing) pairs of electrons from the Lewis bases. Thus the metal–ligand bonds are usually coordinate covalent bonds (see Section 11.4).

The atom in a ligand that is bound directly to the metal atom is known as the **donor atom.** For example, nitrogen is the donor atom in the $[Cu(NH_3)_4]^{2+}$ complex ion. The **coordination number** in coordination compounds is defined as the *number of donor atoms surrounding the central metal atom in a complex ion.* Thus the coordination number of Ag^+ in $[Ag(NH_3)_2]^+$ is 2, that of Cu^{2+} in $[Cu(NH_3)_4]^{2+}$ is 4, and that of Fe^{3+} in $[Fe(CN)_6]^{3-}$ is 6. The most common coordination numbers are 2, 4, and 6, but many complex ions have other coordination numbers.

Depending on the number of donor atoms present in the molecule or ion, ligands can be classified as *monodentate, bidentate,* or *polydentate* (Table 29.3). Ligands such as H_2O and NH_3 are monodentate because there is only one donor atom per ligand. A common bidentate ligand is ethylenediamine:

$$H_2\overset{\cdot\cdot}{N}-CH_2-CH_2-\overset{\cdot\cdot}{N}H_2$$

The two nitrogen atoms can coordinate with a metal atom as shown in Figure 29.3.

Ethylenediaminetetraacetate ion (EDTA) is a polydentate ligand (see Table 29.3) containing six donor atoms—two nitrogen atoms and four oxygen atoms. The four oxygen atoms are in the four —COO⁻ groups that are single bonded to the carbon atoms. *Bi- and polydentate ligands* are also called *chelating agents* because of *their ability to hold the metal atom like a claw* (from the Greek word *chele,* which means "claw").

Oxidation Number of Metals in Coordination Compounds

Another important property that characterizes coordination compounds is the oxidation number of the central metal atom. The net charge of a complex ion is the sum of the charges on the central metal atom and its surrounding ligands. In the $[PtCl_6]^{2-}$ ion, for example, each chloride ion has an oxidation number of -1, so the oxidation number of Pt must be

In a crystal lattice, the *coordination number* of an atom (or ion) is defined as the number of atoms (or ions) surrounding the atom (or ion).

Ethylenediamine is sometimes abreviated *en*; for example, the formula $[Pt(en)_3]^{4+}$ means that three ethylenediamine ligands are coordinated with the Pt(IV) ion.

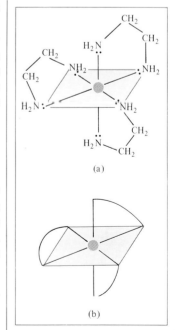

FIGURE 29.3 *(a) Structure of metal-ethylenediamine complex. Each ethylenediamine molecule provides two N donor atoms and is therefore a bidentate ligand. (b) Simplified structure of the same complex.*

TABLE 29.3 Some Common Ligands

	Name	Structure
Monodentate ligands	Ammonia	$H-\ddot{N}-H$ $\quad\quad\;\; \underset{H}{\mid}$
	Pyridine	(pyridine ring structure) $:N$
	Carbon monoxide	$:C\equiv O:$
	Chloride ion	$:\ddot{\underset{..}{Cl}}:^{-}$
	Cyanide ion	$[:C\equiv N:]^{-}$
	Thiocyanate ion	$[:\ddot{\underset{..}{S}}-C\equiv N:]^{-}$
	Triphenylphosphine	$H_5C_6-\overset{..}{P}-C_6H_5$ $\quad\quad\;\; \underset{C_6H_5}{\mid}$
Bidentate ligands	Ethylenediamine	$H_2\ddot{N}-CH_2-CH_2-\ddot{N}H_2$
	Acetylacetonate ion	$\left[H_3C-\overset{\displaystyle :O:}{\overset{\|}{C}}-\underset{\underset{H}{\mid}}{C}=\overset{\displaystyle :\ddot{O}:}{\overset{\|}{C}}-CH_3 \right]^{-}$
	Oxalate ion	$\left[\begin{matrix} :\!\overset{\cdot}{O}\!: & & & :\!\overset{\cdot}{O}\!. \\ & \diagdown & \diagup \\ & C & -C \\ & \diagup & \diagdown \\ :\!\overset{\cdot\cdot}{\underset{..}{O}}\!: & & & :\!\overset{\cdot\cdot}{\underset{..}{O}}\!: \end{matrix} \right]^{2-}$
Polydendate ligands	Ethylenediaminetetraacetate ion (EDTA)	(EDTA structure) $^{4-}$
	Tripolyphosphate ion	$\left[:\ddot{O}-\overset{\displaystyle :O:}{\underset{\displaystyle :\underset{..}{O}:}{\overset{\|}{P}}}-\ddot{O}-\overset{\displaystyle :O:}{\underset{\displaystyle :\underset{..}{O}:}{\overset{\|}{P}}}-\ddot{O}-\overset{\displaystyle :O:}{\underset{\displaystyle :\underset{..}{O}:}{\overset{\|}{P}}}-\ddot{O}: \right]^{5-}$

+4. If the ligands do not bear net charges, the oxidation number of the metal is equal to the charge of the complex ion. Thus, in $[Cu(NH_3)_4]^{2+}$ each NH_3 is neutral, so the oxidation number of Cu is $+2$.

EXAMPLE 29.1

Find the oxidation number of the central metal atom in each of the following compounds: (a) $[Ru(NH_3)_5(H_2O)]Cl_2$, (b) $[Cr(NH_3)_6](NO_3)_3$, (c) $[Fe(CO)_5]$, (d) $K_4[Fe(CN)_6]$.

Answer

(a) Both NH_3 and H_2O are neutral species. Since each chloride ion carries a -1 charge, and there are two Cl^- ions, the oxidation number of Ru must be $+2$.
(b) Each nitrate ion has a charge of -1; therefore the cation must be $[Cr(NH_3)_6]^{3+}$. NH_3 is neutral, so the oxidation number of Cr is $+3$.
(c) Since the CO species are neutral, the oxidation number of Fe is zero.
(d) Each potassium ion has a charge of $+1$; therefore, the anion is $[Fe(CN)_6]^{4-}$. Next, we know that each cyanide group bears a charge of -1, so Fe must have an oxidation number of $+2$.

Similar example: Problem 29.10.

Naming of Coordination Compounds

Having discussed the various types of ligands and the oxidation number of metals, our next step is to learn what to call these coordination compounds. The rules for naming coordination compounds are as follows:

- The cation is named before the anion. This is the same as naming other ionic compounds, and the rule holds regardless of whether the complex ion bears a net positive or a negative charge. For example, in referring to $K_3[Fe(CN)_6]$ and $[Co(NH_3)_4Cl_2]Cl$, we name the K^+ and $[Co(NH_3)_4Cl_2]^+$ cations first, respectively.
- Within a complex ion the ligands are named first, in alphabetical order, and the metal ion is named last.
- The names of anionic ligands end with the letter o, whereas a neutral ligand is usually called by the name of the molecule. The exceptions are H_2O (aquo), CO (carbonyl), and NH_3 (ammine). Table 29.4 lists some common ligands.
- When several ligands of a particular kind are present, we use the Greek prefixes *di-*, *tri-*, *tetra-*, *penta-*, and *hexa-* to name them. Thus the ligands in $[Co(NH_3)_4Cl_2]^+$ are "tetraamminedichloro." If the ligand itself contains a Greek prefix, we use the prefixes *bis* (2), *tris* (3), *tetrakis* (4) to indicate the number of ligands present. For example, the

TABLE 29.4 Names of Common Ligands in Coordination Compounds

Ligand	Name of Ligand in Coordination Compound
Bromide, Br^-	Bromo
Chloride, Cl^-	Chloro
Cyanide, CN^-	Cyano
Hydroxide, OH^-	Hydroxo
Oxide, O^{2-}	Oxo
Carbonate, CO_3^{2-}	Carbonato
Nitrite, NO_2^-	Nitro
Oxalate, $C_2O_4^{2-}$	Oxalato
Ammonia, NH_3	Ammine
Carbon monoxide, CO	Carbonyl
Water, H_2O	Aquo
Ethylenediamine	Ethylenediamine
Ethylenediaminetetraacetate	Ethylenediaminetetraacetato

ligand ethylenediamine already contains *di*; therefore, if two such ligands are present the name is *bis(ethylenediamine)*.

- The oxidation number of the metal is written in Roman numerals following the name of the metal. For example, the Roman numeral III is used to indicate the $+3$ oxidation state of chromium in $[Cr(NH_3)_4Cl_2]^+$, which is called tetraamminedichlorochromium(III) ion.
- If the complex is an anion, its name ends in *-ate*. For example, in $K_4[Fe(CN)_6]$ the anion $[Fe(CN)_6]^{4-}$ is called hexacyanoferrate(II) ion. Note that the Roman numeral II indicates the oxidation state of iron. Table 29.5 gives the names of anions containing metal atoms.

TABLE 29.5 Names of Anions Containing Metal Atoms

Metal	Metal as Named in Anionic Complex
Aluminum	Aluminate
Chromium	Chromate
Cobalt	Cobaltate
Copper	Cuprate
Gold	Aurate
Iron	Ferrate
Lead	Plumbate
Manganese	Manganate
Molybdenum	Molybdate
Nickel	Nickelate
Silver	Argentate
Tin	Stannate
Tungsten	Tungstate
Zinc	Zincate

EXAMPLE 29.2

Give the systematic names of the following compounds: (a) $Ni(CO)_4$, (b) $[Co(NH_3)_4Cl_2]Cl$, (c) $K_3[Fe(CN)_6]$, (d) $[Cr(en)_3]Cl_3$.

Answer

(a) The CO ligands are neutral species and the nickel atom bears no net charge, so the compound is called *tetracarbonylnickel(0)*, or more commonly, *nickel tetracarbonyl*.

(b) Starting with the cation, each of the two chloride ligands bears a negative charge and the ammonia molecules are neutral. Thus the cobalt atom must have an oxidation number of $+3$ (to balance the chloride anion). The compound is called *tetraamminedichlorocobalt(III) chloride*.

(c) The complex ion is the anion and it bears three negative charges. Thus the iron atom must have an oxidation number of $+3$. The compound is *potassium hexacyanoferrate(III)*. This compound is commonly called *potassium ferricyanide*.

(d) As we saw earlier, *en* is the abbreviation for the ligand ethylenediamine. Since there are three en groups present and the name of the ligand already contains *di*, the name of the compound is *tris(ethylenediamine) chromium(III) chloride*.

Similar example: Problem 29.12.

EXAMPLE 29.3

Write the formulas for the following compounds: (a) pentaamminechlorocobalt(III) chloride, (b) dichlorobis(ethylenediamine)platinum(IV) nitrate, (c) sodium hexanitrocobaltate(III).

Answer

(a) The complex cation contains five NH_3 groups, a chloride ion, and a cobalt ion with a $+3$ oxidation number. The net charge on the cation must be $2+$. Therefore, the formula for the compound is $[Co(NH_3)_5Cl]Cl_2$.

(b) There are two chloride ions, two ethylenediamine groups, and a platinum ion with an oxidation number of $+4$ in the complex cation. Therefore, the formula for the compound is $[Pt(en)_2Cl_2](NO_3)_2$.

(c) The complex anion contains six nitro groups and a cobalt ion with an oxidation number of $+3$. Therefore the formula for the compound is $Na_3[Co(NO_2)_6]$.

Similar example: Problem 29.11.

29.3 STEREOCHEMISTRY OF COORDINATION COMPOUNDS

The metal atom and the ligands that make up a complex ion have definite spatial relationships. Figure 29.4 shows four different geometric arrange-

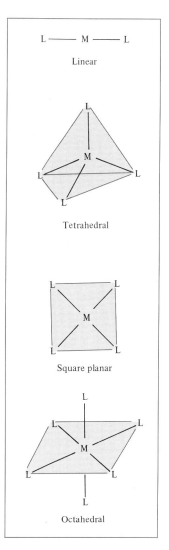

FIGURE 29.4 *Common geometries of complex ions. In each case M is a metal and L is a monodentate ligand.*

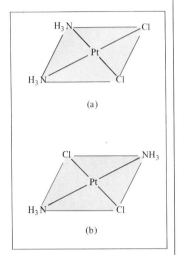

FIGURE 29.5 *The (a) cis and (b) trans isomers of diamminedichloroplatinum (II). Note that the two Cl atoms (and the two NH₃ molecules) are adjacent to each other in the cis isomer and diagonally across from each other in the trans isomer.*

FIGURE 29.6 *The (a) cis and (b) trans isomers of tetraaminedichlorocobalt (III) ion. The structure shown in (c) can be generated by rotating that in (a), and the structure shown in (d) can be generated by rotating that in (b). Therefore the ion has only two geometric isomers, (a) and (b) or (c) and (d).*

ments for metal atoms with monodentate ligands. In these diagrams we see that structure and coordination number of the metal atom relate to each other as follows:

Coordination Number	Structure
2	Linear
4	Tetrahedral or square planar
6	Octahedral

Stereoisomerism

In studying the geometry of coordination compounds, we often find that there is more than one way to arrange ligands around the central atom, thereby generating molecules that have distinctly different physical and chemical properties. The term **stereoisomerism** describes the *occurrence of two or more compounds with the same types and numbers of atoms and the same chemical bonds but different spatial arrangements.* **Stereoisomers** *are compounds possessing the same formula and bonding arrangement but different spatial arrangements of atoms.* There are two types of stereoisomers: geometric isomers and optical isomers. We will look into the importance of these stereoisomers in coordination compounds.

Geometric Isomers. *Geometric isomers are compounds with the same type and number of atoms and the same chemical bonds but different spatial arrangements; such isomers cannot be interconverted without breaking a chemical bond.* We use the terms *cis* and *trans* to distinguish one geometric isomer from another of the same compound: *Cis* means that two particular atoms (or groups of atoms) are adjacent to each other, and *trans* means that the atoms (or groups of atoms) are on opposite sides in the structural formula. *Cis* and *trans* isomers of coordination compounds generally have quite different colors, melting points, dipole moments, and chemical reactivities. Figure 29.5 shows the *cis* and *trans* isomers of diamminedichloroplatinum(II). Note that although the types of bonds are the same in both isomers (two Pt—N and two Pt—Cl bonds), the spatial arrangements are different. Another example is tetraaminedichlorocobalt(III) ion, shown in Figure 29.6.

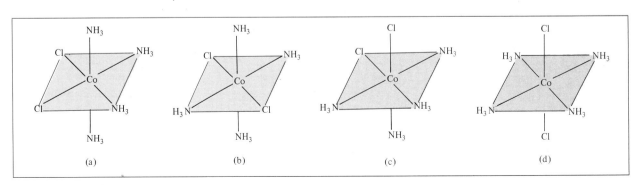

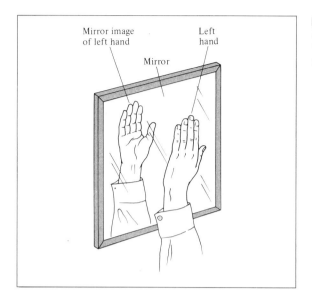

FIGURE 29.7 *A left hand and its mirror image, which looks the same as the right hand.*

Optical Isomers. *Optical isomers* are *compounds that are nonsuperimposable mirror images.* Like geometric isomers, optical isomers of a compound have the same type and number of atoms and the same chemical bonds. Optical isomers differ from geometric isomers, however, in that the former have *identical* physical and chemical properties. The structural relationship between two optical isomers is analogous to the relationship between your left and right hands. If you place your left hand in front of a mirror, you will see that the image of your left hand in the mirror looks the same as your right hand (Figure 29.7); thus we say that your left and right hands are mirror images of each other. However, they are nonsuperimposable, because when you place your left hand over your right hand (with both palms facing down), the hands do not match.

Figure 29.8 shows the *cis* and *trans* isomers of dichlorobis(ethylenediamine)cobalt(III) ion and their images. Careful examination shows that the *trans* isomer and its mirror image are superimposable, but the same

By *superimposable*, we mean that we can place one ion or molecule over the other and be able to match in both the positions of all atoms and groups of atoms.

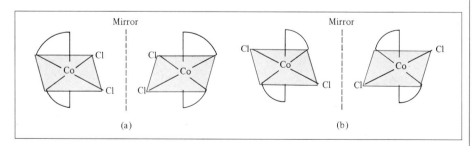

FIGURE 29.8 *The (a) cis and (b) trans isomers of dichlorobis(ethylenediamine)cobalt(III) ion and their mirror images. If you rotate the mirror image of the trans isomer 90° counterclockwise about the axial position and place the ion over the trans isomer, you will find that the two are superimposable. The same does not hold for the cis isomer and its mirror image.*

does not hold for the *cis* isomer and its mirror image. Therefore, the *cis* isomer and its mirror image are optical isomers. *Compounds or ions that are not superimposable with their mirror images* are said to be optically active, or **chiral;** if they are superimposable with their mirror images they are said to be optically inactive, or *achiral.* Thus the *cis* isomer shown in Figure 29.8 is chiral but the *trans* isomer is achiral. *Optical isomers, that is, compounds and their nonsuperimposable mirror images,* are also called **enantiomers.**

Enantiomers differ with respect to their interaction with **plane-polarized light,** that is, *light in which the electric field and magnetic field components are confined to specific planes.* In Section 8.1 we saw that the properties of light can be understood in terms of an oscillating electric field and a magnetic field component. Only the former is important in our discussion here, so let us examine the variation of this component along a given direction, as shown in Figure 29.9. In ordinary or unpolarized light, the waves are not confined to one plane; rather, they vibrate in all planes, so the variation in the amplitude of the wave is represented by Figure 29.9(c). A simple way to produce polarized light is to pass a beam of ordinary light through a Polaroid sheet, which transmits light vibrating only in a particular plane (Figure 29.10).

The difference between a chiral molecule and an achiral molecule is that only the former can rotate the plane of polarization of plane-polarized light when the light is passed through the substance. If this plane is rotated to the right, the substance is *dextrorotatory* (*d*); to the left, *levorotatory* (*l*). (The *d*- and *l*-isomers of the same compound are the two enantiomers.) The

Polaroid sheets are used to make Polaroid sunglasses.

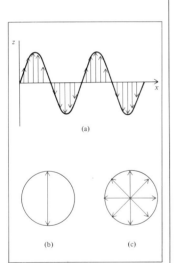

(a)

(b)   (c)

FIGURE 29.9 *(a) The propagation of an electromagnetic wave along the x axis. Only the electric field component is shown. (b) End view (looking along the x axis toward the source of radiation) of the vibration of the electric field component in one plane. (c) End view of the electric field component vibrating in many planes. The double arrows represent the maximum of the amplitude of the wave in the positive and negative directions.*

FIGURE 29.10 *Left: Light passes through a pair of Polaroid sunglasses. Middle: When the axis of polarization in one pair of Polaroid sunglasses is parallel to that in another pair, light passes through both pairs of sunglasses. Right: If the axes of polarization are perpendicular to each other, little or no light passes through both pairs of sunglasses.*

two enantiomers of the same chiral substance always rotate the light by the same amount, but in opposite directions. Thus, in an *equimolar mixture of the two optical isomers,* called a **racemic mixture,** the net rotation is zero.

The *instrument for studying interaction between plane-polarized light and chiral molecules* is known as a **polarimeter** (Figure 29.11). Initially, a beam of unpolarized light is passed through a Polaroid sheet, called the *polarizer,* then through a sample tube. By rotating the *analyzer,* another Polaroid sheet, minimal light transmission can be achieved, as is shown in Figure 29.10. Next, the sample tube is filled with a solution containing an optically active compound. As the plane-polarized light passes through the sample tube this time, its plane of polarization is rotated either to the right or to the left by a certain amount, depending on whether the optical isomer is in the *d-* or *l-*form. This rotation can be measured readily by turning the

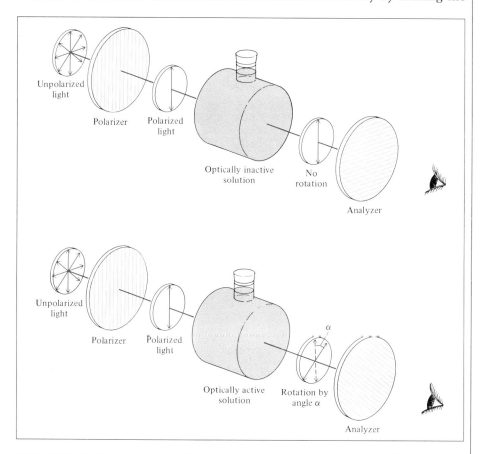

FIGURE 29.11 *An optical rotation experiment. Top: Two Polaroid sheets aligned for minimum light transmission. The achiral solution does not rotate the plane of the polarized light. Bottom: In a chiral medium, the plane of the polarized light is rotated by a certain angle* α *so some light does pass through. To measure this angle, the analyzer is rotated by the same amount (but in the opposite direction) to again achieve minimum light transmission. A compound that rotates the plane of polarized light to the right as one looks toward the light source is called dextrorotatory (d); to the left, levorotatory, (l).*

analyzer in the appropriate direction until minimal light transmission is again achieved. The angle of rotation in degrees, α, depends not only on the nature of the molecules, but also on the concentration of the solution and the length of the sample tube.

29.4   BONDING IN COORDINATION COMPOUNDS

Our understanding of the nature of coordination compounds is the result of the classic work of Alfred Werner (1866–1919). In 1893, at the age of 26, he proposed what is now commonly referred to as *Werner's coordination theory*.

The chemists of that day had been puzzled by a certain class of reactions. For example, it was known that the valences of the elements in cobalt(III) chloride and in ammonia seem to be completely satisfied. Yet these two substances react to form a stable compound having the formula $CoCl_3 \cdot 6NH_3$. To explain this behavior, Werner postulated that most elements exhibit two types of valence: *primary valence* and *secondary valence*, which, in modern terminology, correspond to the oxidation number and coordination number of the element. In $CoCl_3 \cdot 6NH_3$, then, cobalt has a primary valence of three and a secondary valence of six.

Nowadays we write the formula of the compound as $[Co(NH_3)_6]Cl_3$ to indicate that the ammonia molecules and the cobalt atom form a complex ion; the chloride ions are not part of the complex, but are held to it by ionic forces.

Werner also laid the basis for the stereochemistry of coordination compounds. During his relatively short scientific career, Werner synthesized literally hundreds of coordination compounds whose structures and properties were positive proof of his theory.

A satisfactory theory of bonding for coordination compounds must account for properties such as color and magnetism, as well as stereochemistry and bond strengths. No single theory as yet does all of this for us. Rather, several different approaches have been applied to transition metal complexes. We will consider only one of them here—the crystal field theory—because it is straightforward and it accounts satisfactorily for the colors and magnetic properties of many coordination compounds.

Crystal Field Theory

Crystal field theory considers the bonding in complex ions purely in terms of electrostatic interactions between the metal atom and the ligands. We will begin by considering octahedral complex ions.

The first question we will address is: What effect will the surrounding ligands have on the energies of the metal atom's d orbitals? As we observed in Chapter 8, d orbitals have various orientations, but in the absence of external disturbance they all have the same energy (see Figure 8.17). When such a metal ion is in the center of an octahedron surrounded by six lone pairs of electrons (on the six ligands), two types of electrostatic interaction come into play. First, there is the attraction between the lone pairs and the positive metal ion. This is the force that holds the ligands to the metal in the complex. In addition, there is the electrostatic repulsion between the lone pairs on the ligands and the electrons in the metal's d orbitals. However, the magnitude of this repulsion depends on the particular d orbital that is involved. Take the $d_{x^2-y^2}$ orbital as an example. We see that this orbital has its lobes pointing along the x and y axes, where the lone pair electrons are positioned (Figure 29.12). Thus, an electron residing in this orbital would experience a greater repulsion from the ligands than an electron would in, say, the d_{xy} orbital. For this reason, electrons in the $d_{x^2-y^2}$ orbital experience increased energy (they become less stable), whereas electrons in the d_{xy}, d_{yz}, and d_{xz} orbitals lose energy. The d_{z^2} orbital energy is increased, because its lobes are pointed at the ligands along the z axis.

As a result of these metal–ligand interactions, the equality (in energy) of the five d orbitals is nullified, to give two high-lying equal-energy levels ($d_{x^2-y^2}$ and d_{z^2}) and three low-lying equal-energy levels (d_{xy}, d_{yz}, and d_{xz}), shown in Figure 29.13. *The energy difference between these two sets of d orbitals* is called the **crystal field splitting** (Δ); its magnitude depends on

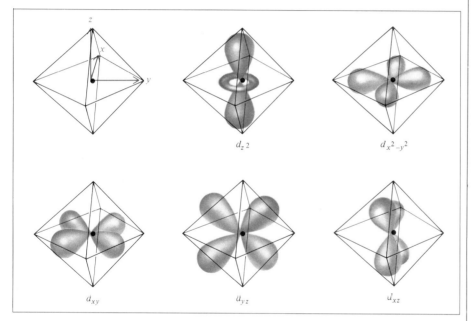

FIGURE 29.12 *The five d orbitals in an octahedral environment. The metal atom (or ion) is at the center of the octahedron, and the six lone pairs on the donor atoms of the ligands are at the corners.*

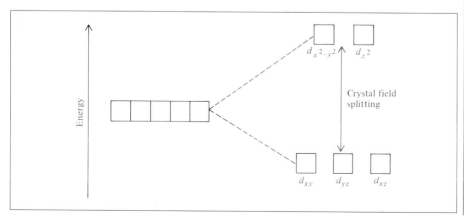

FIGURE 29.13 *Crystal field splitting between d orbitals in an octahedral complex.*

TABLE 29.6   Relationship of Wavelength to Color

Wavelength Absorbed (nm)	Color Observed
400 (violet)	Greenish yellow (570 nm)
450 (blue)	Yellow (600 nm)
490 (blue-green)	Red (700 nm)
570 (yellow-green)	Violet (400 nm)
580 (yellow)	Dark blue (420 nm)
600 (orange)	Blue (450 nm)
650 (red)	Green (510 nm)

the metal and the nature of the ligands. With this basic energy-level scheme in mind, we are ready to discuss two characteristics of octahedral transition metal complex ions: color and magnetic properties.

Color.   A substance appears colored because it absorbs light at one or more wavelengths in the visible part of the electromagnetic spectrum (400 nm–700 nm) and reflects or transmits the others. Each wavelength of light in this region appears as a different color. A combination of all colors appears white (as in sunlight); an absence of lightwaves appears black. Table 29.6 shows the relationship of absorbed wavelength to the observed color. In this table we can see, for example, that the blue hydrated cupric ion $Cu(H_2O)_6^{2+}$ absorbs light in the orange region.

Figure 29.14 shows the quantum mechanical description of the absorption and emission of light. For the sake of simplicity, we consider two particular electron energy levels involved in a transition. Absorption of light may occur when the frequency of the incoming photon, multiplied by the Planck constant, is equal to the difference in energy between these two levels, that is (see Section 8.3)

$$\Delta E = h\nu$$

When we say that the hydrated cupric ion is blue, we mean that each ion is able to absorb a photon whose frequency is about 5×10^{14} Hz, which corresponds to a wavelength of about 600 nm (the orange light). When this component of light is removed, the transmitted light no longer looks white, but appears blue to our eyes. With this knowledge, we can calculate that the electron transition that occurs in the cupric ion must involve an energy change given by

$$\Delta E = (6.63 \times 10^{-34} \text{ J s})(5 \times 10^{14}/\text{s})$$
$$= 3 \times 10^{-19} \text{ J}$$

If the wavelength of the photon absorbed by a molecule lies outside of the visible region, then the transmitted light looks the same (to us) as the incident light—white—and the substance made up of these molecules appears colorless.

Spectroscopic techniques offer the best means for measuring crystal field splitting. The $[Ti(H_2O)_6]^{3+}$ ion provides a particularly simple example, be-

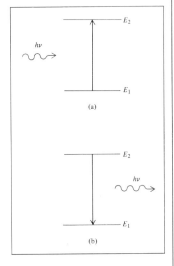

FIGURE 29.14   *(a) A molecule absorbs a photon with energy hν. (b) In the emission process, a molecule gives off a photon with energy hν. The frequency—and hence the wavelength—of the photon involved in either case depends on the difference between the two energy levels.*

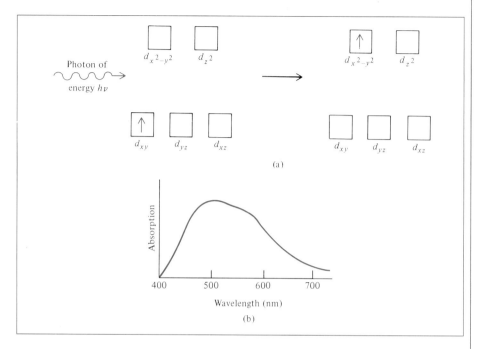

(a)

(b)

FIGURE 29.15 *(a) The absorptive process of a photon and (b) a graph of the absorption spectrum of the $Ti(H_2O)_6^{3+}$ ion. The energy of the incoming photon is equal to the crystal field splitting. The maximum absorption peak in the visible region occurs at 498 nm.*

cause Ti^{3+} has only one $3d$ electron (Figure 29.15). As Figure 29.15b shows, $[Ti(H_2O)_6]^{3+}$ absorbs light in the visible region, and the wavelength corresponding to maximum absorption is 498 nm. This enables us to calculate the crystal field splitting as follows. We start by writing

$$\Delta = h\nu$$

Also

$$\nu = \frac{c}{\lambda}$$

where c is the velocity of light and λ is the wavelength. Therefore

$$\Delta = \frac{hc}{\lambda} = \frac{(6.63 \times 10^{-34} \text{ J s})(3.00 \times 10^8 \text{ m/s})}{(498 \text{ nm})(1 \times 10^{-9} \text{ m/1 nm})}$$
$$= 3.99 \times 10^{-19} \text{ J}$$

This is the energy required to excite one $[Ti(H_2O)_6]^{3+}$ ion. To express this energy difference in the more convenient units of kilojoules per mole, we write

$$\Delta = (3.99 \times 10^{-19} \text{ J/ion})(6.02 \times 10^{23} \text{ ions/mol})$$
$$= 240,000 \text{ J/mol}$$
$$= 240 \text{ kJ/mol}$$

Working with a number of complexes, all having the same metal ion but different ligands, we can calculate the crystal field splitting for each ligand (aided by data from the corresponding absorption spectra), and thus establish a **spectrochemical series,** which is a *list of ligands arranged in order of their abilities to split the d orbital energies.*

The order in the spectrochemical series is the same regardless of what metal atom (or ion) is present.

$$I^- < Br^- < Cl^- < OH^- < F^- < H_2O < NH_3 < en < CN^- < CO$$

These ligands are arranged in the order of increasing value of Δ. Thus, CO and CN^- are called strong-field ligands, because they cause a large splitting of the d orbital energy levels. The halide ions and the hydroxide ion are weak-field ligands, because they split the d orbitals to a lesser extent.

Magnetic Properties. The magnitude of the crystal field splitting also determines the magnetic properties of a complex ion. For $[Ti(H_2O)_6]^{3+}$, the single d electron must be in one of the three lower orbitals, and the ion is always paramagnetic (see Section 8.9). However, in an ion where several d electrons are present, as in Fe^{3+} complexes, the situation becomes more involved. Consider the octahedral complexes $[FeF_6]^{3-}$ and $[Fe(CN)_6]^{3-}$ (Figure 29.16). The electron configuration of Fe^{3+} is $3d^5$, and there are two possibilities for placing the five d electrons in the five d orbitals. According to Hund's rule (see Section 8.9), maximum stability is reached when the electrons enter five separate orbitals with parallel spins. But this arrangement can be achieved only at a cost; two of the five electrons must be energetically promoted to the high-lying $d_{x^2-y^2}$ and d_{z^2} orbitals. No such energy investment is needed if all five electrons enter the d_{xy}, d_{xz}, and d_{yz} orbitals. According to Pauli's exclusion principle (p. 234), there will only be one unpaired electron present in this case.

The actual arrangement of the electrons is determined by the amount of stability gained by having maximum parallel spins versus the investment in energy required to promote electrons to higher d orbitals. Because F^- is a weak-field ligand, the five d electrons enter five separate d orbitals with parallel spins to create a *high-spin* complex. On the other hand, the cyanide ion is a strong-field ligand, so it is energetically preferable for all five electrons to be in the lower orbitals and so a *low-spin* complex is formed.

The actual number of unpaired electrons (or spins) in a complex ion can be found by magnetic measurements, and the general agreement between theory and experiment supports the usefulness of the crystal field theory. A distinction between low- and high-spin complexes can be made only if the metal ion contains more than three and fewer than eight d electrons. Figure 29.17 shows the distribution of electrons among d orbitals that results in low- and high-spin complexes.

The magnetic properties of a complex ion depend on the number of unpaired electrons present. High-spin complexes are more paramagnetic than low-spin complexes.

FIGURE 29.16 *Energy-level diagrams for the Fe^{3+} ion and the FeF_6^{3-} and $Fe(CN)_6^{3-}$ complex ions.*

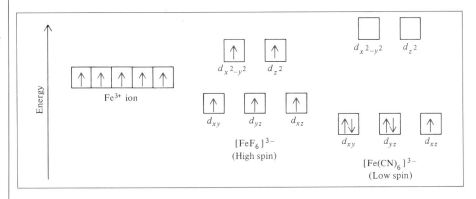

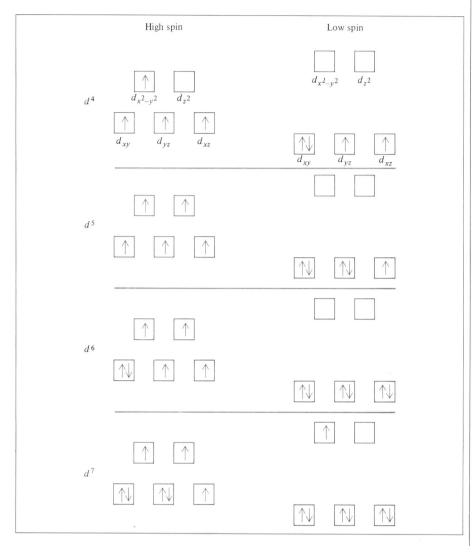

So far we have concentrated on octahedral complexes. The splitting of the d orbital energy levels in two other types of complexes—tetrahedral and square-planar—can also be accounted for satisfactorily by the crystal field theory. In fact, the splitting pattern for a tetrahedral ion is just the reverse of that for octahedral complexes. In this case, the d_{xy}, d_{xz}, and d_{yz} orbitals are more closely directed at the ligands and are therefore raised in energy as compared to the $d_{x^2-y^2}$ and d_{z^2} orbitals (Figure 29.18). Most tetrahedral complexes are high-spin complexes. Presumably, the tetrahedral arrangement reduces the magnitude of metal–ligand interactions, resulting in a smaller Δ value. This is a reasonable assumption since the number of ligands is smaller in a tetrahedral complex.

As Figure 29.19 shows, the splitting pattern for square-planar complexes is the most complicated. Clearly the $d_{x^2-y^2}$ orbital possesses the highest energy (as in the octahedral case), and the d_{xy} orbital the next highest, but

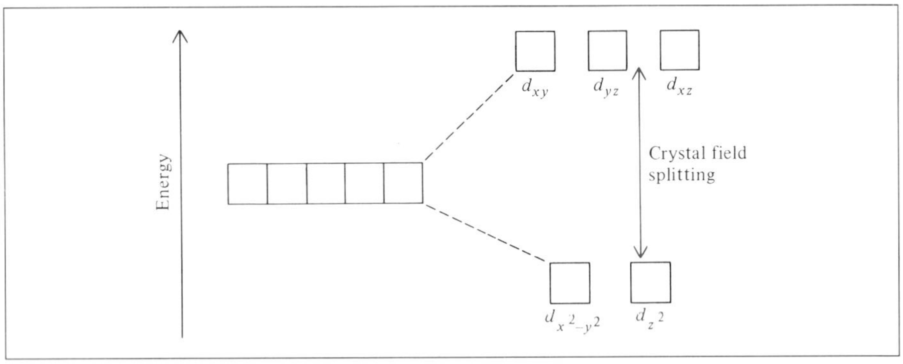

FIGURE 29.18 *Crystal field splitting between d orbitals in a tetrahedral complex.*

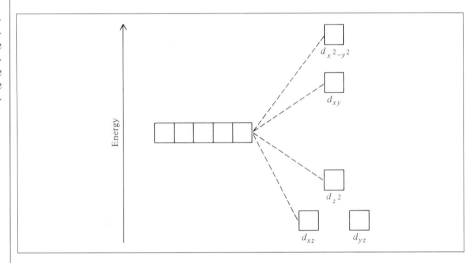

FIGURE 29.19 *Energy-level diagram for a square-planar complex. Because there are more than two energy levels, we cannot define crystal field splitting as we can for octahedral and tetrahedral complexes.*

the relative placement of the d_{z^2} and the d_{xz} and d_{yz} orbitals cannot be decided simply by inspection and must be calculated.

29.5 REACTIONS OF COORDINATION COMPOUNDS

Most complex ions undergo ligand exchange (or ligand substitution) reactions in solution. The rates of ligand exchange reactions vary widely, depending on the nature of the metal ion and the ligands. For example, the reaction between $[Cu(NH_3)_4]^{2+}$ and H^+ takes place rapidly:

$$[Cu(NH_3)_4]^{2+} + 4H^+ + 6H_2O \longrightarrow [Cu(H_2O)_6]^{2+} + 4NH_4^+$$

The $[Cu(NH_3)_4]^{2+}$ ion is thermodynamically less stable in an acid medium. $[Co(NH_3)_6]^{3+}$ ions are also thermodynamically unstable in acid solution. At equilibrium, they are almost completely converted to the hydrated ions as follows:

$$[Co(NH_3)_6]^{3+} + 6H^+ + 6H_2O \longrightarrow [Co(H_2O)_6]^{3+} + 6NH_4^+$$

The slow rate can be attributed to a high activation energy for the reaction.

However, the rate of this reaction is so slow that $[Co(NH_3)_6]^{3+}$ can be kept in acid solution for days at room temperature without noticeable change.

Complex ions such as $[Cu(NH_3)_4]^{2+}$ *that undergo rapid ligand exchange reactions* are said to be **labile complexes.** Those ions such as $[Co(NH_3)_6]^{3+}$ that undergo slow ligand exchange reactions are described as *inert.*

It is important to distinguish the terms *stability of a compound*, as defined by thermodynamics, and *lability of a compound*, as defined by kinetics. Referring to the preceding reaction we see that thermodynamically $[Co(H_2O)_6]^{3+}$ is much more stable than $[Co(NH_3)_6]^{3+}$ in solution. Yet the ligand exchange reaction is very slow for $[Co(NH_3)_6]^{3+}$ because of the high energy of activation (see Section 22.3). Thus an inert complex ion is not necessarily a stable species in the thermodynamic sense because given enough time most or all of it may be converted to the more stable form.

Most complex ions containing Co^{3+}, Cr^{3+}, and Pt^{4+} are kinetically inert. Because they exchange ligands so slowly, they are easier to measure, so our knowledge of the bonding, structure, and isomerism of coordination compounds has been obtained largely from studies of these species.

29.6 APPLICATIONS OF COORDINATION COMPOUNDS

We will now discuss the uses of coordination compounds in households, industry, medicine, and biological systems.

Water Treatment

As we saw in Section 15.9, tripolyphosphate ions and nitrilotriacetic acid (see Figure 15.21) are used to remove Ca^{2+} and Mg^{2+} ions in hard water. These chelating agents form stable, water-soluble complexes with the undesirable ions. EDTA (see Table 29.3) is even better for this purpose because it has a greater affinity for the alkaline earth metal ions.

Metallurgy

The extraction of silver and gold by the formation of cyanide complexes (p. 848) and the purification of nickel (p. 841) by converting the metal to $Ni(CO)_4$ are typical examples of the use of coordination compounds in metallurgical processes.

Dyes

A number of coordination compounds are used as dyes or pigments. A well-known example is phthalocyanine blue. This dark blue pigment, normally called Pigment Blue 15, is prepared by heating phthalonitrile

phthalonitrile

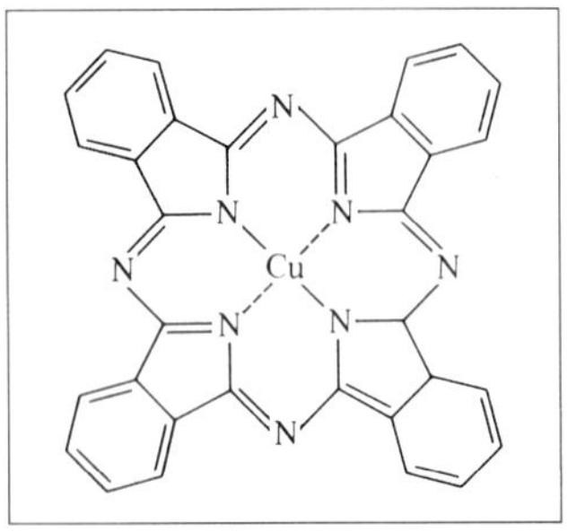

FIGURE 29.20 *Structure of copper phthalocyanine blue. The Cu^{2+} ion is at the center and the four N atoms are at the corners of a square planar complex. The dotted lines indicate the coordinate covalent bonds.*

with copper(I) chloride (CuCl). Phthalocyanine blue (see Figure 29.20) is used in paints, lacquers, papers, colored chalks, and pencils. The copper ion can be replaced by any of a number of other metal ions. Interestingly, the color of the compound is essentially the same, no matter what metal ion is present. This indicates, of course, that the color of the pigment is caused by the organic chelating group, rather than by the metal ion.

Chemical Analysis

Although EDTA has a great affinity for a large number of metal ions (especially 2+ and 3+ ions), other chelates are more selective in binding. For example, dimethylglyoxime

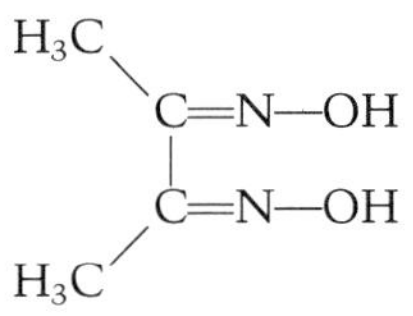

forms an insoluble brick-red solid with Ni^{2+} (Figure 29.21) and an insoluble bright yellow solid with Pd^{2+}. These characteristic colors are used in qualitative analysis to identify nickel and palladium. Further, the quantities of ions present can be determined by gravimetric analysis (see Section 4.5), as follows: To a solution containing Ni^{2+} ions, say, we add an excess of dimethylglyoxime reagent and a brick-red precipitate forms. The precipitate is then filtered, dried, and weighed. Knowing the formula of the complex (see Figure 29.21), we can readily calculate the amount of nickel present in the original solution.

Plant Growth

Plants need various nutrients for healthy growth. Essential nutrients include a number of metals such as iron, zinc, copper, manganese, boron, and molybdenum. Iron in the +3 state in the soil is mostly hydrolyzed to form insoluble iron hydroxides such as $Fe(OH)_3$ which cannot be taken up by plants. Plants deficient in iron are likely to develop a disorder known as

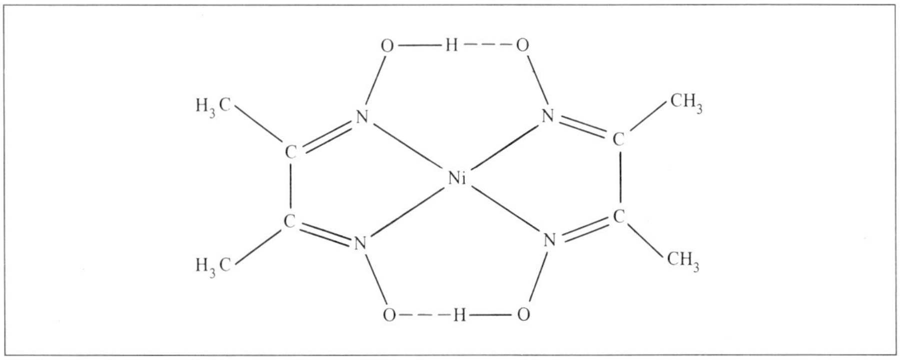

FIGURE 29.21 *Structure of nickel dimethylglyoxime. Note that the overall structure is stabilized by hydrogen bonds (color dotted lines).*

iron chlorosis, evidenced by yellowing leaves. Iron chlorosis particularly affects the yield of fruit from citrus trees. The standard treatment, frequently used in citrus groves, supplies the trees with Fe(III)–EDTA complex. This complex is soluble in water and readily enters the roots of trees where it is eventually converted into a usable form. Certain strains of soybeans growing in alkaline soil, which favors the formation of iron hydroxides, generate and secrete into the soil chelating agents that solubilize the iron needed for plant growth.

Therapeutic Chelating Agents

In a living system, the concentrations of metal ions and their complexes are controlled within rather narrow limits. If the physiological balance of these concentrations is disturbed by internal or external causes, disorders result and the organism may no longer behave normally. Chelating agents are often used to deal with irregularities resulting from an excess of metal ions in our bodies.

In Sections 28.4 and 28.5 we saw that both lead and mercury are toxic substances. Lead poisoning is usually treated with chelating agents such as EDTA and 2,3-dimercaptopropanol, which is more commonly called BAL (British Anti-Lewisite):

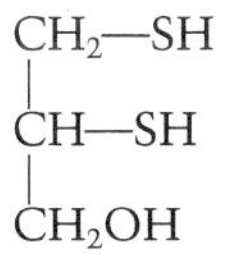

$$CH_2-SH$$
$$CH-SH$$
$$CH_2OH$$

BAL was developed during World War II as an antidote for lewisite, a poison gas containing arsenic. In ionized form, both EDTA and BAL form very stable complexes with Pb^{2+} ions (Figure 29.22). Mercury poisoning is usually treated with EDTA.

Wilson's disease, a liver disorder that affects the central nervous system, results when the amount of copper in tissues such as those of the liver,

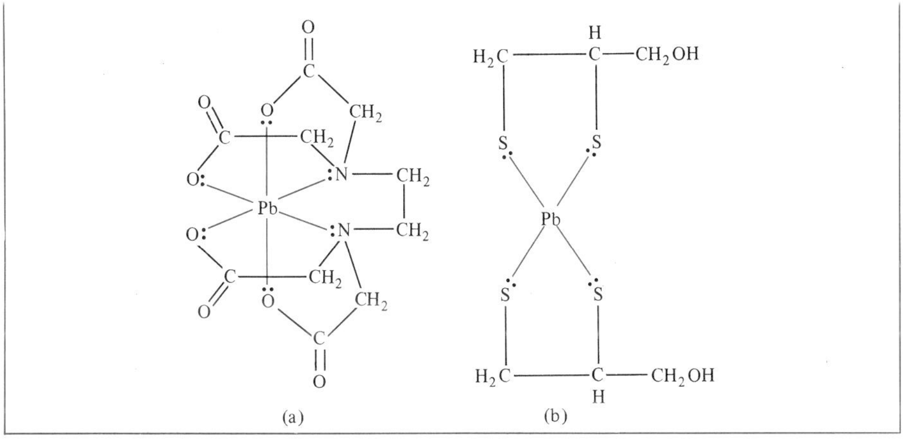

(a)   (b)

FIGURE 29.22 *(a) EDTA complex of lead. The complex ion bears a net charge of 2−, because each O donor atom has one negative charge (there are four of them) and the lead ion carries two positive charges. (b) BAL complex of lead. This complex bears one negative charge. The Pb^{2+} ion is tetrahedrally bonded to the four S atoms.*

brain, and kidneys is much greater than normal. The excess copper can be removed by the chelating agent D-penicillamine (also called 3-mercaptovaline):

$$H_3C-\underset{\underset{SH}{|}}{\overset{\overset{CH_3}{|}}{C}}-\underset{\underset{NH_2}{|}}{\overset{\overset{H}{|}}{C}}-\underset{OH}{\overset{\overset{O}{\|}}{C}}$$

This compound can coordinate with copper ions via sulfur, nitrogen, and oxygen donor atoms. It is also used to treat mercury and lead poisoning.

Recent studies have shown that some of the coordination complexes of platinum effectively inhibit the growth of cancerous cells. It has long been suspected that chelation processes are associated with the development of cancer, as well as with antitumor activity in the living cell. In fact, a compound that has antitumor activity in one environment is generally able to function as a carcinogen in another. Figure 29.23 shows some of the platinum(II) complexes that exhibit antitumor activity. The striking feature of all these complexes is the *cis* arrangement of identical ligands. *Trans*-diamminedichloroplatinum(II)

$$\underset{Cl}{\overset{H_3N}{\diagdown}}Pt\underset{NH_3}{\overset{Cl}{\diagup}}$$

and other *trans* isomers are not active as antitumor agents. Apparently, the antitumor activity is associated in some way with reactions involving chelation. In *cis*-diamminedichloroplatinum(II), the two chlorine atoms are more easily removed by another chelating agent than the ammonia ligands. Replacement of the chlorine atoms in the *trans* isomer by a chelating agent does not occur readily. The mechanism by which the platinum complexes act is not entirely clear.

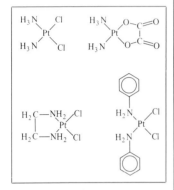

FIGURE 29.23 *Structures of some platinum complexes used in cancer treatment. Only the cis isomers are effective antitumor chelating agents.*

29.7 COORDINATION COMPOUNDS IN LIVING SYSTEMS

In this final section of the chapter we will discuss some of the important coordination compounds in living systems. In particular, we will concentrate on the iron–protein complexes.

Because of its central role as an oxygen carrier for metabolic processes, hemoglobin is probably the most studied of all the proteins. The molecule contains four folded long chains called *subunits*. The main function of hemoglobin is to carry oxygen in the blood from the lungs to the tissues, where it unloads the oxygen molecules to myoglobin. Myoglobin, which is made up of only one subunit, stores oxygen for metabolic processes in muscle.

Figure 29.24 shows the structure of the porphine molecule, which forms an important part of the hemoglobin structure. Upon coordination to a metal, the two H⁺ ions shown bonded to nitrogen atoms are displaced.

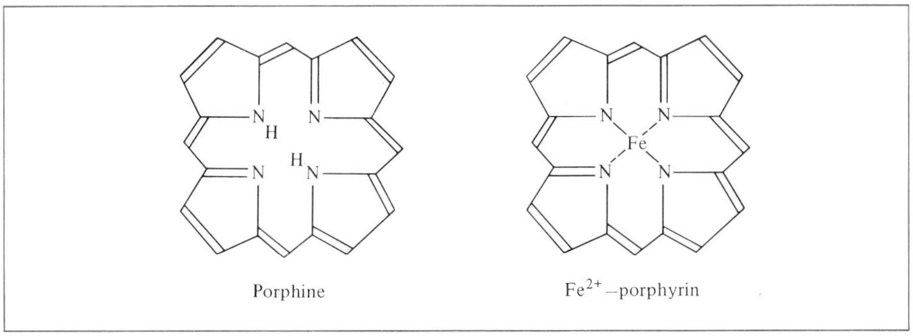

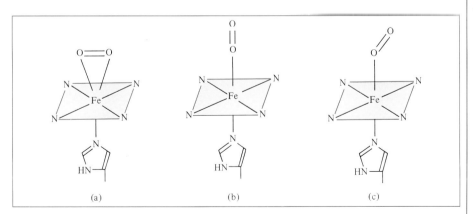

Porphine

Fe²⁺—porphyrin

FIGURE 29.24 *Structures of the porphine molecule and the Fe^{2+}-porphyrin complex. The dotted lines indicate coordinate covalent bonds.*

Complexes derived from porphine are called *porphyrins*, and the iron–porphyrin combination is called the *heme* group. The iron in the heme group has the oxidation number $+2$; it is coordinated to the four nitrogen atoms in the porphine group and also to a nitrogen donor atom in a ligand group which is attached to the protein (Figure 29.25). In the absence of oxygen, the sixth ligand is a water molecule, which binds to the Fe^{2+} ion on the other side of the ring to complete the octahedral complex. Under these conditions, the hemoglobin molecule is called *deoxyhemoglobin* and has a blue color characteristic of venous blood. The water ligand can be replaced readily by molecular oxygen to form the red-colored oxyhemoglobin present in arterial blood. Each subunit contains a heme group, so each hemoglobin molecule can bind up to four O_2 molecules.

For a number of years, the exact arrangement of the oxygen molecule relative to the porphyrin group was not clear. Figure 29.26 shows three possible arrangements in oxyhemoglobin. Figure 29.26(a) would necessitate a coordination number of seven, which is considered unlikely for Fe(II) complexes. Of the remaining two structures, the end-on arrangement shown in Figure 29.26(b) may seem more reasonable; however, evidence points to the structure in Figure 29.26(c) as the most plausible.

Unlike deoxyhemoglobin, oxyhemoglobin is diamagnetic. Because O_2 is a strong-field ligand, the crystal field splitting in oxyhemoglobin is quite

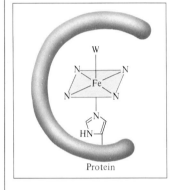

Protein

FIGURE 29.25 *The heme group in hemoglobin. The Fe^{2+} ion is coordinated with the nitrogen atoms of the heme group. The ligand below the porphyrin is the histidine group that is attached to the protein. The sixth ligand is a water molecule. The water molecule is denoted by W because the exact geometry of the bonding of H_2O to Fe^{2+} is not known.*

(a)   (b)   (c)

FIGURE 29.26   *Three possible ways for molecular oxygen to bind to the heme group in hemoglobin. The structure shown in (c) is the most likely arrangement.*

FIGURE 29.27 *Energy level diagrams for (a) paramagnetic deoxyhemoglobin and (b) diamagnetic oxyhemoglobin.*

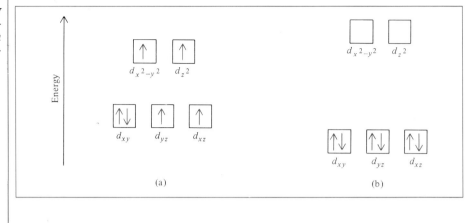

(a)   (b)

large. Consequently, the electron spins are paired, leading to the observed diamagnetism (Figure 29.27).

The porphyrin group is a very effective chelating agent and, not surprisingly, we find it in a number of biological systems. The iron–heme complex is present in another class of proteins called the *cytochromes*. Figure 29.28 shows the structure of cytochrome *c*, which is the most studied member of this class of compounds. The iron forms an octahedral complex, but because both the histidine and the methionine groups are firmly bound to the metal ion, these ligands cannot be displaced by oxygen or other ligands. Instead, the cytochromes act as electron carriers, which also play an essential part in metabolic processes. Unlike the activity in hemoglobin and myoglobin, the iron in cytochromes undergoes rapid reversible redox reactions

$$Fe^{3+} + e^- \rightleftharpoons Fe^{2+}$$

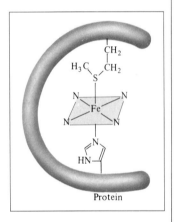

FIGURE 29.28 *The heme group in cytochrome c. The ligands above and below the porphyrin are the methionine group and histidine group, respectively.*

which are coupled to the oxidation of organic molecules such as the carbohydrates.

It is interesting to note that the toxicity of cyanide ion is the result of its binding to *cytochrome oxidase*, an enzyme that catalyzes the oxidation of cytochrome *c*. Cyanide will bind almost irreversibly to the oxidized, Fe(III)-containing form, of cytochrome oxidase, thereby disrupting the redox reactions involving cytochrome *c*. The recommended treatment for cyanide poisoning, if the poisoning is detected early enough, is rapid intravenous administration of an oxidizing agent such as amyl nitrite

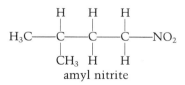

amyl nitrite

to convert some Fe(II) hemoglobin to the Fe(III) form. The latter has a definite tendency to bind to the cyanide ion, and the complex that is formed can then be eliminated.

The porphyrin group is present in two other iron proteins, *catalase* and *peroxidase*. These enzymes catalyze the decomposition of hydrogen peroxide, which is a toxic by-product of oxygen reduction in the cells.

The chlorophyll molecule, which plays an essential role in photosynthesis, also contains the porphyrin ring, but the metal ion there is Mg^{2+} rather than Fe^{2+} (Figure 29.29).

Vitamin B_{12} is another coordination complex of great biological significance. The story of vitamin B_{12} begins with pernicious anemia, a disease that involves the production of abnormal red blood cells and affects the nerve tissues and the gastric mucous membranes. In the 1920s scientists found that pernicious anemia could be controlled by a daily diet of raw liver, a treatment not all patients accepted with enthusiasm. Later it was determined that the ailment is caused by a deficiency of vitamin B_{12}. A *vitamin* is any of a group of organic substances essential in small quantities to normal metabolism and health.

Humans and other animals require small amounts of vitamin B_{12} to perform specific functions. Because it is not synthesized in the human body, it must be obtained from the diet. It is the only metal-containing vitamin. Vitamin B_{12} contains a *corrin* ring (Figure 29.30). The corrin ring bears some resemblance to the porphine ring discussed earlier but, because the ring is nonplanar, it is a less conjugated system. The geometry of the complex is basically octahedral, however. A cobalt atom is coordinated to the four nitrogen atoms and two other ligands above and below the ring. Figure 29.31 shows the structure of vitamin B_{12}.

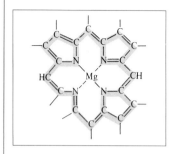

FIGURE 29.29 *The porphyrin structure in chlorophyll. The dotted lines indicate the coordinate covalent bonds. The electron delocalized portion of the molecule is shown in color.*

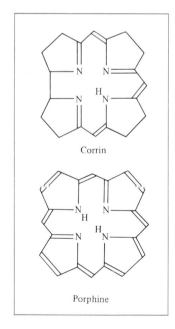

FIGURE 29.30 *Structures of corrin and porphine. Note that corrin does not have as extensive delocalization as porphine. Corrin loses one proton on forming a complex with a metal ion.*

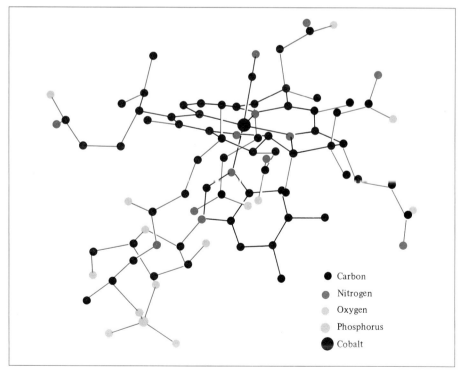

FIGURE 29.31 *The structure of vitamin B_{12} (its formula is $C_{63}H_{84}N_{14}O_{14}PCo$). The H atoms are not shown.*

AN ASIDE ON THE PERIODIC TABLE
Classification of Metals in Coordination Compounds

In Section 21.5 we noted the fact that we can view the formation of complex ions as a set of Lewis acid-base reactions in which the metal ion acts as the acid (the acceptor of lone pairs) and the donor atoms (which donate the lone pairs) of the ligands as the base. In the study of coordination compounds it is possible to classify the metal ions into three categories: the "hard" acids (or class a acceptors), "soft" acids (or class b acceptors), and "borderline" acids (Figure 29.32). You can see that the class a acceptors are the more electropositive elements (see Figure 11.2). Consequently, they tend to form the most stable complex ions by electrostatic interaction with smaller, more compact donor atoms. These donor atoms are the first members of Groups 5A, 6A, and 7A (N, O, and F, respectively). Class b acceptors are less electropositive and tend to form the most stable complexes with donor atoms that are the heavier elements in Groups 5A, 6A, and 7A. The bonding between the metal ion and donor atom in this case is largely covalent in nature. The properties of the borderline acids fall between those of the first two classes.

A nice illustration of the difference between the class a metals in Groups 2A and 2B can be illustrated by the color change of the cobalt complex ions (in an alcoholic solution) as follows:

$$[CoCl_4]^{2-} + 6H_2O \rightleftharpoons [Co(H_2O)_6]^{2+} + 4Cl^-$$

tetrachlorocobaltate(II) ion	hexaaquocobalt(II) ion
Blue	Pink

FIGURE 29.32 Classification of metal ions in coordination compounds as hard acid (class a), soft acid (class b), and borderline acid.

Although both Ca and Zn are class a metals, Ca has a much greater tendency to bind the donor atom (O) in H_2O than does Zn. (Zn's properties are quite close to those of the

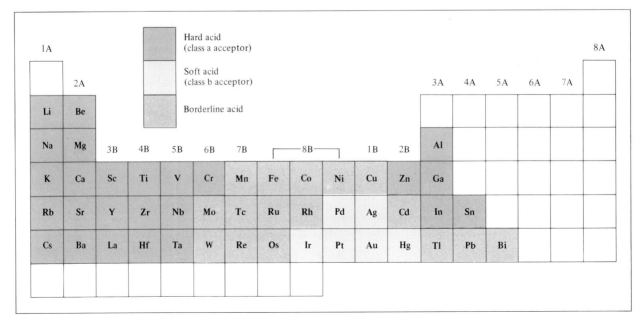

borderline acids.) When Ca^{2+} ions are added to the equilibrium mixture, the color of the solution is that of tetrachlorocobaltate(II) ion because the Ca^{2+} ions take up most of the H_2O molecules. (The equilibrium is shifted to the left.) If Zn^{2+} ions are added to the equilibrium mixture, the color of the solution is that of hexaaquocobalt(II) ion because the Zn^{2+} ion, with its partial class b character, tends to complex with Cl^- ions, thereby freeing the H_2O molecules. (The equilibrium is shifted to the right.)

SUMMARY

1. Transition metals usually have incompletely filled d orbitals and have a pronounced tendency to form complexes. Compounds that contain complex ions are called coordination compounds.
2. The donor atoms in the ligands each contribute an electron pair to the central metal ion in a complex.
3. Coordination compounds may display geometric and/or optical isomerism. All octahedral complexes containing three bidentate ligands are optically active.
4. Crystal field theory explains bonding in complexes in terms of electrostatic interactions. According to crystal field theory, the d orbitals are split into two higher energy and three lower energy orbitals in an octahedral complex. The energy difference between them is the crystal field splitting.
5. Strong-field ligands cause a large splitting, and weak-field ligands cause a small splitting. Electron spins tend to be parallel with weak-field ligands and paired with strong-field ligands, where a greater investment of energy is required to promote electrons into the high-lying d orbitals.
6. Complex ions undergo ligand exchange reactions in solution.
7. Coordination compounds find application in many different areas—for example, in water treatment, in chemical analysis, and as antidotes for metal poisoning.
8. Iron–porphyrin complexes, found in hemoglobin, myoglobin, cytochromes, catalase, and peroxidase, are important in metabolism and other activities. Chlorophyll molecules contain magnesium–porphyrin complex. Vitamin B_{12} contains cobalt complexed to the corrin ring.

KEY WORDS

Chelating agent, p. 887
Chiral, p. 894
Coordination compound, p. 886
Coordination number, p. 887
Crystal field splitting, p. 897
Donor atom, p. 887
Enantiomers, p. 894
Geometric isomers, p. 892
Labile complex, p. 903

Ligand, p. 887
Optical isomers, p. 893
Plane-polarized light, p. 894
Polarimeter, p. 895
Racemic mixture, p. 895
Spectrochemical series, p. 899
Stereoisomerism, p. 892
Stereoisomers, p. 892

PROBLEMS

More challenging problems are marked with an asterisk.

Properties of Transition Metals

29.1 What distinguishes a transition metal from a representative element?

29.2 Why do transition elements have variable oxidation states?

29.3 Explain why atomic radii decrease very gradually from scandium to copper.

29.4 Which ion is a stronger oxidizing agent, Mn^{3+} or Cr^{3+}?

29.5 Why is chromium used to coat other metals?

29.6 Why is zinc not considered a transition metal?

Nomenclature; Oxidation Number

29.7 Give the highest oxidation states from scandium to copper.

29.8 Complete the following statements for the complex ion $[Co(en)_2(H_2O)CN]^{2+}$. (a) en is the abbreviation for _____. (b) The oxidation number of Co is _____. (c) The coordination number of Co is _____. (d) _____ is a bidentate ligand.

29.9 Complete the following statements for the complex ion $[Cr(C_2O_4)_2(H_2O)_2]^-$. (a) The oxidation number of Cr is _____. (b) The coordination number of Cr is _____. (c) _____ is a bidentate ligand.

29.10 Give the oxidation numbers of the metals in the following species:

(a) $K_3[Fe(CN)_6]$ (d) Na_2MoO_4
(b) $K_3[Cr(C_2O_4)_3]$ (e) $MgWO_4$
(c) $[Ni(CN)_4]^{2-}$ (f) $K[Au(OH)_4]$

29.11 Write the formulas for each of the following ions and compounds: (a) tetrahydroxozincate(II), (b) chloropentaaquochromium(III) chloride, (c) tetrabromocuprate(II), (d) ethylenediaminetetraacetatoferrate(II), (e) bis(ethylenediamine)dichlorochromium(III), (f) pentacarbonyliron(0), (g) potassium tetracyanocuprate(II), (h) tetraammineaquochlorocobalt(III) chloride, (i) tris(ethylenediamine)cobalt(III) sulfate.

29.12 What are the systematic names for the following ions and compounds?

(a) $[Co(NH_3)_4Cl_2]^+$ (f) $[cis\text{-}Co(en)_2Cl_2]^+$
(b) $Cr(NH_3)_3Cl_3$ (g) $[Pt(NH_3)_5Cl]Cl_3$
(c) $[Co(en)_2Br_2]^+$ (h) $[Co(NH_3)_6]Cl_3$
(d) $Fe(CO)_5$ (i) $[Co(NH_3)_5Cl]Cl_2$
(e) $trans\text{-}Pt(NH_3)_2Cl_2$ (j) $[Cr(H_2O)_4Cl_2]Cl$

*29.13 What is the systematic name for the compound $[Fe(en)_3][Fe(CO)_4]$? (*Hint:* The oxidation number of Fe in the complex cation is +2.)

Structure

29.14 Specify which of the following structures can exhibit geometric isomerism: (a) linear, (b) square planar, (c) tetrahedral, (d) octahedral.

29.15 How many geometric isomers are in the following species? (a) $[Co(NH_3)_2Cl_4]^-$, (b) $[Co(NH_3)_3Cl_3]$.

29.16 What determines whether a molecule or an ion is chiral?

29.17 Explain the following terms: enantiomers and racemic mixture.

29.18 Draw structures of all the geometric and optical isomers of each of the following cobalt complexes:

(a) $[Co(NH_3)_6]^{3+}$ (d) $[Co(en)_3]^{3+}$
(b) $[Co(NH_3)_5Cl]^{2+}$ (e) $[Co(C_2O_4)_3]^{3-}$
(c) $[Co(NH_3)_4Cl_2]^+$

29.19 The complex ion $[Ni(CN)_2Br_2]^{2-}$ has a square-planar geometry. Draw the structures of the geometric isomers of this complex.

29.20 A student has prepared a cobalt complex which has one of the following three structures: $[Co(NH_3)_6]Cl_3$, $[Co(NH_3)_5Cl]Cl_2$, or $[Co(NH_3)_4Cl_2]Cl$. Explain how the student would distinguish among these possibilities by an electrolytic conductance experiment. At the student's disposal are three strong electrolytes: NaCl, $MgCl_2$, and $FeCl_3$, which may be used for comparison purposes.

Magnetism; Color; Bonding

29.21 What are the factors that determine whether a given complex will be diamagnetic or paramagnetic?

*29.22 For the same type of ligands, explain why the crystal field splitting for an octahedral complex is always greater than that for a tetrahedral complex.

29.23 Transition metal complexes containing CN^- ligands are often yellow in color, whereas those containing H_2O ligands are often green or blue. Explain.

29.24 The $[Ni(CN)_4]^{2-}$ ion, which has a square-planar geometry, is diamagnetic, whereas the $[NiCl_4]^{2-}$ ion, which has a tetrahedral geometry, is paramagnetic. Show the crystal field splitting diagrams for those two complexes.

29.25 Predict the number of unpaired electrons in the following complex ions: (a) $[Cr(CN)_6]^{4-}$, (b) $[Cr(H_2O)_6]^{2+}$.

29.26 The absorption maximum for the complex ion $[Co(NH_3)_6]^{3+}$ occurs at 470 nm. (a) Predict the color of the complex and (b) calculate the crystal field splitting in kJ/mol.

29.27 In each of the following pairs of complexes, choose the one that absorbs light at a shorter wavelength: (a) $[Co(NH_3)_6]^{2+}$, $[Co(H_2O)_6]^{2+}$; (b) $[FeF_6]^{3-}$, $[Fe(CN)_6]^{3-}$; (c) $[Cu(NH_3)_4]^{2+}$, $[CuCl_4]^{2-}$.

29.28 Oxyhemoglobin is bright red, whereas deoxyhemoglobin is purple. Show that the difference in color can be accounted for qualitatively on the basis of high-spin and low-spin complexes.

29.29 A solution made by dissolving 0.875 g of $Co(NH_3)_4Cl_3$ in 25.0 g of water freezes 0.56°C below the freezing point of pure water. Calculate the number of moles of ions produced when one mole of $Co(NH_3)_4Cl_3$ is dissolved in water, and suggest a structure for the complex ion present in this compound.

29.30 A student has prepared three coordination compounds containing chromium, with the following properties:

Formula	Color	Cl^- Ions in Solution per Formula Unit
(a) $CrCl_3 \cdot 6H_2O$	Violet	3
(b) $CrCl_3 \cdot 6H_2O$	Light green	2
(c) $CrCl_3 \cdot 6H_2O$	Dark green	1

Write modern formulas for these compounds and suggest a method for confirming the number of Cl^- ions present in solution in each case. (*Hint:* Some of the compounds may exist as hydrates.)

Reactions

29.31 Oxalic acid, $H_2C_2O_4$, is sometimes used to clean rust stains from sinks and bathtubs. Explain the chemistry involved in this cleaning action.

*29.32 The $[Fe(CN)_6]^{3-}$ complex is more labile than the $[Fe(CN)_6]^{4-}$ complex. Suggest an experiment that would prove the labile nature of the $[Fe(CN)_6]^{3-}$ complex.

*29.33 Aqueous copper(II) sulfate solution is blue in color. When aqueous potassium fluoride is added, a green precipitate is formed. When aqueous potassium chloride is added instead, a bright green solution is formed. Explain what is happening in these two cases.

*29.34 When aqueous potassium cyanide is added to a solution of copper(II) sulfate, a white precipitate, soluble in an excess of potassium cyanide, is formed. No precipitate is formed when hydrogen sulfide is bubbled through the solution. Explain.

*29.35 A concentrated aqueous copper(II) chloride solution is bright green in color. On dilution with water, the solution becomes light blue. Explain.

*29.36 In dilute nitric acid solution, Fe^{3+} reacts with thiocyanate ion (SCN^-) to form a dark red complex:

$$[Fe(H_2O)_6]^{3+} + SCN^- \rightleftharpoons$$
$$H_2O + [Fe(H_2O)_5NCS]^{2+}$$

The equilibrium concentration of $[Fe(H_2O)_5NCS]^{2+}$ may be determined by how darkly colored the solution is (measured by a spectrometer). In one such experiment, 1.0 mL of 0.20 M $Fe(NO_3)_3$ was mixed with 1.0 mL of 1.0×10^{-3} M KSCN and 8.0 mL of dilute HNO_3. The color of the solution indicated that the $[Fe(H_2O)_5NCS]^{2+}$ concentration was 7.3×10^{-5} M. Calculate the formation constant for $[Fe(H_2O)_5NCS]^{2+}$.

Miscellaneous Problems

29.37 Define each of the following terms: (a) ligand, (b) complex, (c) chelating agent, (d) labile complex, (e) coordination number.

29.38 Chemical analysis shows that hemoglobin contains 0.34 percent of Fe by mass. What is the minimum possible molar mass of hemoglobin? The actual molar mass of hemoglobin is about 65,000 g. How do you account for the discrepancy between your minimum value and the actual value?

29.39 Explain the following facts: (a) Copper and iron have several oxidation states, whereas zinc exists in only one. (b) The standard reduction potential of zinc is more negative than that of copper. (c) Copper and iron form colored ions, whereas zinc does not.

29.40 The formation constant for the reaction $Ag^+ + 2NH_3 \rightleftharpoons [Ag(NH_3)_2]^+$ is 1.5×10^7 and that for the reaction $Ag^+ + 2CN^- \rightleftharpoons [Ag(CN)_2]^-$ is 1.0×10^{21} at 25°C (see Table 21.3). Calculate the equilibrium constant and $\Delta G°$ at 25°C for the reaction

$$[Ag(NH_3)_2]^+ + 2CN^- \rightleftharpoons [Ag(CN)_2]^- + 2NH_3$$

29.41 From the standard reduction potentials listed in Table 24.1 for Zn/Zn^{2+} and Cu^{+}/Cu^{2+}, calculate $\Delta G°$ and the equilibrium constant for the reaction

$$Zn(s) + 2Cu^{2+}(aq) \longrightarrow Zn^{2+}(aq) + 2Cu^{+}(aq)$$

*29.42 Using the standard reduction potentials listed in Table 24.1 and the *Handbook of Chemistry and Physics*, show that the following reaction is favorable under standard-state conditions:

$$2Ag(s) + Pt^{2+}(aq) \longrightarrow 2Ag^{+}(aq) + Pt(s)$$

What is the equilibrium constant of this reaction at 25°C?

29.43 Give three examples of coordination compounds in biological systems.

29.44 Discuss, giving suitable examples, the role of chelating agents in medicine.

29.45 The Co^{2+}–porphyrin complex is more stable than the Fe^{2+}–porphyrin complex. Why, then, is iron the metal ion in hemoglobin (and other heme-containing proteins)?

29.46 What are the differences between geometric isomers and optical isomers?

30
ORGANIC CHEMISTRY

Organic chemistry deals with the chemistry of virtually all carbon compounds. The word *organic* was originally used by eighteenth-century chemists to describe substances that originate from living sources—that is, from plants and animals. They believed that nature possessed a certain "vital force" and that it alone could produce organic compounds. This romantic notion was disproved by experiments like those first carried out in 1828 by Friedrich Wohler (1800–1882), who prepared urea, an organic compound, from the reaction between two inorganic compounds—lead cyanate and aqueous ammonia:

$$Pb(OCN)_2 + 2NH_3 + 2H_2O \longrightarrow 2H_2N\overset{\overset{\displaystyle O}{\|}}{C}NH_2 + Pb(OH)_2$$
$$\text{urea}$$

Today well over five million synthetic and natural organic compounds are known. This number is significantly greater than the one hundred thousand or so inorganic compounds known to exist.

30.1   HYDROCARBONS

Recall that the linking of like atoms is called *catenation*.

Carbon can form more compounds than any other element because carbon atoms are able to link up with each other in straight chains and branched chains. Although the number of known organic compounds is enormous, the study of **organic chemistry,** that is, the *branch of chemistry that deals with carbon compounds,* is not as difficult as it may seem. Most organic compounds can be divided into relatively few classes according to the *functional groups* they contain. A **functional group** is *that part of a molecule characterized by a special arrangement of atoms that is largely responsible for the chemical behavior of the parent molecule.* Different molecules containing the same kind of functional group or groups react similarly. Thus, by learning the characteristic properties of a few functional groups, we will be able to study and understand the properties of many organic compounds.

Before discussing functional groups, we will examine a class of compounds that forms the framework of all organic compounds—the hydrocarbons. **Hydrocarbons** are *made up of only two elements, hydrogen and carbon.* On the basis of structure, hydrocarbons are divided into two main classes—aliphatic and aromatic. **Aliphatic hydrocarbons** *do not contain the benzene group or the benzene ring,* whereas **aromatic hydrocarbons** *contain one or more benzene rings.* Aliphatic hydrocarbons are further divided into alkanes, alkenes, and alkynes (Figure 30.1).

Alkanes

Alkanes *have the general formula* C_nH_{2n+2}, *where* $n = 1, 2, \ldots$. The essential characteristic of alkane hydrocarbon molecules is that *only single covalent bonds are present.* In these compounds the bonds are said to be saturated. Thus the alkanes are known as **saturated hydrocarbons.**

FIGURE 30.1 *Classification of hydrocarbons.*

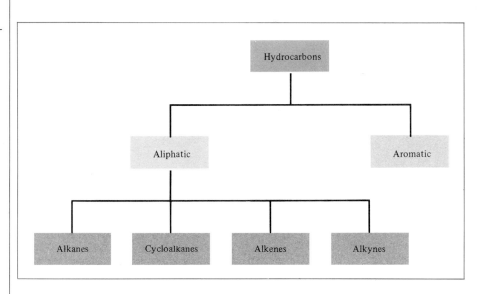

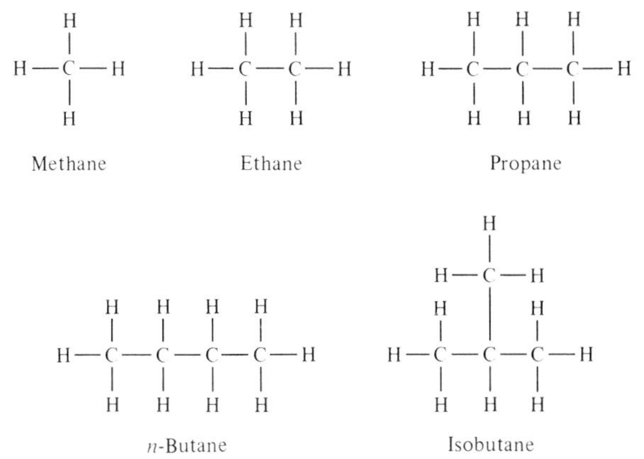

FIGURE 30.2 *The structures of the first four alkanes. Note that butane can exist in two structurally different forms, called structural isomers.*

The simplest member of the family (that is, with $n = 1$) is CH_4, which occurs naturally by the anaerobic bacterial decomposition of vegetable matter under water. Because it was first collected in marshes, methane became known as "marsh gas." Methane is obtained commercially from natural gas. It is also produced in some sewage treatment processes.

Figure 30.2 shows the structures of the first four alkanes (that is, from $n = 1$ to $n = 4$). Natural gas is a mixture of methane, ethane, and a small amount of propane. We discussed the bonding scheme of methane in Chapter 12. Indeed, the carbon atoms in all the alkanes can be assumed to be sp^3-hybridized. The structures of ethane and propane are straightforward, for there is only one way to join the carbon atoms in these molecules. Butane, however, has two possible bonding schemes and can exist as structural isomers called n-butane (where n stands for normal) and isobutane. **Structural isomers** are *molecules that have the same molecular formula, but different structures.*

In the alkane series, as the number of carbon atoms increases, the number of structural isomers increases rapidly. For example, butane has two isomers; decane $(C_{10}H_{22})$ has 75 isomers; and the alkane $C_{30}H_{62}$ has over four hundred million, or 4×10^8 isomers! Obviously, most of these isomers do not exist in nature and have not been synthesized. Nevertheless the numbers help to explain why carbon is found in so many more compounds than any other element.

Bacterial organisms are *anaerobic* if they do not require oxygen. *Aerobic* organisms, on the other hand, cannot survive in an oxygen-free environment.

EXAMPLE 30.1

How many structural isomers can be identified with pentane (C_5H_{12})?

Answer

For small hydrocarbon molecules (eight or fewer carbon atoms), it is relatively easy to determine the number of structural isomers by trial and error. The first step is to write down the straight-chain structure:

(Continued)

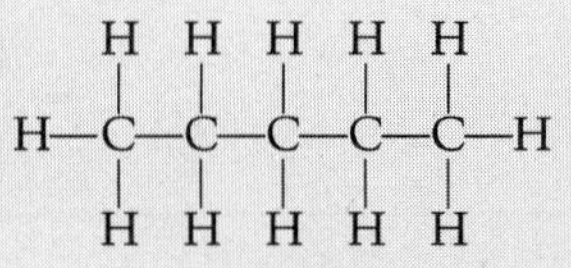

H—C—C—C—C—C—H

n-pentane
(b.p. 36.1°C)

The second structure, by necessity, must be a branched chain:

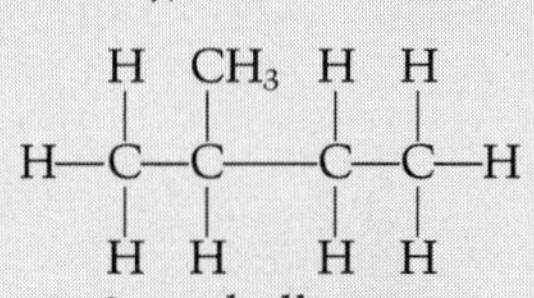

H—C—C—C—C—H

2-methylbutane
(b.p. 27.9°C)

Yet another branched-chain structure is possible:

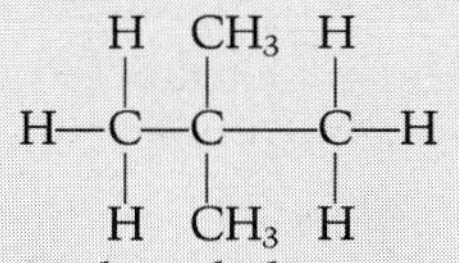

H—C—C—C—H

2,2-dimethylpropane
(b.p. 9.5°C)

We can draw no other structure for an alkane having the molecular formula C_5H_{12}. Thus, pentane has three structural isomers, in which although the structure changes, the numbers of carbon and hydrogen atoms remain unchanged.

Similar example: Problem 30.2.

Table 30.1 shows the melting and boiling points of the straight-chain isomers of the first ten alkanes. The first four are gases at room temperature; and pentane through decane are liquids. As molecular size increases, so does the boiling point, because of the increasing dispersion forces (see Section 14.2).

The system of naming organic compounds is based on the recommendation of the International Union of Pure and Applied Chemistry (IUPAC).

Alkane Nomenclature.   The names of the straight-chain alkanes are fairly easy to remember. As Table 30.1 shows, except for the first four members, the number of carbon atoms in each alkane is indicated by the Greek prefix (see Table 2.3). When one or more hydrogen atoms are replaced by other groups, the name of the compound must indicate the locations of carbon atoms where replacements are made. Let us first consider groups that are themselves derived from alkanes. For example, when a hydrogen atom is removed from methane, we are left with the CH_3 fragment, which is called a *methyl* group. Similarly, removing a hydrogen atom from the ethane molecule gives an *ethyl* group, or C_2H_5. Because they are derived from alkanes, they are also called *alkyl* groups. Table 30.2 lists the names of several common alkyl groups.

TABLE 30.1 The First Ten Straight-Chain Alkanes*

Name of Hydrocarbon	Molecular Formula	Number of Carbon Atoms	Melting Point (°C)	Boiling Point (°C)
Methane	CH_4	1	-182.5	-161.6
Ethane	CH_3—CH_3	2	-183.3	-88.6
Propane	CH_3—CH_2—CH_3	3	-189.7	-42.1
Butane	CH_3—$(CH_2)_2$—CH_3	4	-138.3	-0.5
Pentane	CH_3—$(CH_2)_3$—CH_3	5	-129.8	36.1
Hexane	CH_3—$(CH_2)_4$—CH_3	6	-95.3	68.7
Heptane	CH_3—$(CH_2)_5$—CH_3	7	-90.6	98.4
Octane	CH_3—$(CH_2)_6$—CH_3	8	-56.8	125.7
Nonane	CH_3—$(CH_2)_7$—CH_3	9	-53.5	150.8
Decane	CH_3—$(CH_2)_8$—CH_3	10	-29.7	174.0

* By "straight chain," we mean that the carbon atoms are joined along one line. This does not mean, however, that these molecules are linear. Each carbon atom is sp^3- (not sp-) hybridized.

TABLE 30.2 Several Common Alkyl Groups

Name	Formula
Methyl	—CH_3
Ethyl	—CH_2—CH_3
n-Propyl	—CH_2—CH_2—CH_3
n-Butyl	—CH_2—CH_2—CH_2—CH_3
Isopropyl	$\begin{array}{c} CH_3 \\ | \\ -C-H \\ | \\ CH_3 \end{array}$
t-Butyl*	$\begin{array}{c} CH_3 \\ | \\ -C-CH_3 \\ | \\ CH_3 \end{array}$

* The letter *t* stands for tertiary.

With the aid of Tables 30.1 and 30.2 and the rule that the locations of other groups of carbon atoms are denoted by numbering the atoms along the longest carbon chain, we are now ready to name some substituted alkanes. Note that we always start numbering from the end that is closer to the carbon atom bearing the substituted group. Consider the following example:

$$\begin{array}{c} CH_3 \\ | \\ \overset{1}{H_3C}-\overset{2}{C}-\overset{3}{CH_2}-\overset{4}{CH_3} \\ | \\ H \end{array}$$

Since the longest carbon chain contains four C atoms, the parent name of the compound is butane. We note that a methyl group is attached to carbon

TABLE 30.3
Common Functional Groups

Functional Group	Name
$-NH_2$	Amino
$-F$	Fluoro
$-Cl$	Chloro
$-Br$	Bromo
$-I$	Iodo
$-NO_2$	Nitro
$-CH=CH_2$	Vinyl
⬡	Phenyl

number 2; therefore, the name of the compound is 2-methylbutane. Here is another example:

$$\overset{\text{CH}_3}{\underset{\text{CH}_3}{\overset{|}{\underset{|}{\text{H}_3\overset{1}{\text{C}}-\overset{2}{\text{C}}-\overset{3}{\text{CH}_2}-\overset{\text{CH}_3}{\overset{4}{\text{CH}}}-\overset{5}{\text{CH}_2}-\overset{6}{\text{CH}_3}}}}$$

The parent name of this compound is hexane. Noting that there are two methyl groups attached to carbon number 2 and one methyl group attached to carbon number 4, we call the compound 2,2,4-trimethylhexane.

Of course, we can have many different types of substituents on alkanes. Table 30.3 lists the names of some common functional groups. Thus, the compound

$$\underset{}{\overset{\text{Br}\quad\text{NO}_2}{\underset{}{\overset{|\quad|}{\text{H}_3\overset{1}{\text{C}}-\overset{2}{\text{CH}}-\overset{3}{\text{CH}}-\overset{4}{\text{CH}_3}}}}$$

is called 2-bromo-3-nitrobutane. If the Br is attached to the first carbon atom

$$\underset{}{\overset{\text{NO}_2}{\underset{}{\overset{|}{\text{Br}-\overset{1}{\text{CH}_2}-\overset{2}{\text{CH}_2}-\overset{3}{\text{CH}}-\overset{4}{\text{CH}_3}}}}$$

the compound becomes 1-bromo-3-nitrobutane. As another example, consider the compound with the structure

$$\underset{\text{H}_3\text{C}-\text{CH}-\text{CH}_3}{\text{⬡}}$$

We call this compound 2-phenylpropane.

Reactions of Alkanes. Alkanes are generally not considered to be very reactive substances. However, under suitable conditions they do undergo several kinds of reactions—including combustion. The burning of natural gas, gasoline, and fuel oil all involves the combustion of alkanes. These reactions are all highly exothermic:

$$CH_4(g) + 2O_2(g) \longrightarrow CO_2(g) + 2H_2O(l) \qquad \Delta H° = -890.4 \text{ kJ}$$

$$2C_2H_6(g) + 7O_2(g) \longrightarrow 4CO_2(g) + 6H_2O(l) \qquad \Delta H° = -3119 \text{ kJ}$$

$$C_3H_8(g) + 5O_2(g) \longrightarrow 3CO_2(g) + 4H_2O(l) \qquad \Delta H° = -2299 \text{ kJ}$$

$$2C_4H_{10}(g) + 13O_2(g) \longrightarrow 8CO_2(g) + 10H_2O(l) \qquad \Delta H° = -2874 \text{ kJ}$$

These, and similar reactions, have long been utilized in industrial processes and in domestic heating and cooking.

The *halogenation* of alkanes—that is, the replacement of one or more hydrogen atoms by halogen atoms—is another well-studied kind of reaction. When a mixture of methane and chlorine is heated above 100°C or irradiated with light of a suitable wavelength, methyl chloride is produced:

$$CH_4(g) + Cl_2(g) \longrightarrow CH_3Cl(g) + HCl(g)$$
$$\text{methyl}$$
$$\text{chloride}$$

If chlorine gas is present in sufficient amount, the reaction can proceed further, as follows:

$$CH_3Cl(g) + Cl_2(g) \longrightarrow \quad CH_2Cl_2(l) \quad + HCl(g)$$
$$\text{methylene}$$
$$\text{chloride}$$

$$CH_2Cl_2(g) + Cl_2(g) \longrightarrow \quad CHCl_3(l) \quad + HCl(g)$$
$$\text{chloroform}$$

$$CHCl_3(l) + Cl_2(g) \longrightarrow \quad CCl_4(l) \quad + HCl(g)$$
$$\text{carbon}$$
$$\text{tetrachloride}$$

One of chemistry's principal aims is to understand the mechanism of any reaction at the molecular level. This is not always easy, particularly when complex molecules are involved. Thanks to significant advances made in recent years, we now have clearer ideas about the hows and whys of many reactions. For example, a great deal of experimental evidence suggests that the initial step of the first halogenation reaction occurs as follows:

$$Cl_2 + \text{energy} \longrightarrow Cl\cdot + Cl\cdot$$

Thus the covalent bond breaks and two chlorine atoms form. We know it is the Cl—Cl bond that breaks when the mixture is heated or irradiated because the bond energy of Cl_2 is 242.7 kJ/mol, whereas about 414 kJ/mol is needed to break C—H bonds in CH_4.

A free chlorine atom contains an unpaired electron, as shown by a single dot. These atoms are highly reactive and attack methane molecules according to the equation

$$CH_4 + Cl\cdot \longrightarrow \cdot CH_3 + HCl$$

This reaction produces hydrogen chloride and the methyl radical ($\cdot CH_3$). (We call *any neutral fragment of a molecule containing an unpaired electron* a **radical**.) The methyl radical is another reactive species; it combines with molecular chlorine to give methyl chloride and a chlorine atom:

$$\cdot CH_3 + Cl_2 \longrightarrow CH_3Cl + Cl\cdot$$

The production of methylene chloride from methyl chloride and further reactions can be explained in the same way. The actual mechanism is more complex than the scheme we have shown because "side reactions" that do not lead to the desired products often take place, such as

$$Cl\cdot + Cl\cdot \longrightarrow Cl_2$$
$$\cdot CH_3 + \cdot CH_3 \longrightarrow C_2H_6$$

Alkanes in which one or more hydrogen atoms have been replaced by a halogen atom are called *alkyl halides*. Among the large number of alkyl

The systematic names of methyl chloride, methylene chloride, and chloroform are monochloromethane, dichloromethane, and trichloromethane, respectively.

halides, the best known are chloroform ($CHCl_3$), carbon tetrachloride (CCl_4), methylene chloride (CH_2Cl_2), and the chlorofluorohydrocarbons.

Chloroform is a volatile, sweet-tasting liquid used for many years as an anesthetic. However, because of its toxicity—it can severely damage the liver, kidneys, and heart—it has been replaced by other compounds. Carbon tetrachloride, also a toxic substance, serves as a cleaning liquid, for it effectively removes grease stains from clothing. Methylene chloride is used as a solvent to decaffeinate coffee and as a paint remover.

The chlorofluorohydrocarbons, better known commercially as Freons, have been used as coolants in refrigerators and as propellants in aerosol spray cans. Common Freons include $CFCl_3$, CF_2Cl_2, and CF_3Cl. As we have discussed, recent evidence suggests that these substances may deplete ozone molecules in the stratosphere that protect us from harmful short wavelength (ultraviolet) radiation from the sun. Once released into the atmosphere, the chlorofluorohydrocarbons slowly diffuse upward into the stratosphere and undergo a series of photochemical reactions that consume ozone molecules (see Section 26.6).

Optical Isomerism of Substituted Alkanes.   In Chapter 29 we discussed optical isomerism of coordination compounds (see Section 29.3). Optical isomerism exists also in many organic compounds. Consider the following substituted methanes:

$$\begin{array}{cc}
\begin{array}{c} H \\ | \\ H-C-Br \\ | \\ Cl \\ (a) \end{array}
&
\begin{array}{c} F \\ | \\ H-C-Br \\ | \\ Cl \\ (b) \end{array}
\end{array}$$

Figure 30.3 shows perspective drawings of these two molecules and their mirror images. The two mirror images of (a) are superimposable, but those of (b) are not, no matter how we rotate the molecules. Thus the CHFClBr molecule is chiral. Careful observation shows that most simple chiral molecules contain at least one asymmetric carbon atom—that is, a carbon atom bonded to four different atoms or groups of atoms.

EXAMPLE 30.2

Is the following molecule chiral?

$$\begin{array}{c} Cl \\ | \\ H-C-CH_2-CH_3 \\ | \\ CH_3 \end{array}$$

Answer

We note that the central carbon atom is bonded to a hydrogen atom, a chlorine atom, a —CH_3 group, and a —CH_2—CH_3 group. Therefore the central carbon atom is asymmetric and the molecule is chiral.

Similar examples: Problems 30.14, 30.56.

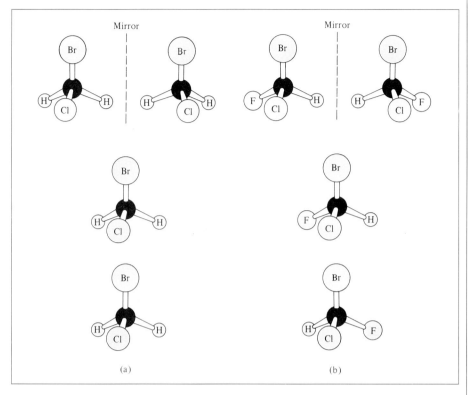

FIGURE 30.3 (a) The CH_2ClBr molecule and its mirror image. Since the molecule and its mirror image are superimposable, it is said to be achiral. (b) The $CHFClBr$ molecule and its mirror image. Since the molecule and its mirror image are not superimposable, no matter how we rotate one with respect to the other, it is said to be chiral.

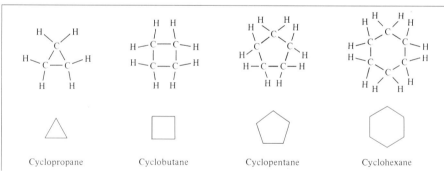

Cyclopropane   Cyclobutane   Cyclopentane   Cyclohexane

FIGURE 30.4 *The structures of the first four cyclic alkanes and their simplified forms.*

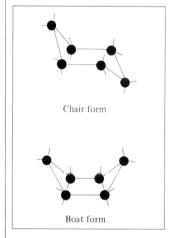

Chair form

Boat form

FIGURE 30.5 *The cyclohexane molecule can exist in various shapes. The most stable shape is the chair form. The boat form is less stable than the chair form. Hydrogen atoms are not shown in these diagrams.*

Cycloalkanes. *Alkanes whose carbon atoms are joined in rings are known as* **cycloalkanes.** They have the general formula C_nH_{2n}, where $n = 3, 4, \ldots$. The structures of some simple cycloalkanes are given in Figure 30.4. Theoretical analysis shows that cyclohexane can assume two different geometries that are relatively free of strain (Figure 30.5). By *strain* we mean that bonds are compressed, stretched, or twisted out of their normal geometric shapes as predicted by hybridization and other considerations. The most stable geometry is the *chair form.*

As Figure 30.6 shows, there are two kinds of hydrogen atoms in the chair form of cyclohexane. Six of these are called *equatorial* hydrogens and the other six are called *axial* hydrogens. The equatorial C—H bonds are in the

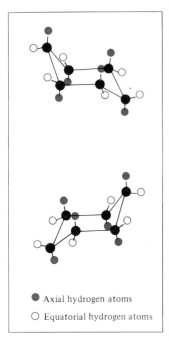

FIGURE 30.6 *Hydrogen atoms in the chair form of cyclohexane can be classified as axial or equatorial. When the chair flips from one form to its mirror image, axial hydrogen atoms become equatorial, and vice versa.*

In the *cis* isomer, the two hydrogen atoms are on the same side of the C═C bond; in the *trans* isomer, the two hydrogen atoms are across from each other.

plane of the "seat" of the chair, whereas the axial C—H bonds are perpendicular to this plane. These atoms differ from each other with respect to their immediate environment in the same molecule.

Study of the various shapes and stability of the cyclohexane ring is important because many biologically significant substances contain one or more such ring systems.

Alkenes

There are two types of ***unsaturated hydrocarbons,*** which are *hydrocarbons that contain carbon–carbon double bonds or carbon–carbon triple bonds.* The ***alkenes*** (also called the *olefins*) *contain at least one carbon–carbon double bond.* (Hydrocarbons containing carbon–carbon triple bonds will be discussed shortly.) *Alkenes have the general formula* C_nH_{2n}, *where n = 2, 3, ...* Thus the simplest alkene is C_2H_4, ethylene, in which both carbon atoms are sp^2-hybridized and the double bond is made up of a sigma bond and a pi bond (see Chapter 12).

Alkene Nomenclature. In naming alkenes it is necessary to indicate the positions of the carbon–carbon double bonds. The names of compounds containing C═C bonds end with *-ene.* The name of the parent compound is still determined by the number of carbon atoms in the longest chain (see Table 30.1), as shown here:

$$CH_2{=}CH{-}CH_2{-}CH_3 \qquad H_3C{-}CH{=}CH{-}CH_3$$
$$\text{1-butene} \qquad\qquad \text{2-butene}$$

The numbers 1 and 2 indicate the first and second carbon atoms where the double bond appears. Butene indicates that they are compounds with four carbon atoms in the longest chain. We must also consider the possibilities of geometric isomers, such as

4-methyl-*cis*-2-hexene 4-methyl-*trans*-2-hexene

Table 30.4 shows the names and structures of some simple alkenes.

Properties and Reactions of Alkenes. Ethylene is an extremely important substance because it is used in large quantities for the manufacture of organic polymers (to be discussed later in this chapter) and in the preparation of many other organic chemicals. Ethylene is prepared industrially by the *cracking* process, that is, the thermal decomposition of a large hydrocarbon into smaller molecules. When ethane is heated to about 800°C, it undergoes the following reaction:

$$C_2H_6(g) \xrightarrow[\text{Pt catalyst}]{\Delta} CH_2{=}CH_2(g) + H_2(g)$$

TABLE 30.4 Some Simple Alkenes

Alkene	Structure	Melting Point (°C)	Boiling Point (°C)
Ethylene	$\begin{array}{c} H \\ \diagdown \\ C=C \\ \diagup \quad \diagdown \\ H \quad\quad H \end{array}$ with H top-left, H top-right, H bottom-left, H bottom-right	−169.4	−102.4
Propene (propylene)	H₃C and H on C=C, H and H below	−185.2	−47.7
2-Methylpropene	H₃C and H on C=C, H₃C and H	−139	−7.0
cis-2-Butene	H₃C and CH₃ on C=C, H and H below	−139	3.7
trans-2-Butene	H₃C and H on C=C, H and CH₃	−105.8	0.9

Other alkenes can be prepared similarly by cracking the higher members of the alkane family.

In Section 26.2 we discussed the hydrogenation of ethylene to produce ethane. Hydrogenation is sometimes described as an "addition reaction." Other addition reactions to the C=C bond include

$$C_2H_4(g) + HX(g) \longrightarrow CH_3-CH_2X(g)$$

$$C_2H_4(g) + X_2(g) \longrightarrow CH_2X-CH_2X(g)$$

where X represents a halogen (Cl, Br, or I). The addition of a hydrogen halide to an unsymmetrical alkene such as propylene is more complicated because two products are possible:

$$\begin{array}{c} H_3C \quad\quad H \\ \diagdown \quad\quad \diagup \\ C=C \\ \diagup \quad\quad \diagdown \\ H \quad\quad\quad H \end{array} + HBr \longrightarrow \begin{array}{c} H \ H \\ | \ | \\ H_3C-C-C-H \\ | \ | \\ H \ Br \end{array} \quad and/or \quad \begin{array}{c} H \ H \\ | \ | \\ H_3C-C-C-H \\ | \ | \\ Br \ H \end{array}$$

propylene   1-bromopropane   2-bromopropane

In practice, however, only one product is formed, and that product is 2-bromopropane. This phenomenon was observed in all reactions between unsymmetrical reagents and alkenes. In 1871, Vladimir Markovnikov (1838–1904) postulated a generalization that enables us to predict the outcome of such an addition reaction. This generalization is now known as *Markovnikov's rule*, which states that: In the addition of unsymmetrical (that is, polar) reagents to alkenes, the positive portion of the reagent (which is usually hydrogen) adds to the carbon atom that already has the most hydrogen atoms.

Geometric Isomers of Alkenes.   In a compound such as ethane (C_2H_6) the rotation of the two methyl groups about the carbon–carbon single bond (which is a sigma bond) is quite free. The situation is different for molecules that contain carbon–carbon double bonds, such as ethylene. In addition to the sigma bond, there is a pi bond between the two carbon atoms. Rotation about the carbon–carbon linkage does not affect the sigma bond, but it does move the two $2p_z$ orbitals out of alignment for overlap and, hence, partially or totally destroys the pi bond (see Figure 12.11). This process requires an input of energy on the order of 270 kJ/mol. For this reason, the rotation of a carbon–carbon double bond is considerably restricted, but not impossible. Consequently molecules containing carbon–carbon double bonds (that is, the alkenes) may have geometric isomers. Dichloroethylene $(ClHC{=}CHCl)$ is an example.

   Generally, *cis* and *trans* isomers have distinctly different physical and chemical properties. The average bond energy for $C{=}C$ is 620 kJ/mol, of which 350 kJ/mol is attributed to the sigma bond and 270 kJ/mol to the pi bond. Thus in many reactions that the alkenes undergo (for example, hydrogenation and addition with hydrogen halides) the weaker pi bond is broken, while the sigma bond remains intact. Heat or irradiation with light is commonly used to bring about the conversion of one geometric isomer to another.

   Dipole moment measurements can be used to distinguish between *cis* and *trans* isomers. For example, we would expect *cis*-dichloroethylene to possess a dipole moment and *trans*-dichloroethylene to be nonpolar, since it has no resultant dipole moment.

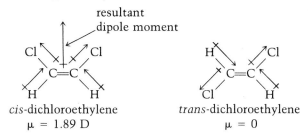

cis-dichloroethylene   *trans*-dichloroethylene

$\mu = 1.89$ D $\mu = 0$

The measured dipole moments (μ) confirm our expectations.

Alkynes

Alkynes have the *general formula* C_nH_{2n-2}, *where* $n = 2, 3, \ldots$. *The alkynes contain at least one carbon–carbon triple bond.*

Alkyne Nomenclature.   Names of compounds containing $C{\equiv}C$ bonds end with *-yne.* The name of the parent compound is still determined by the number of carbon atoms in the longest chain (see Table 30.1). As in the case of alkenes, in naming alkynes it is important to indicate the position of the carbon–carbon triple bond. The name of the parent compound is still determined by the number of carbon atoms in the longest chain, as shown below:

$$HC\equiv C\!-\!CH_2\!-\!CH_3 \qquad H_3C\!-\!C\equiv C\!-\!CH_3$$
$$\text{1-butyne} \qquad\qquad\quad \text{2-butyne}$$

Properties and Reactions of Alkynes. The simplest alkyne is acetylene, C_2H_2, whose structure and bonding were discussed in Section 12.6. Acetylene is a colorless gas (b.p. $-84°C$) prepared by the reaction between calcium carbide and water:

$$CaC_2(s) + 2H_2O(l) \longrightarrow C_2H_2(g) + Ca(OH)_2(aq)$$

Acetylene has many important uses in industry. Because of its high heat of combustion

$$2C_2H_2(g) + 5O_2(g) \longrightarrow 4CO_2(g) + 2H_2O(l) \qquad \Delta H° = -2599.2 \text{ kJ}$$

acetylene burned in an "oxyacetylene torch" gives an extremely hot flame (about 3000°C). Thus oxyacetylene torches are used to weld metals.

The standard free energy of formation of acetylene is positive ($\Delta G_f° = 209.2$ kJ/mol), unlike that of the alkanes. This means that the molecule is unstable (relative to its elements) and has a tendency to decompose:

$$C_2H_2(g) \longrightarrow 2C(s) + H_2(g)$$

In the presence of a suitable catalyst or when the gas is kept under pressure, this reaction can occur with explosive violence. In order for the gas to be transported safely, it must be dissolved in an inert organic solvent such as acetone at moderate pressure. In the liquid state, acetylene is very sensitive to shock, and is highly explosive.

Acetylene can be hydrogenated to yield ethylene:

$$C_2H_2(g) + H_2(g) \longrightarrow C_2H_4(g)$$

It undergoes the following addition reactions with hydrogen halides and halogens:

$$C_2H_2(g) + HX(g) \longrightarrow CH_2\!=\!CHX(g)$$

$$C_2H_2(g) + X_2(g) \longrightarrow CHX\!=\!CHX(g)$$

$$C_2H_2(g) + 2X_2(g) \longrightarrow CHX_2\!-\!CHX_2(g)$$

The CHX=CHX produced generally has the *cis* structure.

Methylacetylene (propyne), $CH_3\!-\!C\equiv C\!-\!H$, is the next member in the acetylene family. It undergoes reactions similar to those of acetylene. The addition reactions of propyne also obey Markovnikov's rule, as in the case of alkenes:

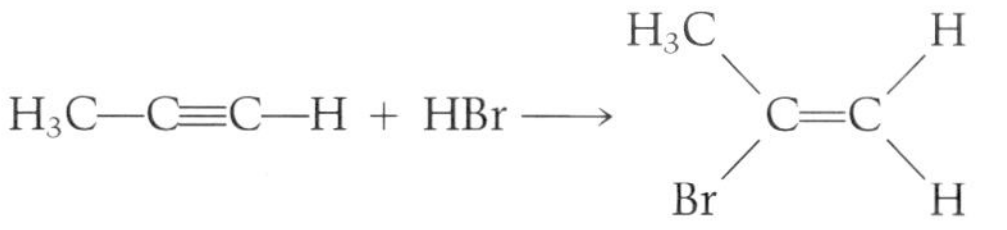

propyne   2-bromopropene

Aromatic Hydrocarbons

Benzene, the parent compound of this large family of organic substances, was discovered by Michael Faraday in 1826. Over the next forty years, chemists were preoccupied with determining its molecular structure. Despite the small number of atoms in the molecule, there are quite a few ways to represent the structure of benzene while maintaining the tetravalency of carbon (Figure 30.7). However, most proposed structures were rejected because they could not explain the known properties of benzene. Finally, in 1865, August Kekulé (1829–1896) deduced that the benzene molecule could be best represented by a ring structure—a cyclic compound consisting of six carbon atoms:

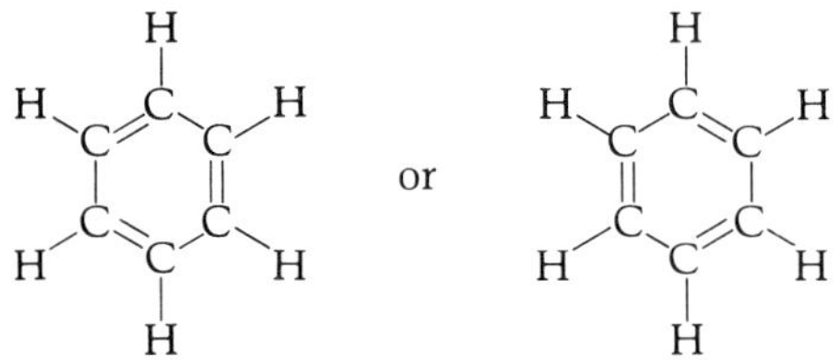

Kekulé's idea did not gain wide acceptance at first, for his formulation suggested that the replacement of two hydrogen atoms by two chlorine atoms on adjacent carbon atoms, say, would produce two *different* molecules:

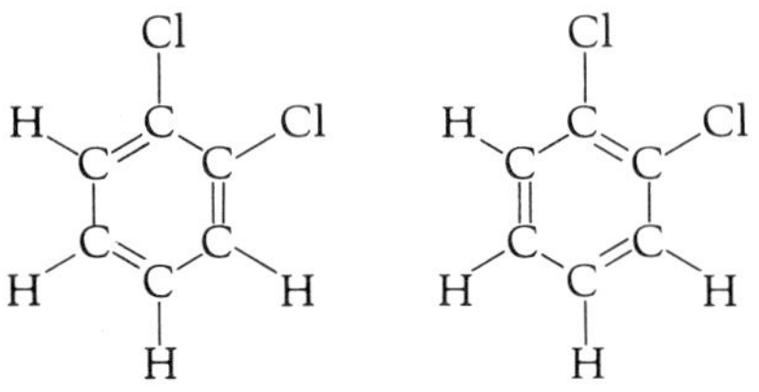

Note that, in the compound on the left, the two chlorine atoms are bonded to carbon atoms joined by a single bond; in the compound on the right, the same two carbon atoms are double-bonded to each other. Yet only one kind of dichlorobenzene with chlorine atoms on adjacent carbon atoms had been prepared. Thus there was a discrepancy between what was predicted by Kekulé's structure and what was known experimentally. This discrepancy was resolved by treating the bonding in benzene in terms of the resonance concept (see Section 11.5) or the delocalization of molecular orbitals (see Section 13.5).

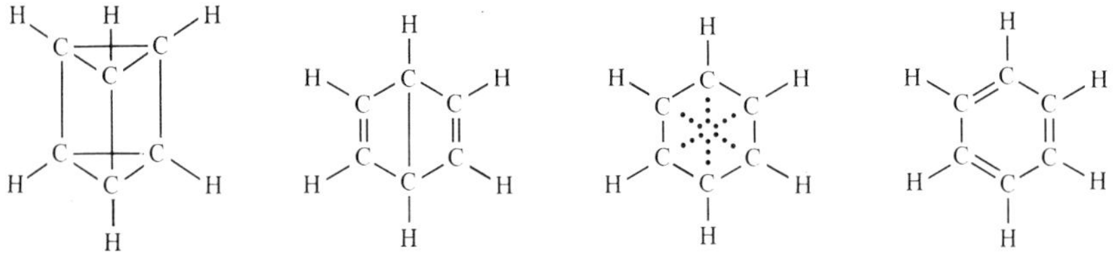

FIGURE 30.7 *Proposed structures of benzene. Molecules corresponding to Ladenburg's formula and Dewar's formula have been synthesized. These compounds bear no resemblance to benzene in their physical and chemical properties.*

Ladenburg's formula   Dewar's formula   Armstrong's formula   Kekulé's formula

Nomenclature of Aromatic Compounds. The naming of monosubstituted benzenes is quite straightforward, as shown below.

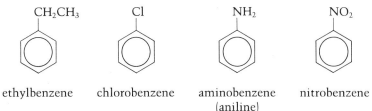

ethylbenzene 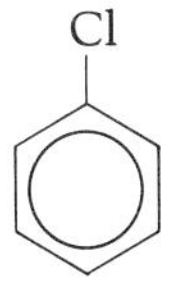   chlorobenzene   aminobenzene   nitrobenzene
                                    (aniline)

If more than one substituent is present, we must indicate the location of the second group relative to the first. The systematic way to accomplish this is to number the carbon atoms as follows:

Thus, there are three different dibromobenzenes:

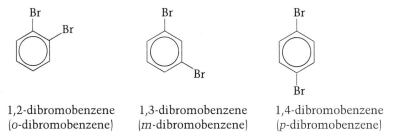

1,2-dibromobenzene   1,3-dibromobenzene   1,4-dibromobenzene
(o-dibromobenzene)   (m-dibromobenzene)   (p-dibromobenzene)

Frequently, the prefixes o- (ortho-), m- (meta-), and p- (para-) are used to denote the relative positions of the two substituted groups. For the dibromobenzenes, the names are as indicated. If the two groups are different, as in

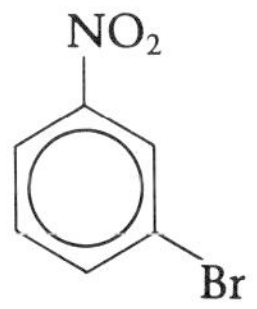

the systematic name is 3-bromonitrobenzene and the common name is *m*-bromonitrobenzene.

Properties and Reactions of Aromatic Compounds. Benzene is a colorless, flammable liquid obtained chiefly from coal tar. Perhaps the most remarkable chemical property of benzene is its relative inertness. Although it has the same empirical formula as acetylene (CH) and a high degree of unsaturation (three C=C bonds), it is much less reactive than either ethylene or acetylene. We now understand that the stability of benzene is the result of electron delocalization. In fact, benzene can be hydrogenated, but only with difficulty. The following reaction is carried out at significantly higher temperatures and pressures than the similar reactions for the alkenes:

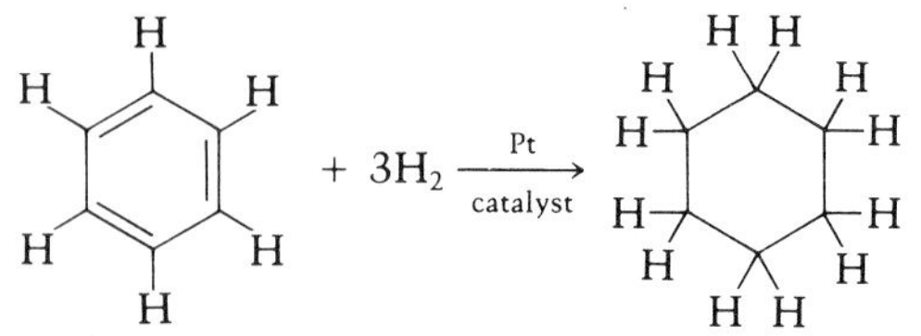

cyclohexane

We saw earlier that alkenes react readily with halogens to form addition products, because the pi bond in C=C can be broken easily. On the other hand, the most common reaction of halogens with benzene is one of *substitution:*

A *substitution reaction* is the replacement of an atom or a group of atoms in a compound by another atom or another group of atoms.

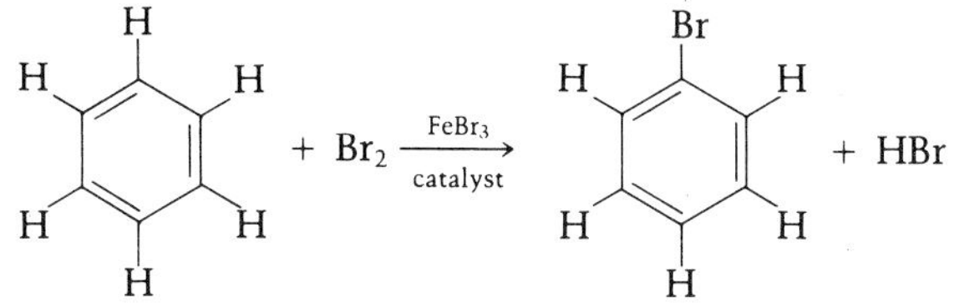

bromobenzene

Note that if the reaction were an addition reaction, electron delocalization would be destroyed in the product

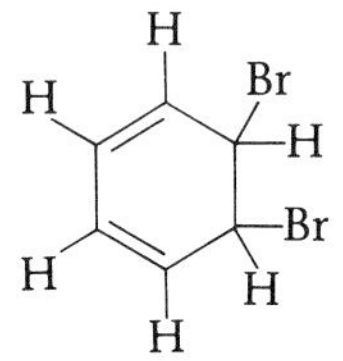

and the molecule would not exhibit the usual aromatic characteristic of chemical unreactivity.

Bromobenzene, C_6H_5Br, contains a benzene ring that has lost one hydrogen atom. The C_6H_5 group is called the *phenyl group. Any group that contains one or more fused benzene rings* is called an **aryl group.** Thus the phenyl group is the simplest aryl group. Naphthalene $(C_{10}H_8)$ less one hydrogen atom $(C_{10}H_7)$ is the next simplest aryl group.

Alkyl groups can be introduced into the ring system by allowing benzene to react with an alkyl halide using $AlCl_3$ as the catalyst:

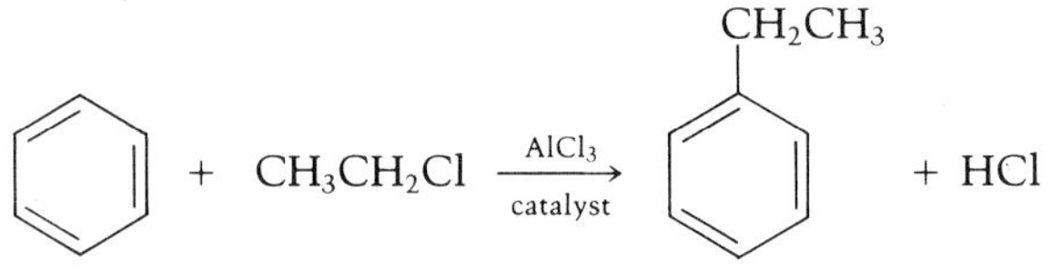

ethyl chloride   ethylbenzene

An enormously large number of compounds can be generated from substances in which benzene rings are fused together. Some of these *polycyclic*

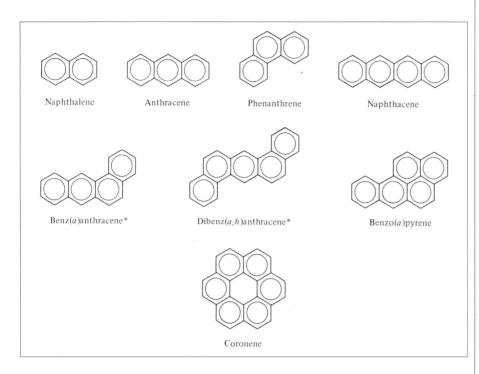

FIGURE 30.8 *Some polycyclic aromatic hydrocarbons. Compounds denoted by * are potent carcinogens. An enormous number of such compounds exist in nature.*

aromatic hydrocarbons are shown in Figure 30.8. The best known of these compounds is naphthalene, an important component of mothballs. These and many other similar compounds are present in coal tar. Some of the compounds with larger numbers of rings are powerful carcinogens—that is, they can cause cancer in humans and other animals. As far back as 1775, the English doctor Percivall Pott treated chimney sweeps for cancer of the scrotum. These workers entered the trade when young and were exposed daily to massive amounts of soot and tar, containing many kinds of carcinogens. Not until some 150 years later, however, were these compounds isolated and demonstrated to be dangerous. Exactly how these substances affect the biological system is still not known.

30.2 FUNCTIONAL GROUPS

We turn our attention now to organic functional groups—groups that are responsible for most of the reactions of the parent compounds. In particular, we will concentrate on oxygen-containing and nitrogen-containing compounds.

Alcohols

*All **alcohols** contain the hydroxyl group —OH.* Some common alcohols are shown in Figure 30.9. Ethyl alcohol, or ethanol, is by far the best known. It is produced biologically by the fermentation of sugar or starch. In the

The functional group in alcohols is the hydroxyl group —OH.

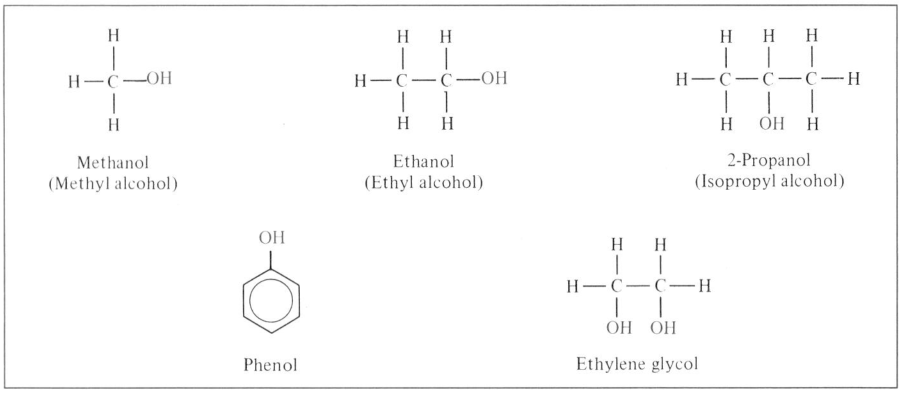

FIGURE 30.9 *Common alcohols. Note that all these compounds contain the —OH group.*

Methanol
(Methyl alcohol)

Ethanol
(Ethyl alcohol)

2-Propanol
(Isopropyl alcohol)

Phenol

Ethylene glycol

absence of oxygen the enzymes present in bacterial culture or yeast catalyze the reaction:

$$C_6H_{12}O_6(aq) \xrightarrow{\text{enzyme}} 2CH_3CH_2OH(aq) + 2CO_2(g)$$

This process gives off energy, which microorganisms use for growth and other functions.

Ethanol is prepared commercially by the addition reaction of water to ethylene at about 280°C and 300 atm:

$$CH_2{=}CH_2(g) + H_2O(g) \xrightarrow[\text{catalyst}]{H_3PO_4} CH_3CH_2OH(g)$$

Ethanol has countless applications as a solvent for organic chemicals and as a starting compound for the manufacture of dyes, synthetic drugs, cosmetics, explosives, and so on. It is familiar as a major beverage constituent. Ethanol is the only nontoxic (more properly, the least toxic) of the straight-chain alcohols; our bodies have an enzyme, called *alcohol dehydrogenase*, that helps to metabolize ethanol by oxidizing it to acetaldehyde:

$$CH_3CH_2OH \xrightarrow{\text{enzyme}} \underset{\text{acetaldehyde}}{CH_3CHO} + H_2$$

This equation is a simplified version of what actually takes place; the H atoms are taken up by other molecules, so that no H_2 gas is evolved.

Ethanol can be oxidized also by inorganic oxidizing agents, such as acidified dichromate, to acetaldehyde and acetic acid:

$$CH_3CH_2OH \xrightarrow[H^+]{Cr_2O_7^{2-}} CH_3CHO \xrightarrow[H^+]{Cr_2O_7^{2-}} CH_3COOH$$

Aliphatic alcohols are derived from alkanes, which are aliphatic hydrocarbons.

The simplest aliphatic alcohol is methanol. Called *wood alcohol*, it was at one time prepared by the dry distillation of wood. It is now synthesized industrially through the reaction of carbon monoxide and molecular hydrogen at high temperatures and pressures:

$$CO(g) + 2H_2(g) \xrightarrow[\text{catalyst}]{Fe_2O_3} CH_3OH(g)$$

Methanol is highly toxic. Ingestion of only a few milliliters can cause nausea and blindness. Ethanol intended for industrial use is often mixed with methanol to prevent people from drinking it. Ethanol containing methanol or other toxic substances is called *denatured alcohol.*

The alcohols are very weakly acidic; they do not react with strong bases, such as NaOH. The alkali metals react with alcohols to produce hydrogen:

$$2CH_3OH + 2Na \longrightarrow 2NaOCH_3 + H_2$$
$$\text{sodium}$$
$$\text{methoxide}$$

However, the reaction is much less violent than that between Na and water:

$$2H_2O + 2Na \longrightarrow 2NaOH + H_2$$

Two other familiar aliphatic alcohols are isopropyl alcohol, commonly known as rubbing alcohol, and ethylene glycol, which is used as an antifreeze. Note that ethylene glycol has two —OH groups and so can form hydrogen bonds with water molecules more effectively than compounds that have only one —OH group (see Figure 30.9). Most alcohols—especially those with low molar masses—are highly flammable.

Ethers

Ethers *contain the R—O—R' linkage, where R and R' are either an alkyl group or an aryl group.* They are formed by the condensation reaction of alcohols. A **condensation reaction** is characterized by the *joining of two molecules and the elimination of a small molecule, usually water.*

$$CH_3OH + HOCH_3 \xrightarrow[\text{catalyst}]{H_2SO_4} CH_3OCH_3 + H_2O$$
$$\text{dimethyl}$$
$$\text{ether}$$

Like alcohols, ethers are extremely flammable. When left standing in air, they have a tendency to slowly form explosive peroxides:

$$\underset{\text{diethyl ether}}{C_2H_5OC_2H_5} + O_2 \longrightarrow C_2H_5O-\overset{\overset{\displaystyle CH_3}{|}}{\underset{\underset{\displaystyle H}{|}}{C}}-O-O-H$$

Peroxides contain the —O—O— linkage; the simplest peroxide is hydrogen peroxide (H_2O_2). Diethyl ether, commonly known as "ether," was used as an anesthetic for many years. It produces unconsciousness by depressing the activity of the central nervous system. The major disadvantages of ether are its irritating effects on the respiratory system and the occurrence of post-anesthetic nausea and vomiting. "Neothyl," or methyl propyl ether, $CH_3OCH_2CH_2CH_3$, is currently favored as an anesthetic because it is relatively free of side effects.

Aldehydes and Ketones

Under mild oxidation conditions, alcohols can be converted to aldehydes and ketones:

$$CH_3OH + \tfrac{1}{2}O_2 \longrightarrow \underset{\text{formaldehyde}}{H_2C{=}O} + H_2O$$

$$C_2H_5OH + \tfrac{1}{2}O_2 \longrightarrow \underset{\text{acetaldehyde}}{\overset{\displaystyle H_3C}{\underset{\displaystyle H}{\diagdown}}}C{=}O + H_2O$$

$$\underset{\underset{\displaystyle OH}{|}}{CH_3{-}\overset{\overset{\displaystyle H}{|}}{C}{-}CH_3} + \tfrac{1}{2}O_2 \longrightarrow \underset{\text{acetone}}{\overset{\displaystyle H_3C}{\underset{\displaystyle H_3C}{\diagdown\!\diagup}}}C{=}O + H_2O$$

The functional group in these compounds is the carbonyl group, $\diagup\!\!\!C{=}O$.

The difference between **aldehydes** and **ketones** is that *in aldehydes at least one hydrogen atom is bonded to the carbon atom in the carbonyl group, whereas in ketones no hydrogen atoms are bonded to this carbon atom.*

Formaldehyde ($H_2C{=}O$) has a tendency to *polymerize;* that is, the individual molecules join together to form a compound of high molar mass. This action gives off much heat and is often explosive, so formaldehyde is usually prepared and stored in aqueous solution (to reduce the concentration). This rather disagreeable-smelling liquid is used as a starting material in the polymer industry (see Section 30.3) and in the laboratory as a preservative for dead animals. Interestingly, the higher molar mass aldehydes, such as cinnamic aldehyde

have a pleasant odor and are used in the manufacture of perfumes.

Ketones generally are less reactive than aldehydes. The simplest ketone, acetone, is a pleasant-smelling liquid that is used mainly as a solvent for organic compounds. Acetone is much more difficult to oxidize than acetaldehyde.

Carboxylic Acids

Under appropriate conditions, both alcohols and aldehydes can be oxidized to **carboxylic acids,** *acids that contain the carboxyl group—COOH:*

$$CH_3CH_2OH + O_2 \longrightarrow CH_3COOH + H_2O$$

$$CH_3CHO + \tfrac{1}{2}O_2 \longrightarrow CH_3COOH$$

The oxidation of ethanol to acetic acid in wine is catalyzed by enzymes.

These reactions occur so readily, in fact, that wine must be protected from atmospheric oxygen while in storage. Otherwise, it would soon taste like vinegar (due to the formation of acetic acid). Figure 30.10 shows the structure of some of the common carboxylic acids.

Carboxylic acids are widely distributed in nature; they are found in both

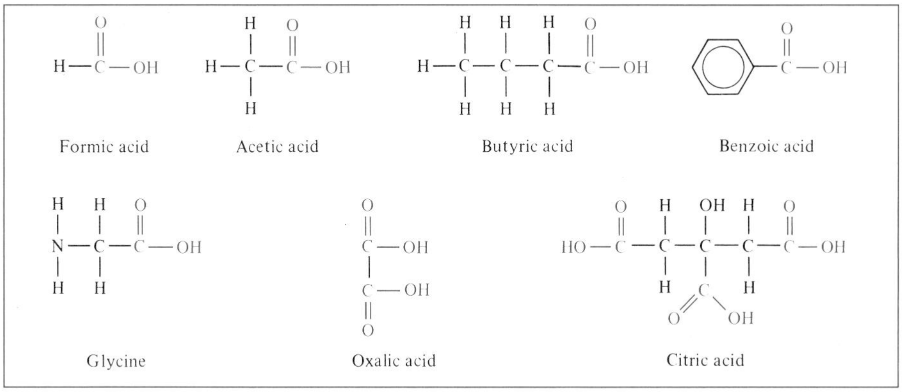

FIGURE 30.10 *Some common carboxylic acids. Note that they all contain the —COOH group. (Glycine is one of the twenty amino acids found in proteins.)*

the plant and animal kingdoms. All protein molecules are made of amino acids, a special kind of carboxylic acid, in which an amino functional group and a carboxylic acid group are present in the same molecule.

The Lewis formula of the carboxyl group (—COOH) is

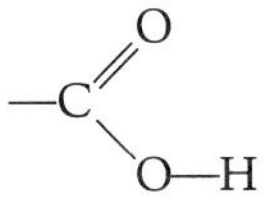

$$-C \diagup\!\!\!\!\diagdown \begin{matrix} O \\ O\!-\!H \end{matrix}$$

The ionizable proton accounts for the acidic properties in carboxylic acids:

$$CH_3COOH \rightleftharpoons CH_3COO^- + H^+$$

acetic   acetate
acid   ion

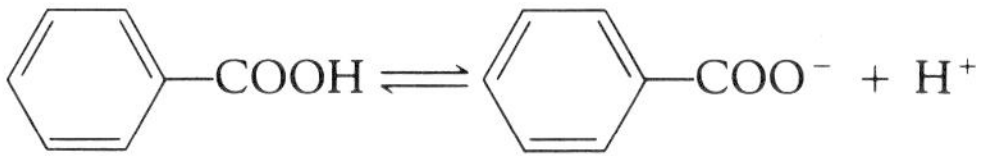

$$\text{benzoic acid} \rightleftharpoons \text{benzoate ion} + H^+$$

Unlike the inorganic acids HCl, HNO_3, and H_2SO_4, carboxylic acids are usually weak. They react with alcohols to form pleasant-smelling esters:

$$CH_3COOH + HOCH_2CH_3 \longrightarrow CH_3\!-\!\overset{\displaystyle O}{\overset{\|}{C}}\!-\!O\!-\!CH_2CH_3 + H_2O$$

ethyl acetate

Other common reactions of carboxylic acids are neutralization

$$CH_3COOH + NaOH \longrightarrow CH_3COONa + H_2O$$

and formation of acid halides, such as acetyl chloride:

$$CH_3COOH + PCl_5 \longrightarrow CH_3COCl + HCl + POCl_3$$

acetyl   phosphoryl
chloride   chloride

Acid halides are reactive compounds used as intermediates in the preparation of many other organic compounds. They hydrolyze in much the same way as many nonmetallic halides, such as $SiCl_4$:

$$CH_3COCl(l) + H_2O(l) \longrightarrow CH_3COOH(aq) + HCl(g)$$

$$SiCl_4(l) + 3H_2O(l) \longrightarrow H_2SiO_3(s) + 4HCl(g)$$
silicic acid

Esters

Esters *have the general formula R'COOR, where R' can be H or an alkyl group or an aryl group and R is an alkyl group or an aryl group.* Esters are used in the manufacture of perfumes and as flavoring agents in the confectionery and soft-drink industries. Many fruits owe their characteristic smell and flavor to the presence of small quantities of esters.

The functional group in esters is the —COOR group. In the presence of an acid, such as HCl, esters undergo hydrolysis to yield a carboxylic acid and an alcohol. For example, in acid solution, ethyl acetate hydrolyzes as follows:

$$CH_3COOC_2H_5 + H_2O \xrightarrow[\text{catalyst}]{H^+} CH_3COOH + C_2H_5OH$$
ethyl acetate $\qquad\qquad\qquad$ acetic acid $\quad$ ethanol

However, this reaction does not go to completion, because the reverse reaction also occurs to an appreciable extent. On the other hand, when NaOH solution is used in hydrolysis:

$$CH_3COOC_2H_5 + NaOH \longrightarrow CH_3COO^-Na^+ + C_2H_5OH$$
ethyl acetate $\qquad\qquad\qquad$ sodium acetate $\quad$ ethanol

The sodium acetate does not react with ethanol, so this reaction does go to completion from left to right. For this reason, ester hydrolysis is usually carried out in basic solutions. Note that NaOH does not act as a catalyst, since it is consumed by the reaction. The term **saponification** (meaning soapmaking) was originally used to describe the alkaline hydrolysis of fatty acid esters to yield soap molecules (sodium stearate):

$$C_{17}H_{35}COOC_2H_5 + NaOH \longrightarrow C_{17}H_{35}COO^-Na^+ + C_2H_5OH$$
ethyl stearate $\qquad\qquad\qquad$ sodium stearate

Saponification has now become *a general term for alkaline hydrolysis of any type of ester.*

Amines

Amines *are organic bases. They have the general formula R_3N, where R may be H, an alkyl group, or an aryl group.* As with ammonia, the reaction of amines with water is

$$RNH_2 + H_2O \rightleftharpoons RNH_3^+ + OH^-$$

where R represents either an alkyl or an aryl group. Like all bases, the amines form salts when reacted with acids:

$$CH_3CH_2NH_2 + HCl \longrightarrow CH_3CH_2NH_3^+Cl^-$$
ethylamine $\qquad\qquad\qquad$ ethylammonium
$\qquad\qquad\qquad\qquad\qquad\qquad$ chloride

These salts are usually colorless, odorless solids.

Aromatic amines are used mainly in the manufacture of dyes. Aniline, the simplest aromatic amine, itself is a toxic compound; a number of other aromatic amines such as 2-naphthylamine and benzidine are potent carcinogens:

aniline   2-naphthylamine   benzidine

Table 30.5 summarizes the common functional groups. A compound can and often does contain more than one functional group. Generally, the reactivity of a compound is determined by the number and types of functional groups in its makeup.

TABLE 30.5 Important Functional Groups and Their Reactions

Functional Group	Name	Typical Reactions
$\diagdown C = C \diagup$	Carbon–carbon double bond	Addition reactions with halogens, hydrogen halides, and water; hydrogenation to yield alkanes
$-C \equiv C-$	Carbon–carbon triple bond	Addition reactions with halogens, hydrogen halides; hydrogenation to yield alkenes and alkanes
$-\ddot{\underset{\cdot\cdot}{X}}:$ (X = F, Cl, Br, I)	Halogen	Exchange reactions: $CH_3CH_2Br + KI \longrightarrow$ $CH_3CH_2I + KBr$
$-\ddot{\underset{\cdot\cdot}{O}}-H$	Hydroxyl	Esterification (formation of an ester) with carboxylic acids; oxidation to aldehydes, ketones, and carboxylic acids
$\diagdown C = \ddot{\underset{\cdot\cdot}{O}}$	Carbonyl	Reduction to yield alcohols; oxidation of aldehydes to yield carboxylic acids
$\overset{:O:}{\underset{}{\overset{\parallel}{-C}}}-\ddot{O}-R$ (R = alkyl or aryl)	Ester	Hydrolysis to yield acids
$\overset{:O:}{\overset{\parallel}{-C}}-\ddot{O}-H$	Carboxyl	Esterification with alcohols; reaction with phosphorus pentachloride to yield acid chlorides
$-\ddot{N} \diagup{\overset{R}{}}_{\diagdown R}$ (R = H, alkyl or aryl)	Amine	Formation of ammonium salts with acids

EXAMPLE 30.3

Cholesterol is a major component of gallstones. It has received much attention because of the suspected correlation between cholesterol level in the blood and certain types of heart disease, for example, atherosclerosis. From the following structure of the compound, predict its reaction with (a) Br_2, (b) H_2 (in the presence of Pt catalyst), (c) CH_3COOH.

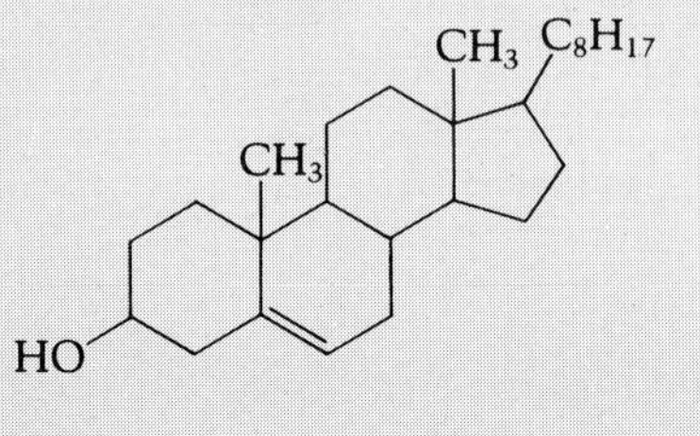

Answer

The first step is to identify the functional groups in cholesterol. There are two: the hydroxyl group and the carbon–carbon double bond. (a) The reaction with bromine results in the addition of bromine to the double-bonded carbons, which become single-bonded. (b) This is a hydrogenation reaction. Again, the carbon–carbon double bond is converted to a carbon–carbon single bond. (c) The acid reacts with the hydroxyl group to form an ester and water. Figure 30.11 shows the products of these reactions.

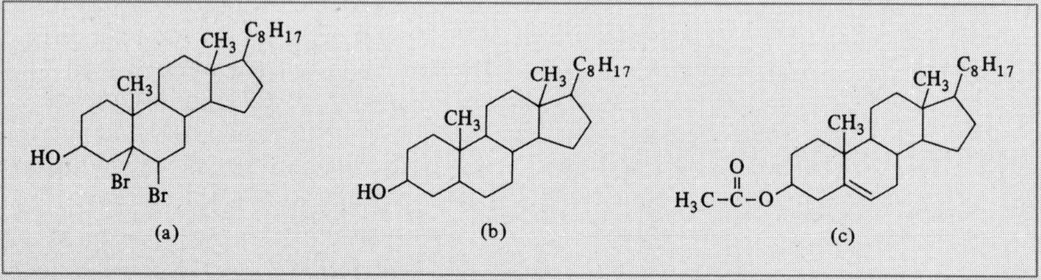

FIGURE 30.11 *The products formed by the reaction of cholesterol with (a) molecular bromine, (b) molecular hydrogen, and (c) acetic acid.*

Similar example: Problem 30.29.

30.3 ORGANIC POLYMERS

A ***polymer*** is a *chemical species distinguished by a high molar mass, ranging into thousands and millions of grams.* Polymers are often called *macromolecules.* The properties of polymers differ greatly from those of small, ordinary molecules, and special techniques are required to study these giant molecules.

Polymers are divided into two classes: natural and synthetic. Examples of natural polymers are proteins, nucleic acids, cellulose (polysaccharides), and rubber (polyisoprene). Most synthetic polymers are organic compounds such as nylon (polyhexamethylene adipamide), Dacron (polyethylene terephthalate), and Lucite or Plexiglas (polymethyl methacrylate). Proteins and nucleic acids will not be discussed in this book. We will focus on rubber and several synthetic organic polymers.

The development of polymer chemistry began in the 1920s, when chemists were making great progress in clarifying the chemical structures of various substances. However, they were generally puzzled by the behavior of certain materials such as wood, gelatin, cotton, and rubber. For example, when rubber, with the known empirical formula of C_5H_8, was dissolved in an organic solvent, the solution displayed several unusual properties—high viscosity, low osmotic pressure, and negligible depression in the freezing point of the solvent.

These observations strongly suggested the presence of very high molar mass solutes, but chemists were not ready at that time to accept the idea that such giant molecules could exist. Instead, they postulated that materials like rubber consist of aggregates of small molecular units, like C_5H_8 or $C_{10}H_{16}$, held together by intermolecular forces. This misconception persisted for a number of years, until Hermann Staudinger (1881–1963) clearly showed that these so-called aggregates are, in fact, enormously large molecules, each of which contains many thousands of atoms held together by covalent bonds.

Once the properties of these macromolecules were understood, the way was open for manufactured polymers, which play such an important role in almost every aspect of our daily lives. About 90 percent of today's chemists, including the biochemists, work with polymers in one respect or another.

Because of their size, polymers may seem inexplicably complex. We might expect molecules containing thousands of carbon and hydrogen atoms to form an almost infinite number of isomers (both structural and geometric). However, *these molecules all contain rather simple repeating units*, called **monomers** (Table 30.6), which severely restrict the number of isomers that can exist.

Consider polyethylene, for example, which is formed by the addition reaction of ethylene molecules. Each step involves the breaking of the pi bond in the C═C linkage in an ethylene molecule, which can then join with two other ethylene molecules whose pi bonds have also been broken. In this fashion, thousands of ethylene units can be joined together to form a very long chain (Figure 30.12). Individual chains pack together well, accounting for the crystal-like properties of the material. Polyethylene is both strong and flexible, but has a rather low melting point.

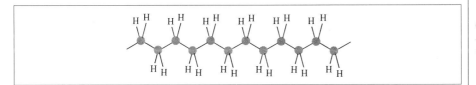

FIGURE 30.12 *The structure of polyethylene. Each carbon atom is sp^3 hybridized.*

TABLE 30.6 Some Monomers and Their Common Synthetic Polymers

Monomer		Polymer	
Formula	Name	Name and Formula	Uses
$H_2C{=}CH_2$	Ethylene	Polyethylene $-(CH_2{-}CH_2)_n$	Plastic piping, bottles, electrical insulation, toys
$H_2C{=}\overset{\displaystyle H}{\underset{\displaystyle CH_3}{C}}$	Propylene	Polypropylene $-(CH{-}CH_2{-}CH{-}CH_2)_n$ with CH_3 CH_3	Packaging film, carpets, crates for soft-drink bottles, lab wares, toys
$H_2C{=}\overset{\displaystyle H}{\underset{\displaystyle Cl}{C}}$	Vinyl chloride	Polyvinyl chloride (PVC) $-(CH_2{-}CH)_n$ with Cl	Piping, siding, gutters, floor tile, clothing, toys
$H_2C{=}\overset{\displaystyle H}{\underset{\displaystyle CN}{C}}$	Acrylonitrile	Polyacrylnitrile (PAN) $-(CH_2{-}CH)_n$ with CN	Carpets, knitwear
$F_2C{=}CF_2$	Tetrafluoroethylene	Teflon $-(CF_2{-}CF_2)_n$	Cooking utensils, electrical insulation, bearings
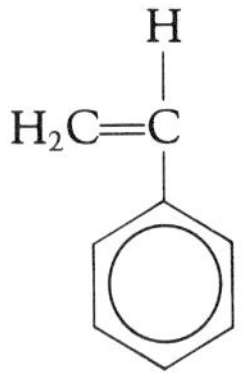 $H_2C{=}\overset{\displaystyle H}{\underset{\displaystyle \bigcirc}{C}}$	Styrene	Polystyrene $-(CH_2{-}CH)_n$ with ⬡ 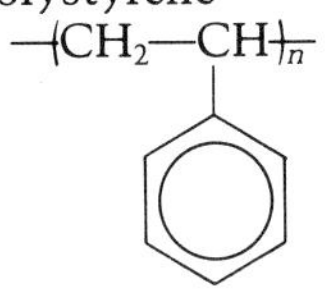	Containers, thermal insulation (ice buckets, water coolers), toys
$H_2C{=}\overset{\displaystyle H}{C}{-}\overset{\displaystyle H}{C}{=}CH_2$	Butadiene	Polybutadiene $-(CH_2CH{=}CHCH_2)_n$	Tire tread, coating resin
See above structures	Butadiene and styrene	Styrene-butadiene rubber 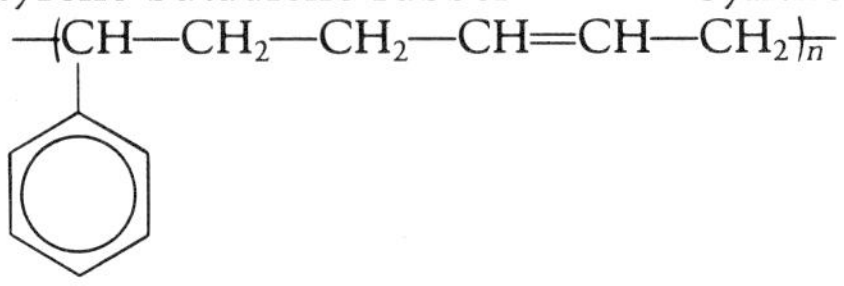 $-(CH{-}CH_2{-}CH_2{-}CH{=}CH{-}CH_2)_n$ with ⬡	Synthetic rubber
$OCN{-}(CH_2)_2{-}NCO$ and $HO{-}(CH_2)_2{-}OH$	Ethylene diisocyanate and 1,2-ethylene glycol	Polyurethane $\left(NH{-}(CH_2)_2{-}NH{-}\overset{\displaystyle O}{\overset{\displaystyle \|}{C}}{-}O{-}(CH_2)_2{-}O\right)_n$	Coating, insulator, adhesive

The chemistry becomes more complex if the starting units are asymmetric; for example

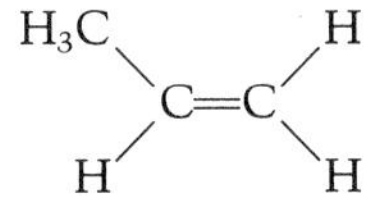

$$H_3C, \quad H$$
$$C=C$$
$$H \quad H$$
propylene

Several structurally different polymers can result from an addition reaction of asymmetrical propylenes. If the additions occur randomly, we obtain the *atactic* polypropylenes, which do not pack together well (Figure 30.13). The result is a rubbery, amorphous polymer of relatively low strength. Two other possibilities yield *isotactic* and *syndiotactic* polypropylenes, also shown in Figure 30.13. Of these, the former has a higher melting point and possesses greater crystallinity and superior mechanical properties.

One of the major problems facing the polymer industry was how to selectively synthesize either the isotactic or syndiotactic polymer without having it contaminated by other products. The answer was provided by the work of Giulio Natta (1903–   ) and Karl Ziegler (1898–   ), who demonstrated that the product with the desired structure can be obtained by using certain "stereo catalysts" such as triethylaluminum $[Al(C_2H_5)_3]$ and titanium trichloride $(TiCl_3)$, which promote the formation only of specific isomers. Their work opened up almost limitless possibilities for chemists to design and synthesize polymers to suit any particular purpose.

Rubber, or poly-*cis*-isoprene, is the best-known natural organic polymer:

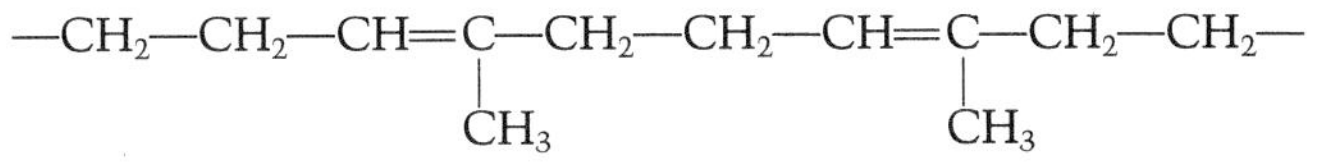

$$-CH_2-CH_2-CH=C-CH_2-CH_2-CH=C-CH_2-CH_2-$$
$$\qquad\qquad\quad CH_3 \qquad\qquad\qquad\quad CH_3$$

It is extracted from the tree *Hevea brasiliensis*. The unusual and useful property of rubber is its elasticity. It can be reversibly extended up to ten times its original length. In contrast, a piece of copper wire can at most be

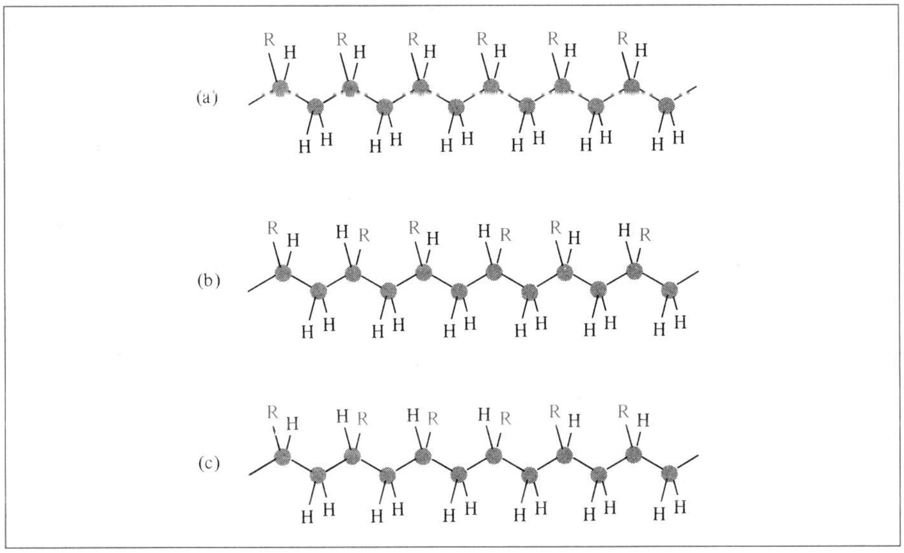

FIGURE 30.13 *Stereoisomers of polymers. When the R group is* —CH_3, *the polymer is polypropylene. (a) When the R groups are all on one side of the chain, the polymer is said to be isotactic. (b) When R groups alternate from side to side, the polymer is said to be syndiotactic. (c) When R groups are disposed at random, the polymer is atactic.*

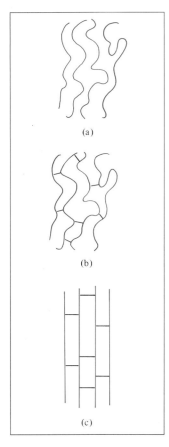

FIGURE 30.14 *Rubber molecules ordinarily are bent and convoluted. (a) and (b) represent the long chains before and after vulcanization, respectively; (c) shows the alignment of molecules when stretched. Without vulcanization, these molecules would slip past one another.*

stretched a few percent of its length. This extraordinary property results from the flexibility of the long-chain molecules. In the bulk state, however, rubber is a tangle of polymeric chains and if the external force is strong enough, individual chains slip past one another, thereby causing the rubber to lose most of its elasticity. In 1839, Charles Goodyear (1800–1860) discovered that natural rubber could be cross-linked with sulfur to prevent chain slippage (Figure 30.14). His process is known as *vulcanization*. This discovery paved the way for many practical and commercial uses of rubber, as in automobile tires.

During World War II there was a shortage of natural rubber in the United States, so an intensive program was launched to produce synthetic rubber. Most synthetic rubbers (called *elastomers*) are formed from products of the petroleum industry such as ethylene, propylene, and butadiene. For example, chloroprene molecules polymerize readily to form polychloroprene, commonly known as *neoprene*, which has properties that are comparable or even superior to those of natural rubber.

$$H_2C=CH-CCl=CH_2 \qquad \left(\begin{array}{c} CH_2 \qquad H \\ \diagdown \qquad \diagup \\ C=C \\ \diagup \qquad \diagdown \\ Cl \qquad CH_2 \end{array} \right)_n$$

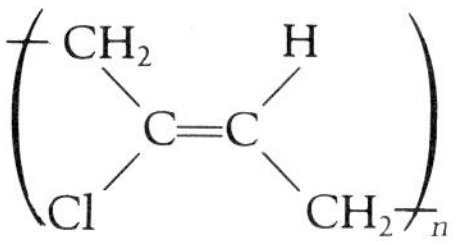

chloroprene   polychloroprene

Another important synthetic rubber is styrene-butadiene rubber (SBR), formed by the addition of butadiene to styrene molecules in 3 to 1 ratio. Because two different types of monomers

$$CH_2=CH-CH=CH_2 \qquad \langle\!\!\!\bigcirc\!\!\!\rangle-CH=CH_2$$

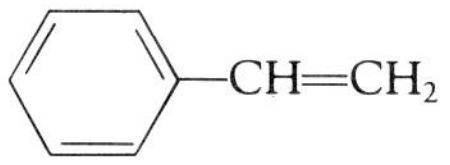

butadiene   styrene

are involved in forming SBR, this process is called *copolymerization*. **Copolymerization** is the *formation of a polymer that contains two or more different monomers.*

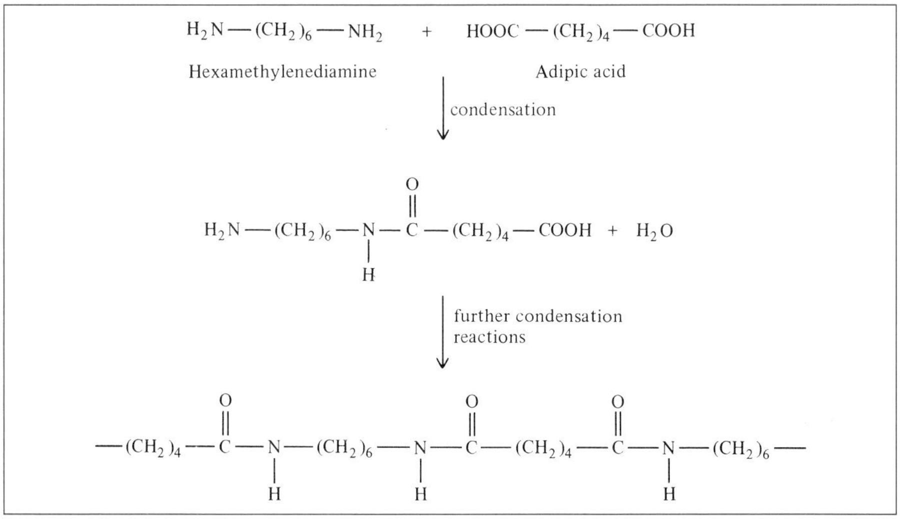

FIGURE 30.15 *The formation of nylon by the condensation reaction between hexamethylenediamine and adipic acid.*

$$H_2N-(CH_2)_6-NH_2 \quad + \quad HOOC-(CH_2)_4-COOH$$

Hexamethylenediamine   Adipic acid

$\downarrow$ condensation

$$H_2N-(CH_2)_6-N-\overset{\overset{\displaystyle O}{\|}}{C}-(CH_2)_4-COOH \ + \ H_2O$$
$\quad\quad\quad\quad\quad\quad\quad\ \ |$
$\quad\quad\quad\quad\quad\quad\quad\ \ H$

$\downarrow$ further condensation reactions

$$-(CH_2)_4-\overset{\overset{\displaystyle O}{\|}}{C}-N-(CH_2)_6-N-\overset{\overset{\displaystyle O}{\|}}{C}-(CH_2)_4-\overset{\overset{\displaystyle O}{\|}}{C}-N-(CH_2)_6-$$
$\quad\quad\quad\quad\quad\quad\ |\quad\quad\quad\quad\quad\ |\quad\quad\quad\quad\quad\quad\quad\quad\ |$
$\quad\quad\quad\quad\quad\quad\ H\quad\quad\quad\quad\quad H\quad\quad\quad\quad\quad\quad\quad\quad H$

FIGURE 30.16 *A Jarvik-7 artificial heart similar to those used on Barney B. Clark, William B. Schroeder, and Murray P. Haydon.*

Polyaddition of individual units is one way to form macromolecules. Another process is called *polycondensation.* We have seen several examples of condensation reactions in this chapter. One of the best-known polycondensation processes is that involving hexamethylenediamine and adipic acid, shown in Figure 30.15. The final product, called *nylon,* was first made by Wallace H. Carothers (1896–1937) at Du Pont in 1931. The versatility of this material is so great that the annual production of nylon and related substances now amounts to several billion pounds.

Figure 30.16 shows an artificial heart and artificial artery sections made of polyurethane supported on an aluminum base.

The first artificial heart (known as the Jarvik-7) implanted in a human (in 1982) was made of polyurethane.

SUMMARY

1. Because carbon atoms can link with other carbon atoms in straight and branched chains, carbon can form more compounds than any other element.
2. Methane, CH_4, is the simplest of the alkanes, a family of hydrocarbons with the general formula C_nH_{2n+2}. Cyclopropane, C_3H_6, is the simplest of the cycloalkanes, a family of alkanes whose carbon atoms are joined in a ring. Alkanes and cycloalkanes are saturated hydrocarbons.
3. Ethylene, $CH_2\!\!=\!\!CH_2$, is the simplest of the olefins, or alkenes, a class of hydrocarbons containing carbon–carbon double bonds and having the general formula C_nH_{2n}. Acetylene, $CH\!\!\equiv\!\!CH$, is the simplest of the alkynes, which are compounds that have the general formula C_nH_{2n-2} and contain carbon–carbon triple bonds.
4. Compounds that contain one or more benzene rings are called aromatic hydrocarbons. These compounds undergo substitution by halogens and alkyl groups.

5. Functional groups impart specific types of chemical reactivity to molecules. Classes of compounds characterized by their functional groups include alcohols, ethers, aldehydes and ketones, carboxylic acids and esters, and amines.
6. The major source of hydrocarbons is the petroleum industry, which furnishes gasoline and heating fuel, as well as raw materials for the chemical industry.
7. Polymers are large molecules formed by the combination of small, repeating units called monomers.
8. Proteins, nucleic acids, cellulose, and rubber are natural polymers. Nylon, Dacron, and Lucite are examples of synthetic polymers.
9. By varying the structure of a polymer from atactic, to isotactic, to syndiotactic, a wide variety of properties can be imparted to the polymer.
10. Synthetic rubbers include polychloroprene and styrene-butadiene rubber, which is made by the copolymerization of styrene and butadiene.

KEY WORDS

Alcohol, p. 931
Aldehyde, p. 934
Aliphatic hydrocarbon, p. 916
Alkane, p. 916
Alkene, p. 924
Alkyne, p. 926
Amine, p. 936
Aromatic hydrocarbon, p. 916
Aryl group, p. 930
Carboxylic acid, p. 934
Condensation reaction, p. 933
Copolymerization, p. 942
Cycloalkane, p. 923

Ester, p. 936
Ether, p. 933
Functional group, p. 916
Hydrocarbon, p. 916
Ketone, p. 934
Monomer, p. 939
Organic chemistry, p. 916
Polymer, p. 938
Radical, p. 921
Saponification, p. 936
Saturated hydrocarbon, p. 916
Structural isomer, p. 917
Unsaturated hydrocarbon, p. 924

PROBLEMS

More challenging problems are marked with an asterisk.

Hydrocarbons

30.1 Describe the differences among the alkanes, alkenes, alkynes, and aromatic hydrocarbons.

30.2 Draw all the possible structural isomers for the following molecules: (a) C_5H_{12}, (b) C_6H_{14}, (c) C_7H_{16}, (d) C_3H_6.

30.3 Draw all the possible isomers for the molecule C_4H_8.

30.4 Discuss how you can determine which of the following compounds might be alkanes, cycloalkanes, alkenes, or alkynes, without drawing their formulas: (a) C_6H_{12}, (b) C_4H_6, (c) C_5H_{12}, (d) C_7H_{14}, (e) C_3H_4.

30.5 The benzene and cyclohexane molecules both contain six-membered rings. Benzene is a planar molecule and cyclohexane is nonplanar. Explain.

30.6 Would you expect cyclobutadiene to be a stable molecule? Explain.

(Hint: Consider the angles between carbon atoms.)

*30.7 Suggest two chemical tests that would help you distinguish between these two compounds:

$$CH_3CH_2CH_2CH_2CH_3 \qquad CH_3CH_2CH_2CH=CH_2$$

*30.8 Draw the structures of *cis*-2-butene and *trans*-2-butene. Which of the two compounds would have the higher heat of hydrogenation? Explain.

30.9 Why do alkynes not exhibit geometric isomerism?

*30.10 Draw the structures of the geometric isomers of the molecule CH_3—$CH=CH$—$CH=CH$—CH_3.

30.11 When a mixture of methane and bromine is exposed to light, the following reaction occurs slowly:

$$CH_4(g) + Br_2(g) \longrightarrow CH_3Br(g) + HBr(g)$$

Suggest a reasonable mechanism for this reaction. *(Hint:* Bromine vapor is deep red; methane is colorless.)

30.12 Sulfuric acid, H_2SO_4 adds across the double bond of alkenes as H^+ and OSO_3H^-. Draw structures for the products of addition of sulfuric acid to (a) ethylene, (b) propylene, (c) 1-butene, and (d) 2-butene.

30.13 Draw all possible isomers for the molecule C_3H_5Br.

30.14 Indicate the asymmetric carbon atoms in the following compounds:

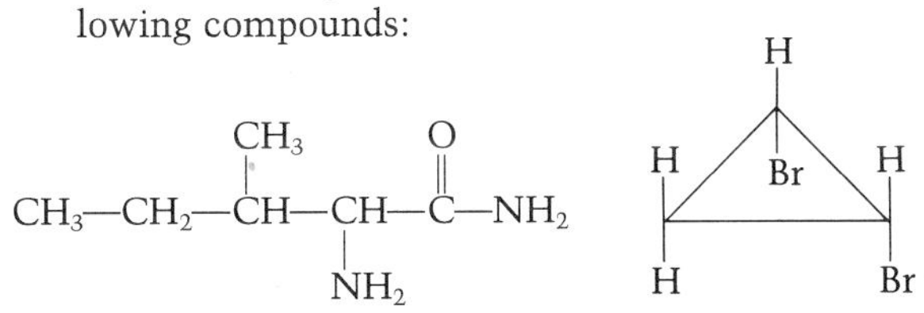

30.15 Ethanol (C_2H_5OH) and dimethyl ether (CH_3OCH_3) are structural isomers. Compare their melting points, boiling points, and solubilities in water.

30.16 Which member of each of the following pairs of compounds is the more reactive? (a) propane and cyclopropane, (b) ethylene and methane, (c) cyclohexane and benzene. Explain.

30.17 How many distinct chloropentanes, $C_5H_{11}Cl$, could be produced in the direct chlorination of *n*-pentane, $CH_3(CH_2)_3CH_3$? Draw the structure of each and identify any that are chiral.

*30.18 The structural isomers of pentane (C_5H_{12}) have substantially different boiling points (see Example 30.1). Explain the observed variation in boiling point, using structure as a reason.

30.19 Under certain conditions acetylene can trimerize to benzene:

$$3C_2H_2(g) \longrightarrow C_6H_6(l)$$

Calculate the standard enthalpy change in kJ for this reaction at 25°C.

30.20 Geometric isomers are not restricted to compounds containing the $C=C$ bond. For example, certain disubstituted cycloalkanes can exist in the *cis* and the *trans* forms. Name the following compounds:

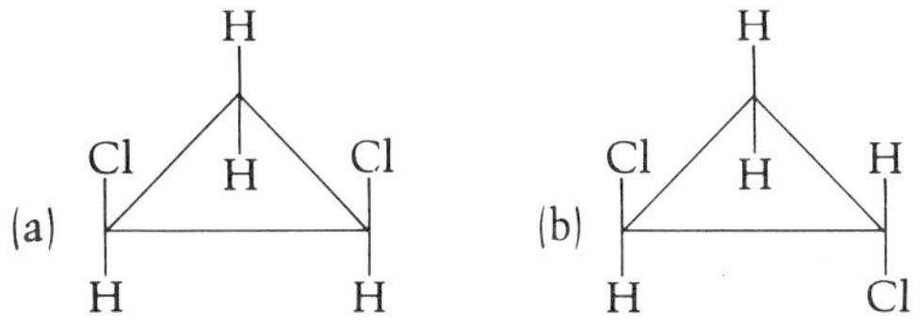

Functional Groups

30.21 Classify each of the following molecules as alcohol, aldehyde, ketone, carboxylic acid, amine, or ether:

(a) H_3C—O—CH_2—CH_3 (b) H_3C—CH_2—NH_2

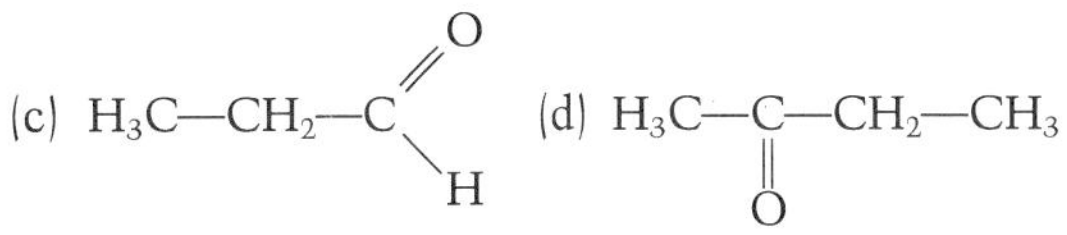

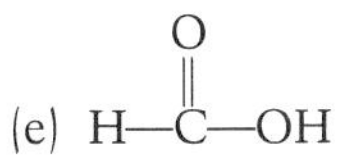

(e) H—C—OH (f) H_3C—CH_2CH_2—OH

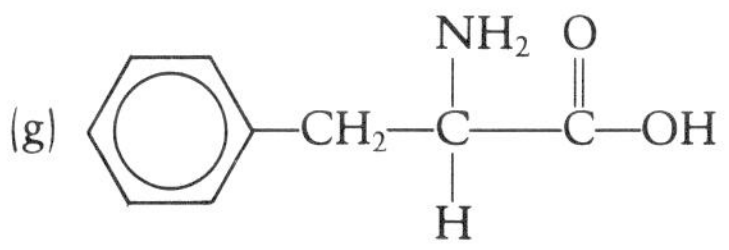

30.22 Draw Lewis formulas of the following functional groups: (a) aldehyde, (b) ketone, (c) carboxyl group, (d) amino group, (e) hydroxyl group.

30.23 Ethanol reacts with (a) gaseous HCl, (b) sodium metal. Predict the products formed in each reaction and write balanced equations for each.

30.24 Complete the following equation and identify the products:

$$CH_3—CH_2—COOH + CH_3OH \longrightarrow$$

*30.25 Generally aldehydes are more susceptible to air oxidation than are ketones. Use acetaldehyde and acetone as examples and show why ketones are more stable than aldehydes in this respect.

30.26 Draw structures for molecules with the following formulas: (a) CH_4O, (b) C_2H_6O, (c) $C_3H_6O_2$, (d) C_3H_8O.

30.27 A compound having the molecular formula $C_4H_{10}O$ does not react with sodium metal. In the presence of light, the compound reacts with Cl_2 to form three compounds having the formula C_4H_9OCl. Draw a structure of the original compound consistent with this information.

30.28 Indicate the classes of compounds having the formulas

(a) $H—C\equiv C—CH_3$
(b) C_4H_9OH
(c) $CH_3OC_2H_5$
(d) C_2H_5CHO
(e) C_6H_5COOH
(f) CH_3NH_2

30.29 Predict the product or products of each of the following reactions:

(a) $C_2H_5OH + HCOOH \longrightarrow$
(b) $H—C\equiv C—CH_3 + H_2 \longrightarrow$

(c) 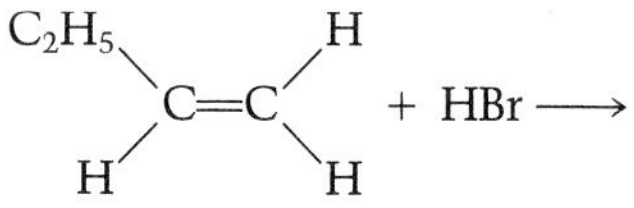
$$\underset{H}{\overset{C_2H_5}{\diagdown}} C=C \underset{H}{\overset{H}{\diagup}} + HBr \longrightarrow$$

30.30 Identify the functional groups in each of the following molecules:

(a) $CH_3CH_2COCH_2CH_2CH_3$
(b) $CH_3COOC_2H_5$
(c) $CH_3CH_2OCH_2CH_2CH_2CH_3$

30.31 Beginning with 3-methyl-1-butyne, show how you would prepare the following compounds:

(a) 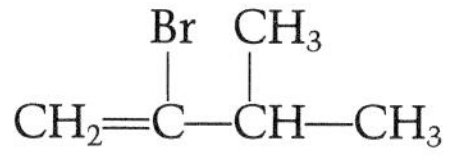
$$\underset{}{CH_2}=\overset{Br}{\underset{}{C}}-\overset{CH_3}{\underset{}{CH}}-CH_3$$

(b)
$$CH_2Br-CBr_2-\overset{CH_3^{\bullet}}{\underset{}{CH}}-CH_3$$

(c)
$$CH_3-\overset{Br}{\underset{}{CH}}-\overset{CH_3}{\underset{}{CH}}-CH_3$$

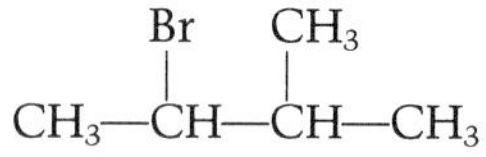

30.32 An organic compound has the empirical formula $C_5H_{12}O$. Upon controlled oxidation, it is converted into a compound of empirical formula $C_5H_{10}O$, which behaves like a ketone. Draw structures of the original compound and the final compound.

Nomenclature

30.33 Name the following compounds:

(a) 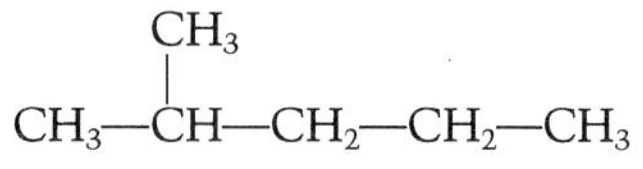
$$CH_3-\overset{CH_3}{\underset{}{CH}}-CH_2-CH_2-CH_3$$

(b) 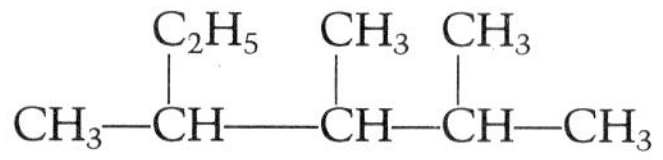
$$CH_3-\overset{C_2H_5}{\underset{}{CH}}-\overset{CH_3}{\underset{}{CH}}-\overset{CH_3}{\underset{}{CH}}-CH_3$$

(c)
$$CH_3-CH_2-\overset{}{\underset{CH_2-CH_2-CH_3}{CH}}-CH_2-CH_3$$

(d) 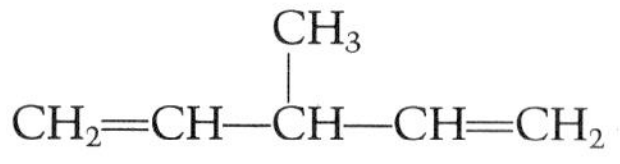
$$CH_2=CH-\overset{CH_3}{\underset{}{CH}}-CH=CH_2$$

(e) $CH_3-C\equiv C-CH_2-CH_3$

(f)

$CH_3-CH_2-CH-CH=CH_2$

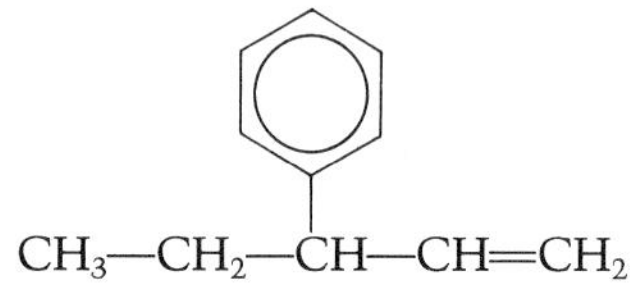

30.34 Name the following compounds:

(a)

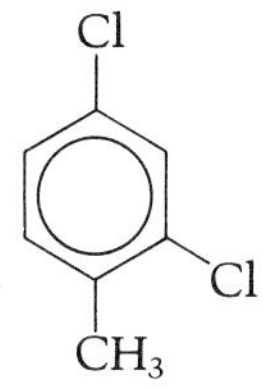

(b)

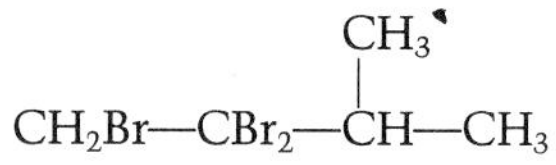

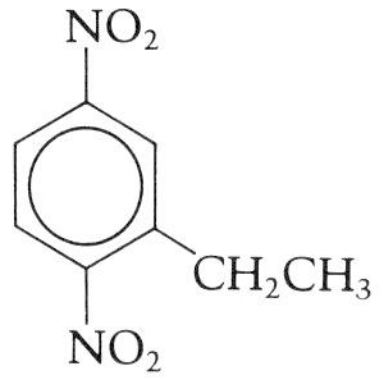

(c)

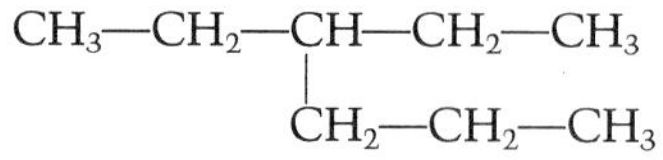

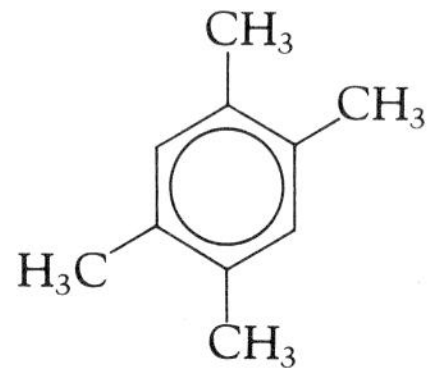

30.35 Write the structural formulas for the following compounds: (a) 3-methylhexane, (b) 1,3,5-trichlorocyclohexane, (c) 2,3-dimethylpentane, (d) 2-phenyl-4-bromopentane, (e) 3,4,5-trimethyloctane.

30.36 Write the structural formulas for the following compounds: (a) *trans*-2-pentene, (b) 2-ethyl-1-butene, (c) 4-ethyl-*trans*-2-heptene, (d) 3-phenylbutyne, (e) 2-nitro-4-bromotoluene.

30.37 Draw structures for the following: (a) cyclopentane, (b) *cis*-2-butene, (c) 2-hexanol, (d) 1,4-dibromobenzene, (e) 2-butyne.

Polymers

30.38 Teflon is a polymer formed by the polymerization of perfluoroethylene ($F_2C=CF_2$). Draw the structure of the polymer.

30.39 Vinyl chloride ($H_2C=CHCl$) undergoes copolymerization with 1,1-dichloroethylene ($H_2C=CCl_2$) to form a polymer commercially known as Saran.

Draw the structure of the polymer, showing the repeating monomer units.

30.40 Dacron is formed by the copolymerization between terephthalic acid

and ethylene glycol. Sketch a portion of the polymer chain showing several monomer units. (*Hint:* The reaction involves condensation.)

*30.41 Nylon can be destroyed easily by strong acids. Explain the chemical basis for the destruction.

30.42 Deduce plausible monomers for polymers with the following repeating units:

(a) $\left[CH_2-CF_2\right]_n$

(b)

(c) $\left[CH_2-CH=CH-CH_2\right]_n$

(d) $\left[CO\left(CH_2\right)_6-NH\right]_n$

Miscellaneous Problems

30.43 Draw all the possible structural isomers for the molecule having the formula C_7H_7Cl. The molecule contains one benzene ring.

30.44 From these data

$$C_2H_4(g) + 3O_2(g) \longrightarrow 2CO_2(g) + 2H_2O(l)$$
$$\Delta H° = -1411 \text{ kJ}$$

$$2C_2H_2(g) + 5O_2(g) \longrightarrow 4CO_2(g) + 2H_2O(l)$$
$$\Delta H° = -2599 \text{ kJ}$$

$$H_2(g) + \tfrac{1}{2}O_2(g) \longrightarrow H_2O(l)$$
$$\Delta H° = -285.8 \text{ kJ}$$

calculate the heat of hydrogenation for acetylene:

$$C_2H_2(g) + H_2(g) \longrightarrow C_2H_4(g)$$

*30.45 An organic compound is found to contain 37.5 percent carbon, 3.2 percent hydrogen, and 59.3 percent fluorine by mass. The following pressure and volume data were obtained for 1.00 g of this substance at 90°C:

P(atm)	V(L)
2.00	0.332
1.50	0.409
1.00	0.564
0.50	1.028

The molecule is known to have no dipole moment. (a) What is the empirical formula of this substance? (b) Does this substance behave like an ideal gas? (c) What is its molecular formula? (d) Draw the Lewis formula of this molecule and describe its geometry. (e) What is the systematic name of this compound?

30.46 Which of the following compounds can form hydrogen bonds with water molecules? (a) carboxylic acids, (b) alkenes, (c) ethers, (d) aldehydes, (e) alkanes, (f) amines.

30.47 Name at least one commercial use for each of the following compounds: (a) 2-propanol (isopropyl alcohol), (b) acetic acid, (c) naphthalene, (d) methanol, (e) ethanol, (f) ethylene glycol, (g) methane, (h) ethylene.

30.48 Explain why carbon is able to form so many more compounds than any other element.

30.49 How many liters of air (78 percent N_2, 22 percent O_2 by volume) at 20°C and 1.00 atm are needed for the complete combustion of 1.0 liter of octane, C_8H_{18}, a typical gasoline component, of density 0.70 g/mL?

30.50 How many carbon–carbon sigma bonds are present in each of the following molecules: (a) benzene, (b) cyclobutane, (c) 2-methyl-3-ethylpentane, (d) 2-butyne, (e) anthracene.

30.51 Draw all the structural isomers of the formula $C_4H_8Cl_2$. Indicate those that are chiral and give them systematic names.

30.52 Combustion of 20.63 mg of compound Y, which contains only C, H, and O, with excess oxygen gave 57.94 mg of CO_2 and 11.85 mg of H_2O. (a) Calculate how many milligrams of C, H, and O were present in the original sample of Y. (b) Derive the empirical formula of Y. (c) Suggest a plausible structure for Y if the empirical formula is the same as the molecular formula.

30.53 Combustion of 3.795 mg of liquid B, which contains only C, H, and O, with excess oxygen gave 9.708 mg CO_2 and 3.969 mg H_2O. In a molar mass determination, 0.205 g of B vaporized at 1.00 atm and 200.0°C occupied a volume of 89.8 mL. Derive the empirical formula, molar mass, and molecular formula of B and draw three plausible structures.

30.54 Discuss what is meant by the term "geometrical isomerism," using N_2F_2 (difluorodiazine) and $C_2H_2Cl_2$ (1,2-dichloroethylene) as examples.

*30.55 Suppose that benzene contained three distinct single bonds and three distinct double bonds. How many different isomers would there be for dichlorobenzene ($C_6H_4Cl_2$)? Draw all your proposed structures.

30.56 How many asymmetric carbon atoms are present
in each of the following compounds?

(a)

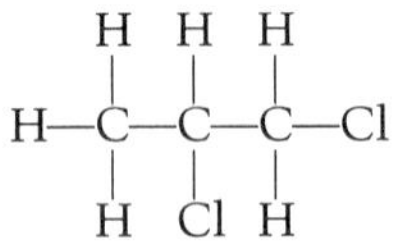

$$H-\underset{\underset{H}{\overset{H}{|}}}{C}-\underset{\underset{Cl}{\overset{H}{|}}}{C}-\underset{\underset{H}{\overset{H}{|}}}{C}-Cl$$

(b)

$$H_3C-\underset{\underset{H}{\overset{OH}{|}}}{C}-\underset{\underset{H}{\overset{CH_3}{|}}}{C}-CH_2OH$$

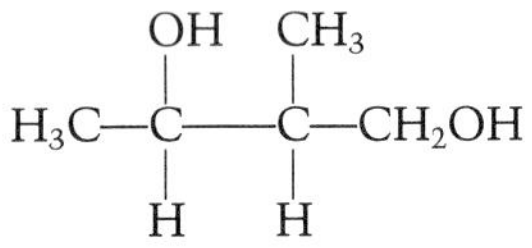

(c)

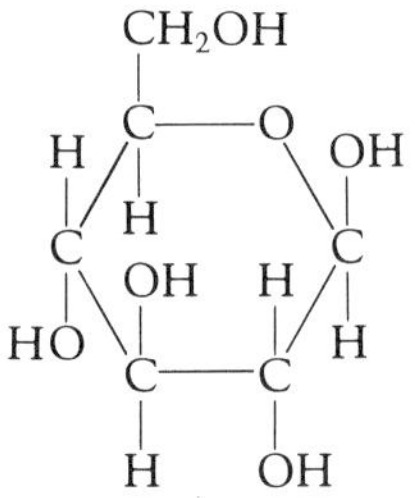

APPENDIXES

APPENDIX 1

Thermodynamic Data for Selected Elements and Their Compounds at 1 atm and 25°C

INORGANIC SUBSTANCES

Substance	State	ΔH_f° (kJ/mol)	ΔG_f° (kJ/mol)	S° (J/K mol)
Ag	s	0	0	42.71
AgCl	s	−127.04	−109.72	96.11
Al	s	0	0	28.32
Al₂O₃	s	−1669.8	−1576.4	50.99
Br₂	l	0	0	152.30
HBr	g	−36.2	−53.22	198.48
C (graphite)	s	0	0	5.69
C (diamond)	s	1.90	2.87	2.44
CO	g	−110.5	−137.3	197.9
CO₂	g	−393.5	−394.4	213.6
Ca	s	0	0	41.63
CaO	s	−635.6	−604.2	39.8
CaCO₃ (calcite)	s	−1206.9	−1128.8	92.9
Cl₂	g	0	0	223.0
HCl	g	−92.3	−95.27	187.0
Cu	s	0	0	33.3
CuO	s	−155.2	−127.2	43.5
F₂	g	0	0	203.34
HF	g	−268.61	−270.71	173.51
H	g	218.2	203.24	114.61
H₂	g	0	0	131.0
H₂O	g	−241.8	−228.6	188.7
H₂O	l	−285.8	−237.2	69.9
H₂O₂	l	−187.6	−118.1	?
Hg	l	0	0	77.40
I₂	s	0	0	116.7
HI	g	25.94	1.30	206.33
Mg	s	0	0	32.51
MgO	s	−601.8	−569.6	26.78
MgCO₃	s	−1112.9	−1029.3	65.69
N₂	g	0	0	192
NH₃	g	−46.3	−16.6	193
NO	g	90.4	86.7	210.62
NO₂	g	33.85	51.8	240.46
N₂O₄	g	9.66	98.29	304.30
N₂O	g	81.56	103.60	219.99
O	g	249.4	230.10	160.95
O₂	g	0	0	205.0
O₃	g	142.2	163.4	237.6
S (rhombic)	s	0	0	31.88
S (monoclinic)	s	0.30	0.10	32.55
SO₂	g	−296.1	−300.4	248.52
SO₃	g	−395.2	−370.4	256.22
H₂S	g	−20.15	−33.02	205.64
ZnO	s	−347.98	−318.19	43.93

Thermodynamic Data for Selected Elements and Their
Compounds at 1 atm and 25°C *(Continued)*

ORGANIC SUBSTANCES

Substance	Formula	State	ΔH_f° (kJ/mol)	ΔG_f° (kJ/mol)	S° (J/K mol)
Acetic acid	CH_3COOH	l	−484.21	−389.45	159.83
Acetaldehyde	Ch_3CHO	g	−166.35	−139.08	264.22
Acetone	CH_3COCH_3	l	−246.81	−153.55	198.74
Acetylene	C_2H_2	g	226.6	209.2	200.8
Benzene	C_6H_6	l	49.04	124.52	124.52
Ethanol	C_2H_5OH	l	−276.98	−174.18	161.04
Ethane	C_2H_6	g	−84.7	−32.89	229.49
Ethylene	C_2H_4	g	52.3	68.12	219.45
Formic acid	$HCOOH$	l	−409.20	−346.02	128.95
Glucose	$C_6H_{12}O_6$	s	−1274.5	−910.56	212.13
Methane	CH_4	g	−74.85	−50.8	186.19
Methanol	CH_3OH	l	−238.66	−166.31	126.78
Sucrose	$C_{12}H_{22}O_{11}$	s	−2221.70	−1544.31	360.24

APPENDIX 2

Units for the Gas Constant

In this appendix we will see how the gas constant R can be expressed in units J/K·mol. Our first step is to derive a relationship between atm and pascal. We start with

$$\begin{aligned}
\text{pressure} &= \frac{\text{force}}{\text{area}} \\
&= \frac{\text{mass} \times \text{acceleration}}{\text{area}} \\
&= \frac{\text{volume} \times \text{density} \times \text{acceleration}}{\text{area}} \\
&= \text{length} \times \text{density} \times \text{acceleration}
\end{aligned}$$

By definition, the standard atmosphere is the pressure exerted by a column of mercury exactly 76 cm high of density 13.5951 g/cm³, in a place where acceleration due to gravity is 980.665 cm/s². However, to express pressure in N/m², it is necessary to write

$$\begin{aligned}
\text{density of mercury} &= 1.35951 \times 10^4 \text{ kg/m}^3 \\
\text{acceleration due to gravity} &= 9.80665 \text{ m/s}^2
\end{aligned}$$

The standard atmosphere is given by

$$\begin{aligned}
1 \text{ atm} &= (0.76 \text{ m Hg})(1.35951 \times 10^4 \text{ kg/m}^3)(9.80665 \text{ m/s}^2) \\
&= 101325 \text{ kg m/m}^2 \cdot \text{s}^2 \\
&= 101325 \text{ N/m}^2 \\
&= 101325 \text{ Pa}
\end{aligned}$$

From Section 6.4 we see that the gas constant R is given by 0.082057 L·atm/K·mol. Using the conversion factors

$$\begin{aligned}
1 \text{ L} &= 1 \times 10^{-3} \text{ m}^3 \\
1 \text{ atm} &= 101325 \text{ N/m}^2
\end{aligned}$$

we write

$$\begin{aligned}
R &= \left(0.082057 \, \frac{\text{L atm}}{\text{K mol}}\right) \left(\frac{1 \times 10^{-3} \text{ m}^3}{1 \text{ L}}\right) \left(\frac{101325 \text{ N/m}^2}{1 \text{ atm}}\right) \\
&= 8.314 \, \frac{\text{N m}}{\text{K mol}} \\
&= 8.314 \, \frac{\text{J}}{\text{K mol}}
\end{aligned}$$

APPENDIX 3

The Elements and the Derivation of Their Names and Symbols*

Element	Symbol	Atomic No.	Atomic Mass†	Date of Discovery	Discoverer and Nationality‡	Derivation
Actinium	Ac	89	(227)	1899	A. Debierne (Fr.)	Gr. *aktis*, beam or ray
Aluminum	Al	13	26.98	1827	F. Woehler (Ge.)	Alum, the aluminum compound in which it was discovered; derived from L *alumen*, astringent taste
Americium	Am	95	(243)	1944	A. Ghiorso (USA) R. A. James (USA) G. T. Seaborg (USA) S. G. Thompson (USA)	The Americas
Antimony	Sb	51	121.8	Ancient		L. *antimonium* (*anti*, opposite of; *monium*, isolated condition), so named because it is a tangible (metallic) substance which combines readily; symbol, L. *stibium*, mark
Argon	Ar	18	39.95	1894	Lord Raleigh (GB) Sir William Ramsay (GB)	Gr. *argon*, inactive
Arsenic	As	33	74.92	1250	Albertus Magnus (Ge.)	Gr. *aksenikon*, yellow pigment; L. *arsenicum*, orpiment; the Greeks once used arsenic trisulfide as a pigment
Astatine	At	85	(210)	1940	D. R. Corson (USA) K. R. MacKenzie (USA) E. Segre (USA)	Gr. *astatos*, unstable
Barium	Ba	56	137.3	1808	Sir Humphry Davy (GB)	Barite, a heavy spar, derived from Gr. *barys*, heavy

SOURCE: Reprinted with permission from "The Elements and Derivation of Their Names and Symbols," G. P. Dinga, *Chemistry* **41** (2), 20-22 (1968). Copyright by the American Chemical Society.
* At the time this table was drawn up, only 103 elements were known to exist.
† The atomic masses given here correspond to the values recommended by the Committee on Teaching of Chemistry, International Union of Pure and Applied Chemistry. Masses in parentheses are those of the most stable or most common isotopes.
‡ The abbreviations are (Ar.) Arabic; (Au.) Austrian; (Du.) Dutch; (Fr.) French; (Ge.) German; (GB) British; (Gr.) Greek; (H.) Hungarian; (I.) Italian; (L.) Latin; (P.) Polish; (R.) Russian; (Sp.) Spanish; (Swe.) Swedish; (USA) American.

The Elements and the Derivation of Their Names and Symbols (Continued)

Element	Symbol	Atomic No.	Atomic Mass	Date of Discovery	Discoverer and Nationality	Derivation
Berkelium	Bk	97	(247)	1950	G. T. Seaborg (USA) S. G. Thompson (USA) A. Ghiorso (USA)	Berkeley, Calif.
Beryllium	Be	4	9.012	1828	F. Woehler (Ge.) A. A. B. Bussy (Fr.)	Fr. L. *beryl*, sweet
Bismuth	Bi	83	209.0	1753	Claude Geoffroy (Fr.)	Ge. *bismuth*, probably a distortion of *weisse masse* (white mass) in which it was found
Boron	B	5	10.81	1808	Sir Humphry Davy (GB) J. L. Gay-Lussac (Fr.) L. J. Thenard (Fr.)	The compound borax, derived from Ar. *buraq*, white
Bromine	Br	35	79.90	1826	A. J. Balard (Fr.)	Gr. *bromos*, stench
Cadmium	Cd	48	112.4	1817	Fr. Stromeyer (Ge.)	Gr. *kadmia*, earth; L. *cadmia*, calamine (because it is found along with calamine)
Calcium	Ca	20	40.08	1808	Sir Humphry Davy (GB)	L. *calx*, lime
Californium	Cf	98	(249)	1950	G. T. Seaborg (USA) S. G. Thompson (USA) A. Ghiorso (USA) K. Street, Jr. (USA)	California
Carbon	C	6	12.01	Ancient		L. *carbo*, charcoal
Cerium	Ce	58	140.1	1803	J. J. Berzelius (Swe.) William Hisinger (Swe.) M. H. Klaproth (Ge.)	Asteroid Ceres
Cesium	Cs	55	132.9	1860	R. Bunsen (Ge.) G. R. Kirchhoff (Ge.)	L. *caesium*, blue (cesium was discovered by its spectral lines, which are blue)
Chlorine	Cl	17	35.45	1774	K. W. Scheele (Swe.)	Gr. *chloros*, light green
Chromium	Cr	24	52.00	1797	L. N. Vauquelin (Fr.)	Gr. *chroma*, color (because it is used in pigments)
Cobalt	Co	27	58.93	1735	G. Brandt (Ge.)	Ge. *Kobold*, goblin (because the ore yielded cobalt instead of the

The Elements and the Derivation of Their Names and Symbols (Continued)

Element	Symbol	Atomic No.	Atomic Mass	Date of Discovery	Discoverer and Nationality	Derivation
Cobalt	Co	27	58.93	1735	G. Brandt (Ge.)	expected metal, copper, it was attributed to goblins)
Copper	Cu	29	63.55	Ancient		L. *cuprum*, copper, derived from *cyprium*, Island of Cyprus, the main source of ancient copper
Curium	Cm	96	(247)	1944	G. T. Seaborg (USA) R. A. James (USA) A. Ghiorso (USA)	Pierre and Marie Curie
Dysprosium	Dy	66	162.5	1886	Lecoq de Boisbaudran (Fr.)	Gr. *dysprositos*, hard to get at
Einsteinium	Es	99	(254)	1952	A. Ghiorso (USA)	Albert Einstein
Erbium	Er	68	167.3	1843	C. G. Mosander (Swe.)	Ytterby, Sweden, where many rare earths were discovered
Europium	Eu	63	152.0	1896	E. Demarcay (Fr.)	Europe
Fermium	Fm	100	(253)	1953	A. Ghiorso (USA)	Enrico Fermi
Fluorine	F	9	19.00	1886	H. Moissan (Fr.)	Mineral fluorspar, from L. *fluere*, flow (because fluorspar was used as a flux)
Francium	Fr	87	(223)	1939	Marguerite Perey (Fr.)	France
Gadolinium	Gd	64	157.3	1880	J. C. Marignac (Fr.)	Johan Gadolin, Finnish rare earth chemist
Gallium	Ga	31	69.72	1875	Lecoq de Boisbaudran (Fr.)	L. *Gallia*, France
Germanium	Ge	32	72.59	1886	Clemens Winkler (Ge.)	L. *Germania*, Germany
Gold	Au	79	197.0	Ancient		L. *aurum*, shining dawn
Hafnium	HF	72	178.5	1923	D. Coster (Du.) G. von Hevesey (H.)	L. *Hafnia*, Copenhagen
Helium	He	2	4.003	1868	P. Janssen (spectr) (Fr.) Sir William Ramsay (isolated) (GB)	Gr. *helios*, sun (because it was first discovered in the sun's spectrum)
Holmium	Ho	67	164.9	1879	P.T. Cleve (Swe.)	L. *Holmia*, Stockholm

The Elements and the Derivation of Their Names and Symbols (Continued)

Element	Symbol	Atomic No.	Atomic Mass	Date of Discovery	Discoverer and Nationality	Derivation
Hydrogen	H	1	1.008	1766	Sir Henry Cavendish (GB)	Gr. *hydro*, water; *genes*, forming (because it produces water when burned with oxygen)
Indium	In	49	114.8	1863	F. Reich (Ge.) T. Richter (Ge.)	Indigo, because of its indigo blue lines in the spectrum
Iodine	I	53	126.9	1811	B. Courtois (Fr.)	Gr. *iodes*, violet
Iridium	Ir	77	192.2	1803	S. Tennant (GB)	L. *iris*, rainbow
Iron	Fe	26	55.85	Ancient		L. *ferrum*, iron
Krypton	Kr	36	83.80	1898	Sir William Ramsay (GB) M. W. Travers (GB)	Gr. *kryptos*, hidden
Lanthanum	La	57	138.9	1839	C. G. Mosander (Swe.)	Gr. *lanthanein*, concealed
Lawrencium	Lr	103	(257)	1961	A. Ghiorso (USA) T. Sikkeland (USA) A. E. Larsh (USA) R. M. Latimer (USA)	E. O. Lawrence (USA), inventor of the cyclotron
Lead	Pb	82	207.2	Ancient		Symbol, L. *plumbum*, lead, meaning heavy
Lithium	Li	3	6.941	1817	A. Arfvedson (Swe.)	Gr. *lithos*, rock (because it occurs in rocks)
Lutetium	Lu	71	175.0	1907	G. Urbain (Fr.) K. Auer von Welsbach (Au.)	*Lutetia*, ancient name for Paris
Magnesium	Mg	12	24.31	1808	Sir Humphry Davy (GB)	*Magnesia*, a district in Thessaly; possibly derived from L. *magnesia*
Manganese	Mn	25	54.94	1774	J. G. Gahn (Swe.)	L. *magnes*, magnet
Mendelevium	Md	101	(256)	1955	A. Ghiorso (USA) G. R. Choppin (USA) G. T. Seaborg (USA) B. G. Harvey (USA) S. G. Thompson (USA)	Mendeleev, Russian chemist who prepared the periodic chart and predicted properties of undiscovered elements

The Elements and the Derivation of Their Names and Symbols (Continued)

Element	Symbol	Atomic No.	Atomic Mass	Date of Discovery	Discoverer and Nationality	Derivation
Mercury	Hg	80	200.6	Ancient		Symbol, L. *hydrargyrum*, liquid silver
Molybdenum	Mo	42	95.94	1778	G. W. Scheele (Swe.)	Gr. *molybdos*, lead
Neodymium	Nd	60	144.2	1885	C. A. von Welsbach (Au.)	Gr. *neos*, new; *didymos*, twin
Neon	Ne	10	20.18	1898	Sir William Ramsay (GB) M. W. Travers (GB)	Gr. *neos*, new
Neptunium	Np	93	(237)	1940	E. M. McMillan (USA) P. H. Abelson (USA)	Planet Neptune
Nickel	Ni	28	58.69	1751	A. F. Cronstedt (Swe.)	Swe. *kopparnickel*, false copper; also Ge. *nickel*, referring to the devil that prevented copper from being extracted from nickel ores
Niobium	Nb	41	92.91	1801	Charles Hatchett (GB)	Gr. *Niobe*, daughter of Tantalus (niobium was considered identical to tantalum, named after *Tantalus*, until 1884; originally called columbium, with symbol Cb)
Nitrogen	N	7	14.01	1772	Daniel Rutherford (GB)	Fr. *nitrogene*, derived from L. *nitrum*, native soda, or Gr. *nitron*, native soda, and Gr. *genes*, forming
Nobelium	No	102	(253)	1958	A. Ghiorso (USA) T. Sikkeland (USA) J. R. Walton (USA) G. T. Seaborg (USA)	Alfred Nobel
Osmium	Os	76	190.2	1803	S. Tennant (GB)	Gr. *osme*, odor

The Elements and the Derivation of Their Names and Symbols (Continued)

Element	Symbol	Atomic No.	Atomic Mass	Date of Discovery	Discoverer and Nationality	Derivation
Oxygen	O	8	16.00	1774	Joseph Priestley (GB) C. W. Scheele (Swe.)	Fr. *oxygene*, generator of acid, derived from Gr. *oxys*, acid, and L. *genes*, forming (because it was once thought to be a part of all acids)
Palladium	Pd	46	106.4	1803	W. H. Wollaston (GB)	Asteroid Pallas
Phosphorus	P	15	30.97	1669	H. Brandt (Ge.)	Gr. *phosphoros*, light bearing
Platinum	Pt	78	195.1	1735 1741	A. de Ulloa (Sp.) Charles Wood (GB)	Sp. *platina*, silver
Plutonium	Pu	94	(242)	1940	G. T. Seaborg (USA) E. M. McMillan (USA) J. W. Kennedy (USA) A. C. Wahl (USA)	Planet Pluto
Polonium	Po	84	(210)	1898	Marie Curie (P.)	Poland
Potassium	K	19	39.10	1807	Sir Humphry Davy (GB)	Symbol, L. *kalium*, potash
Praseodymium	Pr	59	140.9	1885	A. von Welsbach (Aus.)	Gr. *prasios*, green; *didymos*, twin
Promethium	Pm	61	(147)	1945	J. A., Marinsky (USA) L. E. Glendenin (USA) C. D. Coryell (USA)	Gr. mythology, *Prometheus*, the Greek Titan who stole fire from heaven
Protactinium	Pa	91	(231)	1917	O. Hahn (Ge.) L. Meitner (Au.)	Gr. *protos*, first; *actinium* (because it disintegrates into actinium)
Radium	Ra	88	(226)	1898	Pierre and Marie Curie (Fr.; P.)	L. *radius*, ray
Radon	Rn	86	(222)	1900	F. E. Dorn (Ge.)	Derived from radium with suffix "on" common to inert gases (once called nitron, meaning shining, with symbol Nt)
Rhenium	Re	75	186.2	1925	W. Noddack (Ge.) I. Tacke (Ge.) Otto Berg (Ge.)	L. *Rhenus*, Rhine

The Elements and the Derivation of Their Names and Symbols *(Continued)*

Element	Symbol	Atomic No.	Atomic Mass	Date of Discovery	Discoverer and Nationality	Derivation
Rhodium	Rh	45	102.9	1804	W. H. Wollaston (GB)	Gr. *rhodon*, rose (because some of its salts are rose-colored)
Rubidium	Rb	37	85.47	1861	R. W. Bunsen (Ge.) G. Kirchhoff (Ge.)	L. *rubidius*, dark red (discovered with the spectroscope, its spectrum shows red lines)
Ruthenium	Ru	44	101.1	1844	K. K. Klaus (R.)	L. *Ruthenia*, Russia
Samarium	Sm	62	150.4	1879	Lecoq de Boisbaudran (Fr.)	Samarskite, after Samarski, a Russian engineer
Scandium	Sc	21	44.96	1879	L. F. Nilson (Swe.)	Scandinavia
Selenium	Se	34	78.96	1817	J. J. Berzelius (Swe.)	Gr. *selene*, moon (because it resembles tellurium, named for earth)
Silicon	Si	14	28.09	1824	J. J. Berzelius (Swe.)	L. *silex, silicis,* flint
Silver	Ag	47	107.9	Ancient		Symbol, L. *argentum*, silver
Sodium	Na	11	22.99	1807	Sir Humphry Davy (GB)	L. *sodanum,* headache remedy; symbol, L. *natrium,* soda
Strontium	Sr	38	87.62	1808	Sir Humphry Davy (GB)	Strontian, Scotland, derived from mineral strontionite
Sulfur	S	16	32.07	Ancient		L. *sulphurium* (Sanskrit, *sulvere*)
Tantalum	Ta	73	180.9	1802	A. G. Ekeberg (Swe.)	Gr. mythology, *Tantalus,* because of difficulty in isolating it (Tantalus, son of Zeus, was punished by being forced to stand up to his chin in water which receded whenever he tried to drink)

The Elements and the Derivation of Their Names and Symbols (Continued)

Element	Symbol	Atomic No.	Atomic Mass	Date of Discovery	Discoverer and Nationality	Derivation
Technetium	Tc	43	(99)	1937	C. Perrier (I.)	Gr. *technetos*, artificial (because it was the first artificial element)
Tellurium	Te	52	127.6	1782	F. J. Müller (Au.)	L. *tellus*, earth
Terbium	Tb	65	158.9	1843	C. G. Mosander (Swe.)	Ytterby, Sweden
Thallium	Tl	81	204.4	1861	Sir William Crookes (GB)	Gr. *thallos*, a budding twig (because its spectrum shows a bright green line)
Thorium	Th	90	232.0	1828	J. J. Berzelius (Swe.)	Mineral thorite, derived from *Thor*, Norse god of war
Thulium	Tm	69	168.9	1879	P. T. Cleve (Swe.)	*Thule*, early name for Scandinavia
Tin	Sn	50	118.7	Ancient		Symbol, L. *stannum*, tin
Titanium	Ti	22	47.88	1791	W. Gregor (GB)	Gr. giants, the Titans, and L. *titans*, giant deities
Tungsten	W	74	183.9	1783	J. J. and F. de Elhuyar (Sp.)	Swe. *tung sten*, heavy stone; symbol, wolframite, a mineral
Uranium	U	92	238.0	1789 1841	M. H. Klaproth (Ge.) E. M. Peligot (Fr.)	Planet Uranus
Vanadium	V	23	50.94	1801 1830	A. M. del Rio (Sp.) N. G. Sefstrom (Swe.)	*Vanadis*, Norse goddess of love and beauty
Xenon	Xe	54	131.3	1898	Sir William Ramsay (GB) M. W. Travers (GB)	Gr. *xenos*, stranger
Ytterbium	Yb	70	173.0	1907	G. Urbain (Fr.)	Ytterby, Sweden
Yttrium	Y	39	88.91	1843	C. G. Mosander (Swe.)	Ytterby, Sweden
Zinc	Zn	30	65.39	1746	A. S. Marggraf (Ge.)	Ge. *zink*, of obscure origin
Zirconium	Zr	40	91.22	1789	M. H. Klaproth (Ge.)	Zircon, in which it was found, derived from *Ar. zargum*, gold color

APPENDIX 4

Mathematical Operations

Logarithms

Common Logarithms The concept of the logarithm is an extension of the concept of exponents, which is discussed in Chapter 1. The *common*, or base-10, logarithm of any number is the power to which 10 must be raised to equal the number. The following examples illustrate this relationship:

Logarithm	Exponent
$\log 1 = 0$	$10^0 = 1$
$\log 10 = 1$	$10^1 = 10$
$\log 100 = 2$	$10^2 = 100$
$\log 10^{-1} = -1$	$10^{-1} = 0.1$
$\log 10^{-2} = -2$	$10^{-2} = 0.01$

In each case the logarithm of the number can be obtained by inspection.

Since the logarithms of numbers are exponents, they have the same properties as exponents. Thus, we have

Logarithm	Exponent
$\log AB = \log A + \log B$	$10^A \times 10^B = 10^{A+B}$
$\log \dfrac{A}{B} = \log A - \log B$	$\dfrac{10^A}{10^B} = 10^{A-B}$

Furthermore, $\log A^n = n \log A$.

Now, suppose we want to find the common logarithm of 6.75. We write

$$\log 6.75 = x$$

and

$$10^x = 6.75$$

To evaluate x, we can use the table of logarithms in Appendix 5. First we go to the extreme left-hand column of the table to locate the first two digits of the number (that is, 67), and then we look to the top horizontal row to locate the third digit (that is, 5). Thus we find that

$$\log 6.75 = 0.8293$$

The logarithm of any number between 1 and 10 lies between 0 and 1. (Remember that $\log 1 = 0$ and $\log 10 = 1$.) Therefore, our answer must be 0.8293 and not 8.293.

Suppose the number is greater than 10, say 523. We write

$$\log 523 = \log (5.23 \times 10^2)$$
$$= \log 5.23 + \log 10^2$$

As before, we go to Appendix 5 and find

$$\log 5.23 = 0.7185$$

Since $\log 10^2 = 2$, we have

$$\log 523 = 0.7185 + 2$$
$$= 2.7185$$

What if the number is 0.00977? In this case we must write

$$\log 0.00977 = \log (9.77 \times 10^{-3})$$
$$= \log 9.77 + \log 10^{-3}$$

Appendix 5 tells us that

$$\log 9.77 = 0.9899$$

And, since

$$\log 10^{-3} = -3$$

we have

$$\log 0.00977 = 0.9899 - 3$$
$$= -2.0101$$

Sometimes (as in the case of pH calculations) it may be necessary to obtain the number whose logarithm is known. This procedure is known as taking the antilogarithm; it is simply the reverse of taking the logarithm of a number. Suppose we want to evaluate the antilog of 3.455. Many electronic calculators employ a key labeled $\log^{-1}$ or INV log to obtain antilogs. Other calculators have a 10^x or y^x key (where x corresponds to 3.455 in our example and y is 10 for base-10 logarithm). If your calculator does not perform these operations, you can solve the problem by writing

$$\text{antilog } 3.455 = \text{antilog } 0.455 \times \text{antilog } 3$$

From Appendix 5 locate 0.455 in the body of the log table; this is the log of 2.85. Since the antilog of 3 is 10^3, we write

$$\text{antilog } 3.455 = 2.85 \times 10^3$$

Natural Logarithms Logarithms taken to the base e instead of 10 are known as natural logarithms (denoted by ln or $\log_e$); e is equal to 2.7183. The relationship between common logarithms and natural logarithms is as follows:

$$\log 10 = 1 \qquad 10^1 = 10$$
$$\ln 10 = 2.303 \qquad e^{2.303} = 10$$

Thus

$$\ln x = 2.303 \log x$$

The preceding equation allows us to obtain the natural logarithm of any number by using the table of common logarithms. For example, if we want to obtain ln 2.27, first we find

$$\log 2.27 = 0.356$$

and then write

$$\ln 2.27 = 2.303 \times 0.356$$
$$= 0.820$$

The Quadratic Equation

A quadratic equation takes the form

$$ax^2 + bx + c = 0$$

If coefficients a, b, and c are known, then x is given by

$$x = \frac{-b \pm \sqrt{b^2 - 4ac}}{2a}$$

Suppose we have the following quadratic equation:

$$2x^2 + 5x - 12 = 0$$

Solving for x, we write

$$x = \frac{-5 \pm \sqrt{(5)^2 - 4(2)(-12)}}{2(2)}$$

$$= \frac{-5 \pm \sqrt{25 + 96}}{4}$$

Therefore

$$x = \frac{-5 + 11}{4} = \frac{3}{2}$$

and

$$x = \frac{-5 - 11}{4} = -4$$

APPENDIX 5

Table of Logarithms

	0	1	2	3	4	5	6	7	8	9
10	0000	0043	0086	0128	0170	0212	0253	0294	0334	0374
11	0414	0453	0492	0531	0569	0607	0645	0682	0719	0755
12	0792	0828	0864	0899	0934	0969	1004	1038	1072	1106
13	1139	1173	1206	1239	1271	1303	1335	1367	1399	1430
14	1461	1492	1523	1553	1584	1614	1644	1673	1703	1732
15	1761	1790	1818	1847	1875	1903	1931	1959	1987	2014
16	2041	2068	2095	2122	2148	2175	2201	2227	2253	2279
17	2304	2330	2355	2380	2405	2430	2455	2480	2504	2529
18	2553	2577	2601	2625	2648	2672	2695	2718	2742	2765
19	2788	2810	2833	2856	2878	2900	2923	2945	2967	2989
20	3010	3032	3054	3075	3096	3118	3139	3160	3181	3201
21	3222	3243	3263	3284	3304	3324	3345	3365	3385	3404
22	3424	3444	3464	3483	3502	3522	3541	3560	3579	3598
23	3617	3636	3655	3674	3692	3711	3729	3747	3766	3784
24	3802	3820	3838	3856	3874	3892	3909	3927	3945	3962
25	3979	3997	4014	4031	4048	4065	4082	4099	4116	4133
26	4150	4166	4183	4200	4216	4232	4249	4265	4281	4298
27	4314	4330	4346	4362	4378	4393	4409	4425	4440	4456
28	4472	4487	4502	4518	4533	4548	4564	4579	4594	4609
29	4624	4639	4654	4669	4683	4698	4713	4728	4742	4757
30	4771	4786	4800	4814	4829	4843	4857	4871	4886	4900
31	4914	4928	4942	4955	4969	4983	4997	5011	5024	5038
32	5051	5065	5079	5092	5105	5119	5132	5145	5159	5172
33	5185	5198	5211	5224	5237	5250	5263	5276	5289	5302
34	5315	5328	5340	5353	5366	5378	5391	5403	5416	5428
35	5441	5453	5465	5478	5490	5502	5514	5527	5539	5551
36	5563	5575	5587	5599	5611	5623	5635	5647	5658	5670
37	5682	5694	5705	5717	5729	5740	5752	5763	5775	5786
38	5798	5809	5821	5832	5843	5855	5866	5877	5888	5899
39	5911	5922	5933	5944	5955	5966	5977	5988	5999	6010
40	6021	6031	6042	6053	6064	6075	6085	6096	6107	6117
41	6128	6138	6149	6160	6170	6180	6191	6201	6212	6222
42	6232	6243	6253	6263	6274	6284	6294	6304	6314	6325
43	6335	6345	6355	6365	6375	6385	6395	6405	6415	6425
44	6435	6444	6454	6464	6474	6484	6493	6503	6513	6522
45	6532	6542	6551	6561	6571	6580	6590	6599	6609	6618
46	6628	6637	6646	6656	6665	6675	6684	6693	6702	6712
47	6721	6730	6739	6749	6758	6767	6776	6785	6794	6803
48	6812	6821	6830	6839	6848	6857	6866	6875	6884	6893
49	6902	6911	6920	6928	6937	6946	6955	6964	6972	6981
50	6990	6998	7007	7016	7024	7033	7042	7050	7059	7067
51	7076	7084	7093	7101	7110	7118	7126	7135	7143	7152
52	7160	7168	7177	7185	7193	7202	7210	7218	7226	7235
53	7243	7251	7259	7267	7275	7284	7292	7300	7308	7316
54	7324	7332	7340	7348	7356	7364	7372	7380	7388	7396

	0	1	2	3	4	5	6	7	8	9
Table of Logarithms *(Continued)*										
55	7404	7412	7419	7427	7435	7443	7451	7459	7466	7474
56	7482	7490	7497	7505	7513	7520	7528	7536	7543	7551
57	7559	7566	7574	7582	7589	7597	7604	7612	7619	7627
58	7634	7642	7649	7657	7664	7672	7679	7686	7694	7701
59	7709	7716	7723	7731	7738	7745	7752	7760	7767	7774
60	7782	7789	7796	7803	7810	7818	7825	7832	7839	7846
61	7853	7860	7868	7875	7882	7889	7896	7903	7910	7917
62	7924	7931	7938	7945	7952	7959	7966	7973	7980	7987
63	7993	8000	8007	8014	8021	8028	8035	8041	8048	8055
64	8062	8069	8075	8082	8089	8096	8102	8109	8116	8122
65	8129	8136	8142	8149	8156	8162	8169	8176	8182	8189
66	8195	8202	8209	8215	8222	8228	8235	8241	8248	8254
67	8261	8267	8274	8280	8287	8293	8299	8306	8312	8319
68	8325	8331	8338	8344	8351	8357	8363	8370	8376	8382
69	8388	8395	8401	8407	8414	8420	8426	8432	8439	8445
70	8451	8457	8463	8470	8476	8482	8488	8494	8500	8506
71	8513	8519	8525	8531	8537	8543	8549	8555	8561	8567
72	8573	8579	8585	8591	8597	8603	8609	8615	8621	8627
73	8633	8639	8645	8651	8657	8663	8669	8675	8681	8686
74	8692	8698	8704	8710	8716	8722	8727	8733	8739	8745
75	8751	8756	8762	8768	8774	8779	8785	8791	8797	8802
76	8808	8814	8820	8825	8831	8837	8842	8848	8854	8859
77	8865	8871	8876	8882	8887	8893	8899	8904	8910	8915
78	8921	8927	8932	8938	8943	8949	8954	8960	8965	8971
79	8976	8982	8987	8993	8998	9004	9009	9015	9020	9025
80	9031	9036	9042	9047	9053	9058	9063	9069	9074	9079
81	9085	9090	9096	9101	9106	9112	9117	9122	9128	9133
82	9138	9143	9149	9154	9159	9165	9170	9175	9180	9186
83	9191	9196	9201	9206	9212	9217	9222	9227	9232	9238
84	9243	9248	9253	9258	9263	9269	9274	9279	9284	9289
85	9294	9299	9304	9309	9315	9320	9325	9330	9335	9340
86	9345	9350	9355	9360	9365	9370	9375	9380	9385	9390
87	9395	9400	9405	9410	9415	9420	9425	9430	9435	9440
88	9445	9450	9455	9460	9465	9469	9474	9479	9484	9489
89	9494	9499	9504	9509	9513	9518	9523	9528	9533	9538
90	9542	9547	9552	9557	9562	9566	9571	9576	9581	9586
91	9590	9595	9600	9605	9609	9614	9619	9624	9628	9633
92	9638	9643	9647	9652	9657	9661	9666	9671	9675	9680
93	9685	9689	9694	9699	9703	9708	9713	9717	9722	9727
94	9731	9736	9741	9745	9750	9754	9759	9763	9768	9773
95	9777	9782	9786	9791	9795	9800	9805	9809	9814	9818
96	9823	9827	9832	9836	9841	9845	9850	9854	9859	9863
97	9868	9872	9877	9881	9886	9890	9894	9899	9903	9908
98	9912	9917	9921	9926	9930	9934	9939	9943	9948	9952
99	9956	9961	9965	9969	9974	9978	9983	9987	9991	9996

APPENDIX 6

The Top Twenty-Five Chemicals Produced in the United States*

Substance	Formula	Preparation	Uses
1. Sulfuric acid	H_2SO_4	Oxidation of sulfur to SO_3 and further reaction with water (Section 26.6)	Manufacture of fertilizers, explosives, petroleum, detergents, dyestuffs, insecticides, plastics, paints, pharmaceuticals, storage batteries, pickling metal.
2. Molecular nitrogen	N_2	Fractional distillation of liquid air	Manufacture of ammonia, nitric acid, nitrates, etc.; for purging, blanketing, pressurizing, and as a cryogenic agent.
3. Ammonia	NH_3	Haber process (Section 17.6)	Manufacture of fertilizer, explosives, plastics, nitric acid, and others. Used as heat exchanger in some refrigerants.
4. Lime (Calcium oxide)	CaO	Thermal decomposition of limestone ($CaCO_3$)	In glass, cement, waste treatment, insecticides, petroleum refining, food processing, manufacture of calcium carbide.
5. Ethylene	C_2H_4	Cracking of petroleum (Section 30.1)	Manufacture of ethyl alcohol, ethylene glycol, vinyl chloride, ethylene oxide, polyethylene, and other plastics; in agricultural chemistry.
6. Molecular oxygen	O_2	Fractional distillation of liquid air	In metal refining and metal processing (largely in the steel industry), as a rocket fuel, and in medicine (respiration).
7. Caustic soda (Sodium hydroxide)	$NaOH$	Electrolysis of aqueous NaCl solution (Section 26.7)	Manufacture of rayon and cellophane; petroleum refining; pulp and paper; aluminum processing; medicine; hydrolysis of fats to form soaps and detergents.
8. Phosphoric acid	H_3PO_4	From calcium phosphate and sulfuric acid; elemental phosphorus burned and reacted with water (Section 26.5)	Manufacture of fertilizers and detergents; pickling and rustproofing metals; pharmaceuticals; water treatment; flavoring in beverages.

Source: *Chemical Engineering and News*, Volume 63, Number 18, p. 13 (1985).
* The chemicals are ranked in order by the amount produced in the U.S. in 1984.

The Top Twenty-Five Chemicals Produced in the United States (Continued)

Substance	Formula	Preparation	Uses
9. Molecular chlorine	Cl_2	Electrolysis of aqueous NaCl solution (Section 26.7)	Manufacture of chlorinated hydrocarbons, bleach; purification of water.
10. Soda ash (Sodium carbonate)	Na_2CO_3	From CaO, NH_3, and NaCl (Section 28.1)	Manufacture of glass, sodium salts, pulp and paper, soaps and detergents; water treatment; aluminum production; cleaning preparation in medicine; petroleum refining.
11. Nitric acid	HNO_3	From ammonia and oxygen (Section 26.5)	Manufacture of fertilizer and explosives; organic synthesis of dyes, drugs, etc.; etching steel; metallurgy.
12. Propylene	$CH_3CH{=}CH_2$	From catalytic and thermal cracking of hydrocarbons	Manufacture of polypropylene, isopropyl alcohol, propylene oxide, acrylonitrile, acetone.
13. Urea	H_2NCONH_2	From ammonia and carbon dioxide under pressure	Manufacture of fertilizers, animal feeds, plastics, adhesives; softener for cellulose; flameproofing agents; treatment of sickle-cell anemia.
14. Ammonium nitrate	NH_4NO_3	From ammonia and nitric acid	Manufacture of fertilizer and explosives, nitrous oxide, insecticides; ingredient of freezing mixture; matches, catalyst.
15. Ethylene dichloride	$ClCH_2CH_2Cl$	From ethylene and chlorine	Manufacture of vinyl chloride, chlorinated solvents, paint removers, rubber; organic synthesis.
16. Benzene	C_6H_6	By catalytic reforming of petroleum and fractional distillation of coal tar (Section 30.1)	Manufacture of styrene, phenol, synthetic detergents, cyclohexane, nitrobenzene, dyes, paint removers, rubber cement, anti-knock gasoline, and as a solvent.
17. Ethylbenzene	$C_6H_5C_2H_5$	From petroleum refining and by heating benzene and ethylene in presence of aluminum chloride (Section 30.1)	Manufacture of styrene; solvent.

The Top Twenty-Five Chemicals Produced in the United States (Continued)

Substance	Formula	Preparation	Uses
18. Methanol	CH_3OH	By high-pressure catalytic synthesis from carbon monoxide and hydrogen	Manufacture of formaldehyde, dimethyl terephthalate, methyl chloride, acetic acid; aviation fuel; antifreeze; solvent.
19. Carbon dioxide	CO_2	Combustion; reaction of acid with carbonates	Refrigeration, carbonated beverages; fire extinguishing; cloud seeding; lasers.
20. Styrene	$C_6H_5CHCH_2$	From catalytic dehydrogenation of ethylbenzene	Manufacture of polystyrene plastics and other polymers.
21. Vinyl chloride	CH_2CHCl	Pyrolysis of ethylene dichloride	Manufacture of polyvinyl chloride and other polymers, adhesives for plastics.
22. Xylene	$C_6H_4(CH_3)_2$ (all 3 isomers)	Fractional distillation from petroleum and coal tar	Aviation gasoline; solvent; manufacture of phthalic anhydride and terephthalic acid, dyes, etc.
23. Terephthalic acid	$C_6H_4(COOH)_2$	Oxidation of p-xylene	Manufacture of polyester resins and other polymers; additive to poultry feeds.
24. Ethylene oxide	C_2H_4O	Oxidation of ethylene	Manufacture of ethylene glycol; polymers; rocket propellant.
25. Hydrochloric acid	HCl	Chlorination of benzene and other hydrocarbons; reaction between hydrogen and chlorine gases	Food processing; ore reduction; pickling and metal cleaning; alcohol denaturant.

GLOSSARY

Absolute temperature scale—A temperature scale that uses the absolute zero of temperature as the lowest temperature.

Absolute zero—Theoretically the lowest attainable temperature.

Acceptor impurities—Impurities that can accept electrons from semiconductors.

Accuracy—The closeness of a measurement to the true value of the quantity that is measured.

Acid—A substance that yields hydrogen ions (H^+) when dissolved in water.

Acid ionization constant—The equilibrium constant for the acid ionization.

Actinide series—Elements that have incompletely filled $5f$ subshells or readily give rise to cations that have incompletely filled $5f$ subshells.

Activated complex—The species temporarily formed by the reactant molecules as a result of the collision before they form the product.

Activation energy—The minimum amount of energy required to initiate a chemical reaction.

Activity series—A summary of the results of many possible displacement reactions.

Actual yield—The amount of product actually obtained in a reaction.

Adhesion—The attraction between unlike molecules.

Alcohol—An organic compound containing the hydroxyl group —OH.

Aldehydes—Compounds with a carbonyl functional group and the general formula RCHO where R is an H atom, an alkyl, or an aryl group.

Aliphatic hydrocarbons—Hydrocarbons that do not contain the benzene group or the benzene ring.

Alkali metals—The Group 1A elements (Li, Na, K, Rb, Cs, and Fr).

Alkaline earth metals—The Group 2A elements (Be, Mg, Ca, Sr, Ba, and Ra).

Alkanes—Hydrocarbons having the general formula C_nH_{2n+2}, where $n = 1, 2, \ldots$.

Alkenes—Hydrocarbons that contain one or more carbon-carbon double bonds. They have the general formula C_nH_{2n}, where $n = 2, 3, \ldots$.

Alkynes—Hydrocarbons that contain one or more carbon-carbon triple bonds. They have the general formula C_nH_{2n-2}.

Allotropes—Two or more forms of the same element that differ significantly in chemical and physical properties.

Alpha particles—See alpha rays.

Alpha (α) rays—Helium ions with a positive charge of $+2$.

Amines—Organic bases that have the functional group —NR_2 where R may be H, an alkyl group, or an aryl group.

Amorphous solid—A solid that lacks a regular three-dimensional arrangement of atoms or molecules.

Amphoteric oxide—An oxide that exhibits both acidic and basic properties.

Amplitude—The vertical distance from the middle of a wave to the peak or trough.

Anion—An ion with a net negative charge.

Anode—The electrode at which oxidation occurs.

Antibonding molecular orbital—A molecular orbital that is of higher energy and lower stability than the atomic orbitals from which it was formed.

Aromatic hydrocarbon—Hydrocarbons that contain one or more benzene rings.

Aryl group—A group of atoms equivalent to a benzene ring or a set of fused benzene rings, less one hydrogen atom.

Atmospheric pressure—The pressure exerted by Earth's atmosphere.

Atom—The basic unit of an element that can enter into chemical combination.

Atomic mass—The mass of an atom in atomic mass units.

Atomic mass unit—A mass exactly equal to 1/12th the mass of one carbon-12 atom.

Atomic number (Z)—The number of protons in the nucleus of an atom.

Atomic orbital—The probability function that defines the distribution of electron density in space around the atomic nucleus.

Atomic radius—One-half the distance between the two nuclei in two adjacent atoms of the same element in a metal. For elements that exist as diatomic units, the atomic radius is one-half the distance between the nuclei of the two atoms in a particular molecule.

Avogadro's law—At constant pressure and temperature, the volume of a gas is directly proportional to the number of moles of the gas present.

Avogadro's number—6.022×10^{23}; the number of particles in a mole.

Barometer—An instrument that measures atmospheric pressure.

Base—A substance that yields hydroxide ions (OH^-) when dissolved in water.

Base ionization constant—The equilibrium constant for the base ionization.

Battery—An electrochemical cell, or often several electrochemical cells connected in series, that can be used as a source of direct electric current at a constant voltage.

Beta particles—See beta rays.

Beta (β) rays—Electrons.

Bimolecular reaction—An elementary step that involves two molecules.

Binary acid—An acid that contains only two elements.

Binary compounds—A compound formed from just two elements.

Bioamplification—The buildup of any poison along a food chain.

Boiling point—The temperature at which the vapor pressure of a liquid is equal to the external atmospheric pressure.

Bonding molecular orbital—A molecular orbital that is of lower energy and greater stability than the atomic orbitals from which it was formed.

Bond dissociation energy—The enthalpy change required to break a particular bond in a mole of gaseous diatomic molecules.

Bond length—The distance between the centers of two bonded atoms in a molecule.

Bond order—The difference between the numbers of electrons in bonding molecular orbitals and antibonding molecular orbitals, divided by two.

Boranes—Compounds containing only boron and hydrogen.

Born-Haber cycle—The cycle that relates lattice energies of ionic compounds to ionization energies, electron affinities, heats of sublimation and formation, and bond dissociation energies.

Boundary surface diagram—Diagram of the region containing a substantial amount of the electron density (about 90 percent) in an orbital.

Boyle's law—The volume of a fixed amount of gas maintained at constant temperature is inversely proportional to the gas pressure.

Breeder reactor—A nuclear reactor that produces more fissionable materials than it uses.

Brønsted-Lowry acid—A substance capable of donating a proton.

Brønsted-Lowry base—A substance capable of accepting a proton.

Buffer range—The pH range over which a buffer solution is most effective.

Buffer solution—A solution of (a) a weak acid or base and (b) its salt; both components must be present. The solution has the ability to resist changes in pH upon the addition of small amounts of either acid or base.

Calorimetry—The measurement of heat changes.

Carbides—Ionic compounds containing the C_2^{2-} or C^{4-} ion.

Carboxylic acids—Acids that contain the carboxyl group —COOH.

Catalyst—A substance that increases the rate of a chemical reaction without itself being consumed.

Catenation—The ability of the atoms of an element to form bonds with one another.

Cathode—The electrode at which reduction occurs.

Cation—An ion with a net positive charge.

Charles and Gay-Lussac's law—See Charles' law.

Charles' law—The volume of a fixed amount of gas maintained at constant pressure is directly proportional to the absolute temperature of the gas.

Chelating agent—A substance that forms complex ions with metal ions in solution.

Chemical equilibrium—A chemical state in which no net change can be observed.

Chemical formula—An expression showing the chemical composition of a compound in terms of the symbols for the atoms of the elements involved.

Chemical kinetics—The area of chemistry concerned with the speeds, or rates, at which chemical reactions occur.

Chemical property—Any property of a substance that cannot be studied without converting the substance into some other substance.

Chemistry—The science that studies the properties of substances and how substances react with one another.

Chiral—Compounds or ions that are not superimposable with mirror images.

Chlo-alkali process—The production of chlorine gas by the electrolysis of aqueous NaCl solution.

Closed system—A system that allows the exchange of energy (usually in the form of heat) but not mass with its surroundings.

Closest packing—The most efficient arrangements for packing atoms, molecules, or ions in a crystal.

Cohesion—The intermolecular attraction between like molecules.

Colligative properties—Properties of solutions that depend on the number of solute particles in solution and not on the nature of the solute particles.

Combination reaction—Two or more substances react to produce one product.

Common ion effect—The shift in equilibrium caused by the addition of a compound having an ion in common with the dissolved substances.

Complex ion—An ion containing a central metal cation bonded to one or more molecules or ions.

Compound—A substance composed of atoms of two or more elements chemically united in fixed proportions.

Concentration of a solution—The amount of solute dissolved in a given amount of solvent.

Condensation—The phenomenon of going from the gaseous state to the liquid state.

Condensation reaction—A reaction in which two smaller molecules combine to form a larger molecule. Water is invariably one of the products of such a reaction.

Conjugate acid-base pair—An acid and its conjugate base or a base and its conjugate acid.

Coordinate covalent bond—A bond in which the pair of electrons is supplied by one of the two bonded atoms; also called a dative bond.

Coordination compound—A neutral species containing a complex ion.

Coordination number—In a crystal lattice it is defined as the number of atoms (or ions) surrounding an atom (or ion). In coordination compounds it is defined as the number of donor atoms surrounding the central metal atom in a complex.

Copolymerization—The formation of a polymer that contains two or more different monomers.

Corrosion—The deterioration of metals by an electrochemical process.

Covalent bond—A bond in which two electrons are shared by two atoms.

Covalent compounds—Compounds containing only covalent bonds.

Critical mass—The minimum mass of fissionable material required to generate a self-sustaining nuclear chain reaction.

Critical pressure—The minimum pressure that must be applied to bring about liquefaction at the critical temperature.

Critical temperature—The temperature above which a gas will not liquefy.

Crystalline solid—A solid that possesses rigid and long-range order; its atoms, molecules, or ions occupy specific positions.

Crystallization—The process in which dissolved solute comes out of solution and forms crystals.

Crystal field splitting—The energy difference between two sets of d orbitals of a metal atom in the presence of ligands.

Crystal structure—The geometrical order of the lattice points.

Cyanides—Compounds containing the CN^- ion.

Cycloalkanes—Alkanes whose carbon atoms are joined in rings. They have the general formula C_nH_{2n}, where $n = 3, 4, \ldots$.

Dalton's law of partial pressures—The total pressure of a mixture of gases is just the sum of the pressures that each gas would exert if it were present alone.

Decomposition reaction—One substance undergoes a reaction to produce two or more substances.

Delocalized molecular orbitals—Molecular orbitals that are not confined between two adjacent bonding atoms but actually extend over three or more atoms.

Density—The mass of a substance divided by its volume.

Deposition—The process in which the molecules go directly from the vapor into the solid phase.

Desalination—Purification of sea water by the removal of dissolved salts.

Diagonal relationship—Similarities that exist between pairs of elements in different groups and periods of the periodic table.

Diamagnetic—Repelled by a magnet; a diamagnetic substance contains only paired electrons.

Diatomic molecule—A molecule that consists of two atoms.

Diffusion—The gradual mixing of molecules of one gas with the molecules of another by virtue of their kinetic properties.

Dilution—A procedure for preparing a less concentrated solution from a more concentrated solution.

Dimer—A molecule made up of two molecules.

Dipole moment—The product of charge and the distance between the charges in a molecule.

Dipole-dipole forces—Forces that act between polar molecules.

Diprotic acid—Each unit of the acid yields two hydrogen ions.

Dispersion forces—The attractive forces that arise as a result of temporary dipoles induced in the atoms or molecules; also called London forces.

Displacement reaction—An atom or an ion in a compound is replaced by an atom of another element.

Disproportionation reaction—A reaction in which an element in one oxidation state is both oxidized and reduced.

Donor atom—The atom in a ligand that is bonded directly to the metal atom.

Donor impurities—Impurities that provide conduction electrons to semiconductors.

Driving force—Factor (or factors) responsible for converting reactants largely or totally to products.

Dynamic equilibrium—The condition in which the rate of a forward process is exactly balanced by the rate of a reverse process.

Effusion—The process by which a gas under pressure escapes from one compartment of a container to another by passing through a small opening.

Electrolysis—A process in which electrical energy is used to cause a nonspontaneous chemical reaction to occur.

Electrolyte—A substance that, when dissolved in water, results in a solution that can conduct electricity.

Electromagnetic wave—A wave that has an electric field component and a magnetic field component.

Electromotive force (emf)—The voltage difference between electrodes.

Electron—A subatomic particle that has a very low mass and carries a single negative electric charge.

Electronegativity—The ability of an atom to attract electrons toward itself in a chemical bond.

Electron affinity—The energy change when an electron is accepted by an atom (or an ion) in the gaseous state.

Electron configuration—The distribution of electrons among the various orbitals in an atom or molecule.

Electron density—The probability that an electron will be found at a particular region in an atomic orbital.

Element—A substance that cannot be separated into simpler substances by chemical means.

Elementary steps—A series of simple reactions that represent the progress of the overall reaction at the molecular level.

Emission spectra—Continuous or line spectra emitted by substances.

Empirical formula—An expression showing the types of elements present and the ratios of the different kinds of atoms.

Enantiomers—Optical isomers, that is, compounds and their nonsuperimposable mirror images.

Endothermic processes—Processes that absorb heat from the surroundings.

Energy—The capacity to do work or to produce change.

Enthalpy—A thermodynamic quantity used to describe heat changes taking place at constant pressure.

Enthalpy of reaction—The difference between the enthalpies of the products and the enthalpies of the reactants.

Enthalpy of solution—The heat generated or absorbed when a certain amount of solute is dissolved in a certain amount of solvent.

Entropy—A direct measure of the randomness or disorder of a system.

Enzyme—A biological catalyst.

Equilibrium—A state in which there are no observable changes as time goes by.

Equilibrium constant—A number equal to the ratio of the equilibrium concentrations of products to the equilibrium concentrations of reactants, each raised to the power of its stoichiometric coefficient.

Equilibrium vapor pressure—The vapor pressure measured under dynamic equilibrium of condensation and evaporation.

Equivalence point—The point at which the acid is completely reacted with or neutralized by the base.

Equivalent—One equivalent of an oxidizing agent is the amount that gains one mole of electrons. One equivalent of a reducing agent is the amount that loses one mole of electrons.

Equivalent mass—The mass of a substance in grams that gains or loses one mole of electrons in a redox reaction.

Esters—Compounds that have the general formula R'COOR, where R' can be H or an alkyl group or an aryl group and R is an alkyl group or an aryl group.

Ether—An organic compound containing the R—O—R' linkage.

Evaporation—The escape of molecules from the surface of a liquid; also called vaporization.

Excess reagents—One or more reactants present in quantities greater than those needed to react with the quantity of the limiting reagent.

Excited level—See Excited state.

Excited state—A state that has higher energy than the ground state.

Exothermic processes—Processes that give off heat to the surroundings.

Extensive property—A property that depends on how much matter is being considered.

Family—The elements in a vertical column of the periodic table.

First law of thermodynamics—Energy can be converted from one form to another, but cannot be created or destroyed.

First-order reaction—A reaction whose rate depends on reactant concentration raised to the first power.

Formation constant—The equilibrium constant for the complex ion formation.

Fractional crystallization—The separation of a mixture of substances into pure components on the basis of their differing solubilities.

Fractional distillation—A procedure for separating liquid components of a solution that is based on their different boiling points.

Free energy—The energy available to do useful work.

Frequency—The number of waves that pass through a particular point per unit time.

Functional group—That part of a molecule characterized by a special arrangement of atoms that is largely responsible for the chemical behavior of the parent molecule.

Gamma (γ) rays—High energy radiation.

Gas constant (R)—The constant that appears in the ideal gas equation ($PV = nRT$). It is usually expressed as 0.08206 L·atm/K·mol, or 8.314 J/K·mol.

Geometric isomers—Compounds with the same type and number of atoms and the same chemical bonds but different spatial arrangements; such isomers cannot be interconverted without breaking a chemical bond.

Gibbs free energy—See Free energy.

Glass—The optically transparent fusion product of inorganic materials that has cooled to a rigid state without crystallizing.

Graham's law of diffusion—Under the same conditions of temperature and pressure, rates of diffusion for gaseous substances are inversely proportional to the square roots of their molar masses.

Gravimetric analysis—An analytical procedure that involves the measurement of mass.

Ground level—See Ground state.

Ground state—The lowest energy state of a system.

Group—The elements in a vertical column of the periodic table.

Half-cell reactions—Oxidation and reduction reactions at the electrodes.

Half-life—The time required for the concentration of a reactant to decrease to half of its initial concentration.

Half-reaction—A reaction that explicitly shows electrons involved in either oxidation or reduction.

Halogens—The nonmetallic elements in Group 7A (F, Cl, Br, I, and At).

Hard water—Water that contains Ca^{2+} and Mg^{2+} ions.

Heat capacity—The amount of heat required to raise the temperature of a given quantity of the substance by one degree Celsius.

Heat content—See Enthalpy.

Heat of hydration—The heat change (absorbed or released) associated with the hydration process.

Heat of solution—See Enthalpy of solution.

Henry's law—The solubility of a gas in a liquid is proportional to the pressure of the gas over the solution.

Hess's law—The enthalpy change for a reaction is equal to the sum of enthalpy changes for the individual reactions that can be added together to give the desired reaction.

Heterogeneous equilibrium—An equilibrium state in which the reacting species are not all in the same phase.

Heterogeneous mixture—The individual components of a mixture remain physically separated and can be seen as separate components.

Heteronuclear diatomic molecule—A diatomic molecule containing atoms of different elements.

Homogeneous equilibrium—An equilibrium state in which all reacting species are in the same phase.

Homogeneous mixture—The composition of the mixture, after sufficient stirring, is the same throughout the solution.

Homonuclear diatomic molecule—A diatomic molecule containing atoms of the same element.

Hund's rule—The most stable arrangement of electrons in subshells is the one with the greatest number of parallel spins.

Hybridization—The process of mixing the atomic orbitals in an atom (usually the central atom) to generate a set of new atomic orbitals.

Hybrid orbitals—Atomic orbitals obtained when two or more nonequivalent orbitals of the same atom combine.

Hydrates—Compounds that have a specific number of water molecules attached to them.

Hydration—A process in which an ion or a molecule is surrounded by water molecules arranged in a specific manner.

Hydrocarbons—Compounds made up only of carbon and hydrogen.

Hydrogenation—The addition of hydrogen, especially to compounds with double and triple carbon–carbon bonds.

Hydrogen bonding—A special type of dipole–dipole interaction between the hydrogen atom bonded to an atom of a very electronegative element (F, N, O) and another atom of one of the three electronegative elements.

Hydrolysis—The reaction of a substance with water that usually changes the pH of the solution.

Hydrophilic—Water-liking.

Hydrophobic—Water-fearing.

Hydroscopic—Having a tendency to absorb water.

Hypothesis—A tentative explanation for a set of observations.

Ideal gas—A hypothetical gas whose pressure-volume-temperature behavior can be completely accounted for by the ideal gas equation.

Ideal gas equation—An equation expressing the relationships among pressure, volume, temperature, and amount of gas ($PV = nRT$, where R is the gas constant).

Ideal solution—Any solution that obeys Raoult's law.

Indicators—Substances that have distinctly different colors in acidic and basic media.

Induced dipole—The separation of positive and negative charges in a neutral atom (or a nonpolar molecule) caused by the proximity of an ion or a polar molecule.

Inorganic compounds—Compounds other than organic compounds.

Insulator—A substance incapable of conducting electricity.

Intensive property—Properties that do not depend on how much matter is being considered.

Interhalogen compounds—Compounds formed between two halogen elements.

Intermediate—A species that appears in the mechanism of the reaction (that is, the elementary steps) but not in the overall balanced equation.

Intermolecular forces—Attractive forces that exist among molecules.

Intramolecular forces—Forces that hold atoms together in a molecule.

Ion—A charged particle formed when a neutral atom or group of atoms gain or lose one or more electrons.

Ionic bond—The electrostatic force that holds ions together in an ionic compound.

Ionic compound—Any neutral compound containing cations and anions.

Ionic equation—An equation that shows dissolved ionic compounds in terms of their free ions.

Ionic radius—The radius of a cation or an anion as measured in an ionic compound.

Ionization energy—The minimum energy required to remove an electron from an isolated atom (or an ion) in its ground state.

Ion exchange—A process in which undesirable ions are replaced by another type of ions.

Ion pair—A species made up of at least one cation and at least one anion held together by electrostatic forces.

Ion-dipole forces—Forces that operate between an ion and a dipole.

Ion-product constant—Product of hydrogen ion concentration and hydroxide ion concentration (both in molarity) at a particular temperature.

Isoelectronic—Ions, or atoms and ions, that possess the same number of electrons, and hence the same ground-state electron configuration, are said to be isoelectronic.

Isolated system—A system that does not allow the transfer of either mass or energy to or from its surroundings.

Isotopes—Atoms having the same atomic number but different mass numbers.

Joule—Unit of energy given by newtons $\times$ meters.

Kelvin temperature scale—See Absolute temperature scale.

Ketones—Compounds with a carbonyl functional group and the general formula RR'CO, where R and R' are alkyl and/or aryl groups.

Kinetic energy—Energy available because of the motion of an object.

Labile complexes—Complex ions that undergo rapid ligand exchange reactions.

Lanthanide series—Elements that have incompletely filled $4f$ subshells or readily give rise to cations that have incompletely filled $4f$ subshells.

Lattice energy—The energy required to completely separate one mole of a solid ionic compound into gaseous ions.

Lattice points—The positions occupied by atoms, molecules, or ions that define the geometry of a unit cell.

Law—A concise verbal or mathematical statement of a relationship between phenomena that is always the same under the same conditions.

Law of conservation of energy—The total quantity of energy in the universe remains a constant.

Law of conservation of mass—Matter can neither be created nor destroyed.

Law of definite proportions—Different samples of the same compound always contain its constituent elements in the same proportions by mass.

Law of multiple proportions—If two elements can combine to form more than one type of compound, the masses of one element that combine with a fixed mass of the other element are in ratios of small whole numbers.

Leveling effect—The inability of a solvent to differentiate among the relative strengths of all acids stronger than the solvent's conjugate acid.

Lewis acid—A substance that can accept a pair of electrons.

Lewis base—A substance that can donate a pair of electrons.

Lewis dot symbol—The symbol of an element with one or more dots that represent the number of valence electrons in an atom of the element.

Le Chatelier's principle—If an external stress is applied to a system at equilibrium, the system will adjust itself in such a way that the stress is partially offset.

Ligand—A molecule or an ion that is bonded to the metal ion in a complex ion.

Limiting reagent—The reactant used up first in a reaction.

Line spectra—Spectra produced when radiation is absorbed or emitted by substances only at some wavelengths.

Liter—The volume occupied by one cubic decimeter.

Lone pairs—Valence electrons that are not involved in covalent bond formation.

Macroscopic properties—Properties that can be measured directly.

Many-electron atoms—Atoms that contain three or more electrons.

Mass—A measure of the quantity of matter contained in an object.

Mass defect—The difference between the mass of an atom and the sum of the masses of its protons, neutrons, and electrons.

Mass number (A)—The total number of neutrons and protons present in the nucleus of an atom.

Matter—Anything that occupies space and possesses mass.

Melting point—The temperature at which solid and liquid phases coexist in equilibrium.

Metals—Elements that are good conductors of heat and electricity and have the tendency to form positive ions in ionic compounds.

Metalloid—An element with properties intermediate between those of metals and nonmetals.

Metallurgy—The science and technology of separating metals from their ores and of compounding alloys.

Metathesis reaction—A double displacement reaction.

Microscopic properties—Properties that cannot be measured directly without the aid of a microscope or other special instrument.

Mineral—A naturally occurring substance with a characteristic range of chemical composition.

Miscible—Two liquids that are completely soluble in each other in all proportions are said to be miscible.

Mixture—A combination of two or more substances in which the substances retain their identity.

Moderator—A substance that can reduce the kinetic energy of neutrons.

Molality—The number of moles of solute dissolved in 1 kg of solvent.

Molarity—The number of moles of solute in one liter of solution.

Molar heat of fusion—The energy (in kilojoules) required to melt one mole of a solid.

Molar heat of sublimation—the energy (in kilojoules) required to sublime one mole of a solid.

Molar heat of vaporization—The energy (in kilojoules) required to vaporize one mole of a liquid.

Molar mass—The mass (in grams or kilograms) of one mole of atoms, molecules, or other particles.

Molar mass of a chemical compound—The mass (in grams or kilograms) of one mole of the compound.

Molar solution—A solution whose concentration is expressed in molarity.

Mole—The amount of substance that contains as many elementary entities (atoms, molecules, or other particles) as there are atoms in exactly 12 grams (or 0.012 kilograms) of the carbon-12 isotope.

Molecularity of a reaction—The number of molecules reacting in an elementary step.

Molecular equations—Equations in which the formulas of the compounds are written as though all species existed as molecules or whole units.

Molecular formula—An expression showing the exact numbers of atoms of each element in a molecule.

Molecular mass—The sum of the atomic masses (in amu) present in the molecule.

Molecular orbital—An orbital that results from the interaction of the atomic orbitals of the bonding atoms.

Molecule—An aggregate of at least two atoms in a definite arrangement held together by special forces.

Mole fraction—Ratio of the number of moles of one component of a mixture to the total number of moles of all components in the mixture.

Mole method—An approach to determine the amount of product formed in a reaction.

Monomer—The single repeating unit of a polymer.

Monoprotic acid—Each unit of the acid yields one hydrogen ion.

Multiple bonds—Bonds formed when two atoms share two or more pairs of electrons.

Net ionic equation—An equation that indicates only the ionic species that actually take part in the reaction.

Neutralization reaction—A reaction between an acid and a base.

Neutron—A subatomic particle that bears no net electric charge. Its mass is slightly greater than that of the proton.

Newton—The SI unit for force.

Nitrogen fixation—The conversion of molecular nitrogen into nitrogen compounds.

Noble gases—The nonmetallic elements in Group 8A (He, Ne, Ar, Kr, Xe, and Rn). With the exception of helium, these elements all have completely filled p subshells. (The electron configurations are $1s^2$ for helium and ns^2np^6 for the other noble gases, where n is the principal quantum number for the outermost shell.)

Nonbonding electrons—Valence electrons that are not involved in covalent bond formation.

Nonelectrolyte—A substance that, when dissolved in water, gives a solution that is not electrically conducting.

Nonmetals—Elements that are usually poor conductors of heat and electricity.

Nonpolar molecule—A molecule that does not possess a dipole moment.

Nonvolatile—Does not have a measurable vapor pressure.

Normality—The number of equivalents of an oxidizing agent or a reducing agent per liter of solution.

Nuclear chain reaction—A self-sustaining sequence of nuclear fission reactions.

Nuclear fusion—The combining of small nuclei into larger ones.

Nuclear transmutation—The change undergone by a nucleus as a result of bombardment by neutrons or other nuclei.

Nuclear binding energy—The energy required to break up a nucleus into its component protons and neutrons.

Nuclear fission—A heavy nucleus (mass number > 200) divides to form smaller nuclei of intermediate mass and one or more neutrons.

Nucleus—The central core of an atom.

***n*-Type semiconductors**—Semiconductors that contain donor impurities.

Octet rule—An atom other than hydrogen tends to form bonds until it is surrounded by eight valence electrons.

Open system—A system that can exchange mass and energy (usually in the form of heat) with its surroundings.

Optical isomers—Compounds that are nonsuperimposable mirror images.

Orbital—See Atomic orbital and Molecular orbital.

Ore—The material of a mineral deposit concentrated enough to allow economical recovery of a desired metal.

Organic chemistry—The branch of chemistry that deals with carbon compounds.

Organic compounds—Compounds that contain carbon, usually in combination with elements such as hydrogen, oxygen, nitrogen, and sulfur.

Osmosis—The net movement of solvent molecules through a semipermeable membrane from a pure solvent or from a dilute solution to a more concentrated solution.

Osmotic pressure—The pressure required to stop osmosis from pure solvent into the solution.

Overvoltage—The additional voltage required to cause electrolysis.

Oxidation number—The number of charges an atom would have in a molecule if electrons were transferred completely in the direction indicated by the difference in electronegativity.

Oxidation reaction—The half-reaction that involves loss of electrons.

Oxidizing agent—A substance that can accept electrons from another substance or increase the oxidation numbers in another substance.

Oxyacids—Acids containing hydrogen, oxygen, and another element (the central element).

Oxyanions—Anions derived from oxyacids.

Packing efficiency—Percentage of the unit cell space occupied by the spheres.

Paramagnetic—Attracted by a magnet. A paramagnetic substance contains one or more unpaired electrons.

Partial pressure—Pressure of one component in a mixture of gases.

Pascal—A pressure of one newton per square meter (1 N/m^2).

Pauli exclusion principle—No two electrons in an atom can have the same four quantum numbers.

Percent composition by mass—The percent by mass of each element in a compound.

Percent composition by weight—See Percent composition by mass.

Percent yield—The ratio of actual yield to theoretical yield, multiplied by 100%.

Period—A horizontal row of the periodic table.

Periodic table—A tabular arrangement of the elements.

pH—The negative logarithm of the hydrogen ion concentration.

Phase change—Transformation from one phase to another.

Phase diagram—A diagram showing the conditions at which a substance exists as a solid, liquid, or vapor.

Photon—A particle of light.

Physical property—Any property of a substance that can be observed without transforming the substance into some other substance.

Pi bond—A covalent bond formed by sideways overlapping orbitals; it has its electron density concentrated above and below the plane of the nuclei of the bonding atoms.

Pi molecular orbital—A molecular orbital in which the electron density is concentrated above and below the line joining the two nuclei of the bonding atoms.

Plane-polarized light—Light in which the electric field and magnetic field components are confined to specific planes.

Plasma—A state of matter in which a gaseous system consists of positive ions and electrons.

Polarimeter—The instrument for studying interaction between plane-polarized light and chiral molecules.

Polarizability—The ease with which the electron density in a neutral atom (or molecule) can be distorted.

Polar molecule—A molecule that possesses a dipole moment.

Pollutant—A substance that is harmful to the biological environment.

Polyatomic molecule—A molecule that consists of more than two atoms.

Polymer—A chemical species distinguished by a high molar mass, ranging into thousands and millions of grams.

Potential energy—Energy available by virtue of an object's position.

Precipitate—An insoluble solid that separates from the solution.

Precision—The closeness of agreement of two or more measurements of the same quantity.

Pressure—Force applied per unit area.

Product—The substance formed as a result of a chemical reaction.

Proton—A subatomic particle having a single positive electric charge. The mass of a proton is about 1840 times that of an electron.

***p*-Type semiconductors**—Semiconductors that contain acceptor impurities.

Qualitative—Consisting of general observations about the system.

Qualitative analysis—The determination of types of ions present in a solution.

Quantitative—Comprising numbers obtained by various measurements of the system.

Quantum—The smallest quantity of energy that can be emitted (or absorbed) in the form of electromagnetic radiation.

Racemic mixture—An equimolar mixture of the two enantiomers.

Radiation—The emission and transmission of energy through space in the form of waves.

Radical—Any neutral fragment of a molecule containing an unpaired electron.

Radioactivity—The spontaneous breakdown of an atom by emission of particles and/or radiation.

Raoult's law—The partial pressure of the solvent over a solution is given by the product of the vapor pressure of the pure solvent and the mole fraction of the solvent in the solution.

Rare earth series—See Lanthanide series.

Rare gases—The nonmetallic elements in Group 8A (He, Ne, Ar, Kr, Xe, and Rn).

Rate constant—Constant of proportionality between the reaction rate and the concentrations of reactants.

Rate law—An expression relating the rate of a reaction to the rate constant and the concentrations of the reactants.

Rate-determining step—The slowest step in the sequence of steps leading to the formation of products.

Reactants—The starting substances in a chemical reaction.

Reaction mechanism—The sequence of elementary steps that leads to product formation.

Reaction order—The sum of the powers to which all reactant concentrations appearing in the rate law are raised.

Reaction quotient—A number equal to the ratio of product concentrations to reactant concentrations, each raised to the power of its stoichiometric coefficient at some point other than equilibrium.

Reaction rate—The change of the concentration of reactant or product with time.

Redox reaction—A reaction in which there is either a transfer of electrons or a change in the oxidation numbers of the substances taking part in the reaction.

Reducing agent—A substance that can donate electrons to another substance or decrease the oxidation numbers in another substance.

Reduction reaction—The half-reaction that involves gain of electrons.

Representative elements—Elements in Groups 1A through 7A, all of which have incompletely filled s or p subshell of highest principal quantum number.

Resonance—The use of two or more Lewis formulas to represent a particular molecule.

Resonance form—See Resonance structure.

Resonance structure—One of two or more alternative Lewis formulas for a single molecule that cannot be described fully with only one Lewis formula.

Reverse osmosis—A method of desalination using high pressure to force water through a semipermeable membrane from a more concentrated solution to a less concentrated solution.

Reversible reaction—A reaction that can occur in both directions.

Salt—An ionic compound made up of a cation other than H^+ and an anion other than OH^- or O^{2-}.

Salt hydrolysis—The reaction of the anion or cation, or both, of a salt with water.

Saponification—Soapmaking.

Saturated hydrocarbons—Hydrocarbons that contain only single covalent bonds.

Saturated solution—At a given temperature, the solution that results when the maximum amount of a substance has dissolved in a solvent.

Second law of thermodynamics—The entropy of the universe increases in a spontaneous process and remains unchanged in an equilibrium process.

Second-order reaction—A reaction whose rate depends on reactant concentration raised to the second power or on the concentrations of two different reactants, each raised to the first power.

Semiconducting elements—Elements that normally cannot conduct electricity, but can have their conductivity greatly enhanced either by raising the temperature or by the addition of certain impurities.

Semipermeable membrane—A membrane that allows solvent molecules to pass through, but blocks the movement of solute molecules.

Sigma bond—A covalent bond formed by orbitals overlapping end-to-end; it has its electron density concentrated between the nuclei of the bonding atoms.

Sigma molecular orbital—A molecular orbital in which the electron density is concentrated around a line between the two nuclei of the bonding atoms.

Significant figures—The number of meaningful digits in a measured or calculated quantity.

Silanes—Binary compounds containing silicon and hydrogen.

Silicates—Compounds whose anions consist of Si and O atoms.

Soft water—Water that is mostly free of Ca^{2+} and Mg^{2+} ions.

Solubility—The maximum amount of solute that can be dissolved in a given quantity of solvent at a specific temperature.

Solubility product—The product of the molar concentrations of the constituent ions, each raised to the power of its stoichiometric coefficient in the equilibrium equation.

Solute—The substance present in smaller amount in a solution.

Solution—A homogeneous mixture of two or more substances.

Solvent—The substance present in larger amount in a solution.

Specific heat—The amount of heat energy required to raise the temperature of one gram of the substance by one degree Celsius.

Spectator ions—Ions that are not involved in the overall reaction.

Spectrochemical series—A list of ligands arranged in order of their abilities to split the d-orbital energies.

Stability constant—See Formation constant.

Standard emf—The difference between the standard reduction potential of the substance that undergoes reduction and the standard reduction potential of the substance that undergoes oxidation.

Standard enthalpy of formation—The heat change that results when one mole of a compound is formed from its elements in their standard states.

Standard enthalpy of reaction—The enthalpy change when the reaction is carried out under standard state conditions.

Stardard reduction potential—The voltage measured when a reduction reaction occurs at the electrode when all solutes are 1 M and all gases are at 1 atm.

Standard solution—A solution of accurately known concentration.

Standard state—The condition of 1 atm of pressure.

Stardard temperature and pressure (STP)—0°C and 1 atmosphere.

State function—A property that is determined by the state the system is in.

State of a system—The values of all pertinent macroscopic variables (for example, composition, volume, pressure, and temperature) of a system.

Static equilibrium—A state in which no changes are taking place either at the macroscopic, observable level, or at the molecular level.

Stereoisomerism—The occurrence of two or more compounds with the same types and numbers of atoms and the same chemical bonds but different spatial arrangements.

Stereoisomers—Compounds possessing the same formula and bonding arrangement but different spatial arrangements of atoms.

Stoichiometric amounts—The exact molar amounts of reactants and products that appear in the balanced chemical equation.

Stoichiometry—The mass relationships among reactants and products in chemical reactions.

Structural isomers—Molecules that have the same molecular formula but different structures.

Sublimation—The process in which molecules go directly from the solid into the vapor phase.

Substance—A form of matter that has a definite or constant composition (the number and type of basic units present) and distinct properties.

Superconductivity—The loss of electrical resistance in certain metals, alloys, and compounds at low temperatures.

Supercooling—Cooling of a liquid below its freezing point without forming the solid.

Supersaturated solution—A solution that contains more of the solute than is present in a saturated solution.

Surface tension—The amount of energy required to stretch or increase the surface of a liquid by unit area.

Surroundings—The rest of the universe outside the system.

System—Any specific part of the universe that is of interest to us.

Termolecular reaction—An elementary step that involves three molecules.

Ternary acid—An acid that contains three different elements.

Ternary compounds—Compounds consisting of three elements.

Theoretical yield—The amount of product predicted by the balanced equation when all of the limiting reagent has reacted.

Theory—A unifying principle that explains a body of facts and those laws that are based on them.

Thermal pollution—The heating of the environment to temperatures that are harmful to its living inhabitants.

Thermochemistry—The study of heat changes in chemical reactions.

Thermodynamics—The scientific study of the interconversion of heat and other forms of energy.

Thermonuclear reactions—Nuclear fusion reactions that occur at very high temperatures.

Titration—The gradual addition of a solution of accurately known concentration to another solution of unknown concentration until the chemical reaction between the two solutions is complete.

Tracers—Isotopes, especially radioactive isotopes, that are used to trace the path of the atoms of an element in a chemical or biological process.

Transition metals—Elements that have incompletely filled d subshells or readily give rise to cations that have incompletely filled d subshells.

Transuranium elements—Elements with atomic numbers greater than 92.

Triple point—The point at which the vapor, liquid, and solid states of a substance are in equilibrium.

Triprotic acid—An acid that yields three H^+ ions.

Unimolecular reaction—An elementary step in which only one reacting molecule participates.

Unit cell—The basic repeating unit of the arrangement of atoms, molecules, or ions in a crystalline solid.

Unsaturated hydrocarbons—Hydrocarbons that contain carbon–carbon double bonds or carbon–carbon triple bonds.

Unsaturated solution—A solution that contains less solute than it has the capacity to dissolve.

Valence electrons—The outer electrons of an atom, which are those involved in chemical bonding.

Valence shell—The outermost electron-occupied shell of an atom, which holds the electrons that are usually involved in bonding.

Valence-shell electron-pair repulsion (VSEPR) model—A model that accounts for the geometrical arrangements of shared and unshared electron pairs around a central atom in terms of the repulsions between electron pairs.

Valence-shell expansion—The use of d orbitals in addition to s and p orbitals to form covalent bonds.

Van der Waals forces—The dipole–dipole, dipole-induced dipole, and dispersion forces.

Van der Waals radius—One-half the distance between two equivalent nonbonded atoms in their most stable arrangement.

Vaporization—The escape of molecules from the surface of a liquid; also called evaporation.

Viscosity—A measure of a fluid's resistance to flow.

Volatile—Has a measurable vapor pressure.

Wave—A vibrating disturbance by which energy is transmitted.

Wavelength—The distance between identical points on successive waves.

Weight—The force that gravity exerts on an object.

X-ray diffraction—The scattering of X rays by the units of a regular crystalline solid.

Yield of the reaction—The quantity of product obtained from the reaction.

ANSWERS TO SELECTED PROBLEMS

Chapter 1

1.6   (a) Chemical change. (b) Chemical change. (c) Physical change. (d) Physical change. (e) Chemical change.

1.10   (a) Cr. (b) Se. (c) Kr. (d) Zr. (e) Ag. (f) W. (g) K. (h) S.

1.23   105 cm^3.

1.26   (a) 40.6°C. (b) 11.3°F. (c) 1.1×10^4 °F.

1.30   (a) 1.4598×10^2. (b) 3.2×10^2. (c) 6.2×10^{-3}. (d) 9.9×10^{10}. (e) 1.80×10^{-2}. (f) 1.14×10^{10}. (g) -5.0×10^4. (h) 1.3×10^3.

1.32   (a) One. (b) Three. (c) Three. (d) Four. (e) Two. (f) One. (g) One or two (ambiguous).

1.38   (a) 1.8 m; 73.5 kg. (b) 88 km/h. (c) 6.7×10^8 mi. (d) 6.9×10^{-3} g Pb.

1.42   (a) 81 in/s. (b) 1.2×10^2 m/min. (c) 7.4 km/h.

1.45   2.6 g/cm^3.

1.48   (a) 12.6 g/cm^3. (b) 19 g/cm^3. (c) 1.28 g/cm^3.

1.56   F, Cl; Na, K; P, As.

Chapter 2

2.5 4.3×10^{-24} g.

2.8 2.97×10^{22} atoms.

2.15   (a) CH_2O. (b) CH. (c) CH. (d) $NaSO_2$. (e) $AlCl_3$. (f) BH_3. (g) $K_2Cr_2O_7$.

2.18   89 amu; 89 g.

2.19   (a) 16.04 amu. (b) 18.02 amu. (c) 34.02 amu. (d) 179.55 amu. (e) 6.532×10^4 amu.

2.24   1 mol Cl_2 to 3.5 mol O_2.

2.25 3.16×10^{22} C atoms; 5.80×10^{22} H atoms; 2.90×10^{22} O atoms.

2.28   (a) False. (b) False. (c) False.

2.30   (a) Na: 54.75%, F: 45.25%. (b) Na: 39.34%, Cl: 60.66%. (c) Na: 22.34%, Br: 77.66%. (d) Na: 15.34%, 84.66%.

2.31   (a) NaCl. (b) H_2SO_4. (c) $AlCl_3$. (d) CHO. (e) KCN.

2.36 $PtCl_2$ and $PtCl_4$.

2.40   CH.

2.46   The ratios of O atoms in these two compounds is 2:1. The data support the law of multiple proportions.

2.54   (a) Compound. (b) Element. (c) Element. (d) Compound. (e) Element. (f) Element. (g) Element. (h) Compound.

2.57   103.3 amu.

2.61   (a) Inorganic. (b) Organic. (c) Inorganic. (d) Inorganic. (e) Organic. (f) Inorganic. (g) Inorganic. (h) Organic.

Chapter 3

3.5   (a) $CH_4 + 2O_2 \rightarrow CO_2 + 2H_2O$
(b) $CH_4 + 4Br_2 \rightarrow CBr_4 + 4HBr$
(c) $Fe_2O_3 + 3CO \rightarrow 2Fe + 3CO_2$
(d) $S_8 + 8O_2 \rightarrow 8SO_2$
(e) $6HNO_3 + S \rightarrow H_2SO_4 + 2H_2O + 6NO_2$
(f) $Be_2C + 4H_2O \rightarrow 2Be(OH)_2 + CH_4$.

3.8   (a) Weak electrolyte. (b) Strong electrolyte. (c) Strong electrolyte. (d) Weak electrolyte. (e) Nonelectrolyte. (f) Strong electrolyte. (g) Nonelectrolyte. (h) Weak electrolyte. (i) Strong electrolyte.

3.15   (a) No precipitation. (b) $BaSO_4$ precipitate will form.

3.20   Li and Ca.

3.21   (a) No reaction. (b) No reaction.
(c) $Mg(s) + CuSO_4(aq) \rightarrow MgSO_4 (aq) + Cu(s)$
(d) $Cl_2(g) + 2KBr(aq) \rightarrow 2KCl(aq) + Br_2(l)$

Chapter 4

4.1 1.3×10^7 tons S.

4.4   255.9 g; 0.324 L.

4.10   (a) $NH_4NO_3(s) \rightarrow N_2O(g) + 2H_2O(g)$. (b) 20 g N_2O.

4.14 HCl will be used up first; 23.4 g Cl_2.

4.16 65.8%.

4.18 56.0 g.

4.23 (a) 1.16 M. (b) 0.608 M. (c) 1.78 M.

4.30 0.0433 M.

4.31 Dilute 3.00 mL of 4.00 M HNO_3 to 60.0 mL.

4.35 0.215 g AgCl.

4.41 (a) 42.78 mL. (b) 158.5 mL. (c) 79.23 mL.

4.48 $TiCl_4$.

4.52 2.

4.57 $CuO(s) + H_2(g) \rightarrow Cu(s) + H_2O(g)$;Cu: 38%; CuO: 62%.

4.58 744 mL.

Chapter 5

5.8 177.8 kJ.

5.10 (a) -571.6 kJ. (b) -2599 kJ. (c) -1411 kJ. (d) -1124 kJ.

5.14 3.43×10^4 kJ.

5.23 In a nuclear reaction the identity of the element is usually changed from reactant to product. Consequently, the convention of standard states does not apply to these processes.

5.25 -84.6 kJ.

5.29 (a) -65.2 kJ. (b) -9.0 kJ.

5.33 7.28×10^2 kJ.

5.34 3.31 kJ.

5.36 24.8 kJ/g; 603 kJ/mol.

Chapter 6

6.7 1.30×10^3 mL.

6.10 457 mmHg.

6.12 The pressure of the air inside the Ping-Pong ball increases when it is heated. The increased pressure pushes out the dented surface.

6.15 4.7×10^2 °C.

6.22 6.1×10^{-3} atm.

6.24 35.1 g/mol.

6.30 $C_4H_{10}O$.

6.34 3.88 L.

6.36 P_2F_4.

6.39 The gas remaining in NH_3 whose volume is 54.5 L.

6.43 $C_2H_5OH(l) + 3O_2(g) \rightarrow 2CO_2(g) + 3H_2O(l)$; 1.71×10^3 L.

6.45 (a) 0.89 atm. (b) 1.4 L.

6.48 19.8 g.

6.51 On the average, an O_2 molecule moves 3.32 times faster than a UF_6 molecule.

6.56 13 g/mol.

6.58 $HI < Kr < ClO_2 < PH_3 < NH_3$; $r_{NH_3}/r_{HI} = 2.74$.

6.63 From Table 6.4 we see that CH_4 has the largest value for a and Ne has the smallest value.

Ne also has the smallest b value. Therefore, Ne will behave most ideally.

Chapter 7

7.10 2.8×10^{14} g/cm^3.

7.16 $^{14}_{7}$N: 7 protons and 7 neutrons; $^{15}_{7}$N: 7 protons and 8 neutrons; $^{35}_{17}$Cl: 17 protons and 18 neutrons; $^{37}_{17}$Cl: 17 protons and 20 neutrons; $^{51}_{23}$V: 23 protons and 28 neutrons; $^{72}_{32}$Ge: 32 protons and 40 neutrons; $^{231}_{91}$Pa: 91 protons and 140 neutrons.

7.19 (a) $^{23}_{11}$Na. (b) $^{64}_{28}$Ni. (c) $^{186}_{74}$W. (d) $^{201}_{80}$Hg.

7.21 Zn^{2+}.

7.26 (a) 6.30×10^{-12} J; 9.00×10^{-13} J/nucleon. (b) 4.92×10^{-11} J; 1.41×10^{-12} J/nucleon. (c) 4.54×10^{-12} J; 1.14×10^{-12} J/nucleon. (d) 2.63×10^{-10} J; 1.26×10^{-12} J/nucleon.

7.28 35.45 amu.

Chapter 8

8.1 3.5×10^3 nm; 5.30×10^{14} Hz.

8.6 3.19×10^{-19} J.

8.9 1.2×10^2 nm (UV region).

8.24 (a) 1.4×10^2 nm. (b) 5.0×10^{-19} J. (c) 2.0×10^2 nm.

8.25 1.28×10^3 nm.

8.28 3×10^{-22} J.

8.31 9.90 nm.

8.35 $\ell = 0, 1$. $\ell = 0, m_\ell = 0$. $\ell = 1, m_\ell = -1, 0, 1$.

8.44 (a) Two. (b) Six. (c) Ten. (d) Fourteen.

8.47 (a) Three. (b) Six. (c) Zero.

8.52 (a) $2s$. (b) $3p$. (c) $3s$. (d) $4d$.

8.54 (a) m_ℓ cannot be a noninteger. (c) ℓ must be less than n. (e) m_s cannot be 1.

8.58 S^+.

8.61 Al: $1s^2 2s^2 2p^6 3s^2 3p^1$. B: $1s^2 2s^2 2p^1$. F: $1s^2 2s^2 2p^5$.

8.64 (a) $Z = 7$ N: $1s^2 2s^2 2p^3$. (b) $Z = 2$ He: $1s^2$. (c) $Z = 11$ Na: $1s^2 2s^2 2p^6 3s^1$. (d) $Z = 33$ As: $[Ar]4s^2 3d^{10} 4p^3$. (e) $Z = 17$ Cl: $[Ne]3s^2 3p^5$.

Chapter 9

9.6 H_2; C (extensive three-dimensional structure); O_2; P_4; S_8; Ar (monatomic gas); Br_2 (liquid); Ag (metallic state); I_2 (solid); Xe (monatomic gas).

9.12 (a) and (h); (b) and (e); (c) and (g); (d) and (f).

9.14 (a) Group 1A. (b) Group 5A. (c) Group 8A. (d) Group 8B.

9.17 (a) Cs. (b) Ba. (c) Sb. (d) Br. (e) Xe.

9.28 The atom with the electron configuration $1s^2 2s^2 2p^6$ has the first ionization energy of 2080 kJ/mol and that having $1s^2 2s^2 2p^6 3s^1$ has the first ionization energy of 496 kJ/mol.

9.31 5.25×10^3 kJ/mol.

9.34 Cl.

9.44 BaO.

9.50 0.65.

Chapter 10

10.11 (a) Energy decreases. (b) Energy increases. (c) Energy increases. (d) Energy increases.

10.15 (a) IF_7 (iodine heptafluoride): molecular. (b) KBr (potassium bromide): ionic. (c) MgF_2 (magnesium fluoride): ionic. (d) AlI_3 (aluminum iodide): probably molecular.

10.17 (a) $1s^22s^22p^6$ or [Ne]. (b) $1s^22s^22p^6$ or [Ne]. (c) $[Ne]3s^23p^6$ or [Ar]. (d) [Ar]. (e) [Ar]. (f) $[Ar]3d^6$. (g) $[Ar]3d^9$. (h) $[Ar]3d^{10}$.

10.19 (a) Cr^{3+}. (b) Sc^{3+}. (c) Rh^{3+}. (d) Ir^{3+}.

10.22 $Mg^{2+} < Na^+ < F^- < O^{2-} < N^{3-}$.

10.29 Cu is a transition metal and K is a representative element. Therefore, the blue solution contains $CuSO_4$ (the hydrated Cu^{2+} ions are responsible for the blue color).

10.33 (a) MgO. (b) LiF. (c) Mg_3N_2.

10.35 788 kJ.

Chapter 11

11.7 $N_2 < SO_2 < ClF_3 < K_2O < LiF$

11.9 $DG < EG < DF < DE$

11.17

11.19 $:C{\equiv}O:^+$ $:N{\equiv}O:^+$ $:C{\equiv}N:^-$ $:N{\equiv}N:$

11.26

In SeF_4, there are ten valence electrons surrounding the Se atom; in SeF_6, there are twelve electrons surrounding the Se atom. Therefore, the octet rule is not satisfied for Se in both compounds.

11.31

11.33 $^-:\overset{..}{O}{-}C{\equiv}N: \leftrightarrow \overset{..}{O}{=}C{=}\overset{..}{N}{}^- \leftrightarrow :O{\equiv}C{-}\overset{..}{N}:^{2-}$

11.38 $\overset{..}{N}{=}N{=}\overset{..}{O} \leftrightarrow :N{\equiv}N{-}\overset{..}{O}:^- \leftrightarrow :\overset{..}{N}{-}N{\equiv}O:$

11.44 392 kJ.

11.47 (a) -2620 kJ. (b) -2855 kJ.

Chapter 12

12.4 (a) Pyramidal. (b) Tetrahedral. (c) Tetrahedral. (d) Distorted tetrahedron (like SF_4).

12.5 (a) Tetrahedral. (b) Bent. (c) Planar. (d) Linear. (e) Square planar. (f) Tetrahedral. (g) Trigonal bipyramid. (h) Trigonal pyramidal. (i) Tetrahedral. (j) Bent.

12.16 CO_2, $CBr_4 < H_2S < NH_3 < H_2O < HF$

12.24 As is sp^3 hybridized.

12.27 (a) Both C atoms are sp^3 hybridized. (b) C in CH_3 is sp^3 hybridized; the two C atoms double bonded to each other are sp^2 hybridized. (c) Both C atoms are sp^3 hybridized. (d) The C atom in CH_3 is sp^3 hybridized; the C atom that is double bonded to the O atom is sp^2 hybridized. (e) The C atom in CH_3 is sp^3 hybridized; the C atom in COOH is sp^2 hybridized.

12.31 The hybridization of B changes from sp^2 (in BF_3) to sp^3 in (F_3BNH_3) while that of N remains unchanged.

12.36 (a) 180°. (b) 120°. (c) 109.5°. (d) Approximately 109.5°. (e) 180°. (f) Less than 120°. (g) Less than 120°. (h) Less than 120°. (i) 109.5°.

12.38 The molecule is linear: $:\overset{..}{Br}{-}Hg{-}\overset{..}{Br}:$. The two bond moments cancel; the molecule does not possess a dipole moment.

12.44 (a) and (c) are equivalent; (b) and (d) are equivalent.

Chapter 13

13.7 In forming the N_2^+ ion, an electron is removed from the sigma bonding molecular orbital (see Table 13.1). Consequently, the bond order in N_2^+ is 2.5 compared to 3 in N_2. In forming the O_2^+ ion, an electron is removed from the pi antibonding molecular orbital. Consequently, the bond order in O_2^+ is 2.5 compared to 2 in O_2.

13.15 The stability of the benzene molecule (compared to ethylene) is the result of delocalized molecular orbitals.

13.18 (a) We can draw two resonance structures for FNO_2:

(b) The molecule is trigonal planar; the nitrogen atom is sp^2 hybridized.
(c) The bonding consists of sigma bonds joining the nitrogen atom to the fluorine and oxygen atoms. In addition there is a pi delocalized molecular orbital over the molecule.

Chapter 14

14.4 The dispersion forces between the CCl_4 molecules are considerably stronger than the dipole–dipole and dispersion forces between the SO_2 molecules.

14.6 Methane, because it has the lowest boiling point.

14.11 (e).

14.18 (a) Hydrogen bonding in NH_3. (b) Strong ionic bonding in KCl. (c) Both naphthalene and benzene are nonpolar—an example of "like dissolves like." LiBr is an ionic compound.

14.28 Water, because H_2O molecules are held together by hydrogen bonding.

14.30 *Simple cubic cell* 1 sphere/cell; packing efficiency: 52.4%. *Body-centered cubic cell* 2 spheres/cell; packing efficiency: 68.0%. *Face-centered cubic cell* 4 spheres/cell; packing efficiency: 74.0%.

14.31 168 pm.

14.33 19.4 g/cm^3.

14.36 6.17×10^{23} Ba atoms/mol.

14.38 458 pm.

14.41 XY_3.

14.44 45 pm.

14.46 (a) Molecular crystal. (b) Covalent crystal. (c) Molecular crystal. (d) Ionic crystal. (e) Metallic crystal. (f) Covalent crystal. (g) Ionic crystal. (h) Metallic crystal.

14.58 169 kJ.

14.60 47.0 kJ/mol.

14.62 Toluene has the lowest vapor pressure and butane has the highest vapor pressure at $-10°C$.

14.66 The substance that has the higher critical temperature also has stronger intermolecular forces.

14.71 False.

14.87 30.7 kJ/mol.

Chapter 15

15.9 $O_2 < I_2 < LiCl < CH_3OH$.

15.11 (a) $CH_3CH_2CH_2CH_2CH_2OH$. (b) CH_3OH. (c) Same as (a).

15.12 (a) 7.03%. (b) 16.9%. (c) 13%.

15.15 (a) 2.68 m. (b) 8.92 m. (c) 7.82 m.

15.17 3.0×10^2 g.

15.20 (a) 2.41 m. (b) 2.14 M. (c) 58.7 mL. (d) 0.956.

15.23 45.9 g KNO_3.

15.27 During the boiling process, air (oxygen gas) was driven off the water so the goldfish died as a result of the lack of oxygen gas. (It takes some time for water to become saturated with oxygen gas.)

15.30 Both gases react with water.

15.33 2.1×10^3 g.

15.35 Ethanol 30 mmHg; 1-propanol 26 mmHg.

15.39 (a) $CaCl_2$ solution. (b) Urea solution. (c) $CaCl_2$ solution. (d) $CaCl_2$ solution. [In (d), we assume $m \simeq M$.]

15.41 $C_{19}H_{38}O$; 282.5 g/mol.

15.43 0.066 m; 1.3×10^2 g/mol.

15.47 The two solutions have the same molal concentration of solute particles. Other colligative properties will also be the same: boiling point elevation, vapor pressure lowering, and osmotic pressure (if the solutions are dilute

so that molality is approximately equal to molarity.)

15.49 $C_{15}H_{20}O_{10}N_5$.

15.54 0.94 m.

15.56 0.15 m $C_6H_{12}O_6 > 0.15$ m $CH_3COOH > 0.10$ m $Na_3PO_4 > 0.20$ m $MgCl_2 > 0.35$ m $NaCl$.

15.59 2.47

15.65 $\Delta P = 2.05 \times 10^{-5}$ mmHg; $\Delta T_f = 8.91 \times 10^{-5}$ °C; $\Delta T_b = 2.5 \times 10^{-5}$ °C $\pi = 0.889$ mmHg.

15.70 3.5 atm.

Chapter 16

16.3 (a), (f), (g), (i), (j), (k).

16.8 (a) +1, (b) +2, (c) +3, (d) +3, (e) +4, (f) +6, (g) +2, (h) +4, (i) +2, (j) +3, (k) +5, (l) 0, (m) +7, (n) 0, (o) 0.

16.11 (a) +5. (b) +1. (c) +3. (d) +5. (e) +5. (f) +5.

16.12 (a) $C + 2H_2SO_4 \rightarrow CO_2 + 2SO_2 + 2H_2O$
(b) $I_2O_5 + 5CO \rightarrow I_2 + 5CO_2$
(c) $6HI + 2HNO_3 \rightarrow 3I_2 + 2NO + 4H_2O$
(d) $2Ag + 2H_2SO_4 \rightarrow Ag_2SO_4 + SO_2 + 2H_2O$
(e) $6Sb + 10HNO_3 \rightarrow 3Sb_2O_5 + 10NO + 5H_2O$
(f) $3H_2S + 2HNO_3 \rightarrow 3S + 2NO + 4H_2O$
(g) $S + 2H_2SO_4 \rightarrow 3SO_2 + 2H_2O$
(h) $5PbO_2 + 5H_2SO_4 + 2Mn(NO_3)_2 \rightarrow 5PbSO_4 + 2HMnO_4 + 4HNO_3 + 2H_2O$
(i) $Cl_2 + 2KOH \rightarrow KCl + KClO + H_2O$
(j) $H_3AsO_4 + 4Zn + 8HNO_3 \rightarrow AsH_3 + 4Zn(NO_3)_2 + 4H_2O$

16.15 0.156 M

16.17 45.1%

16.22 Fe^{2+}: 0.0920 M; Fe^{3+}: 0.0680 M.

16.28 (a) V_2O_5. (b) FeO. (c) Fe_2O_3. (d) Ti_2O_3. (e) Ni_3N_2. (f) $RuCl_4$.

16.30 (a) $SnCl_2$. (b) $TiCl_2$.

Chapter 17

17.8 1.08×10^7

17.11 0.051

17.13 $K_p = 0.105$; $K_c = 2.05 \times 10^{-3}$.

17.19 6.2×10^{-4}.

17.24 4.7×10^9.

17.28 $[I] = 8.58 \times 10^{-4}$ M; $[I_2] = 0.0194$ M.

17.30 $P_{COCl_2} = 0.408$ atm; $P_{CO} = 0.352$ atm; $P_{Cl_2} = 0.352$ atm.

17.34 0.173 mol.

17.40 (a) K_c increases. (b) K_c decreases. (c) K_c remains unchanged.

17.42 (a) Equilibrium shifted to the right. (b) Equilibrium shifted to the right. (c) Equilibrium shifted to the left.

17.46 No change in the equilibrium position.

17.51 Endothermic.

17.55 (a) 1.7 (b) $P_A = 0.69$ atm; $P_B = 0.8$ atm.

Chapter 18

18.2   (a) Both. (b) Base. (c) Acid. (d) Base. (e) Acid. (f) Base. (g) Base. (h) Base. (i) Acid. (j) Acid.

18.6   (a) H_2S. (b) H_2CO_3. (c) HCO_3^-. (d) H_3PO_4. (e) $H_2PO_4^-$. (f) HPO_4^{2-}. (g) H_2SO_4. (h) HSO_4^-. (i) H_2SO_3. (j) HSO_3^-.

18.9   (a) $3.8 \times 10^{-3} M$. (b) $6.2 \times 10^{-12} M$. (c) $1.1 \times 10^{-7} M$. (d) $1.0 \times 10^{-15} M$.

18.11 (a) 3.00. (b) 13.88. (c) 10.75. (d) 3.28.

18.15 Since the ionization of water is endothermic, according to Le Chatelier's principle $[H^+]$ will be greater at 30°C.

18.19 1.98×10^{-3} mol KOH; 0.444.

18.21 (a) Strong. (b) Weak. (c) Strong. (d) Weak. (e) Weak. (f) Weak. (g) Strong. (h) Weak. (i) Weak.

18.27 (a) H_3O^+. (b) $CH_3COOH_2^+$. (c) NH_4^+.

18.33 (a) $H_2SO_4 > H_2SeO_4$. (b) $H_2SO_3 > H_2SeO_3$. (c) $H_3PO_4 > H_3AsO_4$. (d) $HBrO_4 > HIO_4$.

18.38 (a) Lewis acid. (b) Lewis base. (c) Lewis base. (d) Lewis acid. (e) Lewis base. (f) Lewis base. (g) Lewis acid. (h) Lewis acid.

18.46 0.106 L.

Chapter 19

19.2 2.2×10^{-6}.

19.4 $2.3 \times 10^{-3} M$.

19.6 9.2×10^{-4}.

19.8   (a) 11.11. (b) 8.96. (c) 12.03.

19.10 0.15 M.

19.14 $[H^+] = 0.045 M$; $[SO_4^{2-}] = 0.045 M$. $[HSO_4^-] = 0.16 M$.

19.16 $[H^+] = [HCO_3^-] = 1.0 \times 10^{-4} M$, $[CO_3^{2-}] = 4.8 \times 10^{-11} M$.

19.19 KF, NH_4NO_2, $MgSO_4$, KCN, C_6H_5COONa, $NaHCO_3$, $HCOOK$.

19.21 Basic.

19.27 HZ < HY < HX.

Chapter 20

20.5   (a) $[H^+] = 1.3 \times 10^{-3} M$, $[CH_3COO^-] = 1.3 \times 10^{-3} M$. (b) $[H^+] = 1.00 M$, $[CH_3COO^-] = 1.8 \times 10^{-5} M$.

20.7 NH_3/HN_4NO_3; Na_2HPO_4/NaH_2PO_4; KNO_2/HNO_2; $HCOOK/HCOOH$.

20.9   (a) 4.81. (b) 4.64.

20.12 4.54.

20.14 Before: 9.25; after: 9.18.

20.17 4.3 g

20.18 0.25 M.

20.21 (a) 2.19. (b) 3.95. (c) 8.00. (d) 11.39.

20.30 Red.

20.32 (a) 3.60. (b) 9.69. (c) 6.07.

20.34 $x = 2$.

Chapter 21

21.4   (a) $[I^-] = 6.9 \times 10^{-9} M$. (b) $[Al^{3+}] = 4.4 \times 10^{-9} M$. (c) $[S^{2-}] = 1.3 \times 10^{-20} M$.

21.6   (a) 7.8×10^{-10}. (b) 2.0×10^{-14}. (c) 1.8×10^{-18}.

21.10 Yes.

21.13 9.53.

21.16 (a) Ag_2S first, $AgCl$ next, Ag_3PO_4 last. (b) $[S^{2-}] = 7.0 \times 10^{-32} M$.

21.20 (a) 1.7×10^{-4} g/L. (b) 1.4×10^{-7} g/L.

21.29 $6.5 \times 10^{-3} M$.

21.34 $[Cu^{2+}] = 1.2 \times 10^{-13} M$, $[Cu(NH_3)_4^{2+}] = 1.74 \times 10^{-2} M$, $[NH_3] = 0.23 M$.

21.39 (a) 0.011 M. (b) Yes.

21.44 Add a few drops of concentrated ammonia solution to each solution. Dark blue color due to $Cu(NH_3)_4^{2+}$ indicates presence of Cu^{2+}.

Chapter 22

22.5   (a) 0.049 M/s. (b) 0.025 M/s.

22.11 Reaction first order in A; $k = 0.213$/s.

22.13 (a) 2. (b) 3. (c) $\frac{3}{2}$.

22.16 (a) rate $= k[F_2][ClO_2]$. (b) $1.2/M \cdot s$. (c) $2.4 \times 10^{-4} M$/s.

22.21 0.034 M.

22.23 30 min.

22.24 (a) 0.0198/s. (b) 1.5×10^2 s.

22.30 135 kJ/mol.

22.34 3.0×10^3/s.

22.37 15.1 kJ/mol.

22.46 We need at least one molecule to react.

22.49 1.1×10^{13}.

22.56 (a) rate $= k[H^+][CH_3COCH_3]$. (b) $3.8 \times 10^{-3}/M \cdot s$.

Chapter 23

23.2 -198 J.

23.4 $\Delta H = 0$.

23.6   (a) -336.26 kJ. (b) (i) -1530 kJ. (ii) NH_3 is the better fuel.

23.9 -17 J.

23.11 (a) Spontaneous. (b) Nonspontaneous. (c) Spontaneous. (d) Nonspontaneous.

23.16 $Na(s) < NaCl(s) < Ne(g) < NH_3(g) < SO_2(g)$.

23.18 (a) $\Delta S < 0$. (b) $\Delta S > 0$. (c) $\Delta S > 0$. (d) ΔS is a small positive or negative number. (e) $\Delta S < 0$.

23.20 56.0 J/K.

23.23 Reaction A: Reaction spontaneous above 350 K. Reaction B: Reaction not spontaneous at any temperature. Reaction C: Reaction spontaneous at all temperatures. Reaction D: Reaction spontaneous below 111 K.

23.28 (a) -5.39 kJ. (b) 8.81. (c) -5.23 kJ.

23.34 25°C: 2×10^{-23} atm. 800°C: 0.535 atm.

23.38 79 kJ.

23.48 (a) $\Delta H > 0$, $\Delta S > 0$, $\Delta G < 0$. (b) $\Delta H > 0$, $\Delta S > 0$, $\Delta G = 0$. (c) $\Delta H > 0$, $\Delta S > 0$, $\Delta G > 0$.

Chapter 24

24.5 $Al(s) + 3Ag^+(aq, 1\ M) \rightarrow Al^{3+}(aq, 1\ M) + 3Ag(s)$; $\mathscr{E}° = 2.46$ V.

24.8 (a) Spontaneous. (b) Nonspontaneous. (c) Nonspontaneous. (d) Spontaneous.

24.13 Nonspontaneous.

24.15 (a) Li. (b) H_2. (c) Fe^{2+}. (d) Br^-.

24.20 3×10^{54}.

24.24 0.083 V.

24.29 (a) 0.076 L. (b) 29 L.

24.35 12.2 g.

24.37 0.012 F.

24.41 Cu: 5.33 g; Br_2: 13.4 g.

24.45 1.84 kg.

24.49 27.1 g/mol.

24.52 1.60×10^{-19} C.

24.58 4×10^{-13}.

24.60 (a) 3.14 V. (b) $\mathscr{E} = 3.13$ V.

Chapter 25

25.4 (a) $^{15}_{7}N + ^{1}_{1}p \rightarrow ^{4}_{2}He + ^{12}_{6}C$
(b) $^{27}_{13}Al + ^{2}_{1}d \rightarrow ^{4}_{2}He + ^{25}_{12}Mg$
(c) $^{55}_{25}Mn + ^{1}_{0}n \rightarrow \gamma + ^{56}_{25}Mn$
(d) $^{80}_{34}Se + ^{2}_{1}d \rightarrow ^{1}_{1}p + ^{81}_{34}Se$
(e) $^{9}_{4}Be + ^{2}_{1}d \rightarrow 2^{1}_{1}p + ^{9}_{3}Li$
(f) $^{106}_{46}Pd + ^{4}_{2}He \rightarrow ^{1}_{1}p + ^{109}_{47}Ag$

25.10 $^{18}_{7}N \rightarrow ^{18}_{8}O + ^{0}_{-1}\beta$

25.15 2.7 days.

25.20 (a) $^{40}_{19}K \rightarrow ^{40}_{18}Ar + ^{0}_{+1}\beta$
(b) 3.0×10^9 yr.

25.37 (a) $^{3}_{1}H \rightarrow ^{3}_{2}He + ^{0}_{-1}\beta$. (b) 70.5 disintegrations/min.

25.40 (a) 0.343 millicurie.
(b) $^{237}_{93}Np \rightarrow ^{233}_{91}Pa + ^{4}_{2}He$.

Chapter 26

26.7 Covalent bond (for example, H_2 and CH_4). Ionic bond (for example, LiH and CaH_2). Hydrogen bond (for example, between HF molecules and between H_2O molecules)

26.13 Use palladium metal, which can "dissolve" hydrogen gas but not neon gas.

26.15 11 kg.

26.20 C_2H_6: -84.7 kJ/mol. C_2H_4: 52.3 kJ/mol. Thermodynamic stability decreases as follows: $C_2H_6 > B_2H_6 > C_2H_4$.

26.22 B_2H_6.

26.24 $2B(s) + 6HNO_3(aq) \rightarrow 2H_3BO_3(s) + 6NO_2(g)$

26.28 (a) $Be_2C(s) + 4H_2O(l) \rightarrow 2Be(OH)_2(s) + CH_4(g)$
(b) $CaC_2(s) + 2H_2O(l) \rightarrow Ca(OH)_2(s) + C_2H_2(g)$

26.31 On heating, HCO_3^- decomposes as follows: $2HCO_3^-(aq) \rightarrow CO_3^{2-}(aq) + CO_2(g) + H_2O(l)$ The white precipitate is magnesium carbonate: $Mg^{2+}(aq) + CO_3^{2-}(aq) \rightarrow MgCO_3(s)$

26.36 379 g CO.

26.40 Si_3H_8.

26.42 Molecular formula: Si_2Cl_6. Each Si is sp^3 hybridized.

26.48 (a) $2NaNO_3(s) \rightarrow 2NaNO_2(s) + O_2(g)$
(b) $NaNO_3(s) + C(s) \rightarrow NaNO_2(s) + CO(g)$

26.58 48.0 g KNO_2.

26.59 (a) 86.7 kJ/mol. (b) 4×10^{-31}. (c) 4×10^{-31}.

26.66 125 g/mol; P_4.

26.73 (a) CO_2: discrete molecular units; SiO_2: giant covalent network. (b) H_2O: bent; H_2O_2: nonplanar. (c) MgO: ionic compound—extensive three-dimensional network; NO: discrete molecular units.

26.79 (a) The dark color keeps out light which causes decomposition. (b) 0.372 L.

26.86 2.6×10^2 g.

26.93 Green precipitate is CuF_2. In the presence of Cl^- ions: $Cu^{2+}(aq) + 4Cl^-(aq) \rightarrow CuCl_4^{2-}(aq)$. The complex ion is green.

26.98 2.81 L.

26.106 $XeF_2(+2)$, $XeF_4(+4)$, $XeF_6(+6)$, $XeOF_4(+6)$, $XeO_3(+6)$, $XeO_2F_2(+6)$, $XeO_4(+8)$.

Chapter 27

27.6 4.5×10^5.

27.7 112 h.

27.11 (a) 8.9×10^{12} cm^3. (b) 4.0×10^8 kg SO_2.

27.17 (a) 1.00 V. (b) 1.07×10^4 kJ.

27.25 (i) X-ray diffraction does not support the existence of M_2 species. (ii) It would be difficult to explain the high electrical conductivity in terms of the M_2 species. (iii) Interatomic distance is the same in the metal.

27.29 (a) 250°C. (b) Liquid to vapor transition. (c) For CO, the volume of gases increases. $\Delta G°$ will decrease because the $T\Delta S°$ term is positive. The reverse holds for CO_2 so $\Delta G°$ will increase with temperature.

27.32 1.41×10^3°C.

Chapter 28

28.4 (a) $2K(s) + 2H_2O(l) \rightarrow 2KOH(aq) + H_2(g)$
(b) $2NaOH(aq) + CO_2(g) \rightarrow Na_2CO_3(aq) + H_2O(l)$
(c) $Na_2CO_3(s) + 2HCl(aq) \rightarrow 2NaCl(aq) + CO_2(g) + H_2O(l)$
(d) $NaHCO_3(s) + HCl(aq) \rightarrow NaCl(aq) + H_2O(l) + CO_2(g)$
(e) $2NaHCO_3(s) \rightarrow Na_2CO_3(s) + H_2O(g) + CO_2(g)$
(f) $Na_2CO_3(s) \rightarrow$ No reaction. Unlike $CaCO_3$, Na_2CO_3 is not decomposed by heat.

28.9 (a) 117.6 kJ. (b) 177.8 kJ. $MgCO_3$ is more easily decomposed by heat.

28.14 $BaSO_4 < SrSO_4 < CaSO_4 < MgSO_4$.

28.23 $4Al(NO_3)_3(s) \rightarrow 2Al_2O_3(s) + 12NO_2(g) + 3O_2(g)$.

28.29 17.9%

28.30 (a) White tin. (b) White tin. (c) 2.6 kJ.

28.41 14 kJ; 4×10^{-3}.

28.47 (a) Sulfate ion. $BaSO_4$ is insoluble. (b) H_2S in acid solution. PbS is insoluble. (c) Iodide ions. HgI_2 is insoluble.

Chapter 29

29.4 Mn^{3+}.

29.8 (a) Ethylenediamine. (b) +3. (c) 6. (d) en.

29.12 (a) Tetraamminedichlorocobalt(III).
(b) Triamminetrichlorochromium(III).
(c) Dibromobis(ethylenediamine)cobalt(III).
(d) Pentacarbonyliron(0).
(e) *trans*-Diamminedichloroplatinum(II).
(f) *cis*-Dichlorobis(ethylenediamine)cobalt(III).
(g) Pentaamminechloroplatinum(IV) chloride.
(h) Hexaamminecobalt(III) chloride.
(i) Pentaamminechlorocobalt(III) chloride.
(j) Tetraaquodichlorochromium(III) chloride.

29.15 (a) Two. (b) Two.

29.25 (a) Two. (b) Four.

29.29 Two. $[Co(NH_3)_4Cl_2]^+Cl^-$.

29.32 Use a radioactive label such as $^{14}CN^-$ (in NaCN) to follow the exchange reaction.

Chapter 30

30.4 (a) Cycloalkane or alkene. (b) Alkene or alkyne. (c) Alkane. (d) Cycloalkane or alkene. (e) Alkene or alkyne.

30.8

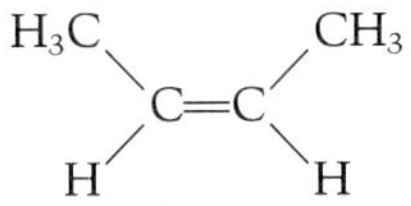

cis-2-Butene 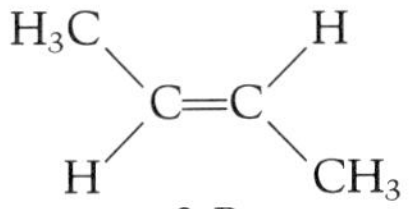   *trans*-2-Butene

The *cis* isomer has a higher heat of hydrogenation.

30.15 Ethanol has a higher boiling point because it can form hydrogen bonds. Both compounds should be soluble in water although ethanol is the more soluble of the two.

30.19 -630.8 kJ.

30.20 (a) *cis*-1,2-dichlorocyclopropane. (b) *trans*-1,2-dichlorocyclopropane.

30.21 (a) Ether. (b) Amine. (c) Aldehyde. (d) Ketone. (e) Carboxylic acid. (f) Alcohol. (g) Amine and Carboxylic acid.

30.30 (a) Ketone. (b) Ester. (c) Ether.

30.33 (a) 2-methylpentane. (b) 2,3,4-trimethylhexane. (c) 3-ethylhexane. (d) 3-methyl-1,4-pentadiene. (e) 2-pentyne. (f) 3-phenylpentene.

30.38 $(CF_2-CF_2-CF_2)_n$.

30.44 -174 kJ.

30.45 (a) CHF. (b) The gas does not behave ideally. (c) $C_2H_2F_2$.
(d)

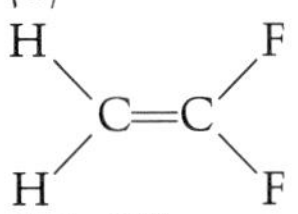

1,1-difluoro-ethylene 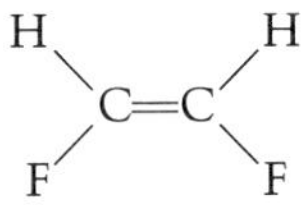   *cis*-difluoro-ethylene 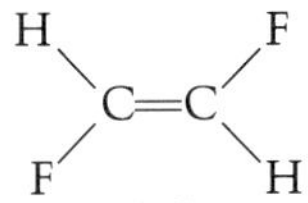   *trans*-difluoro-ethylene

30.49 8.4×10^3 L.

30.56 (a) One. (b) Two. (c) Five.

ACKNOWLEDGMENTS

Line Art

FIGURE 6.13 From *The Chemistry of Oxygen* by James B. Ifft and Julian L. Roberts, Jr., 3d ed. 1975. This is #1228 of the Laboratory Studies in General Chemistry from Franz/Malm's *Essentials of Chemistry in the Laboratory*, 3d ed., 1975, W. H. Freeman and Company. Copyright © (1975).

FIGURE 14.14 From G. Nemethy and H. A. Scheraga, *Journal of Chemical Physics* 36: 3383; 1962.

FIGURE 14.32 From "Glass," J. R. Hutchins, III, and R. V. Harrington from Kirk-Othmer: *Encyclopedia of Chemical Technology*, 2d ed., vol. 10, John Wiley & Sons, Inc., N.Y. 1966.

FIGURE 15.17 Reproduced with permission of Random House, Inc.

FIGURE 22.18 Courtesy, Manufacturing Chemists Association.

FIGURE 26.10 From P. V. Hobbs, H. Harrison, and E. Robinson, *Science* 183: 909–15 (1974). Copyright © 1974 by the American Association for the Advancement of Science

FIGURE 26.11 From "The Carbon Cycle," Bert Bolin. Copyright © 1970 by Scientific American, Inc. All rights reserved.

FIGURE 26.15 From "The Nitrogen Cycle," C. C. Delwiche. Copyright © 1970 by Scientific American, Inc. All rights reserved.

FIGURE 26.22 From "The Oxygen Cycle," Preston Gibor and Aharon Gibor. Copyright © 1970 by Scientific American, Inc. All rights reserved.

FIGURE 28.10 Courtesy of Xerox Corporation.

Text Photo Credits

FIGURE 1.1 Courtesy of the Department of Library Services, American Museum of Natural History.

FIGURE 2.4 Random House photo by Ken Karp.

FIGURE 3.4 Random House photo by Ken Karp.

FIGURE 3.6 (Both) Random House photos by Ken Karp.

FIGURE 3.8 Random House photo by Ken Karp.

FIGURE 3.10 (All) Random House photos by Ken Karp.

FIGURE 3.12 Random House photo by Ken Karp.

FIGURE 4.5 (All) Joel Gordon © 1979.

FIGURE 4.7 Random House photo by Ken Karp.

FIGURE 5.2 UPI/Bettmann Newsphotos.

FIGURE 5.9 (Both) Physicians and Nurses Manufacturing Corporation.

FIGURE 6.11 Random House photo by Ken Karp.

FIGURE 6.19 (Both) Random House photo by Ken Karp

FIGURE 8.10 (Both) by permission of Sargent-Welch Scientific Company, Skokie, Ill.

FIGURE 14.41 Ken Karp.

FIGURE 15.4 (All) Random House photos by Ken Karp.

FIGURE 15.8 Joel Gordon © 1979.

FIGURE 16.2 St. Louis Post-Dispatch.

FIGURE 16.3 Courtesy of Orgo-Thermite, Inc., Lakehurst, N.J.

FIGURE 16.4 (Both) Joel Gordon © 1979.

FIGURE 16.5 National Draeger, Inc.

FIGURE 16.6 Courtesy of Fisher Scientific.

FIGURE 16.7 Courtesy of Fisher Scientific.

FIGURE 16.8 Courtesy of Fisher Scientific.

FIGURE 17.6 USS Agri-Chemicals.

FIGURE 18.2 Random House photo by Ken Karp.

FIGURE 21.1   (All) Random House photos by Ken Karp.
FIGURE 21.3   Satour/Photo Researchers.
FIGURE 21.4   Courtesy of the Permutit Company, Inc.
FIGURE 22.19   Courtesy of General Motors Corp.
FIGURE 23.4   Courtesy of Hedco, Inc.
FIGURE 24.5   Random House photo by Ken Karp.
FIGURE 25.3   Fermi National Accelerator Laboratory.
FIGURE 25.9   Pierre Kopp/West Light.
FIGURE 25.10   (All) courtesy of U.S. Department of Energy, Office of Scientific & Technical Information, Oak Ridge, Tenn.
FIGURE 26.8   General Electric Research & Development Center.
FIGURE 26.9   General Electric.
FIGURE 26.24   Roger Werth/Woodfin Camp & Associates.
FIGURE 26.25   (Both) Landschafts-Verband Westfalen-Lippe, Munich.
FIGURE 26.26   Random House photo by Ken Karp.
FIGURE 26.32   Courtesy of L. C. Clark, Jr.
FIGURE 26.34   Courtesy of Argonne National Laboratory.
FIGURE 27.2   Woods Hole Oceanographic Institution.
FIGURE 27.4   Atlantic Richfield Company.
FIGURE 28.1   Liane Enkelis/Stock, Boston.
FIGURE 28.5   Random House photo by Ken Karp.
FIGURE 28.6   Random House photo by Ken Karp.
FIGURE 28.7   L. V. Bergman & Associates.
FIGURE 29.10   (All) Random House photos by Ken Karp.
FIGURE 30.16   William Strode/Black Star/Humana Inc.

Color Plate Photo Credits

PLATE 1   *Elements:* (Lithium and Sodium) Random House photos by Ken Karp; (Potassium) Life Science Library/MATTER photo by Albert Fenn © 1963 Time-Life Books Inc.; (Rubidium and Cesium) L. V. Bergman & Associates. *Minerals:* (Halite) Ward's Natural Science Establishment; (Spodumene) Richard Megna/Fundamental Photographers.
PLATE 2   *Elements:* (Beryllium, Magnesium, Calcium, Strontium, and Barium) L. V. Bergman & Associates; (Radium) Life Science Library/MATTER photo by Phil Brodatz © 1963 Time-Life Books, Inc. *Minerals:* (All) Ward's Natural Science Establishment. *Mining and Production:* (Both) AMAX, Inc.
PLATE 3   *Elements:* (All) L. V. Bergman & Associates. *Minerals:* (Both) Ward's Natural Science Establishment. Production: (Top left) Aluminum Company of America; (top right) Craig Aurness/West Light; (bottom left and right) Aluminum Company of America.
PLATE 4   *Elements:* (Graphite) Random House photo by Ken Karp; (Diamond) Diamond Information Center; (Silicon) Frank Wing/Stock, Boston; (Germanium and Lead) Random House photos by Ken Karp; (Tin) L. V. Bergman & Associates; (Lead) Random House photo by Ken Karp. *Minerals:* (Sand) Calvin Larsen/Photo Researchers; (Cassiterite and Galena) Ward's Natural Science Establishment. *Mining and Production:* (Top & bottom left) Jeff Smith; (bottom right) Bell Labs.
PLATE 5   *Elements:* (Liquid Nitrogen) Joe McNally/Wheeler Pictures; (Phosphorus) Life Science Library/MATTER photo by Albert Fenn © 1963 Time-Life Books, Inc.; (Arsenic, Antimony, and Bismuth) L. V. Bergman & Associates. *Mining:* (Both) AMAX Inc.
PLATE 6   *Elements:* (Oxygen) Hank Morgan/Photo Researchers; (Sulfur and Selenium) L. V. Bergman & Associates. *Mining and Production:* (Left) Bill Pierce/Rainbow; (right) C. B. Jones/Taurus.
PLATE 7   *Elements:* Random House photo by Joel Gordon. *Production:* (Left) Vulcan Materials Company, photo by Charles Beck; (right) Mula & Haramaty/Phototake.
PLATE 8   (All discharge tubes) Random House photos by Ken Karp; (neon signs) Grant Heilman.
PLATE 9   *Elements:* (All) L. V. Bergman & Associates. *Minerals:* (Both) Ward's Natural Science Establishment. *Mining and Production:* (Left) Taurus; (right) AMAX Inc.
PLATE 10   *Elements:* (Titanium and Chromium) Random House photos by Ken Karp; (Manganese) Jeff Smith; (Iron, Cobalt, and Nickel) Random House photos by Ken Karp; (Copper) L. V. Bergman & Associates; (Ruthenium) Random House photo by Ken Karp; (Silver) Dr. E. R. Degginger, FPSA/Bruce Coleman; (Tungsten and Platinum) Random House photos by Ken Karp; (Gold) Ward's Natural Science Establishment. *Minerals:* (Crocoite, Magnetite, Argentite, and Chalcopyrite) Ward's Natural Science Establishment; (Pyrite) L. V. Bergman & Associates. *Mining and Production:* (Iron mine) Grant Heilman; (Copper mining) Peter Menzel/Stock, Boston; (Copper melting) Jeff Smith; (Copper sheets) Foto du Monde/The Picture Cube; (Molten manganese and steel making) Jeff Smith; (Gold recovery) Paul Logsdon/Phototake.
PLATE 11   Random House photo by Ken Karp.
PLATE 12   Random House photo by Ken Karp.
PLATE 13   (Both) Random House photos by Ken Karp.
PLATE 14   Joel Gordon © 1979.
PLATE 15   (Both) Random House photos by Ken Karp.
PLATE 16   Random House photo by Ken Karp.
PLATE 18   Joel Gordon © 1979.
PLATE 19   Joel Gordon © 1979.
PLATE 20   (All) Random House photos by Ken Karp.
PLATE 21   Joel Gordon © 1979.
PLATE 22   (All) Random House photos by Ken Karp.
PLATE 23   (Both) Random House photos by Ken Karp.
PLATE 24   Random House photo by Ken Karp.
PLATE 25   Random House photo by Ken Karp.
PLATE 26   Random House Photo by Ken Karp.
PLATE 27   Los Alamos National Laboratory.
PLATE 28   Defense Nuclear Agency.
PLATE 29   Courtesy of James L. Dye.
PLATE 30   (Both) Courtesy of Neil Bartlett.

INDEX

FUNDAMENTAL CONSTANTS

Avogadro's number $\quad\quad\quad 6.022 \times 10^{23}$

Electron charge (e) $\quad\quad 1.6022 \times 10^{-19}$ C

Electron mass $\quad\quad\quad 9.1095 \times 10^{-28}$ g

Faraday constant (F) $\quad\quad$ 96,487 C/mol electron

Gas constant (R) $\quad\quad\quad$ 8.314 J/K·mol (0.08206 L·atm/K·mol)

Planck's constant (h) $\quad\quad 6.6256 \times 10^{-34}$ J s

Proton mass $\quad\quad\quad\quad 1.67252 \times 10^{-24}$ g

Speed of light in vacuum $\quad 2.99792458 \times 10^{8}$ m/s

USEFUL CONVERSION FACTORS AND RELATIONSHIPS

1 lb = 453.6 g

1 in = 2.54 cm (exactly)

1 mi = 1.609 km

1 km = 0.6215 mi

1 pm = 1×10^{-12} m = 1×10^{-10} cm

1 atm = 760 mmHg = 760 torr = 101,325 N/m^2 = 101,325 Pa

1 cal = 4.184 J (exactly)

1 L atm = 101.325 J

1 J = 1 C × 1 V

$$?°C = (°F - 32°F) \times \frac{5°C}{9°F}$$

$$?°F = \frac{9°F}{5°C} \times (°C) + 32°F$$

$$K = (°C + 273.15°C)\left(\frac{1K}{1°C}\right)$$

SOME PREFIXES USED WITH SI UNITS

Tera (T) $\quad\quad 10^{12}$

Giga (G) $\quad\quad 10^{9}$

Mega (M) $\quad\quad 10^{6}$

Kilo (k) $\quad\quad 10^{3}$

Deci (d) $\quad\quad 10^{-1}$

Centi (c) $\quad\quad 10^{-2}$

Milli (m) $\quad\quad 10^{-3}$

Micro (μ) $\quad\quad 10^{-6}$

Nano (n) $\quad\quad 10^{-9}$

Pico (p) $\quad\quad 10^{-12}$